Solved Problems in Nonlinear Oscillations

Zeng He · Wen Jiang · Lin Wang

Solved Problems in Nonlinear Oscillations

A Sourcebook for Scientists and Engineers

Springer

Zeng He
Department of Mechanics
Huazhong University of Science
and Technology
Wuhan, China

Wen Jiang
Department of Mechanics
Huazhong University of Science
and Technology
Wuhan, China

Lin Wang
Department of Mechanics
Huazhong University of Science
and Technology
Wuhan, China

ISBN 978-981-97-6115-9 ISBN 978-981-97-6113-5 (eBook)
https://doi.org/10.1007/978-981-97-6113-5

This work was supported by National Natural Science Foundation of China (12325201, 12072119).

This Springer imprint is published by the registered company Springer Nature Singapore Pte Ltd.
The registered company address is: 152 Beach Road, #21-01/04 Gateway East, Singapore 189721, Singapore

Preface

This handbook contains about 200 fully solved problems in analytical and numerical methods for nonlinear oscillations. These comprise all the end-of-chapter problems in Prof. Nayfeh and Prof. Mook's famous book (*Ali H. Nayfeh and Dean T. Mook. Nonlinear Oscillations, Wiley-Interscience, 1979*). Mathematical software is adopted to make those solutions more accessible from a graphical point of view. This book can be adopted as a supplement to course work study for graduates or senior undergraduates. Since many exercise problems are adapted from scientific research papers, this book also has a good reference value for scientists and engineers who work in the area of nonlinear vibration.

We wish to express our appreciation to Jiabiao Yi and Yefeng Pu for their great help in documentation editing and drafting, and to Springer Nature for the opportunity to make available this supplement to *Nonlinear Oscillations*.

Wuhan, China
December 2023

Zeng He
Wen Jiang
Lin Wang

Contents

Chapter 1
Problem List

To make it easier for the reader to access exercise problems, we have listed all problems in this chapter and added a short description for each problem. In addition, in the following chapters, "**the Book**" refers to "*Nonlinear Oscillations. Ali H. Nayfeh & Dean T. Mook, John Wiley and Sons, New York, 1995, ISBN 0–471-12,142–8*"). The solutions are arranged according to the chapter names question-numbering in **the Book**, and our handbook should be viewed as a supplement to **the Book**.

1.1 Chapter 2 Conservative Single-Degree-Of-Freedom Systems

- Exercise 2.1 (Determine solution trajectories from potential energy curves of conservative 1D systems)
- Exercise 2.2 (Solving conservative systems by the method of multiple scales or straightforward expansion combined with reformulation)
- Exercise 2.3 (Show the system transformation and the method of multiple scales)
- Exercise 2.4 (Show the method of averaging)
- Exercise 2.5 (Show the effect of a change in the equilibrium point of a nonlinear spring on the frequency-amplitude relationship)
- Exercise 2.6 (Rods sliding on the smooth walls)
- Exercise 2.7 (Period of the osscillation motion of a single pendulum-type system along separatrices)
- Exercise 2.8 (Particle on a rotating parabola)
- Exercise 2.9 (A single pendulum rotating uniformly around a plumb axis)
- Exercise 2.10 (A single pendulum rotating freely around a plumb axis)
- Exercise 2.11 (Nonlinear oscillation of a slide-slider system)
- Exercise 2.12 (Nonlinear oscillation of a rope-block system)

Z. He et al., *Solved Problems in Nonlinear Oscillations*,
https://doi.org/10.1007/978-981-97-6113-5_1

- Exercise 2.13 (Single pendulum attached with a rolling-without-slipping circular wheel I)
- Exercise 2.14 (Single pendulum attached with a rolling-without-slipping circular wheel II)
- Exercise 2.15 (Single pendulum problem with inelastic collision with an inclined wall)
- Exercise 2.16 (Simplified model for buckling analysis of columns)
- Exercise 2.17 (Rods for pure rolling on a fixed cylindrical surface)
- Exercise 2.18 (Geometrically nonlinear system formed by a linear spring and a mass block)
- Exercise 2.19 (Solving pure cubic nonlinear systems by harmonic balance method)
- Exercise 2.20 (Solving purely fifth-order nonlinear systems by harmonic balance method)
- Exercise 2.21 (Solving pure cubic nonlinear systems by equivalent linearization)
- Exercise 2.22 (Show least residual value method and Galerkin method for solving pure cubic nonlinear systems)
- Exercise 2.23 (Discuss the possibility of solving pure cubic nonlinear systems by the method of multiple scales)
- Exercise 2.24 (Examine the relationship between the period and the amplitude of oscillation)
- Exercise 2.25 (Comparison of equivalent linearization methods and the method of multiple scales for solving different nonlinear equations)
- Exercise 2.26 (Comparison of the Galerkin method with the the method of multiple scales)
- Exercise 2.27 (Equations containing second, third and fourth order nonlinear terms)
- Exercise 2.28 (Solving nonlinear equations with segmentation functions)
- Exercise 2.29 (Nonlinear equations with a single arbitrary subpartial nonlinear term only)
- Exercise 2.30 (Higher order nonlinear equations)
- Exercise 2.31 (Motion of the disk constrained by a linear spring)
- Exercise 2.32 (Rolling without slip rollers restrained by a linear spring)
- Exercise 2.33 (Example of higher-order system)

1.2 Chapter 3 Nonconservative Single-Degree-Of-Freedom Systems

- Exercise 3.1 (Singularity analyses I)
- Exercise 3.2 (Singularity analyses II)
- Exercise 3.3 (Singularity analyses III)
- Exercise 3.4 (Conversion of Rayleigh's equation to van der Pol's equation)
- Exercise 3.5 (Plotting phase trajectories and limit cycle(s) for the van der Pol equation)

- Exercise 3.6 (Plotting phase trajectories and limit cycle(s) for a given equation)
- Exercise 3.7 (Plotting singularity and phase trajectories for a given planar system)
- Exercise 3.8 (Plot the trajectory of a singularity for a given planar system and determine the stability of the singularity)
- Exercise 3.9 (The method of multiple scales for solving a single pendulum with a Coulomb friction torque)
- Exercise 3.10 (Singularities and their stability of a linearly damped pendulum under a constant torque)
- Exercise 3.11 (Singularities and their stability for a single pendulum with a quadratic damping)
- Exercise 3.12 (Analysis of Rayleigh equation singularities, method of averaging of solution)
- Exercise 3.13 (Singularity analysis of the restricted three-body exercise)
- Exercise 3.14 (Singularity analysis of amplitude and phase equations limiting the resonance response of the three-body problem)
- Exercise 3.15 (Modeling of a single pendulum with a damper attached, solving by the method of multiple scales)
- Exercise 3.16 (The method of multiple scales for solving van der Pol oscillators with delay amplitude limiting)
- Exercise 3.17 (Averaging to solve for an oscillator with both Coulomb and viscous damping)
- Exercise 3.18 (Averaging to solve for oscillators with both Coulomb and square damping)
- Exercise 3.19 (Averaging to solve for oscillators with both viscous and square damping)
- Exercise 3.20 (Averaging to solve for an oscillator with both viscous and negative Coulomb damping)
- Exercise 3.21 (Averaging to solve for an oscillator with both squared damping and negative Coulomb damping)
- Exercise 3.22 (Oscillators with simultaneous negative viscous damping, square damping, and cubic nonlinearity)
- Exercise 3.23 (Method of averaging for solving oscillators with simultaneous viscous, quadratic and cubic damping nonlinearities)
- Exercise 3.24 (Averaging to solve and analyze nonlinear vibrations of a pendulum)
- Exercise 3.25 (Viscous damping, negative stiffness, and cubic nonlinear systems)
- Exercise 3.26 (Transformations of variables and three-level number expansion methods for solving a cubic nonlinear system)
- Exercise 3.27 (Application of elliptic functions to represent Solutions of a single pendulum system with a general weak damping function)
- Exercise 3.28 (Masses bounded in definite orbits, analyzing singularities and their bifurcations)
- Exercise 3.29 (Non-autonomous systems with slowly varying functions)

1.3 Chapter 4 Forced Oscillations of Systems Having a Single Degree of Freedom

- Exercise 4.1 (The method of multiple scales for analyzing primary resonances of Coulomb-damped systems with cubic nonlinearities)
- Exercise 4.2 (The method of multiple scales for analyzing primary resonances of cubic nonlinear, quadratic-damped systems)
- Exercise 4.3 (The method of multiple scales for analyzing primary resonances of linear spring and hysteresis damped systems)
- Exercise 4.4 (The method of multiple scales for analyzing primary resonances of linear spring, hysteresis and Coulomb damped systems)
- Exercise 4.5 (The method of multiple scales for analyzing primary resonances of linear springs, hysteresis and linearly damped systems)
- Exercise 4.6 (The method of multiple scales for analyzing primary resonances of linear spring, hysteresis and quadratic damped systems)
- Exercise 4.7 (The method of multiple scales for analyzing primary resonances of quadraticd spring force and quadraticd damped systems)
- Exercise 4.8 (Primary, superharmonic, and subharmonic resonances of cubic nonlinear and cubic damped systems)
- Exercise 4.9 (Method of averaging for analyzing primary resonances of alternating damped systems)
- Exercise 4.10 (Primary, superharmonic, and subharmonic resonances of quadratic and cubic nonlinear systems)
- Exercise 4.11 (Primary and secondary resonances of five nonlinear, linearly damped systems)
- Exercise 4.12 (An iterative method for general undamped nonlinear systems)
- Exercise 4.13 (The method of harmonic balance to analyze viscous damping and primary resonances of cubic nonlinear systems)
- Exercise 4.14 (The method of harmonic balance to analyze the primary resonance of an undamped nonlinear system)
- Exercise 4.15 (Combination resonance for cubic nonlinear systems)
- Exercise 4.16 (The response of cubic nonlinear systems with primary resonance and non-resonance excitation)
- Exercise 4.17 (Primary and combined resonance for cubic nonlinear systems I)
- Exercise 4.18 (Combined primary, subharmonic, and superharmonic resonances of a cubic nonlinear system)
- Exercise 4.19 (Primary and combined resonance for cubic nonlinear systems II)
- Exercise 4.20 (Primary and combined resonance of a cubic nonlinear system)
- Exercise 4.21 (Primary resonance of the system with Coulomb friction and quadratic nonlinearity)
- Exercise 4.22 (Primary resonance problem for the system with quadratic damp and quadratic nonlinear terms)
- Exercise 4.23 (Combined resonance problem for quadratic nonlinear systems)

- Exercise 4.24 (Simultaneous subharmonic and superharmonic resonance of quadratic nonlinear systems)
- Exercise 4.25 (Quadratic nonlinear systems with both subharmonic and combinatorial resonances)
- Exercise 4.26 (Quadratic nonlinear systems with both superharmonic and combinatorial resonances)
- Exercise 4.27 (Quadratic nonlinear systems with superharmonic, subharmonic and combinatorial resonances)
- Exercise 4.28 (Primary and superharmonic resonances of self-excited systems)
- Exercise 4.29 (Primary and subharmonic resonances of self-excited systems)
- Exercise 4.30 (Simultaneous subharmonic and superharmonic resonance of self-excited systems)
- Exercise 4.31 (Self-excited system with both primary and combined resonances I)
- Exercise 4.32 (Self-excited system with both primary and combined resonances II)
- Exercise 4.33 (Combination resonance of self-excited systems)
- Exercise 4.34 (Self-excited system with simultaneous primary, superharmonic, and subharmonic resonances)
- Exercise 4.35 (Self-excited system with simultaneous primary and combinatorial resonances)
- Exercise 4.36 (Simple harmonic response of van der Pol's oscillator)
- Exercise 4.37 (Analysis on the response of a cubic nonlinear system by the slowly varying excitation)
- Exercise 4.38 (Analysis of the response of a self-excited system under slowly varying excitation)
- Exercise 4.39 (Analysis of the response of a self-excited vibrating system under hard excitation at different frequencies)

1.4 Chapter 5 Parametrically Excited Systems

- Exercise 5.1 (The method of strained parameters to determine the transition curves for a Hill equation)
- Exercise 5.2 (The method of strained parameters to determine the transition curves of the Hill's equation for LRC circuits with sinusoidally varying resistance)
- Exercise 5.3 (Stability analysis of a supported movable pendulum)
- Exercise 5.4 (Analyze the linear and nonlinear stability of a rotating pendulum using the method of strained parameters and the method of multiple scales)
- Exercise 5.5 (Parametric excitation of a particle-string system subjected to axial forces at both ends)
- Exercise 5.6 (Linear and nonlinear stability analysis of a pendulum with varying length)

- Exercise 5.7 (Linear and nonlinear stability analysis of a particle sliding on a rotating parabola)
- Exercise 5.8 (Nonlinear solution to the pure rolling of a cylinder on a circular surface)
- Exercise 5.9 (The method of harmonic balance to solve Duffing's equation and stability analysis)
- Exercise 5.10 (The method of harmonic balance for solving pure cubic nonlinear equations)
- Exercise 5.11 (The quadratic parametric excitation of a linear viscous damping system)
- Exercise 5.12 (The cubic parametric excitation of a linear viscous damping system)
- Exercise 5.13 (The high order nonlinear parametric excitation of a linear viscous damping system)
- Exercise 5.14 (The parametric excitation of a system with a nonlinear damping)
- Exercise 5.15 (The nonlinear parametric excitation of a system with a linear viscous damping)
- Exercise 5.16 (The nonlinear parametric excitation of the system with a nonlinear damping)
- Exercise 5.17 (The nonlinear parametric excitation of the system with a nonlinear damping and high order nonlinearities)
- Exercise 5.18 (Cubic nonlinear system subjected to combined parametric and external excitation)
- Exercise 5.19 (Cubic nonlinear system subjected to combined parametric and multi-frequency external excitations)
- Exercise 5.20 (Cubic nonlinear, square-damped system subjected to combined parametric and external excitation)
- Exercise 5.21 (Analysis on a double pendulum with swinging mass attached to springs and moving platform)
- Exercise 5.22 (A two-frequency parametric excitation of a multi degree-of-freedom system with distinct frequencies)
- Exercise 5.23 (A single-frequency parametric excitation of a three-degree-of-freedom system with repeating frequencies)
- Exercise 5.24 (A two-frequency parametric excitation of a three-degree-of-freedom system with repeating frequencies)
- Exercise 5.25 (Parametric excitation analysis of oscillations of a spring pendulum)
- Exercise 5.26 (The buckling of the column with a nonideal energy source)
- Exercise 5.27 (Analysis of two-dimensional sound propagation in a pipe with sinusoidal walls)
- Exercise 5.28 (Analysis of two-dimensional electromagnetic wave propagation in a pipe with sinusoidal walls)

1.5 Chapter 6 Systems Having Finite Degrees of Freedom

- Exercise 6.1 (Internal resonance analysis and nonlinear solution of a double pendulum problem)
- Exercise 6.2 (Internal resonance analysis of a uniform rod hanging from a massless chord)
- Exercise 6.3 (Internal resonance analysis of a disc pendulum)
- Exercise 6.4 (Internal resonance analysis of a spring pendulum)
- Exercise 6.5 (Internal resonance analysis of a uniform rod hanging a spring)
- Exercise 6.6 (Internal resonance analysis of the plane motion of a rigid beam supported by a spring)
- Exercise 6.7 (Analytical solution of a two-degree-of-freedom high-order nonlinear equation)
- Exercise 6.8 (Forced oscillations of a two-degree-of-freedom gyroscopic system with cubic nonlinearity)
- Exercise 6.9 (Forced oscillations of a two-degree-of-freedom system with quadratic and cubic nonlinearities)
- Exercise 6.10 (Analysis of forced oscillations of a spring-slider-pendulum system)
- Exercise 6.11 (Resonance analysis on the cylinder rolling without slip on the circular surface I)
- Exercise 6.12 (Resonance analysis on the cylinder rolling without slip on the circular surface II)
- Exercise 6.13 (Parametric excitation of a stretched wire carrying two particles)
- Exercise 6.14 (Internal resonance, parametric excitation and saturation phenomena for a spring pendulum with a moving support)
- Exercise 6.15 (The response of a ship constrained to pitch and roll only)
- Exercise 6.16 (The method of multiple scales for solving free oscillations of systems with slowly varying frequencies)
- Exercise 6.17 (Analysis of a two-degree-of-freedom system with simultaneous internal, subharmonic, and superharmonic resonances)
- Exercise 6.18 (Nonlinear oscillation analysis of a spherical pendulum with a moving support)
- Exercise 6.19 (Primary resonance analysis of a two-degree-of-freedom self-excited system)
- Exercise 6.20 (Free oscillation analysis of a two-degree-of-freedom self-excited system with heavy eigenvalues)
- Exercise 6.21 (Primary resonance analysis of three-degree-of-freedom systems with repeating frequencies, internal resonance, saturation phenomena)
- Exercise 6.22 (Primary resonance analysis of three-degree-of-freedom systems with internal resonance)
- Exercise 6.23 (Stability analysis of nonlinear forced oscillation of a disk)
- Exercise 6.24 (Forced oscillation of a spherical pendulum)
- Exercise 6.25 (Nonlinear oscillation analysis of a rolling reentry body I)
- Exercise 6.26 (Nonlinear oscillation analysis of a rolling reentry body II)

1.6 Chapter 7 Continuous Systems

- Exercise 7. 1 (Longitudinal oscillation analysis of non-uniform, non-linear elastic rods)
- Exercise 7.2 (Longitudinal oscillation analysis of uniform, nonlinear elastic rods)
- Exercise 7.3 (Longitudinal oscillation analysis of uniform, nonlinear elastic rods)
- Exercise 7.4 (Nonlinear analysis of transverse oscillations of a fixed wire)
- Exercise 7.5 (Nonlinear analysis of transverse oscillations in the plane of an elastic tensioned string)
- Exercise 7.6 (Nonlinear analysis of planar transverse oscillations of elastic tensioned strings under parametric excitation formed by time-varying tension)
- Exercise 7.7 (Transverse oscillation of a hinged-hinged beam excited by first- and second-order primary resonances at $u = O(w^2)$)
- Exercise 7.8 (Forced response of a hinged-hinged beam with a single non-resonant excitation at $u = O(w^2)$)
- Exercise 7.9 (Combined resonance analysis of hinged-hinged beams with two excitations at $u = O(w^2)$)
- Exercise 7.10 (Combined resonance analysis of a hinged-hinged beam with three excitations at $u = O(w^2)$)
- Exercise 7.11 (Parametric resonance analysis on a hinged-hinged beam at $u = O(w^2)$)
- Exercise 7.12 (Transverse oscillation of a hinged-clamped beam under internal resonance and non-resonant excitation at $u = O(w^2)$)
- Exercise 7.13 (Transverse oscillation of a hinged-clamped beam under internal resonance and non-resonance excitation at $u = O(w^2)$)
- Exercise 7.14 (Transverse oscillation of a clamped–clamped beam at internal resonance and resonant excitation at $u = O(w^2)$)
- Exercise 7.15 (Transverse oscillation of a clamped–clamped beam under internal resonance and non-resonance excitation at $u = O(w^2)$)
- Exercise 7.16 (Transverse oscillation of a clamped-supported beam under internal resonance and non-resonance excitation at $u = O(w^2)$)
- Exercise 7.17 (Couple longitudinal and transverse oscillation analysis of hinged-hinged beams at $u = O(w)$)
- Exercise 7.18 (Analysis of internal and primary resonances of cylindrical shells, saturation phenomena)
- Exercise 7.19 (Primary resonance analysis of transverse oscillations of a taut string)
- Exercise 7.20 (Oscillation analysis of a relief valve with boundary nonlinearities)
- Exercise 7.21 (First-order subharmonic resonance analysis of a uniform circular plate clamped along its edge)
- Exercise 7.22 (First-order superharmonic resonance analysis of a uniform circular plate clamped along its edge)
- Exercise 7.23 (Second-order superharmonic resonance analysis of a uniform circular plate clamped along its edge)

- Exercise 7.24 (Combined resonance analysis of a uniform circular plate clamped along its edge I)
- Exercise 7.25 (Combined resonance analysis of a uniform circular plate clamped along its edge II)
- Exercise 7.26 (Combined resonance analysis of a uniform circular plate clamped along its edge III)
- Exercise 7.27 (Axisymmetric response of a uniform circular plate clamped along its edge with 1st to 3rd order modes subjecting to corresponding resonant excitation)
- Exercise 7.28 (Derivation of modal discretization equations for Berger's equation for axisymmetric oscillation analysis of a clamped circular plates)
- Exercise 7.29 (Berger equation for nonlinear oscillation analysis of simply supported rectangular plates)
- Exercise 7.30 (Nonlinear oscillation analysis of rotating circular films)

1.7 Chapter 8 Traveling Waves

- Exercise 8.1 (Determination on dispersive wave)
- Exercise 8.2 (Direct expansion and reformulation analysis of longitudinal waves of semi-infinite uniform rods with material nonlinearity)
- Exercise 8.3 (Direct expansion and reformulation analysis of a nonlinear acoustic equation)
- Exercise 8.4 (Multiscale analysis on a one-dimensional wave equation with cubic nonlinearity)
- Exercise 8.5 (A simplified model for wind-induced oscillation of overhead power lines)
- Exercise 8.6 (Wave propagating along a uniform, initially undeformed nonlinear elastic bar with linear damping)
- Exercise 8.7 (Modeling and multiscale analysis on the high-frequency oscillation of a homogenous visco-elastic rod)
- Exercise 8.8 (Modeling and multiscale analysis on low-frequency oscillation of a homogeneous visco-elastic rod)
- Exercise 8.9 (Transform the Burgers' equation into the heat equation)
- Exercise 8.10 (Stationary solutions of the Burgers' equation)
- Exercise 8.11 (Steady-state solution of the Burgers' equation and their stability analysis)
- Exercise 8.12 (Constant solutions of the KdV equation and their stability analysis)
- Exercise 8.13 (Traveling wave solutions of the sine–Gordon equation)
- Exercise 8.14 (Characteristic transformation and straightforward expansion for waves propagating along a uniform elastic bar with material nonlinearity)
- Exercise 8.15 (Forced excitations of a nonlinear finite elastic bar)
- Exercise 8.16 (Derivation of the Lagrangian form of the wave equation for an inviscid isentropic gas)

- Exercise 8.17 (Exact solution of the one-dimensional wave equation for an inviscid isentropic gas)
- Exercise 8.18 (Approximate solutions of the one-dimensional wave equation for a viscous isentropic gas and its Fourier expansion)
- Exercise 8.19 (Derivation of the Eulerian form of the one-dimensional wave equation for an inviscid isentropic gas)
- Exercise 8.20 (Derivation of linear inviscid acoustic waves in a hardwalled duct)
- Exercise 8.21 (Analysis on the linear waves propagating on the surface of an inviscid liquid of finite depth)
- Exercise 8.22 (The method of multiple scales for wave group propagation governed by Klein–Gordon equation)

Chapter 2
Conservative Single-Degree-of-Freedom Systems

2.1 Exercise 2.1 (Determine Solution Trajectories from Potential Energy Curves of Conservative 1D Systems)

Solution: (**a**) Let $v = \dot{u}$, , then the differential equation for the solution trajectory of (a) is given by

$$\frac{dv}{du} = -\frac{u}{v} \tag{2.1.1}$$

The general solution of (2.1.1) is

$$v^2 + u^2 = 2E \tag{2.1.2}$$

where E is the constant of integration, which is the total mechanical energy of the mechanical system. Therefore, the phase trajectories of (a) are a set of concentric circles centered at the origin (Fig. 2.1a).

(**b**) Let $v = \dot{u}$, then the differential equation for the solution trajectory of (b) is

$$\frac{dv}{du} = \frac{-u + u^3}{v} \tag{2.1.3}$$

$$\text{After integration, } v^2 + u^2 - \frac{1}{2}u^4 = 2E \tag{2.1.4}$$

The equation of the solution trajectory of (b) can be obtained from (2.1.4)

$$v = \pm\sqrt{2E - u^2 + \frac{1}{2}u^4} \tag{2.1.5}$$

Z. He et al., *Solved Problems in Nonlinear Oscillations*,
https://doi.org/10.1007/978-981-97-6113-5_2

The phase trajectories can be sketched directly or from the potential energy, V, of the system:

$$f(u) = u - u^3, \quad V(u) = \int_0^u f(u)du = \frac{1}{2}u^2 - \frac{1}{4}u^4 \tag{2.1.6}$$

One can obtain the singular points of the system s_1 : (0,0), s_2 : (1,0), s_3 : (−1,0) from Eq. (2.1.3), which, together with Eq. (2.1.5), can be adopted to sketch the potential energy curve and phase trajectories of the system (Fig. 2.1b).

(c)

$$\frac{dv}{du} = \frac{u - u^3}{v} \tag{2.1.7}$$

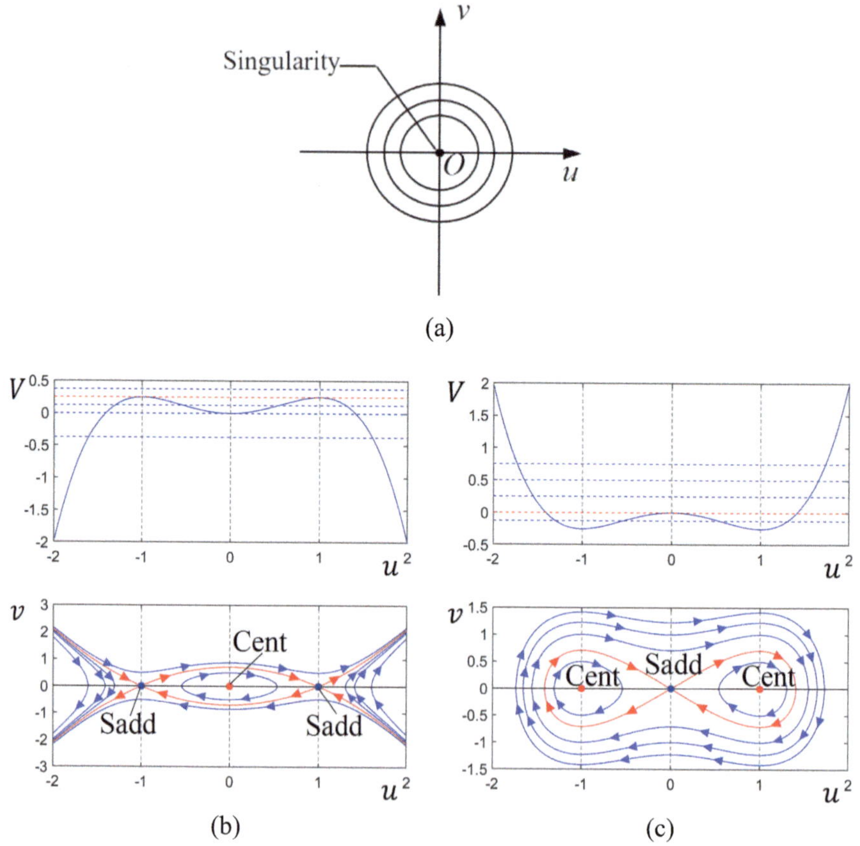

Fig. 2.1 Potential energy curves and solution trajectories for Exercise 2.1 a–g

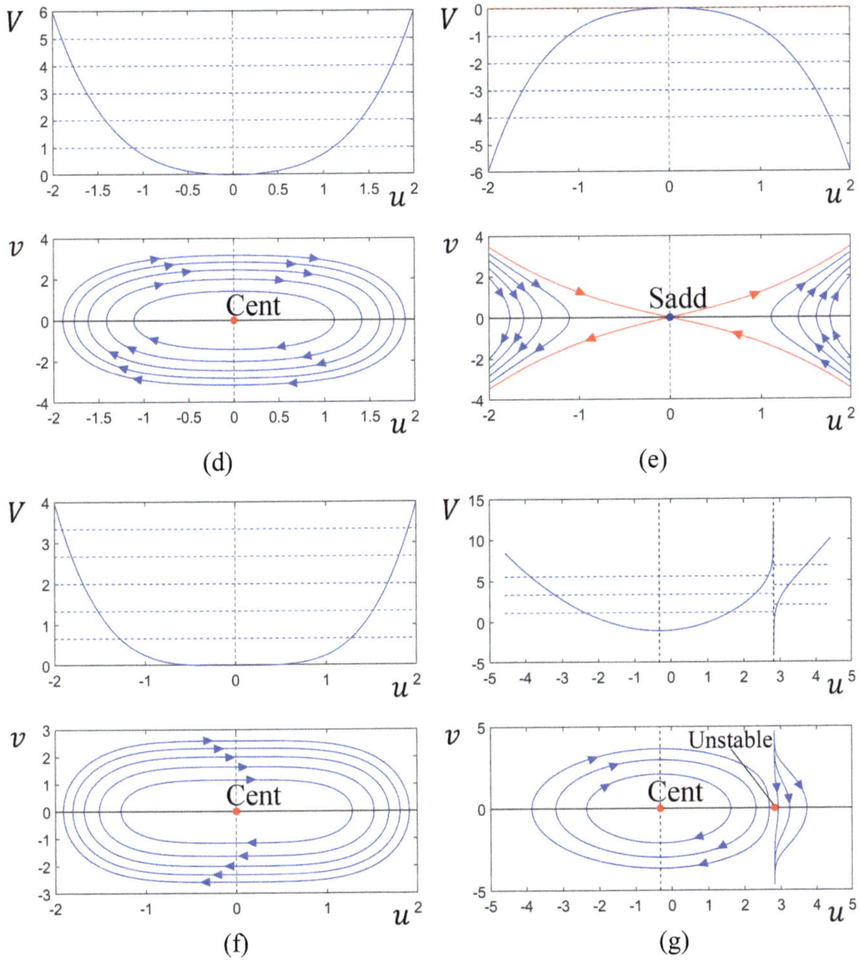

Fig. 2.1 (continued)

$$v^2 - u^2 + \frac{1}{2}u^4 = 2E \tag{2.1.8}$$

$$v = \pm\sqrt{2E + u^2 - \frac{1}{2}u^4} \tag{2.1.9}$$

$$f(u) = -u + u^3, \quad V(u) = \int_0^u f(u)du = -\frac{1}{2}u^2 + \frac{1}{4}u^4 \tag{2.1.10}$$

Singular points of the system:

$$s_1: \ (0,0), \ s_2: \ (1,0), \ s_3: \ (-1,0)$$

The potential energy curve and phase trajectories of the system are shown in Fig. 2.1c.

(d)

$$\frac{dv}{du} = \frac{-u - u^3}{v} \tag{2.1.11}$$

$$v^2 + u^2 + \frac{1}{2}u^4 = 2E \tag{2.1.12}$$

$$v = \pm\sqrt{2E - u^2 - \frac{1}{2}u^4} \tag{2.1.13}$$

$$f(u) = u + u^3, \quad V(u) = \int_0^u f(u)du = \frac{1}{2}u^2 + \frac{1}{4}u^4 \tag{2.1.14}$$

Singular points of the system:

$$s: \ (0,0)$$

The potential energy curve and phase trajectories of the system are shown in Fig. 2.1d.

(e)

$$\frac{dv}{du} = \frac{u + u^3}{v} \tag{2.1.15}$$

$$v^2 - u^2 - \frac{1}{2}u^4 = 2E \tag{2.1.16}$$

$$v = \pm\sqrt{2E + u^2 + \frac{1}{2}u^4} \tag{2.1.17}$$

$$f(u) = -u - u^3, \quad V(u) = \int_0^u f(u)du = -\frac{1}{2}u^2 - \frac{1}{4}u^4 \tag{2.1.18}$$

Singular points of the system:

$$s: \ (0,0)$$

The potential energy curve and phase trajectories of the system are shown in Fig. 2.1e.

(f)

$$\frac{dv}{du} = \frac{-u^3}{v} \tag{2.1.19}$$

$$v^2 + \frac{1}{2}u^4 = 2E \tag{2.1.20}$$

$$v = \pm\sqrt{2E - \frac{1}{2}u^4} \tag{2.1.21}$$

$$f(u) = -u + u^3, \quad V(u) = \int_0^u f(u)du = \frac{1}{4}u^4 \tag{2.1.22}$$

Singular points of the system:

$$s: \ (0,0)$$

The potential energy curve and phase trajectories of the system are shown in Fig. 2.1f.

(g)

$$\frac{dv}{du} = \frac{-u - \frac{\lambda}{u-a}}{v} \tag{2.1.23}$$

This yields that the singular points of the system are

$$s_1: \left(\frac{a + \sqrt{a^2 - 4\lambda}}{2}, 0\right), s_2: \left(\frac{a - \sqrt{a^2 - 4\lambda}}{2}, 0\right) \tag{2.1.24}$$

$$\text{So, only when} a^2 \geq 4\lambda \tag{2.1.25}$$

singular points are real and

$$\frac{1}{2}v^2 + \frac{1}{2}u^2 + \begin{cases} \lambda ln(u-a), & u-a>0 \\ -\lambda ln(a-u), & u-a<0 \end{cases} = E \tag{2.1.26}$$

$$\begin{aligned} v &= \pm\sqrt{2E - u^2 - 2\lambda ln(u-a)}, u-a>0v \\ &= \pm\sqrt{2E - u^2 + 2\lambda ln(a-u)}, u-a<0 \end{aligned} \tag{2.1.27}$$

$$\begin{aligned} f(u) &= u + \frac{\lambda}{u-a}, V(u) = \int_0^u f(u)du \\ &= \begin{cases} \frac{1}{2}u^2 + \lambda ln(u-a), u-a>0 \\ \frac{1}{2}u^2 - \lambda ln(a-u), u-a<0 \end{cases} \end{aligned} \tag{2.1.28}$$

Let $a = \sqrt{8}, \lambda = 1$, then

$$s_1: \quad (\sqrt{2}+1,\ 0), \quad s_2: \quad (\sqrt{2}-1,\ 0) \tag{2.1.29}$$

$$v = \pm\sqrt{2E - u^2 - 2ln(u - \sqrt{8})}, \quad u - \sqrt{8} > 0$$

$$v = \pm\sqrt{2E - u^2 + 2ln(\sqrt{8} - u)}, \quad u - \sqrt{8} < 0 \tag{2.1.30}$$

$$V(u) = \begin{cases} \frac{1}{2}u^2 + ln(u - \sqrt{8}), & u - \sqrt{8} > 0 \\ \frac{1}{2}u^2 - ln(\sqrt{8} - u), & u - \sqrt{8} < 0 \end{cases} \tag{2.1.31}$$

The potential energy curve and phase trajectories of the system are shown in Fig. 2.1g.

2.2 Exercise 2.2 (Solving Conservative Systems by the Method of Multiple Scales or Straightforward Expansion Combined with Reformulation)

Solution: **(a)** Expanding the equation around the singularity $u = 0$ yields

$$\ddot{u} + \omega_0^2 \frac{u + u^3 - u^3}{1 + u^2} = \ddot{u} + \omega_0^2 (u - u^3 + \frac{u^5}{1 + u^2}) \approx \ddot{u} + \omega_0^2 u - \omega_0^2 u^3 = 0$$

$$\text{i.e} \quad \ddot{u} + \omega_0^2 u - \omega_0^2 u^3 = 0 \tag{2.2.1}$$

We introduce a small, dimensionless parameter ε, which is the order of the amplitude of the motion u. Therefore, we can write

$$u = \varepsilon v$$

then Eq. (2.2.1) changes to

$$\ddot{v} + \omega_0^2 v - \varepsilon^2 \omega_0^2 v^3 = 0 \tag{2.2.2}$$

The method of multiple scales. We first solve the Eq. (2.2.2) using the the method of multiple scales. Let the expansion of the solution be

$$v = v_0(T_0, T_1, T_2) + \varepsilon v_1(T_0, T_1, T_2) + \varepsilon^2 v_2(T_0, T_1, T_2) + \cdots \tag{2.2.3}$$

where $T_n = \varepsilon^n t$. Consider

$$\begin{aligned}
\tfrac{d}{dt_2} &= D_0 + \varepsilon D_1 + \varepsilon^2 D_2 + \cdots \\
\tfrac{d^2}{dt^2} &= (D_0 + \varepsilon D_1 + \varepsilon^2 D_2 + \cdots)^2 = D_0^2 + 2\varepsilon D_0 D_1 + \varepsilon^2 (D_1^2 + 2D_0 D_2) + \cdots
\end{aligned} \tag{2.2.4}$$

Substituting (2.2.3) into Eq. (2.2.1), we can obtain

$$\begin{aligned}
0 &= \ddot{v} + \omega_0^2 v - \varepsilon \omega_0^2 v^3 \\
&= [D_0^2 + 2\varepsilon D_0 D_1 + \varepsilon^2 (D_1^2 + 2D_0 D_2)](v_0 + \varepsilon v_1 + \varepsilon^2 v_2) \\
&\quad + \omega_0^2 (v_0 + \varepsilon v_1 + \varepsilon^2 v_2) - \varepsilon^2 \omega_0^2 (v_0 + \varepsilon v_1 + \varepsilon^2 v_2)^3 + \cdots \\
&= D_0^2 v_0 + \omega_0^2 v_0 + \varepsilon (D_0^2 v_1 + \omega_0^2 v_1 + 2D_0 D_1 v_0) \\
&\quad + \varepsilon^2 [D_0^2 v_2 + \omega_0^2 v_2 + 2D_0 D_1 v_1 + (D_1^2 + 2D_0 D_2) v_0 - \omega_0^2 v_0^3] + \cdots
\end{aligned} \tag{2.2.5}$$

Equating the coefficients of ε, $\varepsilon\ ^2$, and $\varepsilon\ ^3$ to zero, we obtain

$$D_0^2 v_0 + \omega_0^2 v_0 = 0 \tag{2.2.6}$$

$$D_0^2 v_1 + \omega_0^2 v_1 = -2D_0 D_1 v_0 \tag{2.2.7}$$

$$D_0^2 v_2 + \omega_0^2 v_2 = -2D_0 D_1 v_1 - (D_1^2 + 2D_0 D_2) v_0 + \omega_0^2 v_0^3 \tag{2.2.8}$$

The solution of Eq. (2.2.6) is

$$v_0 = A(T_1, T_2) e^{i\omega_0 T_0} + cc \tag{2.2.9}$$

Substituing (2.2.9) into the Eq. (2.2.7), we obtain

$$D_0^2 v_1 + \omega_0^2 v_1 = -2i\omega_0 D_1 A e^{i\omega_0 T_0} + cc \tag{2.2.10}$$

where cc denotes the complex conjugate of the preceding terms. In order to eliminate the secular terms from the above equation, there must be

$$D_1 A = 0 \quad \Rightarrow \quad A = A(T_2) \tag{2.2.11}$$

Substituing (2.2.11) into the Eq. (2.2.10), we obtain

$$v_1 = 0 \tag{2.2.12}$$

Substituting (2.2.9) and (2.2.12) into the Eq. (2.2.8), yields

$$vD_0^2 v_2 + \omega_0^2 v_2 = -2D_0 D_2 v_0 + \omega_0^2 v_0^3$$

$$= (-2i\omega_0 D_2 A + 3\omega_0^2 A^2 \overline{A})e^{i\omega_0 T_0} + cc + NST \tag{2.2.13}$$

where *NST* denotes terms which do not generate secular terms. In order to eliminate the secular term from the above equation, there must be

$$-2iD_2 A + 3\omega_0 A^2 \overline{A} = 0 \tag{2.2.14}$$

$$\text{Let} \quad A = \frac{1}{2} a e^{i\beta} \tag{2.2.15}$$

and put (2.2.15) into (2.2.14), we obtain

$$i - a' + a\beta' + \frac{3}{8}\omega_0 a^3 = 0 \tag{2.2.16}$$

Separating the result into real and imaginary parts, we obtain

$$a' = 0, a\beta' + \frac{3}{8}\omega_0 a^3 = 0 \tag{2.2.17}$$

Therefore,

$$a = \text{constant}, \quad \beta = -\frac{3}{8}\omega_0 a^2 t + \beta_0 \tag{2.2.18}$$

From the above results, the two-term approximation of the solution can be obtained as

$$v = a\cos[(1 - \frac{3}{8}a^2)\omega_0 t + \beta_0] + \cdots \tag{2.2.19}$$

Therefore, the frequency-amplitude relationship of the oscillation is

$$\omega = (1 - \frac{3}{8}a^2)\omega_0 \tag{2.2.20}$$

Straightforward Expansion Combined with Reformulation

We then use the straightforward expansion combined with reformulation to solve the Eq. (2.2.2). Let the expansion of the solution be

$$v = v_0(t) + \hat{\varepsilon} v_1(t) + \cdots \tag{2.2.21}$$

where $\hat{\varepsilon} = \varepsilon^2$. Substitute (2.2.21) into the Eq. (2.2.2) and equate the coefficients of the same power of $\hat{\varepsilon}$ in the result, giving

$$\text{i.e.} \quad \ddot{v}_0 + \omega_0^2 v_0 = 0 \tag{2.2.22}$$

$$\ddot{v}_1 + \omega_0^2 v_1 = \omega_0^2 v_0^3 \tag{2.2.23}$$

The solution of Eq. (2.2.22) can be expressed as

$$v_0 = a\cos(\omega_0 t + \beta) \tag{2.2.24}$$

Then the Eq. (2.2.23) can be written as

$$\begin{aligned} \ddot{v}_1 + \omega_0^2 v_1 &= \omega_0^2 a^3 \cos^3(\omega_0 t + \beta) \\ &= \frac{3}{4}\omega_0^2 a^3 \cos(\omega_0 t + \beta) + \frac{1}{4}\omega_0^2 a^3 \cos(3\omega_0 t + 3\beta) \end{aligned} \tag{2.2.25}$$

whose particular solution is

$$v_1 = \frac{3}{8}\omega_0 a^3 t\sin(\omega_0 t + \beta) - \frac{1}{32}a^3 \cos(3\omega_0 t + 3\beta) \tag{2.2.26}$$

Therefore, the solution of Eq. (2.2.2) is

$$v = a\cos(\omega_0 t + \beta) + \hat{\varepsilon}[\frac{3}{8}\omega_0 a^3 t\sin(\omega_0 t + \beta) - \frac{1}{32}a^3 \cos(3\omega_0 t + 3\beta)] + \cdots \tag{2.2.27}$$

This solution is not valid when $.t \geq O(\hat{\varepsilon}^{-1})$

Now we apply the reformulation method to make the above expansion uniformly valid. Let

$$\tau = \Omega t = (\omega_0 + \hat{\varepsilon}\Omega_1 + \cdots)t \quad \Rightarrow \omega_0 t = (1 - \frac{\hat{\varepsilon}\Omega_1}{\omega_0} + \cdots)\tau \tag{2.2.28}$$

put (2.2.28) into (2.2.27), and collect the result according to the power of $\hat{\varepsilon}$:

$$\begin{aligned} lv &= a\cos(\tau + \beta) \\ &+ \hat{\varepsilon}[\frac{\Omega_1 a}{\omega_0}\tau\sin(\tau + \beta) + \frac{3}{8}a^3 \tau\sin(\tau + \beta) - \frac{1}{32}a^3 \cos(3\tau + 3\beta)] + \cdots \end{aligned} \tag{2.2.29}$$

We select Ω_1 to eliminate secular terms of the Eq. (2.2.29) and this gives us

$$\Omega_1 = -\frac{3}{8}a^2 \omega_0 \tag{2.2.30}$$

Therefore, the two-term approximation of the solution of Eq. (2.2.1) is

$$u = \varepsilon a\cos(\tau + \beta) + \cdots = \varepsilon a\cos[(1 - \frac{3}{8}\varepsilon^2 a^2)\omega_0 t + \beta] + \cdots \tag{2.2.31}$$

$$\text{i.e.} \quad u = a\cos(\omega t + \beta) \tag{2.2.32}$$

where the frequency-amplitude relationship of the oscillation is

$$\omega = (1 - \frac{3}{8}a^2)\omega_0 \tag{2.2.33}$$

This result is the same as that of the multiscale method.

(b) Straightforward expansion combined with reformulation. Let

$$u = \varepsilon^{\lambda} v$$

Then the original equation becomes

$$\ddot{v} + \omega_0^2 v + \alpha\varepsilon^{4\lambda} v^5 = 0 \tag{2.2.34}$$

If we take $\lambda = \frac{1}{4}$, then the coefficients of the nonlinear term are of order ε and the Eq. (2.2.34) becomes

$$\ddot{v} + \omega_0^2 v + \varepsilon\alpha v^5 = 0 \tag{2.2.35}$$

We assume that the solution of the above equation can be represented by an expansion having the form

$$v = v_0(t) + \varepsilon v_1(t) + \cdots \tag{2.2.36}$$

Then we substitute (2.2.36) into the Eq. (2.2.35) and set the coefficient of each power of ε equal to zero. This leads to the following set of equations

$$\ddot{v}_0 + \omega_0^2 v_0 = 0 \tag{2.2.37}$$

$$\ddot{v}_1 + \omega_0^2 v_1 = -\alpha v_0^5 \tag{2.2.38}$$

The general solution of Eq. (2.2.37) can be written in the form

$$v_0 = a\cos(\omega_0 t + \beta) \tag{2.2.39}$$

therefore, (2.2.38) becomes

$$\ddot{v}_1 + \omega_0^2 v_1 = -\alpha a^5 \cos^5(\omega_0 t + \beta)$$

$$= -\frac{5}{8}\alpha a^5\cos(\omega_0 t + \beta) - \frac{5}{16}\alpha a^5\cos(3\omega_0 t + 3\beta) - \frac{1}{16}\alpha a^5\cos(5\omega_0 t + 5\beta) \tag{2.2.40}$$

The particular solution of the above equation is

$$u = \varepsilon^{1/4} a\cos(\omega_0 t + \beta) + \varepsilon^{5/4}[-\frac{5}{16\omega_0}\alpha a^5 t\sin(\omega_0 t + \beta) + \frac{5}{128}\alpha a^5\cos(3\omega_0 t + 3\beta) + \frac{1}{384}\alpha a^5\cos(5\omega_0 t + 5\beta)] \tag{2.2.41}$$

Therefore, the solution of the original equation is

$$u = \varepsilon^{1/4} a\cos(\omega_0 t + \beta) + \varepsilon^{5/4}[-\frac{5}{16\omega_0}\alpha a^5 t\sin(\omega_0 t + \beta) + \frac{5}{128}\alpha a^5\cos(3\omega_0 t + 3\beta) + \frac{1}{384}\alpha a^5\cos(5\omega_0 t + 5\beta)] \tag{2.2.42}$$

This solution is not valid when .$t \geq O(\varepsilon^{-1})$

Now we apply the reformulation method to make the above expansions uniformly valid. Let

$$\tau = \Omega t = (\omega_0 + \varepsilon\Omega_1 + \cdots)t \quad \Rightarrow \omega_0 t = (1 - \frac{\varepsilon\Omega_1}{\omega_0} + \cdots)\tau \tag{2.2.43}$$

Substituting (2.2.28) into (2.2.27) leads to

$$u = \varepsilon^{1/4} a\cos(\tau + \beta) + \varepsilon^{5/4}[-\frac{5}{16}\frac{\alpha a^5}{\omega_0^2}\tau\sin(\tau + \beta) + \frac{\Omega_1 a}{\omega_0}\tau\sin(\tau + \beta) + \frac{5}{128}\alpha a^5\cos(3\tau + 3\beta) + \frac{1}{384}\alpha a^5\cos(5\tau + 5\beta) + \cdots] \tag{2.2.44}$$

To eliminate the secular terms from (2.2.29), we must put

$$\Omega_1 = \frac{5}{16\omega_0}\alpha a^4 \tag{2.2.45}$$

Therefore, the first-order approximation of the solution is

$$u = u = \varepsilon^{1/4} a\cos(\tau + \beta) + \cdots = \varepsilon^{1/4} a\cos[(\omega_0 + \frac{5}{16\omega_0}\alpha\varepsilon a^4)t + \beta] + \cdots \tag{2.2.46}$$

Let $\varepsilon = 1$, Eq. (2.2.46) becomes

$$u = a\cos(\omega t + \beta) \tag{2.2.47}$$

where the relationship between the frequency and the amplitude is

$$\omega = (\omega_0 + \frac{5}{16\omega_0}\alpha a^4) \tag{2.2.48}$$

(c) The method of multiple scales. The system described by Exercise 2.1(c) has three singularities (0,0), (−1,0), (1,0), of which (0,0) is the saddle point and the remaining two are the centers. The motion is oscillatory in the neighborhood of a center. Here we discuss the oscillatory motion near the singularity (1,0). It is convenient to shift the origin to the location of the center, $u = 1$. Thus, we let $x = u - 1$, then the equation in Exercise 2.1(c) can be written as

$$\ddot{x} + 2x + 3x^2 + x^3 = 0 \tag{2.2.49}$$

Let

$$u = \varepsilon^{\lambda} v$$

Equation (2.2.49) changes to

$$\ddot{v} + 2v + 3\varepsilon^{\lambda} v^2 + \varepsilon^{2\lambda} v^3 = 0 \tag{2.2.50}$$

If we take $\lambda = 1$ and write the Eq. (2.2.50) in the following form:

$$\ddot{v} + \omega_0^2 v + \varepsilon \delta v^2 + \varepsilon^2 \alpha v^3 = 0 \tag{2.2.51}$$

$$\text{where } \omega_0^2 = 2, \quad \delta = 3, \quad \alpha = 1 \tag{2.2.52}$$

Assume the solution of the above nonlinear equation can be represented by an expansion having the form

$$v = v_0(T_0, T_1, T_2) + \varepsilon v_1(T_0, T_1, T_2) + \varepsilon^2 v_2(T_0, T_1, T_2) + \cdots \tag{2.2.53}$$

where $T_0 = t, \quad T_1 = \varepsilon t$. Substitute (2.2.53) into (2.2.51), we obtain

$$\begin{aligned}
0 &= \ddot{v} + \omega_0^2 v + \varepsilon \delta v^2 + \varepsilon^2 \alpha v^3 \\
&= [D_0^2 + 2\varepsilon D_0 D_1 + \varepsilon^2 (D_1^2 + 2D_0 D_2) + \cdots](v_0 + \varepsilon v_1 + \varepsilon^2 v_2 + \cdots) \\
&\quad + \omega_0^2 (v_0 + \varepsilon v_1 + \varepsilon^2 v_2 + \cdots) + \varepsilon \delta (v_0 + \varepsilon v_1 + \varepsilon^2 v_2 + \cdots)^2 \\
&\quad + \alpha \varepsilon^2 (v_0 + \varepsilon v_1 + \varepsilon^2 v_2 + \cdots)^3 \\
&= D_0^2 v_0 + \omega_0^2 v_0 + \varepsilon (D_0^2 v_1 + 2D_0 D_1 v_0 + \omega_0^2 v_1 + \delta v_0^2) \\
&\quad + \varepsilon^2 [D_0^2 v_2 + 2D_0 D_1 v_1 + (D_1^2 + 2D_0 D_2) v_0 + \omega_0^2 v_2 + 2\delta v_0 v_1 + \alpha v_0^3] + \cdots
\end{aligned} \tag{2.2.54}$$

Set the coefficient of each power of ε equal to zero yield

$$D_0^2 v_0 + \omega_0^2 v_0 = 0 \tag{2.2.55}$$

$$D_0^2 v_1 + \omega_0^2 v_1 = -2D_0 D_1 v_0 - \delta v_0^2 \tag{2.2.56}$$

$$D_0^2 v_2 + \omega_0^2 v_2 = -2D_0 D_1 v_1 - (D_1^2 + 2D_0 D_2) v_0 - 2\delta v_0 v_1 - \alpha v_0^3 \tag{2.2.57}$$

The homogeneous solution of Eq. (2.2.55) is

$$v_0 = A e^{i\omega_0 T_0} + \overline{A} e^{-i\omega_0 T_0} \tag{2.2.58}$$

where $A = A(T_1, T_2)$. Substituting the above equation into (2.2.56) yielding

$$\begin{aligned} &D_0^2 v_1 + \omega_0^2 v_1 = -2D_0 D_1 (A e^{i\omega_0 T_0} + \overline{A} e^{-i\omega_0 T_0}) - \delta (A e^{i\omega_0 T_0} + \overline{A} e^{-i\omega_0 T_0})^2 \\ &= -\delta A\overline{A} - 2i\omega_0 D_1 A e^{i\omega_0 T_0} - \delta A^2 e^{2i\omega_0 T_0} + cc \end{aligned} \tag{2.2.59}$$

where cc denotes the complex conjugate of the preceding terms. To eliminate the secular terms in the above equation, we need $D_1 A = 0$, so $A = A(T_2)$. The particular solution of Eq. (2.2.59) is

$$v_1 = -\frac{\delta A\overline{A}}{\omega_0^2} + \frac{\delta A^2}{3\omega_0^2} e^{2i\omega_0 T_0} + cc \tag{2.2.60}$$

Substituting (2.2.58) and (2.2.60) into (2.2.57), and taking $A = A(T_2)$ into account, we obtain

$$\begin{aligned} &D_0^2 v_2 + \omega_0^2 v_2 = -2D_0 D_1 v_1 - (D_1^2 + 2D_0 D_2) v_0 - 2\delta v_0 v_1 - \alpha v_0^3 \\ &= [-2i\omega_0 D_2 A + (\tfrac{10\delta^2}{3\omega_0^2} - 3\alpha) A^2 \overline{A}] e^{i\omega_0 T_0} + cc + NST \end{aligned} \tag{2.2.61}$$

where cc denotes the complex conjugate of the preceding terms and NST denotes terms which do notgenerate secular terms. In order to eliminate the secular terms in the above equation, we need

$$2i\omega_0 D_2 A - (\frac{10\delta^2}{3\omega_0^2} - 3\alpha) A^2 \overline{A} = 0 \tag{2.2.62}$$

$$\text{Let } A = \frac{1}{2} a e^{i\beta} \tag{2.2.63}$$

Substituting (2.2.63) into (2.2.62) leads to

$$i\omega_0 D_2 a - \omega_0 a D_2 \beta - \frac{1}{8} (\frac{10\delta^2}{3\omega_0^2} - 3\alpha) a^3 = 0 \tag{2.2.64}$$

Separate the above equation into real and imaginary parts and yield

$$D_2 a = 0 \tag{2.2.65}$$

$$D_2\beta + \frac{1}{8\omega_0}(\frac{10\delta^2}{3\omega_0^2} - 3\alpha)a^2 = 0 \tag{2.2.66}$$

Therefore, $a =$ constant and

$$\beta = -\frac{1}{8\omega_0}(\frac{10\delta^2}{3\omega_0^2} - 3\alpha)a^2 T_2 + \beta_0 \tag{2.2.67}$$

where β_0 is a constant.Combining all above results, we can obtain

$$u = \varepsilon a\cos[\omega_0 t - \frac{1}{8\omega_0}(\frac{10\delta^2}{3\omega_0^2} - 3\alpha)a^2\varepsilon^2 t + \beta_0] + \cdots \tag{2.2.68}$$

Let $\varepsilon = 1$ yield

$$u = a\cos[\omega_0 t - \frac{1}{8\omega_0}(\frac{10\delta^2}{3\omega_0^2} - 3\alpha)a^2 t + \beta_0] + \cdots \tag{2.2.69}$$

Therefore, the relation between the oscillation frequency ω and the amplitude a is

$$\omega = \omega_0 - \frac{1}{8\omega_0}(\frac{10\delta^2}{3\omega_0^2} - 3\alpha)a^2 \tag{2.2.70}$$

$$\text{or} \quad \omega^2 = \omega_0^2 + \frac{3}{4}\alpha a^2 - \frac{5}{6\omega_0^2}\delta^2 a^2 \tag{2.2.71}$$

Considering the Eq. (2.2.52), the above equation can be simplified as

$$\omega^2 = 2 - 3a^2 \tag{2.2.72}$$

(d) Let

$$u = \frac{v}{\sqrt{\alpha}} \tag{2.2.73}$$

Then the original equation becomes

$$\ddot{v} + \omega_0^2 v(1 + v^2)^{-1} = 0 \tag{2.2.74}$$

This equation is identical to the equation in subExercise (a), so the first-order approximate solution of the equation is

$$v = a\cos[(1 - \frac{3}{8}a^2)\omega_0 t + \beta] \tag{2.2.75}$$

where

$$\omega = (1 - \frac{3}{8}a^2)\omega_0 \tag{2.2.76}$$

2.3 Exercise 2.3 (Show the System Transformation and the Method of Multiple Scales)

Solution: Since

$$\begin{aligned} \frac{d}{dt}[\frac{m_0\dot{u}}{(1-\dot{u}/c^2)^{1/2}}] &= \frac{m_0\ddot{u}(1-\dot{u}/c^2)^{1/2} + \frac{1}{2}m_0\dot{u}(1-\dot{u}/c^2)^{-1/2}(\ddot{u}/c^2)}{(1-\dot{u}/c^2)} \\ &= \frac{m_0\ddot{u}(1-\frac{1}{2}\dot{u}/c^2)}{(1-\dot{u}/c^2)^{3/2}} = \frac{m_0\ddot{u}c(2c^2-\dot{u})}{2(c^2-\dot{u})^{3/2}} \end{aligned} \tag{2.3.1}$$

the equation can be written as

$$\ddot{u} + \omega_0^2 u \frac{(1-\dot{u}/c^2)^{3/2}}{1-\frac{1}{2}\dot{u}/c^2} = 0 \tag{2.3.2}$$

where $\omega_0^2 = \sqrt{k/m_0}$. We make a Taylor expansion of $(1-\dot{u}/c^2)^{3/2}/(1-\frac{1}{2}\dot{u}/c^2)$ and yield

$$\frac{(1-\dot{u}/c^2)^{3/2}}{1-\frac{1}{2}\dot{u}/c^2} = 1 - \frac{\dot{u}}{c^2} - \frac{1}{8}(\frac{\dot{u}}{c^2})^2 + \cdots \tag{2.3.3}$$

Therefore, the equation can be written as

$$\ddot{u} + \omega_0^2 u - \frac{\omega_0^2}{c^2}u\dot{u} - \frac{\omega_0^2}{8c^4}u\dot{u}^2 = 0 \tag{2.3.4}$$

Let

$$u = \varepsilon^\lambda v$$

then the Eq. (2.3.4) changes to

$$\ddot{v} + \omega_0^2 v - \frac{\omega_0^2}{c^2}\varepsilon^{\lambda} v\dot{v} - \frac{\omega_0^2}{8c^4}\varepsilon^{2\lambda} v\dot{v}^2 = 0 \tag{2.3.5}$$

Let $\lambda = 1$, then the Eq. (2.3.5) becomes

$$\ddot{v} + \omega_0^2 v - \varepsilon\alpha_1 v\dot{v} - \varepsilon^2\alpha_2 v\dot{v}^2 = 0 \tag{2.3.6}$$

where $\alpha_1 = \omega_0^2/c^2$, $\alpha_2 = \omega_0^2/8c^4$.

Assume the solution of the above equation can be represented by an expansion having the form

$$v = v_0(T_0, T_1, T_2) + \varepsilon v_1(T_0, T_1, T_2) + \varepsilon^2 v_2(T_0, T_1, T_2) + \cdots \tag{2.3.7}$$

where $T_0 = t,\quad T_1 = \varepsilon t,\quad T_2 = \varepsilon^2 t$. Substitute (2.3.7) into the Eq. (2.3.6), we obtain

$$\begin{aligned}
0 &= \ddot{v} + \omega_0^2 v - \varepsilon\alpha_1 v\dot{v} - \varepsilon^2\alpha_2 v\dot{v}^2 \\
&= [D_0^2 + 2\varepsilon D_0 D_1 + \varepsilon^2(D_1^2 + 2D_0 D_2) + \cdots](v_0 + \varepsilon v_1 + \varepsilon^2 v_2 + \cdots) \\
&\quad + \omega_0^2(v_0 + \varepsilon v_1 + \varepsilon^2 v_2 + \cdots) \\
&\quad - \alpha_1\varepsilon(v_0 + \varepsilon v_1 + \varepsilon^2 v_2 + \cdots)[(D_0 + \varepsilon D_1 + \cdots)(v_0 + \varepsilon v_1 + \varepsilon^2 v_2 + \cdots)] \\
&\quad - \varepsilon^2\alpha_2(v_0 + \varepsilon v_1 + \varepsilon^2 v_2 + \cdots)[(D_0 + \varepsilon D_1 + \cdots)(v_0 + \varepsilon v_1 + \varepsilon^2 v_2 + \cdots)]^2 \\
&= D_0^2 v_0 + \omega_0^2 v_0 \\
&\quad + \varepsilon(D_0^2 v_1 + \omega_0^2 v_1 + 2D_0 D_1 v_0 - \alpha_1 v_0 D_0 v_0) \\
&\quad + \varepsilon^2[D_0^2 v_2 + \omega_0^2 v_2 + 2D_0 D_1 v_1 + (D_1^2 + 2D_0 D_2)v_0 \\
&\quad - \alpha_1 v_0 D_0 v_1 - \alpha_1 v_1 D_0 v_0 - \alpha_2 v_0 (D_0 v_0)^2] + \cdots
\end{aligned} \tag{2.3.8}$$

Set the coefficient of each power of ε equal to zero yield

$$D_0^2 v_0 + \omega_0^2 v_0 = 0 \tag{2.3.9}$$

$$D_0^2 v_1 + \omega_0^2 v_1 = -2D_0 D_1 v_0 + \alpha_1 v_0 D_0 v_0 \tag{2.3.10}$$

$$D_0^2 v_2 + \omega_0^2 v_2 = -2D_0 D_1 v_1 - (D_1^2 + 2D_0 D_2)v_0 + \alpha_1 v_0 D_0 v_1 + \alpha_1 v_1 D_0 v_0 + \alpha_2 v_0 (D_0 v_0)^2 \tag{2.3.11}$$

The homogeneous solution of Eq. (2.3.9) is

$$v_0 = Ae^{i\omega_0 T_0} + \overline{A}e^{-i\omega_0 T_0} \tag{2.3.12}$$

where $A = A(T_1, T_2)$. Substituting the above equation into (2.3.10), we obtain

$$D_0^2 v_1 + \omega_0^2 v_1 = -2D_0 D_1 (Ae^{i\omega_0 T_0} + \overline{A}e^{-i\omega_0 T_0})$$

$$
+ i\alpha_1\omega_0(Ae^{i\omega_0 T_0} + \overline{A}e^{-i\omega_0 T_0})(Ae^{i\omega_0 T_0} - \overline{A}e^{-i\omega_0 T_0})
$$
$$
= -2i\omega_0 D_1 Ae^{i\omega_0 T_0} + i\alpha_1\omega_0 A^2 e^{2i\omega_0 T_0} + cc \tag{2.3.13}
$$

To eliminate the secular terms in the above equation, we need $D_1 A = 0$, so $A = A(T_2)$. The particular solution of Eq. (2.3.13) is

$$
v_1 = -\frac{i\alpha_1 A^2}{3\omega_0} e^{2i\omega_0 T_0} + cc \tag{2.3.14}
$$

Substituting (2.3.12) and (2.3.14) into (2.3.11), and taking $A = A(T_2)$ into account, we obtain

$$
D_0^2 v_2 + \omega_0^2 v_2 = [-2i\omega_0 D_2 A + (\frac{\alpha_1^2}{3} + \alpha_2\omega_0^2)A^2\overline{A}]e^{i\omega_0 T_0} + cc + NST \tag{2.3.15}
$$

where cc denotes the complex conjugate of its preceding terms and NST denotes terms which do not generate secular terms. In order to eliminate secular terms, we need

$$
2i\omega_0 D_2 A - (\frac{\alpha_1^2}{3} + \alpha_2\omega_0^2)A^2\overline{A} = 0 \tag{2.3.16}
$$

$$
\text{Let} \quad A = \frac{1}{2} a e^{i\beta} \tag{2.3.17}
$$

Substituting (2.3.17) into (2.3.16) leads to

$$
8i\omega_0 D_2 a - 8\omega_0 \sigma D_2 \beta - (\frac{\alpha_1^2}{3} + \alpha_2\omega_0^2)\sigma^3 = 0 \tag{2.3.18}
$$

Separate the above equation into real and imaginary parts and yield

$$
D_2 a = 0 \tag{2.3.19}
$$

$$
-8\omega_0 a D_2 \beta - (\frac{\alpha_1^2}{3} + \alpha_2\omega_0^2)a^3 = 0 \tag{2.3.20}
$$

So a is a constant and

$$
\beta = -\frac{1}{8\omega_0}(\frac{\alpha_1^2}{3} + \alpha_2\omega_0^2)a^2 T_2 + \beta_0 \tag{2.3.21}
$$

Combining the above results, we obtain

$$
u = \varepsilon a \cos\{[\omega_0 - \frac{1}{8\omega_0}(\frac{\alpha_1^2}{3} + \alpha_2\omega_0^2)\varepsilon^2 a^2]t + \beta_0\} + \cdots \tag{2.3.22}
$$

Let $\varepsilon = 1$, the first-order approximate solution of the original equation is

$$u = a\cos(\omega t + \beta_0) + \cdots \tag{2.3.23}$$

where

$$\omega = \omega_0 - \frac{1}{8\omega_0}(\frac{\alpha_1^2}{3} + \alpha_2\omega_0^2)a^2 \tag{2.3.24}$$

Substituting $\alpha_1 = \omega_0^2/c^2$, $\alpha_2 = \omega_0^2/8c^4$ into the above equation, we obtain

$$\omega = \omega_0 - \frac{11}{192}\frac{\omega_0^3}{c^4}a^2 \tag{2.3.25}$$

where $\omega_0 = \sqrt{k/m_0}$.

2.4 Exercise 2.4 (Show the Method of Averaging)

$$\ddot{u} + \omega_0^2 u + u|u| = 0 \tag{2.4.1}$$

Solution: We use the pending power ε^λ, $\lambda > 0$ of the small parameter ε to measure the magnitude of u, i.e.,

$$u = \varepsilon^\lambda v \tag{2.4.2}$$

then the Eq. (2.4.1) changes to

$$\ddot{v} + \omega_0^2 v + \varepsilon^\lambda v|v| = 0 \tag{2.4.3}$$

If we take $\lambda = 1$, then the Eq. (2.3.5) becomes

$$\ddot{v} + \omega_0^2 v + \varepsilon v|v| = 0 \tag{2.4.4}$$

Here we use the method of averaging to solve the Eq. (2.4.4). When $\varepsilon = 0$,

$$v = a\cos(\omega_0 t + \beta) \tag{2.4.5}$$

$$\dot{v} = -\omega_0 a\sin(\omega_0 t + \beta) \tag{2.4.6}$$

here a, β is a constant.

The principle of the method of averaging: when $\varepsilon \neq 0$, the solution of (2.4.4) can still be expressed in the form (2.4.5) and (2.4.6) provided that $a$andβ are considered to be functions of time t rather than constants. Thus, the Eqs. (2.4.5) and (2.4.6) can be viewed as a transformation of $v(t)$ and $\dot{v}(t)$ into the dependent variables $a(t)$ and $\beta(t)$. Differentiating the Eq. (2.4.5) with respect to t gives

$$\dot{v} = \dot{a}\cos\varphi - \omega_0 a\sin\varphi - a\dot{\beta}\sin\varphi, \qquad \varphi = \omega_0 t + \beta \tag{2.4.7}$$

Comparing (2.4.6) and (2.4.7), we find that

$$\dot{a}\cos\varphi - a\dot{\beta}\sin\varphi = 0 \tag{2.4.8}$$

Differentiating (2.4.6) with respect to t yields

$$\ddot{v} = -\omega_0\dot{a}\sin\varphi - \omega_0^2 a\cos\varphi - \omega_0 a\dot{\beta}\cos\varphi \tag{2.4.9}$$

Substituting (2.4.9) and (2.4.5) into the Eq. (2.4.4) yields

$$-\omega_0\dot{a}\sin\varphi - \omega_0 a\dot{\beta}\cos\varphi + \varepsilon a^2\cos\varphi|\cos\varphi| = 0 \tag{2.4.10}$$

Solving (2.4.8) and (2.4.10) for $\dot{a}$ and $\dot{\beta}$, we obtain

$$\dot{a} = \frac{\varepsilon a^2}{\omega_0}\sin\varphi\cos\varphi|\cos\varphi| \tag{2.4.11}$$

$$\dot{\beta} = \frac{\varepsilon a}{\omega_0}\cos^2\varphi|\cos\varphi| \tag{2.4.12}$$

Averaging (2.4.11) and (2.4.12) over the period $2\pi/\omega_0$ and considering $a, \beta, \dot{a}$ and $\dot{\beta}$ to be constants while performing the averaging, we obtain the following equations describing the slow variations of a and β:

$$\begin{aligned}\dot{a} &= \frac{1}{2\pi}\int\limits_0^{2\pi}\frac{\varepsilon a^2}{\omega_0}\sin\phi\cos\phi|\cos\phi|d\phi \\ &= \frac{\varepsilon a^2}{\pi\omega_0}[\int\limits_0^{\pi/2}\sin\phi\cos^2\phi d\phi - \int\limits_{\pi/2}^{\pi}\sin\phi\cos^2\phi d\phi] \\ &= -\frac{\varepsilon a^2}{\pi\omega_0}[\int\limits_1^{0}x^2dx - \int\limits_0^{-1}x^2dx] = 0\end{aligned} \tag{2.4.13}$$

$$\dot{\beta} = \frac{1}{2\pi}\int_0^{2\pi} \frac{\varepsilon a}{\omega_0}\cos^2\phi|\cos\phi|d\phi$$
$$= \frac{\varepsilon a}{\pi\omega_0}[\int_0^{\pi/2} \cos^3\phi d\phi - \int_{\pi/2}^{\pi} \cos^3\phi d\phi]$$
$$= \frac{\varepsilon a}{\pi\omega_0}[\int_0^{1} (1-x^2)dx - \int_{1}^{0} (1-x^2)dx]$$
$$= \frac{2\varepsilon a}{\pi\omega_0}\int_0^1 (1-x^2)dx = \frac{4\varepsilon a}{3\pi\omega_0} \tag{2.4.14}$$

So, a is a constant and

$$\beta = \frac{4\varepsilon a}{3\pi\omega_0}t + \beta_0 \tag{2.4.15}$$

Thus, the first-order approximate solutions of (2.4.1) is

$$u = \varepsilon a\cos[(\omega_0 + \frac{4\varepsilon a}{3\pi\omega_0})t + \beta_0] \tag{2.4.16}$$

Let $\varepsilon = 1$, we can obtain

$$u = a\cos[(\omega_0 + \frac{4a}{3\pi\omega_0})t + \beta_0] \tag{2.4.17}$$

The frequency-amplitude relationship of the oscillation is

$$\omega = \omega_0 + \frac{4a}{3\pi\omega_0} \tag{2.4.18}$$

2.5 Exercise 2.5 (Show the Effect of a Change in the Equilibrium Point of a Nonlinear Spring on the Frequency-Amplitude Relationship)

Solution: (a) Select the position of the mass m when the spring is at rest as the origin of the coordinate, we can obtain the governing equation of the system

$$m\ddot{x} = -f(x), \quad f(x) = k_1 x + k_3 x^3$$

$$\text{i.e.} \quad \ddot{x} + \omega_0^2 x + \alpha x^3 = 0 \tag{2.5.1}$$

$$\text{where } \omega_0^2 = k_1/m, \alpha = k_3/m \tag{2.5.2}$$

Let

$$x = \varepsilon^{\lambda} v \tag{2.5.3}$$

(2.5.1) changes to

$$\ddot{v} + \omega_0^2 v + \alpha \varepsilon^{2\lambda} v^3 = 0 \tag{2.5.4}$$

If we take $\lambda = 1$, then the Eq. (2.5.4) becomes

$$\ddot{v} + \omega_0^2 v + \alpha \varepsilon^2 v^3 = 0 \tag{2.5.5}$$

We use the the method of multiple scales to solve the Eq. (2.5.5). Assume the solution of the above equation can be represented by an expansion having the form

$$v = v_0(T_0, T_1, T_2) + \varepsilon v_1(T_0, T_1, T_2) + \varepsilon^2 v_2(T_0, T_1, T_2) + \cdots \tag{2.5.6}$$

where $T_0 = t, \quad T_1 = \varepsilon t$. Substitute (2.5.6) into (2.5.5), we obtain

$$\begin{aligned}
0 &= \ddot{v} + \omega_0^2 v + \alpha \varepsilon^2 v^3 \\
&= [D_0^2 + 2\varepsilon D_0 D_1 + \varepsilon^2 (D_1^2 + 2D_0 D_2) + \cdots](v_0 + \varepsilon v_1 + \varepsilon^2 v_2 + \cdots) \\
&\quad + \omega_0^2 (v_0 + \varepsilon v_1 + \varepsilon^2 v_2 + \cdots) + \alpha \varepsilon^2 (v_0 + \varepsilon v_1 + \varepsilon^2 v_2 + \cdots)^3 \\
&= D_0^2 v_0 + \omega_0^2 v_0 + \varepsilon (D_0^2 v_1 + 2D_0 D_1 v_0 + \omega_0^2 v_1) \\
&\quad + \varepsilon^2 [D_0^2 v_2 + (D_1^2 + 2D_0 D_2) v_0 + 2D_0 D_1 v_1 + \omega_0^2 v_2 + \alpha v_0^3] + \cdots
\end{aligned} \tag{2.5.7}$$

Let the coefficients of the same power of ε be zero to obtain

$$D_0^2 v_0 + \omega_0^2 v_0 = 0 \tag{2.5.8}$$

$$D_0^2 v_1 + \omega_0^2 v_1 = -2D_0 D_1 v_0 \tag{2.5.9}$$

$$D_0^2 v_2 + \omega_0^2 v_2 = -(D_1^2 + 2D_0 D_2) v_0 - 2D_0 D_1 v_1 - \alpha v_0^3 \tag{2.5.10}$$

The general solution of Eq. (2.5.8) is

$$v_0 = Ae^{i\omega_0 T_0} + \overline{A}e^{-i\omega_0 T_0} \tag{2.5.11}$$

where $A = A(T_1, T_2)$. Substituting it into (2.5.9), we can obtain

$$\begin{aligned} D_0^2 v_1 + \omega_0^2 v_1 &= -2D_0 D_1 (Ae^{i\omega_0 T_0} + \overline{A}e^{-i\omega_0 T_0}) \\ &= -2i\omega_0 D_1 Ae^{i\omega_0 T_0} + cc \end{aligned} \tag{2.5.12}$$

To eliminate the secular term, we need $D_1 A = 0$, so $A = A(T_2)$. Then the particular solution of Eq. (2.5.12) is

$$v_1 = 0 \tag{2.5.13}$$

Substituting (2.5.11) and (2.5.13) into (2.5.10), and taking $A = A(T_2)$ into account, we can obtain

$$D_0^2 v_2 + \omega_0^2 v_2 = (-2i\omega_0 D_2 A - 3\alpha A^2 \overline{A})e^{i\omega_0 T_0} + cc + NST \tag{2.5.14}$$

where cc denotes the complex conjugate term of its preceding terms and NST denotes terms which do not generate secular terms. To eliminate the secular term, we need

$$2i\omega_0 D_2 A + 3\alpha A^2 \overline{A} = 0 \tag{2.5.15}$$

$$\text{Let} \quad A = \frac{1}{2} ae^{i\beta} \tag{2.5.16}$$

and put (2.5.16) into (2.5.15), we can obtain

$$8i\omega_0 D_2 a - 8\omega_0 a D_2 \beta + 3\alpha a^3 = 0 \tag{2.5.17}$$

Separates the real and imaginary parts of the above equation yields

$$D_2 a = 0 \tag{2.5.18}$$

$$-8\omega_0 a D_2 \beta + 3\alpha a^3 = 0 \tag{2.5.19}$$

So a is a constant and

$$\beta = \frac{3\alpha a^2}{8\omega_0} T_2 + \beta_0 \tag{2.5.20}$$

Combining the above results, we can obtain

$$x = \varepsilon a \cos(\omega_0 t + \frac{3\alpha \varepsilon^2 a^2}{8\omega_0} t + \beta_0) + \cdots \tag{2.5.21}$$

Let $\varepsilon = 1$, we can obtain the first-order approximation of the Exercise as the following

$$x = a\cos(\omega_0 t + \frac{3\alpha a^2}{8\omega_0}t + \beta_0) + \cdots \tag{2.5.22}$$

and the the frequency-amplitude relationship of the oscillation is

$$\omega = \omega_0 + \frac{3\alpha a^2}{8\omega_0} \tag{2.5.23}$$

where $\omega_0^2 = k_1/m$, $\alpha = k_3/m$.

(b) Again, choose the position of the mass m when the spring is at rest as the origin of the coordinate, we can obtain the governing equation of the system

$$m\ddot{x} = -f(x) + mg, \quad f(x) = k_1 x + k_3 x^3 \tag{2.5.24}$$

The equilibrium coordinates of the mass Δx are determined by the following equation:

$$k_1 \Delta x + k_3(\Delta x)^3 - mg = 0 \tag{2.5.25}$$

$$\text{Let} \quad x = y + \Delta x \tag{2.5.26}$$

Then the Eq. (2.5.24) becomes

$$m\ddot{y} + [k_1 + 3k_3(\Delta x)^2]y + (3k_3\Delta x)y^2 + k_3 y^3 = 0 \tag{2.5.27}$$

$$\text{i.e} \quad \ddot{y} + \tilde{\omega}_0^2 y + \delta y^2 + \alpha y^3 = 0 \tag{2.5.28}$$

$$\text{where } \tilde{\omega}_0^2 = [k_1 + 3k_3(\Delta x)^2]/m, \quad \delta = 3k_3\Delta x/m, \quad \alpha = k_3/m \tag{2.5.29}$$

Equation (2.5.28) can be changed to the Eq. (2.2.51) in Exercise 2.2(c), the Eq. (2.2.51) the frequency-amplitude relationship of the oscillation is can be obtained from (2.2.70), i.e.,

$$\omega = \tilde{\omega}_0 - \frac{1}{8\tilde{\omega}_0}(\frac{4\delta^2}{3\tilde{\omega}_0^2} - 3\alpha)a^2 \tag{2.5.30}$$

The difference between (2.5.30) and (2.5.23) is caused by the different equilibria of the nonlinear spring in two cases.

2.6 Exercise 2.6 (Rods Sliding on the Smooth Walls)

Solution: Considering the principle of angular impulse and momentum of the rod about the axis of symmetry of the column wall, we can obtain

$$[m\frac{4l^2}{12} + m(R^2 - l^2)]\ddot{\theta} = -mg\sqrt{R^2 - l^2}\sin\theta \tag{2.6.1}$$

$$\text{i.e.}\quad \ddot{\theta} + \frac{g(R^2 - l^2)^{1/2}}{R^2 - \frac{2}{3}l^2}\sin\theta = 0 \tag{2.6.2}$$

When the angle θ is very small, the rod oscillates slightly around the equilibrium position and the Eq. (2.6.2) becomes linear:

$$\ddot{\theta} + \frac{g(R^2 - l^2)^{1/2}}{R^2 - \frac{2}{3}l^2}\theta = 0 \tag{2.6.3}$$

Therefore, the linear natural frequency is

$$\omega_0 = \sqrt{\frac{g(R^2 - l^2)^{1/2}}{R^2 - \frac{2}{3}l^2}} \tag{2.6.4}$$

Expanding $\sin\theta$ in (2.6.2) about $\theta = 0$ and retaining to the third order term, we can obtain

$$\ddot{\theta} + \omega_0^2\theta + \alpha\theta^3 = 0 \tag{2.6.5}$$

where $\alpha = -\omega_0^2/6$. This equation is exactly the same as (2.5.1) in Exercise 2.5(a). According to (2.5.23), the frequency-amplitude relationship of the oscillation is

$$\omega = \omega_0 + \frac{3\alpha a^2}{8\omega_0} = (1 - \frac{a^2}{16})\omega_0 \tag{2.6.6}$$

It can be seen that for a fixed amplitude a, the natural frequency ω of the nonlinear osscillation is proportional to the natural frequency ω_0 of the corresponding linear oscillation. Therefore, the effect of the rod length l on ω can be reflected by the $\omega_0 \sim l$ relationship, shown in Fig. 2.2. Here we rewrite (2.6.4) as

$$\frac{\omega_0}{\sqrt{g/R}} = \sqrt{\frac{[1 - (l/R)^2]^{1/2}}{1 - \frac{2}{3}(l/R)^2}} \tag{2.6.7}$$

It is shown that the natural frequency increases slowly with the increase of the rod length l, and then decreases sharply after reaching the maximum value.

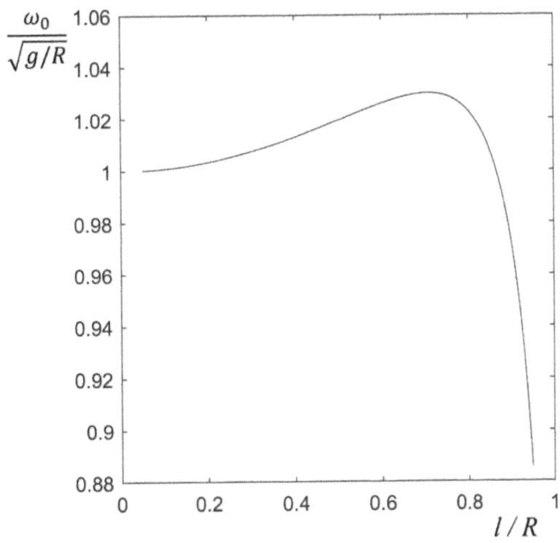

Fig. 2.2 Relationship between the natural frequency and rod length for Exercise 2.6

2.7 Exercise 2.7 (Period of the Osscillation Motion of a Single Pendulum-Type System Along Separatrices)

Solution: (a) The kinetic energy of the column, K, is

$$K = \frac{1}{2}m[(R-r)\dot{\theta}]^2 + \frac{1}{2}\frac{1}{2}mr^2[\frac{(R-r)}{r}\dot{\theta}]^2$$
$$= \frac{1}{2}\frac{3}{2}m(R-r)^2\dot{\theta}^2 \tag{2.7.1}$$

The kinetic energy of the cylinder, V, is

$$V = mg(R-r)(1-\cos\theta) \tag{2.7.2}$$

Substituting them into Lagrange's equation, we can write the differential equation of motion of the cylinder as follows

$$\frac{3}{2}m(R-r)^2\ddot{\theta} + mg(R-r)\sin\theta = 0$$

$$\ddot{\theta} + \frac{2g}{3(R-r)}\sin\theta = 0 \tag{2.7.3}$$

(b) Integrate Eq. (2.7.3) directly, we can obtain

$$\frac{1}{2}\dot{\theta}^2 - \frac{2g}{3(R-r)}\cos\theta = E \tag{2.7.4}$$

where E is the total energy of the cylinder. The minimum angular velocity of the cylinder at point A is $\dot{\theta}(\pi) = 0$, then

$$E = \frac{2g}{3(R - r)} \tag{2.7.5}$$

and Eq. (2.7.4) becomes

$$\frac{1}{2}\dot{\theta}^2 - \frac{2g}{3(R - r)}\cos\theta = \frac{2g}{3(R - r)} \tag{2.7.6}$$

Then the angular velocity of the cylinder at the bottom ($\theta = 0$) can be obtained from the above equation

$$\dot{\theta} = \sqrt{\frac{8g}{3(R - r)}} \tag{2.7.7}$$

(c) The motion of the cylinder is controlled by the Eq. (2.7.6), i.e.,

$$\dot{\theta} = \sqrt{\frac{4g}{3(R - r)}(1 + \cos\theta)} \tag{2.7.8}$$

Therefore, the time required for the cylinder to travel around the circular orbit, T, is

$$\begin{aligned} T &= 2\int_0^{\pi} \frac{d\theta}{\dot{\theta}} = 2\int_0^{\pi} [\frac{4g}{3(R - r)}(1 + \cos\theta)]^{-1/2} d\theta \\ &= \sqrt{\frac{3(R - r)}{g}} \int_0^{\pi} \frac{d\theta}{1 + \cos\theta} = \sqrt{\frac{3(R - r)}{g}} \tan\frac{\theta}{2}\Big|_0^{\pi} \\ &= \infty \end{aligned} \tag{2.7.9}$$

which is infinite because the highest point A of the circular orbit is the saddle point of the cylinder's motion and the cylinder needs to take infinite time to approach the saddle point.

2.8 Exercise 2.8 (Particle on a Rotating Parabola)

Solution: (a) The kinetic energy of the system T is

$$T = \frac{1}{2}m(\dot{x}^2 + \dot{z}^2 + \Omega^2 x^2) = \frac{1}{2}m[(1 + 4p^2x^2)\dot{x}^2 + \dot{\psi}^2 x^2] \tag{2.8.1}$$

where $\dot{\psi} = \Omega$. The potential energy of the system V is

$$V = mgz = mgpx^2 \tag{2.8.2}$$

The differential equation of the motion of the system can be obtained by using Lagrange's equation:

$$\begin{cases} m\ddot{\psi}x^2 + 2m\dot{\psi}\dot{x}x = 0 \\ m(1 + 4p^2x^2)\ddot{x} + 4mp^2x\dot{x}^2 - m\dot{\psi}^2 x = -2mgpx \end{cases}$$

$$\text{i.e.}\dot{\Omega}x + 2\Omega\dot{x} = 0 \tag{2.8.3}$$

$$(1 + 4p^2x^2)\ddot{x} + 4p^2x\dot{x}^2 + (2pg - \Omega^2)x = 0 \tag{2.8.4}$$

(b) Integrate (2.8.3) directly, we can obtain

$$ln\Omega + 2lnx = C \quad \Rightarrow \quad \Omega x^2 = e^C \triangleq \sqrt{H}$$

$$\Omega x^2 = \sqrt{H} \tag{2.8.5}$$

The left side of the above equation represents the angular momentum of the system about the z-axis, so the above equation actually implies the conservation of the angular momentum of the system. Solving Ω from the above equation and substituting it into (2.8.4), we can obtain the governing equation of the system

$$(1 + 4p^2x^2)\ddot{x} + 4p^2x\dot{x}^2 + (2pg - \frac{H}{x^4})x = 0 \tag{2.8.6}$$

(c) The Eq. (2.8.6) can be written as

$$\ddot{x} = -[4p^2x\dot{x}^2 + (2pg - \frac{H}{x^4})x]/(1 + 4p^2x^2) \tag{2.8.7}$$

Let $v = \dot{x}$ (2.8.8)

$$\text{then } \dot{v} = \ddot{x} = -[4p^2xv^2 + (2pg - \frac{H}{x^4})x]/(1 + 4p^2x^2) \tag{2.8.9}$$

from which it follows immediately that

$$\frac{dv}{dx} = -\frac{[4p^2xv^2 + (2pg - \frac{H}{x^4})x]}{v(1 + 4p^2x^2)} \tag{2.8.10}$$

Make a further derivation, we can obtain

$$\frac{dv^2}{dx^2} = -\frac{4p^2v^2 + (2pg - \frac{H}{x^4})}{1 + 4p^2x^2}$$

$$dv^2 + 4p^2x^2dv^2 + 4p^2v^2dx^2 = -(2pg - \frac{H}{x^4})dx^2$$

$$dv^2 + 4p^2d(x^2v^2) = -(2pg - \frac{H}{x^4})dx^2$$

$$d(v^2 + 4p^2x^2v^2) = -d(2pgx^2 + Hx^{-2})$$

$$d(v^2 + 4p^2x^2v^2 + 2pgx^2 + Hx^{-2}) = 0$$

$$v^2 + 4p^2x^2v^2 + 2pgx^2 + Hx^{-2} = h \tag{2.8.11}$$

Now we can discuss the motion of the particle with initial condition $H = H_0$ and $h = h_0$.

(1) $H_0 = 0$

Now Eq. (2.8.11) can be written as

$$v^2 = \frac{h_0 - 2pgx^2}{1 + 4p^2x^2} \tag{2.8.12}$$

The motion of the particle being bounded requires that

$$h_0 - 2pgx^2 \geq 0 \quad \Rightarrow \quad |x| \leq \sqrt{\frac{h_0}{2pg}} \tag{2.8.13}$$

$$\text{So } v^2 \leq h_0 \tag{2.8.14}$$

(2) $H_0 \neq 0$

Now the Eq. (2.8.11) can be written as

$$v^2 = \frac{h_0 - 2pgx^2 - H_0x^{-2}}{1 + 4p^2x^2} \tag{2.8.15}$$

The motion of the particle being bounded requires that

$$h_0 - 2pgx^2 - H_0x^{-2} \geq 0 \tag{2.8.16}$$

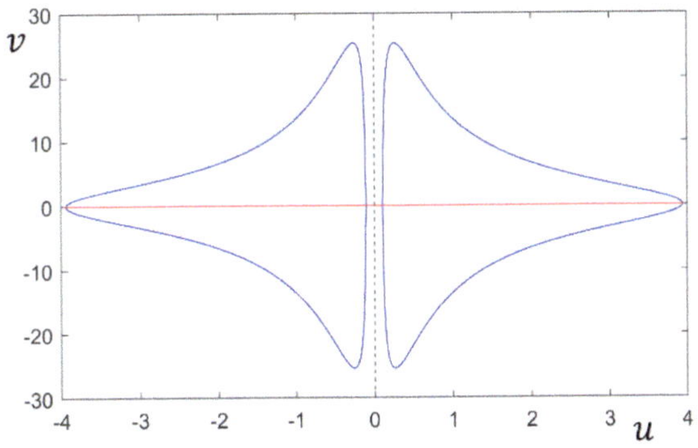

Fig. 2.3 Phase trajectories of the system p = 1, $g = 32.2$, $h = 1000$, $H = 12$ for Exercise 2.8

i.e.,

$$\frac{h_0 - \sqrt{h_0^2 - 8pgH_0}}{4pg} \leq x^2 \leq \frac{h_0 + \sqrt{h_0^2 - 8pgH_0}}{4pg} \tag{2.8.17}$$

(d) Let $p = 1,\ g = 32.2,\ h = 1000,\ H = 12$, we can write the Eq. (2.8.11) as

$$v^2 = \frac{1000 - 64.4x^2 - 12x^{-2}}{1 + 4x^2} \tag{2.8.18}$$

According to (2.8.17), we can obtain

$$-3.9390 \leq x \leq -0.1096 \quad \text{and} \quad 0.1096 \leq x \leq 3.9390 \tag{2.8.19}$$

The trajectories in the phase plane are shown in Fig. 2.3.

2.9 Exercise 2.9 (A Single Pendulum Rotating Uniformly Around a Plumb Axis)

Solution: **(a)** The kinetic energy of the system T is

$$T = \frac{1}{2}m(r^2\dot{\theta}^2 + \Omega^2 r^2 \sin^2\theta) \tag{2.9.1}$$

The potential energy of the system V is

$$V = -mgr\cos\theta \tag{2.9.2}$$

The differential equation of the motion of the pendulum can be obtained by using Lagrange's equation:

$$mr^2\frac{d^2\theta}{dt^2} - m\Omega^2r^2\sin\theta\cos\theta = -mgr\sin\theta$$

$$\text{i.e.}\quad \frac{d^2\theta}{dt^2} - \Omega^2\sin\theta\,\cos\theta + \frac{g}{r}\,\sin\theta = 0 \tag{2.9.3}$$

$$\text{Let}\quad \tau = (g/r)^{1/2}t \tag{2.9.4}$$

(2.9.3) changes to

$$\ddot{\theta} + (1 - \Lambda\cos\theta)\sin\theta = 0 \tag{2.9.5}$$

where $\ddot{\theta} = \partial^2\theta/\partial\tau^2$ and $\Lambda = \Omega^2 r/g$.

(b) Integrate (2.9.5) directly, we can obtain

$$\frac{1}{2}\dot{\theta}^2 + \int(1 - \Lambda\cos\theta)\sin\theta d\theta = E$$

$$\text{i.e.}\quad \frac{1}{2}\dot{\theta}^2 - (1 - \frac{1}{2}\Lambda\cos\theta)\cos\theta = E \tag{2.9.6}$$

Let $\dot{\theta}|\theta = 0 = \dot{\theta}_0$, then

$$E = \frac{1}{2}\dot{\theta}_0^2 - (1 - \frac{1}{2}\Lambda)$$

i.e.$\frac{1}{2}\dot{\theta}^2 = \frac{1}{2}\dot{\theta}_0^2 - [(1 - \frac{1}{2}\Lambda) - (1 - \frac{1}{2}\Lambda\cos\theta)\cos\theta]$

$$\text{Let}\quad F(\theta) = 1 - \frac{1}{2}\Lambda - (1 - \frac{1}{2}\Lambda\cos\theta)\cos\theta \tag{2.9.7}$$

$$\text{then}\ \frac{1}{2}\dot{\theta}^2 = \frac{1}{2}\dot{\theta}_0^2 - F(\theta) \tag{2.9.8}$$

From the Eq. (2.9.5) it is known that the equilibrium point of the system is

$$\Lambda \leq 1; \theta = 0, \quad \pi\ (\text{or}\ -\pi) \tag{2.9.9}$$

$$\Lambda > 1 : \theta = 0, \quad \pi\ (\text{or} - \pi), \quad \pm\theta_\Lambda \tag{2.9.10}$$

where $\theta_\Lambda = arccos(1/\Lambda)$.

The potential energy curve and phase trajectories when $\Lambda \leq 1$ are shown in Fig. 2.4a and b. The equilibrium point $\theta = 0$ is the center, while $\theta = \pi$ (or $-\pi$) is the saddle point. The potential curve and phase trajectories when $\Lambda > 1$ are shown in Fig. 2.4c. The equilibrium points $\theta = 0$ and $\theta = \pi$ (or $-\pi$) are the saddle points, while the equilibrium point $\theta = \pm\theta_\Lambda = \pm arccos(1/\Lambda)$ is the center. From above results, it is clear that $\Lambda = 1$ is the bifurcation point.

(c) Assuming $\Lambda > 1$, the center of the system is

$$\theta = \pm\theta_\Lambda = \pm arccos(1/\Lambda) \tag{2.9.11}$$

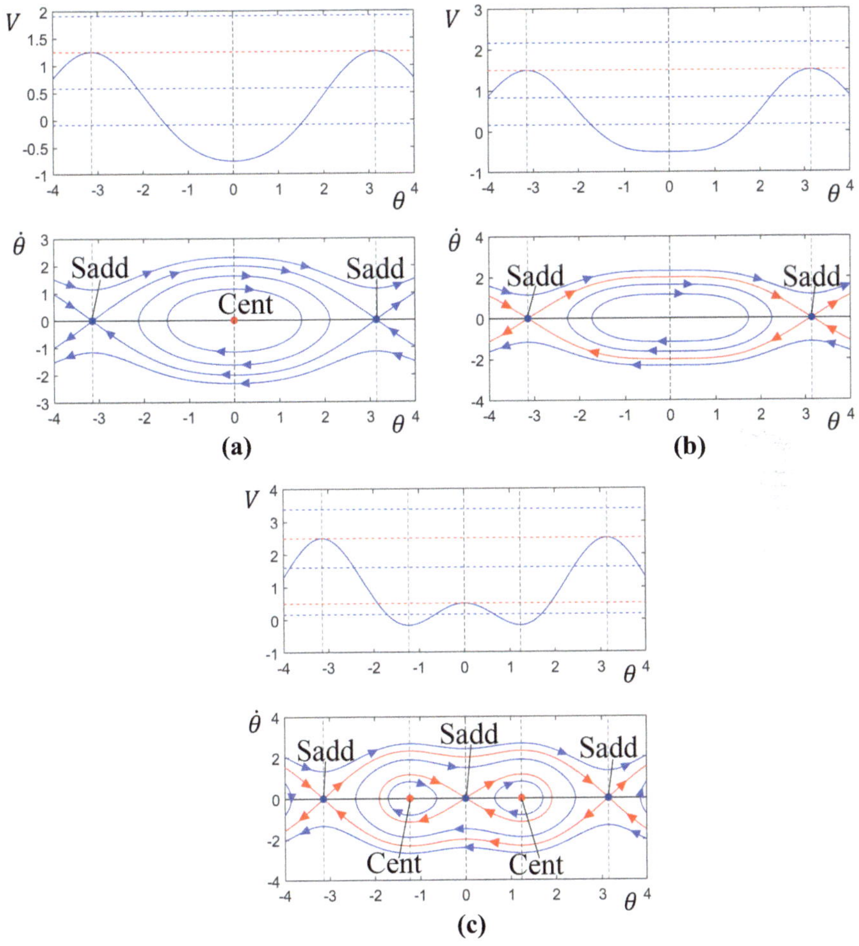

Fig. 2.4 **a** Potential energy curve and phase trajectories when $\Lambda = 0.5$. **b** Potential energy curve and phase trajectories when $\Lambda = 1$ **c** Potential energy curve and phase trajectories when $\Lambda = 3$ for Exercise 2.9

We carry out the expansion of (2.9.5) about $\theta = \theta_\Lambda = arccos(1/\Lambda)$ to $O(\epsilon^3)$ and obtain

$$\begin{aligned}
0 &= \ddot{\theta} + [1 - \Lambda cos(\theta + \theta_\Lambda)]\sin(\theta + \theta_\Lambda) \\
&= \ddot{\theta} + \sin\theta_\Lambda + \theta\cos\theta_\Lambda - \tfrac{1}{2}\theta^2\sin\theta_\Lambda - \tfrac{1}{6}\theta^3\cos\theta_\Lambda \\
&-\tfrac{1}{2}\Lambda[\sin 2\theta_\Lambda + 2\theta\cos 2\theta_\Lambda - \tfrac{1}{2}(2\theta)^2\sin 2\theta_\Lambda - \tfrac{1}{6}(2\theta)^3\cos 2\theta_\Lambda] \\
&= \theta + (\Lambda sin^2\theta_\Lambda)\theta + (\Lambda sin2\theta_\Lambda - \tfrac{1}{2}\sin\theta_\Lambda)\theta^2 \\
&+(\tfrac{2}{3}\Lambda cos2\theta_\Lambda - \tfrac{1}{6}\cos\theta_\Lambda)\theta^3
\end{aligned} \tag{2.9.12}$$

$$\text{Let } \omega_0^2 = \Lambda\sin^2\theta_\Lambda, \quad \alpha = \Lambda\sin 2\theta_\Lambda - \frac{1}{2}\sin\theta_\Lambda, \quad \delta = \frac{2}{3}\Lambda\cos 2\theta_\Lambda - \frac{1}{6}\cos\theta_\Lambda \tag{2.9.13}$$

(2.9.14) becomes

$$\ddot{\theta} + \omega_0^2\theta + \alpha\theta^2 + \delta\theta^3 = 0 \tag{2.9.14}$$

Let

$$\theta = \varepsilon^\lambda v \tag{2.9.15}$$

Equation (2.9.14) changes to $\varepsilon^\lambda\ddot{v} + \omega_0^2\varepsilon^\lambda v + \varepsilon^{2\lambda}\delta v^2 + \varepsilon^{3\lambda}\alpha v^3 = 0$

If we take $\lambda = 1$, the above equation becomes

$$\ddot{v} + \omega_0^2 v + \varepsilon\delta v^2 + \varepsilon^2\alpha v^3 = 0 \tag{2.9.16}$$

Equation (2.9.16) is identical to Eq. (2.2.51), so the frequency-amplitude relationship can be obtained from (2.2.70), i.e.

$$\omega = \omega_0 - \frac{1}{8\omega_0}(\frac{4\delta^2}{3\omega_0^2} - 3\alpha)a^2 \tag{2.9.17}$$

(d) Assuming $\Lambda < 1$, the center of the system is the origin. We carry out the expansion of the Eq. (2.9.5) about the origin to $O(\epsilon^3)$ and obtain

$$\ddot{\theta} + [1 - \Lambda(1 - \frac{1}{2}\theta^2 + \cdots)](\theta - \frac{1}{6}\theta^3) = 0$$

$$\text{i.e.}\ddot{\theta} + (1 - \Lambda)\theta + (\frac{2}{3}\Lambda - \frac{1}{6})\theta^3 = 0 \tag{2.9.18}$$

$$\text{let } \omega_0^2 = 1 - \Lambda, \quad \alpha_1 = \frac{2}{3}\Lambda - \frac{1}{6} \tag{2.9.19}$$

Then this equation is identical to (2.5.1) in Exercise 2.5(a), and therefore, according to (2.5.23), the frequency-amplitude relationship is

$$\omega = \omega_0 + \frac{3\alpha_1 a^2}{8\omega_0} \tag{2.9.20}$$

2.10 Exercise 2.10 (A Single Pendulum Rotating Freely Around a Plumb Axis)

Solution: (a) The kinetic energy of the system T is

$$T = \frac{1}{2}m(r^2\dot{\theta}^2 + \Omega^2 r^2 \sin^2\theta) \tag{2.10.1}$$

where $\Omega = \dot{\psi}$. The kinetic energy of the system V is

$$V = -mgr\cos\theta \tag{2.10.2}$$

The differential equations of the motion of the pendulum can be obtained by using Lagrange's equation:

$$\begin{cases} r^2\sin^2\theta\dot{\Omega} + 2r^2\sin\theta\cos\theta\dot{\theta}\Omega = 0 \\ \ddot{\theta} + (\frac{g}{r} - \Omega^2\cos\theta)\sin\theta = 0 \end{cases} \tag{2.10.3}$$

The first equation can be further integrated as

$$(r\sin\theta)^2\Omega = \sqrt{H} \tag{2.10.4}$$

where $\sqrt{H}$ is associated with the angular momentum of the pendulum about z-axis.

So, the differential equation of motion of the system is

$$(r\sin\theta)^2\Omega = \sqrt{H} \tag{2.10.5}$$

$$\ddot{\theta} + (\frac{g}{r} - \Omega^2\cos\theta)\sin\theta = 0 \tag{2.10.6}$$

(b) Substitute (2.10.5) into (2.10.6) to obtain the equation controlling θ,

$$\ddot{\theta} + \frac{g}{r}(\sin\theta - \Lambda\frac{cot\theta}{\sin^2\theta}) = 0 \tag{2.10.7}$$

where $\Lambda = H/(gr^3)$.

This result is different from the one the question asks us to prove! Therefore, this question is not solved furthermore.

2.11 Exercise 2.11 (Nonlinear Oscillation of a Slide-Slider System)

Solution: **(a)** Let the instantaneous position coordinates of the mass m_2 be y, then there is a constraint relation

$$x^2 + y^2 = l^2 \tag{2.11.1}$$

$$So\ \dot{y} = -\frac{x}{y}\dot{x} \tag{2.11.2}$$

The kinetic energy of the system T is

$$\begin{aligned} T &= \frac{1}{2}m_1\dot{x}^2 + \frac{1}{2}m_2\dot{y}^2 = \frac{1}{2}m_1\dot{x}^2 + \frac{1}{2}m_2\frac{x^2}{y^2}\dot{x}^2 \\ &= \frac{1}{2}m_1\dot{x}^2 + \frac{1}{2}m_2\frac{x^2}{l^2 - x^2}\dot{x}^2 \end{aligned} \tag{2.11.3}$$

The potential energy of the system V is

$$V = m_2 g y + \frac{1}{2}kx^2 = m_2 g\sqrt{l^2 - x^2} + \frac{1}{2}kx^2 \tag{2.11.4}$$

Then the governing equation of the system can be obtained by using Lagrange's equation

$$(m_1 + \frac{m_2 x^2}{l^2 - x^2})\ddot{x} + \frac{m_2 l^2 x\dot{x}^2}{(l^2 - x^2)^2} + kx + m_2 g\frac{x}{(l^2 - x^2)^{1/2}} = 0 \tag{2.11.5}$$

(b) Let $R = m_2/m_1$ and $u = x/l$, from the Eq. (2.11.5), we obtain

$$[1 + \frac{(m_2/m_1)(x/l)^2}{1 - (x/l)^2}]\frac{\ddot{x}}{l} + \frac{m_2 l x\dot{x}^2}{m_1 l^4(1 - x^2/l^2)^2} + \frac{k}{m_1}\frac{x}{l} + \frac{m_2}{m_1}\frac{x}{l^2(1 - x^2/l^2)^{1/2}}g = 0$$

$$(1 + Ru^2)\ddot{u} + Ru\dot{u}^2 + [\frac{k}{m_1} + \frac{Rg}{l(1 - u^2)^{1/2}}]u = 0 \tag{2.11.6}$$

For $|u| << 1$, expanding the last term on the left side of the above equation and retaining it to the third-order yields

$$(1 + Ru^2)\ddot{u} + Ru\dot{u}^2 + \omega_0^2 u + \frac{Rg}{2l}u^3 = 0 \tag{2.11.7}$$

where

$$\omega_0^2 = \frac{k}{m_1} + \frac{Rg}{l}$$

(c) Let

$$u = \varepsilon v \tag{2.11.8}$$

Equation (2.11.7) changes to

$$\ddot{v} + \omega_0^2 v + \varepsilon^2 (Rv^2\ddot{v} + Rv\dot{v}^2 + \frac{Rg}{2l}v^3) = 0 \tag{2.11.9}$$

We seek an expansion of the solution with the following form

$$v = v_0(T_0, T_1, T_2) + \varepsilon v_1(T_0, T_1, T_2) + \varepsilon^2 v_2(T_0, T_1, T_2) + \cdots \tag{2.11.10}$$

where$T_0 = t,\ \ T_1 = \varepsilon t$. Substitute (2.11.10) into (2.11.9), we obtain

$$\begin{aligned}
0 &= \ddot{v} + \omega_0^2 v + \varepsilon^2 (Rv^2\ddot{v} + Rv\dot{v}^2 + \frac{Rg}{2l}v^3) \\
&= [D_0^2 + 2\varepsilon D_0 D_1 + \varepsilon^2 (D_1^2 + 2D_0 D_2) + \cdots](v_0 + \varepsilon v_1 + \varepsilon^2 v_2 + \cdots) \\
&\quad + \omega_0^2 (v_0 + \varepsilon v_1 + \varepsilon^2 v_2 + \cdots) \\
&\quad + \varepsilon^2 [Rv_0^2 D_0^2 v_0 + Rv_0 (D_0 v_0)^2 + \frac{Rg}{2l}v_0^3 + \cdots] \\
&= D_0^2 v_0 + \omega_0^2 v_0 + \varepsilon (D_0^2 v_1 + 2D_0 D_1 v_0 + \omega_0^2 v_1) \\
&\quad + \varepsilon^2 [D_0^2 v_2 + 2D_0 D_1 v_1 + (D_1^2 + 2D_0 D_2) v_0 + \omega_0^2 v_2 \\
&\quad + Rv_0^2 D_0^2 v_0 + Rv_0 (D_0 v_0)^2 + \frac{Rg}{2l}v_0^3] + \cdots
\end{aligned} \tag{2.11.11}$$

Let the coefficients of the same power of ε be zero, we can obtain

$$D_0^2 v_0 + \omega_0^2 v_0 = 0 \tag{2.11.12}$$

$$D_0^2 v_1 + \omega_0^2 v_1 = -2D_0 D_1 v_0 \tag{2.11.13}$$

$$\begin{aligned} D_0^2 v_2 + \omega_0^2 v_2 &= -2D_0 D_1 v_1 - (D_1^2 + 2D_0 D_2) v_0 \\ &-R v_0^2 D_0^2 v_0 - R v_0 (D_0 v_0)^2 - \tfrac{Rg}{2l} v_0^3 \end{aligned} \tag{2.11.14}$$

The general solution of (2.11.12) is

$$v_0 = Ae^{i\omega_0 T_0} + \overline{A}e^{-i\omega_0 T_0} \tag{2.11.15}$$

where $A = A(T_1, T_2)$. Substituting the above equation into (2.11.13), we obtain

$$D_0^2 v_1 + \omega_0^2 v_1 = -2D_0 D_1 (Ae^{i\omega_0 T_0} + \overline{A}e^{-i\omega_0 T_0}) \tag{2.11.16}$$

To eliminate the secular term, we need $D_1 A = 0$. Therefore, $A = A(T_2)$ and the particular solution of Eq. (2.11.13) is

$$v_1 = 0 \tag{2.11.17}$$

Substituting (2.11.15) and (2.11.17) into (2.11.14), and taking $A = A(T_2)$ into account, we obtain

$$\begin{aligned} D_0^2 v_2 + \omega_0^2 v_2 &= -2i\omega_0 D_2 A + 2R\omega_0^2 A^2 \overline{A} e^{i\omega_0 T_0} - \tfrac{3Rg}{2l} A^2 \overline{A} e^{i\omega_0 T_0} \\ &+cc + NST \end{aligned} \tag{2.11.18}$$

where cc denotes the complex conjugate term of its preceding terms and NST denotes terms which do not generate secular terms. In order to eliminate the secular term, we need

$$2i\omega_0 D_2 A - R(2\omega_0^2 - \frac{3g}{2l})A^2\overline{A} = 0 \tag{2.11.19}$$

$$\text{Let} \quad A = \frac{1}{2}ae^{i\beta} \tag{2.11.20}$$

Substitute (2.11.20) into (2.11.19), we obtain

$$i\omega_0 D_2 a - \omega_0 a D_2 \beta - \frac{R}{8}(2\omega_0^2 - \frac{3g}{2l})a^3 = 0 \tag{2.11.21}$$

Separates the real and imaginary parts of the above equation yields

$$D_2 a = 0 \tag{2.11.22}$$

$$D_2 \beta + \frac{R}{8\omega_0}(2\omega_0^2 - \frac{3g}{2l})a^2 = 0 \tag{2.11.23}$$

So a is a constant. Integrate (2.11.23) directly, we obtain

$$\beta = -\frac{R}{8\omega_0}(2\omega_0^2 - \frac{3g}{2l})a^2 T_2 + \beta_0 \tag{2.11.24}$$

The first-order approximation of the solution is then obtained as

$$u = \varepsilon a\cos[\omega_0 t - \frac{R}{8\omega_0}(2\omega_0^2 - \frac{3g}{2l})a^2\varepsilon^2 t + \beta_0] + \cdots \tag{2.11.25}$$

Let $\varepsilon = 1$,

$$u = a\cos[\omega_0 t - \frac{R}{8\omega_0}(2\omega_0^2 - \frac{3g}{2l})a^2 t + \beta_0] + \cdots \tag{2.11.26}$$

Therefore, the frequency-amplitude relationship is

$$\omega = \omega_0 - \frac{R}{8\omega_0}(2\omega_0^2 - \frac{3g}{2l})a^2 \tag{2.11.27}$$

2.12 Exercise 2.12 (Nonlinear Oscillation of a Rope-Block System)

Solution: **(a)** Let the length of the oblique segment of the rope hanging mass m_1 is x, then there is a constraint

$$x^2 = l^2 + y^2 \tag{2.12.1}$$

$$\dot{x} = \frac{y}{x}\dot{y} \tag{2.12.2}$$

The kinetic energy of the system, T, is

$$\begin{aligned} T &= 2 \times \frac{1}{2}m_1\dot{x}^2 + \frac{1}{2}m_2\dot{y}^2 = \frac{1}{2}(2m_1\frac{y^2}{x^2} + m_2)\dot{y}^2 \\ &= \frac{1}{2}(m_2 + \frac{2m_1 y^2}{l^2 + y^2})\dot{y}^2 \end{aligned} \tag{2.12.3}$$

Then calculate the generalized force of the system. For a given virtual displacement δy, we can obtain from (2.12.1) that

$$\delta x = \frac{y}{x}\delta y \tag{2.12.4}$$

The virtual work of the system is

$$\delta W = -2m_1 g\delta x + m_2 g\delta y = (-\frac{2m_1 gy}{x} + m_2 g)\delta y$$

Therefore, the generalized force of the system is

$$Q = -\frac{2m_1 gy}{x} + m_2 g = m_2 g - \frac{2m_1 gy}{(l^2 + y^2)^{1/2}} \tag{2.12.5}$$

Substitute (2.12.3) and (2.12.5) into Lagrange's equation, we can obtain the governing equation of the system

$$m_2 g - \frac{2m_1 gy}{(l^2 + y^2)^{1/2}} = m_2\ddot{y} + \frac{2m_1 y}{l^2 + y^2}(y\ddot{y} + \frac{l^2\dot{y}^2}{l^2 + y^2}) \tag{2.12.6}$$

(**b**) The equilibria of the system can be obtained from (2.12.6) by letting $\dot{y} = \ddot{y} = 0$, i.e.,

$$m_2 g - \frac{2m_1 gy}{(l^2 + y^2)^{1/2}} = 0 \tag{2.12.7}$$

then

$$y_e = \frac{lm_2}{(4m_1^2 - m_2^2)}, \text{ (assume)} 2m_1 > m_2 \tag{2.12.8}$$

(**c**) Let $u = y/l$ and $R = m_1/m_2$, from (2.12.6), we obtain

$$\frac{g}{l} - \frac{2m_1 gy}{m_2 l^2(1 + y^2/l^2)^{1/2}} = \frac{\ddot{y}}{l} + \frac{2m_1 y}{m_2 l^3(1 + y^2/l^2)}(y\ddot{y} + \frac{\dot{y}^2}{1 + y^2/l^2})$$

Finally, the differential equation of motion of the system becomes

$$\frac{g}{l} - \frac{2Rgu}{l(1 + u^2)^{1/2}} = \ddot{u} + \frac{2Ru}{1 + u^2}(u\ddot{u} + \frac{\dot{u}^2}{1 + u^2}) \tag{2.12.9}$$

(**d**) Let

$$u = \frac{y_e}{l} + \eta \text{or } u = u_e + \eta, \quad \text{where } u_e = \frac{y_e}{l} \tag{2.12.10}$$

Substituting this into (2.12.9) and taking (2.12.7) into account, we can obtain

$$-\frac{2Rg\eta}{l[1 + (u_e + \eta)^2]^{1/2}} = \ddot{\eta} + \frac{2R(u_e + \eta)}{1 + (u_e + \eta)^2}[(u_e + \eta)\ddot{\eta} + \frac{\dot{\eta}^2}{1 + (u_e + \eta)^2}]$$

The above equation can be written as

$$-\frac{2Rg}{l}\eta[1+(u_e+\eta)^2]^{3/2} = \ddot{\eta}[1+(u_e+\eta)^2]^2$$

$$+2R(u_e+\eta)\{[1+(u_e+\eta)^2](u_e+\eta)\ddot{\eta}+\dot{\eta}^2\} \tag{2.12.11}$$

Expanding the left side of the equation for the small and finite $|\eta|$, retaining to $O(\eta^3)$ yields

$$\begin{aligned}
&\eta[1+(u_e+\eta)^2]^{3/2}\\
&= b\eta(1+u_e^2+2u_e\eta+\eta^2)[1+u_e^2+2u_e\eta+\eta^2]^{1/2}\\
&= b\eta(b^2+2u_e\eta+\eta^2)[1+b^{-2}(2u_e\eta+\eta^2)]^{1/2}\\
&= b\eta(b^2+2u_e\eta+\eta^2)[1+\frac{1}{2}b^{-2}(2u_e\eta+\eta^2)-\frac{1}{8}b^{-4}(2u_e\eta+\eta^2)^2]\\
&= [b^3\eta+3bu_e\eta^2+\frac{3}{2}b(1+b^{-2}u_e^2)\eta^3]
\end{aligned}$$

where $b^2 = 1+u_e^2$. Expanding the right-hand side of the equation and retaining to $O(\eta^3)$, we obtain

$$\begin{aligned}
&\ddot{\eta}[1+(u_e+\eta)^2]^2+2R(u_e+\eta)\{[1+(u_e+\eta)^2](u_e+\eta)\ddot{\eta}+\dot{\eta}^2\}\\
&= \ddot{\eta}(b^2+2u_e\eta+\eta^2)^2+2R(u_e+\eta)[(b^2+2u_e\eta+\eta^2)(u_e+\eta)\ddot{\eta}+\dot{\eta}^2]\\
&= \ddot{\eta}(b^4+4b^2u_e\eta+4u_e^2\eta^2+2b^2\eta^2)\\
&\quad +2R(u_e+\eta)(b^2u_e\ddot{\eta}+2u_e^2\eta\ddot{\eta}+u_e\eta^2\ddot{\eta}+b^2\eta\ddot{\eta}+2u_e\eta^2\ddot{\eta}+\dot{\eta}^2)\\
&= b^4\ddot{\eta}+4b^2u_e\eta\ddot{\eta}+4u_e^2\eta^2\ddot{\eta}+2b^2\eta^2\ddot{\eta}\\
&\quad +2Rb^2u_e^2\ddot{\eta}+4Ru_e^3\eta\ddot{\eta}+2Ru_e^2\eta^2\ddot{\eta}+2Rb^2u_e\eta\ddot{\eta}+4Ru_e^2\eta^2\ddot{\eta}+2Ru_e\dot{\eta}^2\\
&\quad +2Rb^2u_e\ddot{\eta}\eta+4Ru_e^2\eta^2\ddot{\eta}+2Rb^2\ddot{\eta}\eta^2+2R\dot{\eta}^2\eta\\
&= (b^4+2Rb^2u_e^2)\ddot{\eta}+(4b^2u_e+4Ru_e^3+4Rb^2u_e)\ddot{\eta}\eta+2Ru_e\dot{\eta}^2\\
&\quad +(2b^2+4u_e^2+10Ru_e^2+2Rb^2)\ddot{\eta}\eta^2+2R\dot{\eta}^2\eta
\end{aligned}$$

Substituting these two results into (2.12.11), we obtain

$$\begin{aligned}
&\ddot{\eta}+\frac{2Rb^3g}{l(b^4+2Rb^2u_e^2)}\eta+\frac{6Rbu_eg}{l(b^4+2Rb^2u_e^2)}\eta^2\\
&+\frac{2Ru_e}{(b^4+2Rb^2u_e^2)}\dot{\eta}^2+\frac{3Rbg}{l(b^4+2Rb^2u_e^2)}(1+b^{-2}u_e^2)\eta^3\\
&+\frac{2R}{(b^4+2Rb^2u_e^2)}\dot{\eta}^2\eta+\frac{(4b^2u_e+4Ru_e^3+4Rb^2u_e)}{(b^4+2Rb^2u_e^2)}\ddot{\eta}\eta
\end{aligned}$$

$$+\frac{(2b^2+4u_e^2+10Ru_e^2+2Rb^2)}{(b^4+2Rb^2u_e^2)}\ddot{\eta}\eta^2$$
$$=0$$

Write the above equation as

$$\ddot{\eta}+\omega_0^2\eta+\alpha_1\eta^2+\alpha_2\dot{\eta}^2+\alpha_3\ddot{\eta}\eta+\alpha_4\eta^3+\alpha_5\dot{\eta}^2\eta+\alpha_6\ddot{\eta}\eta^2=0 \tag{2.12.12}$$

where

$$\begin{aligned}
\omega_0^2 &= \frac{2Rb^3g}{l(b^4+2Rb^2u_e^2)}\\
\alpha_1 &= \frac{6Rbu_eg}{l(b^4+2Rb^2u_e^2)}\\
\alpha_2 &= \frac{2Ru_e}{(b^4+2Rb^2u_e^2)}\\
\alpha_3 &= \frac{(4b^2u_e+4Ru_e^3+4Rb^2u_e)}{(b^4+2Rb^2u_e^2)}\\
\alpha_4 &= \frac{3Rbg}{l(b^4+2Rb^2u_e^2)}(1+b^{-2}u_e^2)\\
\alpha_5 &= \frac{2R}{(b^4+2Rb^2u_e^2)}\\
\alpha_6 &= \frac{(2b^2+4u_e^2+10Ru_e^2+2Rb^2)}{(b^4+2Rb^2u_e^2)}
\end{aligned} \tag{2.12.13}$$

Let

$$\eta=\varepsilon v \tag{2.12.14}$$

then Eq. (2.12.12) changes to

$$\ddot{v}+\omega_0^2v+\varepsilon\alpha_1v^2+\varepsilon\alpha_2\dot{v}^2+\varepsilon\alpha_3\ddot{v}v+\varepsilon^2\alpha_4v^3+\varepsilon^2\alpha_5\dot{v}^2v+\varepsilon^2\alpha_6\ddot{v}v^2=0 \tag{2.12.15}$$

We seek an expansion of the following form for the solution of (2.12.15)

$$v=v_0(T_0,T_1,T_2)+\varepsilon v_1(T_0,T_1,T_2)+\varepsilon^2v_2(T_0,T_1,T_2)+\cdots \tag{2.12.16}$$

where $T_0=t,\quad T_1=\varepsilon t$. Substitute (2.12.16) into (2.12.15), we obtain

$$\begin{aligned}
0&=\ddot{v}+\omega_0^2v+\varepsilon\alpha_1v^2+\varepsilon\alpha_2\dot{v}^2+\varepsilon\alpha_3\ddot{v}v+\varepsilon^2\alpha_4v^3+\varepsilon^2\alpha_5\dot{v}^2v+\varepsilon^2\alpha_6\ddot{v}v^2\\
&=[D_0^2+2\varepsilon D_0D_1+\varepsilon^2(D_1^2+2D_0D_2)+\cdots](v_0+\varepsilon v_1+\varepsilon^2v_2+\cdots)\\
&\quad+\omega_0^2(v_0+\varepsilon v_1+\varepsilon^2v_2+\cdots)+\varepsilon\alpha_1(v_0+\varepsilon v_1+\varepsilon^2v_2+\cdots)^2\\
&\quad+\varepsilon\alpha_2[(D_0+\varepsilon D_1+\varepsilon^2D_2+\cdots)(v_0+\varepsilon v_1+\varepsilon^2v_2+\cdots)]^2\\
&\quad+\varepsilon\alpha_3\{[D_0^2+2\varepsilon D_0D_1+\varepsilon^2(D_1^2+2D_0D_2)+\cdots](v_0+\varepsilon v_1+\varepsilon^2v_2+\cdots)\}\\
&\quad\times(v_0+\varepsilon v_1+\varepsilon^2v_2+\cdots)+\varepsilon^2\alpha_4(v_0+\varepsilon v_1+\varepsilon^2v_2+\cdots)^3\\
&\quad+\varepsilon^2\alpha_5[(D_0+\varepsilon D_1+\varepsilon^2D_2+\cdots)(v_0+\varepsilon v_1+\varepsilon^2v_2+\cdots)]^2
\end{aligned}$$

$$
\begin{aligned}
&\times (v_0 + \varepsilon v_1 + \varepsilon^2 v_2 + \cdots) \\
&+ \varepsilon^2 \alpha_6 \{[D_0^2 + 2\varepsilon D_0 D_1 + \varepsilon^2 (D_1^2 + 2D_0 D_2) + \cdots](v_0 + \varepsilon v_1 + \varepsilon^2 v_2 + \cdots)\} \\
&\times (v_0 + \varepsilon v_1 + \varepsilon^2 v_2 + \cdots)^2 \\
= {} & D_0^2 v_0 + \omega_0^2 v_0 \\
&+ \varepsilon [D_0^2 v_1 + 2D_0 D_1 v_0 + \omega_0^2 v_1 + \alpha_1 v_0^2 + \alpha_2 (D_0 v_0)^2 + \alpha_3 v_0 D_0^2 v_0] \\
&+ \varepsilon^2 [D_0^2 v_2 + 2D_0 D_1 v_1 + (D_1^2 + 2D_0 D_2) v_0 + \omega_0^2 v_2 + 2\alpha_1 v_0 v_1 \\
&+ \alpha_2 D_0 v_0 (D_0 v_1 + D_1 v_0) + \alpha_3 v_0 (D_1^2 v_1 + 2D_0 D_2 v_0) \\
&+ \alpha_4 v_0^3 + \alpha_5 v_0 (D_0 v_0)^2 + \alpha_6 v_0^2 D_0^2 v_0] + \cdots
\end{aligned}
\tag{2.12.17}
$$

Let the coefficients of the same power of ε be zero to obtain

$$
D_0^2 v_0 + \omega_0^2 v_0 = 0 \tag{2.12.18}
$$

$$
D_0^2 v_1 + \omega_0^2 v_1 = -2D_0 D_1 v_0 - \alpha_1 v_0^2 - \alpha_2 (D_0 v_0)^2 - \alpha_3 v_0 D_0^2 v_0 \tag{2.12.19}
$$

$$
\begin{aligned}
D_0^2 v_2 + \omega_0^2 v_2 = {} & -2D_0 D_1 v_1 - (D_1^2 + 2D_0 D_2) v_0 - 2\alpha_1 v_0 v_1 \\
& -\alpha_2 D_0 v_0 (D_0 v_1 + D_1 v_0) - \alpha_3 v_0 (D_1^2 v_1 + 2D_0 D_2 v_0) \\
& -\alpha_4 v_0^3 + \alpha_5 v_0 (D_0 v_0)^2 - \alpha_6 v_0^2 D_0^2 v_0
\end{aligned}
\tag{2.12.20}
$$

The general solution of Eq. (2.12.18) is

$$
v_0 = A e^{i\omega_0 T_0} + \overline{A} e^{-i\omega_0 T_0} \tag{2.12.21}
$$

where $A = A(T_1, T_2)$. Substituting the above equation into (2.12.19), we obtain

$$
\begin{aligned}
D_0^2 v_1 + \omega_0^2 v_1 = {} & -2i\omega_0 D_1 A e^{i\omega_0 T_0} - (\alpha_1 - \alpha_2 \omega_0^2 - \alpha_3 \omega_0^2) A\overline{A} \\
& + (-\alpha_1 - \alpha_2 \omega_0^2 + \alpha_3 \omega_0^2) A^2 e^{2i\omega_0 T_0} + cc
\end{aligned}
\tag{2.12.22}
$$

To eliminate secular terms, we need $D_1 A = 0$. Therefore, $A = A(T_2)$ and the particular solution of (2.12.22) is

$$
v_1 = B_1 A\overline{A} + B_2 A^2 e^{2i\omega_0 T_0} + cc \tag{2.12.23}
$$

$$
\text{where } B_1 = -\frac{\alpha_1 - \alpha_2 \omega_0^2 - \alpha_3 \omega_0^2}{\omega_0^2}, \quad B_2 = -\frac{-\alpha_1 - \alpha_2 \omega_0^2 + \alpha_3 \omega_0^2}{3\omega_0^2} \tag{2.12.24}
$$

Substituting (2.12.21) and (2.12.23) into (2.12.20) and taking $A = A(T_2)$ into account, we can obtain

$$D_0^2 v_2 + \omega_0^2 v_2 = (-2i\omega_0 D_2 A - 2\alpha_1 B_1 A^2 \overline{A} - 2\alpha_1 B_2 A^2 \overline{A} - 2\alpha_2 \omega_0^2 B_2 A^2 \overline{A} \\ -3\alpha_4 A^2 \overline{A} - \alpha_5 \omega_0^2 A^2 \overline{A} + 3\alpha_6 \omega_0^2 A^2 \overline{A}) e^{i\omega_0 T_0} + cc + NST \tag{2.12.25}$$

where *cc* denotes the complex conjugate term of its preceding terms and *NST* denotes terms which do not generate secular terms. In order to eliminate secular terms, it is necessary to have

$$2i\omega_0 D_2 A + (2\alpha_1 B_1 + 2\alpha_1 B_2 + 2\alpha_2 \omega_0^2 B_2 + 3\alpha_4 + \alpha_5 \omega_0^2 - 3\alpha_6 \omega_0^2) A^2 \overline{A} = 0 \tag{2.12.26}$$

$$\text{Let} \quad A = \frac{1}{2} a e^{i\beta} \tag{2.12.27}$$

Substitute (2.12.27) into (2.12.26), we obtain

$$i\omega_0 D_2 a - \omega_0 a D_2 \beta \\ + \frac{1}{8}(2\alpha_1 B_1 + 2\alpha_1 B_2 + 2\alpha_2 \omega_0^2 B_2 + 3\alpha_4 + \alpha_5 \omega_0^2 - 3\alpha_6 \omega_0^2) a^3 = 0 \tag{2.12.28}$$

Separate the real and imaginary parts of the above equation yields

$$D_2 a = 0 \tag{2.12.29}$$

$$-\omega_0 a D_2 \beta + \frac{1}{8}(2\alpha_1 B_1 + 2\alpha_1 B_2 + 2\alpha_2 \omega_0^2 B_2 + 3\alpha_4 + \alpha_5 \omega_0^2 - 3\alpha_6 \omega_0^2) a^3 = 0 \tag{2.12.30}$$

So a is a constant. Conduct direct integration of (2.12.30), we obtain

$$\beta = \frac{D}{8\omega_0} a^2 T_2 + \beta_0 \tag{2.12.31}$$

$$\text{where } D = 2\alpha_1 B_1 + 2\alpha_1 B_2 + 2\alpha_2 \omega_0^2 B_2 + 3\alpha_4 + \alpha_5 \omega_0^2 - 3\alpha_6 \omega_0^2 \tag{2.12.32}$$

Combining the above results, the first-order approximation of the solution of the original equation can be written as

$$\eta = \varepsilon a \cos\left(\omega_0 t + \frac{D}{8\omega_0} a^2 \varepsilon^2 t + \beta_0\right) + \cdots \tag{2.12.33}$$

Let $\varepsilon = 1$, we can obtain

$$\eta = a \cos\left(\omega_0 t + \frac{D}{8\omega_0} a^2 t + \beta_0\right) + \cdots \tag{2.12.34}$$

Therefore, the frequency-amplitude relationship of the oscilliation is

$$\omega = \omega_0 + \frac{D}{8\omega_0}a^2 \tag{2.12.35}$$

2.13 Exercise 2.13 (Single Pendulum Attached with a Rolling-Without-Slipping Circular Wheel I)

Solution: (a) The kinetic energy of the system is

$$\begin{aligned} T &= \frac{1}{2}m(l\dot{\theta}\cos\theta - \dot{\theta}r)^2 + \frac{1}{2}m(l\dot{\theta}\sin\theta)^2 \\ &= \frac{1}{2}m(l^2 + r^2 - 2rl\cos\theta)\dot{\theta}^2 \end{aligned} \tag{2.13.1}$$

The potential energy of the system is

$$V = -mgl\cos\theta \tag{2.13.2}$$

Using Lagrange's equation, we can obtain the differential equation of motion of the system

$$(l^2 + r^2 - 2rl\cos\theta)\ddot{\theta} + rl\dot{\theta}^2\sin\theta + gl\sin\theta = 0 \tag{2.13.3}$$

(b) Expand the Eq. (2.13.3) and retain to θ^3 yields

$$\ddot{\theta} + \frac{gl}{(l-r)^2}\theta + \frac{rl}{(l-r)^2}\theta^2\ddot{\theta} + \frac{rl}{(l-r)^2}\dot{\theta}^2\theta - \frac{gl}{6(l-r)^2}\theta^3 = 0 \tag{2.13.4}$$

$$\text{Let } \omega_0^2 = \frac{gl}{(l-r)^2}, \quad \alpha = \frac{rl}{(l-r)^2} \tag{2.13.5}$$

$$\text{then } \ddot{\theta} + \omega_0^2\theta + \alpha\theta^2\ddot{\theta} + \alpha\dot{\theta}^2\theta - \frac{\omega_0^2}{6}\theta^3 = 0 \tag{2.13.6}$$

Let

$$\theta = \varepsilon v \tag{2.13.7}$$

Equation (2.13.6) changes to

$$\ddot{v} + \omega_0^2 v + \varepsilon^2 \alpha v^2 \ddot{v} + \varepsilon^2 \alpha \dot{v}^2 v - \varepsilon^2 \frac{\omega_0^2}{6} v^3 = 0 \tag{2.13.8}$$

We seek an expansion of the following form for the solution of (2.13.8)

$$v = v_0(T_0, T_1, T_2) + \varepsilon v_1(T_0, T_1, T_2) + \varepsilon^2 v_2(T_0, T_1, T_2) + \cdots \tag{2.13.9}$$

where $T_0 = t,\quad T_1 = \varepsilon t$. Substitute (2.13.9) into (2.13.8), we obtain

$$\begin{aligned}
0 &= \ddot{v} + \omega_0^2 v + \varepsilon^2 \alpha v^2 \ddot{v} + \varepsilon^2 \alpha \dot{v}^2 v - \varepsilon^2 \frac{\omega_0^2}{6} v^3 \\
&= [D_0^2 + 2\varepsilon D_0 D_1 + \varepsilon^2 (D_1^2 + 2D_0 D_2) + \cdots](v_0 + \varepsilon v_1 + \varepsilon^2 v_2 + \cdots) \\
&\quad + \omega_0^2 (v_0 + \varepsilon v_1 + \varepsilon^2 v_2 + \cdots) \\
&\quad + \varepsilon^2 [\alpha v_0^2 D_0^2 v_0 + \alpha v_0 (D_0 v_0)^2 - \frac{\omega_0^2}{6} v_0^3 + \cdots] \\
&= D_0^2 v_0 + \omega_0^2 v_0 + \varepsilon (D_0^2 v_1 + 2D_0 D_1 v_0 + \omega_0^2 v_1) \\
&\quad + \varepsilon^2 [D_0^2 v_2 + 2D_0 D_1 v_1 + (D_1^2 + 2D_0 D_2) v_0 + \omega_0^2 v_2 \\
&\quad + \alpha v_0^2 D_0^2 v_0 + \alpha v_0 (D_0 v_0)^2 - \frac{\omega_0^2}{6} v_0^3] + \cdots
\end{aligned} \tag{2.13.10}$$

Let the coefficients of the same power of ε be zeros to obtain

$$D_0^2 v_0 + \omega_0^2 v_0 = 0 \tag{2.13.11}$$

$$D_0^2 v_1 + \omega_0^2 v_1 = -2D_0 D_1 v_0 \tag{2.13.12}$$

$$\begin{aligned}
D_0^2 v_2 + \omega_0^2 v_2 = -2D_0 D_1 v_1 - (D_1^2 + 2D_0 D_2) v_0 \\
-\alpha v_0^2 D_0^2 v_0 - \alpha v_0 (D_0 v_0)^2 + \tfrac{\omega_0^2}{6} v_0^3
\end{aligned} \tag{2.13.13}$$

The general solution of Eq. (2.13.11) is

$$v_0 = A e^{i\omega_0 T_0} + \overline{A} e^{-i\omega_0 T_0} \tag{2.13.14}$$

where$A = A(T_1, T_2)$. Substituting (2.13.14) into (2.13.12), we obtain

$$D_0^2 v_1 + \omega_0^2 v_1 = -2D_0 D_1 (A e^{i\omega_0 T_0} + \overline{A} e^{-i\omega_0 T_0}) \tag{2.13.15}$$

To eliminate secular terms, we need $D_1 A = 0$, so $A = A(T_2)$. And the particular solution of (2.2.59) is

$$v_1 = 0 \tag{2.13.16}$$

Substituting (2.11.15) and (2.11.17) into (2.11.14), and taking $A = A(T_2)$ into account, we can obtain

$$D_0^2 v_2 + \omega_0^2 v_2 = -2i\omega_0 D_2 A e^{i\omega_0 T_0} + 2\alpha\omega_0^2 A^2 \overline{A} e^{i\omega_0 T_0} + \frac{1}{2}\omega_0^2 A^2 \overline{A} e^{i\omega_0 T_0} + cc + NST \tag{2.13.17}$$

where cc denotes the complex conjugate term of its preceding terms and NST denotes terms which do not generate secular terms. In order to eliminate secular terms, it is necessary to have

$$-2i\omega_0 D_2 A + 2\alpha\omega_0^2 A^2 \overline{A} + \frac{1}{2}\omega_0^2 A^2 \overline{A} = 0 \tag{2.13.18}$$

$$\text{Let } A = \frac{1}{2} a e^{i\beta} \tag{2.13.19}$$

Substitute (2.11.20) into (2.11.19), we obtain

$$i\omega_0 D_2 a - \omega_0 a D_2 \beta - \omega_0^2 (\frac{1}{4}\alpha + 4) a^3 = 0 \tag{2.13.20}$$

Separate the real and imaginary parts of the above equation yields

$$D_2 a = 0 \tag{2.13.21}$$

$$D_2 \beta - \omega_0 (\frac{1}{4}\alpha + 4) a^2 = 0 \tag{2.13.22}$$

So a is a constant. By integrating (2.11.23), we can obtain

$$\beta = \omega_0 (\frac{1}{4}\alpha + 4) a^2 T_2 + \beta_0 \tag{2.13.23}$$

Combining the above results, the first-order approximate Solution is obtained as

$$u = \varepsilon a \cos[\omega_0 t + \omega_0 (\frac{1}{4}\alpha + 4) a^2 \varepsilon^2 t + \beta_0] + \cdots \tag{2.13.24}$$

Let $\varepsilon = 1$, we can obtain

$$u = a \cos[\omega_0 t + \omega_0 (\frac{1}{4}\alpha + 4) a^2 t + \beta_0] + \cdots \tag{2.13.25}$$

Then the frequency-amplitude relationship of the oscillation is

$$\omega = \omega_0 + \omega_0(\frac{1}{4}\alpha + 4)a^2 \tag{2.13.26}$$

2.14 Exercise 2.14 (Single Pendulum Attached with a Rolling-Without-Slipping Circular Wheel II)

Solution: (**a**) The kinetic energy of the system is

$$\begin{aligned} T &= \frac{1}{2}m(l\dot{\theta}\cos\theta - \dot{\theta}r)^2 + \frac{1}{2}m(l\dot{\theta}\sin\theta)^2 \\ &= \frac{1}{2}m(l^2 + r^2 - 2rl\cos\theta)\dot{\theta}^2 \end{aligned} \tag{2.14.1}$$

The potential energy of the system is

$$V = -mgl\cos\theta + \frac{1}{2}k(r\theta)^2 \tag{2.14.2}$$

Using Lagrange's equation, we can obtain the differential equation of the motion of the system

$$m(l^2 + r^2 - 2rl\cos\theta)\ddot{\theta} + mrl\dot{\theta}^2\sin\theta + mgl\sin\theta + kr^2\theta = 0 \tag{2.14.3}$$

(**b**) Expand the Eq. (2.13.3) and retain to θ^3 yields

$$\ddot{\theta} + \frac{gl + kr^2/m}{(l-r)^2}\theta + \frac{rl}{(l-r)^2}\theta^2\ddot{\theta} + \frac{rl}{(l-r)^2}\dot{\theta}^2\theta - \frac{gl}{6(l-r)^2}\theta^3 = 0 \tag{2.14.4}$$

$$\text{Let } \omega_0^2 = \frac{gl + kr^2/m}{(l-r)^2}, \quad \alpha = \frac{rl}{(l-r)^2} \tag{2.14.5}$$

$$\text{then}\ddot{\theta} + \omega_0^2\theta + \alpha\theta^2\ddot{\theta} + \alpha\dot{\theta}^2\theta - \frac{1}{6}\omega_0^2\theta^3 = 0 \tag{2.14.6}$$

This equation is the same as the governing equation in Exercise 2.13, except that the value of ω_0^2 is taken differently. Therefore, the relationship between the oscillation frequency ω and the amplitude a is the same as in Exercise 2.13, i.e.,

$$\omega = \omega_0 + \omega_0(\frac{1}{4}\alpha + 4)a^2 \tag{2.14.7}$$

2.15 Exercise 2.15 (Single Pendulum Problem with Inelastic Collision with an Inclined Wall)

Solution: (a) When there is no collision, the whole motion of the system is controlled by the equation $\theta'' + \theta = 0$, so that its energy equation is

$$\theta^{\cdot 2} + \theta^2 = 2E \tag{2.15.1}$$

where E is the total energy. Therefore, the trajectories in the $\theta \sim \dot{\theta}$ phase plane are a family of circles with different initial conditions (E).

(b) When there is an inelastic collision, let the coefficient of restitution be $\mu < 1$, then the governing equation of the system is

$$\begin{array}{rl} \theta \neq \alpha: & \theta^{\cdot\cdot} + \sin\theta = 0 \\ \theta = \alpha,\ \ \dot{\theta}(\alpha) < 0: & \frac{\theta_2^{\cdot}(\alpha)}{-\theta_1^{\cdot}(\alpha)} = \mu \end{array} \tag{2.15.2}$$

where $\theta_1\prime$, $\theta_2\prime$ associated with the angular velocity of the pendulum before and after the collision, respectively. Then the trajectories of the system are governed by

$$\begin{array}{rl} \theta \neq \alpha: & \theta^{\cdot} = \pm\sqrt{2(E + \cos\theta)} \\ \theta = \alpha,\ \ \dot{\theta}(\alpha) < 0: & \theta_2^{\cdot}(\alpha) = -\mu\theta_1^{\cdot}(\alpha) \end{array} \tag{2.15.3}$$

Figure 2.5a–c present the trajectories of the sysstem for (1) $\alpha > 0$, (2) $\alpha = 0$ and (3) $\alpha < 0$, respectively.

Figure 2.5a and b ($\alpha \geq 0$): the ball will always impact the wall with energy loss and velocity decreasing, and finally rest against the wall.

Figure 2.5c ($\alpha < 0$): the ball will impact the wall at the beginning, then the speed will decrease and the energy will be lost, finally the speed will be zero when the ball reaches the wall.

2.16 Exercise 2.16 (Simplified Model for Buckling Analysis of Columns)

Solution: (a) The kinetic energy of the system is

$$T = \frac{1}{2}m\dot{x}^2 \tag{2.16.1}$$

The potential energy of the system is

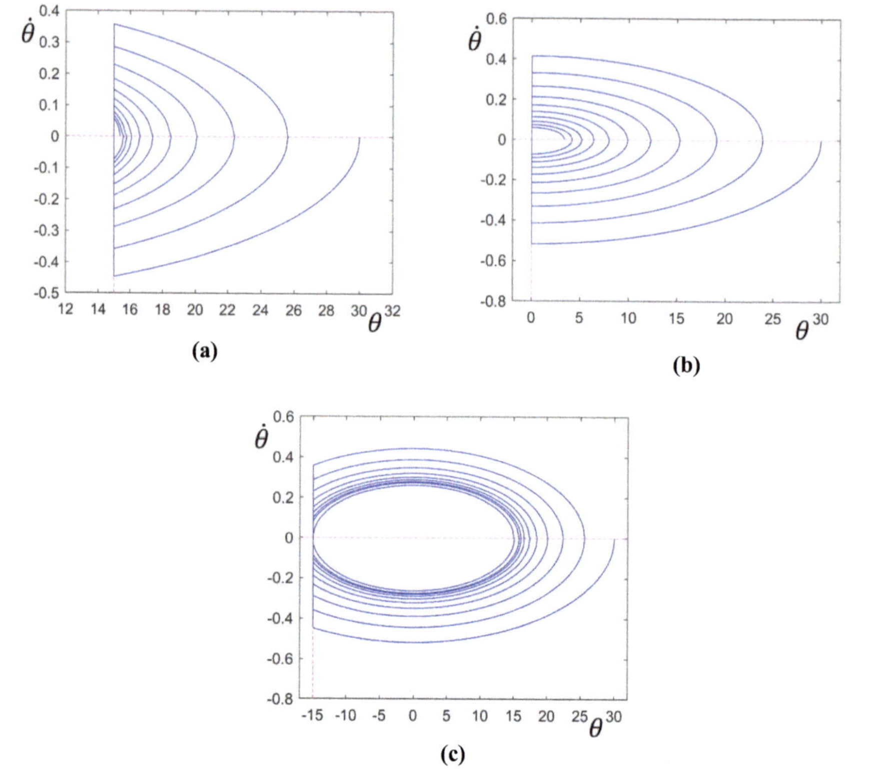

Fig. 2.5 **a** Phase trajectories of the ball at $\alpha > 15°(\theta_0 = 30°, \theta_0\prime = 0, \mu = 0.8)$. **b** Phase trajectories of the ball at $\alpha = 0^{\circ}(\theta_0 = 30°, \theta_0\prime = 0, \mu = 0.8)$. **c** Phase trajectories of the ball at $\alpha > -15°(\theta_0 = 30°, \theta_0\prime = 0, \mu = 0.8)$ for Exercise 2.16

$$
\begin{aligned}
V &= \int_0^x F_{\text{spring}} dx - 2 \times P(l - \sqrt{l^2 - x^2}) \\
&= \int_0^x F_{\text{spring}} dx + 2P\sqrt{l^2 - x^2} - 2Pl
\end{aligned}
\tag{2.16.2}
$$

Substituting the kinetic energy and potential energy into the Lagrange's equation, we obtain

$$
m\ddot{x} = -\frac{\partial V}{\partial x} = -F_{\text{spring}} + \frac{2Px}{\sqrt{l^2 - x^2}}
$$

So $m\ddot{x} + F_{\text{spring}} - \frac{2Px}{\sqrt{l^2 - x^2}} = 0$

$$\text{i.e.}\quad m\ddot{x} + k_1 x + k_3 x^3 - \frac{2Px}{\sqrt{l^2 - x^2}} + \cdots = 0 \tag{2.16.3}$$

Expanding the fourth term on the left-hand side of (2.16.4) and keeping it to $O(x^3)$, we obtain

$$m\ddot{x} + k_1 x + k_3 x^3 - (\frac{2P}{l}x + \frac{P}{l^3}x^3) = 0$$

$$\text{i.e.}\quad m\ddot{x} + (k_1 - \frac{2P}{l})x + (k_3 - \frac{P}{l^3})x^3 = 0 \tag{2.16.4}$$

$$\text{Let } \alpha_1 = \frac{1}{m}(k_1 - \frac{2P}{l}),\quad \alpha_3 = \frac{1}{m}(k_3 - \frac{P}{l^3}) \tag{2.16.5}$$

then (2.16.4) becomes

$$\ddot{x} + \alpha_1 x + \alpha_3 x^3 = 0 \tag{2.16.6}$$

(**b**) In the Eq. (2.16.6) let

$$u = \frac{x}{l},\quad \omega_0 = \begin{cases} \sqrt{\alpha_1}, & \text{when } \alpha_1 \geq 0 \\ \sqrt{-\alpha_1}, & \text{when } \alpha_1 < 0 \end{cases},\quad \tau = \omega_0 t,\quad \alpha = \frac{\alpha_3 l^2}{\omega_0^2} \tag{2.16.7}$$

then (2.16.6) becomes

$$\alpha_1 \geq 0 : u^{''} + u + \alpha u^3 = 0 \tag{2.16.8}$$

$$\alpha_1 < 0 : u^{''} - u + \alpha u^3 = 0 \tag{2.16.9}$$

After integration, the energy equation and the phase trajectories are obtained as

$$\alpha_1 \geq 0 : \begin{cases} \frac{1}{2}u^{'2} + V(u) = E, \text{ where } V(u) = \frac{1}{2}u^2 + \frac{1}{4}\alpha u^4 \\ u^{'} = \pm\sqrt{2(E - V(u))} \end{cases} \tag{2.16.10}$$

$$\alpha_1 < 0 : \begin{cases} \frac{1}{2}u^{'2} + V(u) = E, \text{ where } V(u) = -\frac{1}{2}u^2 + \frac{1}{4}\alpha u^4 \\ u^{'} = \pm\sqrt{2(E - V(u))} \end{cases} \tag{2.16.11}$$

From (2.16.10) and (2.16.11) the potential energy curve $V(u) \sim u$ and the phase trajectories can be calculated and plotted, as shown in Fig. 2.6a and b. The equilibrium point $x = 0$ is the center when $\alpha_1 > 0$ while the saddle point when $\alpha_1 < 0$. Therefore, $\alpha_1 = 0$ is the bifurcation point of the system, i.e., the column is buckled under this condition. Thus, the buckling load P_b can be obtained from Eq. (2.16.5)

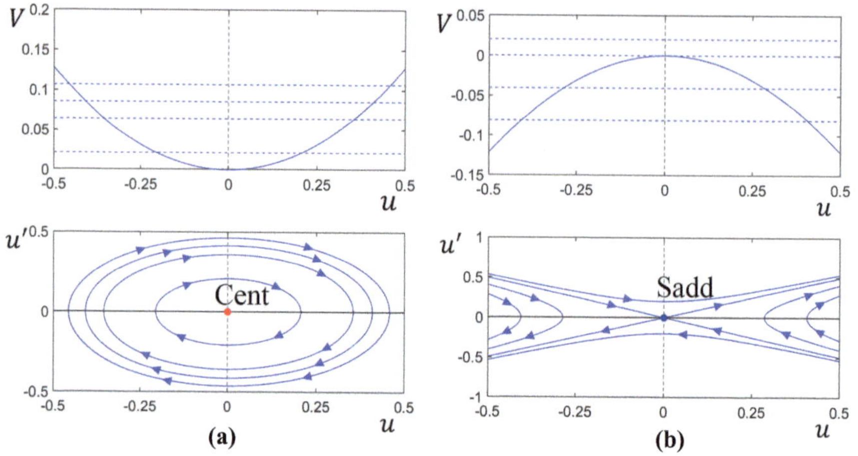

Fig. 2.6 **a** Potential energy curve and phase trajectories ($\alpha_1 > 0, \alpha = 0.2$). **b** Potential energy curve and phase trajectories ($\alpha_1 < 0, \alpha = 0.2$) for Exercise 2.16

$$P_b = \frac{1}{2}k_1 l \tag{2.16.12}$$

2.17 Exercise 2.17 (Rods for Pure Rolling on a Fixed Cylindrical Surface)

Solution: **(a)** Assume that the rod is balanced at the vertex of the cylindrical surface. The kinetic energy of the rod is

$$T = \frac{1}{2}(\frac{1}{12}ml^2 + mr^2\theta^2)\dot{\theta}^2 \tag{2.17.1}$$

The potential energy of the rod is

$$V = mg(r\cos\theta + r\theta\sin\theta) \tag{2.17.2}$$

Substituting the kinetic energy and potential energy into the Lagrange's equation, we obtain

$$(\frac{1}{12}ml^2 + mr^2\theta^2)\ddot{\theta} + mr^2\theta\dot{\theta}^2 = -mgr\theta\cos\theta$$

That is, the differential equation of motion of the rod is

$$(\frac{1}{12}l^2 + r^2\theta^2)\ddot{\theta} + r^2\theta\dot{\theta}^2 + gr\theta\cos\theta = 0 \tag{2.17.3}$$

(b) We expand the Eq. (2.17.3) around the equilibrium point $\theta = 0$ and expand it, retaining to $O(\theta^3)$, to obtain

$$(\frac{1}{12}l^2 + r^2\theta^2)\ddot{\theta} + r^2\theta\dot{\theta}^2 + gr\theta - \frac{1}{2}gr\theta^3 = 0 \tag{2.17.4}$$

$$\text{well sorted } \ddot{\theta} + \omega_0^2\theta + \alpha\theta^2\ddot{\theta} + \alpha\theta\dot{\theta}^2 - \frac{1}{2}\omega_0^2\theta^3 = 0 \tag{2.17.5}$$

$$\text{where } \omega_0^2 = \frac{12rg}{l^2}, \quad \alpha = \frac{12r^2}{l^2} \tag{2.17.6}$$

Let

$$\theta = \varepsilon v \tag{2.17.7}$$

Equation (2.17.4) changes to

$$\ddot{v} + \omega_0^2 v + \varepsilon^2\alpha v^2\ddot{v} + \varepsilon^2\alpha\dot{v}^2 v - \varepsilon^2\frac{\omega_0^2}{2}v^3 = 0 \tag{2.17.8}$$

We seek an expansion of solution with the following form near the singularity $v = 0$

$$v = v_0(T_0, T_1, T_2) + \varepsilon v_1(T_0, T_1, T_2) + \varepsilon^2 v_2(T_0, T_1, T_2) + \cdots \tag{2.17.9}$$

where $T_0 = t, \quad T_1 = \varepsilon t$. Substitute (2.17.9) into (2.17.8), we obtain

$$\begin{aligned}
0 &= \ddot{v} + \omega_0^2 v + \varepsilon^2\alpha v^2\ddot{v} + \varepsilon^2\alpha\dot{v}^2 v - \varepsilon^2\frac{\omega_0^2}{2}v^3 \\
&= [D_0^2 + 2\varepsilon D_0 D_1 + \varepsilon^2(D_1^2 + 2D_0 D_2) + \cdots](v_0 + \varepsilon v_1 + \varepsilon^2 v_2 + \cdots) \\
&\quad + \omega_0^2(v_0 + \varepsilon v_1 + \varepsilon^2 v_2 + \cdots) \\
&\quad + \varepsilon^2[\alpha v_0^2 D_0^2 v_0 + \alpha v_0(D_0 v_0)^2 - \frac{\omega_0^2}{2}v_0^3 + \cdots] \\
&= D_0^2 v_0 + \omega_0^2 v_0 + \varepsilon(D_0^2 v_1 + 2D_0 D_1 v_0 + \omega_0^2 v_1) \\
&\quad + \varepsilon^2[D_0^2 v_2 + 2D_0 D_1 v_1 + (D_1^2 + 2D_0 D_2)v_0 + \omega_0^2 v_2 \\
&\quad + \alpha v_0^2 D_0^2 v_0 + \alpha v_0(D_0 v_0)^2 - \frac{\omega_0^2}{2}v_0^3] + \cdots
\end{aligned} \tag{2.17.10}$$

Retaining the equation to $O(\varepsilon^2)$ and equate the coefficients of the same power of ε to be zero gives

$$D_0^2 v_0 + \omega_0^2 v_0 = 0 \tag{2.17.11}$$

$$D_0^2 v_1 + \omega_0^2 v_1 = -2D_0 D_1 v_0 \tag{2.17.12}$$

$$\begin{aligned} D_0^2 v_2 + \omega_0^2 v_2 = & -2D_0 D_1 v_1 - (D_1^2 + 2D_0 D_2) v_0 \\ & - \alpha v_0^2 D_0^2 v_0 - \alpha v_0 (D_0 v_0)^2 + \frac{\omega_0^2}{2} v_0^3 \end{aligned} \tag{2.17.13}$$

The solution of Eq. (2.17.11) is

$$v_0 = A e^{i\omega_0 T_0} + \overline{A} e^{-i\omega_0 T_0} \tag{2.17.14}$$

where $A = A(T_1, T_2)$. Substituting the above equation into the Eq. (2.17.12), we obtain

$$D_0^2 v_1 + \omega_0^2 v_1 = -2D_0 D_1 (A e^{i\omega_0 T_0} + \overline{A} e^{-i\omega_0 T_0}) \tag{2.17.15}$$

To eliminate the secular term, we need $D_1 A = 0$, so $A = A(T_2)$. Then the solution of (2.17.15) is

$$v_1 = 0 \tag{2.17.16}$$

Substitute (2.17.14) and (2.17.16) into (2.17.13), and take $A = A(T_2)$ into account, we obtain

$$\begin{aligned} D_0^2 v_2 + \omega_0^2 v_2 = & -2i\omega_0 D_2 A e^{i\omega_0 T_0} + 2\alpha\omega_0^2 A^2 \overline{A} e^{i\omega_0 T_0} \\ & + \frac{3\omega_0^2}{2} A^2 \overline{A} e^{i\omega_0 T_0} + cc + NST \end{aligned} \tag{2.17.17}$$

where cc denotes the complex conjugate term of its preceding terms and NST denotes terms which do not generate secular terms. In order to eliminate the secular term, it is necessary to have

$$-2i\omega_0 D_2 A + 2\alpha\omega_0^2 A^2 \overline{A} + \frac{3\omega_0^2}{2} A^2 \overline{A} = 0 \tag{2.17.18}$$

$$\text{Let} \quad A = \frac{1}{2} a e^{i\beta} \tag{2.17.19}$$

Put (2.17.19) into (2.17.18), we obtain

$$i\omega_0 D_2 a - \omega_0 a D_2 \beta - \omega_0^2 (\frac{1}{4}\alpha + 12) a^3 = 0 \tag{2.17.20}$$

Separate the real and imaginary parts of the above equation yields

$$D_2 a = 0 \tag{2.17.21}$$

$$D_2 \beta - \omega_0(\frac{1}{4}\alpha + 12)a^2 = 0 \tag{2.17.22}$$

So a is a constant. Integrate (2.17.22) directly, we obtain

$$\beta = \omega_0(\frac{1}{4}\alpha + 12)a^2 T_2 + \beta_0 \tag{2.17.23}$$

Combining the above results, the first-order approximation of the solution is obtained as

$$u = \varepsilon a\cos[\omega_0 t + \omega_0(\frac{1}{4}\alpha + 12)a^2\varepsilon^2 t + \beta_0] + \cdots \tag{2.17.24}$$

Let $\varepsilon = 1$, we can obtain

$$u = a\cos[\omega_0 t + \omega_0(\frac{1}{4}\alpha + 12)a^2 t + \beta_0] + \cdots \tag{2.17.25}$$

Therefore, the frequency-amplitude relationship of the oscillation is

$$\omega = \omega_0 + \omega_0(\frac{1}{4}\alpha + 12)a^2 \tag{2.17.26}$$

2.18 Exercise 2.18 (Geometrically Nonlinear System Formed by a Linear Spring and a Mass Block)

Solution: **(a)** The kinetic energy of the system is

$$T = \frac{1}{2}m\dot{x}^2 \tag{2.18.1}$$

The potential energy of the system is

$$V = \frac{1}{2}k[(x^2 + l^2)^{1/2} - \hat{l}]^2 \tag{2.18.2}$$

Substituting the kinetic energy and potential energy into the Lagrange's equation, the differential equation of motion of the system is

$$m\ddot{x} + kx(x^2 + l^2)^{-1/2}[(x^2 + l^2)^{1/2} - \hat{l}] = 0 \tag{2.18.3}$$

The equation can be written as

$$\frac{\ddot{x}}{l} + \frac{k}{m}\frac{x}{l}(1 + \frac{x^2}{l^2})^{-1/2}[(1 + \frac{x^2}{l^2})^{1/2} - \frac{\hat{l}}{l}] = 0$$

Let

$$2\omega^2 = \frac{k}{m}, \quad u = \frac{x}{l}, \quad L = \frac{\hat{l}}{l}, \quad \tau = \omega t$$

The equation becomes

$$\ddot{u} + 2u(1 + u^2)^{-1/2}[(1 + u^2)^{1/2} - L] = 0 \tag{2.18.4}$$

$$\text{or } \ddot{u} + 2u - \frac{2Lu}{\sqrt{1 + u^2}} = 0 \tag{2.18.5}$$

Note that the derivatives are obtained for the dimensionless time variable τ, i.e. $\dot{u} = du/d\tau$.

(b) Integrate the Eq. (2.18.5), we can obtain the energy equation

$$\frac{1}{2}\dot{u}^2 + V(u) = E, \quad V(u) = u^2 - 2L\sqrt{1 + u^2} \tag{2.18.6}$$

$$So \dot{u} = \pm\sqrt{2[E - V(u)]} \tag{2.18.7}$$

The potential energy curve and the phase trajectories are shown in Fig. 2.7a–c. The equilibrium point $u = 0$ is a center when $L \leq 1$ while a saddle point when $L > 1$ is unstable; two new equilibrium points are created at the same time.

(c) Expand the Eq. (2.18.5), retaining to $O(u^3)$, to obtain

$$\ddot{u} + 2u - 2Lu(1 - \frac{1}{2}u^2 + \cdots) = 0$$

$$\ddot{u} + 2(1 - L)u + Lu^3 = 0 \tag{2.18.8}$$

For $L = 1$,

$$\ddot{u} + u^3 = 0 \quad \text{when } L = 1 \tag{2.18.9}$$

(d) From (2.18.9), we obtain

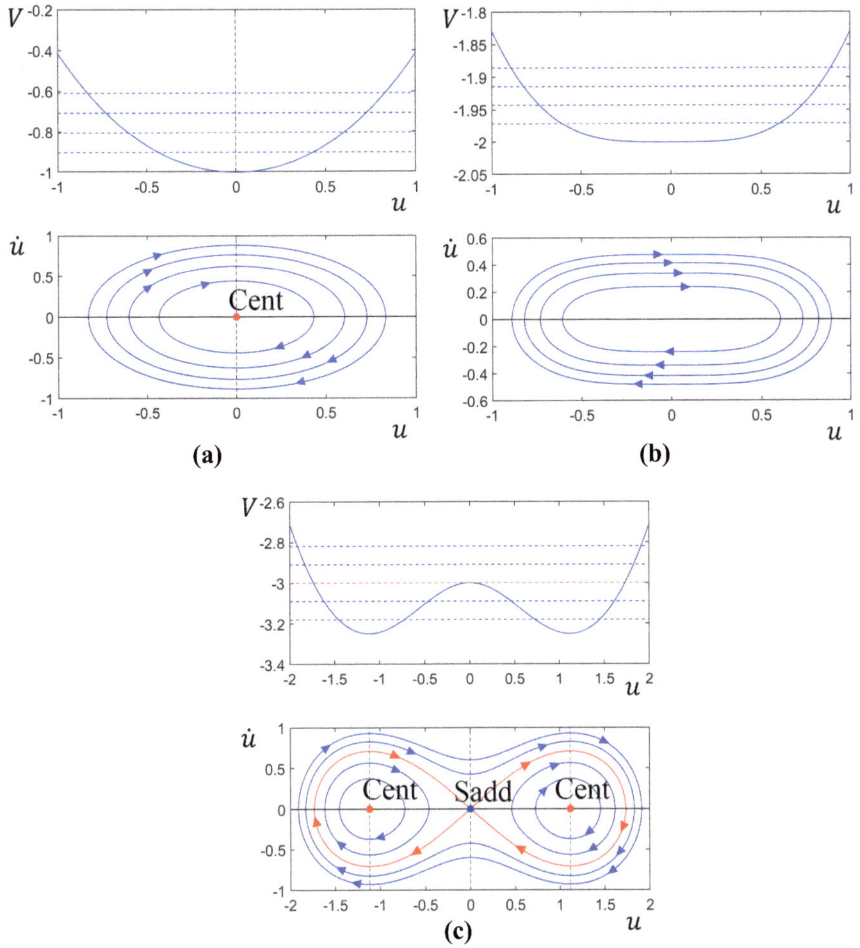

Fig. 2.7 Potential energy curves and solution trajectories when **a** $L = 0.5$, **b** $L = 1$ and **c** $L = 1.5$ for Exercise 2.18(b)

$$d\tau = -\frac{d\dot{u}}{u^3}, \qquad \frac{1}{2}\dot{u}^2 + \frac{1}{4}u^4 = E \tag{2.18.10}$$

If the movement starts at $-u_0$,

$$E = \frac{1}{4}u_0^4, \quad d\dot{u} = -\sqrt{2}\frac{u^3 du}{\sqrt{(u_0^4 - u^4)}}$$

Substitute this into (2.18.10) and make integration of it to get

$$\tau = \sqrt{2}\int_{-u_0}^{u} \frac{du}{\sqrt{(u_0^4 - u^4)}} \tag{2.18.11}$$

$$\text{Let } u = -u_0\cos\varphi \tag{2.18.12}$$

then the Eq. (2.18.11) becomes

$$\tau = \frac{\sqrt{2}}{u_0}\int_0^{\theta} \frac{\sin\varphi d\varphi}{\sqrt{1-\cos^4\varphi}} = \frac{1}{u_0}\int_0^{\theta} \frac{d\varphi}{\sqrt{1-\frac{1}{2}\sin^2\varphi}} \triangleq \frac{1}{u_0}F(\frac{1}{2},\theta) \tag{2.18.13}$$

where $\cos\theta = -u/u_0$;$F(\kappa,\theta)$ are the first class of elliptic integrals defined as

$$F(\kappa,\theta) = \int_0^{\theta} \frac{d\varphi}{\sqrt{1-\kappa^2\sin^2\varphi}}$$

Thus, the period of the system T is

$$T = \frac{4}{u_0}F(\frac{1}{2},\frac{\pi}{2}) = \frac{4}{u_0} \times 1.8612 \approx \frac{7.4448}{u_0} \tag{2.18.14}$$

2.19 Exercise 2.19 (Solving Pure Cubic Nonlinear Systems by Harmonic Balance Method)

Solution: **(a)** Substituting the one-term expansion

$$u = u_0\cos(\omega\tau + \beta) \triangleq u_0\cos\psi$$

into the differential equation of motion of the system, we obtain

$$\begin{aligned} 0 &= -\omega^2 u_0\cos\psi + u_0^3\cos^3\psi \\ &= -\omega^2 u_0\cos\psi + \frac{1}{4}u_0^3(3\cos\psi + \cos3\psi) \\ &= (-\omega^2 + \frac{3}{4}u_0^2)u_0\cos\psi + \frac{1}{4}u_0^3\cos3\psi \end{aligned}$$

Equating the coefficient of $\cos\Psi$ to zero, we obtain

$$\omega = \frac{\sqrt{3}}{2}u_0$$

Therefore, the period of oscillation of the system is

$$T = \frac{2\pi}{\omega} = \frac{4\pi}{u_0\sqrt{3}} \approx \frac{7.2552}{u_0} \tag{2.19.1}$$

From Eq. (2.18.14), the relative error of this result is

$$\text{error} = \frac{7.4448 - 7.2552}{7.4448} = 2.5\%$$

(b) By expanding the integrand integrating term by term, refer to (f) in Exercise 2.18, we obtain

$$\begin{aligned}\frac{1}{(1-\frac{1}{2}\sin^2\phi)^{1/2}} &= 1 + \frac{1}{2}(\frac{1}{2}\sin^2\phi) + \frac{3}{8}(\frac{1}{2}\sin^2\phi)^2 + \frac{5}{16}(\frac{1}{2}\sin^2\phi)^3 + \cdots \\ &= 1 + \frac{1}{4}\sin^2\phi + \frac{3}{32}\sin^4\phi + \frac{5}{128}\sin^6\phi + \cdots\end{aligned}$$

Substituting this into the equation (f) in Exercise 2.18, the period of oscillation of the system can be obtained as

$$T = \frac{4}{u_0}\int_0^{\pi/2}(1 + \frac{1}{4}\sin^2\varphi + \frac{3}{32}\sin^4\varphi + \frac{5}{128}\sin^6\varphi + \cdots)d\varphi \tag{2.19.2}$$

Integrating the first three terms of (2.19.2) gives

$$\begin{aligned}T &= \tfrac{4}{u_0}\int_0^{\pi/2}(1 + \tfrac{1}{4}\sin^2\phi + \tfrac{3}{32}\sin^4\phi)d\phi \\ &= \tfrac{4}{u_0}(\tfrac{\pi}{2} + \tfrac{1}{4}\cdot\tfrac{\pi}{4} + \tfrac{3}{32}\cdot\tfrac{3\pi}{16}) = \tfrac{297\pi}{128u_0} \\ &\approx \tfrac{7.2895}{u_0}\end{aligned} \tag{2.19.3}$$

This result is closer to the exact value (2.18.14) than (2.19.1).

2.20 Exercise 2.20 (Solving Purely Fifth-Order Nonlinear Systems by Harmonic Balance Method)

Solution: Substituting the one-term expansion

$$u = a\cos(\omega t + \beta) \triangleq a\cos\psi$$

Into the differential equation of motion of the system, we obtain

$$\begin{aligned} 0 &= -\omega^2 a\cos\psi + a^5\cos^5\psi \\ &= -\omega^2 a\cos\psi + \frac{1}{16}a^5(\cos 5\psi + 5\cos 3\psi + 10\cos\psi) \\ &= (-\omega^2 + \frac{5}{8}a^4)a\cos\psi + \cdots \end{aligned}$$

Equating the coefficient of $\cos\Psi$ to zero, we obtain

$$\omega = \sqrt{\frac{5}{8}}a^2 \tag{2.20.1}$$

2.21 Exercise 2.21 (Solving Pure Cubic Nonlinear Systems by Equivalent Linearization)

Solution: The average of integrated square of the error over a time interval T is given by

$$e = \frac{1}{T}\int_0^T (\lambda x - x^3)^2 d\tau \triangleq < (\lambda x - x^3)^2 \geq \lambda^2\langle x^2\rangle - 2\lambda\langle x^4\rangle + \langle x^6\rangle$$

Minimize this error with respect to λ yields

$$\frac{de}{d\lambda} = 2\lambda < x^2 > -2 < x^4 > = 0$$

$$\begin{aligned} \lambda &= \frac{<x^4>}{<x^2>} = u_0^2 \frac{\int_0^T \cos^4(\omega\tau+\beta)d\tau}{\int_0^T \cos^2(\omega\tau+\beta)d\tau} = u_0^2 \frac{\int_0^{2\pi} \cos^4\psi d\psi}{\int_0^{2\pi} \cos^2\psi d\psi} \\ &= u_0^2(\tfrac{3}{8}\cdot 2\pi)/((\tfrac{1}{2}\cdot 2\pi)) = \tfrac{3}{4}u_0^2 \end{aligned}$$

So $\omega = \sqrt{\lambda} = \frac{\sqrt{3}}{2}u_0$

2.22 Exercise 2.22 (Show Least Residual Value Method and Galerkin Method for Solving Pure Cubic Nonlinear Systems)

Solution: (a) Substituting the hypothetical solution (b) into (a), the residual can be obtained as

$$\begin{aligned}R &= -\omega^2 u_0 \cos(\omega\tau+\beta) + u_0^3\cos^3(\omega\tau+\beta)\\ &= -\omega^2 u_0\cos(\omega\tau+\beta) + \frac{1}{4}u_0^3[3\cos(\omega\tau+\beta)+\cos(3\omega\tau+3\beta)]\\ &= (\frac{3}{4}u_0^3 - \omega^2 u_0)\cos(\omega\tau+\beta) + \frac{1}{4}u_0^3\cos(3\omega\tau+3\beta)\end{aligned}$$

Thus, the average of integrated square of the error over a time interval $2\pi/\omega$ is

$$\begin{aligned}\langle R^2\rangle &= \frac{1}{T}\{(\frac{3}{4}u_0^3-\omega^2u_0)^2\int_0^{2\pi/\omega}\cos^2(\omega\tau+\beta)d\tau\\ &+\frac{1}{2}u_0^3(\frac{3}{4}u_0^3-\omega^2u_0)\int_0^{2\pi/\omega}\cos(\omega\tau+\beta)\cos(3\omega\tau+3\beta)d\tau\\ &+\frac{1}{16}u_0^6\int_0^{2\pi/\omega}\cos^2(3\omega\tau+3\beta)d\tau\}\\ &=\frac{1}{2}(\frac{3}{4}u_0^3-\omega^2u_0)^2+\frac{1}{32}u_0^6\end{aligned} \tag{2.22.1}$$

If $< R^2 >$ is minimized with respect to u_0 (assuming u_0 is not zero), we have

$$\begin{aligned}\frac{d<R^2>}{du_0} &= (\tfrac{3}{4}u_0^3-\omega^2u_0)(\tfrac{9}{4}u_0^2-\omega^2)+\tfrac{3}{16}u_0^5\\ &=\tfrac{27}{16}u_0^5-\tfrac{9}{4}\omega^2u_0^3-\tfrac{3}{4}\omega^2u_0^3+\omega^4u_0+\tfrac{3}{16}u_0^5\\ &=\tfrac{15}{8}u_0^5-3\omega^2u_0^3+\omega^4u_0=0\end{aligned}$$

So

$$\omega=\frac{1}{2}(\sqrt{6\pm\sqrt{6}})u_0$$

The root with the positive sign maximizes $< R^2 >$ and must be discarded. The other root is not in agreement with the result obtained by the method of harmonic balance and equivalent linearization in Exercises 2.20 & 2.21.

(**b**) Instead of minimizing $< R^2 >$ with respect to u_0, if $< R^2 >$ is minimized with respect to ω, we have

$$\frac{d < R^2 >}{d\omega} = -2\omega u_0(\frac{3}{4}u_0^3 - \omega^2) = 0$$

So

$$\omega = \frac{\sqrt{3}}{2}u_0$$

(**c**) Instead of minimizing $< R^2 >$, R is made orthogonal to the assumed solution, i.e.,

$$0 =< Ru_0\cos(\omega\tau + \beta) >= \frac{1}{T}\int_0^T Ru_0\cos(\omega\tau + \beta)d\tau$$
$$= \frac{1}{T\omega}[u_0(\frac{3}{4}u_0^3 - \omega^2 u_0)\int_0^{2\pi}\cos^2\psi d\psi + \frac{1}{4}u_0^4\int_0^{2\pi}\cos\psi\cos 3\psi d\psi$$
$$= \frac{1}{2}(\frac{3}{4}u_0^3 - \omega^2 u_0)$$

So $\omega = \frac{\sqrt{3}}{2}u_0$

This result is identical to that obtained by the harmonic balance method and the equivalent linearization method shown in Exercises 2.20, 2.21. This method is the Galerkin method.

2.23 Exercise 2.23 (Discuss the Possibility of Solving Pure Cubic Nonlinear Systems by the Method of Multiple Scales)

Solution: Let

$$u = \varepsilon v$$

The equation $u'' + u^3 = 0$ changes to

$$v'' + \varepsilon^2 v^3 = 0 \tag{2.23.1}$$

We seek an expansion of the solution with the following form near the singularity $v = 0$

$$v = v_0(T_0, T_1, T_2) + \varepsilon v_1(T_0, T_1, T_2) + \varepsilon^2 v_2(T_0, T_1, T_2) + \cdots \tag{2.23.2}$$

where $T_n = \varepsilon^n t$. Substitute (2.13.9) into (2.23.1), we obtain

$$\begin{aligned} 0 &= \ddot{v} + \varepsilon^2 v^3 \\ &= [D_0^2 + 2\varepsilon D_0 D_1 + \varepsilon^2 (D_1^2 + 2D_0 D_2) + \cdots](v_0 + \varepsilon v_1 + \varepsilon^2 v_2 + \cdots) \\ &\quad + \varepsilon^2 (v_0 + \varepsilon v_1 + \varepsilon^2 v_2 + \cdots)^3 \\ &= D_0^2 v_0 + \varepsilon (D_0^2 v_1 + 2D_0 D_1 v_0) \\ &\quad + \varepsilon^2 [D_0^2 v_2 + 2D_0 D_1 v_1 + (D_1^2 + 2D_0 D_2) v_0 + v_0^3] + \cdots \end{aligned} \tag{2.23.3}$$

Retaining the above equation to the order of ε^2 and equating coefficients of the same power of ε gives

$$D_0^2 v_0 = 0 \tag{2.23.4}$$

$$D_0^2 v_1 = -2D_0 D_1 v_0 \tag{2.23.5}$$

$$D_0^2 v_2 = -2D_0 D_1 v_1 - (D_1^2 + 2D_0 D_2) v_0 - v_0^3 \tag{2.23.6}$$

In order not to generate secular terms, this set of equations has only zero solutions. Therefore, we do not get the expression for v. The reason is that, according to the above scheme, the ε^0 order in Eq. (2.23.4) is not an oscillational equation, and therefore, the final oscillational solution of v cannot be formed. Therefore, the equation $u'' + u^3 = 0$ cannot be solved by the the method of multiple scales.

2.24 Exercise 2.24 (Examine the Relationship Between the Period and the Amplitude of Oscillation)

Solution: (a) As known from Exercise 2.18, the differential equation of motion of the system is

$$u'' + 2u(1 + u^2)^{-1/2}[(1 + u^2)^{1/2} - L] = 0$$

And from the answer to Exercise 2.18(c), the above equation has been expanded to $O(u^3)$ as

$$u'' + 2(1 - L)u + Lu^3 = 0$$

When $L = 1/2$ is used, the above equation becomes

$$u'' + u + \frac{1}{2}u^3 = 0, \quad \text{when } L = \frac{1}{2} \tag{2.24.1}$$

By integrating (2.24.1), we obtain

$$\frac{1}{2}u'^2 + \frac{1}{2}u^2 + \frac{1}{8}u^4 = E$$

When the initial condition is $u = -u_0, \quad u\prime = 0$

$$E = \frac{1}{2}u_0^2 + \frac{1}{8}u_0^4$$

$$\text{So } u' = \sqrt{(u_0^2 - u^2) + \frac{1}{4}(u_0^4 - u^4)} = \frac{1}{2}\sqrt{(u_0^2 - u^2)(4 + u_0^2 + u^2)} \tag{2.24.2}$$

The time from the initial state of motion to any state is

$$\tau = \int_{-u_0}^{u} \frac{du}{u'} = 2\int_{-u_0}^{u} \frac{du}{\sqrt{(u_0^2 - u^2)(4 + u_0^2 + u^2)}} \tag{2.24.3}$$

(b) Let $u = -u_0\cos\varphi$, the Eq. (2.24.3) becomes

$$\tau = 2\int_0^{\theta} \frac{u_0 \sin\varphi d\varphi}{\sqrt{u_0^2(1 - \cos^2\varphi)(4 + u_0^2 + u_0^2\cos^2\varphi)}} = 2\int_0^{\theta} \frac{d\varphi}{\sqrt{(4 + 2u_0^2 - u_0^2\sin^2\varphi)}}$$

$$= \frac{2}{\sqrt{4 + 2u_0^2}}\int_0^{\theta} \frac{d\varphi}{\sqrt{(1 - k^2\sin^2\varphi)}}$$

$$\text{i.e.} \quad \tau = \frac{2}{\sqrt{4 + 2u_0^2}}\int_0^{\theta} \frac{d\varphi}{\sqrt{(1 - k^2\sin^2\varphi)}} \tag{2.24.4}$$

where $k^2 = u_0^2/(4 + 2u_0^2)$. Therefore, the period of oscillation of the system is

$$T = \frac{8}{\sqrt{4 + 2u_0^2}}\int_0^{\pi/2} \frac{d\varphi}{\sqrt{(1 - k^2\sin^2\varphi)}} \tag{2.24.5}$$

Numerical integration is performed for the Eq. (2.24.5) and $T \sim u_0$ curve is plotted in Fig. 2.8.

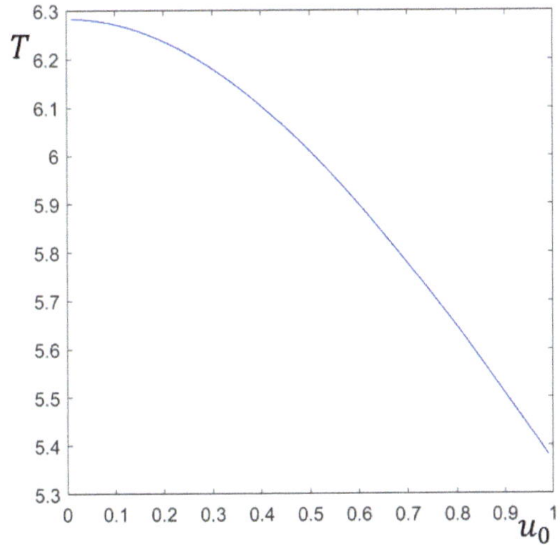

Fig. 2.8 Variation of period T with amplitude u_0 of the oscillation for Exercise 2.24

2.25 Exercise 2.25 (Comparison of Equivalent Linearization Methods and the Method of Multiple Scales for Solving Different Nonlinear Equations)

Solution: (a) The residual is

$$R = \alpha_1 x + \alpha_3 x^3 - \lambda x$$

The mean square value of the residual is

$$\begin{aligned}\langle R^2 \rangle &= \frac{1}{T}\int_0^T (\alpha_1 x + \alpha_3 x^3 - \lambda x)^2 dt \\ &= (\alpha_1 - \lambda)^2 \langle x^2 \rangle + 2\alpha_3(\alpha_1 - \lambda)\langle x^4 \rangle + \alpha_3^2 \langle x^6 \rangle\end{aligned}$$

In order to find the minimum value for $< R^2 >$, we need

$$\frac{d < R^2 >}{d\lambda} = -2(\alpha_1 - \lambda) < x^2 > -2\alpha_3 < x^4 > = 0$$

So

$$\lambda = \frac{\alpha_1 < x^2 > + \alpha_3 < x^4 >}{< x^2 >}$$

Let $x = a\cos(\omega t + \beta)$, then

$$< x^2 > = \frac{1}{T}\int_0^T a^2\cos^2(\omega t+\beta)dt = \frac{a^2}{2\pi}\int_0^{2\pi}\cos^2\psi d\psi = \frac{a^2}{2}$$

$$< x^4 > = \frac{1}{T}\int_0^T a^4\cos^4(\omega t+\beta)dt = \frac{a^4}{2\pi}\int_0^{2\pi}\cos^4\psi d\psi = \frac{3a^4}{8}$$

So $\lambda = \alpha_1 + \frac{3}{4}\alpha_3 a^2$ (2.25.1)

In Exercise 2.5(a), the Eq. (2.5.1) is the same as the nonlinear equation in this Exercise, where the the method of multiple scales has been used to find the frequency-amplitude relationship as in the Eq. (2.5.23). In the equation, let $\omega = \sqrt{\lambda}$, $\omega_1 = \sqrt{\alpha_1}$, $\alpha = \alpha_3$, we can obtain (2.25.1), i.e., the results are the same as those obtained by the the method of multiple scales.

(b) The residual is

$$R = \alpha_1 x + \alpha_2 x^2 + \alpha_3 x^3 - \lambda x$$

The mean square value of the residual is

$$\langle R^2\rangle = \frac{1}{T}\int_0^T [(\alpha_1-\lambda)x + \alpha_2 x^2 + \alpha_3 x^3)^2 dt$$
$$= (\alpha_1-\lambda)^2\langle x^2\rangle + 2\alpha_2(\alpha_1-\lambda)\langle x^3\rangle + 2\alpha_3(\alpha_1-\lambda)\langle x^4\rangle + \alpha_2\alpha_3\langle x^5\rangle + \alpha_2^2\langle x^4\rangle + \alpha_3^2\langle x^6\rangle$$

In order to find the minimum value for $< R^2 >$, we need

$$\frac{d < R^2 >}{d\lambda} = -2(\alpha_1-\lambda) < x^2 > -2\alpha_2 < x^3 > -2\alpha_3 < x^4 > = 0$$

So

$$\lambda = \frac{\alpha_1 < x^2 > + \alpha_2 < x^3 > + \alpha_3 < x^4 >}{< x^2 >}$$

Let $x = a\cos(\omega t+\beta)$, then

$$< x^2 > = \frac{1}{T}\int_0^T a^2\cos^2(\omega t+\beta)dt = \frac{a^2}{2\pi}\int_0^{2\pi}\cos^2\psi d\psi = \frac{a^2}{2}$$

$$< x^4 > = \frac{1}{T}\int_0^T a^4\cos^4(\omega t+\beta)dt = \frac{a^4}{2\pi}\int_0^{2\pi}\cos^4\psi d\psi = \frac{3a^4}{8}$$

$$< x^3 > = \frac{1}{T}\int_0^T a^3\cos^3(\omega t+\beta)dt = 0$$

$$\text{So}\quad \lambda = \alpha_1 + \frac{3}{4}\alpha_3 a^2 \tag{2.25.2}$$

Let $u = \varepsilon v$, we obtain

$$\ddot{v} + \omega_0^2 v + \varepsilon\delta v^2 + \varepsilon^2\alpha v^3 = 0 \tag{2.25.3}$$

where $\omega_0^2 = \alpha_1,\quad \delta = \alpha_2,\quad \alpha = \alpha_3$. Equation (2.25.3) is the same as the Eq. (2.2.51) in Exercise 2.2(c), where the the method of multiple scales has been used to find the nonlinear frequency-amplitude relationship (Eq. (2.2.71)). From (2.2.71), the relationship between the nonlinear frequency and the amplitude of the present Exercise can be obtained by

$$\omega^2 = \alpha_1 + \frac{3}{4}\alpha_3 a^2 - \frac{5}{6}\alpha_1^{-1}\alpha_2^2 a^2 \tag{2.25.4}$$

It can be seen that the result obtained by the equivalent linearization method (2.25.2) is different from that obtained by the the method of multiple scales (2.25.4). The reason is that in the equivalent linearization method, the quadratic nonlinear term has no effect on the mean square error and therefore does not affect the linearization parameter λ; whereas in the the method of multiple scales, the quadratic nonlinear term appears in the equation to eliminate secular terms and thus corrects for the relationship between the frequency and amplitude in nonlinear oscillation.

2.26 Exercise 2.26 (Comparison of the Galerkin Method with the the Method of Multiple Scales)

Solution: **(a)** When $x = a\cos(\omega t + \beta)$, the residual R is

$$\begin{aligned} R &= -a\omega^2\cos(\omega t+\beta) + \alpha_1 a\cos(\omega t+\beta) \\ &\quad + \alpha_2 a^2\cos^2(\omega t+\beta) + \alpha_3 a^3\cos^3(\omega t+\beta) \\ &= (-a\omega^2 + \alpha_1 a + \frac{3}{4}\alpha_3 a^3)\cos(\omega t+\beta) \\ &\quad + \frac{1}{2}\alpha_2 a^2 + \frac{1}{2}\alpha_2 a^2\cos(2\omega t+2\beta) + \frac{1}{4}\alpha_3 a^3\cos(3\omega t+3\beta) \end{aligned}$$

According to the Galerkin method, there are

$$< R a\cos(\omega t+\beta) > = \tfrac{1}{T}\int_0^T R a\cos(\omega t+\beta)dt$$
$$= \tfrac{1}{2}a(-a\omega^2+\alpha_1 a+\tfrac{3}{4}\alpha_3 a^3)=0$$

$$\text{So } \omega^2=\alpha_1+\frac{3}{4}\alpha_3 a^2 \tag{2.26.1}$$

(b) When $x=a\cos(\omega t+\beta)+B$, the residual R is

$$R=-\omega^2 a\cos(\omega t+\beta)+\alpha_1[a\cos(\omega t+\beta)+B]$$
$$+\alpha_2[a\cos(\omega t+\beta)+B]^2+\alpha_3[a\cos(\omega t+\beta)+B]^3$$
$$=[-a\omega^2+\alpha_1 a+2\alpha_2 aB+\frac{3}{4}\alpha_3 a^3+3\alpha_3 aB^3]\cos(\omega t+\beta)$$
$$+\alpha_1 B+\alpha_2(\frac{1}{2}a^2+B^2)+\alpha_3(B^3+\frac{3}{2}a^2 B)$$
$$+[\frac{1}{2}\alpha_2 a^2+\frac{3}{2}\alpha_3 a^2 B]\cos(2\omega t+2\beta)+\frac{1}{4}\alpha_3 a^3\cos(3\omega t+3\beta)$$

(c) Using the condition that

$$< R a\cos(\omega t+\beta) > =0$$

we obtain

$$< R[a\cos(\omega t+\beta)+B] > = \tfrac{1}{T}\int_0^T R[a\cos(\omega t+\beta)+B]dt$$
$$= \tfrac{1}{2}a[-a\omega^2+\alpha_1 a+2\alpha_2 aB+\tfrac{3}{4}\alpha_3 a^3+3\alpha_3 aB^3]=0$$

$$\text{i.e.}\quad -a\omega^2+\alpha_1 a+2\alpha_2 aB+\frac{3}{4}\alpha_3 a^3+3\alpha_3 aB^3=0 \tag{2.26.2}$$

(d) Using the condition that

$$< RB > =0$$

we obtain

$$< RB > = \frac{1}{T}\int_0^T RBdt$$

$$= B\left[\alpha_1 B + \alpha_2\left(\frac{1}{2}a^2 + B^2\right) + \alpha_3\left(B^3 + \frac{3}{2}a^2 B\right)\right] = 0$$

$$\text{i.e.}\quad \alpha_1 B + \alpha_2(\frac{1}{2}a^2 + B^2) + \alpha_3(B^3 + \frac{3}{2}a^2 B) = 0 \tag{2.26.3}$$

From the Eqs. (2.26.2) and (2.26.3), we obtain

$$B = -\frac{1}{2}\alpha_2\alpha_1^{-1}a^2 + \cdots$$

$$\text{and } \omega^2 = \alpha_1 + \frac{3}{4}\alpha_3 a^2 - \frac{5}{6}\alpha_2^2\alpha_1^{-1}a^2 \tag{2.26.4}$$

(e) Compare these results with (2.25.4), we can find that the success of the application of Galerkin procedure depends on the hypothetical solution. In order to obtain the effect of all nonlineat terms on the frequency-amplitude relationship of a nonlinear oscillation, we should choose the hypothetical solution to include all nonlinear coefficients appearing in the odd harmonic terms of the residual R.

2.27 Exercise 2.27 (Equations Containing Second, Third and Fourth Order Nonlinear Terms)

Solution: (a) At the singularity point of the system, there is $\ddot{u} = \dot{u} = 0$, which gives

$$-u + u^4 = 0$$

So $u = 1$ is a singularity of the system and the potential energy of the system is

$$V(u) = -\frac{1}{2}u^2 + \frac{1}{5}u^5$$

Since $.V''(u = 1) = [-1 + 4u^3]_{u=1} = 3 > 0$
the singularity $u = 1$ is the center.
Let $x = u - 1$, which gives

$$\ddot{x} - (x + 1) + (x + 1)^4 = 0$$

$$\text{i.e.}\quad \ddot{x} + 3x + 6x^2 + 4x^3 + x^4 = 0 \tag{2.27.1}$$

Let

$$x = \varepsilon v \tag{2.27.2}$$

Equation (2.27.1) changes to

$$\ddot{v} + 3v + 6\varepsilon v^2 + 4\varepsilon^2 v^3 + \varepsilon^3 v^4 = 0 \tag{2.27.3}$$

The general form of Eq. (2.27.3) is given by

$$\ddot{v} + \omega_0^2 v + \varepsilon \alpha_2 v^2 + \varepsilon^2 \alpha_3 v^3 + \varepsilon^3 \alpha_4 v^4 = 0 \tag{2.27.4}$$

(b) We seek an expansion of the following form for the above equation around the singularity $v = 0$

$$v = v_0(T_0, T_1, T_2) + \varepsilon v_1(T_0, T_1, T_2) + \varepsilon^2 v_2(T_0, T_1, T_2) + \cdots \tag{2.27.5}$$

where $T_n = \varepsilon^n t$. Substitute (2.27.5) into (2.27.4) and retain to $O(\varepsilon^2)$, we obtain

$$\begin{aligned}
0 &= \ddot{v} + \omega_0^2 v + \varepsilon \alpha_2 v^2 + \varepsilon^2 \alpha_3 v^3 + \varepsilon^3 \alpha_4 v^4 \\
&= [D_0^2 + 2\varepsilon D_0 D_1 + \varepsilon^2 (D_1^2 + 2D_0 D_2) + \cdots](v_0 + \varepsilon v_1 + \varepsilon^2 v_2 + \cdots) \\
&+ \omega_0^2 (v_0 + \varepsilon v_1 + \varepsilon^2 v_2 + \cdots) + \varepsilon \alpha_2 (v_0 + \varepsilon v_1 + \varepsilon^2 v_2 + \cdots)^2 \\
&+ \varepsilon^2 \alpha_3 (v_0 + \varepsilon v_1 + \varepsilon^2 v_2 + \cdots)^3 + \varepsilon^3 \alpha_4 (v_0 + \varepsilon v_1 + \varepsilon^2 v_2 + \cdots)^4 \\
&= D_0^2 v_0 + \omega_0^2 v_0 + \varepsilon (D_0^2 v_1 + 2D_0 D_1 v_0 + \omega_0^2 v_1 + \alpha_2 v_0^2) \\
&+ \varepsilon^2 [D_0^2 v_2 + 2D_0 D_1 v_1 + (D_1^2 + 2D_0 D_2) v_0 + \omega_0^2 v_2 + 2\alpha_2 v_0 v_1 + \alpha_3 v_0^3] + \cdots
\end{aligned} \tag{2.27.6}$$

Equate the coefficients of the same power of ε gives

$$D_0^2 v_0 + \omega_0^2 v_0 = 0 \tag{2.27.7}$$

$$D_0^2 v_1 + \omega_0^2 v_1 = -2D_0 D_1 v_0 - \alpha_2 v_0^2 \tag{2.27.8}$$

$$D_0^2 v_2 + \omega_0^2 v_2 = -2D_0 D_1 v_1 - (D_1^2 + 2D_0 D_2) v_0 - 2\alpha_2 v_0 v_1 - \alpha_3 v_0^3 \tag{2.27.9}$$

The general solution of (2.2.55) is

$$v_0 = A e^{i\omega_0 T_0} + \overline{A} e^{-i\omega_0 T_0} \tag{2.27.10}$$

where $A = A(T_1, T_2)$. Substituting the above equation into (2.27.8), we obtain

$$\begin{aligned}
D_0^2 v_1 + \omega_0^2 v_1 &= -2D_0 D_1 (A e^{i\omega_0 T_0} + \overline{A} e^{-i\omega_0 T_0}) - \alpha_2 (A e^{i\omega_0 T_0} + \overline{A} e^{-i\omega_0 T_0})^2 \\
&= -2\alpha_2 A\overline{A} - 2i\omega_0 D_1 A e^{i\omega_0 T_0} - \alpha_2 A^2 e^{2i\omega_0 T_0} + cc
\end{aligned} \tag{2.27.11}$$

To eliminate secular terms, we need $D_1A = 0$. Therefore $A = A(T_2)$ and the particular solution of (2.27.11) is

$$v_1 = -\frac{2\alpha_2 A\overline{A}}{\omega_0^2} + \frac{\alpha_2 A^2}{3\omega_0^2}e^{2i\omega_0 T_0} + cc \tag{2.27.12}$$

Substitute (2.27.10) and (2.27.12) into (2.27.9), and take $A = A(T_2)$ into account, we obtain

$$\begin{aligned} &D_0^2 v_2 + \omega_0^2 v_2 = -2D_0D_1v_1 - (D_1^2 + 2D_0D_2)v_0 - 2\alpha_2 v_0 v_1 - \alpha_3 v_0^3 \\ &= [-2i\omega_0 D_2 A + (\tfrac{10\alpha_2^2}{3\omega_0^2} - 3\alpha_3)A^2\overline{A}]e^{i\omega_0 T_0} + cc + NST \end{aligned} \tag{2.27.13}$$

where cc denotes the complex conjugate term of its preceding terms and NST denotes terms which do not generate secular terms. In order to eliminate the secular term, we need

$$2i\omega_0 D_2 A - (\frac{10\alpha_2^2}{3\omega_0^2} - 3\alpha_3)A^2\overline{A} = 0 \tag{2.27.14}$$

$$\text{Let} \quad A = \frac{1}{2}ae^{i\beta} \tag{2.27.15}$$

Substitute (2.27.15) into (2.27.14), we obtain

$$i\omega_0 D_2 a - \omega_0 a D_2 \beta - \frac{1}{8}(\frac{10\alpha_2^2}{3\omega_0^2} - 3\alpha_3)a^3 = 0 \tag{2.27.16}$$

Separate the real and imaginary parts of the above equation yields

$$D_2 a = 0, D_2\beta + \frac{1}{8\omega_0}(\frac{10\alpha_2^2}{3\omega_0^2} - 3\alpha_3)a^2 = 0 \tag{2.27.17}$$

So a is a constant. Integrating the second equation of (2.27.17), we obtain

$$\beta = -\frac{1}{8\omega_0}(\frac{10\alpha_2^2}{3\omega_0^2} - 3\alpha_3)a^2 T_2 + \beta_0 \tag{2.27.18}$$

Combining the above results, the first-order approximate solution of the Exercise can be written as

$$u = \varepsilon a\cos[\omega_0 t - \frac{1}{8\omega_0}(\frac{10\alpha_2^2}{3\omega_0^2} - 3\alpha_3)a^2\varepsilon^2 t + \beta_0] + \cdots \tag{2.27.19}$$

Let $\varepsilon = 1$ yields

$$u = a\cos[\omega_0 t - \frac{1}{8\omega_0}(\frac{10\alpha_2^2}{3\omega_0^2} - 3\alpha_3)a^2 t + \beta_0] + \cdots \tag{2.27.20}$$

Therefore, the relationship between the frequency and the amplitude of the oscillation is

$$\omega = \omega_0 - \frac{1}{8\omega_0}(\frac{10\alpha_2^2}{3\omega_0^2} - 3\alpha_3)a^2 \tag{2.27.21}$$

$$\text{or } \omega^2 = \omega_0^2 + \frac{3}{4}\alpha_3 a^2 - \frac{5}{6\omega_0^2}\alpha_2^2 a^2 \tag{2.27.22}$$

For this question, $\omega_0^2 = 3,\ \alpha_2 = 6,\ \alpha_3 = 4$, so

$$\omega^2 = 3 - 7a^2 \tag{2.27.23}$$

2.28 Exercise 2.28 (Solving Nonlinear Equations with Segmentation Functions)

Solution: Cases (a) and (b) are special cases of (c), therefore, we solve for case (c).

As seen from the Fig. 2.9a–c, $F(u)$ is an odd function of u, so this Exercise can be solved by harmonic balance method, equivalent linearization method and Galerkin method. Here, we use the Galerkin method.

Let

$$u = a\cos(\omega t + \beta)$$

The residual R is $R = -\omega^2 a\cos(\omega t + \beta) + F[a\cos(\omega t + \beta)]$
From the condition $< Ra\cos(\omega t + \beta) > = 0$, we obtain

$$0 = \langle Ra\cos(\omega t + \beta)\rangle = \frac{1}{T}\int_0^T Ra\cos(\omega t + \beta)dt$$

$$= -\frac{\omega^2 a^2}{2} + \frac{1}{T}\int_0^T a\cos(\omega t + \beta)F[a\cos(\omega t + \beta)]dt$$

$$= -\frac{\omega^2 a^2}{2} + \frac{1}{2\pi}\int_0^{2\pi} a\cos\psi F[a\cos\psi]d\psi$$

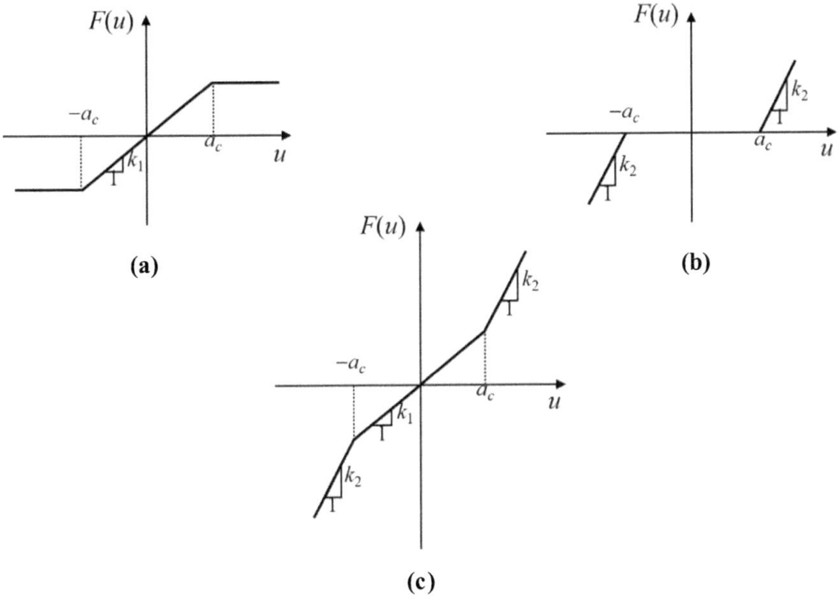

Fig. 2.9 $F(u)$ for three different cases (a)–(c) in Exercise 2.28

$$= -\frac{\omega^2 a^2}{2} + \frac{2}{\pi}\int_0^{\pi/2} a\cos\psi F[a\cos\psi]d\psi$$

Since $d\psi = -du/(a\sin\psi) = -du/\sqrt{a^2-u^2}$, the above equation can be changed to

$$\begin{aligned}
0 &= < R a\cos(\omega t+\beta) > \\
&= -\frac{\omega^2 a^2}{2} - \frac{2}{\pi}\int_a^0 \frac{uF(u)du}{\sqrt{a^2-u^2}} \\
&= -\frac{\omega^2 a^2}{2} + \frac{2}{\pi}\int_0^{a_c} \frac{uF(u)du}{\sqrt{a^2-u^2}} + \frac{2}{\pi}\int_{a_c}^{a} \frac{uF(u)du}{\sqrt{a^2-u^2}} \\
&= -\frac{\omega^2 a^2}{2} + \frac{2}{\pi}\int_0^{a_c} \frac{uF(u)du}{\sqrt{a^2-u^2}} + \frac{2}{\pi}\int_{a_c}^{a} \frac{uF(u)du}{\sqrt{a^2-u^2}}
\end{aligned} \tag{2.28.1}$$

The two integrals on the right-hand side of Eq. (2.28.1) are calculated as follows:

$$\int_0^{a_c} \frac{uF(u)du}{\sqrt{a^2-u^2}} = \int_0^{a_c} \frac{uF(u)}{\sqrt{a^2-u^2}}du = \int_0^{a_c} \frac{k_1u^2}{\sqrt{a^2-u^2}}du$$

$$= k_1(-u\sqrt{a^2-u^2}|_0^{a_c} + \int_0^{a_c} \sqrt{a^2-u^2}du)$$

$$= k_1(-u\sqrt{a^2-u^2}|_0^{a_c} + \frac{u}{2}\sqrt{a^2-u^2}|_0^{a_c} + \frac{a^2}{2}\sin^{-1}\frac{u}{a}|_0^{a_c})$$

$$= k_1(-\frac{u}{2}\sqrt{a^2-u^2}|_0^{a_c} + \frac{a^2}{2}\sin^{-1}\frac{u}{a}|_0^{a_c})$$

$$= -\frac{k_1a_c}{2}\sqrt{a^2-a_c^2} + \frac{k_1a^2}{2}\sin^{-1}\frac{a_c}{a}$$

$$\int_{a_c}^{a} \frac{uF(u)du}{\sqrt{a^2-u^2}} = \int_{a_c}^{a} \frac{uF(u)}{\sqrt{a^2-u^2}}du = \int_{a_c}^{a} \frac{[k_1a_c+k_2(u-a_c)]u}{\sqrt{a^2-u^2}}du$$

$$= \int_{a_c}^{a} \frac{(k_1a_c-k_2a_c)u}{\sqrt{a^2-u^2}}du + \int_{a_c}^{a} \frac{k_2u^2}{\sqrt{a^2-u^2}}du$$

$$= -(k_1a_c-k_2a_c)\sqrt{a^2-u^2}|_{a_c}^{a} - \frac{k_2u}{2}\sqrt{a^2-u^2}|_{a_c}^{a} + \frac{k_2a^2}{2}\sin^{-1}\frac{u}{a}|_{a_c}^{a}$$

$$= (k_1a_c-k_2a_c)\sqrt{a^2-a_c^2} + \frac{k_2a_c}{2}\sqrt{a^2-a_c^2} + \frac{\pi k_2a^2}{4} - \frac{k_2a^2}{2}\sin^{-1}\frac{a_c}{a}$$

Substituting these two results into (2.28.1), we obtain

$$-\frac{\omega^2a^2}{2} + \frac{2}{\pi}[-\frac{k_1a_c}{2}\sqrt{a^2-a_c^2} + \frac{k_1a^2}{2}\sin^{-1}\frac{a_c}{a}$$
$$+ (k_1a_c-k_2a_c)\sqrt{a^2-a_c^2} + \frac{k_2a_c}{2}\sqrt{a^2-a_c^2}$$
$$+\frac{\pi k_2a^2}{4} - \frac{k_2a^2}{2}\sin^{-1}\frac{a_c}{a}] = 0 \tag{2.28.2}$$

$$\text{So } \omega^2 = k_2 - \frac{2}{\pi}(k_2-k_1)[\sin^{-1}\frac{a_c}{a} + \frac{a_c}{a}\sqrt{1-\frac{a_c^2}{a^2}}] \tag{2.28.3}$$

This is the nonlinear frequency-amplitude relationship of the system when $F(u)$ is shown in Fig. 2.9c.

For case (a), $k_2 = 0$ in (2.28.3), so the nonlinear frequency-amplitude relationship can be obtained as

$$\omega^2 = \frac{2k_1}{\pi}[\sin^{-1}\frac{a_c}{a} + \frac{a_c}{a}\sqrt{1 - \frac{a_c^2}{a^2}}] \tag{2.28.4}$$

For case (b), $k_1 = 0$ in (2.28.3), so the nonlinear frequency-amplitude relationship can be obtained as

$$\omega^2 = k_2 - \frac{2k_2}{\pi}[\sin^{-1}\frac{a_c}{a} + \frac{a_c}{a}\sqrt{1 - \frac{a_c^2}{a^2}}] \tag{2.28.5}$$

2.29 Exercise 2.29 (Nonlinear Equations with a Single Arbitrary Subpartial Nonlinear Term Only)

Solution: Let the solution of the equation be

$$u = a\cos(\omega t + \beta)$$

Substituting it for the given nonlinear differential equation, we obtain

$$-\omega^2 a\cos\psi + ka^n\cos^n\psi = 0, \qquad \psi = \omega t + \beta \tag{2.29.1}$$

$\cos^n\psi$ is a periodic function of ψ with a period of 2π, which can be expanded into a Fourier series

$$\cos^n\psi = b_1\cos\psi + \cdots$$

where

$$\begin{aligned} b_1 &= \tfrac{1}{\pi}\int_0^{2\pi} \cos\psi\cos^n\psi d\psi = \tfrac{4}{\pi}\int_0^{\pi/2} \cos^{n+1}\psi d\psi \\ &= \tfrac{4}{\pi}\tfrac{\sqrt{\pi}}{2}\tfrac{\Gamma[\frac{1}{2}(n+2)]}{\Gamma[\frac{1}{2}(n+3)]} = \tfrac{2}{\sqrt{\pi}}\tfrac{\Gamma[\frac{1}{2}(n+2)]}{\Gamma[\frac{1}{2}(n+3)]} \end{aligned}$$

The second equal sign of the above equation holds because n is an odd integer. Substituting this to (2.29.1), we obtain

$$-\omega^2\cos\psi + ka^{n-1}\frac{2}{\sqrt{\pi}}\frac{\Gamma[\frac{1}{2}(n+2)]}{\Gamma[\frac{1}{2}(n+3)]}\cos\psi + \cdots = 0$$

So

$$\omega^2 = \frac{2k}{\sqrt{\pi}}a^{n-1}\frac{\Gamma[\frac{1}{2}(n+2)]}{\Gamma[\frac{1}{2}(n+3)]} \tag{2.29.2}$$

2.30 Exercise 2.30 (Higher Order Nonlinear Equations)

Solution: Let $u = 1 + v$ and substitute it to the governing equation, we obtain

$$
\begin{aligned}
0 &= \ddot{v} - (1+v) + (1+v)^n \\
&= \ddot{v} - (1+v) + 1 + nv + \tfrac{n(n-1)}{2!}v^2 + \tfrac{n(n-1)(n-2)}{3!}v^3 + \cdots
\end{aligned}
$$

i.e.

$$\ddot{v} + (n-1)v + \frac{1}{2}n(n-1)v^2 + \frac{1}{6}n(n-1)(n-2)v^3 + \cdots = 0$$

Let

$$\omega_0^2 = n-1, \quad \alpha_2 = \frac{1}{2}n(n-1), \quad \alpha_3 = \frac{1}{6}n(n-1)(n-2)$$

Retaining to $O(v^3)$, the equation becomes

$$\ddot{v} + \omega_0^2 v + \alpha_2 v^2 + \alpha_3 v^3 = 0 \tag{2.30.1}$$

Equation (2.30.1) is the same as the equation in Exercise 2.27, so the relationship between the oscillation frequency ω and the amplitude a is

$$\omega^2 = \omega_0^2 + \frac{3}{4}\alpha_3 a^2 - \frac{5}{6\omega_0^2}\alpha_2^2 a^2 \tag{2.30.2}$$

$$\text{i.e.}\ \ \omega^2 = (n-1)[1 - \frac{1}{4}(\frac{1}{3}n + 1)na^2] \tag{2.30.3}$$

2.31 Exercise 2.31 (Motion of the Disk Constrained by a Linear Spring)

Solution: **(a)** The kinetic energy of the system is

$$T = \frac{1}{2}mr^2\dot{\theta}^2 \tag{2.31.1}$$

The potential energy of the system is

$$V = \frac{1}{2}k\{[\sqrt{r^2 + (r+l)^2 - 2r(r+l)\cos\theta} - f]^2 - (l-f)^2\}$$

$$= \frac{1}{2}k\{[\sqrt{2r(r+l)(1-\cos\theta)+l^2}-f]^2-(l-f)^2\} \tag{2.31.2}$$

where f is the free length of the spring. Substituting the kinetic and potential energy into the Lagrange's equation, we obtain

$$mr^2\ddot{\theta}+kr(r+l)\{1-\frac{f}{[r^2+(r+l)^2-2r(r+l)\cos\theta]^{1/2}}\}\sin\theta=0$$

$$\text{i.e.,}\quad \ddot{\theta}+\omega^2\{1-\frac{f}{[2r(r+l)(1-\cos\theta)+l^2]^{1/2}}\}\sin\theta=0 \tag{2.31.3}$$

where $\omega^2 = 2k(r+l)/(mr)$.

(b) From (2.31.2), we can write

$$\frac{2V}{kl^2}=[\sqrt{2r_0(r_0+1)(1-\cos\theta)+1}-f_0]^2-(1-f_0)^2 \tag{2.31.4}$$

$$\text{where } f_0=\frac{f}{l},\quad r_0=\frac{r}{l}, \tag{2.31.5}$$

From (2.31.4), we can draw the potential energy curve taking the form of $2V/kl^2 \sim \theta$ in Fig. 2.10a–c, where $r_0 = 1/2$ and f_0 take different constant values. It can be seen that.

(i) When $f \le l$, there is only one equilibrium point $\theta = 0$, which is a center.
(ii) When $l < f < l + 2r$, there are three equilibrium points, where $\theta = 0$ is the saddle point and $\theta = \pm\cos^{-1}[1+\frac{1-f_0^2}{2r_0(r_0+1)}]$ are the centers.
(iii) When $f \ge l + 2r$, there is only one equilibrium point $\theta = 0$, which is the saddle point.

(c) Using (2.31.5), we can write the Eq. (2.31.3) as

$$\ddot{\theta}+\omega^2\{1-\frac{f_0}{[2r_0(r_0+1)(1-\cos\theta)+1]^{1/2}}\}\sin\theta=0 \tag{2.31.6}$$

$$\text{where } f_0=\frac{f}{l},\quad r_0=\frac{r}{l}, \tag{2.31.7}$$

We have already known that there are three equilibrium points when $l < f < l + 2r$:

$$\theta_0=0,\quad \theta_1=\cos^{-1}[1+\frac{1-f_0^2}{2r_0(r_0+1)}],\quad \theta_2=-\cos^{-1}[1+\frac{1-f_0^2}{2r_0(r_0+1)}] \tag{2.31.8}$$

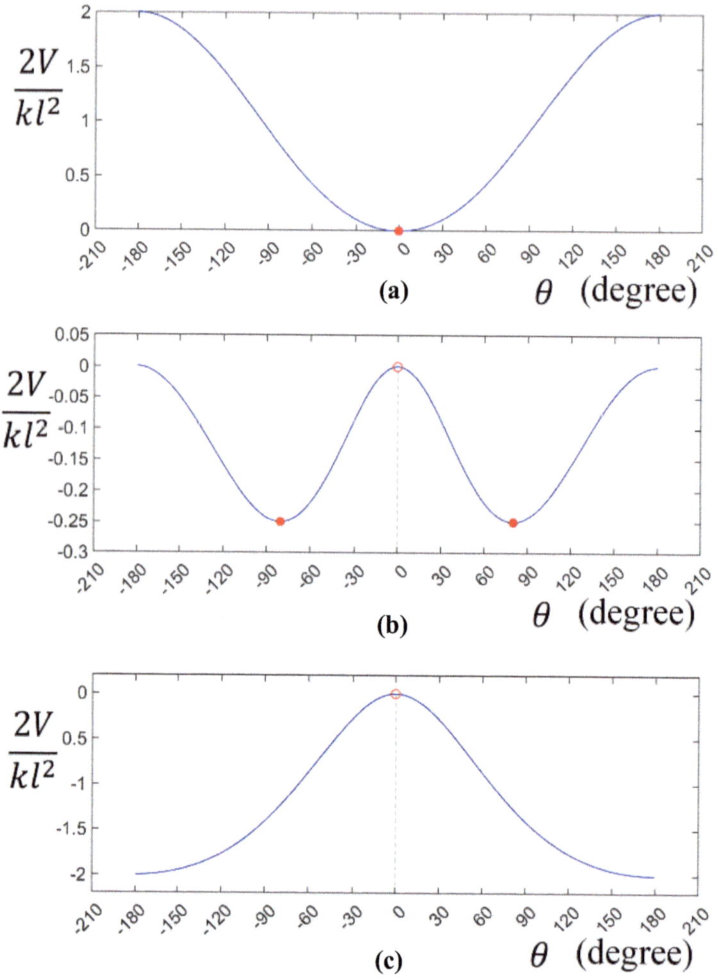

Fig. 2.10 Potential energy curves when **a** $f=0.5l$, **b** $f=l+r$, $r=0.5l$ and **c** $f=l+3r$, $r=0.5l$ for Exercise 2.31

where θ_0 is the saddle point and θ_1 and θ_2 are the centers. Let's study the periodic motion around θ_1. Let

$$\theta=\theta_1+u \tag{2.31.9}$$

Substitute it into (2.31.6) and make an expansion, we obtain

$$\ddot{u}+\omega^2[h'(\theta_1)u+\frac{1}{2}h''(\theta_1)u^2+\frac{1}{6}h'''(\theta_1)u^3+\cdots]=0 \tag{2.31.10}$$

$$\text{where } h(\theta) = \{1 - \frac{f_0}{[2r_0(r_0+1)(1-\cos\theta)+1]^{1/2}}\}\sin\theta \tag{2.31.11}$$

$$\text{and } \omega_0^2 = \omega^2 h^{'}(\theta_1), \quad \alpha_2 = \frac{1}{2}\omega^2 h^{''}(\theta_1), \quad \alpha_3 = \frac{1}{6}\omega^2 h^{'''}(\theta_1) \tag{2.31.12}$$

Retaining to $O(u^3)$, the Eq. (2.31.10) becomes

$$\ddot{u} + \omega_0^2 u + \alpha_2 u^2 + \alpha_3 u^3 = 0 \tag{2.31.13}$$

which is the same type of equation as (2.27.4) in Exercise 2.2(c), so the relationship between the oscillation frequency $\tilde{\omega}$ and the amplitude a is

$$\tilde{\omega} = \omega_0 - \frac{1}{8\omega_0}(\frac{10\alpha_2^2}{3\omega_0^2} - 3\alpha_3)a^2 \tag{2.31.14}$$

2.32 Exercise 2.32 (Rolling Without Slip Rollers Restrained by a Linear Spring)

Solution: (a) The kinetic energy of the system is

$$T = \frac{1}{2}(\frac{3}{2}Mr^2)(\frac{\dot{x}}{r})^2 = \frac{1}{2}\cdot\frac{3}{2}M\dot{x}^2$$

The potential energy of the system is

$$V = \frac{1}{2}k[(\sqrt{x^2+r^2} - f)^2 - (r-f)^2] = \frac{1}{2}k[x^2 + 2f\cdot(r - \sqrt{x^2+r^2})]$$

where f is the free length of the spring. Substituting the kinetic and potential energy into the Lagrange's equation, we obtain

$$\frac{3}{2}M\ddot{x} + kx - kf\frac{x}{\sqrt{x^2+r^2}} = 0$$

$$\text{i.e. } \ddot{x} + \omega^2[1 - (x^2+r^2)^{-1/2}f]x = 0 \tag{2.32.1}$$

where $\omega^2 = 2k/3M$. If x/r and f/r are still be written as x and f in the above equation, the equation controlling the motion of the cylinder can be obtained as

$$\ddot{x} + \omega^2[1 - (1+x^2)^{-1/2}f]x = 0 \tag{2.32.2}$$

Note, however, that x and f are now dimensionless lengths.

(b) Transforming x and f in the potential energy expression to dimensionless lengths, we obtain

$$\frac{2V}{kr^2} = x^2 + 2f \cdot (1 - \sqrt{x^2 + 1}) \tag{2.32.3}$$

From the Eq. (2.32.3), we can draw the potential energy curve taking the form of $2V/kr^2 \sim x$ in Fig. 2.11 for different value of f. It can be seen that: when $f \leq 1$, there is only one equilibrium point $x = 0$, which is a center; when $f > 1$, there are three equilibrium points, of which $x = 0$, is the saddle point, and the remaining two are the centers.

(c) When $f = \sqrt{2}$, as shown in Fig. 2.11, the system has three equilibrium points, of which $x = \pm 1$ are the centers. Let's investigate the periodic motion around $x = 1$ by letting

$$x = 1 + u \tag{2.32.4}$$

Substitute it into the Eq. (2.32.2), we obtain

$$\ddot{u} + \omega^2 [1 - \frac{\sqrt{2}}{\sqrt{1 + (1 + u)^2}}](1 + u) = 0$$

Expanding the above equation and retaining to $O(u^3)$ gives

$$0 = \ddot{u} + \omega^2 [1 - \frac{\sqrt{2}}{\sqrt{2}\sqrt{1 + \frac{u^2 + 2u}{2}}}](1 + u)$$

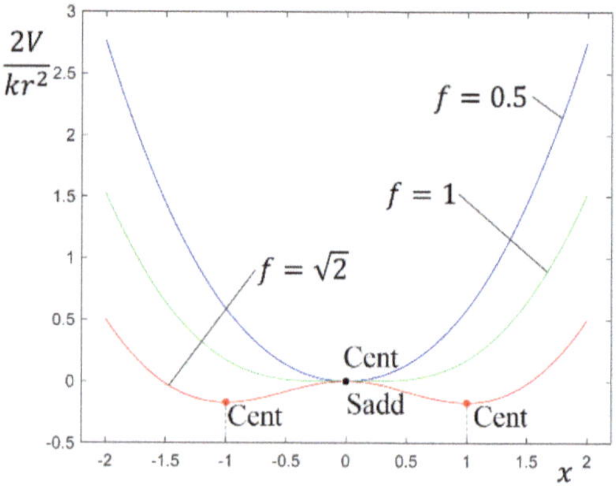

Fig. 2.11 Potential energy curve for different values of f for Exercise 2.32

$$
\begin{aligned}
&= \ddot{u} + \omega^2\{1 - [1 - \frac{u^2 + 2u}{4} + \frac{3}{8}(\frac{u^2 + 2u}{2})^2 + \cdots]\}(1 + u) \\
&= \ddot{u} + \omega^2[\frac{u^2 + 2u}{4} - \frac{3}{32}(u^2 + 2u)^2](1 + u) \\
&= \ddot{u} + \omega^2[\frac{1}{2}u - \frac{1}{8}u^2 - \frac{3}{8}u^3](1 + u) \\
&= \ddot{u} + \omega^2[\frac{1}{2}u - \frac{1}{8}u^2 - \frac{3}{8}u^3 + \frac{1}{2}u^2 - \frac{1}{8}u^3] \\
&= \ddot{u} + \omega^2[\frac{1}{2}u + \frac{3}{8}u^2 - \frac{1}{2}u^3]
\end{aligned}
$$

$$
\text{i.e.} \quad \ddot{u} + \frac{1}{2}\omega^2 u + \frac{3}{8}\omega^2 u^2 - \frac{1}{2}\omega^2 u^3 = 0 \tag{2.32.5}
$$

$$
\text{Let } \omega_0^2 = \frac{1}{2}\omega^2, \quad \alpha_2 = \frac{3}{8}\omega^2, \quad \alpha_3 = -\frac{1}{2}\omega^2 \tag{2.32.6}
$$

(2.32.5) becomes

$$
\ddot{u} + \omega_0^2 u + \alpha_2 u^2 + \alpha_3 u^3 = 0 \tag{2.32.7}
$$

Equation (2.32.7) is the same as the equation in the previous Exercise, so the relationship between the oscillation frequency $\tilde{\omega}$ and the amplitude a is

$$
\tilde{\omega}^2 = \omega_0^2 + \frac{3\alpha_3}{4}a^2 - \frac{5\alpha_2^2}{6\omega_0^2}a^2 \tag{2.32.8}
$$

Substituting (2.32.6) to the above equation, we obtain

$$
\tilde{\omega}^2 = (\frac{1}{2} - \frac{63}{128}a^2)\omega^2 \tag{2.32.9}
$$

2.33 Exercise 2.33 (Example of Higher-Order System)

Solution: (a) When $0 \neq \varepsilon << 1$, the original equation becomes

$$
\dddot{u} - \delta(1 - u^2)\dot{u} = 0 \tag{2.33.1}
$$

Let $x = \dot{u}$, then we have

$$
\begin{cases} \dot{u} = x \\ \ddot{x} - \delta(1 - u^2)x = 0 \end{cases} \tag{2.33.2}
$$

So

$$\frac{d\dot{x}}{du} = \frac{\delta(1-u^2)x}{x} = \delta(1-u^2)$$

After integration, $$\dot{x} - \delta(u - \frac{1}{3}u^3) = b\delta \tag{2.33.3}$$

where b is the integration constant. The above equation is

$$\ddot{u} - \delta(u - \frac{1}{3}u^3) = b\delta$$

Integrate again and obtain $$\dot{u}^2 + 2\delta(\frac{1}{12}u^4 - \frac{1}{2}u^2 + bu) = E \tag{2.33.4}$$

where E is the integration constant.

The first equation of the system of Eqs. (2.33.2), together with the Eq. (2.33.3), form a new system of equations of the first order:

$$\begin{cases} \dot{u} = x \\ \dot{x} = \delta(u - \frac{1}{3}u^3) + b\delta \end{cases} \tag{2.33.5}$$

System of Eqs. (2.33.5) and (2.33.1) are equivalent and, consequently, they have the same equilibrium points. The equilibrium points of (2.33.5) can be determined by

$$\begin{cases} x = 0 \\ \delta(u - \frac{1}{3}u^3) + b\delta = 0 \end{cases} \Rightarrow \begin{cases} x = 0 \\ u^3 - 3u - 3b = 0 \end{cases} \tag{2.33.6}$$

The discriminant of the second equation is

$$\Delta = (\frac{-3b}{2})^2 + (\frac{-3}{3})^3 = \frac{9}{4}b^2 - 1 \tag{2.33.7}$$

So, when $\Delta < 0$, i.e., $b < 2/3$, the system has three equilibrium points. When $\Delta > 0$, i.e., $b > 2/3$, the system has only one equilibrium point.

Here's how to draw the phase trajectories of the system. Write the Eq. (2.33.4) as

$$\begin{cases} V(u) = 2\delta(\frac{1}{12}u^4 - \frac{1}{2}u^2 + bu) \\ \dot{u} = \pm\sqrt{E - V(u)} \end{cases} \tag{2.33.8}$$

For a given value of b, the phase trajectories can be drawn from the Eq. (2.33.8). The potential energy curves and phase trajectories for different values of b are given in Fig. 2.12a–c ($\delta = 1$).

(b) When $0 \neq \varepsilon << 1$, the original equation can be written as

$$\dddot{u} - \delta \dot{u} + \delta u^2 \dot{u} + \varepsilon \ddot{u} + \varepsilon \dot{u} - \varepsilon \delta \dot{u} = 0 \tag{2.33.9}$$

i.e. $\frac{d}{dt}(\ddot{u} - \delta u + \frac{1}{3}\delta u^3 + \varepsilon \dot{u} + \varepsilon u - \varepsilon \delta u) = 0$

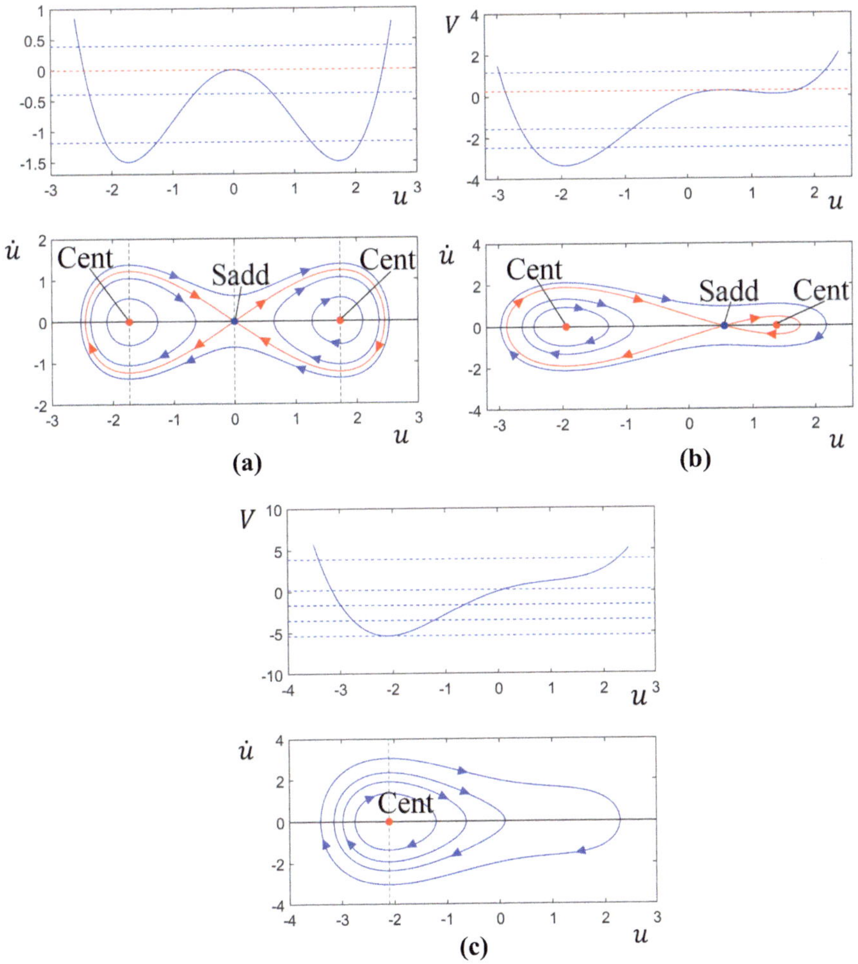

Fig. 2.12 Potential energy curves and solution trajectories when **a** $b = 0$, **b** $b = 0.5$ and **c** $b = 1$ for Exercise 2.33

$$\text{i.e.,}\quad \ddot{u} - \delta u + \frac{1}{3}\delta u^3 + \varepsilon\dot{u} + \varepsilon u - \varepsilon\delta u = h \tag{2.33.10}$$

where h is the integration constant. Let u_s be the singularity of (2.33.10), then it requires

$$\frac{1}{3}\delta u_s^3 - (\delta + \varepsilon\delta - \varepsilon)u_s - h = 0 \tag{2.33.11}$$

$$\text{Let } u = u_s + \varepsilon v \tag{2.33.12}$$

Therefore, the Eq. (2.33.10) becomes

$$\ddot{v} + \delta(u_s^2 - 1)v + \varepsilon(1-\delta)v + \varepsilon\dot{v} + \varepsilon\delta u_s v^2 + \frac{1}{3}\varepsilon^2\delta v^3 = 0 \tag{2.33.13}$$

$$\text{Let } \omega_0^2 = \delta(u_s^2 - 1),\quad \alpha_2 = \delta u_s,\quad \alpha_3 = \frac{1}{3}\delta \tag{2.33.14}$$

then Eq. (2.33.13) changes to

$$\ddot{v} + \omega_0^2 v + \varepsilon(1-\delta)v + \varepsilon\dot{v} + \varepsilon\alpha_2 v^2 + \varepsilon^2\alpha_3 v^3 = 0 \tag{2.33.15}$$

We seek an expansion of the following form for the solution of the above equation around the singularity $v = 0$

$$v = v_0(T_0, T_1, T_2) + \varepsilon v_1(T_0, T_1, T_2) + \varepsilon^2 v_2(T_0, T_1, T_2) + \cdots \tag{2.33.16}$$

where $T_n = \varepsilon^n t$. Substitute (2.33.16) into (2.33.15), we obtain

$$\begin{aligned}
0 &= \ddot{v} + \omega_0^2 v + \varepsilon(1-\delta)v + \varepsilon\dot{v} + \varepsilon\alpha_2 v^2 + \varepsilon^2\alpha_3 v^3 \\
&= [D_0^2 + 2\varepsilon D_0 D_1 + \varepsilon^2(D_1^2 + 2D_0 D_2) + \cdots](v_0 + \varepsilon v_1 + \varepsilon^2 v_2 + \cdots) \\
&\quad + \omega_0^2(v_0 + \varepsilon v_1 + \varepsilon^2 v_2 + \cdots) + \varepsilon(1-\delta)(v_0 + \varepsilon v_1 + \cdots) \\
&\quad + \varepsilon(D_0 + \varepsilon D_1)(v_0 + \varepsilon v_1 + \cdots) + \varepsilon\alpha_2(v_0 + \varepsilon v_1 + \varepsilon^2 v_2 + \cdots)^2 \\
&\quad + \varepsilon^2\alpha_3(v_0 + \varepsilon v_1 + \varepsilon^2 v_2 + \cdots)^3 \\
&= D_0^2 v_0 + \omega_0^2 v_0 + \varepsilon[D_0^2 v_1 + 2D_0 D_1 v_0 + \omega_0^2 v_1 + (1-\delta)v_0 + D_0 v_0 + \alpha_2 v_0^2] \\
&\quad + \varepsilon^2[D_0^2 v_2 + 2D_0 D_1 v_1 + (D_1^2 + 2D_0 D_2)v_0 + \omega_0^2 v_2 + (1-\delta)v_1 \\
&\quad + D_0 v_1 + D_1 v_0 + 2\alpha_2 v_0 v_1 + \alpha_3 v_0^3] + \cdots
\end{aligned} \tag{2.33.17}$$

Retaining to $O(\varepsilon^2)$ and equating coefficients of the same power of ε gives

$$D_0^2 v_0 + \omega_0^2 v_0 = 0 \tag{2.33.18}$$

$$D_0^2 v_1 + \omega_0^2 v_1 = -2D_0 D_1 v_0 - (1-\delta)v_0 - D_0 v_0 - \alpha_2 v_0^2 \tag{2.33.19}$$

$$\begin{aligned} &D_0^2 v_2 + \omega_0^2 v_2 = -2D_0 D_1 v_1 - (D_1^2 + 2D_0 D_2)v_0 - (1-\delta)v_1 \\ &-D_0 v_1 + D_1 v_0 - 2\alpha_2 v_0 v_1 - \alpha_3 v_0^3 \end{aligned} \tag{2.33.20}$$

The general solution of Eq. (2.33.18) is

$$v_0 = Ae^{i\omega_0 T_0} + \overline{A}e^{-i\omega_0 T_0} \tag{2.33.21}$$

where $A = A(T_1, T_2)$. Substituting the above equation into (2.33.19), we obtain

$$\begin{aligned} &D_0^2 v_1 + \omega_0^2 v_1 = -2D_0 D_1 v_0 - (1-\delta)v_0 - D_0 v_0 - \alpha_2 v_0^2 \\ &= -2D_0 D_1 (Ae^{i\omega_0 T_0} + \overline{A}e^{-i\omega_0 T_0}) \\ &-(1-\delta)(Ae^{i\omega_0 T_0} + \overline{A}e^{-i\omega_0 T_0}) \\ &-i\omega_0 (Ae^{i\omega_0 T_0} - \overline{A}e^{-i\omega_0 T_0}) \\ &-\alpha_2 (Ae^{i\omega_0 T_0} + \overline{A}e^{-i\omega_0 T_0})^2 \\ &= -2\alpha_2 A\overline{A} - [2i\omega_0 D_1 A + (1-\delta)A + i\omega_0 A]e^{i\omega_0 T_0} \\ &-\alpha_2 A^2 e^{2i\omega_0 T_0} + cc \end{aligned} \tag{2.33.22}$$

In order to eliminate secular terms, we need

$$2i\omega_0 D_1 A + (1-\delta)A + i\omega_0 A = 0 \tag{2.33.23}$$

$$\text{Let} \quad A = \frac{1}{2}ae^{i\beta} \tag{2.33.24}$$

Substitute (2.33.24) into (2.33.23), we obtain

$$i\omega_0 D_1 a - \omega_0 a D_1 \beta + \frac{1}{2}(1-\delta)a + \frac{1}{2}i\omega_0 a = 0 \tag{2.33.25}$$

Separate the real and imaginary parts of the above equation yields

$$\begin{cases} D_1 a + \frac{1}{2}a = 0 \\ D_1 \beta - \frac{1}{2\omega_0}(1-\delta) = 0 \end{cases} \tag{2.33.26}$$

$$a = a_0 e^{-T_1/2} \tag{2.33.27}$$

$$\beta = \beta_0 exp[\frac{1}{2\omega_0}(1-\delta)T_1] \tag{2.33.28}$$

Combining the above results, we can obtain the first-order approximate solution of the original Eq. (2.33.9)

$$
\begin{aligned}
u &= u_s + \varepsilon v \\
&= u_s + \varepsilon a\cos(\omega_0 T_0 + \beta) \\
&= u_s + \varepsilon a_0 e^{-T_1/2}\cos\{\omega_0 T_0 + \beta_0 \exp[\tfrac{1}{2\omega_0}(1-\delta)T_1]\}
\end{aligned}
$$

$$
\text{i.e.}\quad u = u_s + \varepsilon a_0 e^{-\varepsilon t/2}\cos\{\omega_0 t + \beta_0 exp[\frac{\varepsilon}{2\omega_0}(1-\delta)t]\} \tag{2.33.29}
$$

The oscillation of the system is decaying near the singularity, and therefore, the singularity is a focal point.

Chapter 3
Nonconservative Single-Degree-of-Freedom Systems

3.1 Exercise 3.1 (Singularity Analyses I)

Solution: (a) In order to find the singular points for the equation, we set $\ddot{u} = \dot{u} = 0$ in the original equation and obtain

$$u + u^3 = 0 \tag{3.1.1}$$

therefore, the equation has only one real number solution, i.e., the singular point is

$$u_s = 0 \tag{3.1.2}$$

Let $v = \dot{u}$, then the differential equation for the trajectories is

$$\frac{dv}{du} = -\frac{2\mu v + u + u^3}{v} \tag{3.1.3}$$

The phase trajectories can be sketched in Fig. 3.1a. The singular point is a stable focus.

(b) Set $\ddot{u} = \dot{u} = 0$, we obtain

$$u - u^3 = 0 \tag{3.1.4}$$

therefore, the equation has three real number solutions, i.e., the singularities are

$$u_{s0} = 0, \quad u_{s1,\,s2} = \pm 1 \tag{3.1.5}$$

Let $v = \dot{u}$, then the differential equation for the trajectories is

Z. He et al., *Solved Problems in Nonlinear Oscillations*,
https://doi.org/10.1007/978-981-97-6113-5_3

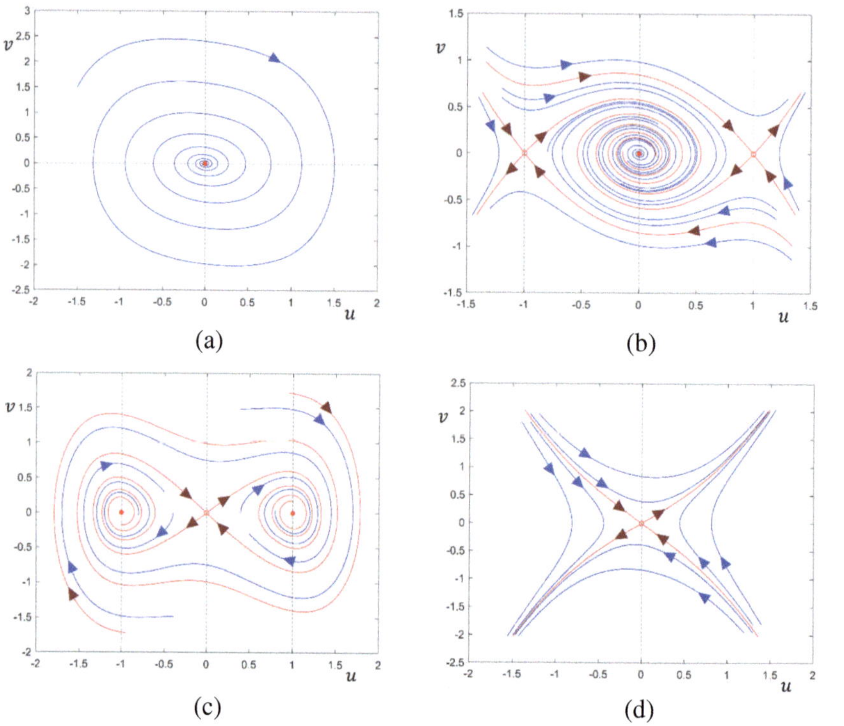

Fig. 3.1 Phase trajectories for Exercise 3.1 (a)–(d) ($\mu = 0.1$)

$$\frac{dv}{du} = -\frac{2\mu v + u - u^3}{v} \tag{3.1.6}$$

The phase trajectories can be sketched in Fig. 3.1b. The singular point $u_{s0} = 0$ is a stable focus, while $u_{s1,s2} = \pm 1$ are two saddle points.

(c) Set $\ddot{u} = \dot{u} = 0$, we obtain

$$-u + u^3 = 0 \tag{3.1.7}$$

therefore, the equation has three real number solutions, i.e., the singularities are

$$u_{s0} = 0,\ u_{s1,s2} = \pm 1 \tag{3.1.8}$$

Let $v = \dot{u}$, then the differential equation for the trajectories is

$$\frac{dv}{du} = -\frac{2\mu v - u + u^3}{v} \tag{3.1.9}$$

The phase trajectories can be sketched in Fig. 3.1c. The singular point $u_{s0} = 0$ is the saddle point, while $u_{s1,s2} = \pm 1$ are two stable foci.

(d) Set $\ddot{u} = \dot{u} = 0$, we obtain

$$-u - u^3 = 0 \tag{3.1.10}$$

therefore, the equation has only one real number solution, i.e., the singular point is

$$u_s = 0 \tag{3.1.11}$$

Let $v = \dot{u}$, then the differential equation for the trajectories is

$$\frac{dv}{du} = -\frac{2\mu v - u - u^3}{v} \tag{3.1.12}$$

The phase trajectories can be sketched in Fig. 3.1d. The singular point $u_{s0} = 0$ is the saddle point.

3.2 Exercise 3.2 (Singularity Analyses II)

Solution: Let $\dot{x} = \dot{y} = 0$, we obtain

$$x^2 - y = 0, x - y = 0$$

and find singularities $s_1(0,\ \ 0)$ and $s_2(1,\ \ 1)$.

Linearizing the original equation around $s_1(0,\ \ 0)$ yields

$$\begin{Bmatrix} \dot{x} \\ \dot{y} \end{Bmatrix} = \begin{bmatrix} 0 & -1 \\ 1 & -1 \end{bmatrix} \begin{Bmatrix} x \\ y \end{Bmatrix}$$

The characteristic equation of the system matrix is

$$\lambda^2 + \lambda + 1 = 0$$

The eigenvalues are

$$\lambda_1 = \frac{-1 + i\sqrt{3}}{2}, \lambda_2 = \frac{-1 - i\sqrt{3}}{2}.$$

Since $\text{Re}(\lambda_{1,2}) < 0$, the singular point $s_1(0,\ \ 0)$ is a stable focus.

Linearizing the original equation around $s_2(1, 1)$ yields

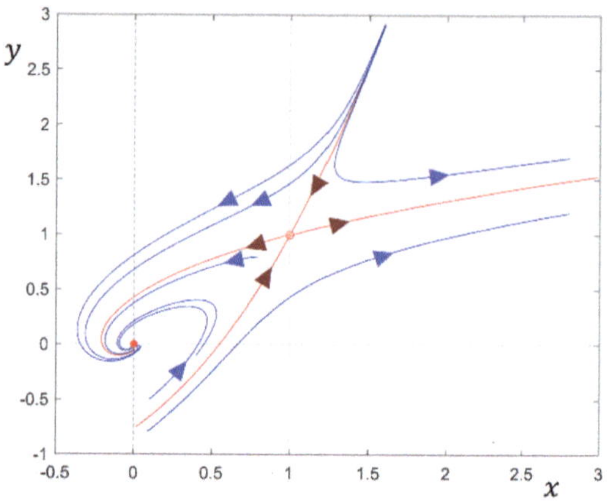

Fig. 3.2 Phase trajectories and separatrices for Exercise 3.2

$$\begin{Bmatrix} \dot{x} \\ \dot{y} \end{Bmatrix} = \begin{bmatrix} 2 & -1 \\ 1 & -1 \end{bmatrix} \begin{Bmatrix} x \\ y \end{Bmatrix}$$

The characteristic equation of the system matrix is

$$\lambda^2 - \lambda - 1 = 0$$

The eigenvalues are

$$\lambda_1 = \frac{1+\sqrt{5}}{2} > 0, \quad \lambda_2 = \frac{1-\sqrt{5}}{2} < 0$$

Therefore, the singular point $s_2(1,\ 1)$ is the saddle point.

We can sketch the trajectories and the separatrices in the state plane by direct numerical integration of the original equations, as in Fig. 3.2.

3.3 Exercise 3.3 (Singularity Analyses III)

Solution: Let $\dot{x} = \dot{y} = 0$, we obtain

$$x^2 + y^2 - 5 = 0, xy - 2 = 0$$

and find singularities $s_1(1,\ 2)$, $s_2(-1,\ -2)$, $s_3(2,\ 1)$ and $s_4(-2,\ -1)$.

Linearizing the original equation around $s_1(1,\ 2)$ yields

$$\begin{Bmatrix} \dot{x} \\ \dot{y} \end{Bmatrix} = \begin{bmatrix} 2 & 4 \\ 2 & 1 \end{bmatrix} \begin{Bmatrix} x \\ y \end{Bmatrix}$$

The characteristic equation of the system matrix is

$$\lambda^2 - 3\lambda - 6 = 0$$

The eigenvalues are

$$\lambda_1 = \frac{3+\sqrt{33}}{2} > 0, \quad \lambda_2 = \frac{3-\sqrt{33}}{2} < 0$$

therefore, the singular point $s_1(1, 2)$ is the saddle point.

Linearizing the original equation around $s_2(-1, -2)$ yields

$$\begin{Bmatrix} \dot{x} \\ \dot{y} \end{Bmatrix} = \begin{bmatrix} -2 & -4 \\ -2 & -1 \end{bmatrix} \begin{Bmatrix} x \\ y \end{Bmatrix}$$

The characteristic equation of the system matrix is

$$\lambda^2 - 3\lambda - 6 = 0$$

The eigenvalues are

$$\lambda_1 = \frac{3+\sqrt{33}}{2} > 0, \quad \lambda_2 = \frac{3-\sqrt{33}}{2} < 0$$

therefore, the singular point $s_2(-1, \ -2)$ is the saddle point.

Linearizing the original equation around $s_3(2, \ 1)$ yields

$$\begin{Bmatrix} \dot{x} \\ \dot{y} \end{Bmatrix} = \begin{bmatrix} 4 & 2 \\ 1 & 2 \end{bmatrix} \begin{Bmatrix} x \\ y \end{Bmatrix}$$

The characteristic equation of the system matrix is

$$\lambda^2 - 6\lambda + 6 = 0$$

The eigenvalues are

$$\lambda_1 = 3 + \sqrt{3} > 0, \quad \lambda_2 = 3 - \sqrt{3} > 0$$

therefore, the singular point $s_3(2, \ 1)$ is an unstable node.

Linearizing the original equation around $s_4(-2, \ -1)$ yields

$$\begin{Bmatrix} \dot{x} \\ \dot{y} \end{Bmatrix} = \begin{bmatrix} -4 & -2 \\ -1 & -2 \end{bmatrix} \begin{Bmatrix} x \\ y \end{Bmatrix}$$

The characteristic equation of the system matrix is

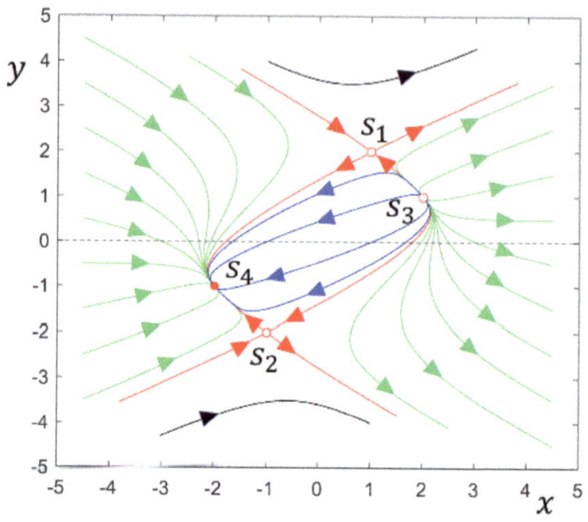

Fig. 3.3 Phase trajectories and separatrices for Exercise 3.3

$$\lambda^2 + 6\lambda + 6 = 0$$

The eigenvalues are

$$\lambda_1 = -3 + \sqrt{3} < 0, \quad \lambda_2 = -3 - \sqrt{3} < 0$$

So, the singular point $s_4(-2, s-1)$ is a stable node.

We can sketch the trajectories and the separatrices in the state plane by direct numerical integration of the original equations, as in Fig. 3.3.

3.4 Exercise 3.4 (Conversion of Rayleigh's Equation to Van Der Pol's Equation)

Solution: Take the derivative of Rayleigh's equation with respect to t to get

$$\dddot{x} - \varepsilon(\ddot{x} - \ddot{x}\dot{x}^2) + \dot{x} = 0$$

Let $u = \dot{x}$, the above equation becomes

$$\ddot{u} - \varepsilon\left(1 - u^2\right)\dot{u} + u = 0$$

This is the van der Pol equation.

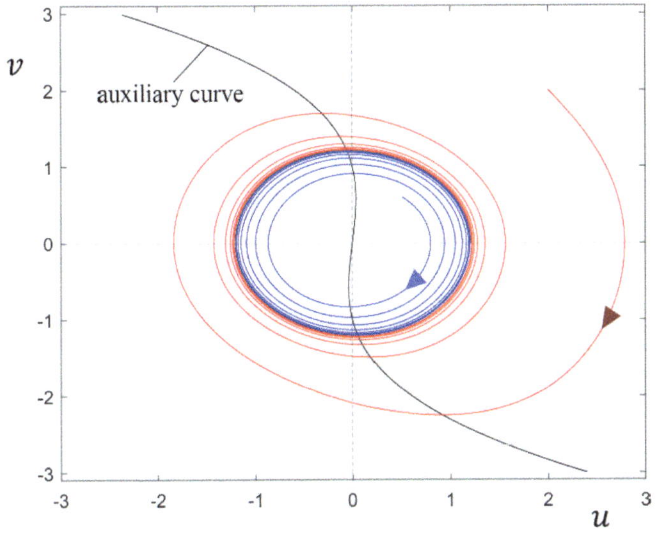

Fig. 3.4 Auxiliary curves, phase trajectories and limit cycle(s) for the given van der Pol equation in Exercise 3.5 ($\varepsilon = 0.1, \quad \beta = 1$)

3.5 Exercise 3.5 (Plotting Phase Trajectories and Limit Cycle(S) for the Van Der Pol Equation)

Solution: Let $v = \dot{u}$, the differential equation for the trajectories is

$$\frac{dv}{du} = -\frac{\varepsilon\left(\beta v^2 - 1\right)v + u}{v}$$

We can sketch the trajectories and the limit cycle of the given van der Pol equation in the state plane by direct numerical integration of the original equations, as in Fig. 3.4. The equation of the auxiliary curve is $\varepsilon\left(\beta v^2 - 1\right)v + u = 0$. The radius of the limit cycle is 1.211 with parameters being set up as $\varepsilon = 0.1$, $\beta = 1$. The reader is also recommended to sketch trajectories by Lienard's method manually and compare with Fig. 3.4.

3.6 Exercise 3.6 (Plotting Phase Trajectories and Limit Cycle(S) for a Given Equation)

Solution: Let $v = \dot{u}$, the differential equation for the trajectories is

$$\frac{dv}{du} = -\frac{\mu \sin v + u}{v}$$

The equation of the auxiliary curve is $\mu \sin v + u = 0$

We can sketch the trajectories and the limit cycle of the given equation in the state plane by direct numerical integration of the original equations, as in Fig. 3.5 ($\mu = 0.8$). The origin of the phase plane is the stable focus. Since the auxiliary curve of this equation varies periodically along the longitudinal axis, the equation actually has an infinite number of limit cycles. As can be seen from the figure, the stable limit cycles (including stable foci) alternate with the unstable limit cycles. It should be pointed out that the phase trajectory from any point in the neighborhood of the unstable limit cycle is divergent, so it is impossible to obtain the exact unstable limit cycle by numerical calculation. We draw the approximate unstable limit cycle in the figure by searching between two neighboring limit cycles and the intersection of auxiliary curves. If the phase trajectories on both sides of a point converge to different stable limit cycles, this point approximately locates on the unstable limit cycle. Two approximate half-limit cycles are obtained by conducting integration on the forward and reverse timeline from this point, respectively. The data will be corrected in a certain way so that they will form a complete approximation of the unstable limit cycle.

The reader is also recommended to sketch trajectories by Lienard's method manually and compare with Fig. 3.5.

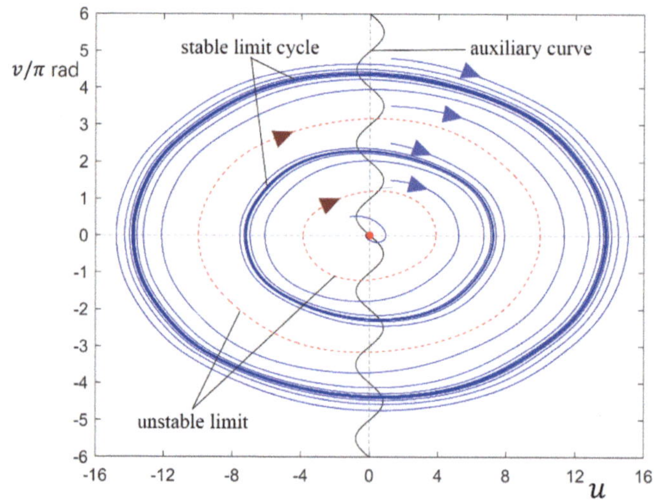

Fig. 3.5 Auxiliary curves, phase trajectories and limit cycle(s) for the given equations ($\mu = 0.8$) in Exercise 3.6

3.7 Exercise 3.7 (Plotting Singularity and Phase Trajectories for a Given Planar System)

Solution: Let $\dot{x}_1 = \dot{x}_2 = 0$, we obtain

$$\begin{cases} -\mu x_1 + k \sin x_2 = 0 \\ \sigma - \alpha x_1^2 + \frac{k}{x_1} \cos x_2 = 0 \end{cases}$$

i.e.,

$$\begin{cases} \mu x_1 = k \sin x_2 \\ \alpha x_1^3 - \sigma x_1 = k \cos x_2 \end{cases} \tag{3.7.1}$$

Squaring and adding these two equations, we obtain

$$\alpha^2 x_1^6 - 2\alpha\sigma x_1^4 + \left(\sigma^2 + \mu^2\right) x_1^2 - k^2 = 0 \tag{3.7.2}$$

x_1 can be obtained from Eq. (3.7.2), x_2 can be obtained by substituting x_1 into the first equation of the system of Eq. (3.7.1).

Find the maximum value of x_1 by differentiating Eq. (3.7.2) with respect to σ and letting $dx_1/d\sigma = 0$, we obtain

$$-2\alpha x_1^4 + 2\sigma x_1^2 = 0 \tag{3.7.3}$$

Join the two Eqs. (3.7.2) and (3.7.3), we obtain

$$\sigma = \frac{\alpha k^2}{\mu^2}, x_{1max} = \frac{k}{\mu} \tag{3.7.4}$$

From (3.7.2), we can solve for σ in terms of x_1

$$\sigma = \alpha x_1^2 \pm \sqrt{\frac{k^2}{x_1^2} - \mu^2} \tag{3.7.5}$$

Then from the system of Eq. (3.7.1), we can obtain

$$\tan x_2 = \frac{\mu x_1}{\alpha x_1^3 - \sigma x_1} \tag{3.7.6}$$

We plot x_1 and x_2 as a function of σ (determined by (3.7.5) and (3.7.6)) for three sets of parameters in the Exercise in Fig. 3.6a, b, respectively. Here, $0 < x_2 < \pi$ can be obtained from the first equation of (3.7.1). In addition, we assume that $x_1 > 0$.

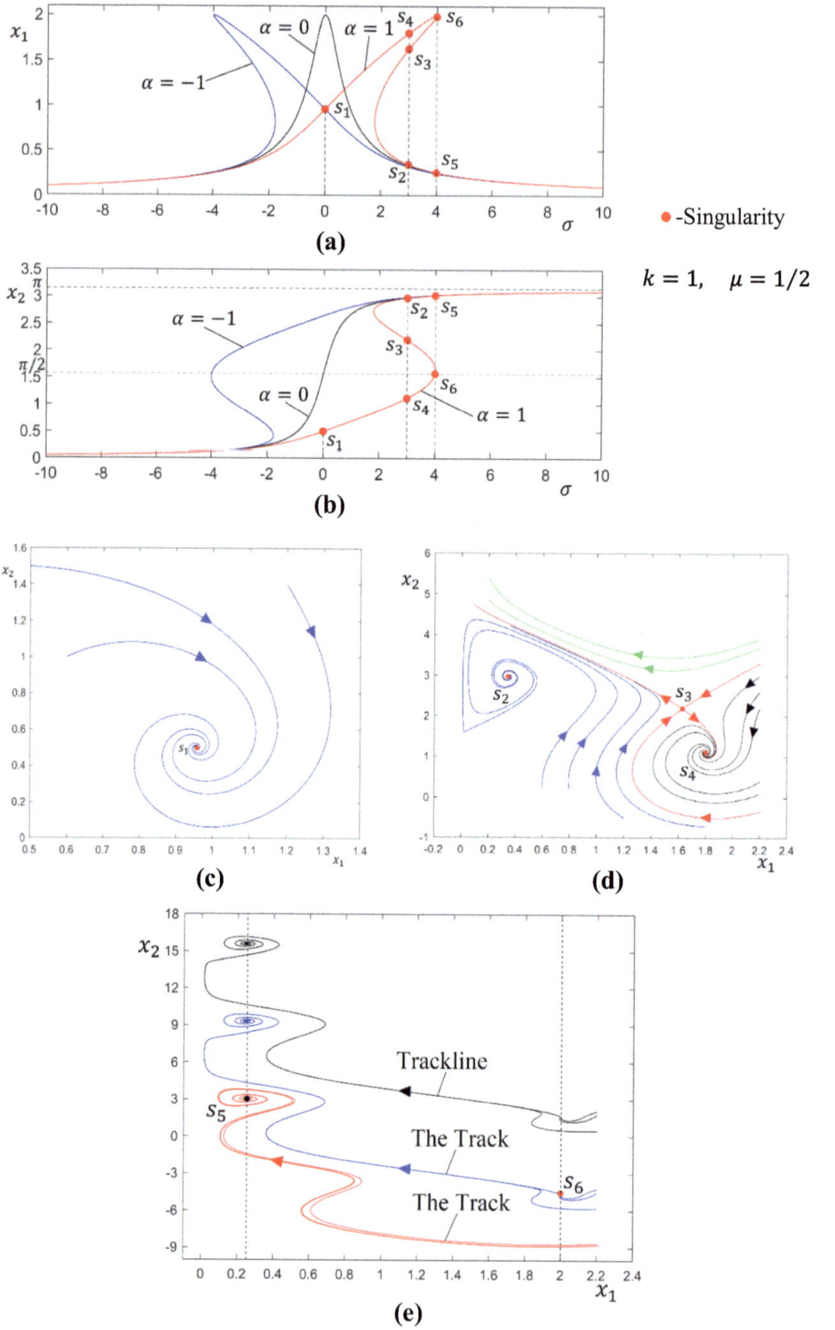

Fig. 3.6 **a** $x_1 \sim \sigma$ and **b** $x_2 \sim \sigma$ curves and trajectories at **c** $\sigma = 0$, **d** $\sigma = 3$ and **e** $\sigma = 4$ for Exercise 3.7

For $\alpha = 1, \quad k = 1$, and$\mu = 1/2$, when $\sigma = 0,3$ and 4, we can obtain the corresponding singular points from the first equation of the system of Eqs. (3.7.1) and (3.7.5):

When $\sigma = 0$, there is only one singular point

$$s_1 : \left(x_1^{(1)}, x_2^{(1)}\right) = (0.9580, 0.5000) \tag{3.7.7}$$

When $\sigma = 3$, there are three singular points

$$s_2 : \left(x_1^{(2)}, x_2^{(2)}\right) = (0.3420, 2.9697)$$

$$s_3 : \left(x_1^{(3)}, x_2^{(3)}\right) = (1.6260, 2.1923)$$

$$s_4 : \left(x_1^{(4)}, x_2^{(4)}\right) = (1.8000, 1.1198) \tag{3.7.8}$$

When $\sigma = 4$, there are two singular points

$$s_5 : \left(x_1^{(5)}, x_2^{(5)}\right) = (0.2520, 3.0153), s_6 : \left(x_1^{(5)}, x_2^{(5)}\right) = \left(2, \frac{\pi}{2}\right) \tag{3.7.9}$$

The coordinate positions of all singular points are shown in Fig. 3.6a, b.

To examine the stability of each singularity, let

$$x_1 = x_1^{(i)} + u, x_2 = x_2^{(i)} + v \tag{3.7.10}$$

where $\left(x_1^{(i)}, x_2^{(i)}\right)$ denotes the coordinate of the singular point s_i. The original equation becomes

$$\begin{cases} \dot{u} = -\mu\left(x_1^{(i)} + u\right) + k\sin(x_2^{(i)} + v) \\ \dot{v} = \sigma - \alpha\left(x_1^{(i)} + u)^2 + \frac{k}{u + x_1^{(i)}}\cos(x_2^{(i)} + v\right) \end{cases}$$

The linearized form of above equations is

$$\begin{Bmatrix} \dot{u} \\ \dot{v} \end{Bmatrix} = \begin{bmatrix} -\mu & k\cos x_2^{(i)} \\ -2\alpha x_1^{(i)} - \frac{k\cos x_2^{(i)}}{(x_1^{(i)})^2} & -\frac{k}{x_1^{(i)}}\sin x_2^{(i)} \end{bmatrix} \begin{Bmatrix} u \\ v \end{Bmatrix} \tag{3.7.11}$$

Considering the system of Eq. (3.7.1), the characteristic equation of coefficient matrix of (3.7.11) can be obtained as

$$\begin{vmatrix} -\mu - \lambda & -x_1^{(i)}\left[\sigma - \left(x_1^{(i)}\right)^2\alpha\right] \\ \frac{1}{x_1^{(i)}}\left[\sigma - 3\left(x_1^{(i)}\right)^2\alpha\right] & -\mu - \lambda \end{vmatrix} = 0 \tag{3.7.12}$$

$$\text{i.e., } \lambda^2 + 2\mu\lambda + \mu^2 + \left[\sigma - \alpha\left(x_1^{(i)}\right)^2\right]\left[\sigma - 3\alpha\left(x_1^{(i)}\right)^2\right] = 0 \tag{3.7.13}$$

$$\text{So, when}\Gamma = \left[\sigma - \alpha\left(x_1^{(i)}\right)^2\right]\left[\sigma - 3\alpha\left(x_1^{(i)}\right)^2\right] + \mu^2 < 0 \tag{3.7.14}$$

the singular point is unstable, otherwise stable.

Substituting the parameter $\alpha = 1, \quad k = 1, \quad \mu = 1/2$ and the coordinates of each singularity into (3.7.14), we can obtain the value of Γ corresponding to each singularity:

	s_1	s_2	s_3	s_4	s_5	s_6
$\Gamma =$	2.777	7.887	−1.506	1.863	15.246	0.250

So, s_3 is an unstable singularity and the others are stable singularities.

We can sketch the trajectories in the state plane for $\sigma = 0, 3,$ and 4 in Fig. 3.6c–e, respectively, when $\alpha = 1, \quad k = 1, \quad \mu = 1/2$.

3.8 Exercise 3.8 (Plot the Trajectory of a Singularity for a Given Planar System and Determine the Stability of the Singularity)

Solution: (a) Let $\dot{x}_1 = \dot{x}_2 = 0$, we can obtain the equations that the singular point should satisfy

$$\begin{cases} x_1\left(1 - x_1^2\right) = -f \cos x_2 \\ x_1\left(\sigma + \nu x_1^2\right) = f \sin x_2 \end{cases} \tag{3.8.1}$$

Squaring these two equations and adding together yields

$$x_1^2\left[(1 - x_1^2)^2 + \left(\sigma + \nu x_1^2\right)^2\right] = f^2$$

i.e.,

$$\rho\left[(1 - \rho)^2 + \left(\sigma + \nu\rho\right)^2\right] = f^2 \tag{3.8.2}$$

where $\rho = x_1^2$.

(b) We can solve for σ from (3.8.2), i.e.,

$$\sigma = -\nu\rho \pm \sqrt{\frac{f^2}{\rho} - (1-\rho)^2} \tag{3.8.3}$$

The $\rho \sim \sigma$ curve can be sketched on the $\rho\sigma$-plane in Figure a for different values of f from Eq. (3.8.3) when $\nu = -0.15$. It can be seen that when $f^2 = 4/27$,ρ is a multi-valued function of σ.

(c) Linearize the original equation around any singularity (x_{1s}, x_{2s}). To do this, let

$$x_1 = x_{1s} + u, \quad x_2 = x_{2s} + v$$

and substitute them into the original equation, we can obtain the linearized equations about u and v:

$$\begin{Bmatrix} \dot{u} \\ \dot{v} \end{Bmatrix} = \begin{bmatrix} 1 - 3x_{1s}^2 & -f\sin x_{2s} \\ 2\nu x_{1s} + \frac{f\sin x_{2s}}{x_{1s}^2} & -\frac{f\cos x_{2s}}{x_{1s}} \end{bmatrix} \begin{Bmatrix} u \\ v \end{Bmatrix}$$

Considering Eq. (3.8.1), the above equation becomes

$$\begin{Bmatrix} \dot{u} \\ \dot{v} \end{Bmatrix} = \begin{bmatrix} 1 - 3x_{1s}^2 & -x_{1s}\left(\sigma + \nu x_{1s}^2\right) \\ 2\nu x_{1s} + \frac{\sigma + \nu x_{1s}^2}{x_{1s}} & 1 - x_{1s}^2 \end{bmatrix} \begin{Bmatrix} u \\ v \end{Bmatrix} \tag{3.8.4}$$

Considering $\rho = x_{1s}^2$, the characteristic equation can be obtained as

$$\lambda^2 - 2(1-2\rho)\lambda + (1-\rho)(1-3\rho) + (\sigma + \nu\rho)(\sigma + 3\nu\rho) = 0 \tag{3.8.5}$$

So, when

$$\Gamma = (1-\rho)(1-3\rho) + (\sigma + \nu\rho)(\sigma + 3\nu\rho) < 0 \tag{3.8.6}$$

the corresponding eigenvalue is an unstable saddle point; when $\Gamma = 0$, the singular point inside the ellipse is an unstable saddle, as shown in Fig. 3.7; when $\Gamma > 0$, the singular point is outside the ellipse and when

$$\begin{aligned} D &= [2(1-2\rho)]^2 - 4\Gamma \\ &= 4\left[\left(1 - 3\nu^2\right)\rho^2 - 4\nu\sigma\rho - \sigma^2\right] > 0 \end{aligned} \tag{3.8.7}$$

the singularity is the node and focus for $D < 0$.

(d) The singular points outside the ellipse are the nodes or foci, and the stability of these nodes or foci depends on the sign of the real part of λ in Eq. (3.8.5). When

$$1 - 2\rho\langle 0 \Rightarrow \rho\rangle \frac{1}{2} \tag{3.8.8}$$

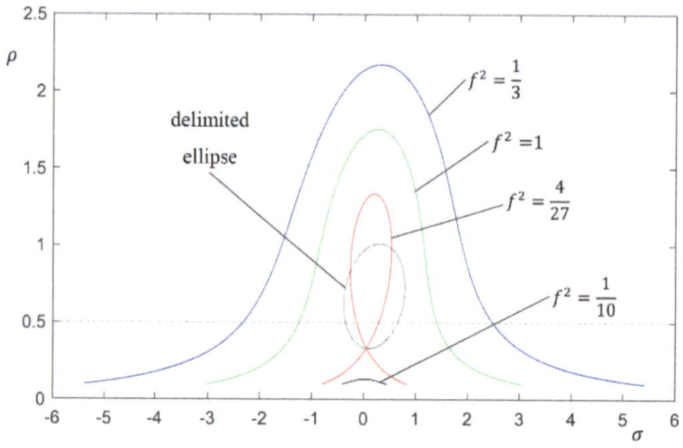

Fig. 3.7 A $\sigma \sim \rho$ curve ($\nu = -0.15$) for Exercise 3.8

these nodes or foci are stable; otherwise, they are unstable.

3.9 Exercise 3.9 (The Method of Multiple Scales for Solving a Single Pendulum with a Coulomb Friction Torque)

Solution: The singular points of the system are $\theta =$ integral multiples of π. The even multiples of π are stable foci, while the odd multiples are saddle points. We seek an approximate solution of the given equations that is uniformly valid near $\theta = 0$, taking both damping and nonlinearity into account. To this end, we expand $\sin\theta$ at $\theta = 0$ and keep only the first two terms; consequently, the above equation becomes

$$\ddot{\theta} + 2\mu \text{sgn}\dot{\theta} + \omega_0^2\theta - \frac{1}{6}\omega_0^2\theta^3 = 0 \tag{3.9.1}$$

Let $\theta = \varepsilon v$, we obtain

$$\ddot{v} + \frac{2\mu}{\varepsilon}\text{sgn}(\dot{v}) + \omega_0^2 v - \frac{1}{6}\varepsilon^2\omega_0^2 v^3 = 0 \tag{3.9.2}$$

We are primarily concerned with lightly damped motions. Consequently, we let

$$\mu = \varepsilon^3\hat{\mu} \tag{3.9.3}$$

where ε is a measure of the amplitude of the motion. Now the nonlinear and damping terms will interact at the same level of approximation.

Substituting this into Eq. (3.9.2) we have

$$\ddot{v} + \omega_0^2 v + 2\varepsilon^2 \hat{\mu}\text{sgn}(\dot{v}) + \varepsilon^2 \alpha_3 v^3 = 0 \tag{3.9.4}$$

where $\alpha_3 = -\omega_0^2/6$. Following the method of multiple scales, we assume

$$v = v_0(T_0, T_1, T_2) + \varepsilon v_1(T_0, T_1, T_2) + \varepsilon^2 v_2(T_0, T_1, T_2) + \cdots \tag{3.9.5}$$

Substituting (3.9.5) into (3.9.4) we have

$$\begin{aligned} 0 &= \ddot{v} + \omega_0^2 v + 2\varepsilon^2 \hat{\mu}\text{sgn}(\dot{v}) + \varepsilon^2 \alpha_3 v^3 \\ &= [D_0^2 + 2\varepsilon D_0 D_1 + \varepsilon^2 (D_1^2 + 2D_0 D_2) + \cdots] \\ &\quad \left(v_0 + \varepsilon v_1 + \varepsilon^2 v_2 + \cdots\right) + \omega_0^2 \left(v_0 + \varepsilon v_1 + \varepsilon^2 v_2 + \cdots\right) \\ &\quad + \varepsilon^2 \left[2\hat{\mu}\text{sgn}(D_0 v_0) + \alpha_3 v_0^3 + \cdots\right] \\ &= D_0^2 v_0 + \omega_0^2 v_0 + \varepsilon\left(D_0^2 v_1 + 2D_0 D_1 v_0 + \omega_0^2 v_1\right) \\ &\quad + \varepsilon^2 \left[D_0^2 v_2 + 2D_0 D_1 v_1 + \left(D_1^2 + 2D_0 D_2\right) v_0 + \omega_0^2 v_2 + 2\hat{\mu}\text{sgn}(D_0 v_0) + \alpha_3 v_0^3\right] + \cdots \end{aligned} \tag{3.9.6}$$

Equalizing coefficients of like powers of ε and retaining to $O(\varepsilon^2)$, we obtain

$$D_0^2 v_0 + \omega_0^2 v_0 = 0 \tag{3.9.7}$$

$$D_0^2 v_1 + \omega_0^2 v_1 = -2D_0 D_1 v_0 \tag{3.9.8}$$

$$D_0^2 v_2 + \omega_0^2 v_2 = -2D_0 D_1 v_1 - \left(D_1^2 + 2D_0 D_2\right) v_0 - 2\hat{\mu}\text{sgn}(D_0 v_0) - \alpha_3 v_0^3 \tag{3.9.9}$$

We can write the solution of Eq. (3.9.7) in the form

$$v_0 = A e^{i\omega_0 T_0} + \overline{A} e^{-i\omega_0 T_0} \tag{3.9.10}$$

where $A = A(T_1, T_2)$. Substituting the above equation into (3.9.8) yields

$$D_0^2 v_1 + \omega_0^2 v_1 = -2i\omega_0 D_1 A e^{i\omega_0 T_0} + cc \tag{3.9.11}$$

where cc denotes the complex conjugate of the preceding terms. In order to eliminate secular terms, we need $D_1 A = 0$, so $A = A(T_2)$. It follows that

$$v_1 = 0 \tag{3.9.12}$$

Substituting (3.9.10) and (3.9.12) into (3.9.9) and taking $A = A(T_2)$ into account, we obtain

$$D_0^2 v_2 + \omega_0^2 v_2 = -2D_0 D_2 v_0 - 2\hat{\mu}\text{sgn}(D_0 v_0) - \alpha_3 v_0^3$$

$$= \left[-2i\omega_0 D_2 A - 3\alpha_3 A^2 \overline{A}\right] e^{i\omega_0 T_0} + cc$$
$$- 2\hat{\mu}\mathrm{sgn}\left(i\omega_0 A e^{i\omega_0 T_0} - i\omega_0 \overline{A} e^{-i\omega_0 T_0}\right) \tag{3.9.13}$$

For a general case, we expand $f(T_0)$ with time period $T_0 = 2\pi/\omega_0$ in a Fourier series as:

$$f(T_0) = \sum_{n=-\infty}^{\infty} f_n e^{in\omega_0 T_0} \tag{3.9.14}$$

$$\text{where} f_n = \frac{\omega_0}{2\pi} \int_0^{2\pi/\omega_0} f(T_0) e^{-in\omega_0 T_0} dT_0 \tag{3.9.15}$$

For this question, let

$$f(T_0) = \mathrm{sgn}(i\omega_0 A e^{i\omega_0 T_0} - i\omega_0 \overline{A} e^{-i\omega_0 T_0}) \tag{3.9.16}$$

In order to eliminate secular terms, we need

$$2i\omega_0 D_2 A + 3\alpha_3 A^2 \overline{A} + 2\hat{\mu} f_1 = 0 \tag{3.9.17}$$

$$\text{Let } A = \frac{1}{2} a e^{i\beta} \tag{3.9.18}$$

$$\text{then } f(T_0) = \mathrm{sgn}(-\omega_0 a \sin\varphi), \quad \text{where } \varphi = \omega_0 T_0 + \beta \tag{3.9.19}$$

Therefore

$$f_1 = \frac{\omega_0}{2\pi} \int_0^{2\pi/\omega_0} f(T_0) e^{-i\omega_0 T_0} dT_0 = \frac{e^{i\beta}}{2\pi} \int_0^{2\pi} \mathrm{sgn}(-\omega_0 a \sin\phi) e^{-i\phi} d\phi$$
$$= -\frac{i\omega_0 a e^{i\beta}}{2\pi} \int_{-\pi}^{\pi} \mathrm{sgn}(-\omega_0 a \sin\phi) \sin\phi d\phi = \frac{ie^{i\beta}}{\pi} \int_0^{\pi} \sin\phi d\phi$$
$$= \frac{2i}{\pi} e^{i\beta} \tag{3.9.20}$$

Substituting (3.9.18) and (3.9.20) into (3.9.17) yields

$$i\omega_0 D_2 a - \omega_0 a D_2 \beta + \frac{3}{8}\alpha_3 a^3 + \frac{4i\hat{\mu}\omega_0 a}{\pi} = 0 \tag{3.9.21}$$

Separate the real and imaginary parts of the above equation yields

$$D_2 a + \frac{4\hat{\mu}}{\pi} = 0 \tag{3.9.22}$$

$$D_2 \beta - \frac{3\alpha_3}{8\omega_0} a^2 = 0 \tag{3.9.23}$$

Therefore

$$a = a_0 - \frac{4\hat{\mu}}{\pi} T_2 = a_0 - \frac{4\mu}{\varepsilon\pi} t \tag{3.9.24}$$

$$\beta = \frac{\pi\alpha_3}{32\hat{\mu}\omega_0}\left(\frac{4\hat{\mu}}{\pi}T_2 - a_0\right)^3 + \beta_0 = \frac{\pi\alpha_3}{32\mu\omega_0}\varepsilon^3 a_0^3\left(\frac{4\mu}{\pi}\frac{t}{\varepsilon a_0} - 1\right)^3 + \beta_0 \tag{3.9.25}$$

Combining the above results, the first order approximate solution of the original equation is

$$\theta = \varepsilon v_0 = \left(\varepsilon a_0 - \frac{4\mu}{\pi} t\right)\cos\left[\omega_0 t + \frac{\pi\alpha_3}{32\mu\omega_0}\varepsilon^3 a_0^3\left(\frac{4\mu}{\pi}\frac{t}{\varepsilon a_0} - 1\right)^3 + \beta_0\right] + \cdots \tag{3.9.26}$$

Let $\varepsilon = 1$, we get

$$\theta = \left(a_0 - \frac{4\mu}{\pi} t\right)\cos\left[\omega_0 t + \frac{\pi\alpha_3}{32\mu\omega_0} a_0^3\left(\frac{4\mu}{\pi}\frac{t}{a_0} - 1\right)^3 + \beta_0\right] + \cdots \tag{3.9.27}$$

where $\alpha_3 = -\omega_0^2/6$.

3.10 Exercise 3.10 (Singularities and Their Stability of a Linearly Damped Pendulum Under a Constant Torque)

Solution: Let $x_1 = \theta$, $x_2 = \dot{x}_1$, the original equation becomes

$$\begin{cases} \dot{x}_1 = x_2 \\ \dot{x}_2 = f - 2\mu x_2 - \omega_0^2 \sin x_1 \end{cases} \tag{3.10.1}$$

For $f = 0$, first Eq. (3.10.1) is linearized near the singular point (0,0) as follows

$$\begin{Bmatrix} \dot{x}_1 \\ \dot{x}_2 \end{Bmatrix} = \begin{bmatrix} 0 & 1 \\ -\omega_0^2 & -2\mu \end{bmatrix}\begin{Bmatrix} x_1 \\ x_2 \end{Bmatrix} \tag{3.10.2}$$

Its characteristic equation is

$$\lambda^2 + 2\mu\lambda + \omega_0^2 = 0 \tag{3.10.3}$$

The eigenvalues are

$$\lambda_1 = -\mu + \sqrt{\mu^2 - \omega_0^2}, \quad \lambda_2 = -\mu - \sqrt{\mu^2 - \omega_0^2} \tag{3.10.4}$$

Since $\mu > 0$, the two eigenvalues are conjugate complex numbers with negative real parts when $\mu < \omega_0$, and the corresponding singularity is a stable focus; when $\mu \geq \omega_0$, both eigenvalues are negative real numbers, and the corresponding singular point is a stable node. In short, the singular point (0,0) is stable.

Second, Eq. (3.10.1) is linearized near the singular point $(\pi, 0)$ as follows

$$\begin{Bmatrix} \dot{x}_1 \\ \dot{x}_2 \end{Bmatrix} = \begin{bmatrix} 0 & 1 \\ \omega_0^2 & -2\mu \end{bmatrix} \begin{Bmatrix} x_1 \\ x_2 \end{Bmatrix} \tag{3.10.5}$$

Its characteristic equation is

$$\lambda^2 + 2\mu\lambda - \omega_0^2 = 0 \tag{3.10.6}$$

The eigenvalues are

$$\lambda_1 = -\mu + \sqrt{\mu^2 + \omega_0^2} > 0, \quad \lambda_2 = -\mu - \sqrt{\mu^2 + \omega_0^2} < 0 \tag{3.10.7}$$

Therefore, the singular point $(\pi, 0)$ is the saddle point and is unstable.

For $f \neq 0$, the equilibrium point should satisfy the following equation

$$\begin{cases} x_2 = 0 \\ f - 2\mu x_2 - \omega_0^2 \sin x_1 = 0 \end{cases} \tag{3.10.8}$$

The coordinates of the equilibrium point are

$$s_1 : x_1 = \sin^{-1}\frac{f}{\omega_0^2}, x_2 = 0; \; s_2 : x_1 = \pi - \sin^{-1}\frac{f}{\omega_0^2}, x_2 = 0 \tag{3.10.9}$$

Therefore, there is no equilibrium point when $f > \omega_0^2$. In that case, the external torque is greater than the maximum gravitational moment of the pendulum, and the pendulum will make a continuous full-circle slewing motion.

When $f \leq \omega_0^2$, first Eq. (3.10.1) is linearized near the singular point s_1

$$\begin{Bmatrix} \dot{x}_1 \\ \dot{x}_2 \end{Bmatrix} = \begin{bmatrix} 0 & 1 \\ -\sqrt{\omega_0^4 - f^2} & -2\mu \end{bmatrix} \begin{Bmatrix} x_1 \\ x_2 \end{Bmatrix} \tag{3.10.10}$$

Its characteristic equation is

$$\lambda^2 + 2\mu\lambda + \sqrt{\omega_0^4 - f^2} = 0 \tag{3.10.11}$$

The eigenvalues are

$$\lambda_1 = -\mu + \sqrt{\mu^2 - \sqrt{\omega_0^4 - f^2}}, \quad \lambda_2 = -\mu - \sqrt{\mu^2 - \omega_0^2} \tag{3.10.12}$$

Since $\mu > 0$, the two eigenvalues are conjugate complex numbers with negative real parts when $\mu^2 < \sqrt{\omega_0^4 - f^2}$ and the corresponding singular point is a stable focus. When $\mu^2 \geq \sqrt{\omega_0^4 - f^2}$, both eigenvalues are negative real numbers and the corresponding singular point is a stable node. In short, the singular point s_1 is stable.

Second, Eq. (3.10.1) is linearized near the singular point s_2

$$\begin{Bmatrix} \dot{x}_1 \\ \dot{x}_2 \end{Bmatrix} = \begin{bmatrix} 0 & 1 \\ \sqrt{\omega_0^4 - f^2} & -2\mu \end{bmatrix} \begin{Bmatrix} x_1 \\ x_2 \end{Bmatrix} \tag{3.10.13}$$

Its characteristic equation is

$$\lambda^2 + 2\mu\lambda - \sqrt{\omega_0^4 - f^2} = 0 \tag{3.10.14}$$

The eigenvalues are

$$\lambda_1 = -\mu + \sqrt{\mu^2 - \sqrt{\omega_0^4 - f^2}}, \quad \lambda_2 = -\mu - \sqrt{\mu^2 - \omega_0^2} \tag{3.10.15}$$

Therefore, the equilibrium point s_2 is the saddle point and is unstable.

$$\text{Let } u = x_1, \quad v = \frac{x_2}{\omega_0}, \quad \tau = \omega_0 t \tag{3.10.16}$$

Equation (3.10.1) change to

$$\begin{cases} u' = v \\ v' = \frac{f}{\omega_0^2} - 2\mu\omega_0 v - \sin u \end{cases} \tag{3.10.17}$$

Taking $f/\omega_0^2 = \sqrt{2}/2$, $\mu\omega_0 = 0.3$, the phase trajectory of the system can be obtained as shown in Fig. 3.8.

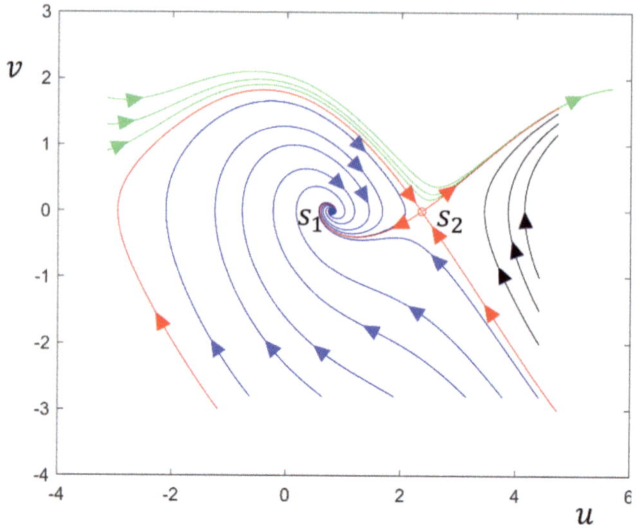

Fig. 3.8 Equilibrium point and phase trajectory line at $f \neq 0$ $\frac{f}{\omega_0^2} = \frac{\sqrt{2}}{2}$, $\mu\omega_0 = 0.3$ for Exercise 3.10

3.11 Exercise 3.11 (Singularities and Their Stability for a Single Pendulum with a Quadratic Damping)

Solution: Let $x_1 = \theta$, $x_2 = \dot{x}_1$, the original equation becomes

$$\begin{cases} \dot{x}_1 = x_2 \\ \dot{x}_2 = -2\mu x_2|x_2| - \omega_0^2 \sin x_1 \end{cases} \tag{3.11.1}$$

Equation (3.11.1) is linearized near the singular point (0,0)

$$\begin{Bmatrix} \dot{x}_1 \\ \dot{x}_2 \end{Bmatrix} = \begin{bmatrix} 0 & 1 \\ -\omega_0^2 & 0 \end{bmatrix} \begin{Bmatrix} x_1 \\ x_2 \end{Bmatrix} \tag{3.11.2}$$

Its characteristic equation is

$$\lambda^2 + \omega_0^2 = 0 \tag{3.11.3}$$

Its eigenvalues are

$$\lambda_1 = i\omega_0, \quad \lambda_2 = -i\omega_0 \tag{3.11.4}$$

Accordingly, the singular point (0,0) is a center, but the original system is not a conservative system, so this result must be incorrect. In order to accurately judge

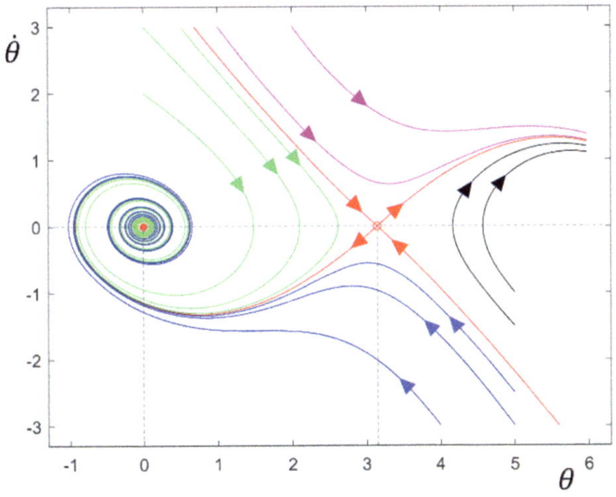

Fig. 3.9 Singular point and phase trajectories of a single pendulum system with quadratic damping $\omega_0 = 1, \quad \mu = 0.2$ for Exercise 3.11

the stability of the singular point (0,0), we need the nonlinear terms in the equation. We directly integrate the nonlinear Eq. (3.11.1) and obtain the phase trajectories as shown in Fig. 3.9. It can be seen that the singular point (0,0) is the stable focus.

3.12 Exercise 3.12 (Analysis of Rayleigh Equation Singularities, Method of Averaging of Solution)

Solution: Let $x_1 = u$, $x_2 = \dot{x}_1$, equation (a) becomes

$$\begin{cases} \dot{x}_1 = x_2 \\ \dot{x}_2 = -\omega_0^2 x_1 + \varepsilon\left(x_2 - x_2^3\right) \end{cases} \tag{3.12.1}$$

Clearly, the singular point is (0,0). Equation (3.12.1) is linearized near this singular point

$$\begin{Bmatrix} \dot{x}_1 \\ \dot{x}_2 \end{Bmatrix} = \begin{bmatrix} 0 & 1 \\ -\omega_0^2 & \varepsilon \end{bmatrix} \begin{Bmatrix} x_1 \\ x_2 \end{Bmatrix} \tag{3.12.2}$$

Its characteristic equation is

$$\lambda^2 - \varepsilon\lambda + \omega_0^2 = 0 \tag{3.12.3}$$

Thus, when $\varepsilon > 0$, the singular point (0,0) is an unstable node or unstable focus, depending on whether $\varepsilon^2 - 4\omega_0^2$ is greater than zero.

When ε is a positive small parameter, equation (a) can be solved by the method of averaging. By setting its solution as

$$u = a\cos(\omega_0 t + \beta) \triangleq a\cos\varphi, \quad \varphi = \omega_0 t + \beta \tag{3.12.4}$$

reintroduce the transformation

$$\dot{u} = -a\omega_0 \sin(\omega_0 t + \beta) \triangleq -a\omega_0 \sin\varphi \tag{3.12.5}$$

Assume a and β are functions of t. Make the derivatives of Eqs. (3.12.4) and (3.12.5) with respect to t yields

$$\dot{u} = \dot{a}\cos\varphi - a\omega_0 \sin\varphi - a\dot{\beta}\sin\varphi \tag{3.12.6}$$

$$\ddot{u} = -\dot{a}\omega_0 \sin\varphi - a\omega_0^2 \cos\varphi - a\omega_0\dot{\beta}\cos\varphi \tag{3.12.7}$$

From (3.12.5) and (3.12.6), we have

$$\dot{a}\cos\varphi - a\dot{\beta}\sin\varphi = 0 \tag{3.12.8}$$

Substitute (3.12.7) and (3.12.5) into equation (a), we obtain

$$\dot{a}\omega_0 \sin\varphi + a\omega_0\dot{\beta}\cos\varphi = -\varepsilon\left(-a\omega_0 \sin\varphi + a^3\omega_0^3 \sin^3\varphi\right) \tag{3.12.9}$$

From Eqs. (3.12.8) and (3.12.9), we can obtain

$$\dot{a} = -\frac{\varepsilon}{\omega_0}\left(-a\omega_0 \sin\varphi + a^3\omega_0^3 \sin^3\varphi\right)\sin\varphi \tag{3.12.10}$$

$$\dot{\beta} = -\frac{\varepsilon}{\omega_0 a}\left(-a\omega_0 \sin\varphi + a^3\omega_0^3 \sin^3\varphi\right)\cos\varphi \tag{3.12.11}$$

When ε is a small parameter, Eqs. (3.12.10) and (3.12.11) imply that a and β vary much more slowly with time t than $\varphi = \omega_0 t + \beta$. In other words,$a$ and β hardly change during the period of oscillation $2\pi/\omega_0$ of φ. This enables us to average out the variations in φ in (3.12.10) and (3.12.11). Averaging these equations over the period $2\pi/\omega_0$ and considering a, β, $\dot{a}$ and $\dot{\beta}$ to be constants while performing the averaging, we obtain the following equations describing the slow variations of a and β:

$$\dot{a} = -\frac{\varepsilon}{2\pi\omega_0}\int_0^{2\pi}\left(-a\omega_0 \sin\varphi + a^3\omega_0^3 \sin^3\varphi\right)\sin\varphi d\varphi \tag{3.12.12}$$

$$\dot{\beta} = -\frac{\varepsilon}{2\pi\omega_0 a}\int_0^{2\pi}\left(-a\omega_0\sin\varphi + a^3\omega_0^3\sin^3\varphi\right)\cos\varphi d\varphi \tag{3.12.13}$$

$$\text{Therefore } \dot{a} = \frac{1}{2}\varepsilon a\left(1 - \frac{3}{4}a^2\omega_0^2\right), \quad \dot{\beta} = 0 \tag{3.12.14}$$

This is equation (b) given in the Exercise. So β is a constant; and the singularity of the equation about a is

$$a_{s1} = 0 \quad \text{or} \quad a_{s2} = \frac{2}{\sqrt{3}\omega_0} \tag{3.12.15}$$

Let $a = a_{si} + v$, the linearized equations in the vicinity of the two singularities are given as

$$\text{Near } a_{s1}: \dot{v} = \frac{1}{2}\varepsilon v \tag{3.12.16}$$

$$\text{Near } a_{s2}: \dot{v} = -\varepsilon v \tag{3.12.17}$$

Therefore, the singularity $a_{s1} = 0$ is unstable while the singularity $a_{s2} = 2/\left(\sqrt{3}\omega_0\right)$ is stable. This implies that the original equation (a) has a steady state constant amplitude $2/\left(\sqrt{3}\omega_0\right)$, i.e., there is a stable limit cycle with amplitude $2/\left(\sqrt{3}\omega_0\right)$.

3.13 Exercise 3.13 (Singularity Analysis of the Restricted Three-Body Exercise)

Solution: (a) Let $\dot{x} = \dot{y} = \ddot{x} = \ddot{y} = 0$, equation (a) becomes

$$\begin{cases} -x = -\frac{m(x-1+m)}{d_1^3} - \frac{(1-m)(x+m)}{d_2^3} \\ -y = -\frac{my}{d_1^3} - \frac{(1-m)y}{d_2^3} \end{cases} \tag{3.13.1}$$

There are 7 sets of solutions, of which we need only real number solutions.

When $x \neq 0, \quad y \neq 0$, the two equations are multiplied by y and x, respectively. Subtraction of the two equations gives

$$\begin{aligned} 0 &= m(m-1)yd_2^3 + m(1-m)yd_1^3 \\ &= my(m-1)\left(d_2^3 - d_1^3\right) \end{aligned}$$

$$= m(m-1)y(d_2 - d_1)\left(d_2^2 + d_1 d_2 + d_1^2\right)$$

$$\text{So } d_2^2 + d_1 d_2 + d_1^2 = 0 \tag{3.13.2}$$

$$\text{or } d_2 - d_1 = 0 \tag{3.13.3}$$

Considering equation (b), we can find no real number solutions for Eq. (3.13.2) but rewrite Eq. (3.13.3) in the following form

$$(1-m-x)^2 = (m+x)^2 \tag{3.13.4}$$

When $x > 1-m$, Eq. (3.13.4) has no solution; while for $x < 1-m$, we can obtain

$$x = \frac{1}{2} - m, \quad d_1^2 = d_2^2 = d^2 = \frac{1}{4} + y^2 \tag{3.13.5}$$

by substituting (3.13.5) into the second equation of (3.13.1), we get

$$(\frac{1}{4} + y^2)^{3/2} = 1$$

From this

$$y = \pm\frac{\sqrt{3}}{2}$$

Therefore, two equilibrium points of the system L_4, L_5 are

$$L_4, \ L_5: \ (x, y) = \left(\frac{1}{2} - m, \ \pm\frac{\sqrt{3}}{2}\right) \tag{3.13.6}$$

When $y = 0$, the system of Eq. (3.13.1) of the second equation is satisfied, and the first equation becomes

$$x = -\frac{m(1-m-x)}{|(1-m-x)|^3} + \frac{(1-m)(x+m)}{|(m+x)|^3} \tag{3.13.7}$$

Previously, we have solved the system for real number solutions in the range $x < 1-m$. Therefore, Eq. (3.13.7) can only have solutions in the range $x > 1-m$, where Eq. (3.13.7) becomes:

$$x - \frac{m}{(1-m-x)^2} - \frac{(1-m)}{(m+x)^2} = 0 \tag{3.13.8}$$

This equation has three real number solutions, which, together with$y = 0$, are the other three equilibrium points of the systemL_1, L_2, L_3. Thus, the system has five equilibrium points L_1, L_2, L_3, L_4, L_5.

(b) Let's analyze the stability of the equilibrium point L_4, L_5. The original equation (a) can be written as

$$\ddot{x} - 2\dot{y} - x = f_1(x, y)$$

$$\ddot{y} + 2\dot{x} - y = f_2(x, y)$$

where

$$f_1(x, y) = -\frac{m(x - 1 + m)}{d_1^3} - \frac{(1 - m)(x + m)}{d_2^3}$$

$$f_2(x, y) = -\frac{my}{d_1^3} - \frac{(1 - m)y}{d_2^3}$$

Linearize the original equation (a) around any equilibrium point (x_s, y_s). For this purpose, let

$$x = x_s + u, \quad y = y_s + v$$

then the linearized equation can be obtained as

$$\ddot{u} - 2\dot{v} - u = \frac{\partial f_1}{\partial x_s}u + \frac{\partial f_1}{\partial y_s}v$$

$$\ddot{v} + 2\dot{u} - v = \frac{\partial f_2}{\partial x_s}u + \frac{\partial f_2}{\partial y_s}v \tag{3.13.9}$$

where

$$\begin{aligned}
\frac{\partial f_1}{\partial x_s} &= -[m - 3m(x_s - 1 + m)^2] - [(1 - m) - 3(1 - m)(x_s + m)^2] \\
&= -1 + 3m + 3(x_s + m)(x_s - m) \\
\frac{\partial f_1}{\partial y_s} &= 3my_s(x_s - 1 + m) + 3y_s(1 - m)(x_s + m) = 3x_s y_s \\
\frac{\partial f_2}{\partial x_s} &= 3my_s(x_s - 1 + m) + 3y_s(1 - m)(x_s + m) = 3x_s y_s \\
\frac{\partial f_2}{\partial y_s} &= -(m - 3my_s^2) - [(1 - m) - 3(1 - m)y_s^2] = 3y_s^2 - 1
\end{aligned} \tag{3.13.10}$$

At the equilibrium point L_4, L_5, the value of the above partial derivative is

$$\frac{\partial f_1}{\partial x_s} = -\tfrac{1}{4}, \qquad \frac{\partial f_1}{\partial y_s} = \pm\tfrac{3\sqrt{3}}{2}\left(\tfrac{1}{2} - m\right)$$
$$\frac{\partial f_2}{\partial x_s} = \pm\tfrac{3\sqrt{3}}{2}\left(\tfrac{1}{2} - m\right), \qquad \frac{\partial f_2}{\partial y_s} = \tfrac{5}{4}$$

Therefore, from Eq. (3.13.9), the linearized equation near the equilibrium point L_4, L_5 is given by

$$\begin{bmatrix} 1 & 0 \\ 0 & 1 \end{bmatrix}\begin{Bmatrix} \ddot{u} \\ \ddot{v} \end{Bmatrix} + \begin{bmatrix} 0 & -2 \\ 2 & 0 \end{bmatrix}\begin{Bmatrix} \dot{u} \\ \dot{v} \end{Bmatrix} + \begin{bmatrix} -\tfrac{3}{4} & \pm\tfrac{3\sqrt{3}}{2}\left(\tfrac{1}{2} - m\right) \\ \pm\tfrac{3\sqrt{3}}{2}\left(\tfrac{1}{2} - m\right) & -\tfrac{9}{4} \end{bmatrix}\begin{Bmatrix} u \\ v \end{Bmatrix} = 0 \tag{3.13.11}$$

Its characteristic equation is

$$\det\begin{bmatrix} \lambda^2 - \tfrac{3}{4} & \pm\tfrac{3\sqrt{3}}{2}\left(\tfrac{1}{2} - m\right) - 2\lambda \\ \pm\tfrac{3\sqrt{3}}{2}\left(\tfrac{1}{2} - m\right) + 2\lambda & \lambda^2 - \tfrac{9}{4} \end{bmatrix} = 0 \tag{3.13.12}$$

$$\text{i.e., } \lambda^4 + \lambda^2 + \frac{27}{4}\left(m - m^2\right) = 0 \tag{3.13.13}$$

$$\text{So } \lambda^2 = \frac{-1 \pm \sqrt{27m^2 - 27m + 1}}{2} \tag{3.13.14}$$

(1) When

$$27m^2 - 27m + 1 > 1 \tag{3.13.15}$$

λ has one positive real root out of four, the system is unstable.

(2) When

$$27m^2 - 27m + 1 = 1 \tag{3.13.16}$$

$m = 1$, and λ has a dual root, $\lambda_{1,\,2} = 0$, then one of the fundamental solutions of the system is

$$\begin{Bmatrix} u \\ v \end{Bmatrix} = c_1\begin{Bmatrix} U_1 \\ V_1 \end{Bmatrix} + c_2\begin{Bmatrix} U_2 \\ V_2 \end{Bmatrix}t \tag{3.13.17}$$

The vectors $\{U_1, V_1\}^T$ and $\{U_2, V_2\}^T$ are the eigenvectors of this dual root, so the system is divergent and unstable.

(3) When

$$27m^2 - 27m + 1 = 0 \tag{3.13.18}$$

there are two dual roots of λ, i.e., $\lambda_{1,\,2} = i/\sqrt{2}$ and$\lambda_{3,\,4} = -i/\sqrt{2}$, then the two fundamental solutions of the system are

$$\begin{Bmatrix} u \\ v \end{Bmatrix}^{(1)} = \left(c_1\begin{Bmatrix} U_1 \\ V_1 \end{Bmatrix} + c_2\begin{Bmatrix} U_2 \\ V_2 \end{Bmatrix}t\right)e^{\frac{i}{\sqrt{2}}t}, \begin{Bmatrix} u \\ v \end{Bmatrix}^{(2)} = \left(c_3\begin{Bmatrix} U_3 \\ V_3 \end{Bmatrix} + c_4\begin{Bmatrix} U_4 \\ V_4 \end{Bmatrix}t\right)e^{-\frac{i}{\sqrt{2}}t} \tag{3.13.19}$$

The vector $\{U_i, V_i\}^T$ is the eigenvector of these two dual roots, so the system is divergent and unstable.

(4) When

$$27m^2 - 27m + 1 < 0 \tag{3.13.20}$$

λ^2 is a pair of conjugate complex numbers with a negative real part, two of the four roots of λ have roots with a positive real part, so the system is unstable.

(5) When

$$0 < 27m^2 - 27m + 1 < 1 \tag{3.13.21}$$

λ^2 are two negative real numbers and the four roots of λ are two pairs of conjugate imaginary numbers, the system is stable. Here

$$0 < m < m_c, \quad \text{where } m_c = \frac{1}{2}\left(1 - \frac{\sqrt{69}}{9}\right) \tag{3.13.22}$$

Combining these results, we clearly know that the equilibrium points of the system L_4, L_5 are stable when $0 < m < m_c$, otherwise it is unstable.

(c) According to (3.13.9) and (3.13.10), the linearized equation of the original equation (a) near the equilibrium point (x_s, y_s) is given by

$$\begin{aligned} &\ddot{u} - 2\dot{v} - [3m + 3(x_s + m)(x_s - m)]u - (3x_s y_s)v = 0 \\ &\ddot{v} + 2\dot{u} - (3x_s y_s)u - \left(3y_s^2\right)v = 0 \end{aligned} \tag{3.13.23}$$

For the equilibrium points $L_1, L_2, L_3, y_s = 0$, so the above equation becomes

$$\begin{aligned} &\ddot{u} - 2\dot{v} - [3m + 3(x_s + m)(x_s - m)]u = 0 \\ &\ddot{v} + 2\dot{u} = 0 \end{aligned} \tag{3.13.24}$$

where x_s can be determined by Eq. (3.13.8). From the second equation of (3.13.24), we have

$$\dot{v} = -2u + \frac{1}{2}c \tag{3.13.25}$$

where c is a constant of integration. Substituting (3.13.25) into the first equation of (3.13.24) yields

$$\ddot{u} - [3m + 3(x_s + m)(x_s - m) - 4]u = c$$

This equation has a constant solution $u = u_0$. Substituting it to (3.13.25), we obtain the general solution of Eq. (3.13.25)

$$v = v_0 + \left(\frac{1}{2}c - 2u_0\right)t \tag{3.13.26}$$

which is divergent. Thus, the equilibrium points L_1, L_2, L_3 are unstable.

3.14 Exercise 3.14 (Singularity Analysis of Amplitude and Phase Equations Limiting the Resonance Response of the Three-Body Problem)

Solution: **(a)** Let $\dot{a}_1 = \dot{a}_2 = \dot{\gamma} = 0$, we obtain

$$\begin{aligned} J_1 a_1 a_2 \sin\gamma &= 0 \\ J_2 a_1^2 \sin\gamma &= 0 \\ \sigma a_2 + \left(J_2 a_1^2 + 2J_1 a_2^2\right)\cos\gamma &= 0 \end{aligned} \tag{3.14.1}$$

The first two equations of this system of equations are satisfied when $\gamma = n\pi$ is satisfied, and then the singular points of this system are given by

$$\gamma = n\pi, \sigma a_2 + \left(J_2 a_1^2 + 2J_1 a_2^2\right)\cos n\pi = 0 \tag{3.14.2}$$

(b) Linearize the original system around a singular point $(a_{1s}, a_{2s}, n\pi)$. For this purpose, let

$$a_1 = a_{1s} + u, a_2 = a_{2s} + v, \gamma = n\pi + w \tag{3.14.3}$$

then the linearized equation can be obtained as

$$\begin{cases} \dot{u} = (J_1 a_{1s} a_{2s} \cos n\pi)w \\ \dot{v} = \left(J_2 a_{1s}^2 \cos n\pi\right)w \\ a_{2s}\dot{w} = \sigma v + (2J_2 a_{1s} u + 4J_1 a_{2s} v)\cos n\pi \end{cases} \tag{3.14.4}$$

Making the derivation of the third equation of (3.14.4) with respect to t and substituting the first two equations into it, we obtain

$$a_{2s}\ddot{w} = \left(\sigma J_2 a_{1s}^2 \cos n\pi + 6J_1 J_2 a_{1s}^2 a_{2s}\right) w \tag{3.14.5}$$

Since (a_{1s}, a_{2s}) satisfies Eq. (3.14.2), we can solve for σa_{2s} from (3.14.2) and substitute it into (3.14.5). Then we have

$$a_{2s}^2 \ddot{w} = a_{1s}^2 \left(4J_1 J_2 a_{2s}^2 - J_2^2 a_{1s}^2\right) w$$

$$\text{i.e., } \ddot{w} + \frac{a_{1s}^2}{a_{2s}^2}\left(J_2^2 a_{1s}^2 - 4J_1 J_2 a_{2s}^2\right) w = 0 \tag{3.14.6}$$

As long as $J_1 J_2 < 0$, $J_2^2 a_{1s}^2 - 4J_1 J_2 a_{2s}^2 > 0$ and then the singular point of the original system is stable.

When $J_1 J_2 > 0$, the singular point of the original system may be unstable. In fact, when the value of $J_1 J_2$ can make $J_2^2 a_{1s}^2 - 4J_1 J_2 a_{2s}^2 < 0$, the singular point of the original system is unstable.

3.15 Exercise 3.15 (Modeling of a Single Pendulum with a Damper Attached, Solving by the Method of Multiple Scales)

Solution: (a) The force and velocity of the pendulum are shown in Fig. 3.10. From the theory of moment of momentum, we have

$$ml_1^2 \ddot{\theta} = -mgl_1 \sin\theta - F_d l_1 \cos(\beta - \theta)$$

where the damping force F_d is

$$F_d = \hat{\mu} mv \cos(\beta - \theta) = \hat{\mu} m\dot{\theta} l_1 \cos(\beta - \theta)$$

Therefore, the differential equation of motion of the pendulum is given by

$$ml_1 \ddot{\theta} = -mg \sin\theta - \hat{\mu} ml_1 \dot{\theta} \cos^2(\beta - \theta) \tag{3.15.1}$$

Since

$$\sin\beta = \frac{l_1 + l_2 - l_1 \cos\theta}{\sqrt{(l_1 + l_2 - l_1 \cos\theta)^2 + l_1^2 \sin^2\theta}}$$

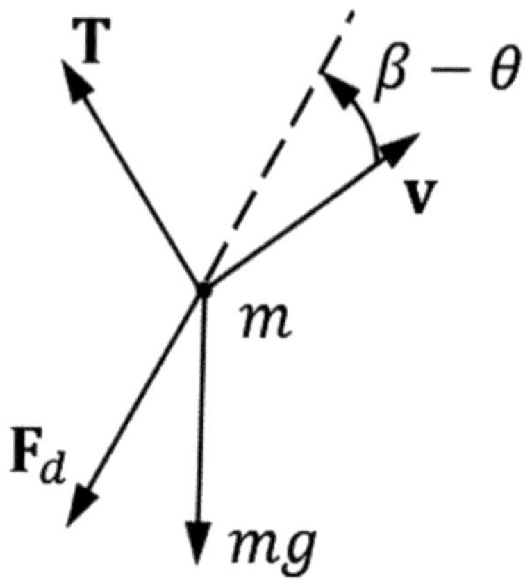

Fig. 3.10 Force and velocity of a mass point for Exercise 3.15

$$= \frac{l_1 + l_2 - l_1 \cos\theta}{\sqrt{(l_1 + l_2)^2 - 2l_1(l_1 + l_2)\cos\theta + l_1^2}}$$

$$\cos\beta = \frac{l_1 \sin\theta}{\sqrt{(l_1 + l_2 - l_1\cos\theta)^2 + l_1^2 \sin^2\theta}}$$
$$= \frac{l_1 \sin\theta}{\sqrt{(l_1 + l_2)^2 - 2l_1(l_1 + l_2)\cos\theta + l_1^2}}$$

then

$$\cos^2(\beta - \theta) = (\cos\beta\cos\theta + \sin\beta\sin\theta)^2$$
$$= \frac{l_1^2 \sin^2\theta\cos^2\theta + (l_1 + l_2 - l_1\cos\theta)^2\sin^2\theta + 2l_1\sin^2\theta\cos\theta(l_1 + l_2 - l_1\cos\theta)}{l_2^2 + 2l_1(l_1 + l_2)(1 - \cos\theta)}$$
$$= \frac{(l_1 + l_2)^2\sin^2\theta}{l_2^2 + 2l_1(l_1 + l_2)(1 - \cos\theta)}$$

Substitute the above equation into (3.15.1), we have

$$\ddot{\theta} + \omega_0^2 \sin\theta + \frac{\hat{\mu}(l_1 + l_2)^2\sin^2\theta}{l_2^2 + 2l_1(l_1 + l_2)(1 - \cos\theta)}\dot{\theta} = 0 \tag{3.15.2}$$

where $\omega_0^2 = g/l_1$.

(b) Expanding the coefficient of $\dot{\theta}$ in (3.15.2) near $\theta = 0$, we obtain

$$\frac{\hat{\mu}(l_1+l_2)^2\sin^2\theta}{l_2^2+2l_1(l_1+l_2)(1-\cos\theta)}=\frac{\hat{\mu}(l_1+l_2)^2(\theta-\frac{1}{6}\theta^3+\cdots)^2}{l_2^2+2l_1(l_1+l_2)\left(\frac{1}{2}\theta^2+\cdots\right)}$$
$$\approx\frac{\hat{\mu}(l_1+l_2)^2\theta^2}{l_2^2}$$

Substituting this into (3.15.2) and keeping to $O(\theta^3)$ yields

$$\ddot{\theta}+\omega_0^2\left(\theta-\frac{1}{6}\theta^3\right)+\frac{\hat{\mu}(l_1+l_2)^2}{l_2^2}\theta^2\dot{\theta}=0$$

$$\text{Let } 2\mu=\frac{\hat{\mu}(l_1+l_2)^2}{l_2^2} \tag{3.15.3}$$

$$\text{then } \ddot{\theta}+\omega_0^2\theta-\frac{1}{6}\omega_0^2\theta^3+2\mu\theta^2\dot{\theta}=0 \tag{3.15.4}$$

The following the method of multiple scales is adopted to solve (3.15.4). Let $\theta=\varepsilon v$, we have

$$\ddot{v}+\omega_0^2 v+\varepsilon^2\alpha_3 v^3+2\varepsilon^2\mu v^2\dot{v}=0 \tag{3.15.5}$$

where

$$\alpha_3=-\frac{1}{6}\omega_0^2. \tag{3.15.6}$$

We seek an approximate solution for (3.15.6) with the following form

$$v=v_0(T_0,T_1,T_2)+\varepsilon v_1(T_0,T_1,T_2)+\varepsilon^2 v_2(T_0,T_1,T_2)+\cdots \tag{3.15.7}$$

Substitute (3.15.7) into (3.15.5), we have

$$\begin{aligned}
0&=\ddot{v}+\omega_0^2 v+\varepsilon^2\alpha_3 v^3+2\varepsilon^2\mu v^2\dot{v}\\
&=\left[D_0^2+2\varepsilon D_0D_1+\varepsilon^2\left(D_1^2+2D_0D_2\right)+\cdots\right]\\
&\quad\left(v_0+\varepsilon v_1+\varepsilon^2 v_2+\cdots\right)+\omega_0^2\left(v_0+\varepsilon v_1+\varepsilon^2 v_2+\cdots\right)\\
&\quad+\varepsilon^2\left(\alpha_3 v_0^3+2\mu v_0^2 D_0 v_0+\cdots\right)\\
&=D_0^2 v_0+\omega_0^2 v_0+\varepsilon\left(D_0^2 v_1+2D_0D_1 v_0+\omega_0^2 v_1\right)\\
&\quad+\varepsilon^2\left[D_0^2 v_2+2D_0D_1 v_1+\left(D_1^2+2D_0D_2\right)v_0+\omega_0^2 v_2+\alpha_3 v_0^3+2\mu v_0^2 D_0 v_0\right]+\cdots
\end{aligned} \tag{3.15.8}$$

Equating the coefficients of the like power of ε and retaining to $O(\varepsilon^2)$ yields

$$D_0^2 v_0+\omega_0^2 v_0=0 \tag{3.15.9}$$

$$D_0^2 v_1 + \omega_0^2 v_1 = -2D_0 D_1 v_0 \tag{3.15.10}$$

$$D_0^2 v_2 + \omega_0^2 v_2 = -2D_0 D_1 v_1 - \left(D_1^2 + 2D_0 D_2\right) v_0 - 2\mu v_0^2 D_0 v_0 - \alpha_3 v_0^3 \tag{3.15.11}$$

The solution of (3.15.9) is

$$v_0 = A e^{i\omega_0 T_0} + \overline{A} e^{-i\omega_0 T_0} \tag{3.15.12}$$

where $A = A(T_1, T_2)$. Substituting the above equation into (3.15.10), we have

$$D_0^2 v_1 + \omega_0^2 v_1 = -2i\omega_0 D_1 A e^{i\omega_0 T_0} + cc \tag{3.15.13}$$

To eliminate secular terms, we need $D_1 A = 0$, so $A = A(T_2)$. And the solution of (3.15.13) is

$$v_1 = 0 \tag{3.15.14}$$

Substituting (3.15.12) and (3.15.14) into (3.15.11), and taking $A = A(T_2)$ into account, we obtain

$$\begin{aligned} D_0^2 v_2 + \omega_0^2 v_2 &= -2D_0 D_2 v_0 - \alpha_3 v_0^3 - 2\mu v_0^2 D_0 v_0 \\ &= \left[-2i\omega_0 D_2 A - 3\alpha_3 A^2 \overline{A} - 2i\mu\omega_0 A^2 \overline{A}\right] e^{i\omega_0 T_0} + cc + NST \end{aligned} \tag{3.15.15}$$

where cc denotes the complex conjugate term of its preceding terms, and NST denotes terms which do not eliminate secular terms. In order to eliminate secular terms, we have

$$2i\omega_0 D_2 A + 3\alpha_3 A^2 \overline{A} + 2i\mu\omega_0 A^2 \overline{A} = 0 \tag{3.15.16}$$

$$\text{Let } A = \frac{1}{2} a e^{i\beta} \tag{3.15.17}$$

Substituting (3.15.17) into (3.15.16) yields

$$i\omega_0 D_2 a - \omega_0 a D_2 \beta + \frac{3}{8}\alpha_3 a^3 + \frac{1}{4} i\mu\omega_0 a^3 = 0 \tag{3.15.18}$$

Separate the real and imaginary parts of the above equation, we obtain

$$D_2 a + \frac{1}{4}\mu a^3 = 0, \ D_2 \beta - \frac{3\alpha_3}{8\omega_0} a^2 = 0 \tag{3.15.19}$$

From (3.15.19), we can obtain

$$a = \frac{a_0}{\sqrt{\frac{1}{2}\mu a_0^2 T_2 + 1}} \tag{3.15.20}$$

$$\beta = \frac{3\alpha_3}{4\mu\omega_0}\ln(\frac{1}{2}\mu a_0^2 T_2 + 1) + \beta_0 \tag{3.15.21}$$

Combining the above results, the first order approximate solution of the original equation is obtained as

$$\theta = \varepsilon v_0 = \frac{\varepsilon a_0}{\sqrt{\frac{1}{2}\mu a_0^2 T_2 + 1}}\cos\left[\omega_0 t + \frac{3\alpha_3}{4\mu\omega_0}\ln\left(\frac{1}{2}\mu a_0^2 T_2 + 1\right) + \beta_0\right] \tag{3.15.22}$$

Considering $\alpha_3 = -\omega_0^2/6, \ \ T_2 = \varepsilon^2 t$, the above equation becomes

$$\theta = \frac{\varepsilon a_0}{\sqrt{\frac{1}{2}\mu a_0^2 \varepsilon^2 t + 1}}\cos[\omega_0 t - \frac{\omega_0}{8\mu}\ln(\frac{1}{2}\mu a_0^2 \varepsilon^2 t + 1) + \beta_0]$$

Let $\varepsilon = 1$, we get

$$\theta = \frac{a_0}{\sqrt{\frac{1}{2}\mu a_0^2 t + 1}}\cos\left\{\omega_0\left[t - \frac{1}{8\mu}\ln\left(\frac{1}{2}\mu a_0^2 t + 1\right)\right] + \beta_0\right\} \tag{3.15.23}$$

3.16 Exercise 3.16 (The Method of Multiple Scales for Solving Van Der Pol Oscillators with Delay Amplitude Limiting)

Solution: (a) Write the original equation as

$$\begin{aligned} &\ddot{v} + \omega_0^2 v = 2\varepsilon\dot{v} - 2\varepsilon z\dot{v} - 2\varepsilon\dot{z}v \\ &\tau\dot{z} + z = v^2 \end{aligned} \tag{3.16.1}$$

We seek an approximate solution for (3.16.1) with the following form

$$v = v_0(T_0, T_1, T_2) + \varepsilon v_1(T_0, T_1, T_2) + \varepsilon^2 v_2(T_0, T_1, T_2) + \cdots \tag{3.16.2}$$

$$z = z_0(T_0, T_1, T_2) + \varepsilon z_1(T_0, T_1, T_2) + \varepsilon^2 z_2(T_0, T_1, T_2) + \cdots \tag{3.16.3}$$

Substitute (3.16.2) and (3.16.3) into the original equation, we get

$$
\begin{aligned}
0 &= \ddot{v} + \omega_0^2 v - 2\varepsilon\dot{v} + 2\varepsilon z\dot{v} + 2\varepsilon\dot{z}v \\
&= \left[D_0^2 + 2\varepsilon D_0 D_1 + \varepsilon^2\left(D_1^2 + 2D_0 D_2\right)\right]\left(v_0 + \varepsilon v_1 + \varepsilon^2 v_2\right) \\
&\quad + \omega_0^2\left(v_0 + \varepsilon v_1 + \varepsilon^2 v_2\right) - 2\varepsilon(D_0 + \varepsilon D_1)(v_0 + \varepsilon v_1) \\
&\quad + 2\varepsilon(z_0 + \varepsilon z_1)(D_0 + \varepsilon D_1)(v_0 + \varepsilon v_1) \\
&\quad + 2\varepsilon(v_0 + \varepsilon v_1)(D_0 + \varepsilon D_1)(z_0 + \varepsilon z_1) + \cdots \\
&= D_0^2 v_0 + \omega_0^2 v_0 \\
&\quad + \varepsilon\left(D_0^2 v_1 + \omega_0^2 v_1 + 2D_0 D_1 v_0 - 2D_0 v_0 + 2z_0 D_0 v_0 + 2v_0 D_0 z_0\right) \\
&\quad + \varepsilon^2[D_0^2 v_2 + \omega_0^2 v_2 + 2D_0 D_1 v_1 + \left(D_1^2 + 2D_0 D_2\right)v_0 - 2D_0 v_1 - 2D_1 v_0 \\
&\quad + 2z_0 D_0 v_1 + 2z_0 D_1 v_0 + 2z_1 D_0 v_0 + 2v_0 D_0 z_1 + 2v_0 D_1 z_0 + 2v_1 D_0 z_0] \\
&\quad + 2z_0 D_0 v_1 + 2z_0 D_1 v_0 + 2z_1 D_0 v_0 + 2v_0 D_0 z_1 + 2v_0 D_1 z_0 + 2v_1 D_0 z_0] + \cdots
\end{aligned}
\tag{3.16.4}
$$

$$
\begin{aligned}
0 &= \tau\dot{z} + z - v^2 \\
&= \tau(D_0 + \varepsilon D_1 + \cdots)(z_0 + \varepsilon z_1 + \cdots) \\
&\quad + (z_0 + \varepsilon z_1 + \cdots) - (v_0 + \varepsilon v_1 + \cdots)^2 \\
&= \tau D_0 z_0 + z_0 - v_0^2 + \varepsilon(\tau D_0 z_1 + z_1 + \tau D_1 z_0 - 2v_0 v_1) + \cdots
\end{aligned}
\tag{3.16.5}
$$

Equating the coefficients of the like power of ε and retaining to $O(\varepsilon)$ yields

$$\varepsilon^0 : D_0^2 v_0 + \omega_0^2 v_0 = 0 \tag{3.16.6}$$

$$\tau D_0 z_0 + z_0 = v_0^2 \tag{3.16.7}$$

$$\varepsilon^1 : D_0^2 v_1 + \omega_0^2 v_1 = -2D_0 D_1 v_0 + 2D_0 v_0 - 2z_0 D_0 v_0 - 2v_0 D_0 z_0 \tag{3.16.8}$$

$$\tau D_0 z_1 + z_1 = -\tau D_1 z_0 + 2v_0 v_1 \tag{3.16.9}$$

The solution of (3.16.6) is

$$v_0 = Ae^{i\omega_0 T_0} + \overline{A}e^{-i\omega_0 T_0} \tag{3.16.10}$$

where $A = A(T_1)$. Substituting the above equation into (3.16.7) yields

$$\tau D_0 z_0 + z_0 = 2A\overline{A} + A^2 e^{2i\omega_0 T_0} + cc \tag{3.16.11}$$

The solution of (3.16.11) is

$$z_0 = be^{-T_0/\tau} + 2A\overline{A} + \frac{A^2(1 - 2i\tau\omega_0)}{1 + 4\omega_0^2\tau^2}e^{2i\omega_0 T_0}$$

$$+\frac{A^2(1+2i\tau\omega_0)}{1+4\omega_0^2\tau^2}e^{-2i\omega_0 T_0} \tag{3.16.12}$$

where $b = b(T_1)$ is a real function. Substituting (3.16.10) and (3.16.12) into (3.16.8), we obtain

$$\begin{aligned}
D_0^2 v_1 + \omega_0^2 v_1 &= -2D_0D_1v_0 + 2D_0v_0 - 2z_0D_0v_0 - 2v_0D_0z_0 \\
&= -2i\omega_0 D_1 A e^{i\omega_0 T_0} + 2i\omega_0 A e^{i\omega_0 T_0} + cc \\
&\quad - 2i\omega_0\left[be^{-T_0/\tau}A + 2A^2\overline{A} - \frac{(1-2i\tau\omega_0)A^2\overline{A}}{1+4\omega_0^2\tau^2}\right]e^{i\omega_0 T_0} \\
&\quad - 2\left[-\frac{b}{\tau}e^{-T_0/\tau}A + \frac{2i\omega_0(1-2i\tau\omega_0)A^2\overline{A}}{1+4\omega_0^2\tau^2}\right]e^{i\omega_0 T_0} \\
&\quad - \frac{6i\omega_0(1-2i\tau\omega_0)A^3}{1+4\omega_0^2\tau^2}e^{3i\omega_0 T_0} + cc \\
&= \Big[-2i\omega_0 D_1 A + 2i\omega_0 A - 2i\omega_0(be^{-T_0/\tau}A + 2A^2\overline{A}) \\
&\quad + \frac{2b}{\tau}e^{-T_0/\tau}A - \frac{2i\omega_0(1-2i\tau\omega_0)A^2\overline{A}}{1+4\omega_0^2\tau^2}\Big]e^{i\omega_0 T_0} \\
&\quad - \frac{6i\omega_0(1-2i\tau\omega_0)A^3}{1+4\omega_0^2\tau^2}e^{3i\omega_0 T_0} + cc
\end{aligned} \tag{3.16.13}$$

where cc denotes the complex conjugate term of its preceding terms. In order to eliminate secular terms, it is necessary to have

$$\begin{aligned}
&2i\omega_0 D_1 A - 2i\omega_0 A + 2i\omega_0\left(be^{-\frac{T_0}{\tau}}A + 2A^2\overline{A}\right) \\
&- \frac{2b}{\tau}e^{-\frac{T_0}{\tau}}A + \frac{2i\omega_0(1-2i\tau\omega_0)A^2\overline{A}}{1+4\omega_0^2\tau^2} = 0
\end{aligned} \tag{3.16.14}$$

$$\text{Let } A = \frac{1}{2}ae^{i\theta} \tag{3.16.15}$$

Substituting (3.16.15) into (3.16.14) yields

$$\begin{aligned}
&i\omega_0 D_1 a - \omega_0 a D_1\theta - i\omega_0 a + i\omega_0\left(be^{-T_0/\tau}a + \frac{1}{2}a^3\right) \\
&- \frac{2b}{\tau}e^{-T_0/\tau}a + \frac{i\omega_0(1-2i\tau\omega_0)a^3}{4\left(1+4\omega_0^2\tau^2\right)} = 0
\end{aligned} \tag{3.16.16}$$

Separating the real and imaginary parts of the above equation yields

$$D_1 a - a + \frac{\left(3 + 8\omega_0^2\tau^2\right)a^3}{4\left(1 + 4\omega_0^2\tau^2\right)} + e^{-T_0/\tau} ab = 0$$

$$D_1\theta - \frac{2\tau\omega_0 a^2}{4\left(1 + 4\omega_0^2\tau^2\right)} + \frac{2b}{\tau\omega_0} e^{-T_0/\tau} = 0$$

Or we can write the above equation as

$$\dot{a} = \varepsilon a\left(1 - \frac{1}{4}\alpha_r a^2\right) + \varepsilon e^{-T_0/\tau} ab, \quad \dot{\theta} = -\frac{1}{4}\varepsilon\alpha_i a^2 + \frac{2\varepsilon b}{\tau\omega_0} e^{-T_0/\tau} \tag{3.16.17}$$

$$\text{where } \alpha_r = \frac{3 + 8\omega_0^2\tau^2}{\left(1 + 4\omega_0^2\tau^2\right)}, \quad \alpha_i = -\frac{2\tau\omega_0}{\left(1 + 4\omega_0^2\tau^2\right)} \tag{3.16.18}$$

Then we can obtain v_1 from (3.16.13)

$$v_1 = \frac{3i(1 - 2i\tau\omega_0)A^3}{4\omega_0\left(1 + 4\omega_0^2\tau^2\right)} e^{3i\omega_0 T_0} + cc \tag{3.16.19}$$

Substituting (3.16.12) and (3.16.19) into (3.16.9) yields

$$\tau D_0 z_1 + z_1 = -\tau D_1 z_0 + 2v_0 v_1$$
$$\tau D_0 z_1 + z_1 = \tau D_1 z_0 + 2v_0 v_1$$

$$\begin{aligned}
&= \tau e^{-T_0/\tau} D_1 b + 2\tau D_1\left(A\overline{A}\right) + \frac{2i\omega_0\tau(1 - 2i\tau\omega_0)A^2}{1 + 4\omega_0^2\tau^2} e^{2i\omega_0 T_0} + cc \\
&\quad + 2\left(Ae^{i\omega_0 T_0} + \overline{A}e^{-i\omega_0 T_0}\right)\left(\frac{3i(1 - 2i\tau\omega_0)A^3}{4\omega_0\left(1 + 4\omega_0^2\tau^2\right)} e^{3i\omega_0 T_0} + cc\right) \\
&= \tau e^{-T_0/\tau} D_1 b + 2\tau D_1\left(A\overline{A}\right) + \frac{2i\omega_0\tau(1 - 2i\tau\omega_0)A^2}{1 + 4\omega_0^2\tau^2} e^{2i\omega_0 T_0} \\
&\quad + \frac{3i(1 - 2i\tau\omega_0)A^3\overline{A}}{2\omega_0\left(1 + 4\omega_0^2\tau^2\right)} e^{2i\omega_0 T_0} + cc + \text{higher order terms}
\end{aligned} \tag{3.16.20}$$

In order to eliminate secular terms, we need

$$D_1 b = 0$$

$$\text{i.e., } \dot{b} = 0 \quad \text{or} \quad b = b_0 = \text{constant} \tag{3.16.21}$$

Note: We were unable to obtain $\dot{b} = 2\left(1 + 2\omega_0^2\tau^2\right)\left(1 + 4\omega_0^2\tau^2\right)^{-1} a^2 b$, which is required by the present Exercise.

Considering b is bounded, The term containing $e^{-T_0/\tau}$ in (3.16.17) decays quickly with time and has no effect on the steady state motion of a and θ, so the governing equations for the steady state motion of a and θ are

$$\dot{a} = \varepsilon a\left(1 - \frac{1}{4}\alpha_r a^2\right), \qquad \dot{\theta} = -\frac{1}{4}\varepsilon\alpha_i a^2 \tag{3.16.22}$$

Substituting (3.16.10) and (3.16.12) into (3.16.2) and (3.16.3), we can obtain the first order steady state approximate solution of the original equation

$$\begin{aligned} v &= a\cos(\omega_0 t + \theta) + O(\varepsilon) \\ z &= b\exp(-t/\tau) + \frac{1}{2}a^2 + \frac{1}{2}a^2(1 + 4\omega_0^2\tau^2)^{-1/2} \\ &\cos(2\omega_0 t + 2\theta - \arctan 2\omega_0\tau) + O(\varepsilon) \end{aligned} \tag{3.16.23}$$

where $a$and θ are given by (3.16.17) and b is given by (3.16.21).

(b) Separating the variables of the first equation of (3.16.22), we obtain

$$\frac{da}{a\left(1 - \frac{1}{4}\alpha_r a^2\right)} = \varepsilon dt \quad \overset{x=a^2}{\Rightarrow} \quad \frac{\frac{1}{2}dx}{x\left(1 - \frac{1}{4}\alpha_r x\right)} = \varepsilon dt$$

Conduct the integration of the above equation, we can obtain

$$\frac{1}{2}\ln\frac{\left(1 - \frac{1}{4}\alpha_r x\right)}{x} = -\varepsilon t + c \quad \Rightarrow \quad \frac{1}{2}\ln\frac{\left(1 - \frac{1}{4}\alpha_r a^2\right)}{a^2} = -\varepsilon t + c$$

where c is a constant of integration. Considering $a = a_0$ when $t = 0$, we can obtain

$$\frac{\left(1 - \frac{1}{4}\alpha_r a^2\right)}{a^2} = \frac{\left(1 - \frac{1}{4}\alpha_r a_0^2\right)}{a_0^2}e^{-2\varepsilon t}$$

$$a^2 = \frac{a_0^2}{\frac{1}{4}\alpha_r a_0^2 + \left(1 - \frac{1}{4}\alpha_r a_0^2\right)e^{-2\varepsilon t}} \tag{3.16.24}$$

Substituting (3.16.24) into the second equation of (3.16.22), we obtain

$$\dot{\theta} = -\frac{\frac{1}{4}\varepsilon\alpha_i a_0^2}{\frac{1}{4}\alpha_r a_0^2 + \left(1 - \frac{1}{4}\alpha_r a_0^2\right)e^{-2\varepsilon t}} \tag{3.16.25}$$

Conduct the integration of the above equation, we can obtain

$$\theta = -\frac{\alpha_i}{2\alpha_r}\{2\varepsilon t + \ln[\frac{1}{4}\alpha_r a_0^2 + \left(1 - \frac{1}{4}\alpha_r a_0^2\right)e^{-2\varepsilon t}]\} + c \tag{3.16.26}$$

where c is a constant of integration. Considering $\theta = \theta_0$ when $t = 0$, we can obtain

$$\theta = -\frac{\alpha_i}{2\alpha_r}\{2\varepsilon t + \ln[\frac{1}{4}\alpha_r a_0^2 + \left(1 - \frac{1}{4}\alpha_r a_0^2\right)e^{-2\varepsilon t}]\} + \theta_0 \tag{3.16.27}$$

(c) Considering the expression of a in (3.16.24), we can find that the system will eventually converge to the steady state amplitude a_∞ for any non-zero initial amplitude a_0:

$$a_\infty = a(t \to \infty) = 2\sqrt{\frac{1}{\alpha_r}} \tag{3.16.28}$$

And considering the expression of θ in (3.16.27), we can find that the system will eventually converge to the steady state phase θ,

$$\theta = -\frac{\alpha_i}{2\alpha_r}\left[2\varepsilon t + \ln(\frac{1}{4}\alpha_r a_0^2\right)] + \theta_0 \tag{3.16.29}$$

Substituting (3.16.28), (3.16.29) and (3.16.21) into (3.16.23), we can obtain the steady state response of the system as

$$v = 2a_\infty \cos[\left(\omega_0 - \frac{\alpha_i}{2\alpha_r}2\varepsilon\right)t - \frac{\alpha_i}{2\alpha_r}\ln(\frac{1}{4}\alpha_r a_0^2) + \theta_0] + O(\varepsilon) \tag{3.16.30}$$

$$\begin{aligned} z &= b_0 e^{-t/\tau} + \tfrac{1}{2}a_\infty^2 + \tfrac{1}{2}a_\infty^2(1 + 4\omega_0^2\tau^2)^{-1/2} \\ &\cdot\cos\left[\left(2\omega_0 - \tfrac{2\alpha_i}{\alpha_r}\varepsilon\right)t - \tfrac{\alpha_i}{\alpha_r}\ln\left(\tfrac{1}{4}\alpha_r a_0^2\right) + 2\theta_0 - \text{arctan}2\omega_0\tau\right] + O(\varepsilon) \end{aligned} \tag{3.16.31}$$

Therefore, v is a stable limit cycle in the phase plane and z is in simple harmonic motion at any time.

3.17 Exercise 3.17 (Averaging to Solve for an Oscillator with Both Coulomb and Viscous Damping)

Solution: **(a)** Write the original equation as

$$\ddot{u} + \omega_0^2 u = \varepsilon f(\dot{u}) \tag{3.17.1}$$

$$\text{where}\, f(\dot{u}) = -(\mu_0 sgn\dot{u} + 2\mu_1\dot{u}) \tag{3.17.2}$$

All the nonlinear parts in (3.17.1) are first-order terms of ε and the nonlinear function $f(\dot{u})$ is an odd function of $\dot{u}$, so it is convenient to solve the equation by the method of averaging. Let its solution be

$$u = a\cos(\omega_0 t + \beta) \triangleq a\cos\varphi, \quad \varphi = \omega_0 t + \beta \tag{3.17.3}$$

reintroduce the transformation$\dot{u} = -a\omega_0\sin(\omega_0 t + \beta) \triangleq -a\omega_0\sin\varphi$ (3.17.4)

Consider a and β as functions of t for (3.17.3) and (3.17.4). The derivative of (3.17.3) and (3.17.4) with respect to t gives

$$\dot{u} = \dot{a}\cos\varphi - a\omega_0\sin\varphi - a\dot{\beta}\sin\varphi \tag{3.17.5}$$

$$\ddot{u} = -\dot{a}\omega_0\sin\varphi - a\omega_0^2\cos\varphi - a\omega_0\dot{\beta}\cos\varphi \tag{3.17.6}$$

By (3.17.4) and (3.17.5), we have

$$\dot{a}\cos\varphi - a\dot{\beta}\sin\varphi = 0 \tag{3.17.7}$$

Substituting (3.17.6), (3.17.3) and (3.17.4) into (3.17.1) yields

$$\dot{a}\omega_0\sin\varphi + a\omega_0\dot{\beta}\cos\varphi = -\varepsilon f(-a\omega_0\sin\varphi) \tag{3.17.8}$$

Solving (3.17.7) and (3.17.8) for $\dot{a}$ and $\dot{\beta}$, we obtain

$$\dot{a} = -\frac{\varepsilon}{\omega_0} f(-a\omega_0\sin\varphi)\sin\varphi \tag{3.17.9}$$

$$\dot{\beta} = -\frac{\varepsilon}{\omega_0 a} f(-a\omega_0\sin\varphi)\cos\varphi \tag{3.17.10}$$

Since ε is a small parameter, (3.17.9) and (3.17.10) imply that a and β vary much more slowly with time t than $\varphi = \omega_0 t + \beta$. In other words, a and β hardly change during the period of oscillation $2\pi/\omega_0$ of $\sin\varphi$ and $\cos\varphi$. This enables us to average out the variations in φ in (3.17.9) and (3.17.10). Averaging these equations over the period $2\pi/\omega_0$ and considering a, β, $\dot{a}$ and $\dot{\beta}$ to be constants while performing the averaging, we obtain the following equations describing the slow variations of a and β:

$$\begin{aligned}\dot{a} &= -\frac{\varepsilon}{2\pi\omega_0}\int_0^{2\pi} f(-a\omega_0\sin\phi)\sin\phi d\phi \\ &= -\frac{\varepsilon}{2\pi\omega_0}\int_0^{2\pi}\left[\mu_0\mathrm{sgn}(a\omega_0\sin\phi) + 2\mu_1 a\omega_0\sin\phi\right]\sin\phi d\phi\end{aligned}$$

$$= -\frac{\varepsilon}{\pi\omega_0}\int_0^{\pi}[\mu_0\sin\phi + 2\mu_1 a\omega_0\sin^2\phi]d\phi$$

$$= -\varepsilon\left(\frac{2\mu_0}{\pi\omega_0} + \mu_1 a\right)$$

$$\dot{\beta} = -\frac{\varepsilon}{2\pi\omega_0 a}\int_0^{2\pi} f(-a\omega_0\sin\phi)\cos\phi d\phi$$

$$= -\frac{\varepsilon}{2\pi\omega_0 a}\int_0^{2\pi}[\mu_0\mathrm{sgn}(a\omega_0\sin\phi) + 2\mu_1 a\omega_0\sin\phi]\cos\phi d\phi$$

$$= 0$$

i.e.,

$$\dot{a} = -\varepsilon\left(\frac{2\mu_0}{\pi\omega_0} + \mu_1 a\right), \quad \dot{\beta} = 0 \tag{3.17.11}$$

(b) Separating the variables of the first equation of (3.17.11) yields

$$\frac{da}{\frac{2\mu_0}{\pi\omega_0} + \mu_1 a} = -\varepsilon dt \quad \Rightarrow \quad \frac{d(\mu_1 a)}{\frac{2\mu_0}{\pi\omega_0} + \mu_1 a} = -d(\varepsilon\mu_1 t)$$

$$\text{So } \ln\left(\frac{2\mu_0}{\pi\omega_0} + \mu_1 a\right) = -\varepsilon\mu_1 t + c \tag{3.17.12}$$

Considering the initial condition that $a = a_0$ when $t = 0$, we obtain

$$\ln\left(\frac{\frac{2\mu_0}{\pi\omega_0} + \mu_1 a}{\frac{2\mu_0}{\pi\omega_0} + \mu_1 a_0}\right) = -\varepsilon\mu_1 t \quad \Rightarrow \quad \frac{2\mu_0}{\pi\omega_0} + \mu_1 a = \left(\frac{2\mu_0}{\pi\omega_0} + \mu_1 a_0\right)\exp(-\varepsilon\mu_1 t)$$

$$\text{i.e., } a = \left(a_0 + \frac{2\mu_0}{\pi\omega_0\mu_1}\right)\exp(-\varepsilon\mu_1 t) - \frac{2\mu_0}{\pi\omega_0\mu_1} \tag{3.17.13}$$

Then from $\dot{\beta} = 0$, we can obtain $\beta = \beta_0$ is a constant. So

$$u = \left[\left(a_0 + \frac{2\mu_0}{\pi\omega_0\mu_1}\right)\exp(-\varepsilon\mu_1 t)\right) - \frac{2\mu_0}{\pi\omega_0\mu_1}]\cos(\omega_0 t + \beta_0) \tag{3.17.14}$$

(c) Assume $u(0) = a_0 > 0$. From (3.17.13) and (3.17.14), we can see that the amplitude a_0 decays exponentially first, then becomes zero and negative, and finally the amplitude reaches to a constant $-2\mu_0/(\pi\omega_0\mu_1)$. The system will oscillate with a constant amplitude $-2\mu_0/(\pi\omega_0\mu_1)$, which is inconsistent with the fact that the given system is dissipative. Therefore, the approximate solution obtained above cannot accurately describe the motion of the system.

Let the motion stops at time t_e, and we follow the above approximate solution to derive the approximate value of t_e. Since $a_0 > 0$, there must be

$$a(t) \geq 0 \tag{3.17.15}$$

when the motion stops, the frictional force is balanced with the restoring force, and from the differential equation of motion of the system, there is

$$|u(t_e)| \leq \frac{\varepsilon\mu_0}{\omega_0^2} \tag{3.17.16}$$

Set $u(0) = a_0 > 0, \quad \dot{u}(0) = 0$, then $\beta_0 = 0$ and

$$u = \left[\left(a_0 + \frac{2\mu_0}{\pi\omega_0\mu_1}\right)\exp(-\varepsilon\mu_1 t)\right) - \frac{2\mu_0}{\pi\omega_0\mu_1}]\cos\omega_0 t \tag{3.17.17}$$

when the motion stops, there must be $\dot{u}(t_e) = 0$, $u(t_e) = a(t_e)$, i.e.

$$-\varepsilon\mu_1\left(a_0 + \frac{2\mu_0}{\pi\omega_0\mu_1}\right)\exp(-\varepsilon\mu_1 t_e)\cos\omega_0 t_e$$

$$+\left[\left(a_0 + \frac{2\mu_0}{\pi\omega_0\mu_1}\right)\exp(-\varepsilon\mu_1 t_e\right) - \frac{2\mu_0}{\pi\omega_0\mu_1}]\omega_0\sin\omega_0 t_e = 0 \tag{3.17.18}$$

$$\text{and } \cos\omega_0 t_e = 1 \tag{3.17.19}$$

From (3.17.19), we have

$$t_e = \frac{2n\pi}{\omega_0}, \quad n = 0, 1, 2, \ldots \tag{3.17.20}$$

In this case, Eqs. (3.17.15), (3.17.16) and (3.17.18) become

$$\left(a_0 + \frac{2\mu_0}{\pi\omega_0\mu_1}\right)\exp(-\varepsilon\mu_1\frac{2n\pi}{\omega_0}) - \frac{2\mu_0}{\pi\omega_0\mu_1} \geq 0 \tag{3.17.21}$$

$$|\left(a_0 + \frac{2\mu_0}{\pi\omega_0\mu_1}\right)\exp(-\varepsilon\mu_1\frac{2n\pi}{\omega_0}) - \frac{2\mu_0}{\pi\omega_0\mu_1}| \leq \frac{\varepsilon\mu_0}{\omega_0^2} \tag{3.17.22}$$

$$\varepsilon\mu_1\left(a_0 + \frac{2\mu_0}{\pi\omega_0\mu_1}\right)\exp(-\varepsilon\mu_1\frac{2n\pi}{\omega_0}) = 0 \tag{3.17.23}$$

These three equations cannot be satisfied at the same time. If (3.17.23) is satisfied, the aforementioned situation where the amplitude $a{=}{-}2\mu_0/(\pi\omega_0\mu_1)$, which is not allowed; therefore, we choose to make Eqs. (3.17.21) and (3.17.22) are satisfied, which yields

$$n = \mathrm{mod}\left\{\frac{\omega_0}{2\varepsilon\mu_1\pi}\ln\left[\left(a_0+\frac{2\mu_0}{\pi\omega_0\mu_1}\right)/\left(\frac{2\mu_0}{\pi\omega_0\mu_1}+\frac{\varepsilon\mu_0}{\omega_0^2}\right)\right]\right\} \tag{3.17.24}$$

where mod(x) denotes the positive integer less than and closest to x. After finding n, we can determine t_e from (3.17.20).

3.18 Exercise 3.18 (Averaging to Solve for Oscillators with Both Coulomb and Square Damping)

Solution: (a) Write the original equation as

$$\ddot{u}+\omega_0^2 u=\varepsilon f(\dot{u}) \tag{3.18.1}$$

$$\text{where } f(\dot{u})=-(\mu_0\mathrm{sgn}\dot{u}+\mu_2\dot{u}|\dot{u}|) \tag{3.18.2}$$

All the nonlinear parts in (3.18.1) are first-order terms of ε and the nonlinear function $f(\dot{u})$ is an odd function of $\dot{u}$. It can be solved by the method of averaging as in the previous Exercise. Let its solution be

$$u=a\cos(\omega_0 t+\beta)\triangleq a\cos\varphi,\ \varphi=\omega_0 t+\beta \tag{3.18.3}$$

Referring to the result of the previous exercise, there is

$$\begin{aligned}
\dot{a} &= -\frac{\varepsilon}{2\pi\omega_0}\int_0^{2\pi} f(-a\omega_0\sin\phi)\sin\phi d\phi \\
&= -\frac{\varepsilon}{2\pi\omega_0}\int_0^{2\pi}\left[\mu_0\mathrm{sgn}(a\omega_0\sin\phi)+\mu_2 a\omega_0\sin\phi|a\omega_0\sin\phi|\right]\sin\phi d\phi \\
&= -\frac{\varepsilon}{\pi\omega_0}\int_0^{\pi}\left[\mu_0\sin\phi+\mu_2 a^2\omega_0^2\sin^3\phi\right]d\phi \\
&= -\frac{\varepsilon}{\pi\omega_0}\left[2\mu_0+\mu_2 a^2\omega_0^2\left(-\cos\phi+\frac{1}{3}\cos^3\phi\right)\Big|_0^{\pi}\right]
\end{aligned}$$

$$= -\varepsilon\left(\frac{2\mu_0}{\pi\omega_0} + \frac{4}{3\pi}\mu_2\omega_0 a^2\right)$$

$$\dot{\beta} = -\tfrac{\varepsilon}{2\pi\omega_0 a}\int_0^{2\pi} f(-a\omega_0 \sin\phi)\cos\phi d\phi$$

$$= -\tfrac{\varepsilon}{2\pi\omega_0 a}\int_0^{2\pi}\left[\mu_0 \mathrm{sgn}(a\omega_0 \sin\phi) + \mu_2 a\omega_0 \sin\phi|a\omega_0 \sin\phi|\right]\cos\phi d\phi$$

$$= 0$$

i.e.,

$$\dot{a} = -\varepsilon\left(\frac{2\mu_0}{\pi\omega_0} + \frac{4}{3\pi}\mu_2\omega_0 a^2\right), \quad \dot{\beta} = 0 \tag{3.18.4}$$

(b) Separating the variables of the first equation of (3.18.4) and integrating it, we obtain

$$\int \frac{da}{\frac{2\mu_0}{\pi\omega_0} + \frac{4}{3\pi}\mu_2\omega_0 a^2} = -\varepsilon\int dt$$

which gives

$$\frac{1}{\sqrt{\frac{4}{3\pi}\mu_2\omega_0 \cdot \frac{2\mu_0}{\pi\omega_0}}}\arctan(a\sqrt{\frac{4}{3\pi}\mu_2\omega_0 / \frac{2\mu_0}{\pi\omega_0}}) = -\varepsilon t + c_0$$

where c_0 is a constant of integration. So

$$a = \frac{1}{\omega_0}\sqrt{\frac{3\mu_0}{2\mu_2}}\tan(c - \frac{\varepsilon}{\pi}\sqrt{\frac{8\mu_0\mu_2}{3}}t) \tag{3.18.5}$$

where c is a constant of integration.

Then from $\dot{\beta} = 0$, we obtain that $\beta = \beta_0$ is a constant value. So

$$u = \left[\frac{1}{\omega_0}\sqrt{\frac{3\mu_0}{2\mu_2}}\tan(c - \frac{\varepsilon}{\pi}\sqrt{\frac{8\mu_0\mu_2}{3}}t\right)]\cos(\omega_0 t + \beta_0) \tag{3.18.6}$$

(c) From (3.18.6), we can find that the initial amplitude of the oscillation is

$$a_0 = \left[\frac{1}{\omega_0}\sqrt{\frac{3\mu_0}{2\mu_2}}\tan(c)\right]\cos\beta_0 \tag{3.18.7}$$

As time increases, the amplitude increases to

$$a(t) = \frac{a_0}{\tan(c)\cos\beta_0}\tan(c - \frac{\varepsilon}{\pi}\sqrt{\frac{8\mu_0\mu_2}{3}t})] \tag{3.18.8}$$

when

$$t_e = \frac{\pi}{\varepsilon}\sqrt{\frac{3}{8\mu_0\mu_2}} \tag{3.18.9}$$

the amplitude becomes zero and the oscillation stops thereafter.

3.19 Exercise 3.19 (Averaging to Solve for Oscillators with Both Viscous and Square Damping)

Solution: (a) Write the original equation as

$$\ddot{u} + \omega_0^2 u = \varepsilon f(\dot{u}) \tag{3.19.1}$$

$$\text{where } f(\dot{u}) = -(2\mu_1\dot{u} + \mu_2\dot{u}|\dot{u}|) \tag{3.19.2}$$

As in the previous Exercise, the method of averaging is still used. Let the solution be

$$u = a\cos(\omega_0 t + \beta) \triangleq a\cos\varphi, \quad \varphi = \omega_0 t + \beta \tag{3.19.3}$$

Referring to the result of the previous exercise, there is

$$\begin{aligned}
\dot{a} &= -\frac{\varepsilon}{2\pi\omega_0}\int_0^{2\pi} f(-a\omega_0\sin\phi)\sin\phi d\phi \\
&= -\frac{\varepsilon}{2\pi\omega_0}\int_0^{2\pi}[2\mu_1 a\omega_0\sin\phi + \mu_2 a\omega_0\sin\phi|a\omega_0\sin\phi|]\sin\phi d\phi \\
&= -\frac{\varepsilon}{\pi\omega_0}\int_0^{\pi}\left[2\mu_1 a\omega_0\sin^2\phi + \mu_2 a^2\omega_0^2\sin^3\phi\right]d\phi \\
&= -\frac{\varepsilon}{\pi\omega_0}\left[\mu_1 a\omega_0\pi + \mu_2 a^2\omega_0^2\left(-\cos\phi + \frac{1}{3}\cos^3\phi\right)\Big|_0^{\pi}\right]
\end{aligned}$$

$$= -\varepsilon\left(\mu_1 a + \frac{4}{3\pi}\mu_2\omega_0 a^2\right)$$

$$\dot{\beta} = -\frac{\varepsilon}{2\pi\omega_0 a}\int_0^{2\pi} f(-a\omega_0\sin\phi)\cos\phi d\phi$$

$$= -\frac{\varepsilon}{2\pi\omega_0 a}\int_0^{2\pi}[2\mu_1 a\omega_0\sin\phi + \mu_2 a\omega_0\sin\phi|a\omega_0\sin\phi|]\cos\phi d\phi$$

$$= 0$$

i.e.,

$$\dot{a} = -\varepsilon\left(\mu_1 a + \frac{4}{3\pi}\mu_2\omega_0 a^2\right),\ \dot{\beta} = 0 \tag{3.19.4}$$

(b) Separating the variables of the first equation of (3.19.4) and integrating it, we obtain

$$\int \frac{da}{\mu_1 a + \frac{4}{3\pi}\mu_2\omega_0 a^2} = -\varepsilon\int dt$$

which gives

$$\frac{1}{\mu_1}\ln\frac{2\cdot\frac{4}{3\pi}\mu_2\omega_0 a + \mu_1 - \mu_1}{2\cdot\frac{4}{3\pi}\mu_2\omega_0 a + \mu_1 + \mu_1} = -\varepsilon t + c_0$$

where c_0 is a constant of integration. So

$$\ln\frac{a}{a + \frac{3\pi}{4\mu_2\omega_0}\mu_1} = -\varepsilon\mu_1 t + c$$

where c is a constant. Let $a(0) = a_0$, we get

$$a = \frac{a_0}{\exp(\varepsilon\mu_1 t) + a_0\frac{4\mu_2\omega_0}{3\pi\mu_1}[\exp(\varepsilon\mu_1 t) - 1]} \tag{3.19.5}$$

Then from $\dot{\beta} = 0$, we obtain that $\beta = \beta_0$ is a constant value. So

$$u = \left[\frac{a_0}{\exp(\varepsilon\mu_1 t) + a_0\frac{4\mu_2\omega_0}{3\pi\mu_1}[\exp(\varepsilon\mu_1 t) - 1]}\right]\cos(\omega_0 t + \beta_0) \tag{3.19.6}$$

3.20 Exercise 3.20 (Averaging to Solve for an Oscillator with Both Viscous and Negative Coulomb Damping)

Solution: **(a)** Write the original equation as

$$\ddot{u} + \omega_0^2 u = \varepsilon f(\dot{u}) \tag{3.20.1}$$

$$\text{where}\, f(\dot{u}) = -(2\mu_1 \dot{u} - \mu_0 \operatorname{sgn} \dot{u}) \tag{3.20.2}$$

As in the previous Exercise, the method of averaging is still used. Let the solution be

$$u = a\cos(\omega_0 t + \beta) \triangleq a\cos\varphi, \quad \varphi = \omega_0 t + \beta \tag{3.20.3}$$

Referring to the result of the previous exercise, there is

$$\begin{aligned}
\dot{a} &= -\frac{\varepsilon}{2\pi\omega_0}\int_0^{2\pi} f(-a\omega_0\sin\phi)\sin\phi d\phi \\
&= -\frac{\varepsilon}{2\pi\omega_0}\int_0^{2\pi} \left[2\mu_1 a\omega_0\sin\phi - \mu_0\operatorname{sgn}(a\omega_0\sin\phi)\right]\sin\phi d\phi \\
&= -\frac{\varepsilon}{\pi\omega_0}\int_0^{\pi} \left[2\mu_1 a\omega_0\sin^2\phi - \mu_0\sin\phi\right] d\phi \\
&= -\frac{\varepsilon}{\pi\omega_0}(\mu_1 a\omega_0\pi - 2\mu_0) \\
&= \varepsilon\left(\frac{2\mu_0}{\pi\omega_0} - \mu_1 a\right)
\end{aligned}$$

$$\begin{aligned}
\dot{\beta} &= -\frac{\varepsilon}{2\pi\omega_0 a}\int_0^{2\pi} f(-a\omega_0\sin\phi)\cos\phi d\phi \\
&= -\frac{\varepsilon}{2\pi\omega_0 a}\int_0^{2\pi} \left[2\mu_1 a\omega_0\sin\phi - \mu_0\operatorname{sgn}(a\omega_0\sin\phi)\right]\cos\phi d\phi \\
&= 0
\end{aligned}$$

i.e.,

$$\dot{a} = \varepsilon\left(\frac{2\mu_0}{\pi\omega_0} - \mu_1 a\right), \dot{\beta} = 0 \tag{3.20.4}$$

(b) Separating the variables of the first equation of (3.20.4) and integrating it, we obtain

$$\int \frac{da}{\frac{2\mu_0}{\pi\omega_0} - \mu_1 a} = \varepsilon \int dt$$

i.e.,

$$\frac{1}{\mu_1}\ln(\frac{2\mu_0}{\pi\omega_0} - \mu_1 a) = -\varepsilon t + c$$

where c is a constant of integration. Let $a(0) = a_0$, we get

$$a = \left(a_0 - \frac{2\mu_0}{\pi\omega_0\mu_1}\right)\exp(-\varepsilon\mu_1 t) + \frac{2\mu_0}{\pi\omega_0\mu_1} \tag{3.20.5}$$

Then from $\dot{\beta} = 0$, we obtain that $\beta = \beta_0$ is a constant value. So

$$u = \left[\left(a_0 - \frac{2\mu_0}{\pi\omega_0\mu_1}\right)\exp(-\varepsilon\mu_1 t\right) + \frac{2\mu_0}{\pi\omega_0\mu_1}]\cos(\omega_0 t + \beta_0) \tag{3.20.6}$$

(c) Since the Coulomb type damping in this system is negative, the system absorbs energy from the external energy source through this damping and on the other hand, the viscous damping is energy dissipating, therefore, the system has the potential to form a limit cycle when the energy absorption is balanced with the energy dissipation.

In fact, from the approximate solution (3.20.6), we can find that the motion of the system will converge to a simple harmonic motion regardless of the value of the initial amplitude:

$$u = \frac{2\mu_0}{\pi\omega_0\mu_1}\cos(\omega_0 t + \beta_0) \tag{3.20.7}$$

In the phase plane, this is a closed trajectory, and since there are no other closed trajectories in its neighborhood, it is a limit cycle and a stable limit cycle.

3.21 Exercise 3.21 (Averaging to Solve for an Oscillator with Both Squared Damping and Negative Coulomb Damping)

Solution: (a) The method of averaging of Solution is still used. Let the Solution be

$$u = a\cos(\omega_0 t + \beta) \triangleq a\cos\varphi, \quad \varphi = \omega_0 t + \beta \tag{3.21.1}$$

Against Exercise 3.18 and its result, we have

$$\dot{a} = \varepsilon\left(\frac{2\mu_0}{\pi\omega_0} - \frac{4\mu_2\omega_0}{3\pi}a^2\right), \quad \dot{\beta} = 0 \tag{3.21.2}$$

(b) Separating the variables of the first equation of (3.21.2) and integrating it, we obtain

$$\int \frac{da}{\frac{2\mu_0}{\pi\omega_0} - \frac{4\mu_2\omega_0}{3\pi}a^2} = \varepsilon \int dt$$

which gives

$$\frac{1}{2\sqrt{\frac{4}{3\pi}\mu_2\omega_0 \cdot \frac{2\mu_0}{\pi\omega_0}}} \ln \frac{a\sqrt{\frac{4\mu_2\omega_0}{3\pi}} + \sqrt{\frac{2\mu_0}{\pi\omega_0}}}{a\sqrt{\frac{4\mu_2\omega_0}{3\pi}} - \sqrt{\frac{2\mu_0}{\pi\omega_0}}} = \varepsilon t + c_0$$

where c_0 is a constant of integration. The above equation can be written as

$$\frac{a\sqrt{\frac{4\mu_2\omega_0}{3\pi}} + \sqrt{\frac{2\mu_0}{\pi\omega_0}}}{a\sqrt{\frac{4\mu_2\omega_0}{3\pi}} - \sqrt{\frac{2\mu_0}{\pi\omega_0}}} = c\exp[\frac{\varepsilon}{\pi}\sqrt{\frac{32\mu_0\mu_2}{3}}t]$$

where c is a constant of integration. Then we can obtain

$$a = \frac{1}{\omega_0}\sqrt{\frac{3\mu_0}{2\mu_2}}\{1 + \frac{2}{c\exp[\frac{\varepsilon}{\pi}\sqrt{\frac{32\mu_0\mu_2}{3}}t] - 1}\} \tag{3.21.3}$$

Then from $\dot{\beta} = 0$, we obtain that $\beta = \beta_0$ is a constant value. So

$$u = \frac{1}{\omega_0}\sqrt{\frac{3\mu_0}{2\mu_2}}\left\{1 + \frac{2}{c\exp[\frac{\varepsilon}{\pi}\sqrt{\frac{32\mu_0\mu_2}{3}}t] - 1}\right\}\cos(\omega_0 t + \beta_0) \tag{3.21.4}$$

(c) Since the Coulomb type damping in this system is negative, the system absorbs energy from the external energy source through this damping and on the other hand, the viscous damping is energy dissipating, therefore, the system has the potential to form a limit cycle when the energy absorption is balanced with the energy dissipation.

In fact, from the approximate solution (3.21.4), we can find that the motion of the system will converge to a simple harmonic motion regardless of the value of the initial amplitude:

$$u = \frac{1}{\omega_0}\sqrt{\frac{3\mu_0}{2\mu_2}}\cos(\omega_0 t + \beta_0) \tag{3.21.5}$$

In the phase plane, this is a closed trajectory, and since there are no other closed trajectories in its neighborhood, it is a limit cycle and a stable limit cycle.

3.22 Exercise 3.22 (Oscillators with Simultaneous Negative Viscous Damping, Square Damping, and Cubic Nonlinearity)

Solution: **(a)** Write the original equation as

$$\ddot{u} + \omega_0^2 u = \varepsilon f(u, \dot{u}) \tag{3.22.1}$$

$$\text{where} f(u, \dot{u}) = -\left[(\mu_2 \dot{u}|\dot{u}| - 2\mu_1 \dot{u}) + \alpha u^3\right] \tag{3.22.2}$$

As in the previous exercise, the method of averaging is still used. Let the solution be

$$u = a\cos(\omega_0 t + \beta) \triangleq a\cos\varphi, \quad \varphi = \omega_0 t + \beta \tag{3.22.3}$$

Referring to Exercise 3.17, there is

$$\begin{aligned}
\dot{a} &= -\frac{\varepsilon}{2\pi\omega_0}\int_0^{2\pi} f(-a\omega_0 \sin\phi)\sin\phi d\phi \\
&= -\frac{\varepsilon}{2\pi\omega_0}\int_0^{2\pi} [2\mu_1 a\omega_0 \sin\phi + \mu_2 a\omega_0 \sin\phi|a\omega_0 \sin\phi|]\sin\phi d\phi \\
&= -\frac{\varepsilon}{\pi\omega_0}\int_0^{\pi} \left[2\mu_1 a\omega_0 \sin^2\phi + \mu_2 a^2\omega_0^2 \sin^3\phi\right] d\phi \\
&= -\frac{\varepsilon}{\pi\omega_0}\left[\mu_1 a\omega_0 \pi + \mu_2 a^2\omega_0^2\left(-\cos\phi + \frac{1}{3}\cos^3\phi\right)\Big|_0^{\pi}\right] \\
&= -\varepsilon\left(\mu_1 a + \frac{4}{3\pi}\mu_2\omega_0 a^2\right)
\end{aligned}$$

$$\dot{\beta} = -\frac{\varepsilon}{2\pi\omega_0 a}\int_0^{2\pi} f(a\cos\phi, -a\omega_0 \sin\phi)\cos\phi d\phi$$

$$= -\frac{\varepsilon}{2\pi\omega_0 a}\int_0^{2\pi}[\mu_2 a\omega_0\sin\phi|a\omega_0\sin\phi| - 2\mu_1 a\omega_0\sin\phi - \alpha(a\cos\phi)^3]\cos\phi d\phi$$

$$= \frac{\varepsilon\alpha a^2}{\pi\omega_0}\int_0^{\pi}\cos^4\phi d\phi = \frac{\varepsilon\alpha a^2}{\pi\omega_0}\cdot\frac{3}{8}\pi$$

$$= \frac{3\varepsilon\alpha}{8\omega_0}a^2$$

i.e.,

$$\dot{a} = \varepsilon\left(\frac{4}{3\pi}\mu_2\omega_0 a^2 - \mu_1 a\right), \quad \dot{\beta} = \frac{3\varepsilon\alpha}{8\omega_0}a^2 \tag{3.22.4}$$

(b) Separating the variables of the first equation of (3.22.4) and integrating it, we obtain

$$\int\frac{da}{\frac{4}{3\pi}\mu_2\omega_0 a^2 - \mu_1 a} = \varepsilon\int dt$$

which gives

$$\frac{1}{\mu_1}\ln\frac{2\cdot\frac{4}{3\pi}\mu_2\omega_0 a + \mu_1 - \mu_1}{2\cdot\frac{4}{3\pi}\mu_2\omega_0 a + \mu_1 + \mu_1} = \varepsilon t + c_0$$

where c_0 is a constant of integration. So

$$\frac{a}{a + \frac{3\pi}{4\mu_2\omega_0}\mu_1} = c\exp(\varepsilon\mu_1 t)$$

where c is a constant. The above equation is reorganized to give

$$a = \frac{3\pi\mu_1}{4\mu_2\omega_0}[1 + \frac{1}{1 + C\exp(-\varepsilon\mu_1 t)}] \tag{3.22.5}$$

where C is a constant.

Separating the variables of the second equation of (3.22.4) and integrating it, we obtain

$$\beta = \frac{3\varepsilon\alpha}{8\omega_0}a^2 t + \beta_0 \tag{3.22.6}$$

Therefore, the approximate solution of the original equation is

$$u = \frac{3\pi\mu_1}{4\mu_2\omega_0}\left[1 + \frac{1}{1 + C\exp(-\varepsilon\mu_1 t)}\right]\cos[\left(\omega_0 + \frac{3\varepsilon\alpha}{8\omega_0}a^2\right)t + \beta_0] \quad (3.22.7)$$

(c) Since the viscous type of damping in this system is negative, the system absorbs energy from the external energy source through this damping, while on the other hand, the viscous damping is energy dissipating, and the system has the potential to form a limit cycle when the energy absorption is balanced with the energy dissipation.

In fact, from the approximate solution (3.22.7), we can find that the motion of the system will converge to a simple harmonic motion regardless of the value of the initial amplitude:

$$u = \frac{3\pi\mu_1}{4\mu_2\omega_0}\cos[\left(\omega_0 + \frac{3\varepsilon\alpha}{8\omega_0}a^2\right)t + \beta_0] \quad (3.22.8)$$

In the phase plane, this is a closed trajectory, and since there are no other closed trajectories in its neighborhood, it is a limit cycle and a stable limit cycle.

From the approximate solution (3.22.7), we can also find that the cubic nonlinear term $\varepsilon\alpha u^3$ affect the frequency of the oscillation but not the amplitude.

3.23 Exercise 3.23 (Method of Averaging for Solving Oscillators with Simultaneous Viscous, Quadratic and Cubic Damping Nonlinearities)

Solution: (a) Write the original equation as

$$\ddot{u} + \omega_0^2 u = \varepsilon f(\dot{u}) \quad (3.23.1)$$

$$\text{where } f(\dot{u}) = -\left(2\mu_1\dot{u} + \mu_2\dot{u}|\dot{u}| + \mu_3\dot{u}^3\right) \quad (3.23.2)$$

The method of averaging is also adopted to solve this problem. Let the solution be

$$u = a\cos(\omega_0 t + \beta) \triangleq a\cos\varphi, \quad \varphi = \omega_0 t + \beta \quad (3.23.3)$$

Referring to the previous exercises, there are

$$\dot{a} = -\frac{\varepsilon}{2\pi\omega_0}\int_0^{2\pi} f(-a\omega_0\sin\phi)\sin\phi d\phi$$

$$
\begin{aligned}
&= -\frac{\varepsilon}{2\pi\omega_0}\int_0^{2\pi}[2\mu_1 a\omega_0\sin\phi + \mu_2 a\omega_0\sin\phi|a\omega_0\sin\phi| + \mu_3(a\omega_0\sin\phi)^3]\sin\phi d\phi \\
&= -\frac{\varepsilon}{\pi\omega_0}\int_0^{\pi}[2\mu_1 a\omega_0\sin^2\phi + \mu_2 a^2\omega_0^2\sin^3\phi + \mu_3 a^3\omega_0^3\sin^4\phi]d\phi \\
&= -\frac{\varepsilon}{\pi\omega_0}\left[\mu_1 a\omega_0\pi + \mu_2 a^2\omega_0^2\left(-\cos\phi + \frac{1}{3}\cos^3\phi\right)\Big|_0^{\pi} + \frac{3\pi}{8}\mu_3 a^3\omega_0^3\right] \\
&= -\varepsilon\left(\mu_1 a + \frac{4}{3\pi}\mu_2\omega_0 a^2 + \frac{3}{8}\mu_3\omega_0^2 a^3\right)
\end{aligned}
$$

$$
\begin{aligned}
\dot{\beta} &= -\frac{\varepsilon}{2\pi\omega_0 a}\int_0^{2\pi} f(-a\omega_0\sin\phi)\cos\phi d\phi \\
&= -\frac{\varepsilon}{2\pi\omega_0 a}\int_0^{2\pi}[2\mu_1 a\omega_0\sin\phi + \mu_2 a\omega_0\sin\phi|a\omega_0\sin\phi| + \mu_3(a\omega_0\sin\phi)^3]\cos\phi d\phi \\
&= 0
\end{aligned}
$$

$$
\text{i.e., } \dot{a} = -\varepsilon\left(\mu_1 a + \frac{4}{3\pi}\mu_2\omega_0 a^2 + \frac{3}{8}\mu_3\omega_0^2 a^3\right) \tag{3.23.4}
$$

$$
\dot{\beta} = 0 \tag{3.23.5}
$$

(b) Let $\dot{a} = 0$, Eq. (3.23.4) becomes

$$
\mu_1 a + \frac{4}{3\pi}\mu_2\omega_0 a^2 + \frac{3}{8}\mu_3\omega_0^2 a^3 = 0 \tag{3.23.6}
$$

Thus, there are three singular points s_1, $s_2$and s_3, where $a_{s1} = 0$ for s_1, and a_s of $s_2$and s_3 satisfies

$$
\frac{3}{8}\mu_3\omega_0^2 a_s^2 + \frac{4}{3\pi}\mu_2\omega_0 a_s + \mu_1 = 0 \tag{3.23.7}
$$

i.e.,

$$
a_{s2,\ s3} = -\frac{16\mu_2}{9\pi\omega_0\mu_3} \pm \frac{4}{3\omega_0\mu_3}\sqrt{\frac{16\mu_2^2}{9\pi^2} - \frac{3}{2}\mu_1\mu_3} \tag{3.23.8}
$$

The '+' number corresponds to the singular point s_2 and the '−' number corresponds to the singular point s_3.

Let's study the stability of the equilibrium point. Linearize Eq. (3.23.4) at the singular point $a = a_s$. For this purpose, let

$$a = a_s + x \tag{3.23.9}$$

then

$$\dot{x} = -\varepsilon\left(\mu_1 + \frac{8}{3\pi}\mu_2\omega_0 a_s + \frac{9}{8}\mu_3\omega_0^2 a_s^2\right)x \tag{3.23.10}$$

Therefore, the condition for the stability of the singular point is

$$\frac{9}{8}\mu_3\omega_0^2 a_s^2 + \frac{8}{3\pi}\mu_2\omega_0 a_s + \mu_1 > 0 \tag{3.23.11}$$

(1) For the singular point $s_1, a_{s1} = 0$, Eq. (3.23.11) becomes

$$\mu_1 > 0 \tag{3.23.12}$$

Therefore, when the above equation is satisfied or the system has positive viscous damping, the stationary state of the system is stable and vice versa.

(2) For the singular point when $a_s \neq 0$, considering (3.23.7), (3.23.11) becomes

$$\frac{6}{8}\mu_3\omega_0^2 a_s^2 + \frac{4}{3\pi}\mu_2\omega_0 a_s > 0 \tag{3.23.13}$$

i.e.,

$$a_s > 0 \quad \text{and} \quad a_s > -\frac{16\mu_2}{9\pi\omega_0\mu_3} \tag{3.23.14}$$

$$\text{or } a_s < 0 \quad \text{and} \quad a_s < -\frac{16\mu_2}{9\pi\omega_0\mu_3} \tag{3.23.15}$$

Since $a_{s2} > 0$,

$$\frac{1}{\mu_3}\sqrt{\frac{16\mu_2^2}{9\pi^2} - \frac{3}{2}\mu_1\mu_3} > \frac{4\mu_2}{3\pi\mu_3} \tag{3.23.16}$$

For the real system, $\mu_3 > 0$, so Eq. (3.23.16) can be written as

$$\sqrt{\frac{16}{9\pi^2}\left(\frac{\mu_2}{\mu_3}\right)^2 - \frac{3}{2}\frac{\mu_1}{\mu_3}} > \frac{4}{3\pi}\frac{\mu_2}{\mu_3} \tag{3.23.17}$$

Therefore, the stability condition for the solution of the singular point s_2 is

$$\frac{\mu_1}{\mu_3}\langle 0 \text{ and } \frac{\mu_2}{\mu_3}\rangle 0$$

or

$$0 < \frac{\mu_1}{\mu_3} < \frac{32}{27\pi^2}\left(\frac{\mu_2}{\mu_3}\right)^2 \text{ and } \frac{\mu_2}{\mu_3} < 0 \tag{3.23.18}$$

Since $a_{s3} < 0$, i.e.

$$\sqrt{\frac{16}{9\pi^2}(\frac{\mu_2}{\mu_3})^2 - \frac{3}{2}\frac{\mu_1}{\mu_3}} > -\frac{4}{3\pi}\frac{\mu_2}{\mu_3} \tag{3.23.19}$$

The stability condition for the solution of the singular point s_3 is

$$\frac{\mu_1}{\mu_3} < 0 \text{ and } \frac{\mu_2}{\mu_3} < 0$$
$$\text{or } 0 < \frac{\mu_1}{\mu_3} < \frac{32}{27\pi^2}\left(\frac{\mu_2}{\mu_3}\right)^2 \text{ and } \frac{\mu_2}{\mu_3} > 0 \tag{3.23.20}$$

However, the second condition of the above equation implies that $\mu_1 > 0,\ \mu_2 > 0,\ \mu_3 > 0$, i.e., all damping is dissipative, which cannot lead to a steady state oscillation with a non-zero amplitude. Therefore, the stability condition for the singular point s_3 is

$$\frac{\mu_1}{\mu_3} < 0 \text{ and } \frac{\mu_2}{\mu_3} < 0 \tag{3.23.21}$$

From (3.23.18) and (3.23.21), the stable regions of singular points s_2 and s_3 can be sketched on the $(\mu_2/\mu_3) \sim (\mu_1/\mu_3)$ plane as shown in Fig. 3.11a. It can be seen that the two stable regions are non-overlapping, so for each set of $\mu_1,\ \mu_2,\ \mu_3$ parameter values that meet the stability condition, only one of the singular points s_2 and s_3 is stable, and correspondingly, only one stable limit cycle can appear on the phase plane.

In order to verify the correctness of the above results, we adopted direct numerical integration of the original equations and drew the phase trajectories and limit cycle(s) in Fig. 3.11b. Here, $\omega_0 = 1,\ \ \varepsilon = 0.1,\ \ \mu_3 = 1$, μ_1 and μ_2 can take the following four different sets of values:

Group 3.1: $\mu_1 = 1$, μ_2 is positive so that $\frac{\mu_1}{\mu_3} = \frac{1}{2}[\frac{32}{27\pi^2}\left(\frac{\mu_2}{\mu_3}\right)^2]$;Group 3.2: $\mu_1 = 1$, μ_2 is negative so that $\frac{\mu_1}{\mu_3} = \frac{1}{2}[\frac{32}{27\pi^2}\left(\frac{\mu_2}{\mu_3}\right)^2]$;Group 3.3: $\mu_1 = -1$,μ_2 is positive so that $\left|\frac{\mu_1}{\mu_3}\right| = \frac{32}{27\pi^2}(\frac{\mu_2}{\mu_3})^2$;Group 3.4: $\mu_1 = -1$, μ_2 is negative so that $\left|\frac{\mu_1}{\mu_3}\right| = \frac{32}{27\pi^2}(\frac{\mu_2}{\mu_3})^2$;

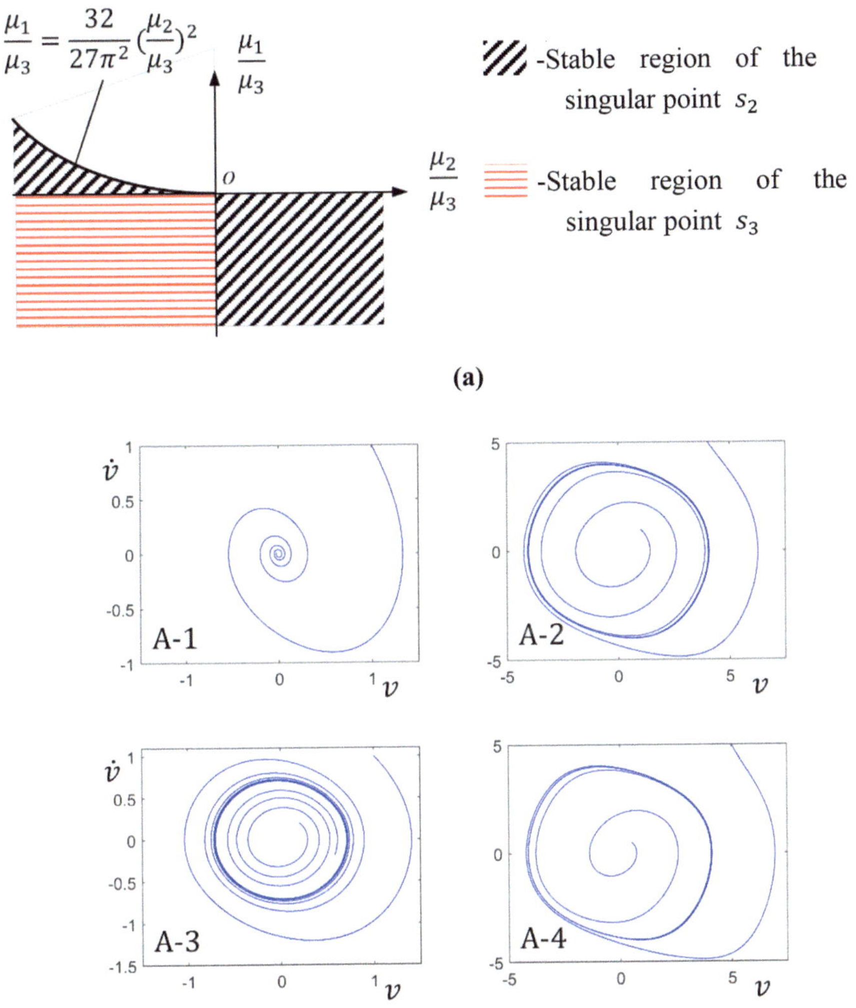

Fig. 3.11 **a** Stable regions of singularities s_2 and s_3. **b** Phase trajectories (including limit cycles) obtained by direct integration of the original equation for Exercise 3.23

According to the second condition of (3.23.20), the case with the first group of parameters should have a limit cycle. But according to the direct integration of the original system, phase trajectories converge to the origin and there is no limit cycle, as shown in Fig. 3.11b A-1. According to the stability conditions (3.23.18) and (3.23.21), each of the cases with the second to the fourth group of parameters has a stable limit cycle, which agrees well with the integration of the original system, as shown in Fig. 3.11b A-2 to A-4.

In summary, the approximate solution and stability conditions obtained above are correct.

3.24 Exercise 3.24 (Averaging to Solve and Analyze Nonlinear Vibrations of a Pendulum)

Solution: (a) We use the small parameter ε to measure the amplitude of oscillation, x, therefore, let $x = \varepsilon u$, and write the original equation as

$$\varepsilon \ddot{u} + \varepsilon \omega_0^2 u + \varepsilon \frac{\lambda}{J} \dot{u} - \varepsilon \frac{I}{2J} (\dot{u} - |\dot{u}|) \delta(u - u_0) = 0 \tag{3.24.1}$$

We study the case where both the friction effect λ / J and the impulsive excitation are small. Let

$$\frac{\lambda}{J} = \varepsilon \lambda_1, \ \frac{I}{2J} = \varepsilon I_1 \tag{3.24.2}$$

Thus, Eq. (3.24.1) becomes

$$\ddot{u} + \omega_0^2 u = \varepsilon f(u, \dot{u}) \tag{3.24.3}$$

$$f(u, \dot{u}) = -[\lambda_1 \dot{u} - I_1 (\dot{u} - |\dot{u}|) \delta(u - u_0)] \tag{3.24.4}$$

Since f is not an even function of the state, it can be solved by the method of averaging. Let its solution be

$$a \cos \psi, \quad \psi = \omega_0 t + \varphi \tag{3.24.5}$$

Referring to Exercise 3.17 or 3.22, there are

$$\begin{aligned}
\dot{a} &= -\frac{\varepsilon}{2\pi \omega_0} \int_0^{2\pi} f(a\cos\psi, -a\omega_0 \sin\psi) \sin\psi d\psi \\
&= -\frac{\varepsilon}{2\pi \omega_0} \int_0^{2\pi} \{\lambda_1 (a\omega_0 \sin\psi) \\
&\quad - I_1 \left[(a\omega_0 \sin\psi) + |(a\omega_0 \sin\psi)|\right] \delta(a\cos\psi - u_0)\} \sin\psi d\psi \\
&= -\frac{\varepsilon a \lambda_1}{2\pi} \int_0^{2\pi} \sin^2\psi d\psi + \frac{\varepsilon I_1 a}{\pi} \int_0^{\pi} \delta(a\cos\psi - u_0) \sin^2\psi d\psi
\end{aligned}$$

$$= -\frac{\varepsilon a\lambda_1}{2} + \frac{\varepsilon I_1 a}{\pi}\int_0^{\pi} \delta(a\cos\psi - u_0)\sin^2\psi d\psi$$

$$\dot{\phi} = -\frac{\varepsilon}{2\pi\omega_0 a}\int_0^{2\pi} f(a\cos\psi, -a\omega_0\sin\psi)\cos\psi d\psi$$

$$= -\frac{\varepsilon}{2\pi\omega_0 a}\int_0^{2\pi} \{\lambda_1(a\omega_0\sin\psi)$$

$$- I_1\left[(a\omega_0\sin\psi) + |(a\omega_0\sin\psi)|\right]\delta(a\cos\psi - u_0)\}\cos\psi d\psi$$

$$= -\frac{\varepsilon\lambda_1}{2\pi}\int_0^{2\pi} \cos\psi\sin\psi d\psi + \frac{\varepsilon I_1}{\pi}\int_0^{\pi} \delta(a\cos\psi - u_0)\cos\psi\sin\psi d\psi$$

Note that the above results are obtained assuming that the impulsive excitation works at the angle within the range of $[0, \ \pi)$ and, therefore, $\delta(a\cos\psi - u_0) = 0$ within the range of $(-\pi, \ 0)$.

Returning to x yields

$$x = a\cos\psi, \quad \psi = \omega_0 t + \varphi \tag{3.24.6}$$

$$\dot{a} = -\frac{a\lambda}{2J} + \frac{Ia}{2J\pi}\int_0^{\pi} \delta(a\cos\psi - x_0)\sin^2\psi d\psi \tag{3.24.7}$$

$$\dot{\varphi} = -\frac{\lambda}{2J\pi}\int_0^{2\pi} \cos\psi\sin\psi d\psi + \frac{I}{2J\pi}\int_0^{\pi} \delta(a\cos\psi - x_0)\cos\psi\sin\psi d\psi \tag{3.24.8}$$

(b) If $a \geq x_0 \geq 0$, according to the property of integration of δ function, we get

$$\dot{a} = -\frac{a\lambda}{2J\pi}\int_0^{2\pi} \sin^2\psi d\psi + \frac{Ia}{2J\pi}\int_0^{\pi} \delta(a\cos\psi - x_0)\sin^2\psi d\psi$$

$$= -\frac{a\lambda}{2J} - \frac{I}{2J\pi}\int_0^{\pi} \delta(a\cos\psi - x_0)\sin\psi d(a\cos\psi)$$

$$= -\frac{a\lambda}{2J} - \frac{I}{2J\pi}\int_0^{\pi} \delta(a\cos\psi - x_0)\sin\psi d(a\cos\psi)$$

$$
\begin{aligned}
&= -\frac{a\lambda}{2J} + \frac{I}{2J\pi}\int_{-a}^{a} \delta(x - x_0)\sin\left(\arccos\frac{x}{a}\right)dx \\
&= -\frac{a\lambda}{2J} + \frac{I}{2J\pi}\int_{0}^{a} \delta(x - x_0)\sin\left(\arccos\frac{x}{a}\right)dx \\
&= -\frac{a\lambda}{2J} + \frac{I}{2J\pi}\sin\left(\arccos\frac{x_0}{a}\right)
\end{aligned}
\tag{3.24.9}
$$

$$
\text{i.e., } \dot{a} = -\frac{a\lambda}{2J} + \frac{I}{2J\pi}\sin\psi_a \tag{3.24.10}
$$

$$
\text{where } \cos\psi_a = \frac{x_0}{a} \quad \Rightarrow \quad a\cos\psi_a - x_0 = 0 \tag{3.24.11}
$$

$$
\begin{aligned}
\dot{\phi} &= -\frac{\lambda}{2J\pi}\int_0^{2\pi} \cos\psi\sin\psi d\psi + \frac{I}{2J\pi}\int_0^{\pi} \delta(a\cos\psi - x_0)\cos\psi\sin\psi d\psi \\
&= -\frac{I}{2J\pi a}\int_0^{\pi} \delta(a\cos\psi - x_0)\cos\psi d(a\cos\psi) \\
&= \frac{I}{2J\pi a}\int_{-a}^{a} \delta(x - x_0)\cos\left(\arccos\frac{x}{a}\right)dx \\
&= \frac{I}{2J\pi a}\int_{0}^{a} \delta(x - x_0)\cos\left(\arccos\frac{x}{a}\right)dx \\
&= \frac{I}{2J\pi a}\cos\left(\arccos\frac{x_0}{a}\right) \\
&= \frac{I}{2J\pi a}\cos\psi_a
\end{aligned}
\tag{3.24.12}
$$

$$
\dot{\varphi} = \frac{I}{2J\pi a}\cos\psi_a \tag{3.24.13}
$$

From (3.24.11), we can obtain

$$
\sin\psi_a = \left(1 - \frac{x_0^2}{a^2}\right)^{1/2} \tag{3.24.14}
$$

Substituting this into (3.24.10) we can obtain

$$
\dot{a} = -\frac{a\lambda}{2J} + \frac{I}{2J\pi}\left(1 - \frac{x_0^2}{a^2}\right)^{1/2} \tag{3.24.15}
$$

(c) If $x_0 > a$, then the domain of definition of the δ function is no longer on $(0, a)$, so the integrals of the last terms in (3.24.9) and (3.24.12) are zeros, which gives

$$\dot{a} = -\frac{a\lambda}{2J}, \quad \dot{\varphi} = 0 \tag{3.24.16}$$

(d) Equation (3.24.16) has the solution $a = \exp(-\lambda/2J)$, which is decaying. Therefore, only (3.24.15) can form a steady state oscillation and the steady state amplitude satisfies

$$-\frac{a_s\lambda}{2J} + \frac{I}{2J\pi}\left(1 - \frac{x_0^2}{a_s^2}\right)^{1/2} = 0 \tag{3.24.17}$$

i.e.,

$$\left(1 - \frac{x_0^2}{a_s^2}\right)^{1/2} = \frac{a_s\pi\lambda}{I} \Rightarrow \left(\frac{\pi\lambda}{I}\right)^2 a_s^4 - a_s^2 + x_0^2 = 0$$

therefore,

$$a_{s1,\,s2}^2 = \frac{1}{2}\left(\frac{I}{\pi\lambda}\right)^2\left[1 \pm \sqrt{1 - 4x_0^2\left(\frac{\pi\lambda}{I}\right)^2}\right] \tag{3.24.18}$$

The real value of a_s exists if

$$4x_0^2\left(\frac{\pi\lambda}{I}\right)^2 \leq 1 \tag{3.24.19}$$

$$\text{i.e., } x_0 \leq \frac{I}{2\pi\lambda} \tag{3.24.20}$$

The stability of the steady state solution is considered below. Linearize Equation (3.24.15) near the steady state amplitude a_s yields

$$\dot{a} = \left[-\frac{\lambda}{2J} + \frac{Ix_0^2}{2J\pi a_s^3}\left(1 - \frac{x_0^2}{a_s^2}\right)^{-1/2} a\right] \tag{3.24.21}$$

Therefore, the stability condition is

$$-\frac{\lambda}{2J} + \frac{Ix_0^2}{2J\pi a_s^3}\left(1 - \frac{x_0^2}{a_s^2}\right)^{-1/2} < 0$$

Taking (3.24.17) into account, the above equation becomes

$$x_0^2 < a_s^2 \tag{3.24.22}$$

Substituting (3.24.18) into the above equation, the stability conditions for the two steady state oscillations become

$$x_0^2 < a_{s1}^2 = 2\left(\frac{I}{2\pi\lambda}\right)^2\left[1+\sqrt{1-4x_0^2\left(\frac{\pi\lambda}{I}\right)^2}\right] \tag{3.24.23}$$

$$x_0^2 < a_{s2}^2 = 2\left(\frac{I}{2\pi\lambda}\right)^2\left[1-\sqrt{1-4x_0^2\left(\frac{\pi\lambda}{I}\right)^2}\right] \tag{3.24.24}$$

It can be seen that when (3.24.20) is satisfied, the stability condition (3.24.23) must also be satisfied. Therefore, the steady state amplitude a_{s1} is stable and it forms a stable limit cycle in the phase plane. When

$$\frac{\sqrt{3}}{2}\frac{I}{2\pi\lambda} \leq x_0 \leq \frac{I}{2\pi\lambda} \tag{3.24.25}$$

the stability condition (3.24.24) is satisfied, then the steady state amplitude a_{s2} is stable and it forms a stable limit cycle in the phase plane.

In summary, when (3.24.25) is satisfied, there are two stable steady state amplitudes; when

$$x_0 < \frac{\sqrt{3}}{2}\frac{I}{2\pi\lambda} \tag{3.24.26}$$

there is only one steady state amplitude a_{s1}.

3.25 Exercise 3.25 (Viscous Damping, Negative Stiffness, and Cubic Nonlinear Systems)

Solution: The singular points of the system are $u_{s1} = 0$, $u_{s2} = \sqrt{\alpha_1/\alpha_2}$ and $u_{s3} = -\sqrt{\alpha_1/\alpha_2}$. Since the stiffness of the system is negative, the singular point u_{s1} is unstable, which is easy to prove. Since the equation remains the same by replacing u with $-u$ in the original equation, the states of motion around the singular points u_{s2} and u_{s3} are the same, and we only need to analyze the motion around the singular point u_{s2}.

Let $u = u_{s2} + v$, we can move the singular point u_{s2} of the original equation to the origin:

$$\ddot{v} + \left(3\alpha_2 u_{s2}^2 - \alpha_1\right)v + \mu\dot{v} + 3\alpha_2 u_{s2} v^2 + \alpha_2 v^3 = 0$$

i.e.,

$$\ddot{v} + 2\alpha_1 v + \mu\dot{v} + 3\sqrt{\alpha_1\alpha_2}v^2 + \alpha_2 v^3 = 0 \tag{3.25.1}$$

The linearization of this equation near the origin is given by

$$\ddot{v} + 2\alpha_1 v + \mu\dot{v} = 0 \tag{3.25.2}$$

This is a viscous damped linear oscillation which decays to simple harmonic oscillation. Therefore, any perturbed motion near the singular point u_{s2} will come back to the singular point u_{s2}.

To summarize, the system diverges around $u = u_{s1} = 0$ and will eventually reach $u = u_{s2} = \sqrt{\alpha_1/\alpha_2}$. For the motion of $u > u_{s2}$, it will eventually converge to $u = u_{s2} = \sqrt{\alpha_1/\alpha_2}$. Therefore, the equilibrium state of the system is $u = u_{s2} = \sqrt{\alpha_1/\alpha_2}$.

3.26 Exercise 3.26 (Transformations of Variables and Three-Level Number Expansion Methods for Solving a Cubic Nonlinear System)

Solution: (a) The linear damping term is not in the same order of the cubic nonlinear term, and therefore, the method of averaging cannot be used.

The independent variables of the original equation transform from t to a and ψ, so

$$\begin{aligned}
\frac{du}{dt} &= \frac{d\psi}{dt}\frac{\partial u}{\partial \psi} + \frac{da}{dt}\frac{\partial u}{\partial a} = \omega\frac{\partial u}{\partial \psi} + \xi\frac{\partial u}{\partial a} \\
\frac{d^2u}{dt^2} &= \frac{d}{dt}\left(\frac{d\psi}{dt}\frac{\partial u}{\partial \psi} + \frac{da}{dt}\frac{\partial u}{\partial a}\right) = \frac{d^2\psi}{dt^2}\frac{\partial u}{\partial \psi} \\
&\quad + \frac{d\psi}{dt}\frac{d}{dt}\left(\frac{\partial u}{\partial \psi}\right) + \frac{d^2a}{dt^2}\frac{\partial u}{\partial a} + \frac{da}{dt}\frac{d}{dt}\left(\frac{\partial u}{\partial a}\right) \\
&= \frac{d^2\psi}{dt^2}\frac{\partial u}{\partial \psi} + \frac{d\psi}{dt}\left(\frac{d\psi}{dt}\frac{\partial^2 u}{\partial \psi^2} + \frac{da}{dt}\frac{\partial^2 u}{\partial \psi \partial a}\right) \\
&\quad + \frac{d^2a}{dt^2}\frac{\partial u}{\partial a} + \frac{da}{dt}\left(\frac{da}{dt}\frac{\partial^2 u}{\partial a^2} + \frac{d\psi}{dt}\frac{\partial^2 u}{\partial \psi \partial a}\right) \\
&= \omega\left(\omega\frac{\partial^2 u}{\partial \psi^2} + \xi\frac{\partial^2 u}{\partial \psi \partial a}\right) + \xi\left(\xi\frac{\partial^2 u}{\partial \psi^2} + \omega\frac{\partial^2 u}{\partial \psi \partial a}\right) \\
&= \frac{da}{dt}\frac{d\omega}{da}\frac{\partial u}{\partial \psi} + \omega^2\frac{\partial^2 u}{\partial \psi^2} + \omega\xi\frac{\partial^2 u}{\partial \psi \partial a} \\
&\quad + \frac{da}{dt}\frac{d\xi}{da}\frac{\partial u}{\partial a} + \xi^2\frac{\partial^2 u}{\partial \psi^2} + \omega\xi\frac{\partial^2 u}{\partial \psi \partial a}
\end{aligned}$$

$$= \omega^2 \frac{\partial^2 u}{\partial \psi^2} + 2\omega\xi \frac{\partial^2 u}{\partial \psi \partial a} + \xi^2 \frac{\partial^2 u}{\partial \psi^2} + \xi \frac{d\omega}{da}\frac{\partial u}{\partial \psi} + \xi \frac{d\xi}{da}\frac{\partial u}{\partial a}$$

$$\text{i.e.,}\ \frac{d}{dt} = \omega \frac{\partial}{\partial \psi} + \xi \frac{\partial}{\partial a} \tag{3.26.1}$$

$$\frac{d^2}{dt^2} = \omega^2 \frac{\partial^2}{\partial \psi^2} + 2\omega\xi \frac{\partial^2}{\partial \psi \partial a} + \xi^2 \frac{\partial^2}{\partial a^2} + \xi \frac{d\xi}{da}\frac{\partial}{\partial a} + \xi \frac{d\omega}{da}\frac{\partial}{\partial \psi} \tag{3.26.2}$$

(b) From (3.26.1) and (3.26.2), the original equation becomes

$$\omega^2 \frac{\partial^2 u}{\partial \psi^2} + 2\omega\xi \frac{\partial^2 u}{\partial \psi \partial a} + \xi^2 \frac{\partial^2 u}{\partial a^2} + \xi \frac{d\xi}{da}\frac{\partial u}{\partial a} + \xi \frac{d\omega}{da}\frac{\partial u}{\partial \psi}$$
$$+ 2\mu\omega \frac{\partial u}{\partial \psi} + 2\mu\xi \frac{\partial u}{\partial a} + u + \varepsilon u^3 = 0 \tag{3.26.3}$$

Substituting the hypothetical solution into (3.26.3) yields

$$(\omega_0 + \varepsilon\omega_1)^2 \frac{\partial^2 (a\cos\psi + \varepsilon u_1)}{\partial \psi^2} + 2(\omega_0 + \varepsilon\omega_1)(-\mu a + \varepsilon\xi_1)\frac{\partial^2 (a\cos\psi + \varepsilon u_1)}{\partial \psi \partial a}$$
$$+ (-\mu a + \varepsilon\xi_1)^2 \frac{\partial^2 (a\cos\psi + \varepsilon u_1)}{\partial a^2} + (-\mu a + \varepsilon\xi_1)\frac{d(-\mu a + \varepsilon\xi_1)}{da}\frac{\partial (a\cos\psi + \varepsilon u_1)}{\partial a}$$
$$+ (-\mu a + \varepsilon\xi_1)\frac{d(\omega_0 + \varepsilon\omega_1)}{da}\frac{\partial (a\cos\psi + \varepsilon u_1)}{\partial \psi}$$
$$+ 2\mu(\omega_0 + \varepsilon\omega_1)\frac{\partial (a\cos\psi + \varepsilon u_1)}{\partial \psi} + 2\mu(-\mu a + \varepsilon\xi_1)\frac{\partial (a\cos\psi + \varepsilon u_1)}{\partial a}$$
$$+ (a\cos\psi + \varepsilon u_1) + \varepsilon(a\cos\psi + \varepsilon u_1)^3 + O(\varepsilon^2) + \cdots = 0$$

Expanding the above equation and retaining to $O(\varepsilon)$ yields

$$0 = \varepsilon[\omega_0^2 \frac{\partial^2 u_1}{\partial \psi^2} - 2\omega_0\omega_1 a\cos\psi - 2\mu\omega_0 a \frac{\partial^2 u_1}{\partial \psi \partial a} + 2\mu a\omega_1 \sin\psi$$
$$- 2\omega_0\xi_1 \sin\psi + \mu^2 a^2 \frac{\partial^2 u_1}{\partial a^2} - \left(\mu\xi_1 + \mu a \frac{d\xi_1}{da}\right)\cos\psi$$
$$+ \mu^2 a \frac{\partial u_1}{\partial a} + \mu a^2 \frac{d\omega_1}{da}\sin\psi + 2\mu\omega_0 \frac{\partial u_1}{\partial \psi}$$
$$- 2\mu\omega_1 a\sin\psi - 2\mu^2 a \frac{\partial u_1}{\partial a}$$
$$+ 2\mu\xi_1\cos\psi + u_1 + a^3\cos^3\psi] + O(\varepsilon^2) + \cdots$$

Equating the coefficient of the like power of ε yields

$$
\begin{aligned}
&\omega_0^2\frac{\partial^2 u_1}{\partial\psi^2} - 2\mu\omega_0 a\frac{\partial^2 u_1}{\partial\psi\partial a} + \mu^2 a^2\frac{\partial^2 u_1}{\partial a^2} \\
&+ 2\mu\omega_0\frac{\partial u_1}{\partial\psi} - \mu^2 a\frac{\partial u_1}{\partial a} + \left(\omega_0^2 + \mu^2\right)u_1 \\
&= 2\omega_0\omega_1 a\cos\psi + 2\omega_0\xi_1\sin\psi - \mu a^2\frac{d\omega_1}{da}\sin\psi \\
&+ \mu\left(a\frac{d\xi_1}{da} - \xi_1\right)\cos\psi \\
&- \frac{3}{4}a^3\cos\psi - \frac{1}{4}a^3\cos 3\psi
\end{aligned} \tag{3.26.4}
$$

where $\omega_0^2 = 1 - \mu^2$.

(c) In order to eliminate secular terms, we need

$$
\begin{aligned}
&2\omega_0\omega_1 a + \mu\left(a\tfrac{d\xi_1}{da} - \xi_1\right) - \tfrac{3}{4}a^3 = 0 \\
&2\omega_0\xi_1 - \mu a^2\tfrac{d\omega_1}{da} = 0
\end{aligned} \tag{3.26.5}
$$

This is a variational coefficient equation of independent variable a. We can observe that the solution of the equation is a polynomial of a, which can be given as

$$
\xi_1 = A_3 a^3 + A_2 a^2 + A_1 a + A_0, \quad \omega_1 = B_2 a^2 + B_1 a + B_0 \tag{3.26.6}
$$

Substituting (3.26.6) into (3.26.5) yields

$$
\begin{aligned}
&\left(2\omega_0 B_2 - \tfrac{3}{4} + 2\mu A_3\right)a^3 + (2\omega_0 B_1 + \mu A_2)a^2 + 2\omega_0 B_0 a - \mu A_0 = 0 \\
&2(\omega_0 A_3 - \mu B_2)a^3 + (2\omega_0 A_2 - \mu B_1)a^2 + 2\omega_0 A_1 a + 2\omega_0 A_0 = 0
\end{aligned} \tag{3.26.7}
$$

Equating the coefficients of the like powers of a yields

$$
A_0 = 0, \quad A_1 = 0, \quad A_2 = 0, \quad A_3 = \frac{3}{8}\mu
$$

$$
B_0 = 0, \quad B_1 = 0, \quad B_2 = \frac{3}{8}\omega_0
$$

Substituting this into (3.26.6), we can obtain

$$
\xi_1 = \frac{3}{8}\mu a^3 \text{ and } \omega_1 = \frac{3}{8}\omega_0 a^2 \tag{3.26.8}
$$

(d) Substituting (3.26.8) into the expressions of $\dot{a}$ and $\dot{\psi}$, and retaining to $O(\varepsilon)$ yields

$$
\dot{a} = -\mu a + \frac{3}{8}\varepsilon\mu a^3 \tag{3.26.9}
$$

$$\dot{\psi} = \omega_0 + \frac{3}{8}\varepsilon\omega_0 a^2 \tag{3.26.10}$$

Integrating (3.26.9) yields

$$\frac{da}{\frac{3}{8}\varepsilon a^3 - a} = d(\mu t) \tag{3.26.11}$$

$$\text{therefore, } \frac{1}{2}\ln\frac{a^2}{\frac{3}{8}\varepsilon a^2 - 1} = -\mu t + c \tag{3.26.12}$$

where c is a constant of integration. When $t = 0$, $a = a_0$, we obtain

$$\frac{a^2}{\frac{3}{8}\varepsilon a^2 - 1} = \frac{a_0^2}{\frac{3}{8}\varepsilon a_0^2 - 1}e^{-2\mu t}$$

$$\text{so } a = a_0 e^{-\mu t}\left[1 + \frac{3}{8}\varepsilon a_0^2\left(e^{-2\mu t} - 1\right)\right]^{-1/2} \tag{3.26.13}$$

Substituting (3.26.13) into (3.26.10) yields

$$\dot{\psi} = \omega_0 + \frac{3}{8}\varepsilon\omega_0 a_0^2 e^{-2\mu t}\left[1 + \frac{3}{8}\varepsilon a_0^2\left(e^{-2\mu t} - 1\right)\right]^{-1} \tag{3.26.14}$$

Integrating the above equation yields

$$\begin{aligned}\psi &= \omega_0 t + \frac{3}{8}\varepsilon\omega_0 a_0^2 \int \frac{e^{-2\mu t}dt}{1 + \frac{3}{8}\varepsilon a_0^2\left(e^{-2\mu t} - 1\right)} \\ &= \omega_0 t - \frac{3}{8}\frac{\varepsilon\omega_0 a_0^2}{2\mu\left(\frac{3}{8}\varepsilon a_0^2\right)}\ln\left[1 + \frac{3}{8}\varepsilon a_0^2\left(e^{-2\mu t} - 1\right)\right] + \psi_0 \\ &= \omega_0 t - \frac{\omega_0}{2\mu}\ln\left[1 + \frac{3}{8}\varepsilon a_0^2\left(e^{-2\mu t} - 1\right)\right] + \psi_0\end{aligned}$$

i.e.,

$$\psi = \psi_0 + \omega_0 t - \frac{\omega_0}{2\mu}\ln\left[1 + \frac{3}{8}\varepsilon a_0^2\left(e^{-2\mu t} - 1\right)\right] \tag{3.26.15}$$

3.27 Exercise 3.27 (Application of Elliptic Functions to Represent Solutions of a Single Pendulum System with a General Weak Damping Function)

Solution: (a) When $\varepsilon = 0$, the original equation becomes

$$\ddot{u} + \omega^2 \sin u = 0 \tag{3.27.1}$$

This is a conservative equation for a single pendulum. After integration, we can obtain

$$\frac{1}{2}\dot{u}^2 - \omega^2 \cos u = E \tag{3.27.2}$$

E is the constant of integration. From (3.27.1) and (3.27.2), we have

$$dt = -\frac{d\dot{u}}{\omega^2 \sin u}, \quad d\dot{u} = -\frac{1}{\sqrt{2}}\frac{\omega^2 \sin u du}{\sqrt{E + \omega^2 \cos u}}$$

Therefore,

$$\begin{aligned} dt &= \frac{1}{\sqrt{2}}\frac{du}{\sqrt{E + \omega^2 \cos u}} = \frac{1}{\sqrt{2}}\frac{du}{\sqrt{E + \omega^2 - 2\omega^2 \sin^2\left(\frac{1}{2}u\right)}} \\ &= \frac{\sqrt{2}d\left(\frac{1}{2}u\right)}{\sqrt{E + \omega^2}\sqrt{1 - (\frac{\sqrt{2}\omega}{\sqrt{E+\omega^2}})^2 \sin^2\left(\frac{1}{2}u\right)}} = \frac{kd\left(\frac{1}{2}u\right)}{\omega\sqrt{1 - k^2 \sin^2\left(\frac{1}{2}u\right)}} \end{aligned} \tag{3.27.3}$$

where $k^2 = 2\omega^2/(E + \omega^2)$. After integration, we can obtain

$$\omega t + \varphi = k \int_0^{u/2} \frac{d\left(\frac{1}{2}u\right)}{\sqrt{1 - k^2 \sin^2(\frac{1}{2}u)}} \tag{3.27.4}$$

Let

$$\psi = \omega t + \varphi, \ \frac{\psi}{k} = \int_0^{u/2} \frac{d\left(\frac{1}{2}u\right)}{\sqrt{1 - k^2 \sin^2(\frac{1}{2}u)}} \tag{3.27.5}$$

From the definition of the Jacobi elliptic function, there is (see Appendix A)

$$\sin\left(\frac{u}{2}\right) = \mathrm{sn}\left(\frac{\psi}{k}, k\right) = k \ \mathrm{sn}(\psi, k) \tag{3.27.6}$$

$$u = 2\sin^{-1}[k\ \mathrm{sn}(\psi, k)] \tag{3.27.7}$$

(b) When $\varepsilon \neq 0$, set the solution of the original equation to be

$$u = 2\sin^{-1}\{k(t)\ [\mathrm{sn}(\psi, k(t)]\} \tag{3.27.8}$$

$$\mathit{where}\psi = \omega t + \varphi(t) \tag{3.27.9}$$

$$\text{At the same time, } \dot{u} = 2k\omega \mathrm{cn}(\psi, k) \tag{3.27.10}$$

Conducting the derivative of (3.27.8) and (3.27.10) with respect to t, respectively, yields

$$\dot{u} = \frac{2}{\mathrm{dn}}\left[\dot{k}\mathrm{sn} + k\mathrm{cndn}(\omega + \dot{\varphi}) + \dot{k}k\frac{d\mathrm{sn}}{dk}\right] \tag{3.27.11}$$

$$\ddot{u} = 2\omega\left[\dot{k}\mathrm{cn} - k\mathrm{sndn}(\omega + \dot{\varphi}) + \dot{k}k\frac{d\mathrm{cn}}{dk}\right] \tag{3.27.12}$$

From (3.27.10) and (3.27.11), we have

$$\dot{k}\mathrm{sn} + k\dot{\varphi}\mathrm{cndn} + \dot{k}k\frac{d\mathrm{sn}}{dk} = 0 \tag{3.27.13}$$

Substitute (3.27.8), (3.27.10) and (3.27.12) into the original equation, we obtain

$$\dot{k}\mathrm{cn} - k\dot{\varphi}\mathrm{sndn} + \dot{k}k\frac{d\mathrm{cn}}{dk} = \frac{\varepsilon}{2\omega}f\left[2\sin^{-1}(k\mathrm{sn}), 2k\omega\mathrm{cn}\right] \tag{3.27.14}$$

Multiply (3.27.13) by sn and (3.27.14) by cn, then add the two together, we can obtain

$$\dot{k} = \frac{\varepsilon\mathrm{cn}}{2\omega}f\left[2\sin^{-1}(k\mathrm{sn}), 2k\omega\mathrm{cn}\right] \tag{3.27.15}$$

Substitute (3.27.15) into (3.27.13), we can obtain

$$\dot{\varphi} = -\frac{\varepsilon}{2\omega k\mathrm{dn}}\left(\mathrm{sn} + k\frac{d\mathrm{sn}}{dk}\right)f\left[2\sin^{-1}(k\mathrm{sn}), 2k\omega\mathrm{cn}\right] \tag{3.27.16}$$

(c) Making the average of (3.27.15) and (3.27.16) over a period $4K$ yields

$$\dot{k} = \frac{\varepsilon}{8\omega K}\int_0^{4K} \mathrm{cn} f\left[2\sin^{-1}(k\mathrm{sn}), 2k\omega\mathrm{cn}\right]d\psi = \varepsilon g_k(k,\varphi) \tag{3.27.17}$$

$$\dot{\varphi} = -\frac{\varepsilon}{8\omega kK}\int_0^{4K} \frac{1}{\mathrm{dn}}\left(\mathrm{sn} + k\frac{d\,\mathrm{sn}}{dk}\right) f\left[2\sin^{-1}(k\,\mathrm{sn}),\right.$$
$$\left. 2k\omega\mathrm{cn}\right]d\psi = \varepsilon g_\varphi(k,\varphi) \tag{3.27.18}$$

When the specific form of the function f is known, it is possible to obtain $g_\varphi(k,\varphi)$ and $g_k(k,\varphi)$. This can be used to solve for $k(t)$ and $\varphi(t)$ and then the first-order approximate solution of the original equation.

3.28 Exercise 3.28 (Masses Bounded in Definite Orbits, Analyzing Singularities and Their Bifurcations)

Solution: **(a)** Let $\dot{u} = \ddot{u} = 0$ in the original equation, we can obtain

$$\mu_2\mu_3 \sin u + \frac{1}{2}\mu_3^2 \sin 2u + \mu_1 = 0 \tag{3.28.1}$$

(b) When $\mu_1 = 0.1$, $\mu_2 = 2.0$, Eq. (3.28.1) becomes

$$\frac{1}{2}\mu_3^2 \sin 2u + 2\mu_3 \sin u + 0.1 = 0 \tag{3.28.2}$$

Solving this equation using numerical methods yields the singularity trajectory shown in Fig. 3.12a.

It can be seen that bifurcation occurs at $\mu_3^{(1)} = 0.04988$ and $\mu_3^{(2)} = 2.24812$. The coordinate values of each singular point of the bifurcation points are:

$$u_{s1} = 4.73673, \quad u_{s2} = 6.27219, \quad u_{s3} = 3.69294, \quad u_{s4} = 2.86984$$

The above results are close to the values given in the question. However, we could not find the bifurcation value of $\mu_3 = 0.3$.

Examine the type of singular point. When $\mu_1 = 0.1$ and $\mu_2 = 2.0$, the original differential equation becomes

$$\ddot{u} + 0.1 \cdot (\dot{u} - 1) - 2\mu_3 \sin u - \frac{1}{2}\mu_3^2 \sin 2u = 0$$

Let $v = \dot{u}$, write this equation as a system of first order differential equations:

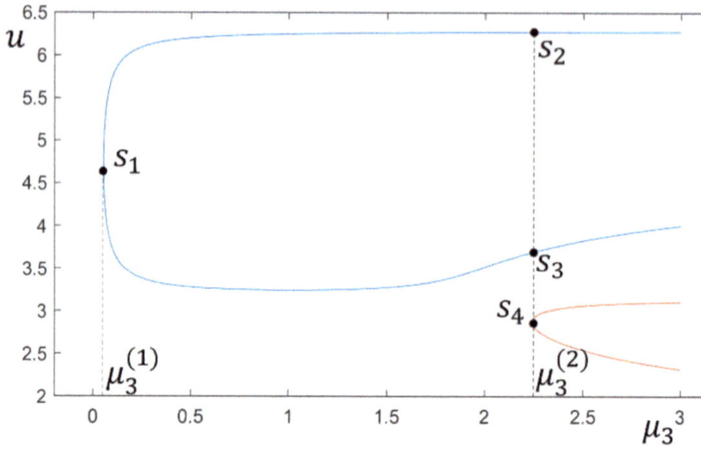

(a)

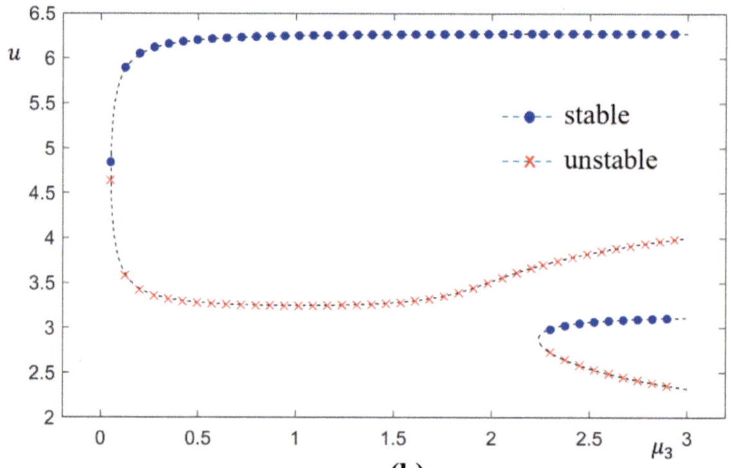

(b)

Fig. 3.12 **a** The bifurcation (choose μ_3 as control parameter) and **b** stability of singular points ($\mu_1 = 0.1$, $\mu_2 = 2.0$) for Exercise 3.28

$$\begin{cases} \dot{u} = v \\ \dot{v} = 2\mu_3 \sin u + \frac{1}{2}\mu_3^2 \sin 2u - 0.1\dot{u} + 0.1 \end{cases} \tag{3.28.3}$$

Linearizing Eq. (3.28.3) around any singular point u_s yields

$$\begin{Bmatrix} \dot{u} \\ \dot{v} \end{Bmatrix} = \begin{bmatrix} 0 & 1 \\ 2\mu_3 \cos u_s + \mu_3^2 \cos 2u_s & -0.1 \end{bmatrix} \begin{Bmatrix} u \\ v \end{Bmatrix} \tag{3.28.4}$$

The characteristic equation is

$$\lambda^2 + 0.1\lambda - \left(2\mu_3 \cos u_s + \mu_3^2 \cos 2u_s\right) = 0 \tag{3.28.5}$$

The eigenvalues are

$$\lambda_{1,\,2} = \frac{-0.1 \pm \sqrt{0.01 + 4\left(2\mu_3 \cos u_s + \mu_3^2 \cos 2u_s\right)}}{2} \tag{3.28.6}$$

Therefore, when

$$2\mu_3 \cos u_s + \mu_3^2 \cos 2u_s < 0 \tag{3.28.7}$$

the corresponding singular point is stable, otherwise it is unstable. According to the singularity trajectory that has been obtained earlier, we can judge the stability of any point on the singularity trajectory by (3.28.7). The stability result is shown in Fig. 3.12b.

3.29 Exercise 3.29 (Non-autonomous Systems with Slowly Varying Functions)

Solution: **(a)** The nonlinear function $f(u, \tau)$ is a function of u and τ, and τ is a slow-varying time variable, so it is hypothesized that the solution u of the system should also contain a slow-varying component with time t. In order to characterize u as slowly varying with time t, we make u as a function of the slow-varying time variable τ and the fast-varying time variable η, but the derivative of η with respect to t is a slow-varying function $g(\tau)$. According to the chain rule for derivatives, we have

$$\begin{aligned}
\frac{du}{dt} &= \frac{\partial u}{\partial \tau}\frac{d\tau}{dt} + \frac{\partial u}{\partial \eta}\frac{d\eta}{dt} = \varepsilon\frac{\partial u}{\partial \tau} + g(\tau)\frac{\partial u}{\partial \eta} \\
\frac{d^2u}{dt^2} &= \frac{d}{dt}\left[\varepsilon\frac{\partial u}{\partial \tau} + g(\tau)\frac{\partial u}{\partial \eta}\right] = \varepsilon\left(\frac{\partial^2 u}{\partial \tau^2}\frac{d\tau}{dt} + \frac{\partial^2 u}{\partial \eta \partial \tau}\frac{d\eta}{dt}\right) \\
&\quad + \frac{\partial u}{\partial \eta}\frac{dg}{d\tau}\frac{d\tau}{dt} + g(\tau)\left(\frac{\partial^2 u}{\partial \eta \partial \tau}\frac{d\tau}{dt} + \frac{\partial^2 u}{\partial \eta^2}\frac{d\eta}{dt}\right) \\
&= \varepsilon^2\frac{\partial^2 u}{\partial \tau^2} + \varepsilon g\frac{\partial^2 u}{\partial \eta \partial \tau} + \varepsilon\frac{dg}{d\tau}\frac{\partial u}{\partial \eta} + \varepsilon g\frac{\partial^2 u}{\partial \eta \partial \tau} \\
&= g^2\frac{\partial^2 u}{\partial \eta^2} + \varepsilon\left(g'\frac{\partial u}{\partial \eta} + 2g\frac{\partial^2 u}{\partial \eta \partial \tau}\right) + O\left(\varepsilon^2\right)
\end{aligned}$$

Written in the form of a differential operator as

$$\frac{d^2}{dt^2} = g^2\frac{\partial^2}{\partial\eta^2} + \varepsilon(g'\frac{\partial}{\partial\eta} + 2g\frac{\partial^2}{\partial\tau\partial\eta}) + O(\varepsilon^2) \tag{3.29.1}$$

The hypothetical solution given in the question is

$$u = u_0(\tau, \eta) + \varepsilon u_1(\tau, \eta) + \cdots \tag{3.29.2}$$

Substituting (3.29.1) and (3.29.2) into the differential equation of the system and retaining to $O(\varepsilon)$ yields

$$g^2\frac{\partial^2(u_0 + \varepsilon u_1)}{\partial\eta^2} + \varepsilon\left[g'\frac{\partial(u_0 + \varepsilon u_1)}{\partial\eta} + 2g\frac{\partial^2(u_0 + \varepsilon u_1)}{\partial\tau\partial\eta}\right] + f[u_0 + \varepsilon u_1, \tau] = 0 \tag{3.29.3}$$

Expanding $f[u_0 + \varepsilon u_1, \tau]$ at u_0 and retaining to $O(\varepsilon)$ yields

$$g^2\frac{\partial^2 u_0}{\partial\eta^2} + f(u_0, \tau) + \varepsilon g^2\frac{\partial^2 u_1}{\partial\eta^2} + \varepsilon g'\frac{\partial u_0}{\partial\eta} + 2\varepsilon g\frac{\partial^2 u_0}{\partial\tau\partial\eta} + \varepsilon\frac{\partial f(u_0, \tau)}{\partial u_0}u_1 = 0$$

Equating the coefficients of the like powers of ε, we obtain

$$g^2\frac{\partial^2 u_0}{\partial\eta^2} + f(u_0, \tau) = 0 \tag{3.29.4}$$

$$g^2\frac{\partial^2 u_1}{\partial\eta^2} + g'\frac{\partial u_0}{\partial\eta} + 2g\frac{\partial^2 u_0}{\partial\tau\partial\eta} + \frac{\partial f(u_0, \tau)}{\partial u_0}u_1 = 0 \tag{3.29.5}$$

Equation (3.29.4) can be written as

$$g^2\frac{\partial u_0}{\partial\eta}\frac{\partial}{\partial\eta}\left(\frac{\partial u_0}{\partial\eta}\right) + f(u_0, \tau)\frac{\partial u_0}{\partial\eta} = 0$$

$$\Rightarrow \frac{1}{2}g^2 d\left[\left(\frac{\partial u_0}{\partial\eta}\right)^2\right] + f(u_0, \tau)du_0 = 0 \tag{3.29.6}$$

After integration

$$g^2\left(\frac{\partial u_0}{\partial\eta}\right)^2 + F(u_0, \tau) = c_1(\tau) \tag{3.29.7}$$

$$\text{where } F(u_0, \tau) = 2\int f(u_0, \tau)du_0 \tag{3.29.8}$$

$c_1(\tau)$ is an integral constant which is an arbitrary function of τ. Equation (3.29.5) can be written as

$$g^2\frac{\partial^2 u_1}{\partial \eta^2}+\frac{\partial f(u_0,\tau)}{\partial u_0}u_1=-g'\frac{\partial u_0}{\partial \eta}-2g\frac{\partial^2 u_0}{\partial \tau \partial \eta} \tag{3.29.9}$$

Multiplying (3.29.9) by $\partial u_0/\partial \eta$ yields

$$g^2\frac{\partial^2 u_1}{\partial \eta^2}\frac{\partial u_0}{\partial \eta}+\frac{\partial f(u_0,\tau)}{\partial u_0}\frac{\partial u_0}{\partial \eta}u_1=-g'\frac{\partial u_0}{\partial \eta}\frac{\partial u_0}{\partial \eta}-2g\frac{\partial^2 u_0}{\partial \tau \partial \eta}\frac{\partial u_0}{\partial \eta}$$

i.e.,

$$g^2\frac{\partial u_0}{\partial \eta}\frac{\partial^2 u_1}{\partial^2 \eta}-u_1 g^2\frac{\partial}{\partial \eta}\left(\frac{\partial^2 u_0}{\partial \eta^2}\right)=-\frac{\partial}{\partial \tau}\left[g\left(\frac{\partial u_0}{\partial \eta}\right)^2\right]$$

Considering (3.29.4) and noting that g is just a function of τ, the above equation can be written as

$$g^2\frac{\partial u_0}{\partial \eta}\frac{\partial^2 u_1}{\partial^2 \eta}-u_1 g^2\frac{\partial}{\partial \eta}\left(\frac{\partial^2 u_0}{\partial \eta^2}\right)=-\frac{\partial}{\partial \tau}[g\left(\frac{\partial u_0}{\partial \eta}\right)^2]$$

i.e.,

$$g^2\left[\frac{\partial u_0}{\partial \eta}d\left(\frac{\partial u_1}{\partial \eta}\right)-u_1 d\left(\frac{\partial^2 u_0}{\partial \eta^2}\right)\right]=-\frac{\partial}{\partial \tau}\left[g\left(\frac{\partial u_0}{\partial \eta}\right)^2 d\eta\right] \tag{3.29.10}$$

The quantity in parentheses on the right-hand side of the above equation reduces to

$$\begin{aligned}
&\frac{\partial u_0}{\partial \eta}d\left(\frac{\partial u_1}{\partial \eta}\right)-u_1 d\left(\frac{\partial^2 u_0}{\partial \eta^2}\right)\\
&=d\left[\frac{\partial u_0}{\partial \eta}\left(\frac{\partial u_1}{\partial \eta}\right)-u_1\left(\frac{\partial^2 u_0}{\partial \eta^2}\right)\right]-\left[\left(\frac{\partial u_1}{\partial \eta}\right)d\left(\frac{\partial u_0}{\partial \eta}\right)-\left(\frac{\partial^2 u_0}{\partial \eta^2}\right)du_1\right]\\
&=d\left[\frac{\partial u_0}{\partial \eta}\left(\frac{\partial u_1}{\partial \eta}\right)-u_1\left(\frac{\partial^2 u_0}{\partial \eta^2}\right)\right]\\
&\quad-\left[\left(\frac{\partial u_1}{\partial \eta}\right)\left(\frac{\partial^2 u_0}{\partial \eta^2}\right)d\eta-\left(\frac{\partial^2 u_0}{\partial \eta^2}\right)\left(\frac{\partial u_1}{\partial \eta}\right)d\eta\right]\\
&=d\left[\frac{\partial u_0}{\partial \eta}\left(\frac{\partial u_1}{\partial \eta}\right)-u_1\left(\frac{\partial^2 u_0}{\partial \eta^2}\right)\right]
\end{aligned}$$

Substituting this into (3.29.10) yields

$$g^2 d\left[\frac{\partial u_0}{\partial \eta}\left(\frac{\partial u_1}{\partial \eta}\right) - u_1\left(\frac{\partial^2 u_0}{\partial \eta^2}\right)\right] = -\frac{\partial}{\partial \tau}\left[g\left(\frac{\partial u_0}{\partial \eta}\right)^2\right] d\eta \tag{3.29.11}$$

Assuming that the motion varies with the fast-varying time variable η by a period of T, and integrating (3.29.11) over $0 \sim T$ yields

$$g^2\left(\frac{\partial u_1}{\partial \eta}\frac{\partial u_0}{\partial \eta} - u_1\frac{\partial^2 u_0}{\partial \eta^2}\right)\Big|_0^T = -\int_0^T \frac{\partial}{\partial \tau}\left[g\left(\frac{\partial u_0}{\partial \eta}\right)^2\right] d\eta \tag{3.29.12}$$

According to the above assumptions, $u_1(\tau, \eta)$ must vary with η by the period T, so the left side of the above equation is zero:

$$\int_0^T \frac{\partial}{\partial \tau}\left[g\left(\frac{\partial u_0}{\partial \eta}\right)^2\right] d\eta = 0$$

i.e.,

$$\frac{\partial}{\partial \tau}\left[g\int_0^T \left(\frac{\partial u_0}{\partial \eta}\right)^2 d\eta\right] = 0 \tag{3.29.13}$$

By the above assumptions, $u_0(\tau, \eta)$ must also vary with η by the period T, so the integral in the above equation is just a function of τ, and hence

$$g\int_0^T \left(\frac{\partial u_0}{\partial \eta}\right)^2 d\eta = 2c \tag{3.29.14}$$

where c is the constant of integration. Assuming that the value of $u_0(\tau, \eta)$ varies antisymmetrically with respect to the midpoint of the period T, the above equation can be written as

$$g\int_0^{T/2} \left(\frac{\partial u_0}{\partial \eta}\right)^2 d\eta = c \tag{3.29.15}$$

From (3.29.7), we can obtain

$$\left(\frac{\partial u_0}{\partial \eta}\right)^2 = \frac{c_1(\tau) - F(u_0, \tau)}{g^2}, \quad \frac{g du_0}{\sqrt{c_1(\tau) - F(u_0, \tau)}} = d\eta$$

Substituting this into (3.29.14), we obtain

$$\int_{0}^{T/2} \sqrt{c_1(\tau) - F(u_0, \tau)} du_0 = c \tag{3.29.16}$$

We replace the integration limit $0 \sim T/2$ of η in the above equation with the integration limit of u_0. Since $u_0(\tau, \eta)$ varies with η periodically and it has been assumed that the value of $u_0(\tau, \eta)$ varies anti-symmetrically with respect to the midpoint of T, the integration of η can be done on virtually any $T/2$. As a result, the integral of Eq. (3.29.16) can be carried out between the very small value y_1 and the very large value y_2 of u_0, so that y_1 and y_2 can make $\partial u_0/\partial \eta = 0$. From (3.29.7), we can obtain

$$c_1(\tau) - F(y_i, \tau) = 0, \quad i = 1, 2 \tag{3.29.17}$$

Therefore, Eq. (3.29.16) can be written as

$$\int_{y_1}^{y_2} \sqrt{c_1(\tau) - F(u_0, \tau)} du_0 = c \tag{3.29.18}$$

(b) If the system is a linear oscillator with slowly varying stiffness, i.e.

$$f(u_0, \tau) = k(\tau) u_0 \tag{3.29.19}$$

by substituting (3.29.19) into (3.29.8) and (3.29.7), we obtain

$$F(u_0, \tau) = 2 \int f(u_0, \tau) du_0 = k(\tau) u_0^2 \tag{3.29.20}$$

$$g^2 (\frac{\partial u_0}{\partial \eta})^2 + k(\tau) u_0^2 = c_1(\tau) \tag{3.29.21}$$

From (3.29.17) and (3.29.20), we have

$$y_1 = -\sqrt{\frac{c_1(\tau)}{k(\tau)}}, \quad y_2 = \sqrt{\frac{c_1(\tau)}{k(\tau)}} \tag{3.29.22}$$

Since y_1 and y_2 are the minimal and maximal values of u_0, we set

$$u_0 = \sqrt{\frac{c_1(\tau)}{k(\tau)}} \sin[\varphi(\eta)] \tag{3.29.23}$$

Substituting this into (3.29.21) yields

$$g^2\frac{c_1(\tau)}{k(\tau)}\cos^2\varphi(\frac{d\varphi}{\partial\eta})^2+\frac{c_1(\tau)}{k(\tau)}k(\tau)\sin^2\varphi=c_1(\tau)$$

$$\text{i.e., } \frac{g^2}{k(\tau)}(\frac{d\varphi}{\partial\eta})^2=1 \tag{3.29.24}$$

Therefore

$$\frac{d\varphi}{\partial\eta}=\frac{\sqrt{k(\tau)}}{g(\tau)} \tag{3.29.25}$$

$$\eta=\frac{g(\tau)}{\sqrt{k(\tau)}}(\varphi+B) \tag{3.29.26}$$

where B is the constant of integration. When $\varphi=-\pi/2$, u_0 reaches the minimum value, so that at this time $\eta=0$, then $B=\pi/2$, so

$$\eta=\frac{g(\tau)}{\sqrt{k(\tau)}}\left(\varphi+\frac{\pi}{2}\right) \tag{3.29.27}$$

When φ changes from 0 to 2π, u_0 also undergoes a period variation. The period of u_0, T, is

$$T=\frac{g(\tau)}{\sqrt{k(\tau)}}\left(2\pi+\frac{\pi}{2}\right)-\frac{g(\tau)}{\sqrt{k(\tau)}}\left(0+\frac{\pi}{2}\right)=\frac{2\pi g(\tau)}{\sqrt{k(\tau)}} \tag{3.29.28}$$

If $T=2\pi$, then we have

$$g(\tau)=\sqrt{k(\tau)} \tag{3.29.29}$$

From (3.29.18) and (3.29.20), we have

$$\begin{aligned} c&=\int_{y_1}^{y_2}\sqrt{c_1(\tau)-k(\tau)u_0^2du_0}\\ &=\sqrt{k(\tau)}\left[\frac{u_0}{2}\sqrt{\frac{c_1(\tau)}{k(\tau)}-u_0^2}+\frac{c_1(\tau)}{2k(\tau)}\arcsin\left(u_0\sqrt{\frac{k(\tau)}{c_1(\tau)}}\right)\right]_{y_1}^{y_2}\\ &=\sqrt{k(\tau)}\left[\frac{c_1(\tau)}{2k(\tau)}\sin\phi\cos\phi+\frac{c_1(\tau)}{2k(\tau)}\phi\right]_0^{\pi}=\frac{\pi c_1(\tau)}{2\sqrt{k(\tau)}} \end{aligned}$$

i.e.,

$$\frac{c_1(\tau)}{\sqrt{k(\tau)}}=\frac{2c}{\pi},\quad \text{a constant} \tag{3.29.30}$$

Substitute (3.29.30) and (3.29.27) into (3.29.23), we obtain

$$u_0 = \sqrt{\frac{2c}{\pi\sqrt{k(\tau)}}} \sin\left[\frac{\sqrt{k(\tau)}}{g(\tau)}\eta - \frac{\pi}{2}\right] \tag{3.29.31}$$

Taking (3.29.29) into account, the above equation becomes

$$u_0 = -\sqrt{\frac{2c}{\pi}} k^{-1/4} \cos\eta \tag{3.29.32}$$

Since

$$\frac{d\eta}{dt} = g(\tau) \quad \Rightarrow \quad d\eta = \frac{1}{\varepsilon} g(\tau) d\tau$$

then

$$\eta = \frac{1}{\varepsilon}\int_0^{\varepsilon t} g(\tau)d\tau + \eta_0 \tag{3.29.33}$$

by substituting (3.29.33) into (3.29.32) and taking c is a constant into account, we obtain

$$u_0 = \frac{a}{[k(\tau)]^{1/4}} \cos\left[\frac{1}{\varepsilon}\int_0^{\varepsilon t} g(\tau)d\tau + \eta_0\right] \tag{3.29.34}$$

Appendix

Introduction to Elliptic Integrals and Jacobi Elliptic Functions

Integral function $F(k, \phi)$

$$z = F(\phi, k) = \int_0^{\sin\phi} \frac{dx}{\sqrt{(1-x^2)(1-k^2x^2)}} = \int_0^{\phi} \frac{d\theta}{\sqrt{(1-k^2\sin^2\theta)}}$$

is called Legendre's elliptic integral of the first kind. The inverse function of this function is usually notated as $\sin\phi = \text{sn}\, z = \text{sn}\,(z, k), \quad \phi = \text{am}\, z$

where $\mathrm{sn}\, z = \mathrm{sn}\,(z, k)$ is called the elliptic sine function. From this, three other related functions can be defined:

$$\mathrm{cn}\, z = \cos\varphi = \sqrt{1 - \mathrm{sn}^2 z}$$
$$\mathrm{tn}\, z = \tan\varphi = \frac{\mathrm{sn}\, n}{\mathrm{cn}\, z} = \frac{\mathrm{sn}\, n}{\sqrt{1 - \mathrm{sn}^2 z}}$$
$$\mathrm{dn}\, z = \sqrt{1 - k^2 \mathrm{sn}^2 z}$$

where $\mathrm{cn}\, z$ is called the elliptic cosine function, $\mathrm{tn}\, z$ is called the elliptic tangent function. $\mathrm{sn}\, z$, $\mathrm{cn}\, z$, $\mathrm{tn}\, z$ and $\mathrm{dn}\, z$ are commonly known as Jacobi elliptic functions.

$\mathrm{sn}\, z$ and $\mathrm{tn}\, z$ are odd functions, and $\mathrm{cn}\, z$ and $\mathrm{dn}\, z$ are even functions. The Jacobi elliptic function is periodic and has a fundamental period of $4K$:

$$4K = 4F\left(\frac{\pi}{2}, k\right) = 4\int_0^{\pi/2} \frac{d\theta}{\sqrt{\left(1 - k^2 \sin^2\theta\right)}}$$

The basic relationships between Jacobi elliptic functions are

$$\mathrm{sn}^2 z + \mathrm{cn}^2 z = 1, \quad \mathrm{dn}^2 z + k^2 \mathrm{sn}^2 z = 1, \quad \mathrm{dn}^2 z - k^2 \mathrm{cn}^2 z = 1 - k^2$$

The formulas for the derivatives of the Jacobi elliptic functions are

$$\frac{d}{dz}\mathrm{sn}\, z = \mathrm{cn}\, z\, \mathrm{dn}\, z, \quad \frac{d}{dz}\mathrm{cn}\, z = -\mathrm{sn}\, z\, \mathrm{dn}\, z, \quad \frac{d}{dz}\mathrm{dn}\, z = -k^2 \mathrm{sn}\, z\, \mathrm{cn}\, z$$

Chapter 4
Forced Oscillations of Systems Having a Single Degree of Freedom

4.1 Exercise 4.1 (The Method of Multiple Scales for Analyzing Primary Resonances of Coulomb-Damped Systems with Cubic Nonlinearities)

Solution: (a) The original differential equation of motion can be written as

$$\ddot{x} + \omega_0^2 x = \frac{d}{m} + \frac{F}{m}\cos\omega t - \frac{k_2}{m}x^3 \tag{4.1.1}$$

where $\omega_0^2 = k_1/m$. Let

$$\tau = \omega_0 t, \quad u = \frac{x}{l} \tag{4.1.2}$$

where l is a reference length; then Eq. (4.1.1) can be written as

$$\frac{d^2u}{d\tau^2} + u = \frac{d}{\omega_0^2 ml} + \frac{F}{\omega_0^2 ml}\cos\frac{\omega}{\omega_0}\tau - \frac{l^2k_2}{\omega_0^2 m}u^3 \tag{4.1.3}$$

Let

$$\overline{f} = \frac{d}{\omega_0^2 ml}, \quad 2K = \frac{F}{\omega_0^2 ml}, \quad \Omega = \frac{\omega}{\omega_0}, \quad \overline{\alpha} = \frac{l^2k_2}{\omega_0^2 m} \tag{4.1.4}$$

then (4.1.3) becomes

$$\frac{d^2u}{d\tau^2} + u = \overline{f} - \overline{\alpha}u^3 + 2K\cos\Omega\tau \tag{4.1.5}$$

Z. He et al., *Solved Problems in Nonlinear Oscillations*,
https://doi.org/10.1007/978-981-97-6113-5_4

For convenience, denote τ as t and $d^2u/d\tau^2$ as $\ddot{u}$, and let the damping force and the cubic nonlinear term are both small quantities and in the same order of ε, thus, let

$$\overline{f} = \varepsilon f, \quad \overline{\alpha} = \varepsilon\alpha \tag{4.1.6}$$

and take l to make it satisfy

$$\frac{mg}{\varepsilon\omega_0^2 ml} = \frac{g}{\varepsilon\omega_0^2 l} = 1 \tag{4.1.7}$$

Then (4.1.5) becomes

$$\ddot{u} + u = \varepsilon\left(f - \alpha u^3\right) + 2K\cos\Omega t \tag{4.1.8}$$

where

$$f = \begin{cases} -1, & \text{if } \dot{u} > 0 \\ 1, & \text{if } \dot{u} < 0 \end{cases}$$

(b) When $\Omega \approx 1$, the solution of (4.1.8) is called the primary resonance response. The the method of multiple scales is used to solve it. First use the detuning parameter σ, which quantitatively describes the nearness of Ω to 1. We write

$$\Omega = 1 + \varepsilon\sigma \tag{4.1.9}$$

We express the solution in terms of different time scales as

$$u(t;\varepsilon) = u_0(T_0, T_1) + \varepsilon u_1(T_0, T_1) + \cdots \tag{4.1.10}$$

where $T_0 = t$ and $T_1 = \varepsilon t$. The linear undamped theory predicts unbounded oscillations when $\sigma = 0$ irrespective of how small the excitation is. In the actual system these large oscillations are limited by the damping and the nonlinearity. Thus, to obtain a uniformly valid approximate solution of this problem, we need to order the excitation so that it appears when the damping and the nonlinearity appear. To accomplish this, we set

$$K = \varepsilon k \tag{4.1.11}$$

This is called soft excitation. Substituting (4.1.9)–(4.1.11) into (4.1.8) and retaining to $O(\varepsilon)$ yields

$$\begin{aligned} 0 &= \ddot{u} + u - \varepsilon\left(f - \alpha u^3\right) - 2\varepsilon k\cos\Omega t \\ &= \left(D_0^2 + 2\varepsilon D_0 D_1\right)(u_0 + \varepsilon u_1) + u_0 + \varepsilon u_1 \\ &\quad + \varepsilon\,\mathrm{sgn}(D_0 u_0) + \varepsilon\alpha(u_0 + \varepsilon u_1)^3 - 2\varepsilon k\cos(T_0 + \sigma T_1) + \cdots \end{aligned}$$

$$
\begin{aligned}
&= D_0^2 u_0 + u_0 + \varepsilon[D_0^2 u_1 + u_1 + 2D_0 D_1 u_0 \\
&\quad + \operatorname{sgn}(D_0 u_0) + \alpha u_0^3 - 2k\cos(T_0 + \sigma T_1)] + \cdots
\end{aligned} \tag{4.1.12}
$$

Equating the coefficient of the like power of ε yields

$$
D_0^2 u_0 + u_0 = 0 \tag{4.1.13}
$$

$$
D_0^2 u_1 + u_1 = -2D_0 D_1 u_0 - \operatorname{sgn}(D_0 u_0) - \alpha u_0^3 + 2k\cos(T_0 + \sigma T_1) \tag{4.1.14}
$$

The general solution of (4.1.13) is

$$
u_0 = Ae^{iT_0} + \overline{A}e^{-iT_0} \tag{4.1.15}
$$

where $A = A(T_1)$. Substitute (4.1.15) into (4.1.14), we obtain

$$
\begin{aligned}
D_0^2 u_1 + u_1 &= -2D_0 D_1 u_0 - \operatorname{sgn}(D_0 u_0) - \alpha u_0^3 + 2k\cos(T_0 + \sigma T_1) \\
&= -2iD_1 Ae^{iT_0} - \operatorname{sgn}(D_0 u_0) - \alpha A^3 e^{3iT_0} \\
&\quad - 3\alpha A^2 \overline{A} e^{iT_0} + ke^{i(T_0 + \sigma T_1)} + cc \\
&= -2iD_1 Ae^{iT_0} - 3\alpha A^2 \overline{A} e^{iT_0} + ke^{i\sigma T_1} e^{iT_0} \\
&\quad - \alpha A^3 e^{3iT_0} + cc - \operatorname{sgn}\left(iAe^{iT_0} - i\overline{A}e^{-iT_0}\right)
\end{aligned} \tag{4.1.16}
$$

In order to further process Eq. (4.1.16), we need to first deal with the symbolic function $\operatorname{sgn}(iAe^{iT_0} - i\overline{A}e^{-iT_0})$, which can be expanded into a Fourier series of complex exponential form:

$$
\operatorname{sgn}(iAe^{iT_0} - i\overline{A}e^{-iT_0}) = \sum_{n=-\infty}^{\infty} f_n e^{inT_0} \tag{4.1.17}
$$

where

$$
f_n = \frac{1}{2\pi} \int_0^{2\pi} \left[\operatorname{sgn}(iAe^{iT_0} - i\overline{A}e^{-iT_0})\right] e^{-inT_0} dT_0 \tag{4.1.18}
$$

From (4.1.16)–(4.1.18), in order to eliminate secular terms, it is necessary to have

$$
-2iD_1 A - 3\alpha A^2 \overline{A} + ke^{i\sigma T_1} - f_1 = 0 \tag{4.1.19}
$$

Let

$$
A = \frac{1}{2} a e^{i\beta} \tag{4.1.20}
$$

By (4.1.18) and (4.1.20), we obtain

$$\begin{aligned} f_1 &= \frac{1}{2\pi}\int_0^{2\pi}\left[\mathrm{sgn}\left(iAe^{iT_0} - i\bar{A}e^{-iT_0}\right)\right]e^{-iT_0}dT_0 \\ &= \frac{e^{i\beta}}{2\pi}\int_0^{2\pi}\mathrm{sgn}(-a\sin\phi)e^{-i\phi}d\phi \\ &= -\frac{ie^{i\beta}}{2\pi}\int_{-\pi}^{\pi}\mathrm{sgn}(-a\sin\phi)\sin\phi d\phi \\ &= \frac{ie^{i\beta}}{\pi}\int_0^{\pi}\sin\phi d\phi = \frac{2i}{\pi}e^{i\beta} \end{aligned} \tag{4.1.21}$$

Substitute (4.1.20) and (4.1.21) into (4.1.19), we obtain

$$ia' - a\beta' + \frac{3}{8}\alpha a^3 - ke^{i(\sigma T_1 - \beta)} + \frac{2i}{\pi} = 0 \tag{4.1.22}$$

The prime denotes the derivative with respect to T_1. Separating the real and imaginary parts of the above equation yields

$$a' = -\frac{2}{\pi} + k\sin(\sigma T_1 - \beta) \tag{4.1.23}$$

$$a\beta' = \frac{3}{8}\alpha a^3 - k\cos(\sigma T_1 - \beta) \tag{4.1.24}$$

Let

$$\gamma = (\sigma T_1 - \beta) \tag{4.1.25}$$

then

$$a' = -\frac{2}{\pi} + k\sin\gamma \tag{4.1.26}$$

$$a\gamma' = a\sigma - \frac{3\alpha a^3}{8} + k\cos\gamma \tag{4.1.27}$$

Therefore, the first order approximate solution of the system is

$$u = a\cos[T_0 + \beta(T_1)] + O(\varepsilon) \tag{4.1.28}$$

(c) Let $a' = \gamma' = 0$ in (4.1.26) and (4.1.27), we can obtain

$$\begin{cases} -\frac{2}{\pi} + k\sin\gamma = 0 \\ a\sigma - \frac{3\alpha a^3}{8} + k\cos\gamma = 0 \end{cases} \Rightarrow \begin{cases} \frac{2}{\pi} = k\sin\gamma \\ \frac{3\alpha a^3}{8} - a\sigma = k\cos\gamma \end{cases} \tag{4.1.29}$$

Squaring and adding these two equations, we obtain the following frequency–response equation:

$$\left(\frac{3\alpha a^3}{8} - a\sigma\right)^2 + \frac{4}{\pi^2} = k^2 \tag{4.1.30}$$

To plot the frequency–response curve of the system $a \sim \sigma$, we can solve for σ in terms of a from (4.1.30)

$$\sigma = \frac{3\alpha a^2}{8} \pm \frac{1}{a}\sqrt{k^2 - \frac{4}{\pi^2}} \tag{4.1.31}$$

Here

$$\sigma = \frac{3\alpha a^2}{8} \tag{4.1.32}$$

defines the backbone curve. From (4.1.29) and (4.1.31), the relationship between the phase γ and σ is

$$\tan\gamma = \frac{2}{\pi} \Big/ \left(\frac{3\alpha a^3}{8} - a\sigma\right) = \frac{\frac{2}{\pi}}{\left[\mp\sqrt{k^2 - \frac{4}{\pi^2}}\right]} \tag{4.1.33}$$

Obviously, when $k = 2/\pi$, the $a \sim \sigma$ curve is the backbone curve. When $k > 2/\pi$, the $a \sim \sigma$ curve has two branches as shown by (4.1.31); the $\gamma \sim \sigma$ curve also has two branches as shown by (4.1.33). The $a \sim \sigma$ curve and $\gamma \sim \sigma$ curve can be sketched from the above formulae, as shown in Fig. 4.1a. It can be seen that the two branches of the $a \sim \sigma$ curve in the backbone curve does not merge; therefore, jump phenomenon of the amplitude occurs only when the frequency was modulated from large to small if $\alpha > 0$, while it occurs only when the frequency modulated from small to large if $\alpha < 0$.

From (4.1.30), we can obtain

$$k = \sqrt{(\frac{3\alpha a^3}{8} - a\sigma)^2 + \frac{4}{\pi^2}} \tag{4.1.34}$$

The excitation-response curves $a \sim k$ can be sketched as shown in Fig. 4.1b. It can be seen that jump phenomenon occurs in the curve $a \sim k$, and there are jumps in both directions of changing k, i.e., from small to large and from large to small.

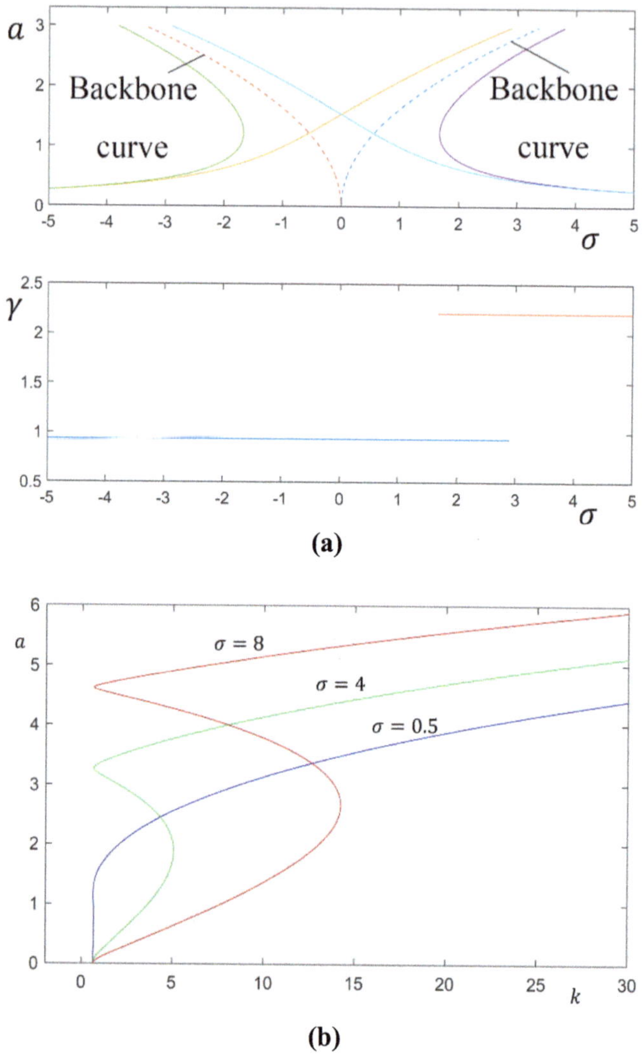

Fig. 4.1 **a** Frequency–response and frequency-phase curves ($k = 1.5$), and **b** Excitation-response curves ($\alpha = 1$) for Exercise 4.1

4.2 Exercise 4.2 (The Method of Multiple Scales for Analyzing Primary Resonances of Cubic Nonlinear, Quadratic-Damped Systems)

Solution: **(a)** Similar to the previous exercise, when $\Omega \approx 1$, the detuning parameter σ is adopted to quantitatively describes the nearness of Ω to 1:

$$\Omega = 1 + \varepsilon\sigma \tag{4.2.1}$$

We express the solution in terms of different time scales as

$$u(t;\varepsilon) = u_0(T_0, T_1) + \varepsilon u_1(T_0, T_1) + \cdots \tag{4.2.2}$$

We set

$$K = \varepsilon k \tag{4.2.3}$$

Substituting (4.2.1)–(4.2.3) into the original differential equation and keeping to $O(\varepsilon)$, we obtain

$$\begin{aligned}
0 &= \ddot{u} + u + \varepsilon\left(|\dot{u}|\dot{u} + \alpha u^3\right) - 2K\cos\Omega t \\
&= \left(D_0^2 + 2\varepsilon D_0 D_1\right)(u_0 + \varepsilon u_1) + u_0 + \varepsilon u_1 \\
&\quad + \varepsilon\,|D_0 u_0|\,D_0 u_0 + \varepsilon\alpha(u_0 + \varepsilon u_1)^3 - 2\varepsilon k\cos(T_0 + \sigma T_1) + \cdots \\
&= D_0^2 u_0 + u_0 + \varepsilon[D_0^2 u_1 + u_1 + 2D_0 D_1 u_0 \\
&\quad + |D_0 u_0|\,D_0 u_0 + \alpha u_0^3 - 2k\cos(T_0 + \sigma T_1)] + \cdots
\end{aligned} \tag{4.2.4}$$

Let the coefficient of the like power of ε be zero, we get

$$D_0^2 u_0 + u_0 = 0 \tag{4.2.5}$$

$$D_0^2 u_1 + u_1 = -2D_0 D_1 u_0 + |D_0 u_0|\,D_0 u_0 - \alpha u_0^3 + 2k\cos(T_0 + \sigma T_1) \tag{4.2.6}$$

The solution of (4.2.5) is

$$u_0 = Ae^{iT_0} + \overline{A}e^{-iT_0} \tag{4.2.7}$$

where $A = A(T_1)$. Substitute (4.2.7) into (4.2.6), we obtain

$$\begin{aligned}
&D_0^2 u_1 + u_1 = -2D_0 D_1 u_0 + |D_0 u_0|\,D_0 u_0 - \alpha u_0^3 + 2k\cos(T_0 + \sigma T_1) \\
&= -2iD_1 Ae^{iT_0} + \left|\left(iAe^{iT_0} - i\overline{A}e^{-iT_0}\right)\right|\left(iAe^{iT_0} - i\overline{A}e^{-iT_0}\right) \\
&-\alpha A^3 e^{3iT_0} - 3\alpha A^2\overline{A}e^{iT_0} + ke^{i(T_0+\sigma T_1)} + cc \\
&= -2iD_1 Ae^{iT_0} - 3\alpha A^2\overline{A}e^{iT_0} + ke^{i\sigma T_1}e^{iT_0} \\
&-\alpha A^3 e^{3iT_0} + cc + \left|\left(iAe^{iT_0} - i\overline{A}e^{-iT_0}\right)\right|\left(iAe^{iT_0} - i\overline{A}e^{-iT_0}\right)
\end{aligned} \tag{4.2.8}$$

In order to further process Eq. (4.2.8), we need to expand the function $\left|\left(iAe^{iT_0} - i\overline{A}e^{-iT_0}\right)\right|\left(iAe^{iT_0} - i\overline{A}e^{-iT_0}\right)$ into a Fourier series of complex exponential form:

$$\left|\left(iAe^{iT_0} - i\bar{A}e^{-iT_0}\right)\right|\left(iAe^{iT_0} - i\bar{A}e^{-iT_0}\right) = \sum_{n=-\infty}^{\infty} f_n e^{inT_0} \tag{4.2.9}$$

where

$$f_n = \frac{1}{2\pi}\int_0^{2\pi}\left[\left|\left(iAe^{iT_0} - i\bar{A}e^{-iT_0}\right)\right|\left(iAe^{iT_0} - i\bar{A}e^{-iT_0}\right)\right]e^{-inT_0}dT_0 \tag{4.2.10}$$

In order to eliminate secular terms, it is necessary to have

$$-2iD_1A - 3\alpha A^2\bar{A} + ke^{i\sigma T_1} + f_1 = 0 \tag{4.2.11}$$

Let

$$A = \frac{1}{2}ae^{i\beta} \tag{4.2.12}$$

By (4.2.10) and (4.2.12),

$$\begin{aligned}
f_1 &= \frac{1}{2\pi}\int_0^{2\pi}\left[\left|\left(iAe^{iT_0} - i\bar{A}e^{-iT_0}\right)\right|\left(iAe^{iT_0} - i\bar{A}e^{-iT_0}\right)\right]e^{-iT_0}dT_0 \\
&= -\frac{1}{2\pi}\int_0^{2\pi}\left[\,|a\sin(T_0+\beta)|\,(a\sin(T_0+\beta)\right]e^{-iT_0}dT_0 \\
&= -\frac{a^2e^{i\beta}}{2\pi}\int_0^{2\pi}|\sin\phi|\,\sin\phi e^{-i\phi}d\phi \\
&= \frac{ia^2e^{i\beta}}{\pi}\int_0^{\pi}\sin^3\phi d\phi = \frac{ia^2e^{i\beta}}{\pi}[-\cos\phi + \frac{1}{3}\cos^3\phi]_0^{\pi} \\
&= \frac{4ia^2e^{i\beta}}{3\pi}
\end{aligned} \tag{4.2.13}$$

Substituting (4.2.12) and (4.2.13) into (4.2.11) yields

$$ia' - a\beta' + \frac{3}{8}\alpha a^3 - ke^{i(\sigma T_1 - \beta)} + \frac{4ia^2}{3\pi} = 0 \tag{4.2.14}$$

The prime s in the above equation denotes the derivative with respect to T_1. Separating the real and imaginary parts of the equation yields

$$a' = -\frac{4a^2}{3\pi} + k\sin(\sigma T_1 - \beta), \quad -a\beta' = -\frac{3\alpha a^3}{8} + k\cos(\sigma T_1 - \beta)$$

Let

$$\gamma = (\sigma T_1 - \beta) \tag{4.2.15}$$

then

$$a' = -\frac{4a^2}{3\pi} + k\sin\gamma \tag{4.2.16}$$

$$a\gamma' = a\sigma - \frac{3\alpha a^3}{8} + k\cos\gamma \tag{4.2.17}$$

So the first order approximate solution of the system is

$$u = a\cos[T_0 + \beta(T_1)] + O(\varepsilon) \tag{4.2.18}$$

(b) Let $a' = \gamma' = 0$ in (4.2.16) and (4.2.17), we obtain:

$$\begin{cases} \frac{4a^2}{3\pi} = k\sin\gamma \\ \frac{3\alpha a^3}{8} - a\sigma = k\cos\gamma \end{cases} \tag{4.2.19}$$

Squaring and adding these two equations, we obtain the following frequency–response equation:

$$\left(\frac{3\alpha a^3}{8} - a\sigma\right)^2 + \frac{16a^4}{9\pi^2} = k^2 \tag{4.2.20}$$

To plot the frequency–response curve of the system $a \sim \sigma$, we can solve for σ in terms of a from (4.2.20):

$$\sigma = \frac{3\alpha a^2}{8} \pm \frac{1}{a}\sqrt{k^2 - \frac{16a^4}{9\pi^2}} \tag{4.2.21}$$

The equation of the backbone curve is

$$\sigma = \frac{3\alpha a^2}{8} \tag{4.2.22}$$

From (4.2.19) and (4.2.21), the relationship between the phase γ and σ is

$$\tan\gamma = \frac{4a^2}{3\pi} / \left(\frac{3\alpha a^3}{8} - a\sigma \right) = \frac{4a^2}{3\pi} / \left[\mp\sqrt{k^2 - \frac{16a^4}{9\pi^2}} \right] \tag{4.2.23}$$

The $a \sim \sigma$ curve has two branches as shown by (4.2.21); the corresponding $\gamma \sim \sigma$ curve also has two branches as shown by (4.2.23). The $a \sim \sigma$ curve and $\gamma \sim \sigma$ curve can be sketched from the above formulae, as shown in Fig. 4.2a. It can be seen that the two branches of the $a \sim \sigma$ curve are stitched together in the backbone curve. When $\alpha > 0$, the response amplitude jumps to a lower amplitude when the frequency changes from small to large, or a larger amplitude when the frequency changes from a large value to a small one. When $\alpha < 0$, the opposite case happens.

From (4.2.20), we can obtain

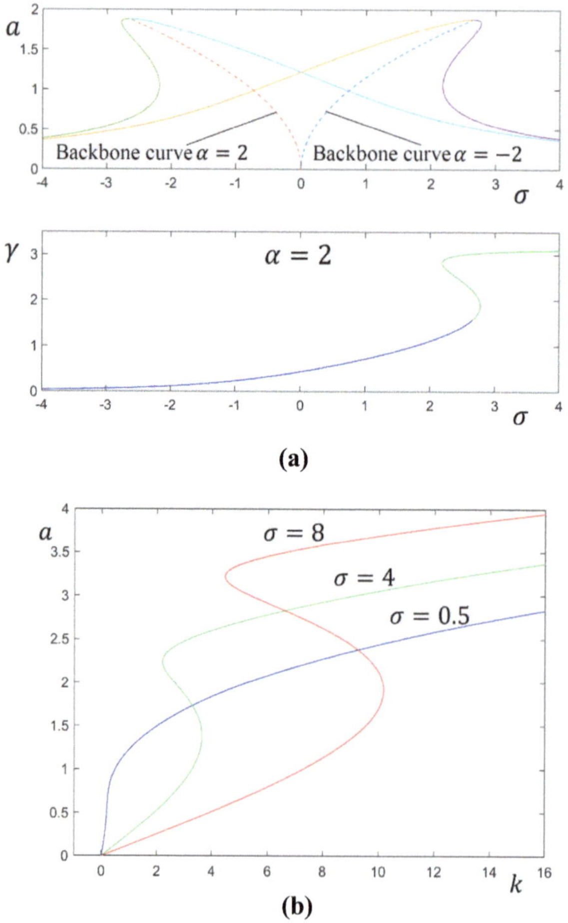

Fig. 4.2 **a** Frequency-response and frequency-phase curves ($k = 1.5$), and **b** Excitation-response curves ($\alpha = 2$) for Exercise 4.2

$$k = \sqrt{(\frac{3\alpha a^3}{8} - a\sigma)^2 + \frac{16a^4}{9\pi^2}} \tag{4.2.24}$$

From this, the excitation-response curve of the system $a \sim k$ can be sketched in Fig. 4.2b, which also demonstrates amplitude jumps in both directions of changing k.

4.3 Exercise 4.3 (The Method of Multiple Scales for Analyzing Primary Resonances of Linear Spring and Hysteresis Damped Systems)

$$-f = \begin{cases} x + x_s - x_b & x_c \le x \le x_b \\ -x_s & x_d \le x \le x_c \\ x - x_s - x_d & x_d \le x \le x_a \\ x_s & x_a \le x \le x_b \end{cases} \tag{4.3.1}$$

$$x_c = x_b - 2x_s \quad \text{and} \quad x_a = x_d + 2x_s \tag{4.3.2}$$

Solution: (a) The differential equation of motion of the system is

$$m\ddot{x} = -f_1 - f_2 + F_0 \cos \omega t$$

where $f_1 = kx$. So

$$\ddot{x} + \omega_0^2 x = -\frac{f_2}{m} + \frac{F_0}{m} \cos \omega t \tag{4.3.3}$$

$$\varepsilon f = -\frac{f_2}{m} \quad \Rightarrow \quad -f = \frac{f_2}{\varepsilon m} \tag{4.3.4}$$

The spring stiffness of f_2 is $k_2 = m\varepsilon$, so $f_2(x_s) = \varepsilon m x_s$ or $-f(x_s) = x_s$. Then the expression for $-f(x)$ can be obtained as (4.3.1). Substituting (4.3.4) into (4.3.3) yields

$$\ddot{x} + \omega_0^2 x = \varepsilon f + \frac{F_0}{m} \cos \omega t$$

Let $X = F_0/(k_1 + k_2)$ and $\tau = \omega_0 t$, the above equation becomes

$$\frac{d^2x}{d\tau^2} + x = \frac{\varepsilon f}{X\omega_0^2} + \frac{F_0}{Xm\omega_0^2} \cos \frac{\omega}{\omega_0}\tau \tag{4.3.5}$$

where

$$K = \frac{F_0}{Xm\omega_0^2}, \quad \Omega = \frac{\omega}{\omega_0} \tag{4.3.6}$$

And, for simplicity, by denoting τ as t, $d^2x/d\tau^2$ as $\ddot{x}$, and $f/X\omega_0^2$ as f, we have

$$\ddot{x} + x = \varepsilon f + K \cos \Omega t \tag{4.3.7}$$

It is important to note that all terms on the right-hand side of (4.3.1) are dimensionless.

(b) When $\Omega \approx 1$, the detuning parameter σ is adopted to quantitatively describes the nearness of Ω to 1:

$$\Omega = 1 + \varepsilon\sigma \tag{4.3.8}$$

We express the solution in terms of different time scales as

$$x(t; \varepsilon) = x_0(T_0, T_1) + \varepsilon x_1(T_0, T_1) + \cdots \tag{4.3.9}$$

We set

$$K = 2\varepsilon k \tag{4.3.10}$$

Substituting (4.3.8)–(4.3.10) into (4.3.7) and retaining to $O(\varepsilon)$ yields

$$\begin{aligned} 0 &= \ddot{x} + x - \varepsilon f(x) - K \cos \Omega t \\ &= \left(D_0^2 + 2\varepsilon D_0 D_1\right)(x_0 + \varepsilon x_1) \\ &\quad + x_0 + \varepsilon x_1 - \varepsilon f(x_0) - 2\varepsilon k \cos(T_0 + \sigma T_1) + \cdots \\ &= D_0^2 x_0 + x_0 + \varepsilon\left[D_0^2 x_1 + x_1 + 2D_0 D_1 x_0 - f(x_0) - 2k \cos(T_0 + \sigma T_1)\right] + \cdots \end{aligned} \tag{4.3.11}$$

Let the coefficient of the like power of ε be zero, we get

$$D_0^2 x_0 + x_0 = 0 \tag{4.3.12}$$

$$D_0^2 x_1 + x_1 = -2D_0 D_1 x_0 + f(x_0) + 2k \cos(T_0 + \sigma T_1) \tag{4.3.13}$$

The solution of (4.3.12) is

$$x_0 = Ae^{iT_0} + \overline{A}e^{-iT_0} \tag{4.3.14}$$

where $A = A(T_1)$. Substitute (4.3.14) into (4.3.13), we obtain

$$
\begin{aligned}
D_0^2 x_1 + x_1 &= -2D_0 D_1 x_0 + f(x_0) + 2k\cos(T_0 + \sigma T_1) \\
&= -2iD_1 A e^{iT_0} + f\left(Ae^{iT_0} + \overline{A}e^{-iT_0}\right) + ke^{i\sigma T_1} e^{iT_0} + cc \qquad (4.3.15)
\end{aligned}
$$

In order to further process the Eq. (4.3.15), we need to deal with the function $f\left(Ae^{iT_0} + \overline{A}e^{-iT_0}\right)$, which can be expanded into a Fourier series of complex exponential form:

$$
f\left(Ae^{iT_0} + \overline{A}e^{-iT_0}\right) = \sum_{n=-\infty}^{\infty} f_n e^{inT_0} \qquad (4.3.16)
$$

Where

$$
f_n = \frac{1}{2\pi}\int_0^{2\pi} \left[f\left(Ae^{iT_0} + \overline{A}e^{-iT_0}\right)\right] e^{-inT_0} dT_0 \qquad (4.3.17)
$$

From (4.3.15)–(4.3.17), in order to eliminate secular terms, it is necessary to have

$$
-2iD_1 A + ke^{i\sigma T_1} + f_1 = 0 \qquad (4.3.18)
$$

Let

$$
A = \frac{1}{2}ae^{i\beta} \qquad (4.3.19)
$$

By (4.3.17) and (4.3.19), we obtain

$$
\begin{aligned}
f_1 &= \frac{1}{2\pi}\int_0^{2\pi} \left[f\left(Ae^{iT_0} + \overline{A}e^{-iT_0}\right)\right] e^{-iT_0} dT_0 \\
&= \frac{1}{2\pi}\int_0^{2\pi} f[a\cos(T_0 + \beta)] e^{-iT_0} dT_0 \\
&= \frac{e^{i\beta}}{2\pi}\int_0^{2\pi} f(a\cos\phi)\, e^{-i\phi} d\phi \qquad (4.3.20)
\end{aligned}
$$

Substituting (4.3.1) into (4.3.20) yields

$$
f_1 = -\frac{e^{i\beta}}{2\pi}\left[\int_{x_b}^{x_c} (x + x_s - x_b)\, e^{-i\phi} d\phi(x)\right.
$$

$$
-\int_{x_c}^{x_d} x_s e^{-i\phi} d\phi(x)
$$

$$
\left.+\int_{x_d}^{x_a} (x - x_s - x_d) e^{-i\phi} d\phi(x) + \int_{x_a}^{x_b} x_s e^{-i\phi} d\phi(x)\right] \tag{4.3.21}
$$

where

$$
x_i = a\cos\varphi_i, \quad i = a,\ b,\ c,\ d \tag{4.3.22}
$$

then (4.3.20) becomes

$$
f_1 = -\frac{e^{i\beta}}{2\pi}\left[\int_{\phi_b}^{\phi_c} (a\cos\phi + x_s - x_b) e^{-i\phi} d\phi - \int_{\phi_c}^{\phi_d} x_s e^{-i\phi} d\phi\right.
$$

$$
\left.+\int_{\phi_d}^{\phi_a} (a\cos\phi - x_s - x_d) e^{-i\phi} d\phi + \int_{\phi_a}^{\phi_b} x_s e^{-i\phi} d\phi\right] \tag{4.3.23}
$$

We note that the period in the variable φ is 2π; and since the motion is periodic, we set $\varphi = 0$ at point B ($\varphi_B = 0$) so that $\varphi = \pi$ at point D ($\varphi_D = 0$). Since

$$
\begin{aligned}
x_b = a, \quad x_c = x_b - 2x_s = a - 2x_s \triangleq a\cos\varphi_1 \\
x_d = -a,\ x_a = x_d + 2x_s = -a + 2x_s \triangleq a\cos\varphi_2
\end{aligned} \tag{4.3.24}
$$

where

$$
\varphi_1 = \cos^{-1}\left(\frac{a - 2x_s}{a}\right), \quad \varphi_2 = \cos^{-1}\left(\frac{2x_s - a}{a}\right) \tag{4.3.25}
$$

Let $f_1 = f_{1R} + i f_{1I}$, and substitute (4.3.24) and (4.3.25) into (4.3.23), we obtain

$$
f_{1R} = -\frac{e^{i\beta}}{2\pi}\left[\int_{0}^{\phi_1} (a\cos\phi + x_s - a)\cos\phi d\phi - \int_{\phi_1}^{\pi} x_s \cos\phi d\phi\right.
$$

$$
+\int_{\pi}^{\phi_2} (a\cos\phi - x_s + a)\cos\phi d\phi + \int_{\phi_2}^{2\pi} x_s \cos\phi d\phi]
$$

$$
= -\frac{e^{i\beta}}{2\pi}[\frac{a\phi_1}{2} + \frac{1}{4}a\sin 2\phi_1 + (x_s - a)\sin\phi_1 + x_s\sin\phi_1
$$

$$
+\frac{a\phi_2}{2} + \frac{1}{4}a\sin 2\phi_2 - \frac{a\pi}{2} - (x_s - a)\sin\phi_2 - x_s\sin\phi_2]
$$

$$= -\frac{ae^{i\beta}}{\pi}[\frac{1}{2}\cos^{-1}\left(\frac{a-2x_s}{a}\right) - \left(1-\frac{2x_s}{a}\right)\left(\frac{x_s}{a}-\frac{x_s^2}{a^2}\right)^{1/2}] \tag{4.3.26}$$

$$f_{1I} = \frac{e^{i\beta}}{2\pi}[\int_0^{\phi_1}(a\cos\phi + x_s - a)\sin\phi d\phi - \int_{\phi_1}^{\pi} x_s \sin\phi d\phi$$

$$+ \int_{\pi}^{\phi_2}(a\cos\phi - x_s + a)\sin\phi d\phi + \int_{\phi_2}^{2\pi} x_s \sin\phi d\phi]$$

$$= \frac{e^{i\beta}}{2\pi}[-\frac{1}{2}a\cos^2\phi_1 + \frac{1}{2}a - (x_s-a)\cos\phi_1 + (x_s-a) - x_s - x_s\cos\phi_1$$

$$-\frac{1}{2}a\cos^2\phi_2 + \frac{1}{2}a + (x_s-a)\cos\phi_2 + (x_s-a) - x_s + x_s\cos\phi_2]$$

$$= \frac{e^{i\beta}}{2\pi}\left[-a\cos^2\phi_1 - 2(2x_s-a)\cos\phi_1 + 2(x_s-a) + a - 2x_s\right]$$

$$= \frac{2x_s e^{i\beta}}{\pi a}(x_s - a) \tag{4.3.27}$$

Substituting the above results into (4.3.18) yields

$$ia' - a\beta' - ke^{i(\sigma T_1 - \beta)} - f_{1R}e^{-i\beta} - if_{1I}e^{-i\beta} = 0 \tag{4.3.28}$$

The prime denotes the derivative with respect to T_1. Separating the real and imaginary parts of the above equation yields

$$\begin{aligned} a' &= k\sin(\sigma T_1 - \beta) + f_{1I}e^{-i\beta} \\ -a\beta' &= k\cos(\sigma T_1 - \beta) + f_{1R}e^{-i\beta} \end{aligned} \tag{4.3.29}$$

i.e.,

$$\begin{aligned} a' &= \frac{2x_s}{\pi a}(x_s - a) + k\sin(\sigma T_1 - \beta) \\ -\beta' &= -\frac{1}{\pi}\left[\frac{1}{2}\cos^{-1}\left(\frac{a-2x_s}{a}\right) - \left(1-\frac{2x_s}{a}\right)\left(\frac{x_s}{a}-\frac{x_s^2}{a^2}\right)^{1/2}\right] \\ &\quad + \frac{k}{a}\cos(\sigma T_1 - \beta) \end{aligned} \tag{4.3.30}$$

Let

$$\gamma = \sigma T_1 - \beta \tag{4.3.31}$$

Then

$$a' = \frac{2x_s}{\pi a}(x_s - a) + k \sin \gamma \tag{4.3.32}$$

$$\gamma' = \sigma - \frac{1}{\pi}\left[\frac{1}{2}\cos^{-1}\left(\frac{a - 2x_s}{a}\right) - \left(1 - \frac{2x_s}{a}\right)\left(\frac{x_s}{a} - \frac{x_s^2}{a^2}\right)^{1/2}\right] + \frac{k}{a}\cos\gamma \tag{4.3.33}$$

Therefore, the first order approximate solution of the system is

$$u = a(T_1)\cos[T_0 + \beta(T_1)] + O(\varepsilon) \tag{4.3.34}$$

(c) Let $a' = \gamma' = 0$ in (4.3.32) and (4.3.33), we can obtain

$$\left.\begin{array}{c} -\frac{2x_s}{\pi a}(x_s - a) = k\sin\gamma \\ \frac{a}{\pi}[\frac{1}{2}\cos^{-1}(\frac{a-2x_s}{a}) - (1 - \frac{2x_s}{a})(\frac{x_s}{a} - \frac{x_s^2}{a^2})^{1/2}] - \sigma a = k\cos\gamma \end{array}\right\} \tag{4.3.35}$$

Squaring and adding these two equations, we obtain the following frequency–response equation:

$$\left\{\frac{a}{\pi}\left[\frac{1}{2}\cos^{-1}\left(\frac{a - 2x_s}{a}\right) - \left(1 - \frac{2x_s}{a}\right)\left(\frac{x_s}{a} - \frac{x_s^2}{a^2}\right)^{1/2}\right] - \sigma a\right\}^2 + \left[\frac{2x_s}{\pi a}(x_s - a)\right]^2 = k^2 \tag{4.3.36}$$

We can solve for σ in terms of a from (4.3.36):

$$\sigma = \frac{1}{\pi}\left[\frac{1}{2}\cos^{-1}\left(\frac{a - 2x_s}{a}\right) - \left(1 - \frac{2x_s}{a}\right)\left(\frac{x_s}{a} - \frac{x_s^2}{a^2}\right)^{1/2}\right] \pm \frac{1}{a}\sqrt{k^2 - \left[\frac{2x_s}{\pi a}(x_s - a)\right]^2} \tag{4.3.37}$$

Since σ is a real number, that requires

$$a > x_s \tag{4.3.38}$$

and

$$k \geq \frac{2x_s}{\pi a}(a - x_s) = \frac{2x_s}{\pi} - \frac{2x_s^2}{\pi a} \quad \Rightarrow \quad a \leq \frac{2x_s^2}{\pi}\frac{1}{\frac{2x_s}{\pi} - k} \tag{4.3.39}$$

From (4.3.38) and (4.3.39), we know that when $0 < k < 2x_s/\pi$,$a > x_s$, and a can reach a maximum value; when $k \geq 2x_s/\pi$, $a > x_s$, but a has no maximum.

Furthermore, it can be seen from (4.3.37) that when $a = x_s$,

$$\sigma_{1,\,2} = \frac{1}{2} \pm \frac{k}{x_s} \tag{4.3.40}$$

It is easy to see from (4.3.37) that as a increases from x_s, $\sigma \in [\sigma_1, \sigma_2]$. Therefore,

$$\frac{1}{2} - \frac{k}{x_s} \leq \sigma \leq \frac{1}{2} + \frac{k}{x_s} \tag{4.3.41}$$

In addition, from (4.3.36), we can obtain

$$k = \sqrt{\begin{array}{c}\left\{\frac{a}{\pi}\left[\frac{1}{2}\cos^{-1}\left(\frac{a-2x_s}{a}\right) - \left(1 - \frac{2x_s}{a}\right)\left(\frac{x_s}{a} - \frac{x_s^2}{a^2}\right)^{1/2}\right] - \sigma a\right\}^2 \\ +\left[\frac{2x_s}{\pi a}(x_s - a)\right]^2\end{array}} \tag{4.3.42}$$

From (4.3.37) and (4.3.42), we can sketch the frequency–response curve of the system $a \sim \sigma$, the excitation-response curve of the system $a \sim k$ in Fig. 4.3a, b, in which no amplitude jump can be found.

4.4 Exercise 4.4 (The Method of Multiple Scales for Analyzing Primary Resonances of Linear Spring, Hysteresis and Coulomb Damped Systems)

Solution: (a) When $\Omega \approx 1$, the detuning parameter σ is adopted to quantitatively describes the nearness of Ω to 1:

$$\Omega = 1 + \varepsilon\sigma \tag{4.4.1}$$

We express the solution in terms of different time scales as

$$x(t;\, \varepsilon) = x_0(T_0, T_1) + \varepsilon x_1(T_0, T_1) + \cdots \tag{4.4.2}$$

In order to keep all damping and excitation terms appearing in the same order, we set

$$K = 2\varepsilon k \tag{4.4.3}$$

Substituting (4.4.1)–(4.4.3) into the differential equation of motion of the system and keeping to $O(\varepsilon)$ yields

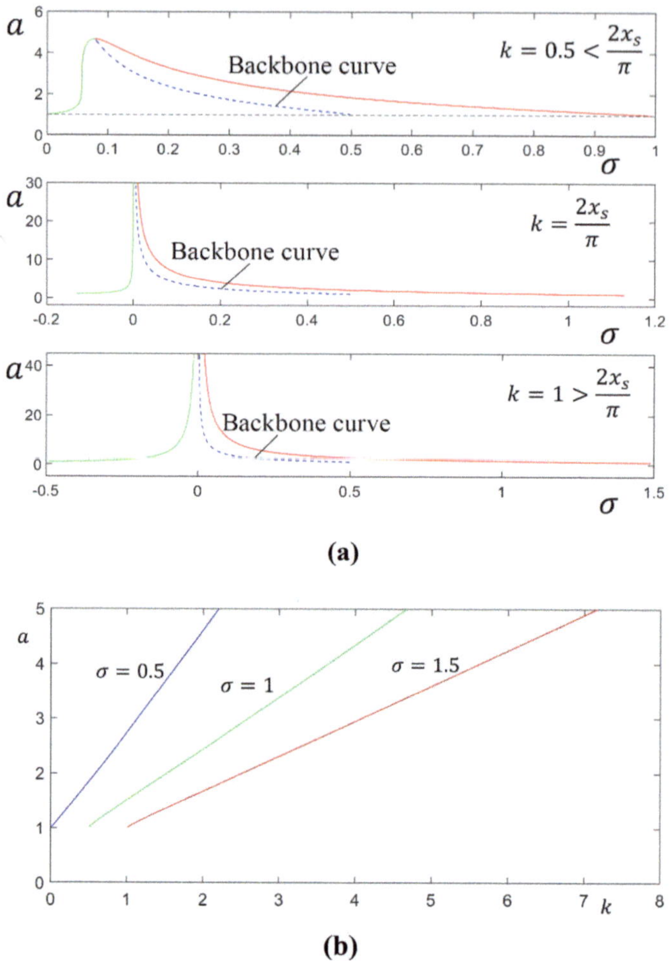

Fig. 4.3 **a** Frequency-response and **b** Excitation-response curves ($x_s = 1$) for Exercise 4.3

$$\begin{aligned}0 &= \ddot{x} + x - \varepsilon f(x) - \varepsilon\mu_0 \mathrm{sgn}\dot{x} - K\cos\Omega t \\ &= \left(D_0^2 + 2\varepsilon D_0 D_1\right)(x_0 + \varepsilon x_1) + x_0 + \varepsilon x_1 \\ &\quad -\varepsilon f(x_0) - \varepsilon\mu_0 \mathrm{sgn}\dot{x}_0 - 2\varepsilon k\cos(T_0 + \sigma T_1) + \cdots \\ &= D_0^2 x_0 + x_0 + \varepsilon[D_0^2 x_1 + x_1 + 2D_0 D_1 x_0 - f(x_0) - \mu_0 \mathrm{sgn}\dot{x}_0 \\ &\quad -2k\cos(T_0 + \sigma T_1)] + \cdots\end{aligned} \tag{4.4.4}$$

Let the coefficient of the like power of ε be zero, we get

$$D_0^2 x_0 + x_0 = 0 \tag{4.4.5}$$

$$D_0^2 x_1 + x_1 = -2D_0 D_1 x_0 + f(x_0) + \mu_0 \mathrm{sgn}\dot{x}_0 + 2k\cos(T_0 + \sigma T_1) \tag{4.4.6}$$

The solution of (4.4.5) is

$$x_0 = Ae^{iT_0} + \overline{A}e^{-iT_0} \tag{4.4.7}$$

where $A = A(T_1)$. Substituting (4.4.7) into (4.4.6) yields

$$\begin{aligned} D_0^2 x_1 + x_1 &= -2D_0 D_1 x_0 + f(x_0) + \mu_0 \mathrm{sgn}\dot{x}_0 + 2k\cos(T_0 + \sigma T_1) \\ &= -2iD_1 Ae^{iT_0} + f\left(Ae^{iT_0} + \overline{A}e^{-iT_0}\right) \\ &\quad + ke^{i\sigma T_1}e^{iT_0} + cc + \mu_0 \mathrm{sgn}\left(iAe^{iT_0} - i\overline{A}e^{-iT_0}\right) \end{aligned} \tag{4.4.8}$$

In order to further process (4.4.8), it is necessary to first deal with the function $f\left(Ae^{iT_0} + \overline{A}e^{-iT_0}\right)$ and the sign function, which can be expanded into a Fourier series of complex exponential form:

$$f\left(Ae^{iT_0} + \overline{A}e^{-iT_0}\right) + \mu_0 \mathrm{sgn}(iAe^{iT_0} - i\overline{A}e^{-iT_0}) = \sum_{n=-\infty}^{\infty} f_n e^{inT_0} \tag{4.4.9}$$

where

$$f_n = \frac{1}{2\pi}\int_0^{2\pi}\left[f\left(Ae^{iT_0} + \overline{A}e^{-iT_0}\right) + \mu_0 \mathrm{sgn}(iAe^{iT_0} - i\overline{A}e^{-iT_0})\right]e^{-inT_0}dT_0 \tag{4.4.10}$$

From (4.4.8)–(4.4.10), in order to eliminate secular terms, it is necessary to have

$$-2iD_1 A + ke^{i\sigma T_1} + f_1 = 0 \tag{4.4.11}$$

Let

$$A = \frac{1}{2}ae^{i\beta} \tag{4.4.12}$$

By (4.4.10) and (4.4.12), we have

$$\begin{aligned} f_1 &= \frac{1}{2\pi}\int_0^{2\pi}\left[f\left(Ae^{iT_0} + \overline{A}e^{-iT_0}\right) + \mu_0 \mathrm{sgn}\left(iAe^{iT_0} - i\overline{A}e^{-iT_0}\right)\right]e^{-iT_0}dT_0 \\ &= \frac{1}{2\pi}\int_0^{2\pi} f[a\cos(T_0+\beta)]e^{-iT_0}dT_0 - \frac{\mu_0}{2\pi}\int_0^{2\pi}\mathrm{sgn}[a\sin(T_0+\beta)]e^{-iT_0}dT_0 \\ &= \frac{e^{i\beta}}{2\pi}\int_0^{2\pi} f(a\cos\phi)e^{-i\phi}d\phi - \frac{\mu_0 e^{i\beta}}{2\pi}\int_0^{2\pi}\mathrm{sgn}(a\sin\phi)e^{-i\phi}d\phi \end{aligned}$$

$$= \frac{e^{i\beta}}{2\pi}\int_0^{2\pi} f(a\cos\phi)\,e^{-i\phi}d\phi + \frac{i\mu_0 e^{i\beta}}{\pi}\int_0^{\pi}\sin\phi d\phi$$

$$= f_{1R} + if_{1I} - \frac{2i\mu_0 e^{i\beta}}{\pi} \tag{4.4.13}$$

where f_{1R}, f_{1I} have already been presented in (4.3.26) and (4.3.27), respectively, in the previous exercise. Substituting the above result into (4.4.11) yields

$$ia' - a\beta' - ke^{i(\sigma T_1-\beta)} - f_{1R}e^{-i\beta} - if_{1I}e^{-i\beta} + \frac{2i\mu_0}{\pi} = 0 \tag{4.4.14}$$

The prime denotes the derivative with respect to T_1. Separating the real and imaginary parts of the above equation yields

$$\begin{aligned} a' &= k\sin(\sigma T_1 - \beta) + f_{1I}e^{-i\beta} - \frac{2\mu_0}{\pi} \\ -a\beta' &= k\cos(\sigma T_1 - \beta) + f_{1R}e^{-i\beta} \end{aligned} \tag{4.4.15}$$

Substituting (4.3.26) and (4.3.27) into the above equation, we get

$$\begin{aligned} a' &= \frac{2x_s}{\pi a}(x_s - a) - \frac{2\mu_0}{\pi} + k\sin(\sigma T_1 - \beta) \\ -\beta' &= -\frac{1}{\pi}\left[\frac{1}{2}\cos^{-1}\left(\frac{a-2x_s}{a}\right) - \left(1 - \frac{2x_s}{a}\right)\left(\frac{x_s}{a} - \frac{x_s^2}{a^2}\right)^{1/2}\right] \\ &\quad + \frac{k}{a}\cos(\sigma T_1 - \beta) \end{aligned} \tag{4.4.16}$$

Let

$$\gamma = (\sigma T_1 - \beta) \tag{4.4.17}$$

$$a' = \frac{2x_s}{\pi a}(x_s - a) - \frac{2\mu_0}{\pi} + k\sin\gamma \tag{4.4.18}$$

$$\gamma' = \sigma - \frac{1}{\pi}\left[\frac{1}{2}\cos^{-1}\left(\frac{a-2x_s}{a}\right) - \left(1 - \frac{2x_s}{a}\right)\left(\frac{x_s}{a} - \frac{x_s^2}{a^2}\right)^{1/2}\right] + \frac{k}{a}\cos\gamma \tag{4.4.19}$$

So the first order approximate solution of the system is

$$u = a(T_1)\cos[T_0 + \beta(T_1)] + O(\varepsilon) \tag{4.4.20}$$

(b) Let $a' = \gamma' = 0$ in (4.4.18) and (4.4.19), we obtain

$$\left.\begin{aligned} &\tfrac{2\mu_0}{\pi} - \tfrac{2x_s}{\pi a}(x_s - a) = k\sin\gamma \\ &\tfrac{a}{\pi}\left[\tfrac{1}{2}\cos^{-1}\left(\tfrac{a-2x_s}{a}\right) - \left(1 - \tfrac{2x_s}{a}\right)\left(\tfrac{x_s}{a} - \tfrac{x_s^2}{a^2}\right)^{1/2}\right] - \sigma a = k\cos\gamma \end{aligned}\right\} \tag{4.4.21}$$

Squaring and adding these two equations, we obtain the following frequency–response equation:

$$\begin{aligned} &\left\{\frac{a}{\pi}\left[\frac{1}{2}\cos^{-1}\left(\frac{a-2x_s}{a}\right) - \left(1 - \frac{2x_s}{a}\right)\left(\frac{x_s}{a} - \frac{x_s^2}{a^2}\right)^{1/2}\right] - \sigma a\right\}^2 \\ &+\left[\frac{2\mu_0}{\pi} - \frac{2x_s}{\pi a}(x_s - a)\right]^2 = k^2 \end{aligned} \tag{4.4.22}$$

We can solve for σ in terms of a from (4.4.22)

$$\begin{aligned} \sigma = &\frac{1}{\pi}\left[\frac{1}{2}\cos^{-1}\left(\frac{a-2x_s}{a}\right) - \left(1 - \frac{2x_s}{a}\right)\left(\frac{x_s}{a} - \frac{x_s^2}{a^2}\right)^{1/2}\right] \\ &\pm \frac{1}{a}\sqrt{k^2 - \left[\frac{2\mu_0}{\pi} - \frac{2x_s}{\pi a}(x_s - a)\right]^2} \end{aligned} \tag{4.4.23}$$

Since σ is a real number, it requires

$$a > x_s \tag{4.4.24}$$

Under this condition, we need

$$\frac{2\mu_0}{\pi} - \frac{2x_s}{\pi a}(x_s - a) > 0 \tag{4.4.25}$$

and

$$k \geq \frac{2\mu_0}{\pi} - \frac{2x_s}{\pi a}(x_s - a) = \frac{2}{\pi}(x_s + \mu_0) - \frac{2x_s^2}{\pi a}$$

From the above equation, we can obtain

$$a \leq \frac{2x_s^2}{\pi}\frac{1}{\frac{2}{\pi}(x_s + \mu_0) - k} \tag{4.4.26}$$

From (4.4.24) and (4.4.26), we know that when $0 < k < k_{\text{crit}} = 2(x_s + \mu_0)/\pi$, $a > x_s$ and a has a maximum value; when $k \geq k_{\text{crit}} = 2(x_s + \mu_0)/\pi$, $a > x_s$ and a has no maximum value.

When $a = x_s$, we can obtain from (4.4.23) that

$$\sigma_{1,2} = \frac{1}{2} \pm \frac{1}{x_s}\sqrt{k^2 - (\frac{2\mu_0}{\pi})^2} \tag{4.4.27}$$

It is easy to see from (4.4.23) that as a increases from x_s, $\sigma \in [\sigma_1, \sigma_2]$. Therefore, the range of values for σ is

$$\frac{1}{2} - \frac{1}{x_s}\sqrt{k^2 - \left(\frac{2\mu_0}{\pi}\right)^2} \le \sigma \le \frac{1}{2} + \frac{1}{x_s}\sqrt{k^2 - \left(\frac{2\mu_0}{\pi}\right)^2} \tag{4.4.28}$$

From (4.4.22), we can obtain

$$k = \sqrt{\begin{array}{c}\left\{\frac{a}{\pi}\left[\frac{1}{2}\cos^{-1}\left(\frac{a-2x_s}{a}\right) - \left(1 - \frac{2x_s}{a}\right)\left(\frac{x_s}{a} - \frac{x_s^2}{a^2}\right)^{1/2}\right] - \sigma a\right\}^2 \\ +\left[\frac{2\mu_0}{\pi} - \frac{2x_s}{\pi a}(x_s - a)\right]^2\end{array}} \tag{4.4.29}$$

The $a \sim \sigma$ curve and $\gamma \sim \sigma$ curve can be sketched from (4.4.23) and (4.4.29) as shown in Fig. 4.4a, b, which confirms the above conclusion about the relationship between $a \sim \sigma$. As can be seen from the figures, there is no amplitude jump.

4.5 Exercise 4.5 (The Method of Multiple Scales for Analyzing Primary Resonances of Linear Springs, Hysteresis and Linearly Damped Systems)

Solution: **(a)** When $\Omega \approx 1$, the detuning parameter σ is adopted to quantitatively describes the nearness of Ω to 1:

$$\Omega = 1 + \varepsilon\sigma \tag{4.5.1}$$

We express the solution in terms of different time scales as

$$x(t; \varepsilon) = x_0(T_0, T_1) + \varepsilon x_1(T_0, T_1) + \cdots \tag{4.5.2}$$

In order to keep all damping and excitation terms appearing simultaneously, we set

$$K = 2\varepsilon k \tag{4.5.3}$$

Substituting (4.5.1)–(4.5.3) into the differential equation of motion of the system and retaining to $O(\varepsilon)$, we obtain

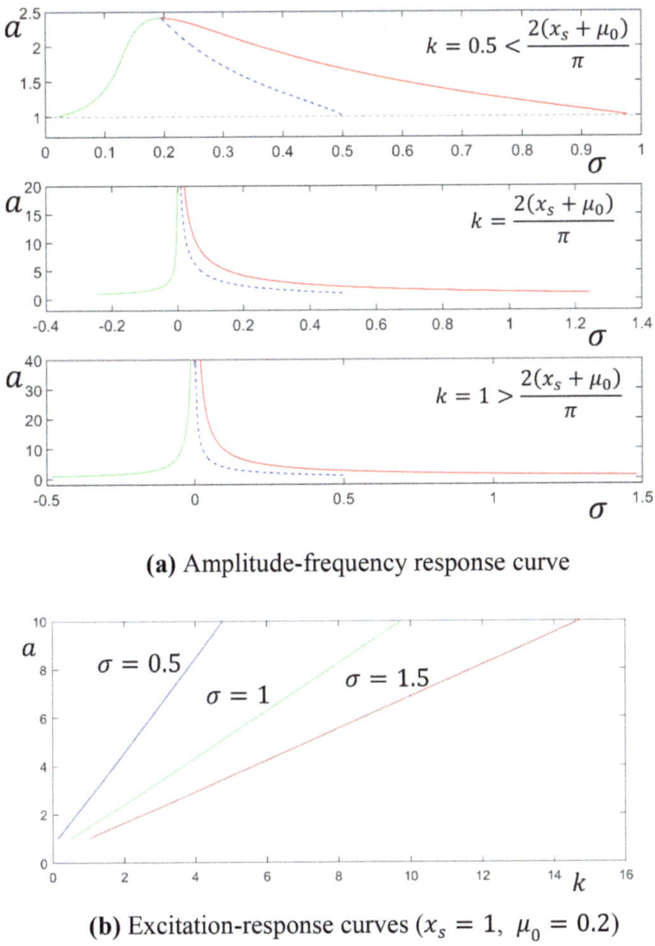

Fig. 4.4 **a** Frequency-response and **b** Excitation-response curves ($x_s = 1$, $\mu_0 = 0.2$) for Exercise 4.4

$$
\begin{aligned}
0 &= \ddot{x} + x - \varepsilon f(x) + 2\varepsilon\mu\dot{x} - K\cos\Omega t \\
&= \left(D_0^2 + 2\varepsilon D_0 D_1\right)(x_0 + \varepsilon x_1) + x_0 + \varepsilon x_1 \\
&\quad - \varepsilon f(x_0) + 2\varepsilon\mu\dot{x}_0 - 2\varepsilon k\cos(T_0 + \sigma T_1) + \cdots \\
&= D_0^2 x_0 + x_0 + \varepsilon[D_0^2 x_1 + x_1 + 2D_0 D_1 x_0 - f(x_0) + 2\mu\dot{x}_0 \\
&\quad - 2k\cos(T_0 + \sigma T_1)] + \cdots
\end{aligned}
\tag{4.5.4}
$$

Equating the coefficient of the like power of ε yields

$$
D_0^2 x_0 + x_0 = 0 \tag{4.5.5}
$$

$$D_0^2x_1 + x_1 = -2D_0D_1x_0 + f(x_0) - 2\mu\dot{x}_0 + 2k\cos(T_0 + \sigma T_1) \quad (4.5.6)$$

The general solution of (4.5.5) is

$$x_0 = Ae^{iT_0} + \overline{A}e^{-iT_0} \quad (4.5.7)$$

where $A = A(T_1)$. Substitute (4.5.7) into (4.5.6), we obtain

$$\begin{aligned} D_0^2x_1 + x_1 &= -2D_0D_1x_0 + f(x_0) - 2\mu\dot{x}_0 + 2k\cos(T_0 + \sigma T_1) \\ &= -2iD_1Ae^{iT_0} + f\left(Ae^{iT_0} + \overline{A}e^{-iT_0}\right) \\ &\quad + ke^{i\sigma T_1}e^{iT_0} + cc - 2\mu\left(iAe^{iT_0} - i\overline{A}e^{-iT_0}\right) \end{aligned} \quad (4.5.8)$$

To further process thc Eqs. (4.5.8), we need to first deal with the functions $f\left(Ae^{iT_0} + \overline{A}e^{-iT_0}\right)$ and $-2\mu\left(iAe^{iT_0} - i\overline{A}e^{-iT_0}\right)$, which can be expanded into a Fourier series of complex exponential form:

$$f\left(Ae^{iT_0} + \overline{A}e^{-iT_0}\right) - 2\mu\left(iAe^{iT_0} - i\overline{A}e^{-iT_0}\right) = \sum_{n=-\infty}^{\infty} f_n e^{inT_0} \quad (4.5.9)$$

where

$$f_n = \frac{1}{2\pi}\int_0^{2\pi} \left[f\left(Ae^{iT_0} + \overline{A}e^{-iT_0}\right) - 2\mu\left(iAe^{iT_0} - i\overline{A}e^{-iT_0}\right)\right]e^{-inT_0}dT_0 \quad (4.5.10)$$

From (4.5.8)–(4.5.10), in order to eliminate secular terms, it is necessary to have

$$-2iD_1A + ke^{i\sigma T_1} + f_1 = 0 \quad (4.5.11)$$

Let

$$A = \frac{1}{2}ae^{i\beta} \quad (4.5.12)$$

By (4.5.10) and (4.5.12), we obtain

$$\begin{aligned} f_1 &= \frac{1}{2\pi}\int_0^{2\pi} \left[f\left(Ae^{iT_0} + \overline{A}e^{-iT_0}\right) - 2\mu\left(iAe^{iT_0} - i\overline{A}e^{-iT_0}\right)\right]e^{-iT_0}dT_0 \\ &= \frac{e^{i\beta}}{2\pi}\int_0^{2\pi} f(a\cos\phi)\, e^{-i\phi}d\phi + \frac{\mu e^{i\beta}}{\pi}\int_0^{2\pi} a\sin\phi e^{-i\phi}d\phi \\ &= f_{1R} + if_{1I} - i\mu ae^{i\beta} \end{aligned} \quad (4.5.13)$$

where $f_{1R}$and f_{1I} have already been presented in (4.3.26) and (4.3.27), respectively. Substituting the above result into (4.5.11) yields

$$ia' - a\beta' - ke^{i(\sigma T_1 - \beta)} - f_{1R}e^{-i\beta} - if_{1I}e^{-i\beta} + i\mu a = 0 \tag{4.5.14}$$

The prime denotes the derivative with respect to T_1. Separating the real and imaginary parts of the above equation yields

$$\begin{aligned} a' &= k\sin(\sigma T_1 - \beta) + f_{1I}e^{-i\beta} - \mu a \\ -a\beta' &= k\cos(\sigma T_1 - \beta) + f_{1R}e^{-i\beta} \end{aligned} \tag{4.5.15}$$

Substituting (4.3.26) and (4.3.27) into the above equation, we get

$$\begin{aligned} a' &= \frac{2x_s}{\pi a}(x_s - a) - \mu a + k\sin(\sigma T_1 - \beta) \\ -\beta' &= -\frac{1}{\pi}\left[\frac{1}{2}\cos^{-1}\left(\frac{a - 2x_s}{a}\right) - \left(1 - \frac{2x_s}{a}\right)\left(\frac{x_s}{a} - \frac{x_s^2}{a^2}\right)^{1/2}\right] \\ &+ \frac{k}{a}\cos(\sigma T_1 - \beta) \end{aligned} \tag{4.5.16}$$

Let

$$\gamma = (\sigma T_1 - \beta) \tag{4.5.17}$$

Then

$$a' = \frac{2x_s}{\pi a}(x_s - a) - \mu a + k\sin\gamma \tag{4.5.18}$$

$$\gamma' = \sigma - \frac{1}{\pi}\left[\frac{1}{2}\cos^{-1}\left(\frac{a - 2x_s}{a}\right) - \left(1 - \frac{2x_s}{a}\right)\left(\frac{x_s}{a} - \frac{x_s^2}{a^2}\right)^{1/2}\right] + \frac{k}{a}\cos\gamma \tag{4.5.19}$$

Therefore, the first order approximate solution of the system is

$$u = a(T_1)\cos[T_0 + \beta(T_1)] + O(\varepsilon) \tag{4.5.20}$$

(b) Let $a' = \gamma' = 0$ in (4.5.18) and (4.5.19), we can obtain

$$\left.\begin{aligned} \mu a - \tfrac{2x_s}{\pi a}(x_s - a) = k\sin\gamma \\ \tfrac{a}{\pi}\left[\tfrac{1}{2}\cos^{-1}\left(\tfrac{a-2x_s}{a}\right) - \left(1 - \tfrac{2x_s}{a}\right)\left(\tfrac{x_s}{a} - \tfrac{x_s^2}{a^2}\right)^{1/2}\right] - \sigma a = k\cos\gamma \end{aligned}\right\} \tag{4.5.21}$$

Squaring and adding these two equations, we obtain the following frequency–response equation:

$$\left\{\frac{a}{\pi}\left[\frac{1}{2}\cos^{-1}\left(\frac{a-2x_s}{a}\right)-\left(1-\frac{2x_s}{a}\right)\left(\frac{x_s}{a}-\frac{x_s^2}{a^2}\right)^{1/2}\right]-\sigma a\right\}^2 + \left[\mu a-\frac{2x_s}{\pi a}(x_s-a)\right]^2 = k^2 \tag{4.5.22}$$

We can solve for σ in terms of a from (4.5.22)

$$\sigma = \frac{1}{\pi}\left[\frac{1}{2}\cos^{-1}\left(\frac{a-2x_s}{a}\right)-\left(1-\frac{2x_s}{a}\right)\left(\frac{x_s}{a}-\frac{x_s^2}{a^2}\right)^{1/2}\right] \pm \frac{1}{a}\sqrt{k^2-\left[\mu a-\frac{2x_s}{\pi a}(x_s-a)\right]^2} \tag{4.5.23}$$

Since σ is a real number, it requires

$$a > x_s \tag{4.5.24}$$

Under this condition, we need

$$\mu a-\frac{2x_s}{\pi a}(x_s-a) > 0 \tag{4.5.25}$$

and

$$k \geq \mu a-\frac{2x_s}{\pi a}(x_s-a) = \mu a-\frac{2x_s^2}{\pi a}+\frac{2x_s}{\pi} \tag{4.5.26}$$

Since the function on the right-hand side of the above equation increases as a increases, when $a = x_s$, k reaches the minimum value

$$k_{min} = \mu x_s-\frac{2x_s^2}{\pi x_s}+\frac{2x_s}{\pi} = \mu x_s \tag{4.5.27}$$

Thus, when $a > a_s$, the steady state motion is possible only when $k \geq k_{min}$. From (4.5.26), we can obtain

$$\mu a^2+\left(\frac{2x_s}{\pi}-k\right)a-\frac{2x_s^2}{\pi} \leq 0 \tag{4.5.28}$$

It can be seen that when $k \geq k_{min}$ and k is finite,a is also finite.

When k is fixed and $k \geq k_{min}$, since the function in the right square bracket on the right side of (4.5.23) is monotonically decreasing, the range of the value of σ is determined by

$$\sigma \in [\sigma_1, \ \sigma_2], \quad \sigma_{1,\,2} = \frac{1}{2} \pm \frac{1}{x_s}\sqrt{k^2 - \mu x_s} \tag{4.5.29}$$

From (4.4.22), we can obtain

$$k = \sqrt{\begin{array}{c} \left\{\frac{a}{\pi}\left[\frac{1}{2}\cos^{-1}\left(\frac{a-2x_s}{a}\right) - \left(1 - \frac{2x_s}{a}\right)\left(\frac{x_s}{a} - \frac{x_s^2}{a^2}\right)^{1/2}\right] - \sigma a\right\}^2 \\ +\left[\mu a - \frac{2x_s}{\pi a}(x_s - a)\right]^2 \end{array}} \tag{4.5.30}$$

The $a \sim \sigma$ curve and $\gamma \sim \sigma$ curve can be sketched from (4.4.23) and (4.4.29) as shown in Fig. 4.5a, b, which confirms the above conclusion about the relationship between $a \sim \sigma$. As can be seen from the figures, there is no amplitude jump.

4.6 Exercise 4.6 (The Method of Multiple Scales for Analyzing Primary Resonances of Linear Spring, Hysteresis and Quadratic Damped Systems)

Solution: (a) When $\Omega \approx 1$, the detuning parameter σ is adopted to quantitatively describes the nearness of Ω to 1:

$$\Omega = 1 + \varepsilon\sigma \tag{4.6.1}$$

We express the solution in terms of different time scales as

$$x(t;\,\varepsilon) = x_0(T_0, T_1) + \varepsilon x_1(T_0, T_1) + \cdots \tag{4.6.2}$$

In order to keep all damping and excitation terms appearing simultaneously, we set

$$K = 2\varepsilon k \tag{4.6.3}$$

Substituting (4.6.1)–(4.6.3) into the differential equation of motion of the system and and keeping to $O(\varepsilon)$ yields

$$\begin{aligned} 0 &= \ddot{x} + x - \varepsilon f(x) + \varepsilon\mu|\dot{x}|\dot{x} - K\cos\Omega t \\ &= \left(D_0^2 + 2\varepsilon D_0 D_1\right)(x_0 + \varepsilon x_1) + x_0 + \varepsilon x_1 \\ &\quad - \varepsilon f(x_0) + \varepsilon\mu\,|\dot{x}_0|\dot{x}_0 - 2\varepsilon k\cos(T_0 + \sigma T_1) + \cdots \end{aligned}$$

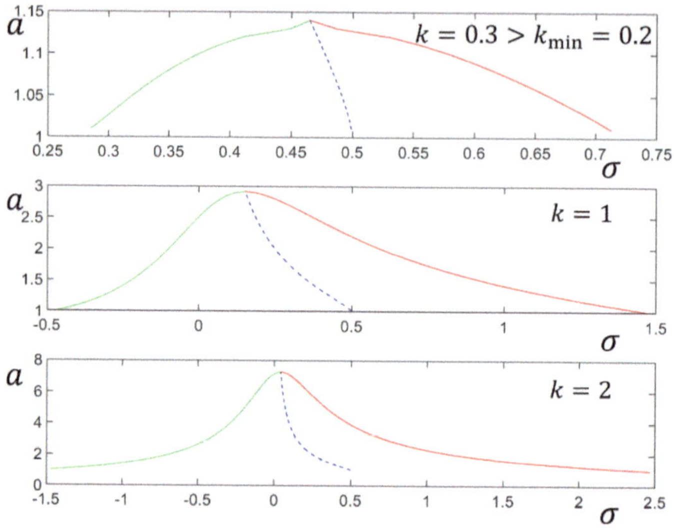

(a) Amplitude-frequency response curve

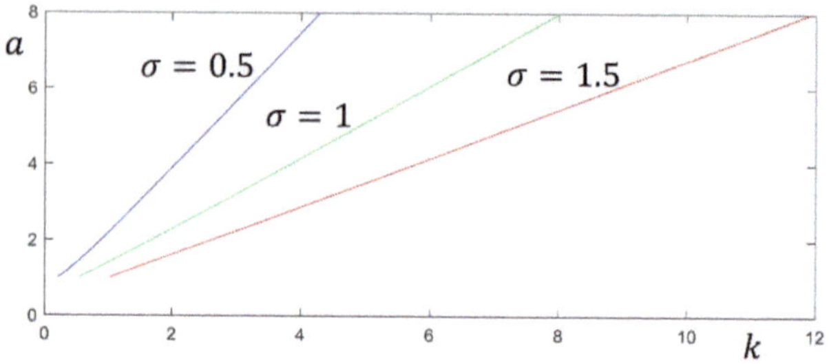

(b) Excitation-response curves ($x_s = 1,\ \mu = 0.2$)

Fig. 4.5 **a** Frequency-response, and **b** Excitation-response curves ($x_s = 1$, $\mu = 0.2$) for Exercise 4.5

$$
\begin{aligned}
&= D_0^2 x_0 + x_0 + \varepsilon[D_0^2 x_1 + x_1 + 2D_0 D_1 x_0 - f(x_0) + \mu\,|\dot{x}_0|\dot{x}_0 \\
&\quad - 2k\cos(T_0 + \sigma T_1)] + \cdots
\end{aligned} \tag{4.6.4}
$$

Let the coefficient of the like power of ε be zero, we get

$$D_0^2 x_0 + x_0 = 0 \tag{4.6.5}$$

$$D_0^2 x_1 + x_1 = -2D_0 D_1 x_0 + f(x_0) - \mu\,|\dot{x}_0|\dot{x}_0 + 2k\cos(T_0 + \sigma T_1) \tag{4.6.6}$$

The solution of (4.6.5) is

$$x_0 = Ae^{iT_0} + \overline{A}e^{-iT_0} \tag{4.6.7}$$

where $A = A(T_1)$. Substituting (4.6.7) into (4.6.6) yields

$$\begin{aligned} D_0^2 x_1 + x_1 &= -2D_0 D_1 x_0 + f(x_0) - \mu\,|\dot{x}_0|\dot{x}_0 + 2k\cos(T_0 + \sigma T_1) \\ &= -2iD_1 A e^{iT_0} + f\left(Ae^{iT_0} + \overline{A}e^{-iT_0}\right) \\ &\quad + ke^{i\sigma T_1}e^{iT_0} + cc - \mu\left|iAe^{iT_0} - i\overline{A}e^{-iT_0}\right|\left(iAe^{iT_0} - i\overline{A}e^{-iT_0}\right) \end{aligned} \tag{4.6.8}$$

In order to further process (4.5.8), we need to first deal with the functions $-\mu\left|iAe^{iT_0} - i\overline{A}e^{-iT_0}\right|\left(iAe^{iT_0} - i\overline{A}e^{-iT_0}\right)$ and $f\left(Ae^{iT_0} + \overline{A}e^{-iT_0}\right)$, which can be expanded into a Fourier series of complex exponential form:

$$f\left(Ae^{iT_0} + \overline{A}e^{-iT_0}\right) - \mu\left|iAe^{iT_0} - i\overline{A}e^{-iT_0}\right|\left(iAe^{iT_0} - i\overline{A}e^{-iT_0}\right) = \sum_{n=-\infty}^{\infty} f_n e^{inT_0} \tag{4.6.9}$$

where $f_n = \frac{1}{2\pi}\int\limits_0^{2\pi} [f\left(Ae^{iT_0} + \overline{A}e^{-iT_0}\right)$

$$-\mu\left|iAe^{iT_0} - i\overline{A}e^{-iT_0}\right|\left(iAe^{iT_0} - i\overline{A}e^{-iT_0}\right)]e^{-inT_0}dT_0 \tag{4.6.10}$$

From (4.6.8)–(4.6.10), in order to eliminate secular terms, it is necessary to have

$$-2iD_1 A + ke^{i\sigma T_1} + f_1 = 0 \tag{4.6.11}$$

Let

$$A = \frac{1}{2}ae^{i\beta} \tag{4.6.12}$$

By (4.6.10) and (4.6.12), we have

$$\begin{aligned} f_1 &= \frac{1}{2\pi}\int\limits_0^{2\pi} [f\left(Ae^{iT_0} + \overline{A}e^{-iT_0}\right) \\ &\quad - \mu\left|iAe^{iT_0} - i\overline{A}e^{-iT_0}\right|\left(iAe^{iT_0} - i\overline{A}e^{-iT_0}\right)]e^{-iT_0}dT_0 \\ &= \frac{e^{i\beta}}{2\pi}\int\limits_0^{2\pi} f(a\cos\phi)\,e^{-i\phi}d\phi + \frac{\mu e^{i\beta}}{2\pi}\int\limits_0^{2\pi} |a\sin\phi|a\sin\phi e^{-i\phi}d\phi \\ &= f_{1R} + if_{1I} - \frac{i\mu a^2 e^{i\beta}}{\pi}\int\limits_0^{\pi}\sin^3\phi d\phi \end{aligned}$$

$$= f_{1R} + if_{1I} - \frac{4i\mu a^2 e^{i\beta}}{3\pi} \tag{4.6.13}$$

where f_{1R}, f_{1I} have already been presented in (4.3.26) and (4.3.27), respectively. Substituting the above result into (4.6.11) yields

$$ia' - a\beta' - ke^{i(\sigma T_1 - \beta)} - f_{1R}e^{-i\beta} - if_{1I}e^{-i\beta} + \frac{4i\mu a^2}{3\pi} = 0 \tag{4.6.14}$$

The prime denotes the derivative with respect to T_1. Separating the real and imaginary parts of the above equation yields

$$\begin{aligned} a' &= k\sin(\sigma T_1 - \beta) + f_{1I}e^{-i\beta} - \tfrac{4\mu a^2}{3\pi} \\ -a\beta' &= k\cos(\sigma T_1 - \beta) + f_{1R}e^{-i\beta} \end{aligned} \tag{4.6.15}$$

Substituting (4.3.26) and (4.3.27) into the above equation, we get

$$\begin{aligned} a' &= \frac{2x_s}{\pi a}(x_s - a) - \frac{4\mu a^2}{3\pi} + k\sin(\sigma T_1 - \beta) \\ -\beta' &= -\frac{1}{\pi}\left[\frac{1}{2}\cos^{-1}\left(\frac{a - 2x_s}{a}\right) - \left(1 - \frac{2x_s}{a}\right)\left(\frac{x_s}{a} - \frac{x_s^2}{a^2}\right)^{1/2}\right] \\ &\quad + \frac{k}{a}\cos(\sigma T_1 - \beta) \end{aligned} \tag{4.6.16}$$

Let

$$\gamma = \sigma T_1 - \beta \tag{4.6.17}$$

Then

$$a' = \frac{2x_s}{a\pi}(x_s - a) - \frac{4\mu}{3\pi}a^2 + k\sin\gamma \tag{4.6.18}$$

$$\begin{aligned} \gamma' &= \sigma - \frac{1}{\pi}\left[\frac{1}{2}\cos^{-1}\left(\frac{a - 2x_s}{a}\right) - \left(1 - \frac{2x_s}{a}\right)\left(\frac{x_s}{a} - \frac{x_s^2}{a^2}\right)^{1/2}\right] \\ &\quad + \frac{k}{a}\cos\gamma \end{aligned} \tag{4.6.19}$$

Therefore, the first order approximate solution of the system is

$$u = a(T_1)\cos[T_0 + \beta(T_1)] + O(\varepsilon) \tag{4.6.20}$$

(b) Let $a' = \gamma' = 0$ in (4.6.18) and (4.6.19), we obtain:

$$\left.\begin{aligned}\tfrac{4\mu}{3\pi}a^2-\tfrac{2x_s}{\pi a}(x_s-a)=k\sin\gamma\\ \tfrac{a}{\pi}\left[\tfrac{1}{2}\cos^{-1}\left(\tfrac{a-2x_s}{a}\right)-\left(1-\tfrac{2x_s}{a}\right)\left(\tfrac{x_s}{a}-\tfrac{x_s^2}{a^2}\right)^{1/2}\right]-\sigma a=k\cos\gamma\end{aligned}\right\} \tag{4.6.21}$$

Squaring and adding these two equations, we obtain the following frequency–response equation:

$$\begin{aligned}&\left\{\frac{a}{\pi}\left[\frac{1}{2}\cos^{-1}\left(\frac{a-2x_s}{a}\right)-\left(1-\frac{2x_s}{a}\right)\left(\frac{x_s}{a}-\frac{x_s^2}{a^2}\right)^{1/2}\right]-\sigma a\right\}^2\\ &+\left[\frac{4\mu}{3\pi}a^2-\frac{2x_s}{\pi a}(x_s-a)\right]^2=k^2\end{aligned} \tag{4.6.22}$$

We can solve for σ in terms of a from (4.6.22)

$$\begin{aligned}\sigma=&\frac{1}{\pi}\left[\frac{1}{2}\cos^{-1}\left(\frac{a-2x_s}{a}\right)-\left(1-\frac{2x_s}{a}\right)\left(\frac{x_s}{a}-\frac{x_s^2}{a^2}\right)^{1/2}\right]\\ &\pm\frac{1}{a}\sqrt{k^2-\left[\frac{4\mu}{3\pi}a^2-\frac{2x_s}{\pi a}(x_s-a)\right]^2}\end{aligned} \tag{4.6.23}$$

Since σ is a real number, it requires

$$a > x_s \tag{4.6.24}$$

Under this condition, we need

$$\frac{4\mu}{3\pi}a^2-\frac{2x_s}{\pi a}(x_s-a)>0 \tag{4.6.25}$$

and

$$k\geq\frac{4\mu}{3\pi}a^2-\frac{2x_s}{\pi a}(x_s-a)=\frac{4\mu}{3\pi}a^2-\frac{2x_s^2}{\pi a}+\frac{2x_s}{\pi} \tag{4.6.26}$$

Since the function on the right-hand side of the above equation increases as a increases, when $a = x_s$, k reaches the minimum value

$$k_{min}=\frac{4\mu}{3\pi}x_s^2 \tag{4.6.27}$$

Thus, when $a > a_s$, the steady state motion is possible only when $k \geq k_{min}$.

When k is fixed and $k \geq k_{min}$, since the function in the right square bracket on the right side of (4.6.23) is monotonically decreasing, the range of the value of σ is determined by

$$\sigma \in [\sigma_1, \sigma_2], \quad \sigma_{1,2} = \frac{1}{2} \pm \frac{1}{x_s}\sqrt{k^2 - \left(\frac{4\mu}{3\pi}x_s^2\right)^2} \tag{4.6.28}$$

From (4.6.22), we can obtain

$$k = \sqrt{\begin{array}{c}\left\{\frac{a}{\pi}\left[\frac{1}{2}\cos^{-1}\left(\frac{a-2x_s}{a}\right) - \left(1 - \frac{2x_s}{a}\right)\left(\frac{x_s}{a} - \frac{x_s^2}{a^2}\right)^{1/2}\right] - \sigma a\right\}^2 \\ +\left[\frac{4\mu}{3\pi}a^2 - \frac{2x_s}{\pi a}(x_s - a)\right]^2\end{array}} \tag{4.6.29}$$

The $a \sim \sigma$ curve and $k \sim a$ curve can be sketched from (4.6.23) and (4.6.29) as shown in Fig. 4.6a, b, which confirms the above conclusion about the relationship between $a \sim \sigma$. As can be seen from the figures, there is no amplitude jump.

4.7 Exercise 4.7 (The Method of Multiple Scales for Analyzing Primary Resonances of Quadraticd Spring Force and Quadraticd Damped Systems)

Solution: (a) Divide the given differential equation of motion by mL, where L is a reference length, we can obtain

$$\frac{\ddot{x}}{L} + \frac{c_1}{m}\frac{\dot{x}}{L} + \frac{c_2 L}{m}\frac{\dot{x}}{L}\left|\frac{\dot{x}}{L}\right| + \frac{k_1}{m}\frac{x}{L} + \frac{k_2 L}{m}\frac{x}{L}\left|\frac{x}{L}\right| = \frac{K}{mL}\cos\omega t \tag{4.7.1}$$

Let

$$\omega_0^2 = \frac{k_1}{m}, \quad \tau = \omega_0 t$$

(4.7.1) becomes

$$\omega_0^2\frac{d^2}{d\tau^2}\left(\frac{x}{L}\right) + \omega_0\frac{c_1}{m}\frac{d}{d\tau}\left(\frac{x}{L}\right) + \omega_0^2\frac{c_2 L}{m}\frac{d}{d\tau}\left(\frac{x}{L}\right)\left|\frac{d}{d\tau}\left(\frac{x}{L}\right)\right|$$
$$+ \omega_0^2\frac{x}{L} + \frac{k_2 L}{m}\frac{x}{L}\left|\frac{x}{L}\right| = \frac{K}{mL}\cos\frac{\omega}{\omega_0}\tau$$

Dividing the above equation by ω_0^2 and, for simplicity, denoting x/L as x, τ as t, $d^2(x/L)/d\tau^2$ and $d(x/L)/d\tau$ as $\ddot{x}$ and $\dot{x}$, we obtain

$$\ddot{x} + \frac{c_1}{m\omega_0}\dot{x} + \frac{c_2 L}{m}\dot{x}|\dot{x}| + x + \frac{k_2 L}{m\omega_0^2}x|x| = \frac{K}{m\omega_0^2 L}\cos\frac{\omega}{\omega_0}\tau$$

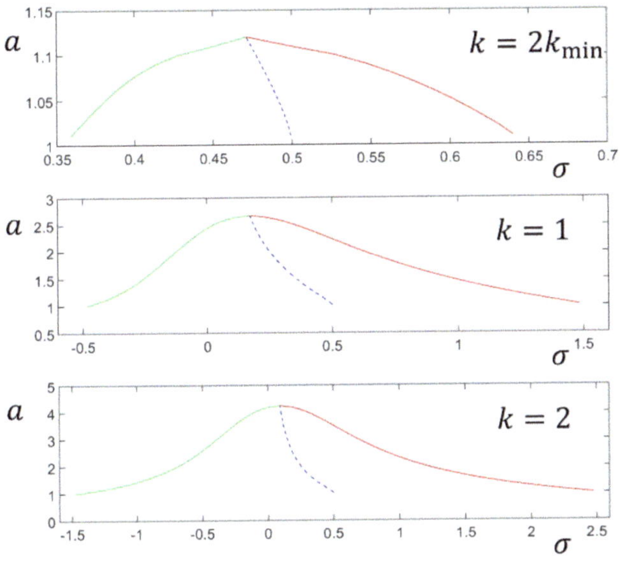

(a) Amplitude-frequency response curve

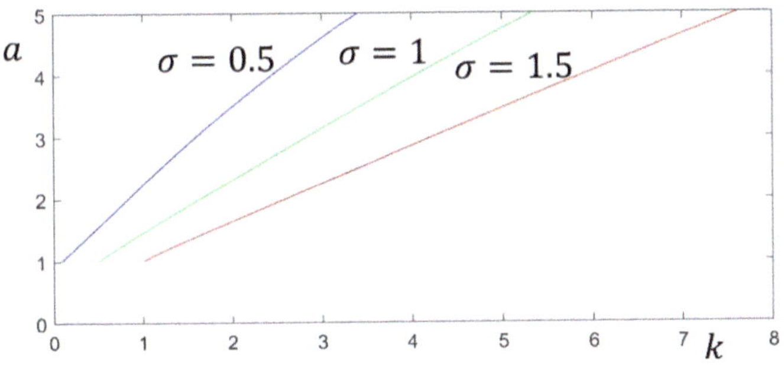

(b) Excitation-response curves ($\boldsymbol{x}_s = 1,\ \mu = 0.2$)

Fig. 4.6 **a** Frequency-response, and **b** Excitation-response curves ($x_s = 1$, $\mu = 0.2$) for Exercise 4.6

Let

$$2\varepsilon\mu_1 = \frac{c_1}{m\omega_0}, \quad \varepsilon\mu_2 = \frac{c_2}{Lm}, \quad \varepsilon\alpha = \frac{k_2 L}{m\omega_0^2}, \quad 2\varepsilon k = \frac{K}{m\omega_0^2 L}, \quad \Omega = \frac{\omega}{\omega_0}$$

Then

$$\ddot{x} + 2\varepsilon\mu_1\dot{x} + \varepsilon\mu_2\dot{x}|\dot{x}| + x + \varepsilon\alpha x|x| = 2\varepsilon k \cos\Omega t \tag{4.7.2}$$

(b) When $\Omega \approx 1$, the detuning parameter σ is adopted to quantitatively describes the nearness of Ω to 1:

$$\Omega = 1 + \varepsilon\sigma \tag{4.7.3}$$

We express the solution in terms of different time scales as

$$x(t;\varepsilon) = x_0(T_0, T_1) + \varepsilon x_1(T_0, T_1) + \cdots \tag{4.7.4}$$

In order to keep all damping and excitation terms appearing simultaneously, we set

$$K = 2\varepsilon k \tag{4.7.5}$$

Substituting (4.7.3)–(4.7.5) into (4.7.2) and retaining to $O(\varepsilon)$, we obtain

$$\begin{aligned} 0 &= \ddot{x} + x + 2\varepsilon\mu_1\dot{x} + \varepsilon\alpha x|x| + \varepsilon\mu_2\dot{x}|\dot{x}| - 2\varepsilon k\cos\Omega t \\ &= \left(D_0^2 + 2\varepsilon D_0 D_1\right)(x_0 + \varepsilon x_1) + x_0 + \varepsilon x_1 + 2\varepsilon\mu_1\dot{x}_0 \\ &\quad + \varepsilon\alpha\,|x_0|x_0 + \varepsilon\mu_2\,|\dot{x}_0|\dot{x}_0 - 2\varepsilon k\cos(T_0 + \sigma T_1) + \cdots \\ &= D_0^2 x_0 + x_0 + \varepsilon[D_0^2 x_1 + x_1 + 2D_0D_1x_0 + 2\mu_1\dot{x}_0 \\ &\quad + \alpha\,|x_0|x_0 + \mu_2\,|\dot{x}_0|\dot{x}_0 - 2k\cos(T_0 + \sigma T_1)] + \cdots \end{aligned} \tag{4.7.6}$$

Let the coefficient of the like power of ε be zero, we get

$$D_0^2 x_0 + x_0 = 0 \tag{4.7.7}$$

$$\begin{aligned} D_0^2 x_1 + x_1 &= -2D_0D_1x_0 - 2\mu_1\dot{x}_0 - \alpha\,|x_0|x_0 - \mu_2\,|\dot{x}_0|\dot{x}_0 \\ &\quad + 2k\cos(T_0 + \sigma T_1) \end{aligned} \tag{4.7.8}$$

The solution of (4.7.7) is

$$x_0 = Ae^{iT_0} + \overline{A}e^{-iT_0} \tag{4.7.9}$$

where $A = A(T_1)$. Substituting (4.7.9) into (4.7.8) yields

$$\begin{aligned} D_0^2 x_1 + x_1 &= -2D_0D_1x_0 - 2\mu_1\dot{x}_0 - \alpha\,|x_0|x_0 - \mu_2\,|\dot{x}_0|\dot{x}_0 + 2k\cos(T_0 + \sigma T_1) \\ &= -2iD_1Ae^{iT_0} - 2i\mu_1 Ae^{iT_0} + ke^{i\sigma T_1}e^{iT_0} + cc \\ &\quad - \alpha\left|Ae^{iT_0} + \bar{A}e^{-iT_0}\right|\left(Ae^{iT_0} + \bar{A}e^{-iT_0}\right) \\ &\quad - \mu_2\left|iAe^{iT_0} - i\bar{A}e^{-iT_0}\right|\left(iAe^{iT_0} - i\bar{A}e^{-iT_0}\right) \end{aligned} \tag{4.7.10}$$

To further process (4.7.10), we need to first deal with the functions $-\mu_2\left|iAe^{iT_0}-i\overline{A}e^{-iT_0}\right|\left(iAe^{iT_0}-i\overline{A}e^{-iT_0}\right)$ and $-\alpha\left|Ae^{iT_0}+\overline{A}e^{-iT_0}\right|\left(Ae^{iT_0}+\overline{A}e^{-iT_0}\right)$, which can be expanded into a Fourier series of complex exponential form:

$$\begin{aligned}&-\alpha\left|Ae^{iT_0}+\overline{A}e^{-iT_0}\right|\left(Ae^{iT_0}+\overline{A}e^{-iT_0}\right)\\&-\mu_2\left|iAe^{iT_0}-i\overline{A}e^{-iT_0}\right|\left(iAe^{iT_0}-i\overline{A}e^{-iT_0}\right)=\sum_{n=-\infty}^{\infty}f_n e^{inT_0}\end{aligned}\tag{4.7.11}$$

where

$$\begin{aligned}f_n=\frac{1}{2\pi}\int_0^{2\pi}&[-\alpha\left|Ae^{iT_0}+\bar{A}e^{-iT_0}\right|\left(Ae^{iT_0}+\bar{A}e^{-iT_0}\right)\\&-\mu_2\left|iAe^{iT_0}-i\bar{A}e^{-iT_0}\right|\left(iAe^{iT_0}-i\bar{A}e^{-iT_0}\right)]e^{-i\,n\;T_0}dT_0\end{aligned}\tag{4.7.12}$$

From (4.7.10)–(4.7.12), in order to eliminate secular terms, it is necessary to have

$$-2iD_1A-2i\mu_1A+ke^{i\sigma T_1}+f_1=0\tag{4.7.13}$$

Let

$$A=\frac{1}{2}ae^{i\beta}\tag{4.7.14}$$

By (4.7.12) and (4.7.14), we obtain

$$\begin{aligned}f_1&=\frac{1}{2\pi}\int_0^{2\pi}[-\alpha\left|Ae^{iT_0}+\overline{A}e^{-iT_0}\right|\left(Ae^{iT_0}+\overline{A}e^{-iT_0}\right)\\&\quad-\mu_2\left|iAe^{iT_0}-i\overline{A}e^{-iT_0}\right|\left(iAe^{iT_0}-i\overline{A}e^{-iT_0}\right)]e^{-iT_0}dT_0\\&=-\frac{\alpha e^{i\beta}}{2\pi}\int_0^{2\pi}|a\cos\theta|a\cos\theta e^{-i\theta}d\theta+\frac{\mu_2e^{i\beta}}{2\pi}\int_0^{2\pi}|a\sin\theta|a\sin\theta e^{-i\theta}d\phi\\&=-\frac{\alpha a^2e^{i\beta}}{\pi}\int_{-\pi/2}^{\pi/2}\cos^3\theta d\theta-\frac{i\mu_2a^2e^{i\beta}}{\pi}\int_0^{\pi}\sin^3\theta d\theta\\&=-\frac{4\alpha a^2e^{i\beta}}{3\pi}-\frac{4i\mu_2a^2e^{i\beta}}{3\pi}\end{aligned}\tag{4.7.15}$$

Substituting the above results into (4.7.10) yields

$$ia' - a\beta' - ke^{i(\sigma T_1 - \beta)} + i\mu_1 a + \frac{4\alpha a^2}{3\pi} + \frac{4i\mu_2 a^2}{3\pi} = 0 \tag{4.7.16}$$

The prime denotes the derivative with respect to T_1. Separating the real and imaginary parts of the above equation yields

$$\begin{aligned} a' &= -\mu_1 a - \frac{4\mu_2 a^2}{3\pi} + k\sin(\sigma T_1 - \beta) \\ -a\beta' &= -\frac{4\alpha a^2}{3\pi} + k\cos(\sigma T_1 - \beta) \end{aligned} \tag{4.7.17}$$

Let

$$\gamma = \sigma T_1 - \beta \tag{4.7.18}$$

Then

$$a' = -\mu_1 a - \frac{4\mu_2}{3\pi}a^2 + k\sin\gamma \tag{4.7.19}$$

$$\gamma' = \sigma - \frac{4\alpha}{3\pi}a + \frac{k}{a}\cos\gamma \tag{4.7.20}$$

So the first order approximate solution of the system is

$$x = a(T_1)\cos\varphi + O(\varepsilon), \quad \varphi = T_0 + \beta(T_1) \tag{4.7.21}$$

(c) Considering (4.7.21), the approximation of $x|x|$ and $\dot{x}|\dot{x}|$ can be written as

$$x|x| = a^2\cos\varphi|\cos\varphi| \tag{4.7.22}$$

$$\dot{x}|\dot{x}| = \frac{\partial x}{\partial\varphi}\frac{\partial\varphi}{\partial T_0} = -a^2\sin\varphi|\sin\varphi| \tag{4.7.23}$$

The curves of $x|x| \sim \varphi$ and $\dot{x}|\dot{x}| \sim \varphi$ can be sketched from (4.7.22) and (4.7.23) as shown in Fig. 4.7a. It can be seen that $x|x|$ is the even function of φ, and $\dot{x}|\dot{x}|$ is the odd function of φ.

The frequency–response equation can be obtained by (4.7.19) and (4.7.20):

$$(\sigma a - \frac{4\alpha}{3\pi}a^2)^2 + (\mu_1 a + \frac{4\mu_2}{3\pi}a^2)^2 = k^2 \tag{4.7.24}$$

We can solve for σ in terms of a from (4.7.24)

$$\sigma = \frac{4\alpha}{3\pi}a \pm \frac{1}{a}\sqrt{k^2 - (\mu_1 a + \frac{4\mu_2}{3\pi}a^2)^2} \tag{4.7.25}$$

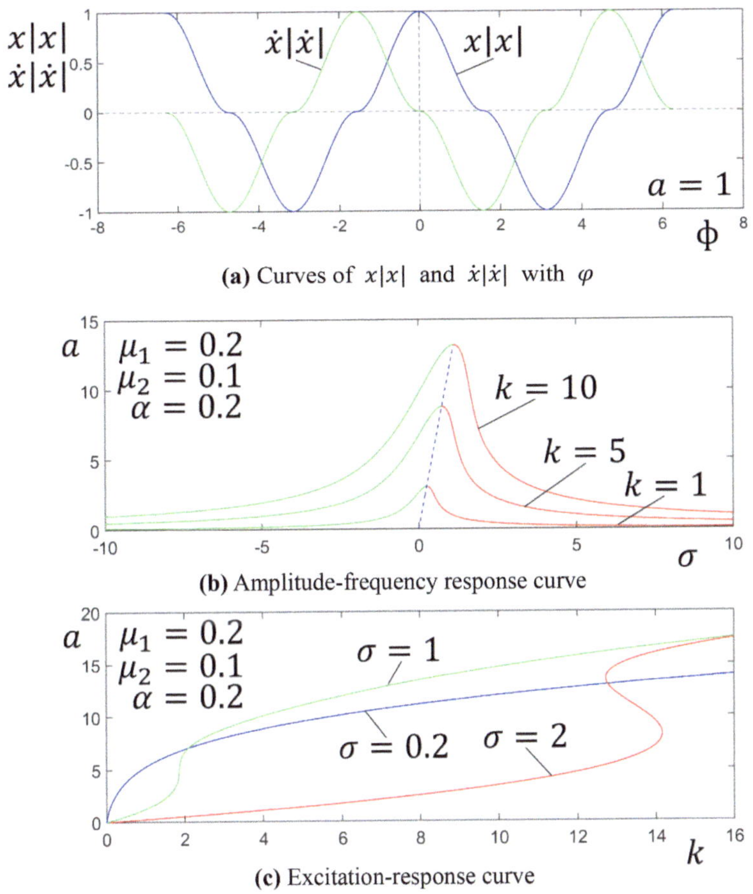

Fig. 4.7 **a** Curves of $x|x|$ and $\dot{x}|\dot{x}|$ with φ, **b** Frequency-response curves, and **c** Excitation-response curves for Exercise 4.7

The excitation-response equation is

$$k = a\sqrt{(\sigma - \frac{4\alpha}{3\pi}a)^2 + (\mu_1 + \frac{4\mu_2}{3\pi}a)^2} \tag{4.7.26}$$

From this, the frequency–response curve of the system $a \sim \sigma$ and the excitation-response curve $a \sim k$ can be drawn as shown in Fig. 4.7b, c. It can be seen that the frequency–response curve of the system has no amplitude jump phenomenon, while the excitation-response curve of the system has amplitude jump phenomenon under some conditions.

It can be seen that there is a maximum value for k in the frequency–response curve of the system. For a given $k > 0$, we can obtain from (4.7.25) that

$$k \geq \mu_1 a + \frac{4\mu_2}{3\pi}a^2 \quad \Rightarrow \quad \frac{4\mu_2}{3\pi}a^2 + \mu_1 a - k \leq 0 \tag{4.7.27}$$

Therefore

$$a \leq \frac{3\pi}{8\mu_2}\left[-\mu_1 + \sqrt{\mu_1^2 + \frac{16k\mu_2}{3\pi}}\right] \tag{4.7.28}$$

As a result, the maximum value of a, a_{max}, is

$$a_{max} = \frac{3\pi\mu_1}{8\mu_2}\left[-1 + \left(1 + \frac{16k\mu_2}{3\pi\mu_1^2}\right)^{1/2}\right] \tag{4.7.29}$$

(d) The effect of the nonlinear spring force $x|x|$ should appear in (4.7.10), (4.7.13) and (4.7.15) (terms containing α), which finally enters into the integral in Eq. (4.7.15). Since the quadrature function value in the following integral

$$\int_0^{2\pi} |a\cos\theta| a\cos\theta e^{-i\theta} d\theta = a^2 \int_0^{2\pi} |\cos\theta| \cos^2\theta d\theta$$

$$= 2a^2 \int_{-\pi/2}^{\pi/2} \cos^3\theta d\theta$$

is greater than zero over the whole period, the integral value is nonzero, so the effect of $x|x|$ can be determined by simply expanding to the first order. If the nonlinear spring force is x^2, then the integral of the above equation becomes

$$\int_0^{2\pi} a^2\cos^2\theta e^{-i\theta} d\theta = a^2 \int_0^{2\pi} \cos^3\theta d\theta = 0$$

When it is expanded to the first order, the effect of the nonlinear spring force x^2 cannot be included, and the second order expansion should be taken into account.

4.8 Exercise 4.8 (Primary, Superharmonic, and Subharmonic Resonances of Cubic Nonlinear and Cubic Damped Systems)

Solution: **(a)** Taylor expansion of $\sin\theta$ is conducted in the original equation near $\theta = 0$, retaining to $O(\theta^3)$:

$$\ddot{\theta} + \theta - \frac{1}{6}\theta^3 + 2\mu\theta^2\dot{\theta} = K\cos\Omega t \tag{4.8.1}$$

We use the dimensionless small parameter ε to measure the magnitude of θ, therefore, setting

$$\theta = \varepsilon u \tag{4.8.2}$$

Equation (4.8.1) becomes

$$\ddot{u} + u - \frac{1}{6}\varepsilon^2 u^3 + 2\varepsilon^2\mu u^2\dot{u} = \frac{K}{\varepsilon}\cos\Omega t \tag{4.8.3}$$

We express the solution in terms of different time scales as

$$u(t;\varepsilon) = u_0(T_0, T_1, T_2) + \varepsilon u_1(T_0, T_1, T_2) + \varepsilon^2 u_2(T_0, T_1, T_2) + \cdots \tag{4.8.4}$$

In order to keep all damping, nonlinear and excitation terms appearing simultaneously, we set

$$K = 2\varepsilon^3 k \tag{4.8.5}$$

Substituting (4.8.4) and (4.8.5) into (4.8.3) and retaining to $O(\varepsilon^2)$ yields

$$\begin{aligned}
0 &= \ddot{u} + u + \frac{1}{6}\varepsilon^2 u^3 - 2\varepsilon^2\mu u^2\dot{u} - 2\varepsilon^2 k\cos\Omega t \\
&= \left[D_0^2 + 2\varepsilon D_0 D_1 + \varepsilon^2\left(D_1^2 + 2D_0 D_2\right)\right]\left(u_0 + \varepsilon u_1 + \varepsilon^2 u_2\right) \\
&\quad + u_0 + \varepsilon u_1 + \varepsilon^2 u_2 + \frac{1}{6}\varepsilon^2(u_0 + \varepsilon u_1 + \varepsilon^2 u_2)^3 \\
&\quad - 2\varepsilon^2\mu(u_0 + \varepsilon u_1 + \varepsilon^2 u_2)^2(D_0 + \varepsilon D_1)\left(u_0 + \varepsilon u_1 + \varepsilon^2 u_2\right) \\
&\quad - 2\varepsilon^2 k\cos\Omega t + \cdots \\
&= D_0^2 u_0 + \varepsilon D_0^2 u_1 + \varepsilon^2 D_0^2 u_2 + 2\varepsilon D_0 D_1 u_0 + 2\varepsilon^2 D_0 D_1 u_1 \\
&\quad + \varepsilon^2\left(D_1^2 + 2D_0 D_2\right)u_0 + u_0 + \varepsilon u_1 + \varepsilon^2 u_2 + \frac{1}{6}\varepsilon^2 u_0^3 \\
&\quad - 2\varepsilon^2\mu u_0^2 D_0 u_0 - 2\varepsilon^2 k\cos\Omega t + \cdots \\
&= D_0^2 u_0 + u_0 + \varepsilon\left(D_0^2 u_1 + u_1 + 2D_0 D_1 u_0\right) \\
&\quad + \varepsilon^2[D_0^2 u_2 + u_2 + 2D_0 D_1 u_1 + \left(D_1^2 + 2D_0 D_2\right)u_0 + \frac{1}{6}u_0^3 \\
&\quad - 2\mu u_0^2 D_0 u_0 - 2k\cos\Omega t] + \cdots
\end{aligned} \tag{4.8.6}$$

Equating the coefficient of the like power of ε yields

$$D_0^2 u_0 + u_0 = 0 \tag{4.8.7}$$

$$D_0^2u_1 + u_1 = -2D_0D_1u_0 \tag{4.8.8}$$

$$\begin{aligned} D_0^2u_2 + u_2 &= -2D_0D_1u_1 - \left(D_1^2 + 2D_0D_2\right)u_0 - \frac{1}{6}u_0^3 \\ &+ 2\mu u_0^2D_0u_0 + 2k\cos\Omega t \end{aligned} \tag{4.8.9}$$

The general solution of (4.8.7) is

$$u_0 = Ae^{iT_0} + \overline{A}e^{-iT_0} \tag{4.8.10}$$

where $A = A(T_1, T_2)$. Substitute (4.8.10) into (4.8.8), we obtain

$$D_0^2u_1 + u_1 = -2D_0D_1u_0 = 0 \tag{4.8.11}$$

Therefore,

$$u_1 = 0, \quad A = A(T_2) \tag{4.8.12}$$

Substituting (4.8.10), (4.8.12) into (4.8.9) yields

$$\begin{aligned} D_0^2u_2 + u_2 &= -2D_0D_2u_0 + \frac{1}{6}u_0^3 - 2\mu u_0^2D_0u_0 + 2k\cos\Omega t \\ &= -2iD_2Ae^{iT_0} + \frac{1}{2}A^2\overline{A}e^{iT_0} - 2i\mu A^2\overline{A}e^{iT_0} \\ &+ ke^{i\Omega T_0} + cc + NST \end{aligned} \tag{4.8.13}$$

When $\Omega \approx 1$, we denote Ω as

$$\Omega = 1 + \hat{\sigma} \tag{4.8.14}$$

where $\hat{\sigma}$ is a small quantity. Therefore, the Eq. (4.8.13) becomes

$$\begin{aligned} D_0^2u_2 + u_2 &= -2iD_2Ae^{iT_0} + \frac{1}{2}A^2\overline{A}e^{iT_0} - 2i\mu A^2\overline{A}e^{iT_0} \\ &+ ke^{i\hat{\sigma}T_0}e^{iT_0} + cc + NST \end{aligned} \tag{4.8.15}$$

In order to eliminate secular terms, one must have

$$-2iD_2A + \frac{1}{2}A^2\overline{A} - 2i\mu A^2\overline{A}e^{iT_0} + ke^{i\hat{\sigma}T_0} = 0 \tag{4.8.16}$$

Since $A = A(T_2)$, $ke^{i\hat{\sigma}T_0}$ must be a function of T_2, so there must be $\hat{\sigma} = \varepsilon^2\sigma$, i.e.,

$$\Omega = 1 + \varepsilon^2\sigma \tag{4.8.17}$$

From this, the Eq. (4.8.16) becomes

$$-2iD_2A + \frac{1}{2}A^2\overline{A} - 2i\mu A^2\overline{A}e^{iT_0} + ke^{i\sigma T_2} = 0 \tag{4.8.18}$$

Let

$$A = \frac{1}{2}ae^{i\beta} \tag{4.8.19}$$

Then

$$ia' - a\beta' - \frac{1}{16}a^3 + \frac{1}{4}i\mu a^3 - ke^{i(\sigma T_2 - \beta)} = 0 \tag{4.8.20}$$

The prime denotes the derivative with respect to T_2. Separating the real and imaginary parts of the above equation yields

$$a' = -\frac{1}{4}\mu a^3 + k\sin(\sigma T_2 - \beta), \quad -a\beta' = \frac{1}{16}a^3 + k\cos(\sigma T_2 - \beta)$$

Let

$$\gamma = (\sigma T_2 - \beta) \tag{4.8.21}$$

Then

$$a' = -\frac{1}{4}\mu a^3 + k\sin\gamma \tag{4.8.22}$$

$$\gamma' = \sigma + \frac{1}{16}a^2 + \frac{k}{a}\cos\gamma \tag{4.8.23}$$

Therefore, the first order approximate solution of the original equation is

$$\theta = \varepsilon u_0 = \varepsilon a\cos[T_0 + \beta(T_2)] + O(\varepsilon^3) \tag{4.8.24}$$

By the Eqs. (4.8.22) and (4.8.23), the frequency–response equation is given by

$$\left(\frac{a^3}{16} + a\sigma\right)^2 + \frac{\mu^2a^6}{16} = k^2 \tag{4.8.25}$$

In order to plot the frequency–response curve of the system $a \sim \sigma$, we can solve for σ and k, respectively, from Eq. (4.8.25), i.e.,

$$\sigma = -\frac{a^2}{16} \pm \frac{1}{a}\sqrt{k^2 - \frac{\mu^2a^6}{16}} \tag{4.8.26}$$

$$k = \sqrt{\left(\frac{a^3}{16} + a\sigma\right)^2 + \frac{\mu^2 a^6}{16}} \tag{4.8.27}$$

From this, the frequency–response curve of the system $a \sim \sigma$ and the excitation-response curve of the system $a \sim k$ can be sketched as shown in Fig. 4.8a, b.

In order to obtain a real number solution for σ from (4.8.26), we need

$$k \geq \frac{\mu a^3}{4} \quad \Rightarrow \quad a^3 \leq \frac{4k}{\mu}$$

Therefore

$$a_{\max} = (4k/\mu)^{1/3}$$

The results in those figures show that there are no amplitude jumps in $a \sim \sigma$ and $a \sim k$ curves.

(b) Let

$$K = \varepsilon k \tag{4.8.28}$$

Equation (4.8.3) change into

$$\ddot{u} + u - \frac{1}{6}\varepsilon^2 u^3 + 2\varepsilon^2 \mu u^2 \dot{u} = k \cos \Omega t \tag{4.8.29}$$

The solution of the Eq. (4.8.29) is given as

$$u(t;\varepsilon) = u_0(T_0, T_1) + \varepsilon u_1(T_0, T_1) + \varepsilon^2 u_2(T_0, T_1) + \cdots \tag{4.8.30}$$

Substituting (4.8.30) into (4.8.29) and retaining to $O(\varepsilon^2)$ yields

$$\begin{aligned}
0 &= \ddot{u} + u + \frac{1}{6}\varepsilon^2 u^3 - 2\varepsilon^2 \mu u^2 \dot{u} - k \cos \Omega T_0 \\
&= \left[D_0^2 + 2\varepsilon D_0 D_1 + \varepsilon^2\left(D_1^2 + 2D_0 D_2\right)\right]\left(u_0 + \varepsilon u_1 + \varepsilon^2 u_2\right) \\
&\quad + u_0 + \varepsilon u_1 + \varepsilon^2 u_2 + \frac{1}{6}\varepsilon^2 (u_0 + \varepsilon u_1 + \varepsilon^2 u_2)^3 \\
&\quad - 2\varepsilon^2 \mu (u_0 + \varepsilon u_1 + \varepsilon^2 u_2)^2 (D_0 + \varepsilon D_1)\left(u_0 + \varepsilon u_1 + \varepsilon^2 u_2\right) - k \cos \Omega T_0 + \cdots \\
&= D_0^2 u_0 + u_0 - k \cos \Omega t + \varepsilon\left(D_0^2 u_1 + u_1 + 2D_0 D_1 u_0\right) \\
&\quad + \varepsilon^2\left[D_0^2 u_2 + u_2 + 2D_0 D_1 u_1 + \left(D_1^2 + 2D_0 D_2\right)u_0 + \frac{1}{6}u_0^3 - 2\mu u_0^2 D_0 u_0\right] + \cdots
\end{aligned} \tag{4.8.31}$$

Let the coefficient of the like power of ε be zero, we get

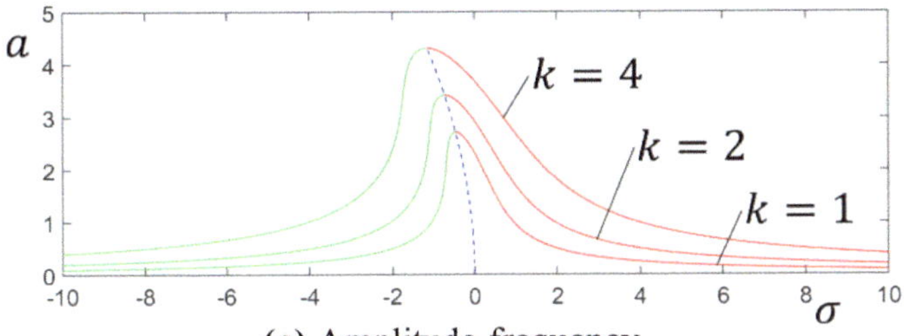

(a) Amplitude-frequency

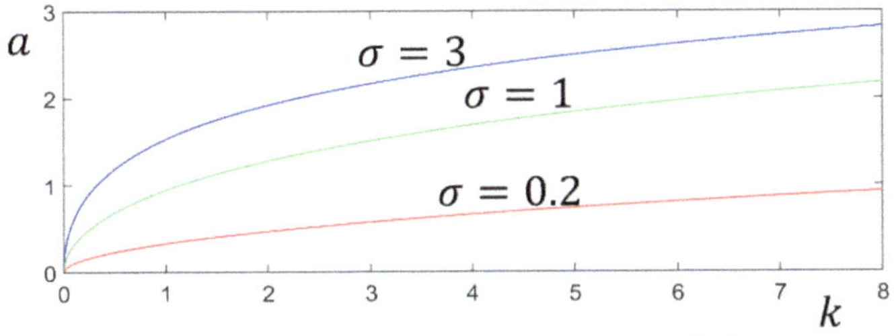

(b) Excitation-response curve ($\mu = 0.2$)

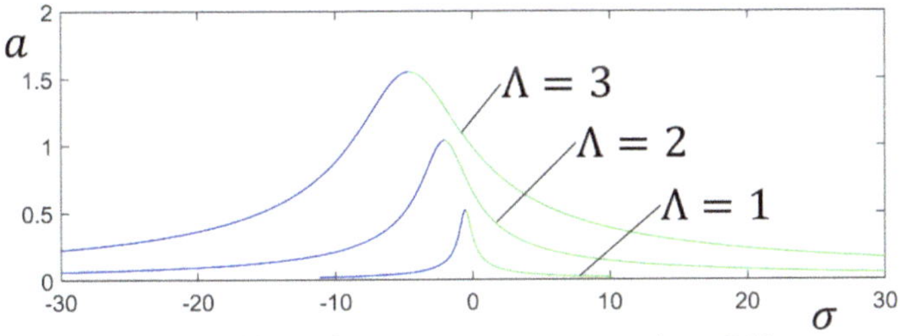

(c) Amplitude-frequency response curve ($\mu = 0.2$)

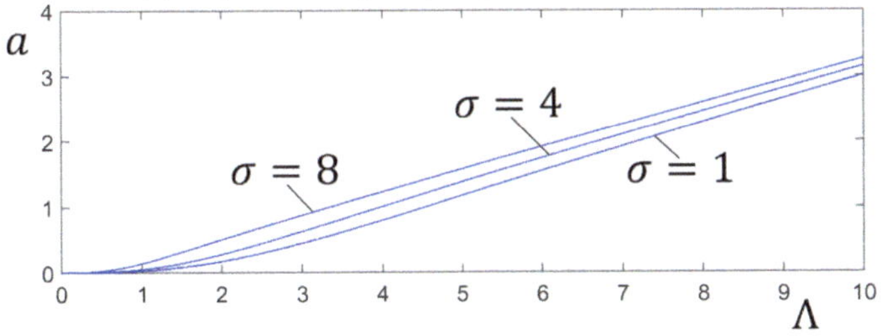

(d) Excitation-response curve ($\mu = 0.2$)

Fig. 4.8 **a** Frequency-response, and **b** excitation-response curves for primary resonance ($\mu = 0.2$); **c** Frequency-response, and **d** excitation-response curves for superharmonic resonance ($\mu = 0.2$); **e** frequency-response, and **f** excitation-response curves for subharmonic resonance ($\mu = \frac{1}{20}$) in Exercise 4.8

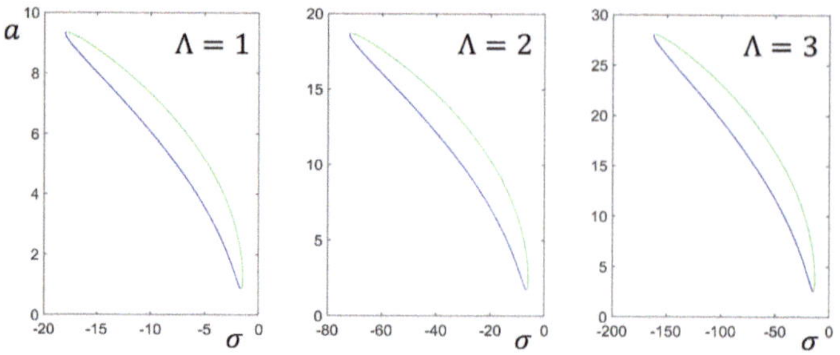

(e) Amplitude-frequency response curve ($\mu = \frac{1}{20}$)

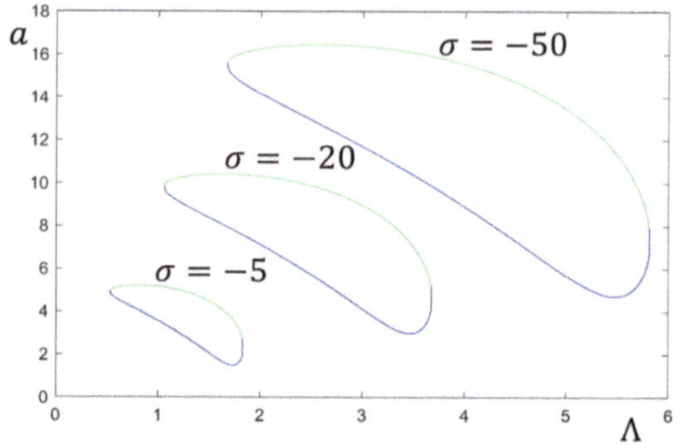

(f) Excitation-response curve ($\mu = \frac{1}{20}$)

Fig. 4.8 (continued)

$$D_0^2 u_0 + u_0 = k \cos \Omega T_0 \tag{4.8.32}$$

$$D_0^2 u_1 + u_1 = -2 D_0 D_1 u_0 \tag{4.8.33}$$

$$\begin{aligned} D_0^2 u_2 + u_2 &= -2 D_0 D_1 u_1 - \left(D_1^2 + 2 D_0 D_2\right) u_0 \\ &\quad - \frac{1}{6} u_0^3 + 2 \mu u_0^2 D_0 u_0 \end{aligned} \tag{4.8.34}$$

The solution for (4.8.32) is

$$u_0 = A e^{i T_0} + \Lambda e^{i \Omega T_0} + cc, \quad \text{where } \Lambda = \frac{1}{2} k \left(1 - \Omega^2\right)^{-1} \tag{4.8.35}$$

where $A = A(T_1, T_2)$. Substituting (4.8.35) into (4.8.33) yields

$$D_0^2 u_1 + u_1 = -2 D_0 D_1 u_0 = 0 \tag{4.8.36}$$

therefore

$$u_1 = 0, \quad A = A(T_2) \tag{4.8.37}$$

Substituting (4.8.35) and (4.8.36) into (4.8.34) yields

$$\begin{aligned} D_0^2 u_2 + u_2 &= -2D_0 D_2 u_0 + \frac{1}{6} u_0^3 - 2\mu u_0^2 D_0 u_0 \\ &= -2iD_2 A e^{iT_0} + \frac{1}{2} A^2 \overline{A} e^{iT_0} + \Lambda^2 A e^{iT_0} + \frac{1}{6} \Lambda^3 e^{3i\Omega T_0} \\ &+ \frac{1}{2} \Lambda \overline{A}^2 e^{i(\Omega-2)T_0} - 2i\mu A^2 \overline{A} e^{iT_0} \\ &- 4i\mu \Lambda^2 A e^{iT_0} - i2\mu\Omega\Lambda^3 e^{3i\Omega T_0} - 2i\mu\Omega\Lambda\overline{A}^2 e^{i(\Omega-2)T_0} \\ &+ 4i\mu\Lambda\overline{A}^2 e^{i(\Omega-2)T_0} + cc + NST \end{aligned} \tag{4.8.38}$$

When $\Omega \approx 1/3$, we denote Ω by the detuning parameter σ in the following equation:

$$\Omega = \frac{1}{3}\left(1 + \varepsilon^2 \sigma\right) \tag{4.8.39}$$

$$\Lambda = \frac{1}{2} k (1 - \Omega^2)^{-1} = \frac{9}{16} k \tag{4.8.40}$$

In order to eliminate secular terms in (4.8.38), we need

$$\begin{aligned} 2iD_2 A - \frac{1}{2} A^2 \overline{A} - \Lambda^2 A + 2i\mu A^2 \overline{A} + 4i\mu \Lambda^2 A \\ - \frac{1}{6} \Lambda^3 e^{i(\sigma T_2 - \beta)} + i2\mu\Omega\Lambda^3 e^{i(\sigma T_2 - \beta)} = 0 \end{aligned} \tag{4.8.41}$$

Let

$$A = \frac{1}{2} a e^{i\beta} \tag{4.8.42}$$

Then

$$\begin{aligned} ia' - a\beta' - \frac{1}{16} a^3 - \frac{1}{2} \Lambda^2 a + \frac{1}{4} i\mu a^3 + 2i\mu\Lambda^2 a \\ - \frac{1}{6} \Lambda^3 \cos(\sigma T_2 - \beta) - \frac{1}{6} i\Lambda^3 \sin(\sigma T_2 - \beta) \\ - \frac{2}{3} \mu\Lambda^3 \sin(\sigma T_2 - \beta) + \frac{2}{3} i\mu\Lambda^3 \cos(\sigma T_2 - \beta) = 0 \end{aligned} \tag{4.8.43}$$

The prime denotes the derivative with respect to T_2. Separating the real and imaginary parts of the above equation yields

$$\begin{aligned} a' &= -2\mu\left(\Lambda^2 + \frac{1}{8}a^2\right)a + \frac{1}{6}\Lambda^3\sin(\sigma T_2 - \beta) \\ &\quad - \frac{2}{3}\mu\Lambda^3\cos(\sigma T_2 - \beta) \\ -a\beta' &= \frac{1}{16}a^3 + \frac{1}{2}\Lambda^2 a + \frac{1}{6}\Lambda^3\cos(\sigma T_2 - \beta) \\ &\quad + \frac{2}{3}\mu\Lambda^3\sin(\sigma T_2 - \beta) \end{aligned}$$

Let

$$\gamma = \sigma T_2 - \beta \tag{4.8.44}$$

Then

$$a' = -2\mu\left(\Lambda^2 + \frac{1}{8}a^2\right)a - \Lambda^3\left(\frac{2}{3}\mu\cos\gamma - \frac{1}{6}\sin\gamma\right) \tag{4.8.45}$$

$$\gamma' = \sigma + \frac{1}{2}\left(\Lambda^2 + \frac{1}{8}a^2\right) + \Lambda^3 a^{-1}\left(\frac{2}{3}\mu\sin\gamma + \frac{1}{6}\cos\gamma\right) \tag{4.8.46}$$

Therefore, the first order approximate solution of the original equation is

$$\begin{aligned} \theta = \varepsilon u_0 &= \varepsilon a(T_2)\cos[T_0 + \beta(T_2)] \\ &\quad + \frac{9}{8}K\cos(\Omega T_0) + O(\varepsilon^3) \end{aligned} \tag{4.8.47}$$

The frequency–response equation can be obtained from Eqs. (4.8.45) and (4.8.46)

$$[\sigma a + \frac{1}{2}\left(\Lambda^2 + \frac{1}{8}a^2\right)a]^2 + [2\mu\left(\Lambda^2 + \frac{1}{8}a^2\right)a]^2 = \Lambda^6\left(\frac{4}{9}\mu^2 + \frac{1}{36}\right) \tag{4.8.48}$$

To plot the frequency–response curve of the system $a \sim \sigma$ (Fig. 4.8c) for a given Λ, we solve for σ in terms of a from (4.8.48)

$$\sigma = -\frac{1}{2}\left(\Lambda^2 + \frac{1}{8}a^2\right) \pm \frac{1}{a}\sqrt{\Lambda^6\left(\frac{4}{9}\mu^2 + \frac{1}{36}\right) - \left[2\mu\left(\Lambda^2 + \frac{1}{8}a^2\right)a\right]^2} \tag{4.8.49}$$

Numerical computation can be conducted to sketch $a \sim \Lambda$ curves (Fig. 4.8d) for a given σ. There is no amplitude jump in the curves $a \sim \sigma$ and $a \sim \Lambda$.

(c) When $\Omega \approx 3$, the detuning parameter σ is adopted to quantitatively describes the nearness of Ω to 3:

$$\Omega = 3 + \varepsilon^2\sigma \tag{4.8.50}$$

$$\Lambda = \frac{1}{2}k(1-\Omega^2)^{-1} = -\frac{1}{16}k \tag{4.8.51}$$

In order to eliminate secular terms from (4.8.38), it is necessary to have

$$\begin{aligned} &2iD_2A - \frac{1}{2}A^2\bar{A} - \Lambda^2 A + 2i\mu A^2\bar{A} + 4i\mu\Lambda^2 A \\ &- \frac{1}{2}\Lambda\bar{A}^2 e^{i\sigma T_2} + 2i\mu\Omega\Lambda\bar{A}^2 e^{i\sigma T_2} - 4i\mu\Lambda\bar{A}^2 e^{i\sigma T_2} = 0 \end{aligned} \tag{4.8.52}$$

Let

$$A = \frac{1}{2}ae^{i\beta} \tag{4.8.53}$$

then

$$\begin{aligned} &ia' - a\beta' - \frac{1}{16}a^3 - \frac{1}{2}\Lambda^2 a + \frac{1}{4}i\mu a^3 \\ &+ 2i\mu\Lambda^2 a - \frac{1}{8}\Lambda a^2\cos(\sigma T_2 - 3\beta) \\ &- \frac{1}{8}i\Lambda a^2\sin(\sigma T_2 - 3\beta) + \frac{1}{2}i\mu\Lambda a^2\cos(\sigma T_2 - 3\beta) \\ &- \frac{1}{2}\mu\Lambda a^2\sin(\sigma T_2 - 3\beta) = 0 \end{aligned} \tag{4.8.54}$$

The prime denotes the derivative with respect to T_2. Separating the real and imaginary parts of the above equation yields

$$\begin{aligned} a' &= -2\mu\left(\Lambda^2 + \tfrac{1}{8}a^2\right)a + \Lambda a^2\left[\tfrac{1}{8}\sin(\sigma T_2 - 3\beta) - \tfrac{1}{2}\mu\cos(\sigma T_2 - 3\beta)\right] \\ -\beta' &= \tfrac{1}{2}\left(\Lambda^2 + \tfrac{1}{8}a^2\right) + \Lambda a\left[\tfrac{1}{8}\cos(\sigma T_2 - 3\beta) + \tfrac{1}{2}\mu\sin(\sigma T_2 - 3\beta)\right] \end{aligned} \tag{4.8.55}$$

Let

$$\gamma = \sigma T_2 - 3\beta \tag{4.8.56}$$

and change Λ to a positive value, i.e., change the Λ in (4.8.55) to $-\Lambda$. Let

$$\Lambda = \frac{1}{16}k \tag{4.8.57}$$

Then

$$a' = -2\mu\left(\Lambda^2 + \frac{1}{8}a^2\right)a - \Lambda a^2\left(\frac{1}{8}\sin\gamma - \frac{1}{2}\mu\cos\gamma\right) \tag{4.8.58}$$

$$\gamma' = \sigma + \frac{3}{2}\left(\Lambda^2 + \frac{1}{8}a^2\right) - 3\Lambda a\left(\frac{1}{8}\cos\gamma + \frac{1}{2}\mu\sin\gamma\right) \tag{4.8.59}$$

So the first order approximate solution of the original equation is

$$\theta = \varepsilon u_0 = \varepsilon a(T_2)\cos[T_0 + \beta(T_2)] - \frac{1}{8}K\cos(\Omega T_0) + O(\varepsilon^3) \tag{4.8.60}$$

The frequency–response equation can be obtained from (4.8.58) and (4.8.59)

$$\{\frac{1}{3}[\sigma + \frac{3}{2}(\Lambda^2 + \frac{1}{8}a^2)]\}^2 + [2\mu(\Lambda^2 + \frac{1}{8}a^2)]^2 = \Lambda^2 a^2(\frac{1}{64} + \frac{1}{4}\mu^2) \tag{4.8.61}$$

In order to obtain the nonzero a^2 from the above equation, one must have

$$\Lambda^2 a^2\left(\frac{1}{64} + \frac{1}{4}\mu^2\right) - \left[2\mu\left(\Lambda^2 + \frac{1}{8}a^2\right)\right]^2 > 0$$

i.e.,

$$4\mu^2 a^4 + \left(48\mu^2\Lambda^2 - \Lambda^2\right)a^2 + 256\mu^2\Lambda^4 < 0 \tag{4.8.62}$$

therefore,

$$0 < \mu^2 < \frac{1}{112} \quad \Rightarrow \quad 0 < \mu < \frac{1}{\sqrt{112}} \tag{4.8.63}$$

Once (4.8.63) is satisfied, we can solve for σ from (4.8.61)

$$\sigma = -\frac{3}{2}(\Lambda^2 + \frac{1}{8}a^2) \pm 3\sqrt{\Lambda^2 a^2(\frac{1}{64} + \frac{1}{4}\mu^2) - [2\mu(\Lambda^2 + \frac{1}{8}a^2)]^2} \tag{4.8.64}$$

For the determined value of Λ, the frequency–response curve of the system of the system $a \sim \sigma$ (Fig. 4.8e) can be plotted.

Numerical computation can be conducted to sketch $a \sim \Lambda$ curves (Fig. 4.8f) for a given σ from (4.8.61).

There are no amplitude jumps in $a{\sim}\sigma$ and $a{\sim}\Lambda$ curves; however, there are multiple values of amplitude in both curves. Which value of the steady state amplitude can be realized is determined by the initial conditions and the stability of the amplitude, which needs to be examined from Eqs. (4.8.58) and (4.8.59).

4.9 Exercise 4.9 (Method of Averaging for Analyzing Primary Resonances of Alternating Damped Systems)

Solution: To find the primary resonance response when $\Omega \approx 1$, let

$$K = 2\varepsilon k, \quad \Omega = 1 + \varepsilon\sigma \tag{4.9.1}$$

where σ is a detuning parameter. The differential equation of motion of the system becomes

$$\ddot{u} + u = -2\varepsilon\mu \sin \dot{u} + 2\varepsilon k \cos(1 + \varepsilon\sigma)t \triangleq \varepsilon f(\dot{u}, t) \tag{4.9.2}$$

$$f(\dot{u}, t) = -2\mu \sin u + 2k \cos(1 + \varepsilon\sigma)t \tag{4.9.3}$$

We try to solve (4.9.2) by the method of averaging.

We suppose that both u and $\dot{u}$ in (4.9.2) have the same forms as the case when $\varepsilon = 0$, i.e.,

$$u = a\cos(t + \beta) \triangleq a\cos\varphi, \quad \varphi = t + \beta \tag{4.9.4}$$

And

$$\dot{u} = -a\sin(t + \beta) \triangleq -a\sin\varphi \tag{4.9.5}$$

where a and β are functions of t.

Differentiating (4.9.4) and (4.9.5) with respect to t, we have

$$\dot{u} = \dot{a}\cos\varphi - a\sin\varphi - a\dot{\beta}\sin\varphi \tag{4.9.6}$$

$$\ddot{u} = -\dot{a}\sin\varphi - a\cos\varphi - a\dot{\beta}\cos\varphi \tag{4.9.7}$$

From (4.9.5) and (4.9.6), we have

$$\dot{a}\cos\varphi - a\dot{\beta}\sin\varphi = 0 \tag{4.9.8}$$

Substituting (4.9.7), (4.9.4) and (4.9.5) into (4.9.2), we get

$$\dot{a}\sin\varphi + a\dot{\beta}\cos\varphi = -\varepsilon f(-a\sin\varphi, t) \tag{4.9.9}$$

Solving (4.9.8) and (4.9.9) for $\dot{a}$ and $\dot{\beta}$, we obtain.

$$\dot{a} = -\varepsilon f(-a\sin\varphi)\sin\varphi, \quad \dot{\beta} = -\frac{\varepsilon}{a} f(-a\sin\varphi)\cos\varphi \tag{4.9.10}$$

Substituting (4.9.3) into the above two equations yields

$$\begin{aligned}\dot{a} &= -\varepsilon[-2\mu\sin(-a\sin\phi) + 2k\cos(1+\varepsilon\sigma)t]\sin\phi \\ &= 2\varepsilon\mu\sin(-a\sin\phi)\sin\phi - 2\varepsilon k[\cos(1+\varepsilon\sigma)t]\sin\phi \\ &= 2\varepsilon\mu\sin(-a\sin\phi)\sin\phi - 2\varepsilon k\cos(\gamma+\phi)\sin\phi \\ &= 2\varepsilon\mu\sin(-a\sin\phi)\sin\phi - 2\varepsilon k(\cos\gamma\cos\phi - \sin\gamma\sin\phi)\sin\phi\end{aligned}$$

$$\begin{aligned}\dot{\beta} &= -\frac{\varepsilon}{a}[-2\mu\sin(-a\sin\phi) + 2k\cos(1+\varepsilon\sigma)t]\cos\phi \\ &= \frac{2\varepsilon\mu}{a}\sin(-a\sin\phi)\cos\phi - \frac{2\varepsilon k}{a}[\cos(1+\varepsilon\sigma)t]\cos\phi \\ &= \frac{2\varepsilon\mu}{a}\sin(-a\sin\phi)\cos\phi - \frac{2\varepsilon k}{a}\cos(\gamma+\phi)\cos\phi \\ &= \frac{2\varepsilon\mu}{a}\sin(-a\sin\phi)\cos\phi - \frac{2\varepsilon k}{a}(\cos\gamma\cos\phi - \sin\gamma\sin\phi)\cos\phi\end{aligned}$$

i.e.,

$$\dot{a} = 2\varepsilon\mu\sin(-a\sin\varphi)\sin\varphi - 2\varepsilon k(\cos\gamma\cos\varphi - \sin\gamma\sin\varphi)\sin\varphi \tag{4.9.11}$$

$$\begin{aligned}\dot{\gamma} = \varepsilon\sigma &- \frac{2\varepsilon\mu}{a}\sin(-a\sin\varphi)\cos\varphi \\ &+ \frac{2\varepsilon k}{a}(\cos\gamma\cos\varphi - \sin\gamma\sin\varphi)\cos\varphi\end{aligned} \tag{4.9.12}$$

Where

$$\gamma = \varepsilon\sigma t - \beta \tag{4.9.13}$$

Averaging these equations over the period 2π, and considering $a, \gamma, \dot{a}$ and $\dot{\gamma}$ to be constants while performing the averaging, we obtain the following equations describing the slow variations of a and γ:

$$\begin{aligned}\dot{a} &= \frac{1}{2\pi}\int_0^{2\pi} 2\varepsilon\mu\sin(-a\sin\phi)\sin\phi - 2\varepsilon k(\cos\gamma\cos\phi - \sin\gamma\sin\phi)\sin\phi d\phi \\ &= -\frac{\varepsilon\mu}{\pi}\int_0^{2\pi} 2\sum_{n=0}^{\infty} J_{2n+1}(a)\sin[(2n+1)\phi]\sin\phi d\phi + \frac{\varepsilon k}{\pi}\sin\gamma\int_0^{2\pi}\sin^2\phi d\phi \\ &= -\frac{2\varepsilon\mu}{\pi}J_1(a)\int_0^{2\pi}\sin^2\phi d\phi + \varepsilon k\sin\gamma\end{aligned}$$

$$= \varepsilon\left[-2\mu J_1(a) + \varepsilon k \sin\gamma\right]$$

$$\dot{\gamma} = \frac{1}{2\pi}\int_0^{2\pi} \varepsilon\sigma - \frac{2\varepsilon\mu}{a}\sin(-a\sin\varphi)\cos\varphi$$

$$+ \frac{2\varepsilon k}{a}(\cos\gamma\cos\varphi - \sin\gamma\sin\varphi)\cos\varphi]d\varphi$$

$$= \varepsilon\sigma + \frac{\varepsilon k}{\pi a}\cos\gamma\int_0^{2\pi}\cos^2\varphi d\varphi$$

$$= \varepsilon\left(\sigma + \frac{k}{a}\cos\gamma\right)$$

i.e.,

$$\dot{a} = \varepsilon\left[-2\mu J_1(a) + k\sin\gamma\right] \tag{4.9.14}$$

$$\dot{\gamma} = \varepsilon\left(\sigma + \frac{k}{a}\cos\gamma\right) \tag{4.9.15}$$

To accomplish the above integrals, we use the following equation:

$$\sin(a\sin\varphi) = 2\sum_{n=0}^{\infty} J_{2n+1}(a)\sin[(2n+1)\varphi] \tag{4.9.16}$$

The frequency–response equation can be obtained from (4.9.14) and (4.9.15):

$$4\mu^2 J_1^2(a) + \sigma^2 a^2 = k^2 \tag{4.9.17}$$

Solving for σ from the above equation, we have

$$\sigma = \pm\frac{1}{a}\sqrt{k^2 - 4\mu^2 J_1^2(a)} \tag{4.9.18}$$

The frequency–response curve of the system $a \sim \sigma$ (Fig. 4.9a) can be plotted from (4.9.18), where $J_1(a)$ can be calculated by the following integral equation

$$J_1(a) = \frac{1}{2\pi}\int_{-\pi}^{\pi}\cos(n\theta - a\sin\theta)d\theta \tag{4.9.19}$$

As can be seen, there are no amplitude jumps; when $\sigma = 0$, the amplitude a becomes infinite.

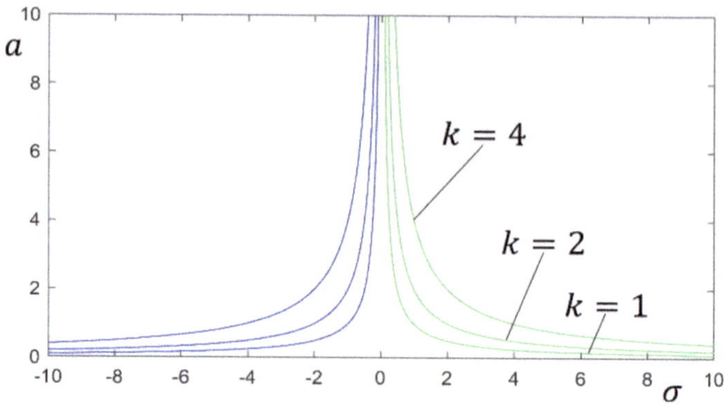

(a) Amplitude-frequency

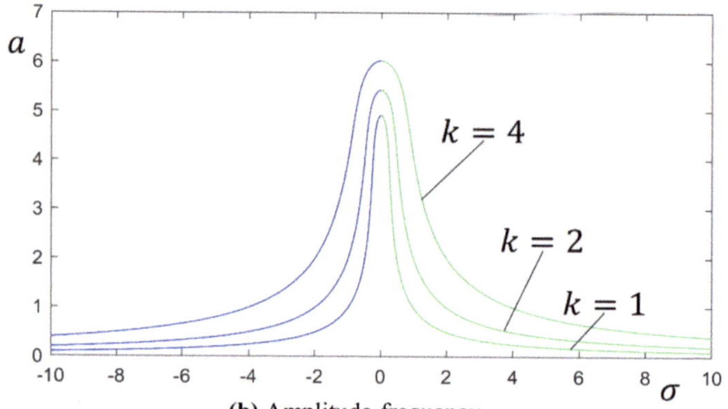

(b) Amplitude-frequency

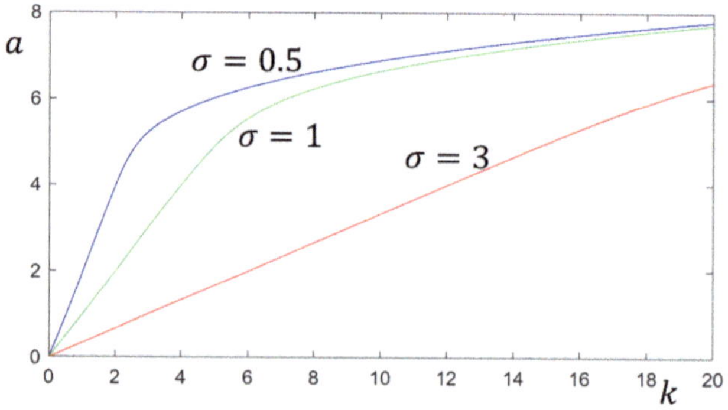

(c) Excitation-response curve

Fig. 4.9 **a** Frequency-response curves for general a, and **b** Frequency-response and **c** Excitation-response curves for a small and finite a in Exercise 4.9

When a is a small and finite, we can make Taylor expansion of $\sin(a \sin \varphi)$ in (4.9.11) and (4.9.12) and retain the first three terms, then (4.9.11) and (4.9.12) become:

$$\dot{a} = -2\varepsilon\mu\left[a\sin\phi - \frac{1}{6}a^3\sin^3\phi + \frac{1}{120}a^5\sin^5\phi\right]\sin\phi \\ - 2\varepsilon k(\cos\gamma\cos\phi - \sin\gamma\sin\phi)\sin\phi \tag{4.9.20}$$

$$\dot{\gamma} = \varepsilon\sigma + 2\varepsilon\mu a^{-1}\left[a\sin\phi - \frac{1}{6}a^3\sin^3\phi + \frac{1}{120}a^5\sin^5\phi\right]\cos\phi \\ + 2\varepsilon k a^{-1}(\cos\gamma\cos\phi - \sin\gamma\sin\phi)\cos\phi \tag{4.9.21}$$

Averaging (4.9.20) and (4.9.21) over one cycle yields

$$\dot{a} = -\varepsilon\left[\mu\left(a - \frac{a^3}{8} + \frac{a^5}{192}\right) + k\sin\gamma\right] \tag{4.9.22}$$

$$\dot{\gamma} = \varepsilon\left(\sigma + \frac{k}{a}\cos\gamma\right) \tag{4.9.23}$$

The corresponding frequency–response equation can be obtained from (4.9.22) and (4.9.23):

$$[\mu\left(a - \frac{1}{8}a^3 + \frac{1}{192}a^5\right)]^2 + \sigma^2 a^2 = k^2 \tag{4.9.24}$$

Therefore,

$$\sigma = \pm\frac{1}{a}\sqrt{k^2 - [\mu\left(a - \frac{1}{8}a^3 + \frac{1}{192}a^5\right)]^2} \tag{4.9.25}$$

$$k = \sqrt{[\mu\left(a - \frac{1}{8}a^3 + \frac{1}{192}a^5\right)]^2 + \sigma^2 a^2} \tag{4.9.26}$$

The frequency–response curve of the system $a \sim \sigma$ and the excitation-response curve of the system $a \sim k$ are sketched in Fig. 4.9b, c, respectively.

4.10 Exercise 4.10 (Primary, Superharmonic, and Subharmonic Resonances of Quadratic and Cubic Nonlinear Systems)

Solution: **(a)** Expanding the differential equation of the system around the center $u = 1$, we get

$$\ddot{u} + u + \frac{3}{2}u^2 + \frac{1}{2}u^3 = K\cos\Omega t \tag{4.10.1}$$

We use the small parameter ε to measure the amplitude of u. In addition, considering searching for the primary resonance response, we let

$$u = \varepsilon x, \quad K = 2\varepsilon^3 k \tag{4.10.2}$$

then Eq. (4.10.1) becomes

$$\ddot{x} + x + \frac{3}{2}\varepsilon x^2 + \frac{1}{2}\varepsilon^2 x^3 = 2\varepsilon^2 k\cos\Omega t \tag{4.10.3}$$

Let the solution of the equation be

$$x(t;\varepsilon) = x_0(T_0, T_1, T_2) + \varepsilon x_1(T_0, T_1, T_2) + \varepsilon^2 x_2(T_0, T_1, T_2) + \cdots \tag{4.10.4}$$

Substituting (4.10.4) into (4.10.3) and retaining to $O(\varepsilon^2)$ yields

$$\begin{aligned}
0 &= \ddot{x} + x + \frac{3}{2}\varepsilon x^2 + \frac{1}{2}\varepsilon^2 x^3 - 2\varepsilon^2 k\cos\Omega t \\
&= \left[D_0^2 + 2\varepsilon D_0 D_1 + \varepsilon^2\left(D_1^2 + 2D_0 D_2\right)\right]\left(x_0 + \varepsilon x_1 + \varepsilon^2 x_2\right) \\
&\quad + \left(x_0 + \varepsilon x_1 + \varepsilon^2 x_2\right) + \frac{3}{2}\varepsilon(x_0 + \varepsilon x_1 + \varepsilon^2 x_2)^2 \\
&\quad + \frac{1}{2}\varepsilon^2(x_0 + \varepsilon x_1 + \varepsilon^2 x_2)^3 - 2\varepsilon^2 k\cos\Omega t \\
&= D_0^2 x_0 + \varepsilon D_0^2 x_1 + \varepsilon^2 D_0^2 x_2 + 2\varepsilon D_0 D_1 x_0 + 2\varepsilon^2 D_0 D_1 x_1 \\
&\quad + \varepsilon^2\left(D_1^2 + 2D_0 D_2\right)x_0 + x_0 + \varepsilon x_1 + \varepsilon^2 x_2 \\
&\quad + \frac{3}{2}\varepsilon x_0^2 + 3\varepsilon^2 x_0 x_1 + \frac{1}{2}\varepsilon^2 x_0^3 - 2\varepsilon^2 k\cos\Omega t + \cdots \\
&= D_0^2 x_0 + x_0 + \varepsilon\left(D_0^2 x_1 + x_1 + 2D_0 D_1 x_0 + \frac{3}{2}x_0^2\right) \\
&\quad + \varepsilon^2[D_0^2 x_2 + x_2 + 2D_0 D_1 x_1 + \left(D_1^2 + 2D_0 D_2\right)x_0 \\
&\quad + 3x_0 x_1 + \frac{1}{2}x_0^3 - 2k\cos\Omega t] + \cdots
\end{aligned} \tag{4.10.5}$$

Let the coefficient of the like power of ε be zero, we get

$$D_0^2x_0 + x_0 = 0 \tag{4.10.6}$$

$$D_0^2x_1 + x_1 = -2D_0D_1x_0 - \frac{3}{2}x_0^2 \tag{4.10.7}$$

$$\begin{aligned} D_0^2x_2 + x_2 = &-2D_0D_1x_1 - \left(D_1^2 + 2D_0D_2\right)x_0 \\ &- 3x_0x_1 - \frac{1}{2}x_0^3 + 2k\cos\Omega t \end{aligned} \tag{4.10.8}$$

The solution of (4.10.6) is

$$x_0 = Ae^{iT_0} + \overline{A}e^{-iT_0} \tag{4.10.9}$$

where $A = A(T_1, T_2)$. Substitute (4.10.9) into (4.10.7), we get

$$D_0^2x_1 + x_1 = -2iD_1Ae^{iT_0} - 3A\overline{A} - \frac{3}{2}A^2e^{2iT_0} + cc \tag{4.10.10}$$

In order to eliminate secular terms, we need $D_1A = 0$, so $A = A(T_2)$. Therefore, the solution of (4.10.10) is

$$x_1 = -3A\overline{A} + \frac{1}{2}A^2e^{2iT_0} + \frac{1}{2}\overline{A}^2e^{-2iT_0} \tag{4.10.11}$$

Since the external excitation term appears in the control equation of x_2(4.10.8), we let

$$\Omega = 1 + \varepsilon^2\sigma \tag{4.10.12}$$

Substitute (4.10.9), (4.10.11) and (4.10.12) into (4.10.8), we obtain

$$D_0^2x_2 + x_2 = \left(-2iD_2A + 6A^2\overline{A} + ke^{i\sigma T_2}\right)e^{iT_0} + cc + NST \tag{4.10.13}$$

In order to eliminate secular terms, there must be

$$2iD_2A - 6A^2\overline{A} - ke^{i\sigma T_2} = 0 \tag{4.10.14}$$

Let

$$A = \frac{1}{2}ae^{i\beta} \tag{4.10.15}$$

Substituting (4.10.15) into (4.10.14) and separating the real and imaginary parts yields

$$a' = k\sin(\sigma T_2 - \beta)$$
$$-a\beta' = 6a^3 + k\cos(\sigma T_2 - \beta)$$

Let

$$\gamma = \sigma T_2 - \beta \tag{4.10.16}$$

$$a' = k \sin\gamma \tag{4.10.17}$$

$$-a\gamma' = a\sigma + 6a^3 + k\cos\gamma \tag{4.10.18}$$

From (4.10.9) and (4.10.11), we obtain the primary resonance response of (4.10.3)

$$x = a\cos(t-\gamma) + \varepsilon a^2[-\frac{3}{4} + \frac{1}{4}\cos(2t-2\gamma)] + O(\varepsilon^2) \tag{4.10.19}$$

Considering $u = \varepsilon x$ and noting εa as a, we obtain the primary resonance response of the original Eq. (4.10.1)

$$u = a\cos(t-\gamma) + a^2[-\frac{3}{4} + \frac{1}{4}\cos(2t-2\gamma)] + O(\varepsilon^3) \tag{4.10.20}$$

(b) For non-primary resonance response, let

$$u = \varepsilon x, \quad K = 2\varepsilon k \tag{4.10.21}$$

Equation (4.10.1) becomes

$$\ddot{x} + x + \frac{3}{2}\varepsilon x^2 + \frac{1}{2}\varepsilon^2 x^3 = 2k\cos\Omega t \tag{4.10.22}$$

Substituting (4.10.4) into (4.10.22) and retaining to $O(\varepsilon^2)$ yields

$$\begin{aligned} 0 &= \ddot{x} + x + \frac{3}{2}\varepsilon x^2 + \frac{1}{2}\varepsilon^2 x^3 - 2k\cos\Omega t \\ &= D_0^2 x_0 + x_0 - 2k\cos\Omega t + \varepsilon\left(D_0^2 x_1 + x_1 + 2D_0 D_1 x_0 + \frac{3}{2}x_0^2\right) \\ &\quad + \varepsilon^2\left[D_0^2 x_2 + x_2 + 2D_0 D_1 x_1 + \left(D_1^2 + 2D_0 D_2\right)x_0 + 3x_0 x_1 + \frac{1}{2}x_0^3\right] + \cdots \end{aligned} \tag{4.10.23}$$

Let the coefficient of the like power of ε be zero, we get

$$D_0^2 x_0 + x_0 = 2k\cos\Omega t \tag{4.10.24}$$

$$D_0^2 x_1 + x_1 = -2D_0 D_1 x_0 - \frac{3}{2} x_0^2 \tag{4.10.25}$$

$$D_0^2 x_2 + x_2 = -2D_0 D_1 x_1 - \left(D_1^2 + 2D_0 D_2\right) x_0 - 3x_0 x_1 - \frac{1}{2} x_0^3 \tag{4.10.26}$$

The solution of (4.10.24) is

$$x_0 = Ae^{iT_0} + \Lambda e^{i\Omega T_0} + cc \tag{4.10.27}$$

where $A = A(T_1, T_2)$ and

$$\Lambda = \frac{k}{1 - \Omega^2} \tag{4.10.28}$$

Substituting (4.10.27) into (4.10.25) yields

$$\begin{aligned} D_0^2 x_1 + x_1 &= -2iD_1 Ae^{iT_0} - \frac{3}{2}(Ae^{iT_0} + \Lambda e^{i\Omega T_0} + cc)^2 \\ &= -3A\overline{A} - 3\Lambda^2 - 2iD_1 Ae^{iT_0} - \frac{3}{2}\Lambda^2 e^{2i\Omega T_0} \\ &\quad - 3\Lambda\overline{A}e^{i(\Omega-1)T_0} - 3\Lambda Ae^{i(\Omega+1)T_0} - \frac{3}{2}A^2 e^{2iT_0} + cc \end{aligned} \tag{4.10.29}$$

The detuning parameter σ is adopted to quantitatively describes the nearness of Ω to 2:

$$\Omega = 2 + \varepsilon\sigma \tag{4.10.30}$$

In order to eliminate secular terms in (4.10.29), we need

$$2iD_1 Ae^{iT_0} + 3\Lambda\overline{A}e^{i(\Omega-1)T_0} = 0$$

i.e.,

$$2iD_1 A + 3\Lambda\overline{A}e^{i\sigma T_1} = 0 \tag{4.10.31}$$

Substituting (4.10.15) into the above equation, we get

$$ia' - a\beta' + \frac{3}{2}\Lambda a e^{i(\sigma T_1 - 2\beta)} = 0 \tag{4.10.32}$$

The prime denotes the derivative with respect to T_1. Separating the real and imaginary parts of the above equation yields

$$a' = -\frac{3}{2}\Lambda a \sin\gamma \tag{4.10.33}$$

$$\gamma' = \sigma - 3\Lambda\cos\gamma \tag{4.10.34}$$

$$\gamma = \sigma T_1 - 2\beta, \quad \Lambda = \frac{k}{1-\Omega^2} = -\frac{k}{3} \tag{4.10.35}$$

The solution of (4.10.29) is

$$\begin{aligned} x_1 = &-3A\overline{A} - 3\Lambda^2 + \frac{1}{2}A^2 e^{2iT_0} + \frac{3}{35}\Lambda A e^{i(\Omega+1)T_0} \\ &+ \frac{1}{10}\Lambda^2 e^{2i\Omega T_0} + cc \end{aligned} \tag{4.10.36}$$

Substituting (4.10.27), (4.10.36), (4.10.30) and (4.10.35) into (4.10.4), we obtain the subharmonic resonance response of (4.10.22)

$$\begin{aligned} x = &a\cos\left(t - \frac{1}{2}\gamma\right) - \frac{2}{3}k\cos 2t \\ &+ \varepsilon[-\frac{3}{4}a^2 - \frac{1}{3}k^2 + \frac{1}{4}a^2\cos(2t+\gamma) \\ &- \frac{1}{35}ka\cos\left(3t - \frac{1}{2}\gamma\right) + \frac{1}{45}k^2\cos 4t] + O\left(\varepsilon^2\right) \end{aligned} \tag{4.10.37}$$

Considering $u = \varepsilon x$, $K = 2\varepsilon k$ and writing εa for a yields the superharmonic resonance response of original Eq. (4.10.1) at $\Omega \approx 2$:

$$\begin{aligned} u = &a\cos\left(t - \frac{1}{2}\gamma\right) - \frac{2}{6}K\cos 2t \\ &+ [-\frac{3}{4}a^2 - \frac{1}{12}K^2 + \frac{1}{4}a^2\cos(2t+\gamma) \\ &- \frac{1}{70}Ka\cos\left(3t - \frac{1}{2}\gamma\right) + \frac{1}{180}K^2\cos 4t] + O\left(\varepsilon^3\right) \end{aligned} \tag{4.10.38}$$

(c) the detuning parameter σ is adopted to quantitatively describes the nearness of Ω to 1/2:

$$\Omega = \frac{1}{2}(1 + \varepsilon\sigma) \tag{4.10.39}$$

In order to eliminate secular terms in (4.10.29), it is necessary to

$$2iD_1 A e^{iT_0} + \frac{3}{2}\Lambda^2 e^{2i\Omega T_0} = 0$$

i.e.,

$$2iD_1A + \frac{3}{2}\Lambda^2 e^{i\sigma T_1} = 0 \tag{4.10.40}$$

Substituting (4.10.15) into the above equation, we get

$$ia' - a\beta' + \frac{3}{2}\Lambda^2 e^{i(\sigma T_1 - \beta)} = 0 \tag{4.10.41}$$

The prime denotes the derivative with respect to T_1. Separating the real and imaginary parts of the above equation yields

$$a' = -\frac{3}{2}\Lambda^2 \sin\gamma \tag{4.10.42}$$

$$a\gamma' = \sigma - \frac{3}{2}\Lambda^2 \cos\gamma \tag{4.10.43}$$

$$\gamma = \sigma T_1 - \beta, \quad \Lambda = \frac{k}{1-\Omega^2} = \frac{4k}{3} \tag{4.10.44}$$

The solution of (4.10.29) is

$$x_1 = -3A\overline{A} - 3\Lambda^2 - 4\Lambda\overline{A}e^{i(\Omega-1)T_0} + \frac{1}{2}A^2e^{2iT_0} + \frac{12}{5}\Lambda A e^{i(\Omega+1)T_0} + cc \tag{4.10.45}$$

Substituting (4.10.27), (4.10.45), (4.10.39) and (4.10.44) into (4.10.4), we obtain the superharmonic resonance response of (4.10.22) at $\Omega \approx 1/2$:

$$\begin{aligned} x &= a\cos(t-\gamma) + \frac{8}{3}k\cos\frac{1}{2}t \\ &+ \quad \varepsilon[-\frac{3}{4}a^2 - \frac{16}{3}k^2 - \frac{16}{3}ka\cos\left(\frac{1}{2}t+\gamma\right) \\ &+ \frac{1}{4}a^2\cos(2t-2\gamma) + \frac{16}{5}ka\cos\left(\frac{3}{2}t-\gamma\right)] + O\left(\varepsilon^2\right) \end{aligned} \tag{4.10.46}$$

Considering $u = \varepsilon x$, $K = 2\varepsilon k$ and writing εa for a yields the superharmonic resonance response of original Eq. (4.10.1) at $\Omega \approx 1/2$:

$$\begin{aligned} u = &\left[a\cos(t-\gamma) + \tfrac{8}{6}k\cos\tfrac{1}{2}t\right] \\ &+[-\tfrac{3}{8}a^2 - \tfrac{8}{3}K^2 - \tfrac{8}{3}Ka\cos\left(\tfrac{1}{2}t+\gamma\right) \\ &+\tfrac{1}{8}a^2\cos(2t-2\gamma) + \tfrac{8}{5}Ka\cos\left(\tfrac{3}{2}t-\gamma\right)] + O\left(\varepsilon^3\right) \end{aligned} \tag{4.10.47}$$

(d), (e): Under non-resonant hard excitation, Eqs. (4.10.21) – (4.10.29) always hold. When $\Omega \approx 3$ or $1/3$, in order to eliminate secular terms in (4.10.29), it is necessary to have $D_1 A = 0$, so $A = A(T_2)$. In this case, the solution of (4.10.29) is

$$x_1 = -3A\overline{A} - 3\Lambda^2 + \frac{1}{2}A^2 e^{2iT_0} + \Lambda_1 e^{2i\Omega T_0} + \Lambda_2 \overline{A} e^{i(\Omega-1)T_0} + \Lambda_3 A e^{i(\Omega+1)T_0} + cc \tag{4.10.48}$$

where

$$\Lambda_1 = -\frac{3\Lambda^2}{2\left(1-4\Omega^2\right)},\quad \Lambda_2 = -\frac{3\Lambda}{1-(\Omega-1)^2},\quad \Lambda_3 = -\frac{3\Lambda}{1-(\Omega+1)^2} \tag{4.10.49}$$

From (4.10.27) and (4.10.48), we can obtain

$$\begin{aligned} x_0 x_1 &= \left(Ae^{iT_0} + \Lambda e^{i\Omega T_0} + cc_1\right)(-3A\overline{A} - 3\Lambda^2 + \frac{1}{2}A^2 e^{2iT_0} \\ &\quad + \Lambda_3 A e^{i(\Omega+1)T_0} + cc + \Lambda_1 e^{2i\Omega T_0} + \Lambda_2 \overline{A} e^{i(\Omega-1)T_0} + \Lambda_3 A e^{i(\Omega+1)T_0} + cc_2) \\ &= -\frac{5}{2}A^2\overline{A}e^{iT_0} - 3\Lambda^2 A e^{iT_0} + \Lambda\Lambda_2 A e^{iT_0} + \Lambda\Lambda_3 A e^{iT_0} \\ &\quad + \Lambda_3 A e^{i(\Omega+1)T_0} + cc + \Lambda\Lambda_1 e^{3i\Omega T_0} + \Lambda_2 \overline{A}^2 e^{i(\Omega-2)T_0} \\ &\quad + \frac{1}{2}\Lambda\overline{A}^2 e^{i(\Omega-2)T_0} + cc + NST \end{aligned} \tag{4.10.50}$$

$$\begin{aligned} x_0^3 &= (Ae^{iT_0} + \Lambda e^{i\Omega T_0} + \overline{A}e^{-iT_0} + \Lambda e^{-i\Omega T_0})^3 \\ &= (Ae^{iT_0} + \Lambda e^{i\Omega T_0})^3 + 3(Ae^{iT_0} + \Lambda e^{i\Omega T_0})^2\left(\overline{A}e^{-iT_0} + \Lambda e^{-i\Omega T_0}\right) + cc \\ &= \Lambda^3 e^{3i\Omega T_0} + 3A^2\overline{A}e^{iT_0} + 6\Lambda^2 A e^{iT_0} + 3\overline{A}^2\Lambda e^{i(\Omega-2)T_0} + cc + NST \end{aligned} \tag{4.10.51}$$

Substituting (4.10.50) and (4.10.51) into (4.10.26), and taking $A = A(T_2)$ into account, we obtain

$$\begin{aligned} D_0^2 x_2 + x_2 &= -2iD_2 A e^{iT_0} - 3x_0 x_1 - \frac{1}{2}x_0^3 \\ &= -2iD_2 A e^{iT_0} + \frac{15}{2}A^2\overline{A}e^{iT_0} + 9\Lambda^2 A e^{iT_0} - 3\Lambda\Lambda_2 A e^{iT_0} \\ &\quad - 3\Lambda\Lambda_3 A e^{iT_0} - 3\Lambda\Lambda_1 e^{3i\Omega T_0} - 3\Lambda_2\overline{A}^2 e^{i(\Omega-2)T_0} - \frac{3}{2}\Lambda\overline{A}^2 e^{i(\Omega-2)T_0} \\ &\quad - \frac{1}{2}\Lambda^3 e^{3i\Omega T_0} - \frac{3}{2}A^2\overline{A}e^{iT_0} - 3\Lambda^2 A e^{iT_0} - \frac{3}{2}\overline{A}^2\Lambda e^{i(\Omega-2)T_0} \\ &\quad + cc + NST \\ &= \left(-2iD_2 A + 6A^2\overline{A} + 6\Lambda^2 A - 3\Lambda\Lambda_2 A - 3\Lambda\Lambda_3 A\right)e^{iT_0} \end{aligned}$$

$$-\left(\frac{1}{2}\Lambda^3+3\Lambda\Lambda_1\right)e^{3i\Omega T_0}-\left(3\Lambda_2\overline{A}^2+3\Lambda\overline{A}^2\right)e^{i(\Omega-2)T_0}+cc+NST \tag{4.10.52}$$

(1) **When $\Omega \approx 3$:**

Let

$$\Omega=3+\varepsilon^2\sigma \tag{4.10.53}$$

In order to eliminate secular terms in (4.10.52), we need

$$2iD_2A-6A^2\overline{A}-6\Lambda^2A+3\Lambda\Lambda_2A+3\Lambda\Lambda_3A+\left(3\Lambda_2\overline{A}^2+3\Lambda\overline{A}^2\right)e^{i\sigma T_2}=0 \tag{4.10.54}$$

Combine (4.10.49) and (4.10.28), we have

$$\Lambda_1=\frac{3\Lambda^2}{70},\quad \Lambda_2=\Lambda,\quad \Lambda_3=\frac{\Lambda}{5},\quad \Lambda=\frac{k}{1-\Omega^2}=-\frac{k}{8} \tag{4.10.55}$$

Substituting (4.10.55) into (4.10.54) yields

$$2iD_2A-6A^2\overline{A}-\frac{12}{5}\Lambda^2A+6\Lambda\overline{A}^2e^{i\sigma T_2}=0 \tag{4.10.56}$$

By making $A=\frac{1}{2}ae^{i\beta}$, the above equation becomes

$$ia'-a\beta'-\frac{3}{4}a^3-\frac{6}{5}\Lambda^2a+6\Lambda a^2e^{i(\sigma T_2-3\beta)}=0 \tag{4.10.57}$$

Separating the real and imaginary parts of the above equation yields

$$a'=\frac{3}{16}ka^2\sin\gamma \tag{4.10.58}$$

$$\gamma'=\sigma+\frac{9}{4}a^2+\frac{9}{160}k^2+\frac{9}{16}ka\cos\gamma \tag{4.10.59}$$

where

$$\gamma=\sigma T_2-3\beta \tag{4.10.60}$$

Substituting (4.10.27), (4.10.53), (4.10.55) and (4.10.60) into (4.10.4), we obtain the superharmonic resonance response of (4.10.22) at $\Omega \approx 3$:

$$x = a\cos(t - \frac{1}{3}\gamma) - \frac{1}{4}k\cos 3t + O(\varepsilon) \tag{4.10.61}$$

Considering $u = \varepsilon x$, $K = 2\varepsilon k$ and writing εa for a yields the superharmonic resonance response of original Eq. (4.10.1) at $\Omega \approx 3$:

$$u = a\cos(t - \frac{1}{3}\gamma) - \frac{1}{8}K\cos 3t + O\left(\varepsilon^2\right) \tag{4.10.62}$$

(2) **When $\Omega \approx \frac{1}{3}$:**

Let

$$\Omega = \frac{1}{3}\left(1 + \varepsilon^2\sigma\right) \tag{4.10.63}$$

In order to eliminate secular terms in (4.10.52), we need

$$2iD_2A - 6A^2\overline{A} - 6\Lambda^2A + 3\Lambda\Lambda_2A + 3\Lambda\Lambda_3A + \left(\frac{1}{2}\Lambda^3 + 3\Lambda\Lambda_1\right)e^{i\sigma T_2} = 0 \tag{4.10.64}$$

Combine (4.10.49) and (4.10.28), we have

$$\Lambda_1 = -\frac{27\Lambda^2}{10}, \quad \Lambda_2 = -\frac{27\Lambda}{5}, \quad \Lambda_3 = \frac{27\Lambda}{7}, \quad \Lambda = \frac{k}{1 - \Omega^2} = \frac{9k}{8} \tag{4.10.65}$$

Substituting (4.10.65) into (4.10.64) yields

$$2iD_2A - 6A^2\overline{A} - 17.68k^2A - 8.4k^3e^{i\sigma T_2} = 0 \tag{4.10.66}$$

By making $A = \frac{1}{2}ae^{i\beta}$, the above equation becomes

$$ia' - a\beta' - \frac{3}{4}a^3 - 8.84k^2a - 8.4k^3e^{i(\sigma T_2 - \beta)} = 0 \tag{4.10.67}$$

Separating the real and imaginary parts of the above equation yields

$$a' = 8.4k^3\sin\gamma \tag{4.10.68}$$

$$\gamma' = \sigma + \frac{3}{4}a^2 + 8.84k^2 + \frac{8.4k^3}{a}\cos\gamma \tag{4.10.69}$$

where

$$\gamma = \sigma T_2 - \beta \tag{4.10.70}$$

Substituting (4.10.27), (4.10.63), (4.10.65) and (4.10.70) into (4.10.4), we obtain the superharmonic resonance response of (4.10.22) at $\Omega \approx 1/3$:

$$x = a\cos(t - \gamma) + \frac{9}{4}k\cos\frac{1}{3}t + O(\varepsilon) \tag{4.10.71}$$

Considering $u = \varepsilon x$, $K = 2\varepsilon k$ and writing εa for a yields the superharmonic resonance response of original Eq. (4.10.1) at $\Omega \approx 1/3$:

$$u = a\cos(t - \gamma) + \frac{9}{8}K\cos 3t + O(\varepsilon^2) \tag{4.10.72}$$

4.11 Exercise 4.11 (Primary and Secondary Resonances of Five Nonlinear, Linearly Damped Systems)

Solution: **(a)** Let us first specify the order of the terms in the original differential equation of motion. Let

$$u = \varepsilon^{\lambda} x \tag{4.11.1}$$

The original equation becomes

$$\ddot{x} + x + \varepsilon^{4\lambda}\alpha x^5 + 2\mu\dot{x} = \frac{K}{\varepsilon^{\lambda}}\cos\Omega t \tag{4.11.2}$$

In order to find the primary resonance response, it is necessary to have the nonlinear term, the damping term and the excitation term are of the same order, so that

$$\lambda = \frac{1}{4}, \quad \mu = \varepsilon\hat{\mu}, \quad K = 2\varepsilon^{5\lambda}k = 2\varepsilon^{5/4}k \tag{4.11.3}$$

Then the Eq. (4.1.1) becomes

$$\ddot{x} + x + \varepsilon\alpha x^5 + 2\varepsilon\hat{\mu}\dot{x} = 2\varepsilon k\cos\Omega t \tag{4.11.4}$$

Let the solution of the Eq. (4.11.4) have the following form:

$$x(t;\varepsilon) = x_0(T_0, T_1, T_2) + \varepsilon x_1(T_0, T_1, T_2) + \varepsilon^2 x_2(T_0, T_1, T_2) + \cdots \tag{4.11.5}$$

Substituting (4.10.4) into (4.11.4) and retaining to $O(\varepsilon)$ yields

$$
\begin{aligned}
0 &= \ddot{x} + x + \varepsilon\alpha x^5 + 2\varepsilon\hat{\mu}\dot{x} - 2\varepsilon k \cos\Omega t \\
&= D_0^2 + 2\varepsilon D_0 D_1)(x_0 + \varepsilon x_1) + (x_0 + \varepsilon x_1) + \varepsilon\alpha(x_0 + \varepsilon x_1)^5 \\
&\quad + 2\varepsilon\hat{\mu}(D_0 + \varepsilon D_1)(x_0 + \varepsilon x_1) - 2\varepsilon k \cos\Omega t + \cdots \\
&= D_0^2 x_0 + \varepsilon D_0^2 x_1 + 2\varepsilon D_0 D_1 x_0 + x_0 + \varepsilon x_1 + \varepsilon\alpha x_0^5 \\
&\quad + 2\varepsilon\hat{\mu} D_0 x_0 - 2\varepsilon k \cos\Omega t + \cdots \\
&= D_0^2 x_0 + x_0 + \varepsilon(D_0^2 x_1 + 2D_0 D_1 x_0 + x_1 + \alpha x_0^5 \\
&\quad + 2\hat{\mu} D_0 x_0 - 2k\cos\Omega t) + \cdots
\end{aligned}
\tag{4.11.6}
$$

Let the coefficient of the like power of ε be zero, we get

$$
D_0^2 x_0 + x_0 = 0 \tag{4.11.7}
$$

$$
D_0^2 x_1 + x_1 = -2D_0 D_1 x_0 - \alpha x_0^5 - 2\hat{\mu} D_0 x_0 + 2k\cos\Omega t \tag{4.11.8}
$$

The solution of (4.10.6) is

$$
x_0 = Ae^{iT_0} + \overline{A}e^{-iT_0} \tag{4.11.9}
$$

where $A = A(T_1)$. Substitute (4.10.9) into (4.10.7), we get

$$
\begin{aligned}
D_0^2 x_1 + x_1 &= -2iD_1 Ae^{iT_0} - 2i\hat{\mu}Ae^{iT_0} \\
&\quad + 2k\cos\Omega t - \alpha(Ae^{iT_0} + \overline{A}e^{-iT_0})^5 + cc \\
&= -2iD_1 Ae^{iT_0} - 2i\hat{\mu}Ae^{iT_0} + 2k\cos\Omega t \\
&\quad - \alpha A^5 e^{5iT_0} - 5\alpha A^4\overline{A}e^{3iT_0} - 10\alpha A^3\overline{A}^2 e^{iT_0} + cc
\end{aligned}
\tag{4.11.10}
$$

When $\Omega \approx 1$, let

$$
\Omega = 1 + \varepsilon\sigma \tag{4.11.11}
$$

Therefore

$$
\begin{aligned}
D_0^2 x_1 + x_1 &= -\left(2iD_1 A + 2i\hat{\mu}A + 10\alpha A^3\overline{A}^2 - ke^{i\sigma T_1}\right)e^{iT_0} \\
&-\alpha A^5 e^{5iT_0} - 5\alpha A^4\overline{A}e^{3iT_0} + cc
\end{aligned}
\tag{4.11.12}
$$

In order to eliminate secular terms, there must be

$$
2iD_1 A + 2i\hat{\mu}A + 10\alpha A^3\overline{A}^2 - ke^{i\sigma T_1} = 0 \tag{4.11.13}
$$

$$
A = \frac{1}{2}ae^{i\beta} \tag{4.11.14}
$$

Then

$$ia' - a\beta' + i\hat{\mu}a + \frac{5}{16}\alpha a^5 - ke^{i(\sigma T_1 - \beta)} = 0 \tag{4.11.15}$$

Separating the real and imaginary parts of the above equation yields

$$a' = -\hat{\mu}a + k\sin\gamma \tag{4.11.16}$$

$$\gamma' = \sigma - \frac{5}{16}\alpha a^4 + \frac{k}{a}\cos\gamma \tag{4.11.17}$$

where

$$\gamma = \sigma T_1 - \beta \tag{4.11.18}$$

The frequency–response equation can be obtained by (4.11.16) and (4.11.17):

$$(\sigma a - \frac{5}{16}\alpha a^5)^2 + \hat{\mu}^2 a^2 = k^2 \tag{4.11.19}$$

Solving for σ and k from the above equation, respectively, yields

$$\sigma = \frac{5}{16}\alpha a^4 \pm \frac{1}{a}\sqrt{k^2 - \hat{\mu}^2 a^2} \tag{4.11.20}$$

$$k = \sqrt{(\sigma a - \frac{5}{16}\alpha a^5)^2 + \hat{\mu}^2 a^2} \tag{4.11.21}$$

The frequency–response curve of the system $a \sim \sigma$ and the excitation-response curve $a \sim k$ can be plotted in Fig. 4.10a, b.

(b) Let

$$\lambda = \frac{1}{4}, \quad \mu = \varepsilon\hat{\mu}, \quad K = 2\varepsilon^{\lambda}k = 2\varepsilon^{1/4}k \tag{4.11.22}$$

Then for this non-resonant hard excitation, Eq. (4.1.1) becomes

$$\ddot{x} + x + \varepsilon\alpha x^5 + 2\varepsilon\hat{\mu}\dot{x} = 2k\cos\Omega t \tag{4.11.23}$$

Substituting (4.10.4) into (4.11.4) and retaining to $O(\varepsilon)$ yields.

$$\begin{aligned} 0 &= \ddot{x} + x + \varepsilon\alpha x^5 + 2\varepsilon\hat{\mu}\dot{x} - 2k\cos\Omega t \\ &= \left(D_0^2 + 2\varepsilon D_0 D_1\right)(x_0 + \varepsilon x_1) + (x_0 + \varepsilon x_1) + \varepsilon\alpha(x_0 + \varepsilon x_1)^5 \\ &\quad + 2\varepsilon\hat{\mu}(D_0 + \varepsilon D_1)(x_0 + \varepsilon x_1) - 2k\cos\Omega t + \cdots \end{aligned}$$

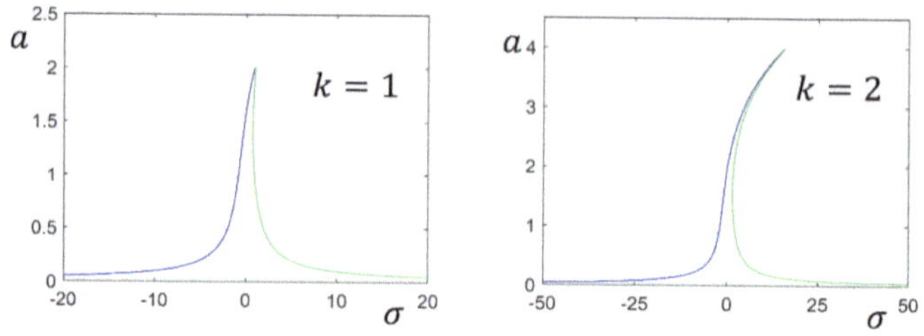

(a) Amplitude-frequency response curve

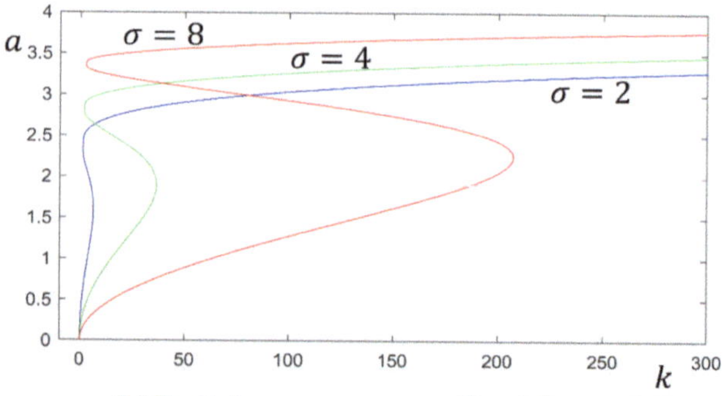

(b) Excitation-response curve ($\hat{\mu} = 0.5,\ \alpha = 0.2$)

Fig. 4.10 **a** Frequency-response and **b** excitation-response curves ($\hat{\mu} = 0.5,\ \alpha = 0.2$) in Exercise 4.11

$$\begin{aligned} &= D_0^2 x_0 + \varepsilon D_0^2 x_1 + 2\varepsilon D_0 D_1 x_0 + x_0 + \varepsilon x_1 + \varepsilon \alpha x_0^5 \\ &\quad + 2\varepsilon \hat{\mu} D_0 x_0 - 2\varepsilon k \cos \Omega t + \cdots \\ &= D_0^2 x_0 + x_0 - 2k \cos \Omega t \\ &\quad + \varepsilon \left(D_0^2 x_1 + 2D_0 D_1 x_0 + x_1 + \alpha x_0^5 + 2\hat{\mu} D_0 x_0 \right) + \cdots \end{aligned} \tag{4.11.24}$$

Let the coefficient of the like power of ε be zero, we get

$$D_0^2 x_0 + x_0 = 2k \cos \Omega t \tag{4.11.25}$$

$$D_0^2 x_1 + x_1 = -2D_0 D_1 x_0 - 2\hat{\mu} D_0 x_0 - \alpha x_0^5 \tag{4.11.26}$$

The solution of (4.11.25) is

$$x_0 = A e^{iT_0} + \Lambda e^{i\Omega T_0} + cc \tag{4.11.27}$$

where $A = A(T_1)$, and

$$\Lambda = \frac{k}{1 - \Omega^2} \tag{4.11.28}$$

Substituting (4.11.27) into (4.11.26) yields

$$
\begin{aligned}
&D_0^2x_1 + x_1 = -2D_0D_1x_0 - 2\hat{\mu}D_0x_0 - \alpha x_0^5 \\
&= -2iD_1Ae^{iT_0} - 2i\hat{\mu}Ae^{iT_0} - 2i\hat{\mu}\Omega\Lambda e^{i\Omega T_0} \\
&-\alpha(Ae^{iT_0} + \Lambda e^{i\Omega T_0})^5 - 5\alpha(Ae^{iT_0} + \Lambda e^{i\Omega T_0})^4(\overline{A}e^{-iT_0} + \Lambda e^{-i\Omega T_0}) \\
&-10\alpha(Ae^{iT_0} + \Lambda e^{i\Omega T_0})^3(\overline{A}e^{-iT_0} + \Lambda e^{-i\Omega T_0})^2 + cc \\
&= -\left(2iD_1A + 2i\hat{\mu}A + 10\alpha A^3\overline{A}^2 + 60\alpha\Lambda^2A^2\overline{A} + 30\alpha\Lambda^4A\right)e^{iT_0} \\
&-5\alpha\Lambda\overline{A}^4e^{i(\Omega-4)T_0} - 30\alpha\Lambda^3\overline{A}^2e^{i(\Omega-2)T_0} - 5\alpha\Lambda A^4e^{i(4-\Omega)T_0} \\
&-10\alpha\Lambda^2\overline{A}^3e^{i(2\Omega-3)T_0} - 5\alpha\Lambda^4\overline{A}e^{i(4\Omega-1)T_0} \\
&-5\alpha\Lambda^5e^{3i\Omega T_0} - 20\alpha A\overline{A}e^{3i\Omega T_0} - 10\alpha A^2\Lambda^3e^{i(2-3\Omega)T_0} \\
&-\alpha\Lambda^5Ae^{5i\Omega T_0} + cc + NST
\end{aligned}
\tag{4.11.29}
$$

We can find, from (4.11.29), that there are secondary resonances when $\Omega \approx 5,\ 3,\ 2,\ \frac{1}{2},\ \frac{1}{3},\ \frac{1}{5}$.

(c) In the above secondary resonances case, we only find the controlling equations for phase and amplitude for the secondary resonance case at $\Omega \approx 5$, and readers are invited to complete the rest of the scenario on their own.

When $\Omega \approx 5$, let

$$\Omega = 5 + \varepsilon\sigma \tag{4.11.30}$$

In order to eliminate secular terms in (4.11.29), there must be

$$2iD_1A + 2i\hat{\mu}A + 10\alpha A^3\overline{A}^2 + 60\alpha\Lambda^2A^2\overline{A} + 30\alpha\Lambda^4A + 5\alpha\Lambda\overline{A}^4e^{i\sigma T_1} = 0 \tag{4.11.31}$$

Let

$$A = \frac{1}{2}ae^{i\beta} \tag{4.11.32}$$

then

$$ia' - a\beta' + i\hat{\mu}a + \frac{5}{16}\alpha a^5 + \frac{15}{2}\alpha\Lambda^2a^3 + 15\alpha\Lambda^4a + \frac{5}{16}\alpha\Lambda a^4e^{i(\sigma T_1 - 5\beta)} = 0 \tag{4.11.33}$$

Separating the real part from the imaginary part of the above equation, we get

$$a' = -\hat{\mu}a - \frac{5}{16}\alpha\Lambda a^4\sin\gamma \tag{4.11.34}$$

$$\gamma' = \sigma - 75\alpha\Lambda^4 - \frac{75}{2}\alpha\Lambda^2 a^2 - \frac{25}{16}\alpha a^4 - \frac{25}{16}\alpha\Lambda a^3 \cos\gamma \tag{4.11.35}$$

where

$$\gamma = \sigma T_1 - 5\beta, \quad \Lambda = \frac{k}{1-\Omega^2} = -\frac{k}{24} \tag{4.11.36}$$

4.12 Exercise 4.12 (An Iterative Method for General Undamped Nonlinear Systems)

Solution: (a) When $K = 0, u(0) = u_0$, $\dot{u}(0) = 0$, making the integration of the original differential equation of motion, we get

$$\frac{1}{2}\dot{u}^2 + \int_{u_0}^{u} f(u)du = 0 \quad \Rightarrow \quad \dot{u}^2 + 2\int_{u_0}^{u} f(u)du = 0$$

Let

$$F(u) = F(u_0) + 2\int_{u_0}^{u} f(u)du$$

then

$$\dot{u} = \pm\sqrt{F(u_0) - F(u)}, \quad \frac{1}{2}F'(u) = f(u) \tag{4.12.1}$$

therefore

$$t = \int_{u_0}^{u} \frac{du}{\dot{u}} = \pm\int_{u_0}^{u} \frac{du}{\sqrt{F(u_0) - F(u)}} = g_1(u) \tag{4.12.2}$$

(b) We can obtain $u(t_1)$, $\dot{u}(t_1)$ from (4.12.1) and (4.12.2) by using the above as the starting solution in an iteration scheme and $u(0) = u_1 = u(t_1)$, $\dot{u}(0) = \dot{u}_1 = \dot{u}(t_1)$ as the initial value for the next calculation. Differentiating (4.12.2) with respect to t twice, we get

$$\ddot{u} + f_1(u) = 0 \tag{4.12.3}$$

$$f_1(u) = \frac{\dot{u}^2 g_1''(u)}{g_1'(u)} = \frac{g_1''(u)}{g_1'(u)}\{\dot{u}[f(u)]\}^2 \tag{4.12.4}$$

Making the integration of (4.12.3) yields

$$\frac{1}{2}\dot{u}^2 + \int_{u_1}^{u} f_1(u)du = \frac{1}{2}\dot{u}_0^2 \tag{4.12.5}$$

Let

$$F_1(u) = F_1(u_0) - \frac{1}{2}\dot{u}_0^2 + 2\int_{u_0}^{u} f_1(u)du$$

then

$$\dot{u} = \pm\sqrt{F_1(u_0) - F_1(u)} \tag{4.12.6}$$

therefore

$$t = \int_{u_0}^{u} \frac{du}{\dot{u}} = \pm\int_{u_0}^{u} \frac{du}{\sqrt{F_1(u_0) - F_1(u)}} \tag{4.12.7}$$

4.13 Exercise 4.13 (The Method of Harmonic Balance to Analyze Viscous Damping and Primary Resonances of Cubic Nonlinear Systems)

Solution: Let the solution of the equation be

$$u = a\cos(\Omega t + \beta) \triangleq a\cos\varphi, \quad \Omega = 1 + \varepsilon\sigma \tag{4.13.1}$$

Here both a and β are constants. Substituting (4.13.1) into the governing equation of the system yields

$$\begin{aligned}
0 &= -a\Omega^2\cos\phi + a\cos\phi - 2\varepsilon\mu\Omega a\sin\phi \\
&\quad + \tfrac{1}{4}\varepsilon\alpha a^3(3\cos\phi + \cos 3\phi) - \varepsilon k\cos\Omega t \\
&= -a\Omega^2\cos\phi + a\cos\phi - 2\varepsilon\mu\Omega a\sin\phi \\
&\quad + \tfrac{1}{4}\varepsilon\alpha a^3(3\cos\phi + \cos 3\phi) - \varepsilon k\cos\beta\cos\phi + \varepsilon k\sin\beta\sin\phi \\
&= \left(a - a\Omega^2 + \tfrac{3}{4}\varepsilon\alpha a^3 - \varepsilon k\cos\beta\right)\cos\phi \\
&\quad + (-2\varepsilon\mu\Omega a + \varepsilon k\sin\beta)\sin\phi + \tfrac{1}{4}\varepsilon\alpha a^3\cos 3\phi
\end{aligned} \tag{4.13.2}$$

According to the method of harmonic balance, we can obtain

$$\begin{aligned} & a - a\Omega^2 + \tfrac{3}{4}\varepsilon\alpha a^3 - \varepsilon k \cos\beta = 0 \\ & -2\varepsilon\mu\Omega a + \varepsilon k \sin\beta = 0 \end{aligned} \tag{4.13.3}$$

i.e.,

$$\begin{aligned} & -2\sigma a + \tfrac{3}{4}\alpha a^3 = k \cos\beta \\ & 2\mu a = k \sin\beta \end{aligned} \tag{4.13.4}$$

The frequency–response equation of the system is

$$(2\sigma a - \frac{3}{4}\alpha a^3)^2 + 4\mu^2 a^2 = k^2 \tag{4.13.5}$$

This result is identical to that obtained by the method of multiscale.

4.14 Exercise 4.14 (The Method of Harmonic Balance to Analyze the Primary Resonance of an Undamped Nonlinear System)

Solution: Let the solution of the equation be

$$u = a\cos(\Omega t + \beta) = a\cos\varphi, \quad \varphi = \Omega t + \beta \tag{4.14.1}$$

Here both a and β are constants. Substituting the above equation into the governing equation of the system yields

$$\begin{aligned} 0 &= \ddot{u} + u^5 - K\cos\Omega t \\ &= -a\Omega^2\cos\phi + a^5\cos^5\phi - K\cos(\phi - \beta) \\ &= -a\Omega^2\cos\phi + \frac{1}{16}a^5(10\cos\phi + \cdots) \\ &\quad - K\cos\beta\cos\phi + K\sin\beta\sin\phi \\ &= \left(-a\Omega^2 + \frac{5}{8}a^5 - K\cos\beta\right)\cos\phi + K\sin\beta\sin\phi + \cdots \end{aligned} \tag{4.14.2}$$

From the method of harmonic balance, we have

$$\begin{aligned} & -a\Omega^2 + \tfrac{5}{8}a^5 - K\cos\beta = 0 \\ & K\sin\beta = 0 \end{aligned} \tag{4.14.3}$$

The frequency–response equation of the system is

$$\frac{5}{8}a^5 - a\Omega^2 = K \tag{4.14.4}$$

4.15 Exercise 4.15 (Combination Resonance for Cubic Nonlinear Systems)

Solution: (a) We assume that both excitations are non-resonant hard excitations. Let the solution of the equation be

$$x(t;\varepsilon) = x_0(T_0, T_1, T_2) + \varepsilon x_1(T_0, T_1, T_2) + \cdots \tag{4.15.1}$$

Substituting (4.15.1) into the original equation and retaining to $O(\varepsilon)$, we obtain

$$\begin{aligned}
0 &= \ddot{u} + \omega_0^2 u + \varepsilon\left(2\mu\dot{u} + \alpha u^3\right) - K_1\cos(\Omega_1 t + \theta_1) - K_2\cos(\Omega_2 t + \theta_2) \\
&= \left(D_0^2 + 2\varepsilon D_0 D_1\right)(x_0 + \varepsilon x_1) + \omega_0^2(x_0 + \varepsilon x_1) \\
&\quad + 2\varepsilon\mu(D_0 + \varepsilon D_1)(x_0 + \varepsilon x_1) + \varepsilon\alpha(x_0 + \varepsilon x_1)^3 \\
&\quad - K_1\cos(\Omega_1 T_0 + \theta_1) - K_2\cos(\Omega_2 T_0 + \theta_2) + \cdots \\
&= D_0^2 x_0 + \varepsilon D_0^2 x_1 + 2\varepsilon D_0 D_1 x_0 + \omega_0^2 x_0 + \varepsilon\omega_0^2 x_1 \\
&\quad + 2\varepsilon\mu D_0 x_0 + \varepsilon\alpha x_0^3 - K_1\cos(\Omega_1 t + \theta_1) - K_2\cos(\Omega_2 t + \theta_2) + \cdots \\
&= D_0^2 x_0 + \omega_0^2 x_0 - K_1\cos(\Omega_1 t + \theta_1) - K_2\cos(\Omega_2 t + \theta_2) \\
&\quad + \varepsilon\left(D_0^2 x_1 + \omega_0^2 x_1 + 2D_0 D_1 x_0 + 2\mu D_0 x_0 + \alpha x_0^3\right) + \cdots
\end{aligned} \tag{4.15.2}$$

Let the coefficient of the like power of ε be zero, we get

$$D_0^2 x_0 + \omega_0^2 x_0 = K_1\cos(\Omega_1 t + \theta_1) + K_2\cos(\Omega_2 t + \theta_2) \tag{4.15.3}$$

$$D_0^2 x_1 + \omega_0^2 x_1 = -2D_0 D_1 x_0 - 2\mu D_0 x_0 - \alpha x_0^3 \tag{4.15.4}$$

The solution of (4.15.3) is

$$x_0 = Ae^{i\omega_0 T_0} + \Lambda_1 e^{i\Omega_1 T_0} + \Lambda_2 e^{i\Omega_2 T_0} + cc \tag{4.15.5}$$

where

$$\Lambda_1 = \frac{K_1 e^{i\theta_1}}{2\left(\omega_0^2 - \Omega_1^2\right)},\quad \Lambda_2 = \frac{K_2 e^{i\theta_2}}{2\left(\omega_0^2 - \Omega_2^2\right)} \tag{4.15.6}$$

and $A = A(T_1)$. Substituting (4.15.5) into (4.15.4) and taking $\Omega_1 + \Omega_2 \approx 2\omega_0$ into account, we can obtain

$$\begin{aligned} & D_0^2 x_1 + \omega_0^2 x_1 = -2D_0 D_1 x_0 - 2\mu D_0 x_0 - \alpha x_0^3 \\ & = -\left(2i\omega_0 D_1 A + 2i\mu\omega_0 A + 3\alpha A^2\overline{A} + 6\alpha\Lambda_1\overline{\Lambda}_1 A + 6\alpha\Lambda_2\overline{\Lambda}_2 A\right)e^{i\omega_0 T_0} \\ & -6\alpha\Lambda_1\Lambda_2\overline{A}e^{i(\Omega_1+\Omega_2-\omega_0)T_0} + cc + NST \end{aligned} \tag{4.15.7}$$

Let

$$\Omega_1 + \Omega_2 = 2\omega_0 + \varepsilon\sigma \tag{4.15.8}$$

In order to eliminate secular terms from (4.15.7), there must be

$$2i\omega_0(D_1 A + \mu A) + \alpha\left(3A\overline{A} + 6\Lambda_1\overline{\Lambda}_1 + 6\Lambda_2\overline{\Lambda}_2\right)A + 6\alpha\Lambda_1\Lambda_2\overline{A}e^{i\sigma T_1} = 0 \tag{4.15.9}$$

(b) Let $A = \frac{1}{2}ae^{i\beta}$ and substitute it into (4.15.9), we have

$$i\omega_0 a' - a\beta' + i\omega_0\mu a + \alpha\left(3a^2 + 6\Lambda_1\overline{\Lambda}_1 + 6\Lambda_2\overline{\Lambda}_2\right)a + 6\alpha\Lambda_1\Lambda_2 a e^{i(\sigma T_1 - 2\beta)} = 0 \tag{4.15.10}$$

i.e.,

$$\begin{aligned} & i\omega_0 a' - a\beta' + i\omega_0\mu a + \alpha\left[\frac{3}{8}a^2 + \frac{3K_1^2}{4(\omega_0^2 - \Omega_1^2)^2} + \frac{3K_2^2}{4(\omega_0^2 - \Omega_2^2)^2}\right]a \\ & + \frac{3\alpha K_1 K_2}{4\left(\omega_0^2 - \Omega_1^2\right)\left(\omega_0^2 - \Omega_2^2\right)} a e^{i(\sigma T_1 - 2\beta + \theta_1 + \theta_2)} = 0 \end{aligned} \tag{4.15.11}$$

Separating the real and imaginary parts of the above equation yields, we get

$$\begin{aligned} & \omega_0 a' + \omega_0\mu a + \frac{3\alpha K_1 K_2}{4\left(\omega_0^2 - \Omega_1^2\right)\left(\omega_0^2 - \Omega_2^2\right)} a\sin(\sigma T_1 - 2\beta + \theta_1 + \theta_2) = 0 \\ & - a\beta' + \alpha\left[\frac{3}{8}a^2 + \frac{3K_1^2}{4(\omega_0^2 - \Omega_1^2)^2} + \frac{3K_2^2}{4(\omega_0^2 - \Omega_2^2)^2}\right]a \\ & + \frac{3\alpha K_1 K_2}{4\left(\omega_0^2 - \Omega_1^2\right)\left(\omega_0^2 - \Omega_2^2\right)} a\cos(\sigma T_1 - 2\beta + \theta_1 + \theta_2) = 0 \end{aligned} \tag{4.15.12}$$

i.e.,

$$a' = -\mu a - \alpha\Gamma_1 a\sin\gamma \tag{4.15.13}$$

$$a\gamma' = (\sigma - 2\alpha\Gamma_2)a - \frac{3\alpha a^3}{4\omega_0} - 2\alpha\Gamma_1 a\cos\gamma \tag{4.15.14}$$

where

$$\begin{aligned}\Gamma_1 &= \frac{3K_1K_2}{4\omega_0(\omega_0^2-\Omega_1^2)(\omega_0^2-\Omega_2^2)}\\ \Gamma_2 &= \frac{3}{4\omega_0}\left[\frac{K_1^2}{(\omega_0^2-\Omega_1^2)}+\frac{K_2^2}{(\omega_0^2-\Omega_2^2)}\right]\\ \gamma &= \sigma T_1 - 2\beta + \theta_1 + \theta_2\end{aligned} \tag{4.15.15}$$

(c) Let $a' = \gamma' = 0$ in (4.15.13) and (4.15.14), we can obtain the amplitude and phase of the steady state

$$-\mu a - \alpha\Gamma_1 a\sin\gamma = 0 \tag{4.15.16}$$

$$(\sigma - 2\alpha\Gamma_2)a - \frac{3\alpha a^3}{4\omega_0} - 2\alpha\Gamma_1 a\cos\gamma = 0 \tag{4.15.17}$$

We can eliminate γ from (4.15.16) and (4.15.17) and obtain

$$\frac{a^2}{4}[(\sigma - 2\alpha\Gamma_2) - \frac{3\alpha a^2}{4\omega_0}]^2 + \mu^2 a^2 = \alpha^2\Gamma_1^2 a^2 \tag{4.15.18}$$

Therefore, the frequency–response equation of the system is

$$a = 0 \tag{4.15.19}$$

or

$$a^2 = \frac{8}{3}\omega_0[\frac{\sigma}{2\alpha} - \Gamma_2 \pm (\Gamma_1^2 - \frac{\mu^2}{\alpha^2})^{1/2}] \tag{4.15.20}$$

(d) To determine the stability of the steady state solution, let

$$a = a_0 + a_1, \quad \gamma = \gamma_0 + \gamma_1 \tag{4.15.21}$$

where a_0 and γ_0 satisfy (4.15.16)–(4.15.20). Substituting (4.15.21) into (4.15.13) and (4.15.14), we can obtain the following linearized equation

$$\begin{Bmatrix} a_1{}' \\ \gamma_1{}' \end{Bmatrix} = \begin{bmatrix} -\mu - \alpha\Gamma_1\sin\gamma_0 & \alpha\Gamma_1 a_0\cos\gamma_0 \\ -\frac{3\alpha a_0}{2\omega_0} & 2\alpha\Gamma_1\sin\gamma_0 \end{bmatrix}\begin{Bmatrix} a_1 \\ \gamma_1 \end{Bmatrix} \tag{4.15.22}$$

Considering (4.15.16) and (4.15.17), the above equation becomes

$$\begin{Bmatrix} a_1{}' \\ \gamma_1{}' \end{Bmatrix} = \begin{bmatrix} 0 & (\frac{1}{2}\sigma - \alpha\Gamma_2)a_0 - \frac{3\alpha a_0^3}{8\omega_0} \\ -\frac{3\alpha a_0}{2\omega_0} & -2\mu a_0 \end{bmatrix} \begin{Bmatrix} a_1 \\ \gamma_1 \end{Bmatrix} \tag{4.15.23}$$

Its characteristic equation is

$$\lambda^2 + 2\mu a_0 \lambda + \frac{3\alpha a_0}{2\omega_0}\left[\left(\frac{1}{2}\sigma - \alpha\Gamma_2\right)a_0 - \frac{3\alpha a_0^3}{8\omega_0}\right] = 0 \tag{4.15.24}$$

Therefore, the stability condition for the steady state solution is

$$\frac{3\alpha a_0}{2\omega_0}\left[\left(\frac{1}{2}\sigma - \alpha\Gamma_2\right)a_0 - \frac{3\alpha a_0^3}{8\omega_0}\right] \leq 0$$

When $\alpha > 0$, the stability condition is

$$\left(\frac{\sigma}{2\alpha} - \Gamma_2\right)a_0 - \frac{3a_0^3}{8\omega_0} \leq 0 \tag{4.15.25}$$

Therefore, there must be

$$a_0 = 0 \quad \text{or} \quad a_0^2 \geq \frac{8\omega_0}{3}\left(\frac{\sigma}{2\alpha} - \Gamma_2\right) \tag{4.15.26}$$

From (4.15.19) and (4.15.20), the frequency–response (amplitude) curve for the steady state is shown in Fig. 4.11, where

$$\sigma_C = 2\alpha\Gamma_2, \quad \sigma_A = 2\alpha\Gamma_2 - 2\alpha(\Gamma_1^2 - \frac{\mu^2}{\alpha^2})^{1/2}, \quad \sigma_B = 2\alpha\Gamma_2 + 2\alpha(\Gamma_1^2 - \frac{\mu^2}{\alpha^2})^{1/2}$$

It can be seen that the steady state response amplitude can be multi-valued. When $\sigma < \sigma_A$, there is only one zero-valued steady state amplitude, which is stable. When $\sigma_A < \sigma < \sigma_C$, there are two steady-state amplitudes, both of which are stable.

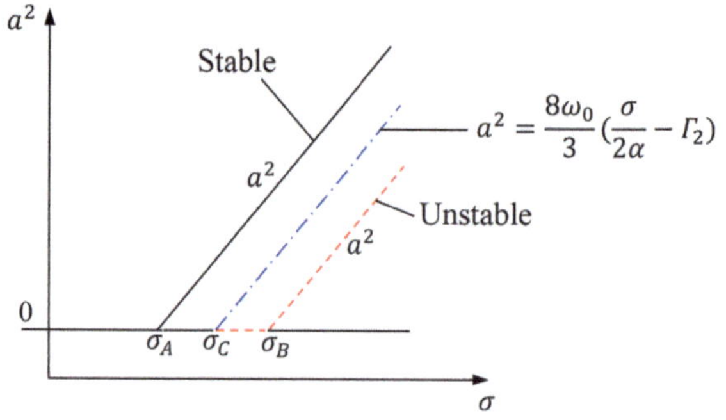

Fig. 4.11 Frequency-response curves and their stability for Exercise 4.15

When $\sigma_C < \sigma < \sigma_B$ and $\sigma > \sigma_B$, there are three steady-state amplitudes, where the response amplitude shown by the dashed line is unstable.

4.16 Exercise 4.16 (The Response of Cubic Nonlinear Systems with Primary Resonance and Non-resonance Excitation)

Solution: **(a)** Since $\Omega_1 \approx 1$ and $\Omega_2 \approx 3$, $K_1 \cos(\Omega_1 t + \theta_1)$ is the primary resonant excitation term, while $K_2 \cos(\Omega_2 t + \theta_2)$ is a non-resonant hard excitation. Let the solution of the equation be

$$x(t;\varepsilon) = x_0(T_0, T_1, T_2) + \varepsilon x_1(T_0, T_1, T_2) + \cdots \tag{4.16.1}$$

and make

$$K_1 = \varepsilon k_1 \tag{4.16.2}$$

Substituting (4.16.1) and (4.16.2) into the original equation and retaining to $O(\varepsilon)$, we get

$$\begin{aligned}
0 &= \ddot{u} + u + 2\varepsilon\mu\dot{u} + \varepsilon\alpha u^3 - K_1\cos(\Omega_1 t + \theta_1) - K_2\cos(\Omega_2 t + \theta_2) \\
&= \left(D_0^2 + 2\varepsilon D_0 D_1\right)(x_0 + \varepsilon x_1) + (x_0 + \varepsilon x_1) \\
&\quad + 2\varepsilon\mu(D_0 + \varepsilon D_1)(x_0 + \varepsilon x_1) + \varepsilon\alpha(x_0 + \varepsilon x_1)^3 \\
&\quad - \varepsilon k_1\cos(\Omega_1 T_0 + \theta_1) - K_2\cos(\Omega_2 T_0 + \theta_2) + \cdots \\
&= D_0^2 x_0 + \varepsilon D_0^2 x_1 + 2\varepsilon D_0 D_1 x_0 + x_0 + \varepsilon x_1 \\
&\quad + 2\varepsilon\mu D_0 x_0 + \varepsilon\alpha x_0^3 - \varepsilon k_1\cos(\Omega_1 t + \theta_1) - K_2\cos(\Omega_2 t + \theta_2) + \cdots \\
&= D_0^2 x_0 + x_0 - K_2\cos(\Omega_2 t + \theta_2) \\
&\quad + \varepsilon\left[D_0^2 x_1 + x_1 + 2D_0 D_1 x_0 + 2\mu D_0 x_0 + \alpha x_0^3 - k_1\cos(\Omega_1 t + \theta_1)\right] + \cdots
\end{aligned} \tag{4.16.3}$$

Let the coefficient of the like power of ε be zero, we get

$$D_0^2 x_0 + x_0 = K_2\cos(\Omega_2 t + \theta_2) \tag{4.16.4}$$

$$D_0^2 x_1 + x_1 = -2D_0 D_1 x_0 - 2\mu D_0 x_0 - \alpha x_0^3 + k_1\cos(\Omega_1 t + \theta_1) \tag{4.16.5}$$

The solution of (4.16.4) is

$$x_0 = Ae^{iT_0} + \Lambda_2 e^{i\theta_2} e^{i\Omega_2 T_0} + cc \tag{4.16.6}$$

where

$$\Lambda_2 = \frac{K_2}{2(1-\Omega_2^2)} = -\frac{1}{16}K_2 \tag{4.16.7}$$

and $A = A(T_1)$. Substitute (4.16.6) into (4.16.5) and take $\Omega_1 \approx 1$ and $\Omega_2 \approx 3$ into account, we get

$$\begin{aligned} D_0^2 x_1 + x_1 &= -2D_0 D_1 x_0 - 2\mu D_0 x_0 - \alpha x_0^3 + k_1 \cos(\Omega_1 t + \theta_1) \\ &= -\left(2iD_1 A + 2i\mu A + 3\alpha A^2 \overline{A} + 6\alpha A \Lambda_2^2\right) e^{iT_0} + k_1 \cos(\Omega_1 t + \theta_1) \\ &\quad - 3\alpha \overline{A}^2 \Lambda_2 e^{i\theta_2} e^{i(\Omega_2 - 2)T_0} + cc + NST \end{aligned} \tag{4.16.8}$$

Let

$$\Omega_1 = 1 + \varepsilon\sigma_1, \quad \Omega_2 = 3 + \varepsilon\sigma_2 \tag{4.16.9}$$

Equation (4.16.8) becomes

$$\begin{aligned} D_0^2 x_1 + x_1 = &-(2iD_1 A + 2i\mu A + 3\alpha A^2 \overline{A} + 6\alpha A \Lambda_2^2 \\ &+ 3\alpha \overline{A}^2 \Lambda_2 e^{i(\sigma_2 T_1 + \theta_2)} - \frac{1}{2} k_1 e^{i(\sigma_1 T_1 + \theta_1)}) e^{iT_0} + cc + NST \end{aligned} \tag{4.16.10}$$

In order to eliminate secular terms, there must be

$$\begin{aligned} &2iD_1 A + 2i\mu A + 3\alpha A^2 \overline{A} + 6\alpha A \Lambda_2^2 \\ &\quad + 3\alpha \overline{A}^2 \Lambda_2 e^{i(\sigma_2 T_1 + \theta_2)} - \frac{1}{2} k_1 e^{i(\sigma_1 T_1 + \theta_1)} = 0 \end{aligned} \tag{4.16.11}$$

Let $A = \frac{1}{2} a e^{i\beta}$, which is given by (4.16.11) we have

$$\begin{aligned} &ia' - a\beta' + i\mu a + \frac{3}{8}\alpha a^3 + 3\alpha \Lambda_2^2 a \\ &\quad + \frac{3}{4}\alpha \Lambda_2 a^2 e^{i(\sigma_2 T_1 + \theta_2 - 3\beta)} - \frac{1}{2} k_1 e^{i(\sigma_1 T_1 + \theta_1 - \beta)} = 0 \end{aligned} \tag{4.16.12}$$

Separating the real and imaginary parts of the above equation yields, we get

$$a' = -\mu a + \frac{1}{2} k_1 \sin(\sigma_1 T_1 - \beta + \theta_1) - \frac{3}{4}\alpha \Lambda_2 a^2 \sin(\sigma_2 T_1 - 3\beta + \theta_2) \tag{4.16.13}$$

$$\begin{aligned} a\beta' = &3\alpha\left(\Lambda_2^2 + \frac{1}{8}a^2\right)a - \frac{1}{2} k_1 \cos(\sigma_1 T_1 - \beta + \theta_1) \\ &+ \frac{3}{4}\alpha \Lambda_2 a^2 \cos(\sigma_2 T_1 - 3\beta + \theta_2) \end{aligned} \tag{4.16.14}$$

(b) It can be seen from Eqs. (4.16.13) and (4.16.14) that the system has a steady state solution when and only when both $\sigma_1 T_1 - \beta$ and $\sigma_2 T_1 - 3\beta$ are constant, hence we have

$$\beta' = \sigma_1 = \frac{1}{3}\sigma_2 \tag{4.16.15}$$

Therefore stead-state motions exist only when $\sigma_2 = 3\sigma_1$, i.e., $\Omega_2 = 3\Omega_1$. That is, steady-state motions occur only when the excitation is periodic.

Let

$$\gamma = \sigma_1 T_1 - \beta \tag{4.16.16}$$

then the amplitude and phase of the system in the steady state should satisfy the following two equations:

$$-\mu a + \frac{1}{2}k_1 \sin(\gamma + \theta_1) - \frac{3}{4}\alpha\Lambda_2 a^2 \sin(3\gamma + \theta_2) = 0 \tag{4.16.17}$$

$$\begin{aligned} &- a\sigma_1 + 3\alpha\left(\Lambda_2^2 + \frac{1}{8}a^2\right)a - \frac{1}{2}k_1 \cos(\gamma + \theta_1) \\ &+ \frac{3}{4}\alpha\Lambda_2 a^2 \cos(3\gamma + \theta_2) = 0 \end{aligned} \tag{4.16.18}$$

We can solve for a and γ from (4.16.17) and (4.16.18), and substitute them into (4.16.6) yields the first order steady state approximate solution

$$x_0 = a\cos(\Omega_1 t - \gamma) + \Lambda_2 \cos(\Omega_2 t + \theta_2) \tag{4.16.19}$$

Therefore steady-state motions, if they exist, are periodic. The effect of the nonlinearity is to adjust the frequency of the free-oscillation term to be perfectly commensurable with the frequencies of the excitation.

(c) For a given set of parameters, Eqs. (4.16.17) and (4.16.18) can be numerically solved to obtain the curves of $a \sim \sigma_1$ and $\gamma \sim \sigma_1$, as shown in Fig. 4.12.

To analyze the stability of the amplitude and phase of the system at any steady state, $(a_0,\ \gamma_0)$, we linearize (4.16.13) and (4.16.14). To do this, we make

$$a = a_0 + a_1, \quad \gamma = \gamma_0 + \gamma_1 \tag{4.16.20}$$

and obtain the linearized forms of (4.16.13) and (4.16.14):

$$\begin{aligned} a_1' = &-\left[\mu + \frac{3}{2}\alpha\ \Lambda_2 a_0 \sin(3\gamma_0 + \theta_2)\right]a_1 + \left[\frac{1}{2}k_1 \cos(\gamma_0 + \theta_1)\right. \\ &- \frac{9}{4}\alpha\ \Lambda_2 a_0^2 \cos(3\gamma_0 + \theta_2)]\ \gamma_1 \end{aligned} \tag{4.16.21}$$

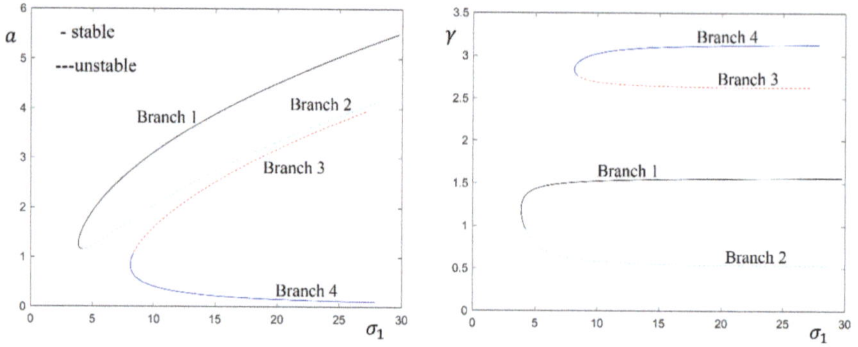

Fig. 4.12 Frequency-response and frequency-phase characteristics of the steady state solution ($\mu = 0.2,\ \alpha = 3,\ \theta_1 = 0,\ \theta_2 = \frac{\pi}{2},\ k_1 = 5,\ K_2 = 10$) for Exercise 4.16

$$\gamma_1' = a_0^{-1}\left[\sigma_1 - 3\alpha\left(\Lambda_2^2 + \frac{3}{8}a_0^2\right) - \frac{3}{2}\alpha\Lambda_2 a_0\cos(3\gamma_0 + \theta_2)\right]a_1 + a_0^{-1}\left[\frac{9}{4}\alpha\Lambda_2 a_0^2\sin(3\gamma_0 + \theta_2) - \frac{1}{2}k_1\sin(\gamma_0 + \theta_1)\right]\gamma_1 \tag{4.16.22}$$

Write the above two equations in matrix form as

$$\begin{Bmatrix} a_1' \\ \gamma_1' \end{Bmatrix} = \begin{bmatrix} l_{11} & l_{12} \\ l_{21} & l_{22} \end{bmatrix}\begin{Bmatrix} a_1 \\ \gamma_1 \end{Bmatrix} \tag{4.16.23}$$

where

$$l_{11} = -[\mu + \frac{3}{2}\alpha\Lambda_2 a_0\sin(3\gamma_0 + \theta_2)]$$

$$l_{12} = \frac{1}{2}k_1\cos(\gamma_0 + \theta_1) - \frac{9}{4}\alpha\Lambda_2 a_0^2\cos(3\gamma_0 + \theta_2)$$

$$l_{21} = a_0^{-1}[\sigma_1 - 3\alpha\left(\Lambda_2^2 + \frac{3}{8}a_0^2\right) - \frac{3}{2}\alpha\Lambda_2 a_0\cos(3\gamma_0 + \theta_2)]$$

$$l_{22} = a_0^{-1}[\frac{9}{4}\alpha\Lambda_2 a_0^2\sin(3\gamma_0 + \theta_2) - \frac{1}{2}k_1\sin(\gamma_0 + \theta_1)]$$

The eigenvalues of Eq. (4.16.23) are

$$\lambda = \frac{1}{2}\left[(l_{11} + l_{22}) \pm \sqrt{(l_{11} + l_{22})^2 - 4(l_{11}l_{22} - l_{12}l_{21})}\right] \tag{4.16.24}$$

The sign of the real part of the eigenvalue determines the stability of the corresponding steady state solution. It follows that branches 2 and 3 in Fig. 4.12 are not stable.

4.17 Exercise 4.17 (Primary and Combined Resonance for Cubic Nonlinear Systems I)

Solution: (a) Since $\Omega_1 \approx 1$, $K_1\cos(\Omega_1 t + \theta_1)$ is the primary resonant excitation term, while the other two excitations $K_2\cos(\Omega_2 t + \theta_2)$ and $K_3\cos(\Omega_3 t + \theta_3)$ are non-resonant hard excitations. Let the solution of the equation be

$$x(t;\varepsilon) = x_0(T_0, T_1, T_2) + \varepsilon x_1(T_0, T_1, T_2) + \cdots \tag{4.17.1}$$

and let the amplitude K_1 is soft, i.e.

$$K_1 = \varepsilon k_1 \tag{4.17.2}$$

Substituting (4.17.1) and (4.17.2) into the original equation, and retaining to $O(\varepsilon)$, we get

$$\begin{aligned}
0 &= \ddot{u} + u + 2\varepsilon\mu\dot{u} + \varepsilon\alpha u^3 - \sum_{n=1}^{3} K_n\cos(\Omega_n t + \theta_n)\\
&= \left(D_0^2 + 2\varepsilon D_0 D_1\right)(x_0 + \varepsilon x_1) + (x_0 + \varepsilon x_1)\\
&\quad + 2\varepsilon\mu(D_0 + \varepsilon D_1)(x_0 + \varepsilon x_1) + \varepsilon\alpha(x_0 + \varepsilon x_1)^3\\
&\quad - \varepsilon k_1\cos(\Omega_1 T_0 + \theta_1) - \sum_{n=2}^{3} K_n\cos(\Omega_n t + \theta_n) + \cdots\\
&= D_0^2 x_0 + \varepsilon D_0^2 x_1 + 2\varepsilon D_0 D_1 x_0 + x_0 + \varepsilon x_1\\
&\quad + 2\varepsilon\mu D_0 x_0 + \varepsilon\alpha x_0^3 - \varepsilon k_1\cos(\Omega_1 t + \theta_1) - \sum_{n=2}^{3} K_n\cos(\Omega_n t + \theta_n) + \cdots\\
&= D_0^2 x_0 + x_0 - \sum_{n=2}^{3} K_n\cos(\Omega_n t + \theta_n)\\
&\quad + \varepsilon\left[D_0^2 x_1 + x_1 + 2D_0 D_1 x_0 + 2\mu D_0 x_0 + \alpha x_0^3 - k_1\cos(\Omega_1 t + \theta_1)\right] + \cdots
\end{aligned} \tag{4.17.3}$$

Let the coefficient of the like power of ε be zero, we get

$$D_0^2 x_0 + x_0 = \sum_{n=2}^{3} K_n \cos(\Omega_n t + \theta_n) \tag{4.17.4}$$

$$D_0^2 x_1 + x_1 = -2D_0 D_1 x_0 - 2\mu D_0 x_0 - \alpha x_0^3 + k_1 \cos(\Omega_1 t + \theta_1) \tag{4.17.5}$$

The solution of (4.17.4) is

$$x_0 = Ae^{iT_0} + \sum_{n=2}^{3} \Lambda_n e^{i\theta_n} e^{i\Omega_n T_0} + cc \tag{4.17.6}$$

where

$$\Lambda_n = \frac{K_n}{2\left(1 - \Omega_n^2\right)} \tag{4.17.7}$$

and $A = A(T_1)$. Substituting (4.17.6) into (4.17.5) and taking $\Omega_1 \approx 1$ and $2\Omega_2 + \Omega_3 \approx 1$ into account, we get

$$\begin{aligned} D_0^2 x_1 + x_1 &= -2D_0 D_1 x_0 - 2\mu D_0 x_0 - \alpha x_0^3 + k_1 \cos(\Omega_1 t + \theta_1) \\ &= -\left(2iD_1 A + 2i\mu A + 3\alpha A^2 \overline{A} + 6\alpha \Lambda_2^2 A + 6\alpha \Lambda_3^2 A\right) e^{iT_0} \\ &\quad - 3\alpha \Lambda_2^2 \Lambda_3 e^{i(2\theta_2 + \theta_3)} e^{i(2\Omega_2 + \Omega_3)T_0} + k_1 \cos(\Omega_1 t + \theta_1) + cc + NST \end{aligned} \tag{4.17.8}$$

Let

$$\Omega_1 = 1 + \varepsilon\sigma_1, \quad 2\Omega_2 + \Omega_3 = 1 + \varepsilon\sigma_2 \tag{4.17.9}$$

Equation (4.16.8) becomes

$$\begin{aligned} D_0^2 x_1 + x_1 &= -2D_0 D_1 x_0 - 2\mu D_0 x_0 - \alpha x_0^3 + k_1 \cos(\Omega_1 t + \theta_1) \\ &= -\left(2iD_1 A + 2i\mu A + 3\alpha A^2 \overline{A} + 6\alpha \Lambda_2^2 A + 6\alpha \Lambda_3^2 A\right) e^{iT_0} \\ &\quad - 3\alpha \Lambda_2^2 \Lambda_3 e^{i(2\theta_2 + \theta_3 + \sigma_2 T_1)} e^{iT_0} + \frac{1}{2} k_1 e^{i(\sigma_1 T_1 + \theta_1)} e^{iT_0} + cc + NST \end{aligned} \tag{4.17.10}$$

In order to eliminate secular terms from (4.17.10), there must be

$$\begin{aligned} &2iD_1 A + 2i\mu A + 3\alpha A^2 \overline{A} + 6\alpha \Lambda_2^2 A + 6\alpha \Lambda_3^2 A \\ &\quad + 3\alpha \Lambda_2^2 \Lambda_3 e^{i(\sigma_2 T_1 + 2\theta_2 + \theta_3)} - \frac{1}{2} k_1 e^{i(\sigma_1 T_1 + \theta_1)} = 0 \end{aligned} \tag{4.17.11}$$

Let

$$A = \frac{1}{2} a e^{i\beta} \tag{4.17.12}$$

Substitute (4.17.12) into (4.16.11), we have

$$\begin{aligned} & ia' - a\beta' + i\mu a + \frac{3}{8}\alpha a^3 + 3\alpha\Lambda_2^2 a + 3\alpha\Lambda_3^2 a \\ & + 3\alpha\Lambda_2^2\Lambda_3 e^{i(\sigma_2 T_1 - \beta + 2\theta_2 + \theta_3)} - \frac{1}{2}k_1 e^{i(\sigma_1 T_1 - \beta + \theta_1)} = 0 \end{aligned} \tag{4.17.13}$$

Separating the real and imaginary parts of the above equation yields

$$\begin{aligned} a' = & -\mu a + \frac{1}{2}k_1 \sin(\sigma_1 T_1 - \beta + \theta_1) \\ & - 3\alpha\Lambda_2^2\Lambda_3 \sin(\sigma_2 T_1 - \beta + 2\theta_2 + \theta_3) \end{aligned}$$

$$\begin{aligned} a\beta' = & \alpha\left(3\Lambda_2^2 + 3\Lambda_3^2 + \frac{3}{8}a^2\right)a - \frac{1}{2}k_1 \cos(\sigma_1 T_1 - \beta + \theta_1) \\ & + 3\alpha\Lambda_2^2\Lambda_3 \cos(\sigma_2 T_1 - \beta + 2\theta_2 + \theta_3) \end{aligned}$$

These two equations can be written as

$$a' = -\mu a + \frac{1}{2}k_1 \sin\gamma_1 - \alpha\Gamma_1 \sin\gamma_2 \tag{4.17.14}$$

$$a\beta' = \alpha\left(\Gamma_2 + \frac{3}{8}a^2\right)a - \frac{1}{2}k_1 \cos\gamma_1 + \alpha\Gamma_1 \cos\gamma_2 \tag{4.17.15}$$

where

$$\gamma_1 = \sigma_1 T_1 - \beta + \theta_1, \quad \gamma_2 = \sigma_2 T_1 - \beta + 2\theta_2 + \theta_3 \tag{4.17.16}$$

$$\Gamma_1 = 3\Lambda_2^2\Lambda_3, \quad \Gamma_2 = 3\left(\Lambda_2^2 + \Lambda_3^2\right) \tag{4.17.17}$$

It can be seen that Γ_1 and Γ_2 are constants. Substitute (4.17.12) into (4.17.6), we obtain the first order approximate solution of the system

$$u = a(T_1)\cos[T_0 + \beta(T_1)] + 2\sum_{n=2}^{3} \Lambda_n \cos(\Omega_n T_0 + \theta_n) + O(\varepsilon) \tag{4.17.18}$$

where a and γ are given by (4.17.14) and (4.17.15).

(b) It can be seen that the system has a steady state solution when and only when both γ_1 and γ_2 are constant, hence

$$\beta' = \sigma_1 = \sigma_2 \tag{4.17.19}$$

Let

$$\gamma = \sigma_1 T_1 - \beta = \sigma_2 T_1 - \beta \tag{4.17.20}$$

The steady state amplitude and phase can be obtained by (4.17.14) and (4.17.15), i.e.,

$$-\mu a + \frac{1}{2}k_1 \sin(\gamma + \theta_1) - \alpha\Gamma_1 \sin(\gamma + 2\theta_2 + \theta_3) = 0 \tag{4.17.21}$$

$$-\sigma_1 a + \alpha\left(\Gamma_2 + \frac{3}{8}a^2\right)a - \frac{1}{2}k_1 \cos(\gamma + \theta_1) + \alpha\Gamma_1 \cos(\gamma + 2\theta_2 + \theta_3) = 0 \tag{4.17.22}$$

We can solve for a and γ from (4.17.21) and (4.17.22) and substitute them into (4.17.18) to obtain the first order steady state solution of the system. The frequencies of these response terms are not necessarily commensurable with each other, so the total response is not necessarily a periodic solution.

(c) For a given set of parameters, Eqs. (4.17.21) and (4.17.22) can be numerically solved to obtain the curves of $a \sim \sigma_1$ and $\gamma \sim \sigma_1$, as shown in Fig. 4.13.

To analyze the stability of any steady state $(a_0,\ \gamma_0)$, we linearize (4.17.14) and (4.17.15) and let

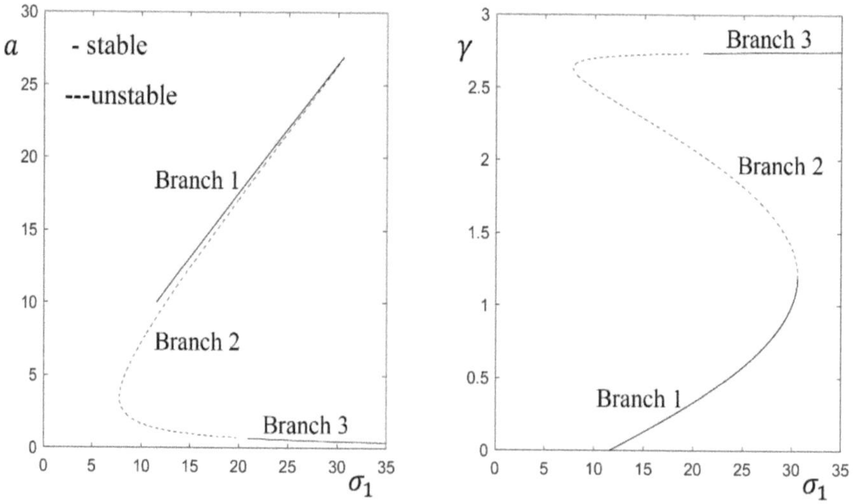

Fig. 4.13 Frequency-response and frequency-phase characteristics of the steady state solution ($\mu = 0.2,\ \alpha = 3,\ \theta_1 = 0,\ \theta_2 = \theta_3 = \frac{\pi}{2},\ \Omega_2 = 0.25,\ \Omega_3 = 0.6)k_1 = 10,\ K_2 = K_3 = 1$) for Exercise 4.17

$$a = a_0 + \tilde{a}, \quad \gamma = \gamma_0 + \tilde{\gamma} \tag{4.17.23}$$

The linearized forms of Eqs. (4.17.14) and (4.17.15) are

$$\tilde{a}\prime = -\mu\tilde{a} + [\frac{1}{2}k_1 \cos(\gamma_0 + \theta_1) - \alpha\Gamma_1 \cos(\gamma_0 + 2\theta_2 + \theta_3)]\tilde{\gamma} \tag{4.17.24}$$

$$\begin{aligned}\tilde{\gamma}\prime = &\left[\frac{\sigma_1}{a_0} - \frac{\alpha}{a_0}\left(\Gamma_2 + \frac{9}{8}a_0^2\right)\right] \\ &\tilde{a} - \frac{1}{a_0}\left[\frac{1}{2}k_1 \sin(\gamma_0 + \theta_1) - \alpha\Gamma_1 \sin(\gamma_0 + 2\theta_2 + \theta_3)\right]\tilde{\gamma}\end{aligned} \tag{4.17.25}$$

Write the above two equations in matrix form as

$$\begin{Bmatrix} \tilde{a}\prime \\ \tilde{\gamma}\prime \end{Bmatrix} = \begin{bmatrix} l_{11} & l_{12} \\ l_{21} & l_{22} \end{bmatrix} \begin{Bmatrix} \tilde{a} \\ \tilde{\gamma} \end{Bmatrix} \tag{4.17.26}$$

where

$$l_{11} = -\mu$$

$$l_{12} = \frac{1}{2}k_1 \cos(\gamma_0 + \theta_1) - \alpha\Gamma_1 \cos(\gamma_0 + 2\theta_2 + \theta_3)$$

$$l_{21} = \frac{1}{a_0}\left[\sigma_1 - \alpha\left(\Gamma_2 + \frac{9}{8}a_0^2\right)\right]$$

$$l_{22} = -\frac{1}{a_0}\left[\frac{1}{2}k_1 \sin(\gamma_0 + \theta_1\right) - \alpha\Gamma_1 \sin(\gamma_0 + 2\theta_2 + \theta_3)]$$

The eigenvalues of (4.17.26) are

$$\lambda = \frac{1}{2}\left[(l_{11} + l_{22}) \pm \sqrt{(l_{11} + l_{22})^2 - 4(l_{11}l_{22} - l_{12}l_{21})}\right] \tag{4.17.27}$$

The stability of the corresponding steady-state solution can be determined from the sign of the real part of the eigenvalue. It follows that branch 2 in Fig. 4.13 is unstable.

4.18 Exercise 4.18 (Combined Primary, Subharmonic, and Superharmonic Resonances of a Cubic Nonlinear System)

Solution: **(a)** Since $\Omega_1 \approx 1$, $K_1\cos(\Omega_1 t + \theta_1)$ is the primary resonant excitation term, while the other two excitations $K_2\cos(\Omega_2 t + \theta_2)$ and $K_3\cos(\Omega_3 t + \theta_3)$ are non-resonant hard excitations. Let the solution of the equation be

$$x(t;\varepsilon) = x_0(T_0, T_1, T_2) + \varepsilon x_1(T_0, T_1, T_2) + \cdots \tag{4.18.1}$$

and let the amplitude K_1 is soft, i.e.

$$K_1 = \varepsilon k_1 \tag{4.18.2}$$

Substituting (4.18.1) and (4.18.2) into the original equation, and retaining to $O(\varepsilon)$, we obtain

$$\begin{aligned}
0 &= \ddot{u} + u + 2\varepsilon\mu\dot{u} + \varepsilon\alpha u^3 - \sum_{n=1}^{3} K_n\cos(\Omega_n t + \theta_n) \\
&= \left(D_0^2 + 2\varepsilon D_0 D_1\right)(x_0 + \varepsilon x_1) + (x_0 + \varepsilon x_1) \\
&\quad + 2\varepsilon\mu(D_0 + \varepsilon D_1)(x_0 + \varepsilon x_1) + \varepsilon\alpha(x_0 + \varepsilon x_1)^3 \\
&\quad - \varepsilon k_1\cos(\Omega_1 T_0 + \theta_1) - \sum_{n=2}^{3} K_n\cos(\Omega_n t + \theta_n) \\
&= D_0^2 x_0 + \varepsilon D_0^2 x_1 + 2\varepsilon D_0 D_1 x_0 + x_0 + \varepsilon x_1 \\
&\quad + 2\varepsilon\mu D_0 x_0 + \varepsilon\alpha x_0^3 - \varepsilon k_1\cos(\Omega_1 t + \theta_1) - \sum_{n=2}^{3} K_n\cos(\Omega_n t + \theta_n) + \cdots \\
&= D_0^2 x_0 + x_0 - \sum_{n=2}^{3} K_n\cos(\Omega_n t + \theta_n) \\
&\quad + \varepsilon\left[D_0^2 x_1 + x_1 + 2D_0 D_1 x_0 + 2\mu D_0 x_0 + \alpha x_0^3 - k_1\cos(\Omega_1 t + \theta_1)\right] + \cdots
\end{aligned} \tag{4.18.3}$$

Let the coefficient of the like power of ε be zero, we get

$$D_0^2 x_0 + x_0 = \sum_{n=2}^{3} K_n\cos(\Omega_n t + \theta_n) \tag{4.18.4}$$

$$D_0^2 x_1 + x_1 = -2D_0 D_1 x_0 - 2\mu D_0 x_0 - \alpha x_0^3 + k_1\cos(\Omega_1 t + \theta_1) \tag{4.18.5}$$

The solution of (4.18.4) is

$$x_0 = Ae^{iT_0} + \sum_{n=2}^{3} \Lambda_n e^{i\theta_n} e^{i\Omega_n T_0} + cc \tag{4.18.6}$$

where

$$\Lambda_n = \frac{K_n}{2\left(1 - \Omega_n^2\right)} \tag{4.18.7}$$

and $A = A(T_1)$. Substituting (4.18.6) into (4.18.5) and taking $\Omega_1 \approx 1$, $\Omega_2 \approx \frac{1}{3}$ and $\Omega_3 \approx 3$ into account, we get

$$\begin{aligned} D_0^2 x_1 + x_1 &= -2D_0 D_1 x_0 - 2\mu D_0 x_0 - \alpha x_0^3 + k_1 \cos(\Omega_1 t + \theta_1) \\ &= -\left(2iD_1 A + 2i\mu A + 3\alpha A^2 \overline{A} + 6\alpha \Lambda_2^2 A + 6\alpha \Lambda_3^2 A\right) e^{iT_0} \\ &\quad - \alpha \Lambda_2^3 e^{i3\theta_2} e^{i3\Omega_2 T_0} - 3\alpha \Lambda_3 \overline{A}^2 e^{i\theta_3} e^{i(\Omega_3 - 2)T_0} \\ &\quad + k_1 \cos(\Omega_1 t + \theta_1) + cc + NST \end{aligned} \tag{4.18.8}$$

Let

$$\Omega_1 = 1 + \varepsilon\sigma_1, \quad 3\Omega_2 = 1 + \varepsilon\sigma_2, \quad \Omega_3 = 3 + \varepsilon\sigma_3 \tag{4.18.9}$$

Equation (4.16.8) becomes

$$\begin{aligned} D_0^2 x_1 + x_1 &= -2D_0 D_1 x_0 - 2\mu D_0 x_0 - \alpha x_0^3 + k_1 \cos(\Omega_1 t + \theta_1) \\ &= -\left(2iD_1 A + 2i\mu A + 3\alpha A^2 \overline{A} + 6\alpha \Lambda_2^2 A + 6\alpha \Lambda_3^2 A\right) e^{iT_0} \\ &\quad - \alpha \Lambda_2^3 e^{i(3\theta_2 + \sigma_2 T_1)} e^{iT_0} - 3\alpha \Lambda_3 \overline{A}^2 e^{i(\theta_3 + \sigma_3 T_1)} e^{iT_0} \\ &\quad + \frac{1}{2} k_1 e^{i(\theta_1 + \sigma_1 T_1)} e^{iT_0} + cc + NST \end{aligned} \tag{4.18.10}$$

In order to eliminate secular terms from, there must be

$$\begin{aligned} &2iD_1 A + 2i\mu A + 3\alpha A^2 \overline{A} + 6\alpha \Lambda_2^2 A + 6\alpha \Lambda_3^2 A + \alpha \Lambda_2^3 e^{i(3\theta_2 + \sigma_2 T_1)} \\ &\quad + 3\alpha \Lambda_3 \overline{A}^2 e^{i(\theta_3 + \sigma_3 T_1)} - \frac{1}{2} k_1 e^{i(\theta_1 + \sigma_1 T_1)} e^{iT_0} = 0 \end{aligned} \tag{4.18.11}$$

Let

$$A = \frac{1}{2} a e^{i\beta} \tag{4.18.12}$$

Substitute (4.18.12) into (4.18.11), we have

$$ia' - a\beta' + i\mu a + \frac{3}{8}\alpha a^3 + 3\alpha \Lambda_2^2 a + 3\alpha \Lambda_3^2 a + \alpha \Lambda_2^3 e^{i(\sigma_2 T_1 - \beta + 3\theta_2)}$$

$$+\frac{3}{4}\alpha\Lambda_3 a^2 e^{i(\sigma_3 T_1 - 3\beta + \theta_3)} - \frac{1}{2}k_1 e^{i(\sigma_1 T_1 - \beta + \theta_1)} = 0 \tag{4.18.13}$$

Separating the real and imaginary parts of the above equation yields

$$a' = -\mu a + \frac{1}{2}k_1 \sin\gamma_1 - \alpha\Lambda_2^3 \sin\gamma_2 - \frac{3}{4}\alpha\Lambda_3 a^2 \sin\gamma_3 \tag{4.18.14}$$

$$a\beta' = 3\alpha\left(\Lambda_2^2 + \Lambda_3^2 + \frac{1}{8}a^2\right)a - \frac{1}{2}k_1\cos\gamma_1 + \alpha\Lambda_2^3\cos\gamma_2 + \frac{3}{4}\alpha\Lambda_3 a^2\cos\gamma_3 \tag{4.18.15}$$

where $\gamma_1 = \sigma_1 T_1 - \beta + \theta_1, \quad \gamma_2 = \sigma_2 T_1 - \beta + 3\theta_2, \quad \gamma_3 = \sigma_3 T_1 - 3\beta + \theta_3$

Substituting (4.18.14) and (4.18.15) into (4.18.16), we can obtain the first-order approximate solution of the system

$$u = a(T_1)\cos[T_0 + \beta(T_1)] + 2\sum_{n=2}^{3}\Lambda_n\cos(\Omega_n T_0 + \theta_n) + O(\varepsilon) \tag{4.18.16}$$

where a and γ are given by (4.18.14) and (4.18.15).

(b) It can be seen from (4.18.14) and (4.18.15) that the system has a steady state solution when and only when γ_1, γ_2 and γ_3 are all constant, hence

$$\beta' = \sigma_1 = \sigma_2 = \frac{1}{3}\sigma_3 \tag{4.18.17}$$

Let

$$\gamma = \sigma_1 T_1 - \beta = \sigma_2 T_1 - \beta = \frac{1}{3}\sigma_3 T_1 - \beta \tag{4.18.18}$$

then the steady state amplitude and phase can be obtained by (4.18.14) and (4.18.15):

$$-\mu a + \frac{1}{2}k_1\sin(\gamma + \theta_1) - \alpha\Lambda_2^3\sin(\gamma + 3\theta_2) - \frac{3}{4}\alpha\Lambda_3 a^2\sin(3\gamma + \theta_3) = 0 \tag{4.18.19}$$

$$-\sigma_1 a + 3\alpha\left(\Lambda_2^2 + \Lambda_3^2 + \frac{1}{8}a^2\right)a - \frac{1}{2}k_1\cos(\gamma + \theta_1) + \alpha\Lambda_2^3\cos(\gamma + 3\theta_2) + \frac{3}{4}\alpha\Lambda_3 a^2\cos(3\gamma + \theta_3) = 0 \tag{4.18.20}$$

We can solve for a and γ from (4.18.20) and (4.18.21), and substitute them into (4.18.17) yields the first order steady state approximate solution. Therefore steady-state motions, if they exist, are periodic. The effect of the nonlinearity is to adjust

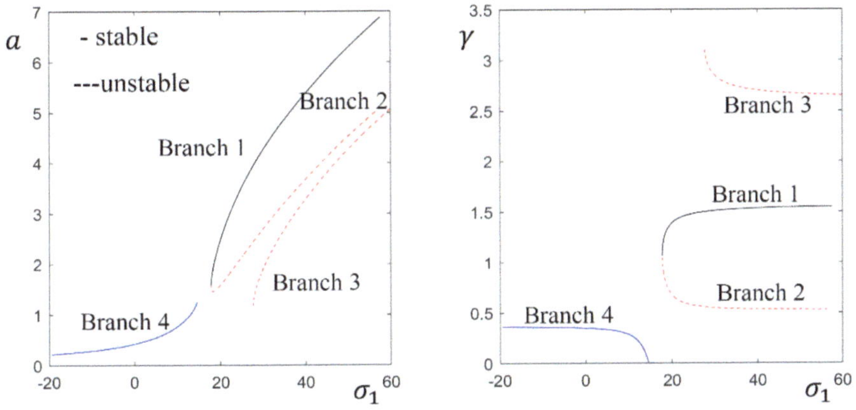

Fig. 4.14 Frequency-response and frequency-phase characteristics of the steady state solution ($\mu = 0.2,\ \alpha = 3,\ \theta_1 = 0,\ \theta_2 = \frac{\pi}{4},\ \theta_3 = \frac{\pi}{2},\ k_1 = 10,\ K_2 = 2,\ K_3 = 15$) for Exercise 4.18

the frequency of the free-oscillation term and other excavation terms to be perfectly commensurable with each other.

(c) For a given set of parameters, Eqs. (4.18.20) and (4.18.21) can be numerically solved to obtain the curves of $a \sim \sigma_1$ and $\gamma \sim \sigma_1$, as shown in Fig. 4.14.

To analyze the stability of any steady state $(a_0,\ \gamma_0)$, we linearize (4.18.14) and (4.18.15) and let

$$a = a_0 + \tilde{a}, \quad \gamma = \gamma_0 + \tilde{\gamma} \tag{4.18.21}$$

The linearized forms of Eqs. (4.18.14) and (4.18.15) are

$$\begin{aligned} \tilde{a}\prime = &-\left[\mu + \frac{3}{2}\alpha\Lambda_3 a_0 \sin(3\gamma + \theta_3)\right]\tilde{a} + \left[\frac{1}{2}k_1 \cos(\gamma_0 + \theta_1)\right.\\ &\left.-\alpha\Lambda_2^3 \cos(\gamma_0 + 3\theta_2) - \frac{9}{4}\alpha\Lambda_3 a^2 \cos(3\gamma_0 + \theta_3)\right]\tilde{\gamma} \\ \tilde{\gamma}\prime = &\, a_0^{-1}\left[\sigma_1 - 3\alpha\left(\Lambda_2^2 + \Lambda_3^2 + \frac{3}{8}a_0^2\right)\frac{3}{2}\alpha\Lambda_3 a_0 \cos(3\gamma_0 + \theta_3)\right]\tilde{a} \\ &+ a_0^{-1}\left[\alpha\Lambda_2^3 \sin(\gamma_0 + 3\theta_2\right] - \frac{1}{2}k_1 \sin(\gamma_0 + \theta_1) + \frac{9}{4}\alpha\Lambda_3 a_0^2 \sin(3\gamma_0 + \theta_3)]\tilde{\gamma} \end{aligned}$$

Write the above two equations in matrix form as

$$\begin{Bmatrix} \tilde{a}' \\ \tilde{\gamma}' \end{Bmatrix} = \begin{bmatrix} l_{11} & l_{12} \\ l_{21} & l_{22} \end{bmatrix} \begin{Bmatrix} \tilde{a} \\ \tilde{\gamma} \end{Bmatrix} \tag{4.18.22}$$

where

$$l_{11} = -\left[\mu + \frac{3}{2}\alpha\Lambda_3 a_0 \sin(3\gamma + \theta_3)\right]$$

$$l_{12} = \frac{1}{2}k_1 \cos(\gamma_0 + \theta_1) - \alpha\Lambda_2^3 \cos(\gamma_0 + 3\theta_2) - \frac{9}{4}\alpha\Lambda_3 a_0^2 \cos(3\gamma_0 + \theta_3)$$

$$l_{21} = a_0^{-1}\left[\sigma_1 - 3\alpha\left(\Lambda_2^2 + \Lambda_3^2 + \frac{3}{8}a_0^2\right) - \frac{3}{2}\alpha\Lambda_3 a_0 \cos(3\gamma_0 + \theta_3)\right]$$

$$l_{22} = a_0^{-1}\left[\alpha\Lambda_2^3 \sin(\gamma_0 + 3\theta_2) - \frac{1}{2}k_1 \sin(\gamma_0 + \theta_1) + \frac{9}{4}\alpha\Lambda_3 a_0^2 \sin(3\gamma_0 + \theta_3)\right]$$

The eigenvalues of Eq. (4.18.23) are

$$\lambda = \frac{1}{2}\left[(l_{11} + l_{22}) \pm \sqrt{(l_{11} + l_{22})^2 - 4(l_{11}l_{22} - l_{12}l_{21})}\right] \tag{4.18.23}$$

The sign of the real part of the eigenvalue determines the stability of the corresponding steady state solution. It follows that branches 2 and 3 in Fig. 4.14 are not stable.

4.19 Exercise 4.19 (Primary and Combined Resonance for Cubic Nonlinear Systems II)

Readers are invited to solve this problem by referring to Exercise 4.17.

4.20 Exercise 4.20 (Primary and Combined Resonance of a Cubic Nonlinear System)

Solution: (a) Since $\Omega_1 \approx 1$, $K_1 \cos(\Omega_1 t + \theta_1)$ is the primary resonant excitation term, while the other two excitations $K_2 \cos(\Omega_2 t + \theta_2)$ and $K_3 \cos(\Omega_3 t + \theta_3)$ are non-resonant hard excitations. Let the solution of the equation be

$$x(t; \varepsilon) = x_0(T_0, T_1, T_2) + \varepsilon x_1(T_0, T_1, T_2) + \cdots \tag{4.20.1}$$

and let the amplitude K_1 is soft, i.e.

$$K_1 = \varepsilon k_1 \tag{4.20.2}$$

Substituting (4.20.1) and (4.20.2) into the original equation, and retaining to $O(\varepsilon)$, we get

$$
\begin{aligned}
0 &= \ddot{u} + u + 2\varepsilon\mu\dot{u} + \varepsilon\alpha u^3 - \sum_{n=1}^{3} K_n \cos(\Omega_n t + \theta_n) \\
&= \left(D_0^2 + 2\varepsilon D_0 D_1\right)(x_0 + \varepsilon x_1) + (x_0 + \varepsilon x_1) \\
&\quad + 2\varepsilon\mu(D_0 + \varepsilon D_1)(x_0 + \varepsilon x_1) + \varepsilon\alpha(x_0 + \varepsilon x_1)^3 \\
&\quad - \varepsilon k_1 \cos(\Omega_1 T_0 + \theta_1) - \sum_{n=2}^{3} K_n \cos(\Omega_n t + \theta_n) + \cdots \\
&= D_0^2 x_0 + \varepsilon D_0^2 x_1 + 2\varepsilon D_0 D_1 x_0 + x_0 + \varepsilon x_1 \\
&\quad + 2\varepsilon\mu D_0 x_0 + \varepsilon\alpha x_0^3 - \varepsilon k_1 \cos(\Omega_1 t + \theta_1) - \sum_{n=2}^{3} K_n \cos(\Omega_n t + \theta_n) + \cdots \\
&= D_0^2 x_0 + x_0 - \sum_{n=2}^{3} K_n \cos(\Omega_n t + \theta_n) \\
&\quad + \varepsilon\left[D_0^2 x_1 + x_1 + 2D_0 D_1 x_0 + 2\mu D_0 x_0 + \alpha x_0^3 - k_1 \cos(\Omega_1 t + \theta_1)\right] + \cdots
\end{aligned}
\tag{4.20.3}
$$

Let the coefficient of the like power of ε be zero, we get

$$
D_0^2 x_0 + x_0 = \sum_{n=2}^{3} K_n \cos(\Omega_n t + \theta_n) \tag{4.20.4}
$$

$$
D_0^2 x_1 + x_1 = -2D_0 D_1 x_0 - 2\mu D_0 x_0 - \alpha x_0^3 + k_1 \cos(\Omega_1 t + \theta_1) \tag{4.20.5}
$$

The solution of (4.20.4) is

$$
x_0 = A e^{iT_0} + \sum_{n=2}^{3} \Lambda_n e^{i\theta_n} e^{i\Omega_n T_0} + cc \tag{4.20.6}
$$

Where

$$
\Lambda_n = \frac{K_n}{2\left(1 - \Omega_n^2\right)} \tag{4.20.7}
$$

and $A = A(T_1)$. Substituting (4.20.6) into (4.20.5) and taking $\Omega_1 \approx 1$ and $\Omega_2 + \Omega_3 \approx 2$ into account, we get

$$
\begin{aligned}
D_0^2 x_1 + x_1 &= -2D_0 D_1 x_0 - 2\mu D_0 x_0 - \alpha x_0^3 + k_1 \cos(\Omega_1 t + \theta_1) \\
&= -\left(2iD_1 A + 2i\mu A + 3\alpha A^2 \bar{A} + 6\alpha\Lambda_2^2 A + 6\alpha\Lambda_3^2 A\right) e^{iT_0} \\
&\quad - 6\alpha\Lambda_2\Lambda_3 \bar{A} e^{i(\theta_2 + \theta_3)} e^{i(\Omega_2 + \Omega_3 - 1)T_0} + k_1 \cos(\Omega_1 t + \theta_1) + cc + NST
\end{aligned}
\tag{4.20.8}
$$

Let

$$\Omega_1 = 1 + \varepsilon\sigma_1, \quad \Omega_2 + \Omega_3 = 2 + \varepsilon\sigma_2 \tag{4.20.9}$$

Equation (4.20.8) becomes

$$\begin{aligned} D_0^2 x_1 + x_1 &= -2D_0D_1x_0 - 2\mu D_0 x_0 - \alpha x_0^3 + k_1\cos(\Omega_1 t + \theta_1) \\ &= -\left(2iD_1A + 2i\mu A + 3\alpha A^2\bar{A} + 6\alpha\Lambda_2^2 A + 6\alpha\Lambda_3^2 A\right)e^{iT_0} \\ &\quad - 6\alpha\Lambda_2\Lambda_3\bar{A}e^{i(\sigma_2 T_1 + \theta_2 + \theta_3)}e^{iT_0} + \frac{1}{2}k_1 e^{i(\sigma_1 T_1 + \theta_1)}e^{iT_0} + cc + NST \end{aligned} \tag{4.20.10}$$

In order to eliminate secular terms from (4.20.10), there must be

$$\begin{aligned} &2iD_1A + 2i\mu A + 3\alpha A^2\overline{A} + 6\alpha\Lambda_2^2 A + 6\alpha\Lambda_3^2 A \\ &\quad + 6\alpha\Lambda_2\Lambda_3\overline{A}e^{i(\sigma_2 T_1 + \theta_2 + \theta_3)} - \frac{1}{2}k_1 e^{i(\sigma_1 T_1 + \theta_1)} = 0 \end{aligned} \tag{4.20.11}$$

Let

$$A = \frac{1}{2}ae^{i\beta} \tag{4.20.12}$$

Substitute (4.20.12) into (4.20.11), we have

$$\begin{aligned} &ia' - a\beta' + i\mu a + \frac{3}{8}\alpha a^3 + 3\alpha\Lambda_2^2 a + 3\alpha\Lambda_3^2 a \\ &\quad + 3\alpha\Lambda_2\Lambda_3 a e^{i(\sigma_2 T_1 - 2\beta + \theta_2 + \theta_3)} - \frac{1}{2}k_1 e^{i(\sigma_1 T_1 - \beta + \theta_1)} = 0 \end{aligned} \tag{4.20.13}$$

Separating the real and imaginary parts of the above equation yields

$$a' = -\mu a + \frac{1}{2}k_1\sin(\sigma_1 T_1 - \beta + \theta_1) - 3\alpha\Lambda_2\Lambda_3 a\sin(\sigma_2 T_1 - 2\beta + \theta_2 + \theta_3)$$

$$\begin{aligned} a\beta' &= \alpha\left(3\Lambda_2^2 + 3\Lambda_3^2 + \frac{3}{8}a^2\right)a - \frac{1}{2}k_1\cos(\sigma_1 T_1 - \beta + \theta_1) \\ &\quad + 3\alpha\Lambda_2\Lambda_3 a\cos(\sigma_2 T_1 - 2\beta + \theta_2 + \theta_3) \end{aligned}$$

These two equations can also be written as

$$a' = -\mu a + \frac{1}{2}k_1\sin\gamma_1 - \alpha\Gamma_1 a\sin\gamma_2 \tag{4.20.14}$$

$$a\beta' = \alpha\left(\Gamma_2 + \frac{3}{8}a^2\right)a - \frac{1}{2}k_1\cos\gamma_1 + \alpha\Gamma_1 a\cos\gamma_2 \tag{4.20.15}$$

where

$$\Gamma_1 = 3\Lambda_2\Lambda_3, \quad \Gamma_2 = 3\left(\Lambda_2^2 + \Lambda_3^2\right) \tag{4.20.16}$$

$$\gamma_1 = \sigma_1 T_1 - \beta + \theta_1, \quad \gamma_2 = \sigma_2 T_1 - 2\beta + \theta_2 + \theta_3 \tag{4.20.17}$$

It can be seen that Γ_1 and Γ_2 are constants. Substitute (4.20.12) into (4.20.6), we obtain the first order approximate solution of the system

$$u = a(T_1)\cos[T_0 + \beta(T_1)] + 2\sum_{n=2}^{3} \Lambda_n \cos(\Omega_n T_0 + \theta_n) + O(\varepsilon) \tag{4.20.18}$$

where a and γ are given by (4.20.14) and (4.20.15).

(b) It can be seen from (4.20.14) and (4.20.15) that the system has a steady state solution when and only when both γ_1 and γ_2 are constant, hence

$$\beta' = \sigma_1 = \frac{1}{2}\sigma_2 \tag{4.20.19}$$

Let

$$\gamma = \sigma_1 T_1 - \beta = \frac{1}{2}(\sigma_2 T_1 - 2\beta) \tag{4.20.20}$$

then the steady state amplitude and phase can be obtained by (4.20.14) and (4.20.15), i.e.,

$$-\mu a + \frac{1}{2}k_1 \sin(\gamma + \theta_1) - \alpha\Gamma_1 a \sin(2\gamma + \theta_2 + \theta_3) = 0 \tag{4.20.21}$$

$$-\sigma_1 a + \alpha\left(\Gamma_2 + \frac{3}{8}a^2\right)a - \frac{1}{2}k_1 \cos(\gamma + \theta_1) + \alpha\Gamma_1 a \cos(2\gamma + \theta_2 + \theta_3) = 0 \tag{4.20.22}$$

We can solve for a and γ from (4.20.21) and (4.20.22) and substitute them into (4.20.18) to obtain the first order steady state solution of the system. The frequencies of these response terms are not necessarily commensurable with each other, so the total response is not necessarily a periodic solution.

(c) For a given set of parameters, Eqs. (4.20.21) and (4.20.22) can be numerically solved to obtain the curves of $a \sim \sigma_1$ and $\gamma \sim \sigma_1$, as shown in Fig. 4.15. To analyze the stability of any steady state $(a_0,\ \gamma_0)$, we linearize (4.20.14) and (4.20.15) and let

$$a = a_0 + \tilde{a}, \quad \gamma = \gamma_0 + \tilde{\gamma} \tag{4.20.23}$$

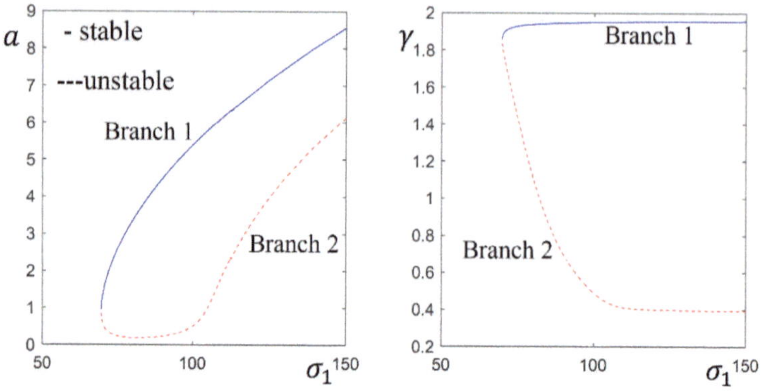

Fig. 4.15 Frequency-response and frequency-phase of the steady state solution ($\mu = 0.2,\ \alpha = 3,\ \theta_1 = 0,\ \theta_2 = \frac{\pi}{4},\ \theta_3 = \frac{\pi}{2}, \Omega_2 = 0.71,\ \Omega_3 = 1.29,\ k_1 = 9,\ K_2 = 3,\ K_3 = 1$) for Exercise 4.20

The linearized form of Eqs. (4.20.14) and (4.20.15) are

$$\begin{aligned}\widetilde{a'} &= -[\mu + \alpha\Gamma_1 \sin(2\gamma_0 + \theta_2 + \theta_3)]\tilde{a}\\ &+[\tfrac{1}{2}k_1 \cos(\gamma_0 + \theta_1) - 2\alpha\Gamma_1 a_0 \cos(2\gamma_0 + \theta_2 + \theta_3)]\tilde{\gamma}\end{aligned} \tag{4.20.24}$$

$$\begin{aligned}\widetilde{\gamma'} &= a_0^{-1}\left[\sigma_1 - \alpha\left(\Gamma_2 + \tfrac{9}{8}a_0^2\right) - \alpha\Gamma_1 \cos(2\gamma_0 + \theta_2 + \theta_3)\right]\tilde{a}\\ &-a_0^{-1}\left[\tfrac{1}{2}k_1 \sin(\gamma_0 + \theta_1) - 2\alpha\Gamma_1 \sin(2\gamma_0 + \theta_2 + \theta_3)\right]\tilde{\gamma}\end{aligned} \tag{4.20.25}$$

Write above two equations in matrix form as

$$\begin{Bmatrix} \tilde{a}\prime \\ \tilde{\gamma}\prime \end{Bmatrix} = \begin{bmatrix} l_{11} & l_{12} \\ l_{21} & l_{22} \end{bmatrix} \begin{Bmatrix} \tilde{a} \\ \tilde{\gamma} \end{Bmatrix} \tag{4.20.26}$$

where $l_{11} = -[\mu + \alpha\Gamma_1 \sin(2\gamma_0 + \theta_2 + \theta_3)]$

$$l_{12} = \frac{1}{2}k_1 \cos(\gamma_0 + \theta_1) - 2\alpha\Gamma_1 a_0 \cos(2\gamma_0 + \theta_2 + \theta_3)$$

$$l_{21} = a_0^{-1}[\sigma_1 - \alpha(\Gamma_2 + \frac{9}{8}a_0^2) - \alpha\Gamma_1 \cos(2\gamma_0 + \theta_2 + \theta_3)]$$

$$l_{22} = -a_0^{-1}[\frac{1}{2}k_1 \sin(\gamma_0 + \theta_1) - 2\alpha\Gamma_1 \sin(2\gamma_0 + \theta_2 + \theta_3)]$$

The eigenvalues of Eq. (4.20.26) are

$$\lambda = \frac{1}{2}\left[(l_{11} + l_{22}) \pm \sqrt{(l_{11} + l_{22})^2 - 4(l_{11}l_{22} - l_{12}l_{21})}\right] \tag{4.20.27}$$

The stability of the corresponding steady-state solution can be determined from the sign of the real part of the eigenvalue. It follows that branch 2 in Fig. 4.15 is unstable.

4.21 Exercise 4.21 (Primary Resonance of the System with Coulomb Friction and Quadratic Nonlinearity)

Solution: (a) Let

$$u = \varepsilon x \tag{4.21.1}$$

The original equation becomes

$$\ddot{x} + x + \frac{\hat{\mu}}{\varepsilon}\mathrm{sgn}(\dot{x}) + \varepsilon\alpha x^2 = \frac{K}{\varepsilon}\cos\Omega t \tag{4.21.2}$$

In order to find the primary resonance solution, it is necessary that the nonlinear, damping and excitation terms are of the same order, for this purpose, the above equation has to be expanded to the order of $O(\varepsilon^2)$, thus letting

$$\hat{\mu} = \varepsilon^3\mu, \quad K = \varepsilon^3 k \tag{4.21.3}$$

Then the Eq. (4.21.2) becomes

$$\ddot{x} + x + \varepsilon^2\mu\mathrm{sgn}(\dot{x}) + \varepsilon\alpha x^2 = \varepsilon^2 k\cos\Omega t \tag{4.21.4}$$

Let the solution of (4.21.4) be

$$x(t;\varepsilon) = x_0(T_0, T_1, T_2) + \varepsilon x_1(T_0, T_1, T_2) + \varepsilon^2 x_1(T_0, T_1, T_2) + \cdots \tag{4.21.5}$$

Substituting (4.21.5) into (4.21.4) and keeping to $O(\varepsilon^2)$, we get

$$\begin{aligned}
0 &= \ddot{x} + x + \varepsilon^2\mu\mathrm{sgn}(\dot{x}) + \varepsilon\alpha x^2 - \varepsilon^2 k\cos\Omega t \\
&= \left[D_0^2 + 2\varepsilon D_0D_1 + \varepsilon^2\left(D_1^2 + 2D_0D_2\right)\right]\left(x_0 + \varepsilon x_1 + \varepsilon^2 x_2\right) \\
&\quad + \left(x_0 + \varepsilon x_1 + \varepsilon^2 x_2\right) + \varepsilon\alpha(x_0 + \varepsilon x_1 + \varepsilon^2 x_2)^2 - \varepsilon^2 k\cos\Omega t \\
&\quad + \varepsilon^2\mu\mathrm{sgn}(D_0x_0) + \cdots \\
&= D_0^2x_0 + \varepsilon D_0^2x_1 + \varepsilon^2 D_0^2x_2 + 2\varepsilon D_0D_1x_0 + 2\varepsilon^2 D_0D_1x_1 \\
&\quad + \varepsilon^2\left(D_1^2 + 2D_0D_2\right)x_0 + x_0 + \varepsilon x_1 + \varepsilon^2 x_2 + \varepsilon\alpha x_0^2 + 2\varepsilon^2\alpha x_0x_1 \\
&\quad - \varepsilon^2 k\cos\Omega t + \varepsilon^2\mu\mathrm{sgn}(D_0x_0) + \cdots \\
&= D_0^2x_0 + x_0 + \varepsilon\left(D_0^2x_1 + x_1 + 2D_0D_1x_0 + \alpha x_0^2\right)
\end{aligned}$$

$$+ \varepsilon^2[D_0^2x_2 + x_2 + 2D_0D_1x_1 + (D_1^2 + 2D_0D_2)x_0 + 2\alpha x_0x_1 + \mu\mathrm{sgn}(D_0x_0) - k\cos\Omega t] + \cdots \tag{4.21.6}$$

Let the coefficient of the like power of ε be zero, we get

$$D_0^2x_0 + x_0 = 0 \tag{4.21.7}$$

$$D_0^2x_1 + x_1 = -2D_0D_1x_0 - \alpha x_0^2 \tag{4.21.8}$$

$$D_0^2x_2 + x_2 = -2D_0D_1x_1 - (D_1^2 + 2D_0D_2)x_0 - 2\alpha x_0x_1 - \mu\mathrm{sgn}(D_0x_0) + k\cos\Omega t \tag{4.21.9}$$

The solution of (4.21.7) is

$$x_0 = Ae^{iT_0} + \overline{A}e^{-iT_0} \tag{4.21.10}$$

where $A = A(T_1, T_2)$. Substituting (4.21.10) into (4.21.8) yields

$$\begin{aligned} D_0^2x_1 + x_1 &= -2D_0D_1x_0 - \alpha x_0^2 \\ &= -2iD_1Ae^{iT_0} - 2\alpha A\bar{A} - \alpha A^2e^{i2T_0} + cc \end{aligned} \tag{4.21.11}$$

In order to eliminate secular terms in the above equation, we need

$$D_1A = 0 \quad \Rightarrow \quad A = A(T_2) \tag{4.21.12}$$

Therefore, Eq. (4.21.11) is solved by

$$x_1 = -2\alpha A\overline{A} + \frac{1}{3}\alpha A^2e^{i2T_0} + \frac{1}{3}\alpha\overline{A}^2e^{-i2T_0} \tag{4.21.13}$$

Substituting x_0 and x_1 into (4.21.9) and letting

$$\Omega = 1 + \varepsilon^2\sigma \tag{4.21.14}$$

then

$$\begin{aligned} D_0^2x_2 + x_2 &= -2D_0D_1x_1 - (D_1^2 + 2D_0D_2)x_0 - 2\alpha x_0x_1 \\ &\quad - \mu\mathrm{sgn}(D_0x_0) + k\cos\Omega t \\ &= -\left[2iD_2A - \frac{10}{3}\alpha^2A^2\overline{A} - \frac{1}{2}ke^{\sigma T_2}\right]e^{iT_0} \\ &\quad + cc - \mu\mathrm{sgn}(iAe^{iT_0} - i\overline{A}e^{-iT_0}) + NST \end{aligned} \tag{4.21.15}$$

In order to further process (4.21.15), we need to first deal with the symbolic function $\mathrm{sgn}(iAe^{iT_0}-i\bar{A}e^{-iT_0})$, which can be expanded into a Fourier series of complex exponential form:

$$\mathrm{sgn}(iAe^{iT_0} - i\bar{A}e^{-iT_0}) = \sum_{n=-\infty}^{\infty} f_n e^{inT_0} \tag{4.21.16}$$

where

$$f_n = \frac{1}{2\pi}\int_0^{2\pi}\left[\mathrm{sgn}(iAe^{iT_0} - i\bar{A}e^{-iT_0})\right]e^{-inT_0}dT_0 \tag{4.21.17}$$

Substituting (4.21.16) into (4.21.11) yields

$$\begin{aligned} D_0^2x_2 + x_2 &= -2D_0D_1x_1 - \left(D_1^2 + 2D_0D_2\right)x_0 - 2\alpha x_0x_1 \\ &\quad - \mu\mathrm{sgn}(D_0x_0) + k\cos\Omega t \\ &= -\left[2iD_2A - \frac{10}{3}\alpha^2A^2\bar{A} - \frac{1}{2}ke^{i\sigma T_2} + \mu f_1\right]e^{iT_0} + cc + NST \end{aligned} \tag{4.21.18}$$

In order to eliminate secular terms, there must be

$$2iD_2A - \frac{10}{3}\alpha^2A^2\bar{A} - \frac{1}{2}ke^{i\sigma T_2} + \mu f_1 = 0 \tag{4.21.19}$$

Let

$$A = \frac{1}{2}ae^{i\beta} \tag{4.21.20}$$

Then

$$\begin{aligned} f_1 &= \frac{1}{2\pi}\int_0^{2\pi}\left[\mathrm{sgn}\left(iAe^{iT_0} - i\bar{A}e^{-iT_0}\right)\right]e^{-iT_0}dT_0 \\ &= -\frac{1}{2\pi}\int^{2\pi}\mathrm{sgn}[a\sin(T_0+\beta)]e^{-iT_0}dT_0 \\ &= -\frac{e^{i\beta}}{2\pi}\int_0^{2\pi}\mathrm{sgn}(a\sin\phi)e^{-i\phi}d\phi = \frac{ie^{i\beta}}{\pi}\int_0^{\pi}\sin\phi d\phi = \frac{2ie^{i\beta}}{\pi} \end{aligned} \tag{4.21.21}$$

Substituting (4.21.20) and (4.21.21) into (4.21.19) yields

$$a' - ia\beta' - \frac{5}{12}\alpha^2a^3 - \frac{1}{2}ke^{i(\sigma T_2-\beta)} + 2i\mu/\pi = 0 \tag{4.21.22}$$

Separating the real and imaginary parts of the above equation yields

$$a' = -2\mu/\pi + \frac{1}{2}k\sin\gamma \tag{4.21.23}$$

$$a\gamma' = a\sigma + \frac{5}{12}\alpha^2 a^3 + \frac{1}{2}k\cos\gamma \tag{4.21.24}$$

where

$$\gamma = \sigma T_2 - \beta \tag{4.21.25}$$

Substituting x_0 and x_1 into (4.21.5), we obtain the first-order approximate solution of (4.21.4):

$$\begin{aligned} x &= a\cos(T_0+\beta) - \frac{1}{2}\varepsilon\alpha a^2 + \frac{1}{6}\varepsilon\alpha a^2\cos(2T_0+2\beta) + O\left(\varepsilon^2\right) \\ &= a\cos(\Omega t-\gamma) + \frac{1}{2}\varepsilon\alpha a^2\left[\frac{1}{3}\alpha\cos(2\Omega t-2\gamma) - 1\right] + O\left(\varepsilon^2\right) \end{aligned} \tag{4.21.26}$$

Then the solution of the original equation is

$$u = \varepsilon a\cos(\Omega t-\gamma) + \frac{1}{2}\alpha\left(\varepsilon^2 a^2\right)\left[\frac{1}{3}\alpha\cos(2\Omega t-2\gamma) - 1\right] + O\left(\varepsilon^3\right) \tag{4.21.27}$$

Denoting εa as a, we obtain the solution of the original equation:

$$u = a\cos(\Omega t-\gamma) + \frac{1}{2}\alpha a^2\left[\frac{1}{3}\alpha\cos(2\Omega t-2\gamma) - 1\right] + O\left(\varepsilon^3\right) \tag{4.21.28}$$

(b) Let $a' = \gamma' = 0$ in (4.21.23) and (4.21.24), we have

$$-\frac{2\mu}{\pi} + \frac{1}{2}k\sin\gamma = 0 \tag{4.21.29}$$

$$a\sigma + \frac{5}{12}\alpha^2 a^3 + \frac{1}{2}k\cos\gamma = 0 \tag{4.21.30}$$

Eliminating γ, we can obtain the frequency–response equation:

$$\left(a\sigma + \frac{5}{12}\alpha^2 a^3\right)^2 + \frac{4\mu^2}{\pi^2} = \frac{1}{4}k^2 \tag{4.21.31}$$

then

$$\sigma = -\frac{5}{12}\alpha^2 a^2 \pm \frac{1}{a}\sqrt{\frac{1}{4}k^2 - \frac{4\mu^2}{\pi^2}} \tag{4.21.32}$$

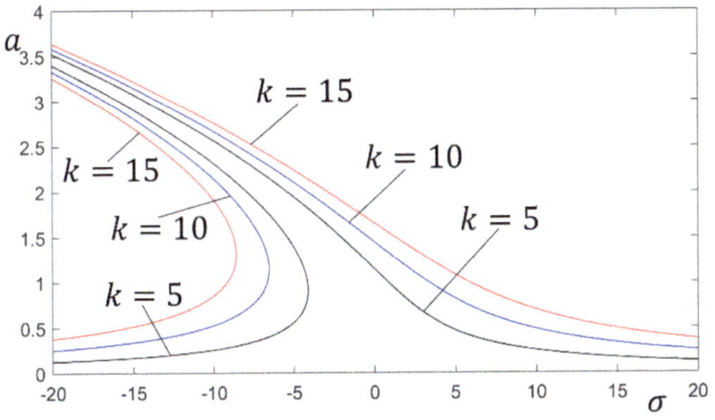

Fig. 4.16 Frequency-response curves ($\mu = 0.2,\ \alpha = 2$) for Exercise 4.21

The $a \sim \sigma$ curve can be sketched from (4.21.32), as shown in Fig. 4.16. As can be seen from the figure, jumps occur when σ changes from a large negative value to a positive σ direction. As σ decreases, a is unbounded.

4.22 Exercise 4.22 (Primary Resonance Problem for the System with Quadratic Damp and Quadratic Nonlinear Terms)

Solution: (a) Let

$$u = \varepsilon x \tag{4.22.1}$$

The original equation becomes

$$\ddot{x} + x + \varepsilon\hat{\mu}\dot{x}|\dot{x}| + \varepsilon\alpha x^2 = \frac{K}{\varepsilon}\cos\Omega t \tag{4.22.2}$$

In order to find the primary resonance solution, it is necessary that the nonlinear, damping and excitation terms are of the same order, for this purpose the above equation has to be expanded to the order of $O(\varepsilon^2)$, thus letting

$$\hat{\mu} = \varepsilon\mu, \quad K = \varepsilon^3 k \tag{4.22.3}$$

Then (4.22.2) becomes

$$\ddot{x} + x + \varepsilon^2\mu\dot{x}|\dot{x}| + \varepsilon\alpha x^2 = \varepsilon^2 k\cos\Omega t \tag{4.22.4}$$

Let the solution of (4.22.4) be

$$x(t;\varepsilon) = x_0(T_0, T_1, T_2) + \varepsilon x_1(T_0, T_1, T_2) + \varepsilon^2 x_1(T_0, T_1, T_2) + \cdots \tag{4.22.5}$$

Substituting (4.22.5) into (4.22.4) and keeping to O(ε^2), we get

$$\begin{aligned}
0 &= \ddot{x} + x + \varepsilon^2 \mu \dot{x}|\dot{x}| + \varepsilon \alpha x^2 - \varepsilon^2 k \cos \Omega t \\
&= \left[D_0^2 + 2\varepsilon D_0 D_1 + \varepsilon^2 \left(D_1^2 + 2D_0 D_2\right)\right]\left(x_0 + \varepsilon x_1 + \varepsilon^2 x_2\right) \\
&\quad + \left(x_0 + \varepsilon x_1 + \varepsilon^2 x_2\right) + \varepsilon \alpha (x_0 + \varepsilon x_1 + \varepsilon^2 x_2)^2 - \varepsilon^2 k \cos \Omega t \\
&\quad + \varepsilon^2 \mu D_0 x_0 \, |D_0 x_0| + \cdots \\
&= D_0^2 x_0 + \varepsilon D_0^2 x_1 + \varepsilon^2 D_0^2 x_2 + 2\varepsilon D_0 D_1 x_0 + 2\varepsilon^2 D_0 D_1 x_1 \\
&\quad + \varepsilon^2 \left(D_1^2 + 2D_0 D_2\right) x_0 + x_0 + \varepsilon x_1 + \varepsilon^2 x_2 + \varepsilon \alpha x_0^2 + 2\varepsilon^2 \alpha x_0 x_1 \\
&\quad - \varepsilon^2 k \cos \Omega t + \varepsilon^2 \mu D_0 x_0 \, |D_0 x_0| + \cdots \\
&= D_0^2 x_0 + x_0 + \varepsilon \left(D_0^2 x_1 + x_1 + 2D_0 D_1 x_0 + \alpha x_0^2\right) \\
&\quad + \varepsilon^2 [D_0^2 x_2 + x_2 + 2D_0 D_1 x_1 + \left(D_1^2 + 2D_0 D_2\right) x_0 \\
&\quad + 2\alpha x_0 x_1 + \mu D_0 x_0 \, |D_0 x_0| - k \cos \Omega t] + \cdots
\end{aligned} \tag{4.22.6}$$

Let the coefficient of the like power of ε be zero, we get

$$D_0^2 x_0 + x_0 = 0 \tag{4.22.7}$$

$$D_0^2 x_1 + x_1 = -2D_0 D_1 x_0 - \alpha x_0^2 \tag{4.22.8}$$

$$D_0^2 x_2 + x_2 = -2D_0 D_1 x_1 - \left(D_1^2 + 2D_0 D_2\right) x_0 - 2\alpha x_0 x_1 - \mu D_0 x_0 \, |D_0 x_0| + k \cos \Omega t \tag{4.22.9}$$

Solving (4.22.7) yields

$$x_0 = A e^{iT_0} + \bar{A} e^{-iT_0} \tag{4.22.10}$$

where $A = A(T_1, T_2)$. Substitute (4.22.10) into (4.22.8), we can obtain

$$\begin{aligned}
D_0^2 x_1 + x_1 &= -2D_0 D_1 x_0 - \alpha x_0^2 \\
&= -2iD_1 A e^{iT_0} - 2\alpha A\bar{A} - \alpha A^2 e^{i2T_0} + cc
\end{aligned} \tag{4.22.11}$$

In order to eliminate secular terms in the above equation, it is necessary to have

$$D_1 A = 0 \quad \Rightarrow \quad A = A(T_2) \tag{4.22.12}$$

Therefore, the Eq. (4.22.8) is solved by

$$x_1 = -2\alpha A\overline{A} + \frac{1}{3}\alpha A^2 e^{i2T_0} + \frac{1}{3}\alpha \overline{A}^2 e^{-i2T_0} \tag{4.22.13}$$

Substitute x_0 and x_1 into (4.22.9) and make

$$\Omega = 1 + \varepsilon^2 \sigma \tag{4.22.14}$$

we can obtain

$$\begin{aligned} D_0^2 x_2 + x_2 &= -2D_0 D_1 x_1 - \left(D_1^2 + 2D_0 D_2\right) x_0 - 2\alpha x_0 x_1 \\ &\quad - \mu \operatorname{sgn}(D_0 x_0) + k\cos\Omega t \\ &= -\left[2iD_2 A - \frac{10}{3}\alpha^2 A^2 \overline{A} - \frac{1}{2} k e^{\sigma T_2}\right] e^{iT_0} + cc \\ &\quad - \mu\left(iAe^{iT_0} - i\overline{A}e^{-iT_0}\right)\left|iAe^{iT_0} - i\overline{A}e^{-iT_0}\right| \operatorname{sgn}\left(iAe^{iT_0} - i\overline{A}e^{-iT_0}\right) + NST \end{aligned} \tag{4.22.15}$$

In order to further process (4.22.15), it is necessary to deal with, in advance, the last function in the above equation, which can be expanded into a Fourier series of complex exponential form:

$$\left(iAe^{iT_0} - i\overline{A}e^{-iT_0}\right)\left|iAe^{iT_0} - i\overline{A}e^{-iT_0}\right| \operatorname{sgn}(iAe^{iT_0} - i\overline{A}e^{-iT_0}) = \sum_{n=-\infty}^{\infty} f_n e^{inT_0} \tag{4.22.16}$$

$$\begin{aligned} f_n = \frac{1}{2\pi} \sum_0^{2\pi} \left(iAe^{iT_0} - i\overline{A}e^{-iT_0}\right)\left|iAe^{iT_0} - i\overline{A}e^{-iT_0}\right| \\ \operatorname{sgn}(iAe^{iT_0} - i\overline{A}e^{-iT_0}) e^{-inT_0} dT_0 \end{aligned} \tag{4.22.17}$$

Substituting (4.22.16) into (4.22.15), we can obtain

$$\begin{aligned} D_0^2 x_2 + x_2 &= -2D_0 D_1 x_1 - \left(D_1^2 + 2D_0 D_2\right) x_0 - 2\alpha x_0 x_1 \\ &\quad - \mu \operatorname{sgn}(D_0 x_0) + k\cos\Omega t \\ &= -\left[2iD_2 A - \frac{10}{3}\alpha^2 A^2 \overline{A} - \frac{1}{2} k e^{i\sigma T_2} + \mu f_1\right] e^{iT_0} + cc + NST \end{aligned} \tag{4.22.18}$$

In order to eliminate secular terms, there must be

$$2iD_2 A - \frac{10}{3}\alpha^2 A^2 \overline{A} - \frac{1}{2} k e^{i\sigma T_2} + \mu f_1 = 0 \tag{4.22.19}$$

Let

$$A = \frac{1}{2}ae^{i\beta} \tag{4.22.20}$$

then

$$\begin{aligned}
f_1 &= \frac{1}{2\pi}\int_0^{2\pi}\left(iAe^{iT_0} - i\bar{A}e^{-iT_0}\right)\left|iAe^{iT_0} - i\bar{A}e^{-iT_0}\right| \\
&\cdot \operatorname{sgn}\left(iAe^{iT_0} - i\bar{A}e^{-iT_0}\right)e^{-inT_0}dT_0 \\
&= -\frac{1}{2\pi}\int_0^{2\pi} a\sin(T_0+\beta)|a\sin(T_0+\beta)|\operatorname{sgn}(a\sin(T_0+\beta))e^{-iT_0}dT_0 \\
&= -\frac{e^{i\beta}}{2\pi}\int_0^{2\pi} a\sin\phi|a\sin\phi|\operatorname{sgn}(a\sin(T_0+\beta))e^{-i\phi}d\phi \\
&= \frac{ia^2e^{i\beta}}{\pi}\int_0^{\pi}\sin^3\phi d\phi = \frac{4ia^2e^{i\beta}}{3\pi}
\end{aligned} \tag{4.22.21}$$

Substituting (4.22.20) and (4.22.21) into (4.22.19), we can obtain

$$a' - ia\beta' - \frac{5}{12}\alpha^2a^3 - \frac{1}{2}ke^{i(\sigma T_2 - \beta)} + \frac{4ia^2\mu e^{i\beta}}{3\pi} = 0 \tag{4.22.22}$$

Separating the real and imaginary parts of the above equation yields

$$a' = -\frac{4\mu a^2}{3\pi} + \frac{1}{2}k\sin\gamma \tag{4.22.23}$$

$$a\gamma' = a\sigma + \frac{5}{12}\alpha^2a^3 + \frac{1}{2}k\cos\gamma \tag{4.22.24}$$

where

$$\gamma = \sigma T_2 - \beta \tag{4.22.25}$$

Substituting x_0 and x_1 into (4.22.5), we can obtain the first-order approximate solution of the system:

$$\begin{aligned}
x &= a\cos(T_0+\beta) - \frac{1}{2}\varepsilon\alpha a^2 + \frac{1}{6}\varepsilon\alpha a^2\cos(2T_0+2\beta) + O\left(\varepsilon^2\right) \\
&= a\cos(\Omega t - \gamma) + \frac{1}{2}\varepsilon\alpha a^2\left[\frac{1}{3}\alpha\cos(2\Omega t - 2\gamma) - 1\right] + O\left(\varepsilon^2\right)
\end{aligned} \tag{4.22.26}$$

Then the solution of the original equation is

$$u = \varepsilon a \cos(\Omega t - \gamma) + \frac{1}{2}\alpha\left(\varepsilon^2 a^2\right)\left[\frac{1}{3}\alpha \cos(2\Omega t - 2\gamma) - 1\right] + O\left(\varepsilon^3\right) \quad (4.22.27)$$

Denoting εa as a, we obtain the solution to the original equation

$$u = a\cos(\Omega t - \gamma) + \frac{1}{2}\alpha a^2\left[\frac{1}{3}\alpha \cos(2\Omega t - 2\gamma) - 1\right] + O\left(\varepsilon^3\right) \quad (4.22.28)$$

(b) Let $a' = \gamma' = 0$ in (4.22.23) and (4.22.24), we have

$$-\frac{4\mu a^2}{3\pi} + \frac{1}{2}k \sin\gamma = 0 \quad (4.22.29)$$

$$a\sigma + \frac{5}{12}\alpha^2 a^3 + \frac{1}{2}k\cos\gamma = 0 \quad (4.22.30)$$

Eliminating γ, we can obtain the frequency–response equation:

$$(a\sigma + \frac{5}{12}\alpha^2 a^3)^2 + \frac{16\mu^2 a^4}{9\pi^2} = \frac{1}{4}k^2 \quad (4.22.31)$$

We can solve σ in terms of a from this

$$\sigma = -\frac{5}{12}\alpha^2 a^2 \pm \frac{1}{a}\sqrt{\frac{1}{4}k^2 - \frac{16\mu^2 a^4}{9\pi^2}} \quad (4.22.32)$$

The $a \sim \sigma$ can be sketched from (4.22.32), as shown in Fig. 4.17. As can be seen from the figure, whenever σ changes from small to large or from large to small, there is a jump. In addition, there is a maximum value for a and the motion is bounded.

4.23 Exercise 4.23 (Combined Resonance Problem for Quadratic Nonlinear Systems)

Solution: **(a)** We assume that $K_n = O(1)(n = 1, 2, 3$, and let the solution of the equation be

$$u(t; \varepsilon) = x_0(T_0, T_1) + \varepsilon x_1(T_0, T_1) + \cdots \quad (4.23.1)$$

Substituting (4.23.1) into the original equation and retaining to $O(\varepsilon)$, we obtain

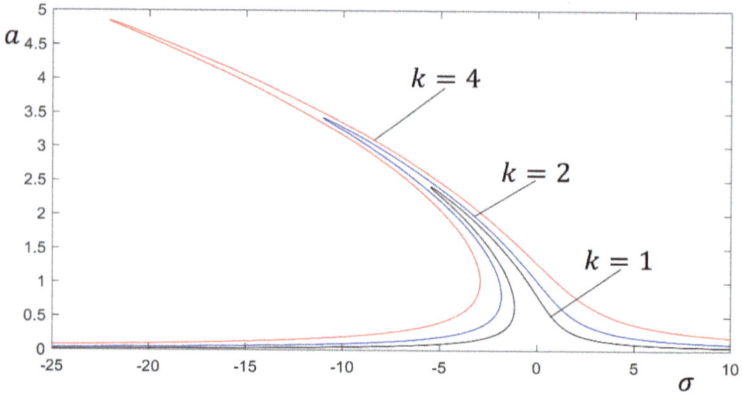

Fig. 4.17 Frequency-response curve ($\mu = 0.2,\ \alpha = 1.5$) for Exercise 4.22

$$
\begin{aligned}
0 &= \ddot{u} + u + 2\varepsilon\mu\dot{u} + \varepsilon\alpha u^2 - \sum_{n=1}^{3} K_n \cos(\Omega_n t + \theta_n) \\
&= \left(D_0^2 + 2\varepsilon D_0 D_1\right)(x_0 + \varepsilon x_1) + (x_0 + \varepsilon x_1) \\
&\quad + 2\varepsilon\mu(D_0 + \varepsilon D_1)(x_0 + \varepsilon x_1) + \varepsilon\alpha(x_0 + \varepsilon x_1)^2 \\
&\quad - \sum_{n=1}^{3} K_n \cos(\Omega_n T_0 + \theta_n) + \cdots \\
&= D_0^2 x_0 + \varepsilon D_0^2 x_1 + 2\varepsilon D_0 D_1 x_0 + x_0 + \varepsilon x_1 \\
&\quad + 2\varepsilon\mu D_0 x_0 + \varepsilon\alpha x_0^2 - \sum_{n=1}^{3} K_n \cos(\Omega_n T_0 + \theta_n) + \cdots \\
&= D_0^2 x_0 + x_0 - \sum_{n=1}^{3} K_n \cos(\Omega_n T_0 + \theta_n) \\
&\quad + \varepsilon\left[D_0^2 x_1 + x_1 + 2D_0 D_1 x_0 + 2\mu D_0 x_0 + \alpha x_0^2\right] + \cdots
\end{aligned} \tag{4.23.2}
$$

Let the coefficient of the like power of ε be zero, we get

$$
D_0^2 x_0 + x_0 = \sum_{n=1}^{3} K_n \cos(\Omega_n T_0 + \theta_n) \tag{4.23.3}
$$

$$
D_0^2 x_1 + x_1 = -2D_0 D_1 x_0 - 2\mu D_0 x_0 - \alpha x_0^2 \tag{4.23.4}
$$

The solution of (4.23.3) is

$$x_0 = Ae^{iT_0} + \sum_{n=1}^{3} \Lambda_n e^{i\theta_n} e^{i\Omega_n T_0} + cc \tag{4.23.5}$$

where $A = A(T_1)$ and

$$\Lambda_n = \frac{K_n}{2\left(1 - \Omega_n^2\right)} \tag{4.23.6}$$

Therefore, the first order approximate solution of the equation is

$$u = A(T_1)e^{iT_0} + cc + 2\sum_{n=1}^{3} \Lambda_n \cos(\Omega_n T_0 + \theta_n) + O(\varepsilon) \tag{4.23.7}$$

Substituting (4.23.5) into (4.23.4) and taking $\Omega_2 - \Omega_1 \approx 1$ and $\Omega_3 - \Omega_2 \approx 1$ into account, we get

$$\begin{aligned} D_0^2 x_1 + x_1 &= -2D_0 D_1 x_0 - 2\mu D_0 x_0 - \alpha x_0^2 \\ &= -(2iD_1 A + 2i\mu A)e^{iT_0} - 2\alpha \Lambda_1 \Lambda_2 e^{i(\theta_2 - \theta_1)} e^{i(\Omega_2 - \Omega_1)T_0} \\ &\quad - 2\alpha \Lambda_2 \Lambda_3 e^{i(\theta_3 - \theta_2)} e^{i(\Omega_3 - \Omega_2)T_0} + cc + NST \end{aligned} \tag{4.23.8}$$

Let

$$\varepsilon\sigma_1 = \Omega_2 - \Omega_1 - 1, \quad \varepsilon\sigma_2 = \Omega_3 - \Omega_2 - 1$$

$$\gamma_1 = \sigma_1 T_1 + \theta_2 - \theta_1, \quad \gamma_2 = \sigma_2 T_1 + \theta_3 - \theta_2 \tag{4.23.9}$$

Equation (4.23.8) becomes

$$\begin{aligned} D_0^2 x_1 + x_1 &= -(2iD_1 A + 2i\mu A)e^{iT_0} - 2\alpha \Lambda_1 \Lambda_2 e^{i(\theta_2 - \theta_1)} e^{i(\Omega_2 - \Omega_1)T_0} \\ &\quad - 2\alpha \Lambda_2 \Lambda_3 e^{i(\theta_3 - \theta_2)} e^{i(\Omega_3 - \Omega_2)T_0} + cc + NST \\ &= -(2iD_1 A + 2i\mu A)e^{iT_0} - 2\alpha \Lambda_1 \Lambda_2 e^{i(\sigma_1 T_1 + \theta_2 - \theta_1)} e^{iT_0} \\ &\quad - 2\alpha \Lambda_2 \Lambda_3 e^{i(\sigma_2 T_1 + \theta_3 - \theta_2)} e^{iT_0} + cc + NST \\ &= -2\left(iA' + i\mu A + \alpha \Lambda_1 \Lambda_2 e^{i\gamma_1} + \alpha \Lambda_2 \Lambda_3 e^{i\gamma_2}\right)e^{iT_0} \\ &\quad + cc + NST \end{aligned} \tag{4.23.10}$$

In order to eliminate secular terms from the above equation, there must be

$$i\left(A' + \mu A\right) + \alpha \Lambda_1 \Lambda_2 e^{i\gamma_1} + \alpha \Lambda_2 \Lambda_3 e^{i\gamma_2} = 0 \tag{4.23.11}$$

The prime denotes the derivative with respect to T_1.

(b) Write the Eq. (4.23.11) as

$$A' + \mu A = i\alpha\Lambda_1\Lambda_2 e^{i\gamma_1} + i\alpha\Lambda_2\Lambda_3 e^{i\gamma_2} \tag{4.23.12}$$

The general solution of the above equation is

$$A_1 = Ce^{-\mu T_1} \tag{4.23.13}$$

and the special solution of the equation is

$$A_2 = \frac{i\alpha\Lambda_1\Lambda_2}{\mu + i\sigma_1} e^{i\gamma_1} + \frac{i\alpha\Lambda_2\Lambda_3}{\mu + i\sigma_2} e^{i\gamma_2} \tag{4.23.14}$$

therefore, the solution of (4.23.12) is

$$A = A_1 + A_2 = Ce^{-\mu T_1} + \frac{i\alpha\Lambda_1\Lambda_2}{\mu + i\sigma_1} e^{i\gamma_1} + \frac{i\alpha\Lambda_2\Lambda_3}{\mu + i\sigma_2} e^{i\gamma_2} \tag{4.23.15}$$

(c) From (4.23.15), we know that the steady state value of A, A_s, is:

$$A_s = \frac{i\alpha\Lambda_1\Lambda_2}{\mu + i\sigma_1} e^{i\gamma_1} + \frac{i\alpha\Lambda_2\Lambda_3}{\mu + i\sigma_2} e^{i\gamma_2} \tag{4.23.16}$$

Substituting A_s into (4.23.7), we can obtain the steady state solution of the original equation as

$$\begin{aligned} x &= \frac{i\alpha\Lambda_1\Lambda_2}{\mu + i\sigma_1} e^{i(T_0+\gamma_1)} + \frac{i\alpha\Lambda_2\Lambda_3}{\mu + i\sigma_2} e^{i(T_0+\gamma_2)} \\ &\quad + cc + 2\sum_{n=1}^{3} \Lambda_n \cos(\Omega_n T_0 + \theta_n) + O(\varepsilon) \\ &= -\frac{2\alpha\Lambda_1\Lambda_2}{\sqrt{\mu^2 + \sigma_1^2}} \sin(T_0 + \gamma_1 - \beta_1) - \frac{2\alpha\Lambda_2\Lambda_3}{\sqrt{\mu^2 + \sigma_2^2}} \sin(T_0 + \gamma_2 + \beta_2) \\ &\quad + 2\sum_{n=1}^{3} \Lambda_n \cos(\Omega_n T_0 + \theta_n) + O(\varepsilon) \end{aligned}$$

i.e.,

$$\begin{aligned} x &= -\frac{2\alpha\Lambda_1\Lambda_2}{\sqrt{\mu^2 + \sigma_1^2}} \sin(T_0 + \gamma_1 - \beta_1) - \frac{2\alpha\Lambda_2\Lambda_3}{\sqrt{\mu^2 + \sigma_2^2}} \sin(T_0 + \gamma_2 + \beta_2) \\ &\quad + 2\sum_{n=1}^{3} \Lambda_n \cos(\Omega_n T_0 + \theta_n) + O(\varepsilon) \end{aligned} \tag{4.23.17}$$

where

$$\tan\beta_1 = \frac{\sigma_1}{\mu}, \quad \tan\beta_2 = \frac{\sigma_2}{\mu} \tag{4.23.18}$$

The frequencies appearing in this solution are not necessarily commensurable with each other, so this solution is not necessarily periodic.

4.24 Exercise 4.24 (Simultaneous Subharmonic and Superharmonic Resonance of Quadratic Nonlinear Systems)

Solution: (a) We assume that $K_n = O(1)(n = 1, 2)$. Let the solution of the equation be

$$u(t;\varepsilon) = x_0(T_0, T_1) + \varepsilon x_1(T_0, T_1) + \cdots \tag{4.24.1}$$

Substituting (4.23.1) into the original equation and retaining to $O(\varepsilon)$ yields

$$\begin{aligned}
0 &= \ddot{u} + u + 2\varepsilon\mu\dot{u} + \varepsilon\alpha u^2 - \sum_{n=1}^{2} K_n\cos(\Omega_n t + \theta_n)\\
&= \left(D_0^2 + 2\varepsilon D_0 D_1\right)(x_0 + \varepsilon x_1) + (x_0 + \varepsilon x_1)\\
&\quad + 2\varepsilon\mu(D_0 + \varepsilon D_1)(x_0 + \varepsilon x_1) + \varepsilon\alpha(x_0 + \varepsilon x_1)^2 - \sum_{n=1}^{2} K_n\cos(\Omega_n T_0 + \theta_n) + \cdots\\
&= D_0^2 x_0 + \varepsilon D_0^2 x_1 + 2\varepsilon D_0 D_1 x_0 + x_0 + \varepsilon x_1\\
&\quad + 2\varepsilon\mu D_0 x_0 + \varepsilon\alpha x_0^2 - \sum_{n=1}^{2} K_n\cos(\Omega_n T_0 + \theta_n) + \cdots\\
&= D_0^2 x_0 + x_0 - \sum_{n=1}^{2} K_n\cos(\Omega_n T_0 + \theta_n)\\
&\quad + \varepsilon\left[D_0^2 x_1 + x_1 + 2D_0 D_1 x_0 + 2\mu D_0 x_0 + \alpha x_0^2\right] + \cdots
\end{aligned} \tag{4.24.2}$$

Let the coefficient of the like power of ε be zero, we get

$$D_0^2 x_0 + x_0 = \sum_{n=1}^{2} K_n\cos(\Omega_n T_0 + \theta_n) \tag{4.24.3}$$

$$D_0^2 x_1 + x_1 = -2D_0 D_1 x_0 - 2\mu D_0 x_0 - \alpha x_0^2 \tag{4.24.4}$$

The solution of (4.23.3) is

$$x_0 = Ae^{iT_0} + \sum_{n=1}^{2} \Lambda_n e^{i\theta_n} e^{i\Omega_n T_0} + cc \tag{4.24.5}$$

where $A = A(T_1)$ and

$$\Lambda_n = \frac{K_n}{2\left(1 - \Omega_n^2\right)} \tag{4.24.6}$$

Therefore, the first order approximate solution of the equation is

$$u = A(T_1)e^{iT_0} + cc + 2\sum_{n=1}^{2} \Lambda_n \cos(\Omega_n T_0 + \theta_n) + O(\varepsilon) \tag{4.24.7}$$

Substituting (4.23.5) into (4.23.4) and taking $\Omega_1 \approx 1/2$ and $\Omega_2 \approx 2$ into account, we get

$$\begin{aligned} D_0^2 x_1 + x_1 &= -2D_0 D_1 x_0 - 2\mu D_0 x_0 - \alpha x_0^2 \\ &= -(2iD_1 A + 2i\mu A)e^{iT_0} - \alpha \Lambda_1^2 e^{i2\theta_1} e^{i2\Omega_1 T_0} - 2\alpha \Lambda_2 \overline{A} e^{i\theta_2} e^{i(\Omega_2 - 1)T_0} \\ &\quad + cc + NST \end{aligned} \tag{4.24.8}$$

Let

$$\varepsilon\sigma_1 = 2\Omega_1 - 1, \quad \varepsilon\sigma_2 = \Omega_2 - 2, \gamma_1 = \sigma_1 T_1 + 2\theta_1, \quad \gamma_2 = \sigma_2 T_1 + \theta_2 \tag{4.24.9}$$

Equation (4.24.8) becomes

$$\begin{aligned} D_0^2 x_1 + x_1 &= -(2iD_1 A + 2i\mu A)e^{iT_0} - \alpha \Lambda_1^2 e^{i(\sigma_1 T_1 + 2\theta_1)} e^{iT_0} \\ &\quad - 2\alpha \Lambda_2 \overline{A} e^{i(\sigma_2 T_1 + \theta_2)} e^{iT_0} + cc + NST \\ &= -\left(2iD_1 A + 2i\mu A + \alpha \Lambda_1^2 e^{i\gamma_1} + 2\alpha \Lambda_2 \overline{A} e^{i\gamma_2}\right)e^{iT_0} \\ &\quad + cc + NST \end{aligned} \tag{4.24.10}$$

In order to eliminate secular terms from the above equation, there must be

$$2i\left(A' + \mu A\right) + \alpha \Lambda_1^2 e^{i\gamma_1} + 2\alpha \Lambda_2 \overline{A} e^{i\gamma_2} = 0 \tag{4.24.11}$$

The prime denotes the derivative with respect to T_1.

(b) Let

$$A = (B_r + iB_i)e^{i\gamma_2/2} \tag{4.24.12}$$

Substitute A into (4.24.11), we get

$$2i\left[(B_{r'} + iB_{i'}) + \frac{i\sigma_2}{2}(B_r + iB_i) + \mu(B_r + iB_i)\right] + \alpha\Lambda_1^2 e^{i(\gamma_1-\gamma_2/2)} + 2\alpha\Lambda_2(B_r - iB_i) = 0$$

i.e.,

$$i[2B_{r'} + 2\mu B_r - (\sigma_2 + 2\alpha\Lambda_2)B_i + \alpha\Lambda_1^2\sin(\gamma_1 - \frac{\gamma_2}{2})] - [2B_{i'} + (\sigma_2 - 2\alpha\Lambda_2)B_r + 2\mu B_i - \alpha\Lambda_1^2\cos(\gamma_1 - \frac{\gamma_2}{2}) = 0 \tag{4.24.13}$$

Separating the real and imaginary parts of the above equation yields

$$2B_{r'} + 2\mu B_r - (\sigma_2 + 2\alpha\Lambda_2)B_i = -\alpha\Lambda_1^2\sin(\gamma_1 - \gamma_2/2)$$

$$2B_{i'} + (\sigma_2 - 2\alpha\Lambda_2)B_r + 2\mu B_i = \alpha\Lambda_1^2\cos(\gamma_1 - \gamma_2/2)$$

The above two equations are written in matrix form as

$$\begin{Bmatrix} B_{r'} \\ B_{i'} \end{Bmatrix} + \begin{bmatrix} \mu & -\frac{1}{2}(\sigma_2 + 2\alpha\Lambda_2) \\ \frac{1}{2}(\sigma_2 - 2\alpha\Lambda_2) & \mu \end{bmatrix}\begin{Bmatrix} B_r \\ B_i \end{Bmatrix} = \frac{1}{2}\alpha\Lambda_1^2\begin{Bmatrix} -\sin(\gamma_1 - \frac{1}{2}\gamma_2) \\ \cos(\gamma_1 - \frac{1}{2}\gamma_2) \end{Bmatrix} \tag{4.24.14}$$

These are the equations controlling B_r and B_i, which are a set of first order linear ordinary differential equations. The eigenvalues can be easily obtained as

$$\lambda = -\mu \pm \frac{1}{2}\sqrt{4\alpha^2\Lambda_2^2 - \sigma_2^2} \tag{4.24.15}$$

Therefore, B_r and B_i are unbounded if

$$\sqrt{4\alpha^2\Lambda_2^2 - \sigma_2^2} > 2\mu \tag{4.24.16}$$

This condition is independent of Λ_1, i.e., independent of the superharmonic excitation term.

4.25 Exercise 4.25 (Quadratic Nonlinear Systems with Both Subharmonic and Combinatorial Resonances)

Solution: **(a)** We assume that $K_n = O(1)(n = 1, 2, 3)$. Let the solution of the equation be

$$u(t; \varepsilon) = x_0(T_0, T_1) + \varepsilon x_1(T_0, T_1) + \cdots \tag{4.25.1}$$

Substituting (4.25.1) into the original equation and retaining to $O(\varepsilon)$ yields

$$\begin{aligned}
0 &= \ddot{u} + u + 2\varepsilon\mu\dot{u} + \varepsilon\alpha u^2 - \sum_{n=1}^{3} K_n \cos(\Omega_n t + \theta_n) \\
&= \left(D_0^2 + 2\varepsilon D_0 D_1\right)(x_0 + \varepsilon x_1) + (x_0 + \varepsilon x_1) \\
&\quad + 2\varepsilon\mu(D_0 + \varepsilon D_1)(x_0 + \varepsilon x_1) + \varepsilon\alpha(x_0 + \varepsilon x_1)^2 \\
&\quad - \sum_{n=1}^{3} K_n \cos(\Omega_n T_0 + \theta_n) + \cdots \\
&= D_0^2 x_0 + \varepsilon D_0^2 x_1 + 2\varepsilon D_0 D_1 x_0 + x_0 + \varepsilon x_1 \\
&\quad + 2\varepsilon\mu D_0 x_0 + \varepsilon\alpha x_0^2 - \sum_{n=1}^{3} K_n \cos(\Omega_n T_0 + \theta_n) + \cdots \\
&= D_0^2 x_0 + x_0 - \sum_{n=1}^{3} K_n \cos(\Omega_n T_0 + \theta_n) \\
&\quad + \varepsilon\left[D_0^2 x_1 + x_1 + 2D_0 D_1 x_0 + 2\mu D_0 x_0 + \alpha x_0^2\right] + \cdots
\end{aligned} \tag{4.25.2}$$

Let the coefficient of the like power of ε be zero, we get

$$D_0^2 x_0 + x_0 = \sum_{n=1}^{3} K_n \cos(\Omega_n T_0 + \theta_n) \tag{4.25.3}$$

$$D_0^2 x_1 + x_1 = -2D_0 D_1 x_0 - 2\mu D_0 x_0 - \alpha x_0^2 \tag{4.25.4}$$

The solution of (4.25.3) is

$$x_0 = Ae^{iT_0} + \sum_{n=1}^{3} \Lambda_n e^{i\theta_n} e^{i\Omega_n T_0} + cc \tag{4.25.5}$$

where $A = A(T_1)$ and

$$\Lambda_n = \frac{K_n}{2(1-\Omega_n^2)} \tag{4.25.6}$$

Therefore, the first order approximate solution of the original equation is

$$u = A(T_1)e^{iT_0} + cc + 2\sum_{n=1}^{3} \Lambda_n \cos(\Omega_n T_0 + \theta_n) + O(\varepsilon) \tag{4.25.7}$$

Substituting (4.25.5) into (4.25.4) and taking $\Omega_1 \approx 2$ and $\Omega_2 + \Omega_3 \approx 1$ into account, we get

$$\begin{aligned} D_0^2 x_1 + x_1 &= -2D_0D_1x_0 - 2\mu D_0 x_0 - \alpha x_0^2 \\ &= -(2iD_1A + 2i\mu A)e^{iT_0} - 2\alpha\Lambda_1\bar{A}e^{i\theta_1}e^{i(\Omega_1-1)T_0} \\ &- 2\alpha\Lambda_2\Lambda_3 e^{i\theta_2}e^{i\theta_3}e^{i(\Omega_2+\Omega_3)T_0} + cc + NST \end{aligned} \tag{4.25.8}$$

Let

$$\varepsilon\sigma_1 = \Omega_1 - 2, \quad \varepsilon\sigma_2 = \Omega_2 + \Omega_3 - 1$$

$$\gamma_1 = \sigma_1 T_1 + \theta_1, \quad \gamma_2 = \sigma_2 T_1 + \theta_2 + \theta_3 \tag{4.25.9}$$

Equation (4.25.8) becomes

$$\begin{aligned} D_0^2 x_1 + x_1 &= -2D_0D_1x_0 - 2\mu D_0 x_0 - \alpha x_0^2 \\ &= -\left(2iD_1A + 2i\mu A + 2\alpha\Lambda_1\bar{A}e^{i\gamma_1} + 2\alpha\Lambda_2\Lambda_3 e^{i\gamma_2}\right)e^{iT_0} \end{aligned} \tag{4.25.10}$$

In order to eliminate secular terms from the above equation, there must be

$$i\left(A' + \mu A\right) + \alpha\Lambda_1\bar{A}e^{i\gamma_1} + \alpha\Lambda_2\Lambda_3 e^{i\gamma_2} = 0 \tag{4.25.11}$$

The prime denotes the derivative with respect to T_1.

(b) Let

$$A = (B_r + iB_i)e^{i\gamma_1/2} \tag{4.25.12}$$

Substitute A into (4.25.11), we obtain

$$\begin{aligned} &i\left[\left(B_r' + iB_i'\right) + \frac{i\sigma_1}{2}(B_r + iB_i) + \mu(B_r + iB_i)\right] \\ &+ \alpha\Lambda_1(B_r - iB_i) + \alpha\Lambda_2\Lambda_3 e^{i(\gamma_2 - \gamma_1/2)} = 0 \end{aligned}$$

i.e., $i[B_r\prime + \mu B_r - (\frac{1}{2}\sigma_1 + \alpha\Lambda_1)B_i + \alpha\Lambda_2\Lambda_3\sin(\gamma_2 - \gamma_1/2)]$

$$-[B_{i'} + \left(\frac{1}{2}\sigma_1 - \alpha\Lambda_1\right)B_r + \mu B_i - \alpha\Lambda_2\Lambda_3\cos(\gamma_2 - \gamma_1/2)] = 0 \tag{4.25.13}$$

Separating the real and imaginary parts of the above equation yields

$$B_{r'} + \mu B_r - \left(\frac{1}{2}\sigma_1 + \alpha\Lambda_1\right)B_i = -\alpha\Lambda_2\Lambda_3\sin(\gamma_2 - \gamma_1/2)$$

$$B_{i'} + \left(\frac{1}{2}\sigma_1 - \alpha\Lambda_1\right)B_r + \mu B_i = \alpha\Lambda_2\Lambda_3\cos(\gamma_2 - \gamma_1/2)$$

The above two equations are written in matrix form as

$$\begin{Bmatrix} B_{r'} \\ B_{i'} \end{Bmatrix} + \begin{bmatrix} \mu & -\left(\frac{1}{2}\sigma_1 + \alpha\Lambda_1\right) \\ \left(\frac{1}{2}\sigma_1 - \alpha\Lambda_1\right) & \mu \end{bmatrix}\begin{Bmatrix} B_r \\ B_i \end{Bmatrix} = \alpha\Lambda_2\Lambda_3\begin{Bmatrix} -\sin(\gamma_2 - \frac{1}{2}\gamma_1) \\ \cos(\gamma_2 - \frac{1}{2}\gamma_1) \end{Bmatrix} \tag{4.25.14}$$

These are the equations controlling B_r and B_i. The eigenvalues can be easily obtained as

$$\lambda = -\mu \pm \frac{1}{2}\sqrt{4\alpha^2\Lambda_1^2 - \sigma_1^2} \tag{4.25.15}$$

Therefore, B_r and B_i are unbounded if

$$\sqrt{4\alpha^2\Lambda_1^2 - \sigma_1^2} > 2\mu \tag{4.25.16}$$

4.26 Exercise 4.26 (Quadratic Nonlinear Systems with Both Superharmonic and Combinatorial Resonances)

Solution: (a) We assume that $K_n = O(1)(n = 1, 2, 3)$. Let the solution of the equation be

$$u(t;\varepsilon) = x_0(T_0, T_1) + \varepsilon x_1(T_0, T_1) + \cdots \tag{4.26.1}$$

Substituting (4.26.1) into the original equation and retaining to $O(\varepsilon)$ yields

$$\begin{aligned} 0 &= \ddot{u} + u + 2\varepsilon\mu\dot{u} + \varepsilon\alpha u^2 - \sum_{n=1}^{3} K_n\cos(\Omega_n t + \theta_n) \\ &= \left(D_0^2 + 2\varepsilon D_0 D_1\right)(x_0 + \varepsilon x_1) + (x_0 + \varepsilon x_1) \end{aligned}$$

$$
\begin{aligned}
&+ 2\varepsilon\mu(D_0 + \varepsilon D_1)(x_0 + \varepsilon x_1) + \varepsilon\alpha(x_0 + \varepsilon x_1)^2 - \sum_{n=1}^{3} K_n \cos(\Omega_n T_0 + \theta_n) + \cdots \\
&= D_0^2 x_0 + \varepsilon D_0^2 x_1 + 2\varepsilon D_0 D_1 x_0 + x_0 + \varepsilon x_1 \\
&+ 2\varepsilon\mu D_0 x_0 + \varepsilon\alpha x_0^2 - \sum_{n=1}^{3} K_n \cos(\Omega_n T_0 + \theta_n) + \cdots \\
&= D_0^2 x_0 + x_0 - \sum_{n=1}^{3} K_n \cos(\Omega_n T_0 + \theta_n) \\
&+ \varepsilon\left[D_0^2 x_1 + x_1 + 2D_0 D_1 x_0 + 2\mu D_0 x_0 + \alpha x_0^2\right] + \cdots
\end{aligned} \tag{4.26.2}
$$

Let the coefficient of the like power of ε be zero, we get

$$D_0^2 x_0 + x_0 = \sum_{n=1}^{3} K_n \cos(\Omega_n T_0 + \theta_n) \tag{4.26.3}$$

$$D_0^2 x_1 + x_1 = -2D_0 D_1 x_0 - 2\mu D_0 x_0 - \alpha x_0^2 \tag{4.26.4}$$

The solution of (4.26.3) is

$$x_0 = Ae^{iT_0} + \sum_{n=1}^{3} \Lambda_n e^{i\theta_n} e^{i\Omega_n T_0} + cc \tag{4.26.5}$$

where $A = A(T_1)$ and

$$\Lambda_n = \frac{K_n}{2\left(1 - \Omega_n^2\right)} \tag{4.26.6}$$

Therefore, the first order approximate solution of the equation is

$$u = A(T_1)e^{iT_0} + cc + 2\sum_{n=1}^{3} \Lambda_n \cos(\Omega_n T_0 + \theta_n) + O(\varepsilon) \tag{4.26.7}$$

Substituting (4.26.5) into (4.26.4) and taking $\Omega_1 \approx 1/2$ and $\Omega_2 + \Omega_3 \approx 1$ into account, we get

$$
\begin{aligned}
D_0^2 x_1 + x_1 &= -2D_0 D_1 x_0 - 2\mu D_0 x_0 - \alpha x_0^2 \\
&= -(2iD_1 A + 2i\mu A)e^{iT_0} - \alpha\Lambda_1^2 e^{i2\theta_1} e^{i2\Omega_1 T_0} \\
&- 2\alpha\Lambda_2\Lambda_3 e^{i\theta_2} e^{i\theta_3} e^{i(\Omega_2+\Omega_3)T_0} + cc + NST
\end{aligned} \tag{4.26.8}
$$

Let

$$\varepsilon\sigma_1 = 2\Omega_1 - 1, \quad \varepsilon\sigma_2 = \Omega_2 + \Omega_3 - 1$$

$$\gamma_1 = \sigma_1 T_1 + 2\theta_1, \quad \gamma_2 = \sigma_2 T_1 + \theta_2 + \theta_3 \tag{4.26.9}$$

Equation (4.26.8) becomes

$$\begin{aligned} D_0^2 x_1 + x_1 &= -2D_0D_1x_0 - 2\mu D_0x_0 - \alpha x_0^2 \\ &= -(2iD_1A + 2i\mu A)e^{iT_0} - \alpha\Lambda_1^2 e^{i(\sigma_1 T_1 + 2\theta_1)}e^{iT_0} \\ &\quad - 2\alpha\Lambda_2\Lambda_3 e^{i(\sigma_2 T_1 + \theta_2 + \theta_3)}e^{iT_0} + cc + NST \\ &= -\left(2iD_1A + 2i\mu A + \alpha\Lambda_1^2 e^{i\gamma_1} + 2\alpha\Lambda_2\Lambda_3 e^{i\gamma_2}\right)e^{iT_0} + cc + NST \end{aligned} \tag{4.26.10}$$

In order to eliminate secular terms from the above equation, there must be

$$2i\left(A' + \mu A\right) + \alpha\Lambda_1^2 e^{i\gamma_1} + 2\alpha\Lambda_2\Lambda_3 e^{i\gamma_2} = 0 \tag{4.26.11}$$

The prime denotes the derivative with respect to T_1.

(b) Write the equation as

$$A' + \mu A = \frac{i\alpha\Lambda_1^2}{2}e^{i\gamma_1} + i\alpha\Lambda_2\Lambda_3 e^{i\gamma_2} \tag{4.26.12}$$

The general solution of (4.26.12) is

$$A_1 = Ce^{-\mu T_1} \tag{4.26.13}$$

The special solution of (4.26.12) is

$$A_2 = \frac{i\alpha\Lambda_1^2}{\mu + i\sigma_1}e^{i\gamma_1} + \frac{i\alpha\Lambda_2\Lambda_3}{\mu + i\sigma_2}e^{i\gamma_2} \tag{4.26.14}$$

Therefore, the solution of (4.26.12) is

$$A = A_1 + A_2 = Ce^{-\mu T_1} + \frac{i\alpha\Lambda_1^2}{\mu + i\sigma_1}e^{i\gamma_1} + \frac{i\alpha\Lambda_2\Lambda_3}{\mu + i\sigma_2}e^{i\gamma_2} \tag{4.26.15}$$

From (4.26.15), we know that the steady state value of A, A_s, is:

$$A_s = \frac{i\alpha\Lambda_1^2}{\mu + i\sigma_1}e^{i\gamma_1} + \frac{i\alpha\Lambda_2\Lambda_3}{\mu + i\sigma_2}e^{i\gamma_2} \tag{4.26.16}$$

Substituting A_s into (4.23.7), we obtain the steady state solution of the original equation as

$$
\begin{aligned}
x &= \frac{i\alpha\Lambda_1^2}{\mu + i\sigma_1} e^{i(T_0+\gamma_1)} + \frac{i\alpha\Lambda_2\Lambda_3}{\mu + i\sigma_2} e^{i(T_0+\gamma_2)} + cc + 2\sum_{n=1}^{3} \Lambda_n \cos(\Omega_n T_0 + \theta_n) + O(\varepsilon) \\
&= -\frac{2\alpha\Lambda_1^2}{\sqrt{\mu^2+\sigma_1^2}} \sin(T_0 + \gamma_1 - \beta_1) - \frac{2\alpha\Lambda_2\Lambda_3}{\sqrt{\mu^2+\sigma_2^2}} \sin(T_0 + \gamma_2 - \beta_2) \\
&\quad + 2\sum_{n=1}^{3} \Lambda_n \cos(\Omega_n T_0 + \theta_n) + O(\varepsilon)
\end{aligned}
$$

i.e.,

$$
\begin{aligned}
x = &-\frac{2\alpha\Lambda_1^2}{\sqrt{\mu^2+\sigma_1^2}} \sin(T_0 + \gamma_1 - \beta_1) \\
&- \frac{2\alpha\Lambda_2\Lambda_3}{\sqrt{\mu^2+\sigma_2^2}} \sin(T_0 + \gamma_2 - \beta_2) \\
&+ 2\sum_{n=1}^{3} \Lambda_n \cos(\Omega_n T_0 + \theta_n) + O(\varepsilon)
\end{aligned}
\tag{4.26.17}
$$

where

$$
\tan\beta_1 = \frac{\sigma_1}{\mu}, \quad \tan\beta_2 = \frac{\sigma_2}{\mu} \tag{4.26.18}
$$

The frequencies appearing in this solution are not necessarily commensurable with each other, so this solution is not necessarily periodic.

4.27 Exercise 4.27 (Quadratic Nonlinear Systems with Superharmonic, Subharmonic and Combinatorial Resonances)

Readers are invited to solve this problem by referring to Exercise 4.25 and Exercise 4.26.

4.28 Exercise 4.28 (Primary and Superharmonic Resonances of Self-Excited Systems) Solution

(a) Let

$$K_1 = 2\varepsilon k_1, \quad K_2 = O(1) \tag{4.28.1}$$

and let the solution of the equation be

$$u(t; \varepsilon) = x_0(T_0, T_1) + \varepsilon x_1(T_0, T_1) + \cdots \tag{4.28.2}$$

Substituting (4.28.1) and (4.28.2) into the original equation and retaining to $O(\varepsilon)$, we get

$$\begin{aligned}
0 &= \left(D_0^2 + 2\varepsilon D_0 D_1\right)(x_0 + \varepsilon x_1) + (x_0 + \varepsilon x_1) \\
&\quad - \varepsilon(D_0 + \varepsilon D_1)(x_0 + \varepsilon x_1) + \frac{1}{3}\varepsilon[(D_0 + \varepsilon D_1)(x_0 + \varepsilon x_1)]^3 \\
&\quad - 2\varepsilon k_n \cos(\Omega_1 T_0 + \theta_{1n}) - K_2 \cos(\Omega_2 T_0 + \theta_2) + \cdots \\
&= D_0^2 x_0 + \varepsilon D_0^2 x_1 + 2\varepsilon D_0 D_1 x_0 + x_0 + \varepsilon x_1 - \varepsilon D_0 x_0 + \frac{1}{3}\varepsilon (D_0 x_0)^3 \\
&\quad - 2\varepsilon k_1 \cos(\Omega_1 T_0 + \theta_1) - K_2 \cos(\Omega_2 T_0 + \theta_2) + \cdots \\
&= D_0^2 x_0 + x_0 - K_2 \cos(\Omega_2 T_0 + \theta_2) \\
&\quad + \varepsilon\left[D_0^2 x_1 + x_1 + 2D_0 D_1 x_0 - D_0 x_0 + \frac{1}{3}(D_0 x_0)^3 - 2k_1 \cos(\Omega_1 T_0 + \theta_1)\right] + \cdots
\end{aligned} \tag{4.28.3}$$

Let the coefficient of the like power of ε be zero, we get

$$D_0^2 x_0 + x_0 = K_2 \cos(\Omega_2 T_0 + \theta_2) \tag{4.28.4}$$

$$D_0^2 x_1 + x_1 = -2D_0 D_1 x_0 + D_0 x_0 - \frac{1}{3}(D_0 x_0)^3 + 2k_1 \cos\left(\Omega_1 T_0 + \theta_1\right) \tag{4.28.5}$$

The solution of (4.28.4) is

$$x_0 = Ae^{iT_0} + \Lambda e^{i\theta_2} e^{i\Omega_2 T_0} + cc \tag{4.28.6}$$

where $A = A(T_1)$ and

$$\Lambda = \frac{K_2}{2\left(1 - \Omega_2^2\right)} = \frac{9}{16} K_2 \tag{4.28.7}$$

Therefore, the first order approximate solution of the equation is

$$u = A(T_1)e^{iT_0} + cc + 2\Lambda \cos(\Omega_2 T_0 + \theta_2) + O(\varepsilon) \tag{4.28.8}$$

Substituting (4.28.6) into (4.28.5) and taking $\Omega_1 \approx 1$ and $\Omega_2 \approx 1/3$ into account, we get

$$D_0^2 x_1 + x_1 = -2D_0 D_1 x_0 + D_0 x_0 - \frac{1}{3}(D_0 x_0)^3 + 2k_1 \cos(\Omega_1 T_0 + \theta_1)$$
$$= -(2iD_1 A - iA)e^{iT_0} - \frac{1}{3}(3iA^2\overline{A}e^{iT_0} + 6i\Omega_2^2\Lambda^2 A e^{iT_0}$$
$$- i\Omega_2^3\Lambda^3 e^{3i\theta_2} e^{3i\Omega_2 T_0}) + 2k_1 \cos(\Omega_1 T_0 + \theta_1) + cc + NST \quad (4.28.9)$$

Let

$$\frac{1}{3}\Omega_2^3\Lambda^3 = k_2, \quad \varepsilon\sigma_1 = \Omega_1 - 1, \quad \varepsilon\sigma_2 = 3\Omega_2 - 1 \quad (4.28.10)$$

Equation (4.28.9) becomes

$$D_0^2 x_1 + x_1 = -(2iD_1 A - iA + iA^2\bar{A} + 2i\Omega_2^2\Lambda^2 A$$
$$- ik_2 e^{i(\sigma_2 T_1 + 3\theta_2)} - k_1 e^{i(\sigma_1 T_1 + \theta_1)})e^{iT_0} + cc + NST \quad (4.28.11)$$

In order to eliminate secular terms from the above equation, there must be

$$2iA' - iA + iA^2\overline{A} + 2i\Omega_2^2\Lambda^2 A - ik_2 e^{i(\sigma_2 T_1 + 3\theta_2)} - k_1 e^{i(\sigma_1 T_1 + \theta_1)} = 0 \quad (4.28.12)$$

where the prime denotes the derivative with respect to T_1. Let

$$A = \frac{1}{2}ae^{i\beta} \quad (4.28.13)$$

$$\gamma_1 = \sigma_1 T_1 - \beta + \theta_1, \quad \gamma_2 = \sigma_2 T_1 - \beta + 3\theta_2 \quad (4.28.14)$$

Then from (4.28.13), we have

$$ia' - a\beta' - i\frac{1}{2}a + i\frac{1}{8}a^3 + i\Omega_2^2\Lambda^2 a - k_1 e^{i\gamma_1} - ik_2 e^{i\gamma_2} = 0 \quad (4.28.15)$$

Separating the real and imaginary parts of the above equation yields

$$a' = \left(\frac{1}{2} - \Omega_2^2\Lambda^2 - \frac{1}{8}a^2\right)a + k_1 \sin\gamma_1 + k_2 \cos\gamma_2 \quad (4.28.16)$$

$$a\beta' = -k_1 \cos\gamma_1 + k_2 \sin\gamma_2 \quad (4.28.17)$$

(b) From (4.28.16) and (4.28.17), we can find that the system has a steady state solution when and only when γ_1 and γ_2 are constant values, hence there are

$$\beta' = \sigma_1 = \sigma_2 \triangleq \sigma \quad (4.28.18)$$

(c) Let $a' = 0$ in (4.28.16) and (4.28.17), we can obtain

$$-\left(\frac{1}{2} - \Omega_2^2\Lambda^2 - \frac{1}{8}a^2\right)a = k_1 \sin\gamma_1 + k_2 \cos\gamma_2 \tag{4.28.19}$$

$$-\sigma a = k_1 \cos\gamma_1 - k_2 \sin\gamma_2 \tag{4.28.20}$$

The two equations are squared and added together to obtain the frequency–response equation as

$$(\frac{1}{2} - \Omega_2^2\Lambda^2 - \frac{1}{8}a^2)^2 a^2 + \sigma^2 a^2 = k_1^2 + k_2^2 + 2k_1k_2 \sin\psi \tag{4.28.21}$$

$$\psi = \gamma_1 - \gamma_2 = \theta_1 - 3\theta_2 \tag{4.28.22}$$

Solving σ from (4.28.21), we have

$$\sigma = \pm[\frac{k_1^2 + 2k_1k_2 \sin\psi + k_2^2}{a^2} - (\frac{1}{2} - \Omega_2^2\Lambda^2 - \frac{1}{8}a^2)^2]^{1/2} \tag{4.28.23}$$

From this, the frequency–response curve of the system can be plotted in Fig. 4.18.

In order to analyze the stability of the steady state solutions, Eqs. (4.28.16) and (4.28.17) are rewritten as

$$a' = \left(\frac{1}{2} - \Omega_2^2\Lambda^2 - \frac{1}{8}a^2\right)a + k_1 \sin(\gamma + \theta_1) + k_2 \cos(\gamma + 3\theta_2) \tag{4.28.24}$$

$$a(\sigma - \gamma') = -k_1 \cos(\gamma + \theta_1) + k_2 \sin(\gamma + 3\theta_2) \tag{4.28.25}$$

where

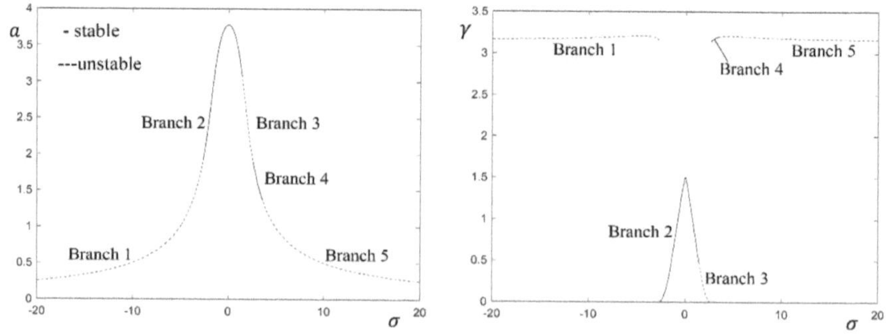

Fig. 4.18 Frequency-response and frequency-phase characteristics of the steady state solution ($\theta_1 = 0,\ \theta_2 = \frac{\pi}{4},\ k_1 = 5,\ K_2 = 1$) for Exercise 4.28

$$\gamma = \sigma_1 T_1 - \beta = \sigma_2 T_1 - \beta = \sigma T_1 - \beta \tag{4.28.26}$$

Let

$$a = a_0 + \tilde{a}, \quad \gamma = \gamma_0 + \tilde{\gamma} \tag{4.28.27}$$

Thus the linearized form of (4.28.24) and (4.28.25) are

$$\tilde{a}\prime = \left(\frac{1}{2} - \Omega_2^2\Lambda^2 - \frac{3}{8}a_0^2\right)\tilde{a} + \left[k_1\cos(\gamma_0 + \theta_1) - k_2\sin(\gamma_0 + 3\theta_2)\right]\tilde{\gamma} \tag{4.28.28}$$

$$\tilde{\gamma}\prime = \frac{\sigma}{a_0}\tilde{a} - \frac{1}{a_0}\left[k_1\sin(\gamma_0 + \theta_1) + k_2\cos(\gamma_0 + 3\theta_2)\right]\tilde{\gamma} \tag{4.28.29}$$

Write the above two equations in matrix form as

$$\begin{Bmatrix} \tilde{a}\prime \\ \tilde{\gamma}\prime \end{Bmatrix} = \begin{bmatrix} l_{11} & l_{12} \\ l_{21} & l_{22} \end{bmatrix} \begin{Bmatrix} \tilde{a} \\ \tilde{\gamma} \end{Bmatrix} \tag{4.28.30}$$

where

$$l_{11} = \frac{1}{2} - \Omega_2^2\Lambda^2 - \frac{3}{8}a_0^2, \quad l_{12} = k_1\cos(\gamma_0 + \theta_1) - k_2\sin(\gamma_0 + 3\theta_2)$$

$$l_{21} = \frac{\sigma}{a_0}, \quad l_{22} = -\frac{1}{a_0}\left[k_1\sin(\gamma_0 + \theta_1) + k_2\cos(\gamma_0 + 3\theta_2)\right]$$

The eigenvalues of Eq. (4.20.26) are

$$\lambda = \frac{(l_{11} + l_{22}) \pm \sqrt{(l_{11} + l_{22})^2 - 4(l_{11}l_{22} - l_{12}l_{21})}}{2} \tag{4.28.31}$$

The sign of the real part of the eigenvalue determines the stability of the corresponding steady state solution. It follows that branches 1, 3, and 5 in Fig. 4.18 are unstable.

4.29 Exercise 4.29 (Primary and Superharmonic Resonances of Self-Excited Systems)

Solution: **(a)** Let

$$K_1 = 2\varepsilon k_1, K_2 = O(1) \tag{4.29.1}$$

and set the solution of the equation be

$$u(t;\varepsilon) = x_0(T_0,T_1) + \varepsilon x_1(T_0,T_1) + \cdots \tag{4.29.2}$$

Substituting (4.28.1) and (4.28.2) into the original equation and retaining to $O(\varepsilon)$, we get

$$\begin{aligned}
0 &= \ddot{u} + u - \varepsilon\left(\dot{u} - \frac{1}{3}\dot{u}^3\right) - \sum_{n=1}^{2} K_n \cos(\Omega_n t + \theta_n) \\
&\quad (D_0^2 + 2\varepsilon D_0 D_1)(x_0 + \varepsilon x_1) + (x_0 + \varepsilon x_1) \\
&\quad - \varepsilon(D_0 + \varepsilon D_1)(x_0 + \varepsilon x_1) + \frac{1}{3}\varepsilon[(D_0 + \varepsilon D_1)(x_0 + \varepsilon x_1)]^3 \\
&\quad - 2\varepsilon k_n \cos(\Omega_1 T_0 + \theta_{1n}) - K_2 \cos(\Omega_2 T_0 + \theta_2) + \cdots \\
&= D_0^2 x_0 + \varepsilon D_0^2 x_1 + 2\varepsilon D_0 D_1 x_0 + x_0 + \varepsilon x_1 - \varepsilon D_0 x_0 + \frac{1}{3}\varepsilon (D_0 x_0)^3 \\
&\quad - 2\varepsilon k_1 \cos(\Omega_1 T_0 + \theta_1) - K_2 \cos(\Omega_2 T_0 + \theta_2) + \cdots \\
&= D_0^2 x_0 + x_0 - K_2 \cos(\Omega_2 T_0 + \theta_2) \\
&\quad + \varepsilon\left[D_0^2 x_1 + x_1 + 2D_0 D_1 x_0 - D_0 x_0 + \frac{1}{3}(D_0 x_0)^3 - 2k_1 \cos(\Omega_1 T_0 + \theta_1)\right] + \cdots
\end{aligned} \tag{4.29.3}$$

Let the coefficient of the like power of ε be zero, we get

$$D_0^2 x_0 + x_0 = K_2 \cos(\Omega_2 T_0 + \theta_2) \tag{4.29.4}$$

$$D_0^2 x_1 + x_1 = -2D_0 D_1 x_0 + D_0 x_0 - \frac{1}{3}(D_0 x_0)^3 + 2k_1 \cos(\Omega_1 T_0 + \theta_1) \tag{4.29.5}$$

The solution of (4.28.4) is

$$x_0 = Ae^{iT_0} + \Lambda e^{i\theta_2} e^{i\Omega_2 T_0} + cc \tag{4.29.6}$$

where $A = A(T_1)$ and

$$\Lambda = \frac{K_2}{2(1 - \Omega_2^2)} = -\frac{1}{16}K_2 \tag{4.29.7}$$

Let

$$A = \frac{1}{2}ae^{i\beta} \tag{4.29.8}$$

Therefore, the first order approximate solution of the equation is

$$u = a(T_1)\cos(T_0 + \beta) + 2\Lambda \cos(\Omega_2 T_0 + \theta_2) + O(\varepsilon) \tag{4.29.9}$$

Substituting (4.28.6) into (4.28.5) and taking $\Omega_1 \approx 1$ and $\Omega_2 \approx 3$ into account, we get

$$\begin{aligned} D_0^2 x_1 + x_1 &= -2D_0 D_1 x_0 + D_0 x_0 - \frac{1}{3}(D_0 x_0)^3 + 2k_1 \cos(\Omega_1 T_0 + \theta_1) \\ &= -(2iD_1 A - iA)e^{iT_0} - \frac{1}{3}(3iA^2\bar{A}e^{iT_0} + 6i\Omega_2^2\Lambda^2 A e^{iT_0} \\ &\quad - 3i\Omega_2\Lambda\bar{A}^2 e^{i\theta_2} e^{i(\Omega_2-2)T_0}) + cc + NST + 2k_1\cos(\Omega_1 T_0 + \theta_1) \end{aligned} \tag{4.29.10}$$

Let

$$k_2 = \frac{1}{4}\Omega_2\Lambda, \quad \varepsilon\sigma_1 = \Omega_1 - 1, \quad \varepsilon\sigma_2 = \Omega_2 - 3 \tag{4.29.11}$$

Equation (4.28.9) becomes

$$\begin{aligned} D_0^2 x_1 + x_1 = &-(2iD_1 A - iA + iA^2\bar{A} + 16ik_2^2 A \\ &- 4ik_2\bar{A}^2 e^{i(\sigma_2 T_1 + \theta_2)} - k_1 e^{i(\sigma_1 T_1 + \theta_1)})e^{iT_0} + cc + NST \end{aligned} \tag{4.29.12}$$

In order to eliminate secular terms from the above equation, there must be

$$2iA' - iA + iA^2\overline{A} + 2i\Omega_2^2\Lambda^2 A - 4ik_2\overline{A}^2 e^{i(\sigma_2 T_1 + \theta_2)} - k_1 e^{i(\sigma_1 T_1 + \theta_1)} = 0 \tag{4.29.13}$$

The prime represents the derivative with respect to T_1. Substituting (4.29.8) into the above equation and let

$$\gamma_1 = \sigma_1 T_1 - \beta + \theta_1, \quad \gamma_2 = \sigma_2 T_1 - 3\beta + \theta_2 \tag{4.29.14}$$

then

$$ia' - a\beta' - i\frac{1}{2}a + i\frac{1}{8}a^3 + i16k_2^2 a - k_1 e^{i\gamma_1} - ik_2 a^2 e^{i\gamma_2} = 0 \tag{4.29.15}$$

Separating the real and imaginary parts of the above equation yields

$$a' = \left(\frac{1}{2} - 16k_2^2 - \frac{1}{8}a^2\right)a + k_1 \sin\gamma_1 + k_2 a^2 \cos\gamma_2 \tag{4.29.16}$$

$$a\beta' = -k_1\cos\gamma_1 + k_2 a^2 \sin\gamma_2 \tag{4.29.17}$$

(b) We can find from (4.28.16) and (4.28.17) that the system has a steady state solution when and only when γ_1 and γ_2 are constants, hence

$$\beta' = \sigma_1 = \frac{1}{3}\sigma_2 \tag{4.29.18}$$

4.30 Exercise 4.30 (Simultaneous Subharmonic and Superharmonic Resonance of Self-Excited Systems)

Solution: **(a)** We assume that $K_n = O(1)$ $(n = 1, 2)$ and the solution of the equation is

$$u(t;\varepsilon) = x_0(T_0, T_1) + \varepsilon x_1(T_0, T_1) + \cdots \tag{4.30.1}$$

Substituting (4.30.1) into the original equation and retaining to $O(\varepsilon)$ yields

$$\begin{aligned}
0 &= \ddot{u} + u - \varepsilon\left(\dot{u} - \frac{1}{3}\dot{u}^3\right) - \sum_{n=1}^{2} K_n \cos(\Omega_n t + \theta_n) \\
&= \left(D_0^2 + 2\varepsilon D_0 D_1\right)(x_0 + \varepsilon x_1) + (x_0 + \varepsilon x_1) \\
&\quad - \varepsilon(D_0 + \varepsilon D_1)(x_0 + \varepsilon x_1) + \frac{1}{3}\varepsilon[(D_0 + \varepsilon D_1)(x_0 + \varepsilon x_1)]^3 \\
&\quad - \sum_{n=1}^{2} K_n \cos(\Omega_n T_0 + \theta_n) + \cdots \\
&= D_0^2 x_0 + \varepsilon D_0^2 x_1 + 2\varepsilon D_0 D_1 x_0 + x_0 + \varepsilon x_1 - \varepsilon D_0 x_0 + \frac{1}{3}\varepsilon (D_0 x_0)^3 \\
&\quad - \sum_{n=1}^{2} K_n \cos(\Omega_n T_0 + \theta_n) + \cdots \\
&= D_0^2 x_0 + x_0 - \sum_{n=1}^{2} K_n \cos(\Omega_n T_0 + \theta_n) \\
&\quad + \varepsilon\left[D_0^2 x_1 + x_1 + 2D_0 D_1 x_0 - D_0 x_0 + \frac{1}{3}(D_0 x_0)^3\right] + \cdots
\end{aligned} \tag{4.30.2}$$

Let the coefficient of the like power of ε be zero, we get

$$D_0^2 x_0 + x_0 = \sum_{n=1}^{2} K_n \cos(\Omega_n T_0 + \theta_n) \tag{4.30.3}$$

$$D_0^2 x_1 + x_1 = -2D_0 D_1 x_0 + D_0 x_0 - \frac{1}{3}(D_0 x_0)^3 \tag{4.30.4}$$

The solution of (4.30.3) is

$$x_0 = Ae^{iT_0} + \sum_{n=1}^{2} \Lambda_n e^{i\theta_n} e^{i\Omega_n T_0} + cc \tag{4.30.5}$$

where $A = A(T_1)$ and

$$\Lambda_1 = \frac{K_1}{2\left(1 - \Omega_1^2\right)} = -\frac{1}{16}K_2, \quad \Lambda_2 = \frac{K_2}{2\left(1 - \Omega_2^2\right)} = \frac{9}{16}K_2 \tag{4.30.6}$$

Let

$$A = \frac{1}{2}ae^{i\beta} \tag{4.30.7}$$

Therefore, the first order approximate solution of the equation is

$$u = a(T_1)\cos(T_0 + \beta) + 2\Lambda\cos(\Omega_2 T_0 + \theta_2) + O(\varepsilon) \tag{4.30.8}$$

Substituting (4.30.5) into (4.30.4) and taking $\Omega_1 \approx 3$ and $\Omega_2 \approx 1/3$ into account, we get

$$\begin{aligned} D_0^2 x_1 + x_1 &= -2D_0 D_1 x_0 + D_0 x_0 - \frac{1}{3}(D_0 x_0)^3 \\ &= -(2iD_1 A - iA)e^{iT_0} - \frac{1}{3}(3iA^2\overline{A}e^{iT_0} \\ &\quad + 6i\Omega_1^2\Lambda_1^2 Ae^{iT_0} + 6i\Omega_2^2\Lambda_2^2 Ae^{iT_0} \\ &\quad - i\Omega_2^3\Lambda_2^3 e^{i3\theta_2} e^{i3\Omega_2 T_0} - 3i\overline{A}^2\Omega_1\Lambda_1 e^{i\theta_1} e^{i(\Omega_1 - 2)T_0}) + cc + NST \end{aligned} \tag{4.30.9}$$

Let

$$\varepsilon\sigma_1 = \Omega_1 - 3, \quad \varepsilon\sigma_2 = 3\Omega_2 - 1 \tag{4.30.10}$$

Equation (4.30.9) becomes

$$\begin{aligned} D_0^2 x_1 + x_1 = &-(2iD_1 A - iA + iA^2\bar{A} + 2i\Omega_1^2\Lambda_1^2 A + 2i\Omega_2^2\Lambda_2^2 A \\ &- i\frac{1}{3}\Omega_2^3\Lambda_2^3 e^{i(\sigma_2 T_1 + 3\theta_2)} - i\bar{A}^2\Omega_1\Lambda_1 e^{i(\sigma_1 T_1 + \theta_1)})e^{iT_0} \\ &+ cc + NST \end{aligned} \tag{4.30.11}$$

In order to eliminate secular terms from the above equation, there must be

$$2A' - A + A^2\overline{A} + 2\Omega_1^2\Lambda_1^2 A + 2\Omega_2^2\Lambda_2^2 A$$

$$-\frac{1}{3}\Omega_2^3\Lambda_2^3 e^{i(\sigma_2 T_1+3\theta_2)} - \overline{A}^2\Omega_1\Lambda_1 e^{i(\sigma_1 T_1+\theta_1)} = 0 \tag{4.30.12}$$

The prime represents the derivative with respect to T_1. Substituting (4.30.7) into the above equation, we get

$$a' + ia\beta' - \frac{1}{2}a + \frac{1}{8}a^3 + \Omega_1^2\Lambda_1^2 a + \Omega_2^2\Lambda_2^2 a$$
$$-\frac{1}{3}\Omega_2^3\Lambda_2^3 e^{i(\sigma_2 T_1-\beta+3\theta_2)} - \frac{1}{4}a^2\Omega_1\Lambda_1 e^{i(\sigma_1 T_1-3\beta+\theta_1)} = 0 \tag{4.30.13}$$

Let

$$k_1 = \frac{1}{4}\Omega_1\Lambda_1, \quad \gamma_1 = \sigma_1 T_1 - 3\beta + \theta_1, \quad \gamma_2 = \sigma_2 T_1 - \beta + 3\theta_2 \tag{4.30.14}$$

then

$$a' - \frac{1}{2}a + \frac{1}{8}a^3 + 16k_1^2 a + \Omega_2^2\Lambda_2^2 a + ia\beta' - k_1 a^2 e^{i\gamma_1} - \frac{1}{3}\Omega_2^3\Lambda_2^3 e^{i\gamma_2} = 0 \tag{4.30.15}$$

Separating the real and imaginary parts of the above equation yields

$$a' = \left(\frac{1}{2} - 16k_1^2 - \Omega_2^2\Lambda_2^2 - \frac{1}{8}a^2\right)a + k_1 a^2\cos\gamma_1 + \frac{1}{3}\Omega_2^3\Lambda_2^3\cos\gamma_2 \tag{4.30.16}$$

$$a\beta' = k_1 a^2\sin\gamma_1 + \frac{1}{3}\Omega_2^3\Lambda_2^3\sin\gamma_2 \tag{4.30.17}$$

(b) It can be found from (4.30.16) and (4.30.17) that the system has a steady state solution when and only when γ_1 and γ_2 are constants, hence

$$\beta' = \frac{1}{3}\sigma_1 = \sigma_2 \tag{4.30.18}$$

4.31 Exercise 4.31 (Self-Excited System with Both Primary and Combined Resonances I)

Solution: **(a)** We assume

$$K_1 = 2\varepsilon k_1, K_2 = O(1) \tag{4.31.1}$$

and let the solution of the equation be

$$u(t;\varepsilon) = x_0(T_0, T_1) + \varepsilon x_1(T_0, T_1) + \cdots \tag{4.31.2}$$

Substituting (4.31.1) and (4.31.2) into the original equation and retaining to $O(\varepsilon)$, we get

$$\begin{aligned}
0 &= \ddot{u} + u - \varepsilon\left(\dot{u} - \frac{1}{3}\dot{u}^3\right) - \sum_{n=1}^{3} K_n \cos(\Omega_n t + \theta_n) \\
&= \left(D_0^2 + 2\varepsilon D_0 D_1\right)(x_0 + \varepsilon x_1) + (x_0 + \varepsilon x_1) \\
&\quad - \varepsilon(D_0 + \varepsilon D_1)(x_0 + \varepsilon x_1) + \frac{1}{3}\varepsilon[(D_0 + \varepsilon D_1)(x_0 + \varepsilon x_1)]^3 \\
&\quad - 2\varepsilon k_1 \cos(\Omega_1 T_0 + \theta_1) - \sum_{n=2}^{3} K_n \cos(\Omega_n T_0 + \theta_n) + \cdots \\
&= D_0^2 x_0 + x_0 - \sum_{n=2}^{3} K_n \cos(\Omega_n T_0 + \theta_n) \\
&\quad + \varepsilon\left[D_0^2 x_1 + x_1 + 2D_0 D_1 x_0 - D_0 x_0 + \frac{1}{3}(D_0 x_0)^3 - 2k_1 \cos(\Omega_1 T_0 + \theta_1)\right] + \cdots
\end{aligned} \tag{4.31.3}$$

Let the coefficient of the like power of ε be zero, we get

$$D_0^2 x_0 + x_0 = \sum_{n=2}^{3} K_n \cos(\Omega_n T_0 + \theta_n) \tag{4.31.4}$$

$$D_0^2 x_1 + x_1 = -2D_0 D_1 x_0 + D_0 x_0 - \frac{1}{3}(D_0 x_0)^3 + 2k_1 \cos(\Omega_1 T_0 + \theta_1) \tag{4.31.5}$$

The solution of (4.31.4) is

$$x_0 = Ae^{iT_0} + \sum_{n=2}^{3} \Lambda_n e^{i\theta_n} e^{i\Omega_n T_0} + cc \tag{4.31.6}$$

where $A = A(T_1)$ and

$$\Lambda_n = \frac{K_n}{2\left(1 - \Omega_n^2\right)} \tag{4.31.7}$$

Let

$$A = \frac{1}{2} a e^{i\beta} \tag{4.31.8}$$

Therefore, the first order approximate solution of the equation is

$$u = a(T_1)\cos[T_0 + \beta(T_1)] + 2\sum_{i=2}^{3} \Lambda_n \cos(\Omega_n T_0 + \theta_n) + O(\varepsilon) \tag{4.31.9}$$

Substituting (4.31.6) into (4.31.5) and taking $\Omega_1 \approx 1$ and $2\Omega_2 + \Omega_3 \approx 1$ into account, we get

$$\begin{aligned} D_0^2 x_1 + x_1 &= -2D_0 D_1 x_0 + D_0 x_0 - \frac{1}{3}(D_0 x_0)^3 \\ &= -(2iD_1 A - iA)e^{iT_0} - \frac{1}{3}(3iA^2\overline{A}e^{iT_0} \\ &\quad + 6iA\Omega_2^2\Lambda_2^2 e^{iT_0} + 6iA\Omega_3^2\Lambda_3^2 e^{iT_0} \\ &\quad - i3\Omega_2^2\Omega_3\Lambda_2^2\Lambda_3 e^{i(2\theta_2+\theta_3)} e^{i(2\Omega_2+\Omega_3)T_0}) + k_1 e^{i(\Omega_1 T_0 + \theta_1)} + cc + NST \end{aligned} \tag{4.31.10}$$

Let

$$\varepsilon\sigma_1 = \Omega_1 - 1, \quad \varepsilon\sigma_2 = 2\Omega_2 + \Omega_3 - 1 \tag{4.31.11}$$

Equation (4.31.10) becomes

$$\begin{aligned} D_0^2 x_1 + x_1 = &-(2iD_1 A - iA + iA^2\bar{A} + 2iA\Omega_2^2\Lambda_2^2 + 2iA\Omega_3^2\Lambda_3^2 \\ &- i\Omega_2^2\Omega_3\Lambda_2^2\Lambda_3 e^{i(\sigma_2 T_1 + 2\theta_2 + \theta_3)} - k_1 e^{i(\sigma_1 T_1 + \theta_1)})e^{iT_0} \\ &+ cc + NST \end{aligned} \tag{4.31.12}$$

In order to eliminate secular terms from the above equation, there must be

$$\begin{aligned} &2A' - A + A^2\overline{A} + 2\Omega_2^2\Lambda_2^2 A + 2\Omega_3^2\Lambda_3^2 A \\ &- \Omega_2^2\Omega_3\Lambda_2^2\Lambda_3 e^{i(\sigma_2 T_1 + 2\theta_2 + \theta_3)} + ik_1 e^{i(\sigma_1 T_1 + \theta_1)} = 0 \end{aligned} \tag{4.31.13}$$

The prime represents the derivative with respect to T_1. Substituting (4.30.7) into the above equation, we get

$$\begin{aligned} &a' + ia\beta' - \frac{1}{2}a + \frac{1}{8}a^3 + \Omega_2^2\Lambda_2^2 a + \Omega_3^2\Lambda_3^2 a \\ &- \Omega_2^2\Omega_3\Lambda_2^2\Lambda_3 e^{i(\sigma_2 T_1 - \beta + 2\theta_2 + \theta_3)} + ik_1 e^{i(\sigma_1 T_1 - \beta + \theta_1)} = 0 \end{aligned} \tag{4.31.14}$$

Let

$$k_2 = \Omega_2^2\Omega_3\Lambda_2^2\Lambda_3, \quad \gamma_1 = \sigma_1 T_1 - \beta + \theta_1, \quad \gamma_2 = \sigma_2 T_1 - \beta + 2\theta_2 + \theta_3 \tag{4.31.15}$$

then

$$a' - \frac{1}{2}a + \frac{1}{8}a^3 + \sum_{n=2}^{3} \Omega_n^2 \Lambda_n^2 + ia\beta' + ik_1 e^{i\gamma_1} - k_2 e^{i\gamma_2} = 0 \tag{4.31.16}$$

Separating the real and imaginary parts of the above equation yields

$$a' = \left(\frac{1}{2} - \sum_{n=2}^{3} \Omega_n^2 \Lambda_n^2 - \frac{1}{8}a^2\right)a + k_1 \sin\gamma_1 + k_2 \cos\gamma_2 \tag{4.31.17}$$

$$a\beta' = -k_1 \cos\gamma_1 + k_2 \sin\gamma_2 \tag{4.31.18}$$

(b) It can be found from (4.30.16) and (4.30.17) that the system has a steady state solution when and only when γ_1 and γ_2 are constants, hence

$$\beta' = \sigma_1 = \sigma_2 \triangleq \sigma \tag{4.31.19}$$

(c) Let $a\prime = 0$and $\beta' = \sigma$ in (4.31.17) and (4.31.18), we can obtain

$$-\left(\frac{1}{2} - \sum_{n=2}^{3} \Omega_n^2 \Lambda_n^2 - \frac{1}{8}a^2\right)a = k_1 \sin\gamma_1 + k_2 \cos\gamma_2 \tag{4.31.20}$$

$$-\sigma a = k_1 \cos\gamma_1 - k_2 \sin\gamma_2 \tag{4.31.21}$$

The two equations are squared and added together to obtain the frequency–response equation as

$$\left(\frac{1}{2} - \sum_{n=2}^{3} \Omega_n^2 \Lambda_n^2 - \frac{1}{8}a^2\right)^2 a^2 + \sigma^2 a^2 = k_1^2 + k_2^2 + 2k_1 k_2 \sin\psi \tag{4.31.22}$$

$$\psi = \gamma_1 - \gamma_2 = \theta_1 - 2\theta_2 - \theta_3 \tag{4.31.23}$$

Solving σ from (4.28.21), we get

$$\sigma = \pm\left[\frac{k_1^2 + 2k_1 k_2 \sin\psi + k_2^2}{a^2} - \left(\frac{1}{2} - \sum_{n=2}^{3} \Omega_n^2 \Lambda_n^2 - \frac{1}{8}a^2\right)^2\right]^{1/2} \tag{4.31.24}$$

From this, the frequency–response curve of the system can be plotted in Fig. 4.19.

In order to analyze the stability of steady state solutions, Eqs. (4.31.17) and (4.31.18) are rewritten as

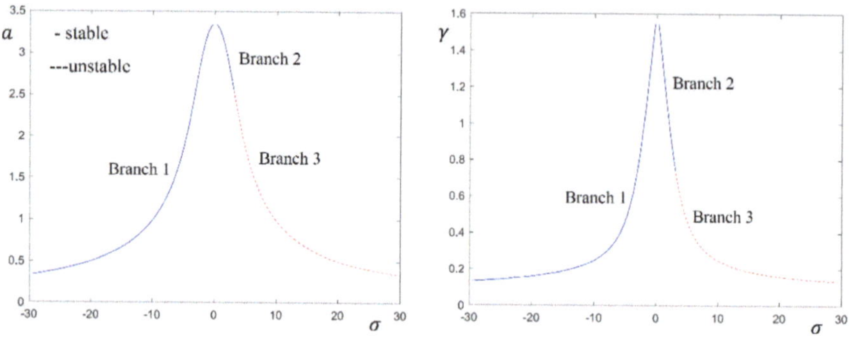

Fig. 4.19 Frequency-response and frequency-phase characteristics of the steady state solution ($\theta_1 = 0,\ \theta_2 = \theta_3 = \frac{\pi}{3},\ k_1 = 10,\ K_2 = K_3 = 5,\ \Omega_2 = 0.3,\ \Omega_3 = 0.4$) for Exercise 4.31

$$a' = \left(\frac{1}{2} - \sum_{n=2}^{3} \Omega_n^2 \Lambda_n^2 - \frac{1}{8}a^2\right)a + k_1 \sin(\gamma + \theta_1) + k_2 \cos(\gamma + 2\theta_2 + \theta_3) \tag{4.31.25}$$

$$a(\sigma - \gamma') = -k_1 \cos(\gamma + \theta_1) + k_2 \sin(\gamma + 2\theta_2 + \theta_3) \tag{4.31.26}$$

where

$$\gamma = \sigma_1 T_1 - \beta = \sigma_2 T_1 - \beta = \sigma T_1 - \beta \tag{4.31.27}$$

Around any set of steady state solution (a_0, γ_0), we let

$$a = a_0 + \tilde{a}, \quad \gamma = \gamma_0 + \tilde{\gamma} \tag{4.31.28}$$

The linearized form of (4.31.25) and (4.31.26) are

$$\tilde{a}\prime = \left(\frac{1}{2} - \sum_{n=2}^{3} \Omega_n^2 \Lambda_n^2 - \frac{3}{8}a_0^2\right)\tilde{a} + [k_1 \cos(\gamma_0 + \theta_1) - k_2 \sin(\gamma_0 + 2\theta_2 + \theta_3)]\tilde{\gamma} \tag{4.31.29}$$

$$\tilde{\gamma}\prime = \frac{\sigma}{a_0}\tilde{a} - \frac{1}{a_0}[k_1 \sin(\gamma_0 + \theta_1) + k_2 \cos(\gamma_0 + 2\theta_2 + \theta_3)]\tilde{\gamma} \tag{4.31.30}$$

Write the above two equations in matrix form as

$$\begin{Bmatrix} \tilde{a}\prime \\ \tilde{\gamma}\prime \end{Bmatrix} = \begin{bmatrix} l_{11} & l_{12} \\ l_{21} & l_{22} \end{bmatrix} \begin{Bmatrix} \tilde{a} \\ \tilde{\gamma} \end{Bmatrix} \tag{4.31.31}$$

where $l_{11} = \frac{1}{2} - \sum_{n=2}^{3} \Omega_n^2 \Lambda_n^2 - \frac{3}{8}a_0^2,\quad l_{12} = k_1 \cos(\gamma_0 + \theta_1) - k_2 \sin(\gamma_0 + 2\theta_2 + \theta_3)$

$$l_{21} = \frac{\sigma}{a_0},\quad l_{22} = -\frac{1}{a_0}\left[k_1 \sin(\gamma_0 + \theta_1) + k_2 \cos(\gamma_0 + 2\theta_2 + \theta_3)\right]$$

The eigenvalues of (4.31.31) are

$$\lambda = \frac{(l_{11} + l_{22}) \pm \sqrt{(l_{11} + l_{22})^2 - 4(l_{11}l_{22} - l_{12}l_{21})}}{2} \tag{4.31.32}$$

The stability of the corresponding steady-state solution can be determined from the sign of the real part of the eigenvalue. It follows that branch 3 in Fig. 4.19 is unstable.

4.32 Exercise 4.32 (Self-Excited System with Both Primary and Combined Resonances II)

Solution: (a) We assume

$$K_1 = 2\varepsilon k_1, K_{2,3} = O(1) \tag{4.32.1}$$

and let the solution of the equation be

$$u(t; \varepsilon) = x_0(T_0, T_1) + \varepsilon x_1(T_0, T_1) + \cdots \tag{4.32.2}$$

Substituting (4.32.1) and (4.32.2) into the original equation and retaining to $O(\varepsilon)$, we get

$$\begin{aligned}
0 &= \ddot{u} + u - \varepsilon\left(\dot{u} - \frac{1}{3}\dot{u}^3\right) - \sum_{n=1}^{3} K_n \cos(\Omega_n t + \theta_n)\\
&= \left(D_0^2 + 2\varepsilon D_0 D_1\right)(x_0 + \varepsilon x_1) + (x_0 + \varepsilon x_1)\\
&\quad - \varepsilon(D_0 + \varepsilon D_1)(x_0 + \varepsilon x_1) + \frac{1}{3}\varepsilon[(D_0 + \varepsilon D_1)(x_0 + \varepsilon x_1)]^3\\
&\quad - 2\varepsilon k_1 \cos(\Omega_1 T_0 + \theta_1) - \sum_{n=2}^{3} K_n \cos(\Omega_n T_0 + \theta_n) + \cdots\\
&= D_0^2 x_0 + \varepsilon D_0^2 x_1 + 2\varepsilon D_0 D_1 x_0 + x_0 + \varepsilon x_1 - \varepsilon D_0 x_0 + \frac{1}{3}\varepsilon (D_0 x_0)^3\\
&\quad - 2\varepsilon k_1 \cos(\Omega_1 T_0 + \theta_1) - \sum_{n=2}^{3} K_n \cos(\Omega_n T_0 + \theta_n) + \cdots
\end{aligned}$$

$$= D_0^2 x_0 + x_0 - \sum_{n=2}^{3} K_n \cos(\Omega_n T_0 + \theta_n)$$
$$+ \varepsilon[D_0^2 x_1 + x_1 + 2D_0 D_1 x_0 - D_0 x_0 + \frac{1}{3}(D_0 x_0)^3 - 2k_1 \cos(\Omega_1 T_0 + \theta_1)] + \cdots \tag{4.32.3}$$

Let the coefficient of the like power of ε be zero, we get

$$D_0^2 x_0 + x_0 = \sum_{n=2}^{3} K_n \cos(\Omega_n T_0 + \theta_n) \tag{4.32.4}$$

$$D_0^2 x_1 + x_1 = -2D_0 D_1 x_0 + D_0 x_0 - \frac{1}{3}(D_0 x_0)^3 + 2k_1 \cos(\Omega_1 T_0 + \theta_1) \tag{4.32.5}$$

The solution of (4.32.4) is

$$x_0 = Ae^{iT_0} + \sum_{n=2}^{3} \Lambda_n e^{i\theta_n} e^{i\Omega_n T_0} + cc \tag{4.32.6}$$

where $A = A(T_1)$ and

$$\Lambda_n = \frac{K_n}{2(1 - \Omega_n^2)} \tag{4.32.7}$$

Let

$$A = \frac{1}{2} a e^{i\beta} \tag{4.32.8}$$

Therefore, the first order approximate solution of the equation is

$$u = a(T_1)\cos[T_0 + \beta(T_1)] + 2\sum_{i=2}^{3} \Lambda_n \cos(\Omega_n T_0 + \theta_n) + O(\varepsilon) \tag{4.32.9}$$

Substituting (4.32.6) into (4.32.5) and taking $\Omega_1 \approx 1$ and $\Omega_2 + \Omega_3 \approx 1$ into account, we get

$$D_0^2 x_1 + x_1 = -2D_0 D_1 x_0 + D_0 x_0 - \frac{1}{3}(D_0 x_0)^3 - 2k_1 \cos(\Omega_1 T_0 + \theta_1))$$
$$= -(2iD_1 A - iA)e^{iT_0} - \frac{1}{3}(3iA^2\overline{A}e^{iT_0}$$
$$+ 6iA\Omega_2^2\Lambda_2^2 e^{iT_0} + 6iA\Omega_3^2\Lambda_3^2 e^{iT_0}$$
$$+ 3i\Omega_2\Omega_3\Lambda_2\Lambda_3\overline{A}e^{i\theta_2}e^{i\theta_3}e^{i(\Omega_2+\Omega_3-1)T_0})$$

$$+ k_1 e^{i(\Omega_1 T_0 + \theta_1)} + cc + NST \tag{4.32.10}$$

Let

$$\varepsilon\sigma_1 = \Omega_1 - 1, \quad \varepsilon\sigma_2 = 2\Omega_2 + \Omega_3 - 1 \tag{4.32.11}$$

Equation (4.32.10) becomes

$$\begin{aligned} D_0^2 x_1 + x_1 = & -(2iD_1 A - iA + iA^2\overline{A} + 2iA\Omega_2^2\Lambda_2^2 + 2iA\Omega_3^2\Lambda_3^2 \\ & + i\Omega_2\Omega_3\Lambda_2\Lambda_3\overline{A}e^{i(\sigma_2 T_1 + \theta_2 + \theta_3)} - k_1 e^{i(\sigma_1 T_1 + \theta_1)})e^{iT_0} \\ & + cc + NST \end{aligned} \tag{4.32.12}$$

In order to eliminate secular terms from the above equation, there must be

$$\begin{aligned} & 2A' - A + A^2\overline{A} + 2\Omega_2^2\Lambda_2^2 A + 2\Omega_3^2\Lambda_3^2 A \\ & \quad + \Omega_2\Omega_3\Lambda_2\Lambda_3\overline{A}e^{i(\sigma_2 T_1 + \theta_2 + \theta_3)} + ik_1 e^{i(\sigma_1 T_1 + \theta_1)} = 0 \end{aligned} \tag{4.32.13}$$

The prime represents the derivative with respect to T_1. Substituting (4.32.8) into the above equation, we get

$$\begin{aligned} & a' + ia\beta' - \frac{1}{2}a + \frac{1}{8}a^3 + \Omega_2^2\Lambda_2^2 a + \Omega_3^2\Lambda_3^2 a \\ & + i\Omega_2\Omega_3\Lambda_2\Lambda_3 a e^{i(\sigma_2 T_1 - 2\beta + \theta_2 + \theta_3)} + ik_1 e^{i(\sigma_1 T_1 - \beta + \theta_1)} = 0 \end{aligned} \tag{4.32.14}$$

Let

$$k_2 = \Omega_2\Omega_3\Lambda_2\Lambda_3, \quad \gamma_1 = \sigma_1 T_1 - \beta + \theta_1, \quad \gamma_2 = \sigma_2 T_1 - 2\beta + \theta_2 + \theta_3 \tag{4.32.15}$$

then

$$a' - \frac{1}{2}a + \frac{1}{8}a^3 + \sum_{n=2}^{3} \Omega_n^2\Lambda_n^2 + ia\beta' + ik_1 e^{i\gamma_1} + k_2 e^{i\gamma_2} = 0 \tag{4.32.16}$$

Separating the real and imaginary parts of the above equation yields

$$a' = \left(\frac{1}{2} - \sum_{n=2}^{3} \Omega_n^2\Lambda_n^2 - \frac{1}{8}a^2\right)a + k_1 \sin\gamma_1 - k_2 \cos\gamma_2 \tag{4.32.17}$$

$$a\beta' = -k_1 \cos\gamma_1 - k_2 \sin\gamma_2 \tag{4.32.18}$$

(b) It can be found from (4.32.17) and (4.32.18) that the system has a steady state solution when and only when γ_1 and γ_2 are constants, hence

$$\beta' = \sigma_1 = \frac{1}{2}\sigma_2 \tag{4.32.19}$$

4.33 Exercise 4.33 (Combination Resonance of Self-Excited Systems)

Solution: **(a)** We assume $K_n = O(1)(n = 1, 2, 3)$ and the solution of the equation is

$$u(t;\varepsilon) = x_0(T_0, T_1) + \varepsilon x_1(T_0, T_1) + \cdots \tag{4.33.1}$$

Substituting (4.31.1) and (4.31.2) into the original equation and retaining to $O(\varepsilon)$, we get

$$\begin{aligned}
0 &= \ddot{u} + u - \varepsilon\left(\dot{u} - \frac{1}{3}\dot{u}^3\right) - \sum_{n=1}^{3} K_n \cos(\Omega_n t + \theta_n) \\
&= \left(D_0^2 + 2\varepsilon D_0 D_1\right)(x_0 + \varepsilon x_1) + (x_0 + \varepsilon x_1) - \varepsilon(D_0 + \varepsilon D_1)(x_0 + \varepsilon x_1 \\
&\quad + \frac{1}{3}\varepsilon[(D_0 + \varepsilon D_1)(x_0 + \varepsilon x_1)]^3 - \sum_{n=1}^{3} K_n \cos(\Omega_n t + \theta_n) \\
&= D_0^2 x_0 + \varepsilon D_0^2 x_1 + 2\varepsilon D_0 D_1 x_0 + x_0 + \varepsilon x_1 - \varepsilon D_0 x_0 + \frac{1}{3}\varepsilon(D_0 x_0)^3 \\
&\quad - \sum_{n=1}^{3} K_n \cos(\Omega_n T_0 + \theta_n) + \cdots \\
&= D_0^2 x_0 + x_0 - \sum_{n=1}^{3} K_n \cos(\Omega_n T_0 + \theta_n) \\
&\quad + \varepsilon[D_0^2 x_1 + x_1 + 2D_0 D_1 x_0 - D_0 x_0 + \frac{1}{3}(D_0 x_0)^3] + \cdots
\end{aligned} \tag{4.33.2}$$

Let the coefficient of the like power of ε be zero, we get

$$D_0^2 x_0 + x_0 = \sum_{n=1}^{3} K_n \cos(\Omega_n T_0 + \theta_n) \tag{4.33.3}$$

$$D_0^2 x_1 + x_1 = -2D_0 D_1 x_0 + D_0 x_0 - \frac{1}{3}(D_0 x_0)^3 \tag{4.33.4}$$

The solution of (4.33.3) is

$$x_0 = Ae^{iT_0} + \sum_{n=1}^{3} \Lambda_n e^{i\theta_n} e^{i\Omega_n T_0} + cc \tag{4.33.5}$$

where $A = A(T_1)$ and

$$\Lambda_n = \frac{K_n}{2(1 - \Omega_n^2)} \tag{4.33.6}$$

Let

$$A = \frac{1}{2} a e^{i\beta} \tag{4.33.7}$$

Therefore, the first order approximate solution of the equation is

$$u = a(T_1)\cos[T_0 + \beta(T_1)] + 2\sum_{i=1}^{3} \Lambda_n \cos(\Omega_n T_0 + \theta_n) + O(\varepsilon) \tag{4.33.8}$$

Substituting (4.33.5) into (4.33.4), and taking $\Omega_1 + \Omega_2 + \Omega_3 \approx 1$ into account, we get

$$\begin{aligned} D_0^2 x_1 + x_1 &= -2D_0 D_1 x_0 + D_0 x_0 - \frac{1}{3}(D_0 x_0)^3 \\ &= -(2iD_1 A - iA)e^{iT_0} - \frac{1}{3}[3iA^2\bar{A}e^{iT_0} + 6iAe^{iT_0}\sum_{n=1}^{3}\Omega_n^2\Lambda_n^2) \\ &\quad - 6i\Omega_1\Omega_2\Omega_3\Lambda_1\Lambda_2\Lambda_3 e^{i(\theta_1+\theta_2+\theta_3)} e^{i(\Omega_1+\Omega_2+\Omega_3)T_0}] + cc + NST \end{aligned} \tag{4.33.9}$$

Let

$$\varepsilon\sigma_2 = \Omega_1 + \Omega_2 + \Omega_3 - 1 \tag{4.33.10}$$

Equation (4.31.10) becomes

$$\begin{aligned} D_0^2 x_1 + x_1 = &-(2iD_1 A - iA + iA^2\bar{A} + 2iA\Omega_1^2\Lambda_1^2 + 2iA\Omega_2^2\Lambda_2^2 + 2iA\Omega_3^2\Lambda_3^2 \\ &- i\Omega_1\Omega_2\Omega_3\Lambda_1\Lambda_2\Lambda_3 e^{i(\sigma T_1+\theta_1+\theta_2+\theta_3)})e^{iT_0} \\ &+ cc + NST \end{aligned} \tag{4.33.11}$$

In order to eliminate secular terms from the above equation, there must be

$$\begin{aligned} &2A' - A + A^2\overline{A} + 2A\Omega_1^2\Lambda_1^2 + 2A\Omega_2^2\Lambda_2^2 + 2A\Omega_3^2\Lambda_3^2 \\ &- 2i\Omega_1\Omega_2\Omega_3\Lambda_1\Lambda_2\Lambda_3 e^{i(\sigma T_1+\theta_1+\theta_2+\theta_3)} = 0 \end{aligned} \tag{4.33.12}$$

The prime represents the derivative with respect to T_1. Substituting (4.33.7) into the above equation, we get

$$\begin{aligned} a' + ia\beta' - \frac{1}{2}a + \frac{1}{8}a^3 + \Omega_1^2\Lambda_1^2 a + \Omega_2^2\Lambda_2^2 a + \Omega_3^2\Lambda_3^2 a \\ - 2\Omega_1\Omega_2\Omega_3\Lambda_1\Lambda_2\Lambda_3 e^{i(\sigma T_1 - \beta + \theta_1 + \theta_2 + \theta_3)} = 0 \end{aligned} \quad (4.33.13)$$

Let

$$k = 2\Omega_1\Omega_2\Omega_3\Lambda_1\Lambda_2\Lambda_3, \quad \gamma = \sigma T_1 - \beta + \theta_1 + \theta_2 + \theta_3 \quad (4.33.14)$$

then

$$a' - \frac{1}{2}a + \frac{1}{8}a^3 + \sum_{n=1}^{3}\Omega_n^2\Lambda_n^2 + ia\beta' - ke^{i\gamma} = 0 \quad (4.33.15)$$

Separating the real and imaginary parts of the above equation yields

$$a' = \left(\frac{1}{2} - \sum_{n=1}^{3}\Omega_n^2\Lambda_n^2 - \frac{1}{8}a^2\right)a + k\cos\gamma \quad (4.33.16)$$

$$a\beta' = k\sin\gamma \quad (4.33.17)$$

(b) Let $a' = 0$and $\beta' = \sigma$ in (4.33.16) and (4.33.17), we obtain

$$-\left(\frac{1}{2} - \sum_{n=1}^{3}\Omega_n^2\Lambda_n^2 - \frac{1}{8}a^2\right)a = k\cos\gamma \quad (4.33.18)$$

$$a\sigma = k\sin\gamma \quad (4.33.19)$$

The two equations are squared and added together to obtain the frequency–response equation as

$$\left(\frac{1}{2} - \sum_{n=1}^{3}\Omega_n^2\Lambda_n^2 - \frac{1}{8}a^2\right)^2 a^2 + \sigma^2 a^2 = k^2 \quad (4.33.20)$$

The solution of (4.33.20) is

$$\sigma = \pm\left[\frac{k^2}{a^2} - \left(\frac{1}{2} - \sum\nolimits_{n=1}^{3}\Omega_n^2\Lambda_n^2 - \frac{1}{8}a^2\right)^2\right]^{1/2} \quad (4.33.21)$$

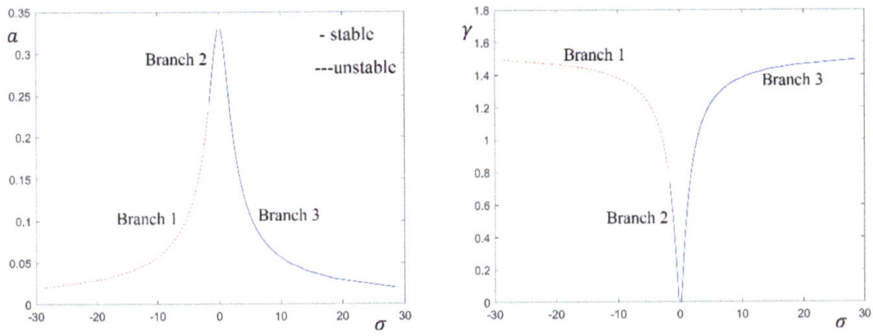

Fig. 4.20 Frequency-response and frequency-phase characteristics of the steady state solution ($K_1 = 5,\ K_2 = 5,\ K_3 = 2,\ \Omega_1 = 0.2,\ \Omega_2 = 0.3,\ \Omega_3 = 0.5$) for Exercise 4.33

From this, the frequency–response curve of the system can be plotted as shown in Fig. 4.20.

In order to analyze the stability of steady state solutions, we can rewrite (4.31.17) and (4.31.18) as

$$a' = \left(\frac{1}{2} - \sum_{n=2}^{3} \Omega_n^2 \Lambda_n^2 - \frac{1}{8}a^2\right)a + k\cos\gamma \tag{4.33.22}$$

$$a(\sigma - \gamma') = k\sin\gamma \tag{4.33.23}$$

Around any set of steady state solution (a_0, γ_0), we let

$$a = a_0 + \tilde{a}, \quad \gamma = \gamma_0 + \tilde{\gamma} \tag{4.33.24}$$

The linearized form of (4.33.22) and (4.33.23) are

$$\tilde{a}\prime = \left(\frac{1}{2} - \sum_{n=1}^{3} \Omega_n^2 \Lambda_n^2 - \frac{3}{8}a_0^2\right)\tilde{a} - (k\sin\gamma_0)\tilde{\gamma} \tag{4.33.25}$$

$$\tilde{\gamma}\prime = \frac{\sigma}{a_0}\tilde{a} - \frac{1}{a_0}(k\cos\gamma_0)\tilde{\gamma} \tag{4.33.26}$$

Write the above two equations in matrix form as

$$\begin{Bmatrix} \tilde{a}\prime \\ \tilde{\gamma}\prime \end{Bmatrix} = \begin{bmatrix} l_{11} & l_{12} \\ l_{21} & l_{22} \end{bmatrix} \begin{Bmatrix} \tilde{a} \\ \tilde{\gamma} \end{Bmatrix} \tag{4.33.27}$$

where

$$l_{11} = \frac{1}{2} - \sum_{n=1}^{3} \Omega_n^2 \Lambda_n^2 - \frac{3}{8} a_0^2, \quad l_{12} = -k \sin \gamma_0$$

$$l_{21} = \frac{\sigma}{a_0}, \quad l_{22} = -\frac{1}{a_0} k \cos \gamma_0$$

The eigenvalues of Eq. (4.31.31) are

$$\lambda = \frac{(l_{11} + l_{22}) \pm \sqrt{(l_{11} + l_{22})^2 - 4(l_{11} l_{22} - l_{12} l_{21})}}{2} \tag{4.33.28}$$

The stability of the corresponding steady-state solution can be determined from the sign of the real part of the eigenvalue. It follows that branch 1 in Fig. 4.20 is unstable.

4.34 Exercise 4.34 (Self-Excited System with Simultaneous Primary, Superharmonic, and Subharmonic Resonances)

Solution: **(a)** We assume

$$K_1 = 2\varepsilon k_1, K_{2,3} = O(1) \tag{4.34.1}$$

and let the solution of the equation be

$$u(t; \varepsilon) = x_0(T_0, T_1) + \varepsilon x_1(T_0, T_1) + \cdots \tag{4.34.2}$$

Substituting (4.34.1) and (4.34.2) into the original equation and retaining to $O(\varepsilon)$, we get

$$\begin{aligned}
0 &= \ddot{u} + u - \varepsilon\left(\dot{u} - \frac{1}{3}\dot{u}^3\right) - \sum_{n=1}^{3} K_n \cos(\Omega_n t + \theta_n) \\
&= \left(D_0^2 + 2\varepsilon D_0 D_1\right)(x_0 + \varepsilon x_1) + (x_0 + \varepsilon x_1) \\
&\quad - \varepsilon(D_0 + \varepsilon D_1)(x_0 + \varepsilon x_1) + \frac{1}{3}\varepsilon[(D_0 + \varepsilon D_1)(x_0 + \varepsilon x_1)]^3 \\
&\quad - 2\varepsilon k_1 \cos(\Omega_1 T_0 + \theta_1) - \sum_{n=2}^{3} K_n \cos(\Omega_n T_0 + \theta_n) + \cdots \\
&= D_0^2 x_0 + \varepsilon D_0^2 x_1 + 2\varepsilon D_0 D_1 x_0 + x_0 + \varepsilon x_1 - \varepsilon D_0 x_0 + \frac{1}{3}\varepsilon (D_0 x_0)^3
\end{aligned}$$

$$
\begin{aligned}
&- 2\varepsilon k_1 \cos(\Omega_1 T_0 + \theta_1) - \sum_{n=2}^{3} K_n \cos(\Omega_n T_0 + \theta_n) + \cdots \\
&= D_0^2 x_0 + x_0 - \sum_{n=2}^{3} K_n \cos(\Omega_n T_0 + \theta_n) \\
&+ \varepsilon[D_0^2 x_1 + x_1 + 2D_0 D_1 x_0 - D_0 x_0 + \frac{1}{3}(D_0 x_0)^3 - 2k_1 \cos(\Omega_1 T_0 + \theta_1)] + \cdots
\end{aligned}
\tag{4.34.3}
$$

Let the coefficient of the like power of ε be zero, we get

$$
D_0^2 x_0 + x_0 = \sum_{n=2}^{3} K_n \cos(\Omega_n T_0 + \theta_n) \tag{4.34.4}
$$

$$
D_0^2 x_1 + x_1 = -2D_0 D_1 x_0 + D_0 x_0 - \frac{1}{3}(D_0 x_0)^3 + 2k_1 \cos(\Omega_1 T_0 + \theta_1) \tag{4.34.5}
$$

The solution of (4.34.4) is

$$
x_0 = Ae^{iT_0} + \sum_{n=2}^{3} \Lambda_n e^{i\theta_n} e^{i\Omega_n T_0} + cc \tag{4.34.6}
$$

where $A = A(T_1)$ and

$$
\Lambda_n = \frac{K_n}{2(1 - \Omega_n^2)} \tag{4.34.7}
$$

Let

$$
A = \frac{1}{2} a e^{i\beta} \tag{4.34.8}
$$

Therefore, the first order approximate solution of the equation is

$$
u = a(T_1) \cos[T_0 + \beta(T_1)] + 2 \sum_{i=2}^{3} \Lambda_n \cos(\Omega_n T_0 + \theta_n) + O(\varepsilon) \tag{4.34.9}
$$

Substituting (4.34.6) into (4.34.5) and taking $\Omega_1 \approx 1$, $\Omega_2 \approx 1/3$ and $\Omega_3 \approx 3$ into account, we get

$$
\begin{aligned}
D_0^2 x_1 + x_1 &= -2D_0 D_1 x_0 + D_0 x_0 - \frac{1}{3}(D_0 x_0)^3 \\
&= -(2iD_1 A - iA)e^{iT_0} - \frac{1}{3}(3iA^2 \overline{A} e^{iT_0} + 6iA\Omega_2^2 \Lambda_2^2 e^{iT_0} + 6iA\Omega_3^2 \Lambda_3^2 e^{iT_0}
\end{aligned}
$$

$$- 3i\bar{A}^2\Omega_3\Lambda_3 e^{i\theta_3}e^{i(\Omega_3-2)T_0} - i\Omega_2^3\Lambda_2^3 e^{i3\theta_2}e^{i3\Omega_2 T_0}) + k_1 e^{i(\Omega_1 T_0+\theta_1)}$$
$$+ cc + NST \tag{4.34.10}$$

Let

$$\varepsilon\sigma_1 = \Omega_1 - 1, \quad \varepsilon\sigma_2 = 3\Omega_2 - 1, \quad \varepsilon\sigma_3 = \Omega_3 - 3 \tag{4.34.11}$$

Equation (4.34.10) becomes

$$\begin{aligned} D_0^2 x_1 + x_1 = &-(2iD_1A - iA + iA^2\bar{A} + 2iA\Omega_2^2\Lambda_2^2 + 2iA\Omega_3^2\Lambda_3^2 \\ &- i\bar{A}^2\Omega_3\Lambda_3 e^{i(\sigma_3 T_1+\theta_3)} - \frac{1}{3}i\Omega_2^3\Lambda_2^3 e^{i(\sigma_2 T_1+3\theta_2)} \\ &- k_1 e^{i(\sigma_1 T_1+\theta_1)})e^{iT_0} + cc + NST \end{aligned} \tag{4.34.12}$$

In order to eliminate secular terms from the above equation, there must be

$$\begin{aligned} &2A' - A + A^2\bar{A} + 2\Omega_2^2\Lambda_2^2 A + 2\Omega_3^2\Lambda_3^2 A - \bar{A}^2\Omega_3\Lambda_3 e^{i(\sigma_3 T_1+\theta_3)} \\ &- \frac{1}{3}\Omega_2^3\Lambda_2^3 e^{i(\sigma_2 T_1+3\theta_2)} + ik_1 e^{i(\sigma_1 T_1+\theta_1)} = 0 \end{aligned} \tag{4.34.13}$$

The prime represents the derivative with respect to T_1. Substituting (4.34.8) into the above equation, we get

$$\begin{aligned} &a' + ia\beta' - \frac{1}{2}a + \frac{1}{8}a^3 + \Omega_2^2\Lambda_2^2 a + \Omega_3^2\Lambda_3^2 a - \frac{1}{4}a^2\Omega_3\Lambda_3 e^{i(\sigma_3 T_1 - 3\beta+\theta_3)} \\ &- \frac{1}{3}\Omega_2^3\Lambda_2^3 e^{i(\sigma_2 T_1-\beta+3\theta_2)} + ik_1 e^{i(\sigma_1 T_1-\beta+\theta_1)} = 0 \end{aligned} \tag{4.34.14}$$

Let

$$\begin{aligned} &k_2 = \frac{1}{3}\Omega_2^3\Lambda_2^3,\, k_3 = \frac{1}{4}\Omega_3\Lambda_3,\, \gamma_1 = \sigma_1 T_1 - \beta + \theta_1, \\ &\gamma_2 = \sigma_2 T_1 - \beta + 3\theta_2,\, \gamma_2 = \sigma_3 T_1 - 3\beta + \theta_3 \end{aligned} \tag{4.34.15}$$

then

$$a' - \frac{1}{2}a + \frac{1}{8}a^3 + \sum_{n=2}^{3}\Omega_n^2\Lambda_n^2 + ia\beta' + ik_1 e^{i\gamma_1} - k_2 e^{i\gamma_2} - k_3 a^2 e^{i\gamma_3} = 0 \tag{4.34.16}$$

Separating the real and imaginary parts of the above equation yields

$$a' = \left(\frac{1}{2} - \sum_{n=2}^{3}\Omega_n^2\Lambda_n^2 - \frac{1}{8}a^2\right)a + k_1\sin\gamma_1 + k_2\cos\gamma_2 + k_3 a^2\cos\gamma_3 \tag{4.34.17}$$

$$a\beta' = -k_1 \cos\gamma_1 + k_2 \sin\gamma_2 + k_3 a^2 \sin\gamma_3 \tag{4.34.18}$$

(b) It can be found from (4.34.17) and (4.34.18) that the system has a steady state solution when and only when γ_1, γ_2 and γ_3 are constants, hence

$$\beta' = \sigma_1 = \sigma_2 = \frac{1}{3}\sigma_3 \tag{4.34.19}$$

4.35 Exercise 4.35 (Self-Excited System with Simultaneous Primary, and Combinatorial Resonances)

Readers are invited to solve this problem by referring to Exercise 4.33 and Exercise 4.34.

4.36 Exercise 4.36 (Simple Harmonic Response of Van Der Pol's Oscillator)

Solution: **(a)** Let the solution to the equation be

$$\begin{aligned} u(t;\varepsilon) &= u_0(T_0, T_1) + \varepsilon u_1(T_0, T_1) + \cdots \\ z(t;\varepsilon) &= z_0(T_0, T_1) + \varepsilon z_1(T_0, T_1) + \cdots \end{aligned} \tag{4.36.1}$$

Substituting (4.36.1) into the original equation and retaining to $O(\varepsilon)$ yields

$$\begin{aligned} 0 &= \ddot{u} + \omega_0^2 u - 2\varepsilon[(1-z)\dot{u} - \dot{z}u] + 2K\Omega \sin\Omega t \\ &= \left(D_0^2 + 2\varepsilon D_0 D_1\right)(u_0 + \varepsilon u_1) + \omega_0^2(u_0 + \varepsilon u_1) - 2\varepsilon(D_0 + \varepsilon D_1)(u_0 + \varepsilon u_1) \\ &\quad + 2\varepsilon(z_0 + \varepsilon z_1)[(D_0 + \varepsilon D_1)(u_0 + \varepsilon u_1)] + \cdots \\ &\quad + 2\varepsilon(u_0 + \varepsilon u_1)[(D_0 + \varepsilon D_1)(z_0 + \varepsilon z_1)] + 2K\Omega \sin\Omega T_0 + \cdots \\ &= D_0^2 u_0 + \varepsilon D_0^2 u_1 + 2\varepsilon D_0 D_1 u_0 + \omega_0^2 u_0 + \varepsilon\omega_0^2 u_1 - 2\varepsilon D_0 u_0 \\ &\quad + 2\varepsilon z_0 D_0 u_0 + 2\varepsilon u_0 D_0 z_0 + 2K\Omega \sin\Omega T_0 + \cdots \\ &= D_0^2 u_0 + \omega_0^2 u_0 + \varepsilon(D_0^2 u_1 + \omega_0^2 u_1 + 2D_0 D_1 u_0 - 2D_0 u_0 \\ &\quad + 2z_0 D_0 u_0 + 2u_0 D_0 z_0) + 2K\Omega \sin\Omega T_0 + \cdots \end{aligned} \tag{4.36.2}$$

$$
\begin{aligned}
0 &= \tau\dot{z} + z - u^2 \\
&= \tau(D_0 + \varepsilon D_1)(z_0 + \varepsilon z_1) + (z_0 + \varepsilon z_1) - (u_0 + \varepsilon u_1)^2 + \cdots \\
&= \tau D_0 z_0 + \tau\varepsilon D_1 z_0 + \tau\varepsilon D_0 z_1 + z_0 + \varepsilon z_1 - u_0^2 - 2\varepsilon u_0 u_1 + \cdots \\
&= \tau D_0 z_0 + z_0 - u_0^2 + \varepsilon(\tau D_0 z_1 + z_1 + \tau D_1 z_0 - 2u_0 u_1) + \cdots
\end{aligned}
\tag{4.36.3}
$$

Considering the primary resonance, we assume.

$$
\Omega = \omega_0 + \varepsilon\sigma, \quad K = \varepsilon k \tag{4.36.4}
$$

Substitute this into (4.36.2) and (4.36.3) and let the coefficient of the like power of ε be zero, we can obtain

$$
\begin{aligned}
D_0^2 u_0 + \omega_0^2 u_0 &= 0 \\
\tau D_0 z_0 + z_0 &= u_0^2
\end{aligned}
\tag{4.36.5}
$$

$$
\begin{aligned}
D_0^2 u_1 + \omega_0^2 u_1 &= -2D_0 D_1 u_0 + 2D_0 u_0 - 2z_0 D_0 u_0 - 2u_0 D_0 z_0 \\
&\quad - 2\Omega k \sin(\omega_0 T_0 + \sigma T_1) \\
D_0 z_1 + \frac{1}{\tau} z_1 &= -D_1 z_0 + \frac{2}{\tau} u_0 u_1
\end{aligned}
\tag{4.36.6}
$$

Solving from (4.36.5) yields

$$
\begin{aligned}
u_0 &= A e^{i\omega_0 T_0} + cc \\
z_0 &= b e^{-T_0/\tau} + 2A\overline{A} + (4\omega_0^2\tau^2 + 1)^{-1/2} A^2 e^{-i\tan^{-1} 2\omega_0\tau} e^{2i\omega_0 T_0} + cc
\end{aligned}
\tag{4.36.7}
$$

where b is the real constant of integration and $A = A(T_1)$. Let

$$
A = \frac{1}{2} a e^{i\beta} \tag{4.36.8}
$$

Therefore, the first order approximate solution of the equation is

$$
\begin{aligned}
u &= a\cos(\omega_0 t + \beta) + O(\varepsilon) \\
z &= b e^{-t/\tau} + \tfrac{1}{2} a^2 + \tfrac{1}{2}\cos\left(2\omega_0 t + 2\beta - \tan^{-1} 2\omega_0\tau\right) + O(\varepsilon)
\end{aligned}
\tag{4.36.9}
$$

Substituting (4.36.7) into (4.36.6) yields

$$
\begin{aligned}
D_0^2 u_1 + \omega_0^2 u_1 &= -2D_0 D_1 u_0 + 2D_0 u_0 - 2z_0 D_0 u_0 - 2u_0 D_0 z_0 - 2\Omega k \sin(\omega_0 T_0 + \sigma T_1) \\
&= -2\omega_0 i(D_1 A - A) e^{i\omega_0 T_0} - i\omega_0 [2bA e^{-T_0/\tau} \\
&\quad + 4A^2\overline{A} - 2\left(4\omega_0^2\tau^2 1\right)^{-1/2} A^2 \overline{A} e^{-i\tan^{-1} 2\omega_0\tau}\Big] e^{i\omega_0 T_0} \\
&\quad - [-\frac{2}{\tau} b e^{-\frac{T_0}{\tau}} A + 4i\omega_0 \left(4\omega_0^2\tau^2 + 1\right)^{-\frac{1}{2}} A^2 \overline{A} e^{-i\tan^{-1} 2\omega_0\tau}\Big] e^{i\omega_0 T_0} \\
&\quad + i\omega_0 k e^{i\sigma T_1} e^{i\omega_0 T_0} + cc + NST
\end{aligned}
$$

$$= -[2\omega_0(iD_1A - iA) + 2i\omega_0 bAe^{-T_0/\tau} + 4i\omega_0 A^2\overline{A} + 2i\omega_0(4\omega_0^2\tau^2+1)^{-1/2}A^2\overline{A}e^{-i\tan^{-1}2\omega_0\tau} - \frac{2}{\tau}be^{-T_0/\tau}A - i\omega_0 ke^{i\sigma T_1}]e^{i\omega_0 T_0} + cc + NST \tag{4.36.10}$$

In order to eliminate secular terms from the above equation, there must be

$$2D_1A - 2A + 4A^2\overline{A} + 2(4\omega_0^2\tau^2+1)^{-\frac{1}{2}}A^2\overline{A}e^{-i\tan^{-1}2\omega_0\tau} + 2bAe^{-\frac{T_0}{\tau}} + i\frac{2}{\tau\omega_0}be^{-\frac{T_0}{\tau}}A - ke^{i\sigma T_1} = 0 \tag{4.36.11}$$

The prime represents the derivative with respect to T_1. Substituting (4.36.8) into the above equation and omitting the fast decay term containing $e^{-T_0/\tau}$, we get

$$a' + ia\beta' - a + \frac{1}{2}a^3 + \frac{1}{4}(4\omega_0^2\tau^2+1)^{-1/2}a^3e^{-i\tan^{-1}2\omega_0\tau} - ke^{i(\sigma T_1-\beta)} = 0 \tag{4.36.12}$$

Separating the real and imaginary parts of the above equation yields

$$\begin{aligned} a' &= \{1 - \tfrac{1}{4}[2 + (4\omega_0^2\tau^2+1)^{-1/2}\cos(\tan^{-1}2\omega_0\tau)]a^2\}a + k\cos(\sigma T_1 - \beta) \\ \beta' &= -\tfrac{1}{4}[-(4\omega_0^2\tau^2+1)^{-1/2}\sin(\tan^{-1}2\omega_0\tau)]a^2 + ka^{-1}\sin(\sigma T_1 - \beta) \end{aligned} \tag{4.36.13}$$

Let

$$\begin{aligned} \alpha_r &= 2 + (4\omega_0^2\tau^2+1)^{-1/2}\cos(\tan^{-1}2\omega_0\tau) \\ \alpha_i &= -(4\omega_0^2\tau^2+1)^{-1/2}\sin(\tan^{-1}2\omega_0\tau) \end{aligned} \tag{4.36.14}$$

and change the time variable T_1 to εt in (4.36.13) we can obtain

$$\begin{aligned} \dot{a} &= \varepsilon(1 - \tfrac{1}{4}\alpha_r a^2)a + \varepsilon k\cos(\varepsilon\sigma t - \beta) \\ \dot{\beta} &= -\tfrac{1}{4}\varepsilon\alpha_i a^2 + \varepsilon ka^{-1}\sin(\varepsilon\sigma t - \beta) \end{aligned} \tag{4.36.15}$$

Let

$$\gamma = \varepsilon\sigma t - \beta \tag{4.36.16}$$

Equation (4.36.15) becomes

$$\begin{aligned} \dot{a} &= \varepsilon(1 - \tfrac{1}{4}\alpha_r a^2)a + \varepsilon k\cos\gamma \\ \dot{\gamma} &= \varepsilon\sigma + \tfrac{1}{4}\varepsilon\alpha_i a^2 - \varepsilon ka^{-1}\sin\gamma \end{aligned} \tag{4.36.17}$$

For the stead-state solutions of (4.36.17), $\dot{a} = \dot{\gamma} = 0$ so that a and γ are the solutions of

$$\begin{aligned} -(1 - \tfrac{1}{4}\alpha_r a^2)a &= k\cos\gamma \\ \sigma a + \tfrac{1}{4}\alpha_i a^3 &= k\sin\gamma \end{aligned} \tag{4.36.18}$$

Squaring and adding two equations in (4.36.18) together yields the frequency–response equation as

$$[\left(1 - \frac{1}{4}\alpha_r a^2\right)a]^2 + (\sigma a + \frac{1}{4}\alpha_i a^3)^2 = k^2 \tag{4.36.19}$$

Solving σ in terms of a from (4.36.19)

$$\sigma = -\frac{1}{4}\alpha_i a^2 \pm \frac{1}{a}\sqrt{k^2 - [\left(1 - \frac{1}{4}\alpha_r a^2\right)a]^2} \tag{4.36.20}$$

To analyze the stability of steady state $(a_0,\ \gamma_0)$, let

$$a = a_0 + a_1, \quad \gamma = \gamma_0 + \gamma_1 \tag{4.36.21}$$

then the linearized form of (4.36.17) is

$$\begin{Bmatrix} \dot{a}_1 \\ \dot{\gamma}_1 \end{Bmatrix} = \varepsilon \begin{bmatrix} l_{11} & l_{12} \\ l_{21} & l_{22} \end{bmatrix} \begin{Bmatrix} a_1 \\ \gamma_1 \end{Bmatrix} \tag{4.36.22}$$

where

$$l_{11} = 1 - \frac{3}{4}\alpha_r a_0^2, \quad l_{12} = -k\sin\gamma_0$$

$$l_{21} = \frac{\sigma}{a_0} + \frac{3}{4}\alpha_i a_0, \quad l_{22} = -\frac{1}{a_0}k\cos\gamma_0$$

The eigenvalues of Eq. (4.36.22) are

$$\frac{\lambda}{\varepsilon} = \frac{(l_{11} + l_{22}) \pm \sqrt{(l_{11} + l_{22})^2 - 4(l_{11}l_{22} - l_{12}l_{21})}}{2} \tag{4.36.23}$$

The stability of the corresponding steady-state solution can be determined from the sign of the real part of the eigenvalue.

(b) For the case of hard nonresonance excitations, i.e., $K = O(1), \quad \Omega - \omega_0 > O(\varepsilon)$, let the coefficient of the like power of ε be zero in (4.36.2) and (4.36.3), we can obtain

$$\begin{aligned} &D_0^2 u_0 + \omega_0^2 u_0 = -2K\Omega \sin \Omega T_0 \\ &\tau D_0 z_0 + z_0 = u_0^2 \end{aligned} \tag{4.36.24}$$

$$\begin{aligned} D_0^2 u_1 + \omega_0^2 u_1 &= -2D_0 D_1 u_0 + 2D_0 u_0 - 2z_0 D_0 u_0 - 2u_0 D_0 z_0 \\ D_0 z_1 + \tfrac{1}{\tau} z_1 &= -D_1 z_0 + \tfrac{2}{\tau} u_0 u_1 \end{aligned} \tag{4.36.25}$$

Solving (u_0, z_0) from (4.36.24) yields

$$u_0 = A e^{i\omega_0 T_0} + iK\Omega(\omega_0^2 - \Omega^2)^{-1} e^{i\Omega T_0} + cc \tag{4.36.26}$$

$$\begin{aligned} z_0 &= b e^{-\frac{T_0}{\tau}} + 2A\overline{A} + 2K^2\Omega^2(\omega_0^2 - \Omega^2)^{-2} \\ &+ (4\omega_0^2\tau^2 + 1)^{-\frac{1}{2}} A^2 e^{-i\tan^{-1} 2\omega_0 \tau} e^{2i\omega_0 T_0} \\ &- K^2\Omega^2(\omega_0^2 - \Omega^2)^{-2}(4\Omega^2\tau^2 + 1)^{-\frac{1}{2}} e^{-i\tan^{-1} 2\Omega\tau} e^{i2\Omega T_0} \\ &+ 2iK\Omega(\omega_0^2 - \Omega^2)^{-1}[(\Omega + \omega_0)^2\tau^2 + 1]^{-\frac{1}{2}} A e^{-i\tan^{-1}(\Omega+\omega_0)\tau} e^{i(\Omega+\omega_0)T_0} \\ &+ 2iK\Omega(\omega_0^2 - \Omega^2)^{-1}[(\Omega - \omega_0)^2\tau^2 + 1]^{-1/2} \overline{A} e^{-i\tan^{-1}(\Omega-\omega_0)\tau} e^{i(\Omega-\omega_0)T_0} + cc \end{aligned} \tag{4.36.27}$$

where b is the real constant of integration and $A = A(T_1)$. Let

$$A = \frac{1}{2} a e^{i\beta} \tag{4.36.28}$$

Therefore, the first order approximate solution of the equation is

$$u = a\cos(\omega_0 t + \beta) - 2K\Omega(\omega_0^2 - \Omega^2)^{-1} \sin \Omega t + O(\varepsilon) \tag{4.36.29}$$

Substituting (4.36.26) and (4.36.27) into (4.36.25) and keeping only the terms containing $e^{i\omega_0 T_0}$ and the terms that can produce subharmonic and superharmonic resonance excitations at $\Omega \approx 3\omega_0$ and $3\Omega \approx \omega_0$, and omitting the fast-decaying term containing $e^{-T_0/\tau}$, we get

$$\begin{aligned} D_0^2 u_1 + \omega_0^2 u_1 &= -2D_0 D_1 u_0 + 2D_0 u_0 - 2D_0(u_0 z_0) \\ &= -2i\omega_0(D_1 A - A)e^{i\omega_0 T_0} - 2i\omega_0[2A\overline{A} + 2K^2\Omega^2(\omega_0^2 - \Omega^2)^{-2}]A e^{i\omega_0 T_0} \\ &\quad - 2i\omega_0(4\omega_0^2\tau^2 + 1)^{-1/2} A^2\overline{A} e^{-i\tan^{-1} 2\omega_0\tau} e^{i\omega_0 T_0} \\ &\quad - 4i\omega_0 K^2\Omega^2(\omega_0^2 - \Omega^2)^{-2}[(\Omega + \omega_0)^2\tau^2 + 1]^{-1/2} A e^{-i\tan^{-1}(\Omega+\omega_0)\tau} e^{i\omega_0 T_0} \\ &\quad - 4i\omega_0 K^2\Omega^2(\omega_0^2 - \Omega^2)^{-2}[(\Omega - \omega_0)^2\tau^2 + 1]^{-1/2} A e^{i\tan^{-1}(\Omega-\omega_0)\tau} e^{i\omega_0 T_0} \\ &\quad - 6K^3\Omega^4(\omega_0^2 - \Omega^2)^{-3}(4\Omega^2\tau^2 + 1)^{-1/2} e^{-i\tan^{-1} 2\Omega\tau} e^{i3\Omega T_0} \\ &\quad + 4K\Omega(\Omega - 2\omega_0)(\omega_0^2 - \Omega^2)^{-1}[(\Omega - \omega_0)^2\tau^2 + 1]^{-1/2} \\ &\quad \overline{A}^2 e^{-i\tan^{-1}(\Omega-\omega_0)\tau} e^{i(\Omega-2\omega_0)T_0} \end{aligned}$$

$$+ 2K\Omega(\Omega - 2\omega_0)(\omega_0^2 - \Omega^2)^{-1}(4\omega_0^2\tau^2 + 1)^{-1/2}$$
$$\overline{A}^2 e^{i\tan^{-1} 2\omega_0\tau} e^{i(\Omega-2\omega_0)T_0} + cc + NST \quad (4.36.30)$$

In order to eliminate secular terms from the above equation, there must be

$$\begin{aligned}
&A' - A + 2K^2\Omega^2(\omega_0^2 - \Omega^2)^{-2}A \\
&+2A^2\overline{A} + (4\omega_0^2\tau^2 + 1)^{-1/2}A^2\overline{A}e^{-i\tan^{-1} 2\omega_0\tau} \\
&+2K^2\Omega^2(\omega_0^2 - \Omega^2)^{-2}[(\Omega+\omega_0)^2\tau^2 + 1]^{-1/2}Ae^{-i\tan^{-1}(\Omega+\omega_0)\tau} \\
&2K^2\Omega^2(\omega_0^2 - \Omega^2)^{-2}[(\Omega-\omega_0)^2\tau^2 + 1]^{-1/2}Ae^{i\tan^{-1}(\Omega-\omega_0)\tau} = 0
\end{aligned} \quad (4.36.31)$$

The prime represents the derivative with respect to T_1. Substituting (4.36.28) into the above equation, we get

$$\begin{aligned}
&a' + ia\beta' - a + \frac{1}{2}a^3 + 2K^2\Omega^2(\omega_0^2 - \Omega^2)^{-2}a \\
&+ \frac{1}{4}(4\omega_0^2\tau^2 + 1)^{-1/2}a^3e^{-i\tan^{-1} 2\omega_0\tau} \\
&+ 2K^2\Omega^2(\omega_0^2 - \Omega^2)^{-2}[(\Omega+\omega_0)^2\tau^2 + 1]^{-1/2}ae^{-i\tan^{-1}(\Omega+\omega_0)\tau} \\
&+ 2K^2\Omega^2(\omega_0^2 - \Omega^2)^{-2}[(\Omega-\omega_0)^2\tau^2 + 1]^{-1/2}ae^{i\tan^{-1}(\Omega-\omega_0)\tau} = 0
\end{aligned} \quad (4.36.32)$$

Separating the real and imaginary parts of the above equation yields

$$\begin{aligned}
a' = &\{1 - 2K^2\Omega^2(\omega_0^2 - \Omega^2)^{-2}[1 + [(\Omega+\omega_0)^2\tau^2 + 1]^{-1/2}\cos(\Omega+\omega_0)\tau \\
&+ [(\Omega-\omega_0)^2\tau^2 + 1]^{-1/2}\cos(\Omega-\omega_0)\tau\}a \\
&- \frac{1}{4}[2 + (4\omega_0^2\tau^2 + 1)^{-1/2}\cos 2\omega_0\tau]a^3
\end{aligned} \quad (4.36.33)$$

$$\begin{aligned}
&\beta' = -[-\frac{1}{4}(4\omega_0^2\tau^2 + 1)^{-1/2}\sin 2\omega_0\tau]a^2 + 2K^2\Omega^2(\omega_0^2 - \Omega^2)^{-2} \\
&\{[(\Omega+\omega_0)^2\tau^2 + 1]^{-1/2}\sin(\Omega+\omega_0)\tau - [(\Omega-\omega_0)^2\tau^2 + 1]^{-1/2}\sin(\Omega-\omega_0)\tau\}
\end{aligned} \quad (4.36.34)$$

Let

$$\begin{aligned}
\eta_r &= 1 - 2K^2\Omega^2(\omega_0^2 - \Omega^2)^{-2}(1 + h_r) \\
\eta_i &= 2K^2\Omega^2(\omega_0^2 - \Omega^2)^{-2}h_i
\end{aligned} \quad (4.36.35)$$

where

$$\begin{aligned}
h_r = &[(\Omega+\omega_0)^2\tau^2 + 1]^{-1/2}\cos(\Omega+\omega_0)\tau \\
&+[(\Omega-\omega_0)^2\tau^2 + 1]^{-1/2}\cos(\Omega-\omega_0)\tau
\end{aligned} \quad (4.36.36)$$

$$h_i = [(\Omega + \omega_0)^2\tau^2 + 1]^{-1/2}\sin(\Omega + \omega_0)\tau \\ -[(\Omega - \omega_0)^2\tau^2 + 1]^{-1/2}\sin(\Omega - \omega_0)\tau \tag{4.36.37}$$

Considering the expressions of α_r and α_i in (4.36.14), (4.23.4) and (4.36.34) become

$$a' = \left(\eta_r - \tfrac{1}{4}\alpha_r a^2\right)a \\ \beta' = \eta_i - \tfrac{1}{4}\alpha_i a^2 \tag{4.36.38}$$

Change the time variable T_1 to εt in (4.36.38), we can obtain

$$\dot{a} = \varepsilon\left(\eta_r - \frac{1}{4}\alpha_r a^2\right)a \\ \dot{\beta} = \varepsilon\left(\eta_i - \frac{1}{4}\alpha_i a^2\right) \tag{4.36.39}$$

Separates the variables in the first equation of (4.36.39) to give

$$\frac{da}{\left(\eta_r - \frac{1}{4}\alpha_r a^2\right)a} = \varepsilon dt \tag{4.36.40}$$

After integration, we can obtain

$$a^2 = \frac{\eta_r e^{2\varepsilon\eta_r t}}{C + \frac{1}{4}\alpha_r e^{2\varepsilon\eta_r t}} \tag{4.36.41}$$

where C is the integration constant.

It can be seen that when $\eta_r > 0$, the amplitude a diverges exponentially; when $\eta_r < 0$, the amplitude a decays exponentially; and when $\eta_r = 0$, the amplitude a is zero, which means that the external excitation does not excite the free vibration.

The result obtained above is different from the result that the Exercise asks us to prove, so the rest two problems are left unsolved.

4.37 Exercise 4.37 (Analysis on the Response of a Cubic Nonlinear System by the Slowly Varying Excitation)

Solution: (a) Let the solution of the equation be

$$u(t;\varepsilon) = x_0(T_0, T_1) + \varepsilon x_1(T_0, T_1) + \cdots \tag{4.37.1}$$

Substituting (4.37.1) into the original equation and retaining to $O(\varepsilon)$ yields

$$\begin{aligned}0 &= \ddot{u} + \omega_0^2 u + 2\varepsilon\mu\dot{u} + \varepsilon\alpha u^3 - \omega_0^2 f(T_1)\\ &= \left(D_0^2 + 2\varepsilon D_0 D_1\right)(u_0 + \varepsilon u_1) + \omega_0^2(u_0 + \varepsilon u_1)\\ &\quad + 2\varepsilon\mu(D_0 + \varepsilon D_1)(u_0 + \varepsilon u_1) + \varepsilon\alpha(u_0 + \varepsilon u_1)^3 - \omega_0^2 f(T_1) + \cdots\\ &= D_0^2 u_0 + \varepsilon D_0^2 u_1 + 2\varepsilon D_0 D_1 u_0 + \omega_0^2 u_0 + \varepsilon\omega_0^2 u_1\\ &\quad + 2\varepsilon\mu D_0 u_0 + \varepsilon\alpha u_0^3 - \omega_0^2 f(T_1) + \cdots\\ &= D_0^2 u_0 + \omega_0^2 u_0 - \omega_0^2 f(T_1)\\ &\quad + \varepsilon\left(D_0^2 u_1 + \omega_0^2 u_1 + 2D_0 D_1 u_0 + 2\mu D_0 u_0 + \alpha u_0^3\right) + \cdots \end{aligned} \tag{4.37.2}$$

Let the coefficient of the like power of ε be zero, we get

$$D_0^2 u_0 + \omega_0^2 u_0 = \omega_0^2 f(T_1) \tag{4.37.3}$$

$$D_0^2 u_1 + \omega_0^2 u_1 = -2D_0 D_1 u_0 - 2\mu D_0 u_0 - \alpha u_0^3 \tag{4.37.4}$$

Since f is a function of T_1, i.e., it is slowly changing with time T_0, it can be considered to be approximately constant with respect to T_0. The solution of (4.37.3) is

$$u_0 = Ae^{i\omega_0 T_0} + cc + f \tag{4.37.5}$$

where $A = A(T_1)$. Therefore, the first order approximate solution of the equation is

$$u = A(T_1)e^{i\omega_0 T_0} + cc + f(T_1) + O(\varepsilon) \tag{4.37.6}$$

Substituting (4.37.5) into (4.37.4) yields

$$\begin{aligned}D_0^2 u_1 + \omega_0^2 u_1 &= -2D_0 D_1 u_0 - 2\mu D_0 u_0 - \alpha u_0^3\\ &= -2i\omega_0(D_1 A + \mu A)e^{i\omega_0 T_0} - 3\alpha A^2 \bar{A} e^{i\omega_0 T_0} - 3\alpha f^2 A e^{i\omega_0 T_0}\\ &\quad + cc + NST \end{aligned} \tag{4.37.7}$$

In order to eliminate secular terms from the above equation, there must be

$$2i\omega_0\left(A' + \mu A\right) + 3\alpha A^2 \bar{A} + 3\alpha f^2 A = 0$$

i.e.,

$$2i\omega_0\left(A' + \mu A\right) + 3\alpha\left(A\bar{A} + f^2\right)A = 0 \tag{4.37.8}$$

The prime denotes the derivative with respect to T_1.

(b) Substituting $A = \frac{1}{2}ae^{i\beta}$ into the above equation, we get

$$i\omega_0\left(a' + ia\beta' + \mu a\right) + \frac{3}{2}\alpha\left(\frac{1}{4}a^2 + f^2\right)a = 0 \tag{4.37.9}$$

Separating the real and imaginary parts of the above equation yields

$$\begin{aligned} a' &= -\mu a \\ \omega_0\beta' &= \tfrac{3}{2}\alpha\left(\tfrac{1}{4}a^2 + f^2\right) \end{aligned} \tag{4.37.10}$$

4.38 Exercise 4.38 (Analysis of the Response of a Self-Excited System Under Slowly Varying Excitation)

Solution: (a) Let the solution be

$$u(t;\varepsilon) = x_0(T_0, T_1) + \varepsilon x_1(T_0, T_1) + \cdots \tag{4.38.1}$$

Substituting (4.38.1) into the original equation and retaining to $O(\varepsilon)$ yields

$$\begin{aligned} 0 &= \ddot{u} + \omega_0^2 u - \varepsilon\left(1 - u^2\right)\dot{u} - \omega_0^2 f(T_1) \\ &= \left(D_0^2 + 2\varepsilon D_0 D_1\right)(u_0 + \varepsilon u_1) \\ &\quad + \omega_0^2(u_0 + \varepsilon u_1) - \varepsilon(D_0 + \varepsilon D_1)(u_0 + \varepsilon u_1) \\ &\quad + \varepsilon(u_0 + \varepsilon u_1)^2(D_0 + \varepsilon D_1)(u_0 + \varepsilon u_1) - \omega_0^2 f(T_1) + \cdots \\ &= D_0^2 u_0 + \varepsilon D_0^2 u_1 + 2\varepsilon D_0 D_1 u_0 + \omega_0^2 u_0 + \varepsilon\omega_0^2 u_1 \\ &\quad - \varepsilon D_0 u_0 + \varepsilon u_0^2 D_0 u_0 - \omega_0^2 f(T_1) + \cdots \\ &= D_0^2 u_0 + \omega_0^2 u_0 - \omega_0^2 f(T_1) \\ &\quad + \varepsilon\left(D_0^2 u_1 + \omega_0^2 u_1 + 2D_0 D_1 u_0 - D_0 u_0 + u_0^2 D_0 u_0\right) + \cdots \end{aligned} \tag{4.38.2}$$

Let the coefficient of the like power of ε be zero, we get

$$D_0^2 u_0 + \omega_0^2 u_0 = \omega_0^2 f(T_1) \tag{4.38.3}$$

$$D_0^2 u_1 + \omega_0^2 u_1 = -2D_0 D_1 u_0 + D_0 u_0 - u_0^2 D_0 u_0 \tag{4.38.4}$$

Since f is a function of T_1, i.e., it is slowly changing with time T_0, it can be considered to be approximately constant with respect to T_0. The solution of (4.38.3) is

$$u_0 = Ae^{i\omega_0 T_0} + cc + f \tag{4.38.5}$$

where $A = A(T_1)$. Therefore, the first order approximate solution of the equation is

$$u = A(T_1)e^{i\omega_0 T_0} + cc + f(T_1) + O(\varepsilon) \tag{4.38.6}$$

Substituting (4.38.5) into (4.38.4) yields

$$\begin{aligned} D_0^2 u_1 + \omega_0^2 u_1 &= -2D_0 D_1 u_0 + D_0 u_0 - u_0^2 D_0 u_0 \\ &= -i\omega_0(2D_1 A - A)e^{i\omega_0 T_0} \\ &\quad - i\omega_0\left(f^2 A + A^2\overline{A}\right)e^{i\omega_0 T_0} + cc + NST \end{aligned} \tag{4.38.7}$$

In order to eliminate secular terms from the above equation, there must be

$$2A' = \left(1 - A\overline{A} - f^2\right)A \tag{4.38.8}$$

The prime denotes the derivative with respect to T_1.

(b) Substituting $A = \frac{1}{2}ae^{i\beta}$ into the above equation, we get

$$a' + ia\beta' = \frac{1}{2}\left(1 - \frac{1}{4}a^2 - f^2\right)a \tag{4.38.9}$$

Separating the real and imaginary parts of the above equation yields

$$\begin{aligned} a' &= \tfrac{1}{2}\left(1 - f^2 - \tfrac{1}{4}a^2\right)a \\ \beta' &= 0 \end{aligned} \tag{4.38.10}$$

4.39 Exercise 4.39 (Analysis of the Response of a Self-Excited Vibrating System Under Hard Excitation at Different Frequencies)

Solution: Let the solution be

$$u(t;\varepsilon) = x_0(T_0, T_1) + \varepsilon x_1(T_0, T_1) + \cdots \tag{4.39.1}$$

Substituting (4.39.1) into the original equation and retaining to $O(\varepsilon)$ yields

$$\begin{aligned} 0 &= \ddot{u} + \omega_0^2 u - \varepsilon\left(1 - u^2\right)\dot{u} - K\cos\Omega t \\ &= \left(D_0^2 + 2\varepsilon D_0 D_1\right)(u_0 + \varepsilon u_1) \\ &\quad + \omega_0^2(u_0 + \varepsilon u_1) - \varepsilon(D_0 + \varepsilon D_1)(u_0 + \varepsilon u_1) \\ &\quad + \varepsilon(u_0 + \varepsilon u_1)^2(D_0 + \varepsilon D_1)(u_0 + \varepsilon u_1) - K\cos\Omega T_0 + \cdots \end{aligned}$$

$$
\begin{aligned}
&= D_0^2 u_0 + \omega_0^2 u_0 - K\cos\Omega T_0 \\
&+ \varepsilon\left(D_0^2 u_1 + \omega_0^2 u_1 + 2D_0 D_1 u_0 - D_0 u_0 + u_0^2 D_0 u_0\right) + \cdots
\end{aligned} \tag{4.39.2}
$$

Let the coefficient of the like power of ε be zero, we get

$$D_0^2 u_0 + \omega_0^2 u_0 = K\cos\Omega T_0 \tag{4.39.3}$$

$$D_0^2 u_1 + \omega_0^2 u_1 = -2D_0 D_1 u_0 + D_0 u_0 - u_0^2 D_0 u_0 \tag{4.39.4}$$

The solution of (4.39.3) is

$$u_0 = A(T_1)e^{i\omega_0 T_0} + \Lambda e^{i\Omega T_0} + cc \tag{4.39.5}$$

where $A = A(T_1), \Lambda = \frac{1}{2}K(\omega_0^2 - \Omega^2)^{-1}$. Therefore, the first order approximate solution of the equation is

$$u = A(T_1)e^{i\omega_0 T_0} + \Lambda e^{i\Omega T_0} + cc + O(\varepsilon) \tag{4.39.6}$$

Substituting (4.39.5) into (4.39.4) yields

$$
\begin{aligned}
D_0^2 u_1 + \omega_0^2 u_1 &= -2D_0 D_1 u_0 + D_0 u_0 - u_0^2 D_0 u_0 \\
&= -i\omega_0(2D_1 A - A)e^{i\omega_0 T_0} - i\omega_0 A^2 \overline{A} e^{i\omega_0 T_0} - 2i\omega_0 \Lambda^2 A e^{i\omega_0 T_0} \\
&- i\Omega\Lambda^3 e^{i3\Omega T_0} - i(\Omega - 2\omega_0)\Lambda\overline{A}^2 e^{i(\Omega - 2\omega_0)T_0} \\
&- i(\omega_0 - 2\Omega)\Lambda^2 A e^{i(\omega_0 - 2\Omega)T_0} - i(2\Omega + \omega_0)\Lambda^2 A e^{i(2\Omega + \omega_0)T_0} \\
&+ cc + NST
\end{aligned} \tag{4.39.7}
$$

(a) For the case of Ω is not close to 0, $3\omega_0$ and $\frac{1}{3}\omega_0$:

In order to eliminate secular terms from the above equation, there must be

$$2A' = A - 2\Lambda^2 A - A^2\overline{A} \tag{4.39.8}$$

The prime denotes the derivative with respect to T_1.

(b) For the case of $\omega_0 = 3\Omega + \varepsilon\sigma$:

In order to eliminate secular terms from (4.39.7), there must be

$$2A' - A + A^2\overline{A} + 2\Lambda^2 A + \omega_0^{-1}\Omega\Lambda^3 e^{-i\sigma T_1} = 0$$

i.e.,

$$2A' = A - 2\Lambda^2 A - A^2\overline{A} - \Omega\omega_0^{-1}\Lambda^3 e^{-i\sigma T_1} \tag{4.39.9}$$

(c) For the case of $\Omega = 3\omega_0 + \varepsilon\sigma$:

In order to eliminate secular terms from (4.39.7), there must be

$$2A' - A + A^2\overline{A} + 2\Lambda^2 A + \omega_0^{-1}(\Omega - 2\omega_0)\Lambda\overline{A}^2 e^{i\sigma T_1} = 0$$

i.e.,

$$2A' = A - 2\Lambda^2 A - A^2\overline{A} + \left(2 - \Omega\omega_0^{-1}\right)\Lambda\overline{A}^2 e^{i\sigma T_1} \tag{4.39.10}$$

(d) For the case of $\Omega = 3\omega_0 + \varepsilon\sigma$:

In order to eliminate secular terms from (4.39.7), there must be

$$2A' - A + A^2\overline{A} + 2\Lambda^2 A + \omega_0^{-1}(\omega_0 - 2\varepsilon\sigma)\Lambda^2 A e^{-i2\sigma T_1}$$
$$+\omega_0^{-1}(2\varepsilon\sigma + \omega_0)\Lambda^2 A e^{i2\sigma T_1} = 0$$

i.e.,

$$2A' = A - 2\Lambda^2 A - A^2\overline{A} - 2\Lambda^2 A\cos 2\sigma T_1 \tag{4.39.11}$$

Chapter 5
Parametrically Excited Systems

5.1 Exercise 5.1 (The Method of Strained Parameters to Determine the Transition Curves for a Hill Equation)

Solution: Based on the Floquet theory, that the characteristic exponent is 0 or i (i.e., the solutions have periods of 2π or $4\ \pi$) and then determines the values of the parameters for which the assumption is true. We seek the solutions of the original equation having periods of 2π and $4\ \pi$ and the equations for the transition curves $a = a(\varepsilon)$ in the form of the following perturbation expansions:

$$u(x;\varepsilon) = u_0(x) + \varepsilon u_1(x) + \varepsilon^2 u_2(x) + \cdots \tag{5.1.1}$$

$$a = a_0 + \varepsilon a_1 + \varepsilon^2 a_2 + \cdots \tag{5.1.2}$$

In addition,

$$\begin{aligned}
(1 - \varepsilon\cos x)^{-1} &= 1 + \varepsilon\cos x + \varepsilon^2\cos^2 x \\
(1 - \varepsilon\cos x)^{-2} &= 1 + 2\varepsilon\cos x + 3\varepsilon^2\cos^2 x \\
(a_0 + \varepsilon a_1 + \varepsilon^2 a_2)^{-1} &= a_0^{-1}(1 + \varepsilon a_0^{-1} a_1 + \varepsilon^2 a_0^{-1} a_2)^{-1} \\
&= a_0^{-1}[1 - (\varepsilon a_0^{-1} a_1 + \varepsilon^2 a_0^{-1} a_2) + (\varepsilon a_0^{-1} a_1 + \varepsilon^2 a_0^{-1} a_2)^2] \\
&= a_0^{-1} - \varepsilon a_0^{-2} a_1 - \varepsilon^2 a_0^{-2} a_2 + \varepsilon^2 a_0^{-3} a_1^2 \\
(a_0 + \varepsilon a_1 + \varepsilon^2 a_2)^{-2} &= a_0^{-2} - 2\varepsilon a_0^{-3} a_1 - 2\varepsilon^2 a_0^{-3} a_2 + 3\varepsilon^2 a_0^{-4} a_1^2
\end{aligned} \tag{5.1.3}$$

Substituting (5.1.1) and (5.1.2) into the original equation and retaining to $O(\varepsilon^2)$, we obtain

Z. He et al., *Solved Problems in Nonlinear Oscillations*,
https://doi.org/10.1007/978-981-97-6113-5_5

$$
\begin{aligned}
0 &= u'' + \frac{1}{4}(1-\varepsilon\cos x)^{-2}a^{-2}\big[2(1-\varepsilon\cos x)\big(2-\varepsilon a^2\cos x\big)+\varepsilon^2 a^2\sin^2 x\big]u \\
&= u'' + (1-\varepsilon\cos x)^{-1}\left(a^{-2}-\frac{1}{2}\varepsilon\cos x\right)u + \frac{1}{4}\varepsilon^2(1-\varepsilon\cos x)^{-2}u\sin^2 x \\
&= u_0'' + \varepsilon u_1'' + \varepsilon^2 u_2'' + \big(1+\varepsilon\cos x+\varepsilon^2\cos^2 x\big) \\
&\quad \times \left(a_0^{-2} - 2\varepsilon a_0^{-3}a_1 - 2\varepsilon^2 a_0^{-3}a_2 + 3\varepsilon^2 a_0^{-4}a_1^2 - \frac{1}{2}\varepsilon\cos x\right)\big(u_0+\varepsilon u_1+\varepsilon^2 u_2\big) \\
&\quad + \frac{1}{4}\varepsilon^2\big(1+2\varepsilon\cos x+3\varepsilon^2\cos^2 x\big)\big(u_0+\varepsilon u_1+\varepsilon^2 u_2\big)\sin^2 x + \cdots \\
&\quad + \frac{1}{4}\varepsilon^2\big(1+2\varepsilon\cos x+3\varepsilon^2\cos^2 x\big)\big(u_0+\varepsilon u_1+\varepsilon^2 u_2\big)\sin^2 x + \cdots \\
&\quad + \varepsilon^2(u_2'' + a_0^{-2}u_2 - 2a_0^{-3}a_2u_0 + 3a_0^{-4}a_1^2u_0 - 2a_0^{-3}a_1u_0\cos x \\
&\quad + \varepsilon^2(u_2'' + a_0^{-2}u_2 - 2a_0^{-3}a_2u_0 + 3a_0^{-4}a_1^2u_0 - 2a_0^{-3}a_1u_0\cos x \\
&\quad - 2a_0^{-3}a_1u_1 - \frac{1}{2}u_1\cos x + a_0^{-2}u_1\cos x) + \cdots
\end{aligned}
\tag{5.1.4}
$$

Equating coefficients of like powers of ε in the above equation yields

$$u_0'' + a_0^{-2}u_0 = 0 \tag{5.1.5}$$

$$u_0'' + a_0^{-2}u_1 = -2a_0^{-3}a_1u_0 + \frac{1}{2}u_0\mathrm{Cos}\,x - a_0^{-2}u_0\mathrm{Cos}\,x \tag{5.1.6}$$

$$
\begin{aligned}
u_2'' + a_0^{-2}u_2 = {} & 2a_0^{-3}a_2u_0 - 3a_0^{-4}a_1^2u_0 + 2a_0^{-3}a_1u_0\cos x \\
& + \frac{1}{2}u_0\cos^2 x - a_0^{-2}u_0\cos^2 x - \frac{1}{4}u_0\sin^2 x \\
& + 2a_0^{-3}a_1u_1 + \frac{1}{2}u_1\cos x - a_0^{-2}u_1\cos x
\end{aligned}
\tag{5.1.7}
$$

The solution of (5.1.5) is

$$u_0 = b\cos a_0^{-1}x + c\sin a_0^{-1}x \tag{5.1.8}$$

CASE I: $\boldsymbol{a_0} = 2$

$$u_0 = b\cos(x/2) + c\sin(x/2) \tag{5.1.9}$$

u_0 is a periodic solution with period $2T = 4\pi$. Substituting (5.1.9) into (5.1.6) yields

$$\begin{aligned}
u_1'' + \frac{1}{4}u_1 &= -\frac{1}{4}a_1\left(b\cos\frac{1}{2}x + c\sin\frac{1}{2}x\right) \\
&\quad + \frac{1}{4}\left(b\cos\frac{1}{2}x + c\sin\frac{1}{2}x\right)\cos x \\
&= -\frac{1}{4}a_1 b\cos\frac{1}{2}x - \frac{1}{4}a_1 c\sin\frac{1}{2}x \\
&\quad + \frac{1}{8}b\cos\frac{1}{2}x - \frac{1}{8}c\sin\frac{1}{2}x + NST \\
&= -\frac{1}{4}b\left(a_1 - \frac{1}{2}\right)\cos\frac{1}{2}x - \frac{1}{4}c\left(a_1 + \frac{1}{2}\right)\sin\frac{1}{2}x + NST
\end{aligned} \tag{5.1.10}$$

In order to eliminate secular terms from the above equation, there must be

$$a_1 = \frac{1}{2} \quad \text{or} \quad a_1 = -\frac{1}{2} \tag{5.1.11}$$

Therefore, the transition curves emanating from $a = 2$ is given by

$$a = 2 \pm \frac{1}{2}\varepsilon + O\left(\varepsilon^2\right) \tag{5.1.12}$$

CASE II: $a_0 = 1$

$$u_0 = b\cos x + c\sin x \tag{5.1.13}$$

u_0 is a periodic solution with period $T = 2\pi$. Substituting (5.1.9) into (5.1.6) yields

$$\begin{aligned}
u_1'' + u_1 &= -2a_1(b\cos x + c\sin x) - \frac{1}{2}(b\cos x + c\sin x)\cos x \\
&= -2ba_1\cos x - 2ca_1\sin x - \frac{1}{4}b - \frac{1}{4}b\cos 2x - \frac{1}{4}c\sin 2x
\end{aligned} \tag{5.1.14}$$

In order to eliminate secular terms from the above equation, there must be

$$a_1 = 0$$

Then the solution of (5.1.14) is

$$u_1 = -\frac{1}{4}b + \frac{1}{12}b\cos 2x + \frac{1}{12}c\sin 2x \tag{5.1.15}$$

Substituting (5.1.13) and (5.1.15) into (5.1.7) yields

$$
\begin{aligned}
u_2'' + u_2 &= 2a_2(b\cos x + c\sin x) - \frac{1}{4}(b\cos x + c\sin x) \\
&\quad - \frac{1}{4}(b\cos x + c\sin x)\cos^2 x \\
&\quad - \frac{1}{2}\left(-\frac{1}{4}b + \frac{1}{12}b\cos 2x + \frac{1}{12}c\sin 2x\right)\cos x \\
&= \left(2a_2 - \frac{1}{3}\right)b\cos x + \left(2a_2 - \frac{1}{3}\right)c\sin x + NST
\end{aligned} \tag{5.1.16}
$$

In order to eliminate secular terms from the above equation, there must be

$$a_2 = 1/6 \tag{5.1.17}$$

Therefore, thc transition curves emanating from $a = 1$ is given by

$$a = 1 + \frac{1}{6}\varepsilon^2 + O(\varepsilon^3) \tag{5.1.18}$$

5.2 Exercise 5.2 (The Method of Strained Parameters to Determine the Transition Curves of the Hill's Equation for LRC Circuits with Sinusoidally Varying Resistance)

Solution: (a) Since

$$\dot{R} = R_0\Omega\alpha\cos\Omega t$$

$$
\begin{aligned}
\dot{I} &= w'\tfrac{d\tau}{dt}\exp\left[-\tfrac{1}{2}L^{-1}\int_0^t R(\xi)d\xi\right] - \tfrac{1}{2}wL^{-1}\text{Rexp}\left[-\tfrac{1}{2}L^{-1}\int_0^t R(\xi)d\xi\right] \\
&= \left(\tfrac{1}{2}\Omega w' - \tfrac{1}{2}wL^{-1}R\right)\exp\left[-\tfrac{1}{2}L^{-1}\int_0^t R(\xi)d\xi\right]
\end{aligned}
$$

$$
\begin{aligned}
\ddot{I} &= [\left(\tfrac{1}{4}\Omega^2 w'' - \tfrac{1}{4}\Omega w' L^{-1}R - \tfrac{1}{2}wL^{-1}\dot{R}\right) \\
&\quad - \left(\tfrac{1}{2}\Omega w' - \tfrac{1}{2}wL^{-1}R\right)\tfrac{1}{2}L^{-1}R]\exp[-\tfrac{1}{2}L^{-1}\int_0^t R(\xi)d\xi] \\
&= \left(\tfrac{1}{4}\Omega^2 w'' - \tfrac{1}{2}\Omega w' L^{-1}R + \tfrac{1}{4}wL^{-2}R^2 - \tfrac{1}{2}w\Omega L^{-1}R_0\alpha\cos\Omega t\right) \\
&\quad \times\exp\left[-\tfrac{1}{2}L^{-1}\int_0^t R(\xi)d\xi\right]
\end{aligned}
$$

Substituting the above results into the original equation

$$L\left(\frac{1}{4}\Omega^2 w'' - \frac{1}{2}\Omega L^{-1} R w' + \frac{1}{4} L^{-2} R^2 w - \frac{1}{2}\Omega L^{-1} R_0 \alpha w \cos \Omega t\right)$$
$$+ R\left(\frac{1}{2}\Omega w' - \frac{1}{2} L^{-1} R w\right) + \left(R_0 \Omega \alpha \cos \Omega t + C^{-1}\right) w = 0$$

i.e.,

$$\begin{aligned} w'' &+ [4\Omega^{-2} L^{-1} C^{-1} - \Omega^{-2} L^{-2} R_0^2 - \frac{1}{2}\Omega^{-2} L^{-2} R_0^2 \alpha^2 \\ &- 2\Omega^{-2} L^{-2} R_0^2 \alpha \sin \Omega t \, 2\Omega^{-1} L^{-1} R_0 \alpha \cos \Omega t \\ &+ \frac{1}{2}\Omega^{-2} L^{-2} R_0^2 \alpha^2 \cos 2\Omega t] \; w = 0 \end{aligned} \tag{5.2.1}$$

Let

$$\varepsilon = \frac{2\alpha R_0}{L\Omega'}, \quad \delta = \frac{4}{\Omega^2}\left[\frac{1}{LC} - \frac{R_0^2}{4L^2}\left(1 + \frac{1}{2}\alpha^2\right)\right] \tag{5.2.2}$$

Considering $\Omega t = 2\tau$, , (5.2.1) becomes

$$w'' + \left[\delta + \varepsilon \cos 2\tau - \frac{1}{2}\varepsilon^2 \alpha^{-1} \sin 2\tau + \frac{1}{8}\varepsilon^2 \cos 4\tau\right] w = 0 \tag{5.2.3}$$

(b) (5.2.3) is a Hill equation with the period of the coefficients $T = \pi$. We use the method of strained parameters to solve this problem. We seek the solutions of the original equation having periods of π and 2 π and the equations for the transition curves $\delta = \delta(\varepsilon)$ in the form of the following perturbation expansions:

$$w(\tau; \varepsilon) = u_0(\tau) + \varepsilon u_1(\tau) + \varepsilon^2 u_2(\tau) + \cdots \tag{5.2.4}$$

$$\delta = \delta_0 + \varepsilon \delta_1 + \varepsilon^2 \delta_2 + \cdots \tag{5.2.5}$$

Substituting the above two equations into (5.2.3) and keeping to $O(\varepsilon^2)$ terms, we get

$$
\begin{aligned}
0 &= w'' + \left[\delta + \varepsilon\cos 2\tau - \frac{1}{2}\varepsilon^2\alpha^{-1}\sin 2\tau + \frac{1}{8}\varepsilon^2\cos 4\tau\right]w \\
&= u_0'' + \varepsilon u_1'' + \varepsilon^2 u_2'' + [\delta_0 + \varepsilon\delta_1 + \varepsilon^2\delta_2 + \varepsilon\cos 2\tau \\
&\quad -\frac{1}{2}\varepsilon^2\alpha^{-1}\sin 2\tau + \frac{1}{8}\varepsilon^2\cos 4\tau]\left(u_0 + \varepsilon u_1 + \varepsilon^2 u_2\right) + \cdots \\
&= u_0'' + \delta_0 u_0 + \varepsilon\left(u_1'' + \delta_0 u_1 + \delta_1 u_0 + u_0\cos 2\tau\right) \\
&\quad +\varepsilon^2(u_2'' + \delta_0 u_2 + \delta_2 u_0 - \frac{1}{2}\alpha^{-1}u_0\sin 2\tau + \frac{1}{8}u_0\cos 4\tau \\
&\quad +\delta_1 u_1 + u_1\cos 2\tau) + \cdots
\end{aligned}
$$

Equating coefficients of like powers of ε in the above equation yields

$$u_0'' + \delta_0 u_0 = 0 \tag{5.2.6}$$

$$u_1'' + \delta_0 u_1 = -\delta_1 u_0 - u_0\cos 2\tau \tag{5.2.7}$$

$$u_2'' + \delta_0 u_2 = -\delta_2 u_0 + \frac{1}{2}\alpha^{-1}u_0\sin 2\tau - \frac{1}{8}u_0\cos 4\tau - \delta_1 u_1 - u_1\cos 2\tau \tag{5.2.8}$$

The solution of (5.2.6) is

$$u_0 = a\cos\sqrt{\delta_0}\tau + b\sin\sqrt{\delta_0}\tau \tag{5.2.9}$$

CASE I: $\delta_0 = 1$

$$u_0 = a\cos\tau + b\sin\tau \tag{5.2.10}$$

u_0 is a periodic solution with period $2T = 2\pi$. Substituting (5.2.10) into (5.2.7) yields

$$
\begin{aligned}
&u''_1 + \delta_0 u_1 = -\delta_1(a\cos\tau + b\sin\tau) - (a\cos\tau + b\sin\tau)\cos 2\tau \\
&= -\left(\delta_1 + \tfrac{1}{2}\right)a\cos\tau - \left(\delta_1\sin\tau - \tfrac{1}{2}\right)b\sin\tau + NST
\end{aligned}
\tag{5.2.11}
$$

In order to eliminate secular terms from the above equation, there must be

$$\delta_1 = \frac{1}{2} \quad \text{or} \quad \delta_1 = -\frac{1}{2} \tag{5.2.12}$$

Therefore, the transition curves emanating from $\delta = 1$ is

$$\delta = 1 \pm \frac{1}{2}\varepsilon + O\left(\varepsilon^2\right) \tag{5.2.13}$$

CASE II: $\delta_0 = 4$

$$u_0 = a\cos 2\tau + b\sin 2\tau \tag{5.2.14}$$

u_0 is a periodic solution with period $T = \pi$. Substituting (5.2.14) into (5.2.7) yields

$$\begin{aligned} u''_1 + \delta_0 u_1 &= -\delta_1(a\cos 2\tau + b\sin 2\tau) \\ &\quad - (a\cos 2\tau + b\sin 2\tau)\cos 2\tau \\ &= -\delta_1 a\cos 2\tau - \delta_1 b\sin 2\tau + NST \end{aligned} \tag{5.2.10}$$

In order to eliminate secular terms from the above equation, there must be

$$\delta_1 = 0 \tag{5.2.11}$$

Therefore, the transition curves emanating from $\delta = 4$ is

$$\delta = 4 + O(\varepsilon^3) \tag{5.2.12}$$

5.3 Exercise 5.3 (Stability Analysis of a Supported Movable Pendulum)

Solution: **(a)** When $Y \equiv 0$, the differential equation of motion of the pendulum becomes

$$\ddot{\theta} + \frac{g}{l}\sin\theta = 0 \tag{5.3.1}$$

Its equilibrium position is given by $sin\theta = 0$, i.e., $\theta = n\pi$, n is an integer. The linearized form of (5.3.1) near the equilibrium position $\theta = 0$ is

$$\ddot{\theta} + \frac{g}{l}\theta = 0 \tag{5.3.2}$$

Its eigenvalues are $\lambda = \pm i\sqrt{g/l}$, so the equilibrium position $\theta = 0$ is the center. The linearized form of (5.3.1) near the equilibrium position $\theta = \pi$ is

$$\ddot{\theta} - \frac{g}{l}\theta = 0 \tag{5.3.3}$$

Its eigenvalues are $\lambda = \pm\sqrt{g/l}$, so the equilibrium position $\theta = \pi$ is the saddle point.

(b) When $Y = \varepsilon g \cos \Omega t$, the differential equation of motion of the pendulum becomes

$$\ddot{\theta} + \delta(1 - \varepsilon \cos \Omega t) \sin \theta = 0, \quad \delta = g/l \tag{5.3.4}$$

The linearized form of (5.3.4) near the equilibrium position $\theta = 0$ is

$$\ddot{\theta} + \delta(1 - \varepsilon \cos \Omega t)\theta = 0, \quad \delta = g/l \tag{5.3.5}$$

This is a Hill equation with a period $T = 2\pi/\Omega$. According to Floquet theory, the transition curves determine the parameters which make the solution of this equation periodic (with a period of T or $2T$). We use the method of strained parameters to determine the transition curves for this problem. We seek the solutions of the original equation having periods of T and $2\,T$ and the equations for the transition curves $\delta = \delta(\varepsilon)$ in the form of the following perturbation expansions:

$$\theta(t; \varepsilon) = u_0(t) + \varepsilon u_1(t) + \varepsilon^2 u_2(t) + \cdots \tag{5.3.6}$$

$$\delta = \delta_0 + \varepsilon \delta_1 + \varepsilon^2 \delta_2 + \cdots \tag{5.3.7}$$

Substituting the above two equations into (5.3.4) and retaining to $O(\varepsilon^2)$, we get

$$\begin{aligned}
0 = \ddot{\theta} + \delta(1 - \varepsilon \cos \Omega t) \\
\theta = \ddot{u}_0 + \varepsilon \ddot{u}_1 + \varepsilon^2 \ddot{u}_2 \\
&+ \left(\delta_0 + \varepsilon \delta_1 + \varepsilon^2 \delta_2\right)(1 - \varepsilon \cos \Omega t)\left(u_0 + \varepsilon u_1 + \varepsilon^2 u_2\right) + \cdots \\
&= \ddot{u}_0 + \varepsilon \ddot{u}_1 + \varepsilon^2 \ddot{u}_2 \\
&+ \left(\delta_0 + \varepsilon \delta_1 - \varepsilon \delta_0 \cos \Omega t + \varepsilon^2 \delta_2 - \varepsilon^2 \delta_1 \cos \Omega t\right)\left(u_0 + \varepsilon u_1 + \varepsilon^2 u_2\right) + \cdots \\
&= \ddot{u}_0 + \delta_0 u_0 + \varepsilon(\ddot{u}_1 + \delta_0 u_1 + \delta_1 u_0 - \delta_0 u_0 \cos \Omega t) \\
&+ \varepsilon^2(\ddot{u}_2 + \delta_0 u_2 + \delta_2 u_0 - \delta_1 u_0 \cos \Omega t + \delta_1 u_1 - \delta_0 u_1 \cos \Omega t) + \cdots
\end{aligned}$$

Equating coefficients of like powers of ε in the above equation yields

$$\ddot{u}_0 + \delta_0 u_0 = 0 \tag{5.3.8}$$

$$\ddot{u}_1 + \delta_0 u_1 = -\delta_1 u_0 + \delta_0 u_0 \cos \Omega t \tag{5.3.9}$$

$$\ddot{u}_2 + \delta_0 u_2 = -\delta_2 u_0 + \delta_1 u_0 \cos \Omega t - \delta_1 u_1 + \delta_0 u_1 \cos \Omega t \tag{5.3.10}$$

The solution of (5.3.8) is

$$u_0 = a \cos \sqrt{\delta_0} t + b \sin \sqrt{\delta_0} t \tag{5.3.11}$$

CASE I: $\delta_0 = \Omega^2$:

$$u_0 = a\cos\Omega t + b\sin\Omega t \tag{5.3.12}$$

$u_0$is a periodic solution with period $T = 2\pi/\Omega$. Substituting (5.3.12) into (5.3.9) yields

$$\begin{aligned}\ddot{u}_1 + \Omega^2 u_1 &= -\delta_1 u_0 + \delta_0 u_0 \cos\Omega t\\ &= -\delta_1(a\cos\Omega t + b\sin\Omega t)\\ &\quad - \Omega^2(a\cos\Omega t + b\sin\Omega t)\cos\Omega t\\ &= -a\delta_1\cos\Omega t - b\delta_1\sin\Omega t\\ &\quad - \frac{1}{2}\Omega^2 a(1+\cos 2\Omega t) - \frac{1}{2}\Omega^2 b\sin 2\Omega t\end{aligned} \tag{5.3.13}$$

In order to eliminate secular terms from the above equation, there must be

$$\delta_1 = 0 \tag{5.3.14}$$

Then the solution of (5.3.13) is

$$u_1 = -\frac{1}{2}a + \frac{1}{6}a\cos 2\Omega t + \frac{1}{6}b\sin 2\Omega t \tag{5.3.15}$$

Substituting (5.3.15) into (5.3.10), we get

$$\begin{aligned}\ddot{u}_2 + \Omega^2 u_2 &= -\delta_2 a\cos\Omega t - \delta_2 b\sin\Omega t\\ &\quad + \delta_0\left(-\frac{1}{2}a + \frac{1}{6}a\cos 2\Omega t + \frac{1}{6}b\sin 2\Omega t\right)\cos\Omega t\\ &= -\left(\delta_2 + \frac{5}{12}\Omega^2\right)a\cos\Omega t\\ &\quad - \left(\delta_2 - \frac{1}{12}\Omega^2\right)b\sin\Omega t + NST\end{aligned} \tag{5.3.16}$$

In order to eliminate secular terms from the above equation, there must be

$$\delta_2 = -\frac{5}{12}\Omega^2 \quad \text{or} \quad \delta_2 = \frac{1}{12}\Omega^2 \tag{5.3.17}$$

Therefore, the transition curves emanating from $\delta = \Omega^2$ is

$$\delta = \frac{g}{l} = \Omega^2\left(1 - \frac{5}{12}\varepsilon^2\right) + O(\varepsilon^3)$$

i.e.,

$$\frac{g}{l} = \Omega^2\left(1 + \frac{1}{12}\varepsilon^2\right) + O(\varepsilon^3) \tag{5.3.18}$$

CASE II: $\delta_0 = \frac{1}{4}\Omega^2$**:**

$$u_0 = a\cos\frac{1}{2}\Omega t + b\sin\frac{1}{2}\Omega t \tag{5.3.19}$$

$u_0$is a periodic solution with period $2T = 4\pi/\Omega$. Substituting (5.3.11) into (5.3.9) yields

$$\begin{aligned}\ddot{u}_1 + \frac{1}{4}\Omega^2 u_1 &= -\delta_1 u_0 + \delta_0 u_0 \cos\Omega t\\ &= -\delta_1\left(a\cos\frac{1}{2}\Omega t + b\sin\frac{1}{2}\Omega t\right)\\ &\quad - \frac{1}{4}\Omega^2\left(a\cos\frac{1}{2}\Omega t + b\sin\frac{1}{2}\Omega t\right)\cos\Omega t\\ &= -\left(\delta_1 + \frac{1}{8}\Omega^2\right)a\cos\frac{1}{2}\Omega t\\ &\quad - \left(\delta_1 - \frac{1}{8}\Omega^2\right)b\sin\frac{1}{2}\Omega t + NST\end{aligned} \tag{5.3.20}$$

In order to eliminate secular terms from the above equation, there must be

$$\delta_1 = \pm\frac{1}{8}\Omega^2 \tag{5.3.21}$$

Therefore, the transition curves emanating from $\delta = \frac{1}{4}\Omega^2$ is

$$\delta = \frac{g}{l} = \Omega^2\left(1 \pm \frac{1}{8}\varepsilon\right) + O(\varepsilon^2) \tag{5.3.22}$$

(c) When $Y = \alpha g\cos\Omega t$, the differential equation of motion of the pendulum becomes

$$\ddot{\theta} + \frac{g}{l}(1 - \alpha\cos\Omega t)\sin\theta = 0 \tag{5.3.23}$$

Let $\theta = \pi + u$, the linearized equation near the equilibrium solution $\theta = \pi$ is

$$\ddot{u} + \delta(1 - \alpha\cos\Omega t)u = 0, \quad \delta = -g/l \tag{5.3.24}$$

This is a Hill equation with a period of $T = 2\pi/\Omega$ and has exactly the same form as (5.3.5). Therefore, the transition curves emanating from $\delta = \Omega^2$ is

$$-\frac{g}{l} = \Omega^2\left(1 - \frac{5}{12}\varepsilon^2\right) + O(\varepsilon^3) \tag{5.3.25}$$

i.e.,

$$-\frac{g}{l} = \Omega^2\left(1 + \frac{1}{12}\varepsilon^2\right) + O(\varepsilon^3) \tag{5.3.26}$$

and the transition curves emanating from $\delta = \frac{1}{4}\Omega^2$ is

$$-\frac{g}{l} = \Omega^2\left(1 \pm \frac{1}{8}\varepsilon\right) + O(\varepsilon^2) \tag{5.3.27}$$

5.4 Exercise 5.4 (Analyze the Linear and Nonlinear Stability of a Rotating Pendulum Using the Method of Strained Parameters and the Method of Multiple Scales)

Solution: (a) The kinetic energy of the system is

$$T = \frac{1}{2}m\left(l^2\dot{\theta}^2 + l^2\Omega^2\sin^2\theta\right) \tag{5.4.1}$$

The potential energy of the system is

$$V = mgl(1 - \cos\theta) \tag{5.4.2}$$

Substituting (5.4.2) and (5.4.1) into the Lagrange's equation, we can obtain the differential equation of motion of the simple pendulum attached to rotating base

$$\ddot{\theta} + \frac{g}{l}\sin\theta - \frac{1}{2}\Omega^2\sin 2\theta = 0 \tag{5.4.3}$$

(b) Let $\Omega = \Omega_0(1 + \varepsilon\cos\omega t)$, where $\varepsilon << 1$, , then (5.4.3) becomes

$$\ddot{\theta} + \frac{g}{l}\sin\theta - \frac{1}{2}\Omega_0^2(1 + \varepsilon\cos\omega t)^2\sin 2\theta = 0 \tag{5.4.4}$$

Linearizing the above equation near $\theta = 0$ yields

$$\ddot{\theta} + \frac{g}{l}\theta - \Omega_0^2(1 + \varepsilon \cos \omega t)^2\theta = 0 \tag{5.4.5}$$

i.e.,

$$\ddot{\theta} + \left(\frac{g}{l} - \Omega_0^2 - 2\Omega_0^2\varepsilon \cos \omega t - \Omega_0^2\varepsilon^2 \cos^2 \omega t\right)\theta = 0 \tag{5.4.6}$$

(c) Denote $\delta = \frac{g}{l} - \Omega_0^2$. We use the method of strained parameters to determine the transition curves for this problem. We seek the solutions of the original equation having periods of T and 2 T and the equations for the transition curves $\delta = \delta(\varepsilon)$ in the form of the following perturbation expansions:

$$\theta(t; \varepsilon) = u_0(t) + \varepsilon u_1(t) + \varepsilon^2 u_2(t) + \cdots \tag{5.4.7}$$

$$\delta = \delta_0 + \varepsilon\delta_1 + \varepsilon^2\delta_2 + \cdots \tag{5.4.8}$$

Substituting the above two equations into (5.4.6) and retaining to $O(\varepsilon^2)$, we get

$$\begin{aligned}
0 &= \ddot{\theta} + \left(\delta - \varepsilon 2\Omega_0^2 \cos \omega t - \varepsilon^2\Omega_0^2 \cos^2 \omega t\right)\theta \\
&= \ddot{u}_0 + \varepsilon\ddot{u}_1 + \varepsilon^2\ddot{u}_2 \\
&\quad + \left(\delta_0 + \varepsilon\delta_1 - 2\varepsilon\Omega_0^2 \cos \omega t + \varepsilon^2\delta_2 - \varepsilon^2\Omega_0^2 \cos^2 \omega t\right) \\
&\quad \times \left(u_0 + \varepsilon u_1 + \varepsilon^2 u_2\right) + \cdots \\
&= \ddot{u}_0 + \delta_0 u_0 + \varepsilon\left(\ddot{u}_1 + \delta_0 u_1 + \delta_1 u_0 - 2\Omega_0^2 u_0 \cos \omega t\right) \\
&\quad + \varepsilon^2(\ddot{u}_2 + \delta_0 u_2 + \delta_2 u_0 - \Omega_0^2 u_0 \cos^2 \omega t \\
&\quad + \delta_1 u_1 - 2\Omega_0^2 u_1 \cos \omega t) + \cdots s
\end{aligned}$$

Equating coefficients of like powers of ε in the above equation yields

$$\ddot{u}_0 + \delta_0 u_0 = 0 \tag{5.4.9}$$

$$\ddot{u}_1 + \delta_0 u_1 = -\delta_1 u_0 + 2\Omega_0^2 u_0 \cos \omega t \tag{5.4.10}$$

$$\begin{aligned}
\ddot{u}_2 + \delta_0 u_2 &= -\delta_2 u_0 + \Omega_0^2 u_0 \cos^2 \omega t \\
&\quad - \delta_1 u_1 + 2\Omega_0^2 u_1 \cos \omega t
\end{aligned} \tag{5.4.11}$$

The solution of (5.4.9) is

$$u_0 = a \cos \sqrt{\delta_0} t + b \sin \sqrt{\delta_0} t \tag{5.4.12}$$

CASE I: $\delta_0 = 0$**:**

$$u_0 = a \tag{5.4.13}$$

u_0 is a constant, which is the periodic solution of any period. Substituting (5.4.13) into (5.4.10) yields

$$\ddot{u}_1 = -\delta_1 a + 2\Omega_0^2 a \cos \omega t \tag{5.4.14}$$

In order to eliminate secular terms from the above equation, there must be

$$\delta_1 = 0 \tag{5.4.15}$$

Then the solution of (5.4.14) is

$$u_1 = -\frac{2\Omega_0^2}{\omega^2} a \cos \omega t \tag{5.4.16}$$

Substituting (5.4.16) and (5.4.13) into (5.4.11), we get

$$\begin{aligned} \ddot{u}_2 &= -\delta_2 a + \Omega_0^2\left(1 - \tfrac{4\Omega_0^2}{\omega^2}\right) a\cos^2\omega t \\ &= -\delta_2 a + 2\Omega_0^2\left(\tfrac{1}{4} - \tfrac{\Omega_0^2}{\omega^2}\right)(1 + \cos 2\omega t) a \end{aligned} \tag{5.4.17}$$

In order to eliminate secular terms from the above equation, there must be

$$\delta_2 = 2\Omega_0^2\left(\frac{1}{4} - \frac{\Omega_0^2}{\omega^2}\right) \tag{5.4.18}$$

Therefore, the transition curves emanating from $\delta = 0$ is

$$\delta = \frac{g}{l} - \Omega_0^2 = 2\Omega_0^2\left(\frac{1}{4} - \frac{\Omega_0^2}{\omega^2}\right)\varepsilon^2 + O(\varepsilon^3)$$

i.e.,

$$\frac{g}{l} = \Omega_0^2\left[1 + 2\left(\frac{1}{4} - \frac{\Omega_0^2}{\omega^2}\right)\varepsilon^2\right] + O(\varepsilon^3) \tag{5.4.19}$$

CASE II: $\delta_0 = \frac{1}{4}\omega^2$**:**

$$u_0 = a \cos \frac{1}{2}\omega t + b \sin \frac{1}{2}\omega t \tag{5.4.20}$$

$u_0$is a periodic solution with period $2T = 4\pi/\omega$. Substituting (5.4.20) into (5.4.10) yields

$$\begin{aligned}\ddot{u}_1 + \frac{1}{4}\omega^2 u_1 &= -\delta_1\left(a\cos\frac{1}{2}\omega t + b\sin\frac{1}{2}\omega t\right)\\ &\quad + 2\Omega_0^2\left(a\cos\frac{1}{2}\omega t + b\sin\frac{1}{2}\omega t\right)\cos\omega t\\ &= -\left(\delta_1 - \Omega_0^2\right)a\cos\frac{1}{2}\omega t\\ &\quad - \left(\delta_1 + \Omega_0^2\right)b\sin\frac{1}{2}\omega t + NST \end{aligned} \tag{5.4.21}$$

In order to eliminate secular terms from the above equation, there must be

$$\delta_1 = \pm\Omega_0^2 \tag{5.4.22}$$

Therefore, the transition curves emanating from $\delta = \frac{1}{4}\omega^2$ is

$$\delta = \frac{g}{l} - \Omega_0^2 = \frac{1}{4}\omega^2 \pm \Omega_0^2\varepsilon + O\left(\varepsilon^2\right)$$

i.e.,

$$\frac{g}{l} = \Omega_0^2 + \frac{1}{4}\omega^2 \pm \Omega_0^2\varepsilon + O\left(\varepsilon^2\right) \tag{5.4.23}$$

CASE III: $\delta_0 = \omega^2$:

$$u_0 = a\cos\omega t + b\sin\omega t \tag{5.4.24}$$

$u_0$is a periodic solution with period $2T = 4\pi/\omega$. Substituting (5.4.24) into (5.4.10) yields

$$\begin{aligned}\ddot{u}_1 + \omega^2 u_1 &= -\delta_1(a\cos\omega t + b\sin\omega t)\\ &\quad + 2\Omega_0^2(a\cos\omega t + b\sin\omega t)\cos\omega t\\ &= -\delta_1 a\cos\omega t - \delta_1 b\sin\omega t\\ &\quad + \Omega_0^2 a(1 + \cos 2\omega t) + \Omega_0^2 b\sin 2\omega \end{aligned} \tag{5.4.25}$$

In order to eliminate secular terms from the above equation, there must be

$$\delta_1 = 0 \tag{5.4.26}$$

The solution of (5.4.25) is

$$u_1 = \frac{\Omega_0^2}{\omega^2}a - \frac{\Omega_0^2}{3\omega^2}a\cos 2\omega t - \frac{\Omega_0^2}{3\omega^2}b\sin 2\omega t \tag{5.4.27}$$

Substituting (5.4.27) and (5.4.24) into (5.4.11), we get

$$\begin{aligned}
\ddot{u}_2 + \delta_0 u_2 &= -\delta_2(a\cos\omega t + b\sin\omega t) \\
&\quad + \Omega_0^2(a\cos\omega t + b\sin\omega t)\cos^2\omega t \\
&\quad + 2\Omega_0^2\left(\frac{\Omega_0^2}{\omega^2}a - \frac{\Omega_0^2}{3\omega^2}a\cos 2\omega t - \frac{\Omega_0^2}{3\omega^2}b\sin 2\omega t\right)\cos\omega t \\
&= -\left(\delta_2 - \frac{3}{4}\Omega_0^2 - \frac{5\Omega_0^4}{3\omega^2}\right)a\cos\omega t \\
&\quad - \left(\delta_2 - \frac{1}{4}\Omega_0^2 + \frac{\Omega_0^4}{3\omega^2}\right)b\sin\omega t + NST
\end{aligned} \tag{5.4.28}$$

In order to eliminate secular terms from the above equation, there must be

$$\delta_2 = \Omega_0^2\left(\frac{3}{4} + \frac{5\Omega_0^2}{3\omega^2}\right) \text{ maybe } \delta_2 = \Omega_0^2\left(\frac{1}{4} - \frac{\Omega_0^2}{3\omega^2}\right) \tag{5.4.29}$$

Therefore, the transition curves emanating from $\delta = \omega^2$ is

$$\delta = \frac{g}{l} - \Omega_0^2 = \omega^2 + \Omega_0^2\left(\frac{3}{4} + \frac{5\Omega_0^2}{3\omega^2}\right)\varepsilon^2 + O(\varepsilon^3)$$

$$\delta = \frac{g}{l} - \Omega_0^2 = \omega^2 + \Omega_0^2\left(\frac{1}{4} - \frac{\Omega_0^2}{3\omega^2}\right)\varepsilon^2 + O(\varepsilon^3)$$

i.e.,

$$\frac{g}{l} = \Omega_0^2 + \omega^2 + \Omega_0^2\left(\frac{3}{4} + \frac{5\Omega_0^2}{3\omega^2}\right)\varepsilon^2 + O(\varepsilon^3) \tag{5.4.30}$$

$$\frac{g}{l} = \Omega_0^2 + \omega^2 + \Omega_0^2\left(\frac{1}{4} - \frac{\Omega_0^2}{3\omega^2}\right)\varepsilon^2 + O(\varepsilon^3) \tag{5.4.31}$$

(d) Expanding the Eq. (5.4.4) near $\theta = 0$ and retaining to $O(\theta^3)$ yields

$$\ddot{\theta} + \left(gl^{-1} - \Omega_0^2 - 2\Omega_0^2\varepsilon\cos\omega t - \Omega_0^2\varepsilon^2\cos^2\omega t\right)\theta + \left(\frac{2}{3}\Omega_0^2 - \frac{1}{6}gl^{-1} + \frac{4}{3}\Omega_0^2\varepsilon\cos\omega t + \frac{2}{3}\Omega_0^2\varepsilon^2\cos^2\omega t\right)\theta^3 = 0 \tag{5.4.32}$$

When $\theta = O\left(\varepsilon^{1/2}\right)$, let

$$\theta = \hat{\varepsilon}u, \quad \hat{\varepsilon} = \varepsilon^{1/2} \tag{5.4.33}$$

Equation (5.4.32) becomes

$$\ddot{u} + \left(\delta - 2\Omega_0^2\hat{\varepsilon}^2\cos\omega t - \Omega_0^2\hat{\varepsilon}^4\cos^2\omega t\right)u + \left(\frac{2}{3}\Omega_0^2\hat{\varepsilon}^2 - \frac{1}{6}gl^{-1}\hat{\varepsilon}^2 + \frac{4}{3}\Omega_0^2\hat{\varepsilon}^4\cos\omega t + \frac{2}{3}\Omega_0^2\hat{\varepsilon}^6\cos^2\omega t\right)u^3 = 0 \tag{5.4.34}$$

where

$$\delta = gl^{-1} - \Omega_0^2 \approx \frac{1}{4}\omega^2 \tag{5.4.35}$$

In order to determine the effect of nonlinear terms on the stability of the equation, we use the method of multiple scales to solve (5.4.34). Setting

$$u(t;\varepsilon) = u_0(T_0, T_1, T_2) + \hat{\varepsilon}u_1(T_0, T_1, T_2) + \hat{\varepsilon}^2u_2(T_0, T_1, T_2) + \cdots \tag{5.4.36}$$

where $T_n = \hat{\varepsilon}^n t$. Substituting the above equation into (5.4.34) and retaining to $O(\hat{\varepsilon}^2)$, we get

$$\begin{aligned}
0 &= \left[D_0^2 + 2\hat{\varepsilon}D_0D_1 + \hat{\varepsilon}^2\left(D_1^2 + 2D_0D_2\right)\right]\left(u_0 + \hat{\varepsilon}u_1 + \hat{\varepsilon}^2u_2\right) \\
&\quad + \left(\delta - 2\Omega_0^2\hat{\varepsilon}^2\cos\omega t - \Omega_0^2\hat{\varepsilon}^4\cos^2\omega t\right)\left(u_0 + \hat{\varepsilon}u_1 + \hat{\varepsilon}^2u_2\right) \\
&\quad + \left(\frac{4}{3}\Omega_0^2\hat{\varepsilon}^2 - \frac{1}{6}gl^{-1}\hat{\varepsilon}^2 + \frac{8}{3}\Omega_0^2\hat{\varepsilon}^4\cos\omega t + \frac{4}{3}\Omega_0^2\hat{\varepsilon}^6\cos^2\omega t\right) \\
&\quad \left(u_0 + \hat{\varepsilon}u_1 + \hat{\varepsilon}^2u_2\right)^3 \\
&= D_0^2u_0 + 2\hat{\varepsilon}D_0D_1u_0 + \hat{\varepsilon}^2\left(D_1^2 + 2D_0D_2\right)u_0 \\
&\quad + \hat{\varepsilon}D_0^2u_1 + 2\hat{\varepsilon}^2D_0D_1u_1 + \hat{\varepsilon}^2D_0^2u_2 + \delta u_0 + \hat{\varepsilon}\delta u_1 + \hat{\varepsilon}^2\delta u_2 \\
&\quad - 2\Omega_0^2\hat{\varepsilon}^2u_0\cos\omega t + \hat{\varepsilon}^2\left(\frac{4}{3}\Omega_0^2 - \frac{1}{6}gl^{-1}\right)u_0^3 + \cdots \\
&= D_0^2u_0 + \delta u_0 + \hat{\varepsilon}\left(D_0^2u_1 + \delta u_1 + 2D_0D_1u_0\right) \\
&\quad + \hat{\varepsilon}^2[D_0^2u_2 + \delta u_2 + \left(D_1^2 + 2D_0D_2\right)u_0 + 2D_0D_1u_1
\end{aligned}$$

$$- 2\Omega_0^2 u_0 \cos \omega t + \left(\frac{4}{3}\Omega_0^2 - \frac{1}{6}gl^{-1}\right)u_0^3] + \cdots$$

Equating coefficients of like powers of $\hat{\varepsilon}$ in the above equation yields

$$D_0^2 u_0 + \omega_0^2 u_0 = 0 \quad (5.4.37)$$

$$D_0^2 u_1 + \omega_0^2 u_1 = -2D_0 D_1 u_0 \quad (5.4.38)$$

$$\begin{aligned} D_0^2 u_2 + \omega_0^2 u_2 = &-\left(D_1^2 + 2D_0 D_2\right)u_0 - 2D_0 D_1 u_1 \\ &+ 2\Omega_0^2 u_0 \cos\omega t + \left(\frac{1}{6}gl^{-1} - \frac{4}{3}\Omega_0^2\right)u_0^3 \end{aligned} \quad (5.4.39)$$

where $\omega_0^2 = \delta$. The solution of (5.4.37) is

$$u_0 = Ae^{i\omega_0 T_0} + \overline{A}\, e^{-i\omega_0 T_0} \quad (5.4.40)$$

where $A = A(T_1, T_2)$. Substituting u_0 into (5.4.38) yields

$$D_0^2 u_1 + \omega_0^2 u_1 = -2i\omega_0 D_1 A e^{i\omega_0 T_0} + cc \quad (5.4.41)$$

In order to eliminate secular terms from the above equation, there must be

$$D_1 A = 0 \quad \Rightarrow \quad A = A(T_2) \quad (5.4.42)$$

Thus

$$u_1 = 0 \quad (5.4.43)$$

Substituting $u_0$and u_1 into (5.4.39) yields

$$\begin{aligned} D_0^2 u_2 + \omega_0^2 u_2 = &-2i\omega_0 D_2 A e^{i\omega_0 T_0} + \Omega_0^2 \overline{A} e^{i(\omega - \omega_0)T_0} \\ &+ \left(\frac{1}{2}gl^{-1} - 4\Omega_0^2\right)A^2 \overline{A} e^{i\omega_0 T_0} + cc + NST \end{aligned} \quad (5.4.44)$$

When $\delta = \omega_0^2 = gl^{-1} - \Omega_0^2 \approx \frac{1}{4}\omega^2$, we use the detuning parameter σ to quantitatively describe the nearness of δ to ω_0, i.e.,

$$\omega = 2\omega_0 + \hat{\varepsilon}^2 \sigma \quad (5.4.45)$$

So (5.4.44) becomes

$$D_0^2 u_2 + \omega_0^2 u_2 = -2i\omega_0 D_2 A e^{i\omega_0 T_0} + \Omega_0^2 \overline{A} e^{i\sigma T_2} e^{i\omega_0 T_0} + \left(\frac{1}{2} g l^{-1} - 4\Omega_0^2\right) A^2 \overline{A} e^{i\omega_0 T_0} + cc + NST \tag{5.4.46}$$

In order to eliminate secular terms from the above equation, there must be

$$-2i\omega_0 D_2 A + \Omega_0^2 e^{i\sigma T_2} \overline{A} + \left(\frac{1}{2} g l^{-1} - 4\Omega_0^2\right) A^2 \overline{A} = 0 \tag{5.4.47}$$

Let $A = \frac{1}{2} a e^{i\beta}$, we get

$$-2i\omega_0 a' + 2\omega_0 a\beta' + \Omega_0^2 a e^{i(\sigma T_2 - 2\beta)} + \frac{1}{4}\left(\frac{1}{2} g l^{-1} - 4\Omega_0^2\right) a^3 = 0 \tag{5.4.48}$$

Separating the real and imaginary parts of the above equation yields

$$a' = \frac{\Omega_0^2}{2\omega_0} a \sin(\sigma T_2 - 2\beta)\beta'$$
$$= -\frac{1}{8\omega_0}\left(\frac{1}{2} g l^{-1} - 4\Omega_0^2\right) a^2 - \frac{\Omega_0^2}{2\omega_0} \cos(\sigma T_2 - 2\beta)$$

Let $\gamma = \sigma T_2 - 2\beta$ and consider $\omega_0^2 \approx \frac{1}{4}\omega^2$, we can write the above equation as

$$a' = \Omega_0^2 \omega^{-1} a \sin\gamma \tag{5.4.49}$$

$$\gamma' = \sigma + \frac{1}{2}\omega^{-1}\left(\frac{1}{2} g l^{-1} - 4\Omega_0^2\right) a^2 + 2\Omega_0^2 \omega^{-1} \cos\gamma \tag{5.4.50}$$

Therefore, the first order approximate solution of (5.4.34) is

$$u_0 = a \cos[\frac{1}{2}(\omega - \varepsilon\sigma)t + \frac{1}{2}\gamma)] + O(\varepsilon^{1/2}) \tag{5.4.51}$$

Let $a' = \gamma' = 0$ in (5.4.49) and (5.4.50), we can obtain

$$\Omega_0^2 \omega^{-1} a \sin\gamma = 0 \tag{5.4.52}$$

$$\sigma + \frac{1}{2}\omega^{-1}\left(\frac{1}{2} g l^{-1} - 4\Omega_0^2\right) a^2 + 2\Omega_0^2 \omega^{-1} \cos\gamma = 0 \tag{5.4.53}$$

Therefore, the steady state solution is

$$a = 0, \quad \cos\gamma = -\frac{1}{2}\sigma \Omega_0^{-2} \omega \tag{5.4.54}$$

The condition for the existence of a steady state solution of γ is

$$|\sigma| \le 2\Omega_0^2\omega^{-1} \tag{5.4.55}$$

Therefore, the critical value of σ for the existence of steady state solution is

$$\sigma = \pm 2\Omega_0^2\omega^{-1} \tag{5.4.56}$$

Substituting this into (5.4.45), we can obtain the critical condition for the existence of a steady state solution

$$gl^{-1} - \Omega_0^2 = \frac{1}{4}\omega^2 \pm \varepsilon\Omega_0^2 + O(\varepsilon^2) \tag{5.4.57}$$

Combining (5.4.23), and (5.4.57), we know that the effect of the nonlinear terms is to limit the unstable linear motion to a finite-amplitude motion, whose frequency is one half the frequency of the excitation. In other words, a subharmonic is generated by the system.

5.5 Exercise 5.5 (Parametric Excitation of a Particle-String System Subjected to Axial Forces at Both Ends)

Solution: (a) The ends of the rope can move only in the horizontal direction. According to Newton's second law, we can obtain the equation of motion of m along x-direction:

$$m\ddot{x} = -2T\frac{x}{\sqrt{x^2 + l^2}} \tag{5.5.1}$$

i.e.,

$$m\ddot{x} + 2T_0(1 + \varepsilon \sin \omega t)x(l^2 + x^2)^{-1/2} = 0 \tag{5.5.2}$$

(b) Linearizing the above equation near the equilibrium position $x = 0$, we get

$$m\ddot{x} + 2T_0(1 + \varepsilon \sin \omega t)l^{-1}x = 0 \tag{5.5.3}$$

i.e.,

$$\ddot{x} + \omega_0^2(1 + \varepsilon \sin \omega t)x = 0, \quad \omega_0^2 = \frac{2T_0}{ml} \tag{5.5.4}$$

(c) (5.5.4) is a Hill equation with a time-varying parameter of period $T = 2\pi/\omega$. We use the method of strained parameters to obtain transition curves. We seek the solutions of the original equation having periods of T and $2\ T$ and the equations for the transition curves $\omega_0^2 = \delta = \delta(\varepsilon)$ in the form of the following perturbation expansions:

$$x(t;\varepsilon) = u_0(t) + \varepsilon u_1(t) + \varepsilon^2 u_2(t) + \cdots \tag{5.5.5}$$

$$\delta = \delta_0 + \varepsilon\delta_1 + \varepsilon^2\delta_2 + \cdots \tag{5.5.6}$$

Substituting the above two equations into (5.5.4) and retaining to $O(\varepsilon^2)$, we get

$$\begin{aligned}
0 &= \ddot{x} + \delta(1 + \varepsilon\sin\omega t)x \\
&= \ddot{u}_0 + \varepsilon\ddot{u}_1 + \varepsilon^2\ddot{u}_2 \\
&\quad + \left(\delta_0 + \varepsilon\delta_1 + \varepsilon^2\delta_2\right)(1 + \varepsilon\sin\omega t)\left(u_0 + \varepsilon u_1 + \varepsilon^2 u_2\right) + \cdots \\
&= \ddot{u}_0 + \varepsilon\ddot{u}_1 + \varepsilon^2\ddot{u}_2 + \delta_0 u_0 + \varepsilon\delta_0 u_1 + \varepsilon\delta_0 u_0\sin\omega t \\
&\quad + \varepsilon^2\delta_0 u_2 + \varepsilon^2\delta_0 u_1\sin\omega t \\
&\quad + \varepsilon\delta_1 u_0 + \varepsilon^2\delta_1 u_1 + \varepsilon^2\delta_1 u_0\sin\omega t + \varepsilon^2\delta_2 u_0 + \cdots \\
&= \ddot{u}_0 + \delta_0 u_0 + \varepsilon(\ddot{u}_1 + \delta_0 u_1 + \delta_1 u_0 + \delta_0 u_0\sin\omega t) \\
&\quad + \varepsilon^2(\ddot{u}_2 + \delta_0 u_2 + \delta_2 u_0 + \delta_1 u_0\sin\omega t + \delta_1 u_1 + \delta_0 u_1\sin\omega t) + \cdots
\end{aligned}$$

Equating coefficients of like powers of ε in the above equation yields

$$\ddot{u}_0 + \delta_0 u_0 = 0 \tag{5.5.7}$$

$$\ddot{u}_1 + \delta_0 u_1 = -\delta_1 u_0 - \delta_0 u_0\sin\omega t \tag{5.5.8}$$

$$\ddot{u}_2 + \delta_0 u_2 = -\delta_2 u_0 - \delta_1 u_0\sin\omega t - \delta_1 u_1 - \delta_0 u_1\sin\omega t \tag{5.5.9}$$

The solution of (5.5.7) is

$$u_0 = a\cos\sqrt{\delta_0}t + b\sin\sqrt{\delta_0}t \tag{5.5.10}$$

CASE I: $\delta_0 = \omega^2$**:**

$$u_0 = a\cos\omega t + b\sin\omega t \tag{5.5.11}$$

u_0 is a periodic solution with period $T = 2\pi/\omega$. Substituting (5.5.11) into (5.5.8) yields

$$\begin{aligned}\ddot{u}_1 + \omega^2 u_1 &= -\left(\delta_1 + \omega^2 \sin \omega t\right)(a \cos \omega t + b \sin \omega t) \\ &= -\delta_1 a \cos \omega t - \delta_1 b \sin \omega t \\ &\quad - \frac{1}{2}\omega^2 a \sin 2\omega t - \frac{1}{2}\omega^2 b(1 - \cos 2\omega t)\end{aligned} \tag{5.5.12}$$

In order to eliminate secular terms from the above equation, there must be

$$\delta_1 = 0 \tag{5.5.13}$$

Then the solution of (5.5.12) is

$$u_1 = -\frac{1}{2}b + \frac{1}{6}a \sin 2\omega t - \frac{1}{6}b \cos 2\omega t \tag{5.5.14}$$

Substituting the above results into (5.5.9), we obtain

$$\begin{aligned}\ddot{u}_2 + \delta_0 u_2 &= -\delta_2 u_0 - \delta_0 u_1 \sin \omega t \\ &= -\delta_2 (a \cos \omega t + b \sin \omega t) \\ &\quad + \omega^2 \left(\frac{1}{2}b - \frac{1}{6}a \sin 2\omega t + \frac{1}{6}b \cos 2\omega t\right) \sin \omega t \\ &= -\left(\delta_2 + \frac{1}{12}\omega^2\right) a \cos \omega t \\ &\quad - \left(\delta_2 - \frac{5}{12}\omega^2\right) b \sin \omega t + NST\end{aligned} \tag{5.5.15}$$

In order to eliminate secular terms from the above equation, there must be

$$\delta_2 = -\frac{1}{12}\omega^2 \quad \text{or} \quad \frac{5}{12}\omega^2 \tag{5.5.16}$$

Therefore, the transition curves emanating from $\delta = \omega^2$ is

$$\delta = \omega^2 - \frac{1}{12}\omega^2 \varepsilon^2 + O\left(\varepsilon^3\right)$$

$$\delta = \omega^2 + \frac{5}{12}\omega^2 \varepsilon^2 + O\left(\varepsilon^3\right)$$

i.e.,

$$\omega_0^2 = \omega^2 \left(1 - \frac{1}{12}\varepsilon^2\right) + O\left(\varepsilon^3\right) \tag{5.5.17}$$

or

$$\omega_0^2 = \omega^2\left(1 + \frac{5}{12}\varepsilon^2\right) + O(\varepsilon^3) \quad (5.5.18)$$

CASE II: $\delta_0 = \frac{1}{4}\omega^2$:

$$u_0 = a\cos\frac{1}{2}\omega t + b\sin\frac{1}{2}\omega t \quad (5.5.19)$$

u_0 is a periodic solution with period $2T = 4\pi/\omega$. Substituting (5.5.19) into (5.5.8) yields

$$\begin{aligned}\ddot{u}_1 + \frac{1}{4}\omega^2 u_1 &= -\delta_1\left(a\cos\frac{1}{2}\omega t + b\sin\frac{1}{2}\omega t\right)\\ &\quad - \frac{1}{4}\omega^2\left(a\cos\frac{1}{2}\omega t + b\sin\frac{1}{2}\omega t\right)\sin\omega t\\ &= -\left(\delta_1 a + \frac{1}{8}\omega^2 b\right)\cos\frac{1}{2}\omega t\\ &\quad - \left(\delta_1 b + \frac{1}{8}\omega^2 a\right)\sin\frac{1}{2}\omega t + NST\end{aligned} \quad (5.5.20)$$

In order to eliminate secular terms from the above equation, there must be

$$\begin{array}{l}\delta_1 a + \frac{1}{8}\omega^2 b = 0\\ \frac{1}{8}\omega^2 a + \delta_1 b = 0\end{array} \Rightarrow \begin{bmatrix}\delta_1 & \frac{1}{8}\omega^2\\ \frac{1}{8}\omega^2 & \delta_1\end{bmatrix}\begin{Bmatrix}a\\ b\end{Bmatrix} = \begin{Bmatrix}0\\ 0\end{Bmatrix} \quad (5.5.21)$$

The condition for having a nontrivial solution is

$$\delta_1 = \pm\frac{1}{8}\omega^2 \quad (5.5.22)$$

Therefore, the transition curves emanating from $\delta = \frac{1}{4}\omega^2$ is

$$\delta = \frac{1}{4}\omega^2 \pm \frac{1}{8}\omega^2\varepsilon + O(\varepsilon^2)$$

i.e.,

$$\omega_0^2 = \frac{1}{4}\omega^2\left(1 \pm \frac{1}{2}\varepsilon\right) + O(\varepsilon^2) \quad (5.5.23)$$

(d) Expand the Eq. (5.5.2) near $x = 0$ and retain to $O(x^3)$, we obtain

$$\frac{\ddot{x}}{l} + \omega_0^2(1 + \varepsilon\sin\omega t)\frac{x}{l}[1 - \frac{1}{2}\left(\frac{x}{l}\right)^2\Big] = 0 \quad (5.5.24)$$

By making $\theta = x/l$, the above equation becomes

$$\ddot{\theta} + \omega_0^2(1 + \varepsilon \sin \omega t)\theta - \frac{1}{2}\omega_0^2(1 + \varepsilon \sin \omega t)\theta^3 = 0 \tag{5.5.25}$$

When $x = O(\varepsilon^{1/2})$, $\theta = O\left(\varepsilon^{\frac{1}{2}}\right)$, let

$$\theta = \hat{\varepsilon} u, \quad \hat{\varepsilon} = \varepsilon^{1/2} \tag{5.5.26}$$

Equation (5.5.25) becomes

$$\ddot{u} + \omega_0^2\left(1 + \hat{\varepsilon}^2 \sin \omega t\right)u - \frac{1}{2}\hat{\varepsilon}^2\omega_0^2\left(1 + \hat{\varepsilon}^2 \sin \omega t\right)u^3 = 0 \tag{5.5.27}$$

In order to determine the effect of the nonlinear terms on the stability of the system, we use the method of multiple scales to solve (5.5.27). Setting

$$u(t; \varepsilon) = u_0(T_0, T_1, T_2) + \hat{\varepsilon} u_1(T_0, T_1, T_2) + \hat{\varepsilon}^2 u_2(T_0, T_1, T_2) + \cdots \tag{5.5.28}$$

where $T_n = \hat{\varepsilon}^n t$. Substituting the above equation into (5.5.27) and retaining to $O(\hat{\varepsilon}^2)$, we obtain

$$\begin{aligned}
0 &= \ddot{u} + \omega_0^2\left(1 + \hat{\varepsilon}^2 \sin \omega t\right)u - \frac{1}{2}\hat{\varepsilon}^2\omega_0^2\left(1 + \hat{\varepsilon}^2 \sin \omega t\right)u^3 \\
&= \left[D_0^2 + 2\hat{\varepsilon} D_0 D_1 + \hat{\varepsilon}^2\left(D_1^2 + 2D_0 D_2\right)\right]\left(u_0 + \hat{\varepsilon} u_1 + \hat{\varepsilon}^2 u_2\right) \\
&\quad + \omega_0^2\left(1 + \hat{\varepsilon}^2 \sin \omega t\right)\left(u_0 + \hat{\varepsilon} u_1 + \hat{\varepsilon}^2 u_2\right) \\
&\quad - \frac{1}{2}\hat{\varepsilon}^2\omega_0^2\left(1 + \hat{\varepsilon}^2 \sin \omega t\right)\left(u_0 + \hat{\varepsilon} u_1 + \hat{\varepsilon}^2 u_2\right)^3 \\
&= D_0^2 u_0 + \hat{\varepsilon} D_0^2 u_1 + \hat{\varepsilon}^2 D_0^2 u_2 + 2\hat{\varepsilon} D_0 D_1 u_0 + 2\hat{\varepsilon}^2 D_0 D_1 u_1 \\
&\quad + \hat{\varepsilon}^2\left(D_1^2 + 2D_0 D_2\right)u_0 + \omega_0^2 u_0 + \hat{\varepsilon}\omega_0^2 u_1 + \hat{\varepsilon}^2\omega_0^2 u_2 \\
&\quad + \hat{\varepsilon}^2\omega_0^2 u_0 \sin \omega t - \frac{1}{2}\hat{\varepsilon}^2\omega_0^2 u_0^3 + \cdots \\
&= D_0^2 u_0 + \omega_0^2 u_0 + \hat{\varepsilon}\left(D_0^2 u_1 + \omega_0^2 u_1 + 2D_0 D_1 u_0\right) \\
&\quad + \hat{\varepsilon}^2[D_0^2 u_2 + \omega_0^2 u_2 + \left(D_1^2 + 2D_0 D_2\right)u_0 + 2D_0 D_1 u_1 \\
&\quad + \omega_0^2 u_0 \sin \omega t - \frac{1}{2}\omega_0^2 u_0^3] + \cdots
\end{aligned}$$

Equating coefficients of like powers of ε in the above equation yields

$$D_0^2 u_0 + \omega_0^2 u_0 = 0 \tag{5.5.29}$$

$$D_0^2 u_1 + \omega_0^2 u_1 = -2D_0 D_1 u_0 \tag{5.5.30}$$

$$\begin{aligned} D_0^2u_2 + \omega_0^2u_2 &= -\left(D_1^2 + 2D_0D_2\right)u_0 - 2D_0D_1u_1 \\ &-\omega_0^2u_0\sin\omega t + \tfrac{1}{2}\omega_0^2u_0^3 \end{aligned} \tag{5.5.31}$$

The solution of (5.5.29) is

$$u_0 = Ae^{i\omega_0 T_0} + \bar{A}\,e^{-i\omega_0 T_0} \tag{5.5.32}$$

where $A = A(T_1, T_2)$. Substituting u_0 into (5.5.30) yields

$$D_0^2u_1 + \omega_0^2u_1 = -2i\omega_0 D_1 A e^{i\omega_0 T_0} + cc \tag{5.5.33}$$

In order to eliminate secular terms from the above equation, there must be

$$D_1 A = 0 \quad \Rightarrow \quad A = A(T_2) \tag{5.5.34}$$

Therefore,

$$u_1 = 0 \tag{5.5.35}$$

Substitute $u_0$and u_1 into (5.5.31) yields

$$\begin{aligned} D_0^2u_2 + \omega_0^2u_2 &= -2i\omega_0 D_2 A e^{i\omega_0 T_0} + \tfrac{1}{2}i\omega_0^2\bar{A}e^{i(\omega-\omega_0)T_0} \\ &+\tfrac{3}{2}\omega_0^2A^2\bar{A}e^{i\omega_0 T_0} + cc + NST \end{aligned} \tag{5.5.36}$$

When $\omega_0^2 \approx \frac{1}{4}\omega^2$, we use the detuning parameter σ to quantitatively describe the nearness of ω to $2\omega_0$, i.e.,

$$\omega = 2\omega_0 + \hat{\varepsilon}^2\sigma \tag{5.5.37}$$

So (5.5.36) becomes

$$\begin{aligned} D_0^2u_2 + \omega_0^2u_2 &= -2i\omega_0 D_2 A e^{i\omega_0 T_0} + \tfrac{1}{2}i\omega_0^2\bar{A}e^{i\sigma T_2}e^{i\omega_0 T_0} \\ &+\tfrac{3}{2}\omega_0^2A^2\bar{A}e^{i\omega_0 T_0} + cc + NST \end{aligned} \tag{5.5.38}$$

In order to eliminate secular terms from the above equation, there must be

$$-2iD_2A + \frac{1}{2}i\omega_0\,\bar{A}\,e^{i\sigma T_2} + \frac{3}{2}\omega_0 A^2\,\bar{A} = 0 \tag{5.5.39}$$

Let $A = \frac{1}{2}ae^{i\beta}$, we get

$$-ia' + a\beta' + \frac{1}{4}i\omega_0 e^{i(\sigma T_2 - 2\beta)}a + \frac{3}{16}\omega_0 a^3 = 0 \tag{5.5.40}$$

Separating the real and imaginary parts of the above equation yields

$$a' = \frac{1}{4}\omega_0 a \cos(\sigma T_2 - 2\beta)$$

$$\beta' = -\frac{3}{16}\omega_0 a^2 + \frac{1}{4}\omega_0 \sin(\sigma T_2 - 2\beta)$$

Let $\gamma = \sigma T_2 - 2\beta$ and considering $\omega_0^2 \approx \frac{1}{4}\omega^2$, the above equation becomes

$$a' = \frac{1}{4}\omega_0 a \cos\gamma \tag{5.5.41}$$

$$\gamma' = \sigma + \frac{3}{8}\omega_0 a^2 - \frac{1}{2}\omega_0 \sin\gamma \tag{5.5.42}$$

Therefore, the first order approximate solution of (5.5.27) is

$$u_0 = a\cos[\frac{1}{2}(\omega - \varepsilon\sigma)t + \frac{1}{2}\gamma)] + O(\varepsilon^{1/2}) \tag{5.5.43}$$

Let $a' = \gamma' = 0$ in (5.5.41) and (5.5.42), we can obtain

$$\frac{1}{4}\omega_0 a \cos\gamma = 0 \tag{5.5.44}$$

$$\sigma + \frac{3}{8}\omega_0 a^2 - \frac{1}{2}\omega_0 \sin\gamma = 0 \tag{5.5.45}$$

Therefore, the steady state solution is

$$a = 0, \quad \sin\gamma = 2\sigma\omega_0^{-1} \tag{5.5.46}$$

The condition for the existence of a steady state solution of γ is

$$|\sigma| \le \frac{1}{2}\omega_0 \tag{5.5.47}$$

Therefore, the critical value of σ for the existence of steady state solution is

$$\sigma = \pm\frac{1}{2}\omega_0 \approx \pm\frac{1}{4}\omega \tag{5.5.48}$$

Substituting this into (5.5.37), we can obtain the critical condition for the existence of a steady state solution

$$\omega_0^2 = \frac{1}{4}\omega^2\left(1 \pm \frac{1}{2}\varepsilon\right) + O(\varepsilon^2) \tag{5.5.49}$$

Combining (5.5.23), and (5.5.49), we know that the effect of the nonlinear terms is to limit the unstable linear motion to a finite-amplitude motion, whose frequency is one half of the frequency of the excitation. In other words, a subharmonic is generated by the system.

5.6 Exercise 5.6 (Linear and Nonlinear Stability Analysis of a Pendulum with Varying Length)

Solution: (**a**) The kinetic energy of the system is

$$\begin{aligned} T &= \frac{1}{2}mv^2 = \frac{1}{2}m\left\{\left[\frac{d(r\sin\theta)}{dt}\right]^2 + \left[\frac{d(r\cos\theta + y)}{dt}\right]^2\right\} \\ &= \frac{1}{2}m[(\dot{r}\sin\theta + r\dot{\theta}\cos\theta)^2 + (\dot{r}\cos\theta - r\dot{\theta}\sin\theta + \dot{y})^2] \\ &= \frac{1}{2}m(\dot{r}^2 + \dot{y}^2 + r^2\dot{\theta}^2 + 2\dot{r}\dot{y}\cos\theta - 2r\dot{y}\dot{\theta}\sin\theta) \\ &= \frac{1}{2}m[2\dot{y}^2 + (l-y)^2\dot{\theta}^2 - 2\dot{y}^2\cos\theta - 2(l-y)\dot{y}\dot{\theta}\sin\theta] \end{aligned} \tag{5.6.1}$$

The potential energy of the system is

$$V = mgr(1-\cos\theta) = mg(l-y)(1-\cos\theta) \tag{5.6.2}$$

Substituting (5.6.1) and (5.6.2) into the Lagrange's equation, i.e.,

$$\frac{d}{dt}\frac{\partial T}{\partial\dot{\theta}} - \frac{\partial T}{\partial\theta} = -\frac{\partial V}{\partial\theta} \tag{5.6.3}$$

we can obtain the differential equation of motion of the system

$$(l-y)\ddot{\theta} - 2\dot{y}\dot{\theta} - \ddot{y}\sin\theta + g\sin\theta = 0 \tag{5.6.4}$$

(**b**) Linearizing the above equation near $\theta = 0$ yields

$$\ddot{\theta} - \frac{2\dot{y}\dot{\theta}}{l-y} + \frac{g-\ddot{y}}{l-y}\theta = 0 \tag{5.6.5}$$

(**c**) When $y = \varepsilon l\cos\Omega t$, (5.6.5) becomes

$$\ddot{\theta} + \frac{2\varepsilon\Omega \sin \Omega t}{1 - \varepsilon \cos \Omega t}\dot{\theta} + \frac{\omega_0^2 + \varepsilon\Omega^2 \cos \Omega t}{1 - \varepsilon \cos \Omega t}\theta = 0, \quad \omega_0^2 = \frac{g}{l} \tag{5.6.6}$$

We use the method of strained parameters to obtain the transition curve. Let the solution of (5.6.6) solution and parameter ω_0^2 have the following expansion:

$$\theta(t; \varepsilon) = u_0(t) + \varepsilon u_1(t) + \varepsilon^2 u_2(t) + \cdots \tag{5.6.7}$$

$$\omega_0^2 = \delta_0 + \varepsilon\delta_1 + \varepsilon^2\delta_2 + \cdots \tag{5.6.8}$$

Substituting the above two equations into (5.6.5) and retaining to $O(\varepsilon^2)$, we get

$$\begin{aligned}
0 &= \ddot{u}_0 + \varepsilon\ddot{u}_1 + \varepsilon^2\ddot{u}_2 + \frac{2\varepsilon\Omega \sin \Omega t}{1 - \varepsilon \cos \Omega t}\left(\dot{u}_0 + \varepsilon\dot{u}_1 + \varepsilon^2\dot{u}_2\right) \\
&\quad + \frac{\omega_0^2 + \varepsilon\Omega^2 \cos \Omega t}{1 - \varepsilon \cos \Omega t}\left(u_0 + \varepsilon u_1 + \varepsilon^2 u_2\right) \\
&= \ddot{u}_0 + \varepsilon\ddot{u}_1 + \varepsilon^2\ddot{u}_2 \\
&\quad + 2\varepsilon\Omega \sin \Omega t(1 + \varepsilon \cos \Omega t)\left(\dot{u}_0 + \varepsilon\dot{u}_1 + \varepsilon^2\dot{u}_2\right) \\
&\quad + \left(\omega_0^2 + \varepsilon\Omega^2 \cos \Omega t\right)\left(1 + \varepsilon \cos \Omega t + \varepsilon^2 \cos^2 \Omega t\right) \\
&\quad \left(u_0 + \varepsilon u_1 + \varepsilon^2 u_2\right) + \cdots \\
&= \ddot{u}_0 + \varepsilon\ddot{u}_1 + \varepsilon^2\ddot{u}_2 + 2\varepsilon\Omega\dot{u}_0 \sin \Omega t \\
&\quad + 2\varepsilon^2\Omega\dot{u}_0 \sin \Omega t \cos \Omega t + 2\varepsilon^2\Omega\dot{u}_1 \sin \Omega t \\
&\quad + \omega_0^2 u_0 + \varepsilon\omega_0^2 u_0 \cos \Omega t + \varepsilon\Omega^2 u_0 \cos \Omega t \\
&\quad + \varepsilon^2\omega_0^2 u_0 \cos^2 \Omega t + \varepsilon^2\Omega^2 u_0 \cos^2 \Omega t \\
&\quad + \varepsilon\omega_0^2 u_1 + \varepsilon^2\omega_0^2 u_1 \cos \Omega t \\
&\quad + \varepsilon^2\Omega^2 u_1 \cos \Omega t + \varepsilon^2\omega_0^2 u_2 + \cdots \\
&= \ddot{u}_0 + \varepsilon\ddot{u}_1 + \varepsilon^2\ddot{u}_2 + 2\varepsilon\Omega\dot{u}_0 \sin \Omega t \\
&\quad + 2\varepsilon^2\Omega\dot{u}_0 \sin \Omega t \cos \Omega t + 2\varepsilon^2\Omega\dot{u}_1 \sin \Omega t \\
&\quad + \left(\delta_0 + \varepsilon\delta_1 + \varepsilon^2\delta_2\right)u_0 + \varepsilon\left(\delta_0 + \varepsilon\delta_1 + \varepsilon^2\delta_2\right)u_0 \cos \Omega t \\
&\quad + \varepsilon\Omega^2 u_0 \cos \Omega t + \varepsilon^2\left(\delta_0 + \varepsilon\delta_1 + \varepsilon^2\delta_2\right)u_0 \cos^2 \Omega t \\
&\quad + \varepsilon^2\Omega^2 u_0 \cos^2 \Omega t \\
&\quad + \varepsilon\left(\delta_0 + \varepsilon\delta_1 + \varepsilon^2\delta_2\right)u_1 + \varepsilon^2\left(\delta_0 + \varepsilon\delta_1 + \varepsilon^2\delta_2\right)u_1 \cos \Omega t \\
&\quad + \varepsilon^2\Omega^2 u_1 \cos \Omega t + \varepsilon^2\left(\delta_0 + \varepsilon\delta_1 + \varepsilon^2\delta_2\right)u_2 + \cdots \\
&= \ddot{u}_0 + \varepsilon\ddot{u}_1 + \varepsilon^2\ddot{u}_2 + 2\varepsilon\Omega\dot{u}_0 \sin \Omega t \\
&\quad + 2\varepsilon^2\Omega\dot{u}_0 \sin \Omega t \cos \Omega t + 2\varepsilon^2\Omega\dot{u}_1 \sin \Omega t \\
&\quad + \delta_0 u_0 + \varepsilon\delta_1 u_0 + \varepsilon^2\delta_2 u_0 \\
&\quad + \varepsilon\delta_0 u_0 \cos \Omega t + \varepsilon^2\delta_1 u_0 \cos \Omega t
\end{aligned}$$

$$
\begin{aligned}
&+\varepsilon\Omega^2 u_0\cos\Omega t+\varepsilon^2\delta_0 u_0\cos^2\Omega t\\
&+\varepsilon^2\Omega^2 u_0\cos^2\Omega t\\
&+\varepsilon\delta_0 u_1+\varepsilon^2\delta_1 u_1+\varepsilon^2\delta_0 u_1\cos\Omega t\\
&+\varepsilon^2\Omega^2 u_1\cos\Omega t+\varepsilon^2\delta_0 u_2+\cdots\\
&=\ddot u_0+\delta_0 u_0\\
&+\varepsilon\big(\ddot u_1+\delta_0 u_1+\delta_1 u_0+2\Omega\dot u_0\sin\Omega t+\delta_0 u_0\cos\Omega t+\Omega^2 u_0\cos\Omega t\big)\\
&+\varepsilon^2[\ddot u_2+\delta_0 u_2+\delta_2 u_0+\delta_1 u_0\cos\Omega t+2\Omega\dot u_0\sin\Omega t\cos\Omega t\\
&+2\Omega\dot u_1\sin\Omega t+\delta_0 u_0\cos^2\Omega t+\Omega^2 u_0\cos^2\Omega t\\
&+\delta_1 u_1+\delta_0 u_1\cos\Omega t+\Omega^2 u_1\cos\Omega t]+\cdots
\end{aligned}
$$

Equating coefficients of like powers of $\hat{\varepsilon}$ in the above equation yields

$$\ddot u_0+\delta_0 u_0=0 \tag{5.6.9}$$

$$
\begin{aligned}
\ddot u_1+\delta_0 u_1&=-\delta_1 u_0-2\Omega\dot u_0\sin\Omega t\\
&\quad-\delta_0 u_0\cos\Omega t-\Omega^2 u_0\cos\Omega t
\end{aligned}
\tag{5.6.10}
$$

$$
\begin{aligned}
\ddot u_2+\delta_0 u_2&=-\delta_2 u_0-\delta_1 u_0\cos\Omega t-\delta_0 u_0\cos^2\Omega t\\
&\quad-\Omega^2 u_0\cos^2\Omega t-\Omega\dot u_0\sin 2\Omega t\\
&\quad-\delta_1 u_1-\delta_0 u_1\cos\Omega t-\Omega^2 u_1\cos\Omega t\\
&\quad-2\Omega\dot u_1\sin\Omega t
\end{aligned}
\tag{5.6.11}
$$

The solution of (5.6.9) is

$$u_0=a\cos\sqrt{\delta_0}t+b\sin\sqrt{\delta_0}t \tag{5.6.12}$$

CASE I: $\sqrt{\delta_0}=\frac{1}{2}\Omega$:

$$u_0=a\cos\frac{1}{2}\Omega t+b\sin\frac{1}{2}\Omega t \tag{5.6.13}$$

Substituting (5.6.13) into (5.6.10) yields

$$
\begin{aligned}
\ddot u_1+\frac{1}{4}\Omega^2 u_1&=-\delta_1 u_0-\frac{5}{4}\Omega^2 u_0\cos\Omega t-2\Omega\dot u_0\sin\Omega t\\
&=-\delta_1\left(a\cos\frac{1}{2}\Omega t+b\sin\frac{1}{2}\Omega t\right)\\
&\quad-\frac{5}{4}\Omega^2\left(a\cos\frac{1}{2}\Omega t+b\sin\frac{1}{2}\Omega t\right)\cos\Omega t
\end{aligned}
$$

$$
\begin{aligned}
&- \Omega^2\left(-a\sin\frac{1}{2}\Omega t + b\cos\frac{1}{2}\Omega t\right)\sin\Omega t \\
&= -\delta_1 a\cos\frac{1}{2}\Omega t - \delta_1 b\sin\frac{1}{2}\Omega t \\
&- \frac{5}{4}\Omega^2 a\cos\frac{1}{2}\Omega t\cos\Omega t \\
&- \frac{5}{4}\Omega^2 b\sin\frac{1}{2}\Omega t\cos\Omega t \\
&+ \Omega^2 a\sin\Omega t\sin\frac{1}{2}\Omega t - \Omega^2 b\sin\Omega t\cos\frac{1}{2}\Omega t \\
&= -\left(\delta_1 + \frac{1}{8}\Omega^2\right)a\cos\frac{1}{2}\Omega t \\
&- \left(\delta_1 - \frac{1}{8}\Omega^2\right)b\sin\frac{1}{2}\Omega t + NST
\end{aligned}
\tag{5.6.14}
$$

In order to eliminate secular terms from the above equation, there must be

$$\delta_1 = \pm\frac{1}{8}\Omega^2 \tag{5.6.15}$$

Therefore, the transition curves emanating from $\omega_0^2 = \frac{1}{4}\Omega^2$ is

$$\omega_0^2 = gl^{-1} = \frac{1}{4}\Omega^2\left(1 \pm \frac{1}{2}\varepsilon\right) + O\left(\varepsilon^2\right) \tag{5.6.16}$$

CASE II: $\sqrt{\delta_0} = \Omega$:

$$u_0 = a\cos\Omega t + b\sin\Omega t \tag{5.6.17}$$

Substituting (5.6.17) into (5.6.10) yields

$$
\begin{aligned}
\ddot{u}_1 + \Omega^2 u_1 &= -\delta_1 u_0 - 2\Omega\dot{u}_0\sin\Omega t \\
&- \Omega^2 u_0\cos\Omega t - \Omega^2 u_0\cos\Omega t \\
&= -\left(\delta_1 + 2\Omega^2\cos\Omega t\right)u_0 - 2\Omega\dot{u}_0\sin\Omega t \\
&= -\left(\delta_1 + 2\Omega^2\cos\Omega t\right)(a\cos\Omega t + b\sin\Omega t) \\
&- 2\Omega^2\sin\Omega t(-a\sin\Omega t + b\cos\Omega t) \\
&= -\delta_1(a\cos\Omega t + b\sin\Omega t) \\
&- \Omega^2 a(1 + \cos 2\Omega t) - \Omega^2 b\sin 2\Omega t \\
&+ \Omega^2 a(1 - \cos 2\Omega t) - \Omega^2 b\sin 2\Omega t \\
&= -\delta_1(a\cos\Omega t + b\sin\Omega t) - 2\Omega^2 a\cos 2\Omega t
\end{aligned}
$$

$$- 2\Omega^2 b \sin 2\Omega t \tag{5.6.18}$$

In order to eliminate secular terms from the above equation, there must be

$$\delta_1 = 0 \tag{5.6.19}$$

Thus, the solution of (5.6.18) is

$$u_1 = \frac{2}{3} a \cos 2\Omega t + \frac{2}{3} b \sin 2\Omega t \tag{5.6.20}$$

Substituting (5.6.17) and (5.6.20) into (5.6.11), we get

$$\begin{aligned}
\ddot{u}_2 + \Omega^2 u_2 &= -\delta_2 u_0 - 2\Omega^2 u_0 \cos^2 \Omega t - \Omega \dot{u}_0 \sin 2\Omega t \\
&\quad - 2\Omega^2 u_1 \cos \Omega t - 2\Omega \dot{u}_1 \sin \Omega t \\
&= -\delta_2 (a \cos \Omega t + b \sin \Omega t) \\
&\quad - 2\Omega^2 (a \cos \Omega t + b \sin \Omega t) \cos^2 \Omega t \\
&\quad - \Omega^2 (-a \sin \Omega t + b \cos \Omega t) \sin 2\Omega t \\
&\quad - 2\Omega^2 \left(\frac{2}{3} a \cos 2\Omega t + \frac{2}{3} b \sin 2\Omega t \right) \cos \Omega t \\
&\quad - 4\Omega^2 \left(-\frac{2}{3} a \sin 2\Omega t + \frac{2}{3} b \cos 2\Omega t \right) \sin \Omega t \\
&= -\left(\delta_2 + \frac{1}{3} \Omega^2 \right) a \cos \Omega t - \left(\delta_2 + 3\Omega^2 \right) b \sin \Omega t + NST
\end{aligned} \tag{5.6.21}$$

In order to eliminate secular terms from the above equation, there must be

$$\delta_2 = -\frac{1}{3} \Omega^2 \quad \text{or} \quad \delta_2 = -3\Omega^2 \tag{5.6.22}$$

Therefore, the transition curves emanating from $\omega_0^2 = \Omega^2$ is

$$\omega_0^2 = g l^{-1} = \Omega^2 \left(1 - \frac{1}{3} \varepsilon^2 \right) + O(\varepsilon^3) \tag{5.6.23}$$

or

$$\omega_0^2 = g l^{-1} = \Omega^2 \left(1 - 3\varepsilon^2 \right) + O(\varepsilon^3) \tag{5.6.24}$$

(d) Expanding (5.6.4) near $\theta = 0$ and retaining to $O(\theta^3)$ yields

$$\ddot{\theta} - \frac{2\dot{y}\dot{\theta}}{l - y} + \frac{g - \ddot{y}}{l - y} \left(\theta - \frac{\theta^3}{6} \right) = 0 \tag{5.6.25}$$

When $y = \varepsilon l \cos \Omega t$, , the nonlinear Eq. (5.6.25) becomes

$$\ddot{\theta} + \frac{2\varepsilon\Omega \sin \Omega t}{1 - \varepsilon \cos \Omega t}\dot{\theta} + \frac{\omega_0^2 + \varepsilon\Omega^2 \cos \Omega t}{1 - \varepsilon \cos \Omega t}\left(\theta - \frac{\theta^3}{6}\right) = 0, \quad \omega_0^2 = \frac{g}{l} \tag{5.6.26}$$

When $\theta = O\left(\varepsilon^{1/2}\right)$, let

$$\theta = \hat{\varepsilon} u, \quad \hat{\varepsilon} = \varepsilon^{1/2} \tag{5.6.27}$$

Equation (5.6.26) becomes

$$\ddot{u} + \frac{2\hat{\varepsilon}^2\Omega \sin \Omega t}{1 - \hat{\varepsilon}^2 \cos \Omega t}\dot{u} + \frac{\omega_0^2 + \hat{\varepsilon}^2\Omega^2 \cos \Omega t}{1 - \hat{\varepsilon}^2 \cos \Omega t}\left(u - \hat{\varepsilon}^2\frac{u^3}{6}\right) = 0 \tag{5.6.28}$$

In order to determine the effect of nonlinear terms on the stability of the equation, we use the method of multiple scales to solve (5.6.28). Setting

$$\begin{aligned} u(t;\varepsilon) = {} & u_0(T_0, T_1, T_2) + \hat{\varepsilon} u_1(T_0, T_1, T_2) \\ & + \hat{\varepsilon}^2 u_2(T_0, T_1, T_2) + \cdots \end{aligned} \tag{5.6.29}$$

where $T_n = \hat{\varepsilon}^n t$. Substituting the above equation into (5.6.28) and retaining to $O(\hat{\varepsilon}^2)$, we get

$$\begin{aligned} 0 &= \ddot{u} + \frac{2\hat{\varepsilon}^2\Omega \sin \Omega t}{1 - \hat{\varepsilon}^2 \cos \Omega t}\dot{u} + \frac{\omega_0^2 + \hat{\varepsilon}^2\Omega^2 \cos \Omega t}{1 - \hat{\varepsilon}^2 \cos \Omega t}\left(u - \hat{\varepsilon}^2\frac{u^3}{6}\right) \\ &= \left[D_0^2 + 2\hat{\varepsilon}D_0D_1 + \hat{\varepsilon}^2\left(D_1^2 + 2D_0D_2\right)\right]\left(u_0 + \hat{\varepsilon}u_1 + \hat{\varepsilon}^2u_2\right) \\ &\quad + 2\hat{\varepsilon}^2\Omega \sin \Omega T_0\left(D_0 + \hat{\varepsilon}D_1\right)\left(u_0 + \hat{\varepsilon}u_1 + \hat{\varepsilon}^2u_2\right) \\ &\quad + \left(\omega_0^2 + \hat{\varepsilon}^2\Omega^2 \cos \Omega T_0\right)\left(1 + \hat{\varepsilon}^2 \cos \Omega T_0\right) \\ &\quad \left(u_0 + \hat{\varepsilon}u_1 + \hat{\varepsilon}^2u_2 - \frac{1}{6}\hat{\varepsilon}^2u_0^3\right) + \cdots \\ &= D_0^2u_0 + \hat{\varepsilon}D_0^2u_1 + \hat{\varepsilon}^2D_0^2u_2 + 2\hat{\varepsilon}D_0D_1u_0 \\ &\quad + 2\hat{\varepsilon}^2D_0D_1u_1 + \hat{\varepsilon}^2\left(D_1^2 + 2D_0D_2\right)u_0 \\ &\quad + 2\hat{\varepsilon}^2\Omega \sin \Omega T_0D_0u_0 + \omega_0^2u_0 + \hat{\varepsilon}\omega_0^2u_1 \\ &\quad + \hat{\varepsilon}^2\omega_0^2u_2 - \frac{1}{6}\hat{\varepsilon}^2\omega_0^2u_0^3 \\ &\quad + \hat{\varepsilon}^2\omega_0^2u_0 \cos \Omega T_0 + \hat{\varepsilon}^2\Omega^2u_0 \cos \Omega T_0 + \cdots \\ &= D_0^2u_0 + \omega_0^2u_0 + \hat{\varepsilon}\left(D_0^2u_1 + \omega_0^2u_1 + 2D_0D_1u_0\right) \\ &\quad + \hat{\varepsilon}^2[D_0^2u_2 + \omega_0^2u_2 + \left(D_1^2 + 2D_0D_2\right)u_0 \\ &\quad + 2D_0D_1u_1 - \frac{1}{6}\omega_0^2u_0^3 + 2\Omega \sin \Omega T_0D_0u_0 \end{aligned}$$

$$+ \omega_0^2 u_0 \cos \Omega T_0 + \Omega^2 u_0 \cos \Omega T_0] + \cdots$$

Equating coefficients of like powers of $\hat{\varepsilon}$ in the above equation yields

$$D_0^2 u_0 + \omega_0^2 u_0 = 0 \tag{5.6.30}$$

$$D_0^2 u_1 + \omega_0^2 u_1 = -2 D_0 D_1 u_0 \tag{5.6.31}$$

$$\begin{aligned} D_0^2 u_2 + \omega_0^2 u_2 = & -\left(D_1^2 + 2 D_0 D_2\right) u_0 - 2 D_0 D_1 u_1 + \frac{1}{6} \omega_0^2 u_0^3 \\ & - 2\Omega \sin \Omega T_0 D_0 u_0 - \omega_0^2 u_0 \cos \Omega T_0 - \Omega^2 u_0 \cos \Omega T_0 \end{aligned} \tag{5.6.32}$$

The solution of (5.6.30) is

$$u_0 = A e^{i\omega_0 T_0} + \bar{A} e^{-i\omega_0 T_0} \tag{5.6.33}$$

where $A = A(T_1, T_2)$. Substituting u_0 into (5.6.31) yields

$$D_0^2 u_1 + \omega_0^2 u_1 = -2 i \omega_0 D_1 A e^{i\omega_0 T_0} + cc \tag{5.6.34}$$

In order to eliminate secular terms from the above equation, there must be

$$D_1 A = 0 \quad \Rightarrow \quad A = A(T_2) \tag{5.6.35}$$

Thus

$$u_1 = 0 \tag{5.6.36}$$

Substitute u_0, u_1 into (5.6.32), we can obtain

$$\begin{aligned} D_0^2 u_2 + \omega_0^2 u_2 & = -\left(D_1^2 + 2 D_0 D_2\right) u_0 - 2\Omega \sin \Omega T_0 D_0 u_0 \\ & \quad - \omega_0^2 u_0 \cos \Omega T_0 - \Omega^2 u_0 \cos \Omega T_0 + \frac{1}{6} \omega_0^2 u_0^3 \\ & = -2 i \omega_0 D_2 A e^{i\omega_0 T_0} + \omega_0 \Omega e^{i(\Omega - \omega_0) T_0} \bar{A} \\ & \quad - \frac{1}{2}\left(\omega_0^2 + \Omega^2\right) e^{i(\Omega - \omega_0) T_0} \bar{A} + \frac{1}{2} \omega_0^2 A^2 \bar{A} e^{i\omega_0 T_0} + cc \end{aligned} \tag{5.6.37}$$

When $\omega_0 = \sqrt{g l^{-1}} \approx \frac{1}{2}\Omega$, we use the detuning parameter σ to quantitatively describe the nearness of Ω to $2\omega_0$, i.e.,

$$\Omega = 2\omega_0 + \hat{\varepsilon}^2 \sigma \tag{5.6.38}$$

So (5.6.37) becomes

$$D_0^2 u_2 + \omega_0^2 u_2 = -i\Omega D_2 A e^{i\omega_0 T_0} - \frac{1}{8}\Omega^2 \bar{A} e^{i\sigma T_2} e^{i\omega_0 T_0} + \frac{1}{8}\Omega^2 A^2 \bar{A} e^{i\omega_0 T_0} \tag{5.6.39}$$

In order to eliminate secular terms from the above equation, there must be

$$-iD_2 A - \frac{1}{8}\Omega \bar{A} e^{i\sigma T_2} + \frac{1}{8}\Omega A^2 \bar{A} = 0 \tag{5.6.40}$$

Let $A = \frac{1}{2} a e^{i\beta}$, we get

$$-ia' + a\beta' - \frac{1}{8}\Omega a e^{i(\sigma T_2 - 2\beta)} + \frac{1}{32}\Omega a^3 = 0 \tag{5.6.41}$$

Separating the real and imaginary parts of the above equation yields

$$a' = -\frac{1}{8}\Omega a \sin(\sigma T_2 - 2\beta)\beta' = -\frac{1}{32}\Omega a^2 + \frac{1}{8}\Omega \cos(\sigma T_2 - 2\beta)$$

Let $\gamma = \sigma T_2 - 2\beta$, the above equation becomes

$$a' = -\frac{1}{8}\Omega a \sin\gamma \tag{5.6.42}$$

$$\gamma' = \sigma + \frac{1}{16}\Omega a^2 - \frac{1}{4}\Omega \cos\gamma \tag{5.6.43}$$

Therefore, the first order approximate solution of (5.6.28) is

$$u_0 = a\cos[\frac{1}{2}\Omega t - \frac{1}{2}\gamma)] + O(\varepsilon^{1/2}) \tag{5.6.44}$$

Let $a' = \gamma' = 0$ in (5.6.42) and (5.6.43), we can obtain

$$\frac{1}{16}\Omega a \sin\gamma = 0 \tag{5.6.45}$$

$$\sigma + \frac{1}{16}\Omega a^2 - \frac{1}{4}\Omega \cos\gamma = 0 \tag{5.6.46}$$

Therefore, the steady state solution is

$$a = 0, \quad \cos\gamma = 4\sigma\Omega^{-1} \tag{5.6.47}$$

The condition for the existence of a steady state solution of γ is

$$|\sigma| \leq \frac{1}{4}\Omega \tag{5.6.48}$$

Therefore, the critical value of σ for the existence of steady state solution is

$$\sigma = \pm\frac{1}{4}\Omega \tag{5.6.49}$$

Substituting this into (5.6.38), we can obtain the critical condition for the existence of a steady state solution

$$\omega_0^2 = \frac{1}{4}\Omega^2\left(1 \pm \frac{1}{2}\hat{\varepsilon}^2\right) \tag{5.6.50}$$

i.e.,

$$gl^{-1} = \frac{1}{4}\Omega^2\left(1 \pm \frac{1}{2}\varepsilon\right) \tag{5.6.51}$$

Combining (5.6.51), we know that the effect of the nonlinear terms is to limit the unstable linear motion to a finite-amplitude motion, whose frequency is one half the frequency of the excitation. In other words, a subharmonic is generated by the system.

5.7 Exercise 5.7 (Linear and Nonlinear Stability Analysis of a Particle Sliding on a Rotating Parabola)

Solution: (a) The kinetic energy of the system is

$$\begin{aligned} T &= \frac{1}{2}m\left(\dot{x}^2 + \dot{z}^2 + \Omega^2 x^2\right) \\ &= \frac{1}{2}m\left[\left(1 + 4p^2x^2\right)\dot{x}^2 + \Omega^2 x^2\right] \end{aligned} \tag{5.7.1}$$

The kinetic energy of the system is

$$V = mgz = mgpx^2 \tag{5.7.2}$$

Substituting (5.7.1) and (5.7.2) into the Lagrange's equation, we can obtain the differential equation of motion of the system

$$m\left(1 + 4p^2x^2\right)\ddot{x} + 4mp^2x\dot{x}^2 - m\Omega^2 x = -2mgpx$$

i.e.,

$$\left(1 + 4p^2x^2\right)\ddot{x} + 4p^2x\dot{x}^2 + \left(2pg - \Omega^2\right)x = 0 \tag{5.7.3}$$

When $\Omega = \Omega_0(1 + \varepsilon\cos\omega t)$, the above equation becomes.
i.e., $\left(1 + 4p^2x^2\right)\ddot{x} + 4p^2x\dot{x}^2 + \left(2pg - \Omega_0^2\right)x - \Omega_0^2\left(2\varepsilon\cos\omega t + \varepsilon^2\cos^2\omega t\right)x = 0$.
Let $\omega_0^2 = 2gp - \Omega_0^2$, we get

$$\left(1 + 4p^2x^2\right)\ddot{x} + 4p^2\dot{x}^2x + \omega_0^2x - \Omega_0^2\left(2\varepsilon\cos\omega t + \varepsilon^2\cos^2\omega t\right)x = 0 \tag{5.7.4}$$

(b) Linearizing the above equation near $x = 0$, we get

$$\ddot{x} + \omega_0^2x - \Omega_0^2\left(2\varepsilon\cos\omega t + \varepsilon^2\cos^2\omega t\right)x = 0 \tag{5.7.5}$$

We use the method of strained parameters to obtain the transition curve. Let the solution of Eq. (5.7.5) and parameter ω_0^2 have the following expansion:

$$x(t;\varepsilon) = u_0(t) + \varepsilon u_1(t) + \varepsilon^2u_2(t) + \cdots \tag{5.7.6}$$

$$\omega_0^2 = \delta_0 + \varepsilon\delta_1 + \varepsilon^2\delta_2 + \cdots \tag{5.7.7}$$

Substituting the above two equations into (5.7.5) and retaining to $O(\varepsilon^2)$, we get

$$\begin{aligned}
0 &= \ddot{x} + \omega_0^2x - \Omega_0^2\left(2\varepsilon\cos\omega t + \varepsilon^2\cos^2\omega t\right)x \\
&= \ddot{u}_0 + \varepsilon\ddot{u}_1 + \varepsilon^2\ddot{u}_2 \\
&\quad + \left(\delta_0 + \varepsilon\delta_1 + \varepsilon^2\delta_2\right)\left(u_0 + \varepsilon u_1 + \varepsilon^2u_2\right) \\
&\quad - \Omega_0^2\left(2\varepsilon\cos\omega t + \varepsilon^2\cos^2\omega t\right)\left(u_0 + \varepsilon u_1 + \varepsilon^2u_2\right) \\
&= \ddot{u}_0 + \varepsilon\ddot{u}_1 + \varepsilon^2\ddot{u}_2 + \delta_0u_0 + \varepsilon\delta_0u_1 \\
&\quad + \varepsilon^2\delta_0u_2 + \varepsilon\delta_1u_0 + \varepsilon^2\delta_1u_1 \\
&\quad + \varepsilon^2\delta_2u_0 - 2\varepsilon\Omega_0^2u_0\cos\omega t \\
&\quad - \varepsilon^2\Omega_0^2u_0\cos^2\omega t - 2\varepsilon^2u_1\Omega_0^2\cos\omega t \\
&= \ddot{u}_0 + \delta_0u_0 + \varepsilon\left(\ddot{u}_1 + \delta_0u_1 + \delta_1u_0 - 2\Omega_0^2u_0\cos\omega t\right) \\
&\quad + \varepsilon^2\left(\ddot{u}_2 + \delta_0u_2 + \delta_2u_0 + \delta_1u_1 - \Omega_0^2u_0\cos^2\omega t - 2\Omega_0^2u_1\cos\omega t\right)
\end{aligned}$$

Equating coefficients of like powers of ε in the above equation yields

$$\ddot{u}_0 + \delta_0u_0 = 0 \tag{5.7.8}$$

$$\ddot{u}_1 + \delta_0u_1 = -\delta_1u_0 + 2\Omega_0^2u_0\cos\omega t \tag{5.7.9}$$

$$\ddot{u}_2 + \delta_0 u_2 = -\delta_2 u_0 - \delta_1 u_1 + \Omega_0^2 u_0 \cos^2 \omega t + 2\Omega_0^2 u_1 \cos \omega t \tag{5.7.10}$$

The solution of (5.7.8) is

$$u_0 = a \cos \sqrt{\delta_0} t + b \sin \sqrt{\delta_0} t \tag{5.7.11}$$

CASE I: $\sqrt{\delta_0} = 0$:

$$u_0 = a \tag{5.7.12}$$

Substituting (5.7.12) into (5.7.9) yields

$$\ddot{u}_1 = -\delta_1 a + 2\Omega_0^2 a \cos \omega t \tag{5.7.13}$$

In order to eliminate secular terms from the above equation, there must be

$$\delta_1 = 0 \tag{5.7.14}$$

Thus

$$u_1 = -2\Omega_0^2 \omega^{-2} a \cos \omega t \tag{5.7.15}$$

Substituting $u_0, \quad u_1$ into (5.7.10) yields

$$\begin{aligned}
\ddot{u}_2 &= -\delta_2 a + \Omega_0^2 a \cos^2 \omega t - 4\Omega_0^4 \omega^{-2} a \cos^2 \omega t \\
&= -\delta_2 a + \frac{1}{2}\Omega_0^2 a (1 + \cos 2\omega t) \\
&\quad - 2\Omega_0^4 \omega^{-2} a (1 + \cos 2\omega t) \\
&= -\left(\delta_2 - \frac{1}{2}\Omega_0^2 + 2\Omega_0^4 \omega^{-2}\right) a \\
&\quad + \left(\frac{1}{2}\Omega_0^2 - 2\Omega_0^4 \omega^{-2}\right) a \cos 2\omega t
\end{aligned} \tag{5.7.16}$$

In order to eliminate secular terms from the above equation, there must be

$$\delta_2 - \frac{1}{2}\Omega_0^2 + 2\Omega_0^4 \omega^{-2} \quad \Rightarrow \quad \delta_2 = \frac{1}{2}\Omega_0^2 - 2\Omega_0^4 \omega^{-2} \tag{5.7.17}$$

Therefore, the transition curves emanating from $\omega_0 = 0$ is

$$\omega_0^2 = \frac{1}{2}\varepsilon^2 \Omega_0^2 (1 - 4\Omega_0^2 \omega^{-2}) + O(\varepsilon^3) \tag{5.7.18}$$

CASE II: $\sqrt{\delta_0} = \frac{1}{2}\omega$:

$$u_0 = a\cos\frac{1}{2}\omega t + b\sin\frac{1}{2}\omega t \tag{5.7.19}$$

Substituting (5.7.19) into (5.7.9) yields

$$\begin{aligned}\ddot{u}_1 + \frac{1}{4}\omega^2 u_1 &= -\delta_1\left(a\cos\frac{1}{2}\omega t + b\sin\frac{1}{2}\omega t\right)\\ &\quad + 2\Omega_0^2\left(a\cos\frac{1}{2}\omega t + b\sin\frac{1}{2}\omega t\right)\cos\omega t\\ &= -\left(\delta_1 - \Omega_0^2\right)a\cos\frac{1}{2}\\ &\quad - \left(\delta_1 + \Omega_0^2\right)b\sin\frac{1}{2}\omega t + NST\end{aligned} \tag{5.7.20}$$

In order to eliminate secular terms from the above equation, there must be

$$\delta_1 = \pm\Omega_0^2 \tag{5.7.21}$$

Therefore, the transition curves emanating from $\omega_0 = \frac{1}{2}\omega$ is

$$\omega_0^2 = \frac{1}{4}\omega^2 \pm \varepsilon\Omega_0^2 + O\left(\varepsilon^2\right) \tag{5.7.22}$$

CASE III: $\sqrt{\delta_0} = \omega$:

$$u_0 = a\cos\frac{1}{2}\omega t + b\sin\frac{1}{2}\omega t \tag{5.7.23}$$

Substituting (5.7.23) into (5.7.9) yields

$$\begin{aligned}\ddot{u}_1 + \frac{1}{4}\omega^2 u_1 &= -\delta_1\left(a\cos\frac{1}{2}\omega t + b\sin\frac{1}{2}\omega t\right)\\ &\quad + 2\Omega_0^2\left(a\cos\frac{1}{2}\omega t + b\sin\frac{1}{2}\omega t\right)\cos\omega t\\ &= -\left(\delta_1 - \Omega_0^2\right)a\cos\frac{1}{2} - \left(\delta_1 + \Omega_0^2\right)b\sin\frac{1}{2}\omega t + NST\end{aligned} \tag{5.7.24}$$

In order to eliminate secular terms from the above equation, there must be

$$\delta_1 = \pm\Omega_0^2 \tag{5.7.25}$$

Therefore, the transition curves emanating from $\omega_0 = \omega$ is

$$\omega_0^2 = \omega^2 \pm \varepsilon\Omega_0^2 + O(\varepsilon^2) \tag{5.7.26}$$

(c) In the neighborhood of $x = 0$, the nonlinear equation of the system (5.7.4) has the same form. When $\theta = O(\varepsilon^{1/2})$, let

$$x = \hat{\varepsilon}u, \quad \hat{\varepsilon} = \varepsilon^{1/2} \tag{5.7.27}$$

Equation (5.7.4) becomes

$$\begin{aligned} &(1 + 4p^2\hat{\varepsilon}^2u^2)\ddot{u} + 4p^2\hat{\varepsilon}^2\dot{u}^2u + \omega_0^2u \\ &- \Omega_0^2(2\hat{\varepsilon}^2\cos\omega t + \hat{\varepsilon}^4\cos^2\omega t)u = 0 \end{aligned} \tag{5.7.28}$$

In order to determine the effect of the nonlinear terms on the stability of the solution, we use the method of multiple scales to solve (5.7.28). Setting

$$u(t;\varepsilon) = u_0(T_0, T_1, T_2) + \hat{\varepsilon}u_1(T_0, T_1, T_2) + \hat{\varepsilon}^2u_2(T_0, T_1, T_2) + \cdots \tag{5.7.29}$$

where $T_n = \hat{\varepsilon}^n t$. Substituting the above equation into (5.7.28) and retaining to $O(\hat{\varepsilon}^2)$, we get

$$\begin{aligned} 0 &= (1 + 4p^2\hat{\varepsilon}^2u^2)\ddot{u} + 4p^2\hat{\varepsilon}^2\dot{u}^2u + \omega_0^2u \\ &\quad - \Omega_0^2(2\hat{\varepsilon}^2\cos\omega t + \hat{\varepsilon}^4\cos^2\omega t)u \\ &= (1 + 4p^2\hat{\varepsilon}^2u_0^2)\left[D_0^2 + 2\hat{\varepsilon}D_0D_1 + \hat{\varepsilon}^2(D_1^2 + 2D_0D_2)\right] \\ &\quad (u_0 + \hat{\varepsilon}u_1 + \hat{\varepsilon}^2u_2) \\ &\quad + 4p^2\hat{\varepsilon}^2(D_0u_0)^2u_0 + \omega_0^2(u_0 + \hat{\varepsilon}u_1 + \hat{\varepsilon}^2u_2) \\ &\quad - \Omega_0^2(2\hat{\varepsilon}^2\cos\omega t + \hat{\varepsilon}^4\cos^2\omega t)(u_0 + \hat{\varepsilon}u_1 + \hat{\varepsilon}^2u_2) + \cdots \\ &= D_0^2u_0 + \hat{\varepsilon}D_0^2u_1 + \hat{\varepsilon}^2D_0^2u_2 + 2\hat{\varepsilon}D_0D_1u_0 + 2\hat{\varepsilon}^2D_0D_1u_1 \\ &\quad + \hat{\varepsilon}^2(D_1^2 + 2D_0D_2)u_0 + 4p^2\hat{\varepsilon}^2u_0^2D_0^2u_0 \\ &\quad + 4p^2\hat{\varepsilon}^2(D_0u_0)^2u_0 \\ &\quad + \omega_0^2u_0 + \hat{\varepsilon}\omega_0^2u_1 + \hat{\varepsilon}^2\omega_0^2u_2 - 2\hat{\varepsilon}^2\Omega_0^2u_0\cos\omega t + \cdots \\ &= D_0^2u_0 + \omega_0^2u_0 + \hat{\varepsilon}(D_0^2u_1 + \omega_0^2u_1 + 2D_0D_1u_0) \\ &\quad + \hat{\varepsilon}^2[D_0^2u_2 + \omega_0^2u_2 + (D_1^2 + 2D_0D_2)u_0 + 2D_0D_1u_1 \\ &\quad + 4p^2u_0^2D_0^2u_0 + 4p^2(D_0u_0)^2u_0 - 2\Omega_0^2u_0\cos\omega t] + \cdots \end{aligned}$$

Equating coefficients of like powers of $\hat{\varepsilon}$ in the above equation yields

$$D_0^2u_0 + \omega_0^2u_0 = 0 \tag{5.7.30}$$

$$D_0^2u_1 + \omega_0^2u_1 = -2D_0D_1u_0 \tag{5.7.31}$$

$$
\begin{aligned}
D_0^2 u_2 + \omega_0^2 u_2 = & -\left(D_1^2 + 2D_0 D_2\right) u_0 - 2D_0 D_1 u_1 \\
& - 4p^2 u_0^2 D_0^2 u_0 - 4p^2 (D_0 u_0)^2 u_0 \\
& + 2\Omega_0^2 u_0 cos\omega t
\end{aligned} \tag{5.7.32}
$$

The solution of (5.7.30) is

$$
u_0 = Ae^{i\omega_0 T_0} + \overline{A}\, e^{-i\omega_0 T_0} \tag{5.7.33}
$$

where $A = A(T_1, T_2)$. Substituting u_0 into (5.7.31) yields

$$
D_0^2 u_1 + \omega_0^2 u_1 = -2i\omega_0 D_1 A e^{i\omega_0 T_0} + cc \tag{5.7.34}
$$

In order to eliminate secular terms from the above equation, there must be

$$
D_1 A = 0 \quad \Rightarrow \quad A = A(T_2) \tag{5.7.35}
$$

Thus

$$
u_1 = 0 \tag{5.7.36}
$$

Substitute u_0, u_1 into (5.7.32) yields

$$
\begin{aligned}
D_0^2 u_2 + \omega_0^2 u_2 = & -\left(D_1^2 + 2D_0 D_2\right) u_0 - 4p^2 u_0^2 D_0^2 u_0 \\
& - 4p^2 (D_0 u_0)^2 u_0 + 2\Omega_0^2 u_0 \cos \omega t \\
= & -2i\omega_0 D_2 A e^{i\omega_0 T_0} + 8p^2 \omega_0^2 A^2 \overline{A} e^{i\omega_0 T_0} \\
& + \Omega_0^2 \overline{A} e^{i(\omega - \omega_0) T_0} + cc + NST
\end{aligned} \tag{5.7.37}
$$

When $\omega_0 \approx \frac{1}{2}\omega$, we use the detuning parameter σ to quantitatively describe the nearness of ω to $2\omega_0$, i.e.,

$$
\omega = 2\omega_0 + \hat{\varepsilon}^2 \sigma \tag{5.7.38}
$$

So the Eq. (5.7.37) becomes

$$
\begin{aligned}
D_0^2 u_2 + \omega_0^2 u_2 = & -2i\omega_0 D_2 A e^{i\omega_0 T_0} + 8p^2 \omega_0^2 A^2 \overline{A} e^{i\omega_0 T_0} \\
& + \Omega_0^2 \overline{A} e^{i\sigma T_2} e^{i\omega_0 T_0} + cc + NST
\end{aligned} \tag{5.7.39}
$$

In order to eliminate secular terms from the above equation, there must be

$$
-2iD_2 A + 8p^2 \omega_0 A^2 \overline{A} + \Omega_0^2 \omega_0^{-1} \overline{A}\, e^{i\sigma T_2} = 0 \tag{5.7.40}
$$

Let $A = \frac{1}{2}ae^{i\beta}$, we get

$$-ia' + a\beta' + p^2\omega_0 a^3 + \frac{1}{2}\Omega_0^2\omega_0^{-1}ae^{i(\sigma T_2 - 2\beta)} = 0 \tag{5.7.41}$$

Separating the real and imaginary parts of the above equation yields

$$a' = \frac{1}{2}\Omega_0^2\omega_0^{-1}a\sin(\sigma T_2 - 2\beta)$$

$$\beta' = p^2\omega_0 a^2 - \frac{1}{2}\Omega_0^2\omega_0^{-1}\cos(\sigma T_2 - 2\beta)$$

Let $\gamma = \sigma T_2 - 2\beta$, the above equation becomes

$$a' = \frac{1}{2}\Omega_0^2\omega_0^{-1}a\sin\gamma \tag{5.7.42}$$

$$\gamma' = \sigma - 2p^2\omega_0 a^2 + \Omega_0^2\omega_0^{-1}\cos\gamma \tag{5.7.43}$$

Therefore, the first order approximate solution of (5.7.28) is

$$u_0 = a\cos[\frac{1}{2}(\Omega - \varepsilon\sigma)t + \frac{1}{2}\gamma)] + O(\varepsilon^{1/2}) \tag{5.7.44}$$

Let $a' = \gamma' = 0$ in (5.7.42) and (5.7.43), we can obtain

$$\frac{1}{2}\Omega_0^2\omega_0^{-1}a\sin\gamma = 0 \tag{5.7.45}$$

$$\sigma - 2p^2\omega_0 a^2 + \Omega_0^2\omega_0^{-1}\cos\gamma = 0 \tag{5.7.46}$$

Therefore, the steady state solution is

$$a = 0, \quad \cos\gamma = -\sigma\Omega_0^2\omega_0 \tag{5.7.47}$$

The condition for the existence of a steady state solution of γ is

$$|\sigma| \le \Omega_0^2\omega_0^{-1} \tag{5.7.48}$$

Therefore, the critical value of σ for the existence of steady state solution is

$$\sigma = \pm\Omega_0^2\omega_0^{-1} \tag{5.7.49}$$

Substituting this into (5.7.38), we can obtain the critical condition for the existence of a steady state solution

$$\omega_0^2 = \frac{1}{4}\omega^2 \pm \varepsilon\Omega_0^2 + O(\varepsilon^2) \tag{5.7.50}$$

Combining (5.7.22) and (5.7.50), we know that the effect of the nonlinear terms is to limit the unstable linear motion to a finite-amplitude motion, whose frequency is one half the frequency of the excitation. In other words, a subharmonic is generated by the system.

5.8 Exercise 5.8 (Nonlinear Solution to the Pure Rolling of a Cylinder on a Circular Surface)

Solution: **(a)** The kinetic energy of the system is

$$\begin{aligned} T &= \frac{1}{2}M\left(\dot{x}^2 + \dot{y}^2\right) \\ &+ \frac{1}{2}m\{[\dot{x} + (R-r)\dot{\theta}\cos\theta]^2 + [\dot{y} + (R-r)\dot{\theta}\sin\theta]^2\} \\ &+ \frac{1}{2}\left(\frac{1}{2}mr^2\right)[\frac{(R-r)\dot{\theta}}{r}]^2 \\ &= \frac{1}{2}M\left(\dot{x}^2 + \dot{y}^2\right) + \frac{1}{2}m\Big\{\left(\dot{x}^2 + \dot{y}^2\right) + \frac{3}{2}(R-r)^2\dot{\theta}^2 \\ &+ 2(R-r)\dot{x}\dot{\theta}\cos\theta + 2(R-r)\dot{y}\dot{\theta}\sin\theta\Big\} \end{aligned} \tag{5.8.1}$$

where M denotes the mass of the cylinder and the block. The kinetic energy of the system is

$$V = Mgy + mg(R-r)(1-\cos\theta) \tag{5.8.2}$$

Substituting (5.8.1) and (5.8.2) into the Lagrange's equation, we can obtain the differential equation of motion of the system

$$m\,[\frac{3}{2}(R-r)^2\ddot{\theta} + (R-r)\ddot{x}\cos\theta + (R-r)\ddot{y}\sin\theta] = -mg(R-r)\sin\theta$$

i.e.,

$$\ddot{\theta} + \frac{2}{3}(R - r)^{-1}(g + \ddot{y}) \sin \theta + \frac{2}{3}(R - r)^{-1}\ddot{x} \cos \theta = 0 \tag{5.8.3}$$

Expanding the above equation near $\theta = 0$ and retaining to $O(\theta^3)$ yields

$$\ddot{\theta} + \frac{2}{3}(R - r)^{-1}(g + \ddot{y})\left(\theta - \frac{1}{6}\theta^3\right) + \frac{2}{3}(R - r)^{-1}\ddot{x}\left(1 - \frac{1}{2}\theta^2\right) = 0 \tag{5.8.4}$$

(b) When $\ddot{x} = \varepsilon K \cos \Omega t$, $\ddot{y} = 0$,, the Eq. (5.8.4) becomes

$$\ddot{\theta} + \frac{2}{3}(R - r)^{-1}g\left(\theta - \frac{1}{6}\theta^3\right) + \frac{2}{3}\varepsilon(R - r)^{-1}K\left(1 - \frac{1}{2}\theta^2\right) \cos \Omega t = 0 \tag{5.8.5}$$

When $\theta = O\left(\varepsilon^{1/2}\right)$, let

$$x = \hat{\varepsilon}u, \quad \hat{\varepsilon} = \varepsilon^{1/2} \tag{5.8.6}$$

Equation (5.8.5) becomes

$$\ddot{u} + \frac{2}{3}(R - r)^{-1}g\left(u - \frac{1}{6}\hat{\varepsilon}^2u^3\right) + \frac{2}{3}\hat{\varepsilon}(R - r)^{-1}K\left(1 - \frac{1}{2}\hat{\varepsilon}^2u^2\right) \cos \Omega t = 0 \tag{5.8.7}$$

Let

$$\omega_0^2 = \frac{2}{3}(R - r)^{-1}g, \quad \hat{K} = K\omega_0^2g^{-1} \tag{5.8.8}$$

Equation (5.8.7) becomes

$$\ddot{u} + \omega_0^2u - \frac{1}{6}\hat{\varepsilon}^2\omega_0^2u^3 - \frac{1}{2}\hat{\varepsilon}^3\hat{K}u^2 \cos \Omega t = -\hat{\varepsilon}\hat{K} \cos \Omega t \tag{5.8.9}$$

We use the method of multiple scales to solve (5.8.9). Let

$$u(t; \varepsilon) = u_0(T_0, T_1, T_2) + \hat{\varepsilon}u_1(T_0, T_1, T_2) + \hat{\varepsilon}^2u_2(T_0, T_1, T_2) + \cdots \tag{5.8.10}$$

where $T_n = \hat{\varepsilon}^n t$.. Substituting the above equation into (5.8.9) and retaining to $O(\hat{\varepsilon}^2)$, we get

$$
\begin{aligned}
0 &= \ddot{u} + \omega_0^2 u - \frac{1}{6}\hat{\varepsilon}^2\omega_0^2 u^3 - \frac{1}{2}\hat{\varepsilon}^3\hat{K}u^2\cos\Omega t + \hat{\varepsilon}\hat{K}\cos\Omega t \\
&= \left[D_0^2 + 2\hat{\varepsilon}D_0D_1 + \hat{\varepsilon}^2\left(D_1^2 + 2D_0D_2\right)\right]\left(u_0 + \hat{\varepsilon}u_1 + \hat{\varepsilon}^2 u_2\right) \\
&\quad + \omega_0^2\left(u_0 + \hat{\varepsilon}u_1 + \hat{\varepsilon}^2 u_2\right) - \frac{1}{6}\hat{\varepsilon}^2\omega_0^2\left(u_0 + \hat{\varepsilon}u_1 + \hat{\varepsilon}^2 u_2\right)^3 \\
&\quad - \frac{1}{2}\hat{\varepsilon}^3\hat{K}\left(u_0 + \hat{\varepsilon}u_1 + \hat{\varepsilon}^2 u_2\right)^2\cos\Omega t + \hat{\varepsilon}\hat{K}\cos\Omega t \\
&= D_0^2 u_0 + \hat{\varepsilon}D_0^2 u_1 + \hat{\varepsilon}^2 D_0^2 u_2 + 2\hat{\varepsilon}D_0D_1u_0 \\
&\quad + 2\hat{\varepsilon}^2 D_0D_1u_1 + \hat{\varepsilon}^2\left(D_1^2 + 2D_0D_2\right)u_0 + \omega_0^2 u_0 \\
&\quad + \hat{\varepsilon}\omega_0^2 u_1 + \hat{\varepsilon}^2\omega_0^2 u_2 - \frac{1}{6}\hat{\varepsilon}^2\omega_0^2 u_0^3 + \hat{\varepsilon}\hat{K}\cos\Omega t + \cdots \\
&= D_0^2 u_0 + \omega_0^2 u_0 + \hat{\varepsilon}\left(D_0^2 u_1 + \omega_0^2 u_1 + 2D_0D_1u_0 + \hat{K}\cos\Omega T_0\right) \\
&\quad + \hat{\varepsilon}^2\left[D_0^2 u_2 + \omega_0^2 u_2 + 2D_0D_1u_1 + \left(D_1^2 + 2D_0D_2\right)u_0 - \frac{1}{6}\omega_0^2 u_0^3\right] + \cdots
\end{aligned}
$$

Equating coefficients of like powers of $\hat{\varepsilon}$ in the above equation yields

$$D_0^2 u_0 + \omega_0^2 u_0 = 0 \tag{5.8.11}$$

$$D_0^2 u_1 + \omega_0^2 u_1 = -2D_0D_1u_0 - \hat{K}\cos\Omega T_0 \tag{5.8.12}$$

$$D_0^2 u_2 + \omega_0^2 u_2 = -\left(D_1^2 + 2D_0D_2\right)u_0 - 2D_0D_1u_1 + \frac{1}{6}\omega_0^2 u_0^3 \tag{5.8.13}$$

The solution of (5.8.11) is

$$u_0 = Ae^{i\omega_0 T_0} + \bar{A}\,e^{-i\omega_0 T_0} \tag{5.8.14}$$

where $A = A(T_1, T_2)$. Substituting u_0 into (5.8.12) yields

$$D_0^2 u_1 + \omega_0^2 u_1 = -2i\omega_0 D_0 D_1 A e^{i\omega_0 T_0} - \frac{1}{2}\hat{K}e^{i\Omega T_0} + cc \tag{5.8.15}$$

CASE I: $\Omega \approx \omega_0$:

We use the detuning parameter σ to quantitatively describe the nearness of Ω to ω_0, i.e.,

$$\Omega = \omega_0 + \hat{\varepsilon}\sigma \tag{5.8.16}$$

therefore,

$$D_0^2u_1 + \omega_0^2u_1 = -2i\omega_0D_0D_1Ae^{i\omega_0T_0} - \frac{1}{2}\hat{K}e^{i\sigma T_1}e^{i\omega_0T_0} + cc \tag{5.8.17}$$

In order to eliminate secular terms from the above equation, there must be

$$-2i\omega_0D_1A - \frac{1}{2}\hat{K}e^{i\sigma T_1} = 0 \tag{5.8.18}$$

then

$$A = \frac{1}{4}\hat{K}\omega_0^{-1}\sigma^{-1}e^{i\sigma T_1} \tag{5.8.19}$$

Substituting this into (5.8.14), we can obtain the first order approximate solution of the system

$$\begin{aligned} u &= \frac{1}{2}\hat{K}\omega_0^{-1}\sigma^{-1}\cos(\omega_0T_0 + \sigma T_1) + O(\hat{\varepsilon}) \\ &= \frac{1}{2}\hat{K}\omega_0^{-1}\sigma^{-1}\cos\left(\omega_0 + \varepsilon^{1/2}\sigma\right)t + O\left(\varepsilon^{1/2}\right) \end{aligned} \tag{5.8.20}$$

CASE II: $\Omega \approx 3\omega_0$**:**

Equation (5.8.15) becomes

$$D_0^2u_1 + \omega_0^2u_1 = -2i\omega_0D_0D_1Ae^{i\omega_0T_0} - \frac{1}{2}\hat{K}e^{3i\omega_0T_0} + cc \tag{5.8.21}$$

In order to eliminate secular terms from the above equation, there must be

$$D_1A = 0 \quad \Rightarrow \quad A = A(T_2) \tag{5.8.22}$$

Then the solution of (5.8.21) is

$$u_1 = \frac{1}{16}\omega_0^{-2}\hat{K}e^{3i\omega_0T_0} + cc \tag{5.8.23}$$

Substituting u_0, u_1 into (5.8.13) yields

$$\begin{aligned} D_0^2u_2 + \omega_0^2u_2 &= -\left(D_1^2 + 2D_0D_2\right)u_0 - 2D_0D_1u_1 + \frac{1}{6}\omega_0^2u_0^3 \\ &= -2i\omega_0D_2Ae^{i\omega_0T_0} + \frac{1}{2}\omega_0^2A^2\overline{A}e^{i\omega_0T_0} + cc + NST \end{aligned} \tag{5.8.24}$$

In order to eliminate secular terms from the above equation, there must be

$$-2iD_2A + \frac{1}{2}\omega_0A^2\overline{A} = 0 \tag{5.8.25}$$

Let $A = \frac{1}{2}ae^{i\beta}$, we get

$$-ia' + a\beta' + \frac{1}{16}\omega_0 a^3 = 0 \tag{5.8.26}$$

Separating the real and imaginary parts of the above equation yields

$$\begin{aligned} a' &= 0 \\ \beta' &= -\frac{1}{16}\omega_0 a^2 \end{aligned} \tag{5.8.27}$$

Near $u = 0$, (5.8.27) has the steady state solution

$$a = 0, \quad \beta = 0 \tag{5.8.28}$$

Therefore, the first order approximate solution of the system is

$$u = 0 + O\left(\varepsilon^{1/2}\right) \tag{5.8.29}$$

This result shows that it is not possible to generate the subharmonic resonance response under soft excitation.

CASE III: $\Omega \approx \frac{1}{3}\omega_0$:
A similar process to the one above leads to the first order approximate solution of the system when $\Omega \approx \frac{1}{3}\omega_0$ is

$$u = 0 + O\left(\varepsilon^{1/2}\right) \tag{5.8.30}$$

(The reader is invited to complete the procedure on their own).

This result shows that it is not possible to generate the super harmonic resonance response under soft excitation.

(c) When $\ddot{x} = \varepsilon K \cos \Omega_1 t, \quad \ddot{y} = \varepsilon \cos \Omega_2 t$, (5.8.4) becomes

$$\begin{aligned} &\ddot{\theta} + \frac{2}{3}(R - r)^{-1}(g + \varepsilon \cos \Omega_2 t)\left(\theta - \frac{1}{6}\theta^3\right) \\ &+ \frac{2}{3}\varepsilon(R - r)^{-1}K\left(1 - \frac{1}{2}\theta^2\right)\cos \Omega_1 t = 0 \end{aligned} \tag{5.8.31}$$

When $\theta = O\left(\varepsilon^{1/2}\right)$, let

$$x = \hat{\varepsilon}u, \quad \hat{\varepsilon} = \varepsilon^{1/2} \tag{5.8.32}$$

Equation (5.8.5) becomes

$$\ddot{u} + \omega_0^2 u - \frac{1}{6}\hat{\varepsilon}^2\omega_0^2 u^3 + \hat{\varepsilon}^2 \alpha u \cos\Omega_2 t = -\hat{\varepsilon}\hat{K}\cos\Omega_1 t \tag{5.8.33}$$

$$\ddot{u} + \frac{2}{3}(R-r)^{-1}\left(g + \hat{\varepsilon}^2\cos\Omega_2 t\right)\left(u - \frac{1}{6}\hat{\varepsilon}^2 u^3\right) + \frac{2}{3}\hat{\varepsilon}(R-r)^{-1}K\left(1 - \frac{1}{2}\hat{\varepsilon}^2 u^2\right)\cos\Omega_1 t = 0 \tag{5.8.34}$$

Let

$$\omega_0^2 = \frac{2}{3}(R-r)^{-1}g, \quad \alpha = \frac{2}{3}(R-r)^{-1}, \quad \hat{K} = \frac{2}{3}(R-r)^{-1}K \tag{5.8.35}$$

Substituting the above equation into (5.8.34) and retaining to $O(\hat{\varepsilon}^2)$, it becomes

$$\ddot{u} + \omega_0^2 u - \frac{1}{6}\hat{\varepsilon}^2\omega_0^2 u^3 + \hat{\varepsilon}^2 \alpha u \cos\Omega_2 t = -\hat{\varepsilon}\hat{K}\cos\Omega_1 t \tag{5.8.36}$$

We use the method of multiple scales to solve the Eq. (5.8.36). Set

$$u(t;\varepsilon) = u_0(T_0, T_1, T_2) + \hat{\varepsilon}u_1(T_0, T_1, T_2) + \hat{\varepsilon}^2 u_2(T_0, T_1, T_2) + \cdots \tag{5.8.37}$$

where $T_n = \hat{\varepsilon}^n t$. Substituting the above equation into (5.8.36) and retaining to $O(\hat{\varepsilon}^2)$, we get

$$\begin{aligned}
0 &= \ddot{u} + \omega_0^2 u - \frac{1}{6}\hat{\varepsilon}^2\omega_0^2 u^3 + \hat{\varepsilon}^2\alpha u\cos\Omega_2 t + \hat{\varepsilon}\hat{K}\cos\Omega_1 t \\
&= \left[D_0^2 + 2\hat{\varepsilon}D_0D_1 + \hat{\varepsilon}^2\left(D_1^2 + 2D_0D_2\right)\right]\left(u_0 + \hat{\varepsilon}u_1 + \hat{\varepsilon}^2 u_2\right) \\
&\quad + \omega_0^2\left(u_0 + \hat{\varepsilon}u_1 + \hat{\varepsilon}^2 u_2\right) - \frac{1}{6}\hat{\varepsilon}^2\omega_0^2\left(u_0 + \hat{\varepsilon}u_1 + \hat{\varepsilon}^2 u_2\right)^3 \\
&\quad + \hat{\varepsilon}^2\alpha\left(u_0 + \hat{\varepsilon}u_1 + \hat{\varepsilon}^2 u_2\right)\cos\Omega_2 T_0 + \hat{\varepsilon}\hat{K}\cos\Omega_1 T_0 \\
&\quad + \hat{\varepsilon}^2\alpha\left(u_0 + \hat{\varepsilon}u_1 + \hat{\varepsilon}^2 u_2\right)\cos\Omega_2 T_0 + \hat{\varepsilon}\hat{K}\cos\Omega_1 T_0 \\
&= D_0^2 u_0 + \hat{\varepsilon}D_0^2 u_1 + \hat{\varepsilon}^2 D_0^2 u_2 + 2\hat{\varepsilon}D_0D_1u_0 + 2\hat{\varepsilon}^2 D_0D_1u_1 \\
&\quad + \hat{\varepsilon}^2\left(D_1^2 + 2D_0D_2\right)u_0 + \omega_0^2 u_0 + \hat{\varepsilon}\omega_0^2 u_1 + \hat{\varepsilon}^2\omega_0^2 u_2 \\
&\quad - \frac{1}{6}\hat{\varepsilon}^2\omega_0^2 u_0^3 + \hat{\varepsilon}^2\alpha u_0\cos\Omega_2 T_0 + \hat{\varepsilon}\hat{K}\cos\Omega_1 T_0 + \cdots \\
&= D_0^2 u_0 + \omega_0^2 u_0 + \hat{\varepsilon}\left(D_0^2 u_1 + \omega_0^2 u_1 + 2D_0D_1u_0 + \hat{K}\cos\Omega_1 T_0\right)
\end{aligned}$$

Equating coefficients of like powers of $\hat{\varepsilon}$ in the above equation yields

$$D_0^2 u_0 + \omega_0^2 u_0 = 0 \tag{5.8.38}$$

$$D_0^2 u_1 + \omega_0^2 u_1 = -2D_0D_1u_0 - \hat{K}\cos\Omega_1 T_0 \tag{5.8.39}$$

$$D_0^2 u_2 + \omega_0^2 u_2 = -\left(D_1^2 + 2D_0 D_2\right)u_0 - 2D_0 D_1 u_1 + \frac{1}{6}\omega_0^2 u_0^3 - \alpha u_0 \cos \Omega_2 T_0 \tag{5.8.40}$$

The solution of (5.8.38) is

$$u_0 = Ae^{i\omega_0 T_0} + \bar{A}\, e^{-i\omega_0 T_0} \tag{5.8.41}$$

where $A = A(T_1, T_2)$. Substituting u_0 into (5.8.39) yields

$$D_0^2 u_1 + \omega_0^2 u_1 = -2i\omega_0 D_0 D_1 A e^{i\omega_0 T_0} - \frac{1}{2}\hat{K} e^{i\Omega T_0} + cc \tag{5.8.42}$$

In order to eliminate secular terms from the above equation, it is necessary to consider different values of Ω_1, and further, if the Eq. (5.8.40) is considered, it is also necessary to consider different values of Ω_2. Two cases are considered here: (i) $\Omega_1 \approx \omega_0,\ \ \Omega_2 \approx 2\omega_0$ and (ii) $\Omega_2 - \Omega_1 \approx \omega_0$.

(i) $\Omega_1 \approx \omega_0,\ \ \Omega_2 \approx 2\omega_0$:

We use the detuning parameter σ to quantitatively describe the nearness of Ω_1 to ω_0, i.e.,

$$\Omega_1 = \omega_0 + \hat{\varepsilon}\sigma \tag{5.8.43}$$

therefore,

$$D_0^2 u_1 + \omega_0^2 u_1 = -2i\omega_0 D_1 A e^{i\omega_0 T_0} - \frac{1}{2}\hat{K} e^{i\sigma T_1} e^{i\omega_0 T_0} + cc \tag{5.8.44}$$

In order to eliminate secular terms from the above equation, there must be

$$-2i\omega_0 D_1 A - \frac{1}{2}\hat{K} e^{i\sigma T_1} = 0 \tag{5.8.45}$$

then

$$A = \frac{1}{4}\hat{K}\omega_0^{-1}\sigma^{-1} e^{i\sigma T_1} \tag{5.8.46}$$

Substituting this into (5.8.14), the first order approximate solution of the system is

$$\begin{aligned} u &= \frac{1}{2}\hat{K}\omega_0^{-1}\sigma^{-1}\cos(\omega_0 T_0 + \sigma T_1) + O(\hat{\varepsilon}) \\ &= \frac{1}{2}\hat{K}\omega_0^{-1}\sigma^{-1}\cos\left(\omega_0 + \varepsilon^{1/2}\sigma\right)t + O\left(\varepsilon^{1/2}\right) \end{aligned} \tag{5.8.47}$$

Therefore, when $\Omega \approx \omega_0$ is adopted, $\Omega_2 \approx 2\omega_0$ is not required.

(ii) $\Omega_2 - \Omega_1 \approx \omega_0$:

Equation (5.8.39) becomes

$$D_0^2 u_1 + \omega_0^2 u_1 = -2i\omega_0 D_1 A e^{i\omega_0 T_0} - \frac{1}{2}\hat{K} e^{i\Omega_1 T_0} + cc \tag{5.8.48}$$

In order to eliminate secular terms from the above equation, there must be

$$D_1 A = 0 \quad \Rightarrow \quad A = A(T_2) \tag{5.8.49}$$

Thus the solution of (5.8.48) is

$$u_1 = \frac{1}{2(\Omega_1^2 - \omega_0^2)} \hat{K} e^{i\Omega_1 T_0} + cc \tag{5.8.50}$$

Substituting $u_0, \quad u_1$ into (5.8.40) yields

$$\begin{aligned} D_0^2 u_2 + \omega_0^2 u_2 &= -(D_1^2 + 2D_0 D_2) u_0 - 2D_0 D_1 u_1 + \frac{1}{6}\omega_0^2 u_0^3 \\ &\quad - \alpha u_0 \cos \Omega_2 T_0 \\ &= -2i\omega_0 D_2 A e^{i\omega_0 T_0} + \frac{1}{2}\omega_0^2 A^2 \overline{A} e^{i\omega_0 T_0} + cc + NST \end{aligned} \tag{5.8.51}$$

$$D_0^2 u_2 + \omega_0^2 u_2 = -(D_1^2 + 2D_0 D_2) u_0 - 2D_0 D_1 u_1 + \frac{1}{6}\omega_0^2 u_0^3 - \alpha u_0 \cos \Omega_2 T_0$$

In order to eliminate secular terms from the above equation, there must be

$$-2iD_2 A + \frac{1}{2}\omega_0 A^2 \overline{A} = 0 \tag{5.8.52}$$

Let $A = \frac{1}{2} a e^{i\beta}$, we get

$$-ia' + a\beta' + \frac{1}{16}\omega_0 a^3 = 0 \tag{5.8.53}$$

Separating the real and imaginary parts of the above equation yields

$$\begin{aligned} a' &= 0 \\ \beta' &= -\frac{1}{16}\omega_0 a^2 \end{aligned} \tag{5.8.54}$$

(5.8.36) has the steady state solution near $u = 0$, i.e.,

$$a = 0, \quad \beta = 0 \tag{5.8.55}$$

Therefore, the first order approximate solution of the system is

$$u = 0 + O\left(\varepsilon^{1/2}\right) \tag{5.8.56}$$

This result shows that it is not possible to generate the combined resonant response of $\Omega_2 - \Omega_1 \approx \omega_0$ under soft excitation.

5.9 Exercise 5.9 (The Method of Harmonic Balance to Solve Duffing's Equation and Stability Analysis)

Solution: (a) Use the method of harmonic balance to find the first order approximate solution of the Duffing equation. Let the solution of the equation be

$$u \approx u_0 = a\cos\omega t \tag{5.9.1}$$

Substituting it into the original equation, we get

$$\begin{aligned}
0 &= \ddot{u} + \omega_0^2 u + \alpha u^3 - K\cos\omega t \\
&= -a\omega^2\cos\omega t + \omega_0^2 a\cos\omega t \\
&\quad + \alpha a^3\cos^3\omega t - K\cos\omega t \\
&= -a\omega^2\cos\omega t + \omega_0^2 a\cos\omega t \\
&\quad + \frac{3}{4}\alpha a^3\cos\omega t + \frac{1}{4}\alpha a^3\cos 3\omega t - K\cos\omega t \\
&= \left(-a\omega^2 + \omega_0^2 a + \frac{3}{4}\alpha a^3 - K\right)\cos\omega t + \cdots
\end{aligned} \tag{5.9.2}$$

Let the coefficient of $\cos\omega t$ be zero, we can obtain the frequency–response equation

$$\left(\omega_0^2 - \omega^2\right)a + \frac{3}{4}\alpha a^3 = K \tag{5.9.3}$$

(b) To examine the stability of the steady state solution, u_0, let

$$u = u_0 + x \tag{5.9.4}$$

where x is a small term. Substituting it into the original Duffing equation and keeping linear terms in x yields

$$\ddot{x} + \omega_0^2 x + 3\alpha u_0^2 x + \ddot{u}_0 + \omega_0^2 u_0 + \alpha u_0^3 = K\cos\omega t$$

Considering $\ddot{u}_0 + \omega_0^2 u_0 + \alpha u_0^3 \approx K\cos\omega t$, we get

$$\ddot{x} + \left(\omega_0^2 + 3\alpha u_0^2\right)x = 0 \tag{5.9.5}$$

(c) Substitute $u_0 = a\cos\omega t$ into (5.9.5), we obtain

$$\ddot{x} + \left(\omega_0^2 + 3\alpha a^2\cos^2\omega t\right)x = 0$$

i.e.,

$$\ddot{x} + \left(\omega_0^2 + \frac{3}{2}\alpha a^2 + \frac{3}{2}\alpha a^2\cos 2\omega t\right)x = 0 \tag{5.9.6}$$

Thus, the stability of the periodic solution is transformed into determining the stability of the solution of the Mathieu Eq. (5.9.6).

(d) We use the method of strained parameters to determine the transition curves of equation. Since $x << u_0$, let the solution and parameters of the equation have the following expansion:

$$x(t;\varepsilon) = \varepsilon x_1(t) + \varepsilon^2 x_2(t) + \varepsilon^3 x_3(t) + \cdots \tag{5.9.7}$$

$$\omega_0^2 + \frac{3}{2}\alpha a^2 = \delta = \delta_0 + \varepsilon\delta_1 + \varepsilon^2\delta_2 + \cdots \tag{5.9.8}$$

In order to let the parametric excitation term appearing in the control equation as the higher order term, let

$$\alpha = \varepsilon\hat{\alpha} \tag{5.9.9}$$

Substituting the above three equations into (5.9.6) and retaining to $O(\varepsilon^3)$, we get

$$
\begin{aligned}
0 &= \ddot{x} + \left(\delta + \frac{3}{2}\varepsilon\hat{\alpha}a^2 \cos 2\omega t\right)x \\
&= \varepsilon\ddot{x}_1 + \varepsilon^2\ddot{x}_2 + \varepsilon^3\ddot{x}_3 + \left(\delta_0 + \varepsilon\delta_1 + \varepsilon^2\delta_2\right)\left(\varepsilon x_1 + \varepsilon^2 x_2 + \varepsilon^3 x_3\right) \\
&+ \frac{3}{2}\varepsilon\hat{\alpha}a^2\left(\varepsilon x_1 + \varepsilon^2 x_2 + \varepsilon^3 x_3\right)\cos 2\omega t + \cdots \\
&= \varepsilon\ddot{x}_1 + \varepsilon^2\ddot{x}_2 + \varepsilon^3\ddot{x}_3 + \varepsilon\delta_0 x_1 + \varepsilon^2\delta_0 x_2 + \varepsilon^3\delta_0 x_3 + \varepsilon^2\delta_1 x_1 + \varepsilon^3\delta_1 x_2 + \varepsilon^3\delta_2 x_1 \\
&+ \frac{3}{2}\varepsilon^2\hat{\alpha}a^2 x_1 \cos 2\omega t + \frac{3}{2}\varepsilon^3\hat{\alpha}a^2 x_2 \cos 2\omega t + \cdots \\
&= \varepsilon(\ddot{x}_1 + \delta_0 x_1) + \varepsilon^2\left(\ddot{x}_2 + \delta_0 x_2 + \delta_1 x_1 + \frac{3}{2}\hat{\alpha}a^2 x_1 \cos 2\omega t\right) \\
&+ \varepsilon^3\left(\ddot{x}_3 + \delta_0 x_3 + \delta_1 x_2 + \delta_2 x_1 + \frac{3}{2}\hat{\alpha}a^2 x_2 \cos 2\omega t\right) + \cdots
\end{aligned}
$$

Equating coefficients of like powers of ε in the above equation yields

$$\ddot{x}_1 + \delta_0 x_1 = 0 \tag{5.9.10}$$

$$\ddot{x}_2 + \delta_0 x_2 = -\delta_1 x_1 - \frac{3}{2}\hat{\alpha}a^2 x_1 \cos 2\omega t \tag{5.9.11}$$

$$\ddot{x}_3 + \delta_0 x_3 = -\delta_1 x_2 - \delta_2 x_1 - \frac{3}{2}\hat{\alpha}a^2 x_2 \cos 2\omega t \tag{5.9.12}$$

The solution of (5.9.10) is

$$x_1 = b\cos\sqrt{\delta_0}t + c\sin\sqrt{\delta_0}t \tag{5.9.13}$$

$\delta_0 = \omega^2$:

$$x_1 = b\cos\omega t + c\sin\omega t \tag{5.9.14}$$

x_1 is a periodic solution with the period of $T = 2\pi/\omega$. Substituting (5.9.14) into (5.9.11) yields

$$
\begin{aligned}
\ddot{x}_2 + \omega^2 x_2 &= -\delta_1(b\text{cos}\omega t + c\text{sin}\omega t) \\
&\quad - \frac{3}{2}\hat{\alpha}a^2(b\text{cos}\omega t + c\text{sin}\omega t)\text{cos}2\omega t \\
&= -\left(\delta_1 + \frac{3}{4}\hat{\alpha}a^2\right)(b\text{cos}\omega t + c\text{sin}\omega t) + NST
\end{aligned} \tag{5.9.15}
$$

In order to eliminate secular terms from the above equation, there must be

$$\delta_1 = -\frac{3}{4}\hat{\alpha}a^2 \tag{5.9.16}$$

Therefore, the transition curves emanating from $\delta = \omega^2$ is

$$\omega_0^2 + \frac{3}{2}\alpha a^2 = \omega^2 - \frac{3}{4}\varepsilon\hat{\alpha}a^2 + O(\varepsilon^2)$$

i.e.,

$$\omega_0^2 = \omega^2 - \frac{9}{4}\varepsilon\hat{\alpha}a^2 + O(\varepsilon^2) \tag{5.9.17}$$

Let

$$\omega = \omega_0 + \varepsilon\sigma$$

then the Eq. (5.9.17) can be written as

$$(\omega - \omega_0)(\omega + \omega_0) - \frac{9}{4}\alpha a^2 \approx 0 \quad \Rightarrow \quad 2\varepsilon\sigma\omega_0(\omega - \omega_0) - \frac{9}{4}\alpha a^2 \approx 0$$

i.e.,

$$\sigma - \frac{9\hat{\alpha}a^2}{8\omega_0} \approx 0 \tag{5.9.18}$$

5.10 Exercise 5.10 (The Method of Harmonic Balance for Solving Pure Cubic Nonlinear Equations)

Solution: (a) Solve the given equation by the method of harmonic balance. Let the approximate solution of the equation be

$$u \approx u_0 = a\cos\omega t + b\cos 3\omega t \tag{5.10.1}$$

Substituting above equation into the original equation, we get.

$$\begin{aligned}
0 &= \ddot{u} + u^3 - K\cos\omega t \\
&= -\omega^2 a\cos\omega t - 9\omega^2 b\cos 3\omega t + a^3\cos^3\omega t \\
&\quad + 3a^2 b\cos^2\omega t\cos 3\omega t \\
&\quad + 3ab^2\cos\omega t\cos^2 3\omega t + b^3\cos^3 3\omega t - K\cos\omega t \\
&= -\omega^2 a\cos\omega t - K\cos\omega t - 9\omega^2 b\cos 3\omega t + \frac{3}{4}a^3\cos\omega t
\end{aligned}$$

$$
\begin{aligned}
&+\frac{1}{4}a^3\cos 3\omega t+\frac{3}{2}a^2b\cos 3\omega t+\frac{3}{4}a^2b\cos\omega t\\
&+\frac{3}{4}a^2b\cos 5\omega t+\frac{3}{2}ab^2\cos\omega t+\frac{3}{4}ab^2\cos 5\omega t\\
&+\frac{3}{4}ab^2\cos 7\omega t+\frac{3}{4}b^3\cos\omega t+\frac{1}{4}b^3\cos 3\omega t\\
&=\left(-\omega^2a-K+\frac{3}{4}a^3+\frac{3}{4}a^2b+\frac{3}{2}ab^2+\frac{3}{4}b^3\right)\cos\omega t\\
&+\left(-9\omega^2b+\frac{1}{4}a^3+\frac{3}{2}a^2b+\frac{1}{4}b^3\right)\cos 3\omega t+\cdots
\end{aligned}
\tag{5.10.2}
$$

Let the coefficients of $\cos\omega t$ and $\cos 3\omega t$ be zero, we get

$$
\begin{aligned}
&-\omega^2a-K+\tfrac{3}{4}a^3+\tfrac{3}{4}a^2b+\tfrac{3}{2}ab^2+\tfrac{3}{4}b^3=0\\
&-9\omega^2b+\tfrac{1}{4}a^3+\tfrac{3}{2}a^2b+\tfrac{1}{4}b^3=0
\end{aligned}
\tag{5.10.3}
$$

(b) To examine the stability of the steady state solution u_0, let

$$u=u_0+x \tag{5.10.4}$$

where x is a small term. Substituting into the original equation and keeping linear terms in x yields

$$\ddot{x}+3u_0^2x+\ddot{u}_0+u_0^3=K\cos\omega t$$

Considering $\ddot{u}_0+u_0^3\approx K\cos\omega t$, we get

$$\ddot{x}+3u_0^2x=0 \tag{5.10.5}$$

Substituting u_0 into (5.10.5) yields

$$\ddot{x}+3(a\cos\omega t+b\cos 3\omega t)^2x=0$$

i.e.,

$$\ddot{x}+\frac{3}{2}\left[a^2+b^2+\left(a^2+2ab\right)\cos 2\omega t+2ab\cos 4\omega t+b^2\cos 6\omega t\right]x=0 \tag{5.10.6}$$

Thus, the stability of the steady state solution u_0 is transformed into determining the stability of the solution of (5.10.6).

(c) We use the method of strained parameters to determine the transition curves of Eq. (5.10.6). Since $x << u_0$, it can be assumed that x is of order $O(\varepsilon)$. In order to make the time-varying parameter terms appear in the higher-order control equations of x, the 3 periodic terms, $\cos 2\omega t$, $\cos 4\omega t$ and $\cos 6\omega t$ are of order $O(\varepsilon)$. So, let the solution and parameters of the equation have the following expansion:

$$x(t;\varepsilon) = \varepsilon x_1(t) + \varepsilon^2 x_2(t) + \varepsilon^3 x_3(t) + \cdots \tag{5.10.7}$$

$$\frac{3}{2}\left(a^2 + b^2\right) = \delta = \delta_0 + \varepsilon\delta_1 + \varepsilon^2\delta_2 + \cdots \tag{5.10.8}$$

$$\frac{3}{2}\left(a^2 + 2ab\right) = \varepsilon\alpha_1, \quad 3ab = \varepsilon\alpha_2, \quad \frac{3}{2}b^2 = \varepsilon\alpha_3 \tag{5.10.9}$$

Substituting the above three equations into (5.10.6) and retaining to $O(\varepsilon^3)$, we get

$$\begin{aligned}
0 &= \ddot{x} + (\delta + \varepsilon\alpha_1\cos 2\omega t + \varepsilon\alpha_2\cos 4\omega t + \varepsilon\alpha_3\cos 6\omega t)x \\
&= \varepsilon\ddot{x}_1 + \varepsilon^2\ddot{x}_2 + \varepsilon^3\ddot{x}_3 \\
&\quad + \left(\delta_0 + \varepsilon\delta_1 + \varepsilon^2\delta_2\right)\left(\varepsilon x_1 + \varepsilon^2 x_2 + \varepsilon^3 x_3\right) \\
&\quad + \varepsilon\alpha_1\left(\varepsilon x_1 + \varepsilon^2 x_2 + \varepsilon^3 x_3\right)\cos 2\omega t \\
&\quad + \varepsilon\alpha_2\left(\varepsilon x_1 + \varepsilon^2 x_2 + \varepsilon^3 x_3\right)\cos 4\omega t \\
&\quad + \varepsilon\alpha_3\left(\varepsilon x_1 + \varepsilon^2 x_2 + \varepsilon^3 x_3\right)\cos 6\omega t \\
&= \varepsilon(\ddot{x}_1 + \delta_0 x_1) \\
&\quad + \varepsilon^2(\ddot{x}_2 + \delta_0 x_2 + \delta_1 x_1 + \alpha_1 x_1\cos 2\omega t + \alpha_2 x_1\cos 4\omega t + \alpha_3 x_1\cos 6\omega t) \\
&\quad + \varepsilon^3(\ddot{x}_3 + \delta_0 x_3 + \delta_1 x_2 + \alpha_1 x_2\cos 2\omega t + \alpha_2 x_2\cos 4\omega t + \alpha_3 x_2\cos 6\omega t)
\end{aligned}$$

Equating coefficients of like powers of ε in the above equation yields

$$\ddot{x}_1 + \delta_0 x_1 = 0 \tag{5.10.10}$$

$$\ddot{x}_2 + \delta_0 x_2 = -\delta_1 x_1 - \alpha_1 x_1\cos 2\omega t - \alpha_2 x_1\cos 4\omega t - \alpha_3 x_1\cos 6\omega t \tag{5.10.11}$$

$$\ddot{x}_3 + \delta_0 x_3 = -\delta_1 x_2 - \alpha_1 x_2\cos 2\omega t - \alpha_2 x_2\cos 4\omega t - \alpha_3 x_2\cos 6\omega t \tag{5.10.12}$$

The solution of (5.10.10) is

$$x_1 = c\cos\sqrt{\delta_0}t + d\sin\sqrt{\delta_0}t \tag{5.10.13}$$

CASE I:$\delta_0 = \omega^2$**:**

$$x_1 = c\cos\omega t + d\sin\omega t \tag{5.10.14}$$

x_1 is a periodic solution with period $2T = 2\pi/\omega$. Substituting (5.10.14) into (5.10.11) yields

$$\begin{aligned}\ddot{x}_2 + 4\omega^2 x_2 &= -\delta_1(c\cos\omega t + d\sin\omega t)\\&\quad - \alpha_1(c\cos\omega t + d\sin\omega t)\cos 2\omega t\\&\quad - \alpha_2(c\cos\omega t + d\sin\omega t)\cos 4\omega t\\&\quad - \alpha_3(c\cos\omega t + d\sin\omega t)\cos 6\omega t\\&= -\left(\delta_1 + \frac{1}{2}\alpha_1\right)c\cos\omega t\\&\quad - \left(\delta_1 - \frac{1}{2}\alpha_1\right)d\sin\omega t + NST\end{aligned} \tag{5.10.15}$$

In order to eliminate secular terms from the above equation, there must be

$$\delta_1 = \pm\frac{1}{2}\alpha_1 \tag{5.10.16}$$

Therefore, the transition curves emanating from $\delta = \omega^2$ is

$$\frac{3}{2}\left(a^2 + b^2\right) = \delta = \omega^2 \pm \frac{1}{2}\varepsilon\alpha_1 + O\left(\varepsilon^2\right) + \cdots$$

i.e.,

$$\omega^2 = \frac{3}{2}\left(a^2 + b^2\right) \pm \frac{3}{4}\left(a^2 + 2ab\right) + O\left(\varepsilon^2\right) \tag{5.10.17}$$

CASE II:$\delta_0 = 4\omega^2$:

$$x_1 = c\cos 2\omega t + d\sin 2\omega t \tag{5.10.18}$$

x_1 is a periodic solution with period $T = \pi/\omega$. Substituting (5.10.18) into (5.10.11) yields

$$\begin{aligned}\ddot{x}_2 + 4\omega^2 x_2 &= -\delta_1(c\cos 2\omega t + d\sin 2\omega t)\\&\quad - \alpha_1(c\cos 2\omega t + d\sin 2\omega t)\cos 2\omega t\\&\quad - \alpha_2(c\cos 2\omega t + d\sin 2\omega t)\cos 4\omega t\\&\quad - \alpha_3(c\cos 2\omega t + d\sin 2\omega t)\cos 6\omega t\\&= -\left(\delta_1 + \frac{1}{2}\alpha_2\right)c\cos 2\omega t\\&\quad - \left(\delta_1 - \frac{1}{2}\alpha_2\right)d\sin 2\omega t + NST\end{aligned} \tag{5.10.19}$$

In order to eliminate secular terms from the above equation, there must be

$$\delta_1 = \pm\frac{1}{2}\alpha_2 \tag{5.10.20}$$

Therefore, the transition curves emanating from $\delta = 4\omega^2$ is

$$\frac{3}{2}\left(a^2 + b^2\right) = \delta = 4\omega^2 \pm \frac{1}{2}\varepsilon\alpha_2 + O\left(\varepsilon^2\right) + \cdots$$

i.e.,

$$\omega^2 = \frac{3}{8}\left(a^2 + b^2 \pm ab\right) + O\left(\varepsilon^2\right) \tag{5.10.21}$$

5.11 Exercise 5.11 (The Quadratic Parametric Excitation of a Linear Viscous Damping System)

Solution: (a) We use the method of multiple scales to solve the first order approximate periodic solution of the system. Let the solution of the equation be

$$u(t;\varepsilon) = u_0(T_0, T_1, T_2) + \varepsilon u_1(T_0, T_1, T_2) + \varepsilon^2 u_2(T_0, T_1, T_2) + \cdots \tag{5.11.1}$$

Substituting the above equation into the original equation and retaining to $O(\varepsilon^2)$, we get

$$\begin{aligned}
0 &= \ddot{u} + \omega_0^2 u + 2\varepsilon\mu\dot{u} + 2\varepsilon u^2 \cos 2t \\
&= \left[D_0^2 + 2\varepsilon D_0 D_1 + \varepsilon^2\left(D_1^2 + 2D_0 D_2\right)\right]\left(u_0 + \varepsilon u_1 + \varepsilon^2 u_2\right) \\
&\quad + \omega_0^2\left(u_0 + \varepsilon u_1 + \varepsilon^2 u_2\right) + 2\varepsilon\mu(D_0 + \varepsilon D_1)\left(u_0 + \varepsilon u_1 + \varepsilon^2 u_2\right) \\
&\quad + 2\varepsilon\left(u_0 + \varepsilon u_1 + \varepsilon^2 u_2\right)^2 \cos 2t \\
&= D_0^2 u_0 + \varepsilon D_0^2 u_1 + \varepsilon^2 D_0^2 u_2 + 2\varepsilon D_0 D_1 u_0 \\
&\quad + 2\varepsilon^2 D_0 D_1 u_1 + \varepsilon^2\left(D_1^2 + 2D_0 D_2\right)u_0 \\
&\quad + \omega_0^2 u_0 + \varepsilon\omega_0^2 u_1 + \varepsilon^2\omega_0^2 u_2 + 2\varepsilon\mu D_0 u_0 \\
&\quad + 2\varepsilon^2\mu D_0 u_1 + 2\varepsilon^2\mu D_1 u_0 \\
&\quad + 2\varepsilon u_0^2 \cos 2t + 4\varepsilon^2 u_0 u_1 \cos 2t + \cdots \\
&= D_0^2 u_0 + \omega_0^2 u_0 \\
&\quad + \varepsilon\left(D_0^2 u_1 + \omega_0^2 u_1 + 2D_0 D_1 u_0 + 2\mu D_0 u_0 + 2u_0^2 \cos 2T_0\right) \\
&\quad + \varepsilon^2\left[D_0^2 u_2 + \omega_0^2 u_2 + \left(D_1^2 + 2D_0 D_2\right)u_0 + 2D_0 D_1 u_1\right. \\
&\quad \left. + 2\mu D_0 u_1 + 2\mu D_1 u_0 + 4u_0 u_1 \cos 2T_0\right] + \cdots
\end{aligned}$$

Equating coefficients of like powers of ε in the above equation yields

$$D_0^2u_0 + \omega_0^2u_0 = 0 \tag{5.11.2}$$

$$D_0^2u_1 + \omega_0^2u_1 = -2D_0D_1u_0 - 2\mu D_0u_0 - 2u_0^2\cos 2t \tag{5.11.3}$$

$$\begin{aligned} D_0^2u_2 + \omega_0^2u_2 &= -\left(D_1^2 + 2D_0D_2\right)u_0 - 2D_0D_1u_1 \\ -2\mu D_0u_1 &- 2\mu D_1u_0 - 4u_0u_1\cos 2t \end{aligned} \tag{5.11.4}$$

The solution of (5.11.2) is

$$u_0 = Ae^{i\omega_0T_0} + \overline{A}\,e^{-i\omega_0T_0} \tag{5.11.5}$$

where $A = A(T_1, T_2)$. Let

$$A = \frac{1}{2}ae^{i\beta} \tag{5.11.6}$$

Substituting this into (5.11.5), we can obtain the first order approximate solution of the given equation

$$u \approx u_0 = a\cos(\omega_0t + \beta) + O(\varepsilon) \tag{5.11.7}$$

Substituting u_0 into (5.11.3) yields

$$\begin{aligned} D_0^2u_1 + \omega_0^2u_1 &= -2D_0D_1u_0 - 2\mu D_0u_0 - 2u_0^2\cos 2T_0 \\ &= -2i\omega_0D_1Ae^{i\omega_0T_0} - 2i\omega_0\mu Ae^{i\omega_0T_0} - 2A\overline{A}e^{2iT_0} \\ &\quad - A^2e^{i(2\omega_0-2)T_0} - \overline{A}^2e^{i(2-2\omega_0)T_0} + cc + NST \end{aligned} \tag{5.11.8}$$

CASE I: When ω_0 is far from 2 or $\frac{2}{3}$:

In order to eliminate secular terms from (5.11.8), we need

$$-A' - \mu A = 0 \tag{5.11.9}$$

The prime denotes the derivative with respect to T_1. Substituting $A = \frac{1}{2}ae^{i\beta}$ into the above equation gives

$$a' + ia\beta' + \mu a = 0 \tag{5.11.10}$$

Separating the real and imaginary parts of the above equation yields

$$a' = -\mu a, \quad \beta' = 0 \tag{5.11.11}$$

CASE II: When $\omega_0 = 2 + \varepsilon\sigma$:

In order to eliminate secular terms from (5.11.8), we need

$$-2i\omega_0 A' - 2i\omega_0\mu A - 2A\bar{A}\,e^{-i\sigma T_1} - A^2 e^{i\sigma T_1} = 0 \tag{5.11.12}$$

The prime denotes the derivative with respect to T_1. Substituting $A = \frac{1}{2}ae^{i\beta}$ into the above equation gives

$$i\omega_0 a' - \omega_0 a\beta' + i\omega_0\mu a + \frac{1}{2}a^2 e^{-i(\sigma T_1+\beta)} + \frac{1}{4}a^2 e^{i(\sigma T_1+\beta)} = 0 \tag{5.11.13}$$

Separating the real and imaginary parts of the above equation yields

$$a' = -\mu a + \frac{1}{4}\omega_0^{-1}a^2\sin(\sigma T_1+\beta)\,a$$

$$\beta' = \frac{3}{4}\omega_0^{-1}a^2\cos(\sigma T_1+\beta)$$

Considering $\omega_0 \approx 2$, the above equation becomes

$$a' = -\mu a + \frac{1}{8}a^2\sin\gamma, \quad a\beta' = \frac{3}{8}a^2\cos\gamma \tag{5.11.14}$$

where

$$\gamma = \sigma T_1 + \beta \tag{5.11.15}$$

CASE III: When $3\omega_0 = 2 + \varepsilon\sigma$:

In order to eliminate secular terms from (5.11.8), we need

$$-2i\omega_0 A' - 2i\omega_0\mu A - \bar{A}^2\,e^{-i\sigma T_1} = 0 \tag{5.11.16}$$

The prime denotes the derivative with respect to T_1. Substituting $A = \frac{1}{2}ae^{i\beta}$ into the above equation gives

$$i\omega_0 a' - \omega_0 a\beta' + i\omega_0\mu a + \frac{1}{4}a^2 e^{-i(\sigma T_1+3\beta)} = 0 \tag{5.11.17}$$

Separating the real and imaginary parts of the above equation yields

$$a' = -\mu a + \frac{1}{4}\omega_0^{-1}a^2\sin(\sigma T_1+3\beta)a\,\beta' = \frac{1}{4}\omega_0^{-1}a^2\cos(\sigma T_1+3\beta)$$

Considering $\omega_0 \approx \frac{2}{3}$, the above equation becomes

$$a' = -\mu a + \frac{3}{8}a^2 \sin\gamma, \quad a\beta' = \frac{3}{8}a^2 \cos\gamma \tag{5.11.18}$$

where

$$\gamma = \sigma T_1 + 3\beta \tag{5.11.19}$$

(b) We only determine the stability of the steady-state solution in the second case above, readers are invited to analyze the rest of the cases on their own. Writing the Eq. (5.11.14) as

$$a' = -\mu a + \frac{3}{8}a^2 \sin\gamma, \quad a\gamma' = a\sigma + \frac{9}{8}a^2 \cos\gamma \tag{5.11.20}$$

Let $a' = \gamma' = 0$ in (5.11.20), we obtain

$$-\mu a + \frac{3}{8}a^2 \sin\gamma = 0, \quad a\sigma + \frac{9}{8}a^2 \cos\gamma = 0 \tag{5.11.21}$$

The steady state solutions can be obtained:

(1) $a = 0$, γ are arbitrary constants;
(2) $a \neq 0$, γ are certain constants.

(1) **$a = 0$ and $\gamma = \gamma_0$ are arbitrary constants.**

Let

$$a = 0 + a_1, \quad \gamma = \gamma_0 + \gamma_1 \tag{5.11.22}$$

Substituting (5.11.22) into (5.11.20), we can obtain the corresponding linearized equation

$$a_1' = -\mu a_1 \tag{5.11.23}$$

So $a_1 = e^{-\mu T_1} = e^{-\varepsilon\mu t} \to 0$, therefore this steady state solution is stable.

(2) **$a = a_0 \neq 0$, $\gamma = \gamma_0$**

For this case, the frequency–response equation is

$$(\mu a)^2 + \frac{1}{9}(a\sigma)^2 = \frac{9}{64}a^4 \tag{5.11.24}$$

Let

$$a = a_0 + a_1, \quad \gamma = \gamma_0 + \gamma_1 \tag{5.11.25}$$

Substituting (5.11.25) into (5.11.20), we can obtain the corresponding linearized equation

$$\begin{Bmatrix} a_1' \\ \gamma_1' \end{Bmatrix} = \begin{bmatrix} -\mu + \frac{3}{4}a_0 \sin\gamma_0 & \frac{3}{8}a_0^2 \cos\gamma_0 \\ \sigma a_0^{-1} + \frac{9}{4}\cos\gamma_0 & -\frac{9}{8}a_0 \sin\gamma_0 \end{bmatrix} \begin{Bmatrix} a_1 \\ \gamma_1 \end{Bmatrix} \tag{5.11.26}$$

For any set of steady state solutions $\boldsymbol{a} = \boldsymbol{a}_0 \neq 0$, $\boldsymbol{\gamma} = \boldsymbol{\gamma}_0$, one can calculate the eigenvalues of Eq. (5.11.26), and the stability of the corresponding steady state solutions can be determined from the sign of the real part of the two eigenvalues.

5.12 Exercise 5.12 (The Cubic Parametric Excitation of a Linear Viscous Damping System)

Solution: (a) We use the method of multiple scales to solve the first order approximate periodic solution of the system. Let the solution of the equation be

$$u(t;\varepsilon) = u_0(T_0, T_1, T_2) + \varepsilon u_1(T_0, T_1, T_2) + \varepsilon^2 u_2(T_0, T_1, T_2) + \cdots \tag{5.12.1}$$

Substituting the above equation into the original equation and retaining to $O(\varepsilon^2)$, we get

$$\begin{aligned}
0 &= \ddot{u} + \omega_0^2 u + 2\varepsilon\mu\dot{u} + 2\varepsilon u^3 \cos 2t \\
&= \left[D_0^2 + 2\varepsilon D_0 D_1 + \varepsilon^2\left(D_1^2 + 2D_0 D_2\right)\right] \\
&\quad \left(u_0 + \varepsilon u_1 + \varepsilon^2 u_2\right) \\
&\quad + \omega_0^2\left(u_0 + \varepsilon u_1 + \varepsilon^2 u_2\right) \\
&\quad + 2\varepsilon\mu(D_0 + \varepsilon D_1)\left(u_0 + \varepsilon u_1 + \varepsilon^2 u_2\right) \\
&\quad + 2\varepsilon\left(u_0 + \varepsilon u_1 + \varepsilon^2 u_2\right)^3 \cos 2t \\
&= D_0^2 u_0 + \varepsilon D_0^2 u_1 + \varepsilon^2 D_0^2 u_2 + 2\varepsilon D_0 D_1 u_0 \\
&\quad + 2\varepsilon^2 D_0 D_1 u_1 + \varepsilon^2\left(D_1^2 + 2D_0 D_2\right)u_0 \\
&\quad + \omega_0^2 u_0 + \varepsilon\omega_0^2 u_1 + \varepsilon^2\omega_0^2 u_2 + 2\varepsilon\mu D_0 u_0 \\
&\quad + 2\varepsilon^2\mu D_0 u_1 + 2\varepsilon^2\mu D_1 u_0 \\
&\quad + 2\varepsilon u_0^3 \cos 2t + 6\varepsilon^2 u_0 u_1 \cos 2t + \cdots \\
&= D_0^2 u_0 + \omega_0^2 u_0 \\
&\quad + \varepsilon\left(D_0^2 u_1 + \omega_0^2 u_1 + 2D_0 D_1 u_0 + 2\mu D_0 u_0 + 2u_0^3 \cos 2T_0\right) \\
&\quad + \varepsilon^2\left[D_0^2 u_2 + \omega_0^2 u_2 + \left(D_1^2 + 2D_0 D_2\right)u_0 + 2D_0 D_1 u_1\right. \\
&\quad \left. + 2\mu D_0 u_1 + 2\mu D_1 u_0 + 6u_0 u_1 \cos 2T_0\right] + \cdots
\end{aligned}$$

Equating coefficients of like powers of ε in the above equation yields

$$D_0^2 u_0 + \omega_0^2 u_0 = 0 \tag{5.12.2}$$

$$D_0^2 u_1 + \omega_0^2 u_1 = -2D_0 D_1 u_0 - 2\mu D_0 u_0 - 2u_0^3 \cos 2T_0 \tag{5.12.3}$$

$$\begin{aligned} D_0^2 u_2 + \omega_0^2 u_2 = &-\left(D_1^2 + 2D_0 D_2\right) u_0 - 2D_0 D_1 u_1 \\ &- 2\mu D_0 u_1 - 2\mu D_1 u_0 - 6u_0 u_1 \cos 2t \end{aligned} \tag{5.12.4}$$

The solution of (5.12.2) is

$$u_0 = Ae^{i\omega_0 T_0} + \bar{A}\, e^{-i\omega_0 T_0} \tag{5.12.5}$$

where $A = A(T_1, T_2)$. Let

$$A = \frac{1}{2} a e^{i\beta} \tag{5.12.6}$$

and substitute this into (5.12.5), we can obtain the first order approximate solution of the given equation

$$u \approx u_0 = a \cos(\omega_0 t + \beta) + O(\varepsilon) \tag{5.12.7}$$

Substituting u_0 into (5.12.3) yields

$$\begin{aligned} D_0^2 u_1 + \omega_0^2 u_1 &= -2D_0 D_1 u_0 - 2\mu D_0 u_0 - 2u_0^3 \cos 2T_0 \\ &= -2i\omega_0 D_1 A e^{i\omega_0 T_0} - 2i\omega_0 \mu A e^{i\omega_0 T_0} \\ &\quad - 3A\overline{A}^2 e^{i(2-\omega_0)T_0} + cc \\ &\quad - \overline{A}^3 e^{i(2-3\omega_0)T_0} - A^3 e^{i(3\omega_0 - 2)T_0} + NST \end{aligned} \tag{5.12.8}$$

CASE I: When ω_0 is far from 1 or $\frac{1}{2}$:
In order to eliminate secular terms from (5.12.8), we need

$$-A' - \mu A = 0 \tag{5.12.9}$$

The prime denotes the derivative with respect to T_1. Substituting $A = \frac{1}{2} a e^{i\beta}$ into the above equation gives

$$a' + ia\beta' + \mu a = 0 \tag{5.12.10}$$

Separating the real and imaginary parts of the above equation yields

$$a' = -\mu a, \quad \beta' = 0 \tag{5.12.11}$$

CASE II: When $\omega_0 = 1 + \varepsilon\sigma$:

In order to eliminate secular terms from (5.12.8), we need

$$-2i\omega_0 A' - 2i\omega_0\mu A - 3A\overset{-2}{A}e^{-i2\sigma T_1} - A^3 e^{i2\sigma T_1} = 0 \tag{5.12.12}$$

The prime denotes the derivative with respect to T_1. Substituting $A = \frac{1}{2}ae^{i\beta}$ into the above equation gives

$$i\omega_0 a' - \omega_0 a\beta' + i\omega_0\mu a + \frac{3}{8}a^3 e^{-i(2\sigma T_1 + 2\beta)}$$
$$+ \frac{1}{8}a^3 e^{i(2\sigma T_1 + 2\beta)} = 0 \tag{5.12.13}$$

Separating the real and imaginary parts of the above equation and taking into account $\omega_0 \approx 1$, we get

$$a' = -\mu a + \frac{1}{4}a^3 \sin\gamma, a\beta' = \frac{1}{2}a^3 \cos\gamma \tag{5.12.14}$$

where

$$\gamma = 2\sigma T_1 + 2\beta \tag{5.12.15}$$

CASE III: When $2\omega_0 = 1 + \varepsilon\sigma$:

In order to eliminate secular terms from (5.12.8), we need

$$-2i\omega_0 D_1 A e^{i\omega_0 T_0} - 2i\omega_0\mu A e^{i\omega_0 T_0} - \overset{-3}{A}e^{-2i\sigma T_1}e^{i\omega_0 T_0} = 0 \tag{5.12.16}$$

Substituting $A = \frac{1}{2}ae^{i\beta}$ into the above equation gives

$$i\omega_0 a' - \omega_0 a\beta' + i\omega_0\mu a + \frac{1}{8}a^3 e^{-i(2\sigma T_1 + 4\beta)} = 0 \tag{5.12.17}$$

The prime denotes the derivative with respect to T_1. Separating the real and imaginary parts of the above equation and taking into account $\omega_0 \approx \frac{1}{2}$ yields

$$a' = -\mu a + \frac{1}{4}a^3 \sin\gamma, a\beta' = \frac{1}{4}a^3 \cos\gamma \tag{5.12.18}$$

where

$$\gamma = 2\sigma T_1 + 4\beta \tag{5.12.19}$$

(b) We only determine the stability of the steady-state solution in the second case above, readers are invited to analyze the rest of the cases on their own. Writing the Eq. (5.12.14) as

$$a' = -\mu a + \frac{1}{4}a^3\sin\gamma,\ a\gamma' = 2\sigma a + a^3\cos\gamma \tag{5.12.20}$$

Let $a' = \gamma' = 0$ in (5.12.20), we obtain

$$-\mu a + \frac{1}{4}a^3\sin\gamma = 0,\ 2\sigma a + a^3\cos\gamma = 0 \tag{5.12.21}$$

The steady state solutions can be obtained:

(1) $a = 0$, γ are arbitrary constants;
(2) $a \neq 0$, γ are certain constants.

(1) $\boldsymbol{a = 0}$ **and** $\boldsymbol{\gamma = \gamma_0}$ **are arbitrary constants.**

Let

$$a = 0 + a_1,\quad \gamma = \gamma_0 + \gamma_1 \tag{5.12.22}$$

Substituting (5.12.22) into (5.12.20), we can obtain the corresponding linearized equation

$$a_1' = -\mu a_1 \tag{5.12.23}$$

So $a_1 = e^{-\mu T_1} = e^{-\varepsilon\mu t} \to 0$, therefore this steady state solution is stable.

(2) $\boldsymbol{a = a_0 \neq 0,\ \gamma = \gamma_0}$

For this case, the frequency–response equation is

$$16(\mu a)^2 + (\sigma a)^2 = a^6 \tag{5.12.24}$$

Let

$$a = a_0 + a_1,\quad \gamma = \gamma_0 + \gamma_1 \tag{5.12.25}$$

Substituting (5.12.25) into (5.12.20), we can obtain the corresponding linearized equation

$$\begin{Bmatrix} a_1' \\ \gamma_1' \end{Bmatrix} = \begin{bmatrix} -\mu + \frac{3}{4}a_0^2\sin\gamma_0 & \frac{1}{4}a_0^3\cos\gamma_0 \\ 2\sigma + 3a_0^2\cos\gamma_0 & -a_0^3\sin\gamma_0 \end{bmatrix} \begin{Bmatrix} a_1 \\ \gamma_1 \end{Bmatrix} \tag{5.12.26}$$

For any set of steady state solutions $\boldsymbol{a} = \boldsymbol{a}_0 \neq 0, \boldsymbol{\gamma} = \boldsymbol{\gamma}_0$, one can calculate the eigenvalues of Eq. (5.12.26), and the stability of the corresponding steady state solutions can be determined from the sign of the real part of the two eigenvalues.

5.13 Exercise 5.13 (The High Order Nonlinear Parametric Excitation of a Linear Viscous Damping System)

Solution: **(a)** We use the method of multiple scales to solve the first order approximate periodic solution of the system. Let the solution of the equation be

$$u(t;\varepsilon) = u_0(T_0, T_1, T_2) + \varepsilon u_1(T_0, T_1, T_2) + \varepsilon^2 u_2(T_0, T_1, T_2) + \cdots \tag{5.13.1}$$

Substituting the above equation into the original equation and retaining to $O(\varepsilon^2)$, we get

$$\begin{aligned}
0 &= \ddot{u} + \omega_0^2 u + 2\varepsilon\mu\dot{u} + 2\varepsilon u^n \cos 2t \\
&= \left[D_0^2 + 2\varepsilon D_0 D_1 + \varepsilon^2\left(D_1^2 + 2D_0 D_2\right)\right] \\
&\quad \left(u_0 + \varepsilon u_1 + \varepsilon^2 u_2\right) \\
&\quad + \omega_0^2\left(u_0 + \varepsilon u_1 + \varepsilon^2 u_2\right) \\
&\quad + 2\varepsilon\mu(D_0 + \varepsilon D_1)\left(u_0 + \varepsilon u_1 + \varepsilon^2 u_2\right) \\
&+ 2\varepsilon(u_0 + \varepsilon u_1 + \varepsilon^2 u_2)^n \cos 2t \\
&= D_0^2 u_0 + \varepsilon D_0^2 u_1 + \varepsilon^2 D_0^2 u_2 + 2\varepsilon D_0 D_1 u_0 \\
&\quad + 2\varepsilon^2 D_0 D_1 u_1 + \varepsilon^2\left(D_1^2 + 2D_0 D_2\right)u_0 \\
&\quad + \omega_0^2 u_0 + \varepsilon\omega_0^2 u_1 + \varepsilon^2\omega_0^2 u_2 + 2\varepsilon\mu D_0 u_0 \\
&\quad + 2\varepsilon^2\mu D_0 u_1 + 2\varepsilon^2\mu D_1 u_0 \\
&\quad + 2\varepsilon u_0^n \cos 2t + 2n\varepsilon^2 u_0^{n-1} u_1 \cos 2t + \cdots \\
&= D_0^2 u_0 + \omega_0^2 u_0 \\
&\quad + \varepsilon\left(D_0^2 u_1 + \omega_0^2 u_1 + 2D_0 D_1 u_0 + 2\mu D_0 u_0 + 2u_0^n \cos 2t\right) \\
&\quad + \varepsilon^2[D_0^2 u_2 + \omega_0^2 u_2 + \left(D_1^2 + 2D_0 D_2\right)u_0 + 2D_0 D_1 u_1 \\
&\quad + 2\mu D_0 u_1 + 2\mu D_1 u_0 + 2n u_0^{n-1} u_1 \cos 2t] + \cdots
\end{aligned}$$

Equating coefficients of like powers of ε in the above equation yields

$$D_0^2 u_0 + \omega_0^2 u_0 = 0 \tag{5.13.2}$$

$$D_0^2 u_1 + \omega_0^2 u_1 = -2D_0 D_1 u_0 - 2\mu D_0 u_0 - 2u_0^n \cos 2T_0 \tag{5.13.3}$$

$$D_0^2 u_2 + \omega_0^2 u_2 = -\left(D_1^2 + 2D_0 D_2\right) u_0 - 2D_0 D_1 u_1 \\ - 2\mu D_0 u_1 - 2\mu D_1 u_0 - 2n u_0^{n-1} u_1 \cos 2t \tag{5.13.4}$$

The solution of (5.13.2) is

$$u_0 = A e^{i\omega_0 T_0} + \overline{A}\, e^{-i\omega_0 T_0} \tag{5.13.5}$$

where $A = A(T_1, T_2)$. Let

$$A = \frac{1}{2} a e^{i\beta} \tag{5.13.6}$$

Substituting this into (5.13.5), we can obtain the first order approximate solution of the given equation

$$u \approx u_0 = a \cos(\omega_0 t + \beta) + O(\varepsilon) \tag{5.13.7}$$

Substituting u_0 into (5.12.3) yields

$$\begin{aligned} D_0^2 u_1 + \omega_0^2 u_1 &= -2D_0 D_1 u_0 - 2\mu D_0 u_0 - 2u_0^n \cos 2T_0 \\ &= -2i\omega_0 D_1 A e^{i\omega_0 T_0} - 2i\omega_0 \mu A e^{i\omega_0 T_0} \\ &\quad - (A e^{i\omega_0 T_0} + \overline{A} e^{-i\omega_0 T_0})^n e^{i2T_0} + cc \\ &= -2i\omega_0 D_1 A e^{i\omega_0 T_0} - 2i\omega_0 \mu A e^{i\omega_0 T_0} \\ &\quad - \sum_{k=0}^{n} C_n^k A^{n-k} \overline{A}^k e^{i[(n-2k)\omega_0 + 2]T_0} + cc \end{aligned} \tag{5.13.8}$$

From the above equation, when the following equation is satisfied

$$(n - 2k)\omega_0 + 2 \approx \omega_0 \tag{5.13.9}$$

i.e.,

$$\omega_0 \approx \frac{2}{1 + 2k - n}, \quad k = 1, 2, \ldots, n \tag{5.13.10}$$

parametric resonance occurs and the first-order solution of Eq. (5.13.8) can be obtained. Since $\omega_0 \geq 0$ is required,$1 + 2k - n \geq 1$, i.e., k is the integer greater than or equal to $\left[\frac{n}{2}\right]$, where $\left[\frac{n}{2}\right]$ denotes the largest integer less than or equal to $\frac{n}{2}$. Therefore, for the case of n is an odd number, parameter resonance occurs when.

$$\omega_0 \approx 1, \ \frac{2}{4}, \ \frac{2}{6}, \ \frac{2}{8}, \ \ldots, \ \frac{2}{n+1} \text{for odd}\, n \tag{5.13.11}$$

For the case of n is an even number, parameter resonance occurs when.
$\omega_0 \approx 2,\ \frac{2}{3},\ \frac{2}{5},\ \frac{2}{7},\ \ldots,\ \frac{2}{n+1}$ for even n.

(b) When $(n+1)\omega_0 = 2 + \varepsilon\sigma$, (5.13.8) can be written as

$$\begin{aligned} D_0^2 u_1 + \omega_0^2 u_1 &= -2i\omega_0 D_1 A e^{i\omega_0 T_0} - 2i\omega_0 \mu A e^{i\omega_0 T_0} \\ &\quad - \sum_{k=0}^{n} C_n^k A^{n-k} \overline{A}^k e^{i[(n-2k)\omega_0 + 2]T_0} + cc \\ &= -2i\omega_0 D_1 A e^{i\omega_0 T_0} - 2i\omega_0 \mu A e^{i\omega_0 T_0} \\ &\quad - \overline{A}^n e^{-i\sigma T_1} e^{i\omega_0 T_0} + NST \end{aligned} \tag{5.13.12}$$

In order to eliminate secular terms from the above equation, there must be

$$2i\omega_0 D_1 A + 2i\omega_0 \mu A + \overline{A}^n e^{-i\sigma T_1} = 0 \tag{5.13.13}$$

Substituting $A = \frac{1}{2} a e^{i\beta}$ into the above equation gives

$$ia' - a\beta' + i\mu a + 2^{-n}\omega_0^{-1} a^n e^{-i[(n+1)\beta + \sigma T_1]} = 0 \tag{5.13.14}$$

The prime denotes the derivative with respect to T_1. Separating the real and imaginary parts of the above equation and considering $(n+1)\omega_0 \approx 2$, we get

$$\begin{aligned} a' &= -\mu a + (n+1)2^{-(n+1)} a^n \sin\gamma \\ a\beta' &= (n+1)2^{-(n+1)} a^n \cos\gamma \end{aligned} \tag{5.13.15}$$

$$\gamma = \sigma T_1 + (n+1)\beta \tag{5.13.16}$$

To analyze the stability of this set of steady-state solution, we can write (5.13.15) as

$$\begin{aligned} a\gamma' &= -\mu a + (n+1)2^{-(n+1)} a^n \sin\gamma \\ a\gamma' &= \sigma a + (n+1)^2 2^{-(n+1)} a^n \cos\gamma \end{aligned} \tag{5.13.17}$$

Let $a' = \gamma' = 0$ in (5.13.17), we obtain

$$- \mu a + (n+1)2^{-(n+1)} a^n \sin \gamma = 0,$$
$$\sigma a + (n+1)^2 2^{-(n+1)} a^n \cos \gamma = 0 \tag{5.13.18}$$

The steady state solutions can be obtained:

(1) $a = 0$, γ are arbitrary constants;
(2) $a \neq 0$, γ are certain constants.

(1) $\boldsymbol{a = 0}$ **and** $\boldsymbol{\gamma = \gamma_0}$ **are arbitrary constants.**

Let

$$a = 0 + a_1, \quad \gamma = \gamma_0 + \gamma_1 \tag{5.13.19}$$

Substituting (5.13.19) into (5.13.17), we can obtain

$$a_1' = -\mu a_1 \tag{5.13.20}$$

So $a_1 = e^{-\mu T_1} = e^{-\varepsilon \mu t} \to 0$, therefore this steady state solution is stable.

(2) $\boldsymbol{a = a_0 \neq 0, \gamma = \gamma_0}$

For this case, the frequency–response equation is

$$\mu^2 a^2 + (\frac{\sigma a}{n+1})^2 = (n+1)^2 2^{-2(n+1)} a^{2n} \tag{5.13.21}$$

Let

$$a = a_0 + a_1, \quad \gamma = \gamma_0 + \gamma_1 \tag{5.13.22}$$

Substituting (5.13.22) into (5.13.17), we can obtain the corresponding linearized equation

$$\begin{Bmatrix} a_1' \\ \gamma_1' \end{Bmatrix} = \begin{bmatrix} -\mu + n(n+1)2^{-(n+1)}a_0^{n-1}\sin\gamma_0 & (n+1)2^{-(n+1)}a_0^n\cos\gamma_0 \\ \sigma a_0^{-1} + n(n+1)^2 2^{-(n+1)}a_0^{n-2}\cos\gamma_0 & -(n+1)^2 2^{-(n+1)}a_0^{n-1}\sin\gamma_0 \end{bmatrix} \begin{Bmatrix} a_1 \\ \gamma_1 \end{Bmatrix} \tag{5.13.23}$$

For any set of steady state solutions $\boldsymbol{a} = \boldsymbol{a}_0 \neq 0$, $\boldsymbol{\gamma} = \boldsymbol{\gamma}_0$, one can calculate the eigenvalues of Eq. (5.13.23), and the stability of the corresponding steady state solutions can be determined from the sign of the real part of the two eigenvalues.

(c) When $(n-1)\omega_0 = 2 + \varepsilon\sigma$, (5.13.8) can be written as

$$\begin{aligned} D_0^2 u_1 + \omega_0^2 u_1 &= -2i\omega_0 D_1 A e^{i\omega_0 T_0} - 2i\omega_0 \mu A e^{i\omega_0 T_0} \\ &\quad - \sum_{k=0}^{n} C_n^k A^{n-k}\overline{A}^k e^{i[(n-2k)\omega_0+2]T_0} + cc \\ &= -2i\omega_0 D_1 A e^{i\omega_0 T_0} - 2i\omega_0 \mu A e^{i\omega_0 T_0} \\ &= -2i\omega_0 D_1 A e^{i\omega_0 T_0} - 2i\omega_0 \mu A e^{i\omega_0 T_0} \\ &\quad - nA\overline{A}^{n-1} e^{-i\sigma T_1} e^{i\omega_0 T_0} - A^n e^{i\sigma T_1} e^{i\omega_0 T_0} + NST \end{aligned} \tag{5.13.24}$$

In order to eliminate secular terms from the above equation, there must be

$$2i\omega_0 D_1 A + 2i\omega_0 \mu A + nA\overline{A}^{n-1} e^{-i\sigma T_1} e^{i\omega_0 T_0} + A^n e^{i\sigma T_1} = 0 \tag{5.13.25}$$

Substituting $A = \frac{1}{2}ae^{i\beta}$ into the above equation gives

$$\begin{aligned} &ia' - a\beta' + i\mu a + 2^{-n} n\omega_0^{-1} a^n e^{-i[(n-1)\beta + \sigma T_1]} \\ &+ 2^{-n}\omega_0^{-1} a^n e^{i[(n-1)\beta + \sigma T_1]} = 0 \end{aligned} \tag{5.13.26}$$

The prime denotes the derivative with respect to T_1. Separating the real and imaginary parts of the above equation and considering $(n-1)\omega_0 \approx 2$, we get

$$\begin{aligned} a' &= -\mu a + (n-1)^2 2^{-(n+1)} a^n \sin\gamma \\ a\beta' &= \left(n^2 - 1\right) 2^{-(n+1)} a^n \cos\gamma \end{aligned} \tag{5.13.27}$$

$$\gamma = \sigma T_1 + (n-1)\beta \tag{5.13.28}$$

To analyze the stability of this set of steady-state solution, (5.13.15) is written as

$$a' = -\mu a + (n-1)^2 2^{-(n+1)} a^n \sin\gamma$$
$$a\gamma' = \sigma a + (n+1)(n-1)^2 2^{-(n+1)} a^n \cos\gamma \quad (5.13.29)$$

Let $a' = \gamma' = 0$ in (5.13.29), we obtain

$$-\mu a + (n-1)^2 2^{-(n+1)} a^n \sin\gamma = 0,$$
$$\sigma a + (n+1)(n-1)^2 2^{-(n+1)} a^n \cos\gamma = 0 \quad (5.13.30)$$

The steady state solutions can be obtained:

(1) $a = 0$, γ are arbitrary constants;
(2) $a \neq 0$, γ are certain constants.
(1) **$a = 0$ and $\gamma = \gamma_0$ are arbitrary constants.**

Let

$$a = 0 + a_1, \quad \gamma = \gamma_0 + \gamma_1 \quad (5.13.31)$$

Substituting (5.13.31) into (5.11.20), we can obtain the corresponding linearized equation

$$a_1' = -\mu a_1 \quad (5.13.32)$$

So $a_1 = e^{-\mu T_1} = e^{-\varepsilon\mu t} \to 0$, therefore this steady state solution is stable.

(2) **$a = a_0 \neq 0$, $\gamma = \gamma_0$**

For this case, the frequency–response equation is

$$\mu^2 a^2 + \left(\frac{\sigma a}{n+1}\right)^2 = (n-1)^4 2^{-2(n+1)} a^{2n} \quad (5.13.33)$$

Let

$$a = a_0 + a_1, \quad \gamma = \gamma_0 + \gamma_1 \quad (5.13.34)$$

Substituting (5.13.34) into (5.13.17), we can obtain the corresponding linearized equation

$$\begin{Bmatrix} a_1' \\ \gamma_1' \end{Bmatrix} = \begin{bmatrix} -\mu + n(n-1)^2 2^{-(n+1)} a_0^{n-1} \sin\gamma_0 & (n-1)^2 2^{-(n+1)} a_0^n \cos\gamma_0 \\ \sigma a_0^{-1} + n(n+1)(n-1)^2 2^{-(n+1)} a_0^{n-2} \cos\gamma_0 & -(n+1)(n-1)^2 2^{-(n+1)} a_0^{n-1} \sin\gamma_0 \end{bmatrix} \begin{Bmatrix} a_1 \\ \gamma_1 \end{Bmatrix} \quad (5.13.35)$$

For any set of steady state solutions $a = a_0 \neq 0, \gamma = \gamma_0$, one can calculate the eigenvalues of Eq. (5.13.36), and the stability of the corresponding steady state solutions can be determined from the sign of the real part of the two eigenvalues.

5.14 Exercise 5.14 (The Parametric Excitation of a System with a Nonlinear Damping)

Solution: (a) We use the method of multiple scales to solve the first order approximate periodic solution of the system. Let the solution of the equation be

$$u(t;\varepsilon) = u_0(T_0, T_1, T_2) + \varepsilon u_1(T_0, T_1, T_2) + \varepsilon^2 u_2(T_0, T_1, T_2) + \cdots \tag{5.14.1}$$

Substituting the above equation into the original equation and retaining to $O(\varepsilon^2)$, we get

$$\begin{aligned}
0 &= \ddot{u} + \omega_0^2 u + \varepsilon\mu|\dot{u}|^n\dot{u} + 2\varepsilon u\cos 2t \\
&= \left(D_0^2 + 2\varepsilon D_0 D_1\right)(u_0 + \varepsilon u_1) + \omega_0^2(u_0 + \varepsilon u_1) \\
&\quad + 2\varepsilon\mu(D_0 + \varepsilon D_1)(u_0 + \varepsilon \dot{u}_1)\left|(D_0 + \varepsilon D_1)(u_0 + \varepsilon u_1)\right|^n \\
&\quad + 2\varepsilon(u_0 + \varepsilon u_1)\cos 2t \\
&= D_0^2 u_0 + \varepsilon D_0^2 u_1 + 2\varepsilon D_0 D_1 u_0 \\
&\quad + \omega_0^2 u_0 + \varepsilon\omega_0^2 u_1 \\
&\quad + \varepsilon\mu D_0 u_0 \left|D_0 u_0\right|^n + 2\varepsilon u_0 \cos 2t + \cdots \\
&= D_0^2 u_0 + \omega_0^2 u_0 + \varepsilon[D_0^2 u_1 + \omega_0^2 u_1 + 2D_0 D_1 u_0 \\
&\quad + \mu D_0 u_0 \left|D_0 u_0\right|^n + 2u_0 \cos 2t] + \cdots
\end{aligned}$$

Equating coefficients of like powers of ε in the above equation yields

$$D_0^2 u_0 + \omega_0^2 u_0 = 0 \tag{5.14.2}$$

$$D_0^2 u_1 + \omega_0^2 u_1 = -2D_0 D_1 u_0 - \mu D_0 u_0 \left|D_0 u_0\right|^n - 2u_0 \cos 2t \tag{5.14.3}$$

The solution of (5.14.2) is

$$u_0 = A e^{i\omega_0 T_0} + \bar{A}\, e^{-i\omega_0 T_0} \tag{5.14.4}$$

where $A = A(T_1, T_2)$. Let

$$A = \frac{1}{2} a e^{i\beta} \tag{5.14.5}$$

Substituting this into(5.14.4), we can obtain the first order approximate solution of the given equation

$$u \approx u_0 = a\cos(\omega_0 t + \beta) + O(\varepsilon) \tag{5.14.6}$$

Substituting u_0 into (5.14.3) yields

$$\begin{aligned}
D_0^2u_1 + \omega_0^2 u_1 &= -2D_0D_1u_0 - \mu D_0u_0\left|D_0u_0\right|^n - 2u_0\cos 2T_0 \\
&= -2i\omega_0 D_1 Ae^{i\omega_0T_0} - \left(Ae^{i\omega_0T_0} + \overline{A}e^{-i\omega_0T_0}\right)e^{i2T_0} + cc \\
&\quad - \mu\left(i\omega_0 Ae^{i\omega_0T_0} - i\omega_0\overline{A}e^{-i\omega_0T_0}\right) \\
&\quad \left|i\omega_0 Ae^{i\omega_0T_0} - i\omega_0\overline{A}e^{-i\omega_0T_0}\right|^n \\
&= -2i\omega_0 D_1Ae^{i\omega_0T_0} - Ae^{i(\omega_0+2)T_0} \\
&\quad - \overline{A}e^{i(2-\omega_0)T_0} + cc \\
&\quad - i\omega_0^{n+1}\mu\left(Ae^{i\omega_0T_0} - \overline{A}e^{-i\omega_0T_0}\right)\left|Ae^{i\omega_0T_0} - \overline{A}e^{-i\omega_0T_0}\right|^n
\end{aligned} \tag{5.14.7}$$

In order to further process the Eq. (5.14.7), we need to deal with the function $\left(Ae^{i\omega_0T_0} - \overline{A}\,e^{-i\omega_0T_0}\right)\left|Ae^{i\omega_0T_0} - \overline{A}\,e^{-i\omega_0T_0}\right|^n$, which can be expanded into a Fourier series of complex exponential form:

$$\left(Ae^{i\omega_0T_0} - \overline{A}\,e^{-i\omega_0T_0}\right)\left|Ae^{i\omega_0T_0} - \overline{A}\,e^{-i\omega_0T_0}\right|^n = \sum_{m=-\infty}^{\infty} f_m e^{im\omega_0T_0} \tag{5.14.8}$$

where

$$f_m = \frac{\omega_0}{2\pi}\int_0^{2\pi/\omega_0}\left[\left(Ae^{i\omega_0T_0} - \overline{A}\,e^{-i\omega_0T_0}\right)\left|Ae^{i\omega_0T_0} - \overline{A}\,e^{-i\omega_0T_0}\right|^n\right]e^{-im\omega_0T_0}dT_0 \tag{5.14.9}$$

Thus, (5.14.7) becomes

$$\begin{aligned}
D_0^2u_1 + \omega_0^2u_1 &= -2i\omega_0D_1Ae^{i\omega_0T_0} - Ae^{i(\omega_0+2)T_0} - \overline{A}e^{i(2-\omega_0)T_0} + cc \\
&\quad - i\omega_0^{n+1}\mu\left(Ae^{i\omega_0T_0} - \overline{A}e^{-i\omega_0T_0}\right)\sum\nolimits_{m=-\infty}^{\infty} f_n e^{im\omega_0T_0} \\
&= -2i\omega_0D_1Ae^{i\omega_0T_0} - Ae^{i(\omega_0+2)T_0} - \overline{A}e^{i(2-\omega_0)T_0} \\
&\quad - i\omega_0^{n+1}\mu f_1 e^{i\omega_0T_0} + cc + NST
\end{aligned} \tag{5.14.10}$$

From the above equation, it can be seen that only when $\omega_0 \approx 1$, parametric resonance can occur.

(b) When ω_0 is far from 1, in order to eliminate secular terms from (5.14.10), there must be

$$2i\omega_0D_1A + 2i\omega_0^{n+1}\mu f_1 = 0 \tag{5.14.11}$$

Let $A = \frac{1}{2}ae^{i\beta}$, then we have

$$
\begin{aligned}
f_1 &= \frac{\omega_0}{2\pi}\int\limits_0^{2\pi/\omega_0}\left[\left(Ae^{i\omega_0 T_0}-\overline{A}e^{-i\omega_0 T_0}\right)\left|Ae^{i\omega_0 T_0}-\overline{A}e^{-i\omega_0 T_0}\right|^n\right]e^{-i\omega_0 T_0}dT_0\\
&= \frac{\omega_0}{2\pi}\int\limits_0^{2\pi/\omega_0} ia^{n+1}e^{i\beta}\sin(\omega_0 T_0+\beta)|\sin(\omega_0 T_0+\beta)|^n e^{-i(\omega_0 T_0+\beta)}dT_0\\
&= \frac{1}{2\pi}\int\limits_0^{2\pi} ia^{n+1}e^{i\beta}\sin\phi|\sin\phi|^n e^{-i\phi}d\phi\\
&= \frac{a^{n+1}e^{i\beta}}{2\pi}\int\limits_0^{2\pi}\sin^2\phi|\sin\phi|^n d\phi\\
&\triangleq ba^{n+1}e^{i\beta}
\end{aligned}
\tag{5.14.12}
$$

Equation (5.14.11) becomes

$$
ia' - a\beta' + i\omega_0^n\mu ba^{n+1} = 0 \tag{5.14.13}
$$

The prime denotes the derivative with respect to T_1. Separating the real and imaginary parts of the above equation gives

$$
a' = -\mu\omega_0^n ba^{n+1},\quad \beta' = 0 \tag{5.14.14}
$$

where

$$
\begin{aligned}
b &= \frac{1}{2\pi}\int\limits_0^{2\pi}\sin^2\varphi|\sin\varphi|^n d\varphi\\
&= \Gamma\left[\frac{1}{2}(n+3)\right]/\sqrt{\pi}\,\Gamma\left[\frac{1}{2}(n+4)\right]
\end{aligned}
\tag{5.14.15}
$$

(c) When $\omega_0 \approx 0$, set

$$
\omega_0 = 1 + \varepsilon\sigma \tag{5.14.16}
$$

In order to eliminate secular terms from the Eq. (5.14.10), there must be

$$
2i\omega_0 D_1 A + \overline{A}\,e^{-2i\sigma T_1} + i\omega_0^{n+1}\mu f_1 = 0 \tag{5.14.17}
$$

Substituting $A = \frac{1}{2}ae^{i\beta}$ and (5.14.12) into the above equation, we have

$$ia' - a\beta' + i\omega_0^n \mu b a^{n+1} + \frac{1}{2}\omega_0^{-1} a e^{-i(2\sigma T_1 + 2\beta)} = 0 \tag{5.14.18}$$

The prime denotes the derivative with respect to T_1. Separating the real and imaginary parts of the above equation and taking into account $\omega_0 \approx 1$, we get

$$\begin{aligned} a' &= -\mu b a^{n+1} + \tfrac{1}{2} a \sin\gamma \\ a\beta' &= \tfrac{1}{2} a \cos\gamma \end{aligned} \tag{5.14.19}$$

where

$$\gamma = 2\sigma T_1 + 2\beta \tag{5.14.20}$$

5.15 Exercise 5.15 (The Nonlinear Parametric Excitation of a System with a Linear Viscous Damping)

Solution: (a) We use the method of multiple scales to solve the first order approximate periodic solution of the system. Let the solution of the equation be

$$u(t;\varepsilon) = u_0(T_0, T_1, T_2) + \varepsilon u_1(T_0, T_1, T_2) + \varepsilon^2 u_2(T_0, T_1, T_2) + \cdots \tag{5.15.1}$$

Substituting the above equation into the original equation and retaining to $O(\varepsilon)$, we get

$$\begin{aligned} 0 &= \ddot{u} + \omega_0^2 u + 2\varepsilon\mu\dot{u} + \varepsilon\alpha_1 u^n + 2\varepsilon\alpha_2 u \cos 2t \\ &= \left(D_0^2 + 2\varepsilon D_0 D_1\right)(u_0 + \varepsilon u_1) + \omega_0^2(u_0 + \varepsilon u_1) \\ &\quad + 2\varepsilon\mu(D_0 + \varepsilon D_1)(u_0 + \varepsilon u_1) + \varepsilon\alpha_1(u_0 + \varepsilon u_1)^n \\ &\quad + 2\varepsilon\alpha_2(u_0 + \varepsilon u_1)\cos 2t + \cdots \\ &= D_0^2 u_0 + \varepsilon D_0^2 u_1 + 2\varepsilon D_0 D_1 u_0 + \omega_0^2 u_0 + \varepsilon\omega_0^2 u_1 \\ &\quad + 2\varepsilon\mu D_0 u_0 + \varepsilon\alpha_1 u_0^n + 2\varepsilon\alpha_2 u_0 \cos 2t + \cdots \\ &= D_0^2 u_0 + \omega_0^2 u_0 + \varepsilon(D_0^2 u_1 + \omega_0^2 u_1 + 2D_0 D_1 u_0 \\ &\quad + 2\mu D_0 u_0 + \alpha_1 u_0^n + 2\alpha_2 u_0 \cos 2t) + \cdots \end{aligned}$$

Equating coefficients of like powers of ε in the above equation yields

$$D_0^2 u_0 + \omega_0^2 u_0 = 0 \tag{5.15.2}$$

$$D_0^2 u_1 + \omega_0^2 u_1 = -2D_0 D_1 u_0 - 2\mu D_0 u_0 - \alpha_1 u_0^n - 2\alpha_2 u_0 \cos 2T_0 \tag{5.15.3}$$

The solution of (5.14.2) is

$$u_0 = Ae^{i\omega_0 T_0} + \bar{A}\,e^{-i\omega_0 T_0} \tag{5.15.4}$$

where $A = A(T_1, T_2)$. Let

$$A = \frac{1}{2}ae^{i\beta} \tag{5.15.5}$$

Substituting this into (5.14.4), we can obtain the first order approximate solution of the given equation

$$u \approx u_0 = a\cos(\omega_0 t + \beta) + O(\varepsilon) \tag{5.15.6}$$

Substituting u_0 into (5.14.3) yields

$$\begin{aligned}
D_0^2 u_1 + \omega_0^2 u_1 &= -2D_0 D_1 u_0 - 2\mu D_0 u_0 - \alpha_1 u_0^n - 2\alpha_2 u_0 \cos 2T_0 \\
&= -2i\omega_0 D_1 A e^{i\omega_0 T_0} - 2i\mu\omega_0 A e^{i\omega_0 T_0} \\
&\quad - \alpha_2\left(Ae^{i\omega_0 T_0} + \bar{A}e^{-i\omega_0 T_0}\right)e^{i2T_0} \\
&\quad + cc - \alpha_1 \sum_{k=0}^{n} C_n^k A^{n-k}\bar{A}^k e^{i(n-2k)\omega_0 T_0} \\
&= -2i\omega_0 D_1 A e^{i\omega_0 T_0} - 2i\mu\omega_0 A e^{i\omega_0 T_0} \\
&\quad - \alpha_2 \bar{A} e^{i(2-\omega_0)T_0} + cc + NST \\
&\quad - \alpha_1 \sum\nolimits_{k=0}^{n} C_n^k A^{n-k}\bar{A}^k e^{i(n-2k)\omega_0 T_0}
\end{aligned} \tag{5.15.7}$$

CASE I: When ω_0 is far from 1:

Let n be an odd number. In order to eliminate secular terms from (5.15.7), there must be

$$2i\omega_0 D_1 A + 2i\mu\omega_0 A + \alpha_1 C_n^{(n+1)/2} A^{(n-1)/2}\bar{A}^{-(n+1)/2} = 0 \tag{5.15.8}$$

Substituting $A = \frac{1}{2}ae^{i\beta}$ into the above equation, we get

$$ia' - a\beta' + i\mu a + 2^{-n}\alpha_1 C_n^{(n+1)/2} a^n = 0 \tag{5.15.9}$$

The prime denotes the derivative with respect to T_1. Separating the real and imaginary parts from the above equation gives

$$a' = -\mu a, \quad \beta' = \frac{b\alpha_1 a^{n-1}}{\omega_0 2^n},$$

$$b = \frac{n!}{\frac{1}{2}(n+1)!\frac{1}{2}(n-1)!} \tag{5.15.10}$$

where

$$b = \frac{n!}{\frac{1}{2}(n+1)!\frac{1}{2}(n-1)!} \tag{5.15.11}$$

CASE II: When $\omega_0 = 1 + \varepsilon\sigma$:

Let n be an odd number. In order to eliminate secular terms from (5.15.7), there must be

$$2i\omega_0 D_1 A + 2i\mu\omega_0 A + \alpha_2 \bar{A} e^{-i2\sigma T_1} + \alpha_1 C_n^{(n+1)/2} A^{(n-1)/2} \bar{A}^{-(n+1)/2} = 0 \tag{5.15.12}$$

Substituting $A = \frac{1}{2}ae^{i\beta}$ into the above equation, we get

$$ia' - a\beta' + i\mu a + 2^{-n}\alpha_1 C_n^{(n+1)/2} a^n + \frac{1}{2}\alpha_2 a e^{-i(2\sigma T_1 + 2\beta)} = 0 \tag{5.15.13}$$

The prime denotes the derivative with respect to T_1. Separating the real and imaginary parts of the above equation gives

$$a' = -\mu a + \frac{1}{2}\alpha_2 a \sin\gamma, \; a\beta' = \alpha_1 b 2^{-n} a^n + \frac{1}{2}\alpha_2 a \cos\gamma \tag{5.15.14}$$

where

$$\gamma = 2\sigma T_1 + 2\beta \tag{5.15.15}$$

(b) We only analyze the stability of the steady state solution in CASE II above. Equation (5.15.14) can be written as

$$\begin{aligned} a' &= -\mu a + \tfrac{1}{2}\alpha_2 a \sin\gamma \\ a\gamma' &= 2\sigma a + \alpha_1 b 2^{-n+1} a^n + \alpha_2 a \cos\gamma \end{aligned} \tag{5.15.16}$$

Let $a' = \gamma' = 0$ in (5.15.16), we can obtain

$$-\mu a + \frac{1}{2}\alpha_2 a \sin\gamma = 0, \; 2\sigma a + \alpha_1 b 2^{-n+1} a^n + \alpha_2 a \cos\gamma = 0 \tag{5.15.17}$$

The steady state solutions can be obtained:

(1) $a = 0$, γ are arbitrary constants;
(2) $a \neq 0$, γ are certain constants.

(1) $\boldsymbol{a = 0}$ **and** $\boldsymbol{\gamma = \gamma_0}$ **are arbitrary constants.**

Let

$$a = 0 + a_1, \quad \gamma = \gamma_0 + \gamma_1 \tag{5.15.18}$$

Substituting (5.15.18) into (5.11.20), the corresponding linearized equation is given by

$$a_1' = -\left(\mu - \frac{1}{2}\alpha_2 \sin\gamma_0\right)a_1 \tag{5.15.19}$$

Therefore, this steady state solution is stable when $\mu \geq \frac{1}{2}\alpha_2 \sin\gamma_0$ and vice versa. Since γ_0 is arbitrary and depends on the initial state of the system, the stability of the system also depends on the initial state of the system.

(2) $\boldsymbol{a = a_0 \neq 0, \gamma = \gamma_0}$

For this case, the frequency–response equation is

$$4\mu^2 + (2\sigma + \alpha_1 b 2^{-n+1} a^{n-1})^2 = \alpha_2^2 \tag{5.15.20}$$

Let

$$a = a_0 + a_1, \quad \gamma = \gamma_0 + \gamma_1 \tag{5.15.21}$$

Substituting (5.15.21) into (5.15.16), the corresponding linearized equation is given by

$$\begin{Bmatrix} a_1' \\ \gamma_1' \end{Bmatrix} = \begin{bmatrix} -\mu + \frac{1}{2}\alpha_2 \sin\gamma_0 & \frac{1}{2}\alpha_2 a_0 \cos\gamma_0 \\ 2\sigma a_0^{-1} + n\alpha_1 b 2^{-n+1} a_0^{n-2} + \alpha_2 a_0^{-1}\cos\gamma_0 & -\alpha_2 \sin\gamma_0 \end{bmatrix} \begin{Bmatrix} a_1 \\ \gamma_1 \end{Bmatrix} \tag{5.15.22}$$

For any set of steady state solutions $\boldsymbol{a = a_0 \neq 0, \gamma = \gamma_0}$, one can calculate the eigenvalues of Eq. (5.15.23), and the stability of the corresponding steady state solutions can be determined from the sign of the real part of the two eigenvalues.

5.16 Exercise 5.16 (The Nonlinear Parametric Excitation of the System with a Nonlinear Damping)

Solution: (a) We use the method of multiple scales to solve the first order approximate periodic solution of the system. Let the solution of the equation be

$$u(t;\varepsilon) = u_0(T_0, T_1, T_2) + \varepsilon u_1(T_0, T_1, T_2) + \varepsilon^2 u_2(T_0, T_1, T_2) + \cdots \tag{5.16.1}$$

Substituting the above equation into the original equation and retaining to $O(\varepsilon^2)$, we get

$$
\begin{aligned}
0 &= \ddot{u} + 2\varepsilon\mu_1\dot{u} + \varepsilon\mu_2|\dot{u}|^{\,n}\dot{u} + \left(\omega_0^2 + \varepsilon\cos 2t\right)\left(u + \varepsilon\alpha u^3\right) \\
&= \ddot{u} + \omega_0^2 u + 2\varepsilon\mu_1\dot{u} + \varepsilon\omega_0^2\alpha u^3 + \varepsilon\mu_2|\dot{u}|^{\,n}\dot{u} \\
&\quad + \varepsilon\left(u + \varepsilon\alpha u^3\right)\cos 2t \\
&= \left(D_0^2 + 2\varepsilon D_0 D_1\right)(u_0 + \varepsilon u_1) + \omega_0^2(u_0 + \varepsilon u_1) \\
&\quad + 2\varepsilon\mu_1(D_0 + \varepsilon D_1)(u_0 + \varepsilon u_1) \\
&\quad + \varepsilon\omega_0^2\alpha(u_0 + \varepsilon u_1)^3 \\
&\quad + \varepsilon\mu_2(D_0 + \varepsilon D_1)(u_0 + \varepsilon u_1)\,|(D_0 + \varepsilon D_1)(u_0 + \varepsilon u_1)|^{\,n} \\
&\quad + \varepsilon\alpha_2(u_0 + \varepsilon u_1)\cos 2t + \cdots \\
&= D_0^2 u_0 + \varepsilon D_0^2 u_1 + 2\varepsilon D_0 D_1 u_0 + \omega_0^2 u_0 \\
&\quad + \varepsilon\omega_0^2 u_1 + 2\varepsilon\mu_1 D_0 u_0 \\
&\quad + \varepsilon\omega_0^2\alpha u_0^3 + \varepsilon\mu_2 D_0 u_0\,|D_0 u_0|^{\,n} + \varepsilon\alpha u_0\cos 2t + \cdots \\
&= D_0^2 u_0 + \omega_0^2 u_0 + \varepsilon(D_0^2 u_1 + \omega_0^2 u_1 \\
&\quad + 2D_0 D_1 u_0 + 2\mu_1 D_0 u_0 + \omega_0^2\alpha u_0^3 \\
&\quad + \mu_2 D_0 u_0\,|D_0 u_0|^{\,n} + \alpha u_0\cos 2t) + \cdots
\end{aligned}
$$

Equating coefficients of like powers of ε in the above equation yields

$$D_0^2 u_0 + \omega_0^2 u_0 = 0 \tag{5.16.2}$$

$$
\begin{aligned}
D_0^2 u_1 + \omega_0^2 u_1 &= -2D_0 D_1 u_0 - 2\mu_1 D_0 u_0 \\
&\quad - \omega_0^2\alpha u_0^3 - \mu_2 D_0 u_0\,|D_0 u_0|^n - \alpha u_0 \cos 2T_0
\end{aligned}
\tag{5.16.3}
$$

The solution of (5.16.2) is

$$u_0 = Ae^{i\omega_0 T_0} + \bar{A}\,e^{-i\omega_0 T_0} \tag{5.16.4}$$

where $A = A(T_1, T_2)$. Let

$$A = \frac{1}{2}ae^{i\beta} \tag{5.16.5}$$

Substituting this into (5.16.4), we can obtain the first order approximate solution of the given equation

$$u \approx u_0 = a\cos(\omega_0 t + \beta) + O(\varepsilon) \tag{5.16.6}$$

Substituting u_0 into (5.16.3) yields

$$\begin{aligned}
D_0^2u_1+\omega_0^2u_1 &= -2D_0D_1u_0-2\mu_1D_0u_0-\omega_0^2\alpha u_0^3\\
&\quad -\mu_2D_0u_0\left|D_0u_0\right|^n-\alpha u_0\cos 2T_0\\
&= -2i\omega_0D_1Ae^{i\omega_0T_0}-2i\omega_0\mu_1Ae^{i\omega_0T_0}\\
&\quad -\frac{1}{2}\alpha\left(Ae^{i\omega_0T_0}+\overline{A}e^{-i\omega_0T_0}\right)e^{i2T_0}\\
&\quad -\omega_0^2\alpha A^3e^{3i\omega_0T_0}-3\omega_0^2\alpha A^2\overline{A}e^{i\omega_0T_0}+cc\\
&\quad -\mu_2\left(i\omega_0Ae^{i\omega_0T_0}-i\omega_0\overline{A}e^{-i\omega_0T_0}\right)\\
&\quad \left|i\omega_0Ae^{i\omega_0T_0}-i\omega_0\overline{A}e^{-i\omega_0T_0}\right|^n\\
&= -2i\omega_0D_1Ae^{i\omega_0T_0}-2i\omega_0\mu_1Ae^{i\omega_0T_0}\\
&\quad -\frac{1}{2}\alpha Ae^{i(\omega_0+2)T_0}-\frac{1}{2}\alpha\overline{A}e^{i(2-\omega_0)T_0}\\
&\quad -3\omega_0^2\alpha A^2\overline{A}e^{i\omega_0T_0}+cc+NST\\
&\quad -i\omega_0^{n+1}\mu_2\left(Ae^{i\omega_0T_0}-\overline{A}e^{-i\omega_0T_0}\right)\\
&\quad \left|Ae^{i\omega_0T_0}-\overline{A}e^{-i\omega_0T_0}\right|^n
\end{aligned}\tag{5.16.7}$$

In order to further process (5.16.7), we need to first deal with the function $\left(Ae^{i\omega_0T_0}-\bar{A}\,e^{-i\omega_0T_0}\right)\left|Ae^{i\omega_0T_0}-\bar{A}\,e^{-i\omega_0T_0}\right|^n$, which can be expanded into a Fourier series of complex exponential form:

$$\left(Ae^{i\omega_0T_0}-\bar{A}\,e^{-i\omega_0T_0}\right)\left|Ae^{i\omega_0T_0}-\bar{A}\,e^{-i\omega_0T_0}\right|^n=\sum_{m=-\infty}^{\infty}f_me^{im\omega_0T_0}\tag{5.16.8}$$

where

$$f_m=\frac{\omega_0}{2\pi}\int_0^{2\pi/\omega_0}\left[\left(Ae^{i\omega_0T_0}-\bar{A}\,e^{-i\omega_0T_0}\right)\left|Ae^{i\omega_0T_0}-\bar{A}\,e^{-i\omega_0T_0}\right|^n\right]e^{-im\omega_0T_0}dT_0\tag{5.16.9}$$

Thus (5.16.7) becomes

$$\begin{aligned}
D_0^2u_1+\omega_0^2u_1 &= -2i\omega_0D_1Ae^{i\omega_0T_0}-2i\omega_0\mu_1Ae^{i\omega_0T_0}\\
&\quad -\frac{1}{2}\alpha Ae^{i(\omega_0+2)T_0}-\frac{1}{2}\alpha\overline{A}e^{i(2-\omega_0)T_0}\\
&\quad -3\omega_0^2\alpha A^2\overline{A}e^{i\omega_0T_0}+cc+NST\\
&\quad -i\omega_0^{n+1}\mu_2\left(Ae^{i\omega_0T_0}-\overline{A}e^{-i\omega_0T_0}\right)\left|Ae^{i\omega_0T_0}-\overline{A}e^{-i\omega_0T_0}\right|^n\\
&= -2i\omega_0D_1Ae^{i\omega_0T_0}-2i\omega_0\mu_1Ae^{i\omega_0T_0}
\end{aligned}$$

$$
\begin{aligned}
&-\frac{1}{2}\alpha Ae^{i(\omega_0+2)T_0}-\frac{1}{2}\alpha\overline{A}e^{i(2-\omega_0)T_0}\\
&-3\omega_0^2\alpha A^2\overline{A}e^{i\omega_0T_0}-i\omega_0^{n+1}\mu_2f_1e^{i\omega_0T_0}+cc+NST
\end{aligned}
\tag{5.16.10}
$$

For the case of $\omega_0=1+\varepsilon\sigma$, in order to eliminate secular terms from the Eq. (5.16.10), there must be

$$
2i\omega_0D_1A+2i\omega_0\mu_1A+\frac{1}{2}\alpha\,\overline{A}\,e^{-2i\sigma T_1}+3\omega_0^2\alpha A^2\,\overline{A}+i\omega_0^{n+1}\mu_2f_1=0 \tag{5.16.11}
$$

Let $A=\frac{1}{2}ae^{i\beta}$, then we have

$$
\begin{aligned}
f_1&=\frac{\omega_0}{2\pi}\int\limits_0^{2\pi/\omega_0}\left[\left(Ae^{i\omega_0T_0}-\overline{A}e^{-i\omega_0T_0}\right)\left|Ae^{i\omega_0T_0}-\overline{A}e^{-i\omega_0T_0}\right|^n\right]e^{-i\omega_0T_0}dT_0\\
&=\frac{\omega_0}{2\pi}\int\limits_0^{2\pi/\omega_0}ia^{n+1}e^{i\beta}\sin(\omega_0T_0+\beta)|\sin(\omega_0T_0+\beta)|^ne^{-i(\omega_0T_0+\beta)}dT_0\\
&=\frac{1}{2\pi}\int\limits_0^{2\pi}ia^{n+1}e^{i\beta}\sin\phi|\sin\phi|^ne^{-i\phi}d\phi\\
&=\frac{a^{n+1}e^{i\beta}}{2\pi}\int\limits_0^{2\pi}\sin^2\phi|\sin\phi|^nd\phi\\
&\triangleq ba^{n+1}e^{i\beta}
\end{aligned}
\tag{5.16.12}
$$

where

$$
b=\frac{1}{2\pi}\int\limits_0^{2\pi}\sin^2\varphi|\sin\varphi|^nd\varphi=\Gamma\left[\frac{1}{2}(n+3)\right]/\sqrt{\pi}\Gamma\left[\frac{1}{2}(n+4)\right] \tag{5.16.13}
$$

Equation (5.16.11) becomes

$$
ia'-a\beta'+2i\mu_1a+\frac{1}{4}\alpha ae^{-i(2\sigma T_1+2\beta)}+\frac{3}{8}\alpha a^3+i\mu_2ba^{n+1}=0 \tag{5.16.14}
$$

The prime denotes the derivative with respect to T_1. Separating the real and imaginary parts of the above equation gives

$$
\begin{aligned}
a'&=-\mu_1a-\mu_2ba^{n+1}+\tfrac{1}{4}a\sin\gamma\\
a\beta'&=\tfrac{3}{8}\alpha a^3+\tfrac{1}{4}a\cos\gamma
\end{aligned}
\tag{5.16.15}
$$

where

$$\gamma = 2\sigma T_1 + 2\beta \tag{5.16.16}$$

(b) Let's analyze the stability of the steady state solution for the above case. Write (5.16.15) as

$$\begin{aligned} a' &= -\mu_1 a - \mu_2 b a^{n+1} + \tfrac{1}{4} a \sin\gamma \\ a\gamma' &= 2\sigma a + \tfrac{3}{4}\alpha a^3 + \tfrac{1}{2} a \cos\gamma \end{aligned} \tag{5.16.17}$$

Let $a' = \gamma' = 0$ in (5.16.17), we obtain

$$-\mu_1 a - \mu_2 b a^{n+1} + \frac{1}{4} a \sin\gamma = 0, \quad 2\sigma a + \frac{3}{4}\alpha a^3 + \frac{1}{2} a \cos\gamma = 0 \tag{5.16.18}$$

The steady state solutions can be obtained:

(1) $a = 0$, γ are arbitrary constants;
(2) $a \neq 0$, γ are certain constants.
(1) $\boldsymbol{a} = 0$ **and** $\boldsymbol{\gamma} = \boldsymbol{\gamma}_0$ **are arbitrary constants.**

Let

$$a = 0 + a_1, \quad \gamma = \gamma_0 + \gamma_1 \tag{5.16.19}$$

Substituting (5.16.19) into (5.16.17), we can obtain the corresponding linearized equation

$$a_1' = -\left(\mu_1 - \frac{1}{4}\sin\gamma_0\right) a_1 \tag{5.16.20}$$

Therefore, this steady state solution is stable when $\mu_1 \geq \frac{1}{4}\sin\gamma_0$ and vice versa. Since γ_0 is arbitrary and depends on the initial state of the system, the stability of the system also depends on the initial state of the system.

(2) $\boldsymbol{a} = \boldsymbol{a}_0 \neq 0, \boldsymbol{\gamma} = \boldsymbol{\gamma}_0$

For this case, the frequency–response equation is

$$16(\mu_1 - \mu_2 b a^n)^2 + (4\sigma + \frac{3}{2}\alpha a^2)^2 = 1 \tag{5.16.21}$$

Let

$$a = a_0 + a_1, \quad \gamma = \gamma_0 + \gamma_1 \tag{5.16.22}$$

Substituting (5.16.22) into (5.16.17), we can obtain the corresponding linearized equation

$$\begin{Bmatrix} a_1' \\ \gamma_1' \end{Bmatrix} = \begin{bmatrix} -\mu_1 - (n+1)\mu_2 b a_0^n + \frac{1}{4}\sin\gamma_0 & \frac{1}{4}a_0\cos\gamma_0 \\ 2\sigma a_0^{-1} + \frac{9}{4}\alpha a_0 + \frac{1}{2}a_0^{-1}\cos\gamma_0 & -\frac{1}{2}\sin\gamma_0 \end{bmatrix} \begin{Bmatrix} a_1 \\ \gamma_1 \end{Bmatrix} \tag{5.16.23}$$

For any set of steady state solutions $a = a_0 \neq 0, \gamma = \gamma_0$, one can calculate the eigenvalues of Eq. (5.16.24), and the stability of the corresponding steady state solutions can be determined from the sign of the real part of the two eigenvalues.

5.17 Exercise 5.17 (The Nonlinear Parametric Excitation of the System with a Nonlinear Damping and High Order Nonlinearities)

Solution: **(a)** We use the method of multiple scales to solve the first order approximate periodic solution of the system. Let the solution of the equation be

$$u(t;\varepsilon) = u_0(T_0, T_1, T_2) + \varepsilon u_1(T_0, T_1, T_2) + \varepsilon^2 u_2(T_0, T_1, T_2) + \cdots \tag{5.17.1}$$

Substituting the above equation into the original equation and retaining to $O(\varepsilon^2)$, we get

$$\begin{aligned} 0 &= \ddot{u} + \omega_0^2 u + \varepsilon\mu|\dot{u}|^n\dot{u} + \varepsilon\alpha_1 u^m + 2\varepsilon\alpha_2 u^k\cos 2t \\ &= \left(D_0^2 + 2\varepsilon D_0 D_1\right)(u_0 + \varepsilon u_1) + \omega_0^2(u_0 + \varepsilon u_1) \\ &\quad + \varepsilon\mu_2(D_0 + \varepsilon D_1)(u_0 + \varepsilon u_1) \\ &\quad |(D_0 + \varepsilon D_1)(u_0 + \varepsilon u_1)|^n \\ &\quad + \varepsilon\alpha_1(u_0 + \varepsilon u_1)^m + 2\varepsilon\alpha_2(u_0 + \varepsilon u_1)^k\cos 2t + \cdots \\ &= D_0^2 u_0 + \varepsilon D_0^2 u_1 + 2\varepsilon D_0 D_1 u_0 + \omega_0^2 u_0 + \varepsilon\omega_0^2 u_1 \\ &\quad + \varepsilon\mu D_0 u_0 \, |D_0 u_0|^n + \varepsilon\alpha_1 u_0^m + 2\varepsilon\alpha_2 u_0^k\cos 2t + \cdots \\ &= D_0^2 u_0 + \omega_0^2 u_0 + \varepsilon(D_0^2 u_1 + \omega_0^2 u_1 + 2D_0 D_1 u_0 \\ &\quad + \mu D_0 u_0 \, |D_0 u_0|^n + \alpha_1 u_0^m + 2\alpha_2 u_0^k\cos 2t) + \cdots \end{aligned}$$

Equating coefficients of like powers of ε in the above equation yields

$$D_0^2 u_0 + \omega_0^2 u_0 = 0 \tag{5.17.2}$$

$$D_0^2 u_1 + \omega_0^2 u_1 = -2D_0 D_1 u_0 - \mu D_0 u_0 \, |D_0 u_0|^n - \alpha_1 u_0^m - 2\alpha_2 u_0^k\cos 2T_0 \tag{5.17.3}$$

The solution of (5.17.3) is

$$u_0 = Ae^{i\omega_0 T_0} + \bar{A}\,e^{-i\omega_0 T_0} \tag{5.17.4}$$

where $A = A(T_1, T_2)$. Substituting u_0 into (5.16.3) yields

$$\begin{aligned}
D_0^2 u_1 + \omega_0^2 u_1 &= -2D_0 D_1 u_0 - \mu D_0 u_0 \left|D_0 u_0\right|^n \\
&\quad - \alpha_1 u_0^m - 2\alpha_2 u_0^k \cos 2T_0 \\
&= -2i\omega_0 D_1 A e^{i\omega_0 T_0} - \alpha_2 (Ae^{i\omega_0 T_0} + \bar{A}e^{-i\omega_0 T_0})^k e^{i2T_0} + cc \\
&\quad - \mu\left(i\omega_0 A e^{i\omega_0 T_0} - i\omega_0 \bar{A} e^{-i\omega_0 T_0}\right)\left|i\omega_0 A e^{i\omega_0 T_0} - i\omega_0 \bar{A} e^{-i\omega_0 T_0}\right|^n \\
&\quad - \alpha_1 (Ae^{i\omega_0 T_0} + \bar{A}e^{-i\omega_0 T_0})^m \\
&= -2i\omega_0 D_1 A e^{i\omega_0 T_0} - \alpha_2 \sum_{j=0}^{k} C_k^j A^{k-j}\bar{A}^j e^{i[(k-2j)\omega_0 + 2]T_0} + cc \\
&\quad - \mu\left(i\omega_0 A e^{i\omega_0 T_0} - i\omega_0 \bar{A} e^{-i\omega_0 T_0}\right)\left|i\omega_0 A e^{i\omega_0 T_0} - i\omega_0 \bar{A} e^{-i\omega_0 T_0}\right|^n \\
&\quad - \alpha_1 \sum_{j=0}^{m} C_m^j A^{m-j}\bar{A}^j e^{i(m-2j)\omega_0 T_0}
\end{aligned} \tag{5.17.5}$$

From the above equation, when the following equation is satisfied

$$(k - 2j)\omega_0 + 2 \approx \omega_0 \tag{5.17.6}$$

i.e.,

$$\omega_0 \approx \frac{2}{1 + 2j - k}, \quad j = 1, 2, \ldots, k \tag{5.17.7}$$

parametric resonance occurs and the first-order solution of Eq. (5.17.5) can be obtained. Since $\omega_0 \geq 0$ is required, $1 + 2k - n \geq 1$, i.e., j is the integer greater than or equal to $\left[\frac{k}{2}\right]$, where $\left[\frac{k}{2}\right]$ denotes the largest integer less than or equal to $\frac{k}{2}$. Therefore, for the case of k is an odd number, parameter resonance occurs when

$$\omega_0 \approx 1, \frac{2}{4}, \frac{2}{6}, \frac{2}{8}, \ldots, \frac{2}{k+1} \text{for odd } k \tag{5.17.8}$$

For the case of k is an even number, parameter resonance occurs when. $\omega_0 \approx 2, \frac{2}{3}, \frac{2}{5}, \frac{2}{7}, \ldots, \frac{2}{k+1}$ for even k.

(b) Let

$$A = \frac{1}{2} a e^{i\beta} \tag{5.17.9}$$

Substituting this into (5.17.4), we can obtain the first order approximate solution of the given equation

$$u \approx u_0 = a\cos(\omega_0 t + \beta) + O(\varepsilon) \tag{5.17.10}$$

where a and β are determined by (5.17.5). The function $\left(Ae^{i\omega_0 T_0} - \overline{A}e^{-i\omega_0 T_0}\right)\left|Ae^{i\omega_0 T_0} - \overline{A}e^{-i\omega_0 T_0}\right|^n$ in (5.17.5) can be expanded into a Fourier series of complex exponential form:

$$\left(Ae^{i\omega_0 T_0} - \overline{A}e^{-i\omega_0 T_0}\right)\left|Ae^{i\omega_0 T_0} - \overline{A}e^{-i\omega_0 T_0}\right|^n = \sum_{l=-\infty}^{\infty} f_l e^{il\omega_0 T_0} \tag{5.17.11}$$

where

$$f_l = \frac{\omega_0}{2\pi}\int_0^{2\pi/\omega_0}\left[\left(Ae^{i\omega_0 T_0} - \overline{A}e^{-i\omega_0 T_0}\right)\left|Ae^{i\omega_0 T_0} - \overline{A}e^{-i\omega_0 T_0}\right|^n\right]e^{-im\omega_0 T_0}dT_0 \tag{5.17.12}$$

Thus, (5.16.7) becomes

$$\begin{aligned} D_0^2 u_1 + \omega_0^2 u_1 &= -2i\omega_0 D_1 A e^{i\omega_0 T_0} - \alpha_2 \sum_{j=0}^{k} C_k^j A^{k-j}\overline{A}^j e^{i[(k-2j)\omega_0 + 2]T_0} \\ &\quad - i\omega_0^{n+1}\mu f_1 e^{i\omega_0 T_0} - \alpha_1 \sum_{j=0}^{m} C_m^j A^{m-j}\overline{A}^j e^{i(m-2j)\omega_0 T_0} \\ &\quad + cc + NST \end{aligned} \tag{5.17.13}$$

where

$$\begin{aligned} f_1 &= \frac{\omega_0}{2\pi}\int_0^{\frac{2\pi}{\omega_0}}\left[\left(Ae^{i\omega_0 T_0} - \overline{A}e^{-i\omega_0 T_0}\right)\left|Ae^{i\omega_0 T_0} - \overline{A}e^{-i\omega_0 T_0}\right|^n\right]e^{-i\omega_0 T_0}dT_0 \\ &= \frac{\omega_0}{2\pi}\int_0^{\frac{2\pi}{\omega_0}} ia^{n+1}e^{i\beta}\sin(\omega_0 T_0 + \beta)\left|\sin(\omega_0 T_0 + \beta)\right|^n e^{-i(\omega_0 T_0 + \beta)}dT_0 \\ &= \frac{1}{2\pi}\int_0^{2\pi} ia^{n+1}e^{i\beta}\sin\phi\left|\sin\phi\right|^n e^{-i\phi}d\phi \\ &= \frac{a^{n+1}e^{i\beta}}{2\pi}\int_0^{2\pi}\sin^2\phi\left|\sin\phi\right|^n d\phi \\ &\triangleq b_1 a^{n+1}e^{i\beta} \end{aligned} \tag{5.17.14}$$

$$\begin{aligned} b_1 &= \frac{1}{2\pi}\int_0^{2\pi} \sin^2\varphi|\sin\varphi|^n d\varphi \\ &= \Gamma\left[\frac{1}{2}(n+3)\right]/\sqrt{\pi}\Gamma\left[\frac{1}{2}(n+4)\right] \end{aligned} \tag{5.17.15}$$

CASE I: $(k-1)\omega_0 = 2 + \varepsilon\sigma$:

For this case, considering m as an odd number, we can write (5.17.13) as

$$\begin{aligned} D_0^2u_1 + \omega_0^2u_1 &= -2i\omega_0 D_1 A e^{i\omega_0 T_0} - \alpha_2 C_k^{k-1} A\overline{A}^{k-1} e^{-i\sigma T_1} e^{i\omega_0 T_0} \\ &- \alpha_2 A^k e^{i(\omega_0+\varepsilon\sigma)T_0} - i\omega_0^{n+1}\mu f_1 e^{i\omega_0 T_0} \\ &- \alpha_1 C_m^{(m-1)/2} A^{(m+1)/2}\overline{A}^{(m-1)/2} e^{i\omega_0 T_0} + cc + NST \end{aligned} \tag{5.17.16}$$

In order to eliminate secular terms from the above equation, there must be

$$\begin{aligned} &2i\omega_0 D_1 A + i\omega_0^{n+1}\mu f_1 + \alpha_1 C_m^{(m-1)/2} A^{(m+1)/2} \overline{A}^{-(m-1)/2} \\ &+ \alpha_2 C_k^{k-1} A \overline{A}^{k-1} e^{-i\sigma T_1} + \alpha_2 A^k e^{i\sigma T_1} = 0 \end{aligned} \tag{5.17.17}$$

Substituting $A = \frac{1}{2}ae^{i\beta}$ into the above equation and considering (5.17.14), we have

$$\begin{aligned} &ia' - a\beta' + i\omega_0^n\mu b_1 a^{n+1} + \alpha_1\omega_0^{-1}2^{-m}\frac{m!}{\frac{1}{2}(m-1)!\frac{1}{2}(m+1)!}a^m \\ &+ \alpha_2\omega_0^{-1}2^{-k}ka^k e^{-i[\sigma T_1+(k-1)\beta]} + \alpha_2\omega_0^{-1}2^{-k}a^k e^{i[\sigma T_1+(k-1)\beta]} = 0 \end{aligned} \tag{5.17.18}$$

where the prime denotes the derivative with respect to T_1 such that

$$b_2 = \frac{2^{-m}m!}{\frac{1}{2}(m-1)!\frac{1}{2}(m+1)!} \tag{5.17.19}$$

Considering $\boldsymbol{\omega_0} \approx 2/(\boldsymbol{k}-1)$, (5.17.18) can be written as

$$\begin{aligned} &ia' - a\beta' + i\omega_0^n\mu b_1 a^{n+1} + b_2\alpha_1\omega_0^{-1}a^m \\ &+ \alpha_2 k(k-1)2^{-(k+1)}a^k e^{-i[\sigma T_1+(k-1)\beta]} \\ &+ \alpha_2(k-1)2^{-(k+1)}a^k e^{i[\sigma T_1+(k-1)\beta]} = 0 \end{aligned} \tag{5.17.20}$$

Separating the real and imaginary parts of the above equation yields

$$\begin{aligned} a' &= -\mu\omega_0^n b_1 a^{n+1} + \alpha_2(k-1)^2 2^{-(k+1)}a^k \sin\gamma \\ a\beta' &= b_2\alpha_1\omega_0^{-1}a^m + \alpha_2\left(k^2-1\right)2^{-(k+1)}a^k\cos\gamma \end{aligned} \tag{5.17.21}$$

where

$$\gamma = \sigma T_1 + (k-1)\beta \tag{5.17.22}$$

CASE II: $(k+1)\omega_0 = 2 + \varepsilon\sigma$**:**
For this case, considering m as an odd number, we can write (5.17.13) as

$$\begin{aligned} D_0^2 u_1 + \omega_0^2 u_1 &= -2i\omega_0 D_1 A e^{i\omega_0 T_0} - \alpha_2 \overline{A}^k e^{-i\sigma T_1} e^{i\omega_0 T_0} \\ &\quad - i\omega_0^{n+1} \mu f_1 e^{i\omega_0 T_0} - \alpha_1 C_m^{(m-1)/2} A^{(m+1)/2} \overline{A}^{(m-1)/2} e^{i\omega_0 T_0} \\ &\quad + cc + NST \end{aligned} \tag{5.17.23}$$

In order to eliminate secular terms from the above equation, there must be

$$2i\omega_0 D_1 A + i\omega_0^{n+1} \mu f_1 + \alpha_1 C_m^{(m-1)/2} A^{(m+1)/2} \overline{A}^{(m-1)/2} + \alpha_2 \overline{A}^k e^{-i\sigma T_1} e^{i\omega_0 T_0} = 0 \tag{5.17.24}$$

Substituting $A = \frac{1}{2} a e^{i\beta}$ into the above equation and considering (5.17.14) and(5.17.19), we have

$$ia' - a\beta' + i\omega_0^n \mu b_1 a^{n+1} + \alpha_1 \omega_0^{-1} b_2 a^m + \alpha_2 \omega_0^{-1} 2^{-k} a^k e^{-i[\sigma T_1 + (k+1)\beta]} = 0 \tag{5.17.25}$$

The prime denotes the derivative with respect to T_1. Considering $\boldsymbol{\omega}_0 \approx 2/(\boldsymbol{k}+1)$, (5.17.25) can be written as

$$ia' - a\beta' + i\omega_0^n \mu b_1 a^{n+1} + b_2 \alpha_1 \omega_0^{-1} a^m + \alpha_2 (k+1) 2^{-(k+1)} a^k e^{-i[\sigma T_1 + (k+1)\beta]} = 0 \tag{5.17.26}$$

Separating the real and imaginary parts of the above equation yields

$$\begin{aligned} a' &= -\mu\omega_0^n b_1 a^{n+1} + \alpha_2 (k+1) 2^{-(k+1)} a^k \sin\gamma \\ a\beta' &= b_2 \alpha_1 \omega_0^{-1} a^m + \alpha_2 (k+1) 2^{-(k+1)} a^k \cos\gamma \end{aligned} \tag{5.17.27}$$

where

$$\gamma = \sigma T_1 + (k+1)\beta \tag{5.17.28}$$

(c) We only analyze the stability of the steady state solution for CASE II above, and the reader is invited to complete the stability analysis for the other case. Write (5.17.27) as

$$\begin{aligned} a' &= -\mu\omega_0^n b_1 a^{n+1} + \alpha_2(k+1)2^{-(k+1)}a^k \sin\gamma \\ a\gamma' &= \sigma a + (k+1)b_2\alpha_1\omega_0^{-1}a^m + \alpha_2(k+1)^2 2^{-(k+1)}a^k \cos\gamma \end{aligned} \tag{5.17.29}$$

Let $a' = \gamma' = 0$ in (5.17.29), we can obtain

$$\begin{aligned} &-\mu\omega_0^n b_1 a^{n+1} + \alpha_2(k+1)2^{-(k+1)}a^k \sin\gamma = 0 \\ &\sigma a + (k+1)b_2\alpha_1\omega_0^{-1}a^m + \alpha_2(k+1)^2 2^{-(k+1)}a^k \cos\gamma = 0 \end{aligned} \tag{5.17.30}$$

The steady state solutions can be obtained:

(1) $a = 0$, γ are arbitrary constants;
(2) $a \neq 0$, γ are certain constants.

(1) $\boldsymbol{a = 0}$ **and** $\boldsymbol{\gamma = \gamma_0}$ **are arbitrary constants.**

Let

$$a = 0 + a_1, \quad \gamma = \gamma_0 + \gamma_1 \tag{5.17.31}$$

Substituting (5.17.31) into (5.16.17), we can obtain

$$a_1' = 0 \quad \Rightarrow \quad a_1 = \text{constant} \tag{5.17.32}$$

Therefore, this steady-state solution is stable.

(2) $\boldsymbol{a = a_0 \neq 0, \gamma = \gamma_0}$

For this case, the frequency–response equation is

$$\left(\frac{\sigma a}{k+1} + b_2\alpha_1\omega_0^{-1}a^m\right)^2 + (\mu\omega_0^n b_1 a^{n+1})^2 = [\alpha_2(k+1)2^{-(k+1)}a^k]^2 \tag{5.17.33}$$

Let

$$a = a_0 + a_1, \quad \gamma = \gamma_0 + \gamma_1 \tag{5.17.34}$$

Substituting (5.17.34) into (5.17.29), the corresponding linearized equation is given by

$$\begin{Bmatrix} a_1' \\ \gamma_1' \end{Bmatrix} = \begin{bmatrix} -\begin{matrix} -\mu(n+1)\omega_0^n b_1 a_0^n \\ +\alpha_2 k(k+1)2^{-(k+1)}a^{k-1}\sin\gamma_0 \end{matrix} & \alpha_2(k+1)2^{-(k+1)}a_0^k \sin\gamma_0 \\ \begin{matrix} \sigma a_0^{-1} + (k+1)b_2\alpha_1\omega_0^{-1}a_0^{m-2}a_1 \\ +\alpha_2 k(k+1)^2 2^{-(k+1)}a_0^{k-2}\cos\gamma_0 \end{matrix} & -\alpha_2(k+1)^2 2^{-(k+1)}a_0^k \sin\gamma_0 \end{bmatrix} \begin{Bmatrix} a_1 \\ \gamma_1 \end{Bmatrix} \tag{5.17.35}$$

For any set of steady state solutions $a = a_0 \neq 0, \gamma = \gamma_0$, one can calculate the eigenvalues of Eq. (5.17.36), and the stability of the corresponding steady state solutions can be determined from the sign of the real part of the two eigenvalues.

(d) When no parametric resonance occurs, the Eq. (5.17.13) can be written as

$$D_0^2u_1 + \omega_0^2u_1 = -2i\omega_0 D_1 A e^{i\omega_0 T_0} - i\omega_0^{n+1}\mu f_1 e^{i\omega_0 T_0} - \alpha_1 \sum_{j=0}^{m} C_m^j A^{m-j}\bar{A}^j e^{i(m-2j)\omega_0 T_0} + cc + NST \tag{5.17.36}$$

In order to eliminate secular terms from the above equation, there must be

$$2i\omega_0 D_1 A + i\omega_0^{n+1}\mu f_1 + \alpha_1 C_m^{(m-1)/2} A^{(m+1)/2} \bar{A}^{-(m-1)/2} = 0 \tag{5.17.37}$$

Substituting $A = \frac{1}{2}ae^{i\beta}$ into the above equation and considering (5.17.14) and(5.17.19), we have

$$ia' - a\beta' + i\omega_0^n \mu b_1 a^{n+1} + \alpha_1 \omega_0^{-1} b_2 a^m = 0 \tag{5.17.38}$$

The prime denotes the derivative with respect to T_1. Separating the real and imaginary parts of the above equation gives

$$a' = -\mu\omega_0^n b_1 a^{n+1}, \quad a\beta' = b_2 \alpha_1 \omega_0^{-1} a^m \tag{5.17.39}$$

5.18 Exercise 5.18 (Cubic Nonlinear System Subjected to Combined Parametric and External Excitation)

Solution: (a) We use the method of multiple scales to obtain the first order approximate periodic solution of the system. When $\Omega = 1 + \varepsilon\sigma_1$ and $\omega_0 = 1 + \varepsilon\sigma_2$, i.e., $\Omega \approx \omega_0$, the external excitation can cause a primary resonance. In order to make the damping, nonlinear, and excitation terms appearing simultaneously in the coefficient of ε, we need to makes

$$K = \varepsilon k \tag{5.18.1}$$

Let the solution of the equation be

$$u(t;\varepsilon) = u_0(T_0, T_1, T_2) + \varepsilon u_1(T_0, T_1, T_2) + \varepsilon^2 u_2(T_0, T_1, T_2) + \cdots \tag{5.18.2}$$

Substituting the above equation into the original equation and retaining to $O(\varepsilon^2)$, we obtain

$$0 = \ddot{u} + \omega_0^2 u + 2\varepsilon\mu\dot{u} + \varepsilon\alpha_1 u^3$$

$$
\begin{aligned}
&+ 2\varepsilon\alpha_2 u \cos 2t - K\cos\Omega t \\
&= \left(D_0^2 + 2\varepsilon D_0 D_1\right)(u_0 + \varepsilon u_1) \\
&+ \omega_0^2(u_0 + \varepsilon u_1) + 2\varepsilon\mu(D_0 + \varepsilon D_1)(u_0 + \varepsilon u_1) \\
&+ \alpha_1(u_0 + \varepsilon u_1)^3 + 2\varepsilon\alpha_2(u_0 + \varepsilon u_1)\cos 2t \\
&- K\cos\Omega t \\
&= D_0^2 u_0 + \varepsilon D_0^2 u_1 + 2\varepsilon D_0 D_1 u_0 \\
&+ \omega_0^2 u_0 + \varepsilon\omega_0^2 u_1 + 2\varepsilon\mu D_0 u_0 \\
&+ \varepsilon\alpha_1 u_0^3 + 2\varepsilon\alpha_2 u_0 \cos 2t \\
&- K\cos\Omega t + \cdots \\
&= D_0^2 u_0 + \omega_0^2 u_0 \\
&+ \varepsilon(D_0^2 u_1 + \omega_0^2 u_1 + 2D_0 D_1 u_0 + 2\mu D_0 u_0 \\
&+ \alpha_1 u_0^3 + 2\alpha_2 u_0 \cos 2t - k\cos\Omega t) + \cdots
\end{aligned} \tag{5.18.3}
$$

Equating coefficients of like powers of ε in the above equation yields

$$
\begin{aligned}
0 &= \ddot{u} + \omega_0^2 u + 2\varepsilon\mu\dot{u} + \varepsilon\alpha_1 u^3 + 2\varepsilon\alpha_2 u\cos 2t - K\cos\Omega t \\
&= \left(D_0^2 + 2\varepsilon D_0 D_1\right)(u_0 + \varepsilon u_1) + \omega_0^2(u_0 + \varepsilon u_1) + 2\varepsilon\mu(D_0 + \varepsilon D_1)(u_0 + \varepsilon u_1) \\
&+\alpha_1(u_0 + \varepsilon u_1)^3 + 2\varepsilon\alpha_2(u_0 + \varepsilon u_1)\cos 2t - K\cos\Omega t \\
&= D_0^2 u_0 + \varepsilon D_0^2 u_1 + 2\varepsilon D_0 D_1 u_0 + \omega_0^2 u_0 + \varepsilon\omega_0^2 u_1 + 2\varepsilon\mu D_0 u_0 \\
&+\varepsilon\alpha_1 u_0^3 + 2\varepsilon\alpha_2 u_0\cos 2t - K\cos\Omega t + \cdots \\
&= D_0^2 u_0 + \omega_0^2 u_0 + \varepsilon(D_0^2 u_1 + \omega_0^2 u_1 + 2D_0 D_1 u_0 + 2\mu D_0 u_0 \\
&+\alpha_1 u_0^3 + 2\alpha_2 u_0\cos 2t - k\cos\Omega t) + \cdots + \alpha_1 u_0^3 + 2\alpha_2 u_0 \cos 2t - k\cos\Omega t) + \cdots
\end{aligned}
$$

$$
D_0^2 u_0 + \omega_0^2 u_0 = 0 \tag{5.18.4}
$$

$$
D_0^2 u_1 + \omega_0^2 u_1 = -2D_0 D_1 u_0 - 2\mu D_0 u_0 - \alpha_1 u_0^3 - 2\alpha_2 u_0 \cos 2T_0 + k\cos\Omega T_0 \tag{5.18.5}
$$

The solution of (5.18.4) is

$$
u_0 = Ae^{i\omega_0 T_0} + \bar{A}\,e^{-i\omega_0 T_0} \tag{5.18.6}
$$

where $A = A(T_1, T_2)$. Let

$$
A = \frac{1}{2} a e^{i\beta} \tag{5.18.7}
$$

Substituting this into (5.18.6), we can obtain the first order approximate solution of the given equation

$$u \approx u_0 = a\cos(\omega_0 t + \beta) + O(\varepsilon) \tag{5.18.8}$$

Substituting u_0 into (5.18.5) yields

$$\begin{aligned} D_0^2 u_1 + \omega_0^2 u_1 =& -2D_0 D_1 u_0 - 2\mu D_0 u_0 - \alpha_1 u_0^3 \\ & -2\alpha_2 u_0 \cos 2T_0 + k\cos\Omega T_0 \\ =& -2i\omega_0 D_1 A e^{i\omega_0 T_0} - 2i\omega_0 \mu A e^{i\omega_0 T_0} \\ & -3\alpha_1 A^2 \overline{A} e^{i\omega_0 T_0} - \alpha_2\left(A e^{i\omega_0 T_0} + \overline{A} e^{-i\omega_0 T_0}\right) e^{i2T_0} \\ & + \frac{1}{2} k e^{i\Omega T_0} + cc + NST \\ =& -2i\omega_0 D_1 A e^{i\omega_0 T_0} - 2i\omega_0 \mu A e^{i\omega_0 T_0} \\ & -3\alpha_1 A^2 \overline{A} e^{i\omega_0 T_0} - \alpha_2 \overline{A} e^{-2i\sigma_2 T_1} e^{i\omega_0 T_0} \\ & + \frac{1}{2} k e^{i(\sigma_1 - \sigma_2)T_1} e^{i\omega_0 T_0} + cc + NST \end{aligned} \tag{5.18.9}$$

In order to eliminate secular terms from the above equation, there must be

$$2i\omega_0 D_1 A + 2i\omega_0 \mu A + 3\alpha_1 A^2 \overline{A} + \alpha_2 \overline{A} e^{-2i\sigma_2 T_1} - \frac{1}{2} k e^{i(\sigma_1 - \sigma_2)T_1} = 0 \tag{5.18.10}$$

Substituting $A = \frac{1}{2} a e^{i\beta}$ into the above equation and taking $\omega_0 \approx 1$ into account, we get

$$ia' - a\beta' + i\mu a + \frac{3}{8}\alpha_1 a^3 + \frac{1}{2}\alpha_2 e^{-i(2\sigma_2 T_1 + 2\beta)} - \frac{1}{2} k e^{i[(\sigma_1 - \sigma_2)T_1 - \beta]} = 0 \tag{5.18.11}$$

The prime denotes the derivative with respect to T_1. Separating the real and imaginary parts of the above equation yields

$$\begin{aligned} a' &= -\mu a + \tfrac{1}{2} k \sin\gamma_1 + \tfrac{1}{2}\alpha_2 a \sin\gamma_2 \\ a\beta' &= \tfrac{3}{8}\alpha_1 a^3 - \tfrac{1}{2} k\cos\gamma_1 + \tfrac{1}{2}\alpha_2 a\cos\gamma_2 \end{aligned} \tag{5.18.12}$$

where

$$\gamma_1 = (\sigma_1 - \sigma_2)T_1 - \beta, \quad \gamma_2 = 2\sigma_2 T_1 + 2\beta \tag{5.18.13}$$

(b) When $\omega_0 = 1 + \varepsilon\sigma_2$, Ω is far from 1, this is a nonresonant case where $K = O(1)$. Therefore, the Eq. (5.18.3) becomes:ss

$$0 = \ddot{u} + \omega_0^2 u + 2\varepsilon\mu\dot{u} + \varepsilon\alpha_1 u^3$$

$$
\begin{aligned}
&+ 2\varepsilon\alpha_2 u \cos 2t - K\cos\Omega t \\
&= D_0^2 u_0 + \varepsilon D_0^2 u_1 + 2\varepsilon D_0 D_1 u_0 \\
&+ \omega_0^2 u_0 + \varepsilon\omega_0^2 u_1 + 2\varepsilon\mu D_0 u_0 \\
&+ \varepsilon\alpha_1 u_0^3 + 2\varepsilon\alpha_2 u_0 \cos 2t - K\cos\Omega t + \cdots \\
&= D_0^2 u_0 + \omega_0^2 u_0 - K\cos\Omega t \\
&+ \varepsilon(D_0^2 u_1 + \omega_0^2 u_1 + 2D_0 D_1 u_0 + 2\mu D_0 u_0 \\
&+ \alpha_1 u_0^3 + 2\alpha_2 u_0 \cos 2t) + \cdots
\end{aligned}
\tag{5.18.14}
$$

Equating coefficients of like powers of ε in the above equation yields

$$D_0^2 u_0 + \omega_0^2 u_0 = K\cos\Omega T_0 \tag{5.18.15}$$

$$D_0^2 u_1 + \omega_0^2 u_1 = -2D_0 D_1 u_0 - 2\mu D_0 u_0 - \alpha_1 u_0^3 - 2\alpha_2 u_0 \cos 2T_0 \tag{5.18.16}$$

The solution of (5.18.15) is

$$u_0 = Ae^{i\omega_0 T_0} + \bar{A}\, e^{-i\omega_0 T_0} + \Lambda\left(e^{i\Omega T_0} + e^{-i\Omega T_0}\right) \tag{5.18.17}$$

where

$$\Lambda = \frac{1}{2\left(\omega_0^2 - \Omega^2\right)} \tag{5.18.18}$$

and $A = A(T_1, T_2)$. Let

$$A = \frac{1}{2} a e^{i\beta} \tag{5.18.19}$$

Substituting this into (5.18.17), we can obtain the first order approximate solution of the given equation.

$$u = a\cos(\omega_0 t + \beta) + 2\Lambda\cos\Omega t + O(\varepsilon) \tag{5.18.20}$$

Substitute u_0 into (5.18.16) yields

$$
\begin{aligned}
D_0^2 u_1 + \omega_0^2 u_1 &= -2D_0 D_1 u_0 - 2\mu D_0 u_0 \\
&- \alpha_1 u_0^3 - 2\alpha_2 u_0 \cos 2T_0 \\
&= -2i\omega_0 D_1 A e^{i\omega_0 T_0} - 2i\omega_0 \mu A e^{i\omega_0 T_0} \\
&- \alpha_2\left[Ae^{i\omega_0 T_0} + \bar{A}e^{-i\omega_0 T_0} + \Lambda\left(e^{i\Omega T_0} + e^{-i\Omega T_0}\right)\right]e^{i2T_0} \\
&+ cc - \alpha_1 [Ae^{i\omega_0 T_0} + \bar{A}e^{-i\omega_0 T_0}
\end{aligned}
$$

$$
\begin{aligned}
&+\Lambda\left(e^{i\Omega T_0}+e^{-i\Omega T_0}\right)]^3 \\
&=-2i\omega_0 D_1 A e^{i\omega_0 T_0}-2i\omega_0\mu A e^{i\omega_0 T_0}-3\alpha_1 A^2\overline{A}e^{i\omega_0 T_0} \\
&-6\alpha_1\Lambda^2 A e^{i\omega_0 T_0}-\alpha_2\overline{A}e^{-2i\sigma_2 T_1}e^{i\omega_0 T_0}-\alpha_2\Lambda e^{i(\Omega-2)T_0} \\
&-3\alpha_1\Lambda\overline{A}^2 e^{i(\Omega-2\omega_0)T_0}-\alpha_1\Lambda^3 e^{3i\Omega T_0}+cc+NST
\end{aligned} \tag{5.18.21}
$$

CASE I: Ω is far from 3 and $\frac{1}{3}$:

In order to eliminate secular terms from (5.18.21), we need

$$
2i\omega_0 D_1 A+2i\omega_0\mu A+3\alpha_1 A^2\,\overline{A}+6\alpha_1\Lambda^2 A+\alpha_2\,\overline{A}\,e^{-2i\sigma_2 T_1}=0 \tag{5.18.22}
$$

Substituting $A=\frac{1}{2}ae^{i\beta}$ into the above equation and taking $\omega_0\approx 1$ into account, we get

$$
ia'-a\beta'+i\mu a+\frac{3}{8}\alpha_1 a^3+3\alpha_1\Lambda^2 a+\frac{1}{2}\alpha_2 a e^{-i(2\sigma_2 T_1+2\beta)}=0 \tag{5.18.23}
$$

The prime denotes the derivative with respect to T_1. Separating the real and imaginary parts of the above equation gives

$$
\begin{aligned}
a'&=-\mu a+\frac{1}{2}\alpha_2 a\sin\gamma_2 \\
a\beta'&=3\alpha_1\left(\Lambda^2+\frac{1}{8}a^2\right)a+\frac{1}{2}\alpha_2 a\cos\gamma_2
\end{aligned} \tag{5.18.24}
$$

where

$$
\gamma_2=2\sigma_2 T_1+2\beta \tag{5.18.25}
$$

CASE II: $3\Omega=1+\varepsilon\sigma_1$:

In order to eliminate secular terms from (5.18.21), we need

$$
\begin{aligned}
&2i\omega_0 D_1 A+2i\omega_0\mu A+3\alpha_1 A^2\overline{A}+6\alpha_1\Lambda^2 A \\
&+\alpha_2\overline{A}e^{-2i\sigma_2 T_1}+\alpha_1\Lambda^3 e^{i(\sigma_1-\sigma_2)T_1}=0
\end{aligned} \tag{5.18.26}
$$

Substituting $A=\frac{1}{2}ae^{i\beta}$ into the above equation and taking $\omega_0\approx 1$ into account, we get

$$
\begin{aligned}
&ia'-a\beta'+i\mu a+\frac{3}{8}\alpha_1 a^3+3\alpha_1\Lambda^2 a+\frac{1}{2}\alpha_2 a e^{-i(2\sigma_2 T_1+2\beta)} \\
&+\alpha_1\Lambda^3 e^{i[(\sigma_1-\sigma_2)T_1-\beta]}=0
\end{aligned} \tag{5.18.27}
$$

The prime denotes the derivative with respect to T_1. Separating the real and imaginary parts of the above equation gives.

$$
\begin{aligned}
a' &= -\mu a - \alpha_1 \Lambda^3 \sin\gamma_1 + \frac{1}{2}\alpha_2 a \sin\gamma_2 \\
a\beta' &= 3\alpha_1\left(\Lambda^2 + \frac{1}{8}a^2\right)a + \alpha_1\Lambda^3\cos\gamma_1 + \frac{1}{2}\alpha_2 a\cos\gamma_2
\end{aligned} \tag{5.18.28}
$$

where

$$
\gamma_1 = (\sigma_1 - \sigma_2)T_1 - \beta, \quad \gamma_2 = 2\sigma_2 T_1 + 2\beta \tag{5.18.29}
$$

CASE III: $\Omega = 3 + \varepsilon\sigma_1$**:**

In order to eliminate secular terms from (5.18.21), we need

$$
\begin{aligned}
&2i\omega_0 D_1 A + 2i\omega_0\mu A + 3\alpha_1 A^2\overline{A} + 6\alpha_1\Lambda^2 A \\
&\quad + \alpha_2\overline{A}e^{-2i\sigma_2 T_1} + 3\alpha_1\Lambda\overline{A}^2 e^{i(\sigma_1 - 3\sigma_2)T_1} + \alpha_2\Lambda e^{i(\sigma_1-\sigma_2)T_1} = 0
\end{aligned} \tag{5.18.30}
$$

Substituting $A = \frac{1}{2}ae^{i\beta}$ into the above equation and taking $\omega_0 \approx 1$ into account, we get

$$
\begin{aligned}
&ia' - a\beta' + i\mu a + \frac{3}{8}\alpha_1 a^3 + 3\alpha_1\Lambda^2 a + \frac{3}{4}\alpha_1\Lambda a^2 e^{i[(\sigma_1 - 3\sigma_2)T_1 - 3\beta]} \\
&\quad + \frac{1}{2}\alpha_2 a e^{-i(2\sigma_2 T_1 + 2\beta)} + \alpha_2\Lambda e^{i[(\sigma_1-\sigma_2)T_1 - \beta]} = 0
\end{aligned} \tag{5.18.31}
$$

The prime denotes the derivative with respect to T_1. Separating the real and imaginary parts of the above equation gives

$$
\begin{aligned}
a' &= -\mu a - \frac{3}{4}\alpha_1\Lambda a^2\sin\gamma_1 + \frac{1}{2}\alpha_2 a\sin\gamma_2 - \alpha_2\Lambda\sin\gamma_3 a\beta' \\
&= 3\alpha_1\left(\Lambda^2 + \frac{1}{8}a^2\right)a + \frac{3}{4}\alpha_1\Lambda a^2\cos\gamma_1 + \frac{1}{2}\alpha_2 a\cos\gamma_2 + \alpha_2\Lambda\cos\gamma_3
\end{aligned} \tag{5.18.32}
$$

where

$$
\begin{aligned}
\gamma_1 &= (\sigma_1 - 3\sigma_2)T_1 - 3\beta, \\
\gamma_2 &= 2\sigma_2 T_1 + 2\beta \\
\gamma_3 &= (\sigma_1 - \sigma_2)T_1 - \beta = \sigma_1 T_1 - \frac{1}{2}\gamma_2
\end{aligned} \tag{5.18.33}
$$

(c) We only analyze the stability of the steady state solution for CASE I above, and readers are invited to complete the stability analysis for other cases. Write (5.18.24) as

$$\begin{aligned} a' &= -\mu a + \frac{1}{2}\alpha_2 a\sin\gamma_2 \\ a\gamma_2' &= 2\sigma_2 a + 6\alpha_1\left(\Lambda^2 + \frac{1}{8}a^2\right)a + \alpha_2 a\cos\gamma_2 \end{aligned} \tag{5.18.34}$$

Let $a' = \gamma' = 0$ in (5.18.34), we obtain

$$\begin{aligned} &-\mu a + \tfrac{1}{2}\alpha_2 a\sin\gamma_2 = 0 \\ &2\sigma_2 a + 6\alpha_1\left(\Lambda^2 + \tfrac{1}{8}a^2\right)a + \alpha_2 a\cos\gamma_2 = 0 \end{aligned} \tag{5.18.35}$$

The steady state solutions can be obtained:

(1) $a = 0$, γ are arbitrary constants;
(2) $a \neq 0$, γ are certain constants.

(1) **$a = 0$ and $\gamma = \gamma_0$ are arbitrary constants.**

Let

$$a = 0 + \tilde{a}, \quad \gamma_2 = \gamma_{20} + \tilde{\gamma} \tag{5.18.36}$$

Substituting (5.18.36) into (5.18.34), we can obtain the corresponding linearized equation

$$\widetilde{a'} = -\left(\mu - \frac{1}{2}\alpha_2\sin\gamma_{20}\right)\tilde{a} \tag{5.18.37}$$

Therefore, this steady state solution is stable when $\mu \geq \frac{1}{2}\alpha_2\sin\gamma_{20}$ and vice versa.

(2) **$a = a_0 \neq 0, \gamma_2 = \gamma_{20}$**

For this case, the frequency–response equation is

$$4\mu^2 a^2 + [2\sigma_2 a + 6\alpha_1\left(\Lambda^2 + \frac{1}{8}a^2\right)a]^2 = \alpha_2^2 a^2 \tag{5.18.38}$$

Let

$$a = a_0 + \tilde{a}, \quad \gamma_2 = \gamma_{20} + \tilde{\gamma} \tag{5.18.39}$$

Substituting (5.18.39) into (5.18.34), we can obtain the corresponding linearized equation

$$\begin{Bmatrix} \tilde{a}' \\ \tilde{\gamma}' \end{Bmatrix} = \begin{bmatrix} -\mu + \frac{1}{2}\alpha_2\sin\gamma_{20} & \frac{1}{2}\alpha_2\cos\gamma_{20} \\ 2\sigma_2 a_0^{-1} + 6\alpha_1 a_0^{-1}\left(\Lambda^2 + \frac{3}{8}a_0^2\right) + \alpha_2 a_0^{-1}\cos\gamma_{20} & -\alpha_2\sin\gamma_{20} \end{bmatrix}\begin{Bmatrix} \tilde{a} \\ \tilde{\gamma} \end{Bmatrix} \tag{5.18.40}$$

For any set of steady state solutions $a = a_0 \neq 0, \gamma = \gamma_0$, one can calculate the eigenvalues of Eq. (5.18.10), and the stability of the corresponding steady state solutions can be determined from the sign of the real part of the two eigenvalues.

(d) When ω_0 is far from 1, this is a nonresonant case where $K = O(1)$. Equations (5.18.3–5.18.18) still hold, and thus the first order approximate solution of the given equation is

$$u = a\cos(\omega_0 t + \beta) + 2\Lambda\cos\Omega t + O(\varepsilon) \tag{5.18.41}$$

Substitute (5.18.17) into (5.18.16) yields

$$\begin{aligned} D_0^2 u_1 + \omega_0^2 u_1 &= -2D_0 D_1 u_0 - 2\mu D_0 u_0 \\ &\quad - \alpha_1 u_0^3 - 2\alpha_2 u_0 \cos 2T_0 \\ &= -2i\omega_0 D_1 A e^{i\omega_0 T_0} - 2i\omega_0 \mu A e^{i\omega_0 T_0} \\ &\quad - \alpha_2\left[A e^{i\omega_0 T_0} + \overline{A} e^{-i\omega_0 T_0} + \Lambda\left(e^{i\Omega T_0} + e^{-i\Omega T_0}\right)\right] e^{i2T_0} \\ &\quad + cc \\ &\quad - \alpha_1 [A e^{i\omega_0 T_0} + \overline{A} e^{-i\omega_0 T_0} + \Lambda\left(e^{i\Omega T_0} + e^{-i\Omega T_0}\right)]^3 \\ &= -2i\omega_0 D_1 A e^{i\omega_0 T_0} - 2i\omega_0 \mu A e^{i\omega_0 T_0} \\ &\quad - 3\alpha_1 A^2 \overline{A} e^{i\omega_0 T_0} - 6\alpha_1 \Lambda^2 A e^{i\omega_0 T_0} \\ &\quad - \alpha_2 \Lambda e^{i(\Omega+2)T_0} - \alpha_2 \Lambda e^{i(\Omega-2)T_0} \\ &\quad - 3\alpha_1 \Lambda \overline{A}^2 e^{i(\Omega - 2\omega_0)T_0} - \alpha_1 \Lambda^3 e^{3i\Omega T_0} \\ &\quad + cc + NST \end{aligned} \tag{5.18.42}$$

From the above equation, it can be seen that the combined resonance of the system occurs when $\Omega \mp 2 \approx \omega_0$, or when $2 - \Omega \approx \omega_0$.

When $\Omega = 2 + \omega_0 + \varepsilon\sigma_1$, in order to eliminate secular terms from (5.18.12), there must be

$$2i\omega_0 D_1 A + 2i\omega_0 \mu A + 3\alpha_1 A^2 \overline{A} + 6\alpha_1 \Lambda^2 A + \alpha_2 \Lambda e^{i\sigma_1 T_1} = 0 \tag{5.18.43}$$

Substituting $A = \frac{1}{2} a e^{i\beta}$ into the above equation gives

$$ia' - a\beta' + i\mu a + \frac{3}{8}\alpha_1 \omega_0^{-1} a^3 + 3\alpha_1 \Lambda^2 \omega_0^{-1} a + \alpha_2 \Lambda \omega_0^{-1} e^{i(\sigma_1 T_1 - \beta)} = 0 \tag{5.18.44}$$

The prime denotes the derivative with respect to T_1. Separating the real and imaginary parts of the above equation gives

$$a' = -\mu a - \Lambda\alpha_2 \omega_0^{-1} \sin\gamma_1$$

$$a\beta' = 3\alpha_1\omega_0^{-1}\left(\Lambda^2 + \frac{1}{8}a^2\right)a + \Lambda\alpha_2\omega_0^{-1}\cos\gamma_1 \tag{5.18.45}$$

where

$$\gamma_1 = \sigma_1 T_1 - \beta \tag{5.18.46}$$

5.19 Exercise 5.19 (Cubic Nonlinear System Subjected to Combined Parametric and Multi-frequency External Excitations)

Solution: (a) We use the method of multiple scales to obtain the first-order approximate solution of the system. Let the solution of the equation be

$$u(t;\varepsilon) = u_0(T_0, T_1, T_2) + \varepsilon u_1(T_0, T_1, T_2) + \varepsilon^2 u_2(T_0, T_1, T_2) + \cdots \tag{5.19.1}$$

Substituting the above equation into the original equation and retaining to $O(\varepsilon^2)$, we obtain

$$\begin{aligned} 0 &= u + \omega_0^2 u + 2\varepsilon\mu u + \varepsilon\alpha_1 u^3 + 2\varepsilon\alpha_2 u\cos 2t \\ &\quad - \sum_{n=1}^{3} K_n\cos(\Omega_n t + \theta_n) \\ &= D_0^2 u_0 + \omega_0^2 u_0 - \sum_{n=1}^{3} K_n\cos(\Omega_n t + \theta_n) \\ &\quad + \varepsilon(D_0^2 u_1 + \omega_0^2 u_1 + 2D_0 D_1 u_0 \\ &\quad + 2\mu D_0 u_0 + \alpha_1 u_0^3 + 2\alpha_2 u_0\cos 2t) + \ldots \end{aligned} \tag{5.19.2}$$

Equating coefficients of like powers of ε in the above equation yields

$$D_0^2 u_0 + \omega_0^2 u_0 = \sum_{n=1}^{3} K_n\cos(\Omega_n T_0 + \theta_n) \tag{5.19.3}$$

$$D_0^2 u_1 + \omega_0^2 u_1 = -2D_0 D_1 u_0 - 2\mu D_0 u_0 - \alpha_1 u_0^3 - 2\alpha_2 u_0\cos 2T_0 \tag{5.19.4}$$

The solution of (5.19.3) is

$$u_0 = Ae^{i\omega_0 T_0} + \bar{A}\,e^{-i\omega_0 T_0} + \sum_{n=1}^{3} \Lambda_n\left[e^{i(\Omega_n T_0+\theta_n)} + e^{-i(\Omega_n T_0+\theta_n)}\right] \tag{5.19.5}$$

where

$$\Lambda_n = \frac{1}{2\left(\omega_0^2 - \Omega_n^2\right)} \tag{5.19.6}$$

and $A = A(T_1, T_2)$. Let

$$A = \frac{1}{2}ae^{i\beta} \tag{5.19.7}$$

Substituting this into (5.19.5), we can obtain the first order approximate solution of the given equation

$$u \approx u_0 = a\cos(\omega_0 t + \beta) + \sum_{n=1}^{3} 2\Lambda_n \cos(\Omega_n T_0 + \theta_n) + O(\varepsilon) \tag{5.19.8}$$

Substituting u_0 into (5.19.4) yields

$$\begin{aligned} D_0^2 u_1 + \omega_0^2 u_1 &= -2D_0 D_1 u_0 - 2\mu D_0 u_0 - \alpha_1 u_0^3 - 2\alpha_2 u_0 \cos 2T_0 \\ &= -2i\omega_0 D_1 A e^{i\omega_0 T_0} - 2i\omega_0 \mu A e^{i\omega_0 T_0} \\ &\quad - \alpha_2 \bar{A} e^{i(2-\omega_0)T_0} - \alpha_2 \sum_{n=1}^{3} \Lambda_n e^{i(\Omega_n T_0 + 2T_0 + \theta_n)} \\ &\quad - \alpha_1 u_0^3 + cc + NST \end{aligned} \tag{5.19.9}$$

where

$$\begin{aligned} u_0^3 = {} & \Lambda_1^3 e^{i(3\Omega_1 T_0 + 3\theta_1)} + \Lambda_2^3 e^{i(3\Omega_2 T_0 + 3\theta_2)} + \Lambda_3^3 e^{i(3\Omega_3 T_0 + 3\theta_3)} \\ & + 3A^2\bar{A}e^{i\omega_0 T_0} + 6\Lambda_1^2 A e^{i\omega_0 T_0} + 6\Lambda_2^2 A e^{i\omega_0 T_0} \\ & + 6\Lambda_3^2 A e^{i\omega_0 T_0} + 3\Lambda_1^2 \bar{A} e^{i(2\Omega_1 T_0 - \omega_0 T_0 + 2\theta_1)} \\ & + 3\Lambda_2^2 \bar{A} e^{i(2\Omega_2 T_0 - \omega_0 T_0 + 2\theta_2)} + 3\Lambda_3^2 \bar{A} e^{i(2\Omega_3 T_0 - \omega_0 T_0 + 2\theta_3)} \\ & + 6\Lambda_1\Lambda_2 \bar{A} e^{i(\Omega_1 T_0 + \Omega_2 T_0 - \omega_0 T_0 + \theta_1 + \theta_2)} \\ & + 6\Lambda_1\Lambda_3 \bar{A} e^{i(\Omega_1 T_0 + \Omega_3 T_0 - \omega_0 T_0 + \theta_1 + \theta_3)} \\ & + 6\Lambda_2\Lambda_3 \bar{A} e^{i(\Omega_2 T_0 + \Omega_3 T_0 - \omega_0 T_0 + \theta_2 + \theta_3)} \\ & + 3\Lambda_1 A^2 e^{i(2\omega_0 T_0 - \Omega_1 T_0 + \theta_1)} + 3\Lambda_2 A^2 e^{i(2\omega_0 T_0 - \Omega_2 T_0 - \theta_2)} \\ & + 3\Lambda_3 A^2 e^{i(2\omega_0 T_0 - \Omega_3 T_0 - \theta_3)} \\ & + 3\Lambda_1\Lambda_2^2 e^{i(2\Omega_2 T_0 - \Omega_1 T_0 + 2\theta_2 - \theta_1)} + 3\Lambda_1^2\Lambda_2 e^{i(2\Omega_1 T_0 - \Omega_2 T_0 + 2\theta_1 - \theta_2)} \end{aligned}$$

$$
\begin{aligned}
&+3\Lambda_1\Lambda_3^2 e^{i(2\Omega_3 T_0-\Omega_1 T_0+2\theta_3-\theta_1)}+3\Lambda_1^2\Lambda_3 e^{i(2\Omega_1 T_0-\Omega_3 T_0+2\theta_1-\theta_3)}\\
&+3\Lambda_2\Lambda_3^2 e^{i(2\Omega_3 T_0-\Omega_2 T_0+2\theta_3-\theta_2)}+3\Lambda_3\Lambda_2^2 e^{i(2\Omega_2 T_0-\Omega_3 T_0+2\theta_2-\theta_3)}\\
&+6\Lambda_1\Lambda_2 A e^{i(\Omega_2 T_0-\Omega_1 T_0+\omega_0 T_0+\theta_2-\theta_1)}+6\Lambda_1\Lambda_2 A e^{i(\Omega_1 T_0-\Omega_2 T_0+\omega_0 T_0+\theta_1-\theta_2)}\\
&+6\Lambda_1\Lambda_3 A e^{i(\Omega_3 T_0-\Omega_1 T_0+\omega_0 T_0+\theta_3-\theta_1)}+6\Lambda_1\Lambda_3 A e^{i(\Omega_1 T_0-\Omega_3 T_0+\omega_0 T_0+\theta_1-\theta_3)}\\
&+6\Lambda_2\Lambda_3 A e^{i(\Omega_3 T_0-\Omega_2 T_0+\omega_0 T_0+\theta_3-\theta_2)}+6\Lambda_2\Lambda_3 A e^{i(\Omega_2 T_0-\Omega_3 T_0+\omega_0 T_0+\theta_2-\theta_3)}\\
&+6\Lambda_1\Lambda_2\Lambda_3 e^{i(\Omega_2 T_0+\Omega_3 T_0-\Omega_1 T_0+\theta_3+\theta_2-\theta_1)}\\
&+6\Lambda_1\Lambda_2\Lambda_3 e^{i(\Omega_1 T_0+\Omega_3 T_0-\Omega_2 T_0+\theta_1+\theta_3-\theta_2)}\\
&+6\Lambda_1\Lambda_2\Lambda_3 e^{i(\Omega_1 T_0+\Omega_2 T_0-\Omega_3 T_0+\theta_1+\theta_2-\theta_3)}\\
&+\Lambda_1^3 e^{i(3\Omega_1 T_0+3\theta_1)}+\Lambda_2^3 e^{i(3\Omega_2 T_0+3\theta_2)}+\Lambda_3^3 e^{i(3\Omega_3 T_0+3\theta_3)}\\
&+3\Lambda_1^2\Lambda_2 e^{i(2\Omega_1 T_0+\Omega_2 T_0+2\theta_1+\theta_2)}+3\Lambda_1\Lambda_2^2 e^{i(\Omega_1 T_0+2\Omega_2 T_0+\theta_1+2\theta_2)}\\
&+3\Lambda_2^2\Lambda_3 e^{i(2\Omega_2 T_0+\Omega_3 T_0+2\theta_2+\theta_3)}+3\Lambda_2\Lambda_3^2 e^{i(\Omega_2 T_0+2\Omega_3 T_0+\theta_2+2\theta_3)}\\
&+3\Lambda_1^2\Lambda_3 e^{i(2\Omega_1 T_0+\Omega_3 T_0+2\theta_1+\theta_3)}+3\Lambda_1\Lambda_3^2 e^{i(\Omega_1 T_0+2\Omega_3 T_0+\theta_1+2\theta_3)}\\
&+6\Lambda_1\Lambda_2\Lambda_3 e^{i(\Omega_1 T_0+\Omega_2 T_0+\Omega_3 T_0+\theta_1+\theta_2+\theta_3)}+cc
\end{aligned}
\tag{5.19.10}
$$

When $\omega_0 = 1+\varepsilon\sigma_2$ and $2\Omega_1+\Omega_2 = 1+\varepsilon\sigma_1$, (5.19.9) becomes

$$
\begin{aligned}
D_0^2 u_1+\omega_0^2 u_1 = &-2i\omega_0 D_1 A e^{i\omega_0 T_0}-2i\omega_0\mu A e^{i\omega_0 T_0}-3\alpha_1 A^2\overline{A} e^{i\omega_0 T_0}\\
&-6\alpha_1\Lambda_1^2 A e^{i\omega_0 T_0}-6\alpha_1\Lambda_2^2 A e^{i\omega_0 T_0}-6\alpha_1\Lambda_3^2 A e^{i\omega_0 T_0}\\
&-6\alpha_1\Lambda_1\Lambda_2\Lambda_3 e^{i(\Omega_1 T_0+\Omega_2 T_0+\Omega_3 T_0+\theta_1+\theta_2+\theta_3)}-\alpha_2\overline{A} e^{i(2-\omega_0)T_0}+cc+NST
\end{aligned}
\tag{5.19.11}
$$

In order to eliminate secular terms from the above equation, there must be

$$
\begin{aligned}
&2i\omega_0 D_1 A+2i\omega_0\mu A+3\alpha_1 A^2\overline{A}+6\alpha_1\sum_{n=1}^{3}\Lambda_n^2 A\\
&+3\alpha_1\Lambda_1^2\Lambda_2 e^{i(2\theta_1+\theta_2)} e^{i(\sigma_1-\sigma_2)T_1}+\alpha_2\overline{A} e^{-i2\sigma_2 T_1}=0
\end{aligned}
\tag{5.19.12}
$$

Substituting $A=\frac{1}{2}ae^{i\beta}$ into the above equation and taking $\omega_0\approx 1$ into account, we get

$$
\begin{aligned}
&ia'-a\beta'+i\mu a+\frac{3}{8}\alpha_1 a^3+3\alpha_1 a\sum_{n=1}^{3}\Lambda_n^2+3\alpha_1\Lambda_1^2\Lambda_2 e^{i[(\sigma_1-\sigma_2)T_1-\beta+2\theta_1+\theta_2]}\\
&+\frac{1}{2}\alpha_2 a e^{-i(2\sigma_2 T_1+2\beta)}=0
\end{aligned}
\tag{5.19.13}
$$

The prime denotes the derivative with respect to T_1. Separating the real and imaginary parts of the above equation gives

$$a' = -\mu a - 3\alpha_1 \Lambda_1^2 \Lambda_2 \sin\gamma_1 + \frac{1}{2}\alpha_2 a \sin\gamma_2$$
$$a\beta' = 3\left[\alpha_1 \sum_{n=1}^{3} \Lambda_n^2 + \frac{1}{8}a^2\right]a + 3\alpha_1 \Lambda_1^2 \Lambda_2 \cos\gamma_1 + \frac{1}{2}\alpha_2 a \cos\gamma_2 \tag{5.19.14}$$

where

$$\gamma_1 = (\sigma_1 - \sigma_2)T_1 - \beta + 2\theta_1 + \theta_2, \quad \gamma_2 = 2\sigma_2 T_1 + 2\beta \tag{5.19.15}$$

(b) CASE I: $\omega_0 = 1 + \varepsilon\sigma_2$, $\Omega_1 + \Omega_2 + \Omega_3 = 1 + \varepsilon\sigma_1$.

In this case, (5.19.9) becomess

$$\begin{aligned} D_0^2 u_1 + \omega_0^2 u_1 &= -2i\omega_0 D_1 A e^{i\omega_0 T_0} - 2i\omega_0 \mu A e^{i\omega_0 T_0} - 3\alpha_1 A^2 \overline{A} e^{i\omega_0 T_0} \\ &\quad - 6\alpha_1 \Lambda_1^2 A e^{i\omega_0 T_0} - 6\alpha_1 \Lambda_2^2 A e^{i\omega_0 T_0} - 6\alpha_1 \Lambda_3^2 A e^{i\omega_0 T_0} \\ &\quad - 6\alpha_1 \Lambda_1 \Lambda_2 \Lambda_3 e^{i(\Omega_1 T_0 + \Omega_2 T_0 + \Omega_3 T_0 + \theta_1 + \theta_2 + \theta_3)} \\ &\quad - \alpha_2 \overline{A} e^{i(2-\omega_0)T_0} + cc + NST \end{aligned} \tag{5.19.16}$$

In order to eliminate secular terms from the above equation, there must be

$$\begin{aligned} &2i\omega_0 D_1 A + 2i\omega_0 \mu A + 3\alpha_1 A^2 \overline{A} + 6\alpha_1 \sum_{n=1}^{3} \Lambda_n^2 A \\ &+ 6\alpha_1 \Lambda_1 \Lambda_2 \Lambda_3 e^{i(\theta_1 + \theta_2 + \theta_3)} e^{i(\sigma_1 - \sigma_2)T_1} + \alpha_2 \overline{A}\, e^{-i2\sigma_2 T_1} = 0 \end{aligned} \tag{5.19.17}$$

Substituting $A = \frac{1}{2} a e^{i\beta}$ into the above equation and taking $\omega_0 \approx 1$ into account, we get

$$\begin{aligned} &ia' - a\beta' + i\mu a + \frac{3}{8}\alpha_1 a^3 + 3\alpha_1 a \sum_{n=1}^{3} \Lambda_n^2 \\ &+ 6\alpha_1 \Lambda_1 \Lambda_2 \Lambda_3 e^{i[(\sigma_1 - \sigma_2)T_1 - \beta + \theta_1 + \theta_2 + \theta_3]} + \frac{1}{2}\alpha_2 a e^{-i(2\sigma_2 T_1 + 2\beta)} = 0 \end{aligned} \tag{5.19.18}$$

The prime denotes the derivative with respect to T_1. Separating the real and imaginary parts of the above equation gives

$$a' = -\mu a - 6\alpha_1 \Lambda_1 \Lambda_2 \Lambda_3 \sin\gamma_1 + \frac{1}{2}\alpha_2 a \sin\gamma_2$$
$$a\beta' = 3\left[\alpha_1 \sum_{n=1}^{3} \Lambda_n^2 + \frac{1}{8}a^2\right]a + 6\alpha_1 \Lambda_1 \Lambda_2 \Lambda_3 \cos\gamma_1 + \frac{1}{2}\alpha_2 a \cos\gamma_2 \tag{5.19.19}$$

where

$$\gamma_1 = (\sigma_1 - \sigma_2)T_1 - \beta + \theta_1 + \theta_2 + \theta_3, \quad \gamma_2 = 2\sigma_2 T_1 + 2\beta \tag{5.19.20}$$

CASE II: $\omega_0 = 1 + \varepsilon\sigma_2$, $\Omega_1 + \Omega_2 = 2 + \varepsilon\sigma_1$

In this case, (5.19.9) becomes

$$\begin{aligned} D_0^2 u_1 + \omega_0^2 u_1 = & -2i\omega_0 D_1 A e^{i\omega_0 T_0} - 2i\omega_0 \mu A e^{i\omega_0 T_0} \\ & - 3\alpha_1 A^2 \overline{A} e^{i\omega_0 T_0} - 6\alpha_1 \Lambda_1^2 A e^{i\omega_0 T_0} \\ & - 6\alpha_1 \Lambda_2^2 A e^{i\omega_0 T_0} - 6\alpha_1 \Lambda_3^2 A e^{i\omega_0 T_0} \\ & - 6\alpha_1 \Lambda_1 \Lambda_2 \overline{A} e^{i(\Omega_1 T_0 + \Omega_2 T_0 - \omega_0 T_0 + \theta_1 + \theta_2)} \\ & - \alpha_2 \overline{A} e^{i(2-\omega_0)T_0} + cc + NST \end{aligned} \tag{5.19.21}$$

In order to eliminate secular terms from the above equation, there must be

$$\begin{aligned} & 2i\omega_0 D_1 A + 2i\omega_0 \mu A + 3\alpha_1 A^2 \overline{A} + 6\alpha_1 \sum_{n=1}^{3} \Lambda_n^2 A \\ & + 6\alpha_1 \Lambda_1 \Lambda_2 \overline{A}\, e^{i[(\sigma_1 - 2\sigma_2)T_1 + \theta_1 + \theta_2]} + \alpha_2 \overline{A}\, e^{-i2\sigma_2 T_1} = 0 \end{aligned} \tag{5.19.22}$$

Substituting $A = \frac{1}{2} a e^{i\beta}$ into the above equation and taking $\omega_0 \approx 1$ into account, we get

$$\begin{aligned} & ia' - a\beta' + i\mu a + \frac{3}{8}\alpha_1 a^3 + 3\alpha_1 a \sum_{n=1}^{3} \Lambda_n^2 \\ & + 3\alpha_1 \Lambda_1 \Lambda_2 a e^{i[(\sigma_1 - 2\sigma_2)T_1 - 2\beta + \theta_1 + \theta_2]} + \frac{1}{2}\alpha_2 a e^{-i(2\sigma_2 T_1 + 2\beta)} = 0 \end{aligned} \tag{5.19.23}$$

The prime denotes the derivative with respect to T_1. Separating the real and imaginary parts of the above equation gives

$$\begin{aligned} a' &= -\mu a - 3\alpha_1 \Lambda_1 \Lambda_2 a \sin\gamma_1 + \frac{1}{2}\alpha_2 a \sin\gamma_2 \\ a\beta' &= 3\left[\alpha_1 \sum_{n=1}^{3} \Lambda_n^2 + \frac{1}{8}a^2\right] a + 3\alpha_1 \Lambda_1 \Lambda_2 a \cos\gamma_1 + \frac{1}{2}\alpha_2 a \cos\gamma_2 \end{aligned} \tag{5.19.24}$$

where

$$\gamma_1 = (\sigma_1 - 2\sigma_2)T_1 - 2\beta + \theta_1 + \theta_2, \quad \gamma_2 = 2\sigma_2 T_1 + 2\beta \tag{5.19.25}$$

CASE III: $\omega_0 = 1 + \varepsilon\sigma_2$, $3\Omega_1 = 1 + \varepsilon\sigma_1$, $\Omega_2 + \Omega_3 = 2 + \varepsilon\sigma_3$

In this case, (5.19.9) becomes

$$\begin{aligned} D_0^2 u_1 + \omega_0^2 u_1 = & -2i\omega_0 D_1 A e^{i\omega_0 T_0} - 2i\omega_0 \mu A e^{i\omega_0 T_0} \\ & - 3\alpha_1 A^2 \overline{A} e^{i\omega_0 T_0} - 6\alpha_1 \Lambda_1^2 A e^{i\omega_0 T_0} \\ & - 6\alpha_1 \Lambda_2^2 A e^{i\omega_0 T_0} - 6\alpha_1 \Lambda_3^2 A e^{i\omega_0 T_0} \\ & - 6\alpha_1 \Lambda_2 \Lambda_3 \overline{A} e^{i(\Omega_2 T_0 + \Omega_3 T_0 - \omega_0 T_0 + \theta_2 + \theta_3)} \\ & - \alpha_1 \Lambda_1^3 e^{i(3\Omega_1 T_0 + 3\theta_1)} \\ & - \alpha_2 \overline{A} e^{i(2-\omega_0)T_0} + cc + NST \end{aligned} \tag{5.19.26}$$

In order to eliminate secular terms from the above equation, there must be

$$\begin{aligned} & 2i\omega_0 D_1 A + 2i\omega_0 \mu A + 3\alpha_1 A^2 \bar{A} + 6\alpha_1 \sum_{n=1}^{3} \Lambda_n^2 A \\ & - 6\alpha_1 \Lambda_2 \Lambda_3 \bar{A} e^{i[(\sigma_3 - 2\sigma_2)T_1 + \theta_2 + \theta_3)} \\ & - \alpha_1 \Lambda_1^3 e^{i[(\sigma_1 - \sigma_2)T_1 + 3\theta_1]} - \alpha_2 \bar{A} e^{-i2\sigma_2 T_1} = 0 \end{aligned} \tag{5.19.27}$$

Substituting $A = \frac{1}{2} a e^{i\beta}$ into the above equation and taking $\omega_0 \approx 1$ into account, we get

$$\begin{aligned} & ia' - a\beta' + i\mu a + \frac{3}{8}\alpha_1 a^3 + 3\alpha_1 a \sum_{n=1}^{3} \Lambda_n^2 + 3\alpha_1 \Lambda_2 \Lambda_3 a e^{i[(\sigma_3 - 2\sigma_2)T_1 - 2\beta + \theta_2 + \theta_3]} \\ & + \alpha_1 \Lambda_1^3 e^{i[(\sigma_1 - \sigma_2)T_1 - \beta + 3\theta_1]} + \frac{1}{2}\alpha_2 a e^{-i(2\sigma_2 T_1 + 2\beta)} = 0 \end{aligned} \tag{5.19.28}$$

The prime denotes the derivative with respect to T_1. Separating the real and imaginary parts of the above equation gives

$$\begin{aligned} a' &= -\mu a - 3\alpha_1 \Lambda_2 \Lambda_3 a \sin\gamma_3 + \frac{1}{2}\alpha_2 a \sin\gamma_2 - \alpha_1 \Lambda_1^3 \sin\gamma_1 \\ a\beta' &= 3\left[\alpha_1 \sum_{n=1}^{3} \Lambda_n^2 + \frac{1}{8}a^2\right] a + 3\alpha_1 \Lambda_2 \Lambda_3 a \cos\gamma_3 \\ &\quad + \frac{1}{2}\alpha_2 a \cos\gamma_2 + \alpha_1 \Lambda_1^3 \cos\gamma_1 \end{aligned} \tag{5.19.29}$$

where

$$\begin{aligned} \gamma_1 &= (\sigma_1 - \sigma_2)T_1 - \beta + 3\theta_1, \\ \gamma_2 &= 2\sigma_2 T_1 2\beta, \\ \gamma_3 &= (\sigma_3 - 2\sigma_2)T_1 - 2\beta + \theta_2 + \theta_3 \end{aligned} \tag{5.19.30}$$

Readers are invited to complete the rest of the cases on their own.

5.20 Exercise 5.20 (Cubic Nonlinear, Square-Damped System Subjected to Combined Parametric and External Excitation)

Solution: **(a)** We use the method of multiple scales to obtain the first order approximate solution of the system. When $\omega_0 = 1 + \varepsilon\sigma_2$, $\Omega = 1 + \varepsilon\sigma_1$, i.e.$\Omega \approx \omega_0$, this can cause a primary resonance, so let

$$K = \varepsilon k \tag{5.20.1}$$

and the solution of the equation be

$$u(t;\varepsilon) = u_0(T_0, T_1, T_2) + \varepsilon u(T_0, T_1, T_2) + \varepsilon^2 u_2(T_0, T_1, T_2) + \tag{5.20.2}$$

Substituting the above equation into the original equation and retaining to $O(\varepsilon^2)$, we get

$$\begin{aligned}
0 &= \ddot{u} + 2\varepsilon\mu_1\dot{u} + \varepsilon\mu_2|\dot{u}|\dot{u} \\
&\quad + \left(\omega_0^2 + \varepsilon\cos 2t\right)\left(u + \varepsilon\alpha u^3\right) - K\cos(\Omega t - \theta) \\
&= \ddot{u} + \omega_0^2 u + 2\varepsilon\mu_1\dot{u} + \varepsilon\mu_2|\dot{u}|\dot{u} + \varepsilon\alpha u^3 \\
&\quad + \varepsilon u\cos 2t - K\cos(\Omega t - \theta) + \cdots \\
&= \left(D_0^2 + 2\varepsilon D_0 D_1\right)(u_0 + \varepsilon u_1) + \omega_0^2(u_0 + \varepsilon u_1) \\
&\quad + 2\varepsilon\mu_1(D_0 + \varepsilon D_1)(u_0 + \varepsilon u_1) \\
&\quad + \varepsilon\mu_2(D_0 + \varepsilon D_1)(u_0 + \varepsilon u_1)\left|(D_0 + \varepsilon D_1)(u_0 + \varepsilon u_1)\right| \\
&\quad + \varepsilon\alpha u_0^3 + \frac{1}{2}\varepsilon(u_0 + \varepsilon u_1)\cos 2T_0 \\
&\quad - K\cos(\Omega T_0 - \theta) + \cdots \\
&= D_0^2 u_0 + \omega_0^2 u_0 + \varepsilon[D_0^2 u_1 + \omega_0^2 u_1 \\
&\quad + 2D_0 D_1 u_0 + 2\mu_1 D_0 u_0 \\
&\quad + \mu_2 D_0 u_0\left|D_0 u_0\right| + \alpha u_0^3 + \frac{1}{2}u_0\cos 2T_0] \\
&\quad - K\cos(\Omega T_0 - \theta) + \cdots \\
&= D_0^2 u_0 + \omega_0^2 u_0 \\
&\quad + \varepsilon[D_0^2 u_1 + \omega_0^2 u_1 + 2D_0 D_1 u_0 + 2\mu_1 D_0 u_0 \\
&\quad + \mu_2 D_0 u_0\left|D_0 u_0\right| + \alpha u_0^3 + \frac{1}{2}u_0\cos 2T_0 \\
&\quad - k\cos(\Omega T_0 - \theta)] + \cdots
\end{aligned} \tag{5.20.3}$$

Equating coefficients of like powers of ε in the above equation yields

$$D_0^2 u_0 + \omega_0^2 u_0 = 0 \tag{5.20.4}$$

$$\begin{aligned} D_0^2 u_1 + \omega_0^2 u_1 &= -2D_0 D_1 u_0 - 2\mu_1 D_0 u_0 \\ &- \mu_2 D_0 u_0 |D_0 u_0| - \alpha u_0^3 - \frac{1}{2} u_0 cos 2T_0 + k cos(\Omega T_0 - \theta) \end{aligned} \tag{5.20.5}$$

The solution of (5.20.4) is

$$u_0 = Ae^{i\omega_0 T_0} + \bar{A}\, e^{-i\omega_0 T_0} \tag{5.20.6}$$

where $A = A(T_1, T_2)$. Let

$$A = \frac{1}{2} a e^{i\beta} \tag{5.20.7}$$

Substituting this into (5.20.6), we can obtain the first order approximate solution of the given equation

$$u \approx u_0 = a\cos(\omega_0 t + \beta) + O(\varepsilon) \tag{5.20.8}$$

Substituting u_0 into (5.20.5) yields

$$\begin{aligned} D_0^2 u_1 + \omega_0^2 u_1 &= -2D_0 D_1 u_0 - 2\mu_1 D_0 u_0 - \mu_2 D_0 u_0 \,|D_0 u_0| \, - \alpha u_0^3 \\ &- \frac{1}{2} u_0 \cos 2T_0 + k \cos(\Omega T_0 - \theta) \\ &= -2i\omega_0 D_1 A e^{i\omega_0 T_0} - 2i\omega_0 \mu_1 A e^{i\omega_0 T_0} \\ &- 3\alpha A^2 \bar{A} e^{i\omega_0 T_0} - \frac{1}{2} \bar{A} e^{i(2-\omega_0)T_0} \\ &+ \frac{1}{2} k e^{i(\Omega T_0 - \theta)} + cc + NST \\ &- i\omega_0^2 \mu_2 \big(Ae^{i\omega_0 T_0} - \bar{A} e^{-i\omega_0 T_0}\big) \left|Ae^{i\omega_0 T_0} - \bar{A} e^{-i\omega_0 T_0}\right| \end{aligned} \tag{5.20.9}$$

In order to further process (5.20.9), we need to deal with $\left(Ae^{i\omega_0 T_0} - \bar{A}\, e^{-i\omega_0 T_0}\right) \left|Ae^{i\omega_0 T_0} - \bar{A}\, e^{-i\omega_0 T_0}\right|$, which can be expanded into a Fourier series of complex exponential form:

$$\left(Ae^{i\omega_0 T_0} - \bar{A}\, e^{-i\omega_0 T_0}\right) \left|Ae^{i\omega_0 T_0} - \bar{A}\, e^{-i\omega_0 T_0}\right| = \sum_{m=-\infty}^{\infty} f_m e^{im\omega_0 T_0} \tag{5.20.10}$$

where

$$f_m = \frac{\omega_0}{2\pi}\int_0^{2\pi/\omega_0}\left[\left(Ae^{i\omega_0 T_0} - \bar{A}\,e^{-i\omega_0 T_0}\right)\left|Ae^{i\omega_0 T_0} - \bar{A}\,e^{-i\omega_0 T_0}\right|\right]e^{-im\omega_0 T_0}dT_0 \quad (5.20.11)$$

Thus, (5.20.9) becomes

$$D_0^2 u_1 + \omega_0^2 u_1 = -2i\omega_0 D_1 Ae^{i\omega_0 T_0} - 2i\omega_0\mu_1 Ae^{i\omega_0 T_0} - 3\alpha A^2\overline{A}e^{i\omega_0 T_0} - \tfrac{1}{2}\overline{A}e^{i(2-\omega_0)T_0} + \tfrac{1}{2}ke^{i(\Omega T_0-\theta)} - i\omega_0^2\mu_2 f_1 e^{i\omega_0 T_0} + cc + NST \quad (5.20.12)$$

In order to eliminate secular terms from (5.20.12), there must be

$$2i\omega_0 D_1 A + 2i\omega_0\mu_1 A + 3\alpha A^2\bar{A} + \frac{1}{2}\bar{A}\,e^{-i2\sigma_2 T_1} - \frac{1}{2}ke^{i(\Omega T_0-\theta)} + i\omega_0^2\mu_2 f_1 = 0 \quad (5.20.13)$$

Let $A = \frac{1}{2}ae^{i\beta}$, then we have.

$$\begin{aligned} f_1 &= \frac{\omega_0}{2\pi}\int_0^{2\pi/\omega_0}\left[\left(Ae^{i\omega_0 T_0} - \overline{A}e^{-i\omega_0 T_0}\right)\left|Ae^{i\omega_0 T_0} - \overline{A}e^{-i\omega_0 T_0}\right|\right]e^{-i\omega_0 T_0}dT_0 \\ &= \frac{\omega_0}{2\pi}\int_0^{2\pi/\omega_0} ia^2e^{i\beta}\sin(\omega_0 T_0+\beta)|\sin(\omega_0 T_0+\beta)|e^{-i(\omega_0 T_0+\beta)}dT_0 \\ &= \frac{1}{2\pi}\int_0^{2\pi} ia^2e^{i\beta}\sin\phi|\sin\phi|e^{-i\phi}d\phi \\ &= \frac{a^2e^{i\beta}}{2\pi}\int_0^{2\pi}\sin^2\phi|\sin\phi|d\phi \\ &= \frac{a^2e^{i\beta}}{\pi}\int_0^{\pi}\sin^3\phi d\phi \\ &= \frac{4a^2e^{i\beta}}{3\pi} \end{aligned} \quad (5.20.14)$$

Considering $\omega_0 \approx 1$, (5.20.13) becomes

$$\begin{aligned} &ia' - a\beta' + i\mu_1 a + \frac{3}{8}\alpha a^3 + \frac{1}{4}ae^{-i(2\sigma_2 T_1+2\beta)} \\ &- \frac{1}{2}ke^{i[(\sigma_1-\sigma_2)T_1-\beta-\theta]} + \frac{4}{3\pi}i\mu_2 a^2 = 0 \end{aligned} \quad (5.20.15)$$

The prime denotes the derivative with respect to T_1. Separating the real and imaginary parts of the above equation gives

$$\begin{aligned} a' &= -\mu_1 a - \tfrac{4\mu_2}{3\pi}a^2 + \tfrac{1}{2}k\sin\gamma_1 + \tfrac{1}{4}a\sin\gamma_2 \\ a\beta' &= \tfrac{3}{8}\alpha a^3 - \tfrac{1}{2}k\cos\gamma_1 + \tfrac{1}{4}a\cos\gamma_2 \end{aligned} \tag{5.20.16}$$

where

$$\gamma_1 = (\sigma_1 - \sigma_2)T_1 - \beta - \theta, \quad \gamma_2 = 2\sigma_2 T_1 + 2\beta \tag{5.20.17}$$

(b) When $\omega_0 = 1 + \varepsilon\sigma_2$ and $\Omega = 2$, only hard excitation can cause resonance, therefore $K = O(1)$. (5.20.3) becomes

$$\begin{aligned} 0 = {} & D_0^2 u_0 + \omega_0^2 u_0 + \varepsilon[D_0^2 u_1 + \omega_0^2 u_1 \\ & + 2D_0 D_1 u_0 + 2\mu_1 D_0 u_0 + \mu_2 D_0 u_0 \left|D_0 u_0\right| \\ & + \alpha u_0^3 + \frac{1}{2}u_0 \cos 2T_0] - K\cos(\Omega T_0 - \theta) + \cdots \end{aligned} \tag{5.20.18}$$

Equating coefficients of like powers of ε in the above equation yields

$$D_0^2 u_0 + \omega_0^2 u_0 = K\cos(\Omega T_0 - \theta) \tag{5.20.19}$$

$$\begin{aligned} D_0^2 u_1 + \omega_0^2 u_1 = {} & -2D_0 D_1 u_0 - 2\mu_1 D_0 u_0 - \mu_2 D_0 u_0 \left|D_0 u_0\right| \\ & - \alpha u_0^3 - \frac{1}{2}u_0 \cos 2T_0 \end{aligned} \tag{5.20.20}$$

The solution of (5.20.19) is

$$\begin{aligned} u_0 &= Ae^{i\omega_0 T_0} + \overline{A}e^{-i\omega_0 T_0} + \Lambda\left[e^{i(\Omega T_0 - \theta)} + e^{-i(\Omega T_0 - \theta)}\right] \\ &= Ae^{i\omega_0 T_0} + \overline{A}e^{-i\omega_0 T_0} + \Lambda\left[e^{i(2T_0 - \theta)} + e^{-i(2T_0 - \theta)}\right] \end{aligned} \tag{5.20.21}$$

where $A = A(T_1, T_2)$ and

$$\Lambda = \frac{K}{2(\omega_0^2 - \Omega^2)} = \frac{K}{2(\omega_0^2 - 4)} = -\frac{K}{6} \tag{5.20.22}$$

Let $A = \frac{1}{2}ae^{i\beta}$, and substitute this into (5.20.6), we obtain the first order approximate solution of the given equation

$$u \approx u_0 = a\cos(\omega_0 t + \beta) + 2\Lambda\cos(2t - \theta) + O(\varepsilon) \tag{5.20.23}$$

Substituting u_0 into (5.20.20) yields

$$
\begin{aligned}
D_0^2 u_1 + \omega_0^2 u_1 &= -2D_0 D_1 u_0 - 2\mu_1 D_0 u_0 - \mu_2 D_0 u_0 \left| D_0 u_0 \right| \\
&\quad - \alpha u_0^3 - \frac{1}{2} u_0 \cos 2T_0 \\
&= -2i\omega_0 D_1 A e^{i\omega_0 T_0} - 2i\omega_0 \mu_1 A e^{i\omega_0 T_0} \\
&\quad - 3\alpha A^2 \overline{A} e^{i\omega_0 T_0} - 6\alpha \Lambda^2 A e^{i\omega_0 T_0} \\
&\quad - \frac{1}{2} \overline{A} e^{i(2-\omega_0)T_0} + cc + NST \\
&\quad - \mu_2 D_0 u_0 \left| D_0 u_0 \right|
\end{aligned} \tag{5.20.24}
$$

In order to further process (5.20.24), we need to deal with the function $D_0 u_0 \left| D_0 u_0 \right|$, which can be expanded into a Fourier series of complex exponential form:

$$
D_0 u_0 \left| D_0 u_0 \right| = \sum_{m=-\infty}^{\infty} g_m e^{im\omega_0 T_0} \tag{5.20.25}
$$

where

$$
g_m = \frac{\omega_0}{2\pi} \int_0^{2\pi/\omega_0} D_0 u_0 \left| D_0 u_0 \right| e^{-im\omega_0 T_0} dT_0 \tag{5.20.26}
$$

Thus, the Eq. (5.20.24) becomes

$$
\begin{aligned}
D_0^2 u_1 + \omega_0^2 u_1 &= -2i\omega_0 D_1 A e^{i\omega_0 T_0} - 2i\omega_0 \mu_1 A e^{i\omega_0 T_0} \\
&\quad - 3\alpha A^2 \overline{A} e^{i\omega_0 T_0} - 6\alpha \Lambda^2 A e^{i\omega_0 T_0} \\
&\quad - \frac{1}{2} \overline{A} e^{i(2-\omega_0)T_0} - \mu_2 g_1 e^{i\omega_0 T_0} + cc + NST
\end{aligned} \tag{5.20.27}
$$

In order to eliminate secular terms from (5.20.27), there must be

$$
\begin{aligned}
&2i\omega_0 D_1 A + 2i\omega_0 \mu_1 A + 6\alpha \Lambda^2 A \\
&\quad + 3\alpha A^2 \overline{A} + \frac{1}{2} \overline{A} e^{-i2\sigma_2 T_1} + \mu_2 g_1 = 0
\end{aligned} \tag{5.20.28}
$$

Substitute $A = \frac{1}{2} a e^{i\beta}$ into the above equation and let

$$
g_1 = e^{i\beta} (g_{1R} + i g_{1I}) \tag{5.20.29}
$$

we obtain

$$ia' - a\beta' + i\mu_1 a + 3\alpha\Lambda^2 a + \frac{3}{8}\alpha a^3 + \frac{1}{4}ae^{-i(2\sigma_2 T_1 + 2\beta)} + \mu_2(g_{1R} + ig_{1I}) = 0 \tag{5.20.30}$$

The prime denotes the derivative with respect to T_1. Separating the real and imaginary parts of the above equation gives

$$\begin{aligned} a' &= -\mu_1 a + \frac{1}{4}a\sin\gamma_2 - \mu_2 g_{1I} \\ a\beta' &= 3\alpha\left(\Lambda^2 + \frac{1}{8}a^2\right)a + \frac{1}{4}a\cos\gamma_2 + \mu_2 g_{1R} \end{aligned} \tag{5.20.31}$$

where

$$\gamma_2 = 2\sigma_2 T_1 + 2\beta \tag{5.20.32}$$

From (5.20.26), and taking $\omega_0 \approx 1$ into account, we have

$$\begin{aligned} g_1 &= \frac{\omega_0}{2\pi}\int\limits_0^{2\pi/\omega_0} D_0 u_0 \left|D_0 u_0\right| e^{-i\omega_0 T_0} dT_0 \\ &= \frac{e^{i\beta}}{2\pi}\int\limits_0^{2\pi} D_0 u_0 \left|D_0 u_0\right| e^{-i(\omega_0 T_0 + \beta)} d\omega_0 T_0 \\ &= \frac{e^{i\beta}}{2\pi}\int\limits_0^{2\pi} G^2(\phi)\mathrm{sgn}[G(\phi)](\cos\phi - i\sin\phi)d\phi \\ &\triangleq e^{i\beta}(g_{iR} + ig_{iI}) \end{aligned} \tag{5.20.33}$$

where $G(\varphi) = D_0 u_0$. From (5.20.21), we can obtain

$$\begin{aligned} G(\phi) = D_0 u_0 &= -\omega_0 a\sin(\omega_0 T_0 + \beta) - 4\Lambda\sin(2T_0 - \theta) \\ &= -\omega_0 a\sin(\omega_0 T_0 + \beta) \\ &\quad - 4\Lambda\sin(2\omega_0 T_0 + 2\beta - 2\sigma_2 T_1 - 2\beta - \theta) \\ &= -\omega_0 a\sin\phi - 4\Lambda\sin(2\phi - \gamma_2 - \theta) \end{aligned}$$

i.e.,

$$G(\varphi) = -\omega_0 a\sin\varphi - 4\Lambda\sin(2\varphi - \gamma_2 - \theta) \tag{5.20.34}$$

Substituting (5.20.33) and (5.20.34) into (5.20.31) yields

$$a' = -\mu_1 a + \frac{1}{4} a\sin\gamma_2 + \frac{\mu_2}{2\pi}\int_0^{2\pi} G^2(\phi)\text{sgn}[G(\phi)]\sin\phi d\phi$$

$$a\beta' = 3\alpha\left(\Lambda^2 + \frac{1}{8}a^2\right)a + \frac{1}{4}a\cos\gamma_2$$

$$+\frac{\mu_2}{2\pi}\int_0^{2\pi} G^2(\phi)\text{sgn}[G(\phi)]\cos\phi d\phi \quad (5.20.35)$$

5.21 Exercise 5.21 (Analysis on a Double Pendulum with Swinging Mass Attached to Springs and Moving Platform)

Solution: **(a)** The double pendulum system oscillates and two springs are in horizontal position and at their original length when $y = 0$and$\theta_1 = \theta_2 = 0$. Since

$$\begin{cases} x_1 = l_1 \sin\theta_1 \\ y_1 = y + l_1\cos\theta_1 \end{cases}; \quad \begin{cases} x_2 = l_1\sin\theta_1 + l_2\sin\theta_2 \\ y_2 = y + l_1\cos\theta_1 + l_2\cos\theta_2 \end{cases} \quad (5.21.1)$$

the velocities of two particles are:

$$v_1^2 = \dot{x}_1^2 + \dot{y}_1^2 = \dot{y}^2 + l_1^2\dot{\theta}_1^2 - 2l_1\dot{y}\dot{\theta}_1\sin\theta_1 \approx \dot{y}^2 + l_1^2\dot{\theta}_1^2 - 2l_1\theta_1\dot{y}\dot{\theta}_1 \quad (5.21.2)$$

$$v_2^2 = \dot{x}_2^2 + \dot{y}_2^2 = \dot{y}^2 + l_1^2\dot{\theta}_1^2 + l_2^2\dot{\theta}_2^2 - 2l_1\dot{y}\dot{\theta}_1\sin\theta_1 - 2l_2\dot{y}\dot{\theta}_2\sin\theta_2$$

$$+2l_1l_2\dot{\theta}_1\dot{\theta}_2\cos\theta_1\cos\theta_2 + 2l_1l_2\dot{\theta}_1\dot{\theta}_2\sin\theta_1\sin\theta_2$$

$$\approx \dot{y}^2 + l_1^2\dot{\theta}_1^2 + l_2^2\dot{\theta}_2^2 - 2l_1\theta_1\dot{y}\dot{\theta}_1 - 2l_2\theta_2\dot{y}\dot{\theta}_2 + 2l_1l_2\dot{\theta}_1\dot{\theta}_2 \quad (5.21.3)$$

The kinetic energy of the system is

$$T = \frac{1}{2}m_1v_1^2 + \frac{1}{2}m_2v_2^2 = \frac{1}{2}m_1\left(\dot{x}_1^2 + \dot{y}_1^2\right)$$

$$+\frac{1}{2}m_2\left(\dot{x}_2^2 + \dot{y}_2^2\right)$$

$$= \frac{1}{2}m_1\left(\dot{y}^2 + l_1^2\dot{\theta}_1^2 - 2l_1\theta_1\dot{y}\dot{\theta}_1\right)$$

$$+\frac{1}{2}m_2$$

$$\left(\dot{y}^2 + l_1^2\dot{\theta}_1^2 + l_2^2\dot{\theta}_2^2 - 2l_1\theta_1\dot{y}\dot{\theta}_1 - 2l_2\theta_2\dot{y}\dot{\theta}_2 + 2l_1l_2\dot{\theta}_1\dot{\theta}_2\right) \quad (5.21.4)$$

The potential energy of the system is

$$\begin{aligned} V &= m_1 g\big[l_1(1-\cos\theta_1) - y\big] \\ &\quad + m_2 g\big[l_1(1-\cos\theta_1) + l_2(1-\cos\theta_2) - y\big] \\ &\quad + \frac{1}{2}k_1 l_1^2\theta_1^2 + \frac{1}{2}k_2(l_1\theta_1 + l_2\theta_2)^2 \\ &\approx \frac{1}{2}m_1 g\big(l_1\theta_1^2 - y\big) + m_2 g\big(l_1\theta_1^2 + l_2\theta_2^2 - y\big) \\ &\quad + \frac{1}{2}k_1 l_1^2\theta_1^2 + \frac{1}{2}k_2(l_1\theta_1 + l_2\theta_2)^2 \end{aligned} \tag{5.21.5}$$

Substitute kinetic and potential energy into the Lagrange's equation:

$$\frac{d}{dt}\frac{\partial T}{\partial\dot{\theta}_i} - \frac{\partial T}{\partial\theta_i} = -\frac{\partial V}{\partial\theta_i}, \quad i = 1,\ 2 \tag{5.21.6}$$

The differential equation of motion of the system can be obtained as

$$\begin{aligned} &(m_1+m_2)l_1\ddot{\theta}_1 + m_2 l_2\ddot{\theta}_2 - (m_1+m_2)\ddot{y}\theta_1 + \big[(m_1+m_2)g + (k_1+k_2)l_1\big]\theta_1 + k_2 l_2\theta_2 \\ &= 0 \\ &\quad m_2 l_2\ddot{\theta}_2 + m_2 l_1\ddot{\theta}_1 - m_2\ddot{y}\theta_2 + k_2 l_1\theta_1 + (m_2 g + k_2 l_2)\theta_2 = 0 \end{aligned} \tag{5.21.7}$$

When $l_1 = l_2 = l,\ \ m_1 = m_2 = m$ and $k_1 = k_2 = 0$, (5.21.7) becomes

$$\ddot{\theta}_2 + \ddot{\theta}_1 - \frac{g-\ddot{y}}{l}\theta_2 = 0$$

A simple transformation of these two equations gives

$$\ddot{\theta}_1 + \frac{1}{2}\ddot{\theta}_2 + \frac{g-\ddot{y}}{l}\theta_1 = 0$$

$$\ddot{\theta}_1 + \frac{g-\ddot{y}}{l}(2\theta_1 - \theta_2) = 0$$

$$\ddot{\theta}_2 + 2\frac{g-\ddot{y}}{l}(\theta_2 - \theta_1) = 0 \tag{5.21.8}$$

(b) When $\ddot{y} \equiv 0$, (5.21.1) becomes

$$\begin{aligned} &(m_1+m_2)l_1\ddot{\theta}_1 + m_2 l_2\ddot{\theta}_2 + \big[(m_1+m_2)g + (k_1+k_2)l_1\big]\theta_1 \\ &\quad + k_2 l_2\theta_2 = 0\, m_2 l_2\ddot{\theta}_2 + m_2 l_1\ddot{\theta}_1 + k_2 l_1\theta_1 + (m_2 g + k_2 l_2)\theta_2 = 0 \end{aligned} \tag{5.21.9}$$

If we make $l_1 = l_2 = l,\ \ m_1 = m_2 = m$ and $k_1 = k_2 = k$, (5.21.3) becomes

$$\begin{Bmatrix} \ddot{\theta}_1 \\ \ddot{\theta}_2 \end{Bmatrix} + \begin{bmatrix} \frac{2g}{l} + \frac{k}{m} & -\frac{g}{l} \\ -\frac{2g}{l} & \frac{2g}{l} + \frac{k}{m} \end{bmatrix} \begin{Bmatrix} \theta_1 \\ \theta_2 \end{Bmatrix} = 0 \tag{5.21.10}$$

The characteristic equation is

$$\left[\omega^2 - \left(\frac{k}{m} + \frac{2g}{l}\right)\right]^2 - 2\left(\frac{g}{l}\right)^2 = 0 \tag{5.21.11}$$

Therefore, the frequency of the system is

$$\omega_{1,2} = \sqrt{\frac{k}{m} + \frac{\left(2 - \sqrt{2}\right)g}{l}}, \quad \sqrt{\frac{k}{m} + \frac{\left(2 + \sqrt{2}\right)g}{l}} \tag{5.21.12}$$

The corresponding eigenmatrix is

$$[\Theta] = \begin{bmatrix} 1 & -1 \\ \sqrt{2} & \sqrt{2} \end{bmatrix} \tag{5.21.13}$$

and

$$[\Theta]^{-1} = \frac{1}{2\sqrt{2}} \begin{bmatrix} \sqrt{2} & 1 \\ -\sqrt{2} & 1 \end{bmatrix} \tag{5.21.14}$$

(c) When $l_1 = l_2 = l, \quad m_1 = m_2 = m$ and $k_1 = k_2 = k$, (5.21.7) becomes

$$\begin{Bmatrix} \ddot{\theta}_1 \\ \ddot{\theta}_2 \end{Bmatrix} + \begin{bmatrix} \frac{2g}{l} + \frac{k}{m} & -\frac{g}{l} \\ -\frac{2g}{l} & \frac{2g}{l} + \frac{k}{m} \end{bmatrix} \begin{Bmatrix} \theta_1 \\ \theta_2 \end{Bmatrix} - \frac{\ddot{y}}{l} \begin{bmatrix} 2 & -1 \\ -2 & 2 \end{bmatrix} \begin{Bmatrix} \theta_1 \\ \theta_2 \end{Bmatrix} = 0 \tag{5.21.15}$$

Make a transformation

$$\{\theta_1, \theta_2\}^T = [\Theta]\{u_1, u_2\}^T \tag{5.21.16}$$

(5.21.9) becomes

$$\begin{Bmatrix} \ddot{u}_1 \\ \ddot{u}_2 \end{Bmatrix} + \begin{bmatrix} \omega_1^2 & 0 \\ 0 & \omega_2^2 \end{bmatrix} \begin{Bmatrix} u_1 \\ u_2 \end{Bmatrix} - \frac{\ddot{y}}{l} \begin{bmatrix} 2 - \sqrt{2} & 0 \\ 0 & 2 + \sqrt{2} \end{bmatrix} \begin{Bmatrix} u_1 \\ u_2 \end{Bmatrix} = 0 \tag{5.21.17}$$

Substituting $y = \varepsilon l \cos \Omega t$ into the above equation yields

$$\begin{aligned} \ddot{u}_1 + \omega_1^2 u_1 + \varepsilon b_1 \Omega^2 u_1 \cos \Omega t = 0 \\ \ddot{u}_2 + \omega_2^2 u_2 + \varepsilon b_2 \Omega^2 u_2 \cos \Omega t = 0 \end{aligned} \tag{5.21.18}$$

where

$$b_1 = 2 - \sqrt{2}, \quad b_2 = 2 + \sqrt{2} \tag{5.21.19}$$

These are two mutually independent, single-degree-of-freedom, single-frequency parametric excitation equations. We use the method of multiple scales to obtain their first-order approximate solutions. Let the solutions of the equations be

$$\begin{aligned} u_n(t;\varepsilon) &= u_{n0}(T_0, T_1, T_2) + \varepsilon u_{n1}(T_0, T_1, T_2) \\ &\quad + \varepsilon^2 u_{n2}(T_0, T_1, T_2) + \cdots, \quad n = 1,\ 2 \end{aligned} \tag{5.21.20}$$

Substituting the above equation(s) into the original equation and retaining to $O(\varepsilon^2)$, we get

$$\begin{aligned} 0 &= \ddot{u}_n + \omega_n^2 u_n + \varepsilon b_n \Omega^2 u_n \cos \Omega t \\ &= \left[D_0^2 + 2\varepsilon D_0 D_1 + \varepsilon^2 \left(D_1^2 + 2D_0 D_2\right)\right]\left(u_{n0} + \varepsilon u_{n1} + \varepsilon^2 u_{n2}\right) \\ &\quad + \omega_n^2 \left(u_{n0} + \varepsilon u_{n1} + \varepsilon^2 u_{n2}\right) \\ &\quad + \varepsilon b_n \Omega^2 \left(u_{n0} + \varepsilon u_{n1} + \varepsilon^2 u_{n2}\right) \cos \Omega T_0 + \cdots \\ &= D_0^2 u_{n0} + \varepsilon D_0^2 u_{n1} + \varepsilon^2 D_0^2 u_{n2} \\ &\quad + 2\varepsilon D_0 D_1 u_{n0} + 2\varepsilon^2 D_0 D_1 u_{n1} \\ &\quad + \varepsilon^2 \left(D_1^2 + 2D_0 D_2\right) u_{n0} + \omega_n^2 u_{n0} + \varepsilon \omega_n^2 u_{n1} + \varepsilon^2 \omega_n^2 u_{n2} \\ &\quad + \varepsilon b_n \Omega^2 u_{n0} \cos \Omega T_0 + \varepsilon^2 b_n \Omega^2 u_{n1} \cos \Omega T_0 + \cdots \\ &= D_0^2 u_{n0} + \omega_n^2 u_{n0} \\ &\quad + \varepsilon \left(D_0^2 u_{n1} + \omega_n^2 u_{n1} + 2D_0 D_1 u_{n0} + b_n \Omega^2 u_{n0} \cos \Omega T_0\right) \\ &\quad + \varepsilon^2 \\ &\quad \left[D_0^2 u_{n2} + \omega_n^2 u_{n2} + 2D_0 D_1 u_{n1} + \left(D_1^2 + 2D_0 D_2\right) u_{n0} + b_n \Omega^2 u_{n1} \cos \Omega T_0\right] + \cdots \end{aligned} \tag{5.21.21}$$

Equating coefficients of like powers of ε in the above equation yields

$$D_0^2 u_{n0} + \omega_n^2 u_{n0} = 0 \tag{5.21.22}$$

$$D_0^2 u_{n1} + \omega_n^2 u_{n1} = -2D_0 D_1 u_{n0} - b_n \Omega^2 u_{n0} \cos \Omega T_0 \tag{5.21.23}$$

$$D_0^2 u_{n2} + \omega_n^2 u_{n2} = -2D_0 D_1 u_{n1} - \left(D_1^2 + 2D_0 D_2\right) u_{n0} - b_n \Omega^2 u_{n1} \cos \Omega T_0 \tag{5.21.24}$$

The solutions of (5.21.22) are

$$u_{n0} = A_n e^{i\omega_n T_0} + \bar{A}_n e^{-i\omega_n T_0} \tag{5.21.25}$$

where $A = A(T_1, T_2)$. Let

$$A_n = \frac{1}{2} a_n e^{i\beta_n} \tag{5.21.26}$$

Substituting this into (5.21.25), we can obtain the first order approximate solutions of the given equations

$$u_n \approx u_{n0} = a_n \cos(\omega_n t + \beta_n) + O(\varepsilon) \tag{5.21.27}$$

Substituting this set of solutions into (5.21.16), we obtain first order approximate solutions to the original equations.

Substitute u_{n0} into (5.21.23) yields

$$\begin{aligned} D_0^2 u_{n1} + \omega_n^2 u_{n1} &= -2D_0 D_1 u_{n0} - b_n \Omega^2 u_{n0} \cos \Omega T_0 \\ &= -2D_0 D_1 A_n e^{i\omega_n T_0} \\ &\quad - \frac{1}{2} b_n \Omega^2 \left(A_n e^{i\omega_n T_0} + \overline{A}_n e^{-i\omega_n T_0}\right) e^{i\Omega T_0} + cc \\ &= -2i\omega_n D_1 A_n e^{i\omega_n T_0} \\ &\quad - \frac{1}{2} b_n \Omega^2 \overline{A}_n e^{i(\Omega - \omega_n) T_0} + cc + NST \end{aligned} \tag{5.21.28}$$

CASE I: $\Omega \approx 2\omega_1$:

Introduce the detuning parameter σ such that

$$\Omega = 2\omega_1 + \varepsilon\sigma \tag{5.21.29}$$

In order to eliminate secular terms from (5.21.28), there must be

$$2i\omega_1 D_1 A_1 + \frac{1}{2} b_1 \Omega^2 \overline{A}_1 e^{i\sigma T_1} = 0 \tag{5.21.30}$$

$$2i\omega_2 D_1 A_2 = 0 \tag{5.21.31}$$

Substituting $A_1 = \frac{1}{2} a_1 e^{i\beta_1}$ into (5.21.30) yields

$$ia_1' - a\beta_1' + \frac{1}{4} \omega_1^{-1} b_1 \Omega^2 a_1 e^{i(\sigma T_1 - 2\beta_1)} = 0 \tag{5.21.32}$$

The prime denotes the derivative with respect to T_1. Separating the real and imaginary parts of the above equation and taking into account $\Omega \approx 2\omega_1$ gives

$$\begin{aligned} a_1' &= -\omega_1 b_1 a_1 \sin\gamma \\ a_1 \gamma' &= a\sigma - 2\omega_1 b_1 a_1 \cos\gamma \end{aligned} \tag{5.21.33}$$

where

$$\gamma = \sigma T_1 - 2\beta_1 \tag{5.21.34}$$

The steady state solution satisfies the following equations:

$$\begin{aligned} -\omega_1 b_1 a_1 sin\gamma &= 0 \\ a_1\sigma - 2\omega_1 b_1 a_1 \cos\gamma &= 0 \end{aligned} \tag{5.21.35}$$

It can be seen that the condition for the existence of a nonzero steady state solution is

$$\sigma = \pm 2\omega_1 b_1 \tag{5.21.36}$$

Substituting this into (5.21.29), we obtain a transition curve separating stability from instability for the system

$$\Omega = 2\omega_1(1 \pm \varepsilon b_1) = 2\omega_1\left[1 \pm \left(2 - \sqrt{2}\right)\varepsilon\right] \tag{5.21.37}$$

From (5.21.31), we can obtain $A_2 = A_2(T_2)$. And then the Eq. (5.21.28) becomes

$$D_0^2 u_{21} + \omega_2^2 u_{21} = -\frac{1}{2} b_2 \Omega^2 \left(A_2 e^{i(\Omega+\omega_2)T_0} + \bar{A}_2\, e^{i(\Omega-\omega_2)T_0} \right) + cc \tag{5.21.38}$$

therefore,

$$u_{21} = -\frac{b_2\Omega^2 A_2 e^{i(\Omega+\omega_2)T_0}}{2[\omega_2^2 - (\Omega+\omega_2)^2]} - \frac{b_2\Omega^2\, \bar{A}_2\, e^{i(\Omega-\omega_2)T_0}}{2[\omega_2^2 - (\Omega-\omega_2)^2]} + cc \tag{5.21.39}$$

Substitute u_{21} into (5.21.24)

$$\begin{aligned} D_0^2 u_{22} + \omega_2^2 u_{22} &= -2D_0 D_2 u_{20} - b_2\Omega^2 u_{21} \cos \Omega T_0 \\ &= -2D_0 D_2 u_{20} - \frac{1}{2} b_2 \Omega^2 u_{21} \left(e^{i\Omega T_0} + cc\right) \\ &= -2i\omega_2 D_2 A_2 e^{i\omega_2 T_0} \\ &\quad + \frac{b_2^2 \Omega^4 A_2}{4[\omega_2^2 - (\Omega-\omega_2)^2]} e^{i\omega_2 T_0} + cc + NST \end{aligned} \tag{5.21.40}$$

In order to eliminate secular terms from the above equation, there must be

$$2i\omega_2 D_2 A_2 - \frac{b_2^2\Omega^4 A_2}{4[\omega_2^2 - (\Omega+\omega_2)^2]} - \frac{b_2^2\Omega^4 A_2}{4[\omega_2^2 - (\Omega-\omega_2)^2]} = 0 \tag{5.21.41}$$

therefore,

$$A_2 = \alpha_2 e^{i\chi_2 T_2} \tag{5.21.42}$$

where α_2 is a complex number and χ_2 is

$$\chi_2 = -\frac{b_2^2\Omega^4}{8\omega_2[\omega_2^2 - (\Omega+\omega_2)^2]} - \frac{b_2^2\Omega^4}{8\omega_2[\omega_2^2 - (\Omega-\omega_2)^2]} \tag{5.21.43}$$

Thus u_{21} is bounded.

CASE II: $\boldsymbol{\Omega \approx \omega_1 + \omega_2}$:

Introduce the detuning parameter σ such that

$$\Omega = \omega_1 + \omega_2 + \varepsilon\sigma \tag{5.21.44}$$

In order to eliminate secular terms from (5.21.28), there must be

$$D_1A_1 = 0, \quad D_1A_2 = 0 \quad \Rightarrow \quad A_1 = A_1(T_2), \quad A_2 = A_2(T_2) \tag{5.21.45}$$

Equation (5.21.28) becomes

$$D_0^2u_{n1} + \omega_2^2u_{n1} = -\frac{1}{2}b_n\Omega^2\left(A_ne^{i(\Omega+\omega_n)T_0} + \bar{A}_n\,e^{i(\Omega-\omega_n)T_0}\right) + cc \tag{5.21.46}$$

Therefore,

$$u_{n1} = -\frac{b_n\Omega^2A_ne^{i(\Omega+\omega_n)T_0}}{2[\omega_n^2 - (\Omega+\omega_n)^2]} - \frac{b_n\Omega^2\,\bar{A}_n\,e^{i(\Omega-\omega_n)T_0}}{2[\omega_n^2 - (\Omega-\omega_n)^2]} + cc \tag{5.21.47}$$

Substitute u_{n1} into (5.21.24)

$$\begin{aligned} D_0^2u_{n2} + \omega_n^2u_{n2} &= -2D_0D_2u_{n0} - b_n\Omega^2u_{n1}\cos\Omega T_0 \\ &= -2D_0D_2u_{n0} - \frac{1}{2}b_n\Omega^2u_{n1}\left(e^{i\Omega T_0} + cc\right) \\ &= -2i\omega_nD_2A_ne^{i\omega_nT_0} + \frac{b_n^2\Omega^4A_n}{4[\omega_n^2 - (\Omega-\omega_n)^2]}e^{i\omega_2T_0} \\ &\quad + \frac{b_n\Omega^2A_n}{2[\omega_n^2 - (\Omega+\omega_n)^2]}e^{i\omega_nT_0} + cc + NST \end{aligned} \tag{5.21.48}$$

In order to eliminate secular terms from the above equation, there must be

$$2i\omega_2D_2A_2 - \frac{b_n^2\Omega^4A_n}{4[\omega_n^2 - (\Omega-\omega_n)^2]} - \frac{b_n\Omega^2A_n}{2[\omega_n^2 - (\Omega+\omega_n)^2]} = 0 \tag{5.21.49}$$

therefore,

$$A_n = \alpha_n e^{i\chi_n T_2} \tag{5.21.50}$$

where α_n are complex numbers and χ_n are

$$\chi_n = -\frac{b_n^2\Omega^4}{8\omega_n[\omega_n^2 - (\Omega+\omega_n)^2]} - \frac{b_n^2\Omega^4}{8\omega_n[\omega_n^2 - (\Omega-\omega_n)^2]} \tag{5.21.51}$$

Thus, u_{n1} are bounded.

In brief, for the case of $\boldsymbol{\Omega} \approx \boldsymbol{\omega}_1 + \boldsymbol{\omega}_2$, the first order approximate solutions of the equations are stable.

Readers are invited to complete the analysis on the case of $\Omega \approx \omega_2 - \omega_1$ on their own.

(d) When $y = \varepsilon l[y_1\cos(\Omega_1 t + \phi_1) + y_2\cos(\Omega_2 t + \phi_2)]$ is substituted into (5.21.31), we get

$$\begin{aligned} &\ddot{u}_1 + \omega_1^2 u_1 + \varepsilon u_1[c_{11}\cos(\Omega_1 t + \phi_1) + c_{12}\cos(\Omega_2 t + \phi_2)] = 0 \\ &\ddot{u}_2 + \omega_2^2 u_2 + \varepsilon u_2[c_{21}\cos(\Omega_1 t + \phi_1) + c_{22}\cos(\Omega_2 t + \phi_2)] = 0 \end{aligned} \tag{5.21.52}$$

where

$$\begin{aligned} &c_{11} = \Omega_1^2 y_1\left(2-\sqrt{2}\right), \quad c_{12} = \Omega_2^2 y_2\left(2-\sqrt{2}\right), \\ &c_{21} = \Omega_1^2 y_1\left(2+\sqrt{2}\right), \quad c_{22} = \Omega_2^2 y_2\left(2+\sqrt{2}\right) \end{aligned} \tag{5.21.53}$$

These are two mutually independent, single-degree-of-freedom, dual-frequency parametric excitation equations, and we use the method of multiple scales to solve their first-order approximate solutions. Let the solutions of the equations be

$$u_n(t;\,\varepsilon) = u_{n0}(T_0, T_1, T_2) + \varepsilon u_{n1}(T_0, T_1, T_2) + \varepsilon^2 u_{n2}(T_0, T_1, T_2) + \cdots, \quad n = 1,\ 2 \tag{5.21.54}$$

Substituting the above equation into (5.21.32) and retaining to $O(\varepsilon^2)$, we get

$$\begin{aligned} 0 &= \ddot{u}_n + \omega_n^2 u_n \\ &\quad + \varepsilon u_n[c_{n1}\cos(\Omega_1 t + \varphi_1) + c_{n2}\cos(\Omega_2 t + \varphi_2)] \\ &= \left[D_0^2 + 2\varepsilon D_0 D_1 + \varepsilon^2\left(D_1^2 + 2D_0 D_2\right)\right]\left(u_{n0} + \varepsilon u_{n1} + \varepsilon^2 u_{n2}\right) \\ &\quad + \omega_n^2\left(u_{n0} + \varepsilon u_{n1} + \varepsilon^2 u_{n2}\right) + \varepsilon\left(u_{n0} + \varepsilon u_{n1} + \varepsilon^2 u_{n2}\right) \\ &\quad \times [c_{n1}\cos(\Omega_1 t + \varphi_1) + c_{n2}\cos(\Omega_2 t + \varphi_2)] + \cdots \\ &= D_0^2 u_{n0} + \varepsilon D_0^2 u_{n1} + \varepsilon^2 D_0^2 u_{n2} + 2\varepsilon D_0 D_1 u_{n0} \\ &\quad + 2\varepsilon^2 D_0 D_1 u_{n1} + \varepsilon^2\left(D_1^2 + 2D_0 D_2\right) u_{n0} \end{aligned}$$

$$
\begin{aligned}
&+\omega_n^2 u_{n0} + \varepsilon\omega_n^2 u_{n1} + \varepsilon^2\omega_n^2 u_{n2}\\
&+\varepsilon u_{n0}[c_{n1}\cos(\Omega_1 t+\varphi_1) + c_{n2}\cos(\Omega_2 t+\varphi_2)]\\
&+\varepsilon^2 u_{n1}[c_{n1}\cos(\Omega_1 t+\varphi_1) + c_{n2}\cos(\Omega_2 t+\varphi_2)] + \cdots\\
&= D_0^2 u_{n0} + \omega_n^2 u_{n0} + \varepsilon\{D_0^2 u_{n1} + \omega_n^2 u_{n1} + 2D_0D_1u_{n0}\\
&+u_{n0}[c_{n1}\cos(\Omega_1 T_0+\varphi_1) + c_{n2}\cos(\Omega_2 T_0+\varphi_2)]\}\\
&+\varepsilon^2\{D_0^2 u_{n2} + \omega_n^2 u_{n2} + 2D_0D_1u_{n1} + \left(D_1^2 + 2D_0D_2\right)u_{n0}\\
&+u_{n1}[c_{n1}\cos(\Omega_1 T_0+\varphi_1) + c_{n2}\cos(\Omega_2 T_0+\varphi_2)]\} + \cdots
\end{aligned}
\tag{5.21.55}
$$

Equating coefficients of like powers of ε in the above equation yields

$$
D_0^2 u_{n0} + \omega_n^2 u_{n0} = 0 \tag{5.21.56}
$$

$$
\begin{aligned}
D_0^2 u_{n1} + \omega_n^2 u_{n1} &= -2D_0D_1u_{n0} - u_{n0}[c_{n1}\cos(\Omega_1 T_0+\phi_1)\\
&\quad + c_{n2}\cos(\Omega_2 T_0+\phi_2)]
\end{aligned}
\tag{5.21.57}
$$

$$
\begin{aligned}
D_0^2 u_{n2} + \omega_n^2 u_{n2} &= -2D_0D_1u_{n1} - \left(D_1^2 + 2D_0D_2\right)u_{n0}\\
&\quad - u_{n1}[c_{n1}\cos(\Omega_1 T_0+\varphi_1) + c_{n2}\cos(\Omega_2 T_0+\varphi_2)]
\end{aligned}
\tag{5.21.58}
$$

The solution of (5.21.56) is

$$
u_{n0} = A_n e^{i\omega_n T_0} + \overline{A}_n e^{-i\omega_n T_0} \tag{5.21.59}
$$

where $A_n = A_n(T_1, T_2)$. Let

$$
A_n = \frac{1}{2}a_n e^{i\beta_n} \tag{5.21.60}
$$

Substituting this into (5.21.59), we can obtain the first order approximate solutions of the given equations

$$
u_n \approx u_{n0} = a_n\cos(\omega_n t + \beta_n) + O(\varepsilon) \tag{5.21.61}
$$

Substituting this set of solutions into (5.21.16), we obtain first order approximate solutions to the original equations.

Substitute u_{n0} into (5.21.57) yields

$$
\begin{aligned}
D_0^2 u_{n1} + \omega_n^2 u_{n1} &= -2D_0D_1u_{n0} - u_{n0}[c_{n1}\cos(\Omega_1 T_0+\varphi_1) + c_{n2}\cos(\Omega_2 T_0+\varphi_2)]\\
&= -2D_0D_1A_n e^{i\omega_n T_0} - \frac{1}{2}c_{n1}\left(A_n e^{i\omega_n T_0} + \overline{A}_n e^{-i\omega_n T_0}\right)e^{i(\Omega_1 T_0+\varphi_1)}\\
&\quad - \frac{1}{2}c_{n2}\left(A_n e^{i\omega_n T_0} + \overline{A}_n e^{-i\omega_n T_0}\right)e^{i(\Omega_2 T_0+\varphi_2)} + cc
\end{aligned}
$$

$$= -2i\omega_n D_1 A_n e^{i\omega_n T_0} - \frac{1}{2}c_{n1}A_n e^{i[(\Omega_1+\omega_n)T_0+\varphi_1]}$$
$$- \frac{1}{2}c_{n1}\overline{A}_n e^{i[(\Omega_1-\omega_n)T_0+\varphi_1]} - \frac{1}{2}c_{n2}A_n e^{i[(\Omega_2+\omega_n)T_0+\varphi_2]}$$
$$- \frac{1}{2}c_{n2}\overline{A}_n e^{i[(\Omega_2-\omega_n)T_0+\varphi_2]} + cc \tag{5.21.62}$$

CASE I: $\Omega_1 \approx \omega_2 - \omega_1$ and $\Omega_2 \approx 2\omega_1$:

Introduce the detuning parameters σ_1 and σ_2 such that

$$\Omega_1 = \omega_2 - \omega_1 + \varepsilon\sigma_1, \quad \Omega_2 = 2\omega_1 + \varepsilon\sigma_2 \tag{5.21.63}$$

In order to eliminate secular terms from (5.21.62), there must be

$$2i\omega_1 D_1 A_1 + \frac{1}{2}c_{12}\overline{A}_1 e^{i(\sigma_2 T_1+\phi_2)} = 0 \tag{5.21.64}$$

$$D_1 A_2 = 0 \tag{5.21.65}$$

Substituting $A_1 = \frac{1}{2}a_1 e^{i\beta_1}$ into (5.21.64) yields

$$ia_1' - a\beta_1' + \frac{1}{4}\omega_1^{-1}c_{12}a_1 e^{i(\sigma_2 T_1-2\beta_1+\phi_2)} = 0 \tag{5.21.66}$$

The prime denotes the derivative with respect to T_1. Separating the real and imaginary parts of the above equation gives

$$\begin{aligned} a_1' &= -\tfrac{1}{4}\omega_1^{-1}c_{12}a_1\sin\gamma_1 \\ a_1\gamma_1' &= \sigma_2 a - \tfrac{1}{2}\omega_1^{-1}c_{12}a_1\cos\gamma_1 \end{aligned} \tag{5.21.67}$$

where

$$\gamma_1 = \sigma_2 T_1 - 2\beta_1 + \phi_2 \tag{5.21.68}$$

Let $a' = \gamma' = 0$ in (5.21.67), we can obtain

$$-\frac{1}{4}\omega_1^{-1}c_{12}a_1\sin\gamma_1 = 0$$
$$\sigma_2 a_1 - \frac{1}{2}\omega_1^{-1}c_{12}a_1\cos\gamma_1 = 0 \tag{5.21.69}$$

It can be seen that the condition for the existence of a nonzero steady state solution is

$$\sigma_2 = \pm\frac{1}{2}\omega_1^{-1}c_{12} \tag{5.21.70}$$

Substituting this into(5.21.63), we obtain a transition curve separating stability from instability for the system

$$\Omega_2 = 2\omega_1 \pm \frac{1}{2}\varepsilon\omega_1^{-1}c_{12} = 2\omega_1\left[1 \pm \left(2-\sqrt{2}\right)\varepsilon y_2\right] \tag{5.21.71}$$

From (5.21.65) we can obtain $A_2 = A_2(T_2)$, then (5.21.62) becomes

$$\begin{aligned} D_0^2u_{21} + \omega_2^2u_{21} = & -\frac{1}{2}c_{21}A_2e^{i[(\Omega_1+\omega_2)T_0+\varphi_1]} - \frac{1}{2}c_{21}\overline{A}_2e^{i[(\Omega_1-\omega_2)T_0+\varphi_1]} \\ & -\frac{1}{2}c_{22}A_2e^{i[(\Omega_2+\omega_2)T_0+\varphi_2]} - \frac{1}{2}c_{22}\overline{A}_2e^{i[(\Omega_2-\omega_2)T_0+\varphi_2]} + cc \end{aligned} \tag{5.21.72}$$

therefore,

$$\begin{aligned} u_{21} = & -\frac{c_{21}A_2e^{i[(\Omega_1+\omega_2)T_0+\varphi_1]}}{2[\omega_2^2-(\Omega_1+\omega_2)^2]} - \frac{c_{21}\overline{A}_2e^{i[(\Omega_1-\omega_2)T_0+\varphi_1]}}{2[\omega_2^2-(\Omega_1-\omega_2)^2]} \\ & -\frac{c_{22}A_2e^{i[(\Omega_2+\omega_2)T_0+\varphi_2]}}{2[\omega_2^2-(\Omega_2+\omega_2)^2]} - \frac{c_{22}\overline{A}_2e^{i[(\Omega_2-\omega_2)T_0+\varphi_2]}}{2[\omega_2^2-(\Omega_2-\omega_2)^2]} + cc \end{aligned} \tag{5.21.73}$$

Substituting u_{21} into (5.21.58) yields

$$\begin{aligned} D_0^2u_{22} + \omega_2^2u_{22} &= -2D_0D_2u_{20} \\ &\quad - u_{21}[c_{21}\cos(\Omega_1T_0+\varphi_1) + c_{22}\cos(\Omega_2T_0+\varphi_2)] \\ &= -2D_0D_2u_{20} \\ &\quad -\frac{1}{2}u_{21}\left[c_{21}\left(e^{i(\Omega_1T_0+\varphi_1)}+cc\right) + c_{22}\left(e^{i(\Omega_2T_0+\varphi_2)}+cc\right)\right] \\ &= -2i\omega_2D_2A_2e^{i\omega_2T_0} \\ &\quad +\{\frac{c_{21}^2}{4[\omega_2^2-(\Omega_1+\omega_2)^2]} + \frac{c_{21}^2}{4[\omega_2^2-(\Omega_1-\omega_2)^2]} \\ &\quad + \frac{c_{22}^2}{4[\omega_2^2-(\Omega_2+\omega_2)^2]} + \frac{c_{22}^2}{4[\omega_2^2-(\Omega_2-\omega_2)^2]}\}A_2e^{i\omega_2T_0} \\ &\quad + cc + NST \end{aligned} \tag{5.21.74}$$

In order to eliminate secular terms from the above equation, there must be

$$2i\omega_2 D_2 A_2 - \left\{ \frac{c_{21}^2}{4[\omega_2^2 - (\Omega_1+\omega_2)^2]} + \frac{c_{21}^2}{4[\omega_2^2 - (\Omega_1-\omega_2)^2]} + \frac{c_{22}^2}{4[\omega_2^2 - (\Omega_2+\omega_2)^2]} + \frac{c_{22}^2}{4[\omega_2^2 - (\Omega_2-\omega_2)^2]} \right\} A_2 = 0(0.21.55) \tag{5.21.75}$$

therefore,

$$A_2 = \alpha_2 e^{i\chi_2 T_2} \tag{5.21.76}$$

where α_2 is a complex number and χ_2 is

$$\chi_2 = -\frac{1}{8\omega_2}\left\{ \frac{c_{21}^2}{\omega_2^2 - (\Omega_1+\omega_2)^2} + \frac{c_{21}^2}{\omega_2^2 - (\Omega_1-\omega_2)^2} + \frac{c_{22}^2}{\omega_2^2 - (\Omega_2+\omega_2)^2} + \frac{c_{22}^2}{\omega_2^2 - (\Omega_2-\omega_2)^2} \right\} \tag{5.21.77}$$

Thus, u_{21} is bounded.

For the case of $\Omega_1 \approx \omega_2 - \omega_1$ and $\Omega_2 \approx \omega_2 + \omega_1$, readers are invited to complete the analysis on their own.

5.22 Exercise 5.22 (A Two-Frequency Parametric Excitation of a Multi Degree-of-Freedom System with Distinct Frequencies)

Solution: These are N mutually coupled, two-frequency parametric excitation linear equations. We use the method of multiple scales to obtain their first-order approximate solutions. Let the solutions of the equations be

$$u_n(t;\varepsilon) = u_{n0}(T_0, T_1, T_2) + \varepsilon u_{n1}(T_0, T_1, T_2) + \varepsilon^2 u_{n2}(T_0, T_1, T_2) + \cdots \tag{5.22.1}$$

Substituting (5.22.1) into the governing equation and retaining to $O(\varepsilon^2)$, we get

$$\begin{aligned} 0 &= \ddot{u}_n + \omega_n^2 u_n \\ &\quad + 2\varepsilon \sum_{m=1}^{N} \left[f_{nm} \cos \Omega_1 t + g_{nm} \cos(\Omega_2 t + \theta) \right] u_m \\ &= \left[D_0^2 + 2\varepsilon D_0 D_1 + \varepsilon^2 \left(D_1^2 + 2D_0 D_2 \right) \right] \left(u_{n0} + \varepsilon u_{n1} + \varepsilon^2 u_{n2} \right) \end{aligned}$$

$$
\begin{aligned}
&+ \omega_n^2\left(u_{n0} + \varepsilon u_{n1} + \varepsilon^2 u_{n2}\right) \\
&+ 2\varepsilon \sum_{m=1}^{N}\left[f_{nm}\cos\Omega_1 T_0 + g_{nm}\cos(\Omega_2 T_0 + \theta)\right] \\
&\times \left(u_{m0} + \varepsilon u_{m1} + \varepsilon^2 u_{m2}\right) + \cdots \\
&= D_0^2 u_{n0} + \varepsilon D_0^2 u_{n1} + \varepsilon^2 D_0^2 u_{n2} \\
&+ 2\varepsilon D_0 D_1 u_{n0} + 2\varepsilon^2 D_0 D_1 u_{n1} \\
&+ \varepsilon^2\left(D_1^2 + 2D_0 D_2\right)u_{n0} + \omega_n^2 u_{n0} + \varepsilon\omega_n^2 u_{n1} + \varepsilon^2\omega_n^2 u_{n2} \\
&+ 2\varepsilon \sum_{m=1}^{N}\left[f_{nm}\cos\Omega_1 T_0 + g_{nm}\cos(\Omega_2 T_0 + \theta)\right]u_{m0} \\
&+ 2\varepsilon^2 \sum_{m=1}^{N}\left[f_{nm}\cos\Omega_1 T_0 + g_{nm}\cos(\Omega_2 T_0 + \theta)\right]u_{m1} + \cdots \\
&= D_0^2 u_{n0} + \omega_n^2 u_{n0} + \varepsilon\{D_0^2 u_{n1} + \omega_n^2 u_{n1} + 2D_0 D_1 u_{n0} \\
&+ 2\sum_{m=1}^{N}\left[f_{nm}\cos\Omega_1 T_0 + g_{nm}\cos(\Omega_2 T_0 + \theta)\right]u_{m0}\} \\
&+ \varepsilon^2\{D_0^2 u_{n2} + \omega_n^2 u_{n2} + 2D_0 D_1 u_{n1} + \left(D_1^2 + 2D_0 D_2\right)u_{n0} \\
&+ 2\sum_{m=1}^{N}\left[f_{nm}\cos\Omega_1 T_0 + g_{nm}\cos(\Omega_2 T_0 + \theta)\right]u_{m1}\} + \cdots
\end{aligned} \tag{5.22.2}
$$

Equating coefficients of like powers of ε in the above equation yields

$$D_0^2 u_{n0} + \omega_n^2 u_{n0} = 0 \tag{5.22.3}$$

$$D_0^2 u_{n1} + \omega_n^2 u_{n1} = -2D_0 D_1 u_{n0} - 2\sum_{m=1}^{N}\left[f_{nm}\cos\Omega_1 T_0 + g_{nm}\cos(\Omega_2 T_0 + \theta)\right]u_{m0} \tag{5.22.4}$$

$$
\begin{aligned}
D_0^2 u_{n2} + \omega_n^2 u_{n2} = &-2D_0 D_1 u_{n1} - \left(D_1^2 + 2D_0 D_2\right)u_{n0} \\
&- 2\sum_{m=1}^{N}\left[f_{nm}\cos\Omega_1 T_0 + g_{nm}\cos(\Omega_2 T_0 + \theta)\right]u_{m1}
\end{aligned} \tag{5.22.5}
$$

From (5.22.3), we can obtain

$$u_{n0} = A_n e^{i\omega_n T_0} + \bar{A}_n e^{-i\omega_n T_0} \tag{5.22.6}$$

where $A_n = A_n(T_1, T_2)$. Substituting u_{n0} into (5.22.4) yields

$$D_0^2 u_{n1} + \omega_n^2 u_{n1} = -2D_0 D_1 u_{n0} - 2\sum_{m=1}^{N}\left[f_{nm}\cos\Omega_1 T_0 + g_{nm}\cos(\Omega_2 T_0 + \theta)\right]u_{m0}$$
$$= -2D_0 D_1 A_n e^{i\omega_n T_0} - \sum_{m=1}^{N} f_{nm}\left(A_m e^{i(\Omega_1+\omega_m)T_0} + \overline{A}_m e^{i(\Omega_1-\omega_m)T_0}\right)$$
$$- \sum_{m=1}^{N} g_{nm}\left(A_m e^{i[(\Omega_2+\omega_m)T_0+\theta]} + \overline{A}_m e^{i[(\Omega_2-\omega_m)T_0+\theta]}\right) + cc \quad (5.22.7)$$

CASE I: $\Omega_1 \approx \omega_2 - \omega_1$ and $\Omega_2 \approx \omega_3 - \omega_2$:

Introduce the detuning parameters σ_1 and σ_2 such that

$$\Omega_1 = \omega_2 - \omega_1 + \varepsilon\sigma_1, \quad \Omega_2 = \omega_3 - \omega_2 + \varepsilon\sigma_2 \quad (5.22.8)$$

then (5.22.7) becomes

$$D_0^2 u_{n1} + \omega_n^2 u_{n1} = -2D_0 D_1 A_n e^{i\omega_n T_0}$$
$$- \sum_{m=1}^{N} f_{nm}\left(A_m e^{i(\omega_2-\omega_1+\varepsilon\sigma_1+\omega_m)T_0} + \overline{A}_m e^{i(\omega_2-\omega_1+\varepsilon\sigma_1-\omega_m)T_0}\right)$$
$$- \sum_{m=1}^{N} g_{nm}\left(A_m e^{i[(\omega_3-\omega_2+\varepsilon\sigma_2+\omega_m)T_0+\theta]} + \overline{A}_m e^{i[(\omega_3-\omega_2+\varepsilon\sigma_2-\omega_m)T_0+\theta]}\right) + cc$$
$$(5.22.9)$$

In order to eliminate secular terms from the above equation, it is necessary to have

$$\begin{aligned} &2i\omega_1 D_1 A_1 + f_{12} A_2 e^{-i\sigma_1 T_1} = 0 \\ &2i\omega_2 D_1 A_2 + f_{21} A_1 e^{i\sigma_1 T_1} + g_{23} A_3 e^{-i(\sigma_2 T_1+\theta)} = 0 \\ &2i\omega_3 D_1 A_3 + g_{32} A_2 e^{i(\sigma_2 T_1+\theta)} = 0 \end{aligned} \quad (5.22.10)$$

$$D_1 A_n = 0 \;\Rightarrow\; A_n = A_n(T_2); \quad \text{for all } n \neq 1,\ 2,\ 3 \quad (5.22.11)$$

The solutions of (5.22.10) are

$$\begin{cases} A_1 = a_1 e^{i(\lambda-\sigma_1)T_1} \\ A_2 = a_2 e^{i\lambda T_1} \\ A_3 = a_3 e^{i[(\lambda+\sigma_2)T_1+\theta]} \end{cases} \quad (5.22.12)$$

where a_1, a_2 and a_3 are complex numbers. Substituting(5.22.12) into(5.22.10) yields

$$\begin{bmatrix} 2\omega_1(\lambda-\sigma_1) & -f_{12} & 0 \\ -f_{21} & 2\omega_2\lambda & -g_{23} \\ 0 & -g_{32} & 2\omega_3(\lambda+\sigma_2) \end{bmatrix} \begin{Bmatrix} a_1 \\ a_2 \\ a_3 \end{Bmatrix} = 0 \tag{5.22.13}$$

So the characteristic equation is

$$\det \begin{bmatrix} 2\omega_1(\lambda-\sigma_1) & -f_{12} & 0 \\ -f_{21} & 2\omega_2\lambda & -g_{23} \\ 0 & -g_{32} & 2\omega_3(\lambda+\sigma_2) \end{bmatrix} = 0$$

i.e.,

$$\lambda - \frac{1}{4}\Lambda_1(\lambda-\sigma_1)^{-1} - \frac{1}{4}\Lambda_2(\lambda+\sigma_2)^{-1} = 0 \tag{5.22.14}$$

where

$$\Lambda_1 = \frac{f_{12}f_{21}}{\omega_1\omega_2}, \quad \Lambda_2 = \frac{g_{23}g_{32}}{\omega_2\omega_3} \tag{5.22.15}$$

The stability of the first order approximate solution can be determined from λ of (5.22.14).

CASE II: $\Omega_1 \approx \omega_3 - \omega_2$ **and** $\Omega_2 \approx \omega_1 + \omega_2$**:**

Introduce the detuning parameters σ_1 and σ_2 such that

$$\Omega_1 = \omega_3 - \omega_2 + \varepsilon\sigma_1, \quad \Omega_2 = \omega_1 + \omega_2 + \varepsilon\sigma_2 \tag{5.22.16}$$

then (5.22.7) becomes

$$\begin{aligned} D_0^2 u_{n1} + \omega_n^2 u_{n1} &= -2D_0D_1A_n e^{i\omega_n T_0} \\ &\quad - \sum_{m=1}^{N} f_{nm}\left(A_m e^{i(\omega_3-\omega_2+\varepsilon\sigma_1+\omega_m)T_0} + \overline{A}_m e^{i(\omega_3-\omega_2+\varepsilon\sigma_1-\omega_m)T_0}\right) \\ &\quad - \sum_{m=1}^{N} g_{nm}\left(A_m e^{i[(\omega_1+\omega_2+\varepsilon\sigma_2+\omega_m)T_0+\theta]} + \overline{A}_m e^{i[(\omega_1+\omega_2+\varepsilon\sigma_2-\omega_m)T_0+\theta]}\right) + cc \end{aligned} \tag{5.22.17}$$

In order to eliminate secular terms from (5.22.17), it is necessary to have.

$$\begin{aligned} &2i\omega_1 D_1 A_1 + g_{12}\overline{A}_2 e^{i(\sigma_2 T_1+\theta)} = 0 \\ &2i\omega_2 D_1 A_2 + f_{23}A_3 e^{-i\sigma_1 T_1} + g_{21}\overline{A}_1 e^{i(\sigma_2 T_1+\theta)} = 0 \\ &2i\omega_3 D_1 A_3 + f_{32}A_2 e^{i\sigma_1 T_1} = 0 \end{aligned} \tag{5.22.18}$$

$$D_1 A_n = 0 \quad \Rightarrow \quad A_n = A_n(T_2); \quad \text{for all } n \neq 1,\ 2,\ 3 \tag{5.22.19}$$

The solutions of (5.22.18) are

$$\begin{cases} A_1 = a_1 e^{i\left[\left(-\bar{\lambda}+\sigma_2\right)T_1\right]} \\ A_2 = a_2 e^{i(\lambda T_1+\theta)} \\ A_3 = a_3 e^{i[(\lambda+\sigma_1)T_1+\theta]} \end{cases} \tag{5.22.20}$$

where a_1, a_2 and a_3 are complex numbers. Substituting (5.22.20) into (5.22.18) yields

$$\begin{bmatrix} 2\omega_1(\lambda-\sigma_2) & g_{12} & 0 \\ -g_{21} & 2\omega_2\lambda & -f_{23} \\ 0 & -f_{32} & 2\omega_3(\lambda+\sigma_1) \end{bmatrix} \begin{Bmatrix} a_1 \\ a_2 \\ a_3 \end{Bmatrix} = 0 \tag{5.22.21}$$

So the characteristic equation is

$$\det \begin{bmatrix} 2\omega_1(\lambda-\sigma_2) & g_{12} & 0 \\ -g_{21} & 2\omega_2\lambda & -f_{23} \\ 0 & -f_{32} & 2\omega_3(\lambda+\sigma_1) \end{bmatrix} = 0$$

i.e.,

$$\lambda - \frac{1}{4}\Lambda_1(\lambda+\sigma_1)^{-1} + \frac{1}{4}\Lambda_2(\lambda-\sigma_2)^{-1} = 0 \tag{5.22.22}$$

where

$$\Lambda_1 = \frac{f_{23}f_{32}}{\omega_2\omega_3}, \quad \Lambda_2 = \frac{g_{12}g_{21}}{\omega_1\omega_2} \tag{5.22.23}$$

The stability of the first order approximate solution can be determined from λ of (5.22.23).

5.23 Exercise 5.23 (A Single-Frequency Parametric Excitation of a Three-Degree-of-Freedom System with Repeating Frequencies)

Solution: This is a linear system having three degrees of freedom, with two of the frequencies being equal. Without parametric excitation, the system is unstable because x_2 contains a secular or resonant term of the form $t\sin(\omega_1 t + \beta)$, where β is a constant. To determine if the parametric excitation can stabilize the system, we assume that all three modes are bounded and then, if possible, determine the values of the parameters which are consistent with this assumption, particularly those values at the boundaries of the region where the assumption is valid.

Although the parametric excitation might stabilize the motion, we still expect the amplitude of the x_2-mode to be much larger than that of the x_1-mode. We do not have any indication of the amplitude of the x_3-mode. To express our expectations systematically, we scale the dependent variables. Without loss of generality, we make

$$x_1 = u_1, \quad x_2 = \varepsilon^{-\lambda 2} u_2, \quad x_3 = \varepsilon^{-\lambda 3} u_3 \tag{5.23.1}$$

where the u_n are $O(1)$ and the λ_n are positive constants to be determined in the solution. Substituting (5.23.1) into the original equations leads to

$$\ddot{u}_1 + \omega_1^2 u_1 + 2\left(\varepsilon f_{11} u_1 + \varepsilon^{1-\lambda 2} f_{12} u_2 + \varepsilon^{1-\lambda 3} f_{13} u_3\right) \cos \omega t = 0 \tag{5.23.2}$$

$$\ddot{u}_2 + \omega_1^2 u_2 + \varepsilon^{\lambda 2} u_1 + 2\left(\varepsilon^{1+\lambda 2} f_{21} u_1 + \varepsilon f_{22} u_2 + \varepsilon^{1-\lambda 3+\lambda 2} f_{23} u_3\right) \cos \omega t = 0 \tag{5.23.3}$$

$$\ddot{u}_3 + \omega_3^2 u_3 + 2\left(\varepsilon^{1+\lambda 3} f_{31} u_1 + \varepsilon^{1-\lambda 2+\lambda 3} f_{32} u_2 + \varepsilon f_{33} u_3\right) \cos \omega t = 0 \tag{5.23.4}$$

We are concerned with various combinations of the frequencies which can lead to a resonant response. We therefore adopt a multiscale approach to determine a uniformly valid approximate solution which exhibits the effects of the repeated frequency and the resonant combinations of frequencies. We write the various time scales as

$$T_0 = t, \quad T_1 = \varepsilon^{\lambda 4} t, \quad T_2 = \varepsilon^{2\lambda 4} t \tag{5.23.5}$$

where $\lambda_4 > 0$ is a constant to be determined. In terms of these scales,

$$\frac{d^2}{dt^2} = D_0^2 + 2\varepsilon^{\lambda 4} D_0 D_1 + \varepsilon^{2\lambda 4}\left(2D_0 D_2 + D_1^2\right) + \cdots \tag{5.23.6}$$

Assume that approximate solutions in the form

$$u_n(t;\varepsilon) = u_{n0}(T_0, T_1, T_2) + \varepsilon^{\lambda 4} u_{n1}(T_0, T_1, T_2) + \varepsilon^{2\lambda 4} u_{n2}(T_0, T_1, T_2) + \cdots \tag{5.23.7}$$

Substituting (5.23.5) ~(5.23.7) into (5.23.2) ~(5.23.4) leads to

$$\begin{aligned} & D_0^2 u_{10} + \omega_1^2 u_{10} + \varepsilon^{\lambda 4}\left(D_0^2 u_{11} + \omega_1^2 u_{11} + 2D_0 D_1 u_{10}\right) \\ & + \varepsilon^{2\lambda 4}\left(D_0^2 u_{12} + \omega_1^2 u_{12} + 2D_0 D_2 u_{10} + D_1^2 u_{10} + 2D_0 D_1 u_{11}\right) \\ & + 2(\varepsilon f_{11} u_{10} + \varepsilon^{1-\lambda 2} f_{12} u_{20} + \varepsilon^{1-\lambda 2+\lambda 4} f_{12} u_{21} \\ & + \varepsilon^{1-\lambda 3} f_{13} u_{30} + \varepsilon^{1-\lambda 3+\lambda 4} f_{13} u_{31}) \cos \omega T_0 + \cdots = 0 \end{aligned} \tag{5.23.8}$$

$$\begin{aligned}&D_0^2u_{20}+\omega_1^2u_{20}+\varepsilon^{\lambda 4}\left(D_0^2u_{21}+\omega_1^2u_{21}+2D_0D_1u_{20}\right)\\&\quad+\varepsilon^{2\lambda 4}\left(D_0^2u_{22}+\omega_1^2u_{22}+2D_0D_1u_{21}+D_1^2u_{20}+2D_0D_2u_{20}\right)\\&\quad+\varepsilon^{\lambda 2}u_{10}+2(\varepsilon^{1+\lambda 2}f_{21}u_{10}+\varepsilon f_{22}u_{20}+\varepsilon^{1+\lambda 4}f_{22}u_{21}\\&\quad+\varepsilon^{1-\lambda 3+\lambda 2}f_{23}u_{30}+\varepsilon^{1-\lambda 3+\lambda 2+\lambda 4}f_{23}u_{31})\cos\omega T_0+\cdots=0\end{aligned}\tag{5.23.9}$$

$$\begin{aligned}&D_0^2u_{30}+\omega_3^2u_{30}+\varepsilon^{\lambda 4}\left(D_0^2u_{31}+\omega_3^2u_{31}+2D_0D_1u_{30}\right)\\&\quad+\varepsilon^{2\lambda 4}\left(D_0^2u_{32}+\omega_3^2u_{32}+2D_0D_1u_{31}+D_1^2u_{30}+2D_0D_2u_{30}\right)\\&\quad+2(\varepsilon^{1+\lambda 3}f_{31}u_{10}+\varepsilon^{1-\lambda 2+\lambda 3}f_{32}u_{20}+\varepsilon^{1-\lambda 2+\lambda 3+\lambda 4}f_{32}u_{21}\\&\quad+\varepsilon f_{33}u_{30}+\varepsilon^{1+\lambda 4}f_{33}u_{31})\cos\omega T_0+\cdots=0\end{aligned}\tag{5.23.10}$$

We note that the governing equations of u_{10}, u_{20} and u_{30} are of order ε^0, i.e.,

$$D_0^2u_{n0}+\omega_n^2u_{n0}=\cdots,\quad n=1,2,3\tag{5.23.11}$$

The right-hand side of the equation is either zero or a non-resonant excitation term. Then, the solutions of u_{n0} will contain free oscillation terms such as $Ae^{\pm i\omega_n T_0}$, and thus u_{k0} cannot enter into the governing equations of u_{n0}, e.g., $\varepsilon^{\lambda 2}u_{10}$ in Eq. (5.23.9) cannot enter into the governing equation of u_{20}, thus

$$\lambda_2\neq 0\tag{5.23.12}$$

Therefore $1+\lambda_2>0$, and the possible forms of the governing equations for u_{10}, $u_{20}$and u_{30} are

$$\begin{aligned}D_0^2u_{10}+\omega_1^2u_{10}&=-2\left(\varepsilon^{1-\lambda 2}f_{12}u_{20}+\varepsilon^{1-\lambda 3}f_{13}u_{30}\right)\cos\omega T_0\\D_0^2u_{20}+\omega_1^2u_{20}&=-2\varepsilon^{1-\lambda 3+\lambda 2}f_{23}u_{30}\cos\omega T_0\\D_0^2u_{30}+\omega_3^2u_{30}&=-2\varepsilon^{1-\lambda 2+\lambda 3}f_{32}u_{20}\cos\omega T_0\end{aligned}\tag{5.23.13}$$

Whether the terms on the right-hand side of the equations are retained or not depends on the values of λ_i and ω. If $u_{n0}\cos\omega T_0$ can produce a resonant excitation term whose frequency is approximately ω_k, then it cannot appear in the governing equation of u_{k0}.

(a) When $\omega\approx 2\omega_1$ and $\omega\approx\omega_3-\omega_1$:

In this case, both $u_{10}\cos\omega T_0$ and $u_{20}\cos\omega T_0$ can produce resonant terms whose frequencies are approximately ω_1 and ω_3, and $u_{30}\cos\omega T_0$ can produce a resonant excitation term whose frequency is approximately ω_1. Therefore, all the terms on the right side of (5.23.12) cannot appear and the governing equations for u_{10}, $u_{20}$and u_{30} are.

Order ε^0:

$$D_0^2u_{10}+\omega_1^2u_{10}=0\tag{5.23.14}$$

$$D_0^2 u_{20} + \omega_1^2 u_{20} = 0 \tag{5.23.15}$$

$$D_0^2 u_{30} + \omega_3^2 u_{30} = 0 \tag{5.23.16}$$

From (5.23.8–5.23.10), we can see that the basic order of the control equations of u_{11}, $u_{21}$and u_{31} is ε^{λ_4} and their basic form is

$$\begin{aligned}&\varepsilon^{\lambda 4}\left(D_0^2 u_{11} + \omega_1^2 u_{11} + 2D_0 D_1 u_{10}\right)\\&+2(\varepsilon f_{11} u_{10} + \varepsilon^{1-\lambda 2} f_{12} u_{20} + \varepsilon^{1-\lambda 2+\lambda 4} f_{12} u_{21}\\&+\varepsilon^{1-\lambda 3} f_{13} u_{30} + \varepsilon^{1-\lambda 3+\lambda 4} f_{13} u_{31}) \cos \omega T_0 = 0\end{aligned} \tag{5.23.17}$$

$$\begin{aligned}&\varepsilon^{\lambda 4}\left(D_0^2 u_{21} + \omega_1^2 u_{21} + 2D_0 D_1 u_{20}\right) + \varepsilon^{\lambda 2} u_{10}\\&+2(\varepsilon^{1+\lambda 2} f_{21} u_{10} + \varepsilon f_{22} u_{20} + \varepsilon^{1-\lambda 3+\lambda 2} f_{23} u_{30}\\&+\varepsilon^{1-\lambda 3+\lambda 2+\lambda 4} f_{23} u_{31}) \cos \omega T_0 = 0\end{aligned} \tag{5.23.18}$$

$$\begin{aligned}&\varepsilon^{\lambda 4}\left(D_0^2 u_{31} + \omega_3^2 u_{31} + 2D_0 D_1 u_{30}\right)\\&+2(\varepsilon^{1+\lambda 3} f_{31} u_{10} + \varepsilon^{1-\lambda 2+\lambda 3} f_{32} u_{20}\\&+\varepsilon^{1-\lambda 2+\lambda 3+\lambda 4} f_{32} u_{21} + \varepsilon f_{33} u_{30}) \cos \omega T_0 = 0\end{aligned} \tag{5.23.19}$$

Here, we are most concerned with the resonant terms that may be involved in these equations, and keep at least the lowest order of those resonant terms. The lowest order resonant term to be retained in (5.23.17) is $\varepsilon^{1-\lambda 2} f_{12} u_{20} \cos \omega T_0$; the lowest-order resonant terms to be retained in (5.23.18) are $\varepsilon^{\lambda 2} u_{10}$ and $\varepsilon f_{22} u_{20} \cos \omega T_0$; and the lowest-order resonant term to be retained in (5.23.19) is $\varepsilon^{1-\lambda 2+\lambda 3} f_{32} u_{20} \cos \omega T_0$. The order for these resonant terms to be preserved is ε^{λ_4}, so

$$\lambda_4 = 1 - \lambda_2 \tag{5.23.20}$$

i.e.,

$$\lambda_4 = \lambda_2 \quad \text{or} \quad \lambda_4 = 1 \tag{5.23.21}$$

and

$$\lambda_4 = 1 - \lambda_2 + \lambda_3 \tag{5.23.22}$$

Since λ_4 is the order of the second term in (5.23.7), we want it to be the lowest possible order. Taking into (5.23.12) account, we have

$$\lambda_4 = \lambda_2 = \frac{1}{2}, \quad \lambda_3 = 0 \tag{5.23.23}$$

Substituting the above results into (5.23.17–5.23.19) and equating coefficients of like powers of ε, we get.

Order $\varepsilon^{1/2}$:

$$D_0^2u_{11} + \omega_1^2u_{11} + 2D_0D_1u_{10} + 2f_{12}u_{20}\cos\omega T_0 = 0 \quad (5.23.24)$$

$$D_0^2u_{21} + \omega_1^2u_{21} + 2D_0D_1u_{20} + u_{10} = 0 \quad (5.23.25)$$

$$D_0^2u_{31} + \omega_3^2u_{31} + 2D_0D_1u_{30} + 2f_{32}u_{20}\cos\omega T_0 = 0 \quad (5.23.26)$$

The solutions of (5.23.14) ~(5.23.16) are

$$u_{10} = A_1(T_1)e^{i\omega_1T_0} + cc \quad (5.23.27)$$

$$u_{20} = A_2(T_1)e^{i\omega_1T_0} + cc \quad (5.23.28)$$

$$u_{30} = A_3(T_1)e^{i\omega_3T_0} + cc \quad (5.23.29)$$

Let

$$\omega - \omega_1 = \omega_1 + 2\varepsilon^{1/2}\sigma, \quad \omega + \omega_1 = \omega_3 + 2\varepsilon^{1/2}\sigma \quad (5.23.30)$$

Substitute u_{01}, $u_{02}$and u_{03} into (5.23.24) ~(5.23.26) and eliminate the permanent term, we get

$$2i\omega_1A_1' + f_{12}\bar{A}_2e^{2i\sigma T_1} = 0 \quad (5.23.31)$$

$$2i\omega_1A_2' + A_1 = 0 \quad (5.23.32)$$

$$2i\omega_3A_3' = 0 \quad (5.23.33)$$

The prime denotes the derivative with respect to T_1. Eliminating A_1 from (5.23.31) and (5.23.32) yields

$$A_2'' + \frac{f_{12}}{4\omega_1^2}\bar{A}_2e^{2i\sigma T_1} = 0 \quad (5.23.34)$$

Let $A_2 = (B_r + iB_i)e^{i\sigma T_1}e^{i\sigma T_1}$,
we have

$$B_r'' - 2\sigma B_i' + \left(\frac{f_{12}}{4\omega_1^2} - \sigma^2\right)B_r = 0 \quad (5.23.35)$$

$$B_i'' - 2\sigma B_r' + \left(\frac{f_{12}}{4\omega_1^2} + \sigma^2\right) B_i = 0 \quad (5.23.36)$$

The nontrivial solutions of Eq. (5.23.35) and(5.23.36) can be expressed as:

$$(B_r, B_i) = (b_r, b_i)e^{\gamma T_1} \quad (5.23.37)$$

where

$$\gamma^2 = -\sigma^2 \pm \frac{f_{12}}{4\omega_1^2} \quad (5.23.38)$$

Therefore, when

$$\sigma^2 > \frac{|f_{12}|}{4\omega_1^2} \quad (5.23.39)$$

the motion is stable, and when

$$\sigma^2 < \frac{|f_{12}|}{4\omega_1^2} \quad (5.23.40)$$

the motion is unstable.

$$\sigma = \pm \frac{|f_{12}|^{1/2}}{2\omega_1} \quad (5.23.41)$$

So, the stability transition curves separating stability from instability for the system are

$$\omega = 2\omega_1 + 2\varepsilon^{1/2}\sigma = 2\omega_1 \pm \varepsilon^{1/2}\frac{|f_{12}|^{1/2}}{\omega_1} + O(\varepsilon) \quad (5.23.42)$$

(b) When $\omega \approx \omega_1$ and $\omega \approx \omega_3 - \omega_1$:

In this case, both $u_{10}\cos\omega T_0$ and $u_{20}\cos\omega T_0$ can produce resonant terms whose frequencies are approximately ω_3, therefore cannot appear in the control equation of u_{30}; $u_{30}\cos\omega T_0$ can produce resonant terms whose frequency is approximately ω_1, therefore cannot appear in the control equations of u_{10} and u_{20}. Therefore, only $\varepsilon^{1-\lambda 2} f_{12} u_{20}\cos\omega T_0$ on the right-hand side of (5.23.17) may be retained.

Thus, if $\lambda_2 \neq 1$, then the governing equations for u_{10}, $u_{20}$and u_{30} are (5.23.14–5.23.16). Similar to the analyses on (5.23.17–5.23.19) in Case (a), the controlling equations of u_{11}, $u_{21}$and u_{31} are still (5.23.24–5.23.26). However, this result omits the resonance case $\omega \approx \omega_1$.

If $\lambda_2 = 1$, the governing equation for u_{10}, $u_{20}$and u_{30} are

$$D_0^2u_{10} + \omega_1^2u_{10} = -2f_{12}u_{20}\cos\omega T_0 \tag{5.23.43}$$

$$D_0^2u_{20} + \omega_1^2u_{20} = 0 \tag{5.23.44}$$

$$D_0^2u_{30} + \omega_3^2u_{30} = 0 \tag{5.23.45}$$

It is required that

$$\lambda_2 = 1 \tag{5.23.46}$$

With this in mind, let's examine the governing equations of u_{11}, u_{21} and u_{31}, which have the possible forms of

$$\begin{aligned}&\varepsilon^{\lambda 4}\left(D_0^2u_{11} + \omega_1^2u_{11} + 2D_0D_1u_{10}\right)\\&+ 2(\varepsilon f_{11}u_{10} + \varepsilon^{1-\lambda 2+\lambda 4}f_{12}u_{21} + \varepsilon^{1-\lambda 3}f_{13}u_{30}\\&+ \varepsilon^{1-\lambda 3+\lambda 4}f_{13}u_{31})\cos\omega T_0 = 0\end{aligned} \tag{5.23.47}$$

$$\begin{aligned}&\varepsilon^{\lambda 4}\left(D_0^2u_{21} + \omega_1^2u_{21} + 2D_0D_1u_{20}\right) + \varepsilon^{\lambda 2}u_{10}\\&+ 2(\varepsilon^{1+\lambda 2}f_{21}u_{10} + \varepsilon f_{22}u_{20} + \varepsilon^{1+\lambda 4}f_{22}u_{21}\\&+ \varepsilon^{1-\lambda 3+\lambda 2}f_{23}u_{30} + \varepsilon^{1-\lambda 3+\lambda 2+\lambda 4}f_{23}u_{31})\cos\omega T_0 = 0\end{aligned} \tag{5.23.48}$$

$$\begin{aligned}&\varepsilon^{\lambda 4}\left(D_0^2u_{31} + \omega_3^2u_{31} + 2D_0D_1u_{30}\right)\\&+ 2(\varepsilon^{1+\lambda 3}f_{31}u_{10} + \varepsilon^{1-\lambda 2+\lambda 3}f_{32}u_{20} + \varepsilon^{1-\lambda 2+\lambda 3+\lambda 4}f_{32}u_{21}\\&+ \varepsilon f_{33}u_{30} + \varepsilon^{1+\lambda 4}f_{33}u_{31})\cos\omega T_0 = 0\end{aligned} \tag{5.23.49}$$

Now (5.23.43) ~ (5.23.45) can be solved by

$$u_{10} = A_1(T_1)e^{i\omega_1T_0} + f_{12}\left[\frac{A_2}{\omega(\omega+2\omega_1)}e^{i(\omega+\omega_1)T_0} + \frac{\bar{A}_2}{\omega(\omega-2\omega_1)}e^{i(\omega-\omega_1)T_0}\right] + cc \tag{5.23.50}$$

$$u_{20} = A_2(T_1)e^{i\omega_1T_0} + cc \tag{5.23.51}$$

$$u_{30} = A_3(T_1)e^{i\omega_3T_0} + cc \tag{5.23.52}$$

If we need to keep $\varepsilon^{\lambda 2}u_{10}$ in (5.23.48), the solution of u_{21} will contain a forced oscillation term of $e^{i(\omega\pm\omega_1)T_0}$. Considering $\omega \approx \omega_1$, we can find that $u_{21}\cos\omega T_0$ can produce a resonant term whose frequency is approximately ω_1; therefore, we may keep $\varepsilon^{1-\lambda 2+\lambda 4}f_{12}u_{21}\cos\omega T_0$ in (5.23.47), and $\varepsilon^{1-\lambda 2+\lambda 3+\lambda 4}f_{32}u_{21}\cos\omega T_0$ in (5.23.49).

Thus, we need

$$\lambda_4 = \lambda_2 = 1, \quad \lambda_3 = 0 \tag{5.23.53}$$

In turn, (5.23.47–5.23.49) become

$$D_0^2 u_{11} + \omega_1^2 u_{11} + 2D_0 D_1 u_{10} + 2(f_{11}u_{10} + f_{12}u_{21} + f_{13}u_{30})\cos\omega T_0 = 0 \tag{5.23.54}$$

$$D_0^2 u_{21} + \omega_1^2 u_{21} + 2D_0 D_1 u_{20} + u_{10} + 2f_{22}u_{20}\cos\omega T_0 = 0 \tag{5.23.55}$$

$$D_0^2 u_{31} + \omega_3^2 u_{31} + 2D_0 D_1 u_{30} + 2(f_{31}u_{10} + f_{32}u_{21} + f_{33}u_{30})\cos\omega T_0 = 0 \tag{5.23.56}$$

Let

$$\omega = \omega_1 + \varepsilon\sigma, \quad \omega + \omega_1 = \omega_3 + \varepsilon\sigma \tag{5.23.57}$$

Substituting u_{01}, $u_{02}$and u_{03} into (5.23.55) and eliminating secular terms yields

$$\begin{aligned} D_0^2 u_{21} + \omega_1^2 u_{21} &= -2D_0 D_1 u_{20} - u_{10} - 2f_{22}u_{20}\cos\omega T_0 \\ &= -2i\omega_1 A_2' e^{i\omega_1 T_0} - A_1 e^{i\omega_1 T_0} \\ &\quad - f_{12}\left[\frac{A_2}{\omega(\omega + 2\omega_1)} e^{i(\omega+\omega_1)T_0} + \frac{\overline{A}_2}{\omega(\omega - 2\omega_1)} e^{i(\omega-\omega_1)T_0}\right] \\ &\quad - f_{22}\left(A_2 e^{i(\omega+\omega_1)T_0} + \overline{A}_2 e^{i(\omega-\omega_1)T_0}\right) + cc \end{aligned} \tag{5.23.58}$$

Eliminating secular terms from the above equation gives

$$2i\omega_1 A_2' + A_1 = 0 \tag{5.23.59}$$

Then the solution for (5.23.58) is

$$\begin{aligned} u_{21} = &\frac{A_2}{\omega(\omega + 2\omega_1)}\left[f_{22} + \frac{f_{12}}{\omega(\omega + 2\omega_1)}\right] e^{i(\omega+\omega_1)T_0} \\ &+ \frac{\overline{A}_2}{\omega(\omega - 2\omega_1)}\left[f_{22} + \frac{f_{12}}{\omega(\omega - 2\omega_1)}\right] e^{i(\omega-\omega_1)T_0} + cc \end{aligned} \tag{5.23.60}$$

Substituting u_{01}, $u_{03}$and u_{21} into (5.23.56) and eliminating secular terms yields

$$\begin{aligned} D_0^2 u_{31} + \omega_3^2 u_{31} &= -2D_0 D_1 u_{30} - 2(f_{31}u_{10} + f_{32}u_{21} + f_{33}u_{30})\cos\omega T_0 \\ &= -2D_0 D_1 u_{30} - (f_{31}u_{10} + f_{32}u_{21} + f_{33}u_{30}) e^{\omega T_0} \\ &= -2i\omega_3 A_3' e^{i\omega_3 T_0} + cc + NST + cc \end{aligned} \tag{5.23.61}$$

Eliminating secular terms from the above equation gives

$$A_3' = 0 \quad \Rightarrow \quad A_3 = A_3(T_2) \tag{5.23.62}$$

Substituting u_{10}, u_{30} and u_{21} into (5.23.54) yields

$$\begin{aligned}
D_0^2u_{11} + \omega_1^2u_{11} &= -2D_0D_1u_{10} - 2(f_{11}u_{10} + f_{12}u_{21} + f_{13}u_{30})\cos\omega T_0 \\
&= -2D_0D_1u_{10} - (f_{11}u_{10} + f_{12}u_{21} + f_{13}u_{30})\left(e^{i\omega T_0} + e^{-i\omega T_0}\right) \\
&= -2i\omega_1A_1'e^{i\omega_1T_0} - f_{11}f_{12}[\frac{\overline{A}_2}{\omega(\omega-2\omega_1)}e^{i(2\omega-\omega_1)T_0} \\
&\quad + \frac{A_2}{\omega(\omega-2\omega_1)}e^{i\omega_1T_0} + \frac{A_2}{\omega(\omega+2\omega_1)}e^{i\omega_1T_0}] \\
&\quad - f_{12}\frac{\overline{A}_2}{\omega(\omega-2\omega_1)}\left[f_{22} + \frac{f_{12}}{\omega(\omega-2\omega_1)}\right]e^{i(2\omega-\omega_1)T_0} \\
&\quad - f_{12}\frac{A_2}{\omega(\omega-2\omega_1)}\left[f_{22} + \frac{f_{12}}{\omega(\omega-2\omega_1)}\right]e^{i\omega_1T_0} \\
&\quad - f_{12}\frac{A_2}{\omega(\omega+2\omega_1)}\left[f_{22} + \frac{f_{12}}{\omega(\omega+2\omega_1)}\right]e^{i\omega_1T_0} \\
&\quad - f_{13}A_3e^{i(\omega_3-\omega)T_0} + cc + NST
\end{aligned} \tag{5.23.63}$$

Taking (5.23.57) into account we have

$$\begin{aligned}
D_0^2u_{11} + \omega_1^2u_{11} &= -2i\omega_1A_1'e^{i\omega_1T_0} - f_{11}f_{12}[\frac{\overline{A}_2}{\omega(\omega-2\omega_1)}e^{2i\sigma T_1}e^{i\omega_1T_0} \\
&\quad + \frac{A_2}{\omega(\omega-2\omega_1)}e^{i\omega_1T_0} + \frac{A_2}{\omega(\omega+2\omega_1)}e^{i\omega_1T_0}] \\
&\quad - f_{12}\frac{\overline{A}_2}{\omega(\omega-2\omega_1)}\left[f_{22} + \frac{f_{12}}{\omega(\omega-2\omega_1)}\right]e^{2i\sigma T_1}e^{i\omega_1T_0} \\
&\quad - f_{12}\frac{A_2}{\omega(\omega-2\omega_1)}\left[f_{22} + \frac{f_{12}}{\omega(\omega-2\omega_1)}\right]e^{i\omega_1T_0} \\
&\quad - f_{12}\frac{A_2}{\omega(\omega+2\omega_1)}\left[f_{22} + \frac{f_{12}}{\omega(\omega+2\omega_1)}\right]e^{i\omega_1T_0} \\
&\quad - f_{13}A_3e^{-i\sigma T_1}e^{i\omega_1T_0} + cc + NST
\end{aligned} \tag{5.23.64}$$

Eliminating secular terms from the above equation gives

$$\begin{aligned}
&2i\omega_1A_1' \\
&+ \left[\frac{f_{11}f_{12}}{\omega(\omega-2\omega_1)} + \frac{f_{11}f_{12}}{\omega(\omega+2\omega_1)} + \frac{f_{12}f_{22}}{\omega(\omega-2\omega_1)} + \frac{f_{12}f_{22}}{\omega(\omega+2\omega_1)} + \frac{f_{12}^2}{\omega^2(\omega-2\omega_1)^2} + \frac{f_{12}^2}{\omega^2(\omega+2\omega_1)^2}\right]A_2
\end{aligned}$$

$$+\left[\frac{f_{11}f_{12}\bar{A}_2}{\omega(\omega-2\omega_1)}+\frac{f_{12}f_{22}\bar{A}_2}{\omega(\omega-2\omega_1)}+\frac{f_{12}^2\bar{A}_2}{\omega^2(\omega-2\omega_1)^2}\right]e^{2i\sigma T_1}+f_{13}A_3e^{-i\sigma T_1}=0$$

i.e.,

$$iA_1'+2\omega_1\alpha_1A_2+2\omega_1\alpha_2\bar{A}_2e^{2i\sigma T_1}+2\omega_1\alpha_3e^{-i\sigma T_1}=0 \tag{5.23.65}$$

where

$$4\omega_1^2\alpha_1=\frac{2f_{12}(f_{11}+f_{22})}{\omega^2-4\omega_1^2}+\frac{f_{12}^2}{\omega^2}\left[\frac{1}{(\omega-2\omega_1)^2}+\frac{1}{(\omega+2\omega_1)^2}\right] \tag{5.23.66}$$

$$4\omega_1^2\alpha_2=\frac{f_{12}(f_{11}+f_{22})}{\omega(\omega-2\omega_1)}+\frac{f_{12}^2}{\omega^2(\omega-2\omega_1)^2} \tag{5.23.67}$$

$$4\omega_1^2\alpha_3=f_{13}A_3 \tag{5.23.68}$$

Eliminating A_1 from (5.23.65) and (5.23.59) yields

$$A_2''+\alpha_1A_2+\alpha_2\bar{A}_2e^{2i\sigma T_1}+\alpha_3e^{-i\sigma T_1}=0 \tag{5.23.69}$$

Let the solution of (5.23.69) be

$$A_2=(B_r+iB_i)e^{i\sigma T_1} \tag{5.23.70}$$

Then we can have

$$\begin{aligned}&\left(B_r''+iB_i''\right)e^{i\sigma T_1}+2i\sigma\left(B_r'+iB_i'\right)e^{i\sigma T_1}-\sigma^2(B_r+iB_i)e^{i\sigma T_1}\\&+\alpha_1(B_r+iB_i)e^{i\sigma T_1}+\alpha_2(B_r-iB_i)e^{i\sigma T_1}+\alpha_3e^{-i\sigma T_1}=0\end{aligned} \tag{5.23.71}$$

Separating the real and imaginary parts of the above equation yields

$$\begin{aligned}&B_r''-2\sigma B_i'+\left(\alpha_1+\alpha_2-\sigma^2\right)B_r\\&=-\alpha_3\cos 2\sigma T_1B_i''+2\sigma B_r'+\left(\alpha_1-\alpha_2-\sigma^2\right)B_i\\&=\alpha_3\sin 2\sigma T_1\end{aligned} \tag{5.23.72}$$

The prime denotes the derivative with respect to T_1. The corresponding homogeneous equations are

$$B_r''-2\sigma B_i'+\left(\alpha_1+\alpha_2-\sigma^2\right)B_r=0B_i''+2\sigma B_r'+\left(\alpha_1-\alpha_2-\sigma^2\right)B_i=0 \tag{5.23.73}$$

The nontrivial solution of (5.23.73) can be expressed as:

$$(B_r, B_i) = (b_r, b_i)\exp(\gamma T_1) \tag{5.23.74}$$

where γ satisfies the characteristic equation

$$\begin{vmatrix} \gamma^2 + \alpha_1 - \sigma^2 + \alpha_2 & -2\sigma\gamma \\ 2\sigma\gamma & \gamma^2 + \alpha_1 - \sigma^2 - \alpha_2 \end{vmatrix} = 0$$

i.e.,

$$(\gamma^2 + \alpha_1 - \sigma^2)^2 - \alpha_2^2 + 4\sigma^2\gamma^2 = 0 \tag{5.23.75}$$

therefore,

$$\gamma^2 = -(\alpha_1 + \sigma^2) \pm \sqrt{4\alpha_1\sigma^2 + \alpha_2^2} \tag{5.23.76}$$

It can be seen from (5.23.70) and (5.23.74) that when the real part of γ is positive, A_2 is unbounded and hence the motion is unstable. Therefore, the critical value of γ^2 changing from stable to unstable is $\gamma^2 = 0$. From (5.23.75), we know that

$$(\alpha_1 - \sigma^2)^2 = \alpha_2^2 \text{ or } \sigma = \pm(\alpha_1 \pm \alpha_2)^{1/2} \tag{5.23.77}$$

Substituting this into (5.23.57), the transition curves separating stability from instability for the system are

$$\omega = \omega_1 \mp \varepsilon(\alpha_1 \pm \alpha_2)^{1/2} + \cdots \tag{5.23.78}$$

5.24 Exercise 5.24 (A Two-Frequency Parametric Excitation of a Three-Degree-of-Freedom System with Repeating Frequencies)

Solution: This is a linear system having three degrees of freedom, with two of the frequencies being equal. Without parametric excitation, the system is unstable because x_2 contains a secular or resonant term of the form $t\sin(\omega_1 t + \beta)$, where β is a constant. To determine if the parametric excitation can stabilize the system, we assume that all three modes are bounded and then, if possible, determine the values of the parameters which are consistent with this assumption, particularly those values at the boundaries of the region where the assumption is valid.

Although the parametric excitation might stabilize the motion, we still expect the amplitude of the x_2-mode to be much larger than that of the x_1-mode. We do not have any indication of the amplitude of the x_3-mode. To express our expectations systematically, we scale the dependent variables. Without loss of generality, we make

$$x_1 = \varepsilon^{-\lambda_1} u_1, \quad x_2 = \varepsilon^{-\lambda_2} u_2, \quad x_3 = \varepsilon^{-\lambda_3} u_3, \quad where\ \lambda_1 = 0 \tag{5.24.1}$$

where the u_n are $O(1)$ and the λ_n are non-negative constants to be determined in the solution. Substituting (5.24.1) into the original equation, we get

$$\ddot{u}_1 + \omega_1^2 u_1 + 2\sum_{n=1}^{3}[f_{1n}\cos\Omega_1 t + g_{1n}\cos(\Omega_2 t + \theta)]\varepsilon^{1-\lambda_n} u_n = 0 \tag{5.24.2}$$

$$\ddot{u}_2 + \omega_1^2 u_2 + \varepsilon^{\lambda_2} u_1 + 2\sum_{n=1}^{3}[f_{2n}\cos\Omega_1 t + g_{2n}\cos(\Omega_2 t + \theta)]\varepsilon^{1-\lambda_n+\lambda_2} u_n = 0 \tag{5.24.3}$$

$$\ddot{u}_3 + \omega_3^2 u_3 + 2\sum_{n=1}^{3}[f_{3n}\cos\Omega_1 t + g_{3n}\cos(\Omega_2 t + \theta)]\varepsilon^{1-\lambda_n+\lambda_3} u_n = 0 \tag{5.24.4}$$

We are concerned with various combinations of the frequencies which can lead to a resonant response. We therefore adopt a multiscale approach to determine a uniformly valid approximate solution which exhibits the effects of the repeated frequency and the resonant combinations of frequencies. We write the various time scales as

$$T_0 = t, \quad T_1 = \varepsilon^{\lambda_4} t, \quad T_2 = \varepsilon^{2\lambda_4} t \tag{5.24.5}$$

where $\lambda_4 > 0$ is a constant to be determined. In terms of these scales,

$$\frac{d^2}{dt^2} = D_0^2 + 2\varepsilon^{\lambda_4} D_0 D_1 + \varepsilon^{2\lambda_4}\left(2D_0 D_2 + D_1^2\right) + \cdots \tag{5.24.6}$$

Assume that approximate solutions in the form

$$\begin{aligned} u_n(t;\varepsilon) = {} & u_{n0}(T_0, T_1, T_2) + \varepsilon^{\lambda_4} u_{n1}(T_0, T_1, T_2) \\ & + \varepsilon^{2\lambda_4} u_{n2}(T_0, T_1, T_2) + \cdots \end{aligned} \tag{5.24.7}$$

Substituting (5.24.5–5.24.7) into (5.24.2–5.24.4) leads to

$$\begin{aligned} & D_0^2 u_{10} + \omega_1^2 u_{10} + \varepsilon^{\lambda_4}\left(D_0^2 u_{11} + \omega_1^2 u_{11} + 2D_0 D_1 u_{10}\right) \\ & + 2(\varepsilon f_{11} u_{10} + \varepsilon^{1-\lambda_2} f_{12} u_{20} + \varepsilon^{1-\lambda_2+\lambda_4} f_{12} u_{21} \\ & + \varepsilon^{1-\lambda_3} f_{13} u_{30} + \varepsilon^{1-\lambda_3+\lambda_4} f_{13} u_{31}) \cos\Omega_1 T_0 \\ & + 2(\varepsilon g_{11} u_{10} + \varepsilon^{1-\lambda_2} g_{12} u_{20} + \varepsilon^{1-\lambda_2+\lambda_4} g_{12} u_{21} \\ & + \varepsilon^{1-\lambda_3} g_{13} u_{30} + \varepsilon^{1-\lambda_3+\lambda_4} g_{13} u_{31}) \cos(\Omega_2 T_0 + \theta) + \cdots = 0 \end{aligned} \tag{5.24.8}$$

$$\begin{aligned}&D_0^2u_{20}+\omega_1^2u_{20}+\varepsilon^{\lambda_4}\left(D_0^2u_{21}+\omega_1^2u_{21}+2D_0D_1u_{20}\right)+\varepsilon^{\lambda_2}u_{10}\\&\quad+2(\varepsilon^{1+\lambda_2}f_{21}u_{10}+\varepsilon f_{22}u_{20}+\varepsilon^{1-\lambda_3+\lambda_2}f_{23}u_{30}+\varepsilon^{1-\lambda_3+\lambda_2+\lambda_4}f_{23}u_{31})\cos\Omega_1T_0\\&\quad+2(\varepsilon^{1+\lambda_2}g_{21}u_{10}+\varepsilon g_{22}u_{20}+\varepsilon^{1-\lambda_3+\lambda_2}g_{23}u_{30}+\varepsilon^{1-\lambda_3+\lambda_2+\lambda_4}g_{23}u_{31})\cos(\Omega_2T_0+\theta)\\&+\cdots=0\end{aligned}\tag{5.24.9}$$

$$\begin{aligned}&D_0^2u_{30}+\omega_3^2u_{30}+\varepsilon^{\lambda_4}\left(D_0^2u_{31}+\omega_3^2u_{31}+2D_0D_1u_{30}\right)\\&\quad+2(\varepsilon^{1+\lambda_3}f_{31}u_{10}+\varepsilon^{1-\lambda_2+\lambda_3}f_{32}u_{20}\\&\quad+\varepsilon^{1-\lambda_2+\lambda_3+\lambda_4}f_{32}u_{21}+\varepsilon f_{33}u_{30})\cos\Omega_1T_0\\&\quad+2(\varepsilon^{1+\lambda_3}g_{31}u_{10}+\varepsilon^{1-\lambda_2+\lambda_3}g_{32}u_{20}\\&\quad+\varepsilon^{1-\lambda_2+\lambda_3+\lambda_4}g_{32}u_{21}+\varepsilon g_{33}u_{30})\cos(\Omega_2T_0+\theta)+\cdots=0\end{aligned}\tag{5.24.10}$$

We note that the governing equations of $u_{10},\quad u_{20}$and u_{30} are of order ε^0, i.e.,

$$D_0^2u_{n0}+\omega_n^2u_{n0}=\cdots,\quad n=1,2,3\tag{5.24.11}$$

The right-hand side of the equation is either zero or a non-resonant excitation term. Then, the solutions of u_{n0} will contain free oscillation terms such as $Ae^{\pm i\omega_nT_0}$, and thus u_{k0} cannot enter into the governing equations of u_{n0}, e.g., $\varepsilon^{\lambda 2}u_{10}$ in Eq. (5.24.9) cannot enter into the governing equation of u_{20}, thus

$$\lambda_2\neq 0\tag{5.24.12}$$

The other resonant terms depend on the values of ω_i and Ω_i, and they are not allowed to enter into the control equations of $u_{10},\quad u_{20}$and u_{30}. On the other hand, in the governing equations of $u_{11},\quad u_{21}$and u_{31}, it is necessary to keep the resonant terms as much as possible. Only in this way, we can solve the problem by eliminating resonant terms.

(a) When $\omega_1\approx\frac{1}{2}$ and $\omega_3-\omega_1\approx\Omega_2$:

From (5.24.8–5.24.10), we can see that the basic order of the control equations of $u_{11},\quad u_{21}$and u_{31} is ε^{λ_4} and their basic form is

$$\begin{aligned}&\varepsilon^{\lambda_4}\left(D_0^2u_{11}+\omega_1^2u_{11}+2D_0D_1u_{10}\right)\\&\quad+2(\varepsilon f_{11}u_{10}+\varepsilon^{1-\lambda_2}f_{12}u_{20}+\varepsilon^{1-\lambda_2+\lambda_4}f_{12}u_{21}\\&\quad+\varepsilon^{1-\lambda_3}f_{13}u_{30}+\varepsilon^{1-\lambda_3+\lambda_4}f_{13}u_{31})\cos\Omega_1T_0\\&\quad+2(\varepsilon g_{11}u_{10}+\varepsilon^{1-\lambda_2}g_{12}u_{20}+\varepsilon^{1-\lambda_2+\lambda_4}g_{12}u_{21}\\&\quad+\varepsilon^{1-\lambda_3}g_{13}u_{30}+\varepsilon^{1-\lambda_3+\lambda_4}g_{13}u_{31})\cos(\Omega_2T_0+\theta)=0\end{aligned}\tag{5.24.13}$$

$$\varepsilon^{\lambda_4}\left(D_0^2u_{21}+\omega_1^2u_{21}+2D_0D_1u_{20}\right)+\varepsilon^{\lambda_2}u_{10}$$

$$
\begin{aligned}
&+2\left(\varepsilon^{1+\lambda_2} f_{21} u_{10}+\varepsilon f_{22} u_{20}+\varepsilon^{1-\lambda_3+\lambda_2} f_{23} u_{30}+\varepsilon^{1-\lambda_3+\lambda_2+\lambda_4} f_{23} u_{31}\right) \cos \Omega_1 T_0 \\
&+2\left(\varepsilon^{1+\lambda_2} g_{21} u_{10}+\varepsilon g_{22} u_{20}+\varepsilon^{1-\lambda_3+\lambda_2} g_{23} u_{30}+\varepsilon^{1-\lambda_3+\lambda_2+\lambda_4} g_{23} u_{31}\right) \cos \left(\Omega_2 T_0+\theta\right) \\
&=0
\end{aligned}
\tag{5.24.14}
$$

$$
\begin{aligned}
&\varepsilon^{\lambda_4}\left(D_0^2 u_{31}+\omega_3^2 u_{31}+2 D_0 D_1 u_{30}\right) \\
&\quad+2\left(\varepsilon^{1+\lambda_3} f_{31} u_{10}+\varepsilon^{1-\lambda_2+\lambda_3} f_{32} u_{20}\right. \\
&\quad\left.+\varepsilon^{1-\lambda_2+\lambda_3+\lambda_4} f_{32} u_{21}+\varepsilon f_{33} u_{30}\right) \cos \Omega_1 T_0 \\
&\quad+2\left(\varepsilon^{1+\lambda_3} g_{31} u_{10}+\varepsilon^{1-\lambda_2+\lambda_3} g_{32} u_{20}\right. \\
&\quad\left.+\varepsilon^{1-\lambda_2+\lambda_3+\lambda_4} g_{32} u_{21}+\varepsilon g_{33} u_{30}\right) \cos \left(\Omega_2 T_0+\theta\right)=0
\end{aligned}
\tag{5.24.15}
$$

Here, we are most concerned with the resonant terms that may be involved in these equations, and keep at least the lowest order of those resonant terms. The lowest-order resonant terms to be retained in (5.24.14) is $\varepsilon^{\lambda 2} u_{10}$, so

$$
\lambda_4=\lambda_2 \tag{5.24.16}
$$

The lowest order resonant term to be retained in (5.24.13) is $\varepsilon^{1-\lambda_3} g_{13} u_{30} \cos \left(\Omega_2 T_0+\theta\right)$, which leads to

$$
\lambda_4=1-\lambda_3 \tag{5.24.17}
$$

The lowest order resonant term to be retained in (5.24.15) is $\varepsilon^{1-\lambda_2+\lambda_3} g_{32} u_{20} \cos \left(\Omega_2 T_0+\theta\right)$, which leads to

$$
\lambda_4=1-\lambda_2+\lambda_3 \tag{5.24.18}
$$

therefore,

$$
\lambda_4=\lambda_2=\frac{2}{3}, \quad \lambda_3=\frac{1}{3} \tag{5.24.19}
$$

Substituting the above results into (5.24.8)~(5.24.10), the governing equations for u_{10}, u_{20}, u_{30} and u_{11}, u_{21}, u_{31} are.

Order ε^0:

$$
D_0^2 u_{10}+\omega_1^2 u_{10}=0 \tag{5.24.20}
$$

$$
D_0^2 u_{20}+\omega_1^2 u_{20}=0 \tag{5.24.21}
$$

$$
D_0^2 u_{30}+\omega_3^2 u_{30}=0 \tag{5.24.22}
$$

Order $\varepsilon^{2/3}$:

$$D_0^2u_{11} + \omega_1^2u_{11} = -2D_0D_1u_{10} - 2f_{13}u_{30}\cos\Omega_1T_0 - 2g_{13}u_{30}\cos(\Omega_2T_0 + \theta) \tag{5.24.23}$$

$$D_0^2u_{21} + \omega_1^2u_{21} = -2D_0D_1u_{20} - u_{10} \tag{5.24.24}$$

$$D_0^2u_{31} + \omega_3^2u_{31} = -2D_0D_1u_{30} - 2f_{32}u_{20}\cos\Omega_1T_0 - 2g_{32}u_{20}\cos(\Omega_2T_0 + \theta) \tag{5.24.25}$$

The solutions of (5.24.20–5.24.22) are

$$u_{10} = A_1(T_1)e^{i\omega_1T_0} + cc \tag{5.24.26}$$

$$u_{20} = A_2(T_1)e^{i\omega_1T_0} + cc \tag{5.24.27}$$

$$u_{30} = A_3(T_1)e^{i\omega_3T_0} + cc \tag{5.24.28}$$

Let

$$\omega_1 = \frac{1}{2} + \varepsilon^{2/3}\sigma, \quad \omega_3 - \omega_1 = \Omega_2 - \varepsilon^{2/3}\sigma \tag{5.24.29}$$

Substitute u_{01}, $u_{02}$and u_{03} into (5.24.23–5.24.25), we obtain

$$\begin{aligned} D_0^2u_{11} + \omega_1^2u_{11} &= -2D_0D_1u_{10} - f_{13}u_{30}\left(e^{i\Omega_1T_0} + e^{-i\Omega_1T_0}\right) \\ &\quad - g_{13}u_{30}\left(e^{i(\Omega_2T_0+\theta)} + e^{-i(\Omega_2T_0+\theta)}\right) \\ &= -2i\omega_1A_1'e^{i\omega_1T_0} - g_{13}A_3e^{-i(\sigma T_1+\theta)}e^{i\omega_1T_0} \\ &\quad - f_{13}A_3e^{i(\omega_3+\Omega_1)T_0} - f_{13}A_3e^{i(\omega_3-\Omega_1)T_0} \\ &\quad - g_{13}A_3e^{i[(\omega_3+\Omega_2)T_0+\theta]} + cc \end{aligned} \tag{5.24.30}$$

$$\begin{aligned} D_0^2u_{21} + \omega_1^2u_{21} &= -2D_0D_1u_{20} - u_{10} \\ &= -2i\omega_1A_2'e^{i\omega_1T_0} - A_1e^{i\omega_1T_0} + cc \end{aligned} \tag{5.24.31}$$

$$\begin{aligned} D_0^2u_{31} + \omega_3^2u_{31} &= -2D_0D_1u_{30} - 2f_{32}u_{20}\left(e^{i\Omega_1T_0} + e^{-i\Omega_1T_0}\right) \\ &\quad - 2g_{32}u_{20}\left(e^{i(\Omega_2T_0+\theta)} + e^{-i(\Omega_2T_0+\theta)}\right) \\ &= -2i\omega_1A_3'e^{i\omega_3T_0} - g_{32}A_2e^{i(\sigma T_1+\theta)}e^{i\omega_3T_0} \\ &\quad - f_{32}A_2e^{i(\omega_1+\Omega_1)T_0} - f_{32}A_2e^{i(\omega_1-\Omega_1)T_0} \\ &\quad - g_{32}A_2e^{i[(\omega_1-\Omega_2)T_0-\theta]} + cc \end{aligned} \tag{5.24.32}$$

The prime denotes the derivative with respect to T_1. Eliminating secular terms in (5.24.30–5.24.32) yields

$$iA_1' + g_{13}A_3e^{-i(\sigma T_1+\theta)} = 0 \tag{5.24.33}$$

$$iA_2' + A_1 = 0 \tag{5.24.34}$$

$$iA_3' + g_{32}A_2e^{i(\sigma T_1+\theta)} = 0 \tag{5.24.35}$$

which has taken into account $\omega_1 \approx \frac{1}{2}$. In the Eqs. (5.24.33) and(5.24.34) eliminating A_1 from the equation yields

$$A_2'' + g_{13}A_3e^{-i(\sigma T_1+\theta)} = 0 \tag{5.24.36}$$

Let

$$\begin{aligned} A_1 &= (B_{1r} + iB_{1i})e^{-i(\sigma T_1+\theta)/2} \\ A_2 &= (B_{2r} + iB_{2i})e^{-i(\sigma T_1+\theta)/2} \\ A_3 &= (B_{3r} + iB_{3i})e^{i(\sigma T_1+\theta)/2} \end{aligned} \tag{5.24.37}$$

Therefore, we can obtain A_2 and A_3 from (5.24.35) and (5.24.36), and then A_1 from (5.24.34). Substituting (5.24.37) into (5.24.26)~(5.24.28), we can obtain the first order approximate solution of the system

$$\begin{aligned} u_1 \approx u_{10} &= 2B_{1r}\cos[(T_0 - \sigma T_1 - \theta)/2] \\ &\quad - 2B_{1i}\sin[(T_0 - \sigma T_1 - \theta)/2] \\ u_2 \approx u_{20} &= 2B_{2r}\cos[(T_0 - \sigma T_1 - \theta)/2] \\ &\quad - 2B_{2i}\sin[(T_0 - \sigma T_1 - \theta)/2] \\ u_3 \approx u_{30} &= 2B_{3r}\cos[(2\omega_3 T_0 + \sigma T_1 + \theta)/2] \\ &\quad - B_{3i}\sin[(2\omega_3 T_0 + \sigma T_1 + \theta)/2] \end{aligned} \tag{5.24.38}$$

(b) When $\omega_1 \approx \Omega_1$ and $\omega_3 - \omega_1 \approx \Omega_2$:

Readers are invited to analyze this case for themselves.

5.25 Exercise 5.25 (Parametric Excitation Analysis of Oscillations of a Spring Pendulum)

Solution: The coordinates of the mass are

$$\begin{aligned} x &= (l + x)\sin\theta \\ y &= Y + (l + x)\cos\theta \end{aligned} \tag{5.25.1}$$

The kinetic energy of the system is

$$\begin{aligned} T &= \frac{1}{2}m\left(\dot{x}^2 + \dot{y}^2\right) \\ &= \frac{1}{2}m\left\{\left[\dot{x}\sin\theta + (l + x)\dot{\theta}\cos\theta\right]^2 + \left[\dot{x}\cos\theta - (l + x)\dot{\theta}\sin\theta - \dot{Y}\right]^2\right\} \\ &= \frac{1}{2}m\left[\dot{x}^2 + (l + x)^2\dot{\theta}^2 + \dot{Y}^2 - 2\dot{x}\dot{Y}\cos\theta + 2(l + x)\dot{\theta}\dot{Y}\sin\theta\right] \end{aligned} \tag{5.25.2}$$

The potential energy of the system is

$$V = \frac{1}{2}kx^2 + mg(l + x)(1 - \cos\theta) \tag{5.25.3}$$

Substitute kinetic and potential energy into the Lagrange's equation:

$$\begin{aligned} \frac{d}{dt}\frac{\partial T}{\partial \dot{x}} - \frac{\partial T}{\partial x} &= -\frac{\partial V}{\partial x} \\ \frac{d}{dt}\frac{\partial T}{\partial \dot{\theta}} - \frac{\partial T}{\partial \theta} &= -\frac{\partial V}{\partial \theta} \end{aligned} \tag{5.25.4}$$

we can obtain

$$m\ddot{x} + kx - m(l + x)\dot{\theta}^2 + mg(1 - \cos\theta) - m\ddot{Y}\cos\theta = 0 \tag{5.25.5}$$

$$m(l + x)\ddot{\theta} + 2m\dot{x}\dot{\theta} + m\ddot{Y}\sin\theta + mg\sin\theta = 0 \tag{5.25.6}$$

Let

$$u = x/l, \quad y = Y/l, \quad \omega_1^2 = k/m, \quad \delta = g/l \tag{5.25.7}$$

(5.25.5) and (5.25.6) becomes

$$\ddot{u} + \omega_1^2 u - (1 + u)\dot{\theta}^2 + \delta(1 - \cos\theta) - \ddot{y}\cos\theta = 0 \tag{5.25.8}$$

$$(1 + u)\ddot{\theta} + 2\dot{u}\dot{\theta} + (\delta + \ddot{y})\sin\theta = 0 \tag{5.25.9}$$

(a) When $y \equiv 0$, the equilibrium position of the system is $x = \theta = 0$. Make the expansion of (5.25.8) and (5.25.9) near this equilibrium point, we can obtain

$$\ddot{u} + \omega_1^2 u - (1 + u)\dot{\theta}^2 + \frac{1}{2}\delta\theta^2 - \ddot{y} + \frac{1}{2}\ddot{y}\theta^2 = 0 \tag{5.25.10}$$

$$(1+u)\ddot{\theta}+2\dot{u}\dot{\theta}+(\delta+\ddot{y})\left(\theta-\frac{1}{6}\theta^3\right)=0 \tag{5.25.11}$$

Omitting the nonlinear term of θ yields

$$\ddot{u}+\omega_1^2 u=\ddot{y} \tag{5.25.12}$$

$$(1+u)\ddot{\theta}+2\dot{u}\dot{\theta}+(\delta+\ddot{y})\theta=0 \tag{5.25.13}$$

(b) When $\ddot{y}=2\varepsilon\cos 2t$, (5.25.12) becomes

$$\ddot{u}+\omega_1^2 u=2\varepsilon\cos 2t \tag{5.25.14}$$

The generalized solution of u is

$$u_e=a\cos(\omega_1 t+\beta)+2\varepsilon(\omega_1^2-4)^{-1}\cos 2t,\quad \omega_1\neq 2 \tag{5.25.15}$$

where a and β are integration constants which are the amplitude and phase of the free oscillation, respectively, and can be determined by initial conditions. Thus, (5.25.13) becomes

$$(1+u_e)\ddot{\theta}+2\dot{u}_e\dot{\theta}+(\delta+2\varepsilon\cos 2t)\theta=0 \tag{5.25.16}$$

This is a parametric excitation equation for θ.
If there is no free oscillation term in u_e, i.e.$a=0$, then (5.25.16) becomes

$$[1+2\varepsilon\left(\omega_1^2-4\right)^{-1}\cos 2t]\ddot{\theta}-8\varepsilon(\omega_1^2-4)^{-1}\dot{\theta}\sin 2t+(\delta+2\varepsilon\cos 2t)\theta=0 \tag{5.25.17}$$

Let the solution of (5.25.17) is

$$\theta(t;\varepsilon)=\theta_0(T_0,T_1,T_2)+\varepsilon\theta_1(T_0,T_1,T_2)+\varepsilon^2\theta_2(T_0,T_1,T_2)+\cdots \tag{5.25.18}$$

Substituting (5.25.18) into (5.25.17) and retains to $O(\varepsilon)$, we get

$$\begin{aligned}
0&=[1+2\varepsilon\kappa\cos 2t]\ddot{\theta}-8\varepsilon\kappa\dot{\theta}\sin 2t+(\delta+2\varepsilon\cos 2t)\theta\\
&=[1+2\varepsilon\kappa\cos 2t]\left(D_0^2+2\varepsilon D_0D_1\right)(\theta_0+\varepsilon\theta_1)\\
&\quad-8\varepsilon\kappa[(D_0+\varepsilon D_1)(\theta_0+\varepsilon\theta_1)]\sin 2T_0\\
&\quad+(\delta+2\varepsilon\cos 2T_0)(\theta_0+\varepsilon\theta_1)+\cdots\\
&=D_0^2\theta_0+\varepsilon D_0^2\theta_1+2\varepsilon D_0D_1\theta_0+2\varepsilon\kappa D_0^2\theta_0\cos 2T_0\\
&\quad-8\varepsilon\kappa D_0\theta_0\sin 2T_0+\delta\theta_0+\varepsilon\delta\theta_1+2\varepsilon\theta_0\cos 2T_0+\cdots\\
&=D_0^2\theta_0+\delta\theta_0+\varepsilon(D_0^2\theta_1+\delta\theta_1+2D_0D_1\theta_0+2\kappa D_0^2\theta_0\cos 2T_0
\end{aligned}$$

$$-8\kappa D_0\theta_0 \sin 2T_0 + 2\theta_0 \cos 2T_0) + \cdots \tag{5.25.19}$$

where $\kappa = (\omega_1^2 - 4)^{-1}$. Equating coefficients of like powers of ε in the above equation yields

$$D_0^2\theta_0 + \omega_0^2\theta_0 = 0 \tag{5.25.20}$$

$$\begin{aligned} D_0^2\theta_1 + \omega_0^2\theta_1 &= -2D_0D_1\theta_0 - 2\kappa D_0^2\theta_0\text{cos}2T_0 \\ &+8\kappa D_0\theta_0\text{sin}2T_0 - 2\theta_0\text{cos}2T_0 \end{aligned} \tag{5.25.21}$$

where $\omega_0 = \sqrt{\delta}$. The solution of (5.25.20) is

$$\theta_0 = Ae^{i\omega_0 T_0} + \overline{A}\, e^{-i\omega_0 T_0} \tag{5.25.22}$$

Substitute θ_0 into (5.25.21), we can obtain

$$\begin{aligned} D_0^2u_1 + \omega_0^2u_1 &= -2D_0D_1\theta_0 - 2\kappa D_0^2\theta_0 \cos 2T_0 \\ &\quad + 8\kappa D_0\theta_0 \sin 2T_0 - 2\theta_0 \cos 2T_0 \\ &= -2i\omega_0 A' e^{i\omega_0 T_0} \\ &\quad + \kappa\omega_0^2\left(Ae^{i\omega_0 T_0} + \overline{A}e^{-i\omega_0 T_0}\right)e^{i2T_0} \\ &\quad + 4i\kappa\omega_0\left(Ae^{i\omega_0 T_0} - \overline{A}e^{-i\omega_0 T_0}\right)e^{i2T_0} \\ &\quad - \left(Ae^{i\omega_0 T_0} + \overline{A}e^{-i\omega_0 T_0}\right)e^{i2T_0} + cc \end{aligned} \tag{5.25.23}$$

The prime denotes the derivative with respect to T_1.
When $\omega_0 \approx 1$, let

$$\omega_0 = 1 + \varepsilon\sigma \tag{5.25.24}$$

then Eq. (5.25.23) becomes

$$\begin{aligned} D_0^2u_1 + \omega_0^2u_1 &= -2i\omega_0 A' e^{i\omega_0 T_0} + \kappa\omega_0^2\overline{A}e^{i(2-\omega_0)T_0} \\ &\quad - 4\kappa i\omega_0\overline{A}e^{i(2-\omega_0)T_0} \\ &\quad - \overline{A}e^{i(2-\omega_0)T_0} + cc + NST \\ &= -2i\omega_0 A' e^{i\omega_0 T_0} - 4i\kappa\omega_0\overline{A}e^{-2i\sigma T_1}e^{i\omega_0 T_0} \\ &\quad + \kappa\omega_0^2\overline{A}e^{-2i\sigma T_1}e^{i\omega_0 T_0} \\ &\quad - \overline{A}e^{-2i\sigma T_1}e^{i\omega_0 T_0} + cc + NST \end{aligned} \tag{5.25.25}$$

In order to eliminate secular terms from the above equation, there must be

$$2i\omega_0 A' + 4i\kappa\omega_0 \bar{A}\, e^{-2i\sigma T_1} - \kappa\omega_0^2 \bar{A}\, e^{-2i\sigma T_1} + \bar{A}\, e^{-2i\sigma T_1} = 0 \tag{5.25.26}$$

Substituting $A = \frac{1}{2}be^{i\beta}$ into the above equation and taking $\omega_0 \approx 1$ into account, we get

$$ib' - b\beta' + 2i\kappa b e^{-i(2\sigma T_1+2\beta)} - \frac{1}{2}\kappa b e^{-i(2\sigma T_1+2\beta)} + \frac{1}{2}b e^{-i(2\sigma T_1+2\beta)} = 0 \tag{5.25.27}$$

Separating the real and imaginary parts of the above equation yields

$$\begin{aligned} b' &= -2\kappa b\cos\gamma - \tfrac{1}{2}(\kappa - 1)b\sin\gamma \\ \gamma' &= 2\sigma b + 4\kappa b\sin\gamma - (\kappa - 1)b\cos\gamma \end{aligned} \tag{5.25.28}$$

where

$$\gamma = 2\sigma T_1 + 2\beta \tag{5.25.29}$$

(5.25.28) are satisfied by the nonzero steady state solutions of

$$2\sigma + \left[4\kappa + \frac{1}{4\kappa}\left(\kappa - 1)^2\right)\right]^2 \sin\gamma = 0 \tag{5.25.30}$$

Thus, the critical value of σ is

$$\sigma = \pm\frac{1}{8\kappa}\left[16\kappa^2 + (\kappa - 1)^2\right] \tag{5.25.31}$$

Substituting this into (5.25.24), we can obtain the transition curves separating stability from instability for the system

$$\omega_0 = \sqrt{\delta} = 1 \pm \varepsilon\frac{1}{8\kappa}\left[16\kappa^2 + (\kappa - 1)^2\right] \tag{5.25.32}$$

(c) Let $\theta = (1 + u)^{-1}\psi$, then

$$\begin{aligned} \dot{\theta} &= (1 + u)^{-1}\dot{\psi} - (1 + u)^{-2}\psi\dot{u}\,\ddot{\theta} \\ &= (1 + u)^{-1}\ddot{\psi} - 2(1 + u)^{-2}\dot{\psi}\dot{u} - (1 + u)^{-2}\psi\ddot{u} + 2(1 + u)^{-3}\psi\dot{u}^2 \end{aligned}$$

Substituting these into (5.25.13), we get

$$\ddot{\psi} + (1 + u)^{-1}(\delta + \ddot{y} - \ddot{u})\psi = 0 \tag{5.25.33}$$

(d) When

$$\ddot{y} = \varepsilon \sum_{n=1}^{N} y_n \cos(\Omega_n t + \theta_n) \tag{5.25.34}$$

we substitute this into (5.25.12) and obtain

$$\ddot{u} + \omega_1^2 u = \varepsilon \sum_{n=1}^{N} y_n \cos(\Omega_n t + \theta_n) \tag{5.25.35}$$

It is assumed that due to the damping of the system, the free oscillation term in (5.25.35) decays very quickly, so that

$$u = \varepsilon \sum_{n=1}^{N} \kappa_n y_n \cos(\Omega_n t + \theta_n) \tag{5.25.36}$$

where

$$\kappa_n = (\omega_1^2 - \Omega_n^2)^{-1} \tag{5.25.37}$$

Substitute (5.25.34) and (5.25.36) into (5.25.33), we get

$$\left[1 + \varepsilon \sum_{n=1}^{N} \kappa_n y_n \cos(\Omega_n t + \theta_n)\right]\ddot{\psi} + \left[\delta + \varepsilon \sum_{n=1}^{N} \alpha_n y_n \cos(\Omega_n t + \theta_n)\right]\psi = 0 \tag{5.25.38}$$

where

$$\alpha_n = 1 + \kappa_n \Omega_n^2 \tag{5.25.39}$$

The method and procedure to find the solution of Eq. (5.25.38) are similar to those in part (b) above. Readers are invited to solve on their own.

5.26 Exercise 5.26 (The Buckling of the Column with a Nonideal Energy Source)

Solution: (a) This is a simply supported beam subjected to axial loads, which are generated by a rotating crankshaft, and the drive motor is considered as a non-ideal energy source.

The Euler–Bernoulli linear elastic beam model is adopted here. Neglecting the longitudinal and rotary inertia and the transverse shear. The diagrams of motion and forces of the beam element are shown in Fig. 5.1. The prime in each variable in the

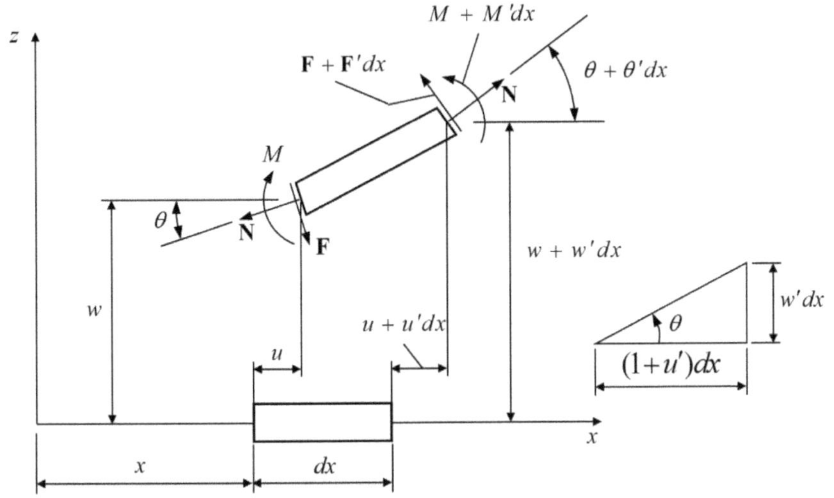

Fig. 5.1 A beam element for Exercise 5.26

figure indicates the partial derivative with respect to x. Applying Newton's second law to the beam element along z direction yields.

$$N\frac{\partial\theta}{\partial x}+\frac{\partial F}{\partial x}=m_1\frac{\partial^2 w}{\partial t^2} \tag{5.26.1}$$

where m_1 is the mass per unit length of the beam.

From the moment of momentum

$$\frac{\partial M}{\partial x}+F=0 \tag{5.26.2}$$

Using

$$\theta\approx\frac{\partial w}{\partial x} \tag{5.26.3}$$

Combining the above three equations,we can have

$$-\frac{\partial^2 M}{\partial x^2}+N\frac{\partial^2 w}{\partial x^2}=m_1\frac{\partial^2 w}{\partial t^2} \tag{5.26.4}$$

and

$$M=EI_x\frac{\partial^2 w}{\partial x^2} \tag{5.26.5}$$

where I_x is the moment of inertia of the beam section; and combining (5.26.4) and (5.26.5), we can obtain the governing equation for a homogeneous isotropic beam under axial force

$$m_1 \frac{\partial^2 w}{\partial t^2} + c \frac{\partial w}{\partial t} + EI_x \frac{\partial^4 w}{\partial x^4} - N \frac{\partial^2 w}{\partial x^2} = 0 \tag{5.26.6}$$

in which a viscous damping term, $c\partial w/\partial t$, is added.

Let the first order mode of the beam be

$$w(x, t) = \psi(t) \sin \frac{\pi x}{l} \tag{5.26.7}$$

Substituting (5.26.7) into (5.26.6) yields

$$\ddot{\psi} + \frac{c}{m_1}\dot{\psi} + \frac{EI_x}{m_1}(\frac{\pi}{l})^4 \psi + \frac{N}{m_1}(\frac{\pi}{l})^2 \psi = 0 \tag{5.26.8}$$

From Fig. 5.1, we can obtain

$$N = -k_1(u_0 - u_B + r \sin \varphi) \tag{5.26.9}$$

where u_0 is the initial displacement of the sliding end B of the beam along the positive x-direction and u_B is the dynamic displacement of the sliding end B of the beam along the negative x-direction, i.e.

$$u_B = -u(l)$$

The axial load N is the first order buckling load of the simply supported beam, which is much less than the elastic ultimate load of the material. Therefore, it can be assumed that the beam axis is inextensible. Furthermore, we can obtain the following relation from the beam element analysis

$$u' = \sqrt{1 - w'^2} - 1 \approx -\frac{1}{2}w'^2$$

So

$$\begin{aligned} u_B &= -u(l) = -\int_0^l u'dx = \tfrac{1}{2}\int_0^l w'^2 dx \\ &= \tfrac{\pi^2\psi^2}{2l^2}\int_0^l \cos^2\tfrac{\pi x}{l}dx = \tfrac{\pi^2}{4l}\psi^2 \end{aligned} \tag{5.26.10}$$

Substituting (5.26.9) and (5.26.10) into (5.26.8) yields

$$\ddot{\psi} + \frac{c}{m_1}\dot{\psi} + \left[\frac{\pi^4}{l^4}\frac{EI_x}{m_1}\left(1 - \frac{k_1 u_0 l^2}{\pi^2 EI_x}\right) - \frac{\pi^2 k_1 r}{l^2 m_1}\sin\varphi\right]\psi + \frac{\pi^4 k_1}{4l^3 m_1}\psi^3 = 0$$

i.e.,

$$\ddot{\psi} + 2\mu\dot{\psi} + \left(\omega^2 - \alpha_1 \sin\varphi\right)\psi + \alpha_2\psi^3 = 0 \tag{5.26.11}$$

where

$$\omega^2 = \frac{\pi^4}{l^4}\frac{EI_x}{m_1}\left(1 - \frac{k_1 u_0 l^2}{\pi^2 EI_x}\right), \quad \alpha_1 = \frac{\pi^2 r k_1}{l^2 m_1}, \quad \alpha_2 = \frac{\pi^4 k_1}{4l^3 m_1}, \quad \mu = \frac{c}{2m_1} \tag{5.26.12}$$

Assuming that the crankshaft rotates at a uniform speed, the driving torque is $L(\dot{\varphi})$ and the damping torque is $H(\dot{\varphi})$, we can obtain the governing equation of the rotating part of the crankshaft by applying the moment of momentum

$$I\ddot{\varphi} = L(\dot{\varphi}) - H(\dot{\varphi}) + Nr\cos\varphi = 0 \tag{5.26.13}$$

where I is the moment of inertia of the rotating part of the crankshaft. Substituting (5.26.9) and (5.26.10) into (5.26.13), we obtain

$$I\ddot{\varphi} = L(\dot{\varphi}) - H(\dot{\varphi}) - k_1\left(u_0 + r\sin\varphi - \frac{\pi^2}{4l}\psi^2\right)r\cos\varphi = 0 \tag{5.26.14}$$

Let

$$M_T(\dot{\varphi}) = L(\dot{\varphi}) - H(\dot{\varphi}) \tag{5.26.15}$$

then

$$I\ddot{\varphi} = M_T(\dot{\varphi}) - k_1\left(u_0 + r\sin\varphi - \frac{\pi^2}{4l}\psi^2\right)r\cos\varphi = 0 \tag{5.26.16}$$

It can be seen from (5.26.16) that the excitation term $\sin\varphi$ or $\cos\varphi$ will be affected by the transverse oscillation response ψ of the beam, therefore, the system is a non-ideal energy system.

(b) Assuming that the crankshaft rotates with an angular velocity Ω, where $\varphi = \Omega t$, (5.26.11) becomes:

$$\ddot{\psi} + 2\mu\dot{\psi} + \left(\omega^2 - \alpha_1 \sin\Omega t\right)\psi + \alpha_2\psi^3 = 0 \tag{5.26.17}$$

Dividing both sides of the equation by l yields

$$\frac{\ddot{\psi}}{l}+2\mu\frac{\dot{\psi}}{l}+\left(\omega^2-\alpha_1\sin\Omega t\right)\frac{\psi}{l}+\alpha_2 l^2\left(\frac{\psi}{l}\right)^3=0 \tag{5.26.18}$$

In addition,

$$\begin{aligned}\frac{\alpha_1}{\omega^2}&\approx\frac{\pi^2 r k_1}{l^2 m_1}\frac{l^4 m_1}{\pi^4 EI_x}=\frac{r k_1}{\pi^2\left(\frac{EI_x}{l^2}\right)}=O\left(\frac{N}{EA}\right)\\ \frac{\alpha_2 l^2}{\omega^2}&\approx\frac{\pi^4 k_1}{4 l m_1}\frac{l^4 m_1}{\pi^4 EI_x}=\frac{l k_1}{4\left(\frac{EI_x}{l^2}\right)}=O\left(\frac{N}{EA}\right)\end{aligned} \tag{5.26.19}$$

where A is the cross-sectional area of the beam, EA is the elastic ultimate load of the beam. Since $N << EA$, both of the above ratios in (5.26.19) are small quantities. Meanwhile, we assume that the system is a small damping system, so we can set

$$\mu=\varepsilon\hat{\mu},\quad \alpha_1=\varepsilon\hat{\alpha}_1,\quad \alpha_2 l^2=\varepsilon\hat{\alpha}_2 \tag{5.26.20}$$

Equation (5.26.18) becomes

$$\ddot{v}+2\varepsilon\hat{\mu}\dot{v}+\left(\omega^2-\varepsilon\hat{\alpha}_1\sin\Omega t\right)v+\varepsilon\hat{\alpha}_2 v^3=0,\quad v=\psi/l \tag{5.26.21}$$

Let the solution of the Eq. (5.26.21) is

$$v(t;\varepsilon)=v_0(T_0,T_1,T_2)+\varepsilon v_1(T_0,T_1,T_2)+\varepsilon^2 v_2(T_0,T_1,T_2)+\cdots \tag{5.26.22}$$

Substituting (5.26.22) into (5.26.21) and retaining to $O(\varepsilon)$, we get

$$\begin{aligned}0&=\ddot{v}+2\varepsilon\hat{\mu}\dot{v}+\left(\omega^2-\varepsilon\hat{\alpha}_1\sin\Omega t\right)v+\varepsilon\hat{\alpha}_2 v^3\\ &=\left(D_0^2+2\varepsilon D_0 D_1\right)(v_0+\varepsilon v_1)+\omega^2(v_0+\varepsilon v_1)\\ &\quad+2\varepsilon\hat{\mu}(D_0+\varepsilon D_1)(v_0+\varepsilon v_1)\\ &\quad-\varepsilon\hat{\alpha}_1(v_0+\varepsilon v_1)\sin\Omega T_0+\varepsilon\hat{\alpha}_2(v_0+\varepsilon v_1)^3+\cdots\\ &=D_0^2 v_0+\varepsilon D_0^2 v_1+2\varepsilon D_0 D_1 v_0+\omega^2 v_0+\varepsilon\omega^2 v_1\\ &\quad+2\varepsilon\hat{\mu}D_0 v_0+\varepsilon\hat{\alpha}_2 v_0^3-\varepsilon\hat{\alpha}_1 v_0\sin\Omega T_0+\cdots\\ &=D_0^2 v_0+\omega^2 v_0+\varepsilon(D_0^2 v_1+\omega^2 v_1+2D_0 D_1 v_0\\ &\quad+2\hat{\mu}D_0 v_0+\hat{\alpha}_2 v_0^3-\hat{\alpha}_1 v_0\sin\Omega T_0)+\cdots\end{aligned} \tag{5.26.23}$$

Equating coefficients of like powers of ε in the above equation yields

$$D_0^2 v_0+\omega^2 v_0=0 \tag{5.26.24}$$

$$D_0^2 v_1+\omega^2 v_1=-2D_0 D_1 v_0-2\hat{\mu}D_0 v_0-\hat{\alpha}_2 v_0^3+\hat{\alpha}_1 v_0\sin\Omega T_0 \tag{5.26.25}$$

The solution of (5.26.24) is

$$v_0 = Ae^{i\omega T_0} + \overline{A}\, e^{-i\omega T_0} \tag{5.26.26}$$

where $A = A(T_1)$. Substituting v_0 into (5.26.25) yields

$$\begin{aligned} D_0^2 v_1 + \omega^2 v_1 &= -2D_0 D_1 v_0 - 2\hat{\mu} D_0 v_0 - \hat{\alpha}_2 v_0^3 + \hat{\alpha}_1 v_0 \sin \Omega T_0 \\ &= -2i\omega A' e^{i\omega T_0} - 2i\hat{\mu}\omega A e^{i\omega T_0} - 3\hat{\alpha}_2 A^2 \overline{A} e^{i\omega T_0} \\ &- \frac{1}{2} i\hat{\alpha}_1 \overline{A} e^{i(\Omega-\omega)T_0} + cc + NST \end{aligned} \tag{5.26.27}$$

The prime denotes the derivative with respect to T_1. It can be seen that the resonance occurs when $\Omega \approx 2\omega$. Let

$$\Omega = 2\omega + \varepsilon\sigma \tag{5.26.28}$$

Equation (5.26.27) becomes

$$\begin{aligned} D_0^2 v_1 + \omega^2 v_1 &= -2D_0 D_1 v_0 - 2\hat{\mu} D_0 v_0 - \hat{\alpha}_2 v_0^3 + \hat{\alpha}_1 v_0 \sin \Omega T_0 \\ &= -2i\omega A' e^{i\omega T_0} - 2i\hat{\mu}\omega A e^{i\omega T_0} - 3\hat{\alpha}_2 A^2 \overline{A} e^{i\omega T_0} \\ &- \frac{1}{2} i\hat{\alpha}_1 \overline{A} e^{i\sigma T_1} e^{i\omega T_0} + cc + NST \end{aligned} \tag{5.26.29}$$

In order to eliminate secular terms from the above equation, there must be

$$2i\omega A' + 2i\hat{\mu}\omega A + 3\hat{\alpha}_2 A^2 \overline{A} + \frac{1}{2} i\hat{\alpha}_1 \overline{A} e^{i\sigma T_1} = 0 \tag{5.26.30}$$

Substituting $A = \frac{1}{2} a e^{i\beta}$ into the above equation and taking $\omega_0 \approx 1$ into account, we get

$$i\omega a' - \omega a\beta' + i\hat{\mu}\omega a + \frac{3}{8}\hat{\alpha}_2 a^3 + \frac{1}{4} i\hat{\alpha}_1 a e^{i(\sigma T_1 - 2\beta)} = 0 \tag{5.26.31}$$

Separating the real and imaginary parts of the above equation yields

$$\begin{aligned} a' &= -\hat{\mu} a - \tfrac{1}{4}\omega^{-1}\hat{\alpha}_1 a \cos\gamma \\ a\gamma' &= a\sigma - \tfrac{3}{4}\omega^{-1}\hat{\alpha}_2 a^3 + \tfrac{1}{2}\omega^{-1}\hat{\alpha}_1 a \sin\gamma \end{aligned} \tag{5.26.32}$$

where

$$\gamma = \sigma T_1 - 2\beta \tag{5.26.33}$$

by substituting(5.26.26) into (5.26.22), we can obtain the first-order approximate solution of the system

$$v \approx v_0 = a\cos(\omega t + \beta) + O(\varepsilon)$$

$$\psi = lv \approx lv_0 = la\cos(\omega t + \beta) + O(\varepsilon) \tag{5.26.34}$$

where a and β are given by(5.26.32) and (5.26.33).

Let the non-trivial steady state solution of (5.26.32) be a_0 and γ_0 and superimpose a little perturbation

$$a = a_0 + a_1, \quad \gamma = \gamma_0 + \gamma_1 \tag{5.26.35}$$

Substituting (5.26.35) into (5.26.32), we can obtain the corresponding linearized equation

$$\begin{Bmatrix} a_1' \\ \gamma_1' \end{Bmatrix} = \begin{bmatrix} -(\hat{\mu} + \frac{1}{4}\omega^{-1}\hat{\alpha}_1\cos\gamma_0) & \frac{1}{4}\omega^{-1}\hat{\alpha}_1 a_0 \sin\gamma_0 \\ a_0^{-1}\sigma - \frac{9}{4}\omega^{-1}\hat{\alpha}_2 a_0 + \frac{1}{2}a_0^{-1}\omega^{-1}\hat{\alpha}_1\sin\gamma_0 & \frac{1}{2}\omega^{-1}\hat{\alpha}_1\cos\gamma_0 \end{bmatrix} \begin{Bmatrix} a_1 \\ \gamma_1 \end{Bmatrix} \tag{5.26.36}$$

For a given set of non-trivial steady state solution a_0 and γ_0, the eigenvalues of (5.26.36) can be found, and the stability of the steady state solution can be determined from the sign of the real part of the eigenvalues.

5.27 Exercise 5.27 (Analysis of Two-Dimensional Sound Propagation in a Pipe with Sinusoidal Walls)

$$\nabla^2\varphi + \omega^2\varphi = 0 \tag{5.27.1}$$

$$\begin{aligned} \frac{\partial\varphi}{\partial y} &= \varepsilon\frac{\partial\varphi}{\partial x}k_w\cos k_w x \text{ at } y = \varepsilon\sin k_w x \\ \frac{\partial\varphi}{\partial y} &= 0 \qquad \text{at} \qquad y = 1 \end{aligned} \tag{5.27.2}$$

$$\begin{aligned} \frac{\partial^2\varphi_1}{\partial y^2} + \frac{\partial^2\varphi_1}{\partial x_0^2} + \omega^2\varphi_1 &= -2ik_m A_m'\cos(m\pi y)e^{ik_m x_0} \\ &\quad - 2ik_n A_n'(x_1)\cos(n\pi y)e^{ik_n x_0} \end{aligned} \tag{5.27.3}$$

$$\begin{aligned} \frac{\partial\varphi_1}{\partial y} &= \frac{1}{2}i\sum_{j=m,\,n} A_j\left(k_j k_w - j^2\pi^2\right)e^{i(k_j + k_w)x_0} \\ &\quad + \frac{1}{2}i\sum_{j=m,\,n} A_j\left(k_j k_w + j^2\pi^2\right)e^{i(k_j - k_w)x_0}, \quad \text{at} \quad y = 0 \end{aligned} \tag{5.27.4}$$

Fig. 5.2 Pipes with sinusoidal walls on one side in Exercise 5.27 and 5.28

$$\frac{\partial \varphi_1}{\partial y} = 0 \quad \text{at} \quad y = 1 \tag{5.27.5}$$

$$\begin{aligned} A'_m &= \tfrac{1}{2} k_m^{-1}\left(k_n k_w + n^2\pi^2\right) A_n e^{-i\sigma x_1} \\ A'_n &= \tfrac{1}{2} k_n^{-1}\left(k_m k_w - m^2\pi^2\right) A_m e^{i\sigma x_1} \end{aligned} \tag{5.27.6}$$

Solution: (a) This is a two-dimensional acoustic wave propagation problem along a pipe with a sinusoidal wave-shaped wall with minor undulations on one side, as shown in the Fig. 5.2.

If both sides of the pipe are smooth, the problem becomes

$$\begin{aligned} &\nabla^2\varphi + \omega^2\varphi = 0 \\ &\tfrac{\partial \varphi}{\partial y} = 0 \text{ at } y = 0 \\ &\tfrac{\partial \varphi}{\partial y} = 0 \text{ at } y = 1 \end{aligned} \tag{5.27.7}$$

Using the method of separation of variables, we can obtain the solution of (5.27.7)

$$\varphi = A\cos(n\pi y)e^{ik_n x}$$

where

$$k_n^2 = \omega^2 - n^2\pi^2$$

Now let the $y = 1$ side is still a smooth wall and the $y = 0$ side is a sinusoidal wall with a slight undulation, so the solution to the Eq. (5.27.1) is assumed to be

$$\varphi = G(y)e^{ik_n x} + \varepsilon E(x, y)e^{ik_n x} + O\left(\varepsilon^2\right) \tag{5.27.8}$$

Near $y = 0$, $\varphi(x, y)$ can be expanded as

$$\varphi(x, y) = \varphi(x, 0) + y\frac{\partial}{\partial y}\varphi(x, 0) + \frac{1}{2}y^2\frac{\partial^2}{\partial y^2}\varphi(x, 0) + \cdots$$

So the boundary condition (5.27.2) can be approximated as

$$\begin{aligned} \frac{\partial\varphi}{\partial y} &= \varepsilon\frac{\partial\varphi}{\partial x}k_w\cos k_w x - \varepsilon\frac{\partial^2\varphi}{\partial y^2}\sin k_w x \text{ at } y = 0 \\ \frac{\partial\varphi}{\partial y} &= 0 \qquad\qquad\qquad\qquad\qquad\quad \text{at } y = 1 \end{aligned} \tag{5.27.9}$$

Substituting (5.27.8) into the boundary conditions of $y = 0$ side in (5.27.9) and equating coefficients of like powers of ε in the above equation yields

$$\begin{aligned} &G'(y) = 0 \\ &E'(x, y) = \frac{1}{2}\big[G(y)k_n k_w + iG''(y)\big]e^{ik_w x};\ \text{at}\, y = 0 \\ &+\frac{1}{2}\big[G(y)k_n k_w - iG''(y)\big]e^{-ik_w x} \end{aligned} \tag{5.27.10}$$

The prime denotes the derivative with respect to y. It appears that $E(x, y)$ can be expressed as a function of the following form:

$$E(x, y) = H_1(y)e^{ik_w x} + H_2(y)e^{-ik_w x} \tag{5.27.11}$$

Therefore, the solution of (5.27.1) can be assumed to be

$$\begin{aligned} \varphi = G(y)e^{ik_n x} &+ \varepsilon H_1(y)e^{i(k_n+k_w)x} \\ &+ \varepsilon H_2(y)e^{i(k_n-k_w)x} + O\big(\varepsilon^2\big) \end{aligned} \tag{5.27.12}$$

Substituting (5.27.12) into (5.27.1) and the boundary conditions (5.27.9) yields

$$\begin{aligned} 0 &= \Delta^2\phi + \omega^2\phi \\ &= -k_n^2 G e^{ik_n x} - \varepsilon(k_n + k_w)^2 H_1 e^{i(k_n+k_w)x} \\ &\quad - \varepsilon(k_n - k_w)^2 H_2 e^{i(k_n-k_w)x} \\ &\quad + G'' e^{ik_n x} + \varepsilon H_1'' e^{i(k_n+k_w)x} + \varepsilon H_2'' e^{i(k_n-k_w)x} \\ &\quad + \omega^2 G e^{ik_n x} + \varepsilon\omega^2 H_1 e^{i(k_n+k_w)x} + \varepsilon\omega^2 H_2 e^{i(k_n-k_w)x} \\ &= \big[G'' + \big(\omega^2 - k_n^2\big)G\big]e^{ik_n x} \\ &\quad + \varepsilon\{H_1'' + [\omega^2 - \big(k_n + k_w\big)^2\big]H_1\}e^{i(k_n+k_w)x} \\ &\quad + \varepsilon\{H_2'' + [\omega^2 - \big(k_n - k_w\big)^2\big]H_2\}e^{i(k_n-k_w)x} + \cdots \end{aligned} \tag{5.27.13}$$

$$\begin{aligned} 0 &= \frac{\partial\varphi}{\partial y} - \varepsilon\frac{\partial\varphi}{\partial x}k_w\cos k_w x + \varepsilon\frac{\partial^2\varphi}{\partial y^2}\sin k_w x \\ &= G' e^{ik_n x} + \varepsilon H_1' e^{i(k_n+k_w)x} + \varepsilon H_2' e^{i(k_n-k_w)x} \quad aty = 0 \\ &\quad - \frac{1}{2}\varepsilon i k_n k_w G\big(e^{i(k_n+k_w)x} + e^{i(k_n-k_w)x}\big) \\ &\quad - \frac{1}{2}\varepsilon i G''\big(e^{i(k_n+k_w)x} - e^{i(k_n-k_w)x}\big) + \cdots \end{aligned} \tag{5.27.14}$$

$$0 = \frac{\partial \varphi}{\partial y} = G' e^{ik_n x} + \varepsilon H_1' e^{i(k_n + k_w)x} + \varepsilon H_2' e^{i(k_n - k_w)x} \text{ at } y = 1 \tag{5.27.15}$$

Equating coefficients of like powers of ε in (5.27.13) ~(5.27.15) yields.
Order ε^0:

$$G'' + (\omega^2 - k_n^2)G = 0$$
$$\begin{cases} G' = 0, \text{ at } y = 0 \\ G' = 0, \text{ at } y = 1 \end{cases} \tag{5.27.16}$$

Order ε^1:

$$H_1'' + [\omega^2 - (k_n + k_w)^2]H_1 = 0$$
$$\begin{cases} H_1' - \frac{1}{2}i(k_n k_w G + G'') = 0, & \text{at } y = 0 \\ H_1' = 0, & \text{at } y = 1 \end{cases} \tag{5.27.17}$$

$$H_2'' + [\omega^2 - (k_n - k_w)^2]H_2 = 0$$
$$\begin{cases} H_2' - \frac{1}{2}i(k_n k_w G - G'') = 0, & \text{at } y = 0 \\ H_2' = 0, & \text{at } y = 1 \end{cases} \tag{5.27.18}$$

The solution of (5.27.16) is

$$G(y) = A\cos(n\pi y) \tag{5.27.19}$$

where A is the constant of integration and the following relation should also be satisfied:

$$k_n^2 = \omega^2 - n^2\pi^2 \tag{5.27.20}$$

By substituting (5.27.19) into (5.27.17) we can get

$$H_1'' + [\omega^2 - (k_n + k_w)^2]H_1 = 0 \tag{5.27.21}$$

$$\begin{cases} H_1' - \frac{1}{2}i(k_n k_w - n^2\pi^2)A = 0, & \text{at } y = 0 \\ H_1' = 0, & \text{at } y = 1 \end{cases} \tag{5.27.22}$$

The solution of (5.27.21) can be expressed as

$$H_1(y) = B_1 \sin \kappa_1 y + C_1 \cos \kappa_1 y \tag{5.27.23}$$

where

$$\kappa_1^2 = \omega^2 - (k_n + k_w)^2 \tag{5.27.24}$$

Substituting (5.27.23) into the boundary conditions (5.27.22) yields

$$\begin{aligned} &B_1\kappa_1 - \tfrac{1}{2}iA\left(k_nk_w - n^2\pi^2\right) = 0 \\ &B_1\cos\kappa_1 - C_1\sin\kappa_1 = 0 \end{aligned} \tag{5.27.25}$$

$$\begin{aligned} B_1 &= \frac{1}{2}iA\left(k_nk_w - n^2\pi^2\right)(\kappa_1\sin\kappa_1)^{-1}\sin\kappa_1 C_1 \\ &= \frac{1}{2}iA\left(k_nk_w - n^2\pi^2\right)(\kappa_1\sin\kappa_1)^{-1} \end{aligned} \tag{5.27.26}$$

$$\begin{aligned} H_1(y) = \frac{1}{2}iA\left(k_nk_w - n^2\pi^2\right)(\kappa_1\sin\kappa_1)^{-1} \\ \left[\sin\kappa_1\sin\kappa_1 y + \cos\kappa_1\cos\kappa_1 y\right] \end{aligned} \tag{5.27.27}$$

Similarly,

$$\begin{aligned} H_2(y) = \frac{1}{2}iA\left(k_nk_w + n^2\pi^2\right)(\kappa_2\sin\kappa_2)^{-1} \\ \left[\sin\kappa_2\sin\kappa_2 y + \cos\kappa_2\cos\kappa_2 y\right] \end{aligned} \tag{5.27.28}$$

where

$$\kappa_2^2 = \omega^2 - (k_n - k_w)^2 \tag{5.27.29}$$

Substituting the above results into (5.27.12), we can obtain an approximate solution to (5.27.12)

$$\begin{aligned} \phi = A\cos(n\pi y)e^{ik_nx} + \frac{1}{2}i\varepsilon A\{\left(k_nk_w - n^2\pi^2\right)\Phi_1(y)e^{i(k_n+k_w)x} \\ + \left(k_nk_w + n^2\pi^2\right)\Phi_2(y)e^{i(k_n-k_w)x}\} + O\left(\varepsilon^2\right) \end{aligned} \tag{5.27.30}$$

where

$$\Phi_m(y) = (\kappa_m\sin\kappa_m)^{-1}\left[\sin\kappa_m\sin\kappa_m y + \cos\kappa_m\cos\kappa_m y\right] \tag{5.27.31}$$

When $\kappa_m - m\pi = O(\varepsilon)$ and $(\kappa_m\sin\kappa_m)^{-1} = O\left(\varepsilon^{-1}\right)$, $\Phi_m(y)$ is large, so the second term on the right-hand side of (5.27.30) cannot be corrected and (5.27.30) is invalid.

From (5.27.24), (5.27.29) and $\kappa_m - m\pi = O(\varepsilon)$, we can obtain

$$(k_n \pm k_w)^2 = \omega^2 - \kappa_m^2 = \omega^2 - m^2\pi^2 + O\left(\varepsilon^2\right) = k_m^2 + O\left(\varepsilon^2\right)$$

therefore,

$$k_w = k_n \pm k_m + O(\varepsilon) \tag{5.27.32}$$

That is, the condition $\kappa_m - m\pi = O(\varepsilon)$ is equivalent to (5.27.32), or when the condition (5.27.32) holds, (5.27.30) fails.

(b) When $k_w = k_n - k_m + \varepsilon\sigma$, condition (5.27.32) holds and (5.27.30) fails. Now we use the method of multiscale to solve (5.27.1) and (5.27.2). To do this, we let

$$\begin{aligned}\varphi(x, y) &= \varphi_0(x_0, x_1, y) + \varepsilon\varphi_1(x_0, x_1, y) + \cdots, \\ x_0 &= x, \quad x_1 = \varepsilon x\end{aligned} \tag{5.27.33}$$

Substituting (5.27.33) into (5.27.1) and (5.27.9), and equating coefficients of like powers of ε yields

$$\begin{aligned}0 &= \nabla^2\phi + \omega^2\phi \\ &= \left(D_0^2 + 2\varepsilon D_0 D_1\right)(\phi_0 + \varepsilon\phi_1) \\ &\quad + \partial^2(\phi_0 + \varepsilon\phi_1)/\partial y^2 + \omega^2(\phi_0 + \varepsilon\phi_1) \\ &= D_0^2\phi_0 + \varepsilon D_0^2\phi_1 + 2\varepsilon D_0 D_1\phi_0 + \partial^2\phi_0/\partial y^2 \\ &\quad + \varepsilon\partial^2\phi_1/\partial y^2 + \omega^2\phi_0 + \varepsilon\omega^2\phi_1 \\ &= D_0^2\phi_0 + \omega^2\phi_0 + \partial^2\phi_0/\partial y^2 \\ &\quad + \varepsilon\left(D_0^2\phi_1 + \omega^2\phi_1 + 2D_0D_1\phi_0 + \partial^2\phi_1/\partial y^2\right)\end{aligned} \tag{5.27.34}$$

$$\begin{aligned}0 &= \tfrac{\partial\phi}{\partial y} - \varepsilon\tfrac{\partial\phi}{\partial x}k_w\text{cos}k_w x_0 + \varepsilon\tfrac{\partial^2\phi}{\partial y^2}\text{sin}k_w x_0 \\ &= \tfrac{\partial\phi_0}{\partial x} + \varepsilon\left(\tfrac{\partial\phi_1}{\partial y} - k_w D_0\phi_0\text{cos}k_w x_0 + \tfrac{\partial^2\phi_0}{\partial y^2}\text{sin}k_w x_0\right), \text{at} y = 0\end{aligned} \tag{5.27.35}$$

$$0 = \frac{\partial\varphi_0}{\partial y} + \varepsilon\frac{\partial\varphi_1}{\partial y}_1, \quad \text{at} \quad y = 1 \tag{5.27.36}$$

where D_0 and D_1 denote the partial derivatives with x_0 and x_1, respectively. Equating coefficients of like powers of ε in (5.27.34–5.27.36) yields

Order ε^0: $\frac{\partial^2\varphi_0}{\partial x_0^2} + \frac{\partial^2\varphi_0}{\partial y^2} + \omega^2\varphi_0 = 0$

$$\begin{aligned}\partial\varphi_0/\partial y &= 0 \text{ at } \quad y = 0 \\ \partial\varphi_0/\partial y &= 0 \text{ at } \quad y = 1\end{aligned} \tag{5.27.37}$$

Order ε^1: $\frac{\partial^2\varphi_1}{\partial x_0^2} + \frac{\partial^2\varphi_1}{\partial y^2} + \omega^2\varphi_1 = -2\frac{\partial^2\varphi_0}{\partial x_0\partial x_1}$

$$\begin{aligned}\tfrac{\partial\varphi_1}{\partial y} &= k_w\tfrac{\partial\varphi_0}{\partial x_0}\cos k_w x_0 - \tfrac{\partial^2\varphi_0}{\partial y^2}\sin k_w x_0 \quad \text{at} \quad y = 0 \\ \tfrac{\partial\varphi_1}{\partial y} &= 0 \quad \text{at} \quad y = 1\end{aligned} \tag{5.27.38}$$

The boundary value problem described by (5.27.37) is the same as (5.27.7). If two modes are taken, the solution is

$$\varphi_0 = A_m(x_1)\cos(m\pi y)e^{ik_m x_0} + A_n(x_1)\cos(n\pi y)e^{ik_n x_0} \tag{5.27.39}$$

where

$$k_m^2 = \omega^2 - m^2\pi^2, \quad k_n^2 = \omega^2 - n^2\pi^2 \tag{5.27.40}$$

Substitute (5.27.39) into (5.27.38), we can obtain the controlling equation of φ_1:

$$\frac{\partial^2 \varphi_1}{\partial x_0^2} + \frac{\partial^2 \varphi_1}{\partial y^2} + \omega^2 \varphi_1 = -2ik_m A_m' \cos(m\pi y)e^{ik_m x_0} - 2ik_n A_n' \cos(n\pi y)e^{ik_n x_0} \tag{5.27.41}$$

$$\begin{aligned} \frac{\partial \varphi_1}{\partial y} &= \frac{1}{2}i \sum_{j=m,\,n} A_j\left(k_j k_w - j^2\pi^2\right)e^{i\left(k_j + k_w\right)x_0} \\ &+ \frac{1}{2}i \sum_{j=m,\,n} A_j\left(k_j k_w + j^2\pi^2\right)e^{i\left(k_j - k_w\right)x_0}, \quad \text{al} \quad y = 0 \end{aligned} \tag{5.27.42}$$

$$\frac{\partial \varphi_1}{\partial y} = 0, \quad \text{at} \quad y = 1 \tag{5.27.43}$$

The prime denotes the derivative with respect to x_1.

It is easy to see that the excitation terms on the right side of Eq. (5.27.41) are the solutions of the corresponding homogeneous equations. So $\cos(m\pi y)$ and $\cos(n\pi y)$ cannot be involved in the solution of φ_1. Let

$$\varphi_1 = B_m(x_1, y)e^{ik_m x_0} + B_n(x_1, y)e^{ik_n x_0} \tag{5.27.44}$$

Substitute (5.27.44) into (5.27.41) yields

$$\frac{\partial^2 B_m}{\partial y^2} + m^2\pi^2 B_m = -2ik_m A_m' \cos(m\pi y) \tag{5.27.45}$$

$$\frac{\partial^2 B_n}{\partial y^2} + n^2\pi^2 B_m = -2ik_n A_n' \cos(n\pi y) \tag{5.27.46}$$

Substitute (5.27.44) into (5.27.42) yields

$$\begin{aligned} \frac{\partial B_m}{\partial y}e^{ik_m x_0} + \frac{\partial B_n}{\partial y}e^{ik_n x_0} &= \frac{1}{2}i \sum_{j=m,\,n} A_j\left(k_j k_w - j^2\pi^2\right)e^{i\left(k_j + k_w\right)x_0} \\ &+ \frac{1}{2}i \sum_{j=m,\,n} A_j\left(k_j k_w + j^2\pi^2\right)e^{i\left(k_j - k_w\right)x_0} \\ &= \frac{1}{2}iA_m\left(k_m k_w - m^2\pi^2\right)e^{i\sigma x_1}e^{ik_n x_0} \\ &+ \frac{1}{2}iA_n\left(k_n k_w - j^2\pi^2\right)e^{i(k_m + k_w)x_0} \end{aligned}$$

$$+\frac{1}{2}iA_m\left(k_mk_w+m^2\pi^2\right)e^{i(k_m-k_w)x_0}$$
$$+\frac{1}{2}iA_n\left(k_nk_w+n^2\pi^2\right)e^{-i\sigma x_1}e^{ik_mx_0},\ \text{at } y=0$$

Making the coefficients of the same harmonic function on both sides of the above equation equal, we get

$$\frac{\partial B_m}{\partial y}=\frac{1}{2}iA_n\left(k_nk_w+n^2\pi^2\right)e^{-i\sigma x_1},\qquad \text{at}\quad y=0 \tag{5.27.47}$$

$$\frac{\partial B_n}{\partial y}=\frac{1}{2}iA_m\left(k_mk_w-m^2\pi^2\right)e^{i\sigma x_1},\qquad \text{at}\quad y=0 \tag{5.27.48}$$

Substitute (5.27.44) into (5.27.43), we have

$$\frac{\partial B_m}{\partial y}=0,\qquad \text{at}\quad y=1 \tag{5.27.49}$$

$$\frac{\partial B_n}{\partial y}=0,\qquad \text{at}\quad y=1 \tag{5.27.50}$$

Since the solution of φ_1 cannot involve terms like $\cos(m\pi y)$ and $\cos(n\pi y)$, neither B_m nor B_n can contain such terms. So we have

$$\int_0^1 B_m\cos(m\pi y)dy=0,\quad \int_0^1 B_n\cos(n\pi y)dy=0 \tag{5.27.51}$$

Multiplying both sides of Eq. (5.27.45) by $\cos(m\pi y)$ and making the integration of y over the interval from 0 to 1, we can obtain

$$\frac{\partial B_m}{\partial y}|y=1-\frac{\partial B_m}{\partial y}|y=0=-2ik_mA'_m\int_0^1\cos^2(m\pi y)dy=-ik_mA'_m \tag{5.27.52}$$

Taking the boundary conditions (5.27.47) and (5.27.49) into account, we have

$$A'_m=\frac{1}{2}k_m^{-1}\left(k_nk_w+n^2\pi^2\right)A_ne^{-i\sigma x_1} \tag{5.27.53}$$

Multiplying both sides of Eq. (5.27.46) by $\cos(n\pi y)$, making the integration of y over the interval from 0 to 1, and taking the boundary conditions (5.27.48) and (5.27.50) into account, we have

$$A'_n=\frac{1}{2}k_n^{-1}\left(k_mk_w-m^2\pi^2\right)A_me^{i\sigma x_1} \tag{5.27.54}$$

(c) Let the solutions of Eqs. (5.27.53) and (5.27.54) be

$$A_m = F_m(x_1)e^{sx_1}, \quad A_n = F_n(x_1)e^{sx_1} \tag{5.27.55}$$

Substituting them into (5.27.53) and (5.27.54), we obtain

$$F'_m + sF_m = \frac{1}{2}k_m^{-1}\left(k_n k_w + n^2\pi^2\right)F_n e^{-i\sigma x_1} \tag{5.27.56}$$

$$F'_n + sF_n = \frac{1}{2}k_n^{-1}\left(k_m k_w - m^2\pi^2\right)F_m e^{i\sigma x_1} \tag{5.27.57}$$

Let $F_m = a_m$ as a constant, and then, by Eq. (5.27.57), we can obtain that $F_n = a_n e^{i\sigma x_1}$. Therefore, by (5.27.55), (5.27.53) and (5.27.54), at least, have the following solutions:

$$A_m = a_m e^{sx_1}, \quad A_n = a_n e^{(s+i\sigma)x_1} \tag{5.27.58}$$

then (5.27.56) and (5.27.57) become

$$\begin{bmatrix} s & -\frac{1}{2}k_m^{-1}\left(k_n k_w + n^2\pi^2\right) \\ -\frac{1}{2}k_n^{-1}\left(k_m k_w - m^2\pi^2\right) & i\sigma + s \end{bmatrix}\begin{Bmatrix} a_m \\ a_n \end{Bmatrix} = 0 \tag{5.27.59}$$

From the non-trivial solution condition, we have

$$\det\begin{bmatrix} s & -\frac{1}{2}k_m^{-1}\left(k_n k_w + n^2\pi^2\right) \\ -\frac{1}{2}k_n^{-1}\left(k_m k_w - m^2\pi^2\right) & i\sigma + s \end{bmatrix} = 0$$

i.e.,

$$s^2 + i\sigma s - \frac{1}{4}k_m^{-1}k_n^{-1}\left(k_n k_w + n^2\pi^2\right)\left(k_m k_w - m^2\pi^2\right) = 0 \tag{5.27.60}$$

therefore,

$$s = \frac{1}{2}\left[-i\sigma \pm \sqrt{-\sigma^2 + k_m^{-1}k_n^{-1}\left(k_m k_w - m^2\pi^2\right)\left(k_n k_w + n^2\pi^2\right)}\right] \tag{5.27.61}$$

When $k_m^{-1}k_n^{-1} > 0$ and if $k_m > 0$and $k_n > 0$, since

$$k_m k_w - m^2\pi^2 = k_m(k_n - k_m + \varepsilon\sigma) - m^2\pi^2 = k_m k_n - \omega^2 + \varepsilon\sigma k_m$$

$$k_n k_m - \omega^2 = -\frac{m^2 k_m^2 + \pi^2 n^2 \omega^2}{k_n k_m + \omega^2} < 0$$

$$\omega^2 - k_n k_m = \frac{m^2 k_m^2 + \pi^2 n^2 \omega^2}{k_n k_m + \omega^2} > 0$$

i.e., $k_m k_w - m^2\pi^2 < 0$

If $k_m < 0$and $k_n < 0$, since

$$k_n k_w + n^2\pi^2 = k_n(k_n - k_m + \varepsilon\sigma) + n^2\pi^2 = \omega^2 - k_m k_n + \varepsilon\sigma k_n$$

i.e., $k_n k_w + n^2\pi^2 > 0$

Anyway

$$k_m^{-1} k_n^{-1}\left(k_m k_w - m^2\pi^2\right)\left(k_n k_w + n^2\pi^2\right) < 0 \tag{5.27.62}$$

It follows that when $k_m^{-1} k_n^{-1} > 0$, s is a pure imaginary root.

When $k_m^{-1} k_n^{-1} < 0$,

$$k_m^{-1} k_n^{-1}\left(k_m k_w - m^2\pi^2\right)\left(k_n k_w + n^2\pi^2\right) > 0 \tag{5.27.63}$$

Thus

$$s = -\frac{1}{2} i\sigma \pm p \tag{5.27.64}$$

where

$$p = \frac{1}{2}\sqrt{k_m^{-1} k_n^{-1}\left(k_m k_w - m^2\pi^2\right)\left(k_n k_w + n^2\pi^2\right) - \sigma^2} > 0 \tag{5.27.65}$$

and then

$$A_m = \left(a_{m1} e^{-p x_1} + a_{m2} e^{p x_1}\right) e^{-i\sigma x_1/2}, \quad A_n = \left(a_{n1} e^{-p x_1} + a_{n2} e^{-p x_1}\right) e^{i\sigma x_1/2} \tag{5.27.66}$$

$e^{-p x_1}$ $e^{p x_1}$quickly diverges, and such divergent acoustic propagation modes cannot occur in linear acoustic systems. Therefore, when $k_m^{-1} k_n^{-1} < 0$, the corresponding acoustic modes cannot propagate, or propagation is cut off.

5.28 Exercise 5.28 (Analysis of Two-Dimensional Electromagnetic Wave Propagation in a Pipe with Sinusoidal Walls)

$$\nabla^2 \varphi + \omega^2 \varphi = 0 \tag{5.28.1}$$

$$\begin{aligned} \frac{\partial^2 \varphi}{\partial x^2} + \omega^2 \varphi &= -\varepsilon k_w \frac{\partial^2 \varphi}{\partial x \partial y} \cos k_w x \text{ at } \quad y = \varepsilon \sin k_w x \\ \frac{\partial^2 \varphi}{\partial x^2} + \omega^2 \varphi &= 0 \qquad\qquad\qquad\qquad \text{at} \quad y = 1 \end{aligned} \tag{5.28.2}$$

$$\begin{aligned} &\frac{\partial^2 \phi_1}{\partial x_0^2} + \frac{\partial^2 \phi_1}{\partial y^2} + \omega^2 \phi_1 = -2ik_m A_m' \sin(m\pi y) e^{ik_m x_0} \\ &-2ik_n A_n'(x_1) \sin(n\pi y) e^{ik_n x_0} \end{aligned} \tag{5.28.3}$$

$$\begin{aligned} \frac{\partial^2 \varphi_1}{\partial x_0^2} + \omega^2 \varphi_1 = \tfrac{1}{2} i\pi &\sum_{j=m,\,n} j\left(j^2\pi^2 - k_j k_w\right) A_j e^{i(k_j + k_w)x_0} \\ -\tfrac{1}{2} i\pi \sum_{j=m,\,n} &j\left(j^2\pi^2 + k_j k_w\right) A_j e^{i(k_j - k_w)x_0}, \qquad \text{at} \quad y = 0 \end{aligned} \tag{5.28.4}$$

$$\frac{\partial^2 \varphi_1}{\partial x_0^2} + \omega^2 \varphi_1 = 0 \quad \text{at} \quad y = 1 \tag{5.28.5}$$

$$\begin{aligned} A_m' &= \frac{1}{2} \frac{n}{m k_m} \left(k_n k_w + n^2\pi^2\right) A_n e^{-i\sigma x_1} \\ A_n' &= \frac{1}{2} \frac{m}{n k_n} \left(k_m k_w - m^2\pi^2\right) A_m e^{i\sigma x_1} \end{aligned} \tag{5.28.6}$$

Solution: (a) This is a two-dimensional electromagnetic wave propagation problem along a pipe with a sinusoidal shaped wall with minor undulations on one side as shown Fig. 5.2

If both sides of the pipe have smooth walls, the problem becomes

$$\nabla^2 \varphi + \omega^2 \varphi = 0$$

$$\begin{aligned} \frac{\partial^2 \varphi}{\partial^2 x} + \omega^2 \varphi = 0 \text{ at } y = 0 \\ \frac{\partial^2 \varphi}{\partial^2 x} + \omega^2 \varphi = 0 \text{ at } y = 1 \end{aligned} \tag{5.28.7}$$

Using the method of separation of variables, we can obtain the solution of (5.28.1)

$$\varphi = A \sin(n\pi y) e^{ik_n x}$$

where

$$k_n^2 = \omega^2 - n^2\pi^2$$

Now let the $y = 1$ side is still a smooth wall and the $y = 0$ side is a sinusoidal wall with a slight undulation, so the solution to the Eq. (5.28.1) with boundary condition (5.28.2) is assumed to be

$$\varphi = G(y) e^{ik_n x} + \varepsilon E(x, y) e^{ik_n x} + O\left(\varepsilon^2\right) \tag{5.28.8}$$

Near $y = 0$, $\varphi(x, y)$ can be expanded as

$$\varphi(x,y) = \varphi(x,0) + y\frac{\partial}{\partial y}\varphi(x,0) + \frac{1}{2}y^2\frac{\partial^2}{\partial y^2}\varphi(x,0) + \cdots$$

Therefore, the boundary condition (5.28.2) can be approximated as

$$\begin{aligned} \frac{\partial^2\varphi}{\partial x^2} + \omega^2\varphi &= -\varepsilon k_w \frac{\partial^2\varphi}{\partial x\partial y}\cos k_w x - \varepsilon\left(\frac{\partial^3\varphi}{\partial x^2\partial y} + \omega^2\frac{\partial\varphi}{\partial y}\right)\sin k_w x \text{ at } \quad y = 0 \\ \frac{\partial^2\varphi}{\partial x^2} + \omega^2\varphi &= 0 \qquad\qquad\qquad\qquad\qquad\qquad\qquad\qquad \text{at } \quad y = 1 \end{aligned} \tag{5.28.9}$$

Substituting (5.28.8) into the boundary conditions of $y = 0$ side in (5.28.9) and equating coefficients of like powers of ε in the above equation yields

$$\begin{aligned} G &= 0 \\ \frac{\partial^2 E}{\partial x^2} + \left(\omega^2 - k_n^2\right)E &= i\left[\tfrac{1}{2}k_n k_w - \left(\omega^2 - k_n^2\right)\right]G' e^{ik_w x} \ ; \ \text{at}\, y = 0 \\ &\quad + i\left[\tfrac{1}{2}k_n k_w + \left(\omega^2 - k_n^2\right)\right]G' e^{ik_w x} \end{aligned} \tag{5.28.10}$$

The prime denotes the derivative with respect to y. It appears that $E(x, y)$ can be expressed as a function of the following form:

$$E(x,y) = H_1(y)e^{ik_w x} + H_2(y)e^{-ik_w x} \tag{5.28.11}$$

Therefore, the solution of (5.28.1) can be assumed to be

$$\begin{aligned} \varphi &= G(y)e^{ik_n x} + \varepsilon H_1(y)e^{i(k_n+k_w)x} \\ &\quad + \varepsilon H_2(y)e^{i(k_n-k_w)x} + O\left(\varepsilon^2\right) \end{aligned} \tag{5.28.12}$$

Substituting (5.28.12) into (5.28.1) and the boundary conditions (5.28.9) yields

$$\begin{aligned} 0 &= \Delta^2\phi + \omega^2\phi \\ &= -k_n^2 G e^{ik_n x} - \varepsilon(k_n + k_w)^2 H_1 e^{i(k_n+k_w)x} \\ &\quad - \varepsilon(k_n - k_w)^2 H_2 e^{i(k_n-k_w)x} \\ &\quad + G'' e^{ik_n x} + \varepsilon H_1'' e^{i(k_n+k_w)x} + \varepsilon H_2'' e^{i(k_n-k_w)x} \\ &\quad + \omega^2 G e^{ik_n x} + \varepsilon\omega^2 H_1 e^{i(k_n+k_w)x} + \varepsilon\omega^2 H_2 e^{i(k_n-k_w)x} \\ &= \left[G'' + \left(\omega^2 - k_n^2\right)G\right]e^{ik_n x} \\ &\quad + \varepsilon\{H_1'' + [\omega^2 - \left(k_n + k_w\right)^2]H_1\}e^{i(k_n+k_w)x} \\ &\quad + \varepsilon\{H_2'' + [\omega^2 - \left(k_n - k_w\right)^2]H_2\}e^{i(k_n-k_w)x} + \cdots \end{aligned} \tag{5.28.13}$$

$$\begin{aligned} 0 &= \frac{\partial^2\varphi}{\partial x^2} + \omega^2\varphi + \varepsilon k_w \frac{\partial^2\varphi}{\partial x\partial y}\cos k_w x \\ &\quad + \varepsilon\left(\frac{\partial^3\varphi}{\partial x^2\partial y} + \omega^2\frac{\partial\varphi}{\partial y}\right)\sin k_w x \end{aligned}$$

$$\begin{aligned}
&= -k_n^2 G e^{ik_n x} - \varepsilon (k_n + k_w)^2 H_1 e^{i(k_n+k_w)x} \\
&\quad - \varepsilon (k_n - k_w)^2 H_2 e^{i(k_n-k_w)x} \\
&\quad + \omega^2 G e^{ik_n x} + \varepsilon \omega^2 H_1 e^{i(k_n+k_w)x} \\
&\quad + \varepsilon \omega^2 H_2 e^{i(k_n-k_w)x} \quad aty = 0 \\
&\quad + \frac{1}{2}\varepsilon i k_n k_w G' \left(e^{i(k_n+k_w)x} + e^{i(k_n-k_w)x}\right) \\
&\quad - \frac{1}{2}\varepsilon i n^2 \pi^2 G' \left(e^{i(k_n+k_w)x} - e^{i(k_n-k_w)x}\right)
\end{aligned} \tag{5.28.14}$$

$$\begin{aligned}
0 &= \frac{\partial^2 \varphi}{\partial x^2} + \omega^2 \varphi \\
&\quad + \varepsilon \left(\frac{\partial^3 \varphi}{\partial x^2 \partial y} + \omega^2 \frac{\partial \varphi}{\partial y} \right) \sin k_w x \\
&= -k_n^2 G e^{ik_n x} - \varepsilon (k_n + k_w)^2 H_1 e^{i(k_n+k_w)x} \\
&\quad - \varepsilon (k_n - k_w)^2 H_2 e^{i(k_n-k_w)x} \quad aty = 1 \\
&\quad + \omega^2 G e^{ik_n x} + \varepsilon \omega^2 H_1 e^{i(k_n+k_w)x} + \varepsilon \omega^2 H_2 e^{i(k_n-k_w)x}
\end{aligned} \tag{5.28.15}$$

Equating coefficients of like powers of ε in (5.28.13–5.28.14) yields

$$\begin{aligned}
0 &= \frac{\partial^2 \varphi}{\partial x^2} + \omega^2 \varphi \\
&\quad + \varepsilon \left(\frac{\partial^3 \varphi}{\partial x^2 \partial y} + \omega^2 \frac{\partial \varphi}{\partial y} \right) \sin k_w x \\
&= -k_n^2 G e^{ik_n x} - \varepsilon (k_n + k_w)^2 H_1 e^{i(k_n+k_w)x} \\
&\quad - \varepsilon (k_n - k_w)^2 H_2 e^{i(k_n-k_w)x} \quad at \quad y = 1 \\
&\quad + \omega^2 G e^{ik_n x} + \varepsilon \omega^2 H_1 e^{i(k_n+k_w)x} + \varepsilon \omega^2 H_2 e^{i(k_n-k_w)x}
\end{aligned}$$

Order ε^0:

$$G'' + \left(\omega^2 - k_n^2\right) G = 0$$
$$\begin{cases} G = 0, \text{ at } y = 0 \\ G = 0, \text{ at } y = 1 \end{cases} \tag{5.28.16}$$

Order ε^1:

$$H_1'' + \kappa_1^2 H_1 = 0$$
$$\begin{cases} \kappa_1^2 H_1 = \frac{1}{2} i \left(n^2 \pi^2 - k_n k_w\right) G', & \text{at } y = 0 \\ H_1 = 0, & \text{at } y = 1 \end{cases} \tag{5.28.17}$$

$$H_2'' + \kappa_2^2 H_2 = 0$$

$$\begin{cases} \kappa_2^2 H_2 = -\frac{1}{2}i\left(n^2\pi^2 + k_n k_w\right)G', & \text{at } y = 0 \\ H_2 = 0, & \text{at } y = 1 \end{cases} \quad (5.28.18)$$

where

$$\kappa_1^2 = \omega^2 - (k_n + k_w)^2, \quad \kappa_2^2 = \omega^2 - (k_n - k_w)^2 \quad (5.28.19)$$

The solution of (5.28.16) is

$$G(y) = A\sin(n\pi y) \quad (5.28.20)$$

where A is the constant of integration and the following relation should also be satisfied

$$k_n^2 = \omega^2 - n^2\pi^2 \quad (5.28.21)$$

By substituting (5.28.20) into (5.28.17) we can get

$$H_1'' + \kappa_1^2 H_1 = 0 \quad (5.28.22)$$

$$\begin{cases} \kappa_1^2 H_1 = \frac{1}{2}in\pi\left(n^2\pi^2 - k_n k_w\right)A, & \text{at } y = 0 \\ H_1 = 0, & \text{at } y = 1 \end{cases} \quad (5.28.23)$$

The solution of (5.28.22) can be expressed as

$$H_1(y) = B_1 \sin\kappa_1 y + C_1 \cos\kappa_1 y \quad (5.28.24)$$

Substituting (5.28.24) into the boundary conditions (5.28.23) yields

$$\kappa_1^2 C_1 = \frac{1}{2}in\pi\left(n^2\pi^2 - k_n k_w\right)A$$

$$B_1 \sin\kappa_1 + C_1 \cos\kappa_1 = 0 \quad (5.28.25)$$

Thus we can obtain

$$C_1 = \frac{1}{2}in\pi(\kappa_1^{-2}\sin\kappa_1)^{-1}A\left(n^2\pi^2 - k_n k_w\right)\sin\kappa_1$$

$$B_1 = -\frac{1}{2}in\pi(\kappa_1^{-2}\sin\kappa_1)^{-1}A\left(n^2\pi^2 - k_n k_w\right)\cos\kappa_1 \quad (5.28.26)$$

Therefore

$$H_1(y) = \frac{1}{2} in\pi A\left(n^2\pi^2 - k_n k_w\right)\left(\kappa_1^{-2}\sin\kappa_1\right)^{-1}$$
$$(\sin\kappa_1\cos\kappa_1 y - \cos\kappa_1\sin\kappa_1 y) \tag{5.28.27}$$

Similarly,

$$H_2(y) = -\frac{1}{2} in\pi A\left(n^2\pi^2 + k_n k_w\right)\left(\kappa_2^{-2}\sin\kappa_2\right)^{-1}$$
$$(\sin\kappa_2\cos\kappa_1 y - \cos\kappa_2\sin\kappa_2 y) \tag{5.28.28}$$

Substituting the above results into (5.28.12), we can obtain an approximate solution to (5.28.12)

$$\phi = A\sin(n\pi y)e^{ik_n x} + \frac{1}{2}\varepsilon in\pi A\{\left(n^2\pi^2 - k_n k_w\right)\Phi_1(y)e^{i(k_n+k_w)x}$$
$$- \left(n^2\pi^2 + k_n k_w\right)\Phi_2(y)e^{i(k_n-k_w)x}\} + O\left(\varepsilon^2\right) \tag{5.28.29}$$

where

$$\Phi_m(y) = \left(\kappa_m^2\sin\kappa_m\right)^{-1}\left[\sin\kappa_m\cos\kappa_m y - \cos\kappa_m\sin\kappa_m y\right] \tag{5.28.30}$$

When $\kappa_m - m\pi = O(\varepsilon)$ and $(\kappa_m\sin\kappa_m)^{-1} = O\left(\varepsilon^{-1}\right)$, $\Phi_m(y)$ is large, so the second term on the right-hand side of (5.28.29) cannot be corrected and (5.28.29) is invalid.

From (5.28.19), (5.28.21) and $\kappa_m - m\pi = O(\varepsilon)$, we can obtain

$$(k_n \pm k_w)^2 = \omega^2 - \kappa_m^2 = \omega^2 - m^2\pi^2 + O\left(\varepsilon^2\right) = k_m^2 + O\left(\varepsilon^2\right)$$

therefore,

$$k_w = k_n \pm k_m + O(\varepsilon) \tag{5.28.31}$$

That is, the condition $\kappa_m - m\pi = O(\varepsilon)$ is equivalent to the condition (5.28.31), or when the condition (5.28.31) holds, (5.28.29) fails.

(b) When $k_w = k_n - k_m + \varepsilon\sigma$, the expansion (5.28.29) fails. Now we use the method of multiscale to solve (5.28.1) and (5.28.2). To do this, we let

$$\varphi(x, y) = \varphi_0(x_0, x_1, y) + \varepsilon\varphi_1(x_0, x_1, y) + \cdots, \quad x_0 = x, \quad x_1 = \varepsilon x \tag{5.28.32}$$

Substituting (5.28.32) into (5.28.1) and (5.28.9), and equating coefficients of like powers of ε yields

$$\begin{aligned} 0 &= \nabla^2\phi + \omega^2\phi \\ &= \left(D_0^2 + 2\varepsilon D_0 D_1\right)(\phi_0 + \varepsilon\phi_1) \end{aligned}$$

$$
\begin{aligned}
&+\frac{\partial^2}{\partial y^2}(\phi_0+\varepsilon\phi_1)+\omega^2(\phi_0+\varepsilon\phi_1)\\
&=D_0^2\phi_0+\varepsilon D_0^2\phi_1+2\varepsilon D_0D_1\phi_0\\
&+\frac{\partial^2\phi_0}{\partial y^2}+\varepsilon\frac{\partial^2\phi_1}{\partial y^2}+\omega^2\phi_0+\varepsilon\omega^2\phi_1\\
&=D_0^2\phi_0+\omega^2\phi_0+\frac{\partial^2\phi_0}{\partial y^2}\\
&+\varepsilon\left(D_0^2\phi_1+\omega^2\phi_1+2D_0D_1\phi_0+\frac{\partial^2\phi_1}{\partial y^2}\right)
\end{aligned}
\tag{5.28.33}
$$

$$
\begin{aligned}
0&=\frac{\partial^2\phi}{\partial x^2}+\omega^2\phi+\varepsilon k_w\frac{\partial^2\phi}{\partial x\partial y}\cos k_wx\\
&+\varepsilon\left(\frac{\partial^3\phi}{\partial x^2\partial y}+\omega^2\frac{\partial\phi}{\partial y}\right)\sin k_wx\\
&=\left(D_0^2+2\varepsilon D_0D_1\right)(\phi_0+\varepsilon\phi_1)+\omega^2(\phi_0+\varepsilon\phi_1)\\
&+\varepsilon k_w\frac{\partial(D_0\phi_0)}{\partial y}\cos k_wx\\
&+\varepsilon\frac{\partial}{\partial y}\left[\left(D_0^2+2\varepsilon D_0D_1\right)(\phi_0+\varepsilon\phi_1)\right]\sin k_wx\\
&+\varepsilon\omega^2\frac{\partial\phi_0}{\partial y}\sin k_wx\\
&=D_0^2\phi_0+\omega^2\phi_0+\varepsilon\{D_0^2\phi_1+\omega^2\phi_1\\
&+2D_0D_1\phi_0+k_w\frac{\partial(D_0\phi_0)}{\partial y}\cos k_wx\\
&+\frac{\partial\left(D_0^2\phi_0\right)}{\partial y}\sin k_wx\\
&+\omega^2\frac{\partial\phi_0}{\partial y}\sin k_wx\},\ aty=0
\end{aligned}
\tag{5.28.34}
$$

$$
\begin{aligned}
0&=\frac{\partial^2\phi}{\partial x^2}+\omega^2\phi\\
&=D_0^2\phi_0+\omega^2\phi_0+\varepsilon\left(D_0^2\phi_1+\omega^2\phi_1+2D_0D_1\phi_0\right),\ aty=1
\end{aligned}
\tag{5.28.35}
$$

where D_0 and D_1 denote the partial derivatives with x_0 and x_1, respectively. Equating coefficients of like powers of ε in (5.28.33–5.28.35) yields

Order ε^0:

$$\begin{aligned} &\frac{\partial^2\varphi_0}{\partial x_0^2}+\frac{\partial^2\varphi_0}{\partial y^2}+\omega^2\varphi_0=0\\ &\frac{\partial^2\varphi_0}{\partial x_0^2}+\omega^2\varphi_0=0 \text{ at } \ y=0\\ &\frac{\partial^2\varphi_0}{\partial x_0^2}+\omega^2\varphi_0=0 \text{ at } \ y=1 \end{aligned} \tag{5.28.36}$$

Order ε^1: $\frac{\partial^2\varphi_1}{\partial x_0^2}+\frac{\partial^2\varphi_1}{\partial y^2}+\omega^2\varphi_1=-2\frac{\partial^2\varphi_0}{\partial x_0\partial x_1}$

$$\begin{aligned} &\frac{\partial^2\varphi_1}{\partial x_0^2}+\omega^2\varphi_1=-2\frac{\partial^2\varphi_0}{\partial x_0\partial x_1}-k_w\frac{\partial^2\varphi_0}{\partial x_0\partial y}\cos k_w x\\ &-\frac{\partial^3\varphi_0}{\partial x_0^2\partial y}\sin k_w x-\omega^2\frac{\partial\varphi_0}{\partial y}\sin k_w x \qquad \text{at} y=0\\ &\frac{\partial^2\varphi_1}{\partial x_0^2}+\omega^2\varphi_1=-2\frac{\partial^2\varphi_0}{\partial x_0\partial x_1} \qquad \text{at} y=1 \end{aligned} \tag{5.28.37}$$

The boundary value problem described by (5.28.36) is the same as (5.28.7). If two modes are taken, the solution is

$$\varphi_0=A_m(x_1)\sin(m\pi y)e^{ik_m x_0}+A_n(x_1)\sin(n\pi y)e^{ik_n x_0} \tag{5.28.38}$$

where

$$k_m^2=\omega^2-m^2\pi^2,\, k_n^2=\omega^2-n^2\pi^2 \tag{5.28.39}$$

Substitute (5.28.38) into (5.28.37), we can obtain the controlling equation of φ_1:

$$\begin{aligned} &\frac{\partial^2\phi_1}{\partial x_0^2}+\frac{\partial^2\phi_1}{\partial y^2}+\omega^2\phi_1=-2ik_m A'_m\sin(m\pi y)e^{ik_m x_0}\\ &-2ik_n A'_n(x_1)\sin(n\pi y)e^{ik_n x_0} \end{aligned} \tag{5.28.40}$$

$$\begin{aligned} &\frac{\partial^2\varphi_1}{\partial x_0^2}+\omega^2\varphi_1=\frac{1}{2}i\pi\sum\nolimits_{j=m,n} j\big(j^2\pi^2-k_jk_w\big)A_je^{i(k_j+k_w)x_0}\\ &-\frac{1}{2}i\pi\sum\nolimits_{j=m,n} j\big(j^2\pi^2+k_jk_w\big)A_je^{i(k_j-k_w)x_0}, \qquad \text{at} y=0 \end{aligned} \tag{5.28.41}$$

$$\frac{\partial^2\varphi_1}{\partial x_0^2}+\omega^2\varphi_1=0, \text{at}\, y=1 \tag{5.28.42}$$

The prime denotes the derivative with respect to x_1.

It is easy to see that the excitation terms on the right side of Eq. (5.28.40) are the solutions of the corresponding homogeneous equations. So $\sin(m\pi y)$ and $\sin(n\pi y)$ cannot be involved in the solution of φ_1. Let

$$\varphi_1=B_m(x_1,y)e^{ik_m x_0}+B_n(x_1,y)e^{ik_n x_0} \tag{5.28.43}$$

Substitute (5.28.43) into (5.28.40) yields

$$\frac{\partial^2 B_m}{\partial y^2} + m^2\pi^2 B_m = -2ik_m A'_m \sin(m\pi y) \tag{5.28.44}$$

$$\begin{aligned}
&m^2\pi^2 B_m e^{ik_m x_0} + n^2\pi^2 B_n e^{ik_n x_0} \\
&= \frac{1}{2} i\pi \sum_{j=m,n} j\big(j^2\pi^2 - k_j k_w\big) A_j e^{i(k_j+k_w)x_0} \\
&-\frac{1}{2} i\pi \sum_{j=m,n} j\big(j^2\pi^2 + k_j k_w\big) A_j e^{i(k_j-k_w)x_0} \\
&= \frac{1}{2} iA_m m\pi\left(m^2\pi^2 - k_m k_w\right) e^{i\sigma x_1} e^{ik_n x_0} + \frac{1}{2} iA_n n\pi\left(n^2\pi^2 - k_n k_w\right) e^{i(k_n+k_w)x_0} \\
&-\frac{1}{2} iA_m m\pi\left(m^2\pi^2 + k_m k_w\right) e^{i(k_m-k_w)x_0} \\
&-\frac{1}{2} iA_n n\pi\left(n^2\pi^2 + k_n k_w\right) e^{-i\sigma x_1} e^{ik_m x_0}, \; at\; y = 0 \\
&\frac{\partial^2 B_n}{\partial y^2} + n^2\pi^2 B_m = -2ik_n A'_n \sin(n\pi y)
\end{aligned} \tag{5.28.45}$$

Substitute (5.28.43) into (5.28.41) yields

$$\begin{aligned}
&m^2\pi^2 B_m e^{ik_m x_0} + n^2\pi^2 B_n e^{ik_n x_0} \\
&= \frac{1}{2} i\pi \sum_{j=m,n} j\big(j^2\pi^2 - k_j k_w\big) A_j e^{i(k_j+k_w)x_0} \\
&-\frac{1}{2} i\pi \sum_{j=m,n} j\big(j^2\pi^2 + k_j k_w\big) A_j e^{i(k_j-k_w)x_0} \\
&= \frac{1}{2} iA_m m\pi\left(m^2\pi^2 - k_m k_w\right) e^{i\sigma x_1} e^{ik_n x_0} + \frac{1}{2} iA_n n\pi\left(n^2\pi^2 - k_n k_w\right) e^{i(k_n+k_w)x_0} \\
&-\frac{1}{2} iA_m m\pi\left(m^2\pi^2 + k_m k_w\right) e^{i(k_m-k_w)x_0} \\
&-\frac{1}{2} iA_n n\pi\left(n^2\pi^2 + k_n k_w\right) e^{-i\sigma x_1} e^{ik_m x_0}, \; at\; y = 0
\end{aligned}$$

Making the coefficients of the same harmonic function on both sides of the above equation equal, we get

$$m^2\pi^2 B_m = -\frac{1}{2} iA_n n\pi\left(k_n k_w + n^2\pi^2\right) e^{-i\sigma x_1}, \text{ at} y = 0 \tag{5.28.46}$$

$$n^2\pi^2 B_n = -\frac{1}{2} iA_m m\pi\left(k_m k_w - m^2\pi^2\right) e^{i\sigma x_1}, \text{ at} y = 0 \tag{5.28.47}$$

Substitute (5.28.43) into (5.28.42), we have

$$B_m = 0, \text{ at } y = 1 \tag{5.28.48}$$

$$B_n = 0, \text{ at } y = 1 \tag{5.28.49}$$

Since the solution of φ_1 cannot involve terms like $\sin(m\pi y)$ and $\sin(m\pi y)$, neither B_m nor B_n can contain such terms. So we have

$$\int_0^1 B_m \cos(m\pi y) dy = 0, \int_0^1 B_n \cos(n\pi y) dy = 0 \tag{5.28.50}$$

Equation (5.28.44) Multiplying both sides by $\sin(m\pi y)$ and integrating overy at[0, 1] yields

$$-m\pi B_m|y = 1 + m\pi B_m|y = 0 = -2ik_m A'_m \int_0^1 \sin^2(m\pi y) dy = -ik_m A'_m \tag{5.28.51}$$

Consider the boundary conditions (5.28.46) and (5.28.46), we have

$$A'_m = \frac{1}{2} \frac{n}{mk_m} \left(k_n k_w + n^2\pi^2\right) A_n e^{-i\sigma x_1} \tag{5.28.52}$$

Multiplying both sides of Eq. (5.28.45) by $\sin(n\pi y)$, making the integration of y over the interval from 0 to 1, and taking into account the boundary conditions (5.28.47) and (5.28.48), we can obtain

$$A'_n = \frac{1}{2} \frac{m}{nk_n} \left(k_m k_w - m^2\pi^2\right) A_m e^{i\sigma x_1} \tag{5.28.53}$$

(c) Let the solutions of Eqs. (5.28.52) and (5.28.53) be

$$A_m = F_m(x_1) e^{sx_1}, A_n = F_n(x_1) e^{sx_1} \tag{5.28.54}$$

Substituting them into (5.28.53) and (5.28.54), we obtain

$$F'_m + sF_m = \frac{1}{2} \frac{n}{mk_m} \left(k_n k_w + n^2\pi^2\right) F_n e^{-i\sigma x_1} \tag{5.28.55}$$

$$F'_n + sF_n = \frac{1}{2} \frac{m}{nk_n} \left(k_m k_w - m^2\pi^2\right) F_m e^{i\sigma x_1} \tag{5.28.56}$$

Let $F_m = a_m$ as a constant, and then, by Eq. (5.28.56), we can obtain that $F_n = a_n e^{i\sigma x_1}$. Therefore, by (5.28.54), (5.28.52) and (5.28.53), at least, have the following solutions:

$$A_m = a_m e^{sx_1}, A_n = a_n e^{(s+i\sigma)x_1} \tag{5.28.57}$$

Then (5.28.52) and (5.28.53) become

$$\begin{bmatrix} s & -\frac{1}{2}\frac{n}{mk_m}\left(k_n k_w + n^2\pi^2\right) \\ -\frac{1}{2}\frac{m}{nk_n}\left(k_m k_w - m^2\pi^2\right) & i\sigma + s \end{bmatrix} \begin{Bmatrix} a_m \\ a_n \end{Bmatrix} = 0 \tag{5.28.58}$$

From the non-trivial solution condition, we have

$$\det \begin{bmatrix} s & -\frac{1}{2}\frac{n}{mk_m}\left(n^2\pi^2 + k_n k_w\right) \\ -\frac{1}{2}\frac{m}{nk_n}\left(k_m k_w - m^2\pi^2\right) & i\sigma + s \end{bmatrix} = 0$$

i.e.,

$$s^2 + i\sigma s - \frac{1}{4}k_m^{-1}k_n^{-1}\left(k_n k_w + n^2\pi^2\right)\left(k_m k_w - m^2\pi^2\right) = 0 \tag{5.28.59}$$

therefore,

$$s = \frac{1}{2}\left[-i\sigma \pm \sqrt{-\sigma^2 + k_m^{-1}k_n^{-1}\left(k_m k_w - m^2\pi^2\right)\left(k_n k_w + n^2\pi^2\right)}\right] \tag{5.28.60}$$

When $k_m^{-1}k_n^{-1} > 0$ and if $k_m > 0$and $k_n > 0$, since

$$k_m k_w - m^2\pi^2 = k_m(k_n - k_m + \varepsilon\sigma) - m^2\pi^2 = k_m k_n - \omega^2 + \varepsilon\sigma k_m$$

$$k_n k_m - \omega^2 = -\frac{m^2 k_m^2 + \pi^2 n^2 \omega^2}{k_n k_m + \omega^2} < 0$$

$$\omega^2 - k_n k_m = \frac{m^2 k_m^2 + \pi^2 n^2 \omega^2}{k_n k_m + \omega^2} > 0$$

therefore, $k_m k_w - m^2\pi^2 < 0$

If $k_m < 0$ and$k_n < 0$, since

$$k_n k_w + n^2\pi^2 = k_n(k_n - k_m + \varepsilon\sigma) + n^2\pi^2 = \omega^2 - k_m k_n + \varepsilon\sigma k_n$$

therefore, $k_n k_w + n^2\pi^2 > 0$

Anyway

$$k_m^{-1}k_n^{-1}\left(k_m k_w - m^2\pi^2\right)\left(k_n k_w + n^2\pi^2\right) < 0 \tag{5.28.61}$$

It follows that when $k_m^{-1}k_n^{-1} > 0$, s is a pure imaginary root.
When $k_m^{-1}k_n^{-1} < 0$,

$$k_m^{-1}k_n^{-1}\left(k_m k_w - m^2\pi^2\right)\left(k_n k_w + n^2\pi^2\right) > 0 \tag{5.28.62}$$

Thus when $k_m^{-1}k_n^{-1}\left(k_m k_w - m^2\pi^2\right)\left(k_n k_w + n^2\pi^2\right) > \sigma^2$,

$$s = -\frac{1}{2}i\sigma \pm p \tag{5.28.63}$$

where

$$p = \frac{1}{2}\sqrt{k_m^{-1}k_n^{-1}\left(k_m k_w - m^2\pi^2\right)\left(k_n k_w + n^2\pi^2\right) - \sigma^2} > 0 \tag{5.28.64}$$

And then

$$\begin{aligned} A_m &= \left(a_{m1}e^{-px_1} + a_{m2}e^{px_1}\right)e^{-i\sigma x_1/2}, \\ A_n &= \left(a_{n1}e^{-px_1} + a_{n2}e^{-px_1}\right)e^{i\sigma x_1/2} \end{aligned} \tag{5.28.65}$$

e^{-px_1} quickly diverges, and such divergent acoustic propagation modes cannot occur in linear acoustic systems. Therefore, when $k_m^{-1}k_n^{-1} < 0$, the corresponding acoustic modes cannot propagate, or propagation is cut off.

Chapter 6
Systems Having Finite Degrees of Freedom

6.1 Exercise 6.1 (Internal Resonance Analysis and Nonlinear Solution of a Double Pendulum Problem)

Solution: (a) The kinetic energy of the system is

$$\begin{aligned} T &= \tfrac{1}{2}m_1 l_1^2\dot{\theta}_1^2 + \tfrac{1}{2}m_2\Big[\left(l_1\dot{\theta}_1\cos\theta_1 + l_2\dot{\theta}_2\cos\theta_2\right)^2 + \left(l_1\dot{\theta}_1\sin\theta_1 + l_2\dot{\theta}_2\sin\theta_2^2\right)\Big] \\ &= \tfrac{1}{2}m_1 l_1^2\dot{\theta}_1^2 + \tfrac{1}{2}m_2 l_1^2\dot{\theta}_1^2 + \tfrac{1}{2}m_2 l_2^2\dot{\theta}_2^2 + m_2 l_1 l_2\dot{\theta}_1\dot{\theta}_2\cos(\theta_2-\theta_1) \end{aligned} \tag{6.1.1}$$

The potential energy of the system is

$$V = -m_1 g l_1\cos\theta_1 - m_2 g(l_1\cos\theta_1 + l_2\cos\theta_2) \tag{6.1.2}$$

Substitute kinetic and potential energy into the Lagrange's equation:

$$\frac{d}{dt}\frac{\partial T}{\partial\dot{\theta}_k} - \frac{\partial T}{\partial\theta_k} = -\frac{\partial V}{\partial\theta_k},\quad k = 1,\ \ 2$$

we can obtain

$$\begin{aligned} &\ddot{\theta}_1 + gl_1^{-1}\sin\theta_1 + \alpha\ddot{\theta}_2\cos(\theta_2-\theta_1) - \alpha\dot{\theta}_2^2\sin(\theta_2-\theta_1) = 0 \\ &\ddot{\theta}_2 + gl_2^{-1}\sin\theta_2 + l_1 l_2^{-1}\ddot{\theta}_1\cos(\theta_2-\theta_1) + l_1 l_2^{-1}\dot{\theta}_1^2\sin(\theta_2-\theta_1) = 0 \end{aligned} \tag{6.1.3}$$

where $\alpha = m_2 l_2 l_1^{-1}(m_1+m_2)^{-1}$.

(b) The equilibrium position of the system is $\theta_1 = \theta_2 = 0$. Linearizing (6.1.3) near the equilibrium point yields

Z. He et al., *Solved Problems in Nonlinear Oscillations*,
https://doi.org/10.1007/978-981-97-6113-5_6

$$\ddot{\theta}_1 + \alpha\ddot{\theta}_2 + gl_1^{-1}\theta_1 = 0$$
$$l_1 l_2^{-1}\ddot{\theta}_1 + \ddot{\theta}_2 + gl_2^{-1}\theta_2 = 0$$

which can be written in the following matrix form:

$$\begin{bmatrix} \alpha^{-1} & 1 \\ 1 & l_1^{-1}l_2 \end{bmatrix} \begin{Bmatrix} \ddot{\theta}_1 \\ \ddot{\theta}_2 \end{Bmatrix} + \begin{bmatrix} \alpha^{-1}gl_1^{-1} & 0 \\ 0 & gl_1^{-1} \end{bmatrix} \begin{Bmatrix} \theta_1 \\ \theta_2 \end{Bmatrix} = 0 \tag{6.1.4}$$

The characteristic equation of this linear system is

$$\left(1 - \alpha l_1 l_2^{-1}\right)\omega^4 - g\left(l_1^{-1} + l_2^{-1}\right)\omega^2 + g^2 l_1^{-1} l_2^{-1} = 0 \tag{6.1.5}$$

The linear natural frequencies of the system, ω_1 and ω_2, are

$$\omega_1 = \left[\frac{g\left(l_1^{-1} + l_2^{-1}\right) - \sqrt{g^2(l_1^{-1} + l_2^{-1})^2 - 4g^2 l_1^{-1} l_2^{-1}\left(1 - \alpha l_1 l_2^{-1}\right)}}{2\left(1 - \alpha l_1 l_2^{-1}\right)}\right]^{1/2} \tag{6.1.6}$$

$$\omega_2 = \left[\frac{g\left(l_1^{-1} + l_2^{-1}\right) + \sqrt{g^2(l_1^{-1} + l_2^{-1})^2 - 4g^2 l_1^{-1} l_2^{-1}\left(1 - \alpha l_1 l_2^{-1}\right)}}{2\left(1 - \alpha l_1 l_2^{-1}\right)}\right]^{1/2} \tag{6.1.7}$$

Assuming that the corresponding normal modal matrix is $[\Psi]$, then using

$$\{\theta_1,\ \theta_2\}^T = [\Psi]\{u_1,\ u_2\}^T \tag{6.1.8}$$

we can transform (6.1.4) into

$$\begin{aligned} \ddot{u}_1 + \omega_1^2 u_1 &= 0 \\ \ddot{u}_2 + \omega_2^2 u_2 &= 0 \end{aligned} \tag{6.1.9}$$

If the system yields a three-to-one internal resonance, i.e., $\omega_2 : \omega_1 = 3 : 1$, there are

$$9 = \frac{\omega_2^2}{\omega_1^2} = \frac{g\left(l_1^{-1} + l_2^{-1}\right) + \sqrt{g^2(l_1^{-1} + l_2^{-1})^2 - 4g^2 l_1^{-1} l_2^{-1}\left(1 - \alpha l_1 l_2^{-1}\right)}}{g\left(l_1^{-1} + l_2^{-1}\right) - \sqrt{g^2(l_1^{-1} + l_2^{-1})^2 - 4g^2 l_1^{-1} l_2^{-1}\left(1 - \alpha l_1 l_2^{-1}\right)}}$$

i.e.,

$$(9 + 100\alpha)l_1^2 - 82 l_1 l_2 + 9 l_2^2 = 0 \tag{6.1.10}$$

When $l_1 = l_2 = l$, we can obtain from (6.1.10) that a three-to-one internal resonance occurs to the system when

$$\alpha = \frac{16}{25} \tag{6.1.11}$$

The linear natural frequency of the system is

$$\omega_1 = \frac{1}{3}\sqrt{5gl^{-1}}, \quad \omega_2 = \sqrt{5gl^{-1}} \tag{6.1.12}$$

The corresponding normal modal matrix is

$$[\Psi] = \begin{bmatrix} \frac{4}{3\sqrt{10}} & -2\sqrt{\frac{2}{5}} \\ \frac{5}{3\sqrt{10}} & \sqrt{\frac{5}{2}} \end{bmatrix} \tag{6.1.13}$$

(c) We use the method of multiple scales to solve (6.1.3). For small but finite amplitudes, we assume $\theta_1, \theta_2 = O(\varepsilon)$. Expanding (6.1.3) and retaining to $O(\varepsilon^3)$ yields

$$\begin{aligned} &\ddot{\theta}_1 + \alpha\ddot{\theta}_2 + gl_1^{-1}\theta_1 - \frac{1}{6}gl_1^{-1}\theta_1^3 - \frac{1}{2}\alpha\ddot{\theta}_2(\theta_2 - \theta_1)^2 - \alpha\dot{\theta}_2^2(\theta_2 - \theta_1) = 0 \\ &l_1 l_2^{-1}\ddot{\theta}_1 + \ddot{\theta}_2 + gl_2^{-1}\theta_2 - \frac{1}{6}gl_2^{-1}\theta_2^3 - \frac{1}{2}l_1 l_2^{-1}\ddot{\theta}_1(\theta_2 - \theta_1)^2 + l_1 l_2^{-1}\dot{\theta}_1^2(\theta_2 - \theta_1) = 0 \end{aligned} \tag{6.1.14}$$

Let the solution of Eq. (6.1.14) be expressed as

$$\begin{aligned} \theta_1 &= \varepsilon\theta_{11}(T_0, T_1, T_2) + \varepsilon^2\theta_{12}(T_0, T_1, T_2) + \varepsilon^3\theta_{13}(T_0, T_1, T_2) + \cdots \\ \theta_2 &= \varepsilon\theta_{21}(T_0, T_1, T_2) + \varepsilon^2\theta_{22}(T_0, T_1, T_2) + \varepsilon^3\theta_{23}(T_0, T_1, T_2) + \cdots \end{aligned} \tag{6.1.15}$$

Substituting (6.1.15) into (6.1.14) and keeping to $O(\varepsilon^3)$, we get

$$
\begin{aligned}
0 &= \ddot{\theta}_1 + \alpha\ddot{\theta}_2 + gl_1^{-1}\theta_1 - \frac{1}{6}gl_1^{-1}\theta_1^3 - \frac{1}{2}\alpha\ddot{\theta}_2(\theta_2-\theta_1)^2 - \alpha\dot{\theta}_2^2(\theta_2-\theta_1) \\
&= \left[D_0^2 + 2\varepsilon D_0D_1 + \varepsilon^2\left(D_1^2 + 2D_0D_2\right)\right]\left(\varepsilon\theta_{11} + \varepsilon^2\theta_{12} + \varepsilon^3\theta_{13}\right) \\
&\quad +\alpha\left[D_0^2 + 2\varepsilon D_0D_1 + \varepsilon^2\left(D_1^2 + 2D_0D_2\right)\right]\left(\varepsilon\theta_{21} + \varepsilon^2\theta_{22} + \varepsilon^3\theta_{23}\right) \\
&\quad +gl_1^{-1}\left(\varepsilon\theta_{11} + \varepsilon^2\theta_{12} + \varepsilon^3\theta_{13}\right) - \frac{1}{6}gl_1^{-1}\left(\varepsilon\theta_{11} + \varepsilon^2\theta_{12} + \varepsilon^3\theta_{13}\right)^3 \\
&\quad -\frac{1}{2}\alpha\left(\varepsilon\theta_{21} - \varepsilon\theta_{11} + \varepsilon^2\theta_{22} - \varepsilon^2\theta_{12} + \varepsilon^3\theta_{23} - \varepsilon^3\theta_{13}\right)^2 \\
&\quad \times\left[D_0^2 + 2\varepsilon D_0D_1 + \varepsilon^2\left(D_1^2 + 2D_0D_2\right)\right]\left(\varepsilon\theta_{21} + \varepsilon^2\theta_{22} + \varepsilon^3\theta_{23}\right) \\
&\quad -\alpha\left(\varepsilon\theta_{21} - \varepsilon\theta_{11} + \varepsilon^2\theta_{22} - \varepsilon^2\theta_{12} + \varepsilon^3\theta_{23} - \varepsilon^3\theta_{13}\right) \\
&\quad \times\left[\left(D_0 + \varepsilon D_1 + \varepsilon^2 D_2\right)\left(\varepsilon\theta_{21} + \varepsilon^2\theta_{22} + \varepsilon^3\theta_{23}\right)\right]^2 \\
&= \varepsilon\left(D_0^2\theta_{11} + \alpha D_0^2\theta_{21} + gl_1^{-1}\theta_{11}\right) + \varepsilon^2\left(D_0^2\theta_{12} + \alpha D_0^2\theta_{22} + gl_1^{-1}\theta_{12}\right. \\
&\quad \left.+2D_0D_1\theta_{11} + 2\alpha D_0D_1\theta_{21}\right) + \varepsilon^3\left[D_0^2\theta_{13} + \alpha D_0^2\theta_{23} + gl_1^{-1}\theta_{13}\right. \\
&\quad +2D_0D_1\theta_{12} + 2\alpha D_0D_1\theta_{22} + \left(D_1^2 + 2D_0D_2\right)\theta_{11} + \alpha\left(D_1^2 + 2D_0D_2\right)\theta_{21} \\
&\quad \left.-\frac{1}{6}gl_1^{-1}\theta_{11}^3 - \frac{1}{2}\alpha(\theta_{21}-\theta_{11})^2 D_0^2\theta_{21} - \alpha(\theta_{21}-\theta_{11})(D_0\theta_{21})^2\right]
\end{aligned}
\tag{6.1.16}
$$

$$
\begin{aligned}
0 &= \mathrm{l}_1\mathrm{l}_2^{-1}\ddot{\theta}_1 + \ddot{\theta}_2 + \mathrm{gl}_2^{-1}\theta_2 - \frac{1}{6}\mathrm{gl}_2^{-1}\theta_2^3 - \frac{1}{2}\mathrm{l}_1\mathrm{l}_2^{-1}\ddot{\theta}_1(\theta_2-\theta_1)^2 + \mathrm{l}_1\mathrm{l}_2^{-1}\dot{\theta}_1^2(\theta_2-\theta_1) \\
&= \left[\mathrm{D}_0^2 + 2\varepsilon\mathrm{D}_0\mathrm{D}_1 + \varepsilon^2\left(\mathrm{D}_1^2 + 2\mathrm{D}_0\mathrm{D}_2\right)\right]\left(\varepsilon\theta_{21} + \varepsilon^2\theta_{22} + \varepsilon^3\theta_{23}\right) \\
&\quad +\mathrm{l}_1\mathrm{l}_2^{-1}\left[\mathrm{D}_0^2 + 2\varepsilon\mathrm{D}_0\mathrm{D}_1 + \varepsilon^2\left(\mathrm{D}_1^2 + 2\mathrm{D}_0\mathrm{D}_2\right)\right]\left(\varepsilon\theta_{11} + \varepsilon^2\theta_{12} + \varepsilon^3\theta_{13}\right) \\
&\quad +\mathrm{gl}_2^{-1}\left(\varepsilon\theta_{21} + \varepsilon^2\theta_{22} + \varepsilon^3\theta_{23}\right) - \frac{1}{6}\mathrm{gl}_2^{-1}\left(\varepsilon\theta_{21} + \varepsilon^2\theta_{22} + \varepsilon^3\theta_{23}\right)^3 \\
&\quad -\frac{1}{2}\mathrm{l}_1\mathrm{l}_2^{-1}\left(\varepsilon\theta_{21} - \varepsilon\theta_{11} + \varepsilon^2\theta_{22} - \varepsilon^2\theta_{12} + \varepsilon^3\theta_{23} - \varepsilon^3\theta_{13}\right)^2 \\
&\quad \times\left[\mathrm{D}_0^2 + 2\varepsilon\mathrm{D}_0\mathrm{D}_1 + \varepsilon^2\left(\mathrm{D}_1^2 + 2\mathrm{D}_0\mathrm{D}_2\right)\right]\left(\varepsilon\theta_{11} + \varepsilon^2\theta_{12} + \varepsilon^3\theta_{13}\right) \\
&\qquad\quad +\mathrm{l}_1\mathrm{I}_2^{-1}\left(\varepsilon\theta_{21} - \varepsilon\theta_{11} + \varepsilon^2\theta_{22} - \varepsilon^2\theta_{12} + \varepsilon^3\theta_{23} - \varepsilon^3\theta_{13}\right) \\
&\quad \times\left[\left(\mathrm{D}_0 + \varepsilon\mathrm{D}_1 + \varepsilon^2\mathrm{D}_2\right)\left(\varepsilon\theta_{11} + \varepsilon^2\theta_{12} + \varepsilon^3\theta_{13}\right)\right]^2 \\
&\quad \times\left[\left(\mathrm{D}_0 + \varepsilon\mathrm{D}_1 + \varepsilon^2\mathrm{D}_2\right)\left(\varepsilon\theta_{11} + \varepsilon^2\theta_{12} + \varepsilon^3\theta_{13}\right)\right]^2 \\
&= \varepsilon\left(\mathrm{l}_1\mathrm{I}_2^{-1}\mathrm{D}_0^2\theta_{11} + \mathrm{D}_0^2\theta_{21} + \mathrm{gl}_2^{-1}\theta_{21}\right) + \varepsilon^2\left(\mathrm{l}_1\mathrm{I}_2^{-1}\mathrm{D}_0^2\theta_{12} + \mathrm{D}_0^2\theta_{22} + \mathrm{gl}_2^{-1}\theta_{22}\right. \\
&\quad \left.+2\mathrm{l}_1\mathrm{l}_2^{-1}\mathrm{D}_0\mathrm{D}_1\theta_{11} + 2\mathrm{D}_0\mathrm{D}_1\theta_{21}\right) + \varepsilon^3\left[\mathrm{I}_1\mathrm{l}_2^{-1}\mathrm{D}_0^2\theta_{13} + \mathrm{D}_0^2\theta_{23} + \mathrm{gl}_2^{-1}\theta_{23}\right. \\
&\quad +2\mathrm{l}_1\mathrm{I}_2^{-1}\mathrm{D}_0\mathrm{D}_1\theta_{12} + 2\mathrm{D}_0\mathrm{D}_1\theta_{22} + \left(\mathrm{D}_1^2 + 2\mathrm{D}_0\mathrm{D}_2\right)\theta_{21} + \mathrm{l}_1\mathrm{l}_2^{-1}\left(\mathrm{D}_1^2 + 2\mathrm{D}_0\mathrm{D}_2\right)\theta_{11} \\
&\quad \left.-\frac{1}{6}\mathrm{gl}_2^{-1}\theta_{21}^3 - \frac{1}{2}\mathrm{l}_1\mathrm{l}_2^{-1}(\theta_{21}-\theta_{11})^2\mathrm{D}_0^2\theta_{11} + \mathrm{l}_1\mathrm{l}_2^{-1}(\theta_{21}-\theta_{11})(\mathrm{D}_0\theta_{11})^2\right]
\end{aligned}
\tag{6.1.17}
$$

Equating coefficients of like powers of ε in (6.1.16) and (6.1.17) yields.

Order ε^1:

$$\begin{aligned} &D_0^2\theta_{11} + \alpha D_0^2\theta_{21} + gl_1^{-1}\theta_{11} = 0 \\ &l_1 l_2^{-1} D_0^2\theta_{11} + D_0^2\theta_{21} + gl_2^{-1}\theta_{21} = 0 \end{aligned} \tag{6.1.18}$$

Order ε^2:

$$\begin{aligned} &D_0^2\theta_{12} + \alpha D_0^2\theta_{22} + gl_1^{-1}\theta_{12} + 2D_0D_1\theta_{11} + 2\alpha D_0D_1\theta_{21} = 0 \\ &l_1 l_2^{-1} D_0^2\theta_{12} + D_0^2\theta_{22} + gl_2^{-1}\theta_{22} + 2l_1 l_2^{-1} D_0D_1\theta_{11} + 2D_0D_1\theta_{21} = 0 \end{aligned} \tag{6.1.19}$$

Order ε^3:

$$\begin{aligned} &D_0^2\theta_{13} + \alpha D_0^2\theta_{23} + gl_1^{-1}\theta_{13} + 2D_0D_1\theta_{12} + 2\alpha D_0D_1\theta_{22} \\ &\quad + \left(D_1^2 + 2D_0D_2\right)\theta_{11} + \alpha\left(D_1^2 + 2D_0D_2\right)\theta_{21} \\ &\quad - \frac{1}{2}\alpha(\theta_{21} - \theta_{11})^2 D_0^2\theta_{21} - \alpha(\theta_{21} - \theta_{11})(D_0\theta_{21})^2 - \frac{1}{6}gl_1^{-1}\theta_{11}^3 = 0 \\ &l_1 l_2^{-1} D_0^2\theta_{13} + D_0^2\theta_{23} + gl_2^{-1}\theta_{23} + 2l_1 l_2^{-1} D_0D_1\theta_{12} + 2D_0D_1\theta_{22} \\ &\quad + \left(D_1^2 + 2D_0D_2\right)\theta_{21} + l_1 l_2^{-1}\left(D_1^2 + 2D_0D_2\right)\theta_{11} \end{aligned} \tag{6.1.20}$$

(6.1.18)–(6.1.20) can also be written in matrix form:

$$\begin{bmatrix} \alpha^{-1} & 1 \\ 1 & l_1^{-1}l_2 \end{bmatrix}\begin{Bmatrix} D_0^2\theta_{11} \\ D_0^2\theta_{21} \end{Bmatrix} + \begin{bmatrix} \alpha^{-1}gl_1^{-1} & 0 \\ 0 & gl_1^{-1} \end{bmatrix}\begin{Bmatrix} \theta_{11} \\ \theta_{21} \end{Bmatrix} = 0 \tag{6.1.21}$$

$$\begin{bmatrix} \alpha^{-1} & 1 \\ 1 & l_1^{-1}l_2 \end{bmatrix}\begin{Bmatrix} D_0^2\theta_{12} \\ D_0^2\theta_{22} \end{Bmatrix} + \begin{bmatrix} \alpha^{-1}gl_1^{-1} & 0 \\ 0 & gl_1^{-1} \end{bmatrix}\begin{Bmatrix} \theta_{12} \\ \theta_{22} \end{Bmatrix} = -2\begin{bmatrix} \alpha^{-1} & 1 \\ 1 & l_1^{-1}l_2 \end{bmatrix} D_0D_1\begin{Bmatrix} \theta_{11} \\ \theta_{21} \end{Bmatrix} \tag{6.1.22}$$

$$\begin{aligned} &\begin{bmatrix} \alpha^{-1} & 1 \\ 1 & l_1^{-1}l_2 \end{bmatrix}\begin{Bmatrix} D_0^2\theta_{13} \\ D_0^2\theta_{23} \end{Bmatrix} + \begin{bmatrix} \alpha^{-1}gl_1^{-1} & 0 \\ 0 & gl_1^{-1} \end{bmatrix}\begin{Bmatrix} \theta_{13} \\ \theta_{23} \end{Bmatrix} = -2\begin{bmatrix} \alpha^{-1} & 1 \\ 1 & l_1^{-1}l_2 \end{bmatrix} D_0D_1\begin{Bmatrix} \theta_{12} \\ \theta_{22} \end{Bmatrix} \\ &- \begin{bmatrix} \alpha^{-1} & 1 \\ 1 & l_1^{-1}l_2 \end{bmatrix}\left(D_1^2 + 2D_0D_2\right)\begin{Bmatrix} \theta_{11} \\ \theta_{21} \end{Bmatrix} + \frac{1}{2}(\theta_{21} - \theta_{11})^2 D_0^2\begin{Bmatrix} \theta_{21} \\ \theta_{11} \end{Bmatrix} \\ &- \begin{Bmatrix} -(\theta_{21} - \theta_{11})(D_0\theta_{21})^2 \\ (\theta_{21} - \theta_{11})(D_0\theta_{11})^2 \end{Bmatrix} + \frac{1}{6}gl_1^{-1}\begin{Bmatrix} \alpha^{-1}\theta_{11}^3 \\ \theta_{21}^3 \end{Bmatrix} \end{aligned} \tag{6.1.23}$$

Apply the transformation to (6.1.21)–(6.1.23), i.e.,

$$\{\theta_{1n},\ \theta_{2n}\}^T = [\Psi]\{u_{1n},\ u_{2n}\}^T,\ n = 1, 2, 3 \tag{6.1.24}$$

we can obtain

$$\begin{Bmatrix} D_0^2 u_{11} \\ D_0^2 u_{21} \end{Bmatrix} + \begin{bmatrix} \omega_1^2 & 0 \\ 0 & \omega_2^2 \end{bmatrix} \begin{Bmatrix} u_{11} \\ u_{21} \end{Bmatrix} = 0 \tag{6.1.25}$$

$$\begin{Bmatrix} D_0^2 u_{12} \\ D_0^2 u_{22} \end{Bmatrix} + \begin{bmatrix} \omega_1^2 & 0 \\ 0 & \omega_2^2 \end{bmatrix} \begin{Bmatrix} u_{12} \\ u_{22} \end{Bmatrix} = -2D_0 D_1 \begin{Bmatrix} u_{11} \\ u_{21} \end{Bmatrix} \tag{6.1.26}$$

$$\begin{Bmatrix} D_0^2 u_{13} \\ D_0^2 u_{23} \end{Bmatrix} + \begin{bmatrix} \omega_1^2 & 0 \\ 0 & \omega_2^2 \end{bmatrix} \begin{Bmatrix} u_{13} \\ u_{23} \end{Bmatrix} = -2D_0 D_1 \begin{Bmatrix} u_{12} \\ u_{22} \end{Bmatrix} - \left(D_1^2 + 2D_0 D_2\right) \begin{Bmatrix} u_{11} \\ u_{21} \end{Bmatrix} + \begin{Bmatrix} F_1 \\ F_2 \end{Bmatrix} \tag{6.1.27}$$

where

$$\begin{aligned} F_1 = {} & \tfrac{1}{2}\Psi_{11}(\theta_{21}-\theta_{11})^2 D_0^2\theta_{21} + \tfrac{1}{2}\Psi_{21}(\theta_{21}-\theta_{11})^2 D_0^2\theta_{11} \\ & +\Psi_{11}(\theta_{21}-\theta_{11})(D_0\theta_{21})^2 - \Psi_{21}(\theta_{21}-\theta_{11})(D_0\theta_{11})^2 \\ & +\tfrac{1}{6}\Psi_{11} g l_1^{-1}\alpha^{-1}\theta_{11}^3 + \tfrac{1}{6}\Psi_{21} g l_1^{-1}\theta_{21}^3 \end{aligned} \tag{6.1.28}$$

$$\begin{aligned} F_2 = {} & \tfrac{1}{2}\Psi_{12}(\theta_{21}-\theta_{11})^2 D_0^2\theta_{21} + \tfrac{1}{2}\Psi_{22}(\theta_{21}-\theta_{11})^2 D_0^2\theta_{11} \\ & +\Psi_{12}(\theta_{21}-\theta_{11})(D_0\theta_{21})^2 - \Psi_{22}(\theta_{21}-\theta_{11})(D_0\theta_{11})^2 \\ & +\tfrac{1}{6}\Psi_{12} g l_1^{-1}\alpha^{-1}\theta_{11}^3 + \tfrac{1}{6}\Psi_{22} g l_1^{-1}\theta_{21}^3 \end{aligned} \tag{6.1.29}$$

And Ψ_{ij} is the element of the normal modal matrix $[\Psi]$.

The solution of (6.1.18) is

$$\begin{Bmatrix} u_{11} \\ u_{21} \end{Bmatrix} = \begin{Bmatrix} A_1 e^{i\omega_1 T_0} + A_1 e^{-i\omega_1 T_0} \\ A_2 e^{i\omega_2 T_0} + A_2 e^{-i\omega_2 T_0} \end{Bmatrix} \tag{6.1.30}$$

where $A_k = A_k(T_1, T_2)$. Substituting (6.1.30) into (6.1.19), we obtain

$$\begin{aligned} D_0^2 u_{12} + \omega_1^2 u_{12} &= -2\left(i\omega_1 D_1 A_1 e^{i\omega_1 t} - i\omega_1 D_1 A_1 e^{i\omega_1 T_0}\right) \\ D_0^2 u_{22} + \omega_2^2 u_{22} &= -2\left(i\omega_2 D_1 A_2 e^{i\omega_2 t} - i\omega_2 D_1 A_2 e^{i\omega_2 T_0}\right) \end{aligned} \tag{6.1.31}$$

In order to eliminate secular terms from the above equation, there must be

$$\mathrm{D_1 A_1 = 0, \quad D_1 A_2 = 0 \quad \Rightarrow \quad A_1 = A_1(T_2), \quad A_2 = A_2(T_2)} \tag{6.1.32}$$

therefore

$$u_{12} = u_{22} = 0 \tag{6.1.33}$$

Substituting (6.1.30) and (6.1.33) into (6.1.27) yields

$$\mathrm{D_0^2 u_{13} + \omega_1^2 u_{13} = -2i\omega_1 A_1' e^{i\omega_1 T_0} + cc + F_1} \tag{6.1.34}$$

$$\mathrm{D_0^2 u_{23} + \omega_2^2 u_{23} = -2i\omega_2 A_2' e^{i\omega_2 T_0} + cc + F_2} \tag{6.1.35}$$

The prime represents the partial derivative with respect to T_2. Substituting the transformations $\{\theta_{11},\ \theta_{21}\}^T = [\Psi]\{u_{11},\ u_{21}\}^T$ and (6.1.30) into (6.1.28) and (6.1.29) yields

$$
\begin{aligned}
F_1 = &-3\omega_1^2\Psi_{11}\Psi_{21}h_1^2A_1^2\overline{A}_1e^{i\omega_1 T_0} - 2\omega_1^2\Psi_{11}\Psi_{21}h_2^2A_1A_2\overline{A}_2e^{i\omega_1 T_0}\\
&-2\omega_2^2\Psi_{11}\Psi_{22}h_1h_2A_1A_2\overline{A}_2e^{i\omega_1 T_0} - 2\omega_2^2\Psi_{12}\Psi_{21}h_1h_2A_1A_2\overline{A}_2e^{i\omega_1 T_0}\\
&-2\omega_1^2\Psi_{11}\Psi_{21}h_1h_2\overline{A}_1^2A_2e^{i(\omega_2-2\omega_1)T_0} - \frac{1}{2}\omega_2^2\Psi_{11}\Psi_{22}h_1^2\overline{A}_1^2A_2e^{i(\omega_2-2\omega_1)T_0}\\
\nu &- \frac{1}{2}\omega_2^2\Psi_{12}\Psi_{21}h_1^2\overline{A}_1^2A_2e^{i(\omega_2-2\omega_1)T_0} - \frac{1}{2}\omega_2^2\Psi_{11}\Psi_{22}h_2^2A_2^3e^{i3\omega_2 T_0}\\
&-\frac{1}{2}\omega_2^2\Psi_{12}\Psi_{21}h_2^2A_2^3e^{i3\omega_2 T_0} + \omega_1^2\Psi_{11}\Psi_{21}^2h_1A_1^2\overline{A}_1e^{i\omega_1 T_0} - \omega_1^2\Psi_{11}^2\Psi_{21}h_1A_1^2\overline{A}_1e^{i\omega_1 T_0}\\
&+2\omega_2^2\Psi_{11}\Psi_{22}^2h_1A_1A_2\overline{A}_2e^{i\omega_1 T_0} - 2\omega_2^2\Psi_{12}^2\Psi_{21}h_1A_1A_2\overline{A}_2e^{i\omega_1 T_0}\\
&+2\omega_1\omega_2\Psi_{11}\Psi_{21}\Psi_{22}h_1\overline{A}_1^2A_2e^{i(\omega_2-2\omega_1)T_0} - \omega_1^2\Psi_{11}\Psi_{21}^2h_2\overline{A}_1^2A_2e^{i(\omega_2-2\omega_1)T_0}\\
&-2\omega_1\omega_2\Psi_{11}\Psi_{12}\Psi_{21}h_1\overline{A}_1^2A_2e^{i(\omega_2-2\omega_1)T_0} + \omega_1^2\Psi_{11}^2\Psi_{21}h_2\overline{A}_1^2A_2e^{i(\omega_2-2\omega_1)T_0}\\
&-\omega_2^2\Psi_{11}\Psi_{22}^2h_2A_2^3e^{i3\omega_2 T_0} + \omega_2^2\Psi_{12}^2\Psi_{21}h_2A_2^3e^{i3\omega_2 T_0}\\
&+\frac{1}{2}gl_1^{-1}\alpha^{-1}\Psi_{11}^4A_1^2\overline{A}_1e^{i\omega_1 T_0} + gl_1^{-1}\alpha^{-1}\Psi_{11}^2\Psi_{12}^2A_1A_2\overline{A}_2e^{i\omega_1 T_0}\\
&+\frac{1}{2}gl_1^{-1}\Psi_{21}^4A_1^2\overline{A}_1e^{i\omega_1 T_0} + gl_1^{-1}\Psi_{21}^2\Psi_{22}^2A_1A_2\overline{A}_2e^{i\omega_1 T_0}\\
&+\frac{1}{2}gl_1^{-1}\alpha^{-1}\Psi_{11}^3\Psi_{12}\overline{A}_1^2A_2e^{i(\omega_2-2\omega_1)T_0} + \frac{1}{2}gl_1^{-1}\Psi_{21}^3\Psi_{22}\overline{A}_1^2A_2e^{i(\omega_2-2\omega_1)T_0}\\
&+\frac{1}{6}gl_1^{-1}\alpha^{-1}\Psi_{11}\Psi_{12}^3A_2^3e^{i3\omega_2 T_0} + \frac{1}{6}gl_1^{-1}\Psi_{21}\Psi_{22}^3A_2^3e^{i3\omega_2 T_0} + cc + NST
\end{aligned}
\tag{6.1.36}
$$

$$
\begin{aligned}
F_2 = &-2\omega_2^2\Psi_{12}\Psi_{22}h_1^2A_1\overline{A}_1A_2e^{i\omega_2 T_0} - 3\omega_2^2\Psi_{12}\Psi_{22}h_2^2A_2^2\overline{A}_2e^{i\omega_2 T_0}\\
&-2\omega_1^2\Psi_{12}\Psi_{21}h_1h_2A_1\overline{A}_1A_2e^{i\omega_2 T_0} - 2\omega_1^2\Psi_{11}\Psi_{22}h_1h_2A_1\overline{A}_1A_2e^{i\omega_2 T_0}\\
&-2\omega_2^2\Psi_{12}\Psi_{22}h_1h_2A_1\overline{A}_2^2e^{i(\omega_1-2\omega_2)T_0} - \frac{1}{2}\omega_1^2\Psi_{12}\Psi_{21}h_2^2A_1\overline{A}_2^2e^{i(\omega_1-2\omega_2)T_0}\\
&-\frac{1}{2}\omega_1^2\Psi_{11}\Psi_{22}h_2^2A_1\overline{A}_2^2e^{i(\omega_1-2\omega_2)T_0} - \frac{1}{2}\omega_1^2\Psi_{12}\Psi_{22}h_1^2A_1^3e^{i3\omega_1 T_0}\\
&-\frac{1}{2}\omega_1^2\Psi_{12}\Psi_{21}h_1^2A_1^3e^{i3\omega_1 T_0} + \omega_2^2\Psi_{12}^2\Psi_{22}h_2A_2^2\overline{A}_2e^{i\omega_2 T_0} - \omega_2^2\Psi_{12}\Psi_{22}^2h_2A_2^2\overline{A}_2e^{i\omega_2 T_0}\\
&-2\omega_1^2\Psi_{12}\Psi_{21}^2h_2A_1\overline{A}_1A_2e^{i\omega_2 T_0} + 2\omega_1^2\Psi_{11}^2\Psi_{22}h_2A_1\overline{A}_1A_2e^{i\omega_2 T_0}
\end{aligned}
$$

$$
\begin{aligned}
&+\omega_2^2\Psi_{12}\Psi_{22}^2 h_1 A_1\overline{A}_2^2 e^{i(\omega_1-2\omega_2)T_0} - 2\omega_1\omega_2\Psi_{12}\Psi_{21}\Psi_{22}h_2 A_1\overline{A}_2^2 e^{i(\omega_1-2\omega_2)T_0} \\
&-\omega_2^2\Psi_{12}^2\Psi_{22}h_1 A_1\overline{A}_2^2 e^{i(\omega_1-2\omega_2)T_0} + 2\omega_1\omega_2\Psi_{11}\Psi_{12}\Psi_{22}h_2 A_1\overline{A}_2^2 e^{i(\omega_1-2\omega_2)T_0} \\
&+\omega_1^2\Psi_{12}\Psi_{21}^2 h_1 A_1^3 e^{i3\omega_1 T_0} - \omega_1^2\Psi_{11}^2\Psi_{22}h_1 A_1^3 e^{i3\omega_1 T_0} \\
&+\frac{1}{2}gl_1^{-1}\alpha^{-1}\Psi_{12}^4 A_2^2\overline{A}_2 e^{i\omega_2 T_0} + gl_1^{-1}\alpha^{-1}\Psi_{11}^2\Psi_{12}^2 A_1\overline{A}_1 A_2 e^{i\omega_2 T_0} \\
&+\frac{1}{2}gl_1^{-1}\Psi_{22}^4 A_2^2\overline{A}_2 e^{i\omega_2 T_0} + gl_1^{-1}\Psi_{21}^2\Psi_{22}^2 A_1\overline{A}_1 A_2 e^{i\omega_2 T_0} \\
&+\frac{1}{2}gl_1^{-1}\alpha^{-1}\Psi_{11}\Psi_{12}^3 A_1\overline{A}_2^2 e^{i(\omega_1-2\omega_2)T_0} + \frac{1}{2}gl_1^{-1}\Psi_{21}\Psi_{22}^3 A_1\overline{A}_2^2 e^{i(\omega_1-2\omega_2)T_0} \\
&+\frac{1}{6}gl_1^{-1}\Psi_{21}^3\Psi_{22}A_1^3 e^{i3\omega_1 T_0} + \frac{1}{6}gl_1^{-1}\alpha^{-1}\Psi_{11}^3\Psi_{12}A_1^3 e^{i3\omega_1 T_0} + cc + NST
\end{aligned}
\tag{6.1.37}
$$

where

$$
h_1 = \psi_{21} - \psi_{11}, \quad h_2 = \psi_{22} - \psi_{12} \tag{6.1.38}
$$

To obtain (6.1.34)–(6.1.37), we did not consider the case of $\omega_1 = \omega_2$. From (6.1.34)–(6.1.37), we can find that there are two possible resonant cases: $\omega_2 \approx 3\omega_1$ and $\omega_1 \approx 3\omega_2$. Without loss of generality, we assume $\omega_2 > \omega_1$. In analyzing the particular solutions of (6.1.34) and (6.1.35). We need to distinguish between the resonant case in which $\omega_2 \approx 3\omega_1$ and the nonresonant case in which ω_2 is away from $3\omega_1$.

(1) The nonresonant case.

In this case the solvability conditions (the conditions for the eliminations of secular terms) become

$$
2A_1' + ib_{11}A_1^2\overline{A}_1 + ib_{12}A_1A_2\overline{A}_2 = 0 \tag{6.1.39}
$$

$$
2A_2' + ib_{21}A_2^2\overline{A}_2 + ib_{22}A_1\overline{A}_1A_2 = 0 \tag{6.1.40}
$$

where

$$
\begin{aligned}
b_{11} &= \omega_1^{-1}\Big(\omega_1^2\Psi_{11}\Psi_{21}^2 h_1 - 3\omega_1^2\Psi_{11}\Psi_{21}h_1^2 \\
&-\omega_1^2\Psi_{11}^2\Psi_{21}h_1 + \frac{1}{2}gl_1^{-1}\Psi_{21}^4 + \frac{1}{2}gl_1^{-1}\alpha^{-1}\Psi_{11}^4\Big) \\
b_{12} &= \omega_1^{-1}\big(2\omega_2^2\Psi_{11}\Psi_{22}^2 h_1 - 2\omega_2^2\Psi_{12}^2\Psi_{21}h_1 - 2\omega_1^2\Psi_{11}\Psi_{21}h_2^2 - 2\omega_2^2\Psi_{11}\Psi_{22}h_1h_2 \\
&-2\omega_2^2\Psi_{12}\Psi_{21}h_1h_2 + gl_1^{-1}\alpha^{-1}\Psi_{11}^2\Psi_{12}^2 + gl_1^{-1}\Psi_{21}^2\Psi_{22}^2\big)A_1A_2\overline{A}_2
\end{aligned}
\tag{6.1.41}
$$

$$b_{21} = \omega_2^{-1}\Big(\omega_2^2\Psi_{12}^2\Psi_{22}h_2 - 3\omega_2^2\Psi_{12}\Psi_{22}h_2^2 - \omega_2^2\Psi_{12}\Psi_{22}^2h_2 + \frac{1}{2}gl_1^{-1}\alpha^{-1}\Psi_{12}^4 + \frac{1}{2}gl_1^{-1}\Psi_{22}^4\Big)$$

$$b_{22} = \omega_2^{-1}\big(2\omega_1^2\Psi_{11}^2\Psi_{22}h_2 - 2\omega_1^2\Psi_{12}\Psi_{21}^2h_2 - 2\omega_2^2\Psi_{12}\Psi_{22}h_1^2 - 2\omega_1^2\Psi_{12}\Psi_{21}h_1h_2 - 2\omega_1^2\Psi_{11}\Psi_{22}h_1h_2 + gl_1^{-1}\alpha^{-1}\Psi_{11}^2\Psi_{12}^2 + gl_1^{-1}\Psi_{21}^2\Psi_{22}^2\big) \tag{6.1.42}$$

Let

$$A_1 = \frac{1}{2}a_1e^{i\beta_1}, \quad A_2 = \frac{1}{2}a_2e^{i\beta_2} \tag{6.1.43}$$

Substituting into (6.1.39) and (6.1.40), we get

$$a_1' + i\beta_1'a_1 + \frac{1}{8}ib_{11}a_1^3 + \frac{1}{8}ib_{12}a_1a_2^2 = 0 \tag{6.1.44}$$

$$a_2' + i\beta_2'a_2 + \frac{1}{8}ib_{21}a_2^3 + \frac{1}{8}ib_{22}a_1^2a_2 = 0 \tag{6.1.45}$$

Separating the real and imaginary parts of (6.1.44) and (6.1.45) yields

$$a_1' = 0$$
$$a_2' = 0$$

$$\begin{aligned} \beta_1'a_1 + \frac{1}{8}b_{11}a_1^3 + \frac{1}{8}b_{12}a_1a_2^2 = 0 \\ \beta_2'a_2 + \frac{1}{8}b_{21}a_2^3 + \frac{1}{8}b_{22}a_1^2a_2 = 0 \end{aligned} \tag{6.1.46}$$

The solution of (6.1.46) is

$$\begin{aligned} &a_1 = a_{10} = \text{constant}, a_2 = a_{20} = \text{constant} \\ &\beta_1 = -\tfrac{1}{8}\big(b_{11}a_{10}^2 + b_{12}a_{20}^2\big)T_2 + \beta_{10}, \quad \beta_2 = -\tfrac{1}{8}\big(b_{21}a_{20}^2 + b_{22}a_{10}^2\big)T_2 + \beta_{20} \end{aligned} \tag{6.1.47}$$

The constants of integration a_{10}, a_{20}, β_{10} and β_{20} are determined by the initial conditions. If εa_1 and εa_2 are written as a_1 and a_2, respectively, then the above equation becomes

$$\begin{aligned} &a_1 = a_{10} = \text{constant}, a_2 = a_{20} = \text{constant} \\ &\beta_1 = -\tfrac{1}{8}\big(b_{11}a_{10}^2 + b_{12}a_{20}^2\big)t + \beta_{10}, \quad \beta_2 = -\tfrac{1}{8}\big(b_{21}a_{20}^2 + b_{22}a_{10}^2\big)t + \beta_{20} \end{aligned} \tag{6.1.48}$$

Substitute (6.1.48), (6.1.43), (6.1.30), (6.1.33) and (6.1.24) into (6.1.15), we can obtain the first order approximate solution of the original equation

$$\begin{Bmatrix} \theta_1 \\ \theta_2 \end{Bmatrix} = [\Psi]\begin{Bmatrix} a_{10}\cos[\omega_1 t + \beta_1(t)] \\ a_{20}\cos[\omega_2 t + \beta_2(t)] \end{Bmatrix} + O(\varepsilon^3) \tag{6.1.49}$$

It can be seen that when there is no internal resonance, the motion of the double pendulum is unconditionally stabilized.

(2) The resonant case (internal resonance $\omega_2 \approx 3\omega_1$).

In this case we introduce a detuning parameter σ according to

$$\omega_2 = 3\omega_1 + \varepsilon^2\sigma \tag{6.1.50}$$

In this case it follows from (6.1.34)–6.1.37) and (6.1.50) that the solvability conditions are

$$2A_1' + ib_{11}A_1^2\overline{A}_1 + ib_{12}A_1A_2\overline{A}_2 + ib_{13}\overline{A}_1^2A_2e^{i\sigma T_2} = 0 \tag{6.1.51}$$

$$2A_2' + ib_{21}A_2^2\overline{A}_2 + ib_{22}A_1\overline{A}_1A_2 + ib_{23}A_1^3e^{-i\sigma T_2} = 0 \tag{6.1.52}$$

where

$$\begin{aligned}
b_{13} &= \omega_1^{-1}\big(\omega_1^2\Psi_{11}^2\Psi_{21}h_2 - \omega_1^2\Psi_{11}\Psi_{21}^2h_2 - 2\omega_1^2\Psi_{11}\Psi_{21}h_1h_2 \\
&+2\omega_1\omega_2\Psi_{11}\Psi_{21}\Psi_{22}h_1 - 2\omega_1\omega_2\Psi_{11}\Psi_{12}\Psi_{21}h_1 \\
&-\frac{1}{2}\omega_2^2\Psi_{11}\Psi_{22}h_1^2 - \frac{1}{2}\omega_2^2\Psi_{12}\Psi_{21}h_1^2 + \frac{1}{2}gl_1^{-1}\alpha^{-1}\Psi_{11}^3\Psi_{12} + \frac{1}{2}gl_1^{-1}\Psi_{21}^3\Psi_{22}\Big) \\
b_{23} &= \omega_2^{-1}\Big(\omega_1^2\Psi_{12}\Psi_{21}^2h_1 - \omega_1^2\Psi_{11}^2\Psi_{22}h_1 - \frac{1}{2}\omega_1^2\Psi_{12}\Psi_{22}h_1^2 - \frac{1}{2}\omega_1^2\Psi_{12}\Psi_{21}h_1^2 \\
&+\frac{1}{6}gl_1^{-1}\Psi_{21}^3\Psi_{22} + \frac{1}{6}gl_1^{-1}\alpha^{-1}\Psi_{11}^3\Psi_{12}\Big)
\end{aligned} \tag{6.1.53}$$

Substituting (6.1.43) into (6.1.51) and (6.1.52) yields

$$a_1' + i\beta_1'a_1 + \frac{1}{8}ib_{11}a_1^3 + \frac{1}{8}ib_{12}a_1a_2^2 + i\frac{1}{8}b_{13}a_1^2a_2e^{i(\sigma T_2 - 3\beta_1 + \beta_2)} = 0 \tag{6.1.54}$$

$$a_2' + i\beta_2'a_2 + \frac{1}{8}ib_{21}a_2^3 + \frac{1}{8}ib_{22}a_1^2a_2 + i\frac{1}{8}b_{23}a_1^3e^{-i(\sigma T_2 - 3\beta_1 + \beta_2)} = 0 \tag{6.1.55}$$

Separating the real and imaginary parts of the above equations yields

$$\begin{aligned} a_1' &= \tfrac{1}{8} b_{13} a_1^2 a_2 \sin\gamma \\ a_2' &= -\tfrac{1}{8} b_{23} a_1^3 \sin\gamma \end{aligned} \tag{6.1.56}$$

$$\begin{aligned} \beta_1' a_1 + \tfrac{1}{8} b_{11} a_1^3 + \tfrac{1}{8} b_{12} a_1 a_2^2 + \tfrac{1}{8} b_{13} a_1^2 a_2 \cos\gamma = 0 \\ \beta_2' a_2 + \tfrac{1}{8} b_{21} a_2^3 + \tfrac{1}{8} b_{22} a_1^2 a_2 + \tfrac{1}{8} b_{23} a_1^3 \cos\gamma = 0 \end{aligned} \tag{6.1.57}$$

where

$$\gamma = \sigma T_2 - 3\beta_1 + \beta_2 \tag{6.1.58}$$

Eliminating γ from (6.1.56), we obtain

$$b_{23} a_1 da_1 + b_{13} a_2 da_2 = 0 \tag{6.1.59}$$

i.e.,

$$\frac{1}{2} d\left(b_{23} a_1^2 + b_{13} a_2^2\right) = 0$$

therefore,

$$b_{23} a_1^2 + b_{13} a_2^2 = E, \quad E = b_{23} a_{10}^2 + b_{13} a_{20}^2 \tag{6.1.60}$$

Substitute,, and into and write εa_1 and εa_2 as a_1 and a_2, respectively (in this case, equations, and remain unchanged, but need to treat the prime as the derivative with respect to time t), we can obtain the first-order approximate solution of the original equation

$$\begin{Bmatrix} \theta_1 \\ \theta_2 \end{Bmatrix} = [\Psi] \begin{Bmatrix} a_1(t)\cos[\omega_1 t + \beta_1(t)] \\ a_2(t)\cos[\omega_2 t + \beta_2(t)] \end{Bmatrix} + O\left(\varepsilon^3\right) \tag{6.1.61}$$

where a_1, a_2, β_1 and β_2 can be determined by,, and.

Next we analyze the equations describing amplitude and phase of the motion, ~ . Eliminating β_1', β_2' from the two equations in, we obtain

$$a_2 \gamma' = a_2 \sigma + \frac{1}{8}(3b_{11} - b_{22}) a_1^2 a_2 + \frac{1}{8}(3b_{12} - b_{21}) a_2^3 + \frac{1}{8}\left(3b_{13} a_1 a_2^2 - b_{23} a_1^3\right)\cos\gamma \tag{6.1.62}$$

Dividing the above equation by the second equation of yields

$$\begin{aligned} -b_{23} a_1^3 a_2 \sin\gamma \tfrac{d\gamma}{da_2} &= 8\sigma a_2 + (3b_{11} - b_{22}) a_1^2 a_2 + (3b_{12} - b_{21}) a_2^3 \\ &+ \left(3b_{13} a_1 a_2^2 - b_{23} a_1^3\right)\cos\gamma \end{aligned}$$

i.e.,

$$b_{23}a_1^3a_2d\cos\gamma + b_{23}a_1^3\cos\gamma da_2 - 3b_{13}a_1a_2^2\cos\gamma da_2$$
$$= 8\sigma a_2da_2 + (3b_{11} - b_{22})a_1^2a_2da_2 + (3b_{12} - b_{21})a_2^3da_2$$

Using, the above equation can be written as

$$\begin{aligned}&b_{23}a_1^3d(a_2\cos\gamma) - 3b_{13}a_1a_2^2\cos\gamma da_2 = \left[8\sigma + (3b_{11} - b_{22})\tfrac{E}{b_{23}}\right]a_2da_2\\&+\left[(3b_{12} - b_{21}) - (3b_{11} - b_{22})\tfrac{b_{13}}{b_{23}}\right]a_2^3da_2\end{aligned} \tag{6.1.63}$$

since

$$a_1^3d(a_2\cos\gamma) = d\left(a_1^3a_2\cos\gamma\right) - 3a_1^2a_2\cos\gamma da_1 \tag{6.1.64}$$

By substituting into the left side of, we can obtain

$$\begin{aligned}&b_{23}d\left(a_1^3a_2\cos\gamma\right) - 3a_1a_2\cos\gamma(b_{23}a_1da_1 + b_{13}a_2da_2)\\&= \left[8\sigma + (3b_{11} - b_{22})\frac{E}{b_{23}}\right]a_2da_2 + \left[(3b_{12} - b_{21}) - (3b_{11} - b_{22})\frac{b_{13}}{b_{23}}\right]a_2^3da_2\end{aligned} \tag{6.1.65}$$

From, we can find that the second term on the left side of equation is zero, so

$$\begin{aligned}&b_{23}d\left(a_1^3a_2\cos\gamma\right) = \left[8\sigma + (3b_{11} - b_{22})\tfrac{E}{b_{23}}\right]a_2da_2\\&+\left[(3b_{12} - b_{21}) - (3b_{11} - b_{22})\tfrac{b_{13}}{b_{23}}\right]a_2^3da_2\end{aligned} \tag{6.1.66}$$

Make the integration of equations, we can obtain

$$\begin{aligned}&b_{23}a_1^3a_2\cos\gamma - \frac{1}{2}\left[8\sigma + (3b_{11} - b_{22})\frac{E}{b_{23}}\right]a_2^2\\&+\frac{1}{4}\left[(3b_{11} - b_{22})\frac{b_{13}}{b_{23}} - (3b_{12} - b_{21})\right]a_2^4 = L\end{aligned} \tag{6.1.67}$$

where L is the constant of integration. Now, let us combine, and into a single variable differential equation. To do this, let

$$a_1^2 = E\xi \tag{6.1.68}$$

Then from, we can obtain

$$a_2^2 = E(1 - b_{23}\xi)/b_{13} \tag{6.1.69}$$

Eliminating γ from and yields

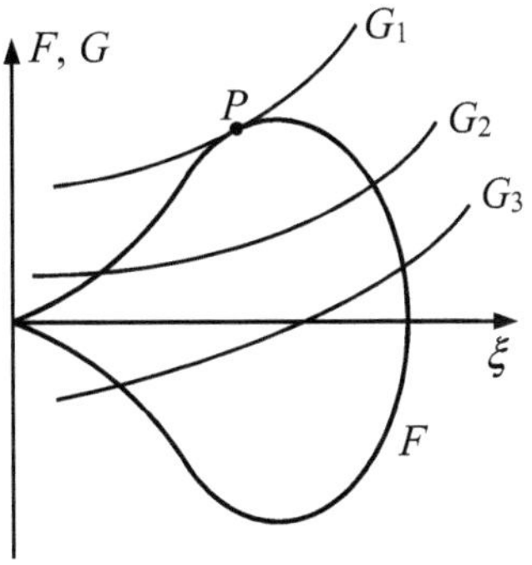

Fig. 6.1 Schematic diagram of $F(\xi)$ and $G(\xi)$ for Exercise 6.1

$$\xi'^2 = F^2(\xi) - G^2(\xi) \tag{6.1.70}$$

where

$$\begin{aligned} &F(\xi) = \pm\tfrac{1}{4}\left[b_{13}E^2\xi^3(1-b_{23}\xi)\right]^{1/2} \\ &G(\xi) = \tfrac{1}{4}\Bigg(\tfrac{(1-b_{23}\xi)^2}{b_{23}^2}\Bigg\{L + \tfrac{1}{2}\Big[8\sigma + (3b_{11}-b_{22})\tfrac{E}{b_{23}}\Big] \\ &+\tfrac{1}{4}\Big[(3b_{12}-b_{21}) - (3b_{11}-b_{22})\tfrac{b_{13}}{b_{23}}\Big]\tfrac{E(1-b_{23}\xi)}{b_{13}}\Bigg\}^2\Bigg)^{1/2} \end{aligned} \tag{6.1.71}$$

The functions $F(\xi)$ and $G(\xi)$ are shown schematically in Fig. 6.1. For real motions, $F^2(\xi) \geq G^2(\xi)$. The points where F meets G corresponds to the vanishing of $\xi\prime$. A curve such as G_2 which meets one branch of F at two different points or a curve G_3 which meets both branches corresponds to a periodic solution. In this case, ξ is periodic and oscillates between two intersection points, and the motion is aperiodic.

Considering the curves G_1 and F are tangent to each other, as shown in Fig. 6.1. In this case, ξ has only one constant-value solution, and the motion of the system is periodic. The nonlinear effect acts to modulate the phase of the motion so that the nonlinear frequencies of the two modes are commensurable.

In order to analyze the steady state solution, let $a_1' = a_2' = \gamma' = 0$, and become.

$$\sin\gamma = 0 \quad \text{or} \quad \gamma = n\pi \tag{6.1.72}$$

$$a_2\sigma + \frac{1}{8}(3b_{11} - b_{22})a_1^2 a_2 + \frac{1}{8}(3b_{12} - b_{21})a_2^3 + \frac{1}{8}\left(3b_{13}a_1 a_2^2 - b_{23}a_1^3\right)\cos n\pi = 0 \tag{6.1.73}$$

Equation is the cubic equation of a_2. For a given set of values of σ, a_1 and $\cos n\pi$, equation has one or three real roots, each of which corresponds to a periodic motion of the system. Thus, the periodic motion of the system is either unique or one of the three possible periodic motions. Figure 6.1 shows that the periodic motion is unstable. With a small perturbation, a curve such as G_1 may become a curve such as G_2 and the motion of the system becomes aperiodic.

For the case of the nonlinear periodic motion of the system with a constant-value solution, we note that the frequency of the motion of the system is

$$\hat{\omega}_1 = \omega_1 + \varepsilon^2 \beta_1', \quad \hat{\omega}_2 = \omega_2 + \varepsilon^2 \beta_2' \tag{6.1.74}$$

therefore,

$$\hat{\omega}_2 - 3\hat{\omega}_1 = \omega_2 - 3\omega_1 + \varepsilon^2(\beta_2' - 3\beta_1') = \varepsilon^2\sigma + \varepsilon^2(\beta_2' - 3\beta_1') = \varepsilon^2\gamma' = 0 \tag{6.1.75}$$

Thus, the nonlinearity of the system modulates the frequency of the motion to exactly 3 to 1, and thus the motion is periodic.

6.2 Exercise 6.2 (Internal Resonance Analysis of a Uniform Rod Hanging from a Massless Chord)

Solution: (a) The kinetic energy of the system is

$$\begin{aligned} T &= \tfrac{1}{2}m\Big[\big(l_1\dot{\theta}_1\cos\theta_1 + \tfrac{1}{2}l_2\dot{\theta}_2\cos\theta_2\big)^2 + \big(-l_1\dot{\theta}_1\sin\theta_1 - \tfrac{1}{2}l_2\dot{\theta}_2\sin\theta_2\big)^2\Big] + \tfrac{1}{2}\big(\tfrac{1}{12}ml_2^2\big)\dot{\theta}_2^2 \\ &= \tfrac{1}{2}ml_1^2\dot{\theta}_1^2 + \tfrac{1}{6}ml_2^2\dot{\theta}_2^2 + \tfrac{1}{2}ml_1l_2\dot{\theta}_1\dot{\theta}_2\cos(\theta_2 - \theta_1) \end{aligned} \tag{6.2.1}$$

The potential energy of the system is

$$V = -mg\left(l_1\cos\theta_1 + \frac{1}{2}l_2\cos\theta_2\right) \tag{6.2.2}$$

Substitute kinetic and potential energy into the Lagrange's equation:

$$\frac{d}{dt}\frac{\partial T}{\partial \dot{\theta}_k} - \frac{\partial T}{\partial \theta_k} = -\frac{\partial V}{\partial \theta_k}, \quad k = 1, \ 2$$

we can obtain

$$\begin{aligned} &l_1\ddot{\theta}_1 + \frac{1}{2}l_2\ddot{\theta}_2\cos(\theta_2 - \theta_1) - \frac{1}{2}l_2\dot{\theta}_2^2\sin(\theta_2 - \theta_1) + g\sin\theta_1 = 0 \\ &\frac{1}{3}l_2\ddot{\theta}_2 + \frac{1}{2}l_1\ddot{\theta}_1\cos(\theta_2 - \theta_1) + \frac{1}{2}l_1\dot{\theta}_1^2\sin(\theta_2 - \theta_1) + \frac{1}{2}g\sin\theta_2 = 0 \end{aligned} \tag{6.2.3}$$

(b) The equilibrium point of the system is $\theta_1 = \theta_2 = 0$. Linearizing (6.2.3) near the equilibrium point yields

$$l_1\ddot{\theta}_1 + \frac{1}{2}l_2\ddot{\theta}_2 + g\theta_1 = 0$$
$$\frac{1}{2}l_1\ddot{\theta}_1 + \frac{1}{3}l_2\ddot{\theta}_2 + \frac{1}{2}g\theta_2 = 0$$

which can be written in the following matrix form:

$$\begin{bmatrix} 2l_1l_2^{-1} & 1 \\ 1 & \frac{2}{3}l_1^{-1}l_2 \end{bmatrix}\begin{Bmatrix} \ddot{\theta}_1 \\ \ddot{\theta}_2 \end{Bmatrix} + \begin{bmatrix} 2gl_2^{-1} & 0 \\ 0 & gl_1^{-1} \end{bmatrix}\begin{Bmatrix} \theta_1 \\ \theta_2 \end{Bmatrix} = 0 \tag{6.2.4}$$

The characteristic equation of this linear system is

$$\omega^4 - 2g\left(2l_1^{-1} + 3l_2^{-1}\right)\omega^2 + 6g^2l_1^{-1}l_2^{-1} = 0 \tag{6.2.5}$$

The linear natural frequencies of the system, ω_1 and ω_2, are

$$\begin{aligned} \omega_1 &= \left[\left(2l_1^{-1} + 3l_2^{-1}\right)g - g\sqrt{\left(2l_1^{-1} + 3l_2^{-1}\right)^2 - 6l_1^{-1}l_2^{-1}}\right]^{1/2} \\ \omega_2 &= \left[\left(2l_1^{-1} + 3l_2^{-1}\right)g + g\sqrt{\left(2l_1^{-1} + 3l_2^{-1}\right)^2 - 6l_1^{-1}l_2^{-1}}\right]^{1/2} \end{aligned} \tag{6.2.6}$$

Assuming that the corresponding normal modal matrix is $[\Psi]$, then using

$$\{\theta_1,\ \theta_2\}^T = [\Psi]\{u_1,\ u_2\}^T \tag{6.2.7}$$

we can transform (6.1.4) into

$$\begin{aligned} \ddot{u}_1 + \omega_1^2 u_1 &= 0 \\ \ddot{u}_2 + \omega_2^2 u_2 &= 0 \end{aligned} \tag{6.2.8}$$

If the system yields an s-to-one internal resonance, i.e., $\omega_2 : \omega_1 = s : 1$, and let $\kappa = l_2/l_1$, there are

$$\left[\left(s^2+1\right)^2 - \left(s^2-1\right)^2\right](2\kappa+3)^2 - 6\left(s^2+1\right)^2\kappa = 0 \tag{6.2.9}$$

Once s is given, $\kappa = l_2/l_1$ can be determined.

For small but finite amplitudes, we assume $\theta_1, \theta_2 = O(\varepsilon)$. Expanding (6.2.3) and retaining to $O\left(\varepsilon^3\right)$ yields

$$\begin{aligned} &\ddot{\theta}_1 + \frac{1}{2}l_1^{-1}l_2\ddot{\theta}_2 + gl_1^{-1}\theta_1 - \frac{1}{6}gl_1^{-1}\theta_1^3 - \frac{1}{4}l_1^{-1}l_2\ddot{\theta}_2(\theta_2-\theta_1)^2 - \frac{1}{2}l_1^{-1}l_2\dot{\theta}_2^2(\theta_2-\theta_1) = 0 \\ &\frac{3}{2}l_1l_2^{-1}\ddot{\theta}_1 + \ddot{\theta}_2 + \frac{3}{2}gl_2^{-1}\theta_2 - \frac{1}{4}gl_2^{-1}\theta_2^3 - \frac{3}{4}l_1l_2^{-1}\ddot{\theta}_1(\theta_2-\theta_1)^2 + \frac{3}{2}l_1l_2^{-1}\dot{\theta}_1^2(\theta_2-\theta_1) = 0 \end{aligned} \tag{6.2.10}$$

Comparing (6.2.10) with (6.1.14) in the previous exercise, we can find that if the following substitutions are made in (6.1.14):

$$\alpha \to \frac{1}{2} l_1^{-1} l_2, \quad l_2^{-1} \to \frac{3}{2} l_2^{-1} \tag{6.2.11}$$

(6.2.10) and (6.1.14) become the same.

We use the method of multiple scales to solve (6.2.10). Therefore, let the solution of Eq. (6.2.10) be expressed as

$$\begin{aligned} \theta_1 &= \varepsilon\theta_{11}(T_0, T_1, T_2) + \varepsilon^2\theta_{12}(T_0, T_1, T_2) + \varepsilon^3\theta_{13}(T_0, T_1, T_2) + \cdots \\ \theta_2 &= \varepsilon\theta_{21}(T_0, T_1, T_2) + \varepsilon^2\theta_{22}(T_0, T_1, T_2) + \varepsilon^3\theta_{23}(T_0, T_1, T_2) + \cdots \end{aligned} \tag{6.2.12}$$

Substituting (6.2.12) into (6.2.10), keeping to $O(\varepsilon^3)$, and equating coefficients of like powers of ε, we can obtain the results similar to (6.1.18)–(6.1.20). The specific result can be obtained by using the substitution (6.2.11) in (6.1.18)–(6.1.20).

Order ε^1:

$$D_0^2\theta_{11} + \frac{1}{2} l_1^{-1} l_2 D_0^2\theta_{21} + g l_1^{-1}\theta_{11} = 0 \, \frac{3}{2} l_1 l_2^{-1} D_0^2\theta_{11} + D_0^2\theta_{21} + \frac{3}{2} g l_2^{-1}\theta_{21} = 0 \tag{6.2.13}$$

Order ε^2:

$$\begin{aligned} & D_0^2\theta_{12} + \frac{1}{2} l_1^{-1} l_2 D_0^2\theta_{22} + g l_1^{-1}\theta_{12} + 2D_0D_1\theta_{11} + l_1^{-1} l_2 D_0 D_1\theta_{21} = 0 \\ & \frac{3}{2} l_1 l_2^{-1} D_0^2\theta_{12} + D_0^2\theta_{22} + \frac{3}{2} g l_2^{-1}\theta_{22} + 3 l_1 l_2^{-1} D_0 D_1\theta_{11} + 2D_0D_1\theta_{21} = 0 \end{aligned} \tag{6.2.14}$$

Order ε^3:

$$\begin{aligned} & D_0^2\theta_{13} + \frac{1}{2} l_1^{-1} l_2 D_0^2\theta_{23} + g l_1^{-1}\theta_{13} + 2D_0D_1\theta_{12} + l_1^{-1} l_2 D_0 D_1\theta_{22} \\ & \quad + (D_1^2 + 2D_0D_2)\theta_{11} + \frac{1}{2} l_1^{-1} l_2 (D_1^2 + 2D_0D_2)\theta_{21} \\ & \quad - \frac{1}{4} l_1^{-1} l_2 (\theta_{21} - \theta_{11})^2 D_0^2\theta_{21} - \frac{1}{2} l_1^{-1} l_2 (\theta_{21} - \theta_{11})(D_0\theta_{21})^2 - \frac{1}{6} g l_1^{-1}\theta_{11}^3 = 0 \\ & \frac{3}{2} l_1 l_2^{-1} D_0^2\theta_{13} + D_0^2\theta_{23} + \frac{3}{2} g l_2^{-1}\theta_{23} + 3 l_1 l_2^{-1} D_0 D_1\theta_{12} + 2D_0D_1\theta_{22} \\ & \quad + (D_1^2 + 2D_0D_2)\theta_{21} + \frac{3}{2} l_1 l_2^{-1} (D_1^2 + 2D_0D_2)\theta_{11} \\ & - \frac{3}{4} l_1 l_2^{-1} (\theta_{21} - \theta_{11})^2 D_0^2\theta_{11} + \frac{3}{2} l_1 l_2^{-1} (\theta_{21} - \theta_{11})(D_0\theta_{11})^2 - \frac{1}{4} g l_2^{-1}\theta_{21}^3 = 0 \end{aligned} \tag{6.2.15}$$

(6.2.13)–(6.2.15) can also be written in matrix form:

$$\begin{bmatrix} 2l_1l_2^{-1} & 1 \\ 1 & \frac{2}{3}l_1^{-1}l_2 \end{bmatrix}\begin{Bmatrix} D_0^2\theta_{11} \\ D_0^2\theta_{21} \end{Bmatrix} + \begin{bmatrix} 2gl_2^{-1} & 0 \\ 0 & gl_1^{-1} \end{bmatrix}\begin{Bmatrix} \theta_{11} \\ \theta_{21} \end{Bmatrix} = 0 \tag{6.2.16}$$

$$\begin{bmatrix} 2l_1l_2^{-1} & 1 \\ 1 & \frac{2}{3}l_1^{-1}l_2 \end{bmatrix}\begin{Bmatrix} D_0^2\theta_{12} \\ D_0^2\theta_{22} \end{Bmatrix} + \begin{bmatrix} 2gl_2^{-1} & 0 \\ 0 & gl_1^{-1} \end{bmatrix}\begin{Bmatrix} \theta_{12} \\ \theta_{22} \end{Bmatrix} = -2\begin{bmatrix} 2l_1l_2^{-1} & 1 \\ 1 & \frac{2}{3}l_1^{-1}l_2 \end{bmatrix}D_0D_1\begin{Bmatrix} \theta_{11} \\ \theta_{21} \end{Bmatrix} \tag{6.2.17}$$

$$\begin{aligned} &\begin{bmatrix} 2l_1l_2^{-1} & 1 \\ 1 & \frac{2}{3}l_1^{-1}l_2 \end{bmatrix}\begin{Bmatrix} D_0^2\theta_{13} \\ D_0^2\theta_{23} \end{Bmatrix} + \begin{bmatrix} 2gl_2^{-1} & 0 \\ 0 & gl_1^{-1} \end{bmatrix}\begin{Bmatrix} \theta_{13} \\ \theta_{23} \end{Bmatrix} = -2\begin{bmatrix} 2l_1l_2^{-1} & 1 \\ 1 & \frac{2}{3}l_1^{-1}l_2 \end{bmatrix}D_0D_1\begin{Bmatrix} \theta_{12} \\ \theta_{22} \end{Bmatrix} \\ &\quad -\begin{bmatrix} 2l_1l_2^{-1} & 1 \\ 1 & \frac{2}{3}l_1^{-1}l_2 \end{bmatrix}\left(D_1^2 + 2D_0D_2\right)\begin{Bmatrix} \theta_{11} \\ \theta_{21} \end{Bmatrix} + \frac{1}{2}(\theta_{21} - \theta_{11})^2D_0^2\begin{Bmatrix} \theta_{21} \\ \theta_{11} \end{Bmatrix} \\ &\quad -\begin{Bmatrix} -(\theta_{21} - \theta_{11})(D_0\theta_{21})^2 \\ (\theta_{21} - \theta_{11})(D_0\theta_{11})^2 \end{Bmatrix} + \frac{1}{6}gl_1^{-1}\begin{Bmatrix} 2l_1l_2^{-1}\theta_{11}^3 \\ \theta_{21}^3 \end{Bmatrix} \end{aligned} \tag{6.2.18}$$

Apply the transformation to (6.2.16)–(6.2.18), i.e.,

$$\{\theta_{1n},\ \theta_{2n}\}^T = [\Psi]\{u_{1n},\ u_{2n}\}^T, \quad n = 1, 2, 3 \tag{6.2.19}$$

we can obtain

$$\begin{Bmatrix} D_0^2u_{11} \\ D_0^2u_{21} \end{Bmatrix} + \begin{bmatrix} \omega_1^2 & 0 \\ 0 & \omega_2^2 \end{bmatrix}\begin{Bmatrix} u_{11} \\ u_{21} \end{Bmatrix} = 0 \tag{6.2.20}$$

$$\begin{Bmatrix} D_0^2u_{12} \\ D_0^2u_{22} \end{Bmatrix} + \begin{bmatrix} \omega_1^2 & 0 \\ 0 & \omega_2^2 \end{bmatrix}\begin{Bmatrix} u_{12} \\ u_{22} \end{Bmatrix} = -2D_0D_1\begin{Bmatrix} u_{11} \\ u_{21} \end{Bmatrix} \tag{6.2.21}$$

$$\begin{Bmatrix} D_0^2u_{13} \\ D_0^2u_{23} \end{Bmatrix} + \begin{bmatrix} \omega_1^2 & 0 \\ 0 & \omega_2^2 \end{bmatrix}\begin{Bmatrix} u_{13} \\ u_{23} \end{Bmatrix} = -2D_0D_1\begin{Bmatrix} u_{12} \\ u_{22} \end{Bmatrix} - \left(D_1^2 + 2D_0D_2\right)\begin{Bmatrix} u_{11} \\ u_{21} \end{Bmatrix} + \begin{Bmatrix} F_1 \\ F_2 \end{Bmatrix} \tag{6.2.22}$$

where

$$\begin{aligned} F_1 &= \tfrac{1}{2}\Psi_{11}(\theta_{21} - \theta_{11})^2D_0^2\theta_{21} + \tfrac{1}{2}\Psi_{21}(\theta_{21} - \theta_{11})^2D_0^2\theta_{11} \\ &+\Psi_{11}(\theta_{21} - \theta_{11})(D_0\theta_{21})^2 - \Psi_{21}(\theta_{21} - \theta_{11})(D_0\theta_{11})^2 \\ &+\tfrac{1}{3}\Psi_{11}gl_2^{-1}\theta_{11}^3 + \tfrac{1}{6}\Psi_{21}gl_1^{-1}\theta_{21}^3 \end{aligned} \tag{6.2.23}$$

$$\begin{aligned} F_2 &= \tfrac{1}{2}\Psi_{12}(\theta_{21} - \theta_{11})^2D_0^2\theta_{21} + \tfrac{1}{2}\Psi_{22}(\theta_{21} - \theta_{11})^2D_0^2\theta_{11} \\ &+\Psi_{12}(\theta_{21} - \theta_{11})(D_0\theta_{21})^2 - \Psi_{22}(\theta_{21} - \theta_{11})(D_0\theta_{11})^2 \\ &+\tfrac{1}{3}\Psi_{12}gl_2^{-1}\theta_{11}^3 + \tfrac{1}{6}\Psi_{22}gl_1^{-1}\theta_{21}^3 \end{aligned} \tag{6.2.24}$$

and Ψ_{ij} is the element of the normal modal matrix $[\Psi]$. Equations (6.2.23) and (6.2.24) can be obtained from (6.1.28) and (6.1.29) using (6.2.11).

The solution of (6.2.20) is

$$\begin{Bmatrix} u_{11} \\ u_{21} \end{Bmatrix} = \begin{Bmatrix} A_1 e^{i\omega_1 T_0} + \overline{A}_1 e^{-i\omega_1 T_0} \\ A_2 e^{i\omega_2 T_0} + \overline{A}_2 e^{-i\omega_2 T_0} \end{Bmatrix} \tag{6.2.25}$$

where $A_k = A_k(T_1, T_2)$. Substituting (6.2.25) int (6.2.21), we obtain

$$\begin{aligned} D_0^2 u_{12} + \omega_1^2 u_{12} &= -2\left(i\omega_1 D_1 A_1 e^{i\omega_1 t} - i\omega_1 D_1 \overline{A}_1 e^{i\omega_1 T_0}\right) \\ D_0^2 u_{22} + \omega_2^2 u_{22} &= -2\left(i\omega_2 D_1 A_2 e^{i\omega_2 t} - i\omega_2 D_1 \overline{A}_2 e^{i\omega_2 T_0}\right) \end{aligned} \tag{6.2.26}$$

In order to eliminate secular terms from the above equation, there must be

$$D_1 A_1 = 0, \quad D_1 A_2 = 0 \quad \Rightarrow \quad A_1 = A_1(T_2), \quad A_2 = A_2(T_2) \tag{6.2.27}$$

therefore

$$u_{12} = u_{12} = 0 \tag{6.2.28}$$

Substituting (6.2.25) and (6.2.28) into (6.2.22) yields

$$D_0^2 u_{13} + \omega_1^2 u_{13} = -2i\omega_1 A_1' e^{i\omega_1 T_0} + cc + F_1 \tag{6.2.29}$$

$$D_0^2 u_{23} + \omega_2^2 u_{23} = -2i\omega_2 A_2' e^{i\omega_2 T_0} + cc + F_2 \tag{6.2.30}$$

The prime represents the partial derivative with respect to T_2. Substituting the transformation $\{\theta_{11},\ \theta_{21}\}^T = [\Psi]\{u_{11},\ u_{21}\}^T$ and (6.2.25) into (6.2.23) and (6.2.24) and the result can also be obtained from Eqs. (6.1.36) and (6.1.37) using (6.2.11)

$$
\begin{aligned}
F_1 = & -3\omega_1^2\Psi_{11}\Psi_{21}h_1^2A_1^2\overline{A}_1e^{i\omega_1T_0} - 2\omega_1^2\Psi_{11}\Psi_{21}h_2^2A_1A_2\overline{A}_2e^{i\omega_1T_0} \\
& -2\omega_2^2\Psi_{11}\Psi_{22}h_1h_2A_1A_2\overline{A}_2e^{i\omega_1T_0} - 2\omega_2^2\Psi_{12}\Psi_{21}h_1h_2A_1A_2\overline{A}_2e^{i\omega_1T_0} \\
& -2\omega_1^2\Psi_{11}\Psi_{21}h_1h_2\overline{A}_1^2A_2e^{i(\omega_2-2\omega_1)T_0} - \frac{1}{2}\omega_2^2\Psi_{11}\Psi_{22}h_1^2\overline{A}_1^2A_2e^{i(\omega_2-2\omega_1)T_0} \\
& -\frac{1}{2}\omega_2^2\Psi_{12}\Psi_{21}h_1^2\overline{A}_1^2A_2e^{i(\omega_2-2\omega_1)T_0} - \frac{1}{2}\omega_2^2\Psi_{11}\Psi_{22}h_2^2A_2^3e^{i3\omega_2T_0} \\
& -\frac{1}{2}\omega_2^2\Psi_{12}\Psi_{21}h_2^2A_2^3e^{i3\omega_2T_0} + \omega_1^2\Psi_{11}\Psi_{21}^2h_1A_1^2\overline{A}_1e^{i\omega_1T_0} - \omega_1^2\Psi_{11}^2\Psi_{21}h_1A_1^2\overline{A}_1e^{i\omega_1T_0} \\
& +2\omega_2^2\Psi_{11}\Psi_{22}^2h_1A_1A_2\overline{A}_2e^{i\omega_1T_0} - 2\omega_2^2\Psi_{12}^2\Psi_{21}h_1A_1A_2\overline{A}_2e^{i\omega_1T_0} \\
& +2\omega_1\omega_2\Psi_{11}\Psi_{21}\Psi_{22}h_1\overline{A}_1^2A_2e^{i(\omega_2-2\omega_1)T_0} - \omega_1^2\Psi_{11}\Psi_{21}^2h_2\overline{A}_1^2A_2e^{i(\omega_2-2\omega_1)T_0} \\
& -2\omega_1\omega_2\Psi_{11}\Psi_{12}\Psi_{21}h_1\overline{A}_1^2A_2e^{i(\omega_2-2\omega_1)T_0} + \omega_1^2\Psi_{11}^2\Psi_{21}h_2\overline{A}_1^2A_2e^{i(\omega_2-2\omega_1)T_0} \\
& -\omega_2^2\Psi_{11}\Psi_{22}^2h_2A_2^3e^{i3\omega_2T_0} + \omega_2^2\Psi_{12}^2\Psi_{21}h_2A_2^3e^{i3\omega_2T_0} \\
& +gl_2^{-1}\Psi_{11}^4A_1^2\overline{A}_1e^{i\omega_1T_0} + 2gl_2^{-1}\Psi_{11}^2\Psi_{12}^2A_1A_2\overline{A}_2e^{i\omega_1T_0} \\
& +\frac{1}{2}gl_1^{-1}\Psi_{21}^4A_1^2\overline{A}_1e^{i\omega_1T_0} + gl_1^{-1}\Psi_{21}^2\Psi_{22}^2A_1A_2\overline{A}_2e^{i\omega_1T_0} \\
& +gl_2^{-1}\Psi_{11}^3\Psi_{12}\overline{A}_1^2A_2e^{i(\omega_2-2\omega_1)T_0} + \frac{1}{2}gl_1^{-1}\Psi_{21}^3\Psi_{22}\overline{A}_1^2A_2e^{i(\omega_2-2\omega_1)T_0} \\
& +\frac{1}{3}gl_2^{-1}\Psi_{11}\Psi_{12}^3A_2^3e^{i3\omega_2T_0} + \frac{1}{6}gl_1^{-1}\Psi_{21}\Psi_{22}^3A_2^3e^{i3\omega_2T_0} + cc + NST
\end{aligned}
\tag{6.2.31}
$$

$$
\begin{aligned}
F_2 = &-2\omega_2^2\Psi_{12}\Psi_{22}h_1^2A_1\overline{A}_1A_2e^{i\omega_2T_0} - 3\omega_2^2\Psi_{12}\Psi_{22}h_2^2A_2^2\overline{A}_2e^{i\omega_2T_0} \\
&-2\omega_1^2\Psi_{12}\Psi_{21}h_1h_2A_1\overline{A}_1A_2e^{i\omega_2T_0} - 2\omega_1^2\Psi_{11}\Psi_{22}h_1h_2A_1\overline{A}_1A_2e^{i\omega_2T_0} \\
&-2\omega_2^2\Psi_{12}\Psi_{22}h_1h_2A_1\overline{A}_2^2e^{i(\omega_1-2\omega_2)T_0} - \frac{1}{2}\omega_1^2\Psi_{12}\Psi_{21}h_2^2A_1\overline{A}_2^2e^{i(\omega_1-2\omega_2)T_0} \\
&-\frac{1}{2}\omega_1^2\Psi_{11}\Psi_{22}h_2^2A_1\overline{A}_2^2e^{i(\omega_1-2\omega_2)T_0} - \frac{1}{2}\omega_1^2\Psi_{12}\Psi_{22}h_1^2A_1^3e^{i3\omega_1T_0} \\
&-\frac{1}{2}\omega_1^2\Psi_{12}\Psi_{21}h_1^2A_1^3e^{i3\omega_1T_0} + \omega_2^2\Psi_{12}^2\Psi_{22}h_2A_2^2\overline{A}_2e^{i\omega_2T_0} - \omega_2^2\Psi_{12}\Psi_{22}^2h_2A_2^2\overline{A}_2e^{i\omega_2T_0} \\
&-2\omega_1^2\Psi_{12}\Psi_{21}^2h_2A_1\overline{A}_1A_2e^{i\omega_2T_0} + 2\omega_1^2\Psi_{11}^2\Psi_{22}h_2A_1\overline{A}_1A_2e^{i\omega_2T_0} \\
&+\omega_2^2\Psi_{12}\Psi_{22}^2h_1A_1\overline{A}_2^2e^{i(\omega_1-2\omega_2)T_0} - 2\omega_1\omega_2\Psi_{12}\Psi_{21}\Psi_{22}h_2A_1\overline{A}_2^2e^{i(\omega_1-2\omega_2)T_0} \\
&-\omega_2^2\Psi_{12}^2\Psi_{22}h_1A_1\overline{A}_2^2e^{i(\omega_1-2\omega_2)T_0} + 2\omega_1\omega_2\Psi_{11}\Psi_{12}\Psi_{22}h_2A_1\overline{A}_2^2e^{i(\omega_1-2\omega_2)T_0} \\
&+\omega_1^2\Psi_{12}\Psi_{21}^2h_1A_1^3e^{i3\omega_1T_0} - \omega_1^2\Psi_{11}^2\Psi_{22}h_1A_1^3e^{i3\omega_1T_0} \\
&+gl_2^{-1}\Psi_{12}^4A_2^2\overline{A}_2e^{i\omega_2T_0} + 2gl_2^{-1}\Psi_{11}^2\Psi_{12}^2A_1\overline{A}_1A_2e^{i\omega_2T_0} \\
&+\frac{1}{2}gl_1^{-1}\Psi_{22}^4A_2^2\overline{A}_2e^{i\omega_2T_0} + gl_1^{-1}\Psi_{21}^2\Psi_{22}^2A_1\overline{A}_1A_2e^{i\omega_2T_0} \\
&+gl_2^{-1}\Psi_{11}\Psi_{12}^3A_1\overline{A}_2^2e^{i(\omega_1-2\omega_2)T_0} + \frac{1}{2}gl_1^{-1}\Psi_{21}\Psi_{22}^3A_1\overline{A}_2^2e^{i(\omega_1-2\omega_2)T_0} \\
&+\frac{1}{6}gl_1^{-1}\Psi_{21}^3\Psi_{22}A_1^3e^{i3\omega_1T_0} + \frac{1}{3}gl_2^{-1}\Psi_{11}^3\Psi_{12}A_1^3e^{i3\omega_1T_0} + cc + NST
\end{aligned} \tag{6.2.32}
$$

where

$$h_1 = \Psi_{21} - \Psi_{11}, \quad h_2 = \Psi_{22} - \Psi_{12} \tag{6.2.33}$$

To obtain (6.2.29)–(6.2.32), we did not consider the case of $\omega_1 = \omega_2$. From (6.2.29)–(6.2.32), we can find that there are two possible resonant cases: $\omega_2 \approx 3\omega_1$ and $\omega_1 \approx 3\omega_2$. Without loss of generality, we assume $\omega_2 > \omega_1$. In analyzing the particular solutions of (6.2.29) and (6.2.30) we need to distinguish between the resonant case in which $\omega_2 \approx 3\omega_1$ and the nonresonant case in which ω_2 is away from $3\omega_1$.

(c) If $s = \omega_2 : \omega_1 = 3 : 1$, then from (6.2.9), we can obtain

$$3(2\kappa + 3)^2 - 50\kappa = 0 \tag{6.2.34}$$

The fact that the equation has no real number solution suggests that there is no length ratio κ that corresponds exactly to $s = 3$, so we examine the $s \sim \kappa$ relationship (Fig. 6.2a), which is given by (6.2.6).

$$s = \left[\frac{(2\kappa + 3) + \sqrt{(2\kappa + 3)^2 - 6\kappa}}{(2\kappa + 3) - \sqrt{(2\kappa + 3)^2 - 6\kappa}}\right]^{1/2} \tag{6.2.35}$$

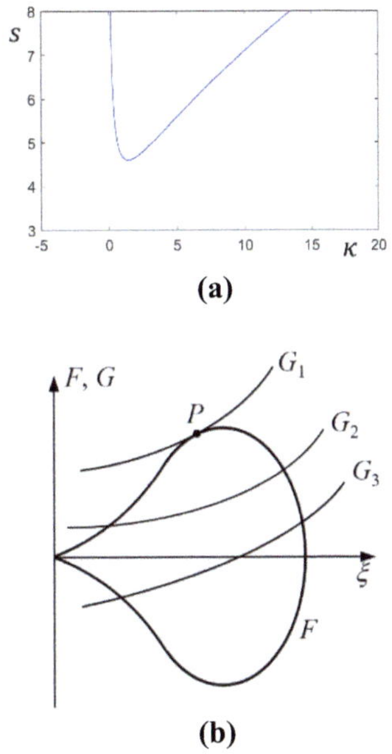

Fig. 6.2 **a** $\kappa \sim s$ curve. **b** Schematic diagram of $F(\xi)$ and $G(\xi)$ in Exercise 6.2

The minimum value of s is 3.7321 at $\kappa = 1.52$. This situation can be considered as the internal resonance of $\omega_2 \approx 3\omega_1$.

The internal resonance of $\omega_2 \approx 3\omega_1$ is investigated below. In this case, the detuning parameter σ is introduced such that

$$\omega_2 = 3\omega_1 + \varepsilon^2\sigma \tag{6.2.36}$$

In order to eliminate secular terms in (6.2.29)–(6.2.32), we need

$$2A_1' + ib_{11}A_1^2\overline{A}_1 + ib_{12}A_1A_2\overline{A}_2 + ib_{13}\overline{A}_1^2A_2e^{i\sigma T_2} = 0 \tag{6.2.37}$$

$$2A_2' + ib_{21}A_2^2\overline{A}_2 + ib_{22}A_1\overline{A}_1A_2 + ib_{23}A_1^3e^{-i\sigma T_2} = 0 \tag{6.2.38}$$

where

$$\begin{aligned}
&b_{11} = \omega_1^{-1}\big(\omega_1^2\Psi_{11}\Psi_{21}^2h_1 - 3\omega_1^2\Psi_{11}\Psi_{21}h_1^2 \\
&-\omega_1^2\Psi_{11}^2\Psi_{21}h_1 + \frac{1}{2}gl_1^{-1}\Psi_{21}^4 + gl_2^{-1}\Psi_{11}^4\Big) \\
&b_{12} = \omega_1^{-1}\big(2\omega_2^2\Psi_{11}\Psi_{22}^2h_1 - 2\omega_2^2\Psi_{12}^2\Psi_{21}h_1 - 2\omega_1^2\Psi_{11}\Psi_{21}h_2^2 - 2\omega_2^2\Psi_{11}\Psi_{22}h_1h_2 \\
&-2\omega_2^2\Psi_{12}\Psi_{21}h_1h_2 + 2gl_2^{-1}\Psi_{11}^2\Psi_{12}^2 + gl_1^{-1}\Psi_{21}^2\Psi_{22}^2\big)A_1A_2\overline{A}_2
\end{aligned} \tag{6.2.39}$$

$$b_{21} = \omega_2^{-1}\big(\omega_2^2\Psi_{12}^2\Psi_{22}h_2 - 3\omega_2^2\Psi_{12}\Psi_{22}h_2^2$$
$$- \omega_2^2\Psi_{12}\Psi_{22}^2h_2 + gl_2^{-1}\Psi_{12}^4 + \frac{1}{2}gl_1^{-1}\Psi_{22}^4\Big)$$
$$b_{22} = \omega_2^{-1}\big(2\omega_1^2\Psi_{11}^2\Psi_{22}h_2 - 2\omega_1^2\Psi_{12}\Psi_{21}^2h_2 - 2\omega_2^2\Psi_{12}\Psi_{22}h_1^2 - 2\omega_1^2\Psi_{12}\Psi_{21}h_1h_2$$
$$- 2\omega_1^2\Psi_{11}\Psi_{22}h_1h_2 + gl_2^{-1}\Psi_{11}^2\Psi_{12}^2 + gl_1^{-1}\Psi_{21}^2\Psi_{22}^2\big) \tag{6.2.40}$$

$$b_{13} = \omega_1^{-1}\big(\omega_1^2\Psi_{11}^2\Psi_{21}h_2 - \omega_1^2\Psi_{11}\Psi_{21}^2h_2 - 2\omega_1^2\Psi_{11}\Psi_{21}h_1h_2$$
$$+2\omega_1\omega_2\Psi_{11}\Psi_{21}\Psi_{22}h_1 - 2\omega_1\omega_2\Psi_{11}\Psi_{12}\Psi_{21}h_1$$
$$- \frac{1}{2}\omega_2^2\Psi_{11}\Psi_{22}h_1^2 - \frac{1}{2}\omega_2^2\Psi_{12}\Psi_{21}h_1^2 + gl_2^{-1}\Psi_{11}^3\Psi_{12} + \frac{1}{2}gl_1^{-1}\Psi_{21}^3\Psi_{22}\Big)$$
$$b_{23} = \omega_2^{-1}\left(\omega_1^2\Psi_{12}\Psi_{21}^2h_1 - \omega_1^2\Psi_{11}^2\Psi_{22}h_1 - \frac{1}{2}\omega_1^2\Psi_{12}\Psi_{22}h_1^2 - \frac{1}{2}\omega_1^2\Psi_{12}\Psi_{21}h_1^2\right.$$
$$\left.+\frac{1}{6}gl_1^{-1}\Psi_{21}^3\Psi_{22} + \frac{1}{3}gl_2^{-1}\Psi_{11}^3\Psi_{12}\right) \tag{6.2.41}$$

Substituting $A_k = \frac{1}{2}ae^{i\beta_k}, \quad k = 1, 2$ into (6.2.36) and (6.2.37), we get

$$a_1' + i\beta_1'a_1 + \frac{1}{8}ib_{11}a_1^3 + \frac{1}{8}ib_{12}a_1a_2^2 + i\frac{1}{8}b_{13}a_1^2a_2e^{i(\sigma T_2 - 3\beta_1 + \beta_2)} = 0 \tag{6.2.38}$$

$$a_2' + i\beta_2'a_2 + \frac{1}{8}ib_{21}a_2^3 + \frac{1}{8}ib_{22}a_1^2a_2 + i\frac{1}{8}b_{23}a_1^3e^{-i(\sigma T_2 - 3\beta_1 + \beta_2)} = 0 \tag{6.2.39}$$

Separating the real and imaginary parts of the above equations yields

$$\begin{aligned} a_1' &= \tfrac{1}{8}b_{13}a_1^2a_2\sin\gamma \\ a_2' &= -\tfrac{1}{8}b_{23}a_1^3\sin\gamma \end{aligned} \tag{6.2.40}$$

$$\begin{aligned} \beta_1'a_1 + \tfrac{1}{8}b_{11}a_1^3 + \tfrac{1}{8}b_{12}a_1a_2^2 + \tfrac{1}{8}b_{13}a_1^2a_2\cos\gamma &= 0 \\ \beta_2'a_2 + \tfrac{1}{8}b_{21}a_2^3 + \tfrac{1}{8}b_{22}a_1^2a_2 + \tfrac{1}{8}b_{23}a_1^3\cos\gamma &= 0 \end{aligned} \tag{6.2.41}$$

where

$$\gamma = \sigma T_2 - 3\beta_1 + \beta_2 \tag{6.2.42}$$

Eliminating γ from (6.2.40), we obtain

$$b_{23}a_1da_1 + b_{13}a_2da_2 = 0 \tag{6.2.43}$$

i.e.,

$$\frac{1}{2}d\left(b_{23}a_1^2+b_{13}a_2^2\right)=0$$

therefore,

$$b_{23}a_1^2+b_{13}a_2^2=E,\quad E=b_{23}a_{10}^2+b_{13}a_{20}^2 \tag{6.2.44}$$

In this case, the Eqs. (6.2.40), (6.2.41) and (6.2.44) remain unchanged, but you need to treat the prime as the derivative with respect to time$t$. Substitute (6.2.19), (6.2.25) and (6.2.28) as well as the polar expression $A_k=\frac{1}{2}ae^{i\beta_k},\quad k=1,2$ into (6.2.12), and write εa_1,εa_2 as a_1,a_2, we can obtain the first order approximate solution of the original equation:

$$\begin{Bmatrix}\theta_1\\ \theta_2\end{Bmatrix}=[\Psi]\begin{Bmatrix}a_1(t)\cos[\omega_1 t+\beta_1(t)]\\ a_2(t)\cos[\omega_2 t+\beta_2(t)]\end{Bmatrix}+O\left(\varepsilon^3\right) \tag{6.2.45}$$

where a_1,a_2,β_1 and β_2 can be determined by (6.2.40), (6.2.41), (6.2.42) and (6.2.44).

Next we analyze the equations describing amplitude and phase of the motion, (6.2.40)–(6.2.42). Eliminating $\beta_1',\quad \beta_2'$ from the two equations in (6.2.41), we obtain

$$a_2\gamma'=a_2\sigma+\frac{1}{8}(3b_{11}-b_{22})a_1^2a_2+\frac{1}{8}(3b_{12}-b_{21})a_2^3+\frac{1}{8}\left(3b_{13}a_1a_2^2-b_{23}a_1^3\right)\cos\gamma \tag{6.2.46}$$

Dividing the above equation by the second equation of (6.2.40) yields

$$\begin{aligned}&-b_{23}a_1^3a_2\sin\gamma\tfrac{d\gamma}{da_2}=8\sigma a_2+(3b_{11}-b_{22})a_1^2a_2+(3b_{12}-b_{21})a_2^3\\&+\left(3b_{13}a_1a_2^2-b_{23}a_1^3\right)\cos\gamma\end{aligned}$$

i.e.,

$$\begin{aligned}&b_{23}a_1^3a_2d\cos\gamma+b_{23}a_1^3\cos\gamma da_2-3b_{13}a_1a_2^2\cos\gamma da_2\\&\quad=8\sigma a_2da_2+(3b_{11}-b_{22})a_1^2a_2da_2+(3b_{12}-b_{21})a_2^3da_2\end{aligned}$$

Using (6.2.44), the above equation can be written as

$$\begin{aligned}&b_{23}a_1^3d(a_2\cos\gamma)-3b_{13}a_1a_2^2\cos\gamma da_2=\left[8\sigma+(3b_{11}-b_{22})\tfrac{E}{b_{23}}\right]a_2da_2\\&+\left[(3b_{12}-b_{21})-(3b_{11}-b_{22})\tfrac{b_{13}}{b_{23}}\right]a_2^3da_2\end{aligned} \tag{6.2.47}$$

since

$$a_1^3d(a_2\cos\gamma)=d\left(a_1^3a_2\cos\gamma\right)-3a_1^2a_2\cos\gamma da_1 \tag{6.2.48}$$

By substituting (6.2.48) into the left side of (6.2.47), we can obtain

$$b_{23}d\left(a_1^3a_2\cos\gamma\right)-3a_1a_2\cos\gamma(b_{23}a_1da_1+b_{13}a_2da_2)$$
$$=\left[8\sigma+(3b_{11}-b_{22})\frac{E}{b_{23}}\right]a_2da_2+\left[(3b_{12}-b_{21})-(3b_{11}-b_{22})\frac{b_{13}}{b_{23}}\right]a_2^3da_2 \tag{6.2.49}$$

From (6.2.43), we can find that the second term on the left side of Eq. (6.2.49) is zero, so

$$\begin{aligned}b_{23}d\left(a_1^3a_2\cos\gamma\right)&=\left[8\sigma+(3b_{11}-b_{22})\tfrac{E}{b_{23}}\right]a_2da_2\\&+\left[(3b_{12}-b_{21})-(3b_{11}-b_{22})\tfrac{b_{13}}{b_{23}}\right]a_2^3da_2\end{aligned} \tag{6.2.50}$$

Make the integration of Eq. (6.2.50), we can obtain

$$\begin{aligned}&b_{23}a_1^3a_2\cos\gamma-\frac{1}{2}\left[8\sigma+(3b_{11}-b_{22})\frac{E}{b_{23}}\right]a_2^2\\&+\frac{1}{4}\left[(3b_{11}-b_{22})\frac{b_{13}}{b_{23}}-(3b_{12}-b_{21})\right]a_2^4=L\end{aligned} \tag{6.2.51}$$

where L is the constant of integration. Now, let us combine (6.2.51), (6.2.44) and (6.2.40) into a single variable differential equation. To do this, let

$$a_1^2=E\xi \tag{6.2.52}$$

Then from (6.2.44), we can obtain

$$a_2^2=E(1-b_{23}\xi)/b_{13} \tag{6.2.53}$$

Eliminating γ from (6.2.51) and (6.2.44) yields

$$\xi'^2=F^2(\xi)-G^2(\xi) \tag{6.2.54}$$

where

$$\begin{aligned}F(\xi)&=\pm\tfrac{1}{4}\left[b_{13}E^2\xi^3(1-b_{23}\xi)\right]^{1/2}\\G(\xi)&=\tfrac{1}{4}\left(\tfrac{(1-b_{23}\xi)^2}{b_{23}^2}\left\{L+\tfrac{1}{2}\left[8\sigma+(3b_{11}-b_{22})\tfrac{E}{b_{23}}\right]\right.\right.\\&\left.\left.+\tfrac{1}{4}\left[(3b_{12}-b_{21})-(3b_{11}-b_{22})\tfrac{b_{13}}{b_{23}}\right]\tfrac{E(1-b_{23}\xi)}{b_{13}}\right\}^2\right)^{1/2}\end{aligned} \tag{6.2.55}$$

The functions $F(\xi)$ and $G(\xi)$ are shown schematically in Fig. 6.2b. For real motions, $F^2(\xi)\geq G^2(\xi)$. The points where F meets G corresponds to the vanishing of ξ'. A curve such as G_2 which meets one branch of F at two different points or a curve G_3 which meets both branches corresponds to a periodic solution. In this case, ξ is periodic and oscillates between two intersection points, and the motion is aperiodic.

Considering the curves G_1 and F are tangent to each other, as shown in Fig. 6.2b. In this case, ξ has only one constant-value solution, and the motion of the system is periodic. The nonlinear effect acts to modulate the phase of the motion so that the nonlinear frequencies of the two modes are commensurable.

In order to analyze the steady state solution, let $a_1' = a_2' = \gamma' = 0$, (6.2.40) and (6.2.51) become.

$$\sin\gamma = 0 \quad \text{or} \quad \gamma = n\pi \tag{6.2.56}$$

$$a_2\sigma + \frac{1}{8}(3b_{11} - b_{22})a_1^2 a_2 + \frac{1}{8}(3b_{12} - b_{21})a_2^3 + \frac{1}{8}\left(3b_{13}a_1 a_2^2 - b_{23}a_1^3\right)\cos n\pi = 0 \tag{6.2.57}$$

Equation (6.2.57) is the cubic equation of a_2. For a given set of values of σ, a_1 and $\cos n\pi$, Eq. (6.2.57) has one or three real roots, each of which corresponds to a periodic motion of the system. Thus, the periodic motion of the system is either unique or one of the three possible periodic motions. Figure 6.2b shows that the periodic motion is unstable. With a small perturbation, a curve such as G_1 may become a curve such as G_2 and the motion of the system becomes aperiodic.

For the case of the nonlinear periodic motion of the system with a constant-value solution, we note that the frequency of the motion of the system is

$$\hat{\omega}_1 = \omega_1 + \varepsilon^2\beta_1', \quad \hat{\omega}_2 = \omega_2 + \varepsilon^2\beta_2' \tag{6.2.58}$$

therefore,

$$\hat{\omega}_2 - 3\hat{\omega}_1 = \omega_2 - 3\omega_1 + \varepsilon^2\left(\beta_2' - 3\beta_1'\right) = \varepsilon^2\sigma + \varepsilon^2\left(\beta_2' - 3\beta_1'\right) = \varepsilon^2\gamma' = 0 \tag{6.2.59}$$

Thus, the nonlinearity of the system modulates the frequency of the motion to exactly 3 to 1, and thus the motion is periodic.

6.3 Exercise 6.3 (Internal Resonance Analysis of a Disc Pendulum)

Solution: (a) The kinetic energy of the system is

$$\begin{aligned} T &= \tfrac{1}{2}I\dot{\theta}_1^2 + \tfrac{1}{2}m\left[\left(R\dot{\theta}_1\cos\theta_1 + r\dot{\theta}_2\cos\theta_2\right)^2 + \left(-R\dot{\theta}_1\sin\theta_1 - r\dot{\theta}_2\sin\theta_2\right)^2\right] \\ &= \tfrac{1}{2}m\left(R^2 + \rho^2\right)\dot{\theta}_1^2 + \tfrac{1}{2}mr^2\dot{\theta}_2^2 + mRr\dot{\theta}_1\dot{\theta}_2\cos(\theta_2 - \theta_1) \end{aligned} \tag{6.3.1}$$

The potential energy of the system is

$$V = -mg(R\cos\theta_1 + r\cos\theta_2) \tag{6.3.2}$$

Substitute kinetic and potential energy into the Lagrange's equation:

$$\frac{d}{dt}\frac{\partial T}{\partial \dot{\theta}_k} - \frac{\partial T}{\partial \theta_k} = -\frac{\partial V}{\partial \theta_k}, \quad k = 1,\ 2$$

we can obtain

$$\begin{aligned}&(R^2+\rho^2)\ddot{\theta}_1 + Rr\ddot{\theta}_2\cos(\theta_2-\theta_1) - Rr\dot{\theta}_2^2\sin(\theta_2-\theta_1) + gR\sin\theta_1 = 0\\ &r\ddot{\theta}_2 + R\ddot{\theta}_1\cos(\theta_2-\theta_1) + R\dot{\theta}_1^2\sin(\theta_2-\theta_1) + g\sin\theta_2 = 0\end{aligned} \tag{6.3.3}$$

(b) The equilibrium position of the system is $\theta_1 = \theta_2 = 0$. Linearizing (6.3.3) near the equilibrium point yields

$$\begin{aligned}&(R^2+\rho^2)\ddot{\theta}_1 + Rr\ddot{\theta}_2 + gR\theta_1 = 0\\ &R\ddot{\theta}_1 + r\ddot{\theta}_2 + g\theta_2 = 0\end{aligned}$$

which can be written in the following matrix form:

$$\begin{bmatrix} m_{11} & 1 \\ 1 & m_{22} \end{bmatrix}\begin{Bmatrix} \ddot{\theta}_1 \\ \ddot{\theta}_2 \end{Bmatrix} + \begin{bmatrix} k_{11} & 0 \\ 0 & k_{22} \end{bmatrix}\begin{Bmatrix} \theta_1 \\ \theta_2 \end{Bmatrix} = 0 \tag{6.3.4}$$

$$m_{11} = (R^2+\rho^2)R^{-1}r^{-1}, \quad m_{22} = R^{-1}r, \quad k_{11} = gr^{-1}, \quad k_{22} = gR^{-1} \tag{6.3.5}$$

The characteristic equation of this linear system is

$$(m_{11}m_{22}-1)\omega^4 - (m_{11}k_{22} + m_{22}k_{11})\omega^2 + k_{11}k_{22} = 0 \tag{6.3.6}$$

The linear natural frequencies of the system, ω_1 and ω_2, are

$$\begin{aligned}\omega_1 &= \frac{(m_{11}k_{22}+m_{22}k_{11}) - \sqrt{(m_{11}k_{22}+m_{22}k_{11})^2 - 4k_{11}k_{22}(m_{11}m_{22}-1)}}{2(m_{11}m_{22}-1)}\\ \omega_2 &= \frac{(m_{11}k_{22}+m_{22}k_{11}) + \sqrt{(m_{11}k_{22}+m_{22}k_{11})^2 - 4k_{11}k_{22}(m_{11}m_{22}-1)}}{2(m_{11}m_{22}-1)}\end{aligned} \tag{6.3.7}$$

Assuming that the corresponding normal modal matrix is $[\Psi]$, then using

$$\{\theta_1,\ \theta_2\}^T = [\Psi]\{u_1,\ u_2\}^T \tag{6.3.8}$$

we can transform (6.3.4) into

$$\begin{aligned} \ddot{u}_1 + \omega_1^2 u_1 &= 0 \\ \ddot{u}_2 + \omega_2^2 u_2 &= 0 \end{aligned} \tag{6.3.9}$$

If the system yields an s-to-one internal resonance, i.e., $\omega_2 : \omega_1 = s : 1$, there are

$$\left[\left(s^2-1\right)^2-\left(s^2+1\right)^2\right](m_{11}k_{22}+m_{22}k_{11})^2+4k_{11}k_{22}(m_{11}m_{22}-1)\left(s^2+1\right)^2=0 \tag{6.3.10}$$

Once s is given, the relation between r, R and ρ can be determined.

For small but finite amplitudes, we assume $\theta_1, \theta_2 = O(\varepsilon)$. Expanding (6.3.3) and retaining to $O\left(\varepsilon^3\right)$ yields

$$\begin{aligned} m_{11}\ddot{\theta}_1+\ddot{\theta}_2+k_{11}\theta_1-\frac{1}{6}k_{11}\theta_1^3-\frac{1}{2}\ddot{\theta}_2(\theta_2-\theta_1)^2-\dot{\theta}_2^2(\theta_2-\theta_1)=0 \\ \ddot{\theta}_1+m_{22}\ddot{\theta}_2+k_{22}\theta_2-\frac{1}{6}k_{22}\theta_2^3-\frac{1}{2}\ddot{\theta}_1(\theta_2-\theta_1)^2+\dot{\theta}_1^2(\theta_2-\theta_1)=0 \end{aligned} \tag{6.3.11}$$

We use the method of multiple scales to solve (6.3.11). Therefore, let the solution of Eq. (6.3.11) be expressed as

$$\begin{aligned} \theta_1 &= \varepsilon\theta_{11}(T_0, T_2)+\varepsilon^3\theta_{13}(T_0, T_2)+\cdots \\ \theta_2 &= \varepsilon\theta_{21}(T_0, T_2)+\varepsilon^3\theta_{23}(T_0, T_2)+\cdots \end{aligned} \tag{6.3.12}$$

where, from the experience of previous two exercises, the second-order term is zero, so the second-order term in (6.3.12) is removed directly. Substituting (6.3.12) into (6.3.11), keeping to O (ε^3), we obtain

$$\begin{aligned} 0 &= m_{11}\ddot{\theta}_1+\ddot{\theta}_2+k_{11}\theta_1-\frac{1}{6}k_{11}\theta_1^3-\frac{1}{2}\ddot{\theta}_2(\theta_2-\theta_1)^2-\dot{\theta}_2^2(\theta_2-\theta_1) \\ &= m_{11}\left(D_0^2+2\varepsilon^2D_0D_2\right)\left(\varepsilon\theta_{11}+\varepsilon^3\theta_{13}\right)+\left(D_0^2+2\varepsilon^2D_0D_2\right)\left(\varepsilon\theta_{21}+\varepsilon^3\theta_{23}\right) \\ &\quad +k_{11}\left(\varepsilon\theta_{11}+\varepsilon^3\theta_{13}\right)-\frac{1}{6}k_{11}\left(\varepsilon\theta_{11}+\varepsilon^3\theta_{13}\right)^3 \\ &\quad -\frac{1}{2}\left(\varepsilon\theta_{21}-\varepsilon\theta_{11}+\varepsilon^3\theta_{23}-\varepsilon^3\theta_{13}\right)^2\left(D_0^2+2\varepsilon^2D_0D_2\right)\left(\varepsilon\theta_{21}+\varepsilon^3\theta_{23}\right) \\ &\quad -\left(\varepsilon\theta_{21}-\varepsilon\theta_{11}+\varepsilon^3\theta_{23}-\varepsilon^3\theta_{13}\right)\left[\left(D_0+\varepsilon^2D_2\right)\left(\varepsilon\theta_{21}+\varepsilon^3\theta_{23}\right)\right]^2+\cdots \\ &= \varepsilon\left(m_{11}D_0^2\theta_{11}+D_0^2\theta_{21}+k_{11}\theta_{11}\right) \\ &\quad +\varepsilon^3\Big[m_{11}D_0^2\theta_{13}+D_0^2\theta_{23}+k_{11}\theta_{13}+2m_{11}D_0D_2\theta_{11}+2D_0D_2\theta_{21} \\ &\quad -\frac{1}{2}(\theta_{21}-\theta_{11})^2D_0^2\theta_{21}-(\theta_{21}-\theta_{11})(D_0\theta_{21})^2-\frac{1}{6}k_{11}\theta_{11}^3\Big]+\cdots \end{aligned} \tag{6.3.13}$$

$$
\begin{aligned}
0 &= \ddot{\theta}_1 + m_{22}\ddot{\theta}_2 + k_{22}\theta_2 - \frac{1}{6}k_{22}\theta_2^3 - \frac{1}{2}\ddot{\theta}_1(\theta_2-\theta_1)^2 + \dot{\theta}_1^2(\theta_2-\theta_1) \\
&= \left(D_0^2 + 2\varepsilon^2 D_0 D_2\right)\left(\varepsilon\theta_{11} + \varepsilon^3\theta_{13}\right) + m_{22}\left(D_0^2 + 2\varepsilon^2 D_0 D_2\right)\left(\varepsilon\theta_{21} + \varepsilon^3\theta_{23}\right) \\
&\quad + k_{22}\left(\varepsilon\theta_{21} + \varepsilon^3\theta_{23}\right) - \frac{1}{6}k_{22}\left(\varepsilon\theta_{21} + \varepsilon^3\theta_{23}\right)^3 \\
&\quad - \frac{1}{2}\left(\varepsilon\theta_{21} - \varepsilon\theta_{11} + \varepsilon^3\theta_{23} - \varepsilon^3\theta_{13}\right)^2\left(D_0^2 + 2\varepsilon^2 D_0 D_2\right)\left(\varepsilon\theta_{11} + \varepsilon^3\theta_{13}\right) \\
&\quad + \left(\varepsilon\theta_{21} - \varepsilon\theta_{11} + \varepsilon^3\theta_{23} - \varepsilon^3\theta_{13}\right)\left[\left(D_0 + \varepsilon^2 D_2\right)\left(\varepsilon\theta_{11} + \varepsilon^3\theta_{13}\right)\right]^2 + \cdots \\
&= \varepsilon\left(D_0^2\theta_{11} + m_{22}D_0^2\theta_{21} + k_{22}\theta_{22}\right) \\
&\quad \varepsilon^3\Big[D_0^2\theta_{13} + m_{22}D_0^2\theta_{23} + k_{22}\theta_{23} + 2D_0D_2\theta_{11} + 2m_{22}D_0D_2\theta_{21} \\
&\quad - \frac{1}{2}(\theta_{21}-\theta_{11})^2 D_0^2\theta_{11} + (\theta_{21}-\theta_{11})(D_0\theta_{11})^2 - \frac{1}{6}k_{22}\theta_{21}^3\Big] + \cdots
\end{aligned}
\tag{6.3.14}
$$

Equating coefficients of like powers of ε in above equations yields.
Order ε^1:

$$
\begin{aligned}
m_{11}D_0^2\theta_{11} + D_0^2\theta_{21} + k_{11}\theta_{11} &= 0 \\
D_0^2\theta_{11} + m_{22}D_0^2\theta_{21} + k_{22}\theta_{22} &= 0
\end{aligned}
\tag{6.3.15}
$$

Order ε^3:

$$
\begin{aligned}
& m_{11}D_0^2\theta_{13} + D_0^2\theta_{23} + k_{11}\theta_{13} + 2m_{11}D_0D_2\theta_{11} + 2D_0D_2\theta_{21} \\
& \quad - \frac{1}{2}(\theta_{21}-\theta_{11})^2 D_0^2\theta_{21} - (\theta_{21}-\theta_{11})(D_0\theta_{21})^2 - \frac{1}{6}k_{11}\theta_{11}^3 = 0 \\
& D_0^2\theta_{13} + m_{22}D_0^2\theta_{23} + k_{22}\theta_{23} + 2D_0D_2\theta_{11} + 2m_{22}D_0D_2\theta_{21} \\
& \quad - \frac{1}{2}(\theta_{21}-\theta_{11})^2 D_0^2\theta_{11} + (\theta_{21}-\theta_{11})(D_0\theta_{11})^2 - \frac{1}{6}k_{22}\theta_{21}^3 = 0
\end{aligned}
\tag{6.3.16}
$$

(6.3.15) and (6.3.16)can also be written in matrix form:

$$
\begin{bmatrix} m_{11} & 1 \\ 1 & m_{22} \end{bmatrix}\begin{Bmatrix} D_0^2\theta_{11} \\ D_0^2\theta_{21} \end{Bmatrix} + \begin{bmatrix} k_{11} & 0 \\ 0 & k_{22} \end{bmatrix}\begin{Bmatrix} \theta_{11} \\ \theta_{21} \end{Bmatrix} = 0
\tag{6.3.17}
$$

$$
\begin{aligned}
& \begin{bmatrix} m_{11} & 1 \\ 1 & m_{22} \end{bmatrix}\begin{Bmatrix} D_0^2\theta_{13} \\ D_0^2\theta_{23} \end{Bmatrix} + \begin{bmatrix} k_{11} & 0 \\ 0 & k_{22} \end{bmatrix}\begin{Bmatrix} \theta_{13} \\ \theta_{23} \end{Bmatrix} = -\begin{bmatrix} m_{11} & 1 \\ 1 & m_{22} \end{bmatrix} 2D_0D_2\begin{Bmatrix} \theta_{11} \\ \theta_{21} \end{Bmatrix} \\
& + \begin{Bmatrix} \frac{1}{2}(\theta_{21}-\theta_{11})^2 D_0^2\theta_{21} + (\theta_{21}-\theta_{11})(D_0\theta_{21})^2 + \frac{1}{6}k_{11}\theta_{11}^3 \\ \frac{1}{2}(\theta_{21}-\theta_{11})^2 D_0^2\theta_{11} - (\theta_{21}-\theta_{11})(D_0\theta_{11})^2 + \frac{1}{6}k_{22}\theta_{21}^3 \end{Bmatrix}
\end{aligned}
\tag{6.3.18}
$$

Apply the transformation to (6.3.17) and (6.3.18), i.e.,

$$
\{\theta_{1n},\ \theta_{2n}\}^T = [\Psi]\{u_{1n},\ u_{2n}\}^T,\quad n = 1, 3
\tag{6.3.19}
$$

we can obtain

$$\begin{Bmatrix} D_0^2 u_{11} \\ D_0^2 u_{21} \end{Bmatrix} + \begin{bmatrix} \omega_1^2 & 0 \\ 0 & \omega_2^2 \end{bmatrix} \begin{Bmatrix} u_{11} \\ u_{21} \end{Bmatrix} = 0 \tag{6.3.20}$$

$$\begin{Bmatrix} D_0^2 u_{13} \\ D_0^2 u_{23} \end{Bmatrix} + \begin{bmatrix} \omega_1^2 & 0 \\ 0 & \omega_2^2 \end{bmatrix} \begin{Bmatrix} u_{13} \\ u_{23} \end{Bmatrix} = -2D_0 D_2 \begin{Bmatrix} u_{11} \\ u_{21} \end{Bmatrix} + \begin{Bmatrix} F_1 \\ F_2 \end{Bmatrix} \tag{6.3.21}$$

where

$$\begin{aligned} F_1 = {} & \tfrac{1}{2}\Psi_{11}(\theta_{21}-\theta_{11})^2 D_0^2\theta_{21} + \tfrac{1}{2}\Psi_{21}(\theta_{21}-\theta_{11})^2 D_0^2\theta_{11} \\ & + \Psi_{11}(\theta_{21}-\theta_{11})(D_0\theta_{21})^2 - \Psi_{21}(\theta_{21}-\theta_{11})(D_0\theta_{11})^2 \\ & + \tfrac{1}{6}\Psi_{11}k_{11}\theta_{11}^3 + \tfrac{1}{6}\Psi_{21}k_{22}\theta_{21}^3 \end{aligned} \tag{6.3.22}$$

$$\begin{aligned} F_2 = {} & \tfrac{1}{2}\Psi_{12}(\theta_{21}-\theta_{11})^2 D_0^2\theta_{21} + \tfrac{1}{2}\Psi_{22}(\theta_{21}-\theta_{11})^2 D_0^2\theta_{11} \\ & + \Psi_{12}(\theta_{21}-\theta_{11})(D_0\theta_{21})^2 - \Psi_{22}(\theta_{21}-\theta_{11})(D_0\theta_{11})^2 \\ & + \tfrac{1}{6}\Psi_{12}k_{11}\theta_{11}^3 + \tfrac{1}{6}\Psi_{22}k_{22}\theta_{21}^3 \end{aligned} \tag{6.3.23}$$

and Ψ_{ij} is the element of the normal modal matrix $[\Psi]$. Equations (6.3.22) and (6.3.23) can be obtained from (6.1.28) and (6.1.29) using the following substitution:

$$g l_1^{-1}\alpha^{-1} \to k_{11}, \quad g l_1^{-1} \to k_{22} \tag{6.3.24}$$

The solution of (6.3.20) is

$$\begin{Bmatrix} u_{11} \\ u_{21} \end{Bmatrix} = \begin{Bmatrix} A_1 e^{i\omega_1 T_0} + \overline{A}_1 e^{-i\omega_1 T_0} \\ A_2 e^{i\omega_2 T_0} + \overline{A}_2 e^{-i\omega_2 T_0} \end{Bmatrix} \tag{6.3.25}$$

where $A_k = A_k(T_2)$. Substituting (6.3.25) into (6.3.21), we obtain

$$D_0^2 u_{13} + \omega_1^2 u_{13} = -2i\omega_1 A_1' e^{i\omega_1 T_0} + cc + F_1 \tag{6.3.26}$$

$$D_0^2 u_{23} + \omega_2^2 u_{23} = -2i\omega_2 A_2' e^{i\omega_2 T_0} + cc + F_2 \tag{6.3.27}$$

The prime represents the partial derivative with respect to T_2. Substituting the transformation $\{\theta_{11},\ \theta_{21}\}^T = [\Psi]\{u_{11},\ u_{21}\}^T$ and (6.3.25) into (6.3.22) and (6.3.23) and the result can also be obtained from Eqs. (6.1.36) and (6.1.37) using (6.3.24):

$$
\begin{aligned}
F_1 = & -3\omega_1^2\Psi_{11}\Psi_{21}h_1^2A_1^2\overline{A}_1e^{i\omega_1T_0} - 2\omega_1^2\Psi_{11}\Psi_{21}h_2^2A_1A_2\overline{A}_2e^{i\omega_1T_0} \\
& -2\omega_2^2\Psi_{11}\Psi_{22}h_1h_2A_1A_2\overline{A}_2e^{i\omega_1T_0} - 2\omega_2^2\Psi_{12}\Psi_{21}h_1h_2A_1A_2\overline{A}_2e^{i\omega_1T_0} \\
& -2\omega_1^2\Psi_{11}\Psi_{21}h_1h_2\overline{A}_1^2A_2e^{i(\omega_2-2\omega_1)T_0} - \frac{1}{2}\omega_2^2\Psi_{11}\Psi_{22}h_1^2\overline{A}_1^2A_2e^{i(\omega_2-2\omega_1)T_0} \\
& -\frac{1}{2}\omega_2^2\Psi_{12}\Psi_{21}h_1^2\overline{A}_1^2A_2e^{i(\omega_2-2\omega_1)T_0} - \frac{1}{2}\omega_2^2\Psi_{11}\Psi_{22}h_2^2A_2^3e^{i3\omega_2T_0} \\
& -\frac{1}{2}\omega_2^2\Psi_{12}\Psi_{21}h_2^2A_2^3e^{i3\omega_2T_0} + \omega_1^2\Psi_{11}\Psi_{21}^2h_1A_1^2\overline{A}_1e^{i\omega_1T_0} - \omega_1^2\Psi_{11}^2\Psi_{21}h_1A_1^2\overline{A}_1e^{i\omega_1T_0} \\
& +2\omega_2^2\Psi_{11}\Psi_{22}^2h_1A_1A_2\overline{A}_2e^{i\omega_1T_0} - 2\omega_2^2\Psi_{12}^2\Psi_{21}h_1A_1A_2\overline{A}_2e^{i\omega_1T_0} \\
& +2\omega_1\omega_2\Psi_{11}\Psi_{21}\Psi_{22}h_1\overline{A}_1^2A_2e^{i(\omega_2-2\omega_1)T_0} - \omega_1^2\Psi_{11}\Psi_{21}^2h_2\overline{A}_1^2A_2e^{i(\omega_2-2\omega_1)T_0} \\
& -2\omega_1\omega_2\Psi_{11}\Psi_{12}\Psi_{21}h_1\overline{A}_1^2A_2e^{i(\omega_2-2\omega_1)T_0} + \omega_1^2\Psi_{11}^2\Psi_{21}h_2\overline{A}_1^2A_2e^{i(\omega_2-2\omega_1)T_0} \\
& -\omega_2^2\Psi_{11}\Psi_{22}^2h_2A_2^3e^{i3\omega_2T_0} + \omega_2^2\Psi_{12}^2\Psi_{21}h_2A_2^3e^{i3\omega_2T_0} \\
& +\frac{1}{2}k_{11}\Psi_{11}^4A_1^2\overline{A}_1e^{i\omega_1T_0} + k_{11}\Psi_{11}^2\Psi_{12}^2A_1A_2\overline{A}_2e^{i\omega_1T_0} \\
& +\frac{1}{2}k_{22}\Psi_{21}^4A_1^2\overline{A}_1e^{i\omega_1T_0} + k_{22}\Psi_{21}^2\Psi_{22}^2A_1A_2\overline{A}_2e^{i\omega_1T_0} \\
& +\frac{1}{2}k_{11}\Psi_{11}^3\Psi_{12}\overline{A}_1^2A_2e^{i(\omega_2-2\omega_1)T_0} + \frac{1}{2}k_{22}\Psi_{21}^3\Psi_{22}\overline{A}_1^2A_2e^{i(\omega_2-2\omega_1)T_0} \\
& +\frac{1}{6}k_{11}\Psi_{11}\Psi_{12}^3A_2^3e^{i3\omega_2T_0} + \frac{1}{6}k_{22}\Psi_{21}\Psi_{22}^3A_2^3e^{i3\omega_2T_0} + cc + NST
\end{aligned}
\tag{6.3.28}
$$

$$\begin{aligned}
F_2 = & -2\omega_2^2\Psi_{12}\Psi_{22}h_1^2A_1\overline{A}_1A_2e^{i\omega_2T_0} - 3\omega_2^2\Psi_{12}\Psi_{22}h_2^2A_2^2\overline{A}_2e^{i\omega_2T_0} \\
& -2\omega_1^2\Psi_{12}\Psi_{21}h_1h_2A_1\overline{A}_1A_2e^{i\omega_2T_0} - 2\omega_1^2\Psi_{11}\Psi_{22}h_1h_2A_1\overline{A}_1A_2e^{i\omega_2T_0} \\
& -2\omega_2^2\Psi_{12}\Psi_{22}h_1h_2A_1\overline{A}_2^2e^{i(\omega_1-2\omega_2)T_0} - \frac{1}{2}\omega_1^2\Psi_{12}\Psi_{21}h_2^2A_1\overline{A}_2^2e^{i(\omega_1-2\omega_2)T_0} \\
& -\frac{1}{2}\omega_1^2\Psi_{11}\Psi_{22}h_2^2A_1\overline{A}_2^2e^{i(\omega_1-2\omega_2)T_0} - \frac{1}{2}\omega_1^2\Psi_{12}\Psi_{22}h_1^2A_1^3e^{i3\omega_1T_0} \\
& -\frac{1}{2}\omega_1^2\Psi_{12}\Psi_{21}h_1^2A_1^3e^{i3\omega_1T_0} + \omega_2^2\Psi_{12}^2\Psi_{22}h_2A_2^2\overline{A}_2e^{i\omega_2T_0} - \omega_2^2\Psi_{12}\Psi_{22}^2h_2A_2^2\overline{A}_2e^{i\omega_2T_0} \\
& -2\omega_1^2\Psi_{12}\Psi_{21}^2h_2A_1\overline{A}_1A_2e^{i\omega_2T_0} + 2\omega_1^2\Psi_{11}^2\Psi_{22}h_2A_1\overline{A}_1A_2e^{i\omega_2T_0} \\
& +\omega_2^2\Psi_{12}\Psi_{22}^2h_1A_1\overline{A}_2^2e^{i(\omega_1-2\omega_2)T_0} - 2\omega_1\omega_2\Psi_{12}\Psi_{21}\Psi_{22}h_2A_1\overline{A}_2^2e^{i(\omega_1-2\omega_2)T_0} \\
& -\omega_2^2\Psi_{12}^2\Psi_{22}h_1A_1\overline{A}_2^2e^{i(\omega_1-2\omega_2)T_0} + 2\omega_1\omega_2\Psi_{11}\Psi_{12}\Psi_{22}h_2A_1\overline{A}_2^2e^{i(\omega_1-2\omega_2)T_0} \\
& +\omega_1^2\Psi_{12}\Psi_{21}^2h_1A_1^3e^{i3\omega_1T_0} - \omega_1^2\Psi_{11}^2\Psi_{22}h_1A_1^3e^{i3\omega_1T_0} \\
& +\frac{1}{2}k_{11}\Psi_{12}^4A_2^2\overline{A}_2e^{i\omega_2T_0} + k_{11}\Psi_{11}^2\Psi_{12}^2A_1\overline{A}_1A_2e^{i\omega_2T_0} \\
& +\frac{1}{2}k_{22}\Psi_{22}^4A_2^2\overline{A}_2e^{i\omega_2T_0} + k_{22}\Psi_{21}^2\Psi_{22}^2A_1\overline{A}_1A_2e^{i\omega_2T_0} \\
& +\frac{1}{2}k_{11}\Psi_{11}\Psi_{12}^3A_1\overline{A}_2^2e^{i(\omega_1-2\omega_2)T_0} + \frac{1}{2}k_{22}\Psi_{21}\Psi_{22}^3A_1\overline{A}_2^2e^{i(\omega_1-2\omega_2)T_0} \\
& +\frac{1}{6}k_{22}\Psi_{21}^3\Psi_{22}A_1^3e^{i3\omega_1T_0} + \frac{1}{6}k_{11}\Psi_{11}^3\Psi_{12}A_1^3e^{i3\omega_1T_0} + cc + NST
\end{aligned} \tag{6.3.29}$$

where

$$h_1 = \Psi_{21} - \Psi_{11}, \quad h_2 = \Psi_{22} - \Psi_{12} \tag{6.3.30}$$

To obtain (6.3.28) and (6.3.29), we did not consider the case of $\omega_1 = \omega_2$. From (6.3.26)–(6.3.29) we can find that there are two possible resonant cases: $\omega_2 \approx 3\omega_1$ and $\omega_1 \approx 3\omega_2$. Without loss of generality, we assume $\omega_2 > \omega_1$. In analyzing the particular solutions of (6.3.26) and (6.3.27) we need to distinguish between the resonant case in which $\omega_2 \approx 3\omega_1$ and the nonresonant case in which ω_2 is away from $3\omega_1$.

(c) If $s = \omega_2 : \omega_1 = 3 : 1$, then from (6.3.10) and (6.3.5), we can obtain

$$-9(m_{11}k_{22} + m_{22}k_{11})^2 + 100k_{11}k_{22}(m_{11}m_{22} - 1) = 0$$

i.e.,

$$9[(R^2 + \rho^2)R^{-1}r^{-1} + 1]^2 - 100\rho^2R^{-1}r^{-1} = 0 \tag{6.3.31}$$

Let

$$x^2 = \frac{\rho^2}{Rr}, \quad y^2 = \frac{R}{r} \tag{6.3.32}$$

Equation (6.3.31) can be written as

$$\left(x - \frac{5}{3}\right)^2 + y^2 = \frac{16}{9} \tag{6.3.33}$$

Therefore, when the three-to-one internal resonance occurs, the parameters of the system are constrained to an arc, as in Fig. 6.3a. For example, one set of parameters can be taken as:

$$x = \frac{\sqrt{7}+5}{3}, \quad y^2 = 1 \quad \Rightarrow \quad r = R, \quad \rho = \frac{\sqrt{7}+5}{3}R \tag{6.3.34}$$

The internal resonance of $\omega_2 \approx 3\omega_1$ is investigated below. In this case, the detuning parameter σ is introduced such that

$$\omega_2 = 3\omega_1 + \varepsilon^2\sigma \tag{6.3.35}$$

In order to eliminate secular terms in (6.3.26)–(6.3.29), we need

$$2A_1' + ib_{11}A_1^2\overline{A}_1 + ib_{12}A_1A_2\overline{A}_2 + ib_{13}\overline{A}_1^2A_2e^{i\sigma T_2} = 0 \tag{6.3.36}$$

$$2A_2' + ib_{21}A_2^2\overline{A}_2 + ib_{22}A_1\overline{A}_1A_2 + ib_{23}A_1^3e^{-i\sigma T_2} = 0 \tag{6.3.37}$$

where

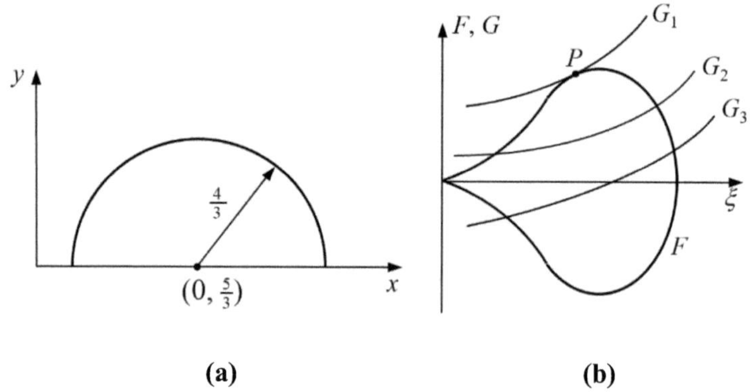

Fig. 6.3 **a** Occurrence of 3:1 internal resonance ($x^2 = \rho^2R^{-1}r^{-1}, y^2 = Rr^{-1}$). **b** Schematic diagram of $F(\xi)$ and $G(\xi)$ for Exercise 6.3

$$
\begin{aligned}
b_{11} &= \omega_1^{-1}\left(\omega_1^2\Psi_{11}\Psi_{21}^2 h_1 - 3\omega_1^2\Psi_{11}\Psi_{21}h_1^2 \right.\\
&\left. -\omega_1^2\Psi_{11}^2\Psi_{21}h_1 + \frac{1}{2}gl_1^{-1}\Psi_{21}^4 + \frac{1}{2}k_{11}\Psi_{11}^4\right)\\
b_{12} &= \omega_1^{-1}\left(2\omega_2^2\Psi_{11}\Psi_{22}^2 h_1 - 2\omega_2^2\Psi_{12}^2\Psi_{21}h_1 - 2\omega_1^2\Psi_{11}\Psi_{21}h_2^2 - 2\omega_2^2\Psi_{11}\Psi_{22}h_1h_2\right.\\
&\left. -2\omega_2^2\Psi_{12}\Psi_{21}h_1h_2 + k_{11}\Psi_{11}^2\Psi_{12}^2 + k_{22}\Psi_{21}^2\Psi_{22}^2\right)A_1A_2\overline{A}_2
\end{aligned}
\tag{6.3.38}
$$

$$
\begin{aligned}
b_{21} &= \omega_2^{-1}\left(\omega_2^2\Psi_{12}^2\Psi_{22}h_2 - 3\omega_2^2\Psi_{12}\Psi_{22}h_2^2 \right.\\
&\left. -\omega_2^2\Psi_{12}\Psi_{22}^2h_2 + \frac{1}{2}k_{11}\Psi_{12}^4 + \frac{1}{2}k_{22}\Psi_{22}^4\right)\\
b_{22} &= \omega_2^{-1}\left(2\omega_1^2\Psi_{11}^2\Psi_{22}h_2 - 2\omega_1^2\Psi_{12}\Psi_{21}^2h_2 - 2\omega_2^2\Psi_{12}\Psi_{22}h_1^2 - 2\omega_1^2\Psi_{12}\Psi_{21}h_1h_2\right.\\
&\left. -2\omega_1^2\Psi_{11}\Psi_{22}h_1h_2 + k_{11}\Psi_{11}^2\Psi_{12}^2 + k_{22}\Psi_{21}^2\Psi_{22}^2\right)
\end{aligned}
\tag{6.3.39}
$$

$$
\begin{aligned}
b_{13} &= \omega_1^{-1}\left(\omega_1^2\Psi_{11}^2\Psi_{21}h_2 - \omega_1^2\Psi_{11}\Psi_{21}^2h_2 - 2\omega_1^2\Psi_{11}\Psi_{21}h_1h_2\right.\\
&+2\omega_1\omega_2\Psi_{11}\Psi_{21}\Psi_{22}h_1 - 2\omega_1\omega_2\Psi_{11}\Psi_{12}\Psi_{21}h_1\\
&\left. -\frac{1}{2}\omega_2^2\Psi_{11}\Psi_{22}h_1^2 - \frac{1}{2}\omega_2^2\Psi_{12}\Psi_{21}h_1^2 + \frac{1}{2}k_{11}\Psi_{11}^3\Psi_{12} + \frac{1}{2}k_{22}\Psi_{21}^3\Psi_{22}\right)\\
b_{23} &= \omega_2^{-1}\left(\omega_1^2\Psi_{12}\Psi_{21}^2h_1 - \omega_1^2\Psi_{11}^2\Psi_{22}h_1 - \frac{1}{2}\omega_1^2\Psi_{12}\Psi_{22}h_1^2 - \frac{1}{2}\omega_1^2\Psi_{12}\Psi_{21}h_1^2\right.\\
&\left. +\frac{1}{6}k_{22}\Psi_{21}^3\Psi_{22} + \frac{1}{6}k_{11}\Psi_{11}^3\Psi_{12}\right)
\end{aligned}
\tag{6.3.40}
$$

Substituting $A_k = \frac{1}{2}ae^{i\beta_k},\ \ k = 1, 2$ into and, we get

$$
a_1' + i\beta_1' a_1 + \frac{1}{8}ib_{11}a_1^3 + \frac{1}{8}ib_{12}a_1a_2^2 + i\frac{1}{8}b_{13}a_1^2a_2e^{i(\sigma T_2 - 3\beta_1 + \beta_2)} = 0 \tag{6.3.41}
$$

$$
a_2' + i\beta_2' a_2 + \frac{1}{8}ib_{21}a_2^3 + \frac{1}{8}ib_{22}a_1^2a_2 + i\frac{1}{8}b_{23}a_1^3e^{-i(\sigma T_2 - 3\beta_1 + \beta_2)} = 0 \tag{6.3.42}
$$

Separating the real and imaginary parts of the above equations yields

$$
\begin{aligned}
a_1' &= \tfrac{1}{8}b_{13}a_1^2a_2\sin\gamma\\
a_2' &= -\tfrac{1}{8}b_{23}a_1^3\sin\gamma
\end{aligned}
\tag{6.3.43}
$$

$$
\begin{aligned}
\beta_1'a_1 + \tfrac{1}{8}b_{11}a_1^3 + \tfrac{1}{8}b_{12}a_1a_2^2 + \tfrac{1}{8}b_{13}a_1^2a_2\cos\gamma = 0\\
\beta_2'a_2 + \tfrac{1}{8}b_{21}a_2^3 + \tfrac{1}{8}b_{22}a_1^2a_2 + \tfrac{1}{8}b_{23}a_1^3\cos\gamma = 0
\end{aligned}
\tag{6.3.44}
$$

where

$$\gamma = \sigma T_2 - 3\beta_1 + \beta_2 \tag{6.3.45}$$

Eliminating γ from (6.3.43), we obtain

$$b_{23}a_1 da_1 + b_{13}a_2 da_2 = 0 \tag{6.3.46}$$

i.e.,

$$\frac{1}{2}d\left(b_{23}a_1^2 + b_{13}a_2^2\right) = 0$$

therefore,

$$b_{23}a_1^2 + b_{13}a_2^2 = E, \quad E = b_{23}a_{10}^2 + b_{13}a_{20}^2 \tag{6.3.47}$$

Combining the above results, the first order approximate solution of the original equation can be obtained as

$$\begin{Bmatrix} \theta_1 \\ \theta_2 \end{Bmatrix} = [\Psi]\begin{Bmatrix} a_1(t)\cos[\omega_1 t + \beta_1(t)] \\ a_2(t)\cos[\omega_2 t + \beta_2(t)] \end{Bmatrix} + O\left(\varepsilon^3\right) \tag{6.3.48}$$

where a_1, a_2, β_1 and β_2 can be determined by (6.3.43), (6.3.44), (6.3.45) and (6.3.47).

Next we analyze the equations describing amplitude and phase of the motion, (6.3.43)–(6.3.45). Eliminating β_1', β_2' from the two equations in (6.3.44), we obtain

$$a_2\gamma' = a_2\sigma + \frac{1}{8}(3b_{11} - b_{22})a_1^2 a_2 + \frac{1}{8}(3b_{12} - b_{21})a_2^3 + \frac{1}{8}\left(3b_{13}a_1 a_2^2 - b_{23}a_1^3\right)\cos\gamma \tag{6.3.49}$$

Dividing the above equation by the second equation of (6.3.43) yields

$$\begin{aligned} -b_{23}a_1^3 a_2 \sin\gamma \tfrac{d\gamma}{da_2} &= 8\sigma a_2 + (3b_{11} - b_{22})a_1^2 a_2 + (3b_{12} - b_{21})a_2^3 \\ &+\left(3b_{13}a_1 a_2^2 - b_{23}a_1^3\right)\cos\gamma \end{aligned}$$

i.e.,

$$\begin{aligned} b_{23}a_1^3 a_2 d\cos\gamma + b_{23}a_1^3\cos\gamma da_2 - 3b_{13}a_1 a_2^2\cos\gamma da_2 \\ = 8\sigma a_2 da_2 + (3b_{11} - b_{22})a_1^2 a_2 da_2 + (3b_{12} - b_{21})a_2^3 da_2 \end{aligned}$$

Using (6.3.47), the above equation can be written as

$$\begin{aligned} b_{23}a_1^3 d(a_2\cos\gamma) - 3b_{13}a_1 a_2^2\cos\gamma da_2 &= \left[8\sigma + (3b_{11} - b_{22})\tfrac{E}{b_{23}}\right]a_2 da_2 \\ +\left[(3b_{12} - b_{21}) - (3b_{11} - b_{22})\tfrac{b_{13}}{b_{23}}\right]a_2^3 da_2 \end{aligned} \tag{6.3.50}$$

since

$$a_1^3 d(a_2\cos\gamma) = d\left(a_1^3 a_2\cos\gamma\right) - 3a_1^2 a_2\cos\gamma\, da_1 \tag{6.3.51}$$

by substituting (6.3.51) into the left side of (6.3.50), we get

$$\begin{aligned} & b_{23} d\left(a_1^3 a_2\cos\gamma\right) - 3a_1 a_2\cos\gamma\left(b_{23}a_1 da_1 + b_{13}a_2 da_2\right) \\ & = \left[8\sigma + (3b_{11} - b_{22})\frac{E}{b_{23}}\right]a_2 da_2 + \left[(3b_{12} - b_{21}) - (3b_{11} - b_{22})\frac{b_{13}}{b_{23}}\right]a_2^3 da_2 \end{aligned} \tag{6.3.52}$$

By substituting (6.3.46) into the left side of (6.3.52), we can obtain

$$\begin{aligned} b_{23} d\left(a_1^3 a_2\cos\gamma\right) &= \left[8\sigma + (3b_{11} - b_{22})\tfrac{E}{b_{23}}\right]a_2 da_2 \\ &+ \left[(3b_{12} - b_{21}) - (3b_{11} - b_{22})\tfrac{b_{13}}{b_{23}}\right]a_2^3 da_2 \end{aligned} \tag{6.3.53}$$

Make the integration of Eq. (6.3.53), we can obtain

$$\begin{aligned} & b_{23}a_1^3 a_2\cos\gamma - \frac{1}{2}\left[8\sigma + (3b_{11} - b_{22})\frac{E}{b_{23}}\right]a_2^2 \\ & + \frac{1}{4}\left[(3b_{11} - b_{22})\frac{b_{13}}{b_{23}} - (3b_{12} - b_{21})\right]a_2^4 = L \end{aligned} \tag{6.3.54}$$

where L is the constant of integration. Now, let us combine (6.3.54), (6.3.47) and (6.3.43) into a single variable differential equation. To do this, let

$$a_1^2 = E\xi \tag{6.3.55}$$

Then from (6.3.47), we can obtain

$$a_2^2 = E(1 - b_{23}\xi)/b_{13} \tag{6.3.56}$$

Eliminating γ from (6.3.54) and (6.3.49) yields

$$\xi'^2 = F^2(\xi) - G^2(\xi) \tag{6.3.57}$$

where

$$\begin{aligned} F(\xi) &= \pm\tfrac{1}{4}\left[b_{13}E^2\xi^3(1 - b_{23}\xi)\right]^{1/2} \\ G(\xi) &= \tfrac{1}{4}\Bigg(\tfrac{(1-b_{23}\xi)^2}{b_{23}^2}\Bigg\{L + \tfrac{1}{2}\left[8\sigma + (3b_{11} - b_{22})\tfrac{E}{b_{23}}\right] \\ &+ \tfrac{1}{4}\left[(3b_{12} - b_{21}) - (3b_{11} - b_{22})\tfrac{b_{13}}{b_{23}}\right]\tfrac{E(1-b_{23}\xi)}{b_{13}}\Bigg\}^2\Bigg)^{1/2} \end{aligned} \tag{6.3.58}$$

The functions $F(\xi)$ and $G(\xi)$ are shown schematically in Fig. 6.3b. For real motions, $F^2(\xi) \geq G^2(\xi)$. The points where F meets G corresponds to the vanishing of ξ'. A curve such as G_2 which meets one branch of F at two different points or a curve G_3 which meets both branches corresponds to a periodic solution. In this case, ξ is periodic and oscillates between two intersection points, and the motion is aperiodic.

Considering the curves G_1 and F are tangent to each other, as shown in Fig. 6.3b. In this case, ξ has only one constant-value solution, and the motion of the system is periodic. The nonlinear effect acts to modulate the phase of the motion so that the nonlinear frequencies of the two modes are commensurable.

In order to analyze the steady state solution, let $a_1' = a_2' = \gamma' = 0$, (6.3.43) and (6.3.49) become

$$\sin\gamma = 0 \quad \text{or} \quad \gamma = n\pi \tag{6.3.59}$$

$$a_2\sigma + \frac{1}{8}(3b_{11} - b_{22})a_1^2 a_2 + \frac{1}{8}(3b_{12} - b_{21})a_2^3 + \frac{1}{8}\left(3b_{13}a_1 a_2^2 - b_{23}a_1^3\right)\cos n\pi = 0 \tag{6.3.60}$$

Equation (6.3.60) is the cubic equation of a_2. For a given set of values of σ, a_1 and $\cos n\pi$, Eq. (6.2.57) has one or three real roots, each of which corresponds to a periodic motion of the system. Thus, the periodic motion of the system is either unique or one of the three possible periodic motions. Figure 6.3b shows that the periodic motion is unstable. With a small perturbation, a curve such as G_1 may become a curve such as G_2 and the motion of the system becomes aperiodic.

For the case of the nonlinear periodic motion of the system with a constant-value solution, we note that the frequency of the motion of the system is

$$\hat{\omega}_1 = \omega_1 + \varepsilon^2\beta_1', \quad \hat{\omega}_2 = \omega_2 + \varepsilon^2\beta_2' \tag{6.3.61}$$

therefore,

$$\hat{\omega}_2 - 3\hat{\omega}_1 = \omega_2 - 3\omega_1 + \varepsilon^2\left(\beta_2' - 3\beta_1'\right) = \varepsilon^2\sigma + \varepsilon^2\left(\beta_2' - 3\beta_1'\right) = \varepsilon^2\gamma' = 0 \tag{6.3.62}$$

Thus, the nonlinearity of the system modulates the frequency of the motion to exactly 3 to 1, and thus the motion is periodic.

6.4 Exercise 6.4 (Internal Resonance Analysis of a Spring Pendulum)

Solution: (a) The kinetic energy of the system is

$$T = \frac{1}{2}m_1 l^2\dot{\theta}^2 + \frac{1}{2}m_2\left(\dot{x}^2 + x^2\dot{\theta}^2\right) \tag{6.4.1}$$

The potential energy of the system is

$$\begin{aligned} V = \tfrac{1}{2}k\left[(x-x_0)^2 - (x_e - x_0)^2\right] + m_1 g l(1-\cos\theta) \\ +m_2 g x(1-\cos\theta) - m_2 g(x - x_e) \end{aligned} \tag{6.4.2}$$

where x_e is the length of the spring when the system is balanced, x_0 is the original length of the spring, and $m_2 g = k(x_e - x_0)$. Here we choose the equilibrium position as the zero potential surface of the system.

Substitute kinetic and potential energy into the Lagrange's equation:

$$\frac{d}{dt}\frac{\partial T}{\partial \dot{x}} - \frac{\partial T}{\partial x} = -\frac{\partial V}{\partial x},\quad \frac{d}{dt}\frac{\partial T}{\partial \dot{\theta}} - \frac{\partial T}{\partial \theta} = -\frac{\partial V}{\partial \theta}$$

we can obtain

$$\begin{aligned} &m_2\ddot{x} + kx - m_2 x\dot{\theta}^2 + m_2 g(1-\cos\theta) = kx_e \\ &\left(m_1 l^2 + m_2 x^2\right)\ddot{\theta} + (m_1 l + m_2 x)g\sin\theta + 2m_2 x\dot{x}\dot{\theta} = 0 \end{aligned} \tag{6.4.3}$$

Let

$$\omega_{10}^2 = k/m_2,\quad \omega_{20}^2 = g/l,\quad m = m_2/m_1,\quad u = x/l,\quad u_e = x_e/l \tag{6.4.4}$$

Equation (6.4.3) changes into

$$\begin{aligned} &\ddot{u} + \omega_{10}^2 u - u\dot{\theta}^2 + \omega_{20}^2(1-\cos\theta) = \omega_{10}^2 u_e \\ &\left(1 + mu^2\right)\ddot{\theta} + (1+mu)\omega_{20}^2\sin\theta + 2mu\dot{u}\dot{\theta} = 0 \end{aligned} \tag{6.4.5}$$

(b) The equilibrium position of the system is $u = u_e,\quad \theta = 0$.

Let

$$u = u_e + u_1,\quad \theta = u_2 \tag{6.4.6}$$

Substituting (6.4.6) into (6.4.5) and retaining to a third-order small quantity yields

$$\begin{aligned} &\ddot{u}_1 + \omega_{10}^2 u_1 - u_1\dot{u}_2^2 + \frac{1}{2}\omega_{20}^2 u_2^2 = 0 \\ &\left(1 + mu_e^2\right)\ddot{u}_2 + (1+mu_e)\omega_{20}^2 u_2 + 2mu_e u_1\ddot{u}_2 + mu_1^2\ddot{u}_2 \\ &+(2mu_e + u_1)\dot{u}_1\dot{u}_2 + \omega_{20}^2 m u_1 u_2 - \frac{1}{6}(1+mu_e)\omega_{20}^2 u_2^3 = 0 \end{aligned} \tag{6.4.7}$$

Let

$$\omega_1^2 = \omega_{10}^2, \quad \omega_2^2 = \frac{1 + mu_e}{1 + mu_e^2}\omega_{20}^2, \quad \alpha = \frac{m}{1 + mu_e^2} \tag{6.4.8}$$

Then the Eq. (6.4.7) can be written as

$$\begin{aligned}
&\ddot{u}_1 + \omega_1^2 u_1 + \frac{1}{2}\omega_{20}^2 u_2^2 - u_1 \dot{u}_2^2 = 0 \\
&\ddot{u}_2 + \omega_2^2 u_2 + \alpha\omega_{20}^2 u_1 u_2 - \frac{1}{6}\omega_2^2 u_2^3 + (2\alpha u_e + u_1)\dot{u}_1\dot{u}_2 + 2\alpha u_e u_1 \ddot{u}_2 + \alpha u_1^2 \ddot{u}_2 = 0
\end{aligned} \tag{6.4.9}$$

where ω_1 and ω_2 are the linear natural frequencies of the system.

For small and finite amplitudes, we use the method of multiple scales to solve (6.4.9). Let the solution of Eq. (6.4.9) be expressed as

$$\begin{aligned}
u_1 &= \varepsilon u_{11}(T_0, T_1) + \varepsilon^2 u_{12}(T_0, T_1) + \cdots \\
u_2 &= \varepsilon u_{21}(T_0, T_1) + \varepsilon^2 u_{22}(T_0, T_1) + \cdots
\end{aligned} \tag{6.4.10}$$

Substituting (6.4.10) into (6.4.9) and keeping to $O\left(\varepsilon^2\right)$, we get

$$\begin{aligned}
0 &= \ddot{u}_1 + \omega_1^2 u_1 + \frac{1}{2}\omega_{20}^2 u_2^2 - u_1 \dot{u}_2^2 \\
&= \left(D_0^2 + 2\varepsilon D_0 D_1\right)\left(\varepsilon u_{11} + \varepsilon^2 u_{12}\right) + \omega_1^2\left(\varepsilon u_{11} + \varepsilon^2 u_{12}\right) + \frac{1}{2}\omega_{20}^2\left(\varepsilon u_{21} + \varepsilon^2 u_{22}\right)^2 \\
&\quad -\left(\varepsilon u_{11} + \varepsilon^2 u_{12}\right)\left[(D_0 + \varepsilon D_1)\left(\varepsilon u_{21} + \varepsilon^2 u_{22}\right)\right]^2 \\
&= \varepsilon\left(D_0^2 u_{11} + \omega_1^2 u_{11}\right) + \varepsilon^2\left(D_0^2 u_{12} + \omega_1^2 u_{12} + 2D_0 D_1 u_{11} + \frac{1}{2}\omega_{20}^2 u_{21}^2\right) + \cdots
\end{aligned} \tag{6.4.11}$$

$$\begin{aligned}
0 &= \ddot{u}_2 + \omega_2^2 u_2 + \alpha\omega_{20}^2 u_1 u_2 - \frac{1}{6}\omega_2^2 u_2^3 + (2\alpha u_e + u_1)\dot{u}_1\dot{u}_2 + 2\alpha u_e u_1 \ddot{u}_2 + \alpha u_1^2 \ddot{u}_2 \\
&= \left(D_0^2 + 2\varepsilon D_0 D_1\right)\left(\varepsilon u_{21} + \varepsilon^2 u_{22}\right) + \omega_2^2\left(\varepsilon u_{21} + \varepsilon^2 u_{22}\right) \\
&\quad +\alpha\omega_{20}^2\left(\varepsilon u_{11} + \varepsilon^2 u_{12}\right)\left(\varepsilon u_{21} + \varepsilon^2 u_{22}\right) - \frac{1}{6}\omega_2^2\left(\varepsilon u_{21} + \varepsilon^2 u_{22}\right)^3 \\
&\quad +2\alpha u_e (D_0 + \varepsilon D_1)\left(\varepsilon u_{11} + \varepsilon^2 u_{12}\right)(D_0 + \varepsilon D_1)\left(\varepsilon u_{21} + \varepsilon^2 u_{22}\right) \\
&\quad +\left(\varepsilon u_{11} + \varepsilon^2 u_{12}\right)(D_0 + \varepsilon D_1)\left(\varepsilon u_{11} + \varepsilon^2 u_{12}\right)(D_0 + \varepsilon D_1)\left(\varepsilon u_{21} + \varepsilon^2 u_{22}\right) \\
&\quad +2\alpha u_e\left(\varepsilon u_{11} + \varepsilon^2 u_{12}\right)\left(D_0^2 + 2\varepsilon D_0 D_1\right)\left(\varepsilon u_{21} + \varepsilon^2 u_{22}\right) \\
&\quad +\alpha\left(\varepsilon u_{11} + \varepsilon^2 u_{12}\right)^2\left(D_0^2 + 2\varepsilon D_0 D_1\right)\left(\varepsilon u_{21} + \varepsilon^2 u_{22}\right) + \cdots \\
&= \varepsilon\left(D_0^2 u_{21} + \omega_2^2 u_{21}\right) + \varepsilon^2\left[D_0^2 u_{22} + \omega_2^2 u_{22} + 2D_0 D_1 u_{21}\right. \\
&\quad \left. +\alpha\omega_{20}^2 u_{11} u_{21} + 2\alpha u_e (D_0 u_{11})(D_0 u_{21}) + 2\alpha u_e u_{11} D_0^2 u_{21}\right] + \cdots
\end{aligned} \tag{6.4.12}$$

Equating coefficients of like powers of ε in (6.4.11) and (6.4.12) yields.
Order ε^1:

$$\begin{aligned} D_0^2 u_{11} + \omega_1^2 u_{11} &= 0 \\ D_0^2 u_{21} + \omega_2^2 u_{21} &= 0 \end{aligned} \tag{6.4.13}$$

Order ε^2:

$$\begin{aligned} &D_0^2 u_{12} + \omega_1^2 u_{12} = -2D_0 D_1 u_{11} - \tfrac{1}{2}\omega_{20}^2 u_{21}^2 \\ &D_0^2 u_{22} + \omega_2^2 u_{22} = -2D_0 D_1 u_{21} - \alpha\omega_{20}^2 u_{11} u_{21} \\ &-2\alpha u_e u_{11} D_0^2 u_{21} - 2\alpha u_e (D_0 u_{11})(D_0 u_{21}) \end{aligned} \tag{6.4.14}$$

The solution of (6.4.13) is

$$\begin{Bmatrix} u_{11} \\ u_{21} \end{Bmatrix} = \begin{Bmatrix} A_1 e^{i\omega_1 T_0} + \overline{A}_1 e^{-i\omega_1 T_0} \\ A_2 e^{i\omega_2 T_0} + \overline{A}_2 e^{-i\omega_2 T_0} \end{Bmatrix} \tag{6.4.15}$$

where $A_k = A_k(T_1, T_2)$. Substituting (6.1.30) into (6.1.19), we get

$$\begin{aligned} &D_0^2 u_{12} + \omega_1^2 u_{12} = -2i\omega_1 D_1 A_1 e^{i\omega_1 t} - \tfrac{1}{2}\omega_{20}^2 A_2^2 e^{2i\omega_2 T_0} + cc + NST \\ &D_0^2 u_{22} + \omega_2^2 u_{22} = -2i\omega_2 D_1 A_2 e^{i\omega_2 t} + \big(2\alpha u_e \omega_0^2 - 2\alpha u_e \omega_1 \omega_2 \\ &-\alpha\omega_{20}^2\big)\overline{A}_2 A_1 e^{i(\omega_1 - \omega_2)T_0} + cc + NST \end{aligned} \tag{6.4.16}$$

The internal resonance case $\omega_1 \approx 2\omega_2$ is investigated below. In this case, the detuning parameter σ is introduced such that

$$\omega_1 = 2\omega_2 + \varepsilon\sigma \tag{6.4.17}$$

In order to eliminate secular terms in (6.4.16), there must be

$$\begin{aligned} &2i\omega_1 A_1' + \frac{1}{2}\omega_{20}^2 A_2^2 e^{-i\sigma T_1} = 0 \\ &2i\omega_2 A_2' + \big(\alpha\omega_{20}^2 + 2\alpha u_e \omega_1 \omega_2 - 2\alpha u_e \omega_0^2\big)\overline{A}_2 A_1 e^{i\sigma T_1} = 0 \end{aligned} \tag{6.4.18}$$

Primes denote the derivative with respect to T_1. Let

$$A_k = \frac{1}{2} a_k e^{i\beta_k}, \quad k = 1, 2 \tag{6.4.19}$$

and substitute (6.4.19) into (6.4.18), we can obtain

$$\begin{aligned} &ia_1' - \beta_1' a_1 + b_1 a_2^2 e^{-i(\sigma T_1 + \beta_1 - 2\beta_2)} = 0 \\ &ia_2' - \beta_2' a_2 + b_2 a_1 a_2 e^{i(\sigma T_1 + \beta_1 - 2\beta_2)} = 0 \end{aligned} \tag{6.4.20}$$

where

$$b_1 = \frac{1}{8}\omega_1^{-1}\omega_{20}^2, \quad b_2 = \frac{1}{4}\omega_2^{-1}\left(\alpha\omega_{20}^2 + 2\alpha u_e\omega_1\omega_2 - 2\alpha u_e\omega_0^2\right) \tag{6.4.21}$$

Separating the real and imaginary parts of the above equations yields

$$\begin{aligned} a_1' &= b_1 a_2^2 \sin\gamma \\ a_2' &= -b_2 a_1 a_2 \sin\gamma \end{aligned} \tag{6.4.22}$$

$$\begin{aligned} \beta_1' a_1 &= b_1 a_2^2 \cos\gamma \\ \beta_2' a_2 &= b_2 a_1 a_2 \cos\gamma \end{aligned} \tag{6.4.23}$$

where

$$\gamma = \sigma T_1 + \beta_1 - 2\beta_2 \tag{6.4.24}$$

Eliminating γ from (6.4.22), we obtain

$$b_2 a_1 da_1 + b_1 a_2 da_2 = 0 \tag{6.4.25}$$

i.e.,

$$b_2 a_1^2 + b_1 a_2^2 = E, \quad E = b_2 a_{10}^2 + b_1 a_{20}^2 \tag{6.4.26}$$

Combine the above results and write εa_1 and εa_2 as a_1 and a_2, respectively (in this case, Eqs. (6.4.22), (6.4.23) and (6.4.26) remain unchanged, but need to treat the prime as the derivative with respect to time t), we can obtain the first-order approximate solution of the original equation

$$\begin{Bmatrix} u \\ \theta \end{Bmatrix} = \begin{Bmatrix} u_e + a_1(t)\cos[\omega_1 t + \beta_1(t)] \\ a_2(t)\cos[\omega_2 t + \beta_2(t)] \end{Bmatrix} + O\left(\varepsilon^2\right) \tag{6.4.27}$$

where a_1, a_2, β_1 and β_2 can be determined by (6.4.22), (6.4.23), (6.4.24) and (6.4.26).

Next we analyze the equations describing amplitude and phase of the motion, (6.4.22)–(6.4.26). Eliminating $\beta_1', \ \beta_2'$ from the two equations in (6.4.23), we obtain

$$a_1\gamma' = \sigma a_1 + b_1 a_2^2 \cos\gamma - 2b_2 a_1^2 \cos\gamma \tag{6.4.28}$$

Dividing the above equation by the second equation of (6.4.22) yields

$$b_1 a_1 a_2^2 \sin\gamma d\gamma = \sigma a_1 da_1 + b_1 a_2^2 \cos\gamma da_1 - 2b_2 a_1^2 \cos\gamma da_1$$

i.e.,

$$-b_1 d\left(a_1 a_2^2 \cos\gamma\right) + 2a_1 \cos\gamma\,(b_1 a_2 da_2 + b_2 a_1 da_1) = \sigma a_1 da_1$$

Using (6.4.25), the above equation becomes

$$b_1 d\left(a_1 a_2^2 \cos\gamma\right) + \frac{1}{2}\sigma da_1^2 = 0 \tag{6.4.29}$$

Make the integration of Eq. (6.4.29), we can obtain

$$b_1 a_1 a_2^2 \cos\gamma + \frac{1}{2}\sigma a_1^2 = L \tag{6.4.30}$$

where L is the constant of integration. Now, let us combine (6.4.30), (6.4.26) and (6.4.22) into a single variable differential equation. To do this, let

$$a_2^2 = E\xi \tag{6.4.31}$$

Then from (6.4.26), we can obtain

$$a_1^2 = \frac{E(1 - b_1\xi)}{b_2} \tag{6.4.32}$$

Eliminating γ from (6.4.30) and (6.4.22) yields

$$\frac{1}{4b_1 b_2 E}\left(\frac{d\xi}{dT_1}\right)^2 = F^2(\xi) - G^2(\xi) \tag{6.4.33}$$

where

$$\begin{aligned} F(\xi) &= \pm\left[\xi^2\left(b_1^{-1} - \xi\right)\right]^{1/2} \\ G(\xi) &= \pm\left(\tfrac{b_2}{E^3 b_1^3}\left[L - \tfrac{\sigma E b_1}{2b_2}\left(b_1^{-1} - \xi\right)\right]^2\right)^{1/2} \end{aligned} \tag{6.4.34}$$

The functions $F(\xi)$ and $G(\xi)$ are shown schematically in Fig. 6.4. For real motions, $F^2(\xi) \geq G^2(\xi)$. The points where F meets G corresponds to the vanishing of ξ'. In general, the curve G meets the branches of F at three intersection points corresponding to the three roots ξ_1, ξ_2 and ξ_3 of the right-hand side of (6.4.33). Let $\xi_1 < \xi_2 < \xi_3$, the motion is confined between ξ_2 and ξ_3 because of $a_2^2 = E\xi$.

When the three roots are distinct corresponding to a curve such as G_1, ξ is periodic and oscillates between ξ_2 and ξ_3, and the motion is aperiodic. In this case, the solution for ξ can be expressed in terms of Jacobi elliptic functions as follows. First, in terms of the ξ_n, we express (6.4.33) as.

$$\frac{1}{4b_1 b_2 E}\left(\frac{d\xi}{dT_1}\right)^2 = (\xi_3 - \xi)(\xi - \xi_2)(\xi - \xi_1) \tag{6.4.35}$$

Fig. 6.4 Schematic diagram of $F(\xi)$ and $G(\xi)$ for Exercise 6.4

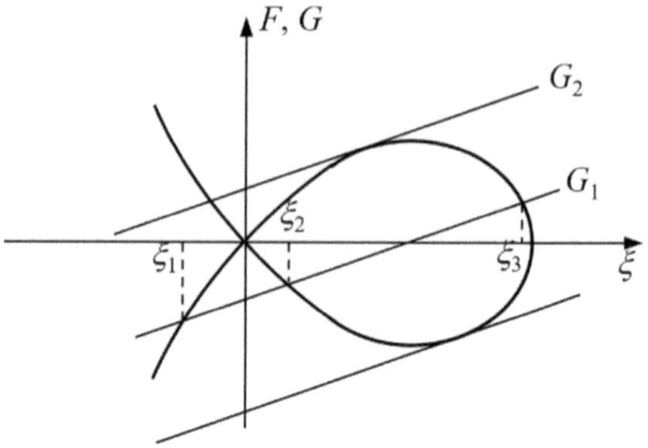

Introducing the transformation

$$\xi_3 - \xi = (\xi_3 - \xi_2)\sin^2\chi \tag{6.4.36}$$

into (6.4.36) we obtain

$$\frac{1}{\sqrt{b_1 b_2 E}}\frac{d\chi}{dT_1} = \pm\sqrt{\xi_3 - \xi_1}\sqrt{1 - \eta^2\sin^2\chi} \tag{6.4.37}$$

where

$$\eta = \sqrt{\frac{\xi_3 - \xi_2}{\xi_3 - \xi_1}} \tag{6.4.38}$$

Separating the variables in (6.4.37), putting $T_1 = \varepsilon t$, and integrating the resulting equation yields

$$(t - t_0)\kappa = \int_0^\chi \frac{d\chi}{\sqrt{1 - \eta^2\sin^2\chi}} \tag{6.4.39}$$

or

$$\sin\chi = \mathrm{sn}[(t - t_0)\kappa;\, \eta] \tag{6.4.40}$$

where t_0 corresponds to $\chi = 0$, sn is a Jacobi elliptic function, and

$$\kappa = \sqrt{E b_1 b_2(\xi_3 - \xi_1)} \tag{6.4.41}$$

Combining (6.4.31), (6.4.36) and (6.4.40) yields

$$\xi = \frac{a_2^2}{E} = \xi_3 - (\xi_3 - \xi_2)\mathrm{sn}^2[(t - t_0)\kappa; \eta] \tag{6.4.42}$$

Thus (6.4.41) and (6.4.42) show that the energy in the system continues to be exchanged between the two modes of oscillation.

When $\xi_2 = \xi_3$ corresponding to the curve G_2, which is tangent to one of the branches of F in Figure 6.4, $\xi = \xi_3$ is a constant according to (6.4.42), and hence $a_2 = \sqrt{E\xi_3}$ and $a_1 = \sqrt{E(1 - b_1\xi)/b_2}$. In this case, the motion of the system is periodic. However, any small disturbance would lead to a curve such as G_1 where the roots are distinct, and hence the motion is aperiodic.

When $\xi_2 = \xi_1$, G coincides with the ξ-axis, and hence $L = \sigma = 0$ according to (6.4.34). Consequently it follows from (6.4.33) and (6.4.34) that $\xi_2 = \xi_1 = 0$, and $\xi_3 = b_1^{-1}$. The solution in this case can be obtained by introducing the transformation

$$\xi = b_1^{-1}\mathrm{sech}^2\varphi \tag{6.4.43}$$

into (6.4.33) with $L = \sigma = 0$. The result is

$$\frac{d\varphi}{dt} = \kappa \tag{6.4.44}$$

whose solution is

$$\varphi = (t - t_0)\kappa \tag{6.4.45}$$

therefore

$$a_2 = \sqrt{E\xi} = \sqrt{Eb_1^{-1}}\mathrm{sech}\varphi = \sqrt{Eb_1^{-1}}\mathrm{sech}[(t - t_0)\kappa] \tag{6.4.46}$$

and

$$a_1 = \sqrt{E(1 - b_1\xi)/b_2} = \sqrt{Eb_2^{-1}}\tanh\varphi = \sqrt{Eb_2^{-1}}\tanh[(t - t_0)\kappa] \tag{6.4.47}$$

We note that $L = \sigma = 0$ demands $\cos\gamma = 0$ according to (6.4.30). Hence it follows from (6.4.23) that $\beta_1' = \beta_2' = 0$; that is, the phases are constant. Therefore, the motion of the system consists of only amplitude-modulated motions (AMM). When $t \to \infty$, $a_2 \to 0$ while $a_1 \to \sqrt{Eb_2^{-1}}$, leading to a motion that is independent of the lower mode. Thus, the internal resonance in this case leads to a complete transfer of energy from the lower mode to the higher mode.

6.5 Exercise 6.5 (Internal Resonance Analysis of a Uniform Rod Hanging a Spring)

Solution: (a) The kinetic energy of the system is

$$\begin{aligned} T &= \tfrac{1}{2}m\Big[\big(\tfrac{1}{2}l\dot{\theta}\cos\theta\big)^2 + \big(\dot{x} - \tfrac{1}{2}l\dot{\theta}\sin\theta\big)^2\Big] + \tfrac{1}{2}\tfrac{1}{12}ml^2\dot{\theta}^2 \\ &= \tfrac{1}{2}m\dot{x}^2 + \tfrac{1}{6}ml^2\dot{\theta}^2 - \tfrac{1}{2}ml\dot{x}\dot{\theta}\sin\theta \end{aligned} \tag{6.5.1}$$

The potential energy of the system is

$$V = \frac{1}{2}k\big[(x - x_0)^2 - (x_e - x_0)^2\big] + \frac{1}{2}mgl(1 - \cos\theta) - mg(x - x_e) \tag{6.5.2}$$

where x_e is the length of the spring when the system is balanced, x_0 is the original length of the spring, and $mg = k(x_e - x_0)$.

Substitute kinetic and potential energy into the Lagrange's equation:

$$\frac{d}{dt}\frac{\partial T}{\partial \dot{x}} - \frac{\partial T}{\partial x} = -\frac{\partial V}{\partial x}, \quad \frac{d}{dt}\frac{\partial T}{\partial \dot{\theta}} - \frac{\partial T}{\partial \theta} = -\frac{\partial V}{\partial \theta}$$

we can obtain

$$\begin{aligned} &\ddot{x} + km^{-1}(x - x_e) = \frac{1}{2}l\ddot{\theta}\sin\theta + \frac{1}{2}l\dot{\theta}^2\cos\theta \\ &\ddot{\theta} + \frac{3}{2}gl^{-1}\sin\theta = \frac{3}{2}\ddot{x}l^{-1}\sin\theta \end{aligned} \tag{6.5.3}$$

Let

$$\omega_1^2 = k/m, \quad \omega_2^2 = 3g/2l, \quad u = (x - x_e)/l \tag{6.5.4}$$

Equation (6.5.3) changes into

$$\begin{aligned} &\ddot{u} + \omega_1^2 u = \frac{1}{2}\ddot{\theta}\sin\theta + \frac{1}{2}\dot{\theta}^2\cos\theta \\ &\ddot{\theta} + \omega_2^2\sin\theta = \frac{3}{2}\ddot{u}\sin\theta \end{aligned} \tag{6.5.5}$$

(b) The equilibrium position of the system is $u = 0, \quad \theta = 0$. Let

$$u = 0 + u_1, \quad \theta = 0 + u_2 \tag{6.5.6}$$

Substituting (6.5.6) into (6.5.5) and retaining to a third-order small quantity yields

$$\begin{aligned} \ddot{u}_1 + \omega_1^2 u_1 &= \tfrac{1}{2}\dot{u}_2^2 + \tfrac{1}{2}u_2\ddot{u}_2 \\ \ddot{u}_2 + \omega_2^2 u_2 &= \tfrac{3}{2}u_2\ddot{u}_1 + \tfrac{1}{6}\omega_2^2 u_2^3 \end{aligned} \tag{6.5.7}$$

where ω_1 and ω_2 are the linear natural frequencies of the system.

For small and finite amplitudes, we use the method of multiple scales to solve (6.5.7). Let the solution of Eq. (6.5.7) be expressed as

$$\begin{aligned} u_1 &= \varepsilon u_{11}(T_0, T_1, T_2) + \varepsilon^2 u_{12}(T_0, T_1, T_2) + \varepsilon^2 u_{13}(T_0, T_1, T_2) + \cdots \\ u_2 &= \varepsilon u_{21}(T_0, T_1, T_2) + \varepsilon^2 u_{22}(T_0, T_1, T_2) + \varepsilon^2 u_{23}(T_0, T_1, T_2) + \cdots \end{aligned} \tag{6.5.8}$$

Substituting (6.5.8) into (6.5.7) and keeping to $O\left(\varepsilon^2\right)$, we get

$$\begin{aligned} 0 &= \ddot{u}_1 + \omega_1^2 u_1 - \frac{1}{2}\dot{u}_2^2 - \frac{1}{2}u_2\ddot{u}_2 \\ &= \left[D_0^2 + 2\varepsilon D_0 D_1 + \varepsilon^2\left(D_1^2 + 2D_0D_2\right)\right]\left(\varepsilon u_{11} + \varepsilon^2 u_{12} + \varepsilon^3 u_{13}\right) \\ &\quad + \omega_1^2\left(\varepsilon u_{11} + \varepsilon^2 u_{12} + \varepsilon^3 u_{13}\right) - \frac{1}{2}\left[\left(D_0 + \varepsilon D_1 + \varepsilon^2 D_2\right)\left(\varepsilon u_{21} + \varepsilon^2 u_{22} + \varepsilon^3 u_{23}\right)\right]^2 \\ &\quad - \frac{1}{2}\left(\varepsilon u_{21} + \varepsilon^2 u_{22} + \varepsilon^3 u_{23}\right)\left[D_0^2 + 2\varepsilon D_0 D_1 + \varepsilon^2\left(D_1^2 + 2D_0D_2\right)\right] \\ &\quad \times\left(\varepsilon u_{21} + \varepsilon^2 u_{22} + \varepsilon^3 u_{23}\right) + \cdots \\ &= \varepsilon\left(D_0^2 u_{11} + \omega_1^2 u_{11}\right) \\ &\quad + \varepsilon^2\left[D_0^2 u_{12} + \omega_1^2 u_{12} + 2D_0D_1u_{11} - \frac{1}{2}(D_0u_{21})^2 - \frac{1}{2}u_{21}D_0^2u_{21}\right] \\ &\quad + \varepsilon^3\left[D_0^2u_{13} + \omega_1^2 u_{13} + 2D_0D_1u_{12} + \left(D_1^2 + 2D_0D_2\right)u_{11} - D_0u_{21}D_1u_{21}\right. \\ &\quad \left. - D_0u_{21}D_0u_{22} - \frac{1}{2}u_{21}D_0^2u_{22} - u_{21}D_0D_1u_{21} - \frac{1}{2}u_{22}D_0^2u_{21}\right] + \cdots \end{aligned} \tag{6.5.9}$$

$$
\begin{aligned}
0 &= \ddot{u}_2 + \omega_2^2 u_2 - \frac{3}{2} u_2 \ddot{u}_1 - \frac{1}{6}\omega_2^2 u_2^3 \\
&= \left[D_0^2 + 2\varepsilon D_0 D_1 + \varepsilon^2\left(D_1^2 + 2D_0 D_2\right)\right]\left(\varepsilon u_{21} + \varepsilon^2 u_{22} + \varepsilon^3 u_{23}\right) \\
&\quad + \omega_2^2\left(\varepsilon u_{21} + \varepsilon^2 u_{22} + \varepsilon^3 u_{23}\right) - \frac{3}{2}\left(\varepsilon u_{21} + \varepsilon^2 u_{22} + \varepsilon^3 u_{23}\right) \\
&\quad \times \left[D_0^2 + 2\varepsilon D_0 D_1 + \varepsilon^2\left(D_1^2 + 2D_0 D_2\right)\right]\left(\varepsilon u_{11} + \varepsilon^2 u_{12} + \varepsilon^3 u_{13}\right) \\
&\quad - \frac{1}{6}\omega_2^2\left(\varepsilon u_{21} + \varepsilon^2 u_{22} + \varepsilon^3 u_{23}\right)^3 + \cdots \\
&= \varepsilon\left(D_0^2 u_{21} + \omega_2^2 u_{21}\right) + \varepsilon^2\left(D_0^2 u_{22} + \omega_2^2 u_{22} + 2D_0 D_1 u_{21} - \frac{3}{2} u_{21} D_0^2 u_{11}\right) \\
&\quad + \varepsilon^3\Big[D_0^2 u_{23} + \omega_2^2 u_{23} + 2D_0 D_1 u_{22} + \left(D_1^2 + 2D_0 D_2\right) u_{21} \\
&\quad - \frac{3}{2} u_{21} D_0^2 u_{12} - 3u_{21} D_0 D_1 u_{11} - \frac{3}{2} u_{22} D_0^2 u_{11} - \frac{1}{6}\omega_2^2 u_{21}^3\Big] + \cdots
\end{aligned}
\tag{6.5.10}
$$

Equating coefficients of like powers of ε in (6.5.9) and (6.5.10) yields.
Order ε^1:

$$
\begin{aligned}
D_0^2 u_{11} + \omega_1^2 u_{11} &= 0 \\
D_0^2 u_{21} + \omega_2^2 u_{21} &= 0
\end{aligned}
\tag{6.5.11}
$$

Order ε^2:

$$
\begin{aligned}
D_0^2 u_{12} + \omega_1^2 u_{12} &= -2D_0 D_1 u_{11} + \tfrac{1}{2}(D_0 u_{21})^2 + \tfrac{1}{2} u_{21} D_0^2 u_{21} \\
D_0^2 u_{22} + \omega_2^2 u_{22} &= -2D_0 D_1 u_{21} + \tfrac{3}{2} u_{21} D_0^2 u_{11}
\end{aligned}
\tag{6.5.12}
$$

Order ε^3:

$$
\begin{aligned}
& D_0^2 u_{13} + \omega_1^2 u_{13} = -2D_0 D_1 u_{12} - \left(D_1^2 + 2D_0 D_2\right) u_{11} + D_0 u_{21} D_1 u_{21} \\
& + D_0 u_{21} D_0 u_{22} + \frac{1}{2} u_{21} D_0^2 u_{22} + u_{21} D_0 D_1 u_{21} + \frac{1}{2} u_{22} D_0^2 u_{21} \\
& D_0^2 u_{23} + \omega_2^2 u_{23} = -2D_0 D_1 u_{22} - \left(D_1^2 + 2D_0 D_2\right) u_{21} + \frac{3}{2} u_{21} D_0^2 u_{12} \\
& + 3u_{21} D_0 D_1 u_{11} + \frac{3}{2} u_{22} D_0^2 u_{11} + \frac{1}{6}\omega_2^2 u_{21}^3
\end{aligned}
\tag{6.5.13}
$$

The solution of (6.5.11) is

$$
\begin{Bmatrix} u_{11} \\ u_{21} \end{Bmatrix} = \begin{Bmatrix} A_1 e^{i\omega_1 T_0} + \overline{A}_1 e^{-i\omega_1 T_0} \\ A_2 e^{i\omega_2 T_0} + \overline{A}_2 e^{-i\omega_2 T_0} \end{Bmatrix}
\tag{6.5.14}
$$

where $A_k = A_k(T_1, T_2)$. Substituting (6.5.14) into (6.5.12), we get

$$\begin{aligned}
&D_0^2u_{12} + \omega_1^2 u_{12} = -2D_0D_1u_{11} + \frac{1}{2}(D_0u_{21})^2 + \frac{1}{2}u_{21}D_0^2u_{21} \\
&= -2i\omega_1 D_1 A_1 e^{i\omega_1 T_0} - \omega_2^2 A_2^2 e^{2i\omega_2 T_0} + cc \\
&D_0^2u_{22} + \omega_2^2 u_{22} = -2D_0D_1u_{21} + \frac{3}{2}u_{21}D_0^2u_{11} \\
&= -2i\omega_2 D_1 A_2 e^{i\omega_2 T_0} - \frac{3}{2}\omega_1^2 A_1 A_2 e^{i(\omega_1+\omega_2)T_0} \\
&-\frac{3}{2}\omega_1^2 A_1 \overline{A}_2 e^{i(\omega_1-\omega_2)T_0} + cc
\end{aligned} \tag{6.5.15}$$

The internal resonance case $\omega_1 \approx 2\omega_2$ is investigated below. In this case, the detuning parameter σ is introduced such that

$$\omega_1 = 2\omega_2 + \varepsilon\sigma \tag{6.5.16}$$

In order to eliminate secular terms in (6.5.15), there must be

$$\begin{aligned}
&2i\omega_1 A_1' + \omega_2^2 A_2^2 e^{-i\sigma T_1} = 0 \\
&2i\omega_2 A_2' + \frac{3}{2}\omega_1^2 A_1 \overline{A}_2 e^{i\sigma T_1} = 0
\end{aligned} \tag{6.5.17}$$

Primes denote the derivative with respect to T_1. Let

$$A_k = \frac{1}{2}a_k e^{i\beta_k}, \quad k = 1, 2 \tag{6.5.18}$$

and substitute (6.5.18) into (6.5.17), we can obtain

$$\begin{aligned}
&ia_1' - \beta_1' a_1 + \frac{1}{4}\omega_1^{-1}\omega_2^2 a_2^2 e^{-i(\sigma T_1 + \beta_1 - 2\beta_2)} = 0 \\
&ia_2' - \beta_2' a_2 + \frac{3}{8}\omega_2^{-1}\omega_1^2 a_1 a_2 e^{i(\sigma T_1 + \beta_1 - 2\beta_2)} = 0
\end{aligned} \tag{6.5.19}$$

Separating the real and imaginary parts of the above equations and taking $\omega_1 = 2\omega_2$ into account yields

$$\begin{aligned}
a_1' &= \tfrac{1}{8}\omega_2 a_2^2 \sin\gamma \\
a_2' &= -\tfrac{3}{2}\omega_2 a_1 a_2 \sin\gamma
\end{aligned} \tag{6.5.20}$$

$$\begin{aligned}
\beta_1' a_1 &= \tfrac{1}{8}\omega_2 a_2^2 \cos\gamma \\
\beta_2' a_2 &= \tfrac{3}{2}\omega_2 a_1 a_2 \cos\gamma
\end{aligned} \tag{6.5.21}$$

where

$$\gamma = \sigma T_1 + \beta_1 - 2\beta_2 \tag{6.5.22}$$

Eliminating γ from (6.5.20), we obtain

$$12a_1da_1 + a_2da_2 = 0 \tag{6.5.23}$$

i.e.,

$$12a_1^2 + a_2^2 = E, \quad E = 12a_{10}^2 + a_{20}^2 \tag{6.5.24}$$

Combine the above results and write εa_1 and εa_2 as a_1 and a_2, respectively, we can obtain the first-order approximate solution of the original equation

$$\begin{Bmatrix} u \\ \theta \end{Bmatrix} = \begin{Bmatrix} a_1(t)\cos[\omega_1 t + \beta_1(t)] \\ a_2(t)\cos[\omega_2 t + \beta_2(t)] \end{Bmatrix} + O(\varepsilon^2) \tag{6.5.25}$$

where a_1, a_2, β_1 and β_2 can be determined by (6.5.20), (6.5.21), (6.5.22), and (6.5.24).

Next we analyze the equations describing amplitude and phase of the motion, (6.5.20)–(6.5.24). Eliminating $\beta_1', \ \beta_2'$ from the two equations in (6.5.21), we obtain

$$a_1\gamma' = \sigma a_1 + \frac{1}{8}\omega_2 a_2^2 \cos\gamma - 3\omega_2 a_1^2 \cos\gamma \tag{6.5.26}$$

Dividing the above equation by the second equation of (6.5.20) yields

$$\frac{1}{8}\omega_2 a_1 a_2^2 \sin\gamma d\gamma = \sigma a_1 da_1 + \frac{1}{8}\omega_2 a_2^2 \cos\gamma da_1 - 3\omega_2 a_1^2 \cos\gamma da_1$$

i.e.,

$$-\frac{1}{8}\omega_2 d\left(a_2^2 a_1 \cos\gamma\right) + \frac{1}{4}\omega_2 a_1 \cos\gamma(12a_1da_1 + a_2da_2) = \sigma a_1 da_1$$

Using (6.5.23), the above equation becomes

$$\frac{1}{8}\omega_2 d\left(a_2^2 a_1 \cos\gamma\right) + \sigma a_1 da_1 = 0 \tag{6.5.27}$$

Make the integration of Eq. (6.5.27), we can obtain

$$\frac{1}{4}\omega_2 a_1 a_2^2 \cos\gamma + \sigma a_1^2 = L \tag{6.5.28}$$

where L is the constant of integration. Now, let us combine (6.5.28), (6.5.24) and (6.5.20) into a single variable differential equation. To do this, let

$$a_2^2 = E\xi \tag{6.5.29}$$

Then from (6.5.24), we can obtain

$$a_1^2 = E(1 - 12\xi) \tag{6.5.30}$$

Eliminating γ from (6.5.28) and (6.5.20) yields

$$\frac{1}{108\omega_2^2 E}\left(\frac{d\xi}{dT_1}\right)^2 = \xi^2\left(\frac{1}{12} - \xi\right) - \frac{4}{3\omega_2^2 E^3}[L - \sigma E(1 - 12\xi)]^2$$

Or

$$\frac{1}{108\omega_2^2 E}\left(\frac{d\xi}{dT_1}\right)^2 = F^2(\xi) - G^2(\xi) \tag{6.5.31}$$

where

$$\begin{aligned} F(\xi) &= \pm\left[\xi^2\left(\tfrac{1}{12} - \xi\right)\right]^{1/2} \\ G(\xi) &= \pm\left(\tfrac{4}{3\omega_2^2 E^3}\left[L - 12\sigma E\left(\tfrac{1}{12} - \xi\right)\right]^2\right)^{1/2} \end{aligned} \tag{6.5.32}$$

The functions $F(\xi)$ and $G(\xi)$ are shown schematically in Fig. 6.5. For real motions, $F^2(\xi) \geq G^2(\xi)$. The points where F meets G corresponds to the vanishing of ξ'. In general, the curve G meets the branches of F at three intersection points corresponding to the three roots ξ_1, ξ_2 and ξ_3 of the right-hand side of (6.5.31). Let $\xi_1 < \xi_2 < \xi_3$, the motion is confined between ξ_2 and ξ_3 because of $a_2^2 = E\xi$. When the three roots are distinct corresponding to a curve such as G_1, ξ is periodic and oscillates between ξ_2 and ξ_3, and the motion is aperiodic. In this case, the solution for ξ can be expressed in terms of Jacobi elliptic functions as follows. First, in terms of the ξ_n, we express (6.5.31) as

$$\frac{1}{108\omega_2^2 E}\left(\frac{d\xi}{dT_1}\right)^2 = (\xi_3 - \xi)(\xi - \xi_2)(\xi - \xi_1) \tag{6.5.33}$$

Introducing the transformation

$$\xi_3 - \xi = (\xi_3 - \xi_2)\sin^2\chi \tag{6.5.34}$$

into (6.5.33) we obtain

$$\frac{1}{\sqrt{27\omega_2^2 E}}\frac{d\chi}{dT_1} = \pm\sqrt{\xi_3 - \xi_1}\sqrt{1 - \eta^2\sin^2\chi} \tag{6.5.35}$$

where

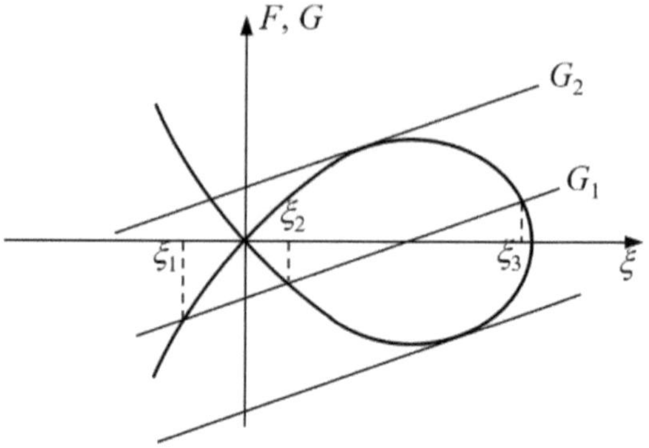

Fig. 6.5 Schematic diagram of $F(\xi)$ and $G(\xi)$ for Exercise 6.5

$$\eta = \sqrt{\frac{\xi_3 - \xi_2}{\xi_3 - \xi_1}} \tag{6.5.36}$$

Separating the variables in (6.5.35), putting $T_1 = \varepsilon t$, and integrating the resulting equation yields

$$(t - t_0)\kappa = \int_0^{\chi} \frac{d\chi}{\sqrt{1 - \eta^2 \sin^2 \chi}} \tag{6.5.37}$$

or

$$\sin \chi = \mathrm{sn}[(t - t_0)\kappa; \eta] \tag{6.5.38}$$

where t_0 corresponds to $\chi = 0$, sn is the Jacobi elliptic function, and

$$\kappa = \sqrt{27\omega_2^2 E(\xi_3 - \xi_1)} \tag{6.5.39}$$

Combining (6.5.34) and (6.5.38) yields

$$\xi = \frac{a_2^2}{E} = \xi_3 - (\xi_3 - \xi_2)\mathrm{sn}^2[(t - t_0)\kappa; \eta] \tag{6.5.40}$$

This equation shows that the energy in the system continues to be exchanged between the two modes of oscillation.

When $\xi_2 = \xi_3$ corresponding to the curve G_2, which is tangent to one of the branches of F in Figure 6.5, $\xi = \xi_3$ is a constant according to (6.5.31), and hence $a_2 = \sqrt{E\xi_3}$ and $a_1 = \sqrt{E(1 - b_1\xi)/b_2}$. In this case, the motion of the system is periodic. However, any small disturbance would lead to a curve such as G_1 where the roots are distinct, and hence the motion is aperiodic.

When $\xi_2 = \xi_1$, G coincides with the ξ-axis, and hence $L = \sigma = 0$ according to (6.5.32). Consequently it follows from (6.5.31) and (6.5.32) that $\xi_2 = \xi_1 = 0$, and

$\xi_3 = \frac{1}{12}$. The solution in this case can be obtained by introducing the transformation

$$\xi = \frac{1}{12}\text{sech}^2\varphi \tag{6.5.41}$$

into (6.5.31) with $L = \sigma = 0$. The result is

$$\frac{d\varphi}{dt} = \kappa \tag{6.5.42}$$

whose solution is

$$\varphi = (t - t_0)\kappa \tag{6.5.43}$$

therefore

$$a_2 = \sqrt{E\xi} = \sqrt{\frac{1}{12}E}\text{sech}\varphi = \sqrt{\frac{1}{12}E}\text{sech}[(t - t_0)\kappa] \tag{6.5.44}$$

and

$$a_1 = \sqrt{E(1 - 12\xi)} = \sqrt{E}\tanh\varphi = \sqrt{E}\tanh[(t - t_0)\kappa] \tag{6.5.45}$$

We note that $L = \sigma = 0$ demands $\cos\gamma = 0$ according to (6.5.28). Hence it follows from (6.5.21) that $\beta_1' = \beta_2' = 0$; that is, the phases are constant. Therefore, the motion of the system consists of only amplitude-modulated motions (AMM). When $t \to \infty$, $a_2 \to 0$ while $a_1 \to \sqrt{E}$, leading to a motion that is independent of the lower mode. Thus, the internal resonance in this case leads to a complete transfer of energy from the lower mode to the higher mode.

(c) When $\omega_1 \approx 2\omega_2$ is not satisfied, no internal resonance occurs in the first order solution. In order to eliminate secular terms in (6.5.15), there must be

$$D_1A_1 = D_1A_2 = 0 \quad \to \quad A_1 = A_1(T_2), \quad A_2 = A_2(T_2) \tag{6.5.46}$$

Equation (6.5.15) changes into

$$\begin{aligned} D_0^2u_{12} + \omega_1^2u_{12} &= -\omega_2^2A_2^2e^{2i\omega_2 T_0} + cc \\ D_0^2u_{22} + \omega_2^2u_{22} &= -\tfrac{3}{2}\omega_1^2A_1A_2e^{i(\omega_1+\omega_2)T_0} - \tfrac{3}{2}\omega_1^2A_1\overline{A}_2e^{i(\omega_1-\omega_2)T_0} + cc \end{aligned} \tag{6.5.47}$$

The solution of (6.5.47) is

$$\begin{aligned} u_{12} &= K_1A_2^2e^{2i\omega_2 T_0} + cc \\ u_{22} &= K_2A_1A_2e^{i(\omega_1+\omega_2)T_0} + K_3A_1\overline{A}_2e^{i(\omega_1-\omega_2)T_0} + cc \end{aligned} \tag{6.5.48}$$

where

$$K_1 = \frac{\omega_2^2}{4\omega_2^2 - \omega_1^2}, \quad K_2 = \frac{3\omega_1^2}{2\left[\left(\omega_1 + \omega_2^2\right) - \omega_2^2\right]}, \quad K_3 = \frac{3\omega_1^2}{2\left[\left(\omega_1 - \omega_2^2\right) - \omega_2^2\right]} \tag{6.5.49}$$

Substituting (6.5.14) and (6.5.48) into (6.5.13) yields

$$\begin{aligned}
D_0^2 u_{13} + \omega_1^2 u_{13} &= -2i\omega_1 D_2 A_1 e^{i\omega_1 T_0} - \frac{1}{2}\omega_2^2 K_2 A_1 A_2 \overline{A}_2 e^{i\omega_1 T_0} - \frac{1}{2}\omega_2^2 K_3 A_1 A_2 \overline{A}_2 e^{i\omega_1 T_0} \\
&+\omega_2(\omega_1 + \omega_2) K_2 A_1 A_2 \overline{A}_2 e^{i\omega_1 T_0} - \omega_2(\omega_1 - \omega_2) K_3 A_1 A_2 \overline{A}_2 e^{i\omega_1 T_0} \\
&-\frac{1}{2}(\omega_1 + \omega_2)^2 K_2 A_1 A_2 \overline{A}_2 e^{i\omega_1 T_0} - \frac{1}{2}(\omega_1 - \omega_2)^2 K_3 A_1 A_2 \overline{A}_2 e^{i\omega_1 T_0} \\
&-\frac{1}{2}\omega_2^2 K_2 A_1 A_2^2 e^{i(\omega_1 + 2\omega_2)T_0} - \omega_2(\omega_1 + \omega_2) K_2 A_1 A_2^2 e^{i(\omega_1 + 2\omega_2)T_0} \\
&-\frac{1}{2}(\omega_1 + \omega_2)^2 K_2 A_1 A_2^2 e^{i(\omega_1 + 2\omega_2)T_0} \\
&-\frac{1}{2}\omega_2^2 K_3 A_1 \overline{A}_2^2 e^{i(\omega_1 - 2\omega_2)T_0} + \omega_2(\omega_1 - \omega_2) K_3 A_1 \overline{A}_2^2 e^{i(\omega_1 - 2\omega_2)T_0} \\
&-\frac{1}{2}(\omega_1 - \omega_2)^2 K_3 A_1 \overline{A}_2^2 e^{i(\omega_1 - 2\omega_2)T_0} + cc
\end{aligned} \tag{6.5.50}$$

$$\begin{aligned}
D_0^2 u_{23} + \omega_2^2 u_{23} &= -2i\omega_2 D_2 A_2 e^{i\omega_2 T_0} - 6\omega_2^2 K_1 A_2^2 \overline{A}_2 e^{i\omega_2 T_0} \\
&-\frac{3}{2}\omega_1^2 K_2 A_1 \overline{A}_1 A_2 e^{i\omega_2 T_0} + \frac{1}{2}\omega_2^2 A_2^2 \overline{A}_2 e^{i\omega_2 T_0} \\
&-\frac{3}{2}\omega_1^2 K_3 \overline{A}_1^2 A_2 e^{i\omega_2 T_0} - \frac{3}{2}\omega_1^2 K_3 A_1 \overline{A}_1 A_2 e^{i\omega_2 T_0} \\
&-6\omega_2^2 K_1 A_2^3 e^{3i\omega_2 T_0} + \frac{1}{6}\omega_2^2 A_2^3 e^{3i\omega_2 T_0} \\
&-\frac{3}{2}\omega_1^2 K_2 A_1^2 A_2 e^{i(2\omega_1 + \omega_2)T_0} + cc
\end{aligned} \tag{6.5.51}$$

From (6.5.50) and (6.5.51), we find no internal resonance excitation. Therefore, the system of this problem does not resonate in the first and second-order approximate solution.

6.6 Exercise 6.6 (Internal Resonance Analysis of the Plane Motion of a Rigid Beam Supported by a Spring)

Solution: (a) The kinetic energy of the system is

$$T = \frac{1}{2}m\dot{x}^2 + \frac{1}{2}I\dot{\theta}^2 \tag{6.6.1}$$

Let $x = \theta = 0$ be the equilibrium position of the system, using the equilibrium position as the zero potential surface of the system, the potential energy of the system is

$$V = \frac{1}{2}k_1(x - l_1 \sin\theta)^2 + \frac{1}{2}k_2(x + l_2 \sin\theta)^2 \tag{6.6.2}$$

Substitute kinetic and potential energy into the Lagrange's equation:

$$\frac{d}{dt}\frac{\partial T}{\partial \dot{x}} - \frac{\partial T}{\partial x} = -\frac{\partial V}{\partial x}, \quad \frac{d}{dt}\frac{\partial T}{\partial \dot{\theta}} - \frac{\partial T}{\partial \theta} = -\frac{\partial V}{\partial \theta}$$

we can obtain

$$\begin{aligned} &m\ddot{x} + (k_1 + k_2)x + (k_2 l_2 - k_1 l_1)\sin\theta = 0 \\ &I\ddot{\theta} + (k_2 l_2 - k_1 l_1)x\cos\theta + \frac{1}{2}\left(k_1 l_1^2 + k_2 l_2^2\right)\sin 2\theta = 0 \end{aligned} \tag{6.6.3}$$

(b) Linearizing (6.6.3) near the equilibrium position of the system $x = 0\,;\theta = 0$, we get

$$\begin{aligned} &m\ddot{x} + (k_1 + k_2)x + (k_2 l_2 - k_1 l_1)\theta = 0 \\ &I\ddot{\theta} + (k_2 l_2 - k_1 l_1)x + \left(k_1 l_1^2 + k_2 l_2^2\right)\theta = 0 \end{aligned} \tag{6.6.4}$$

which can be written in the matrix form as

$$\begin{bmatrix} m & 0 \\ 0 & I \end{bmatrix}\begin{Bmatrix} \ddot{x} \\ \ddot{\theta} \end{Bmatrix} + \begin{bmatrix} (k_1 + k_2) & (k_2 l_2 - k_1 l_1) \\ (k_2 l_2 - k_1 l_1) & \left(k_1 l_1^2 + k_2 l_2^2\right) \end{bmatrix}\begin{Bmatrix} x \\ \theta \end{Bmatrix} = 0 \tag{6.6.5}$$

Therefore, the linear natural frequencies of the system are

$$\begin{aligned} \omega_1 &= \left[\frac{(k_{11} + k_{22}) - \sqrt{(k_{11} - k_{22})^2 + 4k_{12}^2}}{2}\right]^{1/2} \\ \omega_2 &= \left[\frac{(k_{11} + k_{22}) + \sqrt{(k_{11} + k_{22})^2 + 4k_{12}^2}}{2}\right]^{1/2} \end{aligned} \tag{6.6.6}$$

where

$$k_{11} = \frac{k_1 + k_2}{m}, \quad k_{12} = \frac{k_2 l_2 - k_1 l_1}{\sqrt{mI}}, \quad k_{22} = \frac{k_1 l_1^2 + k_2 l_2^2}{I} \tag{6.6.7}$$

Assuming that the corresponding normal modal matrix is

$$[\Psi]=\begin{bmatrix}\Psi_{11} & \Psi_{12}\\ \Psi_{21} & \Psi_{22}\end{bmatrix} \tag{6.6.8}$$

Then using

$$\begin{Bmatrix}x\\ \theta\end{Bmatrix}=[\Psi]\begin{Bmatrix}u_1\\ u_2\end{Bmatrix}=\begin{bmatrix}\Psi_{11} & \Psi_{12}\\ \Psi_{21} & \Psi_{22}\end{bmatrix}\begin{Bmatrix}u_1\\ u_2\end{Bmatrix} \tag{6.6.9}$$

we can transform (6.6.5) into

$$\begin{Bmatrix}\ddot{u}_1\\ \ddot{u}_2\end{Bmatrix}+\begin{bmatrix}\omega_1^2 & 0\\ 0 & \omega_2^2\end{bmatrix}\begin{Bmatrix}u_1\\ u_2\end{Bmatrix}=0 \tag{6.6.10}$$

(c) For small but finite amplitudes, we assume $x,\theta=O(\varepsilon)\cdot x,\theta=O(\varepsilon)$ Expanding (6.6.3) near the equilibrium position of the system and retaining to $O(\varepsilon^3)$ yields

$$\begin{aligned}&m\ddot{x}+(k_1+k_2)x+(k_2l_2-k_1l_1)\theta-\frac{1}{6}(k_2l_2-k_1l_1)\theta^3=0\\ &I\ddot{\theta}+(k_2l_2-k_1l_1)x+\left(k_1l_1^2+k_2l_2^2\right)\theta-\frac{1}{2}(k_2l_2-k_1l_1)x\theta^2-\frac{2}{3}\left(k_1l_1^2+k_2l_2^2\right)\theta^3=0\end{aligned} \tag{6.6.11}$$

which can be written as

$$\begin{aligned}&\begin{bmatrix}m & 0\\ 0 & I\end{bmatrix}\begin{Bmatrix}\ddot{x}\\ \ddot{\theta}\end{Bmatrix}+\begin{bmatrix}(k_1+k_2) & (k_2l_2-k_1l_1)\\ (k_2l_2-k_1l_1) & \left(k_1l_1^2+k_2l_2^2\right)\end{bmatrix}\begin{Bmatrix}x\\ \theta\end{Bmatrix}\\ &-\begin{Bmatrix}\frac{1}{6}(k_2l_2-k_1l_1)\theta^3\\ \frac{1}{2}(k_2l_2-k_1l_1)x\theta^2+\frac{2}{3}\left(k_1l_1^2+k_2l_2^2\right)\theta^3\end{Bmatrix}=0\end{aligned} \tag{6.6.12}$$

Using the transformation (6.6.9), we can transform (6.6.12) into

$$\begin{aligned}&\left[[\Psi]^T\begin{bmatrix}m & 0\\ 0 & I\end{bmatrix}[\Psi]\right]\begin{Bmatrix}\ddot{u}_1\\ \ddot{u}_2\end{Bmatrix}+\left[[\Psi]^T\begin{bmatrix}(k_1+k_2) & (k_2l_2-k_1l_1)\\ (k_2l_2-k_1l_1) & \left(k_1l_1^2+k_2l_2^2\right)\end{bmatrix}[\Psi]\right]\begin{Bmatrix}u_1\\ u_2\end{Bmatrix}\\ &-[\Psi]^T\left\{\begin{matrix}\frac{1}{6}(k_2l_2-k_1l_1)\theta^3\\ \frac{1}{2}(k_2l_2-k_1l_1)x\theta^2+\frac{2}{3}\left(k_1l_1^2+k_2l_2^2\right)\theta^3\end{matrix}\right\}\Bigg|\begin{Bmatrix}x\\ \theta\end{Bmatrix}=[\Psi]\begin{Bmatrix}u_1\\ u_2\end{Bmatrix}=0\end{aligned} \tag{6.6.13}$$

i.e.,

$$\begin{aligned}\ddot{u}_1+\omega_1^2u_1&=\alpha_{11}u_1u_2^2+\alpha_{12}u_1^2u_2+\alpha_{13}u_1^3+\alpha_{14}u_2^3\\ \ddot{u}_2+\omega_2^2u_2&=\alpha_{21}u_1u_2^2+\alpha_{22}u_1^2u_2+\alpha_{23}u_1^3+\alpha_{24}u_2^3\end{aligned} \tag{6.6.14}$$

where

$$\begin{aligned}
\alpha_{11} &= b_{11}\Psi_{11}\Psi_{22}^2 + 2b_{11}\Psi_{12}\Psi_{21}\Psi_{22} + 3b_{12}\Psi_{21}\Psi_{22}^2 \\
\alpha_{12} &= b_{11}\Psi_{12}\Psi_{21}^2 + 2b_{11}\Psi_{11}\Psi_{21}\Psi_{22} + 3b_{12}\Psi_{21}^2\Psi_{22} \\
\alpha_{13} &= b_{11}\Psi_{11}\Psi_{21}^2 + b_{12}\Psi_{21}^3, \quad \alpha_{14} = b_{11}\Psi_{12}\Psi_{22}^2 + b_{12}\Psi_{22}^3 \\
b_{11} &= \frac{1}{2}\Psi_{21}(k_2l_2 - k_1l_1), \quad b_{12} = \frac{1}{6}\Psi_{11}(k_2l_2 - k_1l_1) + \frac{2}{3}\Psi_{21}\left(k_1l_1^2 + k_2l_2^2\right)
\end{aligned} \tag{6.6.15}$$

$$\begin{aligned}
\alpha_{21} &= b_{21}\Psi_{11}\Psi_{22}^2 + 2b_{21}\Psi_{12}\Psi_{21}\Psi_{22} + 3b_{22}\Psi_{21}\Psi_{22}^2 \\
\alpha_{22} &= b_{21}\Psi_{12}\Psi_{21}^2 + 2b_{21}\Psi_{11}\Psi_{21}\Psi_{22} + 3b_{22}\Psi_{21}^2\Psi_{22} \\
\alpha_{23} &= b_{21}\Psi_{11}\Psi_{21}^2 + b_{22}\Psi_{21}^3, \quad \alpha_{24} = b_{21}\Psi_{12}\Psi_{22}^2 + b_{22}\Psi_{22}^3 \\
b_{21} &= \frac{1}{2}\Psi_{22}(k_2l_2 - k_1l_1), \quad b_{22} = \frac{1}{6}\Psi_{12}(k_2l_2 - k_1l_1) + \frac{2}{3}\Psi_{22}\left(k_1l_1^2 + k_2l_2^2\right)
\end{aligned} \tag{6.6.16}$$

We use the method of multiple scales to solve (6.6.14). Let the solution of Eq. (6.6.14) be expressed as

$$\begin{aligned}
u_1 &= \varepsilon u_{11}(T_0, T_1, T_2) + \varepsilon^2 u_{12}(T_0, T_1, T_2) + \varepsilon^3 u_{13}(T_0, T_1, T_2) + \cdots \\
u_2 &= \varepsilon u_{21}(T_0, T_1, T_2) + \varepsilon^2 u_{22}(T_0, T_1, T_2) + \varepsilon^3 u_{23}(T_0, T_1, T_2) + \cdots
\end{aligned} \tag{6.6.17}$$

Substituting (6.6.17) into (6.6.14) and keeping to $O(\varepsilon^3)$, we get

$$\begin{aligned}
0 &= \ddot{u}_1 + \omega_1^2 u_1 - \alpha_{11}u_1u_2^2 - \alpha_{12}u_1^2u_2 - \alpha_{13}u_1^3 - \alpha_{14}u_2^3 \\
&= \left[D_0^2 + 2\varepsilon D_0D_1 + \varepsilon^2\left(D_1^2 + 2D_0D_2\right)\right]\left(\varepsilon u_{11} + \varepsilon^2 u_{12} + \varepsilon^3 u_{13}\right) \\
&+\omega_1^2\left(\varepsilon u_{11} + \varepsilon^2 u_{12} + \varepsilon^3 u_{13}\right) \\
&-\alpha_{11}\left(\varepsilon u_{11} + \varepsilon^2 u_{12} + \varepsilon^3 u_{13}\right)\left(\varepsilon u_{21} + \varepsilon^2 u_{22} + \varepsilon^3 u_{23}\right)^2 \\
&-\alpha_{12}\left(\varepsilon u_{11} + \varepsilon^2 u_{12} + \varepsilon^3 u_{13}\right)^2\left(\varepsilon u_{21} + \varepsilon^2 u_{22} + \varepsilon^3 u_{23}\right) \\
&-\alpha_{13}\left(\varepsilon u_{11} + \varepsilon^2 u_{12} + \varepsilon^3 u_{13}\right)^3 - \alpha_{14}\left(\varepsilon u_{21} + \varepsilon^2 u_{22} + \varepsilon^3 u_{23}\right)^3 + \cdots \\
&= \varepsilon\left(D_0^2 u_{11} + \omega_1^2 u_{11}\right) + \varepsilon^2\left(D_0^2 u_{12} + \omega_1^2 u_{12} + 2D_0D_1u_{11}\right) \\
&+\varepsilon^3\left[D_0^2 u_{13} + \omega_1^2 u_{13} + 2D_0D_1u_{12} + \left(D_1^2 + 2D_0D_2\right)u_{11}\right. \\
&\left.-\alpha_{11}u_{11}u_{21}^2 - \alpha_{12}u_{11}^2u_{21} - \alpha_{13}u_{11}^3 - \alpha_{14}u_{21}^3\right] + \cdots
\end{aligned} \tag{6.6.18}$$

$$\begin{aligned}
0 &= \ddot{u}_1 + \omega_1^2 u_1 - \alpha_{11} u_1 u_2^2 - \alpha_{12} u_1^2 u_2 - \alpha_{13} u_1^3 - \alpha_{14} u_2^3 \\
&= \left[D_0^2 + 2\varepsilon D_0 D_1 + \varepsilon^2 \left(D_1^2 + 2D_0 D_2\right)\right]\left(\varepsilon u_{11} + \varepsilon^2 u_{12} + \varepsilon^3 u_{13}\right) \\
&+\omega_1^2\left(\varepsilon u_{11} + \varepsilon^2 u_{12} + \varepsilon^3 u_{13}\right) \\
&-\alpha_{11}\left(\varepsilon u_{11} + \varepsilon^2 u_{12} + \varepsilon^3 u_{13}\right)\left(\varepsilon u_{21} + \varepsilon^2 u_{22} + \varepsilon^3 u_{23}\right)^2 \\
&-\alpha_{12}\left(\varepsilon u_{11} + \varepsilon^2 u_{12} + \varepsilon^3 u_{13}\right)^2\left(\varepsilon u_{21} + \varepsilon^2 u_{22} + \varepsilon^3 u_{23}\right) \\
&-\alpha_{13}\left(\varepsilon u_{11} + \varepsilon^2 u_{12} + \varepsilon^3 u_{13}\right)^3 - \alpha_{14}\left(\varepsilon u_{21} + \varepsilon^2 u_{22} + \varepsilon^3 u_{23}\right)^3 + \cdots \\
&= \varepsilon\left(D_0^2 u_{11} + \omega_1^2 u_{11}\right) + \varepsilon^2\left(D_0^2 u_{12} + \omega_1^2 u_{12} + 2D_0 D_1 u_{11}\right) \\
&+\varepsilon^3\left[D_0^2 u_{13} + \omega_1^2 u_{13} + 2D_0 D_1 u_{12} + \left(D_1^2 + 2D_0 D_2\right) u_{11}\right. \\
&\left.-\alpha_{11} u_{11} u_{21}^2 - \alpha_{12} u_{11}^2 u_{21} - \alpha_{13} u_{11}^3 - \alpha_{14} u_{21}^3\right] + \cdots
\end{aligned} \tag{6.6.19}$$

Equating coefficients of like powers of ε in (6.6.18) and (6.6.19) yields.
Order ε^1:

$$\begin{aligned}
D_0^2 u_{11} + \omega_1^2 u_{11} &= 0 \\
D_0^2 u_{21} + \omega_2^2 u_{21} &= 0
\end{aligned} \tag{6.6.20}$$

Order ε^2:

$$\begin{aligned}
D_0^2 u_{12} + \omega_1^2 u_{12} &= -2D_0 D_1 u_{11} \\
D_0^2 u_{22} + \omega_2^2 u_{22} &= -2D_0 D_1 u_{21}
\end{aligned} \tag{6.6.21}$$

Order ε^3:

$$\begin{aligned}
&D_0^2 u_{13} + \omega_1^2 u_{13} = -2D_0 D_1 u_{12} - \left(D_1^2 + 2D_0 D_2\right) u_{11} \\
&+\alpha_{11} u_{11} u_{21}^2 + \alpha_{12} u_{11}^2 u_{21} + \alpha_{13} u_{11}^3 + \alpha_{14} u_{21}^3 \\
&D_0^2 u_{23} + \omega_2^2 u_{23} = -2D_0 D_1 u_{22} - \left(D_1^2 + 2D_0 D_2\right) u_{21} \\
&+\alpha_{21} u_{11} u_{21}^2 + \alpha_{22} u_{11}^2 u_{21} + \alpha_{23} u_{11}^3 + \alpha_{24} u_{21}^3
\end{aligned} \tag{6.6.22}$$

The solution of (6.6.20) is

$$\begin{Bmatrix} u_{11} \\ u_{21} \end{Bmatrix} = \begin{Bmatrix} A_1 e^{i\omega_1 T_0} + \overline{A}_1 e^{-i\omega_1 T_0} \\ A_2 e^{i\omega_2 T_0} + \overline{A}_2 e^{-i\omega_2 T_0} \end{Bmatrix} \tag{6.6.23}$$

where $A_k = A_k(T_1, T_2)$. Substituting (6.6.23) into (6.6.21), we obtain

$$\begin{aligned}
D_0^2 u_{12} + \omega_1^2 u_{12} &= -2i\omega_1 D_1 A_1 e^{i\omega_1 T_0} + cc \\
D_0^2 u_{22} + \omega_2^2 u_{22} &= -2i\omega_2 D_1 A_2 e^{i\omega_2 T_0} + cc
\end{aligned} \tag{6.6.24}$$

Eliminating secular terms in the above equation gives

$$D_1 A_1 = D_1 A_2 = 0 \quad \rightarrow \quad A_1 = A_1(T_2), \quad A_2 = A_2(T_2) \tag{6.6.25}$$

therefore

$$u_{12} = 0, \quad u_{22} = 0 \tag{6.6.26}$$

Substituting (6.6.23), (6.6.25) and (6.6.26) into (6.6.22) yields

$$D_0^2 u_{13} + \omega_1^2 u_{13} = -2i\omega_1 A_1' e^{i\omega_1 T_0} + 2\alpha_{11} A_1 A_2 \overline{A}_2 e^{i\omega_1 T_0} + 3\alpha_{13} A_1^2 \overline{A}_1 e^{i\omega_1 T_0} \\ +\alpha_{12} \overline{A}_1^2 A_2 e^{i(\omega_2 - 2\omega_1)T_0} + \alpha_{14} A_2^3 e^{3i\omega_2 T_0} + cc + NST \tag{6.6.27}$$

$$D_0^2 u_{23} + \omega_2^2 u_{23} = -2i\omega_2 A_2' e^{i\omega_2 T_0} + 2\alpha_{22} A_1 \overline{A}_1 A_2 e^{i\omega_2 T_0} + 3\alpha_{24} A_2^2 \overline{A}_2 e^{i\omega_2 T_0} \\ +\alpha_{21} A_1 \overline{A}_2^2 e^{i(\omega_1 - 2\omega_2)T_0} + \alpha_{23} A_1^3 e^{3i\omega_1 T_0} + cc + NST \tag{6.6.28}$$

Primes denote the derivative with respect to T^2. To obtain these two equations, we did not consider the case of $\omega_1 = \omega_2$. From (6.6.27) and (6.6.28), we can find that there are two possible resonant cases:$\omega_2 \approx 3\omega_1$ and $\omega_1 \approx 3\omega_2$. Without loss of generality, we assume $\omega_2 > \omega_1$. In analyzing the particular solutions of (6.6.27) and (6.6.28) we need to distinguish between the resonant case in which $\omega_1 \approx 3\omega_2$ and the nonresonant case in which ω_2 is away from $3\omega_1$.

(1) The nonresonant case.

In this case, $\omega_2 \neq \omega_1, \omega_2 \neq 3\omega_1$, in order to eliminate secular terms from (6.6.27) and (6.6.28), there must be

$$\begin{aligned} 2i\omega_1 A_1' - 2\alpha_{11} A_1 A_2 \overline{A}_2 - 3\alpha_{13} A_1^2 \overline{A}_1 = 0 \\ 2i\omega_2 A_2' - 2\alpha_{22} A_1 \overline{A}_1 A_2 - 3\alpha_{24} A_2^2 \overline{A}_2 = 0 \end{aligned} \tag{6.6.29}$$

Let

$$A_k = \frac{1}{2} a_k e^{i\beta_k}, \quad k = 1, 2 \tag{6.6.30}$$

Substitute (6.6.30) into (6.6.29), we get

$$\begin{aligned} ia_1' - \beta_1' a_1 - \tfrac{1}{4}\alpha_{11}\omega_1^{-1} a_1 a_2^2 - \tfrac{3}{8}\alpha_{13}\omega_1^{-1} a_1^3 = 0 \\ ia_2' - \beta_2' a_2 - \tfrac{1}{4}\alpha_{22}\omega_2^{-1} a_1^2 a_2 - \tfrac{3}{8}\alpha_{24}\omega_2^{-1} a_2^3 = 0 \end{aligned} \tag{6.6.31}$$

Separating the real and imaginary parts of the above equations yields

$$a_1' = 0, \quad a_2' = 0 \tag{6.6.32}$$

$$\begin{aligned} \beta_1' = -\tfrac{1}{4}\alpha_{11}\omega_1^{-1} a_2^2 - \tfrac{3}{8}\alpha_{13}\omega_1^{-1} a_1^2 \\ \beta_2' = -\tfrac{1}{4}\alpha_{22}\omega_2^{-1} a_1^2 - \tfrac{3}{8}\alpha_{24}\omega_2^{-1} a_2^2 \end{aligned} \tag{6.6.33}$$

therefore

$$a_1 = a_{10}, \quad a_2 = a_{20} \tag{6.6.34}$$

$$\begin{aligned} \beta_1 &= \beta_{10} - \left(\tfrac{1}{4}\alpha_{11}\omega_1^{-1}a_{20}^2 + \tfrac{3}{8}\alpha_{13}\omega_1^{-1}a_{10}^2\right)T_2 \\ \beta_2 &= \beta_{20} - \left(\tfrac{1}{4}\alpha_{22}\omega_2^{-1}a_{10}^2 + \tfrac{3}{8}\alpha_{24}\omega_2^{-1}a_{20}^2\right)T_2 \end{aligned} \tag{6.6.35}$$

If εa_1 and εa_2 are written as a_1 and a_2, respectively, then the first order approximate solution can be obtained as

$$\begin{aligned} \begin{Bmatrix} x \\ \theta \end{Bmatrix} &= [\Psi]\begin{Bmatrix} u_1 \\ u_2 \end{Bmatrix} \approx [\Psi]\begin{Bmatrix} u_{11} \\ u_{21} \end{Bmatrix} + O(\varepsilon^3) \\ &= [\Psi]\begin{Bmatrix} a_{10}\cos[\omega_1 t - \left(\frac{1}{4}\alpha_{11}\omega_1^{-1}a_{20}^2 + \frac{3}{8}\alpha_{13}\omega_1^{-1}a_{10}^2\right)t + \beta_{10} \\ a_{20}\cos[\omega_2 t - \left(\frac{1}{4}\alpha_{22}\omega_2^{-1}a_{10}^2 + \frac{3}{8}\alpha_{24}\omega_2^{-1}a_{20}^2\right)t + \beta_{20} \end{Bmatrix} + O(\varepsilon^3) \end{aligned} \tag{6.6.36}$$

(2) The resonant case (internal resonance $\omega_2 \approx 3\omega_1$)

In this case we introduce a detuning parameter σ according to

$$\omega_2 = 3\omega_1 + \varepsilon^2\sigma \tag{6.6.37}$$

Eliminating secular terms from (6.6.27) and (6.6.28), there must be

$$\begin{aligned} &2iA'_1 - 2\alpha_{11}\omega_1^{-1}A_1A_2\overline{A}_2 - 3\alpha_{13}\omega_1^{-1}A_1^2\overline{A}_1 - \alpha_{12}\omega_1^{-1}\overline{A}_1^2A_2e^{i\sigma T_2} = 0 \\ &2iA' - 2\alpha_{22}\omega_2^{-1}A_1\overline{A}_1A_2 - 3\alpha_{24}\omega_2^{-1}A_2^2\overline{A}_2 - \alpha_{23}\omega_2^{-1}A_1^3e^{-i\sigma T_2} = 0 \end{aligned} \tag{6.6.38}$$

Substituting (6.6.30) into (6.6.38) yields

$$\begin{aligned} &ia_1' - a_1\beta_{11}' - \tfrac{1}{4}\alpha_{11}\omega_1^{-1}a_1a_2^2 - \tfrac{3}{8}\alpha_{13}\omega_1^{-1}a_1^3 - \alpha_{12}\omega_1^{-1}a_1^2a_2e^{i(\sigma T_2 - 3\beta_1 + \beta_2)} = 0 \\ &ia_2' - \beta' a_{22} - \tfrac{1}{4}\alpha_{22}\omega_2^{-1}a_1^2a_2 - \tfrac{3}{8}\alpha_{24}\omega_2^{-1}a_2^3 - \alpha_{23}\omega_2^{-1}a_1^3e^{-i(\sigma T_2 - 3\beta_1 + \beta_2)} = 0 \end{aligned} \tag{6.6.39}$$

Separating the real and imaginary parts of the above equations yields

$$\begin{aligned} a_1' &= \alpha_{12}\omega_1^{-1}a_1^2a_2 \sin\gamma \\ a_2' &= -\alpha_{23}\omega_2^{-1}a_1^3 \sin\gamma \end{aligned} \tag{6.6.40}$$

$$\begin{aligned} a_1\beta_1' &= -\frac{1}{4}\alpha_{11}\omega_1^{-1}a_1a_2^2 - \frac{3}{8}\alpha_{13}\omega_1^{-1}a_1^3 - \alpha_{12}\omega_1^{-1}a_1^2a_2\cos\gamma \\ a_2\beta_2' &= -\frac{1}{4}\alpha_{22}\omega_2^{-1}a_1^2a_2 - \frac{3}{8}\alpha_{24}\omega_2^{-1}a_2^3 - \alpha_{23}\omega_2^{-1}a_1^3\cos\gamma \end{aligned} \tag{6.6.41}$$

where

$$\gamma = \sigma T_2 - 3\beta_1 + \beta_2 \tag{6.6.42}$$

Eliminating γ from (6.5.20), we obtain

$$\alpha_{23}a_1da_1 + 3\alpha_{12}a_2da_2 = 0 \tag{6.6.43}$$

therefore,

$$\alpha_{23}a_1^2 + 3\alpha_{12}a_2^2 = E, \quad E = \alpha_{23}a_{10}^2 + 3\alpha_{12}a_{20}^2 \tag{6.6.44}$$

Write εa_1 and εa_2 as a_1 and a_2, respectively, we can obtain the first-order approximate solution of the original equation

$$\begin{aligned} \begin{Bmatrix} x \\ \theta \end{Bmatrix} &= [\Psi]\begin{Bmatrix} u_1 \\ u_2 \end{Bmatrix} \approx [\Psi]\begin{Bmatrix} u_{11} \\ u_{21} \end{Bmatrix} + O(\varepsilon^3) \\ &= [\Psi]\begin{Bmatrix} a_1(t)\cos[\omega_1 t + \beta_1(t)] \\ a_2(t)\cos[\omega_2 t + \beta_2(t)] \end{Bmatrix} + O(\varepsilon^3) \end{aligned} \tag{6.6.45}$$

where a_1, a_2, $\beta_1\beta_1$ and β_2 can be determined by (6.6.40), (6.6.41), (6.6.42) and (6.6.44).

Next we analyze the equations describing amplitude and phase of the motion, (6.6.40)–(6.6.44). Eliminating β_1', β_2' from the two equations of (6.6.41), we get

$$\begin{aligned} a_2\gamma' &= \sigma a_2 + \tfrac{3}{4}\alpha_{11}\omega_1^{-1}a_2^3 + \tfrac{9}{8}\alpha_{13}\omega_1^{-1}a_1^2a_2 + 3\alpha_{12}\omega_1^{-1}a_1a_2^2\cos\gamma \\ &\quad -\tfrac{1}{4}\alpha_{22}\omega_2^{-1}a_1^2a_2 - \tfrac{3}{8}\alpha_{24}\omega_2^{-1}a_2^3 - \alpha_{23}\omega_2^{-1}a_1^3\cos\gamma \end{aligned} \tag{6.6.46}$$

Dividing the above equation with the first equation of (6.6.40) yields

$$\begin{aligned} \alpha_{23}a_1^3d(a_2\cos\gamma) &= 9\alpha_{12}a_1a_2^2\cos\gamma da_2 + \omega_2\sigma a_2da_2 + \tfrac{9}{4}\alpha_{11}a_2^3da_2 \\ &\quad +\tfrac{27}{8}\alpha_{13}a_1^2a_2da_2 - \tfrac{1}{4}\alpha_{22}a_1^2a_2da_2 - \tfrac{3}{8}\alpha_{24}a_2^3da_2 \end{aligned}$$

since

$$\alpha_{23}a_1^3d(a_2\cos\gamma) = \alpha_{23}d\left(a_1^3a_2\cos\gamma\right) - 3\alpha_{23}a_1^2a_2\cos\gamma da_1$$

Thus, the above two equations become

$$\begin{aligned} \alpha_{23}d\left(a_1^3a_2\cos\gamma\right) &= 3a_1a_2\cos\gamma(\alpha_{23}a_1da_1 + 3\alpha_{12}a_2da_2) + \omega_2\sigma a_2da_2 + \tfrac{9}{4}\alpha_{11}a_2^3da_2 \\ &\quad +\tfrac{27}{8}\alpha_{13}a_1^2a_2da_2 - \tfrac{1}{4}\alpha_{22}a_1^2a_2da_2 - \tfrac{3}{8}\alpha_{24}a_2^3da_2 \end{aligned}$$

With Eq. (6.6.43), the above equation becomes

$$\begin{aligned} \alpha_{23}d\left(a_1^3a_2\cos\gamma\right) &= \omega_2\sigma a_2da_2 + \left(\frac{9}{4}\alpha_{11} - \frac{3}{8}\alpha_{24}\right)a_2^3da_2 \\ &\quad + \left(\frac{27}{8}\alpha_{13} - \frac{1}{4}\alpha_{22}\right)a_1^2a_2da_2 \end{aligned}$$

Using (6.6.44) to eliminate a_1^2 on the right-hand side of the above equation, we get

$$\begin{aligned}&\alpha_{23}d\left(a_1^3a_2\cos\gamma\right)=\left[\omega_2\sigma+E\alpha_{23}^{-1}\left(\tfrac{27}{8}\alpha_{13}-\tfrac{1}{4}\alpha_{22}\right)\right]a_2da_2\\&+\left[\left(\tfrac{9}{4}\alpha_{11}-\tfrac{3}{8}\alpha_{24}\right)-3\alpha_{23}^{-1}\alpha_{12}\left(\tfrac{27}{8}\alpha_{13}-\tfrac{1}{4}\alpha_{22}\right)\right]a_2^3da_2\end{aligned}\tag{6.6.47}$$

Make the integration of Eq. (6.6.47), we can obtain

$$a_1^3a_2\cos\gamma-(h_1\sigma+h_2)a_2^2-h_3a_2^4=L\tag{6.6.48}$$

where L is the constant of integration and

$$\begin{aligned}h_1&=\frac{1}{2}\alpha_{23}^{-1}\omega_2,\quad h_2=\frac{1}{2}E\alpha_{23}^{-2}\left(\frac{27}{8}\alpha_{13}-\frac{1}{4}\alpha_{22}\right)\\h_3&=\frac{1}{4}\left[\alpha_{23}^{-1}\left(\frac{9}{4}\alpha_{11}-\frac{3}{8}\alpha_{24}\right)-3\alpha_{23}^{-2}\alpha_{12}\left(\frac{27}{8}\alpha_{13}-\frac{1}{4}\alpha_{22}\right)\right]\end{aligned}\tag{6.6.49}$$

Now, let us combine, and into a single variable differential equation. To do this, let

$$a_1^2=E\xi\tag{6.6.50}$$

Then from (6.6.44), we can obtain

$$a_2^2=\frac{1}{3}\alpha_{12}^{-1}E(1-\alpha_{23}\xi)\tag{6.6.51}$$

Eliminating γ from (6.6.48) and (6.6.40) yields

$$\begin{aligned}&\tfrac{E^2}{4\alpha_{12}^2\omega_1^{-2}}\left(\tfrac{d\xi}{dT_2}\right)^2=\tfrac{1}{3}E^4\alpha_{12}^{-1}\xi^3(1-\alpha_{23}\xi)\\&-\left\{L+\tfrac{1}{3}\alpha_{12}^{-1}E(1-\alpha_{23}\xi)\left[h_1\sigma+h_2+\tfrac{1}{3}Eh_3\alpha_{12}^{-1}(1-\alpha_{23}\xi)\right]\right\}^2\end{aligned}\tag{6.6.52}$$

Or

$$\frac{E^2}{4\alpha_{12}^2\omega_1^{-2}}\left(\frac{d\xi}{dT_2}\right)^2=F^2(\xi)-G^2(\xi)\tag{6.6.53}$$

where

$$\begin{aligned}F(\xi)&=\pm\sqrt{\tfrac{1}{3}E^4\alpha_{12}^{-1}\xi^3(1-\alpha_{23}\xi)}\\G(\xi)&=\sqrt{\left\{L+\tfrac{1}{3}\alpha_{12}^{-1}E(1-\alpha_{23}\xi)\left[h_1\sigma+h_2+\tfrac{1}{3}Eh_3\alpha_{12}^{-1}(1-\alpha_{23}\xi)\right]\right\}^2}\end{aligned}\tag{6.6.54}$$

The functions $F(\xi)$ and $G(\xi)$ are shown schematically in Fig. 6.6. For real motions, $F^2(\xi) \geq G^2(\xi)$. The points where F meets G corresponds to the vanishing of $\xi\prime$. Considering the case of two intersection points of curves F and G, such as the intersection of G_2 and F, as well as the intersection of G_3 and F in Figure 6.6. In this case, ξ is periodic and oscillates between two intersection points, and the motion is aperiodic.

Considering the curves G_1 and F are tangent to each other, as shown in Figure 6.6. In this case, ξ has only one constant-value solution, and the motion of the system is periodic. The nonlinear effect acts to modulate the phase of the motion so that the nonlinear frequencies of the two modes are commensurable.

In order to analyze the steady state solution, let $a_1' = a_2' = \gamma' = 0$, (6.6.40) and (6.6.46) become

$$\sin\gamma = 0 \quad \text{or} \quad \gamma = n\pi \tag{6.6.55}$$

$$\begin{aligned} \sigma a_2 &+ \left(\frac{9}{8}\alpha_{13}\omega_1^{-1} - \frac{1}{4}\alpha_{22}\omega_2^{-1}\right)a_1^2 a_2 + \left(\frac{3}{4}\alpha_{11}\omega_1^{-1} - \frac{3}{8}\alpha_{24}\omega_2^{-1}\right)a_2^3 \\ &+ \left(3\alpha_{12}\omega_1^{-1}a_2^2 - \alpha_{23}\omega_2^{-1}a_1^2\right)a_1\cos n\pi = 0 \end{aligned} \tag{6.6.56}$$

Equation (6.6.56) is the cubic equation of a_2. For a given set of values of σ, a_1 and $\cos n\pi$, Eq. (6.6.56) has one or three real roots, each of which corresponds to a periodic motion of the system. Thus, the periodic motion of the system is either unique or one of the three possible periodic motions. Figure 6.6 shows that the periodic motion is unstable. With a small perturbation, a curve such as G_1 may become a curve such as G_2 and the motion of the system becomes aperiodic.

For the case of the nonlinear periodic motion of the system with a constant-value solution, we note that the frequency of the motion of the system is

$$\hat{\omega}_1 = \omega_1 + \varepsilon^2\beta_1', \quad \hat{\omega}_2 = \omega_2 + \varepsilon^2\beta_2' \tag{6.6.57}$$

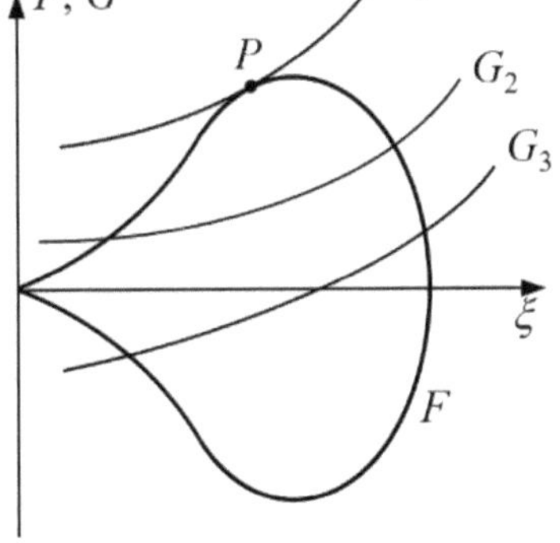

Fig. 6.6 Schematic diagram of $F(\xi)$ and $G(\xi)$ for Exercise 6.6

therefore,

$$\hat{\omega}_2 - 3\hat{\omega}_1 = \omega_2 - 3\omega_1 + \varepsilon^2(\beta_2' - 3\beta_1') = \varepsilon^2\sigma + \varepsilon^2(\beta_2' - 3\beta_1') = \varepsilon^2\gamma' = 0 \tag{6.6.58}$$

Thus, the nonlinearity of the system modulates the frequency of the motion to exactly 3 to 1, and thus the motion is periodic.

6.7 Exercise 6.7 (Analytical Solution of a Two-Degree-of-Freedom High-order Nonlinear Equation)

Solution: (a) From the given equation of the system, we can get

$$\begin{aligned} m_1\ddot{u}_1 &= -\left[k_1 + \alpha_1\left(1 - \frac{u_2}{u_1}\right)^n\right]u_1^n \\ m_2\ddot{u}_2 &= -\left[k_2 - \alpha_1\left(\frac{u_1}{u_2} - 1\right)^n\right]u_2^n \end{aligned} \tag{6.7.1}$$

thus

$$\frac{m_1\ddot{u}_1}{m_2\ddot{u}_2} = \frac{\left[k_1 + \alpha_1\left(1 - \frac{u_2}{u_1}\right)^n\right]}{\left[k_2 - \alpha_1\left(\frac{u_1}{u_2} - 1\right)^n\right]}\left(\frac{u_1}{u_2}\right)^n \tag{6.7.2}$$

It can be seen that there exists a system of solutions u_1, u_2 that satisfy

$$\frac{u_1}{u_2} = c = \text{constant} \tag{6.7.3}$$

In this case, the constant c should satisfy

$$\frac{m_1 c}{m_2} = \frac{\left[k_1 + \alpha_1\left(1 - \frac{1}{c}\right)^n\right]}{\left[k_2 - \alpha_1(c-1)^n\right]}c^n$$

i.e.,

$$m_2\left[k_1 c^n + \alpha_1\left(c^n - 1\right)^n\right] = m_1 c\left[k_2 - \alpha_1(c-1)^n\right] \tag{6.7.4}$$

(b) From the second equation of (6.7.1) and (6.7.3), we can obtain

$$\ddot{u}_2 + \left[k_2 - \alpha_1(c-1)^n\right]m_2^{-1}u_2^n = 0 \tag{6.7.5}$$

Multiplying both sides of the equation by $\dot{u}_2$ gives

$$\dot{u}_2 d\dot{u}_2 + \left[k_2 - \alpha_1(c-1)^n\right]m_2^{-1}u_2^n du_2 = 0 \tag{6.7.6}$$

After integration, we can obtain

$$\frac{1}{2}\dot{u}_2^2 + \frac{1}{n+1}\left[k_2 - \alpha_1(c-1)^n\right]m_2^{-1}u_2^{n+1} = E \tag{6.7.7}$$

E is the constant of integration. Separating the variables for the above equation gives

$$dt = \pm\frac{du_2}{\sqrt{2\left(E - bu_2^{n+1}\right)}} \tag{6.7.8}$$

where

$$b = \frac{1}{n+1}\left[k_2 - \alpha_1(c-1)^n\right]m_2^{-1} \tag{6.7.9}$$

Make the integration of (6.7.8), we can obtain

$$t - t_0 = \pm\int_{u_{20}}^{u_2}\frac{dx}{\sqrt{2\left(E - bx^{n+1}\right)}} \tag{6.7.10}$$

Combining (6.7.3) and (6.7.8), we can obtain

$$dt = \pm\frac{c^{-1}du_1}{\sqrt{2\left(E - bc^{-(n+1)}u_1^{n+1}\right)}} \tag{6.7.11}$$

Make the integration of (6.7.11), we can obtain

$$t - t_0 = \pm c^{-1}\int_{u_{10}}^{u_1}\frac{dx}{\sqrt{2\left(E - bc^{-(n+1)}x^{n+1}\right)}} \tag{6.7.12}$$

6.8 Exercise 6.8 (Forced Oscillations of a Two-Degree-of-Freedom Gyroscopic System with Cubic Nonlinearity)

Solution: (a) The given system equations represent a linear gyroscopic system plus a linear damping term, a cubic nonlinear term, and an excitation term. Its corresponding linear undamped free oscillation equation is:

$$\begin{aligned} \ddot{u}_1 + I\Omega\dot{u}_2 + u_1 &= 0 \\ \ddot{u}_2 - I\Omega\dot{u}_1 + u_2 &= 0 \end{aligned} \tag{6.8.1}$$

Let the solution of Eq. (6.8.1) is

$$u_1 = Ae^{i\omega t}, \quad u_2 = Be^{i\omega t} \tag{6.8.2}$$

The characteristic equation of the linear system can be obtained as

$$\begin{bmatrix} 1-\omega^2 & iI\Omega\omega \\ -iI\Omega\omega & 1-\omega^2 \end{bmatrix} \begin{Bmatrix} A \\ B \end{Bmatrix} = 0 \tag{6.8.3}$$

So

$$\det\begin{bmatrix} -\omega^2+1 & iI\Omega\omega \\ -iI\Omega\omega & -\omega^2+1 \end{bmatrix} = 0 \tag{6.8.4}$$

i.e.,

$$\left(\omega^2-1\right)^2 - I^2\Omega^2\omega^2 = 0 \tag{6.8.5}$$

The natural frequencies of that linear undamped system are:

$$\omega_1 = \frac{\sqrt{I^2\Omega^2+4} - I\Omega}{2}, \quad \omega_2 = \frac{\sqrt{I^2\Omega^2+4} + I\Omega}{2} \tag{6.8.6}$$

Substituting ω_1 and ω_2 into (6.8.3), the corresponding modal vectors can be obtained as

$$\{1,\ i\}^T, \quad \{1,\ -i\}^T \tag{6.8.7}$$

The solution of linear undamped free oscillation Eq. (6.8.1) is

$$\begin{aligned} u_{11} &= A_1 e^{i\omega_1 t} + A_2 e^{i\omega_2 t} + cc \\ u_{21} &= iA_1 e^{i\omega_1 t} - iA_2 e^{i\omega_2 t} + cc \end{aligned} \tag{6.8.8}$$

(b) For small and finite amplitudes, we use the method of multiple scales to solve the original nonlinear equations. Therefore, let the solutions of the equations be

$$\begin{aligned} u_1 &= u_{11}(T_0, T_1) + \varepsilon u_{12}(T_0, T_1) + \cdots \\ u_2 &= u_{21}(T_0, T_1) + \varepsilon u_{22}(T_0, T_1) + \cdots \end{aligned} \tag{6.8.9}$$

Substituting (6.8.9) into the given system equation and retaining to $O(\varepsilon^2)$ yields

$$\begin{aligned} 0 &= \ddot{u}_1 + I\Omega\dot{u}_2 + u_1 + 2\varepsilon\dot{u}_1 + \Omega^2\cos\Omega t \\ &= \left(D_0^2 + 2\varepsilon D_0 D_1\right)(u_{11} + \varepsilon u_{12}) + I\Omega(D_0 + \varepsilon D_1)(u_{21} + \varepsilon u_{22}) \\ &\quad + (u_{11} + \varepsilon u_{12}) + 2\varepsilon(D_0 + \varepsilon D_1)(u_{11} + \varepsilon u_{12}) + \Omega^2\cos\Omega T_0 + \cdots \\ &= D_0^2 u_{11} + I\Omega D_0 u_{21} + u_{11} + \Omega^2\cos\Omega T_0 \\ &\quad + \varepsilon\big(D_0^2 u_{12} + I\Omega D_0 u_{22} + u_{12} \\ &\quad + 2D_0 D_1 u_{11} + I\Omega D_1 u_{21} + 2D_0 u_{11}\big) + \cdots \end{aligned} \tag{6.8.10}$$

$$\begin{aligned} 0 &= \ddot{u}_2 - I\Omega\dot{u}_1 + u_2 + 2\varepsilon\dot{u}_2 + \varepsilon\Omega^2 u_2^3 \\ &= \left(D_0^2 + 2\varepsilon D_0 D_1\right)(u_{21} + \varepsilon u_{22}) - I\Omega(D_0 + \varepsilon D_1)(u_{11} + \varepsilon u_{12}) \\ &\quad + (u_{21} + \varepsilon u_{22}) + 2\varepsilon(D_0 + \varepsilon D_1)(u_{21} + \varepsilon u_{22}) + \varepsilon\Omega^2(u_{21} + \varepsilon u_{22})^3 + \cdots \\ &= D_0^2 u_{21} - I\Omega D_0 u_{11} + u_{21} + \varepsilon\big(D_0^2 u_{22} - I\Omega D_0 u_{12} + u_{22} \\ &\quad - I\Omega D_1 u_{11} + 2D_0 D_1 u_{21} + 2D_0 u_{21} + \Omega^2 u_{21}^3\big) + \cdots \end{aligned} \tag{6.8.11}$$

Equating coefficients of like powers of ε in (6.8.10) and (6.8.11) yields.
Order ε^0:

$$\begin{aligned} D_0^2 u_{11} + I\Omega D_0 u_{21} + u_{11} &= -\Omega^2\cos\Omega T_0 \\ D_0^2 u_{21} - I\Omega D_0 u_{11} + u_{21} &= 0 \end{aligned} \tag{6.8.12}$$

Order ε^1:

$$\begin{aligned} D_0^2 u_{12} + I\Omega D_0 u_{22} + u_{12} &= -2D_0 D_1 u_{11} - I\Omega D_1 u_{21} - 2D_0 u_{11} \\ D_0^2 u_{22} - I\Omega D_0 u_{12} + u_{22} &= -2D_0 D_1 u_{21} + I\Omega D_1 u_{11} - 2D_0 u_{21} - \Omega^2 u_{21}^3 \end{aligned} \tag{6.8.13}$$

The solution of (6.8.12) is

$$\begin{aligned} u_{11} &= A_1 e^{i\omega_1 T_0} + A_2 e^{i\omega_2 T_0} + K_1 e^{i\Omega T_0} + cc \\ u_{21} &= iA_1 e^{i\omega_1 T_0} - iA_2 e^{i\omega_2 T_0} + iK_2 e^{i\Omega T_0} + cc \end{aligned} \tag{6.8.14}$$

where

$$K_1 = \frac{\Omega^2(1-\Omega^2)}{2\left[I^2\Omega^4 - (1-\Omega^2)^2\right]},\quad K_2 = \frac{I\Omega^4}{2\left[I^2\Omega^4 - (1-\Omega^2)^2\right]} \tag{6.8.15}$$

$A_k = A_k(T_1)$. Substituting (6.8.14) into (6.8.13), we can obtain

$$\begin{aligned} &D_0^2 u_{12} + I\Omega D_0 u_{22} + u_{12} = \left(-2i\omega_1 A_1' - iI\Omega A_1' - 2i\omega_1 A_1\right)e^{i\omega_1 T_0} \\ &+\left(-2i\omega_2 A_2' + iI\Omega A_2' - 2i\omega_2 A_2\right)e^{i\omega_2 T_0} \\ &-2i\Omega K_1 e^{i\Omega T_0} + cc \end{aligned} \tag{6.8.16}$$

$$\begin{aligned} D_0^2 u_{22} &- I\Omega D_0 u_{12} + u_{22} = \left(2\omega_1 A_1' + I\Omega A_1' + 2\omega_1 A_1 - 3i\Omega^2 A_1^2\overline{A}_1\right. \\ &\left.- 6i\Omega^2 K_2^2 A_1 - 6i\Omega^2 A_1 A_2 \overline{A}_2\right)e^{i\omega_1 T_0} \\ &+\left(-2\omega_2 A_2' + I\Omega A_2' - 2\omega_2 A_2 + 3i\Omega^2 A_2^2 \overline{A}_2\right. \\ &\left.+6i\Omega^2 K_2^2 A_2 + 6i\Omega^2 A_1 \overline{A}_1 A_2\right)e^{i\omega_2 T_0} \\ &+\left(2\Omega K_2 - 6i\Omega^2 K_2 A_1 \overline{A}_1 - 6i\Omega^2 K_2 A_2 \overline{A}_2\right. \\ &\left.-3i\Omega^2 K_2^3\right)e^{i\Omega T_0} + i\Omega^2 K_2^3 e^{3i\Omega T_0} \\ &+3\Omega^2 K_2 A_1 e^{i(\omega_1+\Omega)T_0} - 3\Omega^2 K_2 A_2 e^{i(\omega_2+\Omega)T_0} \\ &-6i\Omega^2 K_2 A_1 A_2 e^{i(\omega_1+\omega_2+\Omega)T_0} + 6i\Omega^2 K_2 A_1 A_2 e^{i(\omega_1+\omega_2-\Omega)T_0} \\ &-3i\Omega^2 \overline{A}_1 A_2^2 e^{i(2\omega_2-\omega_1)T_0} + 3i\Omega^2 A_1^2 \overline{A}_2 e^{i(2\omega_1-\omega_2)T_0} \\ &-3i\Omega^2 K_2^2 \overline{A}_1 e^{i(2\Omega-\omega_1)T_0} + 3i\Omega^2 K_2^2 \overline{A}_2 e^{i(2\Omega-\omega_2)T_0} \\ &-3i\Omega^2 K_2 A_1^2 e^{i(2\omega_1-\Omega)T_0} - 3i\Omega^2 K_2 A_2^2 e^{i(2\omega_2-\Omega)T_0} \\ &+6i\Omega^2 K_2 A_2 \overline{A}_1 e^{i(\Omega-\omega_1+\omega_2)T_0} + 6i\Omega^2 K_2 A_1 \overline{A}_2 e^{i(\Omega+\omega_1-\omega_2)T_0} \\ &-3\Omega^2 A_1 A_2 e^{i(\omega_1+\omega_2)T_0} + i\Omega^2 A_1^3 e^{3i\omega_1 T_0} - i\Omega^2 A_2^3 e^{3i\omega_2 T_0} + cc \end{aligned} \tag{6.8.17}$$

Primes denote the derivative with respect to T_1. It can be seen from (6.8.16) and (6.8.17) that there are different possibilities: internal resonance, primary resonance and combination resonance, depending on the values of ω_1 and ω_2. In the following, we only present the solution for case (i), i.e.,$\omega_1 + \omega_2 \approx 2\Omega$; for case (ii), i.e.,$3\omega_1 \approx \Omega$ and $\omega_2 \approx 3\Omega$, readers are invited to complete the solution.

(i) $\boldsymbol{\omega_1 + \omega_2 \approx 2\Omega}$**:**

Let

$$\omega_1 + \omega_2 = 2\Omega + \varepsilon\sigma \tag{6.8.18}$$

The corresponding homogeneous equations to (6.8.16) and (6.8.17) are identical to (6.8.1). (6.8.16) and (6.8.17) are coupled with each other, so the excitation terms with frequencies ω_1 and ω_2 can induce the resonances of both $u_{12}$and u_{22}. To make the equations solvable, terms with frequencies ω_1 and ω_2 must be eliminated at the same time. In order to eliminate secular terms of Eqs. (6.8.16) and (6.8.17), we seek the following special solutions for Eqs. (6.8.16) and (6.8.17):

$$\begin{aligned} u_{12} &= P_{11}e^{i\omega_1 T_0} + P_{12}e^{i\omega_2 T_0} \\ u_{22} &= P_{21}e^{i\omega_1 T_0} + P_{22}e^{i\omega_2 T_0} \end{aligned} \tag{6.8.19}$$

Substituting (6.8.19) into the Eqs. (6.8.16) and (6.8.17), taking (6.8.18) into account, and equating the coefficients of $e^{i\omega_1 T_0}$ and $e^{i\omega_2 T_0}$ yields

$$\begin{bmatrix} \left(1-\omega_n^2\right) & i\omega_n I\Omega \\ -i\omega_n I\Omega & \left(1-\omega_n^2\right) \end{bmatrix} \begin{Bmatrix} P_{1n} \\ P_{2n} \end{Bmatrix} = \begin{Bmatrix} R_{1n} \\ R_{2n} \end{Bmatrix}, \quad n = 1,\ 2 \tag{6.8.20}$$

where

$$\begin{aligned} R_{11} &= -2i\omega_1 A_1' - iI\Omega A_1' - 2i\omega_1 A_1 \\ R_{21} &= 2\omega_1 A_1' + I\Omega A_1' + 2\omega_1 A_1 - 3i\Omega^2 A_1^2 \overline{A}_1 \\ &-6i\Omega^2 K_2^2 A_1 - 6i\Omega^2 A_1 A_2 \overline{A}_2 + 3i\Omega^2 K_2^2 \overline{A}_2 e^{-i\sigma T_1} \end{aligned} \tag{6.8.21}$$

$$\begin{aligned} R_{12} &= -2i\omega_2 A_2' + iI\Omega A_2' - 2i\omega_2 A_2 \\ R_{22} &= -2\omega_2 A_2' + I\Omega A_2' - 2\omega_2 A_2 + 3i\Omega^2 A_2^2 \overline{A}_2 \\ &+6i\Omega^2 K_2^2 A_2 + 6i\Omega^2 A_1 \overline{A}_1 A_2 - 3i\Omega^2 K_2^2 \overline{A}_1 e^{-i\sigma T_1} \end{aligned} \tag{6.8.22}$$

According to (6.8.4), the determinant of the coefficient matrix of Eq. (6.8.20) is zero:

$$\det\begin{bmatrix} R_{1n} & i\omega_n I\Omega \\ R_{2n} & \left(1-\omega_n^2\right) \end{bmatrix} = 0$$

i.e.

$$\left(1-\omega_n^2\right) R_{1n} = i\omega_n I\Omega R_{2n} \tag{6.8.23}$$

Substitute (6.8.21) and (6.8.22) into (6.8.23), we get

$$\begin{aligned} \left(\omega_1^2 - \omega_1 I\Omega - 1\right)(2\omega_1 + I\Omega)A_1' &= -2\omega_1\left(\omega_1^2 - 1\right)A_1 + \omega_1 I\Omega\left(2\omega_1 A_1 - 3i\Omega^2 A_1^2 \overline{A}_1\right. \\ &\left.- 6i\Omega^2 K_2^2 A_1 - 6i\Omega^2 A_1 A_2 \overline{A}_2 + 3i\Omega^2 K_2^2 \overline{A}_2 e^{-i\sigma T_1}\right) \end{aligned} \tag{6.8.24}$$

$$\begin{aligned} \left(\omega_2^2 + \omega_2 I\Omega - 1\right)(2\omega_2 - I\Omega)A_2' &= -2\omega_2\left(\omega_2^2 - 1\right)A_2 + \omega_2 I\Omega\left(-2\omega_2 A_2 + 3i\Omega^2 A_2^2 \overline{A}_2\right. \\ &\left.+ 6i\Omega^2 K_2^2 A_2 + 6i\Omega^2 A_1 \overline{A}_1 A_2 - 3i\Omega^2 K_2^2 \overline{A}_1 e^{-i\sigma T_1}\right) \end{aligned} \tag{6.8.25}$$

From (6.8.6), we have

$$\begin{aligned} 2\omega_1 + I\Omega = \omega_1 + \omega_2 \approx 2\Omega, &\quad 2\omega_2 - I\Omega = \omega_1 + \omega_2 \approx 2\Omega \\ \omega_1^2 - \omega_1 I\Omega - 1 = 2\left(\omega_1^2 - 1\right), &\quad \omega_2^2 + \omega_2 I\Omega - 1 = 2\left(\omega_2^2 - 1\right) \\ \omega_1 I\Omega = -\left(\omega_1^2 - 1\right), &\quad \omega_2 I\Omega = \left(\omega_2^2 - 1\right) \end{aligned}$$

Substituting this into (6.8.24) and (6.8.25) yields

$$A_1' = (-s_1 + 2i\alpha_1)A_1 + i\alpha_2 A_1^2\overline{A}_1 + 2i\alpha_2 A_1 A_2 \overline{A}_2 - i\alpha_1 \overline{A}_2 e^{-i\sigma T_1} \tag{6.8.26}$$

$$A_2' = (-s_2 + 2i\alpha_1)A_2 + i\alpha_2 A_2^2\overline{A}_2 + 2i\alpha_2 A_1 \overline{A}_1 A_2 - i\alpha_1 \overline{A}_1 e^{-i\sigma T_1} \tag{6.8.27}$$

where

$$s_1 = \frac{\omega_1}{\Omega}, \quad s_2 = \frac{\omega_2}{\Omega}, \quad \alpha_1 = \frac{3\Omega K_2^2}{4}, \quad \alpha_2 = \frac{3\Omega}{4} \tag{6.8.28}$$

Let

$$A_1 = \frac{1}{2}a_1 e^{i\beta_1}, \quad A_2 = \frac{1}{2}a_2 e^{i\beta_2} \tag{6.8.29}$$

Substituting (6.8.29) into (6.8.26) and (6.8.27), we get

$$a_1' + ia_1\beta_1' = (-s_1 + 2i\alpha_1)a_1 + \frac{1}{4}i\alpha_2 a_1^3 + \frac{1}{2}i\alpha_2 a_1 a_2^2 - i\alpha_1 a_2 e^{-i\gamma} \tag{6.8.30}$$

$$a_2' + ia_2\beta_2' = (-s_2 + 2i\alpha_1)a_2 + \frac{1}{4}i\alpha_2 a_2^3 + \frac{1}{2}i\alpha_2 a_1^2 a_2 - i\alpha_1 a_1 e^{-i\gamma} \tag{6.8.31}$$

where

$$\gamma = \sigma T_1 + \beta_1 + \beta_2 \tag{6.8.32}$$

Separating the real and imaginary parts of (6.8.30) and (6.8.31) yields

$$\begin{aligned} a_1' &= -s_1 a_1 - \alpha_1 a_2 \sin\gamma \\ a_2' &= -s_2 a_2 - \alpha_1 a_1 \sin\gamma \end{aligned} \tag{6.8.33}$$

$$\begin{aligned} a_1\beta_1' &= 2\alpha_1 a_1 + \tfrac{1}{4}\alpha_2 a_1^3 + \tfrac{1}{2}\alpha_2 a_1 a_2^2 - \alpha_1 a_2 \cos\gamma \\ a_2\beta_2' &= 2\alpha_1 a_2 + \tfrac{1}{4}\alpha_2 a_2^3 + \tfrac{1}{2}\alpha_2 a_1^2 a_2 - \alpha_1 a_1 \cos\gamma \end{aligned} \tag{6.8.34}$$

Eliminating β_1 and β_2 in (6.8.34) and (6.8.32) yields

$$a_1 a_2 \gamma' = a_1 a_2 \sigma + 4\alpha_1 a_1 a_2 + \frac{3}{4}\alpha_2\left(a_1^2 + a_2^2\right)a_1 a_2 - \alpha_1\left(a_1^2 + a_2^2\right)\cos\gamma \tag{6.8.35}$$

Let $a_1' = a_2' = \gamma' = 0$ in (6.8.33) and (6.8.35), we can obtain

$$-s_1 a_1 = \alpha_1 a_2 \sin\gamma \tag{6.8.36}$$

$$-s_2 a_2 = \alpha_1 a_1 \sin\gamma \tag{6.8.37}$$

$$a_1 a_2 \sigma + 4\alpha_1 a_1 a_2 + \frac{3}{4}\alpha_2\left(a_1^2 + a_2^2\right) a_1 a_2 = \alpha_1\left(a_1^2 + a_2^2\right)\cos\gamma \tag{6.8.38}$$

Obviously, the following set of solutions can be found

$$a_1 = a_2 = 0, \quad \gamma = \gamma_0 = \text{constant} \tag{6.8.39}$$

When $a_1 \neq$ 0and $a_2 \neq 0$, we can eliminate γ from (6.8.36) and (6.8.37), as well as from (6.8.37) and (6.8.38), and obtain

$$a_2^2 = \frac{s_1}{s_2} a_1^2 \tag{6.8.40}$$

$$\left\{\frac{a_1 a_2}{\alpha_1\left(a_1^2 + a_2^2\right)}\left[\sigma + 4\alpha_1 + \frac{3}{4}\alpha_2\left(a_1^2 + a_2^2\right)\right]\right\}^2 + \left(\frac{s_2 a_2}{\alpha_1 a_1}\right)^2 = 1 \tag{6.8.41}$$

Substituting (6.8.40) into (6.8.41) yields

$$\sigma = -4\alpha_1 - \frac{3}{4}\alpha_2\left(1 + \frac{s_1}{s_2}\right) a_1^2 \pm (s_1 + s_2)\sqrt{\frac{\alpha_1^2 - s_1 s_2}{s_1}} \tag{6.8.42}$$

(6.8.42) and (6.8.40) are the frequency–response equation for nontrivial constant amplitude.

The stability of steady state solution (6.8.39) is analyzed below. The linearized equation of (6.8.33) near the equilibrium point is

$$\begin{Bmatrix} \tilde{a}_1' \\ \tilde{a}_2' \end{Bmatrix} = \begin{bmatrix} -s_1 & -\alpha_1 \sin\gamma_0 \\ -\alpha_1 \sin\gamma_0 & -s_2 \end{bmatrix} \begin{Bmatrix} \tilde{a}_1 \\ \tilde{a}_2 \end{Bmatrix} \tag{6.8.43}$$

Its eigenvalues are

$$\lambda = \frac{-(s_1 + s_2) \pm \sqrt{(s_1 + s_2)^2 - 4\left(s_1 s_2 - \alpha_1^2 \sin^2\gamma_0\right)}}{2} \tag{6.8.44}$$

It can be seen that when

$$s_1 s_2 > \alpha_1^2 \sin^2\gamma_0 \tag{6.8.45}$$

the steady state solution (6.8.39) is stable, otherwise it is unstable. Thus the stability condition is:

$$s_1 s_2 > \alpha_1^2 \text{ or } \omega_1\omega_2 \geq \frac{9\Omega^4 K_2^4}{16} \tag{6.8.46}$$

For a nontrivial constant solution $a_{10},\ \ a_{20},\ \ \gamma_0$, let

$$a_1 = a_{10} + \tilde{a}_1,\quad a_2 = a_{20} + \tilde{a}_2,\quad \gamma = \gamma_0 + \tilde{\gamma} \tag{6.8.47}$$

The linearized equations of (6.8.33) and (6.8.35)

$$\begin{Bmatrix} \tilde{a}_1' \\ \tilde{a}_2' \\ \tilde{\gamma}' \end{Bmatrix} = \begin{bmatrix} -s_1 & -\alpha_1 \sin\gamma_0 & -\alpha_1 a_{20} \cos\gamma_0 \\ -\alpha_1 \sin\gamma_0 & -s_2 & -\alpha_1 a_{10} \cos\gamma_0 \\ \Gamma_1 & \Gamma_2 & \Gamma_3 \end{bmatrix} \begin{Bmatrix} \tilde{a}_1 \\ \tilde{a}_2 \\ \tilde{\gamma} \end{Bmatrix} \tag{6.8.48}$$

where

$$\begin{aligned} \Gamma_1 &= \frac{3}{2}\alpha_2 a_{10} + \alpha_1 a_{20}\left(a_{10}^{-2} - a_{20}^{-2}\right)\cos\gamma_0 \\ \Gamma_2 &= \frac{3}{2}\alpha_2 a_{20} - \alpha_1 a_{10}\left(a_{10}^{-2} - a_{20}^{-2}\right)\cos\gamma_0 \\ \Gamma_3 &= \alpha_1 a_{10} a_{20}\left(a_{10}^{-2} + a_{20}^{-2}\right)\sin\gamma_0 \end{aligned} \tag{6.8.49}$$

The characteristic equation of (6.8.48) is

$$\det\begin{bmatrix} \lambda + s_1 & \alpha_1 \sin\gamma_0 & \alpha_1 a_{20}\cos\gamma_0 \\ \alpha_1 \sin\gamma_0 & \lambda + s_2 & \alpha_1 a_{10}\cos\gamma_0 \\ -\Gamma_1 & -\Gamma_2 & \lambda - \Gamma_3 \end{bmatrix} = 0 \tag{6.8.50}$$

If all three eigenvalues of (6.8.50) have negative real parts, the non-trivial constant solution is stable, otherwise it is unstable.

6.9 Exercise 6.9 (Forced Oscillations of a Two-Degree-of-Freedom System with Quadratic and Cubic Nonlinearities)

Solution: For small and finite amplitudes, we use the method of multiple scales to solve the nonlinear equations. Therefore, let the solution of the equations be

$$\begin{aligned} u_1 &= \varepsilon u_{11}(T_0, T_1) + \varepsilon^2 u_{12}(T_0, T_1) + \cdots \\ u_2 &= \varepsilon u_{21}(T_0, T_1) + \varepsilon^2 u_{22}(T_0, T_1) + \cdots \end{aligned} \tag{6.9.1}$$

In order to make the damping term and nonlinear term appearing in the same order, let

$$\mu_1 = \varepsilon\hat{\mu}_1, \quad \mu_2 = \varepsilon\hat{\mu}_2 \tag{6.9.2}$$

Substituting (6.9.1) and (6.9.2) into the given system equation and retaining to $O(\varepsilon^2)$ yields

$$\begin{aligned}
0 &= \ddot{u}_1 + 2\varepsilon\hat{\mu}_1\dot{u}_1 + \omega_1^2 u_1 + \frac{3}{2}\omega_1^2 u_1^2 + \frac{1}{2}\omega_1^2 u_1^3 + \alpha_1\omega_1^2(u_1+1)u_2^2 - K\cos\Omega t \\
&= \left(D_0^2 + 2\varepsilon D_0 D_1\right)\left(\varepsilon u_{11} + \varepsilon^2 u_{12}\right) + 2\varepsilon\mu_1(D_0 + \varepsilon D_1)\left(\varepsilon u_{11} + \varepsilon^2 u_{12}\right) \\
&+ \omega_1^2\left(\varepsilon u_{11} + \varepsilon^2 u_{12}\right) + \frac{3}{2}\omega_1^2\left(\varepsilon u_{11} + \varepsilon^2 u_{12}\right)^2 + \frac{1}{2}\omega_1^2\left(\varepsilon u_{11} + \varepsilon^2 u_{12}\right)^3 \\
&+ \alpha_1\omega_1^2\left(\varepsilon u_{11} + \varepsilon^2 u_{12} + 1\right)\left(\varepsilon u_{21} + \varepsilon^2 u_{22}\right)^2 - K\cos\Omega T_0 + \cdots \\
&= \varepsilon\left(D_0^2 u_{11} + \omega_1^2 u_{11}\right) + \varepsilon^2\left(D_0^2 u_{12} + \omega_1^2 u_{12} + 2D_0 D_1 u_{11}\right. \\
&\left. + 2\hat{\mu}_1 D_0 u_{11} + \frac{3}{2}\omega_1^2 u_{11}^2 + \alpha_1\omega_1^2 u_{21}^2\right) - K\cos\Omega T_0 + \cdots
\end{aligned} \tag{6.9.3}$$

$$\begin{aligned}
0 &= \ddot{u}_2 + 2\varepsilon\hat{\mu}_2\dot{u}_2 + \omega_2^2 u_2 + \alpha_2 u_2^3 + \alpha_3\left(u_1^2 + 2u_1\right) \\
u_2 &= \left(D_0^2 + 2\varepsilon D_0 D_1\right)\left(\varepsilon u_{21} + \varepsilon^2 u_{22}\right) + 2\varepsilon\hat{\mu}_2(D_0 + \varepsilon D_1)\left(\varepsilon u_{21} + \varepsilon^2 u_{22}\right) \\
&+ \omega_2^2\left(\varepsilon u_{21} + \varepsilon^2 u_{22}\right) + \alpha_2\left(\varepsilon u_{21} + \varepsilon^2 u_{22}\right)^3 \\
&+ \alpha_3\left[\left(\varepsilon u_{11} + \varepsilon^2 u_{12}\right)^2 + 2\left(\varepsilon u_{11} + \varepsilon^2 u_{12}\right)\right]\left(\varepsilon u_{21} + \varepsilon^2 u_{22}\right) + \cdots \\
&\varepsilon\left(D_0^2 u_{21} + \omega_2^2 u_{21}\right) + \varepsilon^2\left(D_0^2 u_{22} + \omega_2^2 u_{22} + 2D_0 D_1 u_{21}\right. \\
&\left. + 2\hat{\mu}_2 D_0 u_{21} + 2\alpha_3 u_{11} u_{21}\right) + \cdots
\end{aligned} \tag{6.9.4}$$

(a) When $\Omega \approx \omega_1$ and ω_2 is away from $\frac{1}{2}\omega_1$.

In this case, we introduce the detuning parameter σ and let

$$\Omega = \omega_1 + \varepsilon\sigma_1 \tag{6.9.5}$$

In this case, the external excitation can cause the primary resonance of the system, which must appear together with the damping and nonlinear terms in the equations of order ε^2, for which let

$$K = \varepsilon^2\hat{K} \tag{6.9.6}$$

Equating coefficients of like powers of ε in (6.9.3) and (6.9.4) yields.
Order ε^1:

$$D_0^2 u_{11} + \omega_1^2 u_{11} = 0 D_0^2 u_{21} + \omega_2^2 u_{21} = 0 \tag{6.9.7}$$

Order ε^2:

$$\begin{aligned} &D_0^2 u_{12} + \omega_1^2 u_{12} = -2D_0 D_1 u_{11} - 2\hat{\mu}_1 D_0 u_{11} - \tfrac{3}{2}\omega_1^2 u_{11}^2 \\ &-\alpha_1 \omega_1^2 u_{21}^2 + \hat{K}\cos\Omega T_0 \\ &D_0^2 u_{22} + \omega_2^2 u_{22} = -2D_0 D_1 u_{21} - 2\hat{\mu}_2 D_0 u_{21} - 2\alpha_3 u_{11} u_{21} \end{aligned} \tag{6.9.8}$$

The solution of (6.9.7) is

$$u_{11} = A_1 e^{i\omega_1 T_0} + cc, \quad u_{21} = A_2 e^{i\omega_2 T_0} + cc \tag{6.9.9}$$

where $A_k = A_k(T_1)$. Substituting (6.9.9) into (6.9.8), we get

$$\begin{aligned} &D_0^2 u_{12} + \omega_1^2 u_{12} = -2i\omega_1 A'_1 e^{i\omega_1 T_0} - 2i\omega_1 \hat{\mu}_1 A_1 e^{i\omega_1 T_0} + \tfrac{1}{2}\hat{K} e^{i\sigma_1 T_1} e^{i\omega_1 T_0} \\ &-\tfrac{3}{2}\omega_1^2 \left(A_1 \overline{A}_1 + A_1^2 e^{2i\omega_1 T_0}\right) - \alpha_1 \omega_1^2 \left(A_2 \overline{A}_2 + A_2^2 e^{2i\omega_2 T_0}\right) + cc \end{aligned} \tag{6.9.10}$$

$$\begin{aligned} &D_0^2 u_{22} + \omega_2^2 u_{22} = -2i\omega_2 A'_2 e^{i\omega_2 T_0} - 2i\omega_2 \hat{\mu}_2 A_2 e^{i\omega_2 T_0} \\ &-2\alpha_3 A_1 A_2 e^{i(\omega_1+\omega_2)T_0} - 2\alpha_3 \overline{A}_1 A_2 e^{i(\omega_2-\omega_1)T_0} + cc \end{aligned} \tag{6.9.11}$$

Primes denote the derivative with respect to T_1. In order to eliminate the secular terms from (6.9.10) and (6.9.11), we need

$$\begin{aligned} &2\omega_1 A'_1 + 2\omega_1 \hat{\mu}_1 A_1 = \tfrac{1}{2}\hat{K} e^{i\sigma_1 T_1} \\ &A'_2 + \hat{\mu}_2 A_2 = 0 \end{aligned} \tag{6.9.12}$$

therefore

$$A_1 = \frac{\hat{K}(i\sigma_1 - \hat{\mu}_1)}{4\omega_1\left(\sigma_1^2 + \hat{\mu}_1^2\right)} e^{i\sigma_1 T_1}, \quad A_2 = A_{20} e^{-\hat{\mu}_2 T_1} \to 0 \tag{6.9.13}$$

Considering $\Omega \approx \omega_1$ and ω_2 is away from $\frac{1}{2}\omega_1$ and substituting (6.9.13) into (6.9.9), we can obtain the first order approximate solution of the system

$$\begin{aligned} u_1 &= \varepsilon u_{11} + O\left(\varepsilon^2\right) \\ &= -\tfrac{\hat{K}}{2\omega_1\left(\sigma_1^2+\hat{\mu}_1^2\right)} [\hat{\mu}_1 \cos(\omega_1 T_0 + \sigma_1 T_1) + \sigma_1 \sin(\omega_1 T_0 + \sigma_1 T_1)] + O\left(\varepsilon^2\right) \\ &= -\tfrac{K}{2\omega_1 \mu_1} \cos(\omega_1 T_0 + \sigma_1 T_1) + O\left(\varepsilon^2\right) \\ u_2 &= \varepsilon u_{21} + O\left(\varepsilon^2\right) = O\left(\varepsilon^2\right) \end{aligned} \tag{6.9.14}$$

(b) When $\Omega \approx \omega_1$ and $\omega_1 \approx 2\omega_2$.

In this case, the detuning parameters σ_1 and σ_2 are introduced such that

$$\Omega = \omega_1 + \varepsilon\sigma_1, \quad \omega_1 = 2\omega_2 + \varepsilon\sigma_2 \tag{6.9.15}$$

Let

$$K = \varepsilon^2 \hat{K} \tag{6.9.16}$$

(6.9.10)–(6.9.11) still hold for the present case. Considering (6.9.15) and eliminating secular terms in (6.9.10) and (6.9.11), we need

$$2\omega_1 A_1' + 2\omega_1 \hat{\mu}_1 A_1 + \frac{1}{2} i\hat{K} e^{i\sigma_1 T_1} - i\alpha_1 \omega_1^2 A_2^2 e^{-i\sigma_2 T_1} = 0 \tag{6.9.17}$$

$$2\omega_2 A_2' + 2\omega_2 \hat{\mu}_2 A_2 - 2i\alpha_3 A_1 \overline{A}_2 e^{i\sigma_2 T_1} = 0 \tag{6.9.18}$$

Let

$$A_1 = \frac{1}{2} a_1 e^{i\beta_1}, \quad A_2 = \frac{1}{2} a_2 e^{i\beta_2} \tag{6.9.19}$$

Substituting (6.9.19) into (6.9.17) and (6.9.18), we have

$$a_1' + ia_1\beta_1' + \hat{\mu}_1 a_1 + \frac{1}{2} i\omega_1^{-1} \hat{K} e^{i(\sigma_1 T_1 - \beta_1)} - \frac{1}{4} i\alpha_1 \omega_1 a_2^2 e^{-i(\sigma_2 T_1 + \beta_1 - 2\beta_2)} = 0 \tag{6.9.20}$$

$$a_2' + ia_2\beta_2' + \hat{\mu}_2 a_2 - \frac{1}{2} i\omega_2^{-1} \alpha_3 a_1 a_2 e^{i(\sigma_2 T_1 + \beta_1 - 2\beta_2)} = 0 \tag{6.9.21}$$

Separating the real and imaginary parts of the above equations yields

$$\begin{aligned}
&a_1' + \hat{\mu}_1 a_1 - \frac{1}{2}\omega_1^{-1} \hat{K} \sin\gamma_1 - \frac{1}{4}\alpha_1 \omega_1 a_2^2 \sin\gamma_2 = 0 \\
&a_1 \beta_1' + \frac{1}{2}\omega_1^{-1} \hat{K} \cos\gamma_1 - \frac{1}{4}\alpha_1 \omega_1 a_2^2 \cos\gamma_2 = 0 \\
&a_2' + \hat{\mu}_2 a_2 + \frac{1}{2}\omega_2^{-1} \alpha_3 a_1 a_2 \sin\gamma_2 = 0 \\
&a_2 \beta_2' - \frac{1}{2}\omega_2^{-1} \alpha_3 a_1 a_2 \cos\gamma_2 = 0
\end{aligned} \tag{6.9.22}$$

where

$$\gamma_1 = \sigma_1 T_1 - \beta_1, \quad \gamma_2 = \sigma_2 T_1 + \beta_1 - 2\beta_2 \tag{6.9.23}$$

Write εa_1 and εa_2 as a_1 and a_2, respectively (6.9.22) can be written in the time variable t as

$$
\begin{aligned}
&\dot{a}_1 + \mu_1 a_1 - \frac{1}{2}\omega_1^{-1} K \sin\gamma_1 - \frac{1}{4}\alpha_1\omega_1 a_2^2 \sin\gamma_2 = 0 \\
&a_1\dot{\beta}_1 + \frac{1}{2}\omega_1^{-1} K \cos\gamma_1 - \frac{1}{4}\alpha_1\omega_1 a_2^2 \cos\gamma_2 = 0 \\
&\dot{a}_2 + \mu_2 a_2 + \frac{1}{2}\omega_2^{-1}\alpha_3 a_1 a_2 \sin\gamma_2 = 0 \\
&a_2\dot{\beta}_2 - \frac{1}{2}\omega_2^{-1}\alpha_3 a_1 a_2 \cos\gamma_2 = 0
\end{aligned}
\tag{6.9.24}
$$

Therefore, the first-order approximate solution of the system is

$$
\begin{aligned}
u_1 &= \varepsilon u_{11} + O(\varepsilon^2) = a_1\cos[\omega_1 t + \beta_1(t)] + O(\varepsilon^2) \\
u_2 &= \varepsilon u_{21} + O(\varepsilon^2) = a_2\cos[\omega_1 t + \beta_2(t)] + O(\varepsilon^2)
\end{aligned}
\tag{6.9.25}
$$

where a_1, a_2, β_1 and β_2 can be determined by (6.9.24) and (6.9.23).

‘ Substituting (6.9.23) into (6.9.24) yields

$$
\begin{aligned}
&\dot{a}_1 + \mu_1 a_1 - \frac{1}{2}\omega_1^{-1} K \sin\gamma_1 - \frac{1}{4}\alpha_1\omega_1 a_2^2 \sin\gamma_2 = 0 \\
&-a_1\dot{\gamma}_1 + \sigma_1 a_1 + \frac{1}{2}\omega_1^{-1} K \cos\gamma_1 - \frac{1}{4}\alpha_1\omega_1 a_2^2 \cos\gamma_2 = 0 \\
&\dot{a}_2 + \mu_2 a_2 + \frac{1}{2}\omega_2^{-1}\alpha_3 a_1 a_2 \sin\gamma_2 = 0 \\
&-a_2\dot{\gamma}_1 - a_2\dot{\gamma}_2 + \sigma_2 a_2 + \sigma_1 a_2 - \omega_2^{-1}\alpha_3 a_1 a_2 \cos\gamma_2 = 0
\end{aligned}
\tag{6.9.26}
$$

From (6.9.26), the steady state solutions of a_1, γ_1 and a_2, γ_2 satisfy the following relations:

$$
\begin{aligned}
&\mu_1 a_1 - \frac{1}{2}\omega_1^{-1} K \sin\gamma_1 - \frac{1}{4}\alpha_1\omega_1 a_2^2 \sin\gamma_2 = 0 \\
&\sigma_1 a_1 + \frac{1}{2}\omega_1^{-1} K \cos\gamma_1 - \frac{1}{4}\alpha_1\omega_1 a_2^2 \cos\gamma_2 = 0 \\
&\mu_2 a_2 + \frac{1}{2}\omega_2^{-1}\alpha_3 a_1 a_2 \sin\gamma_2 = 0 \\
&\sigma_2 a_2 + \sigma_1 a_2 - \omega_2^{-1}\alpha_3 a_1 a_2 \cos\gamma_2 = 0
\end{aligned}
\tag{6.9.27}
$$

From this, we can obtain

$$
a_1^2 = \frac{\omega_2^2}{\alpha_3^2}\left[4\mu_2^2 + (\sigma_2 + \sigma_1)\right] \tag{6.9.28}
$$

$$
(2\omega_1\alpha_3\mu_1 a_1^2 + \omega_1^2\omega_2\alpha_1\mu_2 a_2^2)^2 + [2\omega_1\alpha_3\sigma_1 a_1^2 - \frac{1}{2}\omega_1^2\omega_2\alpha_1(\sigma_2 + \sigma_1)a_2^2]^2 = \alpha_3^2 K a_1^2 \tag{6.9.29}
$$

$$\tan \gamma_1 = -\frac{4\mu_1 a_1 - \alpha_1 \omega_1 a_2^2 \sin \gamma_2}{4\sigma_1 a_1 - \alpha_1 \omega_1 a_2^2 \cos \gamma_2} \tag{6.9.30}$$

$$\tan \gamma_2 = -\frac{2\mu_2}{\sigma_2 + \sigma_1} \tag{6.9.31}$$

Amplitude-frequency and phase-frequency curves can be derived from (6.9.28)–(6.9.31).

Readers are invited to complete the stability analysis of the steady state solution.

(c) When $\Omega \approx \frac{1}{2}\omega_1$ and $\omega_1 \approx \omega_2$.

In this case, the detuning parameters σ_1 and σ_2 are introduced such that

$$2\Omega = \omega_1 + \varepsilon\sigma_1, \quad \omega_1 = 2\omega_2 + \varepsilon\sigma_2 \tag{6.9.32}$$

For this $\Omega \approx \frac{1}{2}\omega_1$ second resonance, this is hard excitation, so let

$$K = \varepsilon \hat{K} \tag{6.9.33}$$

Equating coefficients of like powers of ε in (6.9.3) and (6.9.4) yields.

Order ε_1:

$$\begin{aligned} D_0^2 u_{11} + \omega_1^2 u_{11} &= \hat{K} \cos \Omega T_0 \\ D_0^2 u_{21} + \omega_2^2 u_{21} &= 0 \end{aligned} \tag{6.9.34}$$

Order ε_2:

$$\begin{aligned} D_0^2 u_{12} + \omega_1^2 u_{12} &= -2D_0 D_1 u_{11} - 2\hat{\mu}_1 D_0 u_{11} - \tfrac{3}{2}\omega_1^2 u_{11}^2 - \alpha_1 \omega_1^2 u_{21}^2 \\ D_0^2 u_{22} + \omega_2^2 u_{22} &= -2D_0 D_1 u_{21} - 2\hat{\mu}_2 D_0 u_{21} - 2\alpha_3 u_{11} u_{21} \end{aligned} \tag{6.9.35}$$

The solution of (6.9.34) is

$$u_{11} = A_1 e^{i\omega_1 T_0} + \hat{K}_1 e^{i\Omega T_0} + cc, \quad u_{21} = A_2 e^{i\omega_2 T_0} + cc \tag{6.9.36}$$

where $A_k = A_k(T_1)$, and

$$\hat{K}_1 = \frac{\hat{K}}{2(\omega_1^2 - \Omega^2)} \tag{6.9.37}$$

by substituting (6.9.36) into (6.9.35), we can obtain

$$\begin{aligned}
D_0^2 u_{12} + \omega_1^2 u_{12} &= -2i\omega_1 A_1 e^{i\omega_1 T_0} - 2i\omega_1 \hat{\mu}_1 A_1 e^{i\omega_1 T_0} - 2i\Omega\hat{\mu}_1 \hat{K}_1 e^{i\Omega T_0} \\
&-\frac{3}{2}\omega_1^2 A_1^2 e^{2i\omega_1 T_0} - \frac{3}{2}\omega_1^2 \hat{K}_1^2 e^{2i\Omega T_0} - 3\omega_1^2 \hat{K}_1 A_1 e^{i(\omega_1+\Omega)T_0} \\
&-\frac{3}{2}\omega_1^2 \hat{K}_1 \overline{A}_1 e^{i(\Omega-\omega_1)T_0} - \frac{3}{2}\omega_1^2 \hat{K}_1 A_1 e^{i(\omega_1-\Omega)T_0} - \alpha_1 \omega_1^2 A_2^2 e^{2i\omega_2 T_0} \\
&-\alpha_1\omega_1^2 A_2 \overline{A}_2 - \frac{3}{2}\omega_1^2 A_1 \overline{A}_1 - \hat{K}_1^2 + cc
\end{aligned} \tag{6.9.38}$$

$$\begin{aligned}
D_0^2 u_{22} + \omega_2^2 u_{22} &= -2i\omega_2 A_2' e^{i\omega_2 T_0} - 2i\omega_2 \hat{\mu}_2 A_2 e^{i\omega_2 T_0} \\
&-2\alpha_3 A_1 A_2 e^{i(\omega_1+\omega_2)T_0} - 2\alpha_3 \hat{K}_1 A_2 e^{i(\Omega+\omega_2)T_0} \\
&-2\alpha_3 \overline{A}_1 A_2 e^{i(\omega_2-\omega_1)T_0} - 2\alpha_3 \hat{K}_1 A_2 e^{i(\omega_2-\Omega)T_0} + cc
\end{aligned} \tag{6.9.39}$$

Primes denote the derivative with respect to T_1. Considering (6.9.32) and eliminating secular terms in (6.9.38) and (6.9.39), we need

$$2\omega_1 A_1' + 2\omega_1 \hat{\mu}_1 A_1 - \frac{3}{2} i\omega_1^2 \hat{K}_1^2 e^{i\sigma_1 T_1} - i\alpha_1 \omega_1^2 A_2^2 e^{-i\sigma_2 T_1} = 0 \tag{6.9.40}$$

$$2\omega_2 A_2' + 2\omega_2 \hat{\mu}_2 A_2 - 2i\alpha_3 A_1 \overline{A}_2 e^{i\sigma_2 T_1} = 0 \tag{6.9.41}$$

Let

$$A_1 = \frac{1}{2} a_1 e^{i\beta_1}, \quad A_2 = \frac{1}{2} a_2 e^{i\beta_2} \tag{6.9.42}$$

Substituting (6.9.42) into (6.9.40) and (6.9.41), we have

$$a_1' + i a_1 \beta_1' + \hat{\mu}_1 a_1 - \frac{3}{2} i\omega_1 \hat{K}_1^2 e^{i(\sigma_1 T_1 - \beta_1)} - \frac{1}{4} i\alpha_1 \omega_1 a_2^2 e^{-i(\sigma_2 T_1 + \beta_1 - 2\beta_2)} = 0 \tag{6.9.43}$$

$$a_2' + i a_2 \beta_2' + \hat{\mu}_2 a_2 - \frac{1}{2} i\omega_2^{-1} \alpha_3 a_1 a_2 e^{i(\sigma_2 T_1 + \beta_1 - 2\beta_2)} = 0 \tag{6.9.44}$$

Separating the real and imaginary parts of the above equations yields

$$\begin{aligned}
&a_1' + \hat{\mu}_1 a_1 + \frac{3}{2}\omega_1 \hat{K}_1^2 \sin\gamma_1 - \frac{1}{4}\alpha_1 \omega_1 a_2^2 \sin\gamma_2 = 0 \\
&a_1 \beta_1' - \frac{3}{2}\omega_1 \hat{K}_1^2 \cos\gamma_1 - \frac{1}{4}\alpha_1 \omega_1 a_2^2 \cos\gamma_2 = 0 \\
&a_2' + \hat{\mu}_2 a_2 + \frac{1}{2}\omega_2^{-1} \alpha_3 a_1 a_2 \sin\gamma_2 = 0 \\
&a_2 \beta_2' - \frac{1}{2}\omega_2^{-1} \alpha_3 a_1 a_2 \cos\gamma_2 = 0
\end{aligned} \tag{6.9.45}$$

where

$$\gamma_1 = \sigma_1 T_1 - \beta_1, \quad \gamma_2 = \sigma_2 T_1 + \beta_1 - 2\beta_2 \tag{6.9.46}$$

Write εa_1 and εa_2 as a_1 and a_2, respectively. Then (6.9.45) can be written in the time variable t as

$$\begin{aligned} &\dot{a}_1 + \mu_1 a_1 + \frac{3}{2}\omega_1 K_1^2 \sin\gamma_1 - \frac{1}{4}\alpha_1\omega_1 a_2^2 \sin\gamma_2 = 0 \\ &a_1\dot{\beta}_1 - \frac{3}{2}\omega_1 K_1^2 \cos\gamma_1 - \frac{1}{4}\alpha_1\omega_1 a_2^2 \cos\gamma_2 = 0 \\ &\dot{a}_2 + \mu_2 a_2 + \frac{1}{2}\omega_2^{-1}\alpha_3 a_1 a_2 \sin\gamma_2 = 0 \\ &a_2\dot{\beta}_2 - \frac{1}{2}\omega_2^{-1}\alpha_3 a_1 a_2 \cos\gamma_2 = 0 \end{aligned} \tag{6.9.47}$$

Therefore, the first-order approximate solution of the system is

$$\begin{aligned} u_1 &= \varepsilon u_{11} + O(\varepsilon^2) = a_1\cos[\omega_1 t + \beta_1(t)] + O(\varepsilon^2) \\ u_2 &= \varepsilon u_{21} + O(\varepsilon^2) = a_2\cos[\omega_1 t + \beta_2(t)] + O(\varepsilon^2) \end{aligned} \tag{6.9.48}$$

where α_1, α_2, β_1 and β_2 can be determined by (6.9.24) and (6.9.23).

Substituting (6.9.23) into (6.9.24) yields

$$\begin{aligned} &\dot{a}_1 + \mu_1 a_1 + \frac{3}{2}\omega_1 K_1^2 \sin\gamma_1 - \frac{1}{4}\alpha_1\omega_1 a_2^2 \sin\gamma_2 = 0 \\ &-a_1\dot{\gamma}_1 + \sigma_1 a_1 - \frac{3}{2}\omega_1 K_1^2 \cos\gamma_1 - \frac{1}{4}\alpha_1\omega_1 a_2^2 \cos\gamma_2 = 0 \\ &\dot{a}_2 + \mu_2 a_2 + \frac{1}{2}\omega_2^{-1}\alpha_3 a_1 a_2 \sin\gamma_2 = 0 \\ &-a_2\dot{\gamma}_1 - a_2\dot{\gamma}_2 + \sigma_2 a_2 + \sigma_1 a_2 - \omega_2^{-1}\alpha_3 a_1 a_2 \cos\gamma_2 = 0 \end{aligned} \tag{6.9.49}$$

From(6.9.49), the steady state solutions of α_1, γ_1 and α_2, γ_2 satisfy the following relations:

$$\begin{aligned} &\mu_1 a_1 + \tfrac{3}{2}\omega_1 K_1^2 \sin\gamma_1 - \tfrac{1}{4}\alpha_1\omega_1 a_2^2 \sin\gamma_2 = 0 \\ &\sigma_1 a_1 - \tfrac{3}{2}\omega_1 K_1^2 \cos\gamma_1 - \tfrac{1}{4}\alpha_1\omega_1 a_2^2 \cos\gamma_2 = 0 \\ &\mu_2 a_2 + \tfrac{1}{2}\omega_2^{-1}\alpha_3 a_1 a_2 \sin\gamma_2 = 0 \\ &\sigma_2 a_2 + \sigma_1 a_2 - \omega_2^{-1}\alpha_3 a_1 a_2 \cos\gamma_2 = 0 \end{aligned} \tag{6.9.50}$$

From this, we can obtain

$$a_1^2 = \frac{\omega_2^2}{\alpha_3^2}\left[4\mu_2^2 + (\sigma_2 + \sigma_1)\right] \tag{6.9.51}$$

$$\left(\frac{2}{3}\alpha_3\mu_1 a_1^2 + \frac{1}{3}\omega_1\omega_2\alpha_1\mu_2 a_2^2\right)^2 + \left[\frac{2}{3}\alpha_3\sigma_1 a_1^2 - \frac{1}{6}\alpha_1\omega_1\omega_2 a_2^2(\sigma_2 + \sigma_1)\right]^2 = \omega_1^2\alpha_3^2 K_1^4 a_1^2 \tag{6.9.52}$$

$$\tan\gamma_1 = -\frac{4\mu_1 a_1 - \alpha_1\omega_1 a_2^2 \sin\gamma_2}{4\sigma_1 a_1 - \alpha_1\omega_1 a_2^2 \cos\gamma_2} \tag{6.9.53}$$

$$\tan\gamma_2 = -\frac{2\mu_2}{\sigma_2 + \sigma_1} \tag{6.9.54}$$

Frequency-response and frequency-phase curves can be derived from (6.9.51)–(6.9.54).

Readers are invited to complete the stability analysis of the steady state solution.

(d) When $\Omega \approx 2\omega_1$ and $\omega_1 \approx 2\omega_2$.

Readers are invited to complete the solution for this case.

6.10 Exercise 6.10 (Analysis of Forced Oscillations of a Spring-Slider-Pendulum System)

Solution: (a) The kinetic energy of the system is

$$\begin{aligned} T &= \tfrac{1}{2}m\left[\left(\dot{x} + \tfrac{1}{2}l\dot{\theta}\cos\theta\right)^2 + \left(\tfrac{1}{2}l\dot{\theta}\sin\theta\right)^2\right] + \tfrac{1}{2}\tfrac{1}{12}ml^2\dot{\theta}^2 \\ &= \tfrac{1}{2}m\dot{x}^2 + \tfrac{1}{2}ml\dot{x}\dot{\theta}\cos\theta + \tfrac{1}{6}ml^2\dot{\theta}^2 \end{aligned} \tag{6.10.1}$$

The potential energy of the system is

$$V = \frac{1}{2}kx^2 + \frac{1}{2}mgl(1 - \cos\theta) \tag{6.10.2}$$

The generalized forces are

$$Q_x = F(t), \quad Q_\theta = 0 \tag{6.10.3}$$

Substitute the kinetic energy, potential energy, and generalized forces into the Lagrange's equation:

$$\frac{d}{dt}\frac{\partial T}{\partial \dot{x}} - \frac{\partial T}{\partial x} = -\frac{\partial V}{\partial x} + Q_x, \quad \frac{d}{dt}\frac{\partial T}{\partial \dot{\theta}} - \frac{\partial T}{\partial \theta} = -\frac{\partial V}{\partial \theta} + Q_\theta$$

we can obtain

$$\begin{aligned} & m\ddot{x} + \frac{1}{2}ml\ddot{\theta}\cos\theta - \frac{1}{2}ml\dot{\theta}^2\sin\theta + kx = F_0(t) \\ & \frac{1}{3}ml^2\ddot{\theta} + \frac{1}{2}ml\ddot{x}\cos\theta + \frac{1}{2}mgl\sin\theta = 0 \end{aligned} \tag{6.10.4}$$

Let

$$u = \frac{x}{l}, \omega_{10}^2 = k/m, \omega_{20}^2 = \frac{3g}{2l} \text{ and } \mathrm{F}(t) = F_0(t)/ml \tag{6.10.5}$$

Equation (6.10.4) changes into

$$\begin{aligned} &\ddot{u} + \omega_{10}^2 u + \frac{1}{2}\ddot{\theta}\cos\theta - \frac{1}{2}\dot{\theta}^2\sin\theta = F(t) \\ &\ddot{\theta} + \omega_{20}^2\sin\theta + \frac{3}{2}\ddot{u}\cos\theta = 0 \end{aligned} \tag{6.10.6}$$

(b) Near the equilibrium position of the system $u = \theta = 0$, we can expand (6.10.6) and retain to third-order small quantities yields

$$\begin{aligned} &\ddot{u} + \frac{1}{2}\ddot{\theta} + \omega_{10}^2 u - \frac{1}{4}\theta^2\ddot{\theta} - \frac{1}{2}\theta\dot{\theta}^2 = K\cos\Omega t \\ &\frac{1}{2}\ddot{u} + \frac{1}{3}\ddot{\theta} + \frac{1}{3}\omega_{20}^2\theta - \frac{1}{18}\omega_{20}^2\theta^3 - \frac{1}{4}\theta^2\ddot{u} = 0 \end{aligned} \tag{6.10.7}$$

The corresponding linear free oscillation equation is

$$\begin{bmatrix} 1 & \frac{1}{2} \\ \frac{1}{2} & \frac{1}{3} \end{bmatrix}\begin{Bmatrix} \ddot{u} \\ \ddot{\theta} \end{Bmatrix} + \begin{bmatrix} \omega_{10}^2 & 0 \\ 0 & \frac{1}{3}\omega_{20}^2 \end{bmatrix}\begin{Bmatrix} u \\ \theta \end{Bmatrix} = 0 \tag{6.10.8}$$

Let

$$u = B_1 e^{i\omega t}, \quad \theta = B_2 e^{i\omega t} \tag{6.10.9}$$

and substitute (6.10.9) into (6.10.8), we obtain

$$\begin{bmatrix} \omega_{10}^2 - \omega^2 & -\frac{1}{2}\omega^2 \\ -\frac{1}{2}\omega^2 & \frac{1}{3}\omega_{20}^2 - \frac{1}{3}\omega^2 \end{bmatrix}\begin{Bmatrix} B_1 \\ B_2 \end{Bmatrix} = 0 \tag{6.10.10}$$

Therefore, the characteristic equation is

$$\det\begin{bmatrix} \omega_{10}^2 - \omega^2 & -\frac{1}{2}\omega^2 \\ -\frac{1}{2}\omega^2 & \frac{1}{3}\omega_{20}^2 - \frac{1}{3}\omega^2 \end{bmatrix} = 0 \tag{6.10.11}$$

i.e.,

$$\omega^4 - 4(\omega_{10}^2 + \omega_{20}^2)\omega^2 + 4\omega_{10}^2\omega_{20}^2 = 0 \tag{6.10.12}$$

Therefore, the linear natural frequency is

$$\begin{aligned}\omega_1 &= \left[2\left(\omega_{10}^2+\omega_{20}^2\right)-2\sqrt{\left(\omega_{10}^2+\omega_{20}^2\right)^2-\omega_{10}^2\omega_{20}^2}\right]^{1/2}\\ \omega_2 &= \left[2\left(\omega_{10}^2+\omega_{20}^2\right)+2\sqrt{\left(\omega_{10}^2+\omega_{20}^2\right)^2-\omega_{10}^2\omega_{20}^2}\right]^{1/2}\end{aligned} \tag{6.10.13}$$

Substituting ω_1 and ω_2 into (6.10.10), we can obtain a set of modes of the system as

$$\{B_1, B_2\}_1^T = \{1, h_1\}^T, \quad \{B_1, B_2\}_2^T = \{1, h_2\}^T \tag{6.10.14}$$

where h_1 and h_2 can be obtained from (6.10.10)

$$h_1 = \frac{2\left(\omega_{10}^2-\omega_1^2\right)}{\omega_1^2}, \quad h_2 = \frac{2\left(\omega_{10}^2-\omega_2^2\right)}{\omega_2^2} \tag{6.10.15}$$

Therefore, the solution of (6.10.8) can be expressed as

$$\begin{aligned}u &= A_1 e^{i\omega_1 t} + A_2 e^{i\omega_2 t} + cc\\ \theta &= h_1 A_1 e^{i\omega_1 t} + h_2 A_2 e^{i\omega_2 t} + cc\end{aligned} \tag{6.10.16}$$

For small and finite amplitudes, we use the method of multiple scales to solve (6.10.7). Therefore, let the solution of (6.10.7) be expressed as

$$\begin{aligned}u &= \varepsilon u_{11}(T_0, T_1, T_2) + \varepsilon^2 u_{12}(T_0, T_1, T_2) + \varepsilon^3 u_{13}(T_0, T_1, T_2) + \cdots\\ \theta &= \varepsilon u_{21}(T_0, T_1, T_2) + \varepsilon^2 u_{22}(T_0, T_1, T_2) + \varepsilon^3 u_{23}(T_0, T_1, T_2) + \cdots\end{aligned} \tag{6.10.17}$$

Substituting (6.10.17) into (6.10.7) and keeping to $O\left(\varepsilon^3\right)$, we get

$$
\begin{aligned}
0 &= \ddot{u} + \frac{1}{2}\ddot{\theta} + \omega_{10}^2 u - \frac{1}{4}\theta^2\ddot{\theta} - \frac{1}{2}\theta\dot{\theta}^2 - K\cos\Omega t \\
&= \left[D_0^2 + 2\varepsilon D_0 D_1 + \varepsilon^2\left(D_1^2 + 2D_0 D_2\right)\right]\left(\varepsilon u_{11} + \varepsilon^2 u_{12} + \varepsilon^3 u_{13}\right) \\
&\quad + \frac{1}{2}\left[D_0^2 + 2\varepsilon D_0 D_1 + \varepsilon^2\left(D_1^2 + 2D_0 D_2\right)\right]\left(\varepsilon u_{21} + \varepsilon^2 u_{22} + \varepsilon^3 u_{23}\right) \\
&\quad + \omega_{10}^2\left(\varepsilon u_{11} + \varepsilon^2 u_{12} + \varepsilon^3 u_{13}\right) - \frac{1}{4}\left(\varepsilon u_{21} + \varepsilon^2 u_{22} + \varepsilon^3 u_{23}\right)^2 \\
&\quad \times\left[D_0^2 + 2\varepsilon D_0 D_1 + \varepsilon^2\left(D_1^2 + 2D_0 D_2\right)\right]\left(\varepsilon u_{21} + \varepsilon^2 u_{22} + \varepsilon^3 u_{23}\right) \\
&\quad - \frac{1}{2}\left(\varepsilon u_{21} + \varepsilon^2 u_{22} + \varepsilon^3 u_{23}\right)\left[\left(D_0 + \varepsilon D_1 + \varepsilon^2 D_2\right)\left(\varepsilon u_{21} + \varepsilon^2 u_{22} + \varepsilon^3 u_{23}\right)\right]^2 \\
&\quad - K\cos\Omega T_0 + \cdots \\
&= \varepsilon\left(D_0^2 u_{11} + \frac{1}{2}D_0^2 u_{21} + \omega_{10}^2 u_{11}\right) \\
&\quad + \varepsilon^2\left(D_0^2 u_{12} + \frac{1}{2}D_0^2 u_{22} + \omega_{10}^2 u_{12} + D_0 D_1 u_{21} + 2D_0 D_1 u_{11}\right) \\
&\quad + \varepsilon^3\left[D_0^2 u_{13} + \frac{1}{2}D_0^2 u_{23} + \omega_{10}^2 u_{13} + 2D_0 D_1 u_{12} + \left(D_1^2 + 2D_0 D_2\right)u_{11}\right. \\
&\quad + D_0 D_1 u_{22} + \frac{1}{2}\left(D_1^2 + 2D_0 D_2\right)u_{21} - \frac{1}{4}u_{21}^2 D_0^2 u_{21} \\
&\quad \left. - \frac{1}{2}u_{21}\left(D_0 u_{21}\right)^2\right] - K\cos\Omega T_0 + \cdots
\end{aligned}
\tag{6.10.18}
$$

$$
\begin{aligned}
0 &= \frac{1}{2}\ddot{u} + \frac{1}{3}\ddot{\theta} + \frac{1}{3}\omega_{20}^2\theta - \frac{1}{18}\omega_{20}^2\theta^3 - \frac{1}{4}\theta^2\ddot{u} \\
&= \frac{1}{2}\left[D_0^2 + 2\varepsilon D_0 D_1 + \varepsilon^2\left(D_1^2 + 2D_0 D_2\right)\right]\left(\varepsilon u_{11} + \varepsilon^2 u_{12} + \varepsilon^3 u_{13}\right) \\
&\quad + \frac{1}{3}\left[D_0^2 + 2\varepsilon D_0 D_1 + \varepsilon^2\left(D_1^2 + 2D_0 D_2\right)\right]\left(\varepsilon u_{21} + \varepsilon^2 u_{22} + \varepsilon^3 u_{23}\right) \\
&\quad + \frac{1}{3}\omega_{20}^2\left(\varepsilon u_{21} + \varepsilon^2 u_{22} + \varepsilon^3 u_{23}\right) - \frac{1}{18}\omega_{20}^2\left(\varepsilon u_{21} + \varepsilon^2 u_{22} + \varepsilon^3 u_{23}\right)^3 \\
&\quad - \frac{1}{4}\left(\varepsilon u_{21} + \varepsilon^2 u_{22} + \varepsilon^3 u_{23}\right)^2\left[D_0^2 + 2\varepsilon D_0 D_1 + \varepsilon^2\left(D_1^2 + 2D_0 D_2\right)\right] \\
&\quad \times\left(\varepsilon u_{11} + \varepsilon^2 u_{12} + \varepsilon^3 u_{13}\right) + \cdots \\
&= \varepsilon\left(\frac{1}{2}D_0^2 u_{11} + \frac{1}{3}D_0^2 u_{21} + \frac{1}{3}\omega_{20}^2 u_{21}\right) \\
&\quad + \varepsilon^2\left(\frac{1}{2}D_0^2 u_{12} + \frac{1}{3}D_0^2 u_{22} + \frac{1}{3}\omega_{20}^2 u_{22} + D_0 D_1 u_{11} + \frac{2}{3}D_0 D_1 u_{21}\right) \\
&\quad + \varepsilon^3\left[\frac{1}{2}D_0^2 u_{13} + \frac{1}{3}D_0^2 u_{23} + \frac{1}{3}\omega_{20}^2 u_{23} + D_0 D_1 u_{12}\right. \\
&\quad + \frac{1}{2}\left(D_1^2 + 2D_0 D_2\right)u_{11} + \frac{2}{3}D_0 D_1 u_{22} + \frac{1}{3}\left(D_1^2 + 2D_0 D_2\right)u_{21} \\
&\quad \left. - \frac{1}{18}\omega_{20}^2 u_{21}^3 - \frac{1}{4}u_{21}^2 D_0^2 u_{11}\right] + \cdots
\end{aligned}
\tag{6.10.19}
$$

The order of the term $K \cos \Omega T_0$ is related to the value of Ω.

(i) $\boldsymbol{\Omega} \approx \boldsymbol{\omega}_1$

In this case, $K \cos \Omega T_0$ cannot appear in the order of $O(\varepsilon^1)$. Thus, from (6.10.18) and (6.10.19), we can obtain the order ε^1 equation.

Order ε^1:

$$
\begin{aligned}
&D_0^2 u_{11} + \frac{1}{2}D_0^2 u_{21} + \omega_{10}^2 u_{11} = 0 \\
&\frac{1}{2}D_0^2 u_{11} + \frac{1}{3}D_0^2 u_{21} + \frac{1}{3}\omega_{20}^2 u_{21} = 0
\end{aligned}
\tag{6.10.20}
$$

This set of equations is the same as (6.10.8), and its solution is given by (6.10.16), i.e.,

$$
\begin{aligned}
u_{11} &= A_1 e^{i\omega_1 T_0} + A_2 e^{i\omega_2 T_0} + cc \\
u_{21} &= h_1 A_1 e^{i\omega_1 T_0} + h_2 A_2 e^{i\omega_2 T_0} + cc
\end{aligned}
\tag{6.10.21}
$$

where

$$
A_k = A_k(T_1, T_2), \quad k = 1, 2
$$

If $K\cos\Omega T_0$ appears in the coefficients of ε^2, it is easy to verify that it will resonate with u_{12} and u_{22}, and therefore

$$K = 2\varepsilon^3\hat{K} \tag{6.10.22}$$

Equating coefficients of ε^2 and ε^3 in (6.10.18) and (6.10.19) yields.
Order ε^2:

$$\begin{aligned} &D_0^2u_{12} + \tfrac{1}{2}D_0^2u_{22} + \omega_{10}^2u_{12} = -2D_0D_1u_{11} - D_0D_1u_{21} \\ &\tfrac{1}{2}D_0^2u_{12} + \tfrac{1}{3}D_0^2u_{22} + \tfrac{1}{3}\omega_{20}^2u_{22} = -D_0D_1u_{11} - \tfrac{2}{3}D_0D_1u_{21} \end{aligned} \tag{6.10.23}$$

Order ε^3:

$$\begin{aligned} &D_0^2u_{13} + \frac{1}{2}D_0^2u_{23} + \omega_{10}^2u_{13} = -2D_0D_1u_{12} - \left(D_1^2 + 2D_0D_2\right)u_{11} - D_0D_1u_{22} \\ &-\frac{1}{2}\left(D_1^2 + 2D_0D_2\right)u_{21} + \frac{1}{4}u_{21}^2D_0^2u_{21} \\ &+\frac{1}{2}u_{21}(D_0u_{21})^2 + 2\hat{K}\cos\Omega T_0 \\ &\frac{1}{2}D_0^2u_{13} + \frac{1}{3}D_0^2u_{23} + \frac{1}{3}\omega_{20}^2u_{23} = -D_0D_1u_{12} - \frac{1}{2}\left(D_1^2 + 2D_0D_2\right)u_{11} - \frac{2}{3}D_0D_1u_{22} \\ &-\frac{1}{3}\left(D_1^2 + 2D_0D_2\right)u_{21} + \frac{1}{18}\omega_{20}^2u_{21}^3 + \frac{1}{4}u_{21}^2D_0^2u_{11} \end{aligned} \tag{6.10.24}$$

Substituting (6.10.21) into (6.10.23) yields

$$\begin{aligned} &D_0^2u_{12} + \frac{1}{2}D_0^2u_{22} + \omega_{10}^2u_{12} = -i\omega_1(2 + h_1)D_1A_1e^{i\omega_1T_0} \\ &-i\omega_2(2 + h_2)D_1A_2e^{i\omega_2T_0} + cc \\ &\frac{1}{2}D_0^2u_{12} + \frac{1}{3}D_0^2u_{22} + \frac{1}{3}\omega_{20}^2u_{22} = -i\omega_1\left(1 + \frac{2}{3}h_1\right)D_1A_1e^{i\omega_1T_0} \\ &-i\omega_2\left(1 + \frac{2}{3}h_2\right)D_1A_2e^{i\omega_2T_0} + cc \end{aligned} \tag{6.10.25}$$

In order to eliminate secular terms from the above equations, there must be

$$D_1A_1 = 0, \quad D_1A_2 = 0 \quad \rightarrow \quad A_1 = A_1(T_2), \quad A_2 = A_2(T_2) \tag{6.10.26}$$

from this

$$u_{12} = u_{22} = 0 \tag{6.10.27}$$

Introduce the detuning parameter σ such that

$$\Omega = \omega_1 + \varepsilon^2\sigma \tag{6.10.28}$$

Substituting (6.10.28), (6.10.27), (6.10.26) and (6.10.21) into (6.10.24) yields

$$\begin{aligned}
D_0^2u_{13} + \frac{1}{2}D_0^2u_{23} + \omega_{10}^2u_{13} &= -2D_0D_2u_{11} - D_0D_2u_{21} + \frac{1}{4}u_{21}^2D_0^2u_{21} \\
&\quad + \frac{1}{2}u_{21}(D_0u_{21})^2 + 2\hat{K}\cos\Omega T_0 \\
&= \left(-2i\omega_1A_1' - i\omega_1h_1A_1' - \frac{1}{4}\omega_1^2h_1^3A_1^2\overline{A}_1 - \frac{1}{2}\omega_1^2h_1h_2^2A_1A_2\overline{A}_2 + \hat{K}e^{i\sigma T_2}\right)e^{i\omega_1T_0} \\
&\quad + \left(-2i\omega_2A_2' - i\omega_2h_2A_2' - \frac{1}{4}\omega_2^2h_2^3A_2^2\overline{A}_2 - \frac{1}{2}\omega_2^2h_1^2h_2A_1\overline{A}_1A_2\right)e^{i\omega_2T_0} \\
&\quad - \left(\frac{1}{4}\omega_1^2 + \omega_2^2 + \omega_1\omega_2\right)h_1h_2^2A_1A_2^2e^{i(\omega_1+2\omega_2)T_0} \\
&\quad - \left(\omega_1^2 + \frac{1}{4}\omega_2^2 + \omega_1\omega_2\right)h_1^2h_2A_1^2A_2e^{i(2\omega_1+\omega_2)T_0} \\
&\quad - \left(\frac{1}{4}\omega_1^2 + \omega_2^2 - \omega_1\omega_2\right)h_1h_2^2A_1\overline{A}_2^2e^{i(\omega_1-2\omega_2)T_0} \\
&\quad - \left(\omega_1^2 + \frac{1}{4}\omega_2^2 - \omega_1\omega_2\right)h_1^2h_2\overline{A}_1^2A_2e^{i(\omega_2-2\omega_1)T_0} \\
&\quad - \frac{3}{4}\omega_1^2h_1^3A_1^3e^{3i\omega_1T_0} - \frac{3}{4}\omega_2^2h_2^3A_2^3e^{3i\omega_2T_0} + cc
\end{aligned} \tag{6.10.29}$$

$$
\begin{aligned}
&\frac{1}{2}D_0^2u_{13} + \frac{1}{3}D_0^2u_{23} + \frac{1}{3}\omega_{20}^2u_{23} = -D_0D_2u_{11} - \frac{2}{3}D_0D_2u_{21} + \frac{1}{18}\omega_{20}^2u_{21}^3 + \frac{1}{4}u_{21}^2D_0^2u_{11}\\
&= \left[-i\omega_1A_1' - i\frac{2}{3}\omega_1h_1A_1' + \left(\frac{1}{3}\omega_{20}^2h_1h_2^2 - \omega_2^2h_1h_2 - \frac{1}{2}\omega_1^2h_2^2\right)A_1A_2\overline{A}_2\right.\\
&\quad\left. + \left(\frac{1}{6}\omega_{20}^2h_1^3 - \frac{3}{4}\omega_1^2h_1^2\right)A_1^2\overline{A}_1\right]e^{i\omega_1T_0}\\
&\quad + \left[-i\omega_2A_2' - i\frac{2}{3}\omega_2h_2A_2' + \left(\frac{1}{3}\omega_{20}^2h_1^2h_2 - \omega_1^2h_1h_2 - \frac{1}{2}\omega_2^2h_1^2\right)A_1\overline{A}_1A_2\right.\\
&\quad\left. + \left(\frac{1}{6}\omega_{20}^2h_2^3 - \frac{3}{4}\omega_2^2h_2^2\right)A_2^2\overline{A}_2\right]e^{i\omega_2T_0}\\
&\quad + \left[\frac{1}{6}\omega_{20}^2h_1^2h_2 - \frac{1}{4}(2\omega_1^2h_1h_2 + \omega_2^2h_1^2)\right]A_1^2A_2e^{i(2\omega_1+\omega_2)T_0}\\
&\quad + \left[\frac{1}{6}\omega_{20}^2h_1h_2^2 - \frac{1}{4}(\omega_1^2h_2^2 + 2\omega_2^2h_1h_2)\right]A_1A_2^2e^{i(\omega_1+2\omega_2)T_0}\\
&\quad + \left[\frac{1}{6}\omega_{20}^2h_1h_2^2 - \frac{1}{4}(\omega_1^2h_2^2 + 2\omega_2^2h_1h_2)\right]\overline{A}_1A_2^2e^{i(2\omega_2-\omega_1)T_0}\\
&\quad + \left[\frac{1}{6}\omega_{20}^2h_1^2h_2 - \frac{1}{4}(2\omega_1^2h_1h_2 + \omega_2^2h_1^2)\right]A_1^2\overline{A}_2e^{i(2\omega_1-\omega_2)T_0}\\
&+ \left(\frac{1}{18}\omega_{20}^2h_1^3 - \frac{1}{4}\omega_1^2h_1^2\right)A_1^3e^{3i\omega_1T_0} + \left(\frac{1}{18}\omega_{20}^2h_2^3 - \frac{1}{4}\omega_2^2h_2^2\right)A_2^3e^{3i\omega_2T_0} + cc
\end{aligned}
\tag{6.10.30}
$$

Primes denote the derivative with respect to T_2.

The corresponding homogeneous equations to (6.10.29) and (6.10.30) are the same as those corresponding to the two equations in (6.10.8). Since (6.10.29) and (6.10.30) are coupled with each other, the excitation terms with frequencies ω_1 and ω_2 can induce the resonances of both u_{13} and u_{23}. To make the equations solvable, terms with frequencies ω_1 and ω_2 must be eliminated at the same time. In order to eliminate secular terms of Eqs. (6.10.29) and (6.10.30), we seek the following special solutions for those two equations:

$$
\begin{aligned}
u_{13} &= P_{11}e^{i\omega_1T_0} + P_{12}e^{i\omega_2T_0}\\
u_{23} &= P_{21}e^{i\omega_1T_0} + P_{22}e^{i\omega_2T_0}
\end{aligned}
\tag{6.10.31}
$$

Substituting (6.10.31) into the Eqs. (6.10.29) and (6.10.30), and equating the coefficients of $e^{i\omega_1T_0}$ and $e^{i\omega_2T_0}$ yields

$$
\begin{bmatrix} \omega_{10}^2 - \omega_n^2 & -\frac{1}{2}\omega_n^2 \\ -\frac{1}{2}\omega_n^2 & \frac{1}{3}\omega_{20}^2 - \frac{1}{3}\omega_n^2 \end{bmatrix}\begin{Bmatrix} P_{1n} \\ P_{2n} \end{Bmatrix} = \begin{Bmatrix} R_{1n} \\ R_{2n} \end{Bmatrix}, \quad n = 1,\ 2 \tag{6.10.32}
$$

where

$$R_{11} = -2i\omega_1 A_1' - i\omega_1 h_1 A_1' - \frac{1}{4}\omega_1^2 h_1^3 A_1^2 \overline{A}_1 - \frac{1}{2}\omega_1^2 h_1 h_2^2 A_1 A_2 \overline{A}_2 + \hat{K} e^{i\sigma T_2}$$
$$R_{21} = -i\omega_1 A_1' - i\frac{2}{3}\omega_1 h_1 A_1' + \left(\frac{1}{3}\omega_{20}^2 h_1 h_2^2 - \omega_2^2 h_1 h_2 - \frac{1}{2}\omega_1^2 h_2^2\right) A_1 A_2 \overline{A}_2$$
$$+\left(\frac{1}{6}\omega_{20}^2 h_1^3 - \frac{3}{4}\omega_1^2 h_1^2\right) A_1^2 \overline{A}_1 \tag{6.10.33}$$

$$R_{12} = -2i\omega_2 A_2' - i\omega_2 h_2 A_2' - \frac{1}{4}\omega_2^2 h_2^3 A_2^2 \overline{A}_2 - \frac{1}{2}\omega_2^2 h_1^2 h_2 A_1 \overline{A}_1 A_2$$
$$R_{22} = -i\omega_2 A_2' - i\frac{2}{3}\omega_2 h_2 A_2' + \left(\frac{1}{3}\omega_{20}^2 h_1^2 h_2 - \omega_1^2 h_1 h_2 - \frac{1}{2}\omega_2^2 h_1^2\right) A_1 \overline{A}_1 A_2$$
$$+\left(\frac{1}{6}\omega_{20}^2 h_2^3 - \frac{3}{4}\omega_2^2 h_2^2\right) A_2^2 \overline{A}_2 \tag{6.10.34}$$

According to (6.10.11), the determinant of the coefficient matrix of Eq. (6.10.32) is zero:

$$\det\begin{bmatrix} R_{1n} & -\frac{1}{2}\omega_n^2 \\ R_{2n} & \frac{1}{3}\omega_{20}^2 - \frac{1}{3}\omega_n^2 \end{bmatrix} = 0$$

i.e.,

$$2\left(\omega_n^2 - \omega_{20}^2\right) R_{1n} = 3\omega_n^2 R_{2n} \tag{6.10.35}$$

Substitute (6.10.33) and (6.10.34) into (6.10.35), we get

$$\begin{aligned} &2i\left[\omega_1^2\left(\tfrac{3}{2} + h_1\right) - \left(\omega_1^2 - \omega_{20}^2\right)(2 + h_1)\right] A_1' = \omega_1^3\left(\tfrac{1}{2}h_1^3 - \tfrac{9}{4}h_1^2\right) A_1^2 \overline{A}_1 \\ &+\omega_1\left[h_1 h_2^2 \omega_1^2 - 3\omega_2^2 h_1 h_2 - \tfrac{3}{2}\omega_1^2 h_2^2\right] A_1 A_2 \overline{A}_2 \\ &-2\omega_1^{-1}\left(\omega_1^2 - \omega_{20}^2\right) \hat{K} e^{i\sigma T_2} \end{aligned} \tag{6.10.36}$$

$$\begin{aligned} &2i\left[\omega_2^2\left(\tfrac{3}{2} + h_2\right) - \left(\omega_2^2 - \omega_{20}^2\right)(2 - h_2)\right] A_2' = \omega_2^3\left(\tfrac{1}{2}h_2^3 - \tfrac{9}{4}h_2^2\right) A_2^2 \overline{A}_2 \\ &+\omega_2\left(h_1^2 h_2 \omega_2^2 - 3\omega_1^2 h_1 h_2 - \tfrac{3}{2}\omega_2^2 h_1^2\right) A_1 \overline{A}_1 A_2 \end{aligned} \tag{6.10.37}$$

Simplify the above two equations, we can obtain

$$2iA_1' = 8\alpha_{11} A_1^2 \overline{A}_1 + 8\alpha_{12} A_1 A_2 \overline{A}_2 + K_1 e^{i\sigma T_2} \tag{6.10.38}$$

$$2iA_2' = 8\alpha_{21} A_2^2 \overline{A}_2 + 8\alpha_{22} A_1 \overline{A}_1 A_2 \tag{6.10.39}$$

where

$$\begin{aligned}
\alpha_{11} &= \frac{\omega_1^3\left(\frac{1}{2}h_1^3 - \frac{9}{4}h_1^2\right)}{8\left[\omega_1^2\left(\frac{3}{2}+h_1\right) - \left(\omega_1^2 - \omega_{20}^2\right)(2+h_1)\right]} \\
\alpha_{12} &= \frac{\omega_1\left[h_1 h_2^2 \omega_1^2 - 3\omega_2^2 h_1 h_2 - \frac{3}{2}\omega_1^2 h_2^2\right]}{8\left[\omega_1^2\left(\frac{3}{2}+h_1\right) - \left(\omega_1^2 - \omega_{20}^2\right)(2+h_1)\right]} \\
K_1 &= \frac{-2\omega_1^{-1}\left(\omega_1^2 - \omega_{20}^2\right)}{\left[\omega_1^2\left(\frac{3}{2}+h_1\right) - \left(\omega_1^2 - \omega_{20}^2\right)(2+h_1)\right]}
\end{aligned} \tag{6.10.40}$$

$$\begin{aligned}
\alpha_{21} &= \frac{\omega_2^3\left(\frac{1}{2}h_2^3 - \frac{9}{4}h_2^2\right)}{8\left[\omega_2^2\left(\frac{3}{2}+h_2\right) - \left(\omega_2^2 - \omega_{20}^2\right)(2-h_2)\right]} \\
\alpha_{21} &= \frac{\omega_2\left(h_1^2 h_2 \omega_2^2 - 3\omega_1^2 h_1 h_2 - \frac{3}{2}\omega_2^2 h_1^2\right)}{8\left[\omega_2^2\left(\frac{3}{2}+h_2\right) - \left(\omega_2^2 - \omega_{20}^2\right)(2-h_2)\right]}
\end{aligned} \tag{6.10.41}$$

Let

$$A_1 = \frac{1}{2}a_1 e^{i\beta_1}, \quad A_2 = \frac{1}{2}a_2 e^{i\beta_2} \tag{6.10.42}$$

Substituting (6.10.42) into (6.10.38) and (6.10.39), we get

$$ia_1' + a_1\gamma_1' = \sigma a_1 + \alpha_{11}a_1^3 + \alpha_{12}a_1 a_2^2 + K_1 e^{i\gamma_1} \tag{6.10.43}$$

$$ia_2' - a_2\beta_2' = a_2^3 + \alpha_{22}a_1^2 a_2 \tag{6.10.44}$$

where

$$\gamma_1 = \sigma T_2 - \beta_1 \tag{6.10.45}$$

Separating the real and imaginary parts of (6.10.43) and (6.10.44) yields

$$a_1' = K_1 \sin\gamma_1 \tag{6.10.46}$$

$$a_1\gamma_1' = \sigma a_1 + \alpha_{11}a_1^3 + \alpha_{12}a_1 a_2^2 + K_1\cos\gamma_1 \tag{6.10.47}$$

$$a_2' = 0 \tag{6.10.48}$$

$$\beta_2' = -\left(a_2^2 + \alpha_{22}a_1^2\right) \tag{6.10.49}$$

Let $a_1' = a_2' = \gamma_1' = 0$ in (6.10.46)–(6.10.48), we can obtain

$$K_1 \sin\gamma_1 = 0 \tag{6.10.50}$$

$$\sigma a_1 + \alpha_{11} a_1^3 + \alpha_{12} a_1 a_2^2 + K_1 \cos \gamma_1 = 0 \tag{6.10.51}$$

$$a_2 = a_{20} = \text{constant} \tag{6.10.52}$$

So

$$\gamma_1 = 0, \quad \pm\pi \tag{6.10.53}$$

For $\gamma_1 = 0$, the frequency–response equation is given by

$$\sigma = -\frac{1}{a_1}\left(\alpha_{11} a_1^3 + \alpha_{12} a_1 a_{20}^2 + K_1\right) \tag{6.10.54}$$

and β_2 is given by that

$$\beta_2 = \beta_{20} - \left(a_{20}^2 + \alpha_{22} a_1^2\right) T_2 \tag{6.10.55}$$

Again, analyze the stability of the steady state solution. Let a_{10}, a_{20} and γ_{10} be a set of steady state solution and make a small perturbation of a_{10} and γ_{10}, i.e.,

$$a_1 = a_{10} + \tilde{a}_1, \quad \gamma_1 = \gamma_{10} + \tilde{\gamma}_1 \tag{6.10.56}$$

Substitute (6.10.56) into (6.10.46) and (6.10.47) and make linearization, we can obtain

$$\begin{Bmatrix} \tilde{a}_1' \\ \tilde{\gamma}_1' \end{Bmatrix} = \begin{bmatrix} 0 & K_1 \\ a_{10}^{-1}\left(\sigma + 3\alpha_{11} a_{10}^2 + \alpha_{12} a_{20}^2\right) & 0 \end{bmatrix} \begin{Bmatrix} \tilde{a}_1 \\ \tilde{\gamma}_1 \end{Bmatrix} \tag{6.10.57}$$

It follows that the condition for obtain a stable steady state solution is

$$K_1 a_{10}^{-1}\left(\sigma + 3\alpha_{11} a_{10}^2 + \alpha_{12} a_{20}^2\right) < 0 \tag{6.10.58}$$

From the above results, we can find that the excitation of the first mode of this two-degree-of freedom system does not produce significant responses in the second mode.

(ii) $\Omega \approx \frac{1}{3}\omega_1$

In this case, $K \cos \Omega T_0$ is a non-resonant hard excitation term, so it will appear in the equations of order ε^1. Thus, let

$$K = 2\varepsilon \hat{K} \tag{6.10.59}$$

Equating coefficients of like powers of ε in (6.4.11) and (6.4.12) yields.
Order ε_1:

$$\begin{aligned} & D_0^2 u_{11} + \tfrac{1}{2} D_0^2 u_{21} + \omega_{10}^2 u_{11} = 2\hat{K}\cos\Omega T_0 \\ & \tfrac{1}{2} D_0^2 u_{11} + \tfrac{1}{3} D_0^2 u_{21} + \tfrac{1}{3}\omega_{20}^2 u_{21} = 0 \end{aligned} \tag{6.10.60}$$

Order ε_2:

$$\begin{aligned} & D_0^2 u_{12} + \tfrac{1}{2} D_0^2 u_{22} + \omega_{10}^2 u_{12} = -2D_0 D_1 u_{11} - D_0 D_1 u_{21} \\ & \tfrac{1}{2} D_0^2 u_{12} + \tfrac{1}{3} D_0^2 u_{22} + \tfrac{1}{3}\omega_{20}^2 u_{22} = -D_0 D_1 u_{11} - \tfrac{2}{3} D_0 D_1 u_{21} \end{aligned} \tag{6.10.61}$$

Order ε_3:

$$\begin{aligned} & D_0^2 u_{13} + \frac{1}{2} D_0^2 u_{23} + \omega_{10}^2 u_{13} = -2D_0 D_1 u_{12} - \left(D_1^2 + 2D_0 D_2\right) u_{11} - D_0 D_1 u_{22} \\ & -\frac{1}{2}\left(D_1^2 + 2D_0 D_2\right) u_{21} + \frac{1}{4} u_{21}^2 D_0^2 u_{21} \\ & +\frac{1}{2} u_{21} (D_0 u_{21})^2 \\ & \frac{1}{2} D_0^2 u_{13} + \frac{1}{3} D_0^2 u_{23} + \frac{1}{3}\omega_{20}^2 u_{23} = -D_0 D_1 u_{12} - \frac{1}{2}\left(D_1^2 + 2D_0 D_2\right) u_{11} - \frac{2}{3} D_0 D_1 u_{22} \\ & -\frac{1}{3}\left(D_1^2 + 2D_0 D_2\right) u_{21} + \frac{1}{18}\omega_{20}^2 u_{21}^3 + \frac{1}{4} u_{21}^2 D_0^2 u_{11} \end{aligned} \tag{6.10.62}$$

The solution of (6.10.60) can be expressed as

$$\begin{aligned} u_{11} &= A_1 e^{i\omega_1 T_0} + A_2 e^{i\omega_2 T_0} + \hat{K}_1 e^{i\Omega T_0} + cc \\ u_{21} &= h_1 A_1 e^{i\omega_1 T_0} + h_2 A_2 e^{i\omega_2 T_0} + \hat{K}_2 e^{i\Omega T_0} + cc \end{aligned} \tag{6.10.63}$$

where

$$\hat{K}_1 = \frac{4\left(\omega_{20}^2 - \Omega^2\right)}{4\left(\omega_{10}^2 - \Omega^2\right)\left(\omega_{20}^2 - \Omega^2\right) - 3\Omega^4}\hat{K}, \quad \hat{K}_2 = \frac{6\Omega^2}{4\left(\omega_{10}^2 - \Omega^2\right)\left(\omega_{20}^2 - \Omega^2\right) - 3\Omega^4}\hat{K} \tag{6.10.64}$$

$A_k = A_k(T_1, T_2)$, $k = 1, 2$, h_1, h_2 are given by (6.10.15).
Substituting (6.10.63) into (6.10.61) yields

$$\begin{aligned} & D_0^2 u_{12} + \frac{1}{2} D_0^2 u_{22} + \omega_{10}^2 u_{12} = -i\omega_1 (2 + h_1) D_1 A_1 e^{i\omega_1 T_0} \\ & -i\omega_2 (2 + h_2) D_1 A_2 e^{i\omega_2 T_0} + cc \\ & \frac{1}{2} D_0^2 u_{12} + \frac{1}{3} D_0^2 u_{22} + \frac{1}{3}\omega_{20}^2 u_{22} = -i\omega_1 \left(1 + \frac{2}{3} h_1\right) D_1 A_1 e^{i\omega_1 T_0} \\ & -i\omega_2 \left(1 + \frac{2}{3} h_2\right) D_1 A_2 e^{i\omega_2 T_0} + cc \end{aligned} \tag{6.10.65}$$

In order to eliminate secular terms from the above equations, there must be

$$D_1A_1 = 0, \quad D_1A_2 = 0 \quad \rightarrow \quad A_1 = A_1(T_2), \quad A_2 = A_2(T_2) \tag{6.10.66}$$

therefore

$$u_{12} = u_{22} = 0 \tag{6.10.67}$$

Introduce the detuning parameter σ such that

$$3\Omega = \omega_1 + \varepsilon^2\sigma \tag{6.10.68}$$

Substitute (6.10.68), (6.10.67), (6.10.66) and (6.10.63) into (6.10.62), we can obtain

$$\begin{aligned}
D_0^2u_{13} + \frac{1}{2}D_0^2u_{23} + \omega_{10}^2u_{13} &= -2D_0D_2u_{11} - D_0D_2u_{21} + \frac{1}{4}u_{21}^2D_0^2u_{21} + \frac{1}{2}u_{21}(D_0u_{21})^2 \\
&= \left(-2i\omega_1A_1' - i\omega_1h_1A_1' - \frac{1}{4}\omega_1^2h_1^3A_1^2\overline{A}_1 - \frac{1}{2}\omega_1^2h_1h_2^2A_1A_2\overline{A}_2 \right.\\
&\quad \left. - \frac{1}{2}\omega_1^2\hat{K}_2^2h_1A_1 - \frac{3}{4}\Omega^2\hat{K}_2^3e^{i\sigma T_2}\right)e^{i\omega_1T_0} \\
&\quad + \left(-2i\omega_2A_2' - i\omega_2h_2A_2' - \frac{1}{4}\omega_2^2h_2^3A_2^2\overline{A}_2 - \frac{1}{2}\omega_2^2h_1^2h_2A_1\overline{A}_1A_2 \right.\\
&\quad \left. - \frac{1}{2}\omega_2^2\hat{K}_2^2h_2A_2\right)e^{i\omega_2T_0} \\
&\quad + cc + NST
\end{aligned} \tag{6.10.69}$$

$$
\begin{aligned}
&\frac{1}{2}D_0^2u_{13}+\frac{1}{3}D_0^2u_{23}+\frac{1}{3}\omega_{20}^2u_{23}=-D_0D_2u_{11}-\frac{2}{3}D_0D_2u_{21}+\frac{1}{18}\omega_{20}^2u_{21}^3+\frac{1}{4}u_{21}^2D_0^2u_{11}\\
&=\Bigg[-i\omega_1A_1'-i\frac{2}{3}\omega_1h_1A_1'+\left(\frac{1}{3}\omega_{20}^2h_1h_2^2-\omega_2^2h_1h_2-\frac{1}{2}\omega_1^2h_2^2\right)A_1A_2\bar{A}_2\\
&+\left(\frac{1}{6}\omega_{20}^2h_1^3-\frac{3}{4}\omega_1^2h_1^2\right)A_1^2\bar{A}_1\\
&+\left(\frac{1}{3}\omega_{20}^2\hat{K}_2^2h_1-\frac{1}{2}\omega_1^2\hat{K}_2^2-\Omega^2\hat{K}_1\hat{K}_2h_1\right)A_1\\
&+\left(\frac{1}{18}\omega_{20}^2\hat{K}_2^3-\frac{1}{4}\Omega^2\hat{K}_1\hat{K}_2^2\right)e^{i\sigma T_2}\Bigg]e^{i\omega_1T_0}\\
&+\Bigg[-i\omega_2A_2'-i\frac{2}{3}\omega_2h_2A_2'+\left(\frac{1}{3}\omega_{20}^2h_1^2h_2-\omega_1^2h_1h_2-\frac{1}{2}\omega_2^2h_1^2\right)A_1\bar{A}_1A_2\\
&+\left(\frac{1}{6}\omega_{20}^2h_2^3-\frac{3}{4}\omega_2^2h_2^2\right)A_2^2\bar{A}_2\\
&+\left(\frac{1}{3}\omega_{20}^2\hat{K}_2^2h_2-\frac{1}{2}\omega_2^2\hat{K}_2^2-\Omega^2\hat{K}_1\hat{K}_2h_2\right)A_2\Bigg]e^{i\omega_2T_0}\\
&+cc+NST
\end{aligned}
\tag{6.10.70}
$$

Primes denote the derivative with respect to T_2.

In order to eliminate secular terms in (6.10.69) and (6.10.70), we seek the following special solution for them:

$$
\begin{aligned}
u_{13}&=P_{11}e^{i\omega_1T_0}+P_{12}e^{i\omega_2T_0}\\
u_{23}&=P_{21}e^{i\omega_1T_0}+P_{22}e^{i\omega_2T_0}
\end{aligned}
\tag{6.10.71}
$$

Substituting (6.10.71) into (6.10.69) and (6.10.70) and equating the coefficients of $e^{i\omega_1T_0}$ and $e^{i\omega_2T_0}$, we obtain

$$
\begin{bmatrix}\omega_{10}^2-\omega_n^2 & -\frac{1}{2}\omega_n^2\\ -\frac{1}{2}\omega_n^2 & \frac{1}{3}\omega_{20}^2-\frac{1}{3}\omega_n^2\end{bmatrix}\begin{Bmatrix}P_{1n}\\ P_{2n}\end{Bmatrix}=\begin{Bmatrix}R_{1n}\\ R_{2n}\end{Bmatrix},\quad n=1,\ 2 \tag{6.10.72}
$$

where

$$
\begin{aligned}
R_{11} &= -2i\omega_1 A_1' - i\omega_1 h_1 A_1' - \frac{1}{4}\omega_1^2 h_1^3 A_1^2 \overline{A_1} - \frac{1}{2}\omega_1^2 h_1 h_2^2 A_1 A_2 \overline{A}_2 \\
&\quad -\frac{1}{2}\omega_1^2 \hat{K}_2^2 h_1 A_1 - \frac{3}{4}\Omega^2 \hat{K}_2^3 e^{i\sigma T_2} \\
R_{21} &= -i\omega_1 A_1' - i\frac{2}{3}\omega_1 h_1 A_1' + \left(\frac{1}{3}\omega_{20}^2 h_1 h_2^2 - \omega_2^2 h_1 h_2 - \frac{1}{2}\omega_1^2 h_2^2\right) A_1 A_2 \overline{A}_2 \\
&\quad +\left(\frac{1}{6}\omega_{20}^2 h_1^3 - \frac{3}{4}\omega_1^2 h_1^2\right) A_1^2 \overline{A}_1 + \left(\frac{1}{3}\omega_{20}^2 \hat{K}_2^2 h_1 - \frac{1}{2}\omega_1^2 \hat{K}_2^2 - \Omega^2 \hat{K}_1 \hat{K}_2 h_1\right) A_1 \\
&\quad +\left(\frac{1}{18}\omega_{20}^2 \hat{K}_2^3 - \frac{1}{4}\Omega^2 \hat{K}_1 \hat{K}_2^2\right) e^{i\sigma T_2}
\end{aligned}
\tag{6.10.73}
$$

$$
\begin{aligned}
R_{12} &= -2i\omega_2 A_2' - i\omega_2 h_2 A_2' - \frac{1}{4}\omega_2^2 h_2^3 A_2^2 \overline{A}_2 - \frac{1}{2}\omega_2^2 h_1^2 h_2 A_1 \overline{A}_1 A_2 - \frac{1}{2}\omega_2^2 \hat{K}_2^2 h_2 A_2 \\
R_{22} &= -i\omega_2 A_2' - i\frac{2}{3}\omega_2 h_2 A_2' + \left(\frac{1}{3}\omega_{20}^2 h_1^2 h_2 - \omega_1^2 h_1 h_2 - \frac{1}{2}\omega_2^2 h_1^2\right) A_1 \overline{A}_1 A_2 \\
&\quad +\left(\frac{1}{6}\omega_{20}^2 h_2^3 - \frac{3}{4}\omega_2^2 h_2^2\right) A_2^2 \overline{A}_2 + \left(\frac{1}{3}\omega_{20}^2 \hat{K}_2^2 h_2 - \frac{1}{2}\omega_2^2 \hat{K}_2^2 - \Omega^2 \hat{K}_1 \hat{K}_2 h_2\right) A_2
\end{aligned}
\tag{6.10.74}
$$

According to (6.10.11), the determinant of the coefficient matrix of Eq. (6.10.72) is zero, i.e.,

$$
\det\begin{bmatrix} R_{1n} & -\frac{1}{2}\omega_n^2 \\ R_{2n} & \frac{1}{3}\omega_{20}^2 - \frac{1}{3}\omega_n^2 \end{bmatrix} = 0
$$

i.e.,

$$
2\left(\omega_n^2 - \omega_{20}^2\right) R_{1n} = 3\omega_n^2 R_{2n} \tag{6.10.75}
$$

Substitute (6.10.73) and (6.10.74), respectively, into (6.10.75), we obtain

$$
2iA' = \alpha_{11} A_1 + \alpha_{12} A_1^2 \overline{A}_1 + \alpha_{13} A_1 A_2 \overline{A}_2 + \kappa e^{i\sigma T_2} \tag{6.10.76}
$$

$$
2iA_2' = \alpha_{21} A_2 + \alpha_{22} A_2^2 \overline{A}_2 + \alpha_{23} A_1 \overline{A}_1 A_2 \tag{6.10.77}
$$

where

$$
\begin{aligned}
\alpha_{11} &= \frac{\omega_1^2\left(\omega_1^2\hat{K}_2^2 h_1 - \frac{3}{2}\omega_1^2\hat{K}_2^2 - 3\Omega^2\hat{K}_1\hat{K}_2 h_1\right)}{\omega_1\left[\frac{1}{2}\omega_1^2(3+2h_1) - \left(\omega_1^2 - \omega_{20}^2\right)(2-h_1)\right]} \\
\alpha_{12} &= \frac{\omega_1^4\left(\frac{1}{2}h_1^3 - \frac{9}{4}h_1^2\right)}{\omega_1\left[\frac{1}{2}\omega_1^2(3+2h_1) - \left(\omega_1^2 - \omega_{20}^2\right)(2-h_1)\right]} \\
\alpha_{13} &= \frac{\omega_1^2\left(\omega_1^2 h_1 h_2^2 - 3\omega_2^2 h_1 h_2 - \frac{3}{2}\omega_1^2 h_2^2\right)}{\omega_1\left[\frac{1}{2}\omega_1^2(3+2h_1) - \left(\omega_1^2 - \omega_{20}^2\right)(2-h_1)\right]} \\
\kappa &= \frac{\left(\frac{3}{4}\Omega^2\hat{K}_2^3 + \frac{1}{6}\omega_1^2\omega_{20}^2\hat{K}_2^3 - 3\omega_1^2\frac{1}{4}\Omega^2\hat{K}_1\hat{K}_2^2\right)}{\omega_1\left[\frac{1}{2}\omega_1^2(3+2h_1) - \left(\omega_1^2 - \omega_{20}^2\right)(2-h_1)\right]}
\end{aligned}
\tag{6.10.78}
$$

$$
\begin{aligned}
\alpha_{21} &= \frac{\omega_2^2\left(\omega_2^2\hat{K}_2^2 h_2 - \frac{3}{2}\omega_2^2\hat{K}_2^2 - 3\Omega^2\hat{K}_1\hat{K}_2 h_2\right)}{\omega_2\left[\frac{1}{2}\omega_2^2(3+2h_2) - \left(\omega_2^2 - \omega_{20}^2\right)(2-h_2)\right]} \\
\alpha_{22} &= \frac{\omega_2^4\left(\frac{1}{2}h_2^3 - \frac{9}{4}h_2^2\right)}{\omega_2\left[\frac{1}{2}\omega_2^2(3+2h_2) - \left(\omega_2^2 - \omega_{20}^2\right)(2-h_2)\right]} \\
\alpha_{23} &= \frac{\omega_2^2\left(\omega_2^2 h_1^2 h_2 - 3\omega_1^2 h_1 h_2 - \frac{3}{2}\omega_2^2 h_1^2\right)}{\omega_2\left[\frac{1}{2}\omega_2^2(3+2h_2) - \left(\omega_2^2 - \omega_{20}^2\right)(2-h_2)\right]}
\end{aligned}
\tag{6.10.79}
$$

Let

$$A_1 = \frac{1}{2}a_1 e^{i\beta_1}, \quad A_2 = \frac{1}{2}a_2 e^{i\beta_2} \tag{6.10.80}$$

Substituting (6.10.80) into (6.10.76) and (6.10.77) yields

$$ia_1' + a_1\gamma_1' = \sigma a_1 + \frac{1}{2}\alpha_{11}a_1 + \frac{1}{8}\alpha_{12}a_1^3 + \frac{1}{8}\alpha_{13}a_1a_2^3 + \kappa e^{i\gamma_1} \tag{6.10.81}$$

$$ia_2' - a_2\beta_2' = \frac{1}{2}\alpha_{21}a_2 + \frac{1}{8}\alpha_{22}a_2^3 + \alpha_{23}a_1^2a_2 \tag{6.10.82}$$

where

$$\gamma_1 = \sigma T_2 - \beta_1 \tag{6.10.83}$$

Separating the real and imaginary parts of (6.10.81) and (6.10.82), we obtain

$$a_1' = \kappa \sin\gamma_1 \tag{6.10.84}$$

$$a_1\gamma_1' = \sigma a_1 + \frac{1}{2}\alpha_{11}a_1 + \frac{1}{8}\alpha_{12}a_1^3 + \frac{1}{8}\alpha_{13}a_1a_2^3 + \kappa\cos\gamma_1 \tag{6.10.85}$$

$$a_2' = 0 \tag{6.10.86}$$

$$\beta_2' = -\left(\frac{1}{2}\alpha_{21} + \frac{1}{8}\alpha_{22}a_2^2 + \alpha_{23}a_1^2\right) \tag{6.10.87}$$

The steady state response of the system corresponds to the solution of

$$\kappa \sin \gamma_1 = 0 \tag{6.10.88}$$

$$\sigma a_1 + \frac{1}{2}\alpha_{11}a_1 + \frac{1}{8}\alpha_{12}a_1^3 + \frac{1}{8}\alpha_{13}a_1a_2^3 + \kappa \cos \gamma_1 = 0 \tag{6.10.89}$$

$$a_2 = a_{20} = \text{constant} \tag{6.10.90}$$

So

$$\gamma_1 = 0, \quad \pm\pi \tag{6.10.91}$$

For $\gamma_1 = 0$, the frequency–response equation is given by

$$\sigma = -\frac{1}{a_1}\left(\frac{1}{2}\alpha_{11}a_1 + \frac{1}{8}\alpha_{12}a_1^3 + \frac{1}{8}\alpha_{13}a_1a_{20}^3 + \kappa\right) \tag{6.10.92}$$

From (6.10.87), we can obtain β_2:

$$\beta_2 = \beta_{20} - \left(\frac{1}{2}\alpha_{21} + \frac{1}{8}\alpha_{22}a_{20}^2 + \alpha_{23}a_1^2\right)T_2 \tag{6.10.93}$$

Again, analyze the stability of the steady state solution. Let a_{10}, a_{20} and γ_{10} be a set of steady state solution and make a small perturbation of a_{10} and γ_{10}, i.e.,

$$a_1 = a_{10} + \tilde{a}_1, \quad \gamma_1 = \gamma_{10} + \tilde{\gamma}_1 \tag{6.10.94}$$

Substitute (6.10.94) into (6.10.46) and (6.10.47) and make linearization, we can obtain

$$\begin{Bmatrix} \tilde{a}_1' \\ \tilde{\gamma}_1' \end{Bmatrix} = \begin{bmatrix} 0 & \kappa \\ a_{10}^{-1}\left(\sigma + \frac{1}{2}\alpha_{11} + \frac{3}{8}\alpha_{12}a_{10}^2 + \frac{1}{8}\alpha_{13}a_{20}^3\right) & 0 \end{bmatrix} \begin{Bmatrix} \tilde{a}_1 \\ \tilde{\gamma}_1 \end{Bmatrix} \tag{6.10.95}$$

It follows that the condition for obtain a stable steady state solution is

$$\kappa a_{10}^{-1}\left(\sigma + \frac{1}{2}\alpha_{11} + \frac{3}{8}\alpha_{12}a_{10}^2 + \frac{1}{8}\alpha_{13}a_{20}^3\right) < 0 \tag{6.10.96}$$

From the above results, we can find that the excitation of the first mode of this two-degree-of freedom system does not produce significant responses in the second mode.

(iii) $\Omega \approx 3\omega_1$

Readers are invited to complete the solution for this case.

6.11 Exercise 6.11 (Resonance Analysis on the Cylinder Rolling Without Slip on the Circular Surface I)

Solution: (a) The kinetic energy of the system is

$$\begin{aligned} T &= \tfrac{1}{2}M\dot{x}^2 + \tfrac{1}{2}m_1\{[\dot{x} + (R-r)\dot{\theta}\cos\theta]^2 + [(R-r)\dot{\theta}\sin\theta]^2\} \\ &+ \tfrac{1}{2}\tfrac{1}{2}m_1 r^2[\tfrac{(R-r)\dot{\theta}}{r}]^2 \\ &= \tfrac{1}{2}(m_1 + M)\dot{x}^2 + \tfrac{3}{4}m_1(R-r)^2\dot{\theta}^2 + m_1(R-r)\dot{x}\dot{\theta}\cos\theta \end{aligned} \tag{6.11.1}$$

The potential energy of the system is

$$V = \frac{1}{2}kx^2 - m_1 g(R-r)\cos\theta \tag{6.11.2}$$

The generalized forces are

$$Q_x = F(t), \quad Q_\theta = 0 \tag{6.11.3}$$

Substitute kinetic energy, potential energy, and generalized forces into the Lagrange's equation

$$\frac{d}{dt}\frac{\partial T}{\partial \dot{x}} - \frac{\partial T}{\partial x} = -\frac{\partial V}{\partial x} + Q_x, \quad \frac{d}{dt}\frac{\partial T}{\partial \dot{\theta}} - \frac{\partial T}{\partial \theta} = -\frac{\partial V}{\partial \theta} + Q_\theta$$

we can obtain

$$\begin{aligned} &(m_1 + M)\ddot{x} + m_1(R-r)\ddot{\theta}\cos\theta - m_1(R-r)\dot{\theta}^2\sin\theta + kx = F(t) \\ &\frac{3}{2}m_1(R-r)^2\ddot{\theta} + m_1(R-r)\ddot{x}\cos\theta + m_1 g(R-r)\sin\theta = 0 \end{aligned} \tag{6.11.4}$$

Let

$$\begin{aligned} &u = x/(R-r), \quad \omega_{10}^2 = k/(m_1 + M), \quad \omega_{20}^2 = \frac{2}{3}g/(R-r) \\ &m = m_1/(m_1 + M), \quad f(t) = F(t)/[(R-r)(m_1 + M)] \end{aligned} \tag{6.11.5}$$

Equation (6.11.4) changes into

$$\begin{aligned}&\ddot{u}+m\ddot{\theta}\cos\theta-m\dot{\theta}^2\sin\theta+\omega_{10}^2u=f(t)\\&\ddot{\theta}+\frac{2}{3}\ddot{u}\cos\theta+\omega_{20}^2\sin\theta=0\end{aligned}\tag{6.11.6}$$

(b) Include the viscous damping, which has the form $-cx$, in (6.11.4), we can obtain the equations of motion.

$$\begin{aligned}&(m_1+M)\ddot{x}+m_1(R-r)\ddot{\theta}\cos\theta-m_1(R-r)\dot{\theta}^2\sin\theta+kx=F(t)-c\dot{x}\\&\frac{3}{2}m_1(R-r)^2\ddot{\theta}+m_1(R-r)\ddot{x}\cos\theta+m_1g(R-r)\sin\theta=0\end{aligned}\tag{6.11.7}$$

Let

$$2\omega_{10}\mu=c/(m_1+M)\tag{6.11.8}$$

Considering (6.11.5), Eq. (6.11.7) becomes

$$\begin{aligned}&\ddot{u}+m\ddot{\theta}\cos\theta-m\dot{\theta}^2\sin\theta+\omega_{10}^2u+2\omega_{10}\mu\dot{u}=f(t)\\&\ddot{\theta}+\frac{2}{3}\ddot{u}\cos\theta+\omega_{20}^2\sin\theta=0\end{aligned}\tag{6.11.9}$$

(c) At the equilibrium position of the system $u=\theta=0$, expand (6.11.9) and retain to the third order of smallness yields

$$\begin{aligned}&\ddot{u}+m\ddot{\theta}+\omega_{10}^2u+2\omega_{10}\mu\dot{u}-\frac{1}{2}m\theta^2\,\theta-m\dot{\theta}^2\theta=\sum_{n=1}^{3}K_n\cos(\Omega_nt+\theta_n)\\&\ddot{u}+\frac{3}{2}\ddot{\theta}+\frac{3}{2}\omega_{20}^2\theta-\frac{1}{4}\omega_{20}^2\theta^3-\frac{1}{2}u\,\theta^2=0\end{aligned}\tag{6.11.10}$$

The corresponding equations for the linear undamped free oscillation are

$$\begin{bmatrix}1&m\\1&\frac{3}{2}\end{bmatrix}\begin{Bmatrix}\ddot{u}\\\ddot{\theta}\end{Bmatrix}+\begin{bmatrix}\omega_{10}^2&0\\0&\frac{3}{2}\omega_{20}^2\end{bmatrix}\begin{Bmatrix}u\\\theta\end{Bmatrix}=0\tag{6.11.11}$$

Let

$$u=B_1e^{i\omega t},\quad\theta=B_2e^{i\omega t}\tag{6.11.12}$$

Substitute (6.11.12) into (6.11.11), we can obtain

$$\begin{bmatrix} (\omega_{10}^2 - \omega^2) & -m\omega^2 \\ -\omega^2 & \frac{3}{2}(\omega_{20}^2 - \omega^2) \end{bmatrix} \begin{Bmatrix} B_1 \\ B_2 \end{Bmatrix} = 0 \tag{6.11.13}$$

Therefore, the characteristic equation is

$$\det \begin{bmatrix} (\omega_{10}^2 - \omega^2) & -m\omega^2 \\ -\omega^2 & \frac{3}{2}(\omega_{20}^2 - \omega^2) \end{bmatrix} = 0 \tag{6.11.14}$$

i.e.,

$$(3 - 2m)\omega^4 - 3(\omega_{10}^2 + \omega_{20}^2)\omega^2 + 3\omega_{10}^2\omega_{20}^2 = 0 \tag{6.11.15}$$

Therefore, the linear natural frequencies are

$$\begin{aligned} \omega_1 &= \left[\frac{3(\omega_{10}^2 + \omega_{20}^2) - \sqrt{9(\omega_{10}^2 + \omega_{20}^2)^2 - 12(3 - 2m)\omega_{10}^2\omega_{20}^2}}{2(3 - 2m)} \right]^{1/2} \\ \omega_2 &= \left[\frac{3(\omega_{10}^2 + \omega_{20}^2) + \sqrt{9(\omega_{10}^2 + \omega_{20}^2)^2 - 12(3 - 2m)\omega_{10}^2\omega_{20}^2}}{2(3 - 2m)} \right]^{1/2} \end{aligned} \tag{6.11.16}$$

Substituting ω_1 and ω_2 into (6.11.13), we can obtain

$$\{B_1, B_2\}_1^T = \{1, h_1\}^T, \quad \{B_1, B_2\}_2^T = \{1, h_2\}^T \tag{6.11.17}$$

where h_1 and h_2 can be obtained from (6.11.13)

$$h_1 = \frac{\omega_{10}^2 - \omega_1^2}{m\omega_1^2}, \quad h_2 = \frac{\omega_{10}^2 - \omega_2^2}{m\omega_2^2} \tag{6.11.18}$$

Therefore, the solution for the linear free oscillation (6.11.11) is

$$\begin{aligned} u &= A_1 e^{i\omega_1 t} + A_2 e^{i\omega_2 t} + cc, \\ \theta &= h_1 A_1 e^{i\omega_1 t} + h_2 A_2 e^{i\omega_2 t} + cc \end{aligned} \tag{6.11.19}$$

In order to study the internal resonance of $\omega_2 \approx 3\omega_1$, let

$$s = \omega_2/\omega_1, \quad p = \omega_{20}/\omega_{10} \tag{6.11.20}$$

From (6.11.16),

$$s = \left[\frac{1 + \sqrt{1 - q}}{1 - \sqrt{1 - q}} \right]^{1/2} \tag{6.11.21}$$

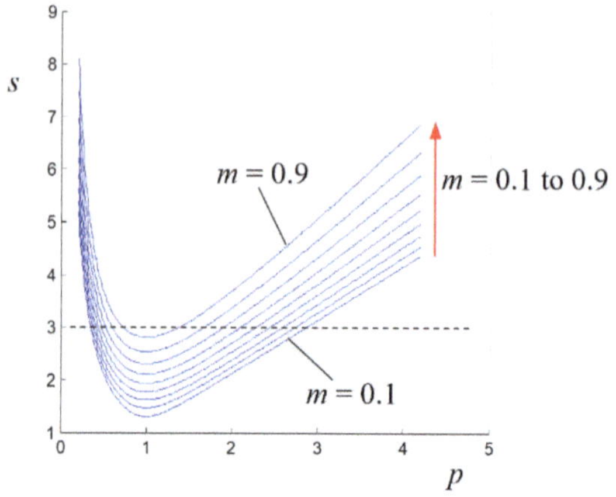

Fig. 6.7 Frequency ratio of p to s for Exercise 6.11

where

$$q = \frac{4(3-2m)p^2}{3(p^2+1)^2} \leq 1, \quad 0 < m < 1 \tag{6.11.22}$$

For different m, the curves of the frequency ratio p versus s are shown in Fig. 6.7. The internal resonance case of $\omega_2 \approx 3\omega_1$ may occur when $m = 0.1$ to 0.9.

(d) For small and finite amplitudes, we use the method of multiple scales to solve (6.11.10). Let the solution of Eq. (6.11.10) be

$$\begin{aligned} u &= \varepsilon u_{11}(T_0, T_2) + \varepsilon^3 u_{13}(T_0, T_2) + \cdots \\ \theta &= \varepsilon u_{21}(T_0, T_2) + \varepsilon^3 u_{23}(T_0, T_2) + \cdots \end{aligned} \tag{6.11.23}$$

In order for the damping and nonlinear terms to appear in the same order equation, we specify that

$$\mu = \varepsilon^2 \hat{\mu} \tag{6.11.24}$$

Substitute (6.11.23) into (6.11.10) and keep to $O(\varepsilon^3)$, we obtain

$$
\begin{aligned}
0 &= \ddot{u} + m\ddot{\theta} + \omega_{10}^2 u + 2\omega_{10}\mu\dot{u} - \frac{1}{2}m\theta^2\,\theta - m\dot{\theta}^2\theta - \sum_{n=1}^{3} K_n\cos(\Omega_n t + \theta_n) \\
&= \left[D_0^2 + 2\varepsilon D_0 D_1 + \varepsilon^2\left(D_1^2 + 2D_0 D_2\right)\right]\left(\varepsilon u_{11} + \varepsilon^3 u_{13}\right) \\
&\quad + m\left[D_0^2 + 2\varepsilon D_0 D_1 + \varepsilon^2\left(D_1^2 + 2D_0 D_2\right)\right]\left(\varepsilon u_{21} + \varepsilon^3 u_{23}\right) \\
&\quad + \omega_{10}^2\left(\varepsilon u_{11} + \varepsilon^3 u_{13}\right) + 2\omega_{10}\varepsilon^2\hat{\mu}\left(D_0 + \varepsilon D_1 + \varepsilon^2 D_2\right)\left(\varepsilon u_{11} + \varepsilon^3 u_{13}\right) \\
&\quad - \frac{1}{2}m\left(\varepsilon u_{21} + \varepsilon^3 u_{23}\right)^2\left[D_0^2 + 2\varepsilon D_0 D_1 + \varepsilon^2\left(D_1^2 + 2D_0 D_2\right)\right]\left(\varepsilon u_{21} + \varepsilon^3 u_{23}\right) \\
&\quad - m\left(\varepsilon u_{21} + \varepsilon^3 u_{23}\right)\left[\left(D_0 + \varepsilon D_1 + \varepsilon^2 D_2\right)\left(\varepsilon u_{21} + \varepsilon^3 u_{23}\right)\right]^2 \\
&\quad - \sum_{n=1}^{3} K_n\cos(\Omega_n T_0 + \theta_n) + \cdots \\
&= \varepsilon\left(D_0^2 u_{11} + mD_0^2 u_{21} + \omega_{10}^2 u_{11}\right) + \varepsilon^3\left[D_0^2 u_{13} + mD_0^2 u_{23} + \omega_{10}^2 u_{13}\right. \\
&\quad + 2D_0 D_2 u_{11} + 2mD_0 D_2 u_{21} + 2\omega_{10}\hat{\mu} D_0 u_{11} \\
&\quad \left. - \frac{1}{2}mu_{21}^2 D_0^2 u_{21} - mu_{21}\left(D_0 u_{21}\right)^2\right] - \sum_{n=1}^{3} K_n\cos(\Omega_n T_0 + \theta_n) + \cdots
\end{aligned}
\tag{6.11.25}
$$

$$
\begin{aligned}
0 &= \ddot{u} + \frac{3}{2}\ddot{\theta} + \frac{3}{2}\omega_{20}^2\theta - \frac{1}{4}\omega_{20}^2\theta^3 - \frac{1}{2}\ddot{u}\theta^2 \\
&= \left[D_0^2 + 2\varepsilon D_0 D_1 + \varepsilon^2\left(D_1^2 + 2D_0 D_2\right)\right]\left(\varepsilon u_{11} + \varepsilon^3 u_{13}\right) \\
&\quad + \frac{3}{2}\left[D_0^2 + 2\varepsilon D_0 D_1 + \varepsilon^2\left(D_1^2 + 2D_0 D_2\right)\right]\left(\varepsilon u_{21} + \varepsilon^3 u_{23}\right) \\
&\quad + \frac{3}{2}\omega_{20}^2\left(\varepsilon u_{21} + \varepsilon^3 u_{23}\right) - \frac{1}{4}\omega_{20}^2\left(\varepsilon u_{21} + \varepsilon^3 u_{23}\right)^3 \\
&\quad - \frac{1}{2}\left(\varepsilon u_{21} + \varepsilon^3 u_{23}\right)^2\left[D_0^2 + 2\varepsilon D_0 D_1 + \varepsilon^2\left(D_1^2 + 2D_0 D_2\right)\right]\left(\varepsilon u_{11} + \varepsilon^3 u_{13}\right) + \cdots \\
&= \varepsilon\left(D_0^2 u_{11} + \frac{3}{2}D_0^2 u_{21} + \frac{3}{2}\omega_{20}^2 u_{21}\right) + \varepsilon^3\left(D_0^2 u_{13} + \frac{3}{2}D_0^2 u_{23} + \frac{3}{2}\omega_{20}^2 u_{23}\right. \\
&\quad \left. + 2D_0 D_2 u_{11} + 3D_0 D_2 u_{21} - \frac{1}{4}\omega_{20}^2 u_{21}^3 - \frac{1}{2}u_{21}^2 D_0^2 u_{11}\right) + \cdots
\end{aligned}
\tag{6.11.26}
$$

We present the answer to only one of 12 cases, (iv), below. Readers are invited to complete the solutions for the rest of cases.

(iv) $\Omega_1 \approx \omega_1$ and $\omega_2 \approx 3\omega_1$.

In this case, $K_1 \cos \Omega_1 T_0$ is a resonant excitation term, which cannot appear in the equations of order ε^1, and the other two excitation terms are non-resonant hard excitations, which should appear in the equations of order ε^1. Thus, let

$$K_1 = 2\varepsilon^3\hat{K}_1, \quad K_2 = 2\varepsilon\hat{K}_2, \quad K_3 = 2\varepsilon\hat{K}_3 \tag{6.11.27}$$

Equating coefficients of like powers of ε in (6.11.25) and (6.11.26) yields.
Order ε^1:

$$\begin{gathered}
D_0^2u_{11} + mD_0^2u_{21} + \omega_{10}^2u_{11} = 2\sum_{n=2}^{3}\hat{K}_n\cos(\Omega_nT_0 + \theta_n) \\
D_0^2u_{11} + \frac{3}{2}D_0^2u_{21} + \frac{3}{2}\omega_{20}^2u_{21} = 0
\end{gathered} \tag{6.11.28}$$

Order ε^3:

$$\begin{aligned}
&D_0^2u_{13} + mD_0^2u_{23} + \omega_{10}^2u_{13} = -2D_0D_2u_{11} - 2mD_0D_2u_{21} \\
&-2\omega_{10}\hat{\mu}D_0u_{11} + \frac{1}{2}mu_{21}^2D_0^2u_{21} \\
&+mu_{21}(D_0u_{21})^2 + 2\hat{K}_1\cos(\Omega_1T_0 + \theta_1) \\
&D_0^2u_{13} + \frac{3}{2}D_0^2u_{23} + \frac{3}{2}\omega_{20}^2u_{23} = -2D_0D_2u_{11} - 3D_0D_2u_{21} + \frac{1}{4}\omega_{20}^2u_{21}^3 + \frac{1}{2}u_{21}^2D_0^2u_{11}
\end{aligned} \tag{6.11.29}$$

The corresponding homogeneous equations to (6.11.28) are the same as (6.11.11), so its solution can be expressed as

$$\begin{aligned}
u_{11} &= A_1e^{i\omega_1T_0} + A_2e^{i\omega_2T_0} + \sum_{n=2}^{3}B_{1n}e^{i(\Omega_nT_0+\theta_n)} + cc \\
u_{21} &= h_1A_1e^{i\omega_1T_0} + h_2A_2e^{i\omega_2T_0} + \sum_{n=2}^{3}B_{2n}e^{i(\Omega_nT_0+\theta_n)} + cc
\end{aligned} \tag{6.11.30}$$

where $A_k = A_k(T_2), \quad k = 1, 2,$ and

$$\left.\begin{aligned}
B_{1n} &= \frac{3(\omega_{20}^2-\Omega_n^2)\hat{K}_n}{3(\omega_{10}^2-\Omega_n^2)(\omega_{20}^2-\Omega_n^2)-2m\Omega_n^2\Omega_n^2} \\
B_{2n} &= \frac{2\Omega_n^2\hat{K}_n}{3(\omega_{10}^2-\Omega_n^2)(\omega_{20}^2-\Omega_n^2)-2m\Omega_n^2\Omega_n^2}
\end{aligned}\right\}; \quad n = 2, 3 \tag{6.11.31}$$

Introduce the detuning parameters σ_1, σ_2 such that

$$\Omega_1 = \omega_1 + \varepsilon^2\sigma_1, \quad \omega_2 = 3\omega_1 + \varepsilon^2\sigma_2 \tag{6.11.32}$$

Substituting (6.11.32) and (6.11.30) into (6.11.29) yields

$$
\begin{aligned}
D_0^2u_{13} + mD_0^2u_{23} + \omega_{10}^2u_{13} = \Big[&-2i\omega_1(1+mh_1)A_1' - 2i\omega_{10}\omega_1\hat{\mu}A_1 \\
&-m\omega_1^2h_1\big(B_{22}^2+B_{23}^2\big)A_1 - \frac{1}{2}m\omega_1^2h_1^3A_1^2\overline{A}_1 - m\omega_1^2h_1h_2^2A_1A_2\overline{A}_2 \\
&+\left(2m\omega_1\omega_2h_1^2h_2 - 2m\omega_1^2h_1^2h_2 - \frac{1}{2}m\omega_2^2h_1^2h_2\right)\overline{A}_1^2A_2e^{i\sigma_2T_2} \\
&+\hat{K}_1e^{i(\sigma_1T_2+\theta_1)}\Big]e^{i\omega_1T_0} \\
&+\big[-2i\omega_2(1+mh_2)A_2' - 2i\omega_{10}\omega_2\hat{\mu}A_2 \\
&-m\omega_2^2h_2\big(B_{22}^2+B_{23}^2\big)A_2 - \frac{1}{2}m\omega_2^2h_2^3A_2^2\overline{A}_2 - m\omega_2^2h_1^2h_2A_1\overline{A}_1A_2 \\
-\frac{3}{2}m\omega_1^2h_1^3A_1^3e^{-i\sigma_2T_2}\Big]&e^{i\omega_2T_0} + cc + NST
\end{aligned}
\tag{6.11.33}
$$

$$
\begin{aligned}
D_0^2u_{13} + \frac{3}{2}D_0^2u_{23} + \frac{3}{2}\omega_{20}^2u_{23} = \Big\{&-i\omega_1(2+3h_1)A_1' \\
&+\left[\left(\frac{3}{2}\omega_{20}^2h_1 - \omega_1^2\right)\big(B_{22}^2+B_{23}^2\big) - 2h_1\big(\Omega_2^2B_{22}^2+\Omega_3^2B_{23}^2\big)\right]A_1 \\
&+\left(\frac{3}{4}\omega_{20}^2h_1^3 - \frac{3}{2}\omega_1^2h_1^2\right)A_1^2\overline{A}_1 + \left(\frac{3}{2}\omega_{20}^2h_1h_2^2 - \omega_1^2h_2^2 - 2\omega_2^2h_1h_2\right)A_1A_2\overline{A}_2 \\
&+\left(\frac{3}{4}\omega_{20}^2h_1^2h_2 - \omega_1^2h_1h_2 - \frac{1}{2}\omega_2^2h_1^2\right)\overline{A}_1^2A_2e^{i\sigma_2T_2}\Big\}e^{i\omega_1T_0} \\
&+\big\{-i\omega_1(2+3h_1)A_2' \\
&+\left[\left(\frac{3}{2}\omega_{20}^2h_2 - \omega_2^2\right)\big(B_{22}^2+B_{23}^2\big) - 2h_2\big(\Omega_2^2B_{22}^2+\Omega_3^2B_{23}^2\big)\right]A_2 \\
&+\left(\frac{3}{4}\omega_{20}^2h_2^3h - \frac{3}{2}\omega_2^2h_2^2\right)A_2^2\overline{A}_2 + \left(\frac{3}{2}\omega_{20}^2h_1^2h_2 - \omega_2^2h_1^2 - 2\omega_1^2h_1h_2\right)A_1\overline{A}_1A_2 \\
+\left(\frac{1}{4}\omega_{20}^2h_1^3 - \frac{1}{2}\omega_1^2h_1^2\right)&A_1^3e^{-i\sigma_2T_2}\Big\}e^{i\omega_2T_0} + cc + NST
\end{aligned}
\tag{6.11.34}
$$

Primes denote the derivative with respect to T_2.

In order to eliminate secular terms of Eqs. (6.11.33) and (6.11.34), we seek the following special solutions for them:

$$
\begin{aligned}
u_{13} &= P_{11}e^{i\omega_1T_0} + P_{12}e^{i\omega_2T_0} \\
u_{23} &= P_{21}e^{i\omega_1T_0} + P_{22}e^{i\omega_2T_0}
\end{aligned}
\tag{6.11.35}
$$

Substituting (6.11.35) into (6.11.33) and (6.11.34), equating the coefficients of $e^{i\omega_1T_0}$ and $e^{i\omega_2T_0}$, we get

$$
\begin{bmatrix} \left(\omega_{10}^2-\omega_n^2\right) & -m\omega_n^2 \\ -\omega_n^2 & \frac{3}{2}\left(\omega_{20}^2-\omega_n^2\right) \end{bmatrix}\begin{Bmatrix} P_{1n} \\ P_{2n} \end{Bmatrix} = \begin{Bmatrix} R_{1n} \\ R_{2n} \end{Bmatrix},\quad n=1,\ 2 \tag{6.11.36}
$$

where

$$
\begin{aligned}
R_{11} &= -2i\omega_1(1+mh_1)A_1' - 2i\omega_{10}\omega_1\hat{\mu}A_1 \\
&-m\omega_1^2h_1\left(B_{22}^2+B_{23}^2\right)A_1 - \frac{1}{2}m\omega_1^2h_1^3A_1^2\overline{A}_1 - m\omega_1^2h_1h_2^2A_1A_2\overline{A}_2 \\
&+\left(2m\omega_1\omega_2h_1^2h_2 - 2m\omega_1^2h_1^2h_2 - \frac{1}{2}m\omega_2^2h_1^2h_2\right)\overline{A}_1^2A_2e^{i\sigma_2T_2} \\
&+\hat{K}_1e^{i(\sigma_1T_2+\theta_1)} \\
R_{21} &= -i\omega_1(2+3h_1)A_1' \\
&+\left[\left(\frac{3}{2}\omega_{20}^2h_1-\omega_1^2\right)\left(B_{22}^2+B_{23}^2\right) - 2h_1\left(\Omega_2^2B_{22}^2+\Omega_3^2B_{23}^2\right)\right]A_1 \\
&+\left(\frac{3}{4}\omega_{20}^2h_1^3 - \frac{3}{2}\omega_1^2h_1^2\right)A_1^2\overline{A}_1 + \left(\frac{3}{2}\omega_{20}^2h_1h_2^2 - \omega_1^2h_2^2 - 2\omega_2^2h_1h_2\right)A_1A_2\overline{A}_2 \\
&+\left(\frac{3}{4}\omega_{20}^2h_1^2h_2 - \omega_1^2h_1h_2 - \frac{1}{2}\omega_2^2h_1^2\right)\overline{A}_1^2A_2e^{i\sigma_2T_2}
\end{aligned}
\tag{6.11.37}
$$

$$
\begin{aligned}
R_{12} &= -2i\omega_2(1+mh_2)A_2' - 2i\omega_{10}\omega_2\hat{\mu}A_2 \\
&-m\omega_2^2h_2\left(B_{22}^2+B_{23}^2\right)A_2 - \frac{1}{2}m\omega_2^2h_2^3A_2^2\overline{A}_2 - m\omega_2^2h_1^2h_2A_1\overline{A}_1A_2 \\
&-\frac{3}{2}m\omega_1^2h_1^3A_1^3e^{-i\sigma_2T_2} \\
R_{22} &= -i\omega_1(2+3h_1)A_2' \\
&+\left[\left(\frac{3}{2}\omega_{20}^2h_2-\omega_2^2\right)\left(B_{22}^2+B_{23}^2\right) - 2h_2\left(\Omega_2^2B_{22}^2+\Omega_3^2B_{23}^2\right)\right]A_2 \\
&+\left(\frac{3}{4}\omega_{20}^2h_2^3h - \frac{3}{2}\omega_2^2h_2^2\right)A_2^2\overline{A}_2 + \left(\frac{3}{2}\omega_{20}^2h_1^2h_2 - \omega_2^2h_1^2 - 2\omega_1^2h_1h_2\right)A_1\overline{A}_1A_2 \\
&+\left(\frac{1}{4}\omega_{20}^2h_1^3 - \frac{1}{2}\omega_1^2h_1^2\right)A_1^3e^{-i\sigma_2T_2}
\end{aligned}
\tag{6.11.38}
$$

According to Eq. (6.11.13), the determinant of the coefficient matrix of Eq. (6.11.36) is zero:

$$
\det\begin{bmatrix} R_{1n} & -m\omega_n^2 \\ R_{2n} & \frac{3}{2}\left(\omega_{20}^2-\omega_n^2\right) \end{bmatrix} = 0
$$

i.e.,

$$
3\left(\omega_n^2-\omega_{20}^2\right)R_{1n} = 2\omega_n^2R_{2n} \tag{6.11.39}
$$

Substitute (6.11.37) and (6.1.20), respectively, into (6.11.39), we obtain

$$\begin{aligned}2iA_1' &= \alpha_{11}A_1 + i\alpha_{12}A_1 + \alpha_{13}A_1A_2\overline{A}_2 + \alpha_{14}A_1^2\overline{A}_1 \\ &+\alpha_{15}\overline{A}_1^2A_2e^{i\sigma_2T_2} + \alpha_{16}e^{i(\sigma_1T_2+\theta_1)}\end{aligned} \tag{6.11.40}$$

$$2iA_2' = \alpha_{21}A_2 + i\alpha_{22}A_2 + \alpha_{23}A_1\overline{A}_1A_2 + \alpha_{24}A_2^2\overline{A}_2 + \alpha_{25}A_1^3e^{-i\sigma_2T_2} \tag{6.11.41}$$

where

$$\begin{aligned}
&\Delta_1 = \omega_1\left[\omega_1^2(2+3h_1) - 3\left(\omega_1^2 - \omega_{20}^2\right)(1+mh_1)\right] \\
&\alpha_{11} = -\Delta_1^{-1}\left\{2\omega_1^2\left[2h_1\left(\Omega_2^2B_{22}^2 + \Omega_3^2B_{23}^2\right) - \left(\frac{3}{2}\omega_{20}^2h_1 - \omega_1^2\right)\left(B_{22}^2 + B_{23}^2\right)\right]\right. \\
&\left.-3\left(\omega_1^2 - \omega_{20}^2\right)m\omega_1^2h_1\left(B_{22}^2 + B_{23}^2\right)\right\} \\
&\alpha_{12} = 6\Delta_1^{-1}\omega_{10}\omega_1\hat{\mu}\left(\omega_1^2 - \omega_{20}^2\right) \\
&\alpha_{13} = \Delta_1^{-1}\left[3\left(\omega_1^2 - \omega_{20}^2\right)m\omega_1^2h_1h_2^2 + 2\omega_1^2\left(\frac{3}{2}\omega_{20}^2h_1h_2^2 - \omega_1^2h_2^2 - 2\omega_2^2h_1h_2\right)\right] \\
&\alpha_{14} = \Delta_1^{-1}\left[\frac{3}{2}\left(\omega_1^2 - \omega_{20}^2\right)m\omega_1^2h_1^3 + 2\omega_1^2\left(\frac{3}{4}\omega_{20}^2h_1^3 - \frac{3}{2}\omega_1^2h_1^2\right)\right] \\
&\alpha_{15} = -\Delta_1^{-1}\left\{3\left(\omega_1^2 - \omega_{20}^2\right)\left(2m\omega_1\omega_2h_1^2h_2 - 2m\omega_1^2h_1^2h_2 - \frac{1}{2}m\omega_2^2h_1^2h_2\right)\right. \\
&\left.-2\omega_1^2\left(\frac{3}{4}\omega_{20}^2h_1^2h_2 - \omega_1^2h_1h_2 - \frac{1}{2}\omega_2^2h_1^2\right)\right\} \\
&\alpha_{16} = -3\Delta_1^{-1}\left(\omega_1^2 - \omega_{20}^2\right)\hat{K}_1e^{i(\sigma_1T_2+\theta_1)}
\end{aligned} \tag{6.11.42}$$

$$\begin{aligned}
&\Delta_2^{-1} = \omega_2\left[\omega_1\omega_2(2+3h_1) - 3\left(\omega_2^2 - \omega_{20}^2\right)(1+mh_2)\right] \\
&\alpha_{21} = -\Delta_2^{-1}\left\{2\omega_2^2\left[2h_2\left(\Omega_2^2B_{22}^2 + \Omega_3^2B_{23}^2\right) - \left(\frac{3}{2}\omega_{20}^2h_2 - \omega_2^2\right)\left(B_{22}^2 + B_{23}^2\right)\right]\right. \\
&\left.-3m\omega_2^2h_2\left(\omega_2^2 - \omega_{20}^2\right)\left(B_{22}^2 + B_{23}^2\right)\right\} \\
&\alpha_{22} = 6\Delta_2^{-1}\omega_{10}\omega_2\hat{\mu}\left(\omega_2^2 - \omega_{20}^2\right) \\
&\alpha_{23} = \Delta_2^{-1}\left[3\left(\omega_2^2 - \omega_{20}^2\right)m\omega_2^2h_1^2h_2 + 2\omega_2^2\left(\frac{3}{2}\omega_{20}^2h_1^2h_2 - \omega_2^2h_1^2 - 2\omega_1^2h_1h_2\right)\right] \\
&\alpha_{24} = \Delta_2^{-1}\left[\frac{3}{2}\left(\omega_2^2 - \omega_{20}^2\right)m\omega_2^2h_2^3A_2^2\overline{A}_2 + 2\omega_2^2\left(\frac{3}{4}\omega_{20}^2h_2^3h - \frac{3}{2}\omega_2^2h_2^2\right)\right] \\
&\alpha_{25} = \Delta_2^{-1}\left[3\left(\omega_2^2 - \omega_{20}^2\right)\frac{3}{2}m\omega_1^2h_1^3 + 2\omega_2^2\left(\frac{1}{4}\omega_{20}^2h_1^3 - \frac{1}{2}\omega_1^2h_1^2\right)\right]
\end{aligned} \tag{6.11.43}$$

Let

$$A_1 = \frac{1}{2} a_1 e^{i\beta_1}, \quad A_2 = \frac{1}{2} a_2 e^{i\beta_2} \tag{6.11.44}$$

Substituting (6.11.44) into (6.11.40) and (6.11.41), we get

$$ia_1' - a_1\beta_1' = \frac{1}{2}\alpha_{11}a_1 + \frac{1}{2}i\alpha_{12}a_1 + \frac{1}{8}\alpha_{13}a_1a_2^2 + \frac{1}{8}\alpha_{14}a_1^3 + \frac{1}{8}\alpha_{15}a_1^2a_2e^{i\gamma_1} + \alpha_{16}e^{i\gamma_2} \tag{6.11.45}$$

$$ia_2' - a_2\beta_2' = \frac{1}{2}\alpha_{21}a_2 + \frac{1}{2}i\alpha_{22}a_2 + \frac{1}{8}\alpha_{23}a_1^2a_2 + \frac{1}{8}\alpha_{24}a_2^3 + \frac{1}{8}\alpha_{25}a_1^3e^{-i\gamma_1} \tag{6.11.46}$$

where

$$\gamma_1 = \sigma_2 T_2 - 3\beta_1 + \beta_2, \quad \gamma_2 = \sigma_1 T_2 - \beta_1 + \theta_1 \tag{6.11.47}$$

Separating the real and imaginary parts of Eqs. (6.11.45) and (6.11.46), we get

$$-a_1\beta_1' = \frac{1}{2}\alpha_{11}a_1 + \frac{1}{8}\alpha_{13}a_1a_2^2 + \frac{1}{8}\alpha_{14}a_1^3 + \frac{1}{8}\alpha_{15}a_1^2a_2\cos\gamma_1 + \alpha_{16}\cos\gamma_2 \tag{6.11.48}$$

$$-a_2\beta_2' = \frac{1}{2}\alpha_{21}a_2 + \frac{1}{8}\alpha_{23}a_1^2a_2 + \frac{1}{8}\alpha_{24}a_2^3 + \frac{1}{8}\alpha_{25}a_1^3\cos\gamma_1 \tag{6.11.49}$$

$$a_1' = \frac{1}{2}\alpha_{12}a_1 + \frac{1}{8}\alpha_{15}a_1^2a_2\sin\gamma_1 + \alpha_{16}\sin\gamma_2 \tag{6.11.50}$$

$$a_2' = \frac{1}{2}\alpha_{22}a_2 - \frac{1}{8}\alpha_{25}a_1^3\sin\gamma_1 \tag{6.11.51}$$

Utilizing (6.11.47) and combining (6.11.48) and (6.11.49), we can obtain

$$\begin{aligned} a_1\gamma_2' &= \sigma_1a_1 + \tfrac{1}{2}\alpha_{11}a_1 + \tfrac{1}{8}\alpha_{13}a_1a_2^2 + \tfrac{1}{8}\alpha_{14}a_1^3 \\ &+ \tfrac{1}{8}\alpha_{15}a_1^2a_2\cos\gamma_1 + \alpha_{16}\cos\gamma_2 \end{aligned} \tag{6.11.52}$$

$$\begin{aligned} -a_2\gamma'_1 + 3a_2\gamma'_2 &= 3\sigma_1a_2 - \sigma_2a_2 + \tfrac{1}{2}\alpha_{21}a_2 + \tfrac{1}{8}\alpha_{23}a_1^2a_2 \\ &+ \tfrac{1}{8}\alpha_{24}a_2^3 + \tfrac{1}{8}\alpha_{25}a_1^3\cos\gamma_1 \end{aligned} \tag{6.11.53}$$

Let $a_1' = \gamma_1' = a_2' = \gamma_2' = 0$ in (6.11.50)–(6.11.53), we can obtain

$$F_1\sin\gamma_1 + \alpha_{16}\sin\gamma_2 = G_1 \tag{6.11.54}$$

$$F_2\sin\gamma_1 = G_2 \tag{6.11.55}$$

$$F_1\cos\gamma_1 + \alpha_{16}\cos\gamma_2 = G_3 - a_1\sigma_1 \tag{6.11.56}$$

$$F_2 \cos\gamma_1 = G_4 - 3\sigma_1 a_2 + \sigma_2 a_2 \tag{6.11.57}$$

where

$$\begin{aligned} F_1 &= \frac{1}{8}\alpha_{15}a_1^2 a_2, \quad F_2 = \frac{1}{8}\alpha_{25}a_1^3 \\ G_1 &= -\frac{1}{2}\alpha_{12}a_1, \quad G_2 = \frac{1}{2}\alpha_{22}a_2 \\ G_3 &= -\left(\frac{1}{2}\alpha_{11}a_1 + \frac{1}{8}\alpha_{13}a_1a_2^2 + \frac{1}{8}\alpha_{14}a_1^3\right) \\ G_4 &= -\left(\frac{1}{2}\alpha_{21}a_2 + \frac{1}{8}\alpha_{23}a_1^2a_2 + \frac{1}{8}\alpha_{24}a_2^3\right) \end{aligned} \tag{6.11.58}$$

Solve for $\sin\gamma_1$ and $\cos\gamma_1$ from (6.11.55) and (6.11.57) and substitute them into (6.11.54) and (6.11.56), we can obtain

$$\alpha_{16}\sin\gamma_2 = G_1 - F_1F_2^{-1}G_2 \tag{6.11.59}$$

$$\alpha_{16}\cos\gamma_2 = G_3 - \sigma_1 a_1 - F_1F_2^{-1}(G_4 - 3\sigma_1 a_2 + \sigma_2 a_2) \tag{6.11.60}$$

Eliminating γ_1 in (6.11.55) and (6.11.57), and γ_2 in (6.11.59) and (6.11.60), we can obtain the frequency–response equations

$$(G_4 - 3\sigma_1 a_2 + \sigma_2 a_2)^2 + G_2^2 = F_2^2 \tag{6.11.61}$$

$$\left[G_3 - \sigma_1 a_1 - F_1F_2^{-1}(G_4 - 3\sigma_1 a_2 + \sigma_2 a_2)\right]^2 + \left(G_1 - F_1F_2^{-1}G_2\right)^2 = \alpha_{16}^2 \tag{6.11.62}$$

Constant amplitudes a_1 and a_2 can be obtained from the above two equations, and the constant phases γ_1 and γ_2 can be obtained through (6.11.54) and (6.11.55).

Again, analyze the stability of the steady state solution. Let a_{10}, a_{20}, γ_{10} and γ_{20} be a set of steady state solution and make a small perturbation of them, i.e.,

$$\begin{aligned} a_1 &= a_{10} + \tilde{a}_1, \quad a_2 = a_{20} + \tilde{a}_2 \\ \gamma_1 &= \gamma_{10} + \tilde{\gamma}_1, \quad \gamma_2 = \gamma_{20} + \tilde{\gamma}_2 \end{aligned} \tag{6.11.63}$$

Substitute (6.11.63) into (6.11.50)–(6.11.53) and make the linearization, we can obtain

$$\begin{Bmatrix} \tilde{a}_1' \\ \tilde{a}_2' \\ \tilde{\gamma}_1' \\ \tilde{\gamma}_2' \end{Bmatrix} = \begin{bmatrix} l_{11} & l_{12} & l_{13} & l_{14} \\ l_{21} & l_{22} & l_{23} & l_{24} \\ l_{31} & l_{32} & l_{33} & l_{34} \\ l_{41} & l_{42} & l_{43} & l_{44} \end{bmatrix} \begin{Bmatrix} \tilde{a}_1 \\ \tilde{a}_2 \\ \tilde{\gamma}_1 \\ \tilde{\gamma}_2 \end{Bmatrix} \tag{6.11.64}$$

where

$$\begin{aligned} l_{11} &= \frac{1}{2}\alpha_{12} + \frac{1}{4}\alpha_{15}a_{10}a_{20}\sin\gamma_{10}, \quad l_{12} = \frac{1}{8}\alpha_{15}a_{10}^2\sin\gamma_{10} \\ l_{13} &= \frac{1}{8}\alpha_{15}a_{10}^2a_{20}\cos\gamma_{10}, \quad l_{14} = \alpha_{16}\sin\gamma_{20} \end{aligned} \tag{6.11.65}$$

$$\begin{aligned} l_{21} &= -\frac{3}{8}\alpha_{25}a_{10}^2\sin\gamma_{10}, \quad l_{22} = \frac{1}{2}\alpha_{22} \\ l_{23} &= -\frac{1}{8}\alpha_{25}a_{10}^3\cos\gamma_{10}, \quad l_{24} = 0 \end{aligned} \tag{6.11.66}$$

$$\begin{aligned} l_{31} &= a_{10}^{-1}\left(3\sigma_1 + \frac{3}{2}\alpha_{11}\tilde{a}_1 + \frac{3}{8}\alpha_{13}a_{20}^2 + \frac{9}{8}\alpha_{14}a_{10}^2 - \frac{1}{4}\alpha_{23}a_{10}^2\right. \\ &\left.+\frac{3}{4}\alpha_{15}a_{10}a_{20}\cos\gamma_{10} - \frac{3}{8}\alpha_{25}a_{10}^3a_{20}^{-1}\cos\gamma_{10}\right) \\ l_{32} &= a_{20}^{-1}\left(\sigma_2 - 3\sigma_1 - \frac{1}{2}\alpha_{21} + \frac{3}{4}\alpha_{13}a_{20}^2 - \frac{1}{8}\alpha_{23}a_{10}^2\right. \\ &\left.+\frac{3}{8}\alpha_{15}a_{10}a_{20}\cos\gamma_{10} - \frac{3}{8}\alpha_{24}a_{20}^2\right) \\ l_{33} &= a_{20}^{-1}\left(\frac{1}{8}\alpha_{25}a_{10}^3 - \frac{3}{8}\alpha_{15}a_{10}a_{20}^2\right)\sin\gamma_{10}, \quad l_{34} = 3\alpha_{16}a_{10}^{-1}\sin\gamma_{20} \end{aligned} \tag{6.11.67}$$

$$\begin{aligned} l_{41} &= a_{10}^{-1}\left(\sigma_1 + \frac{1}{2}\alpha_{11} + \frac{3}{8}\alpha_{14}a_{10}^2 + \frac{1}{8}\alpha_{13}a_{20}^2 + \frac{1}{4}\alpha_{15}a_{10}a_{20}\cos\gamma_{10}\right) \\ l_{42} &= \frac{1}{4}\alpha_{13}a_{20} + \frac{1}{8}\alpha_{15}a_{10}\cos\gamma_{10} \\ l_{43} &= -\frac{1}{8}\alpha_{15}a_{10}a_{20}\sin\gamma_{10}, \quad l_{44} = -\alpha_{16}a_{10}^{-1}\sin\gamma_{20} \end{aligned} \tag{6.11.68}$$

The steady state solutions are stable if the eigenvalues of (6.11.64) have negative real parts.

6.12 Exercise 6.12 (Resonance Analysis on the Cylinder Rolling Without Slip on the Circular Surface II)

Solution: (a) The kinetic energy of the system is

$$\begin{aligned} T &= \tfrac{1}{2}M\dot{y}^2 + \tfrac{1}{2}m_1\left\{\left[(R-r)\dot{\theta}\cos\theta\right]^2 + \left[\dot{y} + (R-r)\dot{\theta}\sin\theta\right]^2\right\} \\ &+\tfrac{1}{2}\tfrac{1}{2}m_1r^2\left[\tfrac{(R-r)\dot{\theta}}{r}\right]^2 \\ &= \tfrac{1}{2}(m_1+M)\dot{y}^2 + \tfrac{3}{4}m_1(R-r)^2\dot{\theta}^2 + m_1(R-r)\dot{y}\dot{\theta}\sin\theta \end{aligned} \tag{6.12.1}$$

Using the equilibrium position as the zero potential surface, we can write the potential energy of the system as

$$\begin{aligned} V = {} & \tfrac{1}{2}k\left[(y - y_0)^2 - (y_c - y_0)^2\right] + \tfrac{1}{2}m_1 g(R - r)(1 - \cos\theta) \\ & +(M + m_1)g(y - y_c) \end{aligned} \tag{6.12.2}$$

where y_0 is the coordinate of the block when the spring is not deformed and y_c is the coordinate of the block when the system is balanced. Therefore, we have

$$k(y_0 - y_c) = (M + m_1)g \tag{6.12.3}$$

The generalized forces are

$$Q_y = F(t), \quad Q_\theta = 0 \tag{6.12.4}$$

Substitute kinetic energy, potential energy, and generalized forces into the Lagrange's equation:

$$\frac{d}{dt}\frac{\partial T}{\partial \dot{y}} - \frac{\partial T}{\partial y} = -\frac{\partial V}{\partial y} + Q_y, \quad \frac{d}{dt}\frac{\partial T}{\partial \dot{\theta}} - \frac{\partial T}{\partial \theta} = -\frac{\partial V}{\partial \theta} + Q_\theta$$

and take (6.12.3) into account, we can obtain

$$\begin{aligned} & (m_1 + M)\ddot{y} + m_1(R - r)\ddot{\theta}\sin\theta + m_1(R - r)\dot{\theta}^2\cos\theta + k(y - y_c) = F(t) \\ & \frac{3}{2}m_1(R - r)^2\ddot{\theta} + m_1(R - r)\ddot{y}\sin\theta + m_1 g(R - r)\sin\theta = 0 \end{aligned} \tag{6.12.5}$$

If we consider that there is viscous damping along y direction, let the drag coefficient be c, and add the drag term directly to the first equation of (6.12.5), we get

$$\begin{aligned} & (m_1 + M)\ddot{y} + m_1(R - r)\ddot{\theta}\sin\theta + m_1(R - r)\dot{\theta}^2\cos\theta + k(y - y_c) + c\dot{y} = F(t) \\ & \frac{3}{2}m_1(R - r)^2\ddot{\theta} + m_1(R - r)\ddot{y}\sin\theta + m_1 g(R - r)\sin\theta = 0 \end{aligned} \tag{6.12.6}$$

Let

$$\begin{aligned} & u = \frac{y - y_c}{R - r}, \quad m = \frac{m_1}{m_1 + M}, \quad \omega_{10}^2 = k/(m_1 + M) \\ & \omega_{20}^2 = 2g/3(R - r), \quad f(t) = F(t)/(m_1 + M)(R - r), \quad 2\omega_{10}\mu = c/(m_1 + M) \end{aligned} \tag{6.12.7}$$

Equation (6.11.4) changes into

$$\begin{aligned} &\ddot{u} + m\ddot{\theta}\sin\theta + m\dot{\theta}^2\cos\theta + \omega_{10}^2 u + 2\omega_{10}\mu\dot{u} = f(t) \\ &\ddot{\theta} + \frac{2}{3}\ddot{u}\sin\theta + \omega_{20}^2\sin\theta = 0 \end{aligned} \tag{6.12.8}$$

where μ is the dimensionless damping ratio.

(b) At the equilibrium position of the system $u = \theta = 0$, we can expand (6.12.8) and retain to third-order small quantities, we get

$$\begin{aligned} &\ddot{u} + m\ddot{\theta} + \omega_{10}^2 u + 2\omega_{10}\mu\dot{u} = \sum_{n=1}^{3} K_n\cos(\Omega_n t + \theta_n) \\ &\ddot{u} + \frac{3}{2}\ddot{\theta} + \frac{3}{2}\omega_{20}^2\theta - \frac{1}{4}\omega_{20}^2\theta^3 = 0 \end{aligned} \tag{6.12.9}$$

The corresponding linear undamped free oscillation equations are

$$\begin{bmatrix} 1 & m \\ 1 & \frac{3}{2} \end{bmatrix}\begin{Bmatrix} \ddot{u} \\ \ddot{\theta} \end{Bmatrix} + \begin{bmatrix} \omega_{10}^2 & 0 \\ 0 & \frac{3}{2}\omega_{20}^2 \end{bmatrix}\begin{Bmatrix} u \\ \theta \end{Bmatrix} = 0 \tag{6.12.10}$$

Let

$$u = B_1 e^{i\omega t}, \quad \theta = B_2 e^{i\omega t} \tag{6.12.11}$$

and substitute them into (6.12.10), we can obtain

$$\begin{bmatrix} \left(\omega_{10}^2 - \omega^2\right) & -m\omega^2 \\ -\omega^2 & \frac{3}{2}\left(\omega_{20}^2 - \omega^2\right) \end{bmatrix}\begin{Bmatrix} B_1 \\ B_2 \end{Bmatrix} = 0 \tag{6.12.12}$$

So the characteristic equation is

$$\det\begin{bmatrix} \left(\omega_{10}^2 - \omega^2\right) & -m\omega^2 \\ -\omega^2 & \frac{3}{2}\left(\omega_{20}^2 - \omega^2\right) \end{bmatrix} = 0 \tag{6.12.13}$$

i.e.,

$$(3 - 2m)\omega^4 - 3\left(\omega_{10}^2 + \omega_{20}^2\right)\omega^2 + 3\omega_{10}^2\omega_{20}^2 = 0 \tag{6.12.14}$$

Therefore, the linear natural frequencies are

$$\begin{aligned}\omega_1 &= \left[\frac{3\left(\omega_{10}^2+\omega_{20}^2\right)-\sqrt{9\left(\omega_{10}^2+\omega_{20}^2\right)^2-12(3-2m)\omega_{10}^2\omega_{20}^2}}{2(3-2m)}\right]^{1/2}\\ \omega_2 &= \left[\frac{3\left(\omega_{10}^2+\omega_{20}^2\right)+\sqrt{9\left(\omega_{10}^2+\omega_{20}^2\right)^2-12(3-2m)\omega_{10}^2\omega_{20}^2}}{2(3-2m)}\right]^{1/2}\end{aligned} \tag{6.12.15}$$

Substituting ω_1 and ω_2 into (6.12.12), we can obtain

$$\{B_1, B_2\}_1^T = \{1, h_1\}^T, \quad \{B_1, B_2\}_2^T = \{1, h_2\}^T \tag{6.12.16}$$

where h_1 and h_2 can be obtained from (6.12.12)

$$h_1 = \frac{\omega_{10}^2 - \omega_1^2}{m\omega_1^2}, \quad h_2 = \frac{\omega_{10}^2 - \omega_2^2}{m\omega_2^2} \tag{6.12.17}$$

Therefore, the solution of the linear free oscillation (6.12.10) can be expressed as

$$\begin{aligned} u &= A_1 e^{i\omega_1 t} + A_2 e^{i\omega_2 t} + cc \\ \theta &= h_1 A_1 e^{i\omega_1 t} + h_2 A_2 e^{i\omega_2 t} + cc \end{aligned} \tag{6.12.18}$$

In order to study the internal resonance of $\omega_2 \approx 3\omega_1$, let.

$$s = \omega_2/\omega_1, \quad p = \omega_{20}/\omega_{10} \tag{6.12.19}$$

From (6.12.15), we can obtain

$$s = \left[\frac{1+\sqrt{1-q}}{1-\sqrt{1-q}}\right]^{1/2} \tag{6.12.20}$$

where

$$q = \frac{4(3-2m)p^2}{3\left(p^2+1\right)^2} \le 1, \quad 0 < m < 1 \tag{6.12.21}$$

For different m, the curves of the frequency ratio p versus s are shown in Fig. 6.8. The internal resonance case of $\omega_2 \approx 3\omega_1$ may occur when $m = 0.1$ to 0.9.

We note that the Eq. (6.12.10) is identical to the equation of the previous exercise, (6.11.11), so the free oscillation solutions and the results for the natural frequencies are identical to those in the previous question.

(c) For small and finite amplitudes, we use the method of multiple scales to solve (6.12.9). Therefore, let the solution of Eq. (6.12.9) be expressed as

Fig. 6.8 Frequency ratio of p to s for Exercise 6.12

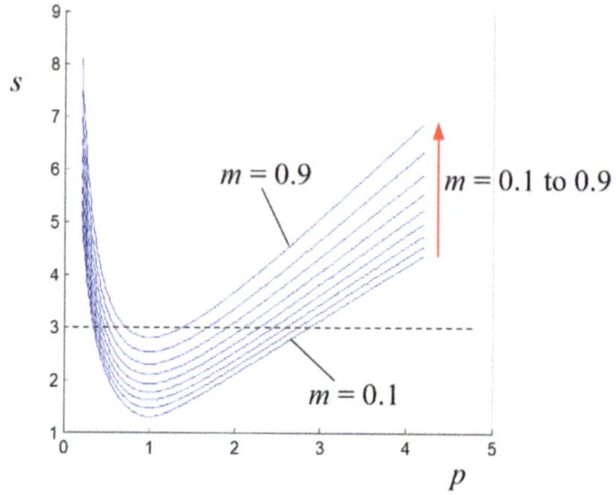

$$
\begin{aligned}
u &= \varepsilon u_{11}(T_0, T_2) + \varepsilon^3 u_{13}(T_0, T_2) + \cdots \\
\theta &= \varepsilon u_{21}(T_0, T_2) + \varepsilon^3 u_{23}(T_0, T_2) + \cdots
\end{aligned}
\tag{6.12.22}
$$

In order to let the damping and nonlinear terms appear in the same order equation, we specify that

$$
\mu = \varepsilon^2 \hat{\mu} \tag{6.12.23}
$$

Substitute (6.12.22) into (6.12.9) and keep to $O(\varepsilon^3)$, we get

$$
\begin{aligned}
0 &= \ddot{u} + m\ddot{\theta} + \omega_{10}^2 u + 2\omega_{10}\mu\dot{u} - \sum_{n=1}^{3} K_n \cos(\Omega_n t + \theta_n) \\
&= \left[D_0^2 + 2\varepsilon D_0 D_1 + \varepsilon^2\left(D_1^2 + 2D_0 D_2\right)\right]\left(\varepsilon u_{11} + \varepsilon^3 u_{13}\right) \\
&\quad + m\left[D_0^2 + 2\varepsilon D_0 D_1 + \varepsilon^2\left(D_1^2 + 2D_0 D_2\right)\right]\left(\varepsilon u_{21} + \varepsilon^3 u_{23}\right) \\
&\quad + \omega_{10}^2\left(\varepsilon u_{11} + \varepsilon^3 u_{13}\right) + 2\omega_{10}\varepsilon^2\hat{\mu}\left(D_0 + \varepsilon D_1 + \varepsilon^2 D_2\right)\left(\varepsilon u_{11} + \varepsilon^3 u_{13}\right) \\
&\quad - \sum_{n=1}^{3} K_n \cos(\Omega_n T_0 + \theta_n) + \cdots \\
&= \varepsilon\left(D_0^2 u_{11} + m D_0^2 u_{21} + \omega_{10}^2 u_{11}\right) + \varepsilon^3\left[D_0^2 u_{13} + m D_0^2 u_{23} + \omega_{10}^2 u_{13}\right. \\
&\quad \left. + 2D_0 D_2 u_{11} + 2m D_0 D_2 u_{21} + 2\omega_{10}\hat{\mu} D_0 u_{11}\right] - \sum_{n=1}^{3} K_n \cos(\Omega_n T_0 + \theta_n) + \cdots
\end{aligned}
\tag{6.12.24}
$$

$$
\begin{aligned}
0 &= \ddot{u} + \frac{3}{2}\ddot{\theta} + \frac{3}{2}\omega_{20}^2\theta - \frac{1}{4}\omega_{20}^2\theta^3 \\
&= \left[D_0^2 + 2\varepsilon D_0 D_1 + \varepsilon^2\left(D_1^2 + 2D_0 D_2\right)\right]\left(\varepsilon u_{11} + \varepsilon^3 u_{13}\right) \\
&+ \frac{3}{2}\left[D_0^2 + 2\varepsilon D_0 D_1 + \varepsilon^2\left(D_1^2 + 2D_0 D_2\right)\right]\left(\varepsilon u_{21} + \varepsilon^3 u_{23}\right) \\
&+ \frac{3}{2}\omega_{20}^2\left(\varepsilon u_{21} + \varepsilon^3 u_{23}\right) - \frac{1}{4}\omega_{20}^2\left(\varepsilon u_{21} + \varepsilon^3 u_{23}\right)^3 + \cdots \\
&= \varepsilon\left(D_0^2 u_{11} + \frac{3}{2}D_0^2 u_{21} + \frac{3}{2}\omega_{20}^2 u_{21}\right) + \varepsilon^3\left(D_0^2 u_{13} + \frac{3}{2}D_0^2 u_{23} + \frac{3}{2}\omega_{20}^2 u_{23}\right. \\
&\quad \left. 2D_0 D_2 u_{11} + 3D_0 D_2 u_{21} - \frac{1}{4}\omega_{20}^2 u_{21}^3\right) + \cdots
\end{aligned}
\tag{6.12.25}
$$

We present the answer to only one of 12 cases, (iv), below. Readers are invited to complete the solutions for the rest of cases.

(iv) $\Omega_1 \approx \omega_1$ and $\omega_2 \approx 3\omega_1$.

In this case, $K_1 \cos \Omega_1 T_0$ is a resonant excitation term, which cannot appear in the equations of order ε^1, and the other two excitation terms are non-resonant hard excitations, which should appear in the equations of order ε^1. Thus, let

$$
K_1 = 2\varepsilon^3 \hat{K}_1, \quad K_2 = 2\varepsilon \hat{K}_2, \quad K_3 = 2\varepsilon \hat{K}_3 \tag{6.12.26}
$$

Equating coefficients of like powers of ε in (6.12.24) and (6.12.25) yields.

Order ε^1:

$$
\begin{aligned}
&D_0^2 u_{11} + mD_0^2 u_{21} + \omega_{10}^2 u_{11} = 2\sum_{n=2}^{3} \hat{K}_n \cos(\Omega_n T_0 + \theta_n) \\
&D_0^2 u_{11} + \tfrac{3}{2}D_0^2 u_{21} + \tfrac{3}{2}\omega_{20}^2 u_{21} = 0
\end{aligned}
\tag{6.12.27}
$$

Order ε^3:

$$
\begin{aligned}
&D_0^2 u_{13} + mD_0^2 u_{23} + \omega_{10}^2 u_{13} = -2D_0 D_2 u_{11} - 2mD_0 D_2 u_{21} \\
&-2\omega_{10}\hat{\mu} D_0 u_{11} + 2\hat{K}_1 \cos(\Omega_1 T_0 + \theta_1) \\
&D_0^2 u_{13} + \tfrac{3}{2}D_0^2 u_{23} + \tfrac{3}{2}\omega_{20}^2 u_{23} = -2D_0 D_2 u_{11} - 3D_0 D_2 u_{21} + \tfrac{1}{4}\omega_{20}^2 u_{21}^3
\end{aligned}
\tag{6.12.28}
$$

The corresponding homogeneous equations to (6.12.27) are the same as (6.12.10), so its solution can be expressed as

$$
\begin{aligned}
u_{11} &= A_1 e^{i\omega_1 T_0} + A_2 e^{i\omega_2 T_0} + \sum_{n=2}^{3} B_{1n} e^{i(\Omega_n T_0 + \theta_n)} + cc \\
u_{21} &= h_1 A_1 e^{i\omega_1 T_0} + h_2 A_2 e^{i\omega_2 T_0} + \sum_{n=2}^{3} B_{2n} e^{i(\Omega_n T_0 + \theta_n)} + cc
\end{aligned}
\tag{6.12.29}
$$

where $A_k = A_k(T_2), \quad k = 1, 2$, and

$$B_{1n} = \frac{3(\omega_{20}^2 - \Omega_n^2)\hat{K}_n}{3(\omega_{10}^2 - \Omega_n^2)(\omega_{20}^2 - \Omega_n^2) - 2m\Omega_n^2\Omega_n^2}; \quad n = 2, 3$$
$$B_{2n} = \frac{2\Omega_n^2\hat{K}_n}{3(\omega_{10}^2 - \Omega_n^2)(\omega_{20}^2 - \Omega_n^2) - 2m\Omega_n^2\Omega_n^2} \tag{6.12.30}$$

Introduce the detuning parameters σ_1, σ_2 such that

$$\Omega_1 = \omega_1 + \varepsilon^2\sigma_1, \quad \omega_2 = 3\omega_1 + \varepsilon^2\sigma_2 \tag{6.12.31}$$

Substituting (6.12.31) and (6.12.29) into (6.12.28) yields

$$\begin{aligned} &D_0^2 u_{13} + mD_0^2 u_{23} + \omega_{10}^2 u_{13} = \left[-2i\omega_1(1 + mh_1)A_1' - 2i\omega_{10}\omega_1\hat{\mu}A_1\right. \\ &\left.+\hat{K}_1 e^{i(\sigma_1 T_2 + \theta_1)}\right]e^{i\omega_1 T_0} \\ &+\left[-2i\omega_2(1 + mh_2)A_2' - 2i\omega_{10}\omega_2\hat{\mu}A_2\right]e^{i\omega_2 T_0} \\ &+cc + NST \end{aligned} \tag{6.12.32}$$

$$\begin{aligned} D_0^2 u_{13} + \frac{3}{2}D_0^2 u_{23} + \frac{3}{2}\omega_{20}^2 u_{23} = \Big\{&-i\omega_1(2 + 3h_1)A_1' \\ &+\frac{3}{2}\omega_{20}^2 h_1\left(B_{22}^2 + B_{23}^2\right)A_1 + \frac{3}{4}\omega_{20}^2 h_1^3 A_1^2\overline{A}_1 + \frac{3}{2}\omega_{20}^2 h_1 h_2^2 A_1 A_2\overline{A} \\ &+\frac{3}{4}\omega_{20}^2 h_1^2 h_2 \overline{A}_1^2 A_2 e^{i\sigma_2 T_2}\Big\}e^{i\omega_1 T_0} \\ &+\Big\{-i\omega_1(2 + 3h_1)A_2' + \frac{3}{2}\omega_{20}^2 h_2\left(B_{22}^2 + B_{23}^2\right)A_2 \\ &+\frac{3}{4}\omega_{20}^2 h_2^3 h A_2^2\overline{A}_2 + \frac{3}{2}\omega_{20}^2 h_1^2 h_2 A_1\overline{A}_1 A_2 \\ &+\frac{1}{4}\omega_{20}^2 h_1^3 A_1^3 e^{-i\sigma_2 T_2}\Big\}e^{i\omega_2 T_0} + cc + NST \end{aligned} \tag{6.12.33}$$

Primes denote the derivative with respect to T_2.

In order to eliminate secular terms of Eqs. (6.12.32) and (6.12.33), we seek the following special solutions for them:

$$\begin{aligned} u_{13} &= P_{11}e^{i\omega_1 T_0} + P_{12}e^{i\omega_2 T_0} \\ u_{23} &= P_{21}e^{i\omega_1 T_0} + P_{22}e^{i\omega_2 T_0} \end{aligned} \tag{6.12.34}$$

Substituting (6.12.34) into (6.12.32) and (6.12.33), equating the coefficients of $e^{i\omega_1 T_0}$ and $e^{i\omega_2 T_0}$, we get

$$\begin{bmatrix} (\omega_{10}^2 - \omega_n^2) & -m\omega_n^2 \\ -\omega_n^2 & \frac{3}{2}(\omega_{20}^2 - \omega_n^2) \end{bmatrix}\begin{Bmatrix} P_{1n} \\ P_{2n} \end{Bmatrix} = \begin{Bmatrix} R_{1n} \\ R_{2n} \end{Bmatrix}, \quad n = 1, \ 2 \tag{6.12.35}$$

where

$$
\begin{aligned}
R_{11} &= -2i\omega_1(1+mh_1)A_1' - 2i\omega_{10}\omega_1\hat{\mu}A_1 + \hat{K}_1 e^{i(\sigma_1 T_2+\theta_1)} \\
R_{21} &= -i\omega_1(2+3h_1)A_1' + \frac{3}{2}\omega_{20}^2 h_1\left(B_{22}^2+B_{23}^2\right)A_1 + \frac{3}{4}\omega_{20}^2 h_1^3 A_1^2\overline{A}_1 \\
&+\frac{3}{2}\omega_{20}^2 h_1 h_2^2 A_1 A_2\overline{A}_2 + \frac{3}{4}\omega_{20}^2 h_1^2 h_2 \overline{A}_1^2 A_2 e^{i\sigma_2 T_2}
\end{aligned}
\tag{6.12.36}
$$

$$
\begin{aligned}
R_{12} &= -2i\omega_2(1+mh_2)A_2' - 2i\omega_{10}\omega_2\mu A_2 \\
R_{22} &= -i\omega_1(2+3h_1)A_2' + \frac{3}{2}\omega_{20}^2 h_2\left(B_{22}^2+B_{23}^2\right)A_2 \\
&+\frac{3}{4}\omega_{20}^2 h_2^3 h A_2^2\overline{A}_2 + \frac{3}{2}\omega_{20}^2 h_1^2 h_2 A_1\overline{A}_1 A_2 + \frac{1}{4}\omega_{20}^2 h_1^3 A_1^3 e^{-i\sigma_2 T_2}
\end{aligned}
\tag{6.12.37}
$$

According to the Eq. (6.12.13), the determinant of the coefficient matrix of Eq. (6.12.35) is zero:

$$
\det\begin{bmatrix} R_{1n} & -m\omega_n^2 \\ R_{2n} & \frac{3}{2}\left(\omega_{20}^2-\omega_n^2\right) \end{bmatrix} = 0
$$

i.e.,

$$
3\left(\omega_n^2-\omega_{20}^2\right)R_{1n} = 2\omega_n^2 R_{2n} \tag{6.12.38}
$$

Substitute (6.12.36) and (6.12.37), respectively, into (6.12.38), we obtain

$$
\begin{aligned}
2iA_1' &= \alpha_{11}A_1 + i\alpha_{12}A_1 + \alpha_{13}A_1A_2\overline{A}_2 + \alpha_{14}A_1^2\overline{A}_1 \\
&+\alpha_{15}\overline{A}_1^2 A_2 e^{i\sigma_2 T_2} + \alpha_{16}e^{i(\sigma_1 T_2+\theta_1)}
\end{aligned}
\tag{6.12.39}
$$

$$
2iA_2' = \alpha_{21}A_2 + i\alpha_{22}A_2 + \alpha_{23}A_1\overline{A}_1A_2 + \alpha_{24}A_2^2\overline{A}_2 + \alpha_{25}A_1^3 e^{-i\sigma_2 T_2} \tag{6.12.40}
$$

where

$$
\begin{aligned}
&\Delta_1 = \omega_1\left[\omega_1^2(2+3h_1) - 3\left(\omega_1^2-\omega_{20}^2\right)(1+mh_1)\right] \\
&\alpha_{11} = 3\Delta_1^{-1}\omega_1^2\omega_{20}^2 h_1\left(B_{22}^2+B_{23}^2\right), \quad \alpha_{12} = 6\Delta_1^{-1}\omega_{10}\omega_1\left(\omega_1^2-\omega_{20}^2\right) \\
&\alpha_{13} = 3\Delta_1^{-1}\omega_1^2\omega_{20}^2 h_1 h_2^2, \quad \alpha_{14} = \frac{3}{2}\Delta_1^{-1}\omega_1^2\omega_{20}^2 h_1^3 \\
&\alpha_{15} = -3\Delta_1^{-1}\left(\omega_1^2-\omega_{20}^2\right)\hat{K}_1, \quad \alpha_{16} = \frac{3}{2}\Delta_1^{-1}\omega_1^2\omega_{20}^2 h_1^2 h_2
\end{aligned}
\tag{6.12.41}
$$

$$\begin{aligned}
&\Delta_2 = \omega_2\left[\omega_1\omega_2(2+3h_1) - 3\left(\omega_2^2 - \omega_{20}^2\right)(1+mh_2)\right] \\
&\alpha_{21} = 3\Delta_2^{-1}\omega_2^2\omega_{20}^2 h_2\left(B_{22}^2 + B_{23}^2\right), \quad \alpha_{22} = 6\Delta_2^{-1}\omega_{10}\omega_2\left(\omega_2^2 - \omega_{20}^2\right)\hat{\mu} \\
&\alpha_{23} = 3\Delta_2^{-1}\omega_2^2\omega_{20}^2 h_1^2 h_2, \quad \alpha_{24} = \frac{3}{2}\Delta_2^{-1}\omega_2^2\omega_{20}^2 h_2^3, \quad \alpha_{25} = \frac{1}{2}\Delta_2^{-1}\omega_2^2\omega_{20}^2 h_1^3
\end{aligned} \tag{6.12.42}$$

Let

$$A_1 = \frac{1}{2}a_1 e^{i\beta_1}, \quad A_2 = \frac{1}{2}a_2 e^{i\beta_2} \tag{6.12.43}$$

Substituting them into (6.12.39) and (6.12.40), we get

$$ia_1' - a_1\beta_1' = \frac{1}{2}\alpha_{11}a_1 + \frac{1}{2}i\alpha_{12}a_1 + \frac{1}{8}\alpha_{13}a_1a_2^2 + \frac{1}{8}\alpha_{14}a_1^3 + \frac{1}{8}\alpha_{15}a_1^2a_2e^{i\gamma_1} + \alpha_{16}e^{i\gamma_2} \tag{6.12.44}$$

$$ia_2' - a_2\beta_2' = \frac{1}{2}\alpha_{21}a_2 + \frac{1}{2}i\alpha_{22}a_2 + \frac{1}{8}\alpha_{23}a_1^2a_2 + \frac{1}{8}\alpha_{24}a_2^3 + \frac{1}{8}\alpha_{25}a_1^3e^{-i\gamma_1} \tag{6.12.45}$$

where

$$\gamma_1 = \sigma_2T_2 - 3\beta_1 + \beta_2, \quad \gamma_2 = \sigma_1T_2 - \beta_1 + \theta_1 \tag{6.12.46}$$

Separating the real and imaginary parts of the above equations yields

$$-a_1\beta_1' = \frac{1}{2}\alpha_{11}a_1 + \frac{1}{8}\alpha_{13}a_1a_2^2 + \frac{1}{8}\alpha_{14}a_1^3 + \frac{1}{8}\alpha_{15}a_1^2a_2\cos\gamma_1 + \alpha_{16}\cos\gamma_2 \tag{6.12.47}$$

$$-a_2\beta_2' = \frac{1}{2}\alpha_{21}a_2 + \frac{1}{8}\alpha_{23}a_1^2a_2 + \frac{1}{8}\alpha_{24}a_2^3 + \frac{1}{8}\alpha_{25}a_1^3\cos\gamma_1 \tag{6.12.48}$$

$$a_1' = \frac{1}{2}\alpha_{12}a_1 + \frac{1}{8}\alpha_{15}a_1^2a_2\sin\gamma_1 + \alpha_{16}\sin\gamma_2 \tag{6.12.49}$$

$$a_2' = \frac{1}{2}\alpha_{22}a_2 - \frac{1}{8}\alpha_{25}a_1^3\sin\gamma_1 \tag{6.12.50}$$

Utilizing (6.12.46) and combining (6.12.47) and (6.12.48), we can obtain

$$\begin{aligned}
&a_1\gamma_2' = \sigma_1a_1 + \tfrac{1}{2}\alpha_{11}a_1 + \tfrac{1}{8}\alpha_{13}a_1a_2^2 + \tfrac{1}{8}\alpha_{14}a_1^3 \\
&+ \tfrac{1}{8}\alpha_{15}a_1^2a_2\cos\gamma_1 + \alpha_{16}\cos\gamma_2
\end{aligned} \tag{6.12.51}$$

$$\begin{aligned}
&-a_2\gamma_1' + 3a_2\gamma_2' = 3\sigma_1a_2 - \sigma_2a_2 + \tfrac{1}{2}\alpha_{21}a_2 + \tfrac{1}{8}\alpha_{23}a_1^2a_2 \\
&+ \tfrac{1}{8}\alpha_{24}a_2^3 + \tfrac{1}{8}\alpha_{25}a_1^3\cos\gamma_1
\end{aligned} \tag{6.12.52}$$

Let $a_1' = \gamma_1' = a_2' = \gamma_2' = 0$ in (6.12.49)–(6.12.52), we can obtain

$$F_1 \sin\gamma_1 + \alpha_{16} \sin\gamma_2 = G_1 \tag{6.12.53}$$

$$F_2 \sin\gamma_1 = G_2 \tag{6.12.54}$$

$$F_1 \cos\gamma_1 + \alpha_{16} \cos\gamma_2 = G_3 - a_1\sigma_1 \tag{6.12.55}$$

$$F_2 \cos\gamma_1 = G_4 - 3\sigma_1 a_2 + \sigma_2 a_2 \tag{6.12.56}$$

where

$$\begin{aligned}
F_1 &= \frac{1}{8}\alpha_{15} a_1^2 a_2, \quad F_2 = \frac{1}{8}\alpha_{25} a_1^3 \\
G_1 &= -\frac{1}{2}\alpha_{12} a_1, \quad G_2 = \frac{1}{2}\alpha_{22} a_2 \\
G_3 &= -\left(\frac{1}{2}\alpha_{11} a_1 + \frac{1}{8}\alpha_{13} a_1 a_2^2 + \frac{1}{8}\alpha_{14} a_1^3\right) \\
G_4 &= -\left(\frac{1}{2}\alpha_{21} a_2 + \frac{1}{8}\alpha_{23} a_1^2 a_2 + \frac{1}{8}\alpha_{24} a_2^3\right)
\end{aligned} \tag{6.12.57}$$

Solve for *sin* γ_1 and *cos* γ_1 from (6.12.54) and (6.12.56) and substitute them into (6.12.53) and (6.12.55), we can obtain

$$\alpha_{16} \sin\gamma_2 = G_1 - F_1 F_2^{-1} G_2 \tag{6.12.58}$$

$$\alpha_{16} \cos\gamma_2 = G_3 - \sigma_1 a_1 - F_1 F_2^{-1}(G_4 - 3\sigma_1 a_2 + \sigma_2 a_2) \tag{6.12.59}$$

Eliminating γ_1 in (6.12.54) and (6.12.56), and γ_2 in (6.12.58) and (6.12.59), we can obtain the frequency–response equations

$$(G_4 - 3\sigma_1 a_2 + \sigma_2 a_2)^2 + G_2^2 = F_2^2 \tag{6.12.60}$$

$$\left[G_3 - \sigma_1 a_1 - F_1 F_2^{-1}(G_4 - 3\sigma_1 a_2 + \sigma_2 a_2)\right]^2 + \left(G_1 - F_1 F_2^{-1} G_2\right)^2 = \alpha_{16}^2 \tag{6.12.61}$$

Constant amplitudes a_1 and a_2 can be obtained from the above two equations, and the constant phases γ_1 and γ_2 can be obtained through (6.12.53) and (6.12.54).

Again, analyze the stability of the steady state solution. Let a_{10}, a_{20}, γ_{10} and γ_{20} be a set of steady state solution and make a small perturbation of them, i.e.,

$$a_1 = a_{10} + \tilde{a}_1, \quad a_2 = a_{20} + \tilde{a}_2$$

$$\gamma_1 = \gamma_{10} + \tilde{\gamma}_1, \quad \gamma_2 = \gamma_{20} + \tilde{\gamma}_2 \tag{6.12.62}$$

Substitute (6.12.62) into (6.12.49)–(6.12.52) and make the linearization, we can obtain

$$\begin{Bmatrix} \tilde{a}_1' \\ \tilde{a}_2' \\ \tilde{\gamma}_1' \\ \tilde{\gamma}_2' \end{Bmatrix} = \begin{bmatrix} l_{11} & l_{12} & l_{13} & l_{14} \\ l_{21} & l_{22} & l_{23} & l_{24} \\ l_{31} & l_{32} & l_{33} & l_{34} \\ l_{41} & l_{42} & l_{43} & l_{44} \end{bmatrix} \begin{Bmatrix} \tilde{a}_1 \\ \tilde{a}_2 \\ \tilde{\gamma}_1 \\ \tilde{\gamma}_2 \end{Bmatrix} \tag{6.12.63}$$

where

$$\begin{aligned} l_{11} &= \frac{1}{2}\alpha_{12} + \frac{1}{4}\alpha_{15}a_{10}a_{20}\sin\gamma_{10}, \quad l_{12} = \frac{1}{8}\alpha_{15}a_{10}^2\sin\gamma_{10} \\ l_{13} &= \frac{1}{8}\alpha_{15}a_{10}^2a_{20}\cos\gamma_{10}, \quad l_{14} = \alpha_{16}\sin\gamma_{20} \end{aligned} \tag{6.12.64}$$

$$\begin{aligned} l_{21} &= -\frac{3}{8}\alpha_{25}a_{10}^2\sin\gamma_{10}, \quad l_{22} = \frac{1}{2}\alpha_{22} \\ l_{23} &= -\frac{1}{8}\alpha_{25}a_{10}^3\cos\gamma_{10}, \quad l_{24} = 0 \end{aligned} \tag{6.12.65}$$

$$\begin{aligned} l_{31} &= a_{10}^{-1}\left(3\sigma_1 + \frac{3}{2}\alpha_{11}\tilde{a}_1 + \frac{3}{8}\alpha_{13}a_{20}^2 + \frac{9}{8}\alpha_{14}a_{10}^2 - \frac{1}{4}\alpha_{23}a_{10}^2\right. \\ &\left.+\frac{3}{4}\alpha_{15}a_{10}a_{20}\cos\gamma_{10} - \frac{3}{8}\alpha_{25}a_{10}^3a_{20}^{-1}\cos\gamma_{10}\right) \\ l_{32} &= a_{20}^{-1}\left(\sigma_2 - 3\sigma_1 - \frac{1}{2}\alpha_{21} + \frac{3}{4}\alpha_{13}a_{20}^2 - \frac{1}{8}\alpha_{23}a_{10}^2\right. \\ &\left.+\frac{3}{8}\alpha_{15}a_{10}a_{20}\cos\gamma_{10} - \frac{3}{8}\alpha_{24}a_{20}^2\right) \\ l_{33} &= a_{20}^{-1}\left(\frac{1}{8}\alpha_{25}a_{10}^3 - \frac{3}{8}\alpha_{15}a_{10}a_{20}^2\right)\sin\gamma_{10}, \quad l_{34} = 3\alpha_{16}a_{10}^{-1}\sin\gamma_{20} \end{aligned} \tag{6.12.66}$$

$$\begin{aligned} l_{41} &= a_{10}^{-1}\left(\sigma_1 + \frac{1}{2}\alpha_{11} + \frac{3}{8}\alpha_{14}a_{10}^2 + \frac{1}{8}\alpha_{13}a_{20}^2 + \frac{1}{4}\alpha_{15}a_{10}a_{20}\cos\gamma_{10}\right) \\ l_{42} &= \frac{1}{4}\alpha_{13}a_{20} + \frac{1}{8}\alpha_{15}a_{10}\cos\gamma_{10} \\ l_{43} &= -\frac{1}{8}\alpha_{15}a_{10}a_{20}\sin\gamma_{10}, \quad l_{44} = -\alpha_{16}a_{10}^{-1}\sin\gamma_{20} \end{aligned} \tag{6.12.67}$$

The steady state solutions are stable if the eigenvalues of (6.12.63) have negative real parts.

6.13 Exercise 6.13 (Parametric Excitation of a Stretched Wire Carrying Two Particles)

Solution: (a) The stretched wire cannot elongate itself, but the two ends of it can move. The magnitude of the tension in the wire is a constant, therefore, applying Newton's second law to each of the mass point gives

$$m_1\ddot{x}_1 = P_1, \quad m_2\ddot{x}_2 = P_2 \tag{6.13.1}$$

where P_1 and P_2 denote the resultant forces exerting on the mass points m_1 and m_2 by the stretched wire along x_1- and x_2- direction, respectively. Therefore,

$$\begin{aligned} P_1 &= -T_1\sin\theta_1 + T_2\sin\theta_2 = -P\sin\theta_1 + P\sin\theta_2 \\ &= -\frac{Px_1}{\left(l_1^2+x_1^2\right)^{1/2}} + \frac{P(x_2-x_1)}{\left[l_2^2+(x_2-x_1)^2\right]^{1/2}} \end{aligned} \tag{6.13.2}$$

$$\begin{aligned} P_2 &= -T_2\sin\theta_2 - T_3\sin\theta_3 = -P\sin\theta_2 - P\sin\theta_3 \\ &= -\frac{Px_2}{\left(l_3^2+x_2^2\right)^{1/2}} - \frac{P(x_2-x_1)}{\left[l_2^2+(x_2-x_1)^2\right]^{1/2}} \end{aligned} \tag{6.13.3}$$

Substitute (6.13.2) and (6.13.3) into (6.13.1), we obtain

$$\begin{aligned} m_1\ddot{x}_1 + Px_1\left(l_1^2+x_1^2\right)^{-1/2} + P(x_1-x_2)\left[l_1^2+(x_2-x_1)^2\right]^{-1/2} &= 0 \\ m_2\ddot{x}_2 + Px_2\left(l_3^2+x_2^2\right)^{-1/2} + P(x_2-x_1)\left[l_2^2+(x_2-x_1)^2\right]^{-1/2} &= 0 \end{aligned} \tag{6.13.4}$$

(b) Write the Eq. (6.13.4) as

$$\begin{aligned} \frac{\ddot{x}_1}{l} + \frac{P}{m_1 l}\frac{x_1/l}{\sqrt{l_1^2/l^2+x_1^2/l^2}} + \frac{P}{m_1 l}\frac{x_1/l - x_2/l}{\sqrt{l_1^2/l^2+(x_2-x_1)^2/l^2}} &= 0 \\ \frac{\ddot{x}_2}{l} + \frac{P}{m_2 l}\frac{x_2/l}{\sqrt{l_3^2/l^2+x_2^2/l^2}} + \frac{P}{m_2 l}\frac{x_2/l - x_1/l}{\sqrt{l_2^2/l^2+(x_2-x_1)^2/l^2}} &= 0 \end{aligned} \tag{6.13.5}$$

where

$$l = l_1 + l_2 + l_3 \tag{6.13.6}$$

Let

$$\begin{aligned} &u_1 = \frac{x_1}{l}, \quad u_2 = \frac{x_2}{l}, \quad r_1 = \frac{l_1}{l}, \quad r_2 = \frac{l_2}{l}, \quad r_3 = \frac{l_3}{l} \\ &P = P_0 p(t), \quad \omega_{10}^2 = \frac{P_0}{m_1 l}, \quad \omega_{20}^2 = \frac{P_0}{m_2 l} \end{aligned} \tag{6.13.7}$$

Then the Eq. (6.13.5) becomes

$$\begin{aligned}
&\ddot{u}_1 + \frac{\omega_{10}^2 p(t) r_1^{-1} u_1}{\sqrt{1 + r_1^{-2} u_1^2}} + \frac{\omega_{10}^2 p(t) r_1^{-1}(u_1 - u_2)}{\sqrt{1 + r_1^{-2}(u_2 - u_1)^2}} = 0 \\
&\ddot{u}_2 + \frac{\omega_{20}^2 p(t) r_3^{-1} u_2}{\sqrt{1 + r_3^{-2} u_2^2}} + \frac{\omega_{20}^2 p(t) r_2^{-1}(u_2 - u_1)}{\sqrt{1 + r_2^{-2}(u_2 - u_1)^2}} = 0
\end{aligned} \tag{6.13.8}$$

Expanding (6.13.8) near the equilibrium position $x_1 = x_2 = 0$ and retaining to third order small quantities yields

$$\begin{aligned}
&\ddot{u}_1 + \left(2\alpha_1 u_1 - \alpha_1 u_2 - 3\alpha_2 u_1 u_2^2 + 3\alpha_2 u_1^2 u_2 - 2\alpha_2 u_1^3 + \alpha_2 u_2^3\right) p(t) = 0 \\
&\ddot{u}_2 + \left(-\alpha_3 u_1 + \alpha_4 u_2 + 3\alpha_5 u_1 u_2^2 - 3\alpha_5 u_1^2 u_2 + \alpha_5 u_1^3 - \alpha_6 u_2^3\right) p(t) = 0
\end{aligned} \tag{6.13.9}$$

where

$$\begin{aligned}
&\alpha_1 = \omega_{10}^2 r_1^{-1}, \quad \alpha_2 = \frac{1}{2}\omega_{10}^2 r_1^{-3}, \quad \alpha_3 = \omega_{20}^2 r_2^{-1} \\
&\alpha_4 = \omega_{20}^2\left(r_2^{-1} + r_3^{-1}\right), \quad \alpha_5 = \frac{1}{2}\omega_{20}^2 r_2^{-3}, \quad \alpha_6 = \frac{1}{2}\omega_{20}^2\left(r_2^{-3} + r_3^{-3}\right)
\end{aligned} \tag{6.13.10}$$

When $P = P_0$, $p(t) = 1$, the linearized form of Eq. (6.13.9) is

$$\begin{aligned}
&\ddot{u}_1 + 2\alpha_1 u_1 - \alpha_1 u_2 = 0 \\
&\ddot{u}_2 - \alpha_3 u_1 + \alpha_4 u_2 = 0
\end{aligned} \tag{6.13.11}$$

Its characteristic equation is

$$\det\begin{bmatrix} 2\alpha_1 - \omega^2 & -\alpha_1 \\ -\alpha_3 & \alpha_4 - \omega^2 \end{bmatrix} = 0 \tag{6.13.12}$$

Therefore,

$$\begin{aligned}
&\omega_1 = \left[\frac{(2\alpha_1 + \alpha_4) - \sqrt{(2\alpha_1 + \alpha_4)^2 - 4(2\alpha_1\alpha_4 - \alpha_1\alpha_3)}}{2}\right]^{1/2} \\
&\omega_2 = \left[\frac{(2\alpha_1 + \alpha_4) + \sqrt{(2\alpha_1 + \alpha_4)^2 - 4(2\alpha_1\alpha_4 - \alpha_1\alpha_3)}}{2}\right]^{1/2}
\end{aligned} \tag{6.13.13}$$

This in turn leads to a set of modes as

$$\{1, h_1\}^T, \quad \{1, h_2\}^T \tag{6.13.14}$$

where

$$h_1 = \frac{2\alpha_1 - \omega_1^2}{\alpha_1}, \quad h_2 = \frac{2\alpha_1 - \omega_2^2}{\alpha_1} \tag{6.13.15}$$

To examine the presence of internal resonance, let the frequency ratio $s = \omega_2/\omega_1$, then.

$$s = \sqrt{\frac{1+\kappa}{1-\kappa}} \tag{6.13.16}$$

where

$$\kappa = \sqrt{1 - \frac{4(2\alpha_1\alpha_4 - \alpha_1\alpha_3)}{(2\alpha_1 + \alpha_4)^2}} = \sqrt{1 - \frac{4m\left(r_1 r_2^{-1} + 2r_1 r_3^{-1}\right)}{\left[2 + m\left(r_1 r_2^{-1} + r_1 r_3^{-1}\right)\right]^2}} \tag{6.13.17}$$

$$m = m_1/m_2, \quad r_1 + r_2 + r_3 = 1$$

The relation between the frequency ratio s and the parameters is shown in Fig. 6.9, as can be seen that an internal resonance of $\omega_2 : \omega_1 = 3 : 1$ occurs.

(c) Solve the nonlinear equation when $P = P_0$.

In this case, $p(t) = 1$ The Eq. (6.13.9) becomes

$$\begin{aligned} &\ddot{u}_1 + 2\alpha_1 u_1 - \alpha_1 u_2 - 3\alpha_2 u_1 u_2^2 + 3\alpha_2 u_1^2 u_2 - 2\alpha_2 u_1^3 + \alpha_2 u_2^3 = 0 \\ &\ddot{u}_2 - \alpha_3 u_1 + \alpha_4 u_2 + 3\alpha_5 u_1 u_2^2 - 3\alpha_5 u_1^2 u_2 + \alpha_5 u_1^3 - \alpha_6 u_2^3 = 0 \end{aligned} \tag{6.13.18}$$

For small and finite amplitudes, we use the method of multiple scales to solve (6.13.18). Therefore, let the solution of Eq. (6.13.18) be expressed as

$$\begin{aligned} u_1 &= \varepsilon u_{11}(T_0, T_2) + \varepsilon^3 u_{13}(T_0, T_2) + \cdots \\ u_2 &= \varepsilon u_{21}(T_0, T_2) + \varepsilon^3 u_{23}(T_0, T_2) + \cdots \end{aligned} \tag{6.13.19}$$

Substitute (6.13.19) into (6.13.18) and keep to $O(\varepsilon^3)$, we get

$$\begin{aligned} 0 &= \ddot{u}_1 + 2\alpha_1 u_1 - \alpha_1 u_2 - 3\alpha_2 u_1 u_2^2 + 3\alpha_2 u_1^2 u_2 - 2\alpha_2 u_1^3 + \alpha_2 u_2^3 \\ &= \left[D_0^2 + 2\varepsilon D_0 D_1 + \varepsilon^2\left(D_1^2 + 2D_0 D_2\right)\right]\left(\varepsilon u_{11} + \varepsilon^3 u_{13}\right) \\ &\quad + 2\alpha_1\left(\varepsilon u_{11} + \varepsilon^3 u_{13}\right) - \alpha_1\left(\varepsilon u_{21} + \varepsilon^3 u_{23}\right) \\ &\quad - 3\alpha_2\left(\varepsilon u_{11} + \varepsilon^3 u_{13}\right)\left(\varepsilon u_{21} + \varepsilon^3 u_{23}\right)^2 + 3\alpha_2\left(\varepsilon u_{11} + \varepsilon^3 u_{13}\right)^2\left(\varepsilon u_{21} + \varepsilon^3 u_{23}\right) \\ &\quad - 2\alpha_2\left(\varepsilon u_{11} + \varepsilon^3 u_{13}\right)^3 + \alpha_2\left(\varepsilon u_{21} + \varepsilon^3 u_{23}\right)^3 + \cdots \\ &= \varepsilon\left(D_0^2 u_{11} + 2\alpha_1 u_{11} - \alpha_1 u_{21}\right) + \varepsilon^3\left(D_0^2 u_{13} + 2\alpha_1 u_{13} - \alpha_1 u_{23} + 2D_0 D_2 u_{11}\right. \\ &\quad \left. - 3\alpha_2 u_{11} u_{21}^2 + 3\alpha_2 u_{11}^2 u_{21} - 2\alpha_2 u_{11}^3 + \alpha_2 u_{21}^3\right) + \cdots \end{aligned} \tag{6.13.20}$$

$$\begin{aligned}
0 &= \ddot{u}_2 - \alpha_3 u_1 + \alpha_4 u_2 + 3\alpha_5 u_1 u_2^2 - 3\alpha_5 u_1^2 u_2 + \alpha_5 u_1^3 - \alpha_6 u_2^3 \\
&= \left[D_0^2 + 2\varepsilon D_0 D_1 + \varepsilon^2\left(D_1^2 + 2D_0 D_2\right)\right]\left(\varepsilon u_{21} + \varepsilon^3 u_{23}\right) \\
&\quad -\alpha_3\left(\varepsilon u_{11} + \varepsilon^3 u_{13}\right) + \alpha_4\left(\varepsilon u_{21} + \varepsilon^3 u_{23}\right) \\
&\quad +3\alpha_5\left(\varepsilon u_{11} + \varepsilon^3 u_{13}\right)\left(\varepsilon u_{21} + \varepsilon^3 u_{23}\right)^2 - 3\alpha_5\left(\varepsilon u_{11} + \varepsilon^3 u_{13}\right)^2\left(\varepsilon u_{21} + \varepsilon^3 u_{23}\right. \\
&\quad +\alpha_5\left(\varepsilon u_{11} + \varepsilon^3 u_{13}\right)^3 - \alpha_6\left(\varepsilon u_{21} + \varepsilon^3 u_{23}\right)^3 + \cdots \\
&= \varepsilon\left(D_0^2 u_{21} - \alpha_3 u_{11} + \alpha_4 u_{21}\right) + \varepsilon^3\left(D_0^2 u_{23} - \alpha_3 u_{13} + \alpha_4 u_{23} + 2D_0 D_2 u_{21}\right. \\
&\quad \left. + 3\alpha_5 u_{11} u_{21}^2 - 3\alpha_5 u_{11}^2 u_{21} + \alpha_5 u_{11}^3 - \alpha_6 u_{12}^3\right) + \cdots
\end{aligned} \tag{6.13.21}$$

Equating coefficients of like powers of ε in (6.13.20) and (6.13.21) yields.
Order ε^1:

$$\begin{aligned}
D_0^2 u_{11} + 2\alpha_1 u_{11} - \alpha_1 u_{21} &= 0 \\
D_0^2 u_{21} - \alpha_3 u_{11} + \alpha_4 u_{21} &= 0
\end{aligned} \tag{6.13.22}$$

Order ε^3:

$$\begin{aligned}
D_0^2 u_{13} + 2\alpha_1 u_{13} - \alpha_1 u_{23} &= -2D_0 D_2 u_{11} + 3\alpha_2 u_{11} u_{21}^2 - 3\alpha_2 u_{11}^2 u_{21} + 2\alpha_2 u_{11}^3 - \alpha_2 u_{21}^3 \\
D_0^2 u_{23} - \alpha_3 u_{13} + \alpha_4 u_{23} &= -2D_0 D_2 u_{21} - 3\alpha_5 u_{11} u_{21}^2 + 3\alpha_5 u_{11}^2 u_{21} - \alpha_5 u_{11}^3 + \alpha_6 u_{12}^3
\end{aligned} \tag{6.13.23}$$

The corresponding homogeneous equations to (6.13.22) are the same as (6.13.11), so its solution can be expressed as

$$\begin{aligned}
u_{11} &= A_1 e^{i\omega_1 T_0} + A_2 e^{i\omega_2 T_0} + cc \\
u_{21} &= h_1 A_1 e^{i\omega_1 T_0} + h_2 A_2 e^{i\omega_2 T_0} + cc
\end{aligned} \tag{6.13.24}$$

where $A_k = A_k(T_2)$, $k = 1, 2, h_1$ and h_2 are given by (6.13.15). Substituting (6.13.24) into (6.13.23) yields

$$\omega_2 = 3\omega_1 + \varepsilon^2 \sigma_2 \tag{6.13.25}$$

$$
\begin{aligned}
D_0^2u_{23} + \alpha_3u_{13} + \alpha_4u_{23} &= -2i\omega_1h_1A_1'e^{i\omega_1T_0} + (6\alpha_6h_1h_2^2 - 6\alpha_5h_2^2 - 12\alpha_5h_1h_2 \\
&+6\alpha_5h_1 + 12\alpha_5h_2 - 6\alpha_5)A_1A_2\overline{A}_2e^{i\omega_1T_0} \\
&+\left(3\alpha_6h_1^3 - 9\alpha_5h_1^2 + 9\alpha_5h_1 - 3\alpha_5\right)A_1^2\overline{A}_1e^{i\omega_1T_0} \\
&+\left(3\alpha_6h_1^2h_2 - 3\alpha_5h_1^2 - 6\alpha_5h_1h_2 + 6\alpha_5h_1 + 3\alpha_5h_2 - 3\alpha_5\right)\overline{A}_1^2A_2e^{i(\omega_2-2\omega_1)T_0} \\
&-2i\omega_2h_2A_2'e^{i\omega_2T_0} + (6\alpha_6h_1^2h_2 - 6\alpha_5h_1^2 - 12\alpha_5h_1h_2 \\
&+6\alpha_5h_2 + 12\alpha_5h_1 - 6\alpha_5)A_1\overline{A}_1A_2e^{i\omega_2T_0} \\
&+\left(3\alpha_6h_2^3 - 9\alpha_5h_2^2 + 9\alpha_5h_2 - 3\alpha_5\right)A_2^2\overline{A}_2e^{i\omega_2T_0} \\
&+\left(\alpha_6h_1^3 - 3\alpha_5h_1^2 + 3\alpha_5h_1 - \alpha_5\right)A_1^3e^{3i\omega_1T_0} + cc + NST
\end{aligned}
\tag{6.13.26}
$$

Primes denote the derivative with respect to T_2. In order to eliminate the secular terms from (6.13.25) and (6.13.26), it is necessary to distinguish between the case where there is internal resonance of $\omega_2 \approx 3\omega_1$ or not. Here we consider the case with three-to-one internal resonance, and readers are invited to investigate the rest of cases.

Introduce the detuning parameter σ_2 such that

$$
\omega_2 = 3\omega_1 + \varepsilon^2\sigma_2 \tag{6.13.27}
$$

Equation (6.13.25) and (6.13.26) become

$$
\begin{aligned}
D_0^2u_{13} + 2\alpha_1u_{13} - \alpha_1u_{23} &= \big[-2i\omega_1A_1' + \big(-6\alpha_2h_1h_2^2 + 6\alpha_2h_2^2 + 12\alpha_2h_1h_2 \\
&-6\alpha_2h_1 - 12\alpha_2h_2 + 12\alpha_2)A_1A_2\overline{A}_2 \\
&+(-3\alpha_2h_1^3 + 9\alpha_2h_1^2 - 9\alpha_2h_1 + 6\alpha_2)A_1^2\overline{A}_1 \\
&+\big(-3\alpha_2h_1^2h_2 + 3\alpha_2h_1^2 + 6\alpha_2h_1h_2 - 6\alpha_2h_1 - 3\alpha_2h_2 + 6\alpha_2\big)\overline{A}_1^2A_2e^{i\sigma_2T_2}\big]e^{i\omega_1T_0} \\
&+\big[-2i\omega_2A_2' + \big(-6\alpha_2h_1^2h_2 + 6\alpha_2h_1^2 + 12\alpha_2h_1h_2 \\
&-6\alpha_2h_2 - 12\alpha_2h_1 + 12\alpha_2)A_1\overline{A}_1A_2 \\
&+\big(-3\alpha_2h_2^3 + 9\alpha_2h_2^2 - 9\alpha_2h_2 + 6\alpha_2\big)A_2^2\overline{A}_2 \\
&+\big(-\alpha_2h_1^3 + 3\alpha_2h_1^2 - 3\alpha_2h_1 + 2\alpha_2\big)A_1^3e^{-i\sigma_2T_2}\big]e^{i\omega_2T_0} + cc + NST
\end{aligned}
\tag{6.13.28}
$$

$$\begin{aligned}
D_0^2u_{23}+\alpha_3u_{13}+\alpha_4u_{23}=&\big[-2i\omega_1h_1A_1'+\big(6\alpha_6h_1h_2^2-6\alpha_5h_2^2-12\alpha_5h_1h_2\\
&+6\alpha_5h_1+12\alpha_5h_2-6\alpha_5)A_1A_2\overline{A}_2\\
&+\big(3\alpha_6h_1^3-9\alpha_5h_1^2+9\alpha_5h_1-3\alpha_5\big)A_1^2\overline{A}_1\\
&+\big(3\alpha_6h_1^2h_2-3\alpha_5h_1^2-6\alpha_5h_1h_2+6\alpha_5h_1+3\alpha_5h_2-3\alpha_5\big)\overline{A}_1^2A_2e^{i\sigma_2T_2}\big]e^{i\omega_1T_0}\\
&+[-2i\omega_2h_2A_2'+(6\alpha_6h_1^2h_2-6\alpha_5h_1^2-12\alpha_5h_1h_2\\
&+6\alpha_5h_2+12\alpha_5h_1-6\alpha_5)A_1\overline{A}_1A_2\\
&+\big(3\alpha_6h_2^3-9\alpha_5h_2^2+9\alpha_5h_2-3\alpha_5\big)A_2^2\overline{A}_2\\
&+\big(\alpha_6h_1^3-3\alpha_5h_1^2+3\alpha_5h_1-\alpha_5\big)A_1^3e^{-i\sigma_2T_2}\big]e^{i\omega_2T_0}+cc+NST
\end{aligned} \tag{6.13.29}$$

In order to eliminate secular terms of Eqs. (6.13.28) and (6.13.29), we seek the following special solutions for them:

$$\begin{aligned}
u_{13}&=P_{11}e^{i\omega_1T_0}+P_{12}e^{i\omega_2T_0}\\
u_{23}&=P_{21}e^{i\omega_1T_0}+P_{22}e^{i\omega_2T_0}
\end{aligned} \tag{6.13.30}$$

Substituting (6.13.30) into (6.13.28) and (6.13.29), equating the coefficients of $e^{i\omega_1T_0}$ and $e^{i\omega_2T_0}$, we get

$$\begin{bmatrix}2\alpha_1-\omega_n^2 & -\alpha_1\\ -\alpha_3 & \alpha_4-\omega_n^2\end{bmatrix}\begin{Bmatrix}P_{1n}\\ P_{2n}\end{Bmatrix}=\begin{Bmatrix}R_{1n}\\ R_{2n}\end{Bmatrix},\quad n=1,\ 2 \tag{6.13.31}$$

where

$$\begin{aligned}
R_{11}&=-2i\omega_1A_1'+\big(-6\alpha_2h_1h_2^2+6\alpha_2h_2^2+12\alpha_2h_1h_2\\
&-6\alpha_2h_1-12\alpha_2h_2+12\alpha_2)A_1A_2\overline{A}_2\\
&+\big(-3\alpha_2h_1^3+9\alpha_2h_1^2-9\alpha_2h_1+6\alpha_2\big)A_1^2\overline{A}_1\\
&+\big(-3\alpha_2h_1^2h_2+3\alpha_2h_1^2+6\alpha_2h_1h_2-6\alpha_2h_1-3\alpha_2h_2+6\alpha_2\big)\overline{A}_1^2A_2e^{i\sigma_2T_2}\\
R_{21}&=-2i\omega_1h_1A_1'+\big(6\alpha_6h_1h_2^2-6\alpha_5h_2^2-12\alpha_5h_1h_2\\
&+6\alpha_5h_1+12\alpha_5h_2-6\alpha_5)A_1A_2\overline{A}_2\\
&+\big(3\alpha_6h_1^3-9\alpha_5h_1^2+9\alpha_5h_1-3\alpha_5\big)A_1^2\overline{A}_1\\
&+\big(3\alpha_6h_1^2h_2-3\alpha_5h_1^2-6\alpha_5h_1h_2+6\alpha_5h_1+3\alpha_5h_2-3\alpha_5\big)\overline{A}_1^2A_2e^{i\sigma_2T_2}
\end{aligned} \tag{6.13.32}$$

$$\begin{aligned}
R_{12} &= -2i\omega_2 A_2' + \left(-6\alpha_2 h_1^2 h_2 + 6\alpha_2 h_1^2 + 12\alpha_2 h_1 h_2 \right.\\
&\quad \left. - 6\alpha_2 h_2 - 12\alpha_2 h_1 + 12\alpha_2\right) A_1 \overline{A}_1 A_2 \\
&\quad + - \left(3\alpha_2 h_2^3 + 9\alpha_2 h_2^2 - 9\alpha_2 h_2 + 6\alpha_2\right) A_2^2 \overline{A}_2 \\
&\quad + \left(-\alpha_2 h_1^3 + 3\alpha_2 h_1^2 - 3\alpha_2 h_1 + 2\alpha_2\right) A_1^3 e^{-i\sigma_2 T_2} \\
R_{22} &= -2i\omega_2 h_2 A_2' + \left(6\alpha_6 h_1^2 h_2 - 6\alpha_5 h_1^2 - 12\alpha_5 h_1 h_2 \right.\\
&\quad \left. + 6\alpha_5 h_2 + 12\alpha_5 h_1 - 6\alpha_5\right) A_1 \overline{A}_1 A_2 \\
&\quad + \left(3\alpha_6 h_2^3 - 9\alpha_5 h_2^2 + 9\alpha_5 h_2 - 3\alpha_5\right) A_2^2 \overline{A}_2 \\
&\quad + \left(\alpha_6 h_1^3 - 3\alpha_5 h_1^2 + 3\alpha_5 h_1 - \alpha_5\right) A_1^3 e^{-i\sigma_2 T_2}
\end{aligned} \tag{6.13.33}$$

According to Eq. (6.13.12), the determinant of the coefficient matrix of Eq. (6.13.31) is zero:

$$\det\begin{bmatrix} R_{1n} & -\alpha_1 \\ R_{2n} & \alpha_4 - \omega_n^2 \end{bmatrix} = 0 \Rightarrow \left(\alpha_4 - \omega_n^2\right) R_{1n} + \alpha_1 R_{2n} = 0 \tag{6.13.34}$$

Substitute (6.13.32) and (6.13.33), respectively, into (6.13.34), we obtain

$$2iA_1' = b_{11} A_1 A_2 \overline{A}_2 + b_{12} A_1^2 \overline{A}_1 + b_{13} \overline{A}_1^2 A_2 e^{i\sigma_2 T_2} \tag{6.13.35}$$

$$2iA_2' = b_{21} A_1 \overline{A}_1 A_2 + b_{22} A_2^2 \overline{A}_2 + b_{23} A_1^3 e^{-i\sigma_2 T_2} \tag{6.13.36}$$

where

$$\begin{aligned}
&\Delta_1 = \omega_1\left(\alpha_4 - \omega_1^2 + \alpha_1 h_1\right) \\
&b_{11} = \Delta_1^{-1}\Big[\left(\alpha_4 - \omega_1^2\right)\left(-6\alpha_2 h_1 h_2^2 + 6\alpha_2 h_2^2 + 12\alpha_2 h_1 h_2 - 6\alpha_2 h_1 - 12\alpha_2 h_2 + 12\alpha_2\right) \\
&+\alpha_1\left(6\alpha_6 h_1 h_2^2 - 6\alpha_5 h_2^2 - 12\alpha_5 h_1 h_2 + 6\alpha_5 h_1 + 12\alpha_5 h_2 - 6\alpha_5\right)\Big] \\
&b_{12} = \Delta_1^{-1}\Big[\left(\alpha_4 - \omega_1^2\right)\left(-3\alpha_2 h_1^3 + 9\alpha_2 h_1^2 - 9\alpha_2 h_1 + 6\alpha_2\right) \\
&+\alpha_1\left(3\alpha_6 h_1^3 - 9\alpha_5 h_1^2 + 9\alpha_5 h_1 - 3\alpha_5\right)\Big] \\
&b_{13} = \Delta_1^{-1}\Big[\left(\alpha_4 - \omega_1^2\right)\left(-3\alpha_2 h_1^2 h_2 + 3\alpha_2 h_1^2 + 6\alpha_2 h_1 h_2 - 6\alpha_2 h_1 - 3\alpha_2 h_2 + 6\alpha_2\right) \\
&+\alpha_1\left(3\alpha_6 h_1^2 h_2 - 3\alpha_5 h_1^2 - 6\alpha_5 h_1 h_2 + 6\alpha_5 h_1 + 3\alpha_5 h_2 - 3\alpha_5\right)\Big]
\end{aligned} \tag{6.13.37}$$

$$
\begin{aligned}
&\Delta_2 = \omega_2\left(\alpha_4 - \omega_2^2 + \alpha_1 h_2\right)\\
&b_{21} = \Delta_2^{-1}\left[\left(\alpha_4 - \omega_2^2\right)\left(-6\alpha_2 h_1^2 h_2 + 6\alpha_2 h_1^2 + 12\alpha_2 h_1 h_2 - 6\alpha_2 h_2 - 12\alpha_2 h_1 + 12\alpha_2\right)\right.\\
&\left.+\alpha_1\left(6\alpha_6 h_1^2 h_2 - 6\alpha_5 h_1^2 - 12\alpha_5 h_1 h_2 + 6\alpha_5 h_2 + 12\alpha_5 h_1 - 6\alpha_5\right)\right]\\
&b_{22} = \Delta_2^{-1}\left[\left(\alpha_4 - \omega_2^2\right)\left(3\alpha_6 h_2^3 - 9\alpha_5 h_2^2 + 9\alpha_5 h_2 - 3\alpha_5\right)\right.\\
&\left.+\alpha_1\left(3\alpha_6 h_2^3 - 9\alpha_5 h_2^2 + 9\alpha_5 h_2 - 3\alpha_5\right)\right]\\
&b_{23} = \Delta_2^{-1}\left[\left(\alpha_4 - \omega_2^2\right)\left(-\alpha_2 h_1^3 + 3\alpha_2 h_1^2 - 3\alpha_2 h_1 + 2\alpha_2\right)\right.\\
&\left.+\alpha_1\left(\alpha_6 h_1^3 - 3\alpha_5 h_1^2 + 3\alpha_5 h_1 - \alpha_5\right)\right]
\end{aligned}
\tag{6.13.38}
$$

Let

$$
A_1 = \frac{1}{2}a_1 e^{i\beta_1}, \quad A_2 = \frac{1}{2}a_2 e^{i\beta_2} \tag{6.13.39}
$$

Substituting (6.13.39) into (6.13.35) and (6.13.36), we get

$$
ia_1' - a_1\beta_1' = \frac{1}{8}b_{11}a_1 a_2^2 + \frac{1}{8}b_{12}a_1^3 + \frac{1}{8}b_{13}a_1^2 a_2 e^{i\gamma} \tag{6.13.40}
$$

$$
ia_2' - a_2\beta_2' = \frac{1}{8}b_{21}a_1^2 a_2 + \frac{1}{8}b_{22}a_2^3 + \frac{1}{8}b_{23}a_1^3 e^{-\gamma} \tag{6.13.41}
$$

where

$$
\gamma = \sigma_2 T_2 - 3\beta_1 + \beta_2 \tag{6.13.42}
$$

Separating the real and imaginary parts of the above equations yields

$$
\begin{aligned}
a_1' &= \tfrac{1}{8}b_{13}a_1^2 a_2 \sin\gamma\\
a_2' &= -\tfrac{1}{8}b_{23}a_1^3 \sin\gamma
\end{aligned}
\tag{6.13.43}
$$

$$
\begin{aligned}
-a_1\beta_1' &= \tfrac{1}{8}b_{11}a_1 a_2^2 + \tfrac{1}{8}b_{12}a_1^3 + \tfrac{1}{8}b_{13}a_1^2 a_2 \cos\gamma\\
-a_2\beta_2' &= \tfrac{1}{8}b_{21}a_1^2 a_2 + \tfrac{1}{8}b_{22}a_2^3 + \tfrac{1}{8}b_{23}a_1^3 \cos\gamma
\end{aligned}
\tag{6.13.44}
$$

Eliminating γ from (6.13.43), we get

$$
b_{23}a_1 da_1 + b_{13}a_2 da_2 = 0 \tag{6.13.45}
$$

i.e.,

$$
b_{23}a_1^2 + b_{13}a_2^2 = E \tag{6.13.46}
$$

Utilizing (6.13.42) and eliminating β_1', β_2' from (6.13.44), we get

$$\begin{aligned} a_2\gamma' = \sigma_2 a_2 + \tfrac{3}{8}b_{11}a_2^3 - \tfrac{1}{8}b_{22}a_2^3 + \tfrac{3}{8}b_{12}a_1^2 a_2 \\ -\tfrac{1}{8}b_{21}a_1^2 a_2 + \tfrac{3}{8}b_{13}a_1 a_2^2 \cos\gamma - \tfrac{1}{8}b_{23}a_1^3 \cos\gamma \end{aligned} \tag{6.13.47}$$

The above equation is divided by the second equation of (6.13.43), then we can obtain

$$\begin{aligned} -b_{23}a_2 a_1^3 \sin\gamma d\gamma = \big(8\sigma_2 a_2 + 3b_{11}a_2^3 - b_{22}a_2^3 + 3b_{12}a_1^2 a_2 \\ - b_{21}a_1^2 a_2 + 3b_{13}a_1 a_2^2 \cos\gamma - b_{23}a_1^3 \cos\gamma\big)da_2 \end{aligned} \tag{6.13.48}$$

since

$$d\big(a_2 a_1^3 \cos\gamma\big) = -a_2 a_1^3 \sin\gamma d\gamma + \cos\gamma a_1^3 da_2 + 3\cos\gamma a_2 a_1^2 da_1$$

the Eq. (6.13.48) becomes

$$\begin{aligned} b_{23}d\big(a_2 a_1^3 \cos\gamma\big) = (8\sigma_2 a_2 + 3b_{11}a_2^3 - b_{22}a_2^3 + 3b_{12}a_1^2 a_2 \\ -b_{21}a_1^2 a_2)da_2 + 3a_1 a_2(b_{13}a_2 da_2 + 3b_{23}a_1 da_1)\cos\gamma \end{aligned} \tag{6.13.49}$$

Using (6.13.45) and (6.13.46), the above equation becomes

$$\begin{aligned} b_{23}d\big(a_2 a_1^3 \cos\gamma\big) = 8\sigma_2 a_2 da_2 + Eb_{23}^{-1}(3b_{12} - b_{21})a_2 da_2 \\ +\big[(3b_{11} - b_{22}) - b_{13}b_{23}^{-1}(3b_{12} - b_{21})\big]a_2^3 da_2 \end{aligned} \tag{6.13.50}$$

Make the integration of (6.13.50), we can obtain

$$\begin{aligned} b_{23}a_1^3 a_2 \cos\gamma - 4\sigma_2 a_2^2 + \frac{1}{2}Eb_{23}^{-1}(3b_{12} - b_{21})a_2^2 \\ -\frac{1}{4}\big[(3b_{11} - b_{22}) - b_{13}b_{23}^{-1}(3b_{12} - b_{21})\big]a_2^4 = L \end{aligned} \tag{6.13.51}$$

where L is the constant of integration.

Now, let us combine (6.13.51), (6.13.46) and (6.13.43) into a single variable differential equation. To do so, let

$$a_1^2 = E\xi \tag{6.13.52}$$

Then from (6.13.46), we can obtain

$$a_2^2 = Eb_{13}^{-1}(1 - b_{23}\xi) \tag{6.13.53}$$

Eliminating γ from (6.13.51) and (6.13.43) yields

$$\xi'^2 = \tfrac{1}{16}E^2 b_{13}^{-1}\xi(1-b_{23}\xi)^3 - \tfrac{1}{16}E^{-2}b_{13}^2 b_{23}^{-2}\{4Eb_{13}^{-1}\sigma_2(1-b_{23}\xi)$$
$$+L - \tfrac{1}{2}E^2 b_{13}^{-1}b_{23}^{-1}(3b_{12}-b_{21})(1-b_{23}\xi)$$
$$+\tfrac{1}{4}\big[(3b_{11}-b_{22}) - b_{13}b_{23}^{-1}(3b_{12}-b_{21})\big]E^2 b_{13}^{-2}(1-b_{23}\xi)^2\}^2$$

Or write as

$$\xi'^2 = F^2(\xi) - G^2(\xi) \tag{6.13.54}$$

where

$$\begin{aligned}
F(\xi) &= \pm\tfrac{1}{4}[E^2 b_{13}^{-1}\xi(1-b_{23}\xi)^3]^{1/2}\\
G(\xi) &= \pm\tfrac{1}{4}E^{-1}b_{13}b_{23}^{-1}\{4Eb_{13}^{-1}\sigma_2(1-b_{23}\xi)\\
&+L - \tfrac{1}{2}E^2 b_{13}^{-1}b_{23}^{-1}(3b_{12}-b_{21})(1-b_{23}\xi)\\
&+\tfrac{1}{4}\big[(3b_{11}-b_{22}) - b_{13}b_{23}^{-1}(3b_{12}-b_{21})\big]E^2 b_{13}^{-2}(1-b_{23}\xi)^2\}
\end{aligned} \tag{6.13.55}$$

The functions $F(\xi)$ and $G(\xi)$ are shown schematically in Fig. 6.9. For real motions, $F^2(\xi) \geq G^2(\xi)$. The points where F meets G corresponds to the vanishing of $\xi\prime$. A curve such as G_2 which meets one branch of F at two different points or a curve G_3 which meets both branches corresponds to a periodic solution. In this case, ξ is periodic and oscillates between two intersection points, and the motion is aperiodic.

Considering the curves G_1 and F are tangent to each other, as shown in Fig. 6.9. In this case, ξ has only one constant-value solution, and the motion of the system is periodic. The nonlinear effect acts to modulate the phase of the motion so that the nonlinear frequencies of the two modes are commensurable.

In order to analyze the steady state solution, let $a_1' = a_2' = \gamma' = 0$, (6.13.47) and (6.13.43) become

$$\sin\gamma = 0 \quad \text{or} \quad \gamma = n\pi \tag{6.13.56}$$

$$\sigma_2 a_2 + \frac{1}{8}(3b_{11}-b_{22})a_2^3 + \frac{1}{8}(3b_{12}-b_{21})a_1^2 a_2 + \frac{1}{8}a_1 a_2^2(3b_{13}-b_{23}a_1)\cos n\pi = 0 \tag{6.13.57}$$

Equation (6.13.57) is the cubic equation of a_2. For a given set of values of σ, a_1 and $\cos n\pi$, Eq. (6.13.57) has one or three real roots, each of which corresponds to a periodic motion of the system. Thus, the periodic motion of the system is either unique or one of the three possible periodic motions. Figure 6.9 shows that the periodic motion is unstable. With a small perturbation, a curve such as G_1 may become a curve such as G_2 and the motion of the system becomes aperiodic.

For the case of the nonlinear periodic motion of the system with a constant-value solution, we note that the frequency of the motion of the system is

$$\hat{\omega}_1 = \omega_1 + \varepsilon^2\beta_1', \quad \hat{\omega}_2 = \omega_2 + \varepsilon^2\beta_2' \tag{6.13.58}$$

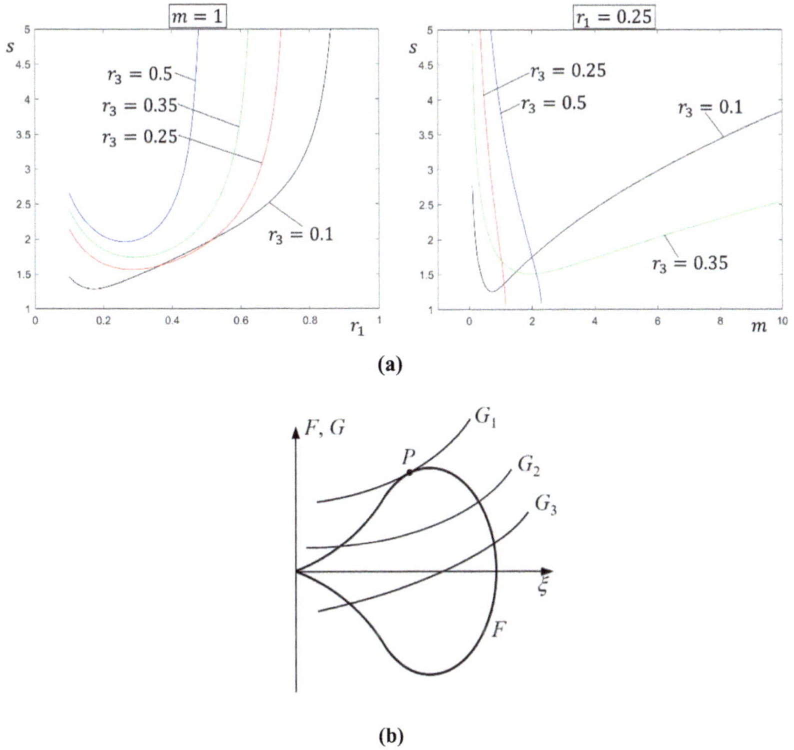

Fig. 6.9 **a** Relations between frequency ratio s and parameters ($m = m_1/m_2$, $r_1 = l_1/l$, $r_2 = l_2/l$, $r_3 = l_3/l$) **b** Schematic diagram of $F(\xi)$ and $G(\xi)$ for Exercise 6.13

therefore,

$$\hat{\omega}_2 - 3\hat{\omega}_1 = \omega_2 - 3\omega_1 + \varepsilon^2(\beta_2' - 3\beta_1') = \varepsilon^2\sigma + \varepsilon^2(\beta_2' - 3\beta_1') = \varepsilon^2\gamma' = 0 \tag{6.13.59}$$

Thus, the nonlinearity of the system modulates the frequency of the motion to exactly 3 to 1, and thus the motion is periodic.

(d) When $P = P_0(1 + \varepsilon^2 \cos\Omega t)$, $p(t) = 1 + \varepsilon^2 \cos\Omega t$, (6.13.9) becomes

$$\begin{aligned} &\ddot{u}_1 + (2\alpha_1 u_1 - \alpha_1 u_2 - 3\alpha_2 u_1 u_2^2 + 3\alpha_2 u_1^2 u_2 - 2\alpha_2 u_1^3 + \alpha_2 u_2^3)(1 + \varepsilon^2 \cos\Omega t) = 0 \\ &\ddot{u}_2 + (-\alpha_3 u_1 + \alpha_4 u_2 + 3\alpha_5 u_1 u_2^2 - 3\alpha_5 u_1^2 u_2 + \alpha_5 u_1^3 - \alpha_6 u_2^3)(1 + \varepsilon^2 \cos\Omega t) = 0 \end{aligned} \tag{6.13.60}$$

Equation (6.13.60) is a two-degree-of-freedom parametric excitation system. We use the method of multiple scales to solve it.

Substituting (6.13.19) into (6.13.60), retaining to $O(\varepsilon^3)$ term, and then equating coefficients of like powers of ε, we get.

Order ε^1:

$$\begin{aligned} D_0^2 u_{11} + 2\alpha_1 u_{11} - \alpha_1 u_{21} &= 0 \\ D_0^2 u_{21} - \alpha_3 u_{11} + \alpha_4 u_{21} &= 0 \end{aligned} \tag{6.13.61}$$

Order ε^3:

$$\begin{aligned} & D_0^2 u_{13} + 2\alpha_1 u_{13} - \alpha_1 u_{23} = -2D_0 D_2 u_{11} + 3\alpha_2 u_{11} u_{21}^2 - 3\alpha_2 u_{11}^2 u_{21} \\ & +2\alpha_2 u_{11}^3 - \alpha_2 u_{21}^3 - 2\alpha_1 u_{11} \cos\Omega t + \alpha_1 u_{21} \cos\Omega t \\ & D_0^2 u_{23} - \alpha_3 u_{13} + \alpha_4 u_{23} = -2D_0 D_2 u_{21} - 3\alpha_5 u_{11} u_{21}^2 + 3\alpha_5 u_{11}^2 u_{21} \\ & -\alpha_5 u_{11}^3 + \alpha_6 u_{12}^3 + \alpha_3 u_{11} \cos\Omega t - \alpha_4 u_{21} \cos\Omega t \end{aligned} \tag{6.13.62}$$

Thc solution of (6.13.61) can be expressed as

$$\begin{aligned} u_{11} &= A_1 e^{i\omega_1 T_0} + A_2 e^{i\omega_2 T_0} + cc \\ u_{21} &= h_1 A_1 e^{i\omega_1 T_0} + h_2 A_2 e^{i\omega_2 T_0} + cc \end{aligned} \tag{6.13.63}$$

where $A_k = A_k(T_2)$, $k = 1, 2$, h_1 and h_2 are given by (6.13.15). Substituting (6.13.63) into (6.13.62) yields

$$\begin{aligned} & D_0^2 u_{13} + 2\alpha_1 u_{13} - \alpha_1 u_{23} = -2i\omega_1 A_1' e^{i\omega_1 T_0} + \big(-6\alpha_2 h_1 h_2^2 + 6\alpha_2 h_2^2 + 12\alpha_2 h_1 h_2 \\ & \quad - 6\alpha_2 h_1 - 12\alpha_2 h_2 + 12\alpha_2\big) A_1 A_2 \overline{A}_2 e^{i\omega_1 T_0} \\ & \quad + \big(-3\alpha_2 h_1^3 + 9\alpha_2 h_1^2 - 9\alpha_2 h_1 + 6\alpha_2\big) A_1^2 \overline{A}_1 e^{i\omega_1 T_0} \\ & \quad + \big(-3\alpha_2 h_1^2 h_2 + 3\alpha_2 h_1^2 + 6\alpha_2 h_1 h_2 - 6\alpha_2 h_1 - 3\alpha_2 h_2 + 6\alpha_2\big) \overline{A}_1^2 A_2 e^{i(\omega_2 - 2\omega_1) T_0} \\ & \quad - 2i\omega_2 A_2' e^{i\omega_2 T_0} + \big(-6\alpha_2 h_1^2 h_2 + 6\alpha_2 h_1^2 + 12\alpha_2 h_1 h_2 \\ & \quad - 6\alpha_2 h_2 - 12\alpha_2 h_1 + 12\alpha_2\big) A_1 \overline{A}_1 A_2 e^{i\omega_2 T_0} \\ & \quad + \big(-3\alpha_2 h_2^3 + 9\alpha_2 h_2^2 - 9\alpha_2 h_2 + 6\alpha_2\big) A_2^2 \overline{A}_2 e^{i\omega_2 T_0} \\ & \quad + \big(-\alpha_2 h_1^3 + 3\alpha_2 h_1^2 - 3\alpha_2 h_1 + 2\alpha_2\big) A_1^3 e^{3i\omega_1 T_0} \\ & \quad + \left(\frac{1}{2}\alpha_1 h_1 - \alpha_1\right) \overline{A}_1 e^{i(\Omega - \omega_1) T_0} + \left(\frac{1}{2}\alpha_1 h_2 - \alpha_1\right) \overline{A}_2 e^{i(\Omega - \omega_2) T_0} \\ & \quad + \left(\frac{1}{2}\alpha_1 h_1 - \alpha_1\right) A_1 e^{i(\Omega + \omega_1) T_0} + cc + NST \end{aligned} \tag{6.13.64}$$

$$
\begin{aligned}
D_0^2 u_{23} - \alpha_3 u_{13} + \alpha_4 u_{23} &= -2i\omega_1 h_1 A_1' e^{i\omega_1 T_0} + \big(6\alpha_6 h_1 h_2^2 - 6\alpha_5 h_2^2 - 12\alpha_5 h_1 h_2 \\
&+ 6\alpha_5 h_1 + 12\alpha_5 h_2 - 6\alpha_5) A_1 A_2 \overline{A}_2 e^{i\omega_1 T_0} \\
&+ \big(3\alpha_6 h_1^3 - 9\alpha_5 h_1^2 + 9\alpha_5 h_1 - 3\alpha_5\big) A_1^2 \overline{A}_1 e^{i\omega_1 T_0} \\
&+ \big(3\alpha_6 h_1^2 h_2 - 3\alpha_5 h_1^2 - 6\alpha_5 h_1 h_2 + 6\alpha_5 h_1 + 3\alpha_5 h_2 - 3\alpha_5\big) \overline{A}_1^2 A_2 e^{i(\omega_2 - 2\omega_1) T_0} \\
&- 2i\omega_2 h_2 A_2' e^{i\omega_2 T_0} + \big(6\alpha_6 h_1^2 h_2 - 6\alpha_5 h_1^2 - 12\alpha_5 h_1 h_2 \\
&+ 6\alpha_5 h_2 + 12\alpha_5 h_1 - 6\alpha_5) A_1 \overline{A}_1 A_2 e^{i\omega_2 T_0} \\
&+ \big(3\alpha_6 h_2^3 - 9\alpha_5 h_2^2 + 9\alpha_5 h_2 - 3\alpha_5\big) A_2^2 \overline{A}_2 e^{i\omega_2 T_0} \\
&+ \big(\alpha_6 h_1^3 - 3\alpha_5 h_1^2 + 3\alpha_5 h_1 - \alpha_5\big) A_1^3 e^{3i\omega_1 T_0} \\
&+ \frac{1}{2}(\alpha_3 - \alpha_4 h_1) \overline{A}_1 e^{i(\Omega - \omega_1) T_0} + \frac{1}{2}(\alpha_3 - \alpha_4 h_2) \overline{A}_2 e^{i(\Omega - \omega_2) T_0} \\
&+ \frac{1}{2}(\alpha_3 - \alpha_4 h_1) A_1 e^{i(\Omega + \omega_1) T_0} + cc + NST
\end{aligned}
\tag{6.13.65}
$$

Primes denote the derivative with respect to T_2. In analyzing the particular solutions of (6.13.64) and (6.13.65) we need to distinguish between the internal resonant case in which $\omega_2 \approx 3\omega_1$ and the other resonant and non-resonant cases. Here, we only present the solution to case (iii) below, and readers are invited to complete the remaining two cases themselves.

Case (iii): $\Omega \approx 2\omega_1$ and $\omega_2 \approx 3\omega_1$.

Introduce the detuning parameter σ_2 such that

$$
\Omega = 2\omega_1 + \varepsilon^2 \sigma_1, \quad \omega_2 = 3\omega_1 + \varepsilon^2 \sigma_2 \tag{6.13.66}
$$

Equation (6.13.64) and (6.13.65) become

$$
\begin{aligned}
D_0^2 u_{13} + 2\alpha_1 u_{13} - \alpha_1 u_{23} = \big[&-2i\omega_1 A_1' + \big(-6\alpha_2 h_1 h_2^2 + 6\alpha_2 h_2^2 + 12\alpha_2 h_1 h_2 \\
&- 6\alpha_2 h_1 - 12\alpha_2 h_2 + 12\alpha_2) A_1 A_2 \overline{A}_2 \\
&+\big(-3\alpha_2 h_1^3 + 9\alpha_2 h_1^2 - 9\alpha_2 h_1 + 6\alpha_2\big) A_1^2 \overline{A}_1 \\
&+\big(-3\alpha_2 h_1^2 h_2 + 3\alpha_2 h_1^2 + 6\alpha_2 h_1 h_2 - 6\alpha_2 h_1 - 3\alpha_2 h_2 + 6\alpha_2\big) \overline{A}_1^2 A_2 e^{i\sigma_2 T_2} \\
&+\left(\frac{1}{2}\alpha_1 h_1 - \alpha_1\right) \overline{A}_1 e^{i\sigma_1 T_2} + \left(\frac{1}{2}\alpha_1 h_2 - \alpha_1\right) A_2 e^{i(-\sigma_1+\sigma_2)T_2} \big] e^{i\omega_1 T_0} \\
&+\big[-2i\omega_2 A_2' + \big(-6\alpha_2 h_1^2 h_2 + 6\alpha_2 h_1^2 + 12\alpha_2 h_1 h_2 \\
&-6\alpha_2 h_2 - 12\alpha_2 h_1 + 12\alpha_2) A_1 \overline{A}_1 A_2 \\
&+\big(-3\alpha_2 h_2^3 + 9\alpha_2 h_2^2 - 9\alpha_2 h_2 + 6\alpha_2\big) A_2^2 \overline{A}_2 \\
&+\big(-\alpha_2 h_1^3 + 3\alpha_2 h_1^2 - 3\alpha_2 h_1 + 2\alpha_2\big) A_1^3 e^{-i\sigma_2 T_2} \\
&+\left(\frac{1}{2}\alpha_1 h_1 - \alpha_1\right) A_1 e^{i(\sigma_1 - \sigma_2)T_2} \Big] e^{i\omega_2 T_0} + cc + NST
\end{aligned}
\tag{6.13.67}
$$

$$
\begin{aligned}
D_0^2 u_{23} + \alpha_3 u_{13} + \alpha_4 u_{23} = \big[&-2i\omega_1 h_1 A_1' + (6\alpha_6 h_1 h_2^2 - 6\alpha_5 h_2^2 - 12\alpha_5 h_1 h_2 \\
&+ 6\alpha_5 h_1 + 12\alpha_5 h_2 - 6\alpha_5) A_1 A_2 \overline{A}_2 \\
&+\big(3\alpha_6 h_1^3 - 9\alpha_5 h_1^2 + 9\alpha_5 h_1 - 3\alpha_5\big) A_1^2 \overline{A}_1 \\
&+\big(3\alpha_6 h_1^2 h_2 - 3\alpha_5 h_1^2 - 6\alpha_5 h_1 h_2 + 6\alpha_5 h_1 + 3\alpha_5 h_2 - 3\alpha_5\big) \overline{A}_1^2 A_2 e^{i\sigma_2 T_2} \\
&+\frac{1}{2}(\alpha_3 - \alpha_4 h_1) \overline{A}_1 e^{i\sigma_1 T_2} + \frac{1}{2}(\alpha_3 - \alpha_4 h_2) A_2 e^{i(\sigma_2 - \sigma_1)T_2} \Big] e^{i\omega_1 T_0} \\
&+\big[-2i\omega_2 h_2 A_2' + (6\alpha_6 h_1^2 h_2 - 6\alpha_5 h_1^2 - 12\alpha_5 h_1 h_2 \\
&+ 6\alpha_5 h_2 + 12\alpha_5 h_1 - 6\alpha_5) A_1 \overline{A}_1 A_2 \\
&+\big(3\alpha_6 h_2^3 - 9\alpha_5 h_2^2 + 9\alpha_5 h_2 - 3\alpha_5\big) A_2^2 \overline{A}_2 \\
&+\big(\alpha_6 h_1^3 - 3\alpha_5 h_1^2 + 3\alpha_5 h_1 - \alpha_5\big) A_1^3 e^{-i\sigma_2 T_2} \\
&+\frac{1}{2}(\alpha_3 - \alpha_4 h_1) A_1 e^{i(\sigma_1 - \sigma_2)T_2} \Big] e^{i\omega_2 T_0} + cc + NST
\end{aligned}
\tag{6.13.68}
$$

In order to eliminate secular terms of Eqs. (6.13.28) and (6.13.29), we seek the following special solutions for them:

$$
\begin{aligned}
u_{13} &= P_{11} e^{i\omega_1 T_0} + P_{12} e^{i\omega_2 T_0} \\
u_{23} &= P_{21} e^{i\omega_1 T_0} + P_{22} e^{i\omega_2 T_0}
\end{aligned}
\tag{6.13.69}
$$

Substituting (6.13.30) into (6.13.28) and (6.13.29), equating the coefficients of $e^{i\omega_1 T_0}$ and $e^{i\omega_2 T_0}$, we get

$$\begin{bmatrix} 2\alpha_1 - \omega_n^2 & -\alpha_1 \\ -\alpha_3 & \alpha_4 - \omega_n^2 \end{bmatrix} \begin{Bmatrix} P_{1n} \\ P_{2n} \end{Bmatrix} = \begin{Bmatrix} R_{1n} \\ R_{2n} \end{Bmatrix}, \quad n = 1, \ 2 \tag{6.13.70}$$

where

$$\begin{aligned} R_{11} &= -2i\omega_1 A_1' + \big(-6\alpha_2 h_1 h_2^2 + 6\alpha_2 h_2^2 + 12\alpha_2 h_1 h_2 \\ &- 6\alpha_2 h_1 - 12\alpha_2 h_2 + 12\alpha_2\big) A_1 A_2 \overline{A}_2 \\ &+ \big(-3\alpha_2 h_1^3 + 9\alpha_2 h_1^2 - 9\alpha_2 h_1 + 6\alpha_2\big) A_1^2 \overline{A}_1 \\ &+ \big(-3\alpha_2 h_1^2 h_2 + 3\alpha_2 h_1^2 + 6\alpha_2 h_1 h_2 - 6\alpha_2 h_1 - 3\alpha_2 h_2 + 6\alpha_2\big) \overline{A}_1^2 A_2 e^{i\sigma_2 T_2} \\ &+ \left(\frac{1}{2}\alpha_1 h_1 - \alpha_1\right) \overline{A}_1 e^{i\sigma_1 T_2} + \left(\frac{1}{2}\alpha_1 h_2 - \alpha_1\right) A_2 e^{i(-\sigma_1 + \sigma_2) T_2} \end{aligned} \tag{6.13.71}$$

$$\begin{aligned} R_{21} &= -2i\omega_1 h_1 A_1' + \big(6\alpha_6 h_1 h_2^2 - 6\alpha_5 h_2^2 - 12\alpha_5 h_1 h_2 \\ &+ 6\alpha_5 h_1 + 12\alpha_5 h_2 - 6\alpha_5\big) A_1 A_2 \overline{A}_2 \\ &+ \big(3\alpha_6 h_1^3 - 9\alpha_5 h_1^2 + 9\alpha_5 h_1 - 3\alpha_5\big) A_1^2 \overline{A}_1 \\ &+ \big(3\alpha_6 h_1^2 h_2 - 3\alpha_5 h_1^2 - 6\alpha_5 h_1 h_2 + 6\alpha_5 h_1 + 3\alpha_5 h_2 - 3\alpha_5\big) \overline{A}_1^2 A_2 e^{i\sigma_2 T_2} \\ &+ \frac{1}{2}(\alpha_3 - \alpha_4 h_1) \overline{A}_1 e^{i\sigma_1 T_2} + \frac{1}{2}(\alpha_3 - \alpha_4 h_2) A_2 e^{i(\sigma_2 - \sigma_1) T_2} \end{aligned} \tag{6.13.72}$$

$$\begin{aligned} R_{12} &= -2i\omega_2 A_2' + \big(-6\alpha_2 h_1^2 h_2 + 6\alpha_2 h_1^2 + 12\alpha_2 h_1 h_2 \\ &- 6\alpha_2 h_2 - 12\alpha_2 h_1 + 12\alpha_2\big) A_1 \overline{A}_1 A_2 \\ &+ \big(-3\alpha_2 h_2^3 + 9\alpha_2 h_2^2 - 9\alpha_2 h_2 + 6\alpha_2\big) A_2^2 \overline{A}_2 \\ &+ \big(-\alpha_2 h_1^3 + 3\alpha_2 h_1^2 - 3\alpha_2 h_1 + 2\alpha_2\big) A_1^3 e^{-i\sigma_2 T_2} + \left(\frac{1}{2}\alpha_1 h_1 - \alpha_1\right) A_1 e^{i(\sigma_1 - \sigma_2) T_2} \end{aligned} \tag{6.13.73}$$

$$\begin{aligned} R_{22} &= -2i\omega_2 h_2 A_2' + \big(6\alpha_6 h_1^2 h_2 - 6\alpha_5 h_1^2 - 12\alpha_5 h_1 h_2 \\ &+ 6\alpha_5 h_2 + 12\alpha_5 h_1 - 6\alpha_5\big) A_1 \overline{A}_1 A_2 \\ &+ \big(3\alpha_6 h_2^3 - 9\alpha_5 h_2^2 + 9\alpha_5 h_2 - 3\alpha_5\big) A_2^2 \overline{A}_2 \\ &+ \big(\alpha_6 h_1^3 - 3\alpha_5 h_1^2 + 3\alpha_5 h_1 - \alpha_5\big) A_1^3 e^{-i\sigma_2 T_2} + \frac{1}{2}(\alpha_3 - \alpha_4 h_1) A_1 e^{i(\sigma_1 - \sigma_2) T_2} \end{aligned} \tag{6.13.74}$$

According to Eq. (6.13.12), the determinant of the coefficient matrix of (6.13.31) is zero:

$$\det \begin{bmatrix} R_{1n} & -\alpha_1 \\ R_{2n} & \alpha_4 - \omega_n^2 \end{bmatrix} = 0 \Rightarrow \big(\alpha_4 - \omega_n^2\big) R_{1n} + \alpha_1 R_{2n} = 0 \tag{6.13.75}$$

Substitute (6.13.71), (6.13.72), (6.13.73) and (6.13.74), respectively, into (6.13.75), we obtain

$$2iA_1' = b_{11}A_1A_2\overline{A}_2 + b_{12}A_1^2\overline{A}_1 + b_{13}\overline{A}_1e^{i\sigma_1T_2} + b_{14}\overline{A}_1^2A_2e^{i\sigma_2T_2} + b_{15}A_2e^{i(\sigma_2-\sigma_1)T_2} \tag{6.13.76}$$

$$2iA_2' = b_{21}A_1\overline{A}_1A_2 + b_{22}A_2^2\overline{A}_2 + b_{23}A_1^3e^{-i\sigma_2T_2} + b_{24}A_1e^{i(\sigma_1-\sigma_2)T_2} \tag{6.13.77}$$

where

$$\begin{aligned}
&\Delta_1 = \omega_1\left(\alpha_4 - \omega_1^2 + \alpha_1h_1\right)\\
&b_{11} = \Delta_1^{-1}\Big[\left(\alpha_4 - \omega_1^2\right)\left(-6\alpha_2h_1h_2^2 + 6\alpha_2h_2^2 + 12\alpha_2h_1h_2 - 6\alpha_2h_1 - 12\alpha_2h_2 + 12\alpha_2\right)\\
&+\alpha_1\left(6\alpha_6h_1h_2^2 - 6\alpha_5h_2^2 - 12\alpha_5h_1h_2 + 6\alpha_5h_1 + 12\alpha_5h_2 - 6\alpha_5\right)\Big]\\
&b_{12} = \Delta_1^{-1}\Big[\left(\alpha_4 - \omega_1^2\right)\left(-3\alpha_2h_1^3 + 9\alpha_2h_1^2 - 9\alpha_2h_1 + 6\alpha_2\right)\\
&+\alpha_1\left(3\alpha_6h_1^3 - 9\alpha_5h_1^2 + 9\alpha_5h_1 - 3\alpha_5\right)\Big]\\
&b_{13} = \Delta_1^{-1}\left[\left(\alpha_4 - \omega_1^2\right)\left(\frac{1}{2}\alpha_1h_1 - \alpha_1\right) + \frac{1}{2}\alpha_1(\alpha_3 - \alpha_4h_1)\right]\\
&b_{14} = \Delta_1^{-1}\Big[\left(\alpha_4 - \omega_1^2\right)\left(-3\alpha_2h_1^2h_2 + 3\alpha_2h_1^2 + 6\alpha_2h_1h_2 - 6\alpha_2h_1 - 3\alpha_2h_2 + 6\alpha_2\right)\\
&+\alpha_1\left(3\alpha_6h_1^2h_2 - 3\alpha_5h_1^2 - 6\alpha_5h_1h_2 + 6\alpha_5h_1 + 3\alpha_5h_2 - 3\alpha_5\right)\Big]\\
&b_{15} = \Delta_1^{-1}\left[\left(\alpha_4 - \omega_1^2\right)\left(\frac{1}{2}\alpha_1h_2 - \alpha_1\right) + \frac{1}{2}\alpha_1(\alpha_3 - \alpha_4h_2)\right]
\end{aligned} \tag{6.13.78}$$

$$\begin{aligned}
&\Delta_2 = \omega_2\left(\alpha_4 - \omega_2^2 + \alpha_1h_2\right)\\
&b_{21} = \Delta_2^{-1}\Big[\left(\alpha_4 - \omega_2^2\right)\left(-6\alpha_2h_1^2h_2 + 6\alpha_2h_1^2 + 12\alpha_2h_1h_2 - 6\alpha_2h_2 - 12\alpha_2h_1 + 12\alpha_2\right)\\
&+\alpha_1\left(6\alpha_6h_1^2h_2 - 6\alpha_5h_1^2 - 12\alpha_5h_1h_2 + 6\alpha_5h_2 + 12\alpha_5h_1 - 6\alpha_5\right)\Big]\\
&b_{22} = \Delta_2^{-1}\Big[\left(\alpha_4 - \omega_2^2\right)\left(-3\alpha_2h_2^3 + 9\alpha_2h_2^2 - 9\alpha_2h_2 + 6\alpha_2\right)\\
&+\alpha_1\left(3\alpha_6h_2^3 - 9\alpha_5h_2^2 + 9\alpha_5h_2 - 3\alpha_5\right)\Big]\\
&b_{23} = \Delta_2^{-1}\Big[\left(\alpha_4 - \omega_2^2\right)\left(-\alpha_2h_1^3 + 3\alpha_2h_1^2 - 3\alpha_2h_1 + 2\alpha_2\right)\\
&+\alpha_1\left(\alpha_6h_1^3 - 3\alpha_5h_1^2 + 3\alpha_5h_1 - \alpha_5\right)\Big]\\
&b_{24} = \Delta_2^{-1}\left[\left(\alpha_4 - \omega_2^2\right)\left(\frac{1}{2}\alpha_1h_1 - \alpha_1\right) + \frac{1}{2}\alpha_1(\alpha_3 - \alpha_4h_1)\right]
\end{aligned} \tag{6.13.79}$$

Let

$$A_1 = \frac{1}{2}a_1e^{i\beta_1}, \quad A_2 = \frac{1}{2}a_2e^{i\beta_2} \tag{6.13.80}$$

Substituting (6.13.80) into (6.13.35) and (6.13.36), we get

$$ia_1' - a_1\beta_1' = \tfrac{1}{8}b_{11}a_1a_2^2 + \tfrac{1}{8}b_{12}a_1^3 + \tfrac{1}{2}b_{13}a_1e^{i(\sigma_1T_2-2\beta_1)} \\ +\tfrac{1}{8}b_{14}a_1^2a_2e^{i(\sigma_2T_2-3\beta_1+\beta_2)} + \tfrac{1}{2}b_{15}a_2e^{i(\sigma_2T_2-\sigma_1T_2-\beta_1+\beta_2)} \tag{6.13.81}$$

$$ia_2' - a_2\beta_2' = \tfrac{1}{8}b_{21}a_1^2a_2 + \tfrac{1}{8}b_{22}a_2^3 + \tfrac{1}{2}b_{23}a_1^3e^{-i(\sigma_2T_2-3\beta_1+\beta_2)} \\ +\tfrac{1}{2}b_{24}a_1e^{-i(\sigma_2T_2-\sigma_1T_2-\beta_1+\beta_2)} \tag{6.13.82}$$

Separating the real and imaginary parts of the above equations yields

$$a_1' = \tfrac{1}{2}b_{13}a_1\sin\gamma_1 + \tfrac{1}{8}b_{14}a_1^2a_2\sin\gamma_2 + \tfrac{1}{2}b_{15}a_2\sin(\gamma_2-\gamma_1) \\ a_1\gamma_1' = \sigma_1a_1 + \tfrac{1}{4}b_{11}a_1a_2^2 + \tfrac{1}{4}b_{12}a_1^3 + b_{13}a_1\cos\gamma_1 + \tfrac{1}{4}b_{14}a_1^2a_2\cos\gamma_2 \\ +b_{15}a_2\cos(\gamma_2-\gamma_1) \tag{6.13.83}$$

$$a_2' = -\tfrac{1}{2}b_{23}a_1^3\sin\gamma_2 - \tfrac{1}{2}b_{24}a_1\sin(\gamma_2-\gamma_1) \\ 3a_2\gamma_1' - 2a_2\gamma_2' = 3\sigma_1a_2 - 2\sigma_2a_2 + \tfrac{1}{4}b_{21}a_1^2a_2 + \tfrac{1}{4}b_{22}a_2^3 + b_{23}a_1^3\cos\gamma_2 \\ +b_{24}a_1\cos(\gamma_2-\gamma_1) \tag{6.13.84}$$

where

$$\gamma_1 = \sigma_1T_2 - 2\beta_1, \quad \gamma_2 = \sigma_2T_2 - 3\beta_1 + \beta_2 \tag{6.13.85}$$

Let $a_1' = a_2' = \gamma_1' = \gamma_2' = 0$, the equation satisfied by the constant solution is given by

$$b_{13}a_1\sin\gamma_1 + \frac{1}{4}b_{14}a_1^2a_2\sin\gamma_2 + b_{15}a_2\sin(\gamma_2-\gamma_1) = 0 \tag{6.13.86}$$

$$\sigma_1a_1 + \frac{1}{4}b_{11}a_1a_2^2 + \frac{1}{4}b_{12}a_1^3 + b_{13}a_1\cos\gamma_1 + \frac{1}{4}b_{14}a_1^2a_2\cos\gamma_2 + b_{15}a_2\cos(\gamma_2-\gamma_1) = 0 \tag{6.13.87}$$

$$b_{23}a_1^3\sin\gamma_2 + b_{24}a_1\sin(\gamma_2-\gamma_1) = 0 \tag{6.13.88}$$

$$3\sigma_1a_2 - 2\sigma_2a_2 + \frac{1}{4}b_{21}a_1^2a_2 + \frac{1}{4}b_{22}a_2^3 + b_{23}a_1^3\cos\gamma_2 + b_{24}a_1\cos(\gamma_2-\gamma_1) = 0 \tag{6.13.89}$$

It is clear that (6.13.86)–(6.13.89) have a set of trivial steady state solution, i.e., $a_1 = a_2 = 0$. The frequency–response equation corresponding to the non-trivial steady state solution is given below. From (6.13.88) and (6.13.89), we have

$$\cos\gamma_1 = \frac{\Gamma_1^2 - b_{23}^2a_1^6 - b_{24}^2a_1^2}{2b_{23}b_{24}a_1^4}, \quad \cos\gamma_2 = \frac{b_{24}^2a_1^2 - \Gamma_1^2 - b_{23}^2a_1^6}{2\Gamma_1b_{23}a_1^3} \tag{6.13.90}$$

where

$$\Gamma_1 = 3\sigma_1 a_2 - 2\sigma_2 a_2 + \frac{1}{4}b_{21}a_1^2 a_2 + \frac{1}{4}b_{22}a_2^3 \tag{6.13.91}$$

By (6.13.86) and (6.13.88), we have

$$b_{13}\sin\gamma_1 = \left(\frac{b_{15}b_{23}a_1a_2}{b_{24}} - \frac{1}{4}b_{14}a_1a_2\right)\sin\gamma_2 \tag{6.13.92}$$

From (6.13.88) and taking the expression of $\cos\gamma_2$ in (6.13.90) into account, we have

$$\frac{b_{23}^2a_1^4}{b_{24}^2}\left[1 - \left(\frac{b_{24}^2a_1^2 - \Gamma_1^2 - b_{23}^2a_1^6}{2\Gamma_1 b_{23}a_1^3}\right)^2\right] = \sin^2(\gamma_2 - \gamma_1) \tag{6.13.93}$$

Substituting (6.13.90) into (6.13.87), we can obtain

$$\frac{1}{b_{15}^2a_2^2}\left[\Gamma_2 + \frac{b_{13}(\Gamma_1^2 - b_{23}^2a_1^6 - b_{24}^2a_1^2)}{2b_{23}b_{24}a_1^3} + \frac{b_{14}a_2(b_{24}^2a_1^2 - \Gamma_1^2 - b_{23}^2a_1^6)}{8\Gamma_1 b_{23}a_1}\right]^2 = \cos^2(\gamma_2 - \gamma_1) \tag{6.13.94}$$

where

$$\Gamma_2 = \sigma_1 a_1 + \frac{1}{4}b_{11}a_1a_2^2 + \frac{1}{4}b_{12}a_1^3 \tag{6.13.95}$$

Squaring both sides of (6.13.92) and substituting (6.13.90) into it, we can obtain a frequency–response equation as the following:

$$\begin{aligned}&\left(\frac{b_{15}b_{23}a_1a_2}{b_{24}} - \frac{1}{4}b_{14}a_1a_2\right)^2\left[1 - \left(\frac{b_{24}^2a_1^2 - \Gamma_1^2 - b_{23}^2a_1^6}{2\Gamma_1 b_{23}a_1^3}\right)^2\right]\\ &= b_{13}^2\left[1 - \left(\frac{\Gamma_1^2 - b_{23}^2a_1^6 - b_{24}^2a_1^2}{2b_{23}b_{24}a_1^4}\right)^2\right]\end{aligned} \tag{6.13.96}$$

Adding (6.13.93) to (6.13.94) yields another frequency–response equation:

$$\begin{aligned}&\frac{b_{23}^2a_1^4}{b_{24}^2}\left[1 - \left(\frac{b_{24}^2a_1^2 - \Gamma_1^2 - b_{23}^2a_1^6}{2\Gamma_1 b_{23}a_1^3}\right)^2\right]\\ &+\frac{1}{b_{15}^2a_2^2}\left[\Gamma_2 + \frac{b_{13}(\Gamma_1^2 - b_{23}^2a_1^6 - b_{24}^2a_1^2)}{2b_{23}b_{24}a_1^3} + \frac{b_{14}a_2(b_{24}^2a_1^2 - \Gamma_1^2 - b_{23}^2a_1^6)}{8\Gamma_1 b_{23}a_1}\right]^2 = 1\end{aligned} \tag{6.13.97}$$

Non-trivial steady state response of the system can be obtained numerically from (6.13.96), (6.13.97) and (6.13.90).

Again, analyze the stability of the steady state solution. Let a_{10}, a_{20}, γ_{10} and γ_{20} be a set of steady state solution and make a small perturbation of them, i.e.,

$$a_1 = a_{10} + \tilde{a}_1, \quad a_2 = a_{20} + \tilde{a}_2, \quad \gamma_1 = \gamma_{10} + \tilde{\gamma}_1, \quad \gamma_2 = \gamma_{20} + \tilde{\gamma}_2 \tag{6.13.98}$$

Substitute (6.13.98) into (6.13.83) and (6.13.84) and make the linearization, we can obtain

$$\begin{bmatrix} 1 & 0 & 0 & 0 \\ 0 & 1 & 0 & 0 \\ 0 & 0 & a_{10} & 0 \\ 0 & 0 & 3a_{20} & -2a_{20} \end{bmatrix} \begin{Bmatrix} \tilde{a}_1' \\ \tilde{a}_2' \\ \tilde{\gamma}_1' \\ \gamma_2' \end{Bmatrix} = \begin{bmatrix} l_{11} & l_{11} & l_{11} & l_{11} \\ l_{11} & l_{11} & l_{11} & l_{11} \\ l_{11} & l_{11} & l_{11} & l_{11} \\ l_{11} & l_{11} & l_{11} & l_{11} \end{bmatrix} \begin{Bmatrix} \tilde{a}_1 \\ \tilde{a}_2 \\ \tilde{\gamma}_1 \\ \gamma_2 \end{Bmatrix} \tag{6.13.99}$$

where

$$\begin{aligned} l_{11} &= \tfrac{1}{2} b_{13} \sin\gamma_{10} + \tfrac{1}{4} b_{14} a_{10} a_{20} \sin\gamma_{20} \\ l_{12} &= \tfrac{1}{8} b_{14} a_{10}^2 \sin\gamma_{20} + \tfrac{1}{2} b_{15} \sin(\gamma_{20} - \gamma_{10}) \\ l_{13} &= \tfrac{1}{2} b_{13} a_{10} \cos\gamma_{10} - \tfrac{1}{2} b_{15} a_{20} \cos(\gamma_{20} - \gamma_{10}) \\ l_{14} &= \tfrac{1}{8} b_{14} a_{10}^2 a_{20} \cos\gamma_{20} + \tfrac{1}{2} b_{15} a_{20} \sin(\gamma_{20} - \gamma_{10}) \end{aligned} \tag{6.13.100}$$

$$\begin{aligned} l_{21} &= -\tfrac{3}{2} b_{23} a_1^2 \sin\gamma_{20} - \tfrac{1}{2} b_{24} \sin(\gamma_{20} - \gamma_{10}) \\ l_{22} &= 0 \\ l_{23} &= \tfrac{1}{2} b_{24} a_{10} \cos(\gamma_{20} - \gamma_{10}) \tilde{\gamma}_1 \\ l_{24} &= -\tfrac{1}{2} b_{23} a_{10}^3 \cos\gamma_{20} - \tfrac{1}{2} b_{24} a_{10} \cos(\gamma_{20} - \gamma_{10}) \end{aligned} \tag{6.13.101}$$

$$\begin{aligned} l_{31} &= \sigma_1 + \frac{1}{4} b_{11} a_{20}^2 + \frac{3}{4} b_{12} a_{10}^2 + b_{13} \cos\gamma_{10} + \frac{1}{2} b_{14} a_{10} a_{20} \cos\gamma_{20} \\ l_{32} &= \frac{1}{2} b_{11} a_{10} a_{20} + \frac{1}{4} b_{14} a_{10}^2 \cos\gamma_{20} + b_{15} \cos(\gamma_{20} - \gamma_{10}) \\ l_{33} &= -b_{13} a_{10} \sin\gamma_{10} + b_{15} a_{20} \sin(\gamma_{20} - \gamma_{10}) \\ l_{34} &= -\frac{1}{4} b_{14} a_{10}^2 a_{20} \sin\gamma_{20} - b_{15} a_{20} \sin(\gamma_{20} - \gamma_{10}) \end{aligned} \tag{6.13.102}$$

$$\begin{aligned} l_{41} &= \tfrac{1}{2} b_{21} a_{10} a_{20} + 3 b_{23} a_{10}^2 \cos\gamma_{20} + b_{24} \cos(\gamma_{20} - \gamma_{10}) \\ l_{42} &= 3\sigma_1 - 2\sigma_2 + \tfrac{1}{4} b_{21} a_{10}^2 + \tfrac{3}{4} b_{22} a_{20}^2 \\ l_{43} &= b_{24} a_{10} \sin(\gamma_{20} - \gamma_{10}) \\ l_{44} &= -b_{23} a_{10}^3 \sin\gamma_{20} - b_{24} a_{10} \sin(\gamma_{20} - \gamma_{10}) \end{aligned} \tag{6.13.103}$$

The steady state solutions are stable if the eigenvalues of (6.13.99) have negative real parts.

6.14 Exercise 6.14 (Internal Resonance, Parametric Excitation and Saturation Phenomena for a Spring Pendulum with a Moving Support)

Solution: (a) The kinetic energy of the system is

$$\begin{aligned} T &= \tfrac{1}{2}m\left\{\left[(l+x)\dot{\theta}\cos\theta + \dot{x}\sin\theta\right]^2 + \left[(l+x)\dot{\theta}\sin\theta - \dot{x}\cos\theta + l\dot{y}\right]^2\right\} \\ &= \tfrac{1}{2}m\dot{x}^2 + \tfrac{1}{2}m(l+x)^2\dot{\theta}^2 + l^2\dot{y}^2 + ml(l+x)\dot{y}\dot{\theta}\sin\theta - ml\dot{y}\dot{x}\cos\theta \end{aligned} \tag{6.14.1}$$

Using the equilibrium position $x = \theta = 0$ as the zero potential surface, the potential energy of the system is

$$V = \frac{1}{2}kx^2 + mg(l+x)(1-\cos\theta) \tag{6.14.2}$$

Substitute kinetic energy, potential energy, and generalized forces into the Lagrange's equation:

$$\frac{d}{dt}\frac{\partial T}{\partial \dot{y}} - \frac{\partial T}{\partial y} = -\frac{\partial V}{\partial y} + Q_y, \quad \frac{d}{dt}\frac{\partial T}{\partial \dot{\theta}} - \frac{\partial T}{\partial \theta} = -\frac{\partial V}{\partial \theta} + Q_\theta$$

The differential equation of motion of the system is

$$\begin{aligned} & m\ddot{x} - ml\ddot{y}\cos\theta - m(l+x)\dot{\theta}^2 = -kx - mg(1-\cos\theta) \\ & m(l+x)\ddot{\theta} + 2m\dot{x}\dot{\theta} + ml\ddot{y}\sin\theta = -mg\sin\theta \end{aligned} \tag{6.14.3}$$

Let

$$u = x/l, \quad \omega_1^2 = k/m, \quad \omega_2^2 = g/l \tag{6.14.4}$$

Equation (6.14.3) changes into

$$\begin{aligned} & \ddot{u} + \omega_1^2 u - (1+u)\dot{\theta}^2 + \omega_2^2(1-\cos\theta) - \ddot{y}\cos\theta = 0 \\ & (1+u)\ddot{\theta} + 2\dot{u}\dot{\theta} + \left(\omega_2^2 + \ddot{y}\right)\sin\theta = 0 \end{aligned} \tag{6.14.5}$$

(b) When $\ddot{y} \equiv 0$, (6.14.5) becomes

$$\begin{aligned} & \ddot{u} + \omega_1^2 u - (1+u)\dot{\theta}^2 + \omega_2^2(1-\cos\theta) = 0 \\ & (1+u)\ddot{\theta} + 2\dot{u}\dot{\theta} + \omega_2^2\sin\theta = 0 \end{aligned} \tag{6.14.6}$$

Expanding (6.14.6) at the equilibrium position $u = \theta = 0$ and retaining to the third order of smallness gives

$$\begin{aligned} &\ddot{u} + \omega_1^2 u - (1+u)\dot{\theta}^2 + \frac{1}{2}\omega_2^2\theta^2 = 0 \\ &\ddot{\theta} + \omega_2^2\theta + u\ddot{\theta} + 2\dot{u}\dot{\theta} - \frac{1}{6}\omega_2^2\theta^3 = 0 \end{aligned} \tag{6.14.7}$$

For small and finite amplitudes, we use the method of multiple scales to solve (6.14.7). Therefore, let the solution be expressed as

$$\begin{aligned} u &= \varepsilon u_{11}(T_0, T_1) + \varepsilon^2 u_{12}(T_0, T_1) + \cdots \\ \theta &= \varepsilon u_{21}(T_0, T_1) + \varepsilon^2 u_{22}(T_0, T_1) + \cdots \end{aligned} \tag{6.14.8}$$

Substitute (6.14.8) into (6.14.7) and keep to $O(\varepsilon^2)$, we get

$$\begin{aligned} 0 &= \ddot{u} + \omega_1^2 u - (1+u)\dot{\theta}^2 + \frac{1}{2}\omega_2^2\theta^2 \\ &= \left(D_0^2 + 2\varepsilon D_0 D_1\right)\left(\varepsilon u_{11} + \varepsilon^2 u_{12}\right) + \omega_1^2\left(\varepsilon u_{11} + \varepsilon^2 u_{12}\right) \\ &\quad -\left(1 + \varepsilon u_{11} + \varepsilon^2 u_{12}\right)\left[(D_0 + \varepsilon D_1)\left(\varepsilon u_{21} + \varepsilon^2 u_{22}\right)\right]^2 \\ &\quad +\frac{1}{2}\omega_2^2\left(\varepsilon u_{21} + \varepsilon^2 u_{22}\right)^2 + \cdots \\ &= \varepsilon\left(D_0^2 u_{11} + \omega_1^2 u_{11}\right) + \varepsilon^2\left[D_0^2 u_{12} + \omega_1^2 u_{12} + 2D_0 D_1 u_{11}\right. \\ &\quad \left. -(D_0 u_{21})^2 + \frac{1}{2}\omega_2^2 u_{21}^2\right] + \cdots \end{aligned} \tag{6.14.9}$$

$$\begin{aligned} 0 &= \ddot{\theta} + \omega_2^2\theta + u\ddot{\theta} + 2\dot{u}\dot{\theta} - \frac{1}{6}\omega_2^2\theta^3 \\ &= \left(D_0^2 + 2\varepsilon D_0 D_1\right)\left(\varepsilon u_{21} + \varepsilon^2 u_{22}\right) + \omega_2^2\left(\varepsilon u_{21} + \varepsilon^2 u_{22}\right) \\ &\quad +\left(\varepsilon u_{11} + \varepsilon^2 u_{12}\right)\left(D_0^2 + 2\varepsilon D_0 D_1\right)\left(\varepsilon u_{21} + \varepsilon^2 u_{22}\right) \\ &\quad +2\left[(D_0 + \varepsilon D_1)\left(\varepsilon u_{11} + \varepsilon^2 u_{12}\right)\right]\left[(D_0 + \varepsilon D_1)\left(\varepsilon u_{21} + \varepsilon^2 u_{22}\right)\right] + \cdots \\ &= \varepsilon\left(D_0^2 u_{21} + \omega_2^2 u_{21}\right) + \varepsilon^2\left[D_0^2 u_{22} + \omega_2^2 u_{22} + 2D_0 D_1 u_{21}\right. \\ &\quad \left. + u_{11} D_0^2 u_{21} + 2(D_0 u_{11})(D_0 u_{21})\right] + \cdots \end{aligned} \tag{6.14.10}$$

Equating the coefficients of the like powers of ε in (6.14.9) and (6.14.10), we get.
Order ε^1:

$$\begin{aligned} D_0^2 u_{11} + \omega_1^2 u_{11} &= 0 \\ D_0^2 u_{21} + \omega_2^2 u_{21} &= 0 \end{aligned} \tag{6.14.11}$$

Order ε^3:

$$\begin{aligned} D_0^2 u_{12} + \omega_1^2 u_{12} &= -2D_0 D_1 u_{11} + (D_0 u_{21})^2 - \tfrac{1}{2}\omega_2^2 u_{21}^2 \\ D_0^2 u_{22} + \omega_2^2 u_{22} &= -2D_0 D_1 u_{21} - u_{11} D_0^2 u_{21} - 2(D_0 u_{11})(D_0 u_{21}) \end{aligned} \tag{6.14.12}$$

The solution of (6.14.11) can be expressed as

$$u_{11} = A_1 e^{i\omega_1 T_0} + cc, \quad u_{21} = A_2 e^{i\omega_2 T_0} + cc \tag{6.14.13}$$

where $A_k = A_k(T_1), \ k = 1, 2$. Substituting (6.14.13) into (6.14.12), we get

$$D_0^2 u_{12} + \omega_1^2 u_{12} = -2i\omega_1 D_1 A_1 e^{i\omega_1 T_0} - \frac{3}{2}\omega_2^2 A_2^2 e^{2i\omega_2 T_0} + cc + \omega_2^2 A_2 \overline{A}_2 \tag{6.14.14}$$

$$\begin{aligned} D_0^2 u_{22} + \omega_2^2 u_{22} &= -2i\omega_2 D_1 A_2 e^{i\omega_2 T_0} + \left(\omega_2^2 + 2\omega_1\omega_2\right) A_1 \overline{A}_2 e^{i(\omega_1 - \omega_2)T_0} \\ &+ \left(\omega_2^2 - 2\omega_1\omega_2\right) A_1 A_2 e^{i(\omega_1 + \omega_2)T_0} + cc \end{aligned} \tag{6.14.15}$$

When $\omega_1 \approx 2\omega_2$, the detuning parameter σ_2 is introduced such that

$$\omega_1 = 2\omega_1 + \varepsilon\sigma \tag{6.14.16}$$

In order to eliminate secular terms from the above two equations, it is necessary to have

$$2i\omega_1 D_1 A_1 + \frac{3}{2}\omega_2^2 A_2^2 e^{-i\sigma T_1} = 0 \tag{6.14.17}$$

$$2i\omega_2 D_1 A_2 + \left(2\omega_1\omega_2 - \omega_2^2\right) A_1 \overline{A}_2 e^{i\sigma T_1} = 0 \tag{6.14.18}$$

Let

$$A_1 = \frac{1}{2} a_1 e^{i\beta_1}, \quad A_2 = \frac{1}{2} a_2 e^{i\beta_2} \tag{6.14.19}$$

$$i\omega_1 a_1' - \omega_1 a_1 \beta_1' + \frac{3}{8}\omega_2^2 a_2^2 e^{-i(\sigma T_1 + \beta_1 - 2\beta_2)} = 0 \tag{6.14.20}$$

$$i\omega_2 a_2' - \omega_2 a_2 \beta_2' + \frac{1}{4}\left(2\omega_1\omega_2 - \omega_2^2\right) a_1 a_2 e^{i(\sigma T_1 + \beta_1 - 2\beta_2)} = 0 \tag{6.14.21}$$

Separating the real and imaginary parts of the above equation and taking into account $\omega_1 \approx 2\omega_2$ yields

$$\begin{aligned} a_1' &= \frac{3}{16}\omega_2 a_2^2 \sin\gamma \\ a_2' &= -\frac{3}{4}\omega_2 a_1 a_2 \sin\gamma \end{aligned} \tag{6.14.22}$$

$$\begin{aligned} a_1 \beta_1' &= \frac{3}{16}\omega_2 a_2^2 \cos\gamma \\ a_2 \beta_2' &= \frac{3}{4}\omega_2 a_1 a_2 \cos\gamma \end{aligned} \tag{6.14.23}$$

where

$$\gamma = \sigma T_1 + \beta_1 - 2\beta_2 \tag{6.14.24}$$

Replacing the independent variable T_1 with t and keeping εa_1 and εa_2 as a_1 and a_2, respectively, we obtain

$$\begin{aligned} \dot{a}_1 &= \frac{3}{16}\omega_2 a_2^2 \sin\gamma \\ \dot{a}_2 &= -\frac{3}{4}\omega_2 a_1 a_2 \sin\gamma \end{aligned} \tag{6.14.25}$$

$$\begin{aligned} a_1\dot{\beta}_1 &= \frac{3}{16}\omega_2 a_2^2 \cos\gamma \\ a_2\dot{\beta}_2 &= \frac{3}{4}\omega_2 a_1 a_2 \cos\gamma \end{aligned} \tag{6.14.26}$$

where

$$\gamma = (\omega_1 - 2\omega_2)t + \beta_1 - 2\beta_2 \tag{6.14.27}$$

The first order approximate solution of (6.14.6) is

$$\begin{aligned} u &= a_1 \sin(\omega_1 t + \beta_1) + \cdots \\ \theta &= a_2 \sin(\omega_2 t + \beta_2) + \cdots \end{aligned}$$

(c) When $\ddot{y} = 2\varepsilon K \cos\Omega t$, (6.14.5) becomes

$$\begin{aligned} &\ddot{u} + \omega_1^2 u - (1+u)\dot{\theta}^2 + \omega_2^2(1-\cos\theta) - 2\varepsilon K \cos\Omega t \cos\theta = 0 \\ &(1+u)\ddot{\theta} + 2\dot{u}\dot{\theta} + \left(\omega_2^2 + 2\varepsilon K\ \cos\Omega t\right)\sin\theta = 0 \end{aligned} \tag{6.14.28}$$

Expanding (6.14.28) at the equilibrium position $u = \theta = 0$ and retaining to the third order of smallness gives

$$\begin{aligned} &\ddot{u} + \omega_1^2 u - (1+u)\dot{\theta}^2 + \frac{1}{2}\omega_2^2\theta^2 - 2\varepsilon K\left(1 - \frac{1}{2}\theta^2\right)\cos\Omega t = 0 \\ &\ddot{\theta} + \omega_2^2\theta + u\ddot{\theta} + 2\dot{u}\dot{\theta} - \frac{1}{6}\omega_2^2\theta^3 + 2\varepsilon K\theta\cos\Omega t = 0 \end{aligned} \tag{6.14.29}$$

For small and finite amplitudes, we use the method of multiple scales to solve (6.14.29). Substituting (6.14.8) into (6.14.29) and retaining to $O(\varepsilon^2)$, we obtain

$$
\begin{aligned}
0 &= \ddot{u} + \omega_1^2 u - (1+u)\dot{\theta}^2 + \frac{1}{2}\omega_2^2\theta^2 - 2\varepsilon K\left(1 - \frac{1}{2}\theta^2\right)\cos\Omega t \\
&= \left(D_0^2 + 2\varepsilon D_0 D_1\right)\left(\varepsilon u_{11} + \varepsilon^2 u_{12}\right) + \omega_1^2\left(\varepsilon u_{11} + \varepsilon^2 u_{12}\right) \\
&\quad -\left(1 + \varepsilon u_{11} + \varepsilon^2 u_{12}\right)\left[\left(D_0 + \varepsilon D_1\right)\left(\varepsilon u_{21} + \varepsilon^2 u_{22}\right)\right]^2 \\
&\quad +\frac{1}{2}\omega_2^2\left(\varepsilon u_{21} + \varepsilon^2 u_{22}\right)^2 - 2\varepsilon K\cos\Omega T_0 + \cdots \\
&= \varepsilon\left(D_0^2 u_{11} + \omega_1^2 u_{11}\right) + \varepsilon^2\left[D_0^2 u_{12} + \omega_1^2 u_{12} + 2D_0 D_1 u_{11}\right. \\
&\quad \left. -(D_0 u_{21})^2 + \frac{1}{2}\omega_2^2 u_{21}^2\right] - 2\varepsilon K\cos\Omega T_0 + \cdots
\end{aligned}
\tag{6.14.30}
$$

$$
\begin{aligned}
0 &= \ddot{\theta} + \omega_2^2\theta + u\ddot{\theta} + 2\dot{u}\dot{\theta} - \frac{1}{6}\omega_2^2\theta^3 + 2\varepsilon K\theta\cos\Omega t \\
&= \left(D_0^2 + 2\varepsilon D_0 D_1\right)\left(\varepsilon u_{21} + \varepsilon^2 u_{22}\right) + \omega_2^2\left(\varepsilon u_{21} + \varepsilon^2 u_{22}\right) \\
&\quad +\left(\varepsilon u_{11} + \varepsilon^2 u_{12}\right)\left(D_0^2 + 2\varepsilon D_0 D_1\right)\left(\varepsilon u_{21} + \varepsilon^2 u_{22}\right) \\
&\quad +2\left[\left(D_0 + \varepsilon D_1\right)\left(\varepsilon u_{11} + \varepsilon^2 u_{12}\right)\right]\left[\left(D_0 + \varepsilon D_1\right)\left(\varepsilon u_{21} + \varepsilon^2 u_{22}\right)\right] \\
&\quad +2\varepsilon K\left(\varepsilon u_{21} + \varepsilon^2 u_{22}\right)\cos\Omega T_0 + \cdots \\
&= \varepsilon\left(D_0^2 u_{21} + \omega_2^2 u_{21}\right) + \varepsilon^2\left[D_0^2 u_{22} + \omega_2^2 u_{22} + 2D_0 D_1 u_{21}\right. \\
&\quad \left. + u_{11} D_0^2 u_{21} + 2(D_0 u_{11})(D_0 u_{21})\right] + 2\varepsilon^2 K u_{21}\cos\Omega T_0 + \cdots
\end{aligned}
\tag{6.14.31}
$$

When $\Omega = 2\omega_2 + \varepsilon\sigma$, and Ω is far from ω_1, we let the coefficients of like power of ε in (6.14.30) and (6.14.31) be zero, we get.

Order ε^1:

$$
\begin{aligned}
D_0^2 u_{11} + \omega_1^2 u_{11} &= 2K\cos\Omega T_0 \\
D_0^2 u_{21} + \omega_2^2 u_{21} &= 0
\end{aligned}
\tag{6.14.32}
$$

Order ε^3:

$$
\begin{aligned}
&D_0^2 u_{12} + \omega_1^2 u_{12} = -2D_0 D_1 u_{11} + (D_0 u_{21})^2 - \tfrac{1}{2}\omega_2^2 u_{21}^2 \\
&D_0^2 u_{22} + \omega_2^2 u_{22} = -2D_0 D_1 u_{21} - u_{11} D_0^2 u_{21} - 2(D_0 u_{11})(D_0 u_{21}) \\
&-2K u_{21}\cos\Omega T_0
\end{aligned}
\tag{6.14.33}
$$

The solution of (6.14.32) can be expressed as

$$
\begin{aligned}
u_{11} &= A_1 e^{i\omega_1 T_0} + K\left(\omega_1^2 - \Omega^2\right)^{-1} e^{i\Omega T_0} + cc \\
u_{21} &= A_2 e^{i\omega_2 T_0} + cc
\end{aligned}
\tag{6.14.34}
$$

where.$A_k = A_k(T_1)$, $k = 1, 2$ Substituting (6.14.34) into (6.14.33), we get

$$
D_0^2 u_{12} + \omega_1^2 u_{12} = -2i\omega_1 D_1 A_1 e^{i\omega_1 T_0} - \frac{3}{2}\omega_2^2 A_2^2 e^{2i\omega_2 T_0} + cc + \omega_2^2 A_2 \overline{A}_2 \tag{6.14.35}
$$

$$
\begin{aligned}
&D_0^2 u_{22} + \omega_2^2 u_{22} = -2i\omega_2 D_1 A_2 e^{i\omega_2 T_0} \\
&-K\left(2\omega_2\Omega - \omega_2^2 + \omega_1^2 - \Omega^2\right)\left(\omega_1^2 - \Omega^2\right)^{-1}\overline{A}_2 e^{i(\Omega-\omega_2)T_0} + cc
\end{aligned} \tag{6.14.36}
$$

When $\Omega = 2\omega_2 + \varepsilon\sigma$, in order to eliminate secular terms from the above two equations, there must be

$$
D_1 A_1 = 0 \tag{6.14.37}
$$

$$
2i\omega_2 D_1 A_2 + K\left(2\omega_2\Omega - \omega_2^2 + \omega_1^2 - \Omega^2\right)\left(\omega_1^2 - \Omega^2\right)^{-1}\overline{A}_2 e^{i\sigma T_1} = 0 \tag{6.14.38}
$$

Replacing the independent variable T_1 with t, (6.14.37) and (6.14.38) become

$$
\dot{A}_1 = 0 \tag{6.14.39}
$$

$$
2i\omega_2 \dot{A}_2 + \varepsilon K\left(2\omega_2\Omega - \omega_2^2 + \omega_1^2 - \Omega^2\right)\left(\omega_1^2 - \Omega^2\right)^{-1}\overline{A}_2 e^{i\varepsilon\sigma t} = 0 \tag{6.14.40}
$$

Let

$$
A_1 = \frac{1}{2} a_1 e^{i\beta_1}, \quad A_2 = \frac{1}{2} a_2 e^{i\beta_2} \tag{6.14.41}
$$

Substituting (6.14.41) into (6.14.34), we can obtain the first-order approximate solution of (6.14.6):

$$
\begin{aligned}
u &= \varepsilon A_1 e^{i\omega_1 T_0} + 2K\left(\omega_1^2 - \Omega^2\right)^{-1}\cos\Omega t + cc + O\left(\varepsilon^2\right) \\
\theta &= \varepsilon A_2 e^{i\omega_2 T_0} + cc + O\left(\varepsilon^2\right)
\end{aligned}
$$

By substituting (6.14.41) into (6.14.40), we get

$$
i\omega_2 \dot{a}_2 - \omega_2 a_2 \dot{\beta}_2 + \frac{1}{2}\varepsilon K\left(2\omega_2\Omega - \omega_2^2 + \omega_1^2 - \Omega^2\right)\left(\omega_1^2 - \Omega^2\right)^{-1} a_2 e^{i(\varepsilon\sigma t - 2\beta_2)} = 0 \tag{6.14.42}
$$

Separating the real and imaginary parts of the above equations yields

$$
\begin{aligned}
\omega_2 \dot{a}_2 &= -\tfrac{1}{2}\varepsilon K\left(2\omega_2\Omega - \omega_2^2 + \omega_1^2 - \Omega^2\right)\left(\omega_1^2 - \Omega^2\right)^{-1} a_2 \sin\gamma \\
\omega_2 a_2 \dot{\gamma} &= \varepsilon\omega_2 a_2 \sigma - \varepsilon K\left(2\omega_2\Omega - \omega_2^2 + \omega_1^2 - \Omega^2\right)\left(\omega_1^2 - \Omega^2\right)^{-1} a_2 \cos\gamma
\end{aligned} \tag{6.14.43}
$$

where

$$
\gamma = \varepsilon\sigma t - 2\beta_2 \tag{6.14.44}
$$

Clearly, $a_2 = 0$ (i.e.,$\theta = 0$) is a constant solution of the system. In this case, (6.14.43) is linearized as

$$\begin{aligned}&\omega_2\dot{\tilde{a}}_2 = \left[-\frac{1}{2}\varepsilon K\left(2\omega_2\Omega - \omega_2^2 + \omega_1^2 - \Omega^2\right)\left(\omega_1^2 - \Omega^2\right)^{-1}\sin\gamma_0\right]\tilde{a}_2\\&\left[\varepsilon\omega_2\sigma - \varepsilon K\left(2\omega_2\Omega - \omega_2^2 + \omega_1^2 - \Omega^2\right)\left(\omega_1^2 - \Omega^2\right)^{-1}\cos\gamma_0\right]\tilde{a}_2 = 0\end{aligned} \tag{6.14.45}$$

So the stability condition is

$$\varepsilon\omega_2\sigma - \varepsilon K\left(2\omega_2\Omega - \omega_2^2 + \omega_1^2 - \Omega^2\right)\left(\omega_1^2 - \Omega^2\right)^{-1}\cos\gamma_0 = 0 \tag{6.14.46}$$

To make γ_0 exist, there must be

$$\left|\omega_2\sigma\left[K\left(2\omega_2\Omega - \omega_2^2 + \omega_1^2 - \Omega^2\right)\left(\omega_1^2 - \Omega^2\right)^{-1}\right]\right| \le 1 \tag{6.14.47}$$

The critical condition separating stability from instability is

$$K\left(2\omega_2\Omega - \omega_2^2 + \omega_1^2 - \Omega^2\right) = \pm\omega_2\left(\omega_1^2 - \Omega^2\right)\sigma \tag{6.14.48}$$

(d) When $\ddot{y} = 2\varepsilon K\cos\Omega t$, (6.14.30) and (6.14.31) still hold, but the excitation frequency takes a different value. When $= \omega_1 + \varepsilon\sigma_1, \omega_1 = 2\omega_2 + \varepsilon\sigma_2$ there are both parametric excitation resonances and internal resonances. Therefore, it is necessary to specify $K = \varepsilon k$.

Let the coefficients of the same power of ε in (6.14.9) and (6.14.10) be zero, we can obtain.

Order ε^1:

$$\begin{aligned}D_0^2 u_{11} + \omega_1^2 u_{11} = 0\\D_0^2 u_{21} + \omega_2^2 u_{21} = 0\end{aligned} \tag{6.14.49}$$

Order ε^3:

$$\begin{aligned}D_0^2 u_{12} + \omega_1^2 u_{12} &= -2D_0D_1u_{11} + (D_0u_{21})^2 - \tfrac{1}{2}\omega_2^2 u_{21}^2 + 2k\cos\Omega T_0\\D_0^2 u_{22} + \omega_2^2 u_{22} &= -2D_0D_1u_{21} - u_{11}D_0^2u_{21} - 2(D_0u_{11})(D_0u_{21})\end{aligned} \tag{6.14.50}$$

The solution of (6.14.49) is still (6.14.13). Substituting (6.14.13) into (6.14.50), we get

$$D_0^2 u_{12} + \omega_1^2 u_{12} = -2i\omega_1 D_1 A_1 e^{i\omega_1 T_0} - \frac{3}{2}\omega_2^2 A_2^2 e^{2i\omega_2 T_0} + k e^{i\Omega T_0} + cc + \omega_2^2 A_2 \overline{A}_2 \tag{6.14.51}$$

$$D_0^2 u_{22} + \omega_2^2 u_{22} = -2i\omega_2 D_1 A_2 e^{i\omega_2 T_0} + \left(\omega_2^2 + 2\omega_1\omega_2\right) A_1 \overline{A}_2 e^{i(\omega_1-\omega_2)T_0} \\ +\left(\omega_2^2 - 2\omega_1\omega_2\right) A_1 A_2 e^{i(\omega_1+\omega_2)T_0} + cc \tag{6.14.52}$$

When $\Omega = \omega_1 + \varepsilon\sigma_1$ and $\omega_1 = 2\omega_2 + \varepsilon\sigma_2$, in order to eliminate secular terms from the above two equations, there must be

$$2i\omega_1 D_1 A_1 + \frac{3}{2}\omega_2^2 A_2^2 e^{-i\sigma_2 T_1} - k e^{i\sigma_1 T_1} = 0 \tag{6.14.53}$$

$$2i\omega_2 D_1 A_2 + \left(2\omega_1\omega_2 - \omega_2^2\right) A_1 \overline{A}_2 e^{i\sigma_2 T_1} = 0 \tag{6.14.54}$$

Let

$$A_1 = \frac{1}{2} a_1 e^{i\beta_1}, \quad A_2 = \frac{1}{2} a_2 e^{i\beta_2} \tag{6.14.55}$$

The first order approximate solution of the original equation is

$$u = \varepsilon a_1 \sin(\omega_1 t + \beta_1) + O\left(\varepsilon^2\right)$$
$$\theta = \varepsilon a_2 \sin(\omega_2 t + \beta_2) + O\left(\varepsilon^2\right)$$

Substituting (6.14.55) into (6.14.53) and (6.14.54) yields

$$i\omega_1 a_1' - \omega_1 a_1 \beta_1' + \frac{3}{8}\omega_2^2 a_2^2 e^{-i(\sigma_2 T_1 + \beta_1 - 2\beta_2)} - k e^{i(\sigma_1 T_1 - \beta_1)} = 0 \tag{6.14.56}$$

$$i\omega_2 a_2' - \omega_2 a_2 \beta_2' + \frac{1}{4}\left(2\omega_1\omega_2 - \omega_2^2\right) a_1 a_2 e^{i(\sigma_2 T_1 + \beta_1 - 2\beta_2)} = 0 \tag{6.14.57}$$

Separating the real and imaginary parts of the above equation and taking into account $\omega_1 \approx 2\omega_2$ yields

$$\begin{aligned} a_1' &= \frac{3}{16}\omega_2 a_2^2 \sin\gamma_1 + \frac{k}{\omega_1}\sin\gamma_2 \\ a_2' &= -\frac{3}{4}\omega_2 a_1 a_2 \sin\gamma_1 \end{aligned} \tag{6.14.58}$$

$$\begin{aligned} a_1\beta_1' &= \frac{3}{16}\omega_2 a_2^2 \cos\gamma_1 - \frac{k}{\omega_1}\cos\gamma_2 \\ a_2\beta_2' &= \frac{3}{4}\omega_2 a_1 a_2 \cos\gamma_1 \end{aligned} \tag{6.14.59}$$

where

$$\gamma_1 = \sigma_2 T_1 + \beta_1 - 2\beta_2, \quad \gamma_2 = \sigma_1 T_1 - \beta_1 \tag{6.14.60}$$

Substituting (6.14.60) into (6.14.59) yields

$$\begin{aligned} a_1\gamma_2' &= \sigma_1 a_1 - \tfrac{3}{16}\omega_2 a_2^2 \cos\gamma_1 + \tfrac{k}{\omega_1}\cos\gamma_2 \\ a_2(\gamma_1' + \gamma_2') &= (\sigma_1 + \sigma_2)a_2 - \tfrac{3}{2}\omega_2 a_1 a_2 \cos\gamma_1 \end{aligned} \tag{6.14.61}$$

From (6.14.58) and (6.14.61), we can find that the steady state solution should satisfy the following equations:

$$\begin{aligned} &\frac{3}{4}\omega_2 a_1 a_2 \sin\gamma_1 = 0 \\ &(\sigma_1 + \sigma_2)a_2 - \frac{3}{2}\omega_2 a_1 a_2 \cos\gamma_1 = 0 \end{aligned} \tag{6.14.62}$$

$$\begin{aligned} &\frac{3}{16}\omega_2 a_2^2 \sin\gamma_1 + \frac{k}{\omega_1}\sin\gamma_2 = 0 \\ &\sigma_1 a_1 - \frac{3}{16}\omega_2 a_2^2 \cos\gamma_1 + \frac{k}{\omega_1}\cos\gamma_2 = 0 \end{aligned} \tag{6.14.63}$$

It is easy to see that there is no trivial steady state response. From (6.14.62), we can obtain that

$$\sin\gamma_1 = 0, \quad \cos\gamma_1 = \frac{2(\sigma_1 + \sigma_2)}{3\omega_2 a_1} \tag{6.14.64}$$

Thus (6.14.63) changes into

$$\begin{aligned} &\frac{k}{\omega_1}\sin\gamma_2 = 0 \\ &\sigma_1 a_1 - \frac{3}{8}(\sigma_1 + \sigma_2)a_1^{-1}a_2^2 + \frac{k}{\omega_1}\cos\gamma_2 = 0 \end{aligned} \tag{6.14.65}$$

Eliminating γ_1 from (6.14.62) an γ_2 from (6.14.65), we can obtain

$$\begin{aligned} a_1 &= \pm\frac{2}{3}(\sigma_1 + \sigma_2) \\ a_2 &= 4\sqrt{\frac{2}{3}\sigma_1(\sigma_1 + \sigma_2) \pm \omega_1^{-1}k} \end{aligned} \tag{6.14.66}$$

As can be seen, a_1 is independent of the excitation amplitude k, so the amplitude a_1 is saturated.

Readers are invited to complete the stability analysis of steady state response.

6.15 Exercise 6.15 (The Response of a Ship Constrained to Pitch and Roll Only)

Solution: (a) For small and finite amplitudes, we use the method of multiple scales to solve the nonlinear equations presented in this exercise. Let the solution of the equations be

$$\begin{aligned}\varphi &= \varepsilon u_{11}(T_0, T_1) + \varepsilon^2 u_{12}(T_0, T_1) + \cdots \\ \theta &= \varepsilon u_{21}(T_0, T_1) + \varepsilon^2 u_{22}(T_0, T_1) + \cdots\end{aligned} \tag{6.15.1}$$

In order to make the damping and nonlinear terms appear in the second-order equation, let

$$\mu_1 = \varepsilon\hat{\mu}_1, \quad \mu_2 = \varepsilon\hat{\mu}_2 \tag{6.15.2}$$

Substituting (6.15.1) and (6.15.2) into the given nonlinear equation and retaining to $O(\varepsilon^2)$ yields

$$\begin{aligned}0 &= \ddot{\phi} + \omega_1^2\phi - 2\phi\theta + 2\mu_1\dot{\phi} - \sum_{n=1}^{N} K_n\cos(\Omega_n t + \theta_n) \\ &= \left(D_0^2 + 2\varepsilon D_0 D_1\right)\left(\varepsilon u_{11} + \varepsilon^2 u_{12}\right) + \omega_1^2\left(\varepsilon u_{11} + \varepsilon^2 u_{12}\right) \\ &\quad -2\left(\varepsilon u_{11} + \varepsilon^2 u_{12}\right)\left(\varepsilon u_{21} + \varepsilon^2 u_{22}\right) + 2\varepsilon\hat{\mu}_1(D_0 + \varepsilon D_1)\left(\varepsilon u_{11} + \varepsilon^2 u_{12}\right) \\ &\quad - \sum_{n=1}^{N} K_n\cos(\Omega_n t + \theta_n) + \cdots \\ &= \varepsilon\left(D_0^2 u_{11} + \omega_1^2 u_{11}\right) + \varepsilon^2\left(D_0^2 u_{12} + \omega_1^2 u_{12} + 2D_0 D_1 u_{11}\right. \\ &\quad \left. + 2\hat{\mu}_1 D_0 u_{11} - 2u_{11}u_{21}\right) - \sum_{n=1}^{N} K_n\cos(\Omega_n t + \theta_n) + \cdots\end{aligned} \tag{6.15.3}$$

$$\begin{aligned}0 &= \ddot{\theta} + \omega_2^2\theta - \phi^2 + 2\mu_2\dot{\theta} - \sum_{n=1}^{N} M_n\cos(\Omega_n t + \tau_n) \\ &= \left(D_0^2 + 2\varepsilon D_0 D_1\right)\left(\varepsilon u_{21} + \varepsilon^2 u_{22}\right) + \omega_2^2\left(\varepsilon u_{21} + \varepsilon^2 u_{22}\right) \\ &\quad -\left(\varepsilon u_{11} + \varepsilon^2 u_{12}\right)^2 + 2\varepsilon\hat{\mu}_2(D_0 + \varepsilon D_1)\left(\varepsilon u_{21} + \varepsilon^2 u_{22}\right) \\ &\quad - \sum_{n=1}^{N} M_n\cos(\Omega_n t + \tau_n) + \cdots \\ &= \varepsilon\left(D_0^2 u_{21} + \omega_2^2 u_{21}\right) + \varepsilon^2\left(D_0^2 u_{22} + \omega_2^2 u_{22} + 2D_0 D_1 u_{21}\right. \\ &\quad \left. + 2\hat{\mu}_2 D_0 u_{21} - u_{11}^2\right) - \sum_{n=1}^{N} M_n\cos(\Omega_n t + \tau_n) + \cdots\end{aligned} \tag{6.15.4}$$

When $\Omega_1 \approx \omega_2$, $M_1 \cos(\Omega_1 t + \tau_1)$ becomes a resonant excitation term of the Eq. (6.15.4) The resonant excitation term needs to appear in the second order equation, and the rest of excitation terms are non-resonant hard excitations and they should appear in the first order equation. For this purpose, let

$$\begin{aligned} K_n &= \varepsilon k_n, n = 1, 2, \ldots, N \\ M_1 &= \varepsilon^2 m_1, M_n = \varepsilon m_n, n = 2, \ldots, N \end{aligned} \tag{6.15.5}$$

Let the coefficients of the same power of ε in (6.15.3) and (6.15.4) be zero, we obtain.

Order ε^1:

$$\begin{aligned} D_0^2 u_{11} + \omega_1^2 u_{11} &= \sum_{n=1}^{N} k_n \cos(\Omega_n t + \theta_n) \\ D_0^2 u_{21} + \omega_2^2 u_{21} &= \sum_{n=2}^{N} m_n \cos(\Omega_n t + \tau_n) \end{aligned} \tag{6.15.6}$$

Order ε^2:

$$\begin{aligned} D_0^2 u_{12} + \omega_1^2 u_{12} &= -2D_0 D_1 u_{11} - 2\hat{\mu}_1 D_0 u_{11} + 2u_{11} u_{21} \\ D_0^2 u_{22} + \omega_2^2 u_{22} &= -2D_0 D_1 u_{21} - 2\hat{\mu}_2 D_0 u_{21} + u_{11}^2 + m_1 \cos(\Omega_1 t + \tau_1) \end{aligned} \tag{6.15.7}$$

The solution of (6.15.6) can be expressed as

$$\begin{aligned} u_{11} &= A_1 e^{i\omega_1 T_0} + \tfrac{1}{2} \sum_{n=1}^{N} \left(\omega_1^2 - \Omega_n^2\right)^{-1} k_n e^{i(\Omega_n T_0 + \theta_n)} + cc \\ u_{21} &= A_2 e^{i\omega_2 T_0} + \tfrac{1}{2} \sum_{n=2}^{N} \left(\omega_2^2 - \Omega_n^2\right)^{-1} m_n e^{i(\Omega_n T_0 + \tau_n)} + cc \end{aligned} \tag{6.15.8}$$

where $A_k = A_k(T_1), \quad k = 1, 2$. Substituting (6.15.8) into (6.15.7), we obtain

$$\begin{aligned} D_0^2 u_{12} + \omega_1^2 u_{12} &= -2i\omega_1 D_1 A_1 e^{i\omega_1 T_0} - 2i\omega_1 \hat{\mu}_1 A_1 e^{i\omega_1 T_0} \\ &+ 2\overline{A}_1 A_2 e^{i(\omega_2 - \omega_1) T_0} + cc + NST \end{aligned} \tag{6.15.9}$$

$$\begin{aligned} D_0^2 u_{22} + \omega_2^2 u_{22} &= -2i\omega_2 D_1 A_2 e^{i\omega_2 T_0} - 2i\omega_2 \hat{\mu}_2 A_2 e^{i\omega_2 T_0} \\ &+ A_1^2 e^{2i\omega_1 T_0} + \tfrac{1}{2} m_1 e^{i(\Omega_1 T_0 + \tau_1)} + cc + NST \end{aligned} \tag{6.15.10}$$

Introduce the detuning parameters σ_1, σ_2 such that

$$\Omega_1 = \omega_2 + \varepsilon\sigma_1, \quad \omega_2 = 2\omega_1 + \varepsilon\sigma_2 \tag{6.15.11}$$

In order to eliminate secular terms from and, we need

$$2i\omega_1 A_1' + 2i\omega_1 \hat{\mu}_1 A_1 - 2\overline{A}_1 A_2 e^{i\sigma_2 T_1} = 0 \tag{6.15.12}$$

$$2i\omega_2 D_1 A_2 + 2i\omega_2 \hat{\mu}_2 A_2 - \frac{1}{2} m_1 e^{i(\sigma_1 T_1 + \tau_1)} - A_1^2 e^{-i\sigma_2 T_1} = 0 \tag{6.15.13}$$

Let

$$A_1 = \frac{1}{2} a_1 e^{i\beta_1}, A_2 = \frac{1}{2} a_2 e^{i\beta_2} \tag{6.15.14}$$

$$i\omega_1 a_1' - \omega_1 a_1 \beta_1' + i\omega_1 \hat{\mu}_1 a_1 - \frac{1}{2} a_1 a_2 e^{i(\sigma_2 T_1 - 2\beta_1 + \beta_2)} = 0 \tag{6.15.15}$$

$$i\omega_2 a_2' - \omega_2 a_2 \beta_2' + i\omega_2 \hat{\mu}_2 a_2 - \frac{1}{2} m_1 e^{i(\sigma_1 T_1 - \beta_2 + \tau_1)} - \frac{1}{4} a_1^2 e^{-i(\sigma_2 T_1 - 2\beta_1 + \beta_2)} = 0 \tag{6.15.16}$$

Separating the real and imaginary parts of the above equations yields

$$\begin{aligned} \omega_1 a_1' + \omega_1 \hat{\mu}_1 a_1 - \frac{1}{2} a_1 a_2 \sin\gamma_1 = 0 \\ -\omega_1 a_1 \beta_1' - \frac{1}{2} a_1 a_2 \cos\gamma_1 = 0 \end{aligned} \tag{6.15.17}$$

$$\begin{aligned} \omega_2 a_2' + \omega_2 \hat{\mu}_2 a_2 + \frac{1}{4} a_1^2 \sin\gamma_1 - \frac{1}{2} m_1 \sin\gamma_2 = 0 \\ -\omega_2 a_2 \beta_2' - \frac{1}{4} a_1^2 \cos\gamma_1 - \frac{1}{2} m_1 \cos\gamma_2 = 0 \end{aligned} \tag{6.15.18}$$

where

$$\gamma_1 = \sigma_2 T_1 - 2\beta_1 + \beta_2, \quad \gamma_2 = \sigma_1 T_1 - \beta_2 + \tau_1 \tag{6.15.19}$$

Utilizing (6.15.19), we can write (6.15.17) and (6.15.18) as

$$\begin{aligned} \omega_1 a_1' &= -\omega_1 \hat{\mu}_1 a_1 + \tfrac{1}{2} a_1 a_2 \sin\gamma_1 \\ \omega_2 a_2' &= -\omega_2 \hat{\mu}_2 a_2 - \tfrac{1}{4} a_1^2 \sin\gamma_1 + \tfrac{1}{2} m_1 \sin\gamma_2 \\ \omega_2 a_2 \gamma_2' &= \omega_2 \sigma_1 a_2 + \tfrac{1}{4} a_1^2 \cos\gamma_1 + \tfrac{1}{2} m_1 \cos\gamma_2 \\ \omega_1 a_1 (\gamma_1' + \gamma_2') &= \omega_1 a_1 (\sigma_1 + \sigma_2) + a_1 a_2 \cos\gamma_1 \end{aligned} \tag{6.15.20}$$

Therefore, the steady state solution should satisfy the following equations

$$-\omega_1 \hat{\mu}_1 a_1 + \frac{1}{2} a_1 a_2 \sin\gamma_1 = 0 \tag{6.15.21}$$

$$\omega_1 a_1 (\sigma_1 + \sigma_2) + a_1 a_2 \cos\gamma_1 = 0 \tag{6.15.22}$$

$$-\omega_2\hat{\mu}_2 a_2 - \frac{1}{4}a_1^2 \sin\gamma_1 + \frac{1}{2}m_1 \sin\gamma_2 = 0 \tag{6.15.23}$$

$$\omega_2\sigma_1 a_2 + \frac{1}{4}a_1^2 \cos\gamma_1 + \frac{1}{2}m_1 \cos\gamma_2 = 0 \tag{6.15.24}$$

From (6.15.21) and (6.15.22), we have

$$\sin\gamma_1 = \frac{2\omega_1\hat{\mu}_1}{a_2}, \quad \cos\gamma_1 = -\frac{\omega_1(\sigma_1+\sigma_2)}{a_2} \tag{6.15.25}$$

Substituting (6.15.25) into (6.15.23) and (6.15.24) yields

$$\omega_2\hat{\mu}_2 a_2 + \frac{1}{2}\omega_1\hat{\mu}_1 a_1^2 a_2^{-1} = \frac{1}{2}m_1 \sin\gamma_2 \tag{6.15.26}$$

$$\omega_2\sigma_1 a_2 - \frac{1}{4}\omega_1(\sigma_1+\sigma_2)a_1^2 a_2^{-1} = -\frac{1}{2}m_1 \cos\gamma_2 \tag{6.15.27}$$

Eliminating γ_1 from (6.15.21) and (6.15.22) and γ_2 from (6.15.26) and (6.15.27), we can obtain

$$a_2^2 = 4\omega_1^2\hat{\mu}_1^2 + \omega_1^2(\sigma_1+\sigma_2)^2 \tag{6.15.28}$$

$$\begin{gathered}\left(\omega_2\hat{\mu}_2 a_2 + \frac{1}{2}\omega_1\hat{\mu}_1 a_1^2 a_2^{-1}\right)^2 \\ +\left[\omega_2\sigma_1 a_2 - \frac{1}{4}\omega_1(\sigma_1+\sigma_2)a_1^2 a_2^{-1}\right]^2 = \frac{1}{4}m_1^2\end{gathered} \tag{6.15.29}$$

From (6.15.28) and (6.15.29), we can obtain

$$\begin{aligned} a_1 &= 2\left[-\Gamma_1 \pm \sqrt{\frac{1}{4}m_1^2 - \Gamma_2^2}\right]^{1/2} \\ a_2 &= 2\omega_1\sqrt{\hat{\mu}_1^2 + (\sigma_1+\sigma_2)^2} \end{aligned} \tag{6.15.30}$$

where

$$\begin{aligned} \Gamma_1 &= \omega_1\omega_2[2\hat{\mu}_1\hat{\mu}_2 - \sigma_1(\sigma_1+\sigma_2)] \\ \Gamma_2 &= \omega_1\omega_2[2\hat{\mu}_1\sigma_1 + \hat{\mu}_2(\sigma_1+\sigma_2)] \end{aligned} \tag{6.15.31}$$

The steady state response can be obtained from (6.15.30) and (6.15.25).

Below we discuss when does Eq. (6.15.30) have real roots. It is clear that a_2 is always a real number and does not vary with the excitation amplitude. Let

$$\zeta_1 = 2|\Gamma_2|, \quad \zeta_2 = 2\sqrt{\left(\Gamma_1^2 + \Gamma_2^2\right)} \tag{6.15.32}$$

Clearly there is $\zeta_2 \geq \zeta_1$. So.
$\Gamma_1 \geq 0$:

$$\text{when } m_1 > \zeta_2,\ a_1 \text{ has one positive real root} \tag{6.15.33}$$

$\Gamma_1 < 0$:

$$\text{when } \zeta_1 < m_1 < \zeta_2,\ a_1 \text{ has two positive real roots} \tag{6.15.34}$$

$$\text{when } m_1 > \zeta_2,\ a_1 \text{ has one positive real root} \tag{6.15.35}$$

To analyze the stability of any set of steady state response a_{10}, a_{20}, γ_{10}, γ_{20}, let

$$\begin{aligned} a_1 &= a_{10} + \tilde{a}_1, a_2 = a_{20} + \tilde{a}_2 \\ \gamma_1 &= \gamma_{10} + \tilde{\gamma}_1, \gamma_2 = \gamma_{20} + \tilde{\gamma}_2 \end{aligned} \tag{6.15.36}$$

Substituting (6.15.36) into (6.15.20) and linearizing it, we obtain

$$\begin{bmatrix} \omega_1 & 0 & 0 & 0 \\ 0 & \omega_2 & 0 & 0 \\ 0 & 0 & 0 & \omega_2 a_{20} \\ 0 & 0 & \omega_1 a_{10} & \omega_1 a_{10} \end{bmatrix} \begin{Bmatrix} \tilde{a}_1' \\ \tilde{a}_2' \\ \tilde{\gamma}_1' \\ \tilde{\gamma}_2' \end{Bmatrix} = \begin{bmatrix} l_{11} & l_{12} & l_{13} & l_{14} \\ l_{21} & l_{22} & l_{23} & l_{24} \\ l_{31} & l_{32} & l_{33} & l_{34} \\ l_{41} & l_{42} & l_{43} & l_{44} \end{bmatrix} \begin{Bmatrix} \tilde{a}_1 \\ \tilde{a}_2 \\ \tilde{\gamma}_1 \\ \tilde{\gamma}_2 \end{Bmatrix} \tag{6.15.37}$$

where

$$\begin{aligned} l_{13} &= \frac{1}{2} a_{10} a_{20} \cos \gamma_{10}, \quad l_{14} = 0 \\ l_{21} &= -\frac{1}{2} a_{10} \sin \gamma_{10}, \quad l_{22} = -\omega_2 \hat{\mu}_2 \\ l_{23} &= -\frac{1}{4} a_{10}^2 \cos \gamma_{10}, \quad l_{24} = \frac{1}{2} m_1 \cos \gamma_{20} \\ l_{31} &= \frac{1}{4} a_{10} \cos \gamma_{10}, \quad l_{32} = \omega_2 \sigma_1 \\ l_{33} &= -\frac{1}{4} a_{10}^2 \sin \gamma_{10}, \quad l_{34} = -\frac{1}{2} m_1 \sin \gamma_{20} \\ l_{41} &= \omega_1 (\sigma_1 + \sigma_2) + a_{20} \cos \gamma_{10}, \quad l_{42} = a_{10} \cos \gamma_{10} \\ l_{43} &= -a_{10} a_{20} \sin \gamma_{10}, \quad l_{44} = 0 \end{aligned} \tag{6.15.38}$$

if all four eigenvalues of the Eq. (6.15.37) have negative real parts, the corresponding steady state responses are stable, otherwise they are unstable.

(b) When $\Omega_1 \approx \omega_1$ and $\omega_2 \approx 2\omega_1$, (6.15.3) and (6.15.4) still hold. However, $K_1\cos(\Omega_1 t+\theta_1)$ becomes the resonant excitation term of Eq. (6.15.3) and should appear in the second order equation; and the rest of the excitation terms are non-resonant hard excitations and they should appear in the first order equation. For this purpose, we make

$$\begin{aligned} &K_1=\varepsilon^2 k_1, K_n=\varepsilon k_n, n=2,\ldots,N\\ &M_n=\varepsilon m_n, n=1,2,\ldots,N \end{aligned} \tag{6.15.39}$$

Let the coefficients of the like powers of ε in (6.15.3) and (6.15.4) be zero, we can obtain.

Order ε^1:

$$\begin{aligned} D_0^2u_{11}+\omega_1^2u_{11}&=\sum_{n=2}^{N}k_n\cos(\Omega_n t+\theta_n)\\ D_0^2u_{21}+\omega_2^2u_{21}&=\sum_{n=1}^{N}m_n\cos(\Omega_n t+\tau_n) \end{aligned} \tag{6.15.40}$$

Order ε^2:

$$\begin{aligned} D_0^2u_{12}+\omega_1^2u_{12}&=-2D_0D_1u_{11}-2\hat{\mu}_1D_0u_{11}+2u_{11}u_{21}+k_1\cos(\Omega_1 t+\theta_1)\\ D_0^2u_{22}+\omega_2^2u_{22}&=-2D_0D_1u_{21}-2\hat{\mu}_2D_0u_{21}+u_{11}^2 \end{aligned} \tag{6.15.41}$$

The solution of the Eq. (6.15.40) can be expressed as

$$\begin{aligned} u_{11}&=A_1e^{i\omega_1T_0}+\tfrac{1}{2}\sum_{n=2}^{N}\left(\omega_1^2-\Omega_n^2\right)^{-1}k_ne^{i(\Omega_nT_0+\theta_n)}+cc\\ u_{21}&=A_2e^{i\omega_2T_0}+\tfrac{1}{2}\sum_{n=1}^{N}\left(\omega_2^2-\Omega_n^2\right)^{-1}m_ne^{i(\Omega_nT_0+\tau_n)}+cc \end{aligned} \tag{6.15.42}$$

where $A_k=A_k(T_1),\ \ k=1,2$. Substituting (6.15.42) into (6.15.41), we get

$$\begin{aligned} &D_0^2u_{12}+\omega_1^2u_{12}=-2i\omega_1A_1'e^{i\omega_1T_0}-2i\omega_1\hat{\mu}_1A_1e^{i\omega_1T_0}+2\overline{A}_1A_2e^{i(\omega_2-\omega_1)T_0}\\ &+\tfrac{1}{2}k_1e^{i(\Omega_1T_0+\theta_1)}+cc+NST \end{aligned} \tag{6.15.43}$$

$$D_0^2u_{22}+\omega_2^2u_{22}=-2i\omega_2A_2'e^{i\omega_2T_0}-2i\omega_2\hat{\mu}_2A_2e^{i\omega_2T_0}+A_1^2e^{2i\omega_1T_0}+cc+NST \tag{6.15.44}$$

Introduce the detuning parameters σ_1,σ_2 so that

$$\Omega_1=\omega_1+\varepsilon\sigma_1,\quad \omega_2=2\omega_1+\varepsilon\sigma_2 \tag{6.15.45}$$

Therefore, in order to eliminate the secular terms in (6.15.43) and (6.15.44), there must be

$$2i\omega_1 A_1' + 2i\omega_1 \hat{\mu}_1 A_1 - 2\overline{A}_1 A_2 e^{i\sigma_2 T_1} - \frac{1}{2}k_1 e^{i(\sigma_1 T_1 + \theta_1)} = 0 \tag{6.15.46}$$

$$2i\omega_2 A_2' e^{i\omega_2 T_0} + 2i\omega_2 \hat{\mu}_2 A_2 e^{i\omega_2 T_0} - A_1^2 e^{-i\sigma_2 T_1} = 0 \tag{6.15.47}$$

Let

$$A_1 = \frac{1}{2}a_1 e^{i\beta_1}, \quad A_2 = \frac{1}{2}a_2 e^{i\beta_2} \tag{6.15.48}$$

By substituting (6.15.48) into (6.15.46) and (6.15.47), we obtain

$$i\omega_1 a_1' - \omega_1 a_1 \beta_1' + i\omega_1 \hat{\mu}_1 a_1 - \frac{1}{2}a_1 a_2 e^{i(\sigma_2 T_1 - 2\beta_1 + \beta_2)} - \frac{1}{2}k_1 e^{i(\sigma_1 T_1 - \beta_1 + \theta_1)} = 0 \tag{6.15.49}$$

$$i\omega_2 a_2' - \omega_2 a_2 \beta_2' + i\omega_2 \hat{\mu}_2 a_2 - \frac{1}{4}a_1^2 e^{-i(\sigma_2 T_1 - 2\beta_1 + \beta_2)} = 0 \tag{6.15.50}$$

Separating the real and imaginary parts of the above equations yields

$$\begin{aligned} \omega_1 a_1' &= -\omega_1 \hat{\mu}_1 a_1 + \tfrac{1}{2}a_1 a_2 \sin\gamma_1 + \tfrac{1}{2}k_1 \sin\gamma_2 \\ \omega_2 a_2' &= -\omega_2 \hat{\mu}_2 a_2 - \tfrac{1}{4}a_1^2 \sin\gamma_1 \end{aligned} \tag{6.15.51}$$

$$\begin{aligned} \omega_1 a_1 \beta_1' &= -\tfrac{1}{2}a_1 a_2 \cos\gamma_1 - \tfrac{1}{2}k_1 \cos\gamma_2 \\ \omega_2 a_2 \beta_2' &= -\tfrac{1}{4}a_1^2 \cos\gamma_1 \end{aligned} \tag{6.15.52}$$

where

$$\gamma_1 = \sigma_2 T_1 - 2\beta_1 + \beta_2, \quad \gamma_2 = \sigma_1 T_1 - \beta_1 + \theta_1 \tag{6.15.53}$$

Utilizing (6.15.53), we can write (6.15.52) as

$$\begin{aligned} \omega_1 a_1 \gamma_2' &= \omega_1 a_1 \sigma_1 + \tfrac{1}{2}a_1 a_2 \cos\gamma_1 + \tfrac{1}{2}k_1 \cos\gamma_2 \\ \omega_2 a_2 (\gamma_1' - 2\gamma_2') &= \omega_2 a_2 (\sigma_2 - 2\sigma_1) - \tfrac{1}{4}a_1^2 \cos\gamma_1 \end{aligned} \tag{6.15.54}$$

Therefore, the steady state response should satisfy the following equations

$$\omega_2 \hat{\mu}_2 a_2 + \frac{1}{4}a_1^2 \sin\gamma_1 = 0 \tag{6.15.55}$$

$$\omega_2(\sigma_2 - 2\sigma_1)a_2 - \frac{1}{4}a_1^2 \cos\gamma_1 = 0 \tag{6.15.56}$$

$$-\omega_1\hat{\mu}_1 a_1 + \frac{1}{2}a_1 a_2 \sin\gamma_1 + \frac{1}{2}k_1 \sin\gamma_2 = 0 \tag{6.15.57}$$

$$\omega_1\sigma_1 a_1 + \frac{1}{2}a_1 a_2 \cos\gamma_1 + \frac{1}{2}k_1 \cos\gamma_2 = 0 \tag{6.15.58}$$

From (6.15.55) and (6.15.56), we have

$$\sin\gamma_1 = -4\omega_2\hat{\mu}_2 a_1^{-2} a_2, \quad \cos\gamma_1 = 4\omega_2 a_1^{-2} a_2(\sigma_2 - 2\sigma_1) \tag{6.15.59}$$

Substituting (6.15.59) into (6.15.57) and (6.15.58), we get

$$\omega_1\hat{\mu}_1 a_1 + 2\omega_2\hat{\mu}_2 a_1^{-1} a_2^2 = \frac{1}{2}k_1 \sin\gamma_2 \tag{6.15.60}$$

$$\omega_1 a_1\sigma_1 + 2\omega_2(\sigma_2 - 2\sigma_1)a_1^{-1} a_2^2 = -\frac{1}{2}k_1 \cos\gamma_2 \tag{6.15.61}$$

Eliminating γ_1 from (6.15.55) and (6.15.56) and γ_2 from (6.15.60) and (6.15.61), we can obtain the frequency–response equation

$$\omega_2^2\left[\hat{\mu}_2^2 + (\sigma_2 - 2\sigma_1)^2\right]a_2^2 = \frac{1}{16}a_1^4 \tag{6.15.62}$$

$$\left(\omega_1\hat{\mu}_1 a_1^2 + 2\omega_2\hat{\mu}_2 a_2^2\right)^2 + \left[\omega_1\sigma_1 a_1^2 + 2\omega_2(\sigma_2 - 2\sigma_1)a_2^2\right]^2 = \frac{1}{4}k_1^2 a_1^2 \tag{6.15.63}$$

Steady state responses of the system can be obtained numerically from (6.15.59)–(6.15.63). Since the amplitude a_1 and a_2 of steady state response are related to the excitation amplitude k_1, saturation does not occur.

Again, analyze the stability of the steady state solution. Let a_{10}, a_{20}, γ_{10} and γ_{20} be a set of steady state solution and make a small perturbation of them, i.e.,

$$\begin{aligned} a_1 &= a_{10} + \tilde{a}_1, a_2 = a_{20} + \tilde{a}_2 \\ \gamma_1 &= \gamma_{10} + \tilde{\gamma}_1, \gamma_2 = \gamma_{20} + \tilde{\gamma}_2 \end{aligned} \tag{6.15.64}$$

Substitute (6.15.64) into (6.15.51) and (6.15.54), we obtain the linearized equation as

$$\begin{bmatrix} \omega_1 & 0 & 0 & 0 \\ 0 & \omega_2 & 0 & 0 \\ 0 & 0 & 0 & \omega_1 a_{10} \\ 0 & 0 & \omega_2 a_{20} & -2\omega_2 a_{20} \end{bmatrix} \begin{Bmatrix} \tilde{a}_1' \\ \tilde{a}_2' \\ \tilde{\gamma}_1' \\ \tilde{\gamma}_2' \end{Bmatrix} = \begin{bmatrix} l_{11} & l_{12} & l_{13} & l_{14} \\ l_{21} & l_{22} & l_{23} & l_{24} \\ l_{31} & l_{32} & l_{33} & l_{34} \\ l_{41} & l_{42} & l_{43} & l_{44} \end{bmatrix} \begin{Bmatrix} \tilde{a}_1 \\ \tilde{a}_2 \\ \tilde{\gamma}_1 \\ \tilde{\gamma}_2 \end{Bmatrix} \tag{6.15.65}$$

where

$$
\begin{aligned}
l_{11} &= -\omega_1\hat{\mu}_1 + \frac{1}{2}a_{20}\sin\gamma_{10},\ l_{12} = \frac{1}{2}a_{10}\sin\gamma_{10} \\
l_{13} &= \frac{1}{2}a_{10}a_{20}\cos\gamma_{10},\ l_{14} = \frac{1}{2}k_1\cos\gamma_{20} \\
l_{21} &= -\frac{1}{2}a_{10}\sin\gamma_{10},\ l_{22} = -\omega_2\hat{\mu}_2 \\
l_{23} &= -\frac{1}{4}a_{10}^2\cos\gamma_{10},\ l_{24} = 0 \\
l_{31} &= \omega_1\sigma_1 + \frac{1}{2}a_{20}\cos\gamma_{10},\ l_{32} = \frac{1}{2}a_{10}\cos\gamma_{10} \\
l_{33} &= -\frac{1}{2}a_{10}a_{20}\sin\gamma_{10},\ l_{34} = -\frac{1}{2}k_1\sin\gamma_{20}\tilde{\gamma}_2 \\
l_{41} &= -\frac{1}{2}a_{10}\cos\gamma_{10},\ l_{42} = \omega_2(\sigma_2 - 2\sigma_1) \\
l_{43} &= \frac{1}{4}a_{10}^2\sin\gamma_{10},\ l_{44} = 0
\end{aligned}
\tag{6.15.66}
$$

The steady state solutions are stable if all four eigenvalues of (6.15.66) have negative real parts.

(c) We only analyze case (ii) $\Omega_1 + \Omega_2 \approx \omega_1$ and $\omega_2 \approx 2\omega_1$, and readers are invited to complete the rest of those cases. Equations (6.15.3) and (6.15.4) still hold. However, all excitation terms are non-resonant hard excitations. Let

$$
K_n = \varepsilon k_n, \quad M_n = \varepsilon m_n, \quad n = 1, 2, \ldots, N \tag{6.15.67}
$$

Equating coefficients of like powers of ε in (6.15.3) and (6.15.4) yields.
Order ε^1:

$$
\begin{aligned}
D_0^2u_{11} + \omega_1^2u_{11} &= \sum_{n=1}^{N} k_n\cos(\Omega_n t + \theta_n) \\
D_0^2u_{21} + \omega_2^2u_{21} &= \sum_{n=1}^{N} m_n\cos(\Omega_n t + \tau_n)
\end{aligned}
\tag{6.15.68}
$$

Order ε^2:

$$
\begin{aligned}
D_0^2u_{12} + \omega_1^2u_{12} &= -2D_0D_1u_{11} - 2\hat{\mu}_1D_0u_{11} + 2u_{11}u_{21} \\
D_0^2u_{22} + \omega_2^2u_{22} &= -2D_0D_1u_{21} - 2\hat{\mu}_2D_0u_{21} + u_{11}^2
\end{aligned}
\tag{6.15.69}
$$

The solution of (6.15.68) can be expressed as

$$
\begin{aligned}
u_{11} &= A_1e^{i\omega_1T_0} + \tfrac{1}{2}\sum_{n=1}^{N}\left(\omega_1^2 - \Omega_n^2\right)^{-1}k_ne^{i(\Omega_nT_0+\theta_n)} + cc \\
u_{21} &= A_2e^{i\omega_2T_0} + \tfrac{1}{2}\sum_{n=1}^{N}\left(\omega_2^2 - \Omega_n^2\right)^{-1}m_ne^{i(\Omega_nT_0+\tau_n)} + cc
\end{aligned}
\tag{6.15.70}
$$

where $A_k = A_k(T_1),\quad k = 1, 2$. Substituting (6.15.70) into (6.15.69), we get

$$D_0^2 u_{12} + \omega_1^2 u_{12} = -2i\omega_1 A_1' e^{i\omega_1 T_0} - 2i\omega_1 \hat{\mu}_1 A_1 e^{i\omega_1 T_0} + 2\overline{A}_1 A_2 e^{i(\omega_2-\omega_1)T_0} + f_1 e^{i(\Omega_1 T_0+\Omega_2 T_0+\theta_2+\tau_1)} + f_2 e^{i(\Omega_1 T_0+\Omega_2 T_0+\theta_1+\tau_2)} + cc + NST \tag{6.15.71}$$

$$D_0^2 u_{22} + \omega_2^2 u_{22} = -2i\omega_2 A_2' e^{i\omega_2 T_0} - 2i\omega_2 \hat{\mu}_2 A_2' e^{i\omega_2 T_0} + A_1^2 e^{2i\omega_1 T_0} \tag{6.15.72}$$

where

$$\begin{aligned} f_1 &= \tfrac{1}{2} k_2 m_1 (\omega_1^2 - \Omega_2^2)^{-1} (\omega_2^2 - \Omega_1^2)^{-1} \\ f_2 &= \tfrac{1}{2} k_1 m_2 (\omega_1^2 - \Omega_1^2)^{-1} (\omega_2^2 - \Omega_2^2)^{-1} \end{aligned} \tag{6.15.73}$$

Introduce the detuning parameters σ_1, σ_2 so that

$$\Omega_1 + \Omega_2 = \omega_1 + \varepsilon\sigma_1, \quad \omega_2 = 2\omega_1 + \varepsilon\sigma_2 \tag{6.15.74}$$

In order to eliminate secular terms of Eqs. (6.15.71) and (6.15.72), there must be

$$2i\omega_1 A_1' + 2i\omega_1 \hat{\mu}_1 A_1 - 2\overline{A}_1 A_2 e^{i\sigma_2 T_1} - f_1 e^{i(\sigma_1 T_1+\theta_2+\tau_1)} - f_2 e^{i(\sigma_1 T_1+\theta_1+\tau_2)} = 0 \tag{6.15.75}$$

$$2i\omega_2 D_1 A_2 + 2i\omega_2 \hat{\mu}_2 A_2 - A_1^2 e^{-i\sigma_2 T_1} = 0 \tag{6.15.76}$$

Let

$$A_1 = \frac{1}{2} a_1 e^{i\beta_1}, \quad A_2 = \frac{1}{2} a_2 e^{i\beta_2} \tag{6.15.77}$$

$$\begin{aligned} &i\omega_1 a_1' - \omega_1 a_1 \beta_1' + i\omega_1 \hat{\mu}_1 a_1 - \frac{1}{2} a_1 a_2 e^{i(\sigma_2 T_1 - 2\beta_1 + \beta_2)} \\ &-f_1 e^{i(\sigma_1 T_1 - \beta_1 + \theta_2 + \tau_1)} - f_2 e^{i(\sigma_1 T_1 - \beta_1 + \theta_1 + \tau_2)} = 0 \end{aligned} \tag{6.15.78}$$

$$i\omega_2 a_2' - \omega_2 a_2 \beta_2' + i\omega_2 \hat{\mu}_2 a_2 - \frac{1}{4} a_1^2 e^{-i(\sigma_2 T_1 - 2\beta_1 + \beta_2)} = 0 \tag{6.15.79}$$

Separating the real and imaginary parts of the above equations yields

$$\begin{aligned} &\omega_1 a_1' + \omega_1 \hat{\mu}_1 a_1 - \frac{1}{2} a_1 a_2 \sin\gamma_1 - f_1 \sin(\gamma_2 + \theta_2 + \tau_1) - f_2 \sin(\gamma_2 + \theta_1 + \tau_2) = 0 \\ &-\omega_1 a_1 \beta_1' - \frac{1}{2} a_1 a_2 \cos\gamma_1 - f_1 \cos(\gamma_2 + \theta_2 + \tau_1) - f_2 \cos(\gamma_2 + \theta_1 + \tau_2) = 0 \end{aligned} \tag{6.15.80}$$

$$\begin{aligned} &\omega_2 a_2' + \omega_2 \hat{\mu}_2 a_2 + \frac{1}{4} a_1^2 \sin\gamma_1 = 0 \\ &-\omega_2 a_2 \beta_2' - \frac{1}{4} a_1^2 \cos\gamma_1 = 0 \end{aligned} \tag{6.15.81}$$

where

$$\gamma_1 = \sigma_2 T_1 - 2\beta_1 + \beta_2, \quad \gamma_2 = \sigma_1 T_1 - \beta_1 \tag{6.15.82}$$

Utilizing (6.15.82), we can write (6.15.80) and (6.15.81) as

$$\begin{aligned}
&\omega_1 a_1' = -\omega_1 \hat{\mu}_1 a_1 + \frac{1}{2} a_1 a_2 \sin\gamma_1 + f_1 \sin(\gamma_2 + \theta_2 + \tau_1) \\
&+ f_2 \sin(\gamma_2 + \theta_1 + \tau_2) \\
&\omega_2 a_2' = -\omega_2 \hat{\mu}_2 a_2 - \frac{1}{4} a_1^2 \sin\gamma_1 \\
&\omega_1 a_1 \gamma_2' = \omega_1 a_1 \sigma_1 + \frac{1}{2} a_1 a_2 \cos\gamma_1 + f_1 \cos(\gamma_2 + \theta_2 + \tau_1) \\
&+ f_2 \cos(\gamma_2 + \theta_1 + \tau_2) \\
&\omega_2 a_2 \left(\gamma_1' - 2\gamma_2'\right) = \omega_2 a_2 (\sigma_2 - 2\sigma_1) - \frac{1}{4} a_1^2 \cos\gamma_1
\end{aligned} \tag{6.15.83}$$

Therefore, the steady state response should satisfy the following equations

$$\begin{aligned}
&\omega_2 \hat{\mu}_2 a_2 + \frac{1}{4} a_1^2 \sin\gamma_1 = 0 \\
&\omega_2 a_2 (\sigma_2 - 2\sigma_1) - \frac{1}{4} a_1^2 \cos\gamma_1 = 0
\end{aligned} \tag{6.15.84}$$

$$\begin{aligned}
&-\omega_1 \hat{\mu}_1 a_1 + \frac{1}{2} a_1 a_2 \sin\gamma_1 + f_1 \sin(\gamma_2 + \theta_2 + \tau_1) + f_2 \sin(\gamma_2 + \theta_1 + \tau_2) = 0 \\
&\omega_1 a_1 \sigma_1 + \frac{1}{2} a_1 a_2 \cos\gamma_1 + f_1 \cos(\gamma_2 + \theta_2 + \tau_1) + f_2 \cos(\gamma_2 + \theta_1 + \tau_2) = 0
\end{aligned} \tag{6.15.85}$$

From, we can obtain

$$\sin\gamma_1 = -4\omega_2 \hat{\mu}_2 a_1^{-2} a_2, \quad \cos\gamma_1 = 4\omega_2 a_1^{-2} a_2 (\sigma_2 - 2\sigma_1) \tag{6.15.86}$$

Substituting into yields

$$\begin{aligned}
&-\omega_1 \hat{\mu}_1 a_1 - 2\omega_2 \hat{\mu}_2 a_1^{-1} a_2^2 \\
&\quad + f_1 \sin(\gamma_2 + \theta_2 + \tau_1) + f_2 \sin(\gamma_2 + \theta_1 + \tau_2) = 0 \\
&\omega_1 a_1 \sigma_1 + 2\omega_2 (\sigma_2 - 2\sigma_1) a_1^{-1} a_2^2 \\
&+ f_1 \cos(\gamma_2 + \theta_2 + \tau_1) + f_2 \cos(\gamma_2 + \theta_1 + \tau_2) = 0
\end{aligned} \tag{6.15.87}$$

Eliminating γ_1 from (6.15.84) and γ_2 from (6.15.87), we can obtain the frequency–response equations

$$\omega_2^2\left[\hat{\mu}_2^2 + (\sigma_2 - 2\sigma_1)^2\right]a_2^2 = \frac{1}{16}a_1^4 \tag{6.15.88}$$

$$\begin{aligned}&\left(\omega_1\hat{\mu}_1 a_1^2 + 2\omega_2\hat{\mu}_2 a_2^2\right)^2 + \left[\omega_1\sigma_1 a_1^2 + 2\omega_2(\sigma_2 - 2\sigma_1)a_2^2\right]^2 \\ &= \left[f_1^2 + f_2^2 + 2f_1 f_2\cos(\theta_2 - \theta_1 + \tau_1 - \tau_2)\right] a_1^2\end{aligned} \tag{6.15.89}$$

Steady state responses of the system can be obtained numerically from (6.15.85), (6.15.86), (6.15.88) and (6.15.89). To analyze the stability of any set of constant solutions a_{10}, a_{20}, γ_{10}, γ_{20}, let

$$\begin{aligned}a_1 &= a_{10} + \tilde{a}_1, a_2 = a_{20} + \tilde{a}_2 \\ \gamma_1 &= \gamma_{10} + \tilde{\gamma}_1, \gamma_2 = \gamma_{20} + \tilde{\gamma}_2\end{aligned} \tag{6.15.90}$$

Substitute (6.15.90) into (6.15.51) and (6.15.54) and make the linearization, we can obtain

$$\begin{bmatrix}\omega_1 & 0 & 0 & 0 \\ 0 & \omega_2 & 0 & 0 \\ 0 & 0 & 0 & \omega_1 a_{10} \\ 0 & 0 & \omega_2 a_{20} & -2\omega_2 a_{20}\end{bmatrix}\begin{Bmatrix}\tilde{a}_1' \\ \tilde{a}_2' \\ \tilde{\gamma}_1' \\ \tilde{\gamma}_2'\end{Bmatrix} = \begin{bmatrix}l_{11} & l_{12} & l_{13} & l_{14} \\ l_{21} & l_{22} & l_{23} & l_{24} \\ l_{31} & l_{32} & l_{33} & l_{34} \\ l_{41} & l_{42} & l_{43} & l_{44}\end{bmatrix}\begin{Bmatrix}\tilde{a}_1 \\ \tilde{a}_2 \\ \tilde{\gamma}_1 \\ \tilde{\gamma}_2\end{Bmatrix} \tag{6.15.91}$$

where

$$\begin{aligned}&l_{11} = -\omega_1\hat{\mu}_1 + \frac{1}{2}a_{20}\sin\gamma_{10}, \quad l_{12} = \frac{1}{2}a_{10}\sin\gamma_{10}, \quad l_{13} = \frac{1}{2}a_{10}a_{20}\cos\gamma_{10} \\ &l_{14} = f_1\cos(\gamma_{20} + \theta_2 + \tau_1) + f_2\cos(\gamma_{20} + \theta_1 + \tau_2) \\ &l_{21} = -\frac{1}{2}a_{10}\sin\gamma_{10}, \quad l_{22} = -\omega_2\hat{\mu}_2, \quad l_{23} = -\frac{1}{4}a_{10}^2\cos\gamma_{10}, \quad l_{24} = 0 \\ &l_{31} = \omega_1\sigma_1 + \frac{1}{2}a_{20}\cos\gamma_{10}, \quad l_{32} = \frac{1}{2}a_{10}\cos\gamma_{10}, \quad l_{33} = -\frac{1}{2}a_{10}a_{20}\sin\gamma_{10} \\ &l_{34} = -f_1\sin(\gamma_{20} + \theta_2 + \tau_1) - f_2\sin(\gamma_{20} + \theta_1 + \tau_2) \\ &l_{41} = -\frac{1}{2}a_{10}\cos\gamma_{10}, \quad l_{42} = \omega_2(\sigma_2 - 2\sigma_1), \quad l_{43} = \frac{1}{4}a_{10}^2\sin\gamma_{10}, \quad l_{44} = 0\end{aligned} \tag{6.15.92}$$

The steady state solutions are stable if the eigenvalues of (6.15.91) have negative real parts.

6.16 Exercise 6.16 (The Method of Multiple Scales for Solving Free Oscillations of Systems with Slowly Varying Frequencies)

Solution: (a) In the generalized multiscale format, the time derivatives of the variables are

$$\begin{aligned}
\tfrac{d}{dt} &= \omega_1 D_1 + \omega_2 D_2 + \varepsilon D_\tau + \cdots \\
\tfrac{d^2}{dt^2} &= \tfrac{d}{dt}(\omega_1 D_1 + \omega_2 D_2 + \varepsilon D_\tau + \cdots) \\
&= (\omega_1 D_1 + \omega_2 D_2)^2 + 2\varepsilon(\omega_1 D_1 + \omega_2 D_2)D_\tau \\
&+\varepsilon(D_\tau\omega_1)D_1 + \varepsilon(D_\tau\omega_2)D_2 + \varepsilon^2 D_\tau^2 + \cdots
\end{aligned} \tag{6.16.1}$$

where

$$D_1 = \partial/\partial\eta_1, \quad D_2 = \partial/\partial\eta_2, \quad D_\tau = \partial/\partial\tau \tag{6.16.2}$$

Thus, substituting the expression of the solution into the equation of the system and retaining to $O\left(\varepsilon^1\right)$ yields

$$\begin{aligned}
0 &= \ddot{u}_1 + \omega_1^2 u_1 - \varepsilon\alpha_1 u_1 u_2 \\
&= \big[(\omega_1 D_1 + \omega_2 D_2)^2 + 2\varepsilon(\omega_1 D_1 + \omega_2 D_2)D_\tau \\
&\quad + \varepsilon(D_\tau\omega_1)D_1 + \varepsilon(D_\tau\omega_2)D_2\big](u_{10} + \varepsilon u_{11}) \\
&\quad + \omega_1^2(u_{10} + \varepsilon u_{11}) - \varepsilon\alpha_1(u_{10} + \varepsilon u_{11})(u_{20} + \varepsilon u_{21}) + \cdots \\
&= (\omega_1 D_1 + \omega_2 D_2)^2 u_{10} + \omega_1^2 u_{10} + \varepsilon\big[(\omega_1 D_1 + \omega_2 D_2)^2 u_{11} + \omega_1^2 u_{11} \\
&\quad 2(\omega_1 D_1 + \omega_2 D_2)D_\tau u_{10} + (D_\tau\omega_1)D_1 u_{10} + (D_\tau\omega_2)D_2 u_{10} - \alpha_1 u_{10} u_{20}\big] + \cdots
\end{aligned} \tag{6.16.3}$$

$$\begin{aligned}
0 &= \ddot{u}_2 + \omega_2^2 u_2 - \varepsilon\alpha_2 u_1^2 \\
&= \big[(\omega_1 D_1 + \omega_2 D_2)^2 + 2\varepsilon(\omega_1 D_1 + \omega_2 D_2)D_\tau \\
&\quad + \varepsilon(D_\tau\omega_1)D_1 + \varepsilon(D_\tau\omega_2)D_2\big](u_{20} + \varepsilon u_{21}) \\
&\quad + \omega_2^2(u_{20} + \varepsilon u_{21}) - \varepsilon\alpha_2(u_{10} + \varepsilon u_{11})^2 + \cdots \\
&= (\omega_1 D_1 + \omega_2 D_2)^2 u_{20} + \omega_2^2 u_{20} + \varepsilon\big[(\omega_1 D_1 + \omega_2 D_2)^2 u_{21} + \omega_2^2 u_{21} \\
&\quad + 2(\omega_1 D_1 + \omega_2 D_2)D_\tau u_{20} + (D_\tau\omega_1)D_1 u_{20} + (D_\tau\omega_2)D_2 u_{20} - \alpha_2 u_{10}^2\big] + \cdots
\end{aligned} \tag{6.16.4}$$

Equating coefficients of like powers of ε in the above two equations yields.
Order ε^0:

$$\begin{aligned}
D_1^2 u_{10} + u_{10} &= 0 \\
D_2^2 u_{20} + u_{20} &= 0
\end{aligned} \tag{6.16.5}$$

Order ε^1:

$$(\omega_1 D_1 + \omega_2 D_2)^2 u_{11} + \omega_\tau^2 u_{11} = -2\omega_1 D_1 D_\tau u_{10} - (D_\tau \omega_1) D_1 u_{10} + \alpha_1 u_{10} u_{20}$$
$$(\omega_1 D_1 + \omega_2 D_2)^2 u_{21} + \omega_\tau^2 u_{21} = -2\omega_2 D_2 D_\tau u_{20} - (D_\tau \omega_2) D_2 u_{20} + \alpha_2 u_{10}^2 \tag{6.16.6}$$

in which $D_2 u_{10} = 0$ and $D_1 u_{20} = 0$ are taken into account. The solution of (6.16.5) is

$$\begin{aligned} u_{10} &= A_1(\tau) e^{i\eta_1} + cc \\ u_{20} &= A_2(\tau) e^{i\eta_2} + cc \end{aligned} \tag{6.16.7}$$

(b) Substituting (6.16.7) into (6.16.6) yields

$$\begin{aligned} (\omega_1 D_1 + \omega_2 D_2)^2 u_{11} &= -2i\omega_1 A_1' e^{i\eta_1} - i\omega_1' A_1 e^{i\eta_1} + \alpha_1 A_1 A_2 e^{i\eta_1} e^{i\eta_2} \\ &+ \alpha_1 A_2 \overline{A}_1 e^{-i\eta_1} e^{i\eta_2} + cc \end{aligned} \tag{6.16.8}$$

$$(\omega_1 D_1 + \omega_2 D_2)^2 u_{21} = -2i\omega_2 A_2' e^{i\eta_2} - i\omega_2' A_2 e^{i\eta_2} + \alpha_2 A_1^2 e^{2i\eta_1} + cc + 2A_1\overline{A}_1 \tag{6.16.9}$$

Primes denote the derivative with respect to τ. When $\omega_2 \approx 2\omega_1$, we can induce the detuning parameter $\sigma(\tau)$ such that

$$\dot{\sigma} = \omega_2 - 2\omega_1 \text{ or } \sigma = \int(\omega_2 - 2\omega_1)dt = \eta_2 - 2\eta_1 \tag{6.16.10}$$

Therefore, in order to eliminate the secular terms from (6.16.8) and (6.16.9), we need

$$\begin{aligned} 2i\omega_1 A_1' + i\omega'_1 A_1 - \alpha_1 A_2 \overline{A}_1 e^{i[\sigma(\tau)]} &= 0 \\ 2i\omega_2 A_2' + i\omega'_2 A_2 - \alpha_2 A_1^2 e^{-i[\sigma(\tau)]} &= 0 \end{aligned} \tag{6.16.11}$$

(c) Let

$$A_1 = \frac{1}{2} a_1 e^{i\beta_1}, \quad A_2 = \frac{1}{2} a_2 e^{i\beta_2} \tag{6.16.12}$$

and substitute (6.16.12) into (6.16.11), we can obtain

$$i\omega_1 a_1' - \omega_1 a_1 \beta_1' + \frac{1}{2} i\omega_1' a_1 - \frac{1}{4}\alpha_1 a_2 a_1 e^{i[\sigma(\tau) - 2\beta_1 + \beta_2]} = 0 \tag{6.16.13}$$

$$i\omega_2 a_2' - \omega_2 a_2 \beta_2' + \frac{1}{2} i\omega_2' a_2 - \frac{1}{4}\alpha_2 a_1^2 e^{-i[\sigma(\tau) - 2\beta_1 + \beta_2]} = 0 \tag{6.16.14}$$

Separating the real and imaginary parts of the above equations yields

$$\begin{aligned}\omega_1 a_1' + \frac{1}{2}\omega_1' a_1 &= \frac{1}{4}\alpha_1 a_1 a_2 \sin[\sigma(\tau) - 2\beta_1 + \beta_2] \\ \omega_2 a_2' + \frac{1}{2}\omega_2' a_2 &= -\frac{1}{4}\alpha_2 a_1^2 \sin[\sigma(\tau) - 2\beta_1 + \beta_2]\end{aligned} \quad (6.16.15)$$

$$\begin{aligned}\omega_2 a_2 \beta_2' &= -\frac{1}{4}\alpha_2 a_1^2 \cos[\sigma(\tau) - 2\beta_1 + \beta_2] \\ \omega_1 a_1 \beta_1' &= -\frac{1}{4}\alpha_1 a_1 a_2 \cos[\sigma(\tau) - 2\beta_1 + \beta_2]\end{aligned} \quad (6.16.16)$$

Substituting (6.16.12) into (6.16.7), we can obtain the first order approximate solution of the system

$$u_1 \approx a_1 \cos(\eta_1 + \beta_1), \quad u_2 \approx a_2 \cos(\eta_2 + \beta_2) \quad (6.16.17)$$

To illustrate the numerical procedure, we assume that

$$\omega_1 = \omega_{10} + \nu_1 \tau, \quad \omega_2 = \omega_{20} + \nu_2 \tau \quad (6.16.18)$$

then

$$\begin{aligned}\sigma &= \sigma_0 + (\omega_{20} - 2\omega_{10})t + \frac{1}{2}\varepsilon(\nu_2 - 2\nu_1)t^2 \\ \eta_1 &= \omega_{10} t + \frac{1}{2}\varepsilon \nu_1 t^2, \quad \eta_2 = \omega_{20} t + \frac{1}{2}\varepsilon \nu_2 t^2\end{aligned} \quad (6.16.19)$$

Equation (6.16.15) and (6.16.16) become

$$\begin{aligned}\omega_1 a_1' &= -\tfrac{1}{2}\nu_1 a_1 + \tfrac{1}{4}\alpha_1 a_1 a_2 \sin[\sigma(\tau) - 2\beta_1 + \beta_2] \\ \omega_2 a_2' &= -\tfrac{1}{2}\nu_2 a_2 - \tfrac{1}{4}\alpha_2 a_1^2 \sin[\sigma(\tau) - 2\beta_1 + \beta_2]\end{aligned} \quad (6.16.20)$$

$$\begin{aligned}\omega_1 a_1 \beta_1' &= -\tfrac{1}{4}\alpha_1 a_1 a_2 \cos[\sigma(\tau) - 2\beta_1 + \beta_2] \\ \omega_2 a_2 \beta_2' &= -\tfrac{1}{4}\alpha_2 a_1^2 \cos[\sigma(\tau) - 2\beta_1 + \beta_2]\end{aligned} \quad (6.16.21)$$

Substituting (6.16.18) and (6.16.19) into (6.16.15) and (6.16.16), and replacing the time variable with t, we get

$$\begin{aligned}\dot{a}_1 &= -\frac{1}{2}\varepsilon\omega_1^{-1}\nu_1 a_1 + \frac{1}{4}\varepsilon\omega_1^{-1}\alpha_1 a_1 a_2 \sin[\sigma(t) - 2\beta_1 + \beta_2] \\ \dot{a}_2 &= -\frac{1}{2}\varepsilon\omega_2^{-1}\nu_2 a_2 - \frac{1}{4}\varepsilon\omega_2^{-1}\alpha_2 a_1^2 \sin[\sigma(t) - 2\beta_1 + \beta_2] \\ a_1\dot{\beta}_1 &= -\frac{1}{4}\varepsilon\omega_1^{-1}\alpha_1 a_1 a_2 \cos[\sigma(t) - 2\beta_1 + \beta_2] \\ a_2\dot{\beta}_2 &= -\frac{1}{4}\varepsilon\omega_1^{-1}\alpha_2 a_1^2 \cos[\sigma(t) - 2\beta_1 + \beta_2]\end{aligned} \quad (6.16.22)$$

where

$$\begin{aligned} &\omega_1 = \omega_{10} + \varepsilon \nu_1 t, \quad \omega_2 = \omega_{20} + \varepsilon \nu_2 t \\ &\sigma(t) = \sigma_0 + (\omega_{20} - 2\omega_{10})t + \frac{1}{2}\varepsilon(\nu_2 - 2\nu_1)t^2 \end{aligned} \tag{6.16.23}$$

Equation (6.16.17) changes into

$$u_1 \approx a_1 \cos\left(\omega_{10} t + \frac{1}{2}\varepsilon \nu_1 t^2 + \beta_1\right), \quad u_2 \approx a_2 \cos\left(\omega_{20} t + \frac{1}{2}\varepsilon \nu_2 t^2 + \beta_2\right) \tag{6.16.24}$$

From the given initial conditions and ~, we have

$$\begin{aligned} &a_1(0)\cos[\beta_1(0)] = 1 \\ &a_2(0)\cos[\beta_2(0)] = 0 \\ &\dot{a}_1(0)\cos[\beta_1(0)] - a_1(0)\left[\omega_{10} + \dot{\beta}_1(0)\right]\sin[\beta_1(0)] = 0 \\ &\dot{a}_2(0)\cos[\beta_2(0)] - a_2(0)\left[\omega_{20} + \dot{\beta}_2(0)\right]\sin[\beta_2(0)] = 0 \end{aligned} \tag{6.16.25}$$

$$\begin{aligned} \dot{a}_1(0) &= -\frac{1}{2}\varepsilon\omega_{10}^{-1}\nu_1 a_1(0) + \frac{1}{4}\varepsilon\omega_{10}^{-1}\alpha_1 a_1(0)a_2(0)\sin[\sigma_0 - 2\beta_1(0) + \beta_2(0)] \\ \dot{a}_2(0) &= -\frac{1}{2}\varepsilon\omega_{20}^{-1}\nu_2 a_2(0) - \frac{1}{4}\varepsilon\omega_{20}^{-1}\alpha_2 a_1^2(0)\sin[\sigma_0 - 2\beta_1(0) + \beta_2(0)] \\ a_1(0)\dot{\beta}_1(0) &= -\frac{1}{4}\varepsilon\omega_{10}^{-1}\alpha_1 a_1(0)a_2(0)\cos[\sigma_0 - 2\beta_1(0) + \beta_2(0)] \\ a_2(0)\dot{\beta}_2(0) &= -\frac{1}{4}\varepsilon\omega_{10}^{-1}\alpha_2 a_1^2(0)\cos[\sigma_0 - 2\beta_1(0) + \beta_2(0)] \end{aligned} \tag{6.16.26}$$

From the above two sets of equations, we obtain

$$\begin{aligned} &\cos[\beta_2(0)] = 0 \\ &a_1(0)\cos[\beta_1(0)] = 1 \\ &\frac{1}{4}\varepsilon\omega_{10}^{-1}\alpha_2 a_1^2(0)\cos[\sigma_0 - 2\beta_1(0) + \beta_2(0)] = \omega_{20} a_2(0) \\ &\left\{\frac{1}{4}\varepsilon\omega_{20}^{-1}\alpha_2 a_1^2(0)\sin[\sigma_0 - 2\beta_1(0) + \beta_2(0)] + \frac{1}{2}\varepsilon\omega_{20}^{-1}\nu_2 a_2(0)\right\}\cos[\beta_1(0)] + \\ &\left\{-\frac{1}{4}\varepsilon\omega_{10}^{-1}\alpha_1 a_1(0)a_2(0)\cos[\sigma_0 - 2\beta_1(0) + \beta_2(0)] + a_1(0)\omega_{10}\right\}\sin[\beta_1(0)] = 0 \end{aligned} \tag{6.16.27}$$

We can solve for $a_1(0),\ a_2(0),\ \beta_1(0),\ \beta_2(0)$ from the system of Eq. (6.16.27), and then obtain $a_1(t),\ a_2(t),\ \beta_1(t),\ \beta_2(t)$ from the step-by-step integration of the system of Eq. (6.16.22).

6.17 Exercise 6.17 (Analysis of a Two-Degree-of-Freedom System with Simultaneous Internal, Subharmonic, and Superharmonic Resonances)

Solution: Solve by the method of multiple scale. Let the solution of the equation be

$$\begin{aligned} x &= u_{11}(T_0, T_1) + \varepsilon u_{12}(T_0, T_1) + \cdots \\ y &= u_{21}(T_0, T_1) + \varepsilon u_{22}(T_0, T_1) + \cdots \end{aligned} \tag{6.17.1}$$

Substituting (6.17.1) into the equation of the system and retaining to $O(\varepsilon^1)$ yields

$$\begin{aligned} 0 &= \ddot{x} + x + 4\cos 2t - \varepsilon\left[\frac{4}{3}\dot{y} - \left(1 - x^2\right)\dot{x}\right] \\ &= \left(D_0^2 + 2\varepsilon D_0 D_1\right)(u_{11} + \varepsilon u_{12}) + u_{11} + \varepsilon u_{12} + 4\cos 2t \\ &\quad -\frac{4}{3}\varepsilon(D_0 + \varepsilon D_1)(u_{21} + \varepsilon u_{22}) + \varepsilon(D_0 + \varepsilon D_1)(u_{11} + \varepsilon u_{12}) \\ &\quad -\varepsilon(u_{11} + \varepsilon u_{12})^2(D_0 + \varepsilon D_1)(u_{11} + \varepsilon u_{12}) + \cdots \\ &= D_0^2 u_{11} + u_{11} + 4\cos 2t + \varepsilon\left(D_0^2 u_{12} + u_{12}\right. \\ &\quad \left. + 2D_0 D_1 u_{11} - \frac{4}{3}D_0 u_{21} + D_0 u_{11} - u_{11}^2 D_0 u_{11}\right) + \cdots \end{aligned} \tag{6.17.2}$$

$$\begin{aligned} 0 &= \ddot{y} + \frac{1}{4}y - 5\cos 2t - \frac{4}{3}\varepsilon\left[\left(1 - x^2\right)\dot{x} - \dot{y}\right] \\ &= \left(D_0^2 + 2\varepsilon D_0 D_1\right)(u_{21} + \varepsilon u_{22}) + \frac{1}{4}(u_{21} + \varepsilon u_{22}) - 5\cos 2t \\ &\quad -\frac{4}{3}\varepsilon(D_0 + \varepsilon D_1)(u_{11} + \varepsilon u_{12}) + \frac{4}{3}\varepsilon(u_{11} + \varepsilon u_{12})^2(D_0 + \varepsilon D_1)(u_{11} + \varepsilon u_{12}) \\ &\quad +\frac{4}{3}\varepsilon(D_0 + \varepsilon D_1)(u_{21} + \varepsilon u_{22}) + \cdots \\ &= D_0^2 u_{21} + \frac{1}{4}u_{21} - 5\cos 2t + \varepsilon\left(D_0^2 u_{22} + \frac{1}{4}u_{22} + 2D_0 D_1 u_{21}\right. \\ &\quad \left. -\frac{4}{3}D_0 u_{11} + \frac{4}{3}u_{11}^2 D_0 u_{11} + \frac{4}{3}D_0 u_{21}\right) + \cdots \end{aligned} \tag{6.17.3}$$

Making the coefficients of each power of ε zero in the above two equations, we obtain.

Order ε^0:

$$\begin{aligned} D_0^2 u_{11} + u_{11} &= -4\cos 2t \\ D_0^2 u_{21} + \tfrac{1}{4} u_{21} &= 5\cos 2t \end{aligned} \tag{6.17.4}$$

Order ε^1:

$$\begin{aligned} D_0^2 u_{12} + u_{12} &= -2D_0D_1u_{11} - D_0u_{11} + \tfrac{4}{3}D_0u_{21} + u_{11}^2 D_0 u_{11} \\ D_0^2 u_{22} + \tfrac{1}{4}u_{22} &= -2D_0D_1u_{21} + \tfrac{4}{3}D_0u_{11} - \tfrac{4}{3}D_0u_{21} - \tfrac{4}{3}u_{11}^2 D_0 u_{11} \end{aligned} \tag{6.17.5}$$

The solution of (6.17.4) can be expressed as

$$\begin{aligned} u_{11} &= A_1 e^{iT_0} + \frac{2}{3} e^{i2T_0} + cc \\ u_{21} &= A_2 e^{iT_0/2} - \frac{2}{3} e^{i2T_0} + cc \end{aligned} \tag{6.17.6}$$

where $A_k = A_k(T_1), \quad k = 1, 2$. Substitute (6.17.6) into (6.17.5), we get

$$\begin{aligned} &D_0^2 u_{12} + u_{12} = -2D_0D_1u_{11} - D_0u_{11} + \tfrac{4}{3}D_0u_{21} + u_{11}^2 D_0u_{11} \\ &= -2iA_1' e^{iT_0} + \tfrac{23}{9} iA_1 e^{iT_0} + 3iA_1^2\overline{A}_1 e^{iT_0} + cc + NST \end{aligned} \tag{6.17.7}$$

$$\begin{aligned} &D_0^2 u_{22} + \tfrac{1}{4}u_{22} = -2D_0D_1u_{21} + \tfrac{4}{3}D_0u_{11} - \tfrac{4}{3}D_0u_{21} - \tfrac{4}{3}u_{11}^2 D_0u_{11} \\ &= -iA_2' e^{iT_0/2} - \tfrac{2}{3} iA_2 e^{iT_0/2} + cc + NST \end{aligned} \tag{6.17.8}$$

In order to eliminate secular terms from (6.17.7) and (6.17.8), we need

$$2A_1' - \frac{23}{9}A_1 - 3A_1^2\overline{A}_1 = 0 \tag{6.17.9}$$

$$A_2' + \frac{2}{3}A_2 = 0 \tag{6.17.10}$$

therefore,

$$A_2 = A_{20} e^{-2t/3} \quad \to \quad 0 \tag{6.17.11}$$

Let

$$A_1 = \frac{1}{2} a_1 e^{i\beta_1} \tag{6.17.12}$$

Substituting (6.17.12) into (6.17.9), we can obtain

$$\omega_1 a_1' + i\omega_1 a_1 \beta_1' - \frac{23}{18} a_1 - \frac{3}{8} a_1^3 = 0 \tag{6.17.13}$$

Separating the real and imaginary parts of the above equations yields

$$\begin{aligned} \omega_1 a_1' &= \frac{23}{18} a_1 + \frac{3}{8} a_1^3 \\ \beta_1' &= 0 \end{aligned} \tag{6.17.14}$$

So, the steady state response is $\beta_1 = \beta_{10}$; $a_{10} = 0$. The linearized equation of a_1 in the above equation at $a_{10} = 0$ is

$$\omega_1 \widetilde{a_1'} = \frac{23}{18} \tilde{a}_1 \tag{6.17.15}$$

Obviously, the perturbation $\tilde{a}_1$ is divergent and hence the steady state solution $a_{10} = 0$ is unstable. From the above results, the first order approximate solution of the system is

$$x = \frac{4}{3} \cos 2t + O(\varepsilon), \quad y = -\frac{4}{3} \cos 2t + O(\varepsilon) \tag{6.17.16}$$

This solution is unstable.

6.18 Exercise 6.18 (Nonlinear Oscillation Analysis of a Spherical Pendulum with a Moving Support)

Solution: (a) The coordinates of the absolute motion of the mass are

$$\begin{aligned} x &= l \sin\theta \cos\varphi \\ y &= l \sin\theta \sin\varphi \\ z &= l \cos\theta \end{aligned} \tag{6.18.1}$$

The kinetic energy of the system is

$$
\begin{aligned}
T &= \frac{1}{2}m\left\{\left[\dot{x}^2+\dot{y}^2\right]+\left[\frac{d}{dt}(z-\varepsilon l\sin\Omega t)\right]^2\right\}\\
&= \frac{1}{2}m\left\{\left[\frac{d}{dt}(l\sin\theta\cos\phi)\right]^2\left[\frac{d}{dt}(l\sin\theta\sin\phi)\right]^2\right.\\
&\quad\left.+\left[\frac{d}{dt}(l\cos\theta)-\varepsilon l\Omega\cos\Omega t\right]^2\right\}\\
&= \frac{1}{2}ml^2\left[\left(\dot{\theta}\cos\theta\cos\phi-\dot{\phi}\sin\theta\sin\phi\right)^2+\left(\dot{\theta}\cos\theta\sin\phi+\dot{\phi}\sin\theta\cos\phi\right)^2\right.\\
&\quad\left.+\left(\dot{\theta}\sin\theta+\varepsilon\Omega\cos\Omega t\right)^2\right]\\
&= \frac{1}{2}ml^2\left(\dot{\theta}^2+\dot{\phi}^2\sin^2\theta+2\varepsilon\Omega\dot{\theta}\sin\theta\cos\Omega t+\varepsilon^2\Omega^2\cos^2\Omega t\right)
\end{aligned}
\tag{6.18.2}
$$

Using the static equilibrium position as the zero potential position, the potential energy of the system can be written as

$$
V = mgl(1-\cos\theta) \tag{6.18.3}
$$

Substitute kinetic energy, potential energy, and generalized forces into the Lagrange's equation:

$$
\frac{d}{dt}\frac{\partial T}{\partial\dot{\theta}}-\frac{\partial T}{\partial\theta}=-\frac{\partial V}{\partial\theta},\quad \frac{d}{dt}\frac{\partial T}{\partial\dot{\varphi}}-\frac{\partial T}{\partial\varphi}=-\frac{\partial V}{\partial\varphi}
$$

we can obtain

$$
\begin{aligned}
&\ddot{\theta}+\left(\frac{g}{l}-\varepsilon\Omega^2\sin\Omega t-\dot{\varphi}^2\cos\theta\right)\sin\theta=0\\
&\ddot{\varphi}\sin\theta+2\dot{\varphi}\dot{\theta}\cos\theta=0
\end{aligned}
\tag{6.18.4}
$$

(b) Separating the variables for the second equation of (6.18.4) yields

$$
\frac{d\dot{\varphi}}{\dot{\varphi}}=-\frac{2\cos\theta}{\sin\theta}d\theta
$$

After integration

$$
\dot{\varphi}\sin^2\theta=P \tag{6.18.5}
$$

where P is the constant of integration, determined from initial conditions. Substituting (6.18.5) into the first equation in (6.18.4), we obtain

$$
\ddot{\theta}+\left(\frac{g}{l}-\varepsilon\Omega^2\sin\Omega t\right)\sin\theta-\frac{P^2\cos\theta}{\sin^3\theta}=0 \tag{6.18.6}
$$

(c) When $\varepsilon = 0$, (6.18.6) becomes

$$\ddot{\theta} + \frac{g}{l}\sin\theta - \frac{P^2\cos\theta}{\sin^3\theta} = 0 \tag{6.18.7}$$

therefore, the equilibrium position of the system satisfies

$$\frac{g}{l}\sin\theta - \frac{P^2\cos\theta}{\sin^3\theta} = 0 \tag{6.18.8}$$

(I) When $\dot{\varphi} \equiv 0 : P \equiv 0$, then the equilibrium θ can be obtained from:

$$\theta_s = 0, \quad \pi \tag{6.18.9}$$

(II) When $\dot{\varphi}(0) \neq 0,\ \theta(0) \neq 0 : P \neq 0$, hence, the equilibrium θ should satisfy

$$\sin^3\theta_s \tan\theta_s = \frac{P^2}{\omega_0^2} \tag{6.18.10}$$

where

$$\omega_0^2 = g/l \tag{6.18.11}$$

(d) Different types of equilibrium positions need to be analyzed separately.
(I) $\theta_s = 0, \quad \pi$:

In this case, due to $\dot{\varphi} \equiv 0$ and $P \equiv 0$, the system becomes a single pendulum with a movable support. If there is no support excitation, near the equilibrium position, the linearized equation of (6.18.6) is

$$\theta = 0 : \ \ddot{\theta} + \omega_0^2\theta = 0 \tag{6.18.12}$$

$$\theta = \pi : \ \ddot{\theta} - \omega_0^2\theta = 0 \tag{6.18.13}$$

It is clear that the motion described by (6.18.12) is stable, while the motion by (6.18.13) is divergent.

When there is a support excitation, (6.18.6) becomes

$$\ddot{\theta} + \left(\omega_0^2 - \varepsilon\Omega^2 \sin\Omega t\right)\sin\theta = 0 \tag{6.18.14}$$

The stability of the steady state response of (6.18.14) is left to readers to complete.

(II) the equilibrium is given by (6.18.10).

If there is no support excitation, linearized equation of (6.18.6) near the equilibrium position is

$$\ddot{\theta}+\left[\omega_0^2\cos\theta_s+P^2\left(3\sin^{-4}\theta_s\cos^2\theta_s+\sin^{-2}\theta_s\right)\right]\theta=0 \tag{6.18.15}$$

By (6.18.10), we have

$$-\frac{1}{2}\pi<\theta_s<\frac{1}{2}\pi \quad \text{and} \quad \theta_s\neq 0$$

therefore,

$$\omega_0^2\cos\theta_s+P^2\left(3\sin^{-4}\theta_s\cos^2\theta_s+\sin^{-2}\theta_s\right)>0 \tag{6.18.16}$$

Therefore, the motion determined by (6.18.15) is stable.

When there is a support excitation, expand (6.18.6) near the equilibrium position $\theta=\theta_s$ and retain to the third order of smallness yields

$$\ddot{\theta}+\omega^2\theta+\alpha_1\theta^2+\alpha_2\theta^3+\varepsilon\left(\delta_1\theta+\delta_2\theta^2\right)\sin\Omega t=0 \tag{6.18.17}$$

where

$$\begin{aligned}
\omega^2&=\omega_0^2\cos\theta_s+P^2\left(3\sin^{-4}\theta_s\cos^2\theta_s+\sin^{-2}\theta_s\right)\\
\alpha_1&=-\frac{1}{2}\left[\omega_0^2\sin\theta_s+P^2\left(12\sin^{-5}\theta_s\cos^3\theta_s+8\sin^{-3}\theta_s\cos\theta_s\right)\right]\\
\alpha_2&=\frac{1}{6}\left[-\omega_0^2\cos\theta_s+P^2\left(60\sin^{-6}\theta_s\cos^4\theta_s\right.\right.\\
&\left.\left.+60\sin^{-4}\theta_s\cos^2\theta_s+8\sin^{-2}\theta_s\right)\right]\\
\delta_1&=-\Omega^2\cos\theta_s\\
\delta_2&=\frac{1}{2}\Omega^2\sin\theta_s
\end{aligned} \tag{6.18.18}$$

We use the method of multiple scales to solve (6.18.17), and let the solution of the equation be

$$\theta=u_1(T_0,T_1,T_2)+\varepsilon u_2(T_0,T_1,T_2)+\varepsilon^2u_3(T_0,T_1,T_2)+\cdots \tag{6.18.19}$$

Substituting (6.18.19) into (6.18.17) and retaining to $\mathrm{O}\left(\varepsilon^3\right)$, we get

$$
\begin{aligned}
0 &= \ddot{\theta} + \omega^2\theta + \alpha_1\theta^2 + \alpha_2\theta^3 + \varepsilon\left(\delta_1\theta + \delta_2\theta^2\right)\sin\Omega t \\
&= \left[D_0^2 + 2\varepsilon D_0 D_1 + \varepsilon^2\left(D_1^2 + 2D_0 D_2\right)\right]\left(\varepsilon u_1 + \varepsilon^2 u_2 + \varepsilon^3 u_3\right) \\
&\quad + \omega^2\left(\varepsilon u_1 + \varepsilon^2 u_2 + \varepsilon^3 u_3\right) + \alpha_1\left(\varepsilon u_1 + \varepsilon^2 u_2 + \varepsilon^3 u_3\right)^2 \\
&\quad + \alpha_2\left(\varepsilon u_1 + \varepsilon^2 u_2 + \varepsilon^3 u_3\right)^3 + \varepsilon\delta_1\left(\varepsilon u_1 + \varepsilon^2 u_2 + \varepsilon^3 u_3\right)\sin\Omega T_0 \\
&\quad + \varepsilon\delta_2\left(\varepsilon u_1 + \varepsilon^2 u_2 + \varepsilon^3 u_3\right)^2\sin\Omega T_0 + \cdots \\
&= \varepsilon\left(D_0^2 u_1 + \omega^2 u_1\right) + \varepsilon^2\left(D_0^2 u_2 + \omega^2 u_2 + 2D_0 D_1 u_1\right. \\
&\quad \left. + \alpha_1 u_1^2 + \delta_1 u_1 \sin\Omega T_0\right) + \varepsilon^3\left[D_0^2 u_3 + \omega^2 u_3 + \left(D_1^2 + 2D_0 D_2\right)u_1\right. \\
&\quad \left. + 2D_0 D_1 u_2 + 2\alpha_1 u_1 u_2 + \alpha_2 u_1^3 + \delta_1 u_2 \sin\Omega T_0 + \delta_2 u_1^2 \sin\Omega T_0\right] + \cdots
\end{aligned}
\tag{6.18.20}
$$

Equating coefficients of the like power of ε in the above equation yields.
Order ε^1:

$$D_0^2 u_1 + \omega^2 u_1 = 0 \tag{6.18.21}$$

Order ε^2:

$$D_0^2 u_2 + \omega^2 u_2 = -2D_0 D_1 u_1 - \alpha_1 u_1^2 - \delta_1 u_1 \sin\Omega T_0 \tag{6.18.22}$$

Order ε^3:

$$
\begin{aligned}
D_0^2 u_3 + \omega^2 u_3 &= -\left(D_1^2 + 2D_0 D_2\right)u_1 - 2D_0 D_1 u_2 - 2\alpha_1 u_1 u_2 \\
&\quad -\alpha_2 u_1^3 - \delta_2 u_1^2 \sin\Omega T_0 - \delta_1 u_2 \sin\Omega T_0
\end{aligned}
\tag{6.18.23}
$$

The solution of (6.18.21) can be expressed as

$$u_1 = Ae^{i\omega T_0} + cc \tag{6.18.24}$$

where $A = A(T_1, T_2)$. Substituting (6.18.24) into (6.18.22), we get

$$
\begin{aligned}
D_0^2 u_2 + \omega^2 u_2 &= -2D_0 D_1 u_1 - \alpha_1 u_1^2 - \delta_1 u_1 \sin\Omega T_0 \\
&= -2i\omega D_1 A e^{i\omega T_0} - 2\alpha_1 A\overline{A} - \alpha_1 A^2 e^{2i\omega T_0} \\
&\quad + \tfrac{1}{2} i\delta_1 A e^{i(\Omega+\omega)T_0} + \tfrac{1}{2} i\delta_1 \overline{A} e^{i(\Omega-\omega)T_0} + cc
\end{aligned}
\tag{6.18.25}
$$

In order to eliminate secular terms from (6.18.25), the different values of Ω need to be discussed.

CASE (1): $\Omega \approx 2\omega$.

In this case, induce the detuning parameter σ such that

$$\Omega = 2\omega + \varepsilon\sigma \tag{6.18.26}$$

In order to remove secular terms from (6.18.25), we need

$$2\omega D_1 A - \frac{1}{2}\delta_1 \bar{A} e^{i\sigma T_1} = 0 \tag{6.18.27}$$

Let

$$A = \frac{1}{2} a e^{i\beta} \tag{6.18.28}$$

Substituting (6.18.28) into (6.18.27), we can obtain

$$\omega a' + i\omega a\beta' - \frac{1}{2}\delta_1 a e^{i(\sigma T_1 - 2\beta)} = 0 \tag{6.18.29}$$

Separating the real and imaginary parts of the above equations yields

$$\begin{aligned} a' &= \tfrac{1}{2}\omega^{-1}\delta_1 a\cos(\sigma T_1 - 2\beta) \\ a\beta' &= \tfrac{1}{2}\omega^{-1}\delta_1 a\sin(\sigma T_1 - 2\beta) \end{aligned} \tag{6.18.30}$$

$$\gamma = \sigma T_1 - 2\beta \tag{6.18.31}$$

Equation (6.18.30) can be written as

$$\begin{aligned} a' &= \tfrac{1}{2}\omega^{-1}\delta_1 a\cos\gamma \\ a\gamma' &= \sigma a - \tfrac{1}{2}\omega^{-1}\delta_1 a\sin\gamma \end{aligned} \tag{6.18.32}$$

The necessary condition for the existence of a non-trivial steady state response of (6.18.32) is

$$\sigma = \pm\frac{1}{2}\omega^{-1}\delta_1 \tag{6.18.33}$$

Substituting the above equation into (6.18.26), we can obtain the transition curve separating the stability from instability is

$$\Omega = 2\omega \pm \varepsilon\frac{1}{2}\omega^{-1}\delta_1 \tag{6.18.34}$$

Therefore, the equilibrium position θ_s is conditionally stabilized when the system has a support excitation and $\Omega \approx 2\omega$.

CASE (2): Ω Stay away from 2ω.

In this case, in order to remove secular terms from (6.18.25), we need

$$D_1 A = 0 \quad \Rightarrow \quad A = A(T_2) \tag{6.18.35}$$

Then the solution of (6.18.25) can be obtained:

$$u_2 = -2\omega^{-2}\alpha_1 A\overline{A} + \frac{1}{3}\omega^{-2}\alpha_1 A^2 e^{2i\omega T_0} + ik_1 A e^{i(\Omega+\omega)T_0} + ik_2\overline{A}e^{i(\Omega-\omega)T_0} + cc \tag{6.18.36}$$

where

$$k_1 = \frac{1}{2}\delta_1\left[\omega^2 - (\Omega+\omega)^2\right],\ k_2 = \frac{1}{2}\delta_1\left[\omega^2 - (\Omega+\omega)^2\right] \tag{6.18.37}$$

Substitute (6.18.36) and (6.18.24) into (6.18.23), we get

$$\begin{aligned} D_0^2 u_3 + \omega^2 u_3 &= -2D_0D_2u_1 - 2\alpha_1 u_1 u_2 - \alpha_2 u_1^3 - \delta_2 u_1^2 \sin\Omega T_0 - \delta_1 u_2 \sin\Omega T_0 \\ &= \left[-2\omega^{-2}\alpha_1 A' + \left(\frac{10}{3}\omega^{-2}\alpha_1^2 - 3\alpha_2\right)A^2\overline{A} - \frac{1}{2}\delta_1(k_1+k_2)A\right]e^{i\omega T_0} \\ &\quad + i\left(2\alpha_1 k_1 - 2\alpha_1 k_2 - \omega^{-2}\alpha_1\delta_1 + \delta_2\right)A\overline{A}e^{i\Omega T_0} \\ &\quad - i\left(2\alpha_1 k_2 + \frac{1}{2}\delta_2 + \frac{1}{6}\omega^{-2}\alpha_1\delta_1\right)A^2 e^{i(2\omega-\Omega)T_0} \\ &\quad - \frac{1}{2}\delta_1 k_2 \overline{A}e^{i(2\Omega-\omega)T_0} + cc + NST \end{aligned} \tag{6.18.38}$$

The prime represents the derivative with respect to T_2. In order to eliminate secular terms from the above equation, it is necessary to distinguish between two cases where Ω is far away from ω and $\Omega \approx \omega$. Only the solution for the primary resonance $\Omega \approx \omega$ is given below, and readers are invited to complete the solution for the other case.

When $\Omega \approx \omega$, let

$$\Omega = \omega + \varepsilon\sigma \tag{6.18.39}$$

In order to remove secular terms from (6.18.38), we need

$$\begin{aligned} &2\omega^{-2}\alpha_1 A' - \left(\frac{10}{3}\omega^{-2}\alpha_1^2 - 3\alpha_2\right)A^2\overline{A} - \frac{1}{2}\delta_1(k_1+k_2)A \\ &- i\left(2\alpha_1 k_1 - 2\alpha_1 k_2 - \omega^{-2}\alpha_1\delta_1 + \delta_2\right)A\overline{A}e^{i\sigma T_1} \\ &+ i\left(2\alpha_1 k_2 + \frac{1}{2}\delta_2 + \frac{1}{6}\omega^{-2}\alpha_1\delta_1\right)A^2 e^{-i\sigma T_1} + \frac{1}{2}\delta_1 k_2 \overline{A}e^{2i\sigma T_1} = 0 \end{aligned} \tag{6.18.40}$$

Let

$$A = \frac{1}{2}ae^{i\beta} \tag{6.18.41}$$

Substituting (6.18.41) into (6.18.40), we obtain

$$
\begin{aligned}
& a' + ia\beta' - \frac{1}{8}\omega^2\alpha_1^{-1}\left(\frac{10}{3}\omega^{-2}\alpha_1^2 - 3\alpha_2\right)a^3 - \frac{1}{4}\omega^2\alpha_1^{-1}\delta_1(k_1 + k_2)a \\
& -\frac{1}{4}i\omega^2(2k_1 - 2k_2 - \omega^{-2}\delta_1 + \delta_2)a^2 e^{i(\sigma T_1 - \beta)} \\
& +\frac{1}{4}i\omega^2\left(2k_2 + \frac{1}{6}\omega^{-2}\delta_1 + \frac{1}{2}\alpha_1^{-1}\delta_2\right)a^2 e^{-i(\sigma T_1 - \beta)} \\
& +\frac{1}{4}\omega^2\alpha_1^{-1}\delta_1 k_2 a e^{2i(\sigma T_1 - \beta)} = 0
\end{aligned} \tag{6.18.42}
$$

Separating the real and imaginary parts of the above equations yields

$$
\begin{aligned}
& a' - \frac{1}{8}\omega^2\alpha_1^{-1}\left(\frac{10}{3}\omega^{-2}\alpha_1^2 - 3\alpha_2\right)a^3 - \frac{1}{4}\omega^2\alpha_1^{-1}\delta_1(k_1 + k_2)a \\
& \quad +\frac{1}{4}\omega^2(2k_1 - 2k_2 - \omega^{-2}\delta_1 + \delta_2)a^2 \sin(\sigma T_1 - \beta) \\
& \quad +\frac{1}{4}\omega^2\left(2k_2 + \frac{1}{6}\omega^{-2}\delta_1 + \frac{1}{2}\alpha_1^{-1}\delta_2\right)a^2 \sin(\sigma T_1 - \beta) \\
& \quad +\frac{1}{4}\omega^2\alpha_1^{-1}\delta_1 k_2 a \cos[2(\sigma T_1 - \beta)] = 0 \qquad (6.18.43) \\
& a\beta' - \frac{1}{4}\omega^2(2k_1 - 2k_2 - \omega^{-2}\delta_1 + \delta_2)a^2 \cos(\sigma T_1 - \beta) \\
& \quad -\frac{1}{4}\omega^2\left(2k_2 + \frac{1}{6}\omega^{-2}\delta_1 + \frac{1}{2}\alpha_1^{-1}\delta_2\right)a^2 \cos(\sigma T_1 - \beta) \\
& \quad +\frac{1}{4}\omega^2\alpha_1^{-1}\delta_1 k_2 a \sin[2(\sigma T_1 - \beta)] = 0
\end{aligned}
$$

Let

$$
\gamma = \sigma T_1 - \beta \tag{6.18.44}
$$

Equation (6.18.43) can be written as

$$
\begin{aligned}
& a' + \Gamma_1 a^3 + \Gamma_3 a + \Gamma_2 a^2 \sin\gamma + \Gamma_4 a \cos 2\gamma = 0 \\
& a\gamma' - \sigma a + \Gamma_2 a^2 \cos\gamma - \Gamma_4 a \sin 2\gamma = 0
\end{aligned} \tag{6.18.45}
$$

where

$$
\begin{aligned}
\Gamma_1 &= -\frac{1}{8}\omega^2\alpha_1^{-1}\left(\frac{10}{3}\omega^{-2}\alpha_1^2 - 3\alpha_2\right) \\
\Gamma_2 &= \frac{1}{4}\omega^2\left(2k_1 - 2k_2 - \omega^{-2}\delta_1 + \delta_2\right)a^2 \\
&\quad + \frac{1}{4}\omega^2\left(2k_2 + \frac{1}{6}\omega^{-2}\delta_1 + \frac{1}{2}\alpha_1^{-1}\delta_2\right) \\
\Gamma_3 &= -\frac{1}{4}\omega^2\alpha_1^{-1}\delta_1(k_1 + k_2) \\
\Gamma_4 &= \frac{1}{4}\omega^2\alpha_1^{-1}\delta_1 k_2
\end{aligned}
\tag{6.18.46}
$$

The steady state response should satisfy the following equations

$$
\begin{aligned}
\Gamma_1 a^3 + \Gamma_3 a &= -\Gamma_2 a^2 \sin\gamma - \Gamma_4 a\cos 2\gamma \\
\sigma a &= \Gamma_2 a^2 \cos\gamma - \Gamma_4 a \sin 2\gamma
\end{aligned}
\tag{6.18.47}
$$

The steady state responses a_0, γ_0 can be obtained by solving (6.18.47) numerically. To further analyze the stability of the steady state response, let

$$
a = a_0 + \tilde{a}, \quad \gamma = \gamma_0 + \tilde{\gamma} \tag{6.18.48}
$$

Substitute (6.18.48) into (6.18.45), we can obtain the linearized equation

$$
\begin{Bmatrix} \tilde{a}' \\ \tilde{\gamma}' \end{Bmatrix} + \begin{bmatrix} l_{11} & l_{12} \\ l_{21} & l_{22} \end{bmatrix} \begin{Bmatrix} \tilde{a} \\ \tilde{\gamma} \end{Bmatrix} = 0 \tag{6.18.49}
$$

where

$$
\begin{aligned}
l_{11} &= 3\Gamma_1 a_0^2 + \Gamma_3 + 2\Gamma_2 a_0 \sin\gamma_0 + \Gamma_4 \cos 2\gamma_0 \\
l_{12} &= \Gamma_2 a_0^2 \cos\gamma_0 - 2\Gamma_4 a_0 \sin 2\gamma_0 = 0 \\
l_{21} &= -\sigma + 2\Gamma_2 a_0 \cos\gamma_0 - \Gamma_4 \sin 2\gamma_0 \\
l_{22} &= -\left(\Gamma_2 a_0^2 \sin\gamma_0 + 2\Gamma_4 a_0 \cos 2\gamma_0\right)
\end{aligned}
\tag{6.18.50}
$$

The steady state solutions are stable if all eigenvalues of (6.18.49) have negative real parts, the steady state response is stable, otherwise it is unstable.

6.19 Exercise 6.19 (Primary Resonance Analysis of a Two-Degree-of-Freedom Self-Excited System)

Solution: Let the solution of the equation be

$$\begin{aligned} x &= u_{11}(T_0, T_1) + \varepsilon u_{12}(T_0, T_1) + \cdots \\ y &= u_{21}(T_0, T_1) + \varepsilon u_{22}(T_0, T_1) + \cdots \end{aligned} \tag{6.19.1}$$

Substituting (6.19.1) into the given system equation and retaining to $O(\varepsilon)$ yields

$$\begin{aligned} 0 &= \ddot{x} + x - \varepsilon\left[\lambda_0 \sin t + \alpha_1\left(1 - x^2\right)\dot{x} + \alpha_2\dot{y}\right] \\ &= \left(D_0^2 + 2\varepsilon D_0 D_1\right)(u_{11} + \varepsilon u_{12}) + u_{11} + \varepsilon u_{12} \\ &-\varepsilon\lambda_0 \sin T_0 - \varepsilon\alpha_1\left[1 - (u_{11} + \varepsilon u_{12})^2\right](D_0 + \varepsilon D_1)(u_{11} + \varepsilon u_{12}) \\ &-\varepsilon\alpha_2(D_0 + \varepsilon D_1)(u_{21} + \varepsilon u_{22}) + \cdots \\ &= D_0^2 u_{11} + u_{11} + \varepsilon\left(D_0^2 u_{12} + u_{12} + 2D_0 D_1 u_{11}\right. \\ &\left.-\alpha_1 D_0 u_{11} + \alpha_1 u_{11}^2 D_0 u_{11} - \alpha_2 D_0 u_{21} - \lambda_0 \sin T_0\right) + \cdots \end{aligned} \tag{6.19.2}$$

$$\begin{aligned} 0 &= \ddot{y} + \frac{1}{4}y - \varepsilon\left[-\frac{1}{4}\lambda_0 \sin t + \alpha_3\left(1 - x^2\right)\dot{x} + \alpha_4\dot{y}\right] \\ &= \left(D_0^2 + 2\varepsilon D_0 D_1\right)(u_{21} + \varepsilon u_{22}) + \frac{1}{4}(u_{21} + \varepsilon u_{22}) \\ &+\frac{1}{4}\varepsilon\lambda_0 \sin T_0 - \varepsilon\alpha_3\left[1 - (u_{11} + \varepsilon u_{12})^2\right](D_0 + \varepsilon D_1)(u_{11} + \varepsilon u_{12}) \\ &-\varepsilon\alpha_4(D_0 + \varepsilon D_1)(u_{21} + \varepsilon u_{22}) + \cdots \\ &= D_0^2 u_{21} + \frac{1}{4}u_{21} + \varepsilon\left(D_0^2 u_{22} + \frac{1}{4}u_{22} + 2D_0 D_1 u_{21}\right. \\ &\left.-\alpha_3 D_0 u_{11} + \alpha_3 u_{11}^2 D_0 u_{11} - \alpha_4 D_0 u_{21} + \frac{1}{4}\lambda_0 \sin T_0\right) + \cdots \end{aligned} \tag{6.19.3}$$

Making the coefficients of the like power of ε equal to zero in the above two equations, we get.

Order ε^0:

$$\begin{aligned} D_0^2 u_{11} + u_{11} &= 0 \\ D_0^2 u_{21} + \tfrac{1}{4}u_{21} &= 0 \end{aligned} \tag{6.19.4}$$

Order ε^1:

$$\begin{aligned} D_0^2 u_{12} + u_{12} &= -2D_0 D_1 u_{11} + \alpha_1 D_0 u_{11} - \alpha_1 u_{11}^2 D_0 u_{11} + \alpha_2 D_0 u_{21} + \lambda_0 \sin T_0 \\ D_0^2 u_{22} + \tfrac{1}{4}u_{22} &= -2D_0 D_1 u_{21} + \alpha_3 D_0 u_{11} - \alpha_3 u_{11}^2 D_0 u_{11} + \alpha_4 D_0 u_{21} - \tfrac{1}{4}\lambda_0 \sin T_0 \end{aligned} \tag{6.19.5}$$

The solution of (6.19.4) can be expressed as

$$u_{11} = A_1 e^{iT_0} + cc, \quad u_{21} = A_2 e^{iT_0/2} + cc \tag{6.19.6}$$

where $A_k = A_k(T_1), \quad k = 1, 2$. Substituting this into (6.19.5), we obtain

$$D_0^2u_{12} + u_{12} = -2D_0D_1u_{11} + \alpha_1 D_0u_{11} - \alpha_1 u_{11}^2 D_0u_{11} + \alpha_2 D_0u_{21} + \lambda_0 \sin T_0$$
$$= -2iA_1' e^{iT_0} + i\alpha_1 A_1 e^{iT_0} - i\alpha_1 A_1^2\overline{A}_1 e^{iT_0} - \tfrac{1}{2} i\lambda_0 e^{iT_0} + cc + NST \tag{6.19.7}$$

$$D_0^2u_{22} + \tfrac{1}{4}u_{22} = -2D_0D_1u_{21} + \alpha_3 D_0u_{11} - \alpha_3 u_{11}^2 D_0u_{11} + \alpha_4 D_0u_{21} - \tfrac{1}{4}\lambda_0 \sin T_0$$
$$= -iA_2' e^{iT_0/2} + \tfrac{1}{2} i\alpha_4 A_2 e^{iT_0/2} + cc + NST \tag{6.19.8}$$

In order to eliminate secular terms from the above two equations, we need

$$2A_1' - \alpha_1 A_1 + \alpha_1 A_1^2\overline{A}_1 + \frac{1}{2}\lambda_0 = 02A_1' - \alpha_1 A_1 + \alpha_1 A_1^2\overline{A}_1 + \frac{1}{2}\lambda_0 = 0 \tag{6.19.9}$$

$$A_2' - \frac{1}{2}\alpha_4 A_2 = 0 \tag{6.19.10}$$

can be solved directly, i.e.,

$$A_2 = \frac{1}{2}a_{20}e^{\alpha_4 T_1/2} \tag{6.19.11}$$

Let

$$A_1 = \frac{1}{2}a_1 e^{i\beta_1} \tag{6.19.12}$$

Substituting it into (6.19.9) and get

$$a_1' + ia_1\beta_1' - \alpha_1 a_1 + \alpha_1 a_1^3 + \frac{1}{2}\lambda_0 e^{-i\beta_1} = 0 \tag{6.19.13}$$

Separating the real and imaginary parts of the above equations yields

$$\begin{aligned} &a_1' - \alpha_1 a_1 + \alpha_1 a_1^3 + \frac{1}{2}\lambda_0 \cos\beta_1 = 0 \\ &a_1\beta_1' - \frac{1}{2}\lambda_0 \sin\beta_1 = 0 \end{aligned} \tag{6.19.14}$$

The steady state response should satisfy the following equations

$$\begin{aligned} &\alpha_1 a_1^3 - \alpha_1 a_1 + \frac{1}{2}\lambda_0 \cos\beta_1 = 0 \\ &\sin\beta_1 = 0 \end{aligned} \tag{6.19.15}$$

Therefore,

$$\beta_{10} = n\pi \tag{6.19.16}$$

$$\alpha_1 a_1^3 - \alpha_1 a_1 + \frac{1}{2}\lambda_0 \cos n\pi = 0 \tag{6.19.17}$$

The steady state response a_{10} can be found from (6.19.17). Substituting the above results into (6.19.6), we can obtain the first order approximate solution of the system

$$\begin{aligned} x &= a_1 \cos(t + n\pi) + O(\varepsilon) \\ y &= a_{20} e^{\varepsilon\alpha_4 t/2} \cos \frac{1}{2}t + O(\varepsilon) \end{aligned} \tag{6.19.18}$$

In order to analyze the stability of the steady state responses a_{10},β_{10}, let

$$a_1 = a_{10} + \tilde{a}_1,^\circ\ \beta_1 = n\pi + \tilde{\beta}_1 \tag{6.19.19}$$

Substituting it into (6.19.14) and linearizing the equation, we can obtain

$$\begin{aligned} \tilde{a}_1' &= \alpha_1\left(1 - 3a_{10}^2\right)\tilde{a}_1 \\ \tilde{\beta}_1' &= \frac{1}{2}a_{10}^{-1}\lambda_0 \cos n\pi\, \tilde{\beta}_1 \end{aligned} \tag{6.19.20}$$

So, the stability conditions for the steady state responses a_{10}, β_{10} are

$$\alpha_1\left(1 - 3a_{10}^2\right) < 0^\circ \text{ and}^\circ a_{10}^{-1}\lambda_0 \cos n\pi < 0 \tag{6.19.21}$$

In addition, one can find from (6.19.18) that the stability condition of y is

$$\alpha_4 < 0 \tag{6.19.22}$$

6.20 Exercise 6.20 (Free Oscillation Analysis of a Two-Degree-of-Freedom Self-Excited System with Heavy Eigenvalues)

Solution: (a) Let the solution of the equation be

$$\begin{aligned} u_1 &= u_{11}(T_0, T_1) + \varepsilon u_{12}(T_0, T_1) + \cdots \\ u_2 &= u_{21}(T_0, T_1) + \varepsilon u_{22}(T_0, T_1) + \cdots \end{aligned} \tag{6.20.1}$$

Substituting them into the control equation of the system and retaining to $O(\varepsilon)$ yields

$$
\begin{aligned}
0 &= \ddot{u}_1 + \omega_1^2 u_1 - \varepsilon\left(\alpha_1 - \alpha_2 u_1^2\right)\dot{u}_1 - \varepsilon\alpha_3 u_2 \\
&= \left(D_0^2 + 2\varepsilon D_0 D_1\right)(u_{11} + \varepsilon u_{12}) + \omega_1^2(u_{11} + \varepsilon u_{12}) \\
&\quad -\varepsilon\left[\alpha_1 - \alpha_2(u_{11} + \varepsilon u_{12})^2\right](D_0 + \varepsilon D_1)(u_{11} + \varepsilon u_{12}) \\
&\quad -\varepsilon\alpha_3(u_{21} + \varepsilon u_{22}) + \cdots \\
&= D_0^2 u_{11} + \omega_1^2 u_{11} + \varepsilon\left(D_0^2 u_{12} + \omega_1^2 u_{12} + 2D_0 D_1 u_{11}\right. \\
&\quad \left. - \alpha_1 D_0 u_{11} + \alpha_2 u_{11}^2 D_0 u_{11} - \alpha_3 u_{21}\right) + \cdots
\end{aligned}
\tag{6.20.2}
$$

$$
\begin{aligned}
0 &= \ddot{u}_2 + \omega_2^2 u_2 - \varepsilon\left(\alpha_1 - \alpha_2 u_2^2\right)\dot{u}_2 - \varepsilon\alpha_4 u_1 \\
&= \left(D_0^2 + 2\varepsilon D_0 D_1\right)(u_{21} + \varepsilon u_{22}) + \omega_2^2(u_{21} + \varepsilon u_{22}) \\
&\quad -\varepsilon\left[\alpha_1 - \alpha_2(u_{21} + \varepsilon u_{22})^2\right](D_0 + \varepsilon D_1)(u_{21} + \varepsilon u_{22}) \\
&\quad -\varepsilon\alpha_4(u_{11} + \varepsilon u_{12}) + \cdots \\
&= D_0^2 u_{21} + \omega_2^2 u_{21} + \varepsilon\left(D_0^2 u_{22} + \omega_2^2 u_{22} + 2D_0 D_1 u_{21}\right. \\
&\quad \left. - \alpha_1 D_0 u_{21} + \alpha_2 u_{21}^2 D_0 u_{21} - \alpha_4 u_{11}\right) + \cdots
\end{aligned}
\tag{6.20.3}
$$

Making the coefficient of the like power of ε equal to zero in the above two equations, we get.

Order ε^0:

$$
\begin{aligned}
D_0^2 u_{11} + \omega_1^2 u_{11} &= 0 \\
D_0^2 u_{21} + \omega_2^2 u_{21} &= 0
\end{aligned}
\tag{6.20.4}
$$

Order ε^1:

$$
\begin{aligned}
D_0^2 u_{12} + \omega_1^2 u_{12} &= -2D_0 D_1 u_{11} + \alpha_1 D_0 u_{11} - \alpha_2 u_{11}^2 D_0 u_{11} + \alpha_3 u_{21} \\
D_0^2 u_{22} + \omega_2^2 u_{22} &= -2D_0 D_1 u_{21} + \alpha_1 D_0 u_{21} - \alpha_2 u_{21}^2 D_0 u_{21} + \alpha_4 u_{11}
\end{aligned}
\tag{6.20.5}
$$

The solution of (6.20.4) can be expressed as

$$
u_{11} = A_1 e^{i\omega_1 T_0} + cc, \quad u_{21} = A_2 e^{i\omega_2 T_0} + cc \tag{6.20.6}
$$

where.$A_k = A_k(T_1), \quad k = 1, 2$ Substituting this into (6.20.5), we get

$$
\begin{aligned}
&D_0^2 u_{12} + \omega_1^2 u_{12} = -2D_0 D_1 u_{11} + \alpha_1 D_0 u_{11} - \alpha_2 u_{11}^2 D_0 u_{11} + \alpha_3 u_{21} \\
&= \left(-2i\omega_1 A_1' + i\omega_1\alpha_1 A_1 - 3i\omega_1\alpha_2 A_1^2 \overline{A}_1\right)e^{i\omega_1 T_0} \\
&+\alpha_3 A_2 e^{i\omega_2 T_0} + cc + NST
\end{aligned}
\tag{6.20.7}
$$

$$
\begin{aligned}
&D_0^2 u_{22} + \omega_2^2 u_{22} = -2D_0 D_1 u_{21} + \alpha_1 D_0 u_{21} - \alpha_2 u_{21}^2 D_0 u_{21} + \alpha_4 u_{11} \\
&= \left(-2i\omega_2 A_2' + i\omega_2\alpha_1 A_2 - 3i\omega_2\alpha_2 A_2^2 \overline{A}_2\right)e^{i\omega_2 T_0} \\
&+\alpha_4 A_1 e^{i\omega_1 T_0} + cc + NST
\end{aligned}
\tag{6.20.8}
$$

In order to eliminate secular terms from the above two equations when $\omega_2 = \omega_1 + \varepsilon\sigma$, we need

$$\begin{aligned} 2i\omega_1 A'_1 &= i\omega_1(\alpha_1 - 3\alpha_2 A_1\overline{A}_1)A_1 + \alpha_3 A_2 e^{i\sigma T_1} \\ 2i\omega_2 A'_2 &= i\omega_2(\alpha_1 - 3\alpha_2 A_2\overline{A}_2)A_2 + \alpha_4 A_1 e^{-i\sigma T_1} \end{aligned} \tag{6.20.9}$$

(b) Set

$$A_1 = \frac{1}{2}a_1 e^{i\beta_1}, \quad A_2 = \frac{1}{2}a_2 e^{i\beta_2} \tag{6.20.10}$$

Substituting them into (6.20.9) and get

$$i\omega_1 a_1' - \omega_1 a_1 \beta_1' = i\omega_1(\alpha_1 - 3\alpha_2 a_1^2)a_1 + \alpha_3 a_2 e^{i(\sigma T_1 - \beta_1 + \beta_2)} \tag{6.20.11}$$

$$i\omega_2 a_2' - \omega_2 a_2 \beta_2' = i\omega_2(\alpha_1 - 3\alpha_2 a_2^2)a_2 + \alpha_4 a_1 e^{-i(\sigma T_1 - \beta_1 + \beta_2)} \tag{6.20.12}$$

Separating the real and imaginary parts of the above equations yields

$$\begin{aligned} \omega_1 a_1' &= \omega_1(\alpha_1 - 3\alpha_2 a_1^2)a_1 + \alpha_3 a_2 \sin\gamma \\ \omega_2 a_2' &= \omega_2(\alpha_1 - 3\alpha_2 a_2^2)a_2 - \alpha_4 a_1 \sin\gamma \end{aligned} \tag{6.20.13}$$

$$\begin{aligned} \omega_1 a_1 \beta_1' &= -\alpha_3 a_2 \cos\gamma \\ \omega_2 a_2 \beta_2' &= -\alpha_4 a_1 \cos\gamma \end{aligned} \tag{6.20.14}$$

where

$$\gamma = \sigma T_1 - \beta_1 + \beta_2 \tag{6.20.15}$$

Eliminating β_1 and β_2 from (6.20.14), we get

$$\omega_1\omega_2 a_1 a_2 \gamma' = \omega_1\omega_2 a_1 a_2 \sigma + (\omega_2\alpha_3 a_2^2 - \omega_1\alpha_4 a_1^2)\cos\gamma \tag{6.20.16}$$

By (6.20.13) and (6.20.16), the steady state response should satisfy the following equations

$$\begin{aligned} &\omega_1(\alpha_1 - 3\alpha_2 a_1^2)a_1 + \alpha_3 a_2 \sin\gamma = 0 \\ &\omega_2(\alpha_1 - 3\alpha_2 a_2^2)a_2 - \alpha_4 a_1 \sin\gamma = 0 \\ &\omega_1\omega_2 a_1 a_2 \sigma + (\omega_2\alpha_3 a_2^2 - \omega_1\alpha_4 a_1^2)\cos\gamma = 0 \end{aligned} \tag{6.20.17}$$

This set of equations has trivial solution $a_1 = a_2 = 0$. For non-trivial solutions, the frequency–response equation can be obtained as

$$\omega_2\left(\alpha_1 - 3\alpha_2 a_2^2\right)a_2^2 = -\frac{\alpha_4}{\alpha_3}\omega_1\left(\alpha_1 - 3\alpha_2 a_1^2\right)a_1^2 \tag{6.20.18}$$

$$\left[\frac{\omega_1\left(\alpha_1 - 3\alpha_2 a_1^2\right)a_1}{\alpha_3 a_2}\right]^2\left[\frac{\omega_1\omega_2 a_1 a_2\sigma}{\left(\omega_2\alpha_3 a_2^2 - \omega_1\alpha_4 a_1^2\right)}\right]^2 = 1 \tag{6.20.19}$$

From this, the non-trivial steady state response, a_{10}, a_{20} and γ_0, can be obtained. To analyze the stability of the steady state response, let

$$a_1 = a_{10} + \tilde{a}_1, \quad a_2 = a_{20} + \tilde{a}_2, \quad \gamma = \gamma_0 + \tilde{\gamma} \tag{6.20.20}$$

Substituting (6.20.21) into (6.20.13) and (6.20.16) and linearizing the equation, we can obtain

$$\begin{bmatrix} 1\,0\,0 & \\ 0\,1\,0 & \\ 0\,0\,a_{10}a_{20} & \end{bmatrix}\begin{Bmatrix} \widetilde{a'}_1 \\ \widetilde{a'}_2 \\ \widetilde{\gamma'} \end{Bmatrix} = \begin{bmatrix} l_{11}\ l_{12}\ l_{13} \\ l_{21}\ l_{22}\ l_{23} \\ l_{31}\ l_{32}\ l_{33} \end{bmatrix}\begin{Bmatrix} \tilde{a}_1 \\ \tilde{a}_2 \\ \tilde{\gamma} \end{Bmatrix} \tag{6.20.21}$$

where

$$\begin{aligned} l_{11} &= \alpha_1 - 9\alpha_2 a_{10}^2, \quad l_{12} = \omega_1^{-1}\alpha_3 \sin\gamma_0, \quad l_{13} = \omega_1^{-1}\alpha_3 a_{20}\cos\gamma_0 \\ l_{21} &= -\omega_2^{-1}\alpha_4 \sin\gamma_0, \quad l_{22} = \alpha_1 - 9\alpha_2 a_{20}^2, \quad l_{23} = -\omega_2^{-1}\alpha_4 a_{10}\cos\gamma_0 \\ l_{31} &= a_{20}\sigma - 2\omega_2^{-1}\alpha_4 a_{10}\cos\gamma_0, \quad l_{32} = a_{10}\sigma + 2\omega_1^{-1}\alpha_3 a_{20}\cos\gamma_0 \\ l_{33} &= \omega_2^{-1}\alpha_4 a_{10}^2 \sin\gamma_0 - \omega_1^{-1}\alpha_3 a_{20}^2 \sin\gamma_0 \end{aligned} \tag{6.20.22}$$

For the trivial steady state response $a_{10} = a_{20} = 0$, the coefficient matrix on the right-hand side (6.20.22) is singular, so the trivial steady state response is unstable. For the non-trivial solution, we can obtain three eigenvalues from (6.20.22). if all three eigenvalues have negative real parts, then the non-trivial solution is stable, otherwise it is unstable.

6.21 Exercise 6.21 (Primary Resonance Analysis of Three-Degree-of-Freedom Systems with Repeating Frequencies, Internal Resonance, Saturation Phenomena)

Solution: (a) Let the solution of the equation be

$$\begin{aligned} u_1 &= \varepsilon u_{11}(T_0, T_1) + \varepsilon^2 u_{12}(T_0, T_1) + \cdots \\ u_2 &= \varepsilon u_{21}(T_0, T_1) + \varepsilon^2 u_{22}(T_0, T_1) + \cdots \\ u_3 &= \varepsilon u_{31}(T_0, T_1) + \varepsilon^2 u_{32}(T_0, T_1) + \cdots \end{aligned} \tag{6.21.1}$$

Substituting them into the control equation of the system and retaining to $O(\varepsilon^2)$ yields

$$\begin{aligned}
0 &= \ddot{u}_1 + \omega_1^2 u_1 + 2\varepsilon\mu_1\dot{u}_1 + \alpha_1 u_1 u_3 + \frac{1}{2}\alpha_2 u_1 \ddot{u}_3 - 2K_1 \cos \Omega t \\
&= \left(D_0^2 + 2\varepsilon D_0 D_1\right)\left(\varepsilon u_{11} + \varepsilon^2 u_{12}\right) + \omega_1^2\left(\varepsilon u_{11} + \varepsilon^2 u_{12}\right) \\
&+ 2\varepsilon\mu_1(D_0 + \varepsilon D_1)\left(\varepsilon u_{11} + \varepsilon^2 u_{12}\right) + \alpha_1\left(\varepsilon u_{11} + \varepsilon^2 u_{12}\right)\left(\varepsilon u_{31} + \varepsilon^2 u_{32}\right) \\
&+ \frac{1}{2}\alpha_2\left(\varepsilon u_{11} + \varepsilon^2 u_{12}\right)\left(D_0^2 + 2\varepsilon D_0 D_1\right)\left(\varepsilon u_{31} + \varepsilon^2 u_{32}\right) - 2K_1 \cos \Omega T_0 \\
&= \varepsilon\left(D_0^2 u_{11} + \omega_1^2 u_{11}\right) + \varepsilon^2\Big(D_0^2 u_{12} + \omega_1^2 u_{12} + 2D_0 D_1 u_{11} + 2\mu_1 D_0 u_{11} \\
&+ \alpha_1 u_{11} u_{31} + \frac{1}{2}\alpha_2 u_{11} D_0^2 u_{31}\Big) - 2K_1 \cos \Omega T_0 + \cdots
\end{aligned} \tag{6.21.2}$$

$$\begin{aligned}
0 &= \ddot{u}_2 + \omega_2^2 u_2 + 2\varepsilon\mu_2\dot{u}_2 + \alpha_1 u_2 u_3 + \frac{1}{2}\alpha_2 u_2 \ddot{u}_3 \\
&= \left(D_0^2 + 2\varepsilon D_0 D_1\right)\left(\varepsilon u_{21} + \varepsilon^2 u_{22}\right) + \omega_2^2\left(\varepsilon u_{21} + \varepsilon^2 u_{22}\right) \\
&+ 2\varepsilon\mu_2(D_0 + \varepsilon D_1)\left(\varepsilon u_{21} + \varepsilon^2 u_{22}\right) + \alpha_1\left(\varepsilon u_{21} + \varepsilon^2 u_{22}\right)\left(\varepsilon u_{31} + \varepsilon^2 u_{32}\right) \\
&+ \frac{1}{2}\alpha_2\left(\varepsilon u_{21} + \varepsilon^2 u_{22}\right)\left(D_0^2 + 2\varepsilon D_0 D_1\right)\left(\varepsilon u_{31} + \varepsilon^2 u_{32}\right) \\
&= \varepsilon\left(D_0^2 u_{21} + \omega_2^2 u_{21}\right) + \varepsilon^2\Big(D_0^2 u_{22} + \omega_1^2 u_{22} + 2D_0 D_1 u_{21} + 2\mu_2 D_0 u_{21} \\
&+ \alpha_1 u_{21} u_{31} + \frac{1}{2}\alpha_2 u_{21} D_0^2 u_{31}\Big) + \cdots
\end{aligned} \tag{6.21.3}$$

$$\begin{aligned}
0 &= \ddot{u}_3 + \omega_3^2 u_3 + 2\varepsilon\mu_3\dot{u}_3 + \frac{1}{2}\alpha_1\left(u_1^2 + u_2^2\right) + \frac{1}{2}\alpha_2\frac{d}{dt}(u_1\dot{u}_1 + u_2\dot{u}_2) - 2K_3 \cos \Omega t \\
&= \left(D_0^2 + 2\varepsilon D_0 D_1\right)\left(\varepsilon u_{31} + \varepsilon^2 u_{32}\right) + \omega_3^2\left(\varepsilon u_{31} + \varepsilon^2 u_{32}\right) \\
&+ 2\varepsilon\mu_3(D_0 + \varepsilon D_1)\left(\varepsilon u_{31} + \varepsilon^2 u_{32}\right) + \frac{1}{2}\alpha_1\left[\left(\varepsilon u_{11} + \varepsilon^2 u_{12}\right)^2 + \left(\varepsilon u_{21} + \varepsilon^2 u_{22}\right)^2\right] \\
&+ \frac{1}{2}\alpha_2(D_0 + \varepsilon D_1)\left[\left(\varepsilon u_{11} + \varepsilon^2 u_{12}\right)(D_0 + \varepsilon D_1)\left(\varepsilon u_{11} + \varepsilon^2 u_{12}\right)\right. \\
&+ \left.\left(\varepsilon u_{21} + \varepsilon^2 u_{22}\right)(D_0 + \varepsilon D_1)\left(\varepsilon u_{21} + \varepsilon^2 u_{22}\right)\right] - 2K_3 \cos \Omega T_0 \\
&= \varepsilon\left(D_0^2 u_{31} + \omega_3^2 u_{31}\right) + \varepsilon^2\Big[D_0^2 u_{32} + \omega_3^2 u_{32} + 2D_0 D_1 u_{31} + 2\mu_3 D_0 u_{31} + \frac{1}{2}\alpha_1 u_{11}^2 \\
&+ \frac{1}{2}\alpha_1 u_{21}^2 + \frac{1}{2}\alpha_2 D_0(u_{11} D_0 u_{11}) + \frac{1}{2}\alpha_2 D_0(u_{21} D_0 u_{21})\Big] - 2K_3 \cos \Omega T_0 + \cdots
\end{aligned} \tag{6.21.4}$$

It can be seen that we need to discuss on the value of Ω to eliminate secular terms from (6.21.2) and (6.21.4).

(i) $\Omega = \omega_3 + \varepsilon\sigma_3$

Let

$$K_1 = \varepsilon k_1, \quad K_3 = \varepsilon^2 k_3 \tag{6.21.5}$$

Equating the coefficients of the like power of ε in (6.21.2)–(6.21.4), we get.
Order ε^1:

$$\begin{aligned} D_0^2 u_{11} + \omega_1^2 u_{11} &= 2k_1 \cos\Omega T_0 \\ D_0^2 u_{21} + \omega_2^2 u_{21} &= 0 \\ D_0^2 u_{31} + \omega_3^2 u_{31} &= 0 \end{aligned} \tag{6.21.6}$$

Order ε^2:

$$\begin{aligned} D_0^2 u_{12} + \omega_1^2 u_{12} &= -2D_0 D_1 u_{11} - 2\mu_1 D_0 u_{11} - \alpha_1 u_{11} u_{31} - \frac{1}{2}\alpha_2 u_{11} D_0^2 u_{31} \\ D_0^2 u_{22} + \omega_1^2 u_{22} &= -2D_0 D_1 u_{21} - 2\mu_2 D_0 u_{21} - \alpha_1 u_{21} u_{31} - \frac{1}{2}\alpha_2 u_{21} D_0^2 u_{31} \\ D_0^2 u_{32} + \omega_3^2 u_{32} &= -2D_0 D_1 u_{31} - 2\mu_3 D_0 u_{31} - \frac{1}{2}\alpha_1 u_{11}^2 - \frac{1}{2}\alpha_1 u_{21}^2 \\ -\frac{1}{2}\alpha_2 D_0 (u_{11} D_0 u_{11}) &- \frac{1}{2}\alpha_2 D_0 (u_{21} D_0 u_{21}) + 2k_3 \cos\Omega T_0 \end{aligned} \tag{6.21.7}$$

The solution of (6.21.6) can be expressed as

$$u_{11} = A_1 e^{i\omega_1 T_0} + \hat{k}_1 e^{i\Omega T_0} + cc, \quad u_{21} = A_2 e^{i\omega_2 T_0} + cc, \quad u_{31} = A_3 e^{i\omega_3 T_0} + cc \tag{6.21.8}$$

where

$$\hat{k}_1 = \frac{k_1}{\omega_1^2 - \Omega^2} \tag{6.21.9}$$

$A_k = A_k(T_1), \quad k = 1, 2, 3$. Substituting (6.21.8) into (6.21.7), we obtain

$$\begin{aligned} &D_0^2 u_{12} + \omega_1^2 u_{12} = -2D_0 D_1 u_{11} - 2\mu_1 D_0 u_{11} - \alpha_1 u_{11} u_{31} - \tfrac{1}{2}\alpha_2 u_{11} D_0^2 u_{31} \\ &= -2i\omega_1 A_1' e^{i\omega_1 T_0} - 2i\omega_1 \mu_1 A_1 e^{i\omega_1 T_0} - \alpha_1 \overline{A}_1 A_3 e^{i(\omega_3 - \omega_1) T_0} \\ &+ \tfrac{1}{2}\omega_3^2 \alpha_2 \overline{A}_1 A_3 e^{i(\omega_3 - \omega_1) T_0} + cc + NST \end{aligned} \tag{6.21.10}$$

$$\begin{aligned} &D_0^2 u_{22} + \omega_1^2 u_{22} = -2D_0 D_1 u_{21} - 2\mu_2 D_0 u_{21} - \alpha_1 u_{21} u_{31} - \tfrac{1}{2}\alpha_2 u_{21} D_0^2 u_{31} \\ &= -2i\omega_2 A'_2 e^{i\omega_2 T_0} - 2i\omega_2 \mu_2 A_2 e^{i\omega_2 T_0} - \alpha_1 \overline{A}_2 A_3 e^{i(\omega_3 - \omega_2) T_0} \\ &+ \tfrac{1}{2}\omega_3^2 \alpha_2 \overline{A}_2 A_3 e^{i(\omega_3 - \omega_2) T_0} + cc + NST \end{aligned} \tag{6.21.11}$$

$$
\begin{aligned}
D_0^2u_{32} + \omega_3^2u_{32} &= -2D_0D_1u_{31} - 2\mu_3D_0u_{31} - \frac{1}{2}\alpha_1u_{11}^2 - \frac{1}{2}\alpha_1u_{21}^2 \\
&-\frac{1}{2}\alpha_2D_0(u_{11}D_0u_{11}) - \frac{1}{2}\alpha_2D_0(u_{21}D_0u_{21}) + 2k_3\cos\Omega T_0 \\
&= -2i\omega_3A'e^{i\omega_3T_0} - 2i\omega_3\mu_3A_3e^{i\omega_3T_0} - \frac{1}{2}\alpha_1A_1^2e^{2i\omega_1T_0} \\
&-\frac{1}{2}\alpha_1A_2^2e^{2i\omega_2T_0} + \alpha_2\omega_1^2A_1^2e^{2i\omega_1T_0} + \alpha_2\omega_2^2A_2^2e^{2i\omega_2T_0} \\
&+k_3e^{i\Omega T_0} + cc + NST
\end{aligned}
\tag{6.21.12}
$$

In order to eliminate secular terms from the above three equations when $\omega_2 = \omega_1, \quad \omega_3 = 2\omega_1 + \varepsilon\sigma, \quad \Omega = \omega_3 + \varepsilon\sigma_3$, we need

$$
\begin{aligned}
&2i\omega_1\left(A_1' + \mu_1A_1\right) + \left(\alpha_1 - \frac{1}{2}\omega_3^2\alpha_2\right)\overline{A}_1A_3e^{i\sigma T_1} = 0 \\
&2i\omega_2\left(A_2' + \mu_2A_2\right) + \left(\alpha_1 - \frac{1}{2}\omega_3^2\alpha_2\right)\overline{A}_2A_3e^{i\sigma T_1} = 0 \\
&2i\omega_3\left(A_3' + \mu_3A_3\right) + \left(\frac{1}{2}\alpha_1 - \omega_1^2\alpha_2\right)\left(A_1^2 + A_2^2\right)e^{-i\sigma T_1} - k_3e^{i\sigma_3T_1} = 0
\end{aligned}
\tag{6.21.13}
$$

(ii) $\Omega = \omega_1 + \varepsilon\sigma_1$

Let

$$
K_1 = \varepsilon^2k_1, \quad K_3 = \varepsilon k_3 \tag{6.21.14}
$$

Equating the coefficients of the like power of ε in (6.21.2)–(6.21.4), we get.
Order ε^1:

$$
\begin{aligned}
D_0^2u_{11} + \omega_1^2u_{11} &= 0 \\
D_0^2u_{21} + \omega_2^2u_{21} &= 0 \\
D_0^2u_{31} + \omega_3^2u_{31} &= 2k_1\cos\Omega T_0
\end{aligned}
\tag{6.21.15}
$$

Order ε^2:

$$D_0^2u_{12}+\omega_1^2u_{12}=-2D_0D_1u_{11}-2\mu_1D_0u_{11}-\alpha_1u_{11}u_{31}-\frac{1}{2}\alpha_2u_{11}D_0^2u_{31}$$
$$+2k_1\cos\Omega T_0$$
$$D_0^2u_{22}+\omega_1^2u_{22}=-2D_0D_1u_{21}-2\mu_2D_0u_{21}-\alpha_1u_{21}u_{31}-\frac{1}{2}\alpha_2u_{21}D_0^2u_{31}$$
$$D_0^2u_{32}+\omega_3^2u_{32}=-2D_0D_1u_{31}-2\mu_3D_0u_{31}-\frac{1}{2}\alpha_1u_{11}^2-\frac{1}{2}\alpha_1u_{21}^2$$
$$-\frac{1}{2}\alpha_2D_0(u_{11}D_0u_{11})-\frac{1}{2}\alpha_2D_0(u_{21}D_0u_{21}) \tag{6.21.16}$$

The solution of (6.21.15) can be expressed as

$$u_{11}=A_1e^{i\omega_1T_0}+cc,\quad u_{21}=A_2e^{i\omega_2T_0}+cc,\quad u_{31}=A_3e^{i\omega_3T_0}+\hat{k}_3e^{i\Omega T_0}+cc \tag{6.21.17}$$

where

$$\hat{k}_3=\frac{k_3}{\omega_3^2-\Omega^2} \tag{6.21.18}$$

$A_k=A_k(T_1),\quad k=1,2,3$. Substituting (6.21.17) into (6.21.16), we obtain

$$D_0^2u_{12}+\omega_1^2u_{12}=-2D_0D_1u_{11}-2\mu_1D_0u_{11}-\alpha_1u_{11}u_{31}-\frac{1}{2}\alpha_2u_{11}D_0^2u_{31}$$
$$=-2i\omega_1A_1'e^{i\omega_1T_0}-2i\omega_1\mu_1A_1e^{i\omega_1T_0}-\alpha_1\overline{A}_1A_3e^{i(\omega_3-\omega_1)T_0}$$
$$+\frac{1}{2}\omega_3^2\alpha_2\overline{A}_1A_3e^{i(\omega_3-\omega_1)T_0}+k_1e^{i\Omega T_0}+cc+NST \tag{6.21.19}$$

$$D_0^2u_{22}+\omega_1^2u_{22}=-2D_0D_1u_{21}-2\mu_2D_0u_{21}-\alpha_1u_{21}u_{31}-\tfrac{1}{2}\alpha_2u_{21}D_0^2u_{31}$$
$$=-2i\omega_2A'_2e^{i\omega_2T_0}-2i\omega_2\mu_2A_2e^{i\omega_2T_0}-\alpha_1\overline{A}_2A_3e^{i(\omega_3-\omega_2)T_0}$$
$$+\tfrac{1}{2}\omega_3^2\alpha_2\overline{A}_2A_3e^{i(\omega_3-\omega_2)T_0}+cc+NST \tag{6.21.20}$$

$$D_0^2u_{32}+\omega_3^2u_{32}=-2D_0D_1u_{31}-2\mu_3D_0u_{31}-\frac{1}{2}\alpha_1u_{11}^2-\frac{1}{2}\alpha_1u_{21}^2$$
$$-\frac{1}{2}\alpha_2D_0(u_{11}D_0u_{11})-\frac{1}{2}\alpha_2D_0(u_{21}D_0u_{21})+2k_3\cos\Omega T_0$$
$$=-2i\omega_3A_3'e^{i\omega_3T_0}-2i\omega_3\mu_3A_3e^{i\omega_3T_0}-\frac{1}{2}\alpha_1A_1^2e^{2i\omega_1T_0} \tag{6.21.21}$$
$$-\frac{1}{2}\alpha_1A_2^2e^{2i\omega_2T_0}+\alpha_2\omega_1^2A_1^2e^{2i\omega_1T_0}+\alpha_2\omega_2^2A_2^2e^{2i\omega_2T_0}$$
$$+cc+NST$$

In order to eliminate secular terms from the above three equations when $\omega_2 = \omega_1, \ \ \omega_3 = 2\omega_1 + \varepsilon\sigma, \ \ \ \Omega = \omega_1 + \varepsilon\sigma_1$, we need

$$
\begin{aligned}
&2i\omega_1\left(A_1' + \mu_1 A_1\right) + \left(\alpha_1 - \frac{1}{2}\omega_3^2\alpha_2\right)\overline{A}_1 A_3 e^{i\sigma T_1} - k_1 e^{i\sigma_1 T_1} = 0\\
&2i\omega_2\left(A_2' + \mu_2 A_2\right) + \left(\alpha_1 - \frac{1}{2}\omega_3^2\alpha_2\right)\overline{A}_2 A_3 e^{i\sigma T_1} = 0\\
&2i\omega_3\left(A_3' + \mu_3 A_3\right) + \left(\frac{1}{2}\alpha_1 - \omega_1^2\alpha_2\right)\left(A_1^2 + A_2^2\right)e^{-i\sigma T_1} = 0
\end{aligned}
\tag{6.21.22}
$$

Equation (6.21.13) and (6.21.22) can be combined and written as

$$
\begin{aligned}
&2i\omega_1\left(A_1' + \mu_1 A_1\right) + \left(\alpha_1 - \frac{1}{2}\omega_3^2\alpha_2\right)A_3\overline{A}_1 e^{i\sigma T_1} - k_1\delta_{1s} e^{i\sigma_1 T_1} = 0\\
&2i\omega_2\left(A_2' + \mu_2 A_2\right) + \left(\alpha_1 - \frac{1}{2}\omega_3^2\alpha_2\right)A_3\overline{A}_2 e^{i\sigma T_1} = 0\\
&2i\omega_3\left(A_3' + \mu_3 A_3\right) + \left(\frac{1}{2}\alpha_1 - \omega_1^2\alpha_2\right)\left(A_1^2 + A_2^2\right)e^{-i\sigma T_1} - k_3\delta_{3s} e^{i\sigma_3 T_1} = 0
\end{aligned}
\tag{6.21.23}
$$

where δ_{ns} is the Kronecker symbol.

(b) Set

$$
A_1 = \frac{1}{2}a_1 e^{i\beta_1}, A_2 = \frac{1}{2}a_2 e^{i\beta_2}, A_3 = \frac{1}{2}a_3 e^{i\beta_3}
\tag{6.21.24}
$$

Substituting (6.21.24) into (6.21.23) when $s = 3$ and $\Omega = \omega_3 + \varepsilon\sigma_3$, we get

$$
\begin{aligned}
&i\omega_1 a_1' - \omega_1 a_1 \beta_1' + i\omega_1\mu_1 a_1 + \frac{1}{4}\left(\alpha_1 - \frac{1}{2}\omega_3^2\alpha_2\right)a_1 a_3 e^{i(\sigma T_1 - 2\beta_1 + \beta_3)} = 0\\
&i\omega_2 a_2' - \omega_2 a_2 \beta_2' + i\omega_2\mu_2 a_2 + \frac{1}{4}\left(\alpha_1 - \frac{1}{2}\omega_3^2\alpha_2\right)a_2 a_3 e^{i(\sigma T_1 - 2\beta_2 + \beta_3)} = 0\\
&i\omega_3 a_3' - \omega_3 a_3 \beta_3' + i\omega_3\mu_3 a_3 + \frac{1}{4}\left(\frac{1}{2}\alpha_1 - \omega_1^2\alpha_2\right)a_1^2 e^{-i(\sigma T_1 - 2\beta_1 + \beta_3)}\\
&+\frac{1}{4}\left(\frac{1}{2}\alpha_1 - \omega_1^2\alpha_2\right)a_2^2 e^{-i(\sigma T_1 - 2\beta_2 + \beta_3)} - k_3 e^{i(\sigma_3 T_1 - \beta_3)} = 0
\end{aligned}
\tag{6.21.25}
$$

Separating the real and imaginary parts of the above equations yields

$$
\begin{aligned}
a_1' &= -\mu_1 a_1 - \frac{1}{4}\omega_1^{-1}\left(\alpha_1 - \frac{1}{2}\omega_3^2\alpha_2\right)a_1 a_3 \sin\gamma_1 \\
a_2' &= -\mu_2 a_2 - \frac{1}{4}\omega_2^{-1}\left(\alpha_1 - \frac{1}{2}\omega_3^2\alpha_2\right)a_2 a_3 \sin\gamma_2 \\
a_3' &= -\mu_3 a_3 + \frac{1}{4}\omega_3^{-1}\left(\frac{1}{2}\alpha_1 - \omega_1^2\alpha_2\right)a_1^2 \sin\gamma_1 \\
&+\frac{1}{4}\omega_3^{-1}\left(\frac{1}{2}\alpha_1 - \omega_1^2\alpha_2\right)a_2^2 \sin\gamma_2 + \omega_3^{-1}k_3 \sin\gamma_3
\end{aligned}
\tag{6.21.26}
$$

$$
\begin{aligned}
a_1\beta_1' &= \frac{1}{4}\omega_1^{-1}\left(\alpha_1 - \frac{1}{2}\omega_3^2\alpha_2\right)a_1 a_3 \cos\gamma_1 \\
a_2\beta_2' &= \frac{1}{4}\omega_2^{-1}\left(\alpha_1 - \frac{1}{2}\omega_3^2\alpha_2\right)a_2 a_3 \cos\gamma_2 \\
a_3\beta_3' &= \frac{1}{4}\omega_3^{-1}\left(\frac{1}{2}\alpha_1 - \omega_1^2\alpha_2\right)a_1^2 \cos\gamma_1 \\
&+\frac{1}{4}\omega_3^{-1}\left(\frac{1}{2}\alpha_1 - \omega_1^2\alpha_2\right)a_2^2 \cos\gamma_2 - \omega_3^{-1}k_3 \cos\gamma_3
\end{aligned}
\tag{6.21.27}
$$

where

$$
\gamma_1 = \sigma T_1 - 2\beta_1 + \beta_3,\ \gamma_2 = \sigma T_1 - 2\beta_2 + \beta_3,\ \gamma_3 = \sigma_3 T_1 - \beta_3 \tag{6.21.28}
$$

From (6.21.28), we have

$$
2\beta_1 = \sigma T_1 + \sigma_3 T_1 - \gamma_1 - \gamma_3,\ 2\beta_2 = \sigma T_1 + \sigma_3 T_1 - \gamma_2 - \gamma_3,\ \beta_3 = \sigma_3 T_1 - \gamma_3 \tag{6.21.29}
$$

Substituting this into (6.21.27), we can obtain

$$
\begin{aligned}
a_1\gamma_1' + a_1\gamma_3' &= \sigma a_1 + \sigma_3 a_1 - \frac{1}{2}\omega_1^{-1}\left(\alpha_1 - \frac{1}{2}\omega_3^2\alpha_2\right)a_1 a_3 \cos\gamma_1 \\
a_2\gamma_2' + a_2\gamma_3' &= \sigma a_2 + \sigma_3 a_2 - \frac{1}{2}\omega_2^{-1}\left(\alpha_1 - \frac{1}{2}\omega_3^2\alpha_2\right)a_2 a_3 \cos\gamma_2 \\
a_3\gamma_3' &= \sigma_3 a_3 - \frac{1}{4}\omega_3^{-1}\left(\frac{1}{2}\alpha_1 - \omega_1^2\alpha_2\right)a_1^2 \cos\gamma_1 \\
&-\frac{1}{4}\omega_3^{-1}\left(\frac{1}{2}\alpha_1 - \omega_1^2\alpha_2\right)a_2^2 \cos\gamma_2 + \omega_3^{-1}k_3 \cos\gamma_3
\end{aligned}
\tag{6.21.30}
$$

Let $a_n' = 0$ and $\gamma_n' = 0$ in (6.21.26) and (6.21.30), the steady state response should satisfy the following equations

$$
\begin{aligned}
0 &= -\mu_1 a_1 - \frac{1}{4}\omega_1^{-1}\left(\alpha_1 - \frac{1}{2}\omega_3^2\alpha_2\right)a_1 a_3 \sin\gamma_1 \\
0 &= -\mu_2 a_2 - \frac{1}{4}\omega_2^{-1}\left(\alpha_1 - \frac{1}{2}\omega_3^2\alpha_2\right)a_2 a_3 \sin\gamma_2 \\
0 &= -\mu_3 a_3 + \frac{1}{4}\omega_3^{-1}\left(\frac{1}{2}\alpha_1 - \omega_1^2\alpha_2\right)a_1^2 \sin\gamma_1 \\
&\quad + \frac{1}{4}\omega_3^{-1}\left(\frac{1}{2}\alpha_1 - \omega_1^2\alpha_2\right)a_2^2 \sin\gamma_2 + \omega_3^{-1}k_3 \sin\gamma_3
\end{aligned}
\tag{6.21.31}
$$

$$
\begin{aligned}
0 &= \sigma a_1 + \sigma_3 a_1 - \frac{1}{2}\omega_1^{-1}\left(\alpha_1 - \frac{1}{2}\omega_3^2\alpha_2\right)a_1 a_3 \cos\gamma_1 \\
0 &= \sigma a_2 + \sigma_3 a_2 - \frac{1}{2}\omega_2^{-1}\left(\alpha_1 - \frac{1}{2}\omega_3^2\alpha_2\right)a_2 a_3 \cos\gamma_2 \\
0 &= \sigma_3 a_3 - \frac{1}{4}\omega_3^{-1}\left(\frac{1}{2}\alpha_1 - \omega_1^2\alpha_2\right)a_1^2 \cos\gamma_1 \\
&\quad - \frac{1}{4}\omega_3^{-1}\left(\frac{1}{2}\alpha_1 - \omega_1^2\alpha_2\right)a_2^2 \cos\gamma_2 + \omega_3^{-1}k_3 \cos\gamma_3
\end{aligned}
\tag{6.21.32}
$$

Eliminating γ_1, γ_2 from the first and second equations in (6.21.31) and (6.21.32) yields

$$
\begin{aligned}
16\mu_1^2 a_1^2 + 4a_1^2(\sigma + \sigma_3)^2 &= \omega_1^{-2}\left(\alpha_1 - \frac{1}{2}\omega_3^2\alpha_2\right)^2 a_1^2 a_3^2 \\
16\mu_2^2 a_2^2 + 4a_2^2(\sigma + \sigma_3)^2 &= \omega_2^{-2}\left(\alpha_1 - \frac{1}{2}\omega_3^2\alpha_2\right)^2 a_2^2 a_3^2
\end{aligned}
\tag{6.21.33}
$$

Since $\omega_1 = \omega_2$ when $\mu_1 \neq \mu_2$, there must be $a_1 = a_2 = 0$. When $\mu_1 = \mu_2$, a_1 and a_2 can be either zero or non-zeros. When $a_1 = a_2 = 0$, we can eliminate γ_3 from the third equation of (6.21.31) and (6.21.32), i.e.,

$$
\left(\mu_3^2 + \sigma_3^2\right)a_3^2 = \omega_3^{-2}k_3^2, \quad a_1 = a_2 = 0 \tag{6.21.34}
$$

When $\mu_1 = \mu_2, a_1 \neq 0, a_2 \neq 0$ and $a_3 \neq 0$, there is $a_1 = a_2$, $\gamma_1 = \gamma_2$, (6.21.33) degenerates to the following equation:

$$
a_3^2 = \omega_1^2\left(\alpha_1 - \frac{1}{2}\omega_3^2\alpha_2\right)^{-2}\left[16\mu_1^2 + 4(\sigma + \sigma_3)^2\right] \tag{6.21.35}
$$

Equations (6.21.31) and (6.21.32) degenerate into

$$\begin{aligned}
0 &= -\mu_1 a_1 - \frac{1}{4}\omega_1^{-1}\left(\alpha_1 - \frac{1}{2}\omega_3^2\alpha_2\right)a_1 a_3 \sin\gamma_1 \\
0 &= \sigma a_1 + \sigma_3 a_1 - \frac{1}{2}\omega_1^{-1}\left(\alpha_1 - \frac{1}{2}\omega_3^2\alpha_2\right)a_1 a_3 \cos\gamma_1 \\
0 &= -\mu_3 a_3 + \frac{1}{2}\omega_3^{-1}\left(\frac{1}{2}\alpha_1 - \omega_1^2\alpha_2\right)a_1^2 \sin\gamma_1 + \omega_3^{-1}k_3\sin\gamma_3 \\
0 &= \sigma_3 a_3 - \frac{1}{2}\omega_3^{-1}\left(\frac{1}{2}\alpha_1 - \omega_1^2\alpha_2\right)a_1^2 \cos\gamma_1 + \omega_3^{-1}k_3\cos\gamma_3
\end{aligned} \tag{6.21.36}$$

Eliminating $\gamma_1,\ \ \gamma_3$ in (6.21.36), we get

$$\begin{aligned}
&\left[\mu_3 a_3^2 + 2\omega_1\omega_3^{-1}\mu_1\left(\frac{1}{2}\alpha_1 - \omega_1^2\alpha_2\right)\left(\alpha_1 - \frac{1}{2}\omega_3^2\alpha_2\right)^{-1} a_1^2\right]^2 \\
&+\left[\sigma_3 a_3^2 - \omega_1\omega_3^{-1}\left(\frac{1}{2}\alpha_1 - \omega_1^2\alpha_2\right)\left(\alpha_1 - \frac{1}{2}\omega_3^2\alpha_2\right)^{-1}(\sigma+\sigma_3)a_1^2\right]^2 \\
&= \omega_3^{-2}k_3^2 a_3^2
\end{aligned} \tag{6.21.37}$$

The steady state responses can be determined by (6.21.35) and (6.21.37). From (6.21.35), we know that a_3 is a constant and independent of the excitation k_3, therefore, a_3 is saturated. The stability of the steady state response depends on the eigenvalues of the linearized equations of (6.21.26) and (6.21.30).

(c) When $s = 1$, $\Omega = \omega_1 + \varepsilon\sigma_1$, substitute (6.21.24) into (6.21.22), we get

$$\begin{aligned}
&i\omega_1 a_1' - \omega_1 a_1\beta_1' + i\omega_1\mu_1 a_1 + \frac{1}{4}\left(\alpha_1 - \frac{1}{2}\omega_3^2\alpha_2\right)a_1 a_3 e^{i(\sigma T_1 - 2\beta_1 + \beta_3)} - k_1 e^{i(\sigma_1 T_1 - \beta_1)} = 0 \\
&i\omega_2 a_2' - \omega_2 a_2\beta_2' + i\omega_2\mu_2 a_2 + \frac{1}{4}\left(\alpha_1 - \frac{1}{2}\omega_3^2\alpha_2\right)a_2 a_3 e^{i(\sigma T_1 - 2\beta_2 + \beta_3)} = 0 \\
&i\omega_3 a_3' - \omega_3 a_3\beta_3' + i\omega_3\mu_3 a_3 + \frac{1}{4}\left(\frac{1}{2}\alpha_1 - \omega_1^2\alpha_2\right)a_1^2 e^{-i(\sigma T_1 - 2\beta_1 + \beta_3)} \\
&+\frac{1}{4}\left(\frac{1}{2}\alpha_1 - \omega_1^2\alpha_2\right)a_2^2 e^{-i(\sigma T_1 - 2\beta_2 + \beta_3)} = 0
\end{aligned} \tag{6.21.38}$$

$$\begin{aligned}
&i\omega_1 a_1' - \omega_1 a_1\beta_1' + i\omega_1\mu_1 a_1 - k_1 e^{i\gamma_1} + \frac{1}{4}\left(\alpha_1 - \frac{1}{2}\omega_3^2\alpha_2\right)a_1 a_3 e^{i\gamma_2} = 0 \\
&i\omega_2 a_2' - \omega_2 a_2\beta_2' + i\omega_2\mu_2 a_2 + \frac{1}{4}\left(\alpha_1 - \frac{1}{2}\omega_3^2\alpha_2\right)a_2 a_3 e^{i\gamma_3} = 0 \\
&i\omega_3 a_3' - \omega_3 a_3\beta_3' + i\omega_3\mu_3 a_3 + \frac{1}{4}\left(\frac{1}{2}\alpha_1 - \omega_1^2\alpha_2\right)a_1^2 e^{-i\gamma_2} \\
&+\frac{1}{4}\left(\frac{1}{2}\alpha_1 - \omega_1^2\alpha_2\right)a_2^2 e^{-i\gamma_3} = 0
\end{aligned} \tag{6.21.39}$$

Separating the real and imaginary parts of the above equations yields

$$\begin{aligned}
a_1' &= \omega_1^{-1}k_1\sin\gamma_1 - \mu_1 a_1 - \frac{1}{4}\omega_1^{-1}\left(\alpha_1 - \frac{1}{2}\omega_3^2\alpha_2\right)a_1a_3\sin\gamma_2\\
a_2' &= -\mu_2 a_2 - \frac{1}{4}\omega_2^{-1}\left(\alpha_1 - \frac{1}{2}\omega_3^2\alpha_2\right)a_2a_3\sin\gamma_3\\
a_3' &= -\mu_3 a_3 + \frac{1}{4}\omega_3^{-1}\left(\frac{1}{2}\alpha_1 - \omega_1^2\alpha_2\right)a_1^2\sin\gamma_2 + \frac{1}{4}\omega_3^{-1}\left(\frac{1}{2}\alpha_1 - \omega_1^2\alpha_2\right)a_2^2\sin\gamma_3
\end{aligned} \tag{6.21.40}$$

$$\begin{aligned}
a_1\beta_1' &= -\omega_1^{-1}k_1\cos\gamma_1 + \frac{1}{4}\omega_1^{-1}\left(\alpha_1 - \frac{1}{2}\omega_3^2\alpha_2\right)a_1a_3\cos\gamma_2\\
a_2\beta_2' &= \frac{1}{4}\omega_2^{-1}\left(\alpha_1 - \frac{1}{2}\omega_3^2\alpha_2\right)a_2a_3\cos\gamma_3\\
a_3\beta_3' &= \frac{1}{4}\omega_3^{-1}\left(\frac{1}{2}\alpha_1 - \omega_1^2\alpha_2\right)a_1^2\cos\gamma_2 + \frac{1}{4}\omega_3^{-1}\left(\frac{1}{2}\alpha_1 - \omega_1^2\alpha_2\right)a_2^2\cos\gamma_3
\end{aligned} \tag{6.21.41}$$

where

$$\gamma_1 = \sigma_1 T_1 - \beta_1,\quad \gamma_2 = \sigma T_1 - 2\beta_1 + \beta_3,\quad \gamma_3 = \sigma T_1 - 2\beta_2 + \beta_3 \tag{6.21.42}$$

From (6.21.42), we can obtain

$$\begin{aligned}
\beta_1 &= \sigma_1 T_1 - \gamma_1\\
\beta_3 &= -2\gamma_1 + \gamma_2 - (\sigma - 2\sigma_1)T_1\\
2\beta_2 &= -2\gamma_1 + \gamma_2 - \gamma_3 + 2\sigma_1 T_1
\end{aligned} \tag{6.21.43}$$

Substituting (6.21.43) into (6.21.41) yields

$$\begin{aligned}
&a_1\gamma_1' = \sigma_1 a_1 + \omega_1^{-1}k_1\cos\gamma_1 - \frac{1}{4}\omega_1^{-1}\left(\alpha_1 - \frac{1}{2}\omega_3^2\alpha_2\right)a_1a_3\cos\gamma_2\\
&2a_2\gamma_1' - a_2\gamma_2' + a_2\gamma_3' = 2\sigma_1 a_2 - \frac{1}{2}\omega_2^{-1}\left(\alpha_1 - \frac{1}{2}\omega_3^2\alpha_2\right)a_2a_3\cos\gamma_3\\
&2a_3\gamma_1 - a_3\gamma_2 = -(\sigma - 2\sigma_1)a_3 - \frac{1}{4}\omega_3^{-1}\left(\frac{1}{2}\alpha_1 - \omega_1^2\alpha_2\right)a_1^2\cos\gamma_2\\
&-\frac{1}{4}\omega_3^{-1}\left(\frac{1}{2}\alpha_1 - \omega_1^2\alpha_2\right)a_2^2\cos\gamma_3
\end{aligned} \tag{6.21.44}$$

For brevity, (6.21.40) and (6.21.44) are written as:

$$
\begin{aligned}
a_1' &= \omega_1^{-1}k_1\sin\gamma_1 - \mu_1 a_1 - \omega_1^{-1}\Gamma_1 a_1 a_3 \sin\gamma_2 \\
a_2' &= -\mu_2 a_2 - \omega_2^{-1}\Gamma_1 a_2 a_3 \sin\gamma_3 \\
a_3' &= -\mu_3 a_3 + \omega_3^{-1}\Gamma_2 a_1^2 \sin\gamma_2 + \omega_3^{-1}\Gamma_2 a_2^2 \sin\gamma_3
\end{aligned}
\tag{6.21.45}
$$

$$
\begin{aligned}
&a_1\gamma_1' = \sigma_1 a_1 + \omega_1^{-1}k_1\cos\gamma_1 - \omega_1^{-1}\Gamma_1 a_1 a_3\cos\gamma_2 \\
&2a_2\gamma_1' - a_2\gamma_2' + a_2\gamma_3' = 2\sigma_1 a_2 - 2\omega_2^{-1}\Gamma_1 a_2 a_3 \cos\gamma_3 \\
&2a_3\gamma_1 - a_3\gamma_2 = -(\sigma - 2\sigma_1)a_3 - \omega_3^{-1}\Gamma_2 a_1^2\cos\gamma_2 - \omega_3^{-1}\Gamma_2 a_2^2 \cos\gamma_3
\end{aligned}
\tag{6.21.46}
$$

where

$$
\Gamma_1 = \frac{1}{4}\left(\alpha_1 - \frac{1}{2}\omega_3^2\alpha_2\right), \quad \Gamma_2 = \frac{1}{4}\left(\frac{1}{2}\alpha_1 - \omega_1^2\alpha_2\right)
\tag{6.21.47}
$$

Let $a_n' = 0$, $\gamma_n' = 0$ in (6.21.45) and (6.21.46), the steady state response should satisfy the following equations

$$
\begin{aligned}
0 &= \omega_1^{-1}k_1\sin\gamma_1 - \mu_1 a_1 - \omega_1^{-1}\Gamma_1 a_1 a_3 \sin\gamma_2 \\
0 &= -\mu_2 a_2 - \omega_2^{-1}\Gamma_1 a_2 a_3 \sin\gamma_3 \\
0 &= -\mu_3 a_3 + \omega_3^{-1}\Gamma_2 a_1^2 \sin\gamma_2 + \omega_3^{-1}\Gamma_2 a_2^2 \sin\gamma_3
\end{aligned}
\tag{6.21.48}
$$

$$
\begin{aligned}
0 &= \sigma_1 a_1 + \omega_1^{-1}k_1\cos\gamma_1 - \omega_1^{-1}\Gamma_1 a_1 a_3\cos\gamma_2 \\
0 &= 2\sigma_1 a_2 - 2\omega_2^{-1}\Gamma_1 a_2 a_3 \cos\gamma_3 \\
0 &= -(\sigma - 2\sigma_1)a_3 - \omega_3^{-1}\Gamma_2 a_1^2\cos\gamma_2 - \omega_3^{-1}\Gamma_2 a_2^2 \cos\gamma_3
\end{aligned}
\tag{6.21.49}
$$

It is not obvious from (6.21.48) and (6.21.49) that whether the system yields a trivial steady state response or not. For non-trivial solutions, (6.21.48) and (6.21.49) can be written as:

$$
\begin{aligned}
&\omega_1\mu_1 a_1^2 + \omega_2\mu_2 a_2^2 + \omega_3\mu_3\Gamma_1\Gamma_2^{-1}a_3^2 = k_1 a_1\sin\gamma_1 \\
&-\omega_2\mu_2 = \Gamma_1 a_3 \sin\gamma_3 \\
&\omega_3\mu_3 a_3^2 + \omega_2\mu_2\Gamma_1^{-1}\Gamma_2 a_2^2 = \Gamma_2 a_1^2 a_3 \sin\gamma_2
\end{aligned}
\tag{6.21.50}
$$

$$
\begin{aligned}
&-\omega_1\sigma_1 a_1^2 - \omega_2\sigma_1 a_2^2 - \omega_3(\sigma - 2\sigma_1)\Gamma_1\Gamma_2^{-1}a_3^2 = k_1 a_1\cos\gamma_1 \\
&\omega_2\sigma_1 = \Gamma_1 a_3\cos\gamma_3 \\
&-\omega_3(\sigma - 2\sigma_1)a_3^2 - \omega_2\Gamma_1^{-1}\Gamma_2\sigma_1 a_2^2 = \Gamma_2 a_1^2 a_3\cos\gamma_2
\end{aligned}
\tag{6.21.51}
$$

Eliminating γ_1, γ_2 and γ_3 in (6.21.50) and (6.21.51), we get

$$
\begin{aligned}
&\Gamma_1^2 a_3^2 = \omega_2^2\left(\mu_2^2 + \sigma_1^2\right) \\
&k_1^2 a_1^2 = \left(\omega_1\mu_1 a_1^2 + \omega_2\mu_2 a_2^2 + \omega_3\mu_3\Gamma_1\Gamma_2^{-1}a_3^2\right)^2 \\
&+\left[\omega_1\sigma_1 a_1^2 + \omega_2\sigma_1 a_2^2 + \omega_3(\sigma - 2\sigma_1)\Gamma_1\Gamma_2^{-1}a_3^2\right]^2 \\
&\Gamma_2^2 a_1^4 a_3^2 = \left(\omega_3\mu_3 a_3^2 + \omega_2\mu_2\Gamma_1^{-1}\Gamma_2 a_2^2\right)^2 + \left[\omega_3(\sigma - 2\sigma_1)a_3^2 + \omega_2\Gamma_1^{-1}\Gamma_2\sigma_1 a_2^2\right]^2
\end{aligned}
\tag{6.21.52}
$$

The steady state responses for a_1, a_2 and a_3 can be determined from (6.21.52). From the first equation of (6.21.52), we can find that a_3 is constant and independent of the excitation k_1, therefore, a_3 is saturated. The stability of the steady state response depends on the eigenvalues of the linearized equations of (6.21.45) and (6.21.46).

6.22 Exercise 6.22 (Primary Resonance Analysis of Three-Degree-of-Freedom Systems with Internal Resonance)

Readers are invited to complete this problem by referring to Exercise 6.21.

6.23 Exercise 6.23 (Stability Analysis of Nonlinear Forced Oscillation of a Disk)

Solution: (a) Let the solution of the equation be

$$\begin{aligned} u &= u_0(T_0, T_1) + \varepsilon u_1(T_0, T_1) + \cdots \\ v &= v_0(T_0, T_1) + \varepsilon v_1(T_0, T_1) + \cdots \end{aligned} \tag{6.23.1}$$

Substituting them into the control equation of the system and retaining to $O(\varepsilon)$ yields

$$\begin{aligned} 0 &= \ddot{v} + v + 8\varepsilon\left(v^3 + u^2 v\right) \\ &= \left(D_0^2 + 2\varepsilon D_0 D_1\right)(v_0 + \varepsilon v_1) + v_0 + \varepsilon v_1 \\ &\quad + 8\varepsilon(v_0 + \varepsilon v_1)^3 + 8\varepsilon(u_0 + \varepsilon u_1)^2(v_0 + \varepsilon v_1) + \cdots \\ &= D_0^2 v_0 + v_0 + \varepsilon\left(D_0^2 v_1 + v_1 + 2D_0 D_1 v_0 + 8v_0^3 + 8u_0^2 v_0\right) + \cdots \end{aligned} \tag{6.23.2}$$

$$\begin{aligned} 0 &= \ddot{v} + v + 8\varepsilon\left(v^3 + u^2 v\right) \\ &= \left(D_0^2 + 2\varepsilon D_0 D_1\right)(v_0 + \varepsilon v_1) + v_0 + \varepsilon v_1 \\ &\quad + 8\varepsilon(v_0 + \varepsilon v_1)^3 + 8\varepsilon(u_0 + \varepsilon u_1)^2(v_0 + \varepsilon v_1) + \cdots \\ &= D_0^2 v_0 + v_0 + \varepsilon\left(D_0^2 v_1 + v_1 + 2D_0 D_1 v_0 + 8v_0^3 + 8u_0^2 v_0\right) + \cdots \end{aligned} \tag{6.23.3}$$

Making the coefficient of the like power of ε zero in the above two equations, we get

Order ε^0:

$$\begin{aligned} D_0^2 u_0 + u_0 &= 0 \\ D_0^2 v_0 + v_0 &= 0 \end{aligned} \tag{6.23.4}$$

Order ε^1:

$$\begin{aligned} D_0^2u_1 + u_1 &= -2D_0D_1u_0 - 8u_0^3 - 8u_0v_0^2 + 2k\cos\omega T_0 \\ D_0^2v_1 + v_1 &= -2D_0D_1v_0 - 8v_0^3 - 8u_0^2v_0 \end{aligned} \tag{6.23.5}$$

The solution of (6.23.4) is

$$\begin{aligned} u_0 &= Ae^{iT_0} + \overline{A}e^{-iT_0} \\ v_0 &= Be^{iT_0} + \overline{B}e^{-iT_0} \end{aligned} \tag{6.23.6}$$

Let

$$A = \frac{1}{2}ae^{i\alpha}, \quad B = \frac{1}{2}be^{i\beta} \tag{6.23.7}$$

The first order approximate solution of the original equation is given by (6.23.1), i.e.,

$$\begin{aligned} u &= a\cos(t+\alpha) + O(\varepsilon) \\ v &= b\cos(t+\beta) + O(\varepsilon) \end{aligned} \tag{6.23.8}$$

Substituting (6.23.6) into (6.23.5) and taking into account $\omega = 1 + \varepsilon\sigma$, we obtain

$$\begin{aligned} &D_0^2u_1 + u_1 = -2D_0D_1u_0 - 8u_0^3 - 8u_0v_0^2 + 2k\cos\omega T_0 \\ &= -\left(2iA' + 24A^2\overline{A} + 16AB\overline{B} + 8\overline{A}B^2 - ke^{i\sigma T_1}\right)e^{iT_0} + cc + NST \\ &D_0^2v_1 + v_1 = -\left(2iB\prime + 24B^2\overline{B} + 16A\overline{A}B + 8A^2\overline{B}\right)e^{iT_0} + cc + NST \end{aligned} \tag{6.23.9}$$

In order to eliminate secular terms from the above equation, we need

$$\begin{aligned} &2iA' + 24A^2\overline{A} + 16AB\overline{B} + 8\overline{A}B^2 - ke^{i\sigma T_1} = 0 \\ &2iB\prime + 24B^2\overline{B} + 16A\overline{A}B + 8A^2\overline{B} = 0 \end{aligned} \tag{6.23.10}$$

Substituting (6.23.7) into (6.23.10) yields

$$\begin{aligned} &ia' - a\alpha' + 3a^3 + 2ab^2 + ab^2e^{2i(\beta-\alpha)} - ke^{i(\sigma T_1 - \alpha)} = 0 \\ &ib\prime - b\beta' + 3b^3 + 2a^2b + a^2be^{-2i(\beta-\alpha)} = 0 \end{aligned} \tag{6.23.11}$$

Separating the real and imaginary parts of the above equations yields

$$\begin{aligned} a' &= -ab^2\sin\gamma + k\sin\upsilon \\ a\alpha' &= 3a^3 + 2ab^2 + ab^2\cos\gamma - k\cos\upsilon \\ b\prime &= a^2b\sin\gamma \\ b\beta' &= 3b^3 + 2a^2b + a^2b\cos\gamma \end{aligned} \tag{6.23.12}$$

where

$$\gamma = 2\beta - 2\alpha, \quad \nu = \sigma T_1 - \alpha \tag{6.23.13}$$

Utilizing (6.23.13), we can write (6.23.12) as

$$\begin{aligned} a' &= -ab^2\sin\gamma + k\sin\nu \\ a\nu\prime &= \sigma a - 3a^3 - 2ab^2 - ab^2\cos\gamma + k\cos\nu \\ b\prime &= a^2 b\sin\gamma \\ 2b\nu\prime + b\gamma' &= -2\sigma b + 6b^3 + 4a^2 b + 2a^2 b\cos\gamma \\ b\prime &= \nu\prime \end{aligned} \tag{6.23.14}$$

(b) Let $a' = b\prime = \nu\prime = \gamma' = 0$ in (6.23.14), the steady state response should satisfy the following equations:

$$\begin{aligned} 0 &= -ab^2\sin\gamma + k\sin\nu \\ 0 &= \sigma a - 3a^3 - 2ab^2 - ab^2\cos\gamma + k\cos\nu \\ 0 &= a^2 b\sin\gamma \\ 0 &= -2\sigma b + 6b^3 + 4a^2 b + 2a^2 b\cos\gamma \end{aligned} \tag{6.23.15}$$

From the third equation in (6.23.15), we can obtain

$$b = 0 \quad \text{or} \quad a = 0 \quad \text{or} \quad \sin\gamma = 0 \tag{6.23.16}$$

When $b = 0$, the fourth equation of (6.23.15) is automatically satisfied, and the first and second Eq. (6.23.15) become

$$\sin\nu = 0, \quad a\sigma - 3a^3 + k\cos\nu = 0 \tag{6.23.17}$$

So $a \neq 0$.

When $a = 0$, the fourth equation of (6.23.15) requires $b = 0$, but this makes the first and second equations of (6.23.15) do not hold, so $a = 0$ is not possible.

When $\sin\gamma = 0, a \neq 0$, otherwise there are no meaningful solutions can be found for the first and second equations of (6.23.15). Then we can find from the second and fourth equations of (6.23.15) that there exists a non-trivial solution to b.

To sum up, it is clear that either (i) $b = 0$ and $a \neq 0$ or (ii) $b \neq 0$ and $a \neq 0$ can be found for the steady state response of the system.

(c) $b = 0, a \neq 0$.

From the first equation of (6.23.15), we get $\nu = 0$ or π. And the frequency–response equation can be obtained from the second equation of (6.23.15)

$$\sigma = 3a^2 - \frac{k}{a}\cos\nu, \quad \nu = 0 \quad \text{or} \quad \pi \tag{6.23.18}$$

To analyze the stability of an arbitrary steady state response $a_0,\ b_0,\ \nu_0,\ \gamma_0$, let

$$a = a_0 + \tilde{a}, \quad b = b_0 + \tilde{b}, \quad \nu = \nu_0 + \tilde{\nu}, \quad \gamma = \gamma_0 + \tilde{\gamma} \tag{6.23.19}$$

Substituting (6.23.19) into (6.23.15), we can obtain the corresponding linearized equation:

$$\begin{bmatrix} 1\,0 & 0\,0 \\ 0\,a_0 & 0\,0 \\ 0\,0 & 1\,0 \\ 0\,2b_0 & 0\,b_0 \end{bmatrix} \begin{Bmatrix} \tilde{a}' \\ \tilde{\nu}' \\ \tilde{b}' \\ \tilde{\gamma}' \end{Bmatrix} = \begin{bmatrix} l_{11}\, l_{12}\, l_{13}\, l_{14} \\ l_{21}\, l_{22}\, l_{23}\, l_{24} \\ l_{31}\, l_{32}\, l_{33}\, l_{34} \\ l_{41}\, l_{42}\, l_{43}\, l_{44} \end{bmatrix} \begin{Bmatrix} \tilde{a} \\ \tilde{\nu} \\ \tilde{b} \\ \tilde{\gamma} \end{Bmatrix} \tag{6.23.20}$$

where

$$\begin{aligned}
&l_{11} = -b_0^2 \sin\gamma_0,\, l_{12} = k\cos\nu_0 \\
&l_{13} = -2a_0 b_0 \sin\gamma_0,\, l_{14} = -a_0 b_0^2 \cos\gamma_0 \\
&l_{21} = \sigma - 9a_0^2 - (2+\cos\gamma_0) b_0^2,\, l_{22} = -k\sin\nu_0 \\
&l_{23} = -2(2+\cos\gamma_0) a_0 b_0,\, l_{24} = a_0 b_0^2 \sin\gamma_0 \\
&l_{31} = 2a_0 b_0 \sin\gamma_0,\, l_{32} = 0,\, l_{33} = a_0^2 \sin\gamma_0,\, l_{34} = a_0^2 b_0 \cos\gamma_0 \\
&l_{41} = 2(4+2\cos\gamma_0) a_0 b_0,\, l_{42} = 0 \\
&l_{43} = -\left[2\sigma - 18b_0^2 - (4+2\cos\gamma_0) a_0^2\right],\, l_{44} = -2a_0^2 b_0 \sin\gamma_0
\end{aligned} \tag{6.23.21}$$

Since $b_0 = 0$ and $a_0 \neq 0$, (6.23.20) becomes

$$\begin{aligned}
&\begin{bmatrix} 1\,0 & 0\,0 \\ 0\,a_0 & 0\,0 \\ 0\,0 & 1\,0 \\ 0\,0 & 0\,0 \end{bmatrix} \begin{Bmatrix} \tilde{a}' \\ \tilde{\nu}' \\ \tilde{b}' \\ \tilde{\gamma}' \end{Bmatrix} \\
&= \begin{bmatrix} 0 & k\cos\nu_0 & 0 & 0 \\ a_0^{-1}\left(\sigma - 9a_0^2\right) & -a_0^{-1} k \sin\nu_0 & 0 & 0 \\ 0 & 0 & a_0^2 \sin\gamma_0 & 0 \\ 0 & 0 & -2\sigma + (4+2\cos\gamma_0) a_0^2 & 0 \end{bmatrix} \begin{Bmatrix} \tilde{a} \\ \tilde{\nu} \\ \tilde{b} \\ \tilde{\gamma} \end{Bmatrix}
\end{aligned} \tag{6.23.22}$$

Considering $\nu = 0$ or π and noting the steady state responses a_0, ν_0, γ_0 as a, ν, γ, we can obtain the eigenvalues of the first three equations of (6.23.22)

$$\lambda_{1,2} = \pm\sqrt{a^{-1}\left(\sigma - 9a^2\right) k \cos\nu}, \quad \lambda_3 = a^2 \sin\gamma \tag{6.23.23}$$

From the fourth equation of (6.23.22), we have

$$-\sigma + (2 + \cos\gamma) a^2 = 0 \tag{6.23.24}$$

Substituting (6.23.18) into (6.23.23) and (6.23.24), we get

$$\lambda_{1,2} = \pm\sqrt{-a^{-2}(6a^3 + k\cos\nu)k\cos\nu}, \quad \lambda_3 = a^2\sin\gamma \tag{6.23.25}$$

$$\frac{k}{a}\cos\nu - a^2 + a^2\cos\gamma = 0 \tag{6.23.26}$$

Eliminating γ in (6.23.23) and (6.23.26), we get

$$\begin{aligned} \lambda_{1,2} &= \pm\sqrt{-a^{-2}(6a^3 + k\cos\nu)k\cos\nu} \\ \lambda_3 &= \pm a^2\sqrt{\frac{k}{a^3}\left(2 - \frac{k}{a^3}\cos\nu\right)\cos\nu} \end{aligned} \tag{6.23.27}$$

Therefore,

$$\begin{cases} \lambda_{1,2} = \pm\sqrt{-a^{-2}(6a^3 + k)k} \\ \lambda_3 = \pm a^2\sqrt{\frac{k}{a^3}\left(2 - \frac{k}{a^3}\right)} \end{cases} \quad \nu = 0 \tag{6.23.28}$$

$$\begin{cases} \lambda_{1,2} = \pm\sqrt{a^{-2}(6a^3 - k)k} \\ \lambda_3 = \pm a^2\sqrt{-\frac{k}{a^3}\left(2 + \frac{k}{a^3}\right)} \end{cases} \quad \nu = \pi \tag{6.23.29}$$

Assume $a > 0, \quad k > 0$. If $\nu = 0$, λ_1 and λ_2 are two conjugate pure imaginary roots and the steady state response is stable. If $a > a_A = (\frac{1}{2}k)^{1/3}$, λ_3 are two real roots which are equal but with opposite sign, and the steady state response is not stable.

If $\nu = \pi$, λ_3 are two conjugate pure imaginary roots and the steady state response is stable. If $a > a_C = \left(\frac{1}{6}k\right)^{1/3}$, λ_1 and λ_2 are two real roots which are equal but with opposite sign, and the steady state response is not stable.

(d) $b \neq 0, a \neq 0$.

From the third equation of (6.23.15), we get $\gamma = 0$ or π. From the first equation of (6.23.15), we get $\nu = 0$ or π. If $\gamma = \pi$, we can obtain from the second and fourth equation of (6.23.15) that

$$\begin{aligned} 0 &= \sigma a - 3a^3 - ab^2 + k\cos\nu \\ 0 &= -\sigma + 3b^2 + a^2 \end{aligned}$$

which leads to

$$\sigma = 4a^2 - \frac{3k}{2a}\cos\nu, \quad b^2 = a^2 - \frac{k}{2a}\cos\nu \tag{6.23.30}$$

where $\nu = 0$ or π.

Considering the steady state responses $b_0 \neq 0, a_0 \neq 0$, $\gamma_0 = \pi$ and $\nu_0 = 0$ or π and denoting the steady state response as $a, \quad b, \quad \nu$, we can obtain from (6.23.20) that

$$\begin{aligned}
\widetilde{a'} &= k\cos\nu\tilde{\nu} - ab^2\tilde{\gamma} \\
\widetilde{b'} &= -a^2 b\tilde{\gamma} \\
\widetilde{\nu'} &= a^{-1}\left(\sigma - 9a^2 - b^2\right)\tilde{a} - 2b\tilde{b} \\
\widetilde{\gamma'} &= \left[22a - 2a^{-1}\left(\sigma - b^2\right)\right]\tilde{a} + \left[22b - 2b^{-1}\left(\sigma - a^2\right)\right]\tilde{b}
\end{aligned}$$

Substitute (6.23.30) into the above equation, we get

$$\begin{Bmatrix} \hat{a}' \\ \hat{b}' \\ \hat{\nu}' \\ \hat{\gamma}' \end{Bmatrix} = \begin{bmatrix} 0 & 0 & k\cos\nu & -\Gamma_1 \\ 0 & 0 & 0 & -a \\ -\Gamma_2 & -2\Gamma_1 & 0 & 0 \\ \Gamma_3 & 16\Gamma_1 & 0 & 0 \end{bmatrix} \begin{Bmatrix} \hat{a} \\ \hat{b} \\ \hat{\nu} \\ \hat{\gamma} \end{Bmatrix} \tag{6.23.32}$$

where

$$\begin{aligned}
&\hat{a} = a\tilde{a}, \quad \hat{b} = b^{-1}\tilde{b}, \quad \hat{\nu} = a\tilde{\nu}, \quad \hat{\gamma} = a\tilde{\gamma} \\
&\Gamma_1 = a^3 - \frac{k}{2}\cos\nu, \quad \Gamma_2 = 6a + \frac{k}{a^2}\cos\nu, \quad \Gamma_3 = 16a + \frac{2k}{a^2}\cos\nu
\end{aligned} \tag{6.23.33}$$

The characteristic equation of (6.23.32)

$$\det\begin{bmatrix} -\lambda & 0 & k\cos\nu & -\Gamma_1 \\ 0 & -\lambda & 0 & -a \\ -\Gamma_2 & -2\Gamma_1 & -\lambda & 0 \\ \Gamma_3 & 16\Gamma_1 & 0 & -\lambda \end{bmatrix} = 0 \tag{6.23.34}$$

i.e.,

$$\lambda^4 + 8\left(4a^4 - ka\cos\nu\right)\lambda^2 + ak\left(64a^4 - 20ka\cos\nu - \frac{6k^2}{a^2}\right)\cos\nu = 0 \tag{6.23.35}$$

therefore,

$$\nu = 0: \lambda^4 + 8a\left(4a^3 - k\right)\lambda^2 + 2a^{-1}k\left(32a^6 - 10ka^3 - 3k^2\right) = 0 \tag{6.23.36}$$

$$\nu = \pi: \lambda^4 + 8a\left(4a^3 + k\right)\lambda^2 - 2a^{-1}k\left(32a^6 + 10ka^3 - 3k^2\right) = 0 \tag{6.23.37}$$

Assume $a > 0, \quad k > 0$. Thus, when $\nu = 0$, from (6.23.36), we have

$$4a^3 - k > 0, \quad 32a^6 - 10ka^3 - 3k^2 > 0 \tag{6.23.38}$$

Then all four eigenvalues are pure imaginary roots and the steady state response is stable. From (6.23.38), we have $a > a_A = (\frac{1}{2}k)^{1/3}$.

When $\nu = \pi$, from (6.23.37), we know that if

$$\begin{aligned} &32a^6 + 10ka^3 - 3k^2 < 0 \\ &\left[8a\left(4a^3 + k\right)\right]^2 + 8a^{-1}k\left(32a^6 + 10ka^3 - 3k^2\right) > 0 \end{aligned} \tag{6.23.39}$$

then all four eigenvalues are pure imaginary roots and the steady state response is stable. From the first inequality of (6.23.39), we can obtain $a \le a_D = \left(\frac{3}{16}k\right)^{1/3}$. The second inequality of (6.23.39) can be written as

$$f\left(a^3, k\right) = \left(\frac{a^3}{k}\right)^3 + \frac{3}{4}\left(\frac{a^3}{k}\right)^2 + \frac{9}{64}\left(\frac{a^3}{k}\right) - \frac{3}{128} > 0 \tag{6.23.40}$$

This leads to $a^3/k > 0.10275$; or $a > a_E, a_E \approx 0.468k^{1/3}$.

Thus, when $\nu = \pi$, if $\cdot\, a_E < a \le a_D = \left(\frac{3}{16}k\right)^{1/3}$, $a_E \approx 0.468k^{1/3}$, then the corresponding steady state response is stable.

6.24 Exercise 6.24 (Forced Oscillation of a Spherical Pendulum)

Readers are invited to complete this problem.

6.25 Exercise 6.25 (Nonlinear Oscillation Analysis of a Rolling Reentry Body I)

Solution: (a) when $\varepsilon = 0$ the corresponding linear equation of oscillation of the system (1) is

$$\ddot{\xi} - ip\dot{\xi} + \omega_0^2\xi = 0 \tag{6.25.1}$$

This is the free oscillation equation for a gyroscopic system. Let the free oscillation solution of the equation be

$$\xi = Ae^{i\omega t} \tag{6.25.2}$$

The characteristic equation can be obtained as

$$\omega^2 - p\omega - \omega_0^2 = 0 \tag{6.25.3}$$

Therefore, the linear natural frequencies of the system are

$$\omega_{1,\,2} = \frac{1}{2}p \pm \sqrt{\frac{1}{4}p^2 + \omega_0^2} \tag{6.25.4}$$

(b) Let the solution of the system Eq. (1) be

$$\xi = \varepsilon\xi_1(T_0, T_1, T_2) + \varepsilon^2\xi_2(T_0, T_1, T_2) + \varepsilon^3\xi_3(T_0, T_1, T_2) + \cdots \tag{6.25.5}$$

Substituting (6.25.5) into the governing equation of the system Eq. (1) and retaining to $O(\varepsilon^3)$ yields

$$\begin{aligned}
0 &= \ddot{\xi} - ip\dot{\xi} + \omega_0^2\xi - \gamma|\xi^2|\xi - \varepsilon^2\mu_1\dot{\xi} - \varepsilon K e^{i(pt+\phi_0)} \\
&= \left[D_0^2 + 2\varepsilon D_0 D_1 + \varepsilon^2\left(D_1^2 + 2D_0D_2\right)\right]\left(\varepsilon\xi_1 + \varepsilon^2\xi_2 + \varepsilon^3\xi_3\right) \\
&\quad -ip\left(D_0 + \varepsilon D_1 + \varepsilon^2 D_2\right)\left(\varepsilon\xi_1 + \varepsilon^2\xi_2 + \varepsilon^3\xi_3\right) \\
&\quad -\varepsilon^2\mu_1\left(D_0 + \varepsilon D_1 + \varepsilon^2 D_2\right)\left(\varepsilon\xi_1 + \varepsilon^2\xi_2 + \varepsilon^3\xi_3\right) \\
&\quad +\omega_0^2\left(\varepsilon\xi_1 + \varepsilon^2\xi_2 + \varepsilon^3\xi_3\right) \\
&\quad -\gamma\left|\left(\varepsilon\xi_1 + \varepsilon^2\xi_2 + \varepsilon^3\xi_3\right)^2\right|\left(\varepsilon\xi_1 + \varepsilon^2\xi_2 + \varepsilon^3\xi_3\right) + \cdots \\
&= \varepsilon\left(D_0^2\xi_1 - ipD_0\xi_1 + \omega_0^2\xi_1\right) + \varepsilon^2\left(D_0^2\xi_2 - ipD_0\xi_2 + \omega_0^2\xi_2\right. \\
&\quad \left.+2D_0D_1\xi_1 - ipD_1\xi_1\right) + \varepsilon^3\left[D_0^2\xi_3 - ipD_0\xi_3 + \omega_0^2\xi_3 - ipD_2\xi_1 - \mu_1 D_0\xi_1\right. \\
&\quad \left.-\gamma\left|\xi_1^2\right|\xi_1 + \left(D_1^2 + 2D_0D_2\right)\xi_1 + 2D_0D_1\xi_2 - ipD_1\xi_2\right] - \varepsilon K e^{i(pt+\phi_0)} + \cdots
\end{aligned} \tag{6.25.6}$$

When $K \equiv 0$, making the coefficient of the like powers of ε zero in the above equation, we get.

Order ε^1:

$$D_0^2\xi_1 - ipD_0\xi_1 + \omega_0^2\xi_1 = 0 \tag{6.25.7}$$

Order ε^2:

$$D_0^2\xi_2 - ipD_0\xi_2 + \omega_0^2\xi_2 = ipD_1\xi_1 - 2D_0D_1\xi_1 \tag{6.25.8}$$

Order ε^3:

$$\begin{aligned}
&D_0^2\xi_3 - ipD_0\xi_3 + \omega_0^2\xi_3 = ipD_2\xi_1 + \mu_1 D_0\xi_1 + \gamma\left|\xi_1^2\right|\xi_1 \\
&-\left(D_1^2 + 2D_0D_2\right)\xi_1 - 2D_0D_1\xi_2 + ipD_1\xi_2
\end{aligned} \tag{6.25.9}$$

According to (6.25.1), the solution of (6.25.7) can be expressed as

$$\xi_1 = A_1 e^{i\omega_1 T_0} + A_2 e^{i\omega_2 T_0} \tag{6.25.10}$$

where $A_1 = A_1(T_1, T_2), A_2 = A_2(T_1, T_2)$. Substituting (6.25.10) into (6.25.8), we get

$$D_0^2\xi_2 - ipD_0\xi_2 + \omega_0^2\xi_2 = (ip - 2i\omega_1)D_1A_1e^{i\omega_1T_0} + (ip - 2i\omega_2)D_1A_2e^{i\omega_2T_0} \tag{6.25.11}$$

In order to eliminate secular terms from the above equation, we need

$$D_1A_1 = D_1A_2 = 0 \quad \Rightarrow \quad A_1 = A_1(T_2), \quad A_2 = A_2(T_2) \tag{6.25.12}$$

From (6.25.11), we obtain

$$\xi_2 = 0 \tag{6.25.13}$$

Substituting (6.25.10), (6.25.12) and (6.25.13) into (6.25.9), we obtain

$$\begin{aligned}
&D_0^2\xi_3 - ipD_0\xi_3 + \omega_0^2\xi_3 = -2D_0D_2\xi_1 + ipD_2\xi_1 + \mu_1D_0\xi_1 + \gamma\left|\xi_1^2\right|\xi_1 \\
&\quad = -2i\omega_1A_1'e^{i\omega_1T_0} + ipA_1'e^{i\omega_1T_0} + i\omega_1\mu_1A_1e^{i\omega_1T_0} \\
&\quad -2i\omega_2A_2'e^{i\omega_2T_0} + ipA_2'e^{i\omega_2T_0} + i\omega_2\mu_1A_2e^{i\omega_2T_0} \\
&\quad +\gamma\left|\left(A_1e^{i\omega_1T_0} + A_2e^{i\omega_2T_0}\right)^2\right|\left(A_1e^{i\omega_1T_0} + A_2e^{i\omega_2T_0}\right) \\
&\quad = -2i\omega_1A_1'e^{i\omega_1T_0} + ipA_1'e^{i\omega_1T_0} + i\omega_1\mu_1A_1e^{i\omega_1T_0} \\
&\quad -2i\omega_2A_2'e^{i\omega_2T_0} + ipA_2'e^{i\omega_2T_0} + i\omega_2\mu_1A_2e^{i\omega_2T_0} \\
&\quad +\gamma\left(A_1e^{i\omega_1T_0} + A_2e^{i\omega_2T_0}\right)^2\left(\overline{A}_1e^{-i\omega_1T_0} + \overline{A}_2e^{-i\omega_2T_0}\right) \\
&\quad = \left(-2i\omega_1A_1' + ipA_1' + i\omega_1\mu_1A_1 + \gamma A_1^2\overline{A}_1 + 2\gamma A_1A_2\overline{A}_2\right)e^{i\omega_1T_0} \\
&\quad +\left(-2i\omega_2A_2' + ipA_2' + i\omega_2\mu_1A_2 + \gamma A_2^2\overline{A}_2 + 2\gamma A_1\overline{A}_1A_2\right)e^{i\omega_2T_0} \\
&\quad +\gamma\overline{A}_1A_2^2e^{i(2\omega_2-\omega_1)T_0} + \gamma A_1^2\overline{A}_2e^{i(2\omega_1-\omega_2)T_0}
\end{aligned} \tag{6.25.14}$$

Primes denote the derivative with respect to T_2. In order to eliminate secular terms from the above equation, we need

$$\begin{aligned}
-2i\omega_1A_1' + ipA_1' + i\omega_1\mu_1A_1 + \gamma A_1^2\overline{A}_1 + 2\gamma A_1A_2\overline{A}_2 = 0 \\
-2i\omega_2A_2' + ipA_2' + i\omega_2\mu_1A_2 + \gamma A_2^2\overline{A}_2 + 2\gamma A_1\overline{A}_1A_2 = 0
\end{aligned} \tag{6.25.15}$$

Let

$$A_1 = a_1e^{i\theta_1}, \quad A_2 = a_2e^{i\theta_2} \tag{6.25.16}$$

Substituting (6.25.16) into (6.25.15), we obtain

$$
\begin{aligned}
&-2i\omega_1 a_1' + 2\omega_1 a_1\theta_1' + ipa_1' - pa_1\theta_1' + i\omega_1\mu_1 a_1 + \gamma a_1^3 + 2\gamma a_1 a_2^2 = 0 \\
&-2i\omega_2 a_2' + 2\omega_2 a_2\theta_2' + ipa_2' - pa_2\theta_2' + i\omega_2\mu_1 a_2 + \gamma a_2^3 + 2\gamma a_2 a_1^2 = 0
\end{aligned}
\tag{6.25.17}
$$

Separating the real and imaginary parts of the above equations yields

$$
a_1' = (\omega_1 - \omega_2)^{-1}\omega_1\mu_1 a_1 \triangleq \Lambda_1 \tag{6.25.18}
$$

$$
a_2' = -(\omega_1 - \omega_2)^{-1}\omega_2\mu_1 a_2 \triangleq \Lambda_2 \tag{6.25.19}
$$

$$
(\omega_1 - \omega_2)(\theta_1', -\theta_2') = -\gamma\left(a_1^2 + 2a_2^2,\ 2a_1^2 + a_2^2\right) \triangleq (\Lambda_3,\ \Lambda_4) \tag{6.25.20}
$$

From the above results, we can obtain the solution of the given system equation when $K = 0$:

$$
\xi = \varepsilon a_1(T_2)e^{i\omega_1 T_0 + i\theta_1(T_2)} + \varepsilon a_2(T_2)e^{i\omega_2 T_0 + i\theta_2(T_2)} + O(\varepsilon^3) \tag{6.25.21}
$$

We can solve a_1 and a_2 from (6.25.18) and (6.25.19):

$$
a_1 = a_{10}\exp\left[(\omega_1 - \omega_2)^{-1}\omega_1\mu_1 T_2\right],\quad a_2 = a_{20}\exp\left[-(\omega_1 - \omega_2)^{-1}\omega_2\mu_1 T_2\right] \tag{6.25.22}
$$

$$
a_2' = -(\omega_1 - \omega_2)^{-1}\omega_2\mu_1 a_2 \tag{6.25.23}
$$

Therefore, one of a_1 and a_2 is always divergent, i.e. the free oscillation of the given system is unstable.

(c) When $K \equiv \varepsilon^2 k$, (6.25.5)–(6.25.8), (6.25.10)–(6.25.13) still hold and (6.25.9) becomes.

Order ε^3:

$$
\begin{aligned}
&D_0^2\xi_3 - ipD_0\xi_3 + \omega_0^2\xi_3 = ipD_2\xi_1 + \mu_1 D_0\xi_1 + \gamma\left|\xi_1^2\right|\xi_1 - \left(D_1^2 + 2D_0D_2\right)\xi_1 \\
&-2D_0D_1\xi_2 + ipD_1\xi_2 + ke^{i(pT_0+\phi_0)}
\end{aligned}
\tag{6.25.24}
$$

Substitute (6.25.10), (6.25.12) and (6.25.13) into (6.25.9), we get

$$
\begin{aligned}
D_0^2\xi_3 - ipD_0\xi_3 + \omega_0^2\xi_3 = &\left(-2i\omega_1 A_1' + ipA_1' + i\omega_1\mu_1 A_1 + \gamma A_1^2\overline{A}_1 + 2\gamma A_1 A_2\overline{A}_2\right)e^{i\omega_1 T_0} \\
&+\left(-2i\omega_2 A_2' + ipA_2' + i\omega_2\mu_1 A_2 + \gamma A_2^2\overline{A}_2 + 2\gamma A_1\overline{A}_1 A_2\right)e^{i\omega_2 T_0} \\
&+\gamma\overline{A}_1 A_2^2 e^{i(2\omega_2-\omega_1)T_0} + \gamma A_1^2\overline{A}_2 e^{i(2\omega_1-\omega_2)T_0} + ke^{i(pT_0+\phi_0)}
\end{aligned}
\tag{6.25.25}
$$

Primes denote the derivative with respect to T_2. When $p = \omega_1 + \varepsilon^2\sigma$, in order to eliminate secular terms from the above equation, one must have

$$\begin{aligned}&-2i\omega_1 A_1' + ipA_1' + i\omega_1\mu_1 A_1 + \gamma A_1^2\overline{A}_1 + 2\gamma A_1 A_2\overline{A}_2 + ke^{i(\sigma T_2+\phi_0)} = 0\\&-2i\omega_2 A_2' + ipA_2' + i\omega_2\mu_1 A_2 + \gamma A_2^2\overline{A}_2 + 2\gamma A_1\overline{A}_1 A_2 = 0\end{aligned} \tag{6.25.26}$$

Let

$$A_1 = a_1 e^{i\theta_1}, \quad A_2 = a_2 e^{i\theta_2} \tag{6.25.27}$$

Substituting (6.25.27) into (6.25.15), we can obtain

$$\begin{aligned}&-2i\omega_1 a_1' + 2\omega_1 a_1\theta_1' + ipa_1' - pa_1\theta_1' + i\omega_1\mu_1 a_1 + \gamma a_1^3 + 2\gamma a_1 a_2^2 + ke^{i(\sigma T_2-\theta_1+\varphi_0)} = 0\\&-2i\omega_2 a_2' + 2\omega_2 a_2\theta_2' + ipa_2' - pa_2\theta_2' + i\omega_2\mu_1 a_2 + \gamma a_2^3 + 2\gamma a_2 a_1^2 = 0\end{aligned} \tag{6.25.28}$$

Separating the real and imaginary parts of the above equations yields

$$\begin{aligned}&a_1' = (\omega_1-\omega_2)^{-1}[\omega_1\mu_1 a_1 + k\sin(\sigma T_2 - \theta_1 + \phi_0)]\\&a_1\theta_1' = -(\omega_1-\omega_2)^{-1}\left[\gamma\left(a_1^2+2a_2^2\right)a_1 + k\cos(\sigma T_2-\theta_1+\phi_0)\right]\\&a_2' = -(\omega_1-\omega_2)^{-1}\omega_2\mu_1 a_2\\&a_2\theta_2' = (\omega_1-\omega_2)^{-1}\gamma\left(2a_1^2+a_2^2\right)a_2\end{aligned} \tag{6.25.29}$$

From the above results, the solution of the given system when $K \equiv \varepsilon^2 k$ is

$$\xi = \varepsilon a_1(T_2)e^{i\omega_1 T_0 + i\theta_1(T_2)} + \varepsilon a_2(T_2)e^{i\omega_2 T_0 + i\theta_2(T_2)} + O\left(\varepsilon^3\right) \tag{6.25.30}$$

where a_1, a_2, θ_1 and θ_2 are given by (6.25.29).
Rewrite (6.25.29) as

$$\begin{aligned}&a_1' = (\omega_1-\omega_2)^{-1}\omega_1\mu_1 a_1 + (\omega_1-\omega_2)^{-1}k\sin\beta_1\\&a_1\beta_1' = \sigma a_1 + (\omega_1-\omega_2)^{-1}\gamma\left(a_1^2+2a_2^2\right)a_1 + (\omega_1-\omega_2)^{-1}k\cos\beta_1\\&a_2' = -(\omega_1-\omega_2)^{-1}\omega_2\mu_1 a_2\\&a_2\beta_2' = (\omega_1-\omega_2)^{-1}\gamma\left(2a_1^2+a_2^2\right)a_2\end{aligned} \tag{6.25.31}$$

where

$$\beta_1 = \sigma T_2 - \theta_1 + \varphi_0, \quad \beta_2 = \theta_2 \tag{6.25.32}$$

Let $a_1' = a_2' = \beta_1' = \beta_2' = 0$, the steady state response should satisfy the following equations

$$\begin{aligned}&0 = (\omega_1-\omega_2)^{-1}\omega_1\mu_1 a_1 + (\omega_1-\omega_2)^{-1}k\sin\beta_1\\&0 = \sigma a_1 + (\omega_1-\omega_2)^{-1}\gamma\left(a_1^2+2a_2^2\right)a_1 + (\omega_1-\omega_2)^{-1}k\cos\beta_1\\&0 = -(\omega_1-\omega_2)^{-1}\omega_2\mu_1 a_2\\&0 = (\omega_1-\omega_2)^{-1}\gamma\left(2a_1^2+a_2^2\right)a_2\end{aligned} \tag{6.25.33}$$

When $\omega_1 < \omega_2$, the motion of the system is unstable as can be seen by the third equation of (6.25.31) in which a_2 is diverge.

When $\omega_1 > \omega_2$, from the third equation of (6.25.31), the steady state solution of a_2 decays and its steady state solution is $a_2 = 0$, so $a_2 = 0$ is an asymptotically stable steady state response. In this case, the fourth equation of (6.25.33) is automatically satisfied; and $a_1 \neq 0$ from the first and second equations of (6.25.33). Eliminating β_1 from the first and second equations of (6.25.33), we obtain

$$\omega_1^2\mu_1^2a_1^2 + \left[(\omega_1 - \omega_2)\sigma + \gamma a_1^2\right]^2 a_1^2 = k^2 \tag{6.25.34}$$

The steady state response of a_1 can be determined from (6.25.34), and then the steady state response of β_1 can be obtained from the first or second equation of (6.25.33).

In order to analyze the stability of an arbitrary non-trivial steady state response a_{10}, β_{10}, let

$$a_1 = a_{10} + \tilde{a}_1, \quad \beta_1 = \beta_{10} + \tilde{\beta}_1 \tag{6.25.35}$$

Substituting (6.25.35) into (6.25.31), we can obtain the corresponding linearized equation for $a_2 = 0$.

$$\begin{Bmatrix} \tilde{a}'_1 \\ \tilde{\beta}'_1 \end{Bmatrix} = \begin{bmatrix} (\omega_1 - \omega_2)^{-1}\omega_1\mu_1 & (\omega_1 - \omega_2)^{-1}k\cos\beta_{10} \\ a_{10}^{-1}\sigma + 3(\omega_1 - \omega_2)^{-1}\gamma a_{10} & -(\omega_1 - \omega_2)^{-1}a_{10}^{-1}k\sin\beta_{10} \end{bmatrix} \begin{Bmatrix} \tilde{a}_1 \\ \tilde{\beta}_1 \end{Bmatrix} \tag{6.25.36}$$

From this, we can calculate the eigenvalues of (6.25.36). a_{10}, β_{10} are stable if both eigenvalues have negative real parts, otherwise they are unstable.

(d) When $K = O(1)$, equating coefficients of like powers of ε in (6.25.6) yields.

Order ε^1:

$$D_0^2\xi_1 - ipD_0\xi_1 + \omega_0^2\xi_1 = Ke^{i(pT_0+\varphi_0)} \tag{6.25.37}$$

Order ε^2:

$$D_0^2\xi_2 - ipD_0\xi_2 + \omega_0^2\xi_2 = ipD_1\xi_1 - 2D_0D_1\xi_1 \tag{6.25.38}$$

Order ε^3:

$$\begin{aligned} D_0^2\xi_3 - ipD_0\xi_3 + \omega_0^2\xi_3 &= ipD_2\xi_1 + \mu_1D_0\xi_1 + \gamma\left|\xi_1^2\right|\xi_1 \\ &\quad -\left(D_1^2 + 2D_0D_2\right)\xi_1 - 2D_0D_1\xi_2 + ipD_1\xi_2 \end{aligned} \tag{6.25.39}$$

The solution of (6.25.37) can be expressed as

$$\xi_1 = A_1 e^{i\omega_1 T_0} + A_2 e^{i\omega_2 T_0} + \hat{K} e^{i(pT_0+\varphi_0)}, \quad \hat{K} = \frac{K}{\omega_0^2 - 2p^2} \tag{6.25.40}$$

where $A_1 = A_1(T_1, T_2)$, $A_2 = A_2(T_1, T_2)$. Substituting (6.25.40) into (6.25.38), we obtain

$$D_0^2\xi_2 - ipD_0\xi_2 + \omega_0^2\xi_2 = (ip - 2i\omega_1)D_1A_1e^{i\omega_1 T_0} + (ip - 2i\omega_2)D_1A_2e^{i\omega_2 T_0} \tag{6.25.41}$$

In order to eliminate secular terms from the above equation, we need

$$D_1A_1 = D_1A_2 = 0 \Rightarrow A_1 = A_1(T_2), A_2 = A_2(T_2) \tag{6.25.42}$$

From (6.25.41), we can obtain

$$\xi_2 = 0 \tag{6.25.43}$$

Substitute (6.25.40), (6.25.42) and (6.25.43) into (6.25.39), we obtain

$$\begin{aligned}
&D_0^2\xi_3 - ipD_0\xi_3 + \omega_0^2\xi_3 = -2D_0D_2\xi_1 + ipD_2\xi_1 + \mu_1 D_0\xi_1 + \gamma\left|\xi_1^2\right|\xi_1 \\
&= -2i\omega_1 A_1' e^{i\omega_1 T_0} + ipA_1' e^{i\omega_1 T_0} + i\omega_1\mu_1 A_1 e^{i\omega_1 T_0} \\
&\quad -2i\omega_2 A_2' e^{i\omega_2 T_0} + ipA_2' e^{i\omega_2 T_0} + i\omega_2\mu_1 A_2 e^{i\omega_2 T_0} + \hat{K}p\mu_1 e^{i(pT_0+\varphi_0)} \\
&\quad +\gamma(A_1 e^{i\omega_1 T_0} + A_2 e^{i\omega_2 T_0} + \hat{K}e^{i(pT_0+\varphi_0)})^2\left(\overline{A}_1 e^{-i\omega_1 T_0} + \overline{A}_2 e^{-i\omega_2 T_0} + \hat{K}e^{-i(pT_0+\varphi_0)}\right) \\
&= \left(-2i\omega_1 A_1' + ipA_1' + i\omega_1\mu_1 A_1 + \gamma A_1^2\overline{A}_1 + 2\gamma A_1 A_2\overline{A}_2 + 2\gamma\hat{K}^2 A_1\right)e^{i\omega_1 T_0} \\
&\quad +\left(-2i\omega_2 A_2' + ipA_2' + i\omega_2\mu_1 A_2 + \gamma A_2^2\overline{A}_2 + 2\gamma A_1\overline{A}_1 A_2 + 2\gamma\hat{K}^2 A_2\right)e^{i\omega_2 T_0} \\
&\quad +\left(\hat{K}p\mu_1 + 2\gamma\hat{K}A_1\overline{A}_1 + 2\gamma\hat{K}A_2\overline{A}_2 + \gamma\hat{K}^3\right)e^{i(pT_0+\varphi_0)} \\
&\quad +2\gamma\hat{K}A_1\overline{A}_2 e^{i(\omega_1 T_0 - \omega_2 T_0 + pT_0 + \varphi_0)} + 2\gamma\overline{A}_1\hat{K}A_2 e^{i(\omega_2 T_0 - \omega_1 T_0 + pT_0 + \varphi_0)} \\
&\quad +\gamma\hat{K}^2\overline{A}_2 e^{i(2pT_0 - \omega_2 T_0 + 2\varphi_0)} + \gamma\hat{K}^2\overline{A}_1 e^{i(2pT_0 - \omega_1 T_0 + 2\varphi_0)} \\
&\quad +\gamma\hat{K}A_1^2 e^{i(2\omega_1 T_0 - pT_0 - \varphi_0)} + \gamma\hat{K}A_2^2 e^{i(2\omega_2 T_0 - pT_0 - \varphi_0)} \\
&\quad +2\gamma\hat{K}A_1A_2 e^{i(\omega_1 T_0 + \omega_2 T_0 - pT_0 - \varphi_0)} + \gamma\overline{A}_1 A_2^2 e^{i(2\omega_2 - \omega_1)T_0} + \gamma A_1^2\overline{A}_2 e^{i(2\omega_1 - \omega_2)T_0}
\end{aligned} \tag{6.25.44}$$

Primes denote the derivative with respect to T_2. In order to eliminate secular terms from the above equation, it is necessary to consider the value of p. Below we give the solution for the case $p \approx 0$. Readers are invited to complete the solution for the remaining two cases.

(i) $p \approx 0$:

In this case, let

$$p = \varepsilon^2\sigma \tag{6.25.45}$$

Also considering $p = \omega_1 + \omega_2$, we obtain

$$\begin{aligned} e^{i(2pT_0-\omega_2 T_0+2\varphi_0)} &= e^{i(pT_0+\omega_1 T_0+2\varphi_0)} = e^{i(\sigma T_2+2\varphi_0)} e^{i\omega_1 T_0} \\ e^{i(2pT_0-\omega_1 T_0+2\varphi_0)} &= e^{i(pT_0+\omega_2 T_0+2\varphi_0)} = e^{i(\sigma T_2+2\varphi_0)} e^{i\omega_2 T_0} \end{aligned} \quad (6.25.46)$$

Therefore, in order to eliminate secular terms from (6.25.44), we need

$$\begin{aligned} &-2i\omega_1 A_1' + ipA_1' + i\omega_1\mu_1 A_1 + \gamma A_1^2\overline{A}_1 + 2\gamma A_1 A_2 \overline{A}_2 \\ &+2\gamma\hat{K}^2 A_1 + \gamma\hat{K}^2\overline{A}_2 e^{i(\sigma T_2+2\varphi_0)} = 0 \\ &-2i\omega_2 A_2' + ipA_2' + i\omega_2\mu_1 A_2 + \gamma A_2^2\overline{A}_2 + 2\gamma A_1\overline{A}_1 A_2 \\ &+2\gamma\hat{K}^2 A_2 + \gamma\hat{K}^2\overline{A}_1 e^{i(\sigma T_2+2\varphi_0)} = 0 \end{aligned} \quad (6.25.47)$$

Let

$$A_1 = a_1 e^{i\theta_1}, \quad A_2 = a_2 e^{i\theta_2} \quad (6.25.48)$$

Substituting (6.25.48) into (6.25.47), we obtain

$$\begin{aligned} &-2i\omega_1 a_1' + 2\omega_1 a_1\theta_1' + ipa_1' - pa_1\theta_1' + i\omega_1\mu_1 a_1 \\ &+\gamma a_1^3 + 2\gamma a_1 a_2^2 + 2\gamma\hat{K}^2 a_1 + \gamma\hat{K}^2 a_2 e^{i(\sigma T_2-\theta_1-\theta_2+2\varphi_0)} = 0 \end{aligned} \quad (6.25.49)$$

$$\begin{aligned} &-2i\omega_2 a_2' + 2\omega_2 a_2\theta_2' + ipa_2' - pa_2\theta_2' + i\omega_2\mu_1 a_2 \\ &+\gamma a_2^3 + 2\gamma a_2 a_1^2 + 2\gamma\hat{K}^2 a_2 + \gamma\hat{K}^2 a_1 e^{i(\sigma T_2-\theta_1-\theta_2+2\varphi_0)} = 0 \end{aligned} \quad (6.25.50)$$

Separating the real and imaginary parts of the above equations yields

$$\begin{aligned} (\omega_1-\omega_2)a_1' &= \omega_1\mu_1 a_1 + \gamma\hat{K}^2 a_2 \sin\beta \\ (\omega_1-\omega_2)a_2' &= -\omega_2\mu_1 a_2 - \gamma\hat{K}^2 a_1 \sin\beta \\ (\omega_1-\omega_2)a_1\theta_1' &= -\gamma\left[a_1^3 + 2a_1 a_2^2 + 2\hat{K}^2 a_1 + \hat{K}^2 a_2\cos\beta\right] \\ (\omega_1-\omega_2)a_2\theta_2' &= \gamma\left[a_2^3 + 2a_2 a_1^2 + 2\hat{K}^2 a_2 + \hat{K}^2 a_1\cos\beta\right] \end{aligned} \quad (6.25.51)$$

where

$$\beta = \sigma T_2 - \theta_1 - \theta_2 + 2\varphi_0 \quad (6.25.52)$$

From the above result, the first order approximate solution of the original equation is

$$\xi = a_1 e^{i(\omega_1 T_0+\theta_1)} + a_2 e^{i(\omega_2 T_0+\theta_2)} + \hat{K} e^{i(pT_0+\varphi_0)} + O(\varepsilon^3) \quad (6.25.53)$$

where $a_1,\ a_2,\ \theta_1,\ \theta_2$ are given by (6.25.51).

Readers are invited to complete the analysis of the steady state response of the system and its stability.

6.26 Exercise 6.26 (Nonlinear Oscillation Analysis of a Rolling Reentry Body II)

Readers are invited to complete this exercise referring to Exercise 6.25.

Chapter 7
Continuous Systems

7.1 Exercise 7.1 (Longitudinal Oscillation Analysis of Non-uniform, Non-linear Elastic Rods)

Solution: Applying Newton's second law to the rod element, as in Fig. 7.1, yields

$$\rho(x)A(x)\frac{\partial^2 u}{\partial t^2}dx = A(x)\sigma + \frac{\partial}{\partial x}[A(x)\sigma]dx - A(x)\sigma + F(x,t)dx - 2\mu\frac{\partial u}{\partial t}dx$$

where $-2\mu(\partial u/\partial t)dx$ is the linear damping involved in the rod element. The above equation can be written as

$$\rho(x)A(x)\frac{\partial^2 u}{\partial t^2} = \frac{\partial}{\partial x}[A(x)\sigma] - 2\mu\frac{\partial u}{\partial t} + F(x,t) \tag{7.1.1}$$

The nonlinear constitutive relation is adopted in the present problem, i.e.

$$\sigma = E(x)e\big[1 + E_1(x)e + E_2(x)e^2 + \cdots\big], \quad e = \frac{\partial u}{\partial x} \tag{7.1.2}$$

and the boundary conditions at the both ends of the rod are

$$u(0,t) = u(1,t) = 0 \tag{7.1.3}$$

(**a**) Substituting (7.1.2) into (7.1.1) yields

$$\begin{aligned}\rho(x)A(x)\frac{\partial^2 u}{\partial t^2} &= \frac{\partial}{\partial x}\left\{A(x)E(x)\frac{\partial u}{\partial x}\left[1 + E_1(x)\frac{\partial u}{\partial x} + E_2(x)\left(\frac{\partial u}{\partial x}\right)^2 + \cdots\right]\right\}\\ &\quad - 2\mu\frac{\partial u}{\partial t} + F(x,t)\end{aligned}$$

Z. He et al., *Solved Problems in Nonlinear Oscillations*,
https://doi.org/10.1007/978-981-97-6113-5_7

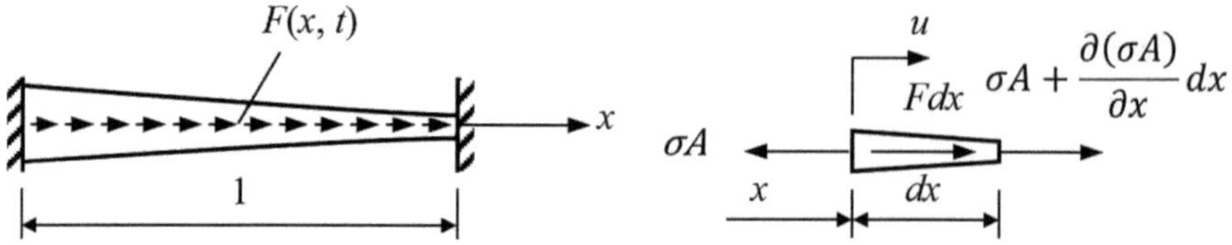

Fig. 7.1 Longitudinal oscillating rod and rod element for Exercise 7.1

i.e.,

$$\rho(x)A(x)\frac{\partial^2 u}{\partial t^2} = \frac{\partial}{\partial x}\left[E(x)A(x)\frac{\partial u}{\partial x}\right] + \frac{\partial}{\partial x}\left[E(x)A(x)E_1(x)\left(\frac{\partial u}{\partial x}\right)^2 + E(x)A(x)E_2(x)\left(\frac{\partial u}{\partial x}\right)^3\right] - 2\mu\frac{\partial u}{\partial t} + F(x,t) \tag{7.1.4}$$

(**b**) Without damping, nonlinear and excitation terms, Eq. (7.1.4) becomes

$$\rho(x)A(x)\frac{\partial^2 u}{\partial t^2} = \frac{\partial}{\partial x}\left[E(x)A(x)\frac{\partial u}{\partial x}\right] \tag{7.1.5}$$

Let

$$u(x,t) = \phi(x)e^{i\omega t} \tag{7.1.6}$$

and substitute it into (7.1.5), we can obtain

$$\frac{d}{dx}\left[E(x)A(x)\frac{d\phi}{dx}\right] + \omega^2\rho(x)A(x)\phi = 0 \tag{7.1.7}$$

From (7.1.6) and the boundary condition (7.1.3), we have

$$\phi(0) = \phi(1) = 1 \tag{7.1.8}$$

(7.1.7) and (7.1.8) are the governing equation for $\phi(x)$, which is the mode shape corresponding to the frequency ω, and its natural frequency ω.

Let $\omega^2 = \lambda$ and assume that λ and ϕ are complex numbers and complex functions. $\overline{\lambda}$ and $\overline{\phi}$ are also the eigenvalue and mode shape function of (7.1.7) and (7.1.8), therefore

$$\overline{\phi}(0) = \overline{\phi}(1) = 1 \tag{7.1.9}$$

Multiplying Eq. (7.1.7) by $\overline{\phi}$ and integrating over the domain [0, 1] yields

$$\lambda\int\limits_0^1 \rho A\phi\overline{\phi}dx = -\int\limits_0^1 \overline{\phi}\frac{d}{dx}\left(EA\frac{d\phi}{dx}\right)dx$$

Applying the integration by part to the right-hand side of the above equation and taking boundary conditions into account yields

$$\lambda\int\limits_0^1 \rho A\phi\overline{\phi}dx = \int\limits_0^1 EA\frac{d\phi}{dx}\frac{d\overline{\phi}}{dx}dx$$

therefore

$$\lambda = \frac{\int_0^1 EA\frac{d\phi}{dx}\frac{d\overline{\phi}}{dx}dx}{\int_0^1 \rho A\phi\overline{\phi}dx} \tag{7.1.10}$$

Both integrals in Eq. (7.1.10) are positive real, so λ is positive real, and thus $\omega = \pm\sqrt{\lambda}$ is real.

Let $\phi = \phi_R + i\phi_I$, we have

$$\frac{d}{dx}\left[E(x)A(x)\frac{d\phi_R}{dx}\right] + \omega^2\rho(x)A(x)\phi_R = 0, \quad \phi_R(0) = \phi_R(1) = 0 \tag{7.1.11}$$

$$\frac{d}{dx}\left[E(x)A(x)\frac{d\phi_I}{dx}\right] + \omega^2\rho(x)A(x)\phi_I = 0, \quad \phi_I(0) = \phi_I(1) = 0 \tag{7.1.12}$$

It can be seen that ϕ_R and ϕ_I satisfy the same boundary-value problem, so $\phi_I = c\phi_R$, where c is an arbitrary nonzero real number; and $\phi = \phi_R(1 + ic)$. Since the mode shape ϕ corresponding to the natural frequency ω is still the mode shape corresponding to ω of the equation multiplying by any non-zero constant, ϕ_R is also the mode shape corresponding to the natural frequency ω. Therefore, the eigenfrequencies are real.

Let $\phi_n(x)$ and $\phi_m(x)$ be the mode shape functions corresponding to natural frequencies ω_n and ω_m, respectively, so we have

$$\omega_n^2\rho A\phi_n = -\frac{d}{dx}\left(EA\frac{d\phi_n}{\partial x}\right), \quad \phi_n(0) = \phi_n(1) = 0$$

$$\omega_m^2\rho A\phi_m = -\frac{d}{dx}\left(EA\frac{d\phi_m}{\partial x}\right), \quad \phi_m(0) = \phi_m(1) = 0$$

Multiplying these two equations by $\phi_m(x)$ and $\phi_n(x)$, respectively, integrating over [0, 1], applying the integration by part and considering boundary conditions, yields

$$\omega_n^2 \int_0^1 \rho A\phi_n\phi_m dx = -\int_0^1 \phi_m \frac{d}{dx}\left(EA\frac{d\phi_n}{\partial x}\right)dx = \int_0^1 EA\frac{d\phi_n}{dx}\frac{d\phi_m}{dx}dx \tag{7.1.13}$$

$$\omega_m^2 \int_0^1 \rho A\phi_n\phi_m dx = -\int_0^1 \phi_n \frac{d}{dx}\left(EA\frac{d\phi_m}{dx}\right)dx = \int_0^1 EA\frac{d\phi_n}{dx}\frac{d\phi_m}{dx}dx \tag{7.1.14}$$

(7.1.13)–(7.1.14) yields

$$\left(\omega_n^2 - \omega_m^2\right)\int_0^1 \rho A\phi_n\phi_m dx = 0$$

When $\omega_n \neq \omega_m$, we have

$$\int_0^1 \rho(x)A(x)\phi_n(x)\phi_m(x)dx = 0$$

When $\omega_n \neq \omega_m$, $\phi_n(x)$ can be normalized such that

$$\int_0^1 \rho(x)A(x)\phi_n^2(x)dx = 1$$

therefore

$$\int_0^1 \rho(x)A(x)\phi_n(x)\phi_m(x)dx = \delta_{mn} \tag{7.1.15}$$

Then from (7.1.13) or (7.1.14), we have

$$\int_0^1 E(x)A(x)\phi_n'(x)\phi_m'(x)dx = \begin{cases} 0, & n \neq m \\ \omega_n^2, & n = m \end{cases} \tag{7.1.16}$$

where the prime denotes the derivative with respect to x. Equations (7.1.15) and (7.1.16) show that the mode shape functions $\phi_n(x)$ and $\phi_m(x)$ are orthogonal and can be made orthonormal.

Let the nonlinear Eq. (7.1.4) be solved by

$$u(x,t)=\sum_{n=1}^{\infty}\psi_n(t)\phi_n(x) \tag{7.1.17}$$

Substituting (7.1.17) into (7.1.4) yields

$$\begin{aligned}
&\rho(x)A(x)\sum_{m=1}^{\infty}\ddot{\psi}_m(t)\phi_m(x)\\
&=\frac{\partial}{\partial x}\left[E(x)A(x)\sum_{m=1}^{\infty}\psi_m(t)\phi_m'(x)\right]\\
&+\frac{\partial}{\partial x}\left\{E(x)A(x)E_1(x)\left[\sum_{m=1}^{\infty}\psi_m(t)\phi_m'(x)\right]^2\right.\\
&\left.+E(x)A(x)E_2(x)\left[\sum_{n=1}^{\infty}\psi_m(t)\phi_m'(x)\right]^3\right\}-2\mu\sum_{m=1}^{\infty}\dot{\psi}_m(t)\phi_m(x)+F(x,t)
\end{aligned}$$

Multiplying the above equation by $\phi_n(x)$ and integrating over [0, 1] yields

$$\begin{aligned}
&\sum_{m=1}^{\infty}\int_0^1\rho(x)A(x)\ddot{\psi}_m(t)\phi_n(x)\phi_m(x)dx\\
&=\int_0^1\phi_n(x)\frac{\partial}{\partial x}\left[E(x)A(x)\sum_{m=1}^{\infty}\psi_m(t)\phi_m'(x)\right]dx\\
&+\int_0^1\phi_n(x)\frac{\partial}{\partial x}\left\{E(x)A(x)E_1(x)\left[\sum_{m=1}^{\infty}\psi_m(t)\phi_m'(x)\right]^2\right.\\
&\left.+E(x)A(x)E_2(x)\left[\sum_{m=1}^{\infty}\psi_m(t)\phi_m'(x)\right]^3\right\}dx\\
&-2\int_0^1\sum_{m=1}^{\infty}\dot{\psi}_m(t)\mu\phi_m(x)\phi_n(x)dx+\int_0^1F(x,t)\phi_n(x)dx\\
&=-\sum_{m=1}^{\infty}\psi_m(t)\int_0^1E(x)A(x)\phi_m'(x)\phi_n'(x)dx\\
&+\int_0^1\phi_n(x)\frac{\partial}{\partial x}\left\{E(x)A(x)E_1(x)\left[\sum_{m=1}^{\infty}\psi_m(t)\phi_m'(x)\right]^2\right.
\end{aligned}$$

$$+ E(x)A(x)E_2(x)\left[\sum_{m=1}^{\infty}\psi_m(t)\phi_m'(x)\right]^3\Bigg\}dx$$

$$-2\sum_{m=1}^{\infty}\int_0^1\dot{\psi}_m(t)\mu\phi_m(x)\phi_n(x)dx+\int_0^1 F(x,t)\phi_n(x)dx$$

Considering the orthogonality of the mode shape functions, we can obtain

$$\begin{aligned}\ddot{\psi}_n = &-\omega_n^2\psi_n+\int_0^1\phi_n\frac{\partial}{\partial x}\left[EAE_1\sum_{p=1}^{\infty}\psi_p\phi_p'\sum_{q=1}^{\infty}\psi_q\phi_q'\right]dx\\ &+\int_0^1\phi_n\frac{\partial}{\partial x}\left[EAE_2\sum_{p=1}^{\infty}\psi_p\phi_p'\sum_{q=1}^{\infty}\psi_q\phi_q'\sum_{r=1}^{\infty}\psi_r\phi_r'\right]dx\\ &-2\sum_{m=1}^{\infty}\left(\dot{\psi}_m\int_0^1\mu\phi_m\phi_n dx\right)+\int_0^1 F(x,t)\phi_n(x)dx\\ =&-\omega_n^2\psi_n+\int_0^1\phi_n\frac{\partial}{\partial x}\left[\sum_{p,q}\psi_p\psi_q EAE_1\phi_p'\phi_q'\right]dx\\ &+\int_0^1\phi_n\frac{\partial}{\partial x}\left[\sum_{p,q,r}\psi_p\psi_q\psi_r EAE_2\phi_p'\phi_q'\phi_r'\right]dx-2\mu_n\dot{\psi}_n+f_n(t)\\ =&-\omega_n^2\psi_n+\sum_{p,q}\psi_p\psi_q\int_0^1\phi_n\left(EAE_1\phi_p'\phi'\right)'dx_q\\ &+\sum_{p,q,r}\psi_p\psi_q\psi_r\int_0^1\phi_n\left(EAE_2\phi_p'\phi_q'\phi_r'\right)'dx-2\mu_n\dot{\psi}_n+f_n(t)\end{aligned}$$

Write the above equation as

$$\ddot{\psi}_n+\omega_n^2\psi_n=-2\mu_n\dot{\psi}_n+N_n(\psi_1,\psi_2,\ldots,\psi_m,\ldots)+f_n(t) \tag{7.1.18}$$

where

$$N_n=\sum_{p,q}\psi_p\psi_q\int_0^1\phi_n\frac{d}{dx}\left[EAE_1\frac{d\phi_p}{dx}\frac{d\phi_q}{dx}\right]dx$$

$$+\sum_{p,q,r}\psi_p\psi_q\psi_r\int_0^1\phi_n\frac{d}{dx}\left[EAE_2\frac{d\phi_p}{dx}\frac{d\phi_q}{dx}\frac{d\phi_r}{dx}\right]dx \tag{7.1.19}$$

$$f_n(t)=\int_0^1 F(x,t)\phi_n(x)dx \tag{7.1.20}$$

In addition, the present problem assumes that the damping is assumed to be associated with mode shape functions, i.e.

$$\mu_n=\int_0^1 \mu\phi_n^2(x)dx \tag{7.1.21}$$

and when $m\neq n$, we have

$$\int_0^1 \mu\phi_m(x)\phi_n(x)dx=0$$

In the following, we find the solution for the primary resonance case of $\Omega\approx\omega_m$ (no internal resonance) only.

When $f_n(t)=K_n\cos\Omega t$, we can write the equation (7.1.18) as:

$$\ddot{\psi}_n+\omega_n^2\psi_n=-2\mu_n\dot{\psi}_n+\sum_{p,q}g_{npq}\psi_p\psi_q+\sum_{p,q,r}h_{npqr}\psi_p\psi_q\psi_r+K_n\cos\Omega \tag{7.1.22}$$

where

$$g_{npq}=\int_0^1\phi_n\frac{d}{dx}\left[EAE_1\frac{d\phi_p}{dx}\frac{d\phi_q}{dx}\right]dx=-\int_0^1 EAE_1\frac{d\phi_n}{dx}\frac{d\phi_p}{dx}\frac{d\phi_q}{dx}dx \tag{7.1.23}$$

$$h_{npqr}=\int_0^1\phi_n\frac{d}{dx}\left[EAE_2\frac{d\phi_p}{dx}\frac{d\phi_q}{dx}\frac{d\phi_r}{dx}\right]dx=-\int_0^1 EAE_2\frac{d\phi_n}{dx}\frac{d\phi_p}{dx}\frac{d\phi_q}{dx}\frac{d\phi_r}{dx}dx \tag{7.1.24}$$

Let the solution of the equation (7.1.22) be

$$\psi_n=\varepsilon\psi_{n1}(T_0,T_1,T_2)+\varepsilon^2\psi_{n2}(T_0,T_1,T_2)+\varepsilon^3\psi_{n3}(T_0,T_1,T_2)+\cdots \tag{7.1.25}$$

In order to let the damping, nonlinear and excitation terms appearing simultaneously in the third-order equation, we let

$$\mu_n = \varepsilon^2 \hat{\mu}_n, K_n = 2\varepsilon^3 k_n \tag{7.1.26}$$

Substituting (7.1.25) and (7.1.26) into (7.1.22) and retaining to $O(\varepsilon^3)$, we obtain

$$\begin{aligned}
0 &= \ddot{\psi}_n + \omega_n^2 \psi_n + 2\varepsilon^2 \hat{\mu}_n \dot{\psi}_n - \sum_{p,q} g_{npq} \psi_p \psi_q - \sum_{p,q,r} h_{npqr} \psi_p \psi_q \psi_r - K_n \cos \Omega t \\
&= \left[D_0^2 + 2\varepsilon D_0 D_1 + \varepsilon^2 \left(D_1^2 + 2D_0 D_2\right)\right]\left(\varepsilon \psi_{n1} + \varepsilon^2 \psi_{n2} + \varepsilon^3 \psi_{n3}\right) \\
&\quad + \omega_n^2 \left(\varepsilon \psi_{n1} + \varepsilon^2 \psi_{n2} + \varepsilon^3 \psi_{n3}\right) \\
&\quad + 2\varepsilon^2 \hat{\mu}_n \left(D_0 + \varepsilon D_1 + \varepsilon^2 D_2\right)\left(\varepsilon \psi_{n1} + \varepsilon^2 \psi_{n2} + \varepsilon^3 \psi_{n3}\right) \\
&\quad - \sum_{p,q} g_{npq} \left(\varepsilon \psi_{p1} + \varepsilon^2 \psi_{p2} + \varepsilon^3 \psi_{p3}\right)\left(\varepsilon \psi_{q1} + \varepsilon^2 \psi_{q2} + \varepsilon^3 \psi_{q3}\right) \\
&\quad - \sum_{p,q,r} h_{npqr} \left(\varepsilon \psi_{p1} + \varepsilon^2 \psi_{p2} + \varepsilon^3 \psi_{p3}\right)\left(\varepsilon \psi_{q1} + \varepsilon^2 \psi_{q2} + \varepsilon^3 \psi_{q3}\right) \\
&\quad \times \left(\varepsilon \psi_{r1} + \varepsilon^2 \psi_{r2} + \varepsilon^3 \psi_{r3}\right) - 2\varepsilon^3 k_n \cos \Omega T_0 + \cdots \\
&= \varepsilon \left(D_0^2 \psi_{n1} + \omega_n^2 \psi_{n1}\right) + \varepsilon^2 \left(D_0^2 \psi_{n2} + \omega_n^2 \psi_{n2} + 2D_0 D_1 \psi_{n1} - \sum_{p,q} g_{npq} \psi_{p1} \psi_{q1}\right) \\
&\quad + \varepsilon^3 [D_0^2 \psi_{n3} + \omega_n^2 \psi_{n3} + \left(D_1^2 + 2D_0 D_2\right) \psi_{n1} + 2D_0 D_1 \psi_{n2} \\
&\quad + 2\hat{\mu}_n D_0 \psi_{n1} - 2 \sum\nolimits_{p,q} g_{npq} \psi_{p1} \psi_{q2} \\
&\quad - \sum_{p,q,r} h_{npqr} \psi_{p1} \psi_{q1} \psi_{r1} - 2k_n \cos \Omega T_0] + \cdots
\end{aligned} \tag{7.1.27}$$

Let the coefficient of the same power of ε in (7.1.27) be zero, we obtain.
Order ε:

$$D_0^2 \psi_{n1} + \omega_n^2 \psi_{n1} = 0 \tag{7.1.28}$$

Order ε^2:

$$D_0^2 \psi_{n2} + \omega_n^2 \psi_{n2} = -2D_0 D_1 \psi_{n1} + \sum_{p,q} g_{npq} \psi_{p1} \psi_{q1} \tag{7.1.29}$$

Order ε^3:

$$\begin{aligned}
D_0^2 \psi_{n3} + \omega_n^2 \psi_{n3} &= -\left(D_1^2 + 2D_0 D_2\right) \psi_{n1} - 2D_0 D_1 \psi_{n2} - 2\hat{\mu}_n D_0 \psi_{n1} \\
&\quad + 2 \sum_{p,q} g_{npq} \psi_{p1} \psi_{q2} + \sum_{p,q,r} h_{npqr} \psi_{p1} \psi_{q1} \psi_{r1} + 2k_n \cos \Omega T_0
\end{aligned} \tag{7.1.30}$$

The solution of (7.1.28) is

$$\psi_{n1} = A_n e^{i\omega_n T_0} + cc \tag{7.1.31}$$

Substituting (7.1.31) into (7.1.29) yields

$$D_0^2\psi_{n2} + \omega_n^2\psi_{n2} = -2i\omega_n A_n' e^{i\omega_n T_0} + \sum_{p,q} g_{npq}\left(A_p A_q e^{i(\omega_q+\omega_p)T_0} + \overline{A}_p A_q e^{i(\omega_q-\omega_p)T_0}\right) + cc \tag{7.1.32}$$

In order to eliminate secular terms from the above equation, we need

$$A_n' = 0 \rightarrow \quad A_n = A_n(T_2) \tag{7.1.33}$$

From (7.1.32), we can obtain

$$\psi_{n2} = \sum_{p,q} g_{npq}\left(\Lambda_{npq}^{(1)} A_p A_q e^{i(\omega_q+\omega_p)T_0} + \Lambda_{npq}^{(2)} \overline{A}_p A_q e^{i(\omega_q-\omega_p)T_0}\right) + cc \tag{7.1.34}$$

where

$$\Lambda_{npq}^{(1)} = \left[\omega_n^2 - \left(\omega_q + \omega_p\right)^2\right]^{-1}, \quad \Lambda_{npq}^{(2)} = \left[\omega_n^2 - \left(\omega_q - \omega_p\right)^2\right]^{-1} \tag{7.1.35}$$

Substitute (7.1.31), (7.1.33) and (7.1.34) into (7.1.30), we obtain

$$\begin{aligned} D_0^2\psi_{n3} + \omega_n^2\psi_{n3} &= -2i\omega_n A_n' e^{i\omega_n T_0} - 2i\omega_n \hat{\mu}_n A_n e^{i\omega_n T_0} + k_n e^{i\Omega T_0} \\ &\quad + \sum_{p,q,j,k} 2g_{npq} g_{qjk}\left(\Lambda_{qjk}^{(1)} \overline{A}_p A_j A_k e^{i(\omega_j+\omega_k-\omega_p)T_0}\right. \\ &\quad \left. + \Lambda_{qjk}^{(2)} A_p \overline{A}_j A_k e^{i(\omega_k+\omega_p-\omega_j)T_0}\right) \\ &\quad + \sum_{p,q,r} 3h_{npqr} A_p \overline{A}_q A_r e^{i(\omega_p+\omega_r-\omega_q)T_0} + cc + NST \\ &= -2i\omega_n A_n' e^{i\omega_n T_0} - 2i\omega_n \hat{\mu}_n A_n e^{i\omega_n T_0} + k_n e^{i\Omega T_0} \\ &\quad + \sum_{p,q,j,k} 2g_{npq} g_{qjk} \Lambda_{qjk}^{(1)} \overline{A}_p A_j A_k e^{i(\omega_j+\omega_k-\omega_p)T_0} \\ &\quad + \sum_{p,q,j,k} 2g_{nkq} g_{qjp} \Lambda_{qjp}^{(2)} \overline{A}_p A_j A_k e^{i(\omega_j+\omega_k-\omega_p)T_0} \\ &\quad + \sum_{p,j,k} 3h_{npjk} \overline{A}_p A_j A_k e^{i(\omega_j+\omega_k-\omega_p)T_0} + cc + NST \\ &= -2i\omega_n A_n' e^{i\omega_n T_0} - 2i\omega_n \hat{\mu}_n A_n e^{i\omega_n T_0} + k_n e^{i\Omega T_0} \\ &\quad + \sum_{p,q,j,k} \Gamma_{npqjk} \overline{A}_p A_j A_k e^{i(\omega_j+\omega_k-\omega_p)T_0} + cc + NST \end{aligned} \tag{7.1.36}$$

where

$$\Gamma_{npqjk} = 2g_{npq}g_{qjk}\Lambda^{(1)}_{qjk} + 2g_{nkq}g_{qjp}\Lambda^{(2)}_{qjp} + 3h_{npjk} \tag{7.1.37}$$

Note that g_{npq} and h_{npqr} remain unchanged after permutation of subscripts, while $\Lambda^{(1)}_{qjk}$ and $\Lambda^{(2)}_{qjk}$ remain unchanged after the exchange of the second and third subscripts.

Introduce the detuning parameter σ such that

$$\Omega = \omega_m + \varepsilon^2\sigma \tag{7.1.38}$$

In order to eliminate secular terms in (7.1.36), we need

$$2i\omega_m A'_m + 2i\omega_m\hat{\mu}_m A_m - A_m \sum_{p,q} \left(\Gamma_{mpqpm} + \Gamma_{mpqmp}\right)\overline{A}_p A_p - k_m e^{i\sigma T_2} = 0 \quad (n = m) \tag{7.1.39}$$

$$2i\omega_n A'_n + 2i\omega_n\hat{\mu}_n A_n - A_n \sum_{p,q} \left(\Gamma_{npqpn} + \Gamma_{npqnp}\right)\overline{A}_p A_p = 0 \quad (n \neq m) \tag{7.1.40}$$

Let

$$A_n = \frac{1}{2}a_n e^{i\beta_n}, \quad A_m = \frac{1}{2}a_m e^{i\beta_m} \tag{7.1.41}$$

and substitute them into (7.1.40), we have

$$i\omega_n a'_n - \omega_n a_n \beta'_n + i\omega_n\hat{\mu}_n a_n - \frac{1}{8}a_n \sum_{p,q} \left(\Gamma_{npqpn} + \Gamma_{npqnp}\right)a_p^2 = 0 \tag{7.1.42}$$

Separating the real and imaginary parts of the above equation yields

$$a'_n + \hat{\mu}_n a_n = 0 \tag{7.1.43}$$

$$\omega_n a_n \beta'_n + \frac{1}{8}a_n \sum_{p,q} \left(\Gamma_{npqpn} + \Gamma_{npqnp}\right)a_p^2 = 0 \tag{7.1.44}$$

From (7.1.43), we have

$$a_n = a_{n0}e^{-\hat{\mu}_n T_2}, \ n \neq m \tag{7.1.45}$$

Since $\hat{\mu}_n > 0$, $a_n \to 0$ for all $n \neq m$.

Substituting (7.1.41) into (7.1.39) and taking (7.1.45) into account, we obtain

$$i\omega_m a'_m - \omega_m a_m \beta'_m + i\omega_m\hat{\mu}_m a_m - \frac{1}{4}a_m^3\Gamma_m - k_m e^{i(\sigma T_2 - \beta_m)} = 0 \tag{7.1.46}$$

where

$$\Gamma_m = 2g_m^2\Lambda_m^{(1)} + 2g_m^2\Lambda_m^{(2)} + 3h_m$$
$$g_m = -\int_0^1 EAE_1\left(\frac{d\phi_m}{dx}\right)^3 dx, \quad h_m = -\int_0^1 EAE_2\left(\frac{d\phi_m}{dx}\right)^4 dx$$
$$\Lambda_m^{(1)} = -\frac{1}{3}\omega_m^{-2}, \ \Lambda_m^{(2)} = \omega_m^{-2} \tag{7.1.47}$$

Separating the real and imaginary parts of Eq. (7.1.46), we obtain

$$\begin{aligned} a_m' &= -\hat{\mu}_m a_m + \omega_m^{-1} k_m \sin(\sigma T_2 - \beta_m) \\ a_m\beta_m' &= -\frac{1}{4}\omega_m^{-1}\Gamma_m a_m^3 - \omega_m^{-1} k_m \cos(\sigma T_2 - \beta_m) \end{aligned} \tag{7.1.48}$$

or

$$\begin{aligned} a_m' &= -\hat{\mu}_m a_m + \omega_m^{-1} k_m \sin\gamma_m \\ a_m\gamma_m' &= \sigma a_m + \frac{1}{4}\omega_m^{-1}\Gamma_m a_m^3 + \omega_m^{-1} k_m \cos\gamma_m \end{aligned} \tag{7.1.49}$$

where

$$\gamma_m = \sigma T_2 - \beta_m \tag{7.1.50}$$

Therefore, the solution to (7.1.22) is

$$\psi_n = \begin{cases} a_m\cos(\omega_m t + \beta_m) + \frac{1}{2}a_m^2 g_m\left[\Lambda_m^{(1)}\cos 2(\omega_m T_0 + \beta_m) + \frac{1}{2}\Lambda_m^{(2)}\right] + O(\varepsilon^3), & n = m \\ 0, & n \neq m \end{cases} \tag{7.1.51}$$

where a_m and β_m are given by (7.1.49) and (7.1.50).

Readers are invited to complete Exercise 7.1(g).

7.2 Exercise 7.2 (Longitudinal Oscillation Analysis of Uniform, Nonlinear Elastic Rods)

Solution: (a) Substituting the expressions for axial stress σ and axial strain e into the governing equation of the rod, we obtain

$$\rho\frac{\partial^2 u}{\partial t^2} = E\frac{\partial}{\partial x}\left[e\left(1 + E_1 e + E_2 e^2 + \cdots\right)\right] - 2\rho\mu\frac{\partial u}{\partial t} + \rho F(x, t)$$

$$= E\frac{\partial^2 u}{\partial x^2} + 2EE_1\frac{\partial u}{\partial x}\frac{\partial^2 u}{\partial x^2} + 3EE_2\left(\frac{\partial u}{\partial x}\right)^2\frac{\partial^2 u}{\partial x^2} - 2\rho\mu\frac{\partial u}{\partial t} + \rho F(x,t)$$

i.e.,

$$\frac{\partial^2 u}{\partial t^2} - c^2\frac{\partial^2 u}{\partial x^2} = 2c^2E_1\frac{\partial u}{\partial x}\frac{\partial^2 u}{\partial x^2} + 3c^2E_2\left(\frac{\partial u}{\partial x}\right)^2\frac{\partial^2 u}{\partial x^2} - 2\mu\frac{\partial u}{\partial t} + F(x,t) \quad (7.2.1)$$

where

$$c^2 = \frac{E}{\rho} \quad (7.2.2)$$

The corresponding linear undamped free oscillation equation to (7.2.1) is given by

$$\frac{\partial^2 u}{\partial t^2} - c^2\frac{\partial^2 u}{\partial x^2} = 0 \quad (7.2.3)$$

Let the solution of (7.2.3) be

$$u = \phi(x)e^{i\omega t} \quad (7.2.4)$$

Substituting (7.2.4) into (7.2.3) yields

$$\phi'' + \frac{\omega^2}{c^2}\phi = 0 \quad (7.2.5)$$

therefore,

$$\phi(x) = A\sin kx + B\cos kx, \quad k = \omega/c \quad (7.2.6)$$

From the boundary conditions, we have

$$\left.\begin{array}{r} B = 0 \\ A\sin k + hkA\cos kx = 0 \end{array}\right\} \xrightarrow{\text{Non-trivial solution condition}} \sin k + hk\cos kx = 0 \quad (7.2.7)$$

So

$$\tan k = -hk \quad (7.2.8)$$

The mode shape function is

$$\phi(x) = \sin kx, \quad k = \omega/c \tag{7.2.9}$$

When $h = 1$, the first 5 positive roots of the Eq. (7.2.8) are:

$$k_1 = 2.0288, \quad k_2 = 4.9132, \quad k_3 = 7.9786, \quad k_4 = 11.0855, \quad k_5 = 14.2075$$

Let the solution of nonlinear Eq. (7.2.1) be

$$u = \sqrt{2}\sum_{n=1}^{N} \psi_n(t)\sin k_n x \tag{7.2.10}$$

Substituting this into (7.2.1), we obtain

$$\begin{aligned}
&\sqrt{2}\sum_{p=1}^{N} \ddot{\psi}_p \sin k_p x + \sqrt{2}c^2 \sum_{p=1}^{N} k_p^2 \psi_p \sin k_p x \\
&= -4c^2 E_1 \sum_{p,q}^{N} k_p k_q^2 \psi_p \psi_q \sin k_q x \cos k_p x \\
&\quad - 6\sqrt{2}c^2 E_2 \sum_{p,q,r}^{N} k_p k_q^2 k_r \psi_p \psi_q \psi_r \cos k_p x \sin k_q x \cos k_r x \\
&\quad - 2\sqrt{2}\mu \sum_{p=1}^{N} \dot{\psi}_p(t) \sin k_p x + F(x,t)
\end{aligned}$$

Multiplying both sides of the above equation by $\sqrt{2}\sin k_n x$ and making the integration of x over [0, 1] yields

$$\ddot{\psi}_n + \omega_n^2 \psi_n = -2\mu_n \dot{\psi}_n + c^2 \sum_{p,q}^{N} \Gamma_{npq} \psi_p \psi_q + c^2 \sum_{p,q,r}^{N} \Gamma_{npqr} \psi_p \psi_q \psi_r + f_n(t) \tag{7.2.11}$$

where

$$\Gamma_{npq} = -4\sqrt{2}E_1 k_p k_q^2 \int_0^1 \sin k_n x \sin k_q x \cos k_p x dx$$

$$\Gamma_{npqr} = -12 E_2 k_p k_q^2 k_r \int_0^1 \sin k_n x \sin k_q x \cos k_p x \cos k_r x dx$$

$$f_n(t) = \sqrt{2}\int_0^1 F(x,t)\sin k_n x dx \tag{7.2.12}$$

When $f_n(t) = K_n\cos\Omega t$, the Eq. (7.2.11) becomes

$$\ddot{\psi}_n + \omega_n^2\psi_n = -2\mu_n\dot{\psi}_n + c^2\sum_{p,q}^{N}\Gamma_{npq}\psi_p\psi_q + c^2\sum_{p,q,r}^{N}\Gamma_{npqr}\psi_p\psi_q\psi_r + K_n\cos\Omega t \tag{7.2.13}$$

In the following, we find the primary resonance solution of (7.2.13). Thus let $\Omega \approx \omega_m$ and assume that this equation does not have any internal resonances. Let the solution of (7.2.13) be

$$\psi_n = \varepsilon\psi_{n1}(T_0, T_1, T_2) + \varepsilon^2\psi_{n2}(T_0, T_1, T_2) + \varepsilon^3\psi_{n3}(T_0, T_1, T_2) + \cdots \tag{7.2.14}$$

In order to make the damping, nonlinear and excitation terms appearing simultaneously in the third-order equation, we let

$$\mu_n = \varepsilon^2\hat{\mu}_n,\ K_n = 2\varepsilon^3 k_n \tag{7.2.15}$$

Substituting (7.2.14) and (7.2.15) into (7.2.13) and retaining to $O(\varepsilon^3)$, we obtain

$$\begin{aligned}
0 &= \ddot{\psi}_n + \omega_n^2\psi_n + 2\mu_n\dot{\psi}_n - c^2\sum_{p,q}^{N}\Gamma_{npq}\psi_p\psi_q - c^2\sum_{p,q,r}^{N}\Gamma_{npqr}\psi_p\psi_q\psi_r - K_n\cos\Omega t \\
&= \left[D_0^2 + 2\varepsilon D_0D_1 + \varepsilon^2\left(D_1^2 + 2D_0D_2\right)\right]\left(\varepsilon\psi_{n1} + \varepsilon^2\psi_{n2} + \varepsilon^3\psi_{n3}\right) \\
&\quad + \omega_n^2\left(\varepsilon\psi_{n1} + \varepsilon^2\psi_{n2} + \varepsilon^3\psi_{n3}\right) \\
&\quad + 2\varepsilon^2\hat{\mu}_n\left(D_0 + \varepsilon D_1 + \varepsilon^2 D_2\right)\left(\varepsilon\psi_{n1} + \varepsilon^2\psi_{n2} + \varepsilon^3\psi_{n3}\right) \\
&\quad - c^2\sum_{p,q}\Gamma_{npq}\left(\varepsilon\psi_{p1} + \varepsilon^2\psi_{p2} + \varepsilon^3\psi_{p3}\right)\left(\varepsilon\psi_{q1} + \varepsilon^2\psi_{q2} + \varepsilon^3\psi_{q3}\right) \\
&- c^2\sum_{p,q,r}\Gamma_{npqr}\left(\varepsilon\psi_{p1} + \varepsilon^2\psi_{p2} + \varepsilon^3\psi_{p3}\right)\left(\varepsilon\psi_{q1} + \varepsilon^2\psi_{q2} + \varepsilon^3\psi_{q3}\right) \\
&\times\left(\varepsilon\psi_{r1} + \varepsilon^2\psi_{r2} + \varepsilon^3\psi_{r3}\right) - 2\varepsilon^3 k_n\cos\Omega T_0 + \cdots \\
&= \varepsilon\left(D_0^2\psi_{n1} + \omega_n^2\psi_{n1}\right) + \varepsilon^2\left(D_0^2\psi_{n2} + \omega_n^2\psi_{n2} + 2D_0D_1\psi_{n1} - c^2\sum_{p,q}\Gamma_{npq}\psi_{p1}\psi_{q1}\right) \\
&+ \varepsilon^3\left[D_0^2\psi_{n3} + \omega_n^2\psi_{n3} + \left(D_1^2 + 2D_0D_2\right)\psi_{n1} + 2\hat{\mu}_n D_0\psi_{n1} + 2D_0D_1\psi_{n2}\right. \\
&- c^2\sum_{p,q}\Gamma_{npq}\left(\psi_{p1}\psi_{q2} + \psi_{p2}\psi_{q1}\right)
\end{aligned}$$

$$-c^2\sum_{p,q,r}\Gamma_{npqr}\psi_{p1}\psi_{q1}\psi_{r1}-2k_n\cos\Omega T_0\Bigg]+\cdots \tag{7.2.16}$$

Let the coefficient of the same power of ε in the equation (7.2.16) be zero, we obtain.

Order ε:

$$D_0^2\psi_{n1}+\omega_n^2\psi_{n1}=0 \tag{7.2.17}$$

Order ε^2:

$$D_0^2\psi_{n2}+\omega_n^2\psi_{n2}=-2D_0D_1\psi_{n1}+c^2\sum_{p,q}\Gamma_{npq}\psi_{p1}\psi_{q1} \tag{7.2.18}$$

Order ε^3:

$$\begin{aligned}D_0^2\psi_{n3}+\omega_n^2\psi_{n3}=&-\left(D_1^2+2D_0D_2\right)\psi_{n1}-2D_0D_1\psi_{n2}-2\hat{\mu}_nD_0\psi_{n1}\\&+c^2\sum\nolimits_{p,\ q}\Gamma_{npq}\left(\psi_{p1}\psi_{q2}+\psi_{p2}\psi_{q1}\right)\\&+c^2\sum\nolimits_{p,\ q,\ r}\Gamma_{npqr}\psi_{p1}\psi_{q1}\psi_{r1}+2k_n\cos\Omega T_0\end{aligned} \tag{7.2.19}$$

The solution of the Eq. (7.2.17) is

$$\psi_{n1}=A_ne^{i\omega_nT_0}+cc \tag{7.2.20}$$

where $A_n=A_n(T_1,T_2)$. Substituting (7.2.20) into (7.2.18), we obtain

$$\begin{aligned}D_0^2\psi_{n2}+\omega_n^2\psi_{n2}=&-2i\omega_nD_1A_ne^{i\omega_nT_0}\\&+c^2\sum_{p,q}\Gamma_{npq}\left(A_pA_qe^{i(\omega_q+\omega_p)T_0}+\overline{A}_pA_qe^{i(\omega_q-\omega_p)T_0}\right)+cc\end{aligned} \tag{7.2.21}$$

In order to eliminate secular terms from the above equation, we need

$$D_1A_n=0\ \rightarrow\quad A_n=A_n(T_2) \tag{7.2.22}$$

Furthermore, we can obtain from (7.2.21) that

$$\psi_{n2}=c^2\sum_{p,\,q}\Gamma_{npq}\left(\Lambda_{npq}^{(1)}A_pA_qe^{i(\omega_q+\omega_p)T_0}+\Lambda_{npq}^{(2)}\overline{A}_pA_qe^{i(\omega_q-\omega_p)T_0}\right)+cc \tag{7.2.23}$$

where

$$\Lambda_{npq}^{(1)} = \left[\omega_n^2 - \left(\omega_q + \omega_p\right)^2\right]^{-1}, \quad \Lambda_{npq}^{(2)} = \left[\omega_n^2 - \left(\omega_q - \omega_p\right)^2\right]^{-1} \tag{7.2.24}$$

Substituting (7.2.20), (7.2.22) and (7.2.23) into (7.2.19) yields

$$\begin{aligned}
&D_0^2\psi_{n3} + \omega_n^2\psi_{n3} \\
&= -2i\omega_n A_n' e^{i\omega_n T_0} - 2i\omega_n\hat{\mu}_n A_n e^{i\omega_n T_0} \\
&\quad + c^4 \sum_{p,q,j,k} \Gamma_{npq}\Gamma_{qjk}\left(\Lambda_{qjk}^{(1)}\bar{A}_p A_j A_k e^{i(\omega_k+\omega_j-\omega_p)T_0} + \Lambda_{qjk}^{(2)} A_p\bar{A}_j A_k e^{i(\omega_k-\omega_j+\omega_p)T_0}\right) \\
&\quad + c^4 \sum_{p,q,j,k} \Gamma_{npq}\Gamma_{pjk}\left(\Lambda_{pjk}^{(1)}\bar{A}_q A_j A_k e^{i(\omega_k+\omega_j-\omega_q)T_0} + \Lambda_{pjk}^{(2)} A_q\bar{A}_j A_k e^{i(\omega_k-\omega_j+\omega_q)T_0}\right) \\
&\quad + c^2 \sum_{p,q,r} \Gamma_{npqr}\left(\bar{A}_p A_q A_r e^{i(\omega_q-\omega_p+\omega_r)T_0} + A_p\bar{A}_q A_r e^{i(\omega_p-\omega_q+\omega_r)T_0}\right. \\
&\quad \left. + A_p A_q\bar{A}_r e^{i(\omega_p+\omega_q-\omega_r)T_0}\right) + k_n e^{i\Omega T_0} + cc + NST \\
&= -2i\omega_n A_n' e^{i\omega_n T_0} - 2i\omega_n\hat{\mu}_n A_n e^{i\omega_n T_0} + c^4 \sum_{p,q} g_{pq}\bar{A}_p A_p A_n e^{i\omega_n T_0} \\
&\quad + c^2 \sum\nolimits_p h_p\bar{A}_p A_p A_n e^{i\omega_n T_0} + k_n e^{i\Omega T_0} + cc + NST
\end{aligned} \tag{7.2.25}$$

where the prime denotes the derivative with respect to T_2 and

$$\begin{aligned}
g_{pq} &= \Gamma_{npq}\Gamma_{qnp}\Lambda_{qnp}^{(1)} + \Gamma_{nqp}\Gamma_{qnp}\Lambda_{qnp}^{(1)} + \Gamma_{npq}\Gamma_{qpn}\Lambda_{qpn}^{(1)} + \Gamma_{nqp}\Gamma_{qpn}\Lambda_{qpn}^{(1)} \\
&\quad + \Gamma_{npq}\Gamma_{qpn}\Lambda_{qpn}^{(2)} + \Gamma_{nnq}\Gamma_{qpp}\Lambda_{qpp}^{(2)} + \Gamma_{nqn}\Gamma_{qpp}\Lambda_{qpp}^{(2)} + \Gamma_{nqp}\Gamma_{qpn}\Lambda_{qpn}^{(2)} \\
h_p &= 2\Gamma_{nppn} + 2\Gamma_{nnpp} + 2\Gamma_{npnp}
\end{aligned} \tag{7.2.26}$$

Introduce the detuning parameter σ such that

$$\Omega = \omega_m + \varepsilon^2\sigma \tag{7.2.27}$$

Therefore, in order to eliminate secular terms in (7.2.25), we need

$$2i\omega_m A_m' + 2i\omega_m\hat{\mu}_m A_m - c^2 A_m \sum_{p,q}\left(c^2 g_{pq} + h_p\right)\bar{A}_p A_p - k_m e^{i\sigma T_2} = 0 \quad (n = m) \tag{7.2.28}$$

$$2i\omega_n A_n' + 2i\omega_n\hat{\mu}_n A_n - c^2 A_n \sum_{p,q}\left(c^2 g_{pq} + h_p\right)\bar{A}_p A_p = 0 \quad (n \neq m) \tag{7.2.29}$$

Let

$$A_n = \frac{1}{2}a_n e^{i\beta_n}, \quad A_m = \frac{1}{2}a_m e^{i\beta_m} \tag{7.2.30}$$

and substitute (7.2.30) into (7.2.29), we have

$$i\omega_n a_n' - \omega_n a_n \beta_n' + i\omega_n \hat{\mu}_n a_n - \frac{1}{8} a_n \sum_{p,q} (c^2 g_{pq} + h_p) a_p^2 = 0 \tag{7.2.31}$$

Separating the real and imaginary parts of the above equation yields

$$a_n' + \hat{\mu}_n a_n = 0 \tag{7.2.32}$$

$$\omega_n a_n \beta_n' + \frac{1}{8} a_n \sum_{p,q} \left(c^2 g_{pq} + h_p\right) a_p^2 = 0 \tag{7.2.33}$$

The solution of (7.2.32) is

$$a_n = a_{n0} e^{-\hat{\mu}_n T_2}, \quad n \neq m \tag{7.2.34}$$

Since $\hat{\mu}_n > 0$, $a_n \to 0$ for all $n \neq m$.

Substituting (7.2.32) into the Eq. (7.2.28) and taking (7.2.34) into account, we obtain

$$i\omega_m a_m' - \omega_m a_m \beta_m' + i\omega_m \hat{\mu}_m a_m - \frac{1}{8} c^2 (c^2 g_{mm} + h_m) a_m^3 - k_m e^{i(\sigma T_2 - \beta_m)} = 0 \tag{7.2.35}$$

Separating the real and imaginary parts of the above equation yields

$$\begin{aligned} a_m' &= -\hat{\mu}_m a_m + \omega_m^{-1} k_m \sin(\sigma T_2 - \beta_m) \\ a_m \beta_m' &= -\frac{1}{8} c^2 (c^2 g_{mm} + h_m) a_m^3 - \omega_m^{-1} k_m \cos(\sigma T_2 - \beta_m) \end{aligned} \tag{7.2.36}$$

or

$$\begin{aligned} a_m' &= -\hat{\mu}_m a_m + \omega_m^{-1} k_m \sin\gamma_m \\ a_m \gamma_m' &= \sigma a_m + \frac{1}{8} c^2 (c^2 g_{mm} + h_m) a_m^3 - \omega_m^{-1} k_m \cos\gamma_m \end{aligned} \tag{7.2.37}$$

where

$$\gamma_m = \sigma T_2 - \beta_m \tag{7.2.38}$$

The solution to the Eq. (7.2.13) is

$$\psi_n = \begin{cases} a_m \cos(\omega_m t + \beta_m) + \frac{1}{2} c^2 \Gamma_{mmm} \left[\Lambda^{(1)}_{mmm} \cos 2(\omega_m T_0 + \beta_m) \right. \\ \left. + \Lambda^{(2)}_{mmm} \right] a_m^2 + O\left(\varepsilon^3\right) & , \quad n = m \\ 0 & , \quad n \neq m \end{cases} \tag{7.2.39}$$

where a_m and β_m are given by (7.2.38) and (7.2.39).

Readers are invited to complete Exercise 7.2(e).

7.3 Exercise 7.3 (Longitudinal Oscillation Analysis of Uniform, Nonlinear Elastic Rods)

Solution: **(a)** The corresponding linear undamped free oscillation equation to the Eq. (7.2.1) is

$$\frac{\partial^2 u}{\partial t^2} - c^2 \frac{\partial^2 u}{\partial x^2} = 0 \tag{7.3.1}$$

Let the solution of the Eq. (7.2.1) be

$$u = \phi(x) e^{i\omega t} \tag{7.3.2}$$

Substituting the above equation into (7.3.1), we obtain

$$\phi'' + \frac{\omega^2}{c^2} \phi = 0 \tag{7.3.3}$$

therefore,

$$\phi(x) = A \sin kx + B \cos kx, \quad k = \omega / c \tag{7.3.4}$$

From the boundary conditions, we have

$$\begin{aligned} & - h_1 k A + B = 0 \\ & (\sin k + h_2 k \cos k) A + (\cos k - h_2 k \sin k) B = 0 \end{aligned} \tag{7.3.5}$$

From the non-trivial solution condition, we can obtain the eigen equation

$$\left(1 - h_1 h_2 k^2\right) \tan k + (h_1 + h_2) k = 0 \tag{7.3.6}$$

From the first equation of (7.3.5), we have

$$B = h_1 k A \tag{7.3.7}$$

So the mode shape function can be written as

$$\phi(x) = A(\sin kx + h_1 k \cos kx) \tag{7.3.8}$$

After finding the eigenvalue k_m from the Eq. (7.3.6), we can write the corresponding mode shape function as

$$\phi_m(x) = c_m(\sin k_m x + h_1 k_m \cos k_m x), \quad \omega_m = ck_m \tag{7.3.9}$$

where c_m is chosen to normalize ϕ_m.

Let k and $2k$ be the two roots of the Eq. (7.3.6), then we have

$$\begin{aligned} &\left(1 - h_1 h_2 k^2\right) \tan k + (h_1 + h_2)k = 0 \\ &\left(1 - 4h_1 h_2 k^2\right) \tan 2k + 2(h_1 + h_2)k = 0 \end{aligned} \tag{7.3.10}$$

Combine those two equations, we have

$$\left(1 - h_1 h_2 k^2\right)\tan^2 k = 3h_1 h_2 k^2 \tag{7.3.11}$$

Eliminating $\tan k$ from the first equation of (7.3.10) and (7.3.11) yields

$$3h_1^2 h_1^2 k^4 + \left(h_1^2 + h_1^2 - h_1 h_2\right)k^2 = 0 \tag{7.3.12}$$

therefore

$$k = 0 \quad k^2 = -\frac{h_1^2 + h_1^2 - h_1 h_2}{3h_1^2 h_1^2} < 0 \tag{7.3.13}$$

Therefore, the Eq. (7.3.6) does not have non-trivial real roots such as k and $2k$, and thus does not have a natural frequency that is twice of another natural frequency.

When $h_1 = 1$ and $h_2 = 1.51883$, the eigen Eq. (7.3.6) becomes

$$\left(1 - 1.51883k^2\right)\tan k = -2.51883k \tag{7.3.14}$$

The first five roots of the Eq. (7.3.14) are

$$k_0 = 0, \quad k_1 = 1.19812, \quad k_2 = 3.59437, \quad k_3 = 6.53570, \quad k_4 = 9.59717, \quad k_5 = 12.69714$$

It can be seen that the non-trivial root $k_2/k_1 = 3.59437/1.19812 = 3.000008 \approx 3$, and hence $\omega_2/\omega_1 = k_2/k_1 \approx 3$.

Rewrite the nonlinear Eq. (7.3.1) as follows:

$$\frac{\partial^2 u}{\partial t^2} - c^2 \frac{\partial^2 u}{\partial x^2} = 2c^2 E_1 \frac{\partial u}{\partial x}\frac{\partial^2 u}{\partial x^2} + 3c^2 E_2 \left(\frac{\partial u}{\partial x}\right)^2 \frac{\partial^2 u}{\partial x^2} - 2\mu \frac{\partial u}{\partial t} + F(x, t) \tag{7.3.15}$$

With the present boundary conditions, we let the solution of Eq. (7.3.15) be

$$u=\sum_{n=0}^{N}\psi_n(t)\phi_n(x)=\sum_{n=0}^{N}c_n\psi_n(t)(\sin k_nx+h_1k_n\cos k_nx) \tag{7.3.16}$$

Substituting this into (7.3.15), we obtain

$$\begin{aligned}
&\sum_{p}\ddot{\psi}_p\phi_p(x)+c^2\sum_{p=0}^{N}k_p^2\psi_p\phi_p(x)\\
&=-2c^2E_1\sum_{p,q=0}^{N}k_q^2\psi_p\psi_q\phi_p'(x)\phi_q(x)\\
&\quad-3c^2E_2\sum_{p,\,q,\,r=0}^{N}k_r^2\psi_p\psi_q\psi_r\phi_p'(x)\phi_q'(x)\phi_r(x)\\
&\quad-2\mu\sum_{p=0}^{N}\dot{\psi}_p\phi_p(x)+F(x,t)
\end{aligned} \tag{7.3.17}$$

Multiplying both sides of the above equation by $\phi_n(x)$ and making the integration of x over [0, 1] yields

$$\ddot{\psi}_n+\omega_n^2\psi_n=-2\mu_n\dot{\psi}_n+c^2\sum_{p,q}^{N}\Gamma_{npq}\psi_p\psi_q+c^2\sum_{p,q,r}^{N}\Gamma_{npqr}\psi_p\psi_q\psi_r+f_n(t) \tag{7.3.18}$$

where

$$\begin{aligned}
\Gamma_{npq}&=-2E_1\int_0^1k_q^2\phi_n(x)\phi_p'(x)\phi_q(x)dx\\
\Gamma_{npqr}&=-3E_2\int_0^1k_r^2\phi_n(x)\phi_p'(x)\phi_q'(x)\phi_r(x)dx\\
f_n(x,t)&=\int_0^1\phi_n(x)F(x,t)dx
\end{aligned} \tag{7.3.19}$$

When $f_n(t)=K_n\cos\Omega t$, the Eq. (7.2.18) becomes

$$\ddot{\psi}_n + \omega_n^2 \psi_n = -2\mu_n \dot{\psi}_n + c^2 \sum_{p,q}^{N} \Gamma_{npq} \psi_p \psi_q + c^2 \sum_{p,q,r}^{N} \Gamma_{npqr} \psi_p \psi_q \psi_r + K_n \cos \Omega t \tag{7.3.20}$$

In the following part, we find the solution of Eq. (7.3.20) for the case of $\Omega \approx \omega_1,\ \omega_2 \approx 3\omega_1$. Readers are invited to find the solution for the case of $\Omega \approx \omega_2,\ \omega_2 \approx 3\omega_1$.

Since $\omega_2 \approx 3\omega_1$, the internal resonance excitation will be generated by a third order nonlinear term, and Eq. (7.3.20) needs to be expanded to the third order. Let the solution to Eq. (7.3.20) be

$$\psi_n = \varepsilon \psi_{n1}(T_0, T_1, T_2) + \varepsilon^2 \psi_{n2}(T_0, T_1, T_2) + \varepsilon^3 \psi_{n3}(T_0, T_1, T_2) + \cdots \tag{7.3.21}$$

In order to make the damping, nonlinear and excitation terms appearing simultaneously in the third-order equation, we let

$$\mu_n = \varepsilon^2 \hat{\mu}_n, \quad K_n = 2\varepsilon^3 k_n \tag{7.3.22}$$

Substituting (7.3.21) and (7.3.22) into (7.3.20) and retaining to $O(\varepsilon^3)$, we obtain

$$\begin{aligned}
0 &= \ddot{\psi}_n + \omega_n^2 \psi_n + 2\mu_n \dot{\psi}_n - c^2 \sum_{p,q}^{N} \Gamma_{npq} \psi_p \psi_q - c^2 \sum_{p,q,r}^{N} \Gamma_{npqr} \psi_p \psi_q \psi_r - K_n \cos \Omega t \\
&= \left[D_0^2 + 2\varepsilon D_0 D_1 + \varepsilon^2 \left(D_1^2 + 2D_0 D_2\right)\right]\left(\varepsilon \psi_{n1} + \varepsilon^2 \psi_{n2} + \varepsilon^3 \psi_{n3}\right) \\
&\quad + \omega_n^2 \left(\varepsilon \psi_{n1} + \varepsilon^2 \psi_{n2} + \varepsilon^3 \psi_{n3}\right) \\
&\quad + 2\varepsilon^2 \hat{\mu}_n \left(D_0 + \varepsilon D_1 + \varepsilon^2 D_2\right)\left(\varepsilon \psi_{n1} + \varepsilon^2 \psi_{n2} + \varepsilon^3 \psi_{n3}\right) \\
&\quad - c^2 \sum_{p,q} \Gamma_{npq} \left(\varepsilon \psi_{p1} + \varepsilon^2 \psi_{p2} + \varepsilon^3 \psi_{p3}\right)\left(\varepsilon \psi_{q1} + \varepsilon^2 \psi_{q2} + \varepsilon^3 \psi_{q3}\right) \\
&\quad - c^2 \sum_{p,q,r} \Gamma_{npqr} \left(\varepsilon \psi_{p1} + \varepsilon^2 \psi_{p2} + \varepsilon^3 \psi_{p3}\right)\left(\varepsilon \psi_{q1} + \varepsilon^2 \psi_{q2} + \varepsilon^3 \psi_{q3}\right) \\
&\quad \times \left(\varepsilon \psi_{r1} + \varepsilon^2 \psi_{r2} + \varepsilon^3 \psi_{r3}\right) - 2\varepsilon^3 k_n \cos \Omega T_0 + \cdots \\
&= \varepsilon \left(D_0^2 \psi_{n1} + \omega_n^2 \psi_{n1}\right) + \varepsilon^2 \left(D_0^2 \psi_{n2} + \omega_n^2 \psi_{n2} + 2D_0 D_1 \psi_{n1} - c^2 \sum_{p,q} \Gamma_{npq} \psi_{p1} \psi_{q1}\right) \\
&\quad + \varepsilon^3 [D_0^2 \psi_{n3} + \omega_n^2 \psi_{n3} + \left(D_1^2 + 2D_0 D_2\right)\psi_{n1} + 2\hat{\mu}_n D_0 \psi_{n1} + 2D_0 D_1 \psi_{n2} \\
&\quad - c^2 \sum_{p,q} \Gamma_{npq} \left(\psi_{p1} \psi_{q2} + \psi_{p2} \psi_{q1}\right) - c^2 \sum_{p,q,r} \Gamma_{npqr} \psi_{p1} \psi_{q1} \psi_{r1} - 2k_n \cos \Omega T_0] + \cdots
\end{aligned} \tag{7.3.23}$$

Let the coefficient of the same power of ε in Eq. (7.3.23) be zero, we obtain
Order ε:

$$D_0^2\psi_{n1} + \omega_n^2\psi_{n1} = 0 \tag{7.3.24}$$

Order ε^2:

$$D_0^2\psi_{n2} + \omega_n^2\psi_{n2} = -2D_0D_1\psi_{n1} + c^2\sum_{p,q}\Gamma_{npq}\psi_{p1}\psi_{q1} \tag{7.3.25}$$

Order ε^3:

$$\begin{aligned} D_0^2\psi_{n3} + \omega_n^2\psi_{n3} = & -\left(D_1^2 + 2D_0D_2\right)\psi_{n1} - 2D_0D_1\psi_{n2} - 2\hat{\mu}_nD_0\psi_{n1} \\ & + c^2\sum_{p,q}\Gamma_{npq}\left(\psi_{p1}\psi_{q2} + \psi_{p2}\psi_{q1}\right) \\ & + c^2\sum_{p,q,r}\Gamma_{npqr}\psi_{p1}\psi_{q1}\psi_{r1} + 2k_n\cos\Omega T_0 \end{aligned} \tag{7.3.26}$$

The solution of Eq. (7.3.24) is

$$\psi_{n1} = A_ne^{i\omega_nT_0} + cc \tag{7.3.27}$$

where $A_n = A_n(T_1, T_2)$. Substituting (7.3.27) into (7.3.25), we obtain

$$\begin{aligned} D_0^2\psi_{n2} + \omega_n^2\psi_{n2} = & -2i\omega_nD_1A_ne^{i\omega_nT_0} \\ & + c^2\sum_{p,q}\Gamma_{npq}\left(A_pA_qe^{i(\omega_q+\omega_p)T_0} + \overline{A}_pA_qe^{i(\omega_q-\omega_p)T_0}\right) + cc \end{aligned} \tag{7.3.28}$$

In order to eliminate secular terms from the above equation, we need

$$D_1A_n = 0 \rightarrow \quad A_n = A_n(T_2) \tag{7.3.29}$$

From (7.3.28), we can obtain

$$\psi_{n2} = c^2\sum_{p,q}\Gamma_{npq}\left(\Lambda_{npq}^{(1)}A_pA_qe^{i(\omega_q+\omega_p)T_0} + \Lambda_{npq}^{(2)}\overline{A}_pA_qe^{i(\omega_q-\omega_p)T_0}\right) + cc \tag{7.3.30}$$

where

$$\Lambda_{npq}^{(1)} = \left[\omega_n^2 - \left(\omega_q + \omega_p\right)^2\right]^{-1}, \quad \Lambda_{npq}^{(2)} = \left[\omega_n^2 - \left(\omega_q - \omega_p\right)^2\right]^{-1} \tag{7.3.31}$$

Substituting (7.3.27), (7.3.29) and (7.3.30) into (7.3.26) yields

$$\begin{aligned} & D_0^2\psi_{n3} + \omega_n^2\psi_{n3} \\ & = -2i\omega_nA_n'e^{i\omega_nT_0} - 2i\omega_n\hat{\mu}_nA_ne^{i\omega_nT_0} \end{aligned}$$

$$+c^4\sum_{p,q}\Gamma_{npq}\sum_{j,k}\Gamma_{qjk}\Big(\Lambda^{(1)}_{qjk}A_pA_jA_ke^{i(\omega_p+\omega_k+\omega_j)T_0}+\Lambda^{(2)}_{qjk}A_p\overline{A}_jA_ke^{i(\omega_p+\omega_k-\omega_j)T_0}\Big)$$
$$+c^4\sum_{p,q}\Gamma_{npq}\sum_{j,k}\Gamma_{qjk}\Big(\Lambda^{(1)}_{qjk}\overline{A}_pA_jA_ke^{i(\omega_k+\omega_j-\omega_p)T_0}+\Lambda^{(2)}_{qjk}\overline{A}_p\overline{A}_jA_ke^{i(\omega_k-\omega_j-\omega_p)T_0}\Big)$$
$$+c^4\sum_{p,q}\Gamma_{npq}\sum_{j,k}\Gamma_{pjk}\Big(\Lambda^{(1)}_{pjk}A_qA_jA_ke^{i(\omega_q+\omega_k+\omega_j)T_0}+\Lambda^{(2)}_{pjk}A_q\overline{A}_jA_ke^{i(\omega_q+\omega_k-\omega_j)T_0}\Big)$$
$$+c^4\sum_{p,q}\Gamma_{npq}\sum_{j,k}\Gamma_{pjk}\Big(\Lambda^{(1)}_{pjk}\overline{A}_qA_jA_ke^{i(\omega_k+\omega_j-\omega_q)T_0}+\Lambda^{(2)}_{pjk}\overline{A}_q\overline{A}_jA_ke^{i(\omega_k-\omega_j-\omega_q)T_0}\Big)$$
$$+c^2\sum_{p,q,r}\Gamma_{npqr}\Big(A_pA_qA_re^{i(\omega_p+\omega_q+\omega_r)T_0}+\overline{A}_pA_qA_re^{i(\omega_q+\omega_r-\omega_p)T_0}\Big)$$
$$+c^2\sum_{p,q,r}\Gamma_{npqr}\Big(A_p\overline{A}_qA_re^{i(\omega_p-\omega_q+\omega_r)T_0}+\overline{A}_p\overline{A}_qA_re^{i(\omega_r-\omega_p-\omega_q)T_0}\Big)+k_ne^{i\Omega T_0}+cc$$

The prime represents the derivative with respect to T_2. The above equation can be rewritten as

$$\begin{aligned}D_0^2\psi_{n3}+\omega_n^2\psi_{n3}&=-2i\omega_nA_n'e^{i\omega_nT_0}-2i\omega_n\hat{\mu}_nA_ne^{i\omega_nT_0}+A_ne^{i\omega_nT_0}\sum_pE_{np}A_p\overline{A}_p\\&\quad+G_n\overline{A}_n^2A_{n+1}e^{i(\omega_{n+1}-2\omega_n)T_0}+H_{n-1}A_{n-1}^3e^{3i\omega_{n-1}T_0}+k_ne^{i\Omega T_0}+cc+NST\end{aligned}\tag{7.3.32}$$

where

$$\begin{aligned}E_{np}&=c^2\big(\Gamma_{nppn}+\Gamma_{npnp}+\Gamma_{npqp}+\Gamma_{nnpp}+\Gamma_{nppn}+\Gamma_{nqpp}\big)\\&\quad+c^4\sum_q\Big(\Gamma_{npq}\Gamma_{qnp}\Lambda^{(1)}_{qnp}+\Gamma_{npq}\Gamma_{qpn}\Lambda^{(1)}_{qpn}+\Gamma_{nqp}\Gamma_{qnp}\Lambda^{(1)}_{qnp}+\Gamma_{nqp}\Gamma_{qpn}\Lambda^{(1)}_{qpn}\\&\quad+\Gamma_{npq}\Gamma_{qpn}\Lambda^{(2)}_{qpn}+\Gamma_{nnq}\Gamma_{qpp}\Lambda^{(2)}_{qpp}+\Gamma_{nnq}\Gamma_{qpp}\Lambda^{(2)}_{qpp}+\Gamma_{npq}\Gamma_{qnp}\Lambda^{(2)}_{qnp}\\&\quad+\Gamma_{nqp}\Gamma_{qpn}\Lambda^{(2)}_{qpn}+\Gamma_{nqn}\Gamma_{qpp}\Lambda^{(2)}_{qpp}+\Gamma_{nqp}\Gamma_{qnp}\Lambda^{(2)}_{qnp}+\Gamma_{nqn}\Gamma_{qpp}\Lambda^{(2)}_{qpp}\Big)\\G_n&=c^2\big(\Gamma_{n(n+1)nn}+\Gamma_{nn(n+1)n}+\Gamma_{nn(n+1)}\big)\\&\quad+c^4\sum_p\Big(\Gamma_{nnp}\Gamma_{p(n+1)n}\Lambda^{(2)}_{p(n+1)n}+\Gamma_{n(n+1)p}\Gamma_{pnn}\Lambda^{(1)}_{pnn}+\Gamma_{nnp}\Gamma_{pn(n+1)}\Lambda^{(2)}_{pn(n+1)}\\&\quad+\Gamma_{npn}\Gamma_{p(n+1)n}\Lambda^{(2)}_{p(n+1)n}+\Gamma_{np(n+1)}\Gamma_{pnn}\Lambda^{(1)}_{pnn}+\Gamma_{npn}\Gamma_{pn(n+1)}\Lambda^{(2)}_{pn(n+1)}\Big)\\H_n&=c^2\Gamma_{nnnn}+c^4\sum_p\Big(\Gamma_{nnp}\Gamma_{pnn}\Lambda^{(1)}_{pnn}+\Gamma_{npn}\Gamma_{pnn}\Lambda^{(1)}_{pnn}\Big)\end{aligned}\tag{7.3.33}$$

Introduce the detuning parameters σ_1 and σ_2 such that

$$\Omega=\omega_1+\varepsilon^2\sigma_1,\quad\omega_2=3\omega_1+\varepsilon^2\sigma_2\tag{7.3.34}$$

Therefore, in order to eliminate secular terms in Eq. (7.3.32), we need

$$2i\omega_1 A_1' + 2i\omega_1\hat{\mu}_1 A_1 - A_1\sum_p E_{1p}A_p\overline{A}_p - G_1\overline{A}_1^2 A_2 e^{i\sigma_2 T_2} - k_1 e^{i\sigma_1 T_2} = 0 \quad (n=1) \tag{7.3.35}$$

$$2i\omega_2 A_2' + 2i\omega_2\hat{\mu}_2 A_2 - A_2\sum_p E_{2p}A_p\overline{A}_p - H_1 A_1^3 e^{-i\sigma_2 T_2} = 0 \quad (n=2) \tag{7.3.36}$$

$$2i\omega_n A_n' + 2i\omega_n\hat{\mu}_n A_n - A_n\sum_p E_{np}A_p\overline{A}_p = 0 \quad (n \geq 3) \tag{7.3.37}$$

Let

$$A_n = \frac{1}{2}a_n e^{i\beta_n} \tag{7.3.38}$$

and substitute (7.3.38) into (7.3.37), we can obtain

$$i\omega_n a_n' - \omega_n a_n\beta_n' + i\omega_n\hat{\mu}_n a_n - \frac{1}{8}a_n\sum_p E_{np}a_p^2 = 0 \tag{7.3.39}$$

Separating the real and imaginary parts of the above equation yields

$$a_n' + \hat{\mu}_n a_n = 0 \tag{7.3.40}$$

$$\omega_n a_n\beta_n' + \frac{1}{8}a_n\sum_p E_{np}a_p^2 = 0 \tag{7.3.41}$$

Equation (7.3.4) can be solved by

$$a_n = a_{n0}e^{-\hat{\mu}_n T_2}, \quad n \geq 3 \tag{7.3.42}$$

Since $\hat{\mu}_n > 0$, $a_n \to 0$ for all $n \geq 3$.

Substituting (7.3.38) into (7.3.37) and (7.3.36), and taking (7.3.42) into account, we obtain

$$\begin{aligned} &i\omega_1 a_1' - \omega_1 a_1\beta_1' + i\omega_1\hat{\mu}_1 a_1 - \frac{1}{8}a_1\left(E_{11}a_1^2 + E_{12}a_2^2\right) \\ &-\frac{1}{8}G_1 a_1^2 a_2 e^{i(\sigma_2 T_2 - 3\beta_1 + \beta_2)} - k_1 e^{i(\sigma T_2 - \beta_1)} = 0 \end{aligned} \tag{7.3.43}$$

$$\begin{aligned} &i\omega_2 a_2' - \omega_2 a_2\beta_2' + i\omega_2\hat{\mu}_2 a_2 - \frac{1}{8}a_2\left(E_{21}a_1^2 + E_{22}a_2^2\right) \\ &-\frac{1}{8}H_1 a_1^3 e^{-i(\sigma_2 T_2 - 3\beta_1 + \beta_2)} = 0 \end{aligned} \tag{7.3.44}$$

Let

$$\gamma_1 = \sigma_1 T_2 - \beta_1, \quad \gamma_2 = \sigma_2 T_2 - 3\beta_1 + \beta_2 \tag{7.3.45}$$

and separate the real and imaginary parts from (7.3.43) and (7.3.44), we obtain

$$\begin{aligned} a_1' &= -\hat{\mu}_1 a_1 + \frac{1}{8}\omega_1^{-1} G_1 a_1^2 a_2 \sin\gamma_2 + \omega_1^{-1} k_1 \sin\gamma_1 \\ a_1\gamma_1' &= \sigma_1 a_1 + \frac{1}{8}\omega_1^{-1} G_1 a_1^2 a_2 \cos\gamma_2 + \omega_1^{-1} k_1 \cos\gamma_1 \end{aligned} \tag{7.3.46}$$

$$\begin{aligned} a_2' &= -\hat{\mu}_2 a_2 - \frac{1}{8}\omega_2^{-1} H_1 a_1^3 \sin\gamma_2 \\ 3a_2\gamma_1' - a_2\gamma_2' &= -a_2\sigma_2 + 3a_2\sigma_1 + \frac{1}{8}\omega_2^{-1} a_2\left(E_{21}a_1^2 + E_{22}a_2^2\right) + \frac{1}{8}\omega_2^{-1} H_1 a_1^3 \sin\gamma_2 \end{aligned} \tag{7.3.47}$$

From the above results, the approximate solution of Eq. (7.3.20) for the case of $\Omega \approx \omega_1, \omega_2 \approx 3\omega_1$ is

$$\begin{aligned} \psi_1 = {} & a_1\cos(\omega_1 t + \beta_1) + b_{11} + b_{12}\cos(2\omega_1 T_0 + 2\beta_1) \\ & + b_{13}\cos(2\omega_1 T_0 - \beta_1 + \beta_2) + b_{14}\cos(4\omega_1 T_0 + \beta_1 + \beta_2) \\ & + b_{15}\cos(6\omega_1 T_0 + 2\beta_2) \end{aligned} \tag{7.3.48}$$

$$\begin{aligned} \psi_2 = {} & a_2\cos(\omega_2 t + \beta_2) + b_{21} + b_{22}\cos(2\omega_1 T_0 + 2\beta_1) \\ & + b_{23}\cos(2\omega_1 T_0 - \beta_1 + \beta_2) + b_{24}\cos(4\omega_1 T_0 + \beta_1 + \beta_2) \\ & + b_{25}\cos(6\omega_1 T_0 + 2\beta_2) \end{aligned} \tag{7.3.49}$$

$$\psi_n = 0, n \geq 3 \tag{7.3.50}$$

where $a_1, a_2, \beta_1, \beta_2$ are given by (7.3.46) and (7.3.47) and

$$\begin{aligned} b_{n1} &= \frac{1}{2}c^2\left(\Gamma_{n11}\Lambda_{n11}^{(2)}a_1^2 + \Gamma_{n22}\Lambda_{n22}^{(2)}a_2^2\right) \\ b_{n2} &= \frac{1}{2}c^2\Gamma_{n11}\Lambda_{n11}^{(1)}a_1^2 \\ b_{n3} &= \frac{1}{2}c^2\left(\Gamma_{n12}\Lambda_{n12}^{(2)} + \Gamma_{n21}\Lambda_{n2}^{(2)}\right)a_1 a_2 \\ b_{n4} &= \frac{1}{2}c^2\left(\Gamma_{n12}\Lambda_{n12}^{(1)} + \Gamma_{n21}\Lambda_{n21}^{(1)}\right)a_1 a_2 \\ b_{n5} &= \frac{1}{2}c^2\Gamma_{n22}\Lambda_{n22}^{(1)}a_2^2(6\omega_1 T_0 + 2\beta_2); \quad n = 1, 2 \end{aligned} \tag{7.3.51}$$

7.4 Exercise 7.4 (Nonlinear Analysis of Transverse Oscillations of a Fixed Wire)

Solution: **(a)** The deformed configuration of a tensioned string and the initial coordinate system x_1x_2 are given in Fig. 7.2(a). Point P_0 denotes the unloaded position of the observed string mass. Point $\hat{P}$ denotes the post-deformation position reached by P_0 with static pre-tension, which changes the undeformed length of the string l to $l + \Delta l$. Point P denotes the post-deformation position reached by P_0 under a combination of static pre-tension and dynamic loading.

Here, we denote by s the undeformed length of the string from the point O to P_0, by x the length of the same part after the static deformation, and by $\tilde{s}$ the dynamic arc length at the same point. Also, we denote the unloaded mass density of the string by $\rho_0(s)$ and the cross-sectional area of the unloaded string by $A_0(s)$, such that $()' = \partial()/\partial s$. The coordinates of the point $\hat{P}$ are $(x, 0)$, where $x = s + \int_0^s e_0 ds$, e_0 is the static axial strain from pre-tension, so $\Delta l = \int_0^l e_0 ds$. We use (x_1, x_2) to denote the coordinates of the point P, and u, w to denote the dynamic displacements along the direction x_1 and x_2, respectively. Therefore, we have

$$x_1 = x + u, \quad x_2 = w \tag{7.4.1}$$

According to the free-body diagram shown in Fig. 7.2(b), we can obtain the axial strain of the string

$$e = \sqrt{(1 + e_0 + u')^2 + w'^2} - 1 \tag{7.4.2}$$

Due to Poisson effect, the deformed cross-sectional area is

$$A = (1 - \nu e)^2 A_0 \tag{7.4.3}$$

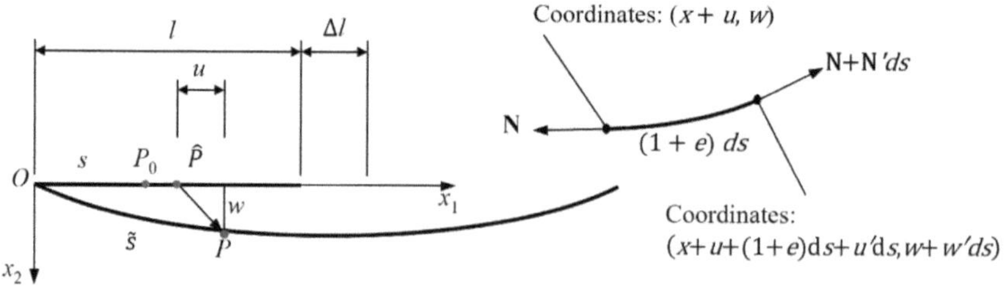

Fig. 7.2 String model: **a** Configuration and coordinate of a tensioned string; **b** free-body diagram of a string element for Exercise 7.4

where ν is the Poisson ratio. Since the string is very thin, we can assume that the stresses of each cross section are approximately uniform. Furthermore, the string is assumed to be linearly elastic. Applying Hooke's law and (7.4.3), we can obtain the internal tension of the string

$$N(s,t) = EAe = EA_0(1-\nu e)^2 e \tag{7.4.4}$$

where E is the Young's modulus.

From Fig. 7.2(b) and Newton's second law, we have

$$m_0(\ddot{u}\mathbf{i}_1 + \ddot{w}\mathbf{i}_2) + \tilde{\mu}_1\dot{u}\mathbf{i}_1 + \tilde{\mu}_2\dot{w}\mathbf{i}_2 = \mathbf{N}' + \tilde{f}_1\mathbf{i}_1 + \tilde{f}_2\mathbf{i}_2 \tag{7.4.5}$$

where $m_0 = \rho_0 A_0$, $\tilde{\mu}_i$ is the damping coefficient per unit length of the undeformed string along the direction of x_i, $\tilde{f}_i$ is the dynamic load per unit length of the undeformed string along the direction of x_i, and $\mathbf{N}$ is the internal force in the cross-section, which has the following form

$$\mathbf{N} = \frac{N}{1+e}\left[\left(1+e_0+u'\right)\mathbf{i}_1 + w'\mathbf{i}_2\right] \tag{7.4.6}$$

Substituting (7.4.6) into (7.4.5), we can obtain the governing equation of motion for the string.

$$m_0\ddot{u} + \tilde{\mu}_1\dot{u} = \frac{\partial}{\partial s}\left[\frac{N}{1+e}\left(1+e_0+u'\right)\right] + \tilde{f}_1 \tag{7.4.7}$$

$$m_0\ddot{w} + \tilde{\mu}_2\dot{w} = \frac{\partial}{\partial s}\left[\frac{N}{1+e}w'\right] + \tilde{f}_2 \tag{7.4.8}$$

Boundary conditions are specified at $s = 0$ and l for the displacement or tension:

$$u \quad \text{or} \quad \frac{N\left(1+e_0+u'\right)}{1+e} \tag{7.4.9}$$

$$w \quad or \quad \frac{Nw'}{1+e} \tag{7.4.10}$$

In static equilibrium, $u = v = w = 0, \quad \tilde{f}_i = 0, \quad e = e_0$, from (7.4.7) we have

$$\frac{dN_0}{ds} = 0 \tag{7.4.11}$$

where N_0 is the pre-tension in the string, which can be determined by (7.4.11):

$$N_0 = EA_0(1-\nu e_0)^2 e_0 = \text{constant} \tag{7.4.12}$$

We note that A_0 and e_0 may be functions of s and e_0 may be large strain.

Equations (7.4.7)–(7.4.10) form a complete set of governing equations for the boundary value problem for strings, and they are completely nonlinear. Ignoring the Poisson effect, the Eq. (7.4.4) becomes $N(s,t) = EA_0 e$, and thus the Eqs. (7.4.7) and (7.4.8) become

$$m_0\ddot{u} + \tilde{\mu}_1\dot{u} = \frac{\partial}{\partial s}\left[\frac{EA_0 e\left(1 + e_0 + u'\right)}{1+e}\right] + \tilde{f}_1 \tag{7.4.13}$$

$$m_0\ddot{w} + \tilde{\mu}_2\dot{w} = \frac{\partial}{\partial s}\left[\frac{EA_0 e w'}{1+e}\right] + \tilde{f}_2 \tag{7.4.14}$$

The boundary conditions are specified at $s = 0$ and l for the displacement or tension:

$$u \quad \text{or} \quad \frac{EA_0 e\left(1 + e_0 + u'\right)}{1+e} \tag{7.4.15}$$

$$w \quad \text{or} \quad \frac{EA_0 e w'}{1+e} \tag{7.4.16}$$

In the following, we transfer the independent variable from the undeformed coordinate s to the static equilibrium coordinate x. Because

$$x = s + \int_0^s e_0 ds$$

then

$$dx = (1 + e_0)ds = \alpha ds$$

where $\alpha = 1 + e_0$. From this, Eqs. (7.4.13) and (7.4.14) become

$$m_0\ddot{u} + \tilde{\mu}_1\dot{u} = \alpha\frac{\partial}{\partial x}\left[\frac{\alpha EA_0 e}{1+e}\left(1 + \frac{\partial u}{\partial x}\right)\right] + \tilde{f}_1 \tag{7.4.17}$$

$$m_0\ddot{w} + \tilde{\mu}_2\dot{w} = \alpha\frac{\partial}{\partial x}\left[\frac{\alpha EA_0 e}{1+e}\frac{\partial w}{\partial x}\right] + \tilde{f}_2 \tag{7.4.18}$$

From (7.4.2), we have

$$e = \alpha\sqrt{(1 + u_x)^2 + w_x^2} - 1 \tag{7.4.19}$$

therefore

$$\frac{\alpha e}{1+e} = \alpha - 1 + u_x - u_x^2 + \frac{1}{2}w_x^2 + u_x^3 - \frac{3}{2}u_x w_x^2 + \cdots \tag{7.4.20}$$

Substituting (7.4.20) into (7.4.17) and (7.4.18) and only keeping to the third-order small quantities yields

$$m_0\ddot{u} + \tilde{\mu}_1\dot{u} - \alpha^2 EA_0 u_{xx} = \alpha EA_0 \frac{\partial}{\partial x}\left[\left(\frac{1}{2} - u_x\right)w_x^2\right] + \tilde{f}_1 \tag{7.4.21}$$

$$m_0\ddot{w} + \tilde{\mu}_2\dot{w} - \alpha(\alpha-1)EA_0 w_{xx} = \alpha EA_0 \frac{\partial}{\partial x}\left[w_x\left(u_x - u_x^2 + \frac{1}{2}w_x^2\right)\right] + \tilde{f}_2 \tag{7.4.22}$$

Dividing the above two equations by $m_0 = \rho_0 A_0$, we can obtain the approximate equations ignoring the Poisson effect

$$u_{tt} + \mu_1 u_t - c_1^2 u_{xx} = \left(c_1^2 - c_2^2\right)\frac{\partial}{\partial x}\left[\left(\frac{1}{2} - u_x\right)w_x^2\right] + f_1 \tag{7.4.23}$$

$$w_{tt} + \mu_2 w_t - c_2^2 w_{xx} = \left(c_1^2 - c_2^2\right)\frac{\partial}{\partial x}\left[w_x\left(u_x - u_x^2 + \frac{1}{2}w_x^2\right)\right] + f_2 \tag{7.4.24}$$

where $\mu_n = \tilde{\mu}_n/m_0$, $f_n = \tilde{f}_n/m_0$.

$$c_1^2 = \frac{EA_0\alpha^2}{m_0} = \frac{EA_0\alpha^2}{\rho_0 A_0} = \frac{E\alpha^2}{\rho_0} \tag{7.4.25}$$

$$c_2^2 = \frac{EA_0\alpha(\alpha-1)}{m_0} = \frac{EA_0\alpha e_0}{\rho_0 A_0} = \frac{N_0\alpha}{\rho_0 A_0} \tag{7.4.26}$$

Substituting (7.4.20) into (7.4.15) and (7.4.16) yields the boundary conditions for the approximate equations specified at $s = 0$ and l for the displacement or tension:

$$u \text{ or } EA_0\left[\alpha u_x + \left(\frac{1}{2} - u_x\right)\left(v_x^2 + w_x^2\right)\right] \tag{7.4.27}$$

$$w \text{ or } EA_0 w_x\left[\alpha - 1 + u_x - u_x^2 + \frac{1}{2}\left(v_x^2 + w_x^2\right)\right] \tag{7.4.28}$$

In the following, we will combine two approximate Eqs. (7.4.23) and (7.4.24) into one equation. To do this, we need to first consider the corresponding linear undamped free oscillation problem.

Omitting the damping, excitation and nonlinear terms in (7.4.23)–(7.4.24), we obtain the following uncoupled equations

$$u_{tt} - c_1^2 u_{xx} = 0 \tag{7.4.29}$$

$$w_{tt} - c_2^2 w_{xx} = 0 \tag{7.4.30}$$

We use the boundary conditions

$$u = w = 0 \text{ at } x = 0 \text{ and } l \tag{7.4.31}$$

Combining (7.4.29) and (7.4.31), we can obtain the mode shape of the motion along axial direction

$$\varphi_m(x) = \sin\frac{m\pi x}{l} \tag{7.4.32}$$

The corresponding natural frequency is

$$\omega_m = \frac{m\pi c_1}{l} \tag{7.4.33}$$

where m is a positive integer. Similarly, the normalized mode shape functions for transverse oscillation of the string are

$$\phi_n(x) = \sqrt{2}\sin\frac{n\pi x}{l} \tag{7.4.34}$$

The corresponding natural frequency is

$$\omega_n = \frac{n\pi c_2}{l} \tag{7.4.35}$$

where m and k are positive integers. Since

$$\frac{c_1}{c_2} = \sqrt{\frac{\alpha}{\alpha - 1}} = \sqrt{\frac{1 + e_0}{e_0}} \tag{7.4.36}$$

For typical metallic wires within the elastic range, this is a dimensionless large value. So, for a given frequency order (i.e., $m = n$), the transverse frequency is much smaller than the longitudinal frequency of the wire from (7.4.33) to (7.4.35). Therefore, if the excitation frequency is much smaller than the longitudinal fundamental frequency $\pi c_1/l$, then the longitudinal inertia in Eq. (7.4.23) can be omitted. To illustrate this point, we apply a displacement excitation, whose frequency is close to the mth order transverse fundamental frequency, at the right end of the wire. We normalize the displacement and time by l and $l/n\pi c_2$ in Eq. (7.4.23). Substitute

$$\xi = \frac{x}{l}, \quad \tau = \frac{n\pi c_2 t}{l}, \quad U = \frac{u}{l}, \quad W = \frac{w}{l}$$

into the Eq. (7.4.23), we obtain

$$\frac{n^2\pi^2c_2^2}{c_1^2}U_{\tau\tau} + \frac{\mu_1 n\pi c_2}{c_1^2}U_\tau - U_{\xi\xi} = \frac{c_1^2 - c_2^2}{c_1^2}\frac{\partial}{\partial\xi}\left[\left(\frac{1}{2} - U_\xi\right)W_\xi^2\right] \quad (7.4.37)$$

where the longitudinally distributed excitation $f_1(x, t) = 0$ is assumed. For a typical metallic wire within the elastic range, $c_2^2/c_1^2 << 1.$, so the terms containing $U_{\tau\,\tau}$ and U_τ in the above equation can be omitted, and the equation becomes

$$U_{\xi\xi} \approx -\frac{\partial}{\partial\xi}\left[\left(\frac{1}{2} - U_\xi\right)W_\xi^2\right] \quad (7.4.38)$$

After integration, we obtain

$$U_\xi = -\left(\frac{1}{2} - U_\xi\right)W_\xi^2 + b(\tau) \quad (7.4.39)$$

where $b(\tau)$ is the integration constant. Omitting the third order term, we obtain

$$U_\xi = -\frac{1}{2}W_\xi^2 + b(\tau) \quad (7.4.40)$$

Since it was assumed earlier that the displacement excitation is applied to the right end of the wire, the boundary condition can be assumed to be

$$U(0, \tau) = 0, U(1, \tau) = P(\tau) \quad (7.4.41)$$

where $P(t)$ is the kinematic function of the right support. Integrate (7.4.40) and apply (7.4.41), we can obtain

$$U = -\frac{1}{2}\int_0^\xi W_\xi^2 d\xi + b(\tau)\xi \quad (7.4.42)$$

where

$$b(\tau) = \frac{1}{2}\int_0^1 W_\xi^2 d\xi + P(\tau) \quad (7.4.43)$$

Then (7.4.40) and (7.4.41) become

$$u_x = -\frac{1}{2}w_x^2 + b(t) \quad (7.4.44)$$

where

$$b(t) = \frac{1}{2l}\int_0^l w_x^2 dx + P(t) \tag{7.4.45}$$

Substituting (7.4.45) into (7.4.44), we obtain

$$u_x + \frac{1}{2}w_x^2 = \frac{1}{2l}\int_0^l w_x^2 dx + P(t) \tag{7.4.46}$$

Substituting (7.4.46) into (7.4.24), omitting the c_2^2 term from the terms of the same order containing c_1^2, c_2^2, and keeping the result to the third order, we obtain

$$w_{tt} + 2\hat{\mu}w_t - c_2^2 w_{xx} = c_1^2 w_{xx}P(t) + \frac{c_1^2}{2l}w_{xx}\int_0^l w_x^2 dx + F(x, t) \tag{7.4.47}$$

where

$$\mu_2 = 2\hat{\mu}, \quad f_2(x, t) = F(x, t) \tag{7.4.48}$$

When $P(t) = 0$, (7.4.46) becomes

$$u_x + \frac{1}{2}w_x^2 = \frac{1}{2l}\int_0^l w_x^2 dx \tag{7.4.49}$$

It can be seen that $u_x = O\left(w_x^2\right)$, then the Eq. (7.4.47) becomes

$$w_{tt} - c_2^2 w_{xx} = \frac{c_1^2}{2l}w_{xx}\int_0^l w_x^2 dx - 2\hat{\mu}w_t + F(x, t) \tag{7.4.50}$$

Equation (7.4.50) is the equation governing the transverse oscillations of a fixed wire.

Let the solution of Eq. (7.4.50) be

$$w(x, t) = \sqrt{2}\sum_{n=1}^{\infty}\psi_n(t)\sin\frac{n\pi x}{l} \tag{7.4.51}$$

Substituting (7.4.51) into(7.4.50), we obtain

$$\sum_{p=1}^{\infty}\sqrt{2}\,\ddot{\psi}_p(t)\sin\frac{p\pi x}{l}+c_2^2\sum_{p=1}^{\infty}\sqrt{2}\frac{p^2\pi^2}{l^2}\psi_p(t)\sin\frac{p\pi x}{l}$$

$$=-\frac{c_1^2}{2l}\sum_{p=1}^{\infty}\sqrt{2}\frac{p^2\pi^2}{l^2}\psi_p(t)\sin\frac{p\pi x}{l}\int_0^l\left[\sum_{q=1}^{\infty}\sqrt{2}\frac{q\pi}{l}\psi_p(t)\cos\frac{q\pi x}{l}\right]^2 dx$$

$$-2\hat{\mu}\sum_{p=1}^{\infty}\sqrt{2}\dot{\psi}_p(t)\sin\frac{p\pi x}{l}+F(x,t) \tag{7.4.52}$$

Multiplying both sides of the above equation by $\sqrt{2}\sin(n\pi x/l)$, integrating over $[0, l]$ and taking into account the orthogonality of the mode shape function, we obtain

$$\ddot{\psi}_n+\omega_n^2\psi_n=-\frac{c_1^2n^2\pi^4}{2l^4}\psi_n\sum_{m=1}^{\infty}m^2\psi_m^2-2\hat{\mu}_n\dot{\psi}_n+f_n(t) \tag{7.4.53}$$

where

$$\omega_n^2=\frac{c_2^2n^2\pi^2}{l^2},\quad f_n(t)=\frac{\sqrt{2}}{l}\int_0^l F(x,t)\sin\frac{n\pi x}{l}dx \tag{7.4.54}$$

When $f_n(t)=\delta_{1n}K_1\cos\Omega t$, the Eq. (7.4.54) becomes

$$\ddot{\psi}_n+\omega_n^2\psi_n=-\lambda_n^2\psi_n\sum_{m=1}^{\infty}m^2\psi_m^2-2\hat{\mu}_n\dot{\psi}_n+\delta_{1n}K_1\cos\Omega t \tag{7.4.55}$$

where

$$\lambda_n^2=\frac{c_1^2n^2\pi^4}{2l^4} \tag{7.4.56}$$

Since the Eq. (7.4.55) contains only cubic nonlinear terms, the solution of the equation needs to be expanded to the third order to find a first-order approximate solution. Let the solution of the equation be

$$\psi_n=\varepsilon\psi_{n1}(T_0,T_1,T_2)+\varepsilon^2\psi_{n2}(T_0,T_1,T_2)+\varepsilon^3\psi_{n3}(T_0,T_1,T_2)+\cdots \tag{7.4.57}$$

In order to make both the damping and nonlinear terms appearing in the third order equation, we let

$$\hat{\mu}_n=\varepsilon^2\tilde{\mu}_n \tag{7.4.58}$$

Substituting (7.4.57) and (7.4.58) into (7.4.55) and retaining to $O(\varepsilon^3)$, we obtain

$$
\begin{aligned}
0 &= \ddot{\psi}_n + \omega_n^2\psi_n + 2\hat{\mu}_n\dot{\psi}_n + \lambda_n^2\psi_n\sum_{m=1}^{\infty} m^2\psi_m^2 - \delta_{1n}K_1\cos\Omega t \\
&= \left[D_0^2 + 2\varepsilon D_0D_1 + \varepsilon^2\left(D_1^2 + 2D_0D_2\right)\right]\left(\varepsilon\psi_{n1} + \varepsilon^2\psi_{n2} + \varepsilon^3\psi_{n3}\right) \\
&\quad + \omega_n^2\left(\varepsilon\psi_{n1} + \varepsilon^2\psi_{n2} + \varepsilon^3\psi_{n3}\right) \\
&\quad + 2\varepsilon^2\tilde{\mu}_n\left(D_0 + \varepsilon D_1 + \varepsilon^2 D_2\right)\left(\varepsilon\psi_{n1} + \varepsilon^2\psi_{n2} + \varepsilon^3\psi_{n3}\right) \\
&\quad + \lambda_n^2\left(\varepsilon\psi_{n1} + \varepsilon^2\psi_{n2} + \varepsilon^3\psi_{n3}\right)\sum_{m=1}^{\infty} m^2\left(\varepsilon\psi_{m1} + \varepsilon^2\psi_{m2} + \varepsilon^3\psi_{m3}\right)^2 \\
&\quad - \delta_{1n}K_1\cos\Omega t + \cdots \\
&= \varepsilon\left(D_0^2\psi_{n1} + \omega_n^2\psi_{n1}\right) + \varepsilon^2\left(D_0^2\psi_{n2} + \omega_n^2\psi_{n2} + 2D_0D_1\psi_{n1}\right) \\
&\quad + \varepsilon^3\Bigg[D_0^2\psi_{n3} + \omega_n^2\psi_{n3} + \left(D_1^2 + 2D_0D_2\right)\psi_{n1} + 2D_0D_1\psi_{n2} \\
&\quad + 2\tilde{\mu}_n D_0\psi_{n1} + \lambda_n^2\psi_{n1}\sum_{m=1}^{\infty} m^2\psi_{m1}^2\Bigg] - \delta_{1n}K_1\cos\Omega t + \cdots
\end{aligned}
\tag{7.4.59}
$$

The order of the external excitation K_1 cosΩt in the above equation needs to be determined by Ω, i.e.,

$$
K_1 = \begin{cases} 2\varepsilon^3 k_1, & \Omega \approx \omega_1 \\ 2\varepsilon k_1, & \Omega \neq \omega_1 \end{cases}
\tag{7.4.60}
$$

$\boldsymbol{\Omega \approx \omega_1}$**:**

Let the coefficient of the same power of ε in Eq. (7.4.59) be zero, we obtain

Order ε:

$$
D_0^2\psi_{n1} + \omega_n^2\psi_{n1} = 0
\tag{7.4.61}
$$

Order ε^2:

$$
D_0^2\psi_{n2} + \omega_n^2\psi_{n2} = -2D_0D_1\psi_{n1}
\tag{7.4.62}
$$

Order ε^3:

$$
\begin{aligned}
D_0^2\psi_{13} + \omega_1^2\psi_{13} &= -\left(D_1^2 + 2D_0D_2\right)\psi_{11} - 2D_0D_1\psi_{12} \\
&\quad - 2\tilde{\mu}_1 D_0\psi_{11} - \lambda_1^2\psi_{11}\sum_{m=1}^{\infty} m^2\psi_{m1}^2 + 2k_1\cos\Omega T_0
\end{aligned}
\tag{7.4.63}
$$

$$
\begin{aligned}
D_0^2\psi_{n3} + \omega_n^2\psi_{n3} &= -\left(D_1^2 + 2D_0D_2\right)\psi_{n1} - 2D_0D_1\psi_{n2} \\
&\quad - 2\tilde{\mu}_n D_0\psi_{n1} - \lambda_n^2\psi_{n1}\sum_{m=1}^{\infty} m^2\psi_{m1}^2\,,\ n \geq 2
\end{aligned}
\tag{7.4.64}
$$

The solution of Eq. (7.4.61) is

$$\psi_{n1} = A_n e^{i\omega_n T_0} + cc \tag{7.4.65}$$

where $A_n = A_n(T_1, T_2)$. Substituting (7.4.65) into (7.4.62), we obtain

$$D_0^2\psi_{n2} + \omega_n^2\psi_{n2} = -2i\omega_n D_1 A_n e^{i\omega_n T_0} + cc \tag{7.4.66}$$

In order to eliminate secular terms from the above equation, there must be

$$D_1 A_n = 0 \Rightarrow A_n = A_n(T_2) \tag{7.4.67}$$

Then the solution to (7.4.66) is

$$\psi_{n2} = 0 \tag{7.4.68}$$

Substitute (7.4.65), (7.4.67) and (7.4.68) into (7.4.63) and (7.4.64), we can obtain

$$\begin{aligned} D_0^2\psi_{n3} + \omega_n^2\psi_{n3} &= -\left(D_1^2 + 2D_0D_2\right)\psi_{n1} - 2\tilde{\mu}_n D_0\psi_{n1} - \lambda_n^2\psi_{n1}\sum_{m=1}^{\infty} m^2\psi_{m1}^2 \\ &= -2i\omega_n A' e^{i\omega_n T_0} - 2i\omega_n\tilde{\mu}_n A_n e^{i\omega_n T_0} \\ &\quad - \lambda_n^2 A_n e^{i\omega_n T_0}\sum_{m=2}^{\infty} 2m^2 A_m\overline{A}_m - \lambda_n^2 n^2 A_n^2\overline{A}_n e^{i\omega_n T_0} \\ &\quad + cc + NST, \quad n \geq 2 \end{aligned} \tag{7.4.69}$$

The prime denotes the derivative with respect to T_2. In order to eliminate secular terms from the above equation, we need

$$2i\omega_n A_n' + 2i\omega_n\tilde{\mu}_n A_n + 2\lambda_n^2 A_n\sum_{m=2}^{\infty} m^2 A_m\overline{A}_m + \lambda_n^2 n^2 A_n^2\overline{A}_n = 0, \quad n \geq 2 \tag{7.4.70}$$

Substituting $A_n = \frac{1}{2}a_n e^{i\beta_n}$ into the above equation yields

$$i\omega_n a_n' - \omega_n a_n\beta_n' + i\omega_n\tilde{\mu}_n a_n + \frac{1}{4}\lambda_n^2 a_n\sum_{m=2}^{\infty} m^2 a_m^2 + \frac{1}{8}\lambda_n^2 n^2 a_n^3 = 0, \quad n \geq 2 \tag{7.4.71}$$

Separating the real and imaginary parts of the above equation yields

$$a_n' = -\tilde{\mu}_n a_n, \quad n \geq 2 \tag{7.4.72}$$

therefore

$$a_n = a_{n0}e^{-\tilde{\mu}_n T_2} \to 0, \quad n \geq 2 \tag{7.4.73}$$

and the first-order approximate solution of the equation is

$$\psi_n = a_n \cos(\omega_n t + \beta_n) \to 0, n \geq 2 \tag{7.4.74}$$

$\Omega \neq \omega_1$:

Let the coefficient of the same power of ε in (7.4.59) be zero, we obtain.

Order ε:

$$D_0^2 \psi_{11} + \omega_1^2 \psi_{11} = 2k_1 \cos\Omega T_0 \tag{7.4.75}$$

$$D_0^2 \psi_{n1} + \omega_n^2 \psi_{n1} = 0, n \geq 2 \tag{7.4.76}$$

Order ε^2:

$$D_0^2 \psi_{n2} + \omega_n^2 \psi_{n2} = -2D_0 D_1 \psi_{n1} \tag{7.4.77}$$

Order ε^3:

$$\begin{aligned} D_0^2 \psi_{n3} + \omega_n^2 \psi_{n3} = &-\left(D_1^2 + 2D_0 D_2\right)\psi_{n1} - 2D_0 D_1 \psi_{n2} \\ &- 2\tilde{\mu}_n D_0 \psi_{n1} - \lambda_n^2 \psi_{n1} \sum_{m=1}^{\infty} m^2 \psi_{m1}^2 \end{aligned} \tag{7.4.78}$$

The solution of (7.4.75) is

$$\psi_{11} = A_1 e^{i\omega_1 T_0} + \hat{k}_1 e^{i\Omega T_0} + cc \tag{7.4.79}$$

where

$$\hat{k}_1 = k_1 / \left(\omega_1^2 - \Omega^2\right) \tag{7.4.80}$$

The solution of (7.4.76) is

$$\psi_{n1} = A_n e^{i\omega_n T_0} + cc, n \geq 2 \tag{7.4.81}$$

where $A_n = A_n(T_1, T_2)$. Substitute (7.4.79) and (7.4.81) into (7.4.77), we obtain

$$D_0^2 \psi_{n2} + \omega_n^2 \psi_{n2} = -2i\omega_n D_1 A_n e^{i\omega_n T_0} + cc \tag{7.4.82}$$

In order to eliminate secular terms from the above equation, we need

$$D_1 A_n = 0 \Rightarrow A_n = A_n(T_2) \tag{7.4.83}$$

The solution to (7.4.82) is

$$\psi_{n2} = 0 \tag{7.4.84}$$

Substitute (7.4.79), (7.4.81), (7.4.83) and (7.4.84) into (7.4.78), we obtain

$$\begin{aligned} D_0^2\psi_{n3} + \omega_n^2\psi_{n3} &= -2D_0D_2\psi_{n1} - 2\tilde{\mu}_n D_0\psi_{n1} - \lambda_n^2\psi_{n1}\sum_{m=1}^{\infty} m^2\psi_{m1}^2 \\ &= -2i\omega_n A_n' e^{i\omega_n T_0} - 2i\omega_n \tilde{n}_n A_n e^{i\omega_n T_0} \\ &\quad - \lambda_n^2 n^2 A_n^2 \overline{A}_n e^{i\omega_n T_0} - 2\lambda_n^2 A_n e^{i\omega_n T_0}\sum_{m=2}^{\infty} m^2 A_m \overline{A}_m \\ &\quad - \lambda_n^2 A_n e^{i\omega_n T_0}\left(2A_1\overline{A}_1 + 2\hat{k}_1^2\right) + cc + NST, \quad n \geq 2 \end{aligned} \tag{7.4.85}$$

In order to eliminate secular terms from the above equation, we need

$$\begin{aligned} &2i\omega_n A_n' + 2i\omega_n\tilde{\mu}_n A_n + \lambda_n^2 n^2 A_n^2 \overline{A}_n + 2\lambda_n^2 A_n \sum_{m=2}^{\infty} m^2 A_m \overline{A}_m \\ &+ \lambda_n^2 A_n\left(2A_1\overline{A}_1 + 2\hat{k}_1^2\right) = 0, \quad n \geq 2 \end{aligned} \tag{7.4.86}$$

Substituting $A_n = \frac{1}{2}a_n e^{i\beta_n}$ into the above equation yields

$$\begin{aligned} &i\omega_n a_n' - \omega_n a_n \beta_n' + i\omega_n\tilde{\mu}_n a_n + \frac{1}{4}\lambda_n^2 n^2 a_n^3 \\ &+ \frac{1}{8}\lambda_n^2 a_n \sum_{m=2}^{\infty} m^2 a_m^2 + \frac{1}{2}\lambda_n^2\left(\frac{1}{2}a_1^2 + 2\hat{k}_1^2\right)a_n = 0, \quad n \geq 2 \end{aligned} \tag{7.4.87}$$

Separating the real and imaginary parts of the above equation yields

$$a_n' = -\tilde{\mu}_n a_n, \; n \geq 2 \tag{7.4.88}$$

therefore

$$a_n = a_{n0}e^{-\tilde{\mu}_n T_2} \to 0, n \geq 2 \tag{7.4.89}$$

and the first-order approximate solution of the equation is

$$\psi_n = a_n \cos(\omega_n t + \beta_n) \to 0, n \geq 2 \tag{7.4.90}$$

By now, we have proved that for both $\Omega \approx \omega_1$ and $\Omega \neq \omega_1$, $\psi_n \to 0, \; n \geq 2$.

$\Omega \approx \omega_1$:

From (7.4.65), we can obtain

$$\psi_{11} = A_1 e^{i\omega_1 T_0} + cc \tag{7.4.91}$$

Introduce the detuning parameter σ such that

$$\Omega = \omega_1 + \varepsilon^2 \sigma \tag{7.4.92}$$

Substituting (7.4.91) into (7.4.63), taking into account the results obtained earlier, yields

$$\begin{aligned} D_0^2\psi_{13} + \omega_1^2\psi_{13} &= -2D_0D_2\psi_{11} - 2\tilde{\mu}_1 D_0\psi_{11} - \lambda_n^2\psi_{11}^3 + 2k_1\cos\Omega T_0 \\ &= \left(-2i\omega_1 A_1' - 2i\omega_1\tilde{\mu}_1 A_1 - \lambda_1^2 A_1^2\overline{A}_1 + k_1 e^{i\sigma T_2}\right)e^{i\omega_1 T_0} \\ &\quad + cc + NST \end{aligned} \tag{7.4.93}$$

In order to eliminate secular terms from the above equation, there must be

$$2i\omega_1 A_1' + 2i\omega_1\tilde{\mu}_1 A_1 + \lambda_1^2 A_1^2\overline{A}_1 - k_1 e^{i\sigma T_2} = 0 \tag{7.4.94}$$

Substituting $A_1 = \frac{1}{2}a_1 e^{i\beta_1}$ into the above equation yields

$$i\omega_1 a_1' - \omega_1 a_1\beta_1' + i\omega_1\tilde{\mu}_1 a_1 + \frac{1}{8}\lambda_1^2 a_1^3 - k_1 e^{i(\sigma T_2 - \beta_1)} = 0 \tag{7.4.95}$$

Separating the real and imaginary parts of the above equation yields

$$\begin{aligned} a_1' &= -\tilde{\mu}_1 a_1 + \omega_1^{-1} k_1 \sin\gamma_1 \\ \gamma_1' a_1 &= \sigma a_1 - \frac{1}{8}\omega_1^{-1}\lambda_1^2 a_1^3 + \omega_1^{-1} k_1 \cos\gamma_1 \end{aligned} \tag{7.4.96}$$

where

$$\gamma_1 = \sigma T_2 - \beta_1 \tag{7.4.97}$$

Let $a_1' = \gamma_1' = 0$, we can obtain that the steady state solution should satisfy the following equations

$$\begin{aligned} \tilde{\mu}_1 a_1 &= \omega_1^{-1} k_1 \sin\gamma_1 \\ -\left(\sigma a_1 - \frac{1}{8}\omega_1^{-1}\lambda_1^2 a_1^3\right) &= \omega_1^{-1} k_1 \cos\gamma_1 \end{aligned} \tag{7.4.98}$$

Therefore, the frequency–response equation is

$$\left(\sigma a_1 - \frac{1}{8}\omega_1^{-1}\lambda_1^2 a_1^3\right)^2 + \tilde{\mu}_1^2 a_1^2 = \omega_1^{-2} k_1^2 \tag{7.4.99}$$

The steady state solution of the system can be found by (7.4.99) and (7.4.100). To analyze the stability of any steady state $(a_{10},\ \gamma_{10})$, we let

$$a_1 = a_{10} + \tilde{a}_1, \gamma_1 = \gamma_{10} + \tilde{\gamma}_1 \tag{7.4.100}$$

and substitute (7.4.100) into (7.4.96). After linearization, we can obtain

$$\begin{Bmatrix} \tilde{a}_1' \\ \tilde{\gamma}_1' \end{Bmatrix} = \begin{bmatrix} -\tilde{\mu}_1 & \omega_1^{-1} k_1 \cos\gamma_{10} \\ \sigma a_{10}^{-1} - \frac{3}{8}\omega_1^{-1}\lambda_1^2 a_{10} & -a_{10}^{-1}\omega_1^{-1} k_1 \sin\gamma_{10} \end{bmatrix} \begin{Bmatrix} \tilde{a}_1 \\ \tilde{\gamma}_1 \end{Bmatrix} \tag{7.4.101}$$

The stability of the steady state solutions a_{10},γ_{10} can be determined by the eigenvalues of (7.4.101).

$\boldsymbol{\Omega \approx 3\omega_1}$**:**

ψ_{11} is given by (7.4.79). Considering the results obtained earlier and (7.4.78), we can obtain the governing equation for ψ_{13}:

$$D_0^2\psi_{13} + \omega_1^2\psi_{13} = -2D_0D_2\psi_{11} - 2\tilde{\mu}_1 D_0\psi_{11} - \lambda_1^2\psi_{11}^3 \tag{7.4.102}$$

Substituting (7.4.79) into (7.4.102) yields

$$\begin{aligned} D_0^2\psi_{13} + \omega_1^2\psi_{13} &= -2D_0D_2\psi_{11} - 2\tilde{\mu}_1 D_0\psi_{11} - \lambda_1^2\psi_{11}^3 \\ &= -\left(2i\omega_1 A_1' + 2i\omega_1\tilde{\mu}_1 A_1 + 6\lambda_1^2\hat{k}_1^2 A_1 + 3\lambda_1^2 A_1^2\overline{A}_1\right)e^{i\omega_1 T_0} \\ &\quad - 3\lambda_1^2\hat{k}_1\overline{A}_1^2 e^{i(\Omega-2\omega_1)T_0} - \lambda_1^2\hat{k}_1^3 e^{3i\Omega T_0} + cc + NST \end{aligned} \tag{7.4.103}$$

Introduce the detuning parameter σ such that

$$\Omega = 3\omega_1 + \varepsilon^2\sigma \tag{7.4.104}$$

In order to eliminate secular terms from (7.4.103), there must be

$$2i\omega_1 A_1' + 2i\omega_1\tilde{\mu}_1 A_1 + 6\lambda_1^2\hat{k}_1^2 A_1 + 3\lambda_1^2 A_1^2\overline{A}_1 + 3\lambda_1^2\hat{k}_1\overline{A}_1^2 e^{i\sigma T_2} = 0 \tag{7.4.105}$$

Substituting $A_1 = \frac{1}{2}a_1 e^{i\beta_1}$ into the above equation yields

$$i\omega_1 a_1' - \omega_1 a_1\beta_1' + i\omega_1\tilde{\mu}_1 a_1 + 3\lambda_1^2\hat{k}_1^2 a_1 + \frac{3}{8}\lambda_1^2 a_1^3 + \frac{3}{4}\lambda_1^2\hat{k}_1 a_1^2 e^{i(\sigma T_2 - 3\beta_1)} = 0 \tag{7.4.106}$$

Separating the real and imaginary parts of the above equation yields

$$\begin{aligned} a_1' &= -\tilde{\mu}_1 a_1 + \frac{3}{4}\omega_1^{-1}\lambda_1^2\hat{k}_1 a_1^2 \sin\gamma_1 \\ a_1\gamma_1' &= \sigma a_1 - 9\omega_1^{-1}\lambda_1^2\hat{k}_1^2 a_1 - \frac{9}{8}\omega_1^{-1}\lambda_1^2 a_1^3 - \frac{9}{4}\omega_1^{-1}\lambda_1^2\hat{k}_1 a_1^2 \cos\gamma_1 \end{aligned} \tag{7.4.107}$$

Where

$$\gamma_1 = \sigma T_2 - 3\beta_1 \tag{7.4.108}$$

Let $a_1' = \gamma_1' = 0$, we can obtain that the steady state solution should satisfy the following equations

$$\begin{aligned} 0 &= -\tilde{\mu}_1 a_1 + \frac{3}{4}\omega_1^{-1}\lambda_1^2\hat{k}_1 a_1^2 \sin\gamma_1 \\ 0 &= \sigma a_1 - 9\omega_1^{-1}\lambda_1^2\hat{k}_1^2 a_1 - \frac{9}{8}\omega_1^{-1}\lambda_1^2 a_1^3 - \frac{9}{4}\omega_1^{-1}\lambda_1^2\hat{k}_1 a_1^2 \cos\gamma_1 \end{aligned} \tag{7.4.109}$$

Therefore, the frequency–response equation is

$$\left(\sigma a_1 - 9\omega_1^{-1}\lambda_1^2\hat{k}_1^2 a_1 - \frac{9}{8}\omega_1^{-1}\lambda_1^2 a_1^3\right)^2 + 27\tilde{\mu}_1^2 a_1^2 = \frac{81}{16}\omega_1^{-2}\lambda_1^4\hat{k}_1^2 a_1^4 \tag{7.4.110}$$

The steady state solution of the system can be found by (7.4.110) and (7.4.109). To analyze the stability of any steady state $(a_{10}, \ \gamma_{10})$, we let

$$a_1 = a_{10} + \tilde{a}_1, \gamma_1 = \gamma_{10} + \tilde{\gamma}_1 \tag{7.4.111}$$

and substitute (7.4.111) into (7.4.107). After linearization, we can obtain

$$\begin{aligned} \tilde{a}_1' &= \left(-\tilde{\mu}_1 + \frac{3}{2}\omega_1^{-1}\lambda_1^2\hat{k}_1 a_{10}\sin\gamma_{10}\right)\tilde{a}_1 + \left(\frac{3}{4}\omega_1^{-1}\lambda_1^2\hat{k}_1 a_{10}^2\cos\gamma_{10}\right)\tilde{\gamma}_1 \\ a_{10}\tilde{\gamma}_1' &= \left(\sigma - 9\omega_1^{-1}\lambda_1^2\hat{k}_1^2 - \frac{27}{8}\omega_1^{-1}\lambda_1^2 a_{10}^2 - \frac{18}{4}\omega_1^{-1}\lambda_1^2\hat{k}_1 a_{10}\cos\gamma_1\right)\tilde{a}_1 \\ &\quad + \left(\frac{9}{4}\omega_1^{-1}\lambda_1^2\hat{k}_1 a_{10}^2\cos\gamma_{10}\right)\tilde{\gamma}_1 \end{aligned} \tag{7.4.112}$$

Obviously, there exists a steady state solution $a_{10} = 0$, at which point the Eq. (7.4.112) becomes

$$\widetilde{a_1'} = -\tilde{\mu}_1\tilde{a}_1, \quad (\sigma - 9\omega_1^{-1}\lambda_1^2\hat{k}_1^2)\tilde{a}_1 = 0 \tag{7.4.113}$$

therefore

$$\tilde{a}_1 = Ce^{-\tilde{\mu}_1 T_2} \to 0 \tag{7.4.114}$$

Thus, the steady state solution $a_{10} = 0$ is stable.
For a non-trivial steady state solution $a_{10} \neq 0$, the Eq. (7.4.112) becomes

$$\begin{Bmatrix} \tilde{a}_1' \\ \tilde{\gamma}_1' \end{Bmatrix} = \begin{bmatrix} -\tilde{\mu}_1 + \frac{3}{2}\omega_1^{-1}\lambda_1^2\hat{k}_1 a_{10}\sin\gamma_{10} & \frac{3}{4}\omega_1^{-1}\lambda_1^2\hat{k}_1 a_{10}^2\cos\gamma_{10} \\ a_{10}^{-1}\big(\sigma - 9\omega_1^{-1}\lambda_1^2\hat{k}_1^2 - \frac{27}{8}\omega_1^{-1}\lambda_1^2 a_{10}^2 & \frac{9}{4}\omega_1^{-1}\lambda_1^2\hat{k}_1 a_{10}\cos\gamma_{10} \\ -\frac{18}{4}\omega_1^{-1}\lambda_1^2\hat{k}_1 a_{10}\cos\gamma_1\big) & \end{bmatrix} \begin{Bmatrix} \tilde{a}_1 \\ \tilde{\gamma}_1 \end{Bmatrix} \tag{7.4.114}$$

The stability of the steady state solutions a_{10},γ_{10} can be determined by the eigenvalues of (7.4.101).

$\boldsymbol{\Omega} \approx \frac{1}{3}\boldsymbol{\omega}_1$:

Readers are invited to complete this case.

When $f_1(t) = K_1\cos\Omega t$, $f_3(t) = K_3\cos(\Omega t + \tau)$ and $f_n = 0,\ \ n \neq 1, 3$, Eq. (7.4.59) becomes

$$\begin{aligned} 0 &= \varepsilon\big(D_0^2\psi_{n1} + \omega_n^2\psi_{n1}\big) + \varepsilon^2\big(D_0^2\psi_{n2} + \omega_n^2\psi_{n2} + 2D_0D_1\psi_{n1}\big) \\ &\quad + \varepsilon^3\begin{bmatrix} D_0^2\psi_{n3} + \omega_n^2\psi_{n3} + \big(D_1^2 + 2D_0D_2\big)\psi_{n1} \\ + 2D_0D_1\psi_{n2} + 2\tilde{\mu}_n D_0\psi_{n1} + \lambda_n^2\psi_{n1}\sum_{m=1}^{\infty} m^2\psi_{m1}^2 \end{bmatrix} \\ &\quad - \delta_{1n}K_1\cos\Omega t - \delta_{3n}K_3\cos(\Omega t + \tau) + \cdots \end{aligned} \tag{7.4.116}$$

When $\Omega \approx 3\omega_1$ is adopted, the excitation term $K_3\cos(\Omega t + \tau)$ in the above equation becomes a resonant excitation term of ψ_3 due to $\omega_3 = 3\omega_1$. Therefore we specify:

$$K_1 = 2\varepsilon k_1, \quad K_3 = 2\varepsilon^3 k_3 \tag{7.4.117}$$

Let the coefficient of the same power of ε in the Eq. (7.4.116) be zero, we obtain.

Order ε:

$$D_0^2\psi_{11} + \omega_1^2\psi_{11} = 2k_1\cos\Omega T_0 \tag{7.4.118}$$

$$D_0^2\psi_{n1} + \omega_n^2\psi_{n1} = 0, n \geq 2 \tag{7.4.119}$$

Order ε^2:

$$D_0^2\psi_{n2} + \omega_n^2\psi_{n2} = -2D_0D_1\psi_{n1} \tag{7.4.120}$$

Order ε^3:

$$D_0^2\psi_{33} + \omega_3^2\psi_{33} = -\left(D_1^2 + 2D_0D_2\right)\psi_{31} - 2D_0D_1\psi_{32} - 2\tilde{\mu}_3D_0\psi_{31}$$
$$- \lambda_3^2\psi_{31}\sum_{m=1}^{\infty} m^2\psi_{m1}^2 + 2k_3\cos(\Omega T_0 + \tau) \tag{7.4.121}$$

$$D_0^2\psi_{n3} + \omega_n^2\psi_{n3} = -\left(D_1^2 + 2D_0D_2\right)\psi_{n1} - 2D_0D_1\psi_{n2}$$
$$- 2\tilde{\mu}_nD_0\psi_{n1} - \lambda_n^2\psi_{n1}\sum_{m=1}^{\infty} m^2\psi_{m1}^2,\quad n \neq 3 \tag{7.4.122}$$

The solution of the Eq. (7.4.118) is

$$\psi_{11} = A_1e^{i\omega_1T_0} + \hat{k}_1e^{i\Omega T_0} + cc \tag{7.4.123}$$

where

$$\hat{k}_1 = k_1/\left(\omega_1^2 - \Omega^2\right) \tag{7.4.124}$$

The solution of the Eq. (7.4.119) is

$$\psi_{n1} = A_ne^{i\omega_nT_0} + cc, n \geq 2 \tag{7.4.125}$$

where $A_n = A_n(T_1, T_2)$. Substitute (7.4.123) and (7.4.125) into (7.4.120), we obtain

$$D_0^2\psi_{n2} + \omega_n^2\psi_{n2} = -2i\omega_nD_1A_ne^{i\omega_nT_0} + cc \tag{7.4.126}$$

In order to eliminate secular terms from the above equation, there must be

$$D_1A_n = 0 \Rightarrow A_n = A_n(T_2) \tag{7.4.127}$$

The solution to (7.4.126) is

$$\psi_{n2} = 0 \tag{7.4.128}$$

Substitute (7.4.123), (7.4.125), (7.4.127) and (7.4.128) into (7.4.122), we obtain

$$D_0^2\psi_{n3} + \omega_n^2\psi_{n3} = -2D_0D_2\psi_{n1} - 2\tilde{\mu}_nD_0\psi_{n1} - \lambda_n^2\psi_{n1}\sum_{m=1}^{\infty} m^2\psi_{m1}^2$$
$$= -2i\omega_nA_n'e^{i\omega_nT_0} - 2i\omega_n\tilde{\mu}_nA_ne^{i\omega_nT_0}$$
$$- \lambda_n^2n^2A_n^2\overline{A}_ne^{i\omega_nT_0} - 2\lambda_n^2A_ne^{i\omega_nT_0}\sum_{m=2}^{\infty} m^2A_m\overline{A}_m$$
$$- \lambda_n^2A_ne^{i\omega_nT_0}\left(2A_1\overline{A}_1 + 2\hat{k}_1^2\right) + cc + NST,\quad n \neq 1 \text{ and } 3 \tag{7.4.129}$$

In order to eliminate secular terms from the above equation, we need

$$\begin{aligned}&2i\omega_n A_n' + 2i\omega_n \tilde{\mu}_n A_n + \lambda_n^2 n^2 A_n^2 \overline{A}_n + 2\lambda_n^2 A_n \sum_{m=2}^{\infty} m^2 A_m \overline{A}_m \\ &\quad + \lambda_n^2 A_n \left(2A_1\overline{A}_1 + 2\hat{k}_1^2\right) = 0, \quad n \neq 1 \text{ and } 3\end{aligned} \tag{7.4.130}$$

Substituting $A_n = \frac{1}{2}a_n e^{i\beta_n}$ into the above equation yields

$$\begin{aligned}&i\omega_n a_n' - \omega_n a_n \beta_n' + i\omega_n \tilde{\mu}_n a_n + \frac{1}{4}\lambda_n^2 n^2 a_n^3 \\ &\quad + \frac{1}{8}\lambda_n^2 a_n \sum_{m=2}^{\infty} m^2 a_m^2 + \frac{1}{2}\lambda_n^2\left(\frac{1}{2}a_1^2 + 2\hat{k}_1^2\right)a_n = 0, \quad n \neq 1 \text{ and } 3\end{aligned} \tag{7.4.131}$$

Separating the real and imaginary parts of the above equation yields

$$a_n' = -\tilde{\mu}_n a_n, \; n \neq 1 \text{ and } 3 \tag{7.4.132}$$

Therefore,

$$a_n = a_{n0} e^{-\tilde{\mu}_n T_2} \to 0, \quad n \neq 1 \text{ and } 3 \tag{7.4.133}$$

and the first-order approximate solution of the equation is

$$\psi_n = a_n \cos(\omega_n t + \beta_n) \to 0, \; n \neq 1 \text{ and } \; 3 \tag{7.4.134}$$

From the above results, the governing equations of ψ_1 and ψ_3 can be rewritten as:

$$\begin{aligned}&D_0^2 \psi_{11} + \omega_1^2 \psi_{11} = 2k_1 \cos\Omega T_0 \\ &D_0^2 \psi_{31} + \omega_3^2 \psi_{31} = 0\end{aligned} \tag{7.4.135}$$

$$\begin{aligned}&D_0^2 \psi_{13} + \omega_1^2 \psi_{13} = -2D_0 D_2 \psi_{11} - 2\tilde{\mu}_1 D_0 \psi_{11} - \lambda_1^2 \psi_{11} \sum_{m=1}^{\infty} m^2 \psi_{m1}^2 \\ &D_0^2 \psi_{33} + \omega_3^2 \psi_{33} = -2D_0 D_2 \psi_{31} - 2\tilde{\mu}_3 D_0 \psi_{31} \\ &\qquad - \lambda_3^2 \psi_{31} \sum_{m=1}^{\infty} m^2 \psi_{m1}^2 + 2k_3 \cos(\Omega T_0 + \tau)\end{aligned} \tag{7.4.136}$$

The solution of the Eq. (7.4.135) has been obtained earlier as

$$\begin{aligned}&\psi_{11} = A_1 e^{i\omega_1 T_0} + \hat{k}_1 e^{i\Omega T_0} + cc \\ &\psi_{31} = A_3 e^{i\omega_3 T_0} + cc\end{aligned} \tag{7.4.137}$$

Substituting (7.4.137) into (7.4.136) and taking into account $\Omega \approx 3\omega_1$ and $\omega_3 = 3\omega_1$, we obtain

$$\begin{aligned}D_0^2\psi_{13} + \omega_1^2\psi_{13} &= -2D_0D_2\psi_{11} - 2\tilde{\mu}_1 D_0\psi_{11} - \lambda_1^2\psi_{11}\sum_{m=1}^{\infty} m^2\psi_{m1}^2 \\ &= -2i\omega_1 A_1' e^{i\omega_1 T_0} - 2i\omega_1\tilde{\mu}_1 A_1 e^{i\omega_1 T_0} - \lambda_1^2\psi_{11}^3 - 9\lambda_1^2\psi_{11}\psi_{31}^2 \\ &= -\big(2i\omega_1 A_1' + 2i\omega_1\tilde{\mu}_1 A_1 + 3\lambda_1^2 A_1^2\overline{A}_1 \\ &\quad + 6\lambda_1^2\hat{k}_1^2 A_1 + 18A_1\lambda_1^2 A_3\overline{A}_3\big)e^{i\omega_1 T_0} \\ &\quad - 3\lambda_1^2\hat{k}_1\overline{A}_1^2 e^{i(\Omega-2\omega_1)T_0} + cc + NST \end{aligned} \tag{7.4.138}$$

$$\begin{aligned}D_0^2\psi_{33} + \omega_3^2\psi_{33} &= -2D_0D_2\psi_{31} - 2\tilde{\mu}_3 D_0\psi_{31} \\ &\quad - \lambda_3^2\psi_{31}\psi_{11}^2 - 9\lambda_3^2\psi_{31}^3 + 2k_3\cos(\Omega T_0 + \tau) \\ &= -\big(2i\omega_3 A_3' + 2i\omega_3\tilde{\mu}_3 A_3 + 2\lambda_3^2 A_1\overline{A}_1 A_3 \\ &\quad + 2\lambda_3^2\hat{k}_1^2 A_3 + 9\lambda_3^2 A_3\overline{A}_3^2\big)e^{i\omega_3 T_0} \\ &\quad - \lambda_3^2\hat{k}_1^2\overline{A}_3 e^{i(2\Omega-\omega_3)T_0} + k_3 e^{i(\Omega T_0+\tau)} + cc + NST \end{aligned} \tag{7.4.139}$$

We introduce the detuning parameters σ_1 and σ_3 and make

$$\Omega = 3\omega_1 + \varepsilon^2\sigma_1, \quad \Omega = \omega_3 + \varepsilon^2\sigma_3 \tag{7.4.140}$$

Thus, in order to eliminate secular terms from (7.4.138) and (7.4.139), we need

$$\begin{aligned}&2i\omega_1 A_1' + 2i\omega_1\tilde{\mu}_1 A_1 + 6\lambda_1^2\hat{k}_1^2 A_1 \\ &\quad + 3\lambda_1^2 A_1^2\overline{A}_1 + 18\lambda_1^2 A_1 A_3\overline{A}_3 + 3\lambda_1^2\hat{k}_1\overline{A}_1^2 e^{i\sigma_1 T_2} = 0\end{aligned} \tag{7.4.141}$$

$$\begin{aligned}&2i\omega_3 A_3' + 2i\omega_3\tilde{\mu}_3 A_3 + 2\lambda_3^2\hat{k}_1^2 A_3 + 2\lambda_3^2 A_1\overline{A}_1 A_3 \\ &\quad + 9\lambda_3^2 A_3\overline{A}_3^2 + \lambda_3^2\hat{k}_1^2\overline{A}_3 e^{2i\sigma_3 T_2} + k_3 e^{i(\sigma_3 T_2+\tau)} = 0\end{aligned} \tag{7.4.142}$$

Let

$$A_1 = \frac{1}{2}a_1 e^{i\beta_1}, \quad A_3 = \frac{1}{2}a_3 e^{i\beta_3} \tag{7.4.143}$$

Substituting (7.4.143) into (7.4.141) and (7.4.142) yields

$$\begin{aligned}&i\omega_1 a_1' - \omega_1 a_1\beta_1' + i\omega_1\tilde{\mu}_1 a_1 + 3\lambda_1^2\hat{k}_1^2 a_1 \\ &\quad + \frac{3}{8}\lambda_1^2 a_1^3 + \frac{9}{4}\lambda_1^2 a_1 a_3^2 + \frac{3}{4}\lambda_1^2\hat{k}_1 a_1^2 e^{i(\sigma_1 T_2 - 3\beta_1)} = 0\end{aligned} \tag{7.4.144}$$

$$i\omega_3 a_3' - \omega_3 a_3 \beta_3' + i\omega_3 \tilde{\mu}_3 a_3 + \lambda_3^2 \hat{k}_1^2 a_3 + \frac{1}{4}\lambda_3^2 a_1^2 a_3 + \frac{9}{8}\lambda_3^2 a_3^3 + \frac{1}{2}\lambda_3^2 \hat{k}_1^2 a_3 e^{i(2\sigma_3 T_2 - 2\beta_3)} + k_3 e^{i(\sigma_3 T_2 - \beta_3 + \tau)} = 0 \tag{7.4.145}$$

Separating the real and imaginary parts of the above equation yields

$$\begin{aligned} a_1' &= -\tilde{\mu}_1 a_1 - \frac{3}{4}\omega_1^{-1}\lambda_1^2 \hat{k}_1 a_1^2 \sin\gamma_1 \\ a_1\gamma_1' &= \sigma_1 a_1 - 9\omega_1^{-1}\lambda_1^2 \hat{k}_1^2 a_1 - \frac{9}{8}\omega_1^{-1}\lambda_1^2 a_1^3 \\ &\quad - \frac{27}{4}\omega_1^{-1}\lambda_1^2 a_1 a_3^2 - \frac{9}{4}\omega_1^{-1}\lambda_1^2 \hat{k}_1 a_1^2 \cos\gamma_1 \end{aligned} \tag{7.4.146}$$

$$\begin{aligned} a_3' &= -\tilde{\mu}_3 a_3 - \frac{1}{2}\omega_3^{-1}\lambda_3^2 \hat{k}_1^2 a_3 \sin 2\gamma_3 - \omega_3^{-1} k_3 \sin(\gamma_3 + \tau) \\ a_3\gamma_3' &= \sigma_3 a_3 - \omega_3^{-1}\lambda_3^2 \hat{k}_1^2 a_3 - \frac{1}{4}\omega_3^{-1}\lambda_3^2 a_1^2 a_3 - \frac{9}{8}\omega_3^{-1}\lambda_3^2 a_3^3 \\ &\quad - \frac{1}{2}\omega_3^{-1}\lambda_3^2 \hat{k}_1^2 a_3 \cos 2\gamma_3 - \omega_3^{-1} k_3 \cos(\gamma_3 + \tau) \end{aligned} \tag{7.4.147}$$

$$\gamma_1 = \sigma_1 T_2 - 3\beta_1, \gamma_3 = \sigma_3 T_2 - \beta_3 \tag{7.4.148}$$

Let $a_1\prime = \gamma_1\prime = a_3\prime = \gamma_3\prime = 0$, we can obtain that the steady state solution should satisfy the following equations

$$\begin{aligned} &\frac{3}{4}\omega_1^{-1}\lambda_1^2 \hat{k}_1 a_1^2 \sin\gamma_1 = -\tilde{\mu}_1 a_1 \\ &\frac{9}{4}\omega_1^{-1}\lambda_1^2 \hat{k}_1 a_1^2 \cos\gamma_1 = \sigma_1 a_1 - 9\omega_1^{-1}\lambda_1^2 \hat{k}_1^2 a_1 - \frac{9}{8}\omega_1^{-1}\lambda_1^2 a_1^3 - \frac{27}{4}\omega_1^{-1}\lambda_1^2 a_1 a_3^2 \end{aligned} \tag{7.4.149}$$

$$\begin{aligned} &\frac{1}{2}\omega_3^{-1}\lambda_3^2 \hat{k}_1^2 a_3 \sin 2\gamma_3 + \omega_3^{-1} k_3 \sin(\gamma_3 + \tau) = -\tilde{\mu}_3 a_3 \\ &\frac{1}{2}\omega_3^{-1}\lambda_3^2 \hat{k}_1^2 a_3 \cos 2\gamma_3 + \omega_3^{-1} k_3 \cos(\gamma_3 + \tau) = \sigma_3 a_3 - \omega_3^{-1}\lambda_3^2 \hat{k}_1^2 a_3 \\ &\quad - \frac{1}{4}\omega_3^{-1}\lambda_3^2 a_1^2 a_3 - \frac{9}{8}\omega_3^{-1}\lambda_3^2 a_3^3 \end{aligned} \tag{7.4.150}$$

Eliminating γ_1 from the system of Eq. (7.4.149) yields

$$\frac{81}{16}\omega_1^{-2}\lambda_1^4 \hat{k}_1^2 a_1^4 = 9\tilde{\mu}_1^2 a_1^2 + \left(\sigma_1 a_1 - 9\omega_1^{-1}\lambda_1^2 \hat{k}_1^2 a_1 - \frac{9}{8}\omega_1^{-1}\lambda_1^2 a_1^3 - \frac{27}{4}\omega_1^{-1}\lambda_1^2 a_1 a_3^2\right)^2 \tag{7.4.151}$$

To analyze the stability of any steady state solution $(a_{10}, \gamma_{10}, a_{30}, \gamma_{30})$, we let

$$\begin{aligned} a_1 &= a_{10} + \tilde{a}_1, \quad \gamma_1 = \gamma_{10} + \tilde{\gamma}_1 \\ a_3 &= a_{30} + \tilde{a}_3, \quad \gamma_3 = \gamma_{30} + \tilde{\gamma}_3 \end{aligned} \tag{7.4.152}$$

Substituting this into (7.4.146) and (7.4.147). After linearization, we can obtain

$$\begin{bmatrix} 1 & 0 & 0 & 0 \\ 0 & 1 & 0 & 0 \\ 0 & 0 & a_{10} & 0 \\ 0 & 0 & 0 & a_{30} \end{bmatrix} \begin{Bmatrix} \tilde{a}_1' \\ \tilde{a}_3' \\ \tilde{\gamma}_1{}' \\ \tilde{\gamma}_3{}' \end{Bmatrix} = \begin{bmatrix} l_{11} & l_{12} & l_{13} & l_{14} \\ l_{21} & l_{22} & l_{23} & l_{24} \\ l_{31} & l_{32} & l_{33} & l_{34} \\ l_{41} & l_{42} & l_{43} & l_{44} \end{bmatrix} \begin{Bmatrix} \tilde{a}_1 \\ \tilde{a}_3 \\ \tilde{\gamma}_1 \\ \tilde{\gamma}_3 \end{Bmatrix} \tag{7.4.153}$$

where

$$\begin{aligned} l_{11} &= -\left(\tilde{\mu}_1 + \frac{3}{2}\omega_1^{-1}\lambda_1^2\hat{k}_1 a_{10}\sin\gamma_{10}\right), \quad l_{12} = 0, \quad l_{13} = -\frac{3}{4}\omega_1^{-1}\lambda_1^2\hat{k}_1 a_{10}^2\cos\gamma_{10}, \quad l_{14} = 0 \\ l_{21} &= 0, \quad l_{22} = -\left(\tilde{\mu}_3 + \frac{1}{2}\omega_3^{-1}\lambda_3^2\hat{k}_1^2\sin 2\gamma_{30}\right), \quad l_{23} = 0 \\ l_{24} &= -\left[\omega_3^{-1}\lambda_3^2\hat{k}_1^2 a_{30}\cos 2\gamma_{30} + \omega_3^{-1}k_3\cos(\gamma_{30}+\tau)\right] \\ l_{31} &= \left(\sigma_1 - 9\omega_1^{-1}\lambda_1^2\hat{k}_1^2 - \frac{27}{8}\omega_1^{-1}\lambda_1^2 a_{10}^2 - \frac{27}{4}\omega_1^{-1}\lambda_1^2 a_{30}^2 - \frac{18}{4}\omega_1^{-1}\lambda_1^2\hat{k}_1 a_{10}\cos\gamma_{10}\right) \\ l_{32} &= 0, \quad l_{33} = -\frac{54}{4}\omega_1^{-1}\lambda_1^2 a_{10}a_{30}\tilde{a}_3 + \frac{9}{4}\omega_1^{-1}\lambda_1^2\hat{k}_1 a_1^2\sin\gamma_{10}, \quad l_{34} = 0 \\ l_{41} &= -\frac{1}{2}\omega_3^{-1}\lambda_3^2 a_{10}a_{30} \\ l_{42} &= \left(\sigma_3 - \omega_3^{-1}\lambda_3^2\hat{k}_1^2 - \frac{1}{4}\omega_3^{-1}\lambda_3^2 a_{10}^2 - \frac{27}{8}\omega_3^{-1}\lambda_3^2 a_{30}^2 - \frac{1}{2}\omega_3^{-1}\lambda_3^2\hat{k}_1^2\cos 2\gamma_{30}\right) \\ l_{43} &= 0, \quad l_{44} = \left[\omega_3^{-1}\lambda_3^2\hat{k}_1^2 a_{30}\sin 2\gamma_{30} + \omega_3^{-1}k_3\sin(\gamma_{30}+\tau)\right] \end{aligned} \tag{7.4.154}$$

The stability of the steady state solution $(a_{10}, \gamma_{10}, a_{30}, \gamma_{30})$ can be determined by the eigenvalues of (7.4.153).

Readers are invited to complete the analysis on Exercise 7.4(*e*).

7.5 Exercise 7.5 (Nonlinear Analysis of Transverse Oscillations in the Plane of an Elastic Tensioned String)

Solution: **(a)** The governing equations for the corresponding linear undamped free oscillation of the system are

$$w_{tt} - c_2^2 w_{xx} = 0 \tag{7.5.1}$$

Let

$$w(x,t) = \phi(x)e^{i\omega t} \tag{7.5.2}$$

and substitute it into (7.5.1), we obtain

$$\frac{d^2\phi}{dx^2} + k^2\phi = 0, k = \frac{\omega}{c_2} \tag{7.5.3}$$

The boundary condition becomes

$$\phi(0) - h_1\phi'(0) = 0, \ \phi(1) + h_2\phi\prime(1) = 0 \tag{7.5.4}$$

The solution of (7.5.3) is

$$\phi(x) = C\sin kx + D\cos kx \tag{7.5.5}$$

Applying the boundary conditions (7.5.4), we can obtain

$$\begin{aligned} &-kh_1C + D = 0 \\ &(\sin k + Ch_2k\cos k)C + (\cos k - h_2k\sin k)D = 0 \end{aligned} \tag{7.5.6}$$

From the non-trivial solution condition, we can obtain the eigen equation

$$\left(1 - h_1h_2k^2\right)\tan k + (h_1 + h_2)k = 0 \tag{7.5.7}$$

The equation consists of an infinite number of solutions. Let the nth solution be k_n, then the natural frequency $\omega_n = c_2k_n$. The first equation of (7.5.6) is

$$D_n = k_nh_1C_n \tag{7.5.8}$$

Therefore, the nth order mode shape function is

$$\phi_n(x) = C_n(\sin k_nx + h_1k_n\cos k_nx) \tag{7.5.9}$$

Since C_n can be any non-zero constant, we have $C_n = \delta_n$, where δ_n is the modal (mode) normalization factor, i.e.

$$\delta_n^2\int_0^1 (\sin k_nx + h_1k_n\cos k_nx)^2dx = 1 \tag{7.5.10}$$

Therefore, the mode shape function is

$$\phi_n(x) = \delta_n(\sin k_nx + h_1k_n\cos k_nx) \tag{7.5.11}$$

The orthogonality of the mode shape functions is proved below. From (7.5.3), we have

$$k_n^2\phi_n = \phi_n'', \quad k_m^2\phi_m = \phi_m''$$

The prime denotes the derivative with respect to x. Multiplying these two equations by ϕ_m and ϕ_n, respectively, and integrating over [0, 1] yields

$$k_n^2\int_0^1 \phi_n\phi_m dx = \int_0^1 \phi_m\phi_n'' dx = -\int_0^1 \phi_n'\phi_m' dx + \phi_n'\phi_m\big|_0^1 \tag{7.5.12}$$

$$k_m^2\int_0^1 \phi_n\phi_m dx = \int_0^1 \phi_n\phi_m'' dx = -\int_0^1 \phi_n'\phi_m' dx + \phi_m'\phi_n\big|_0^1 \tag{7.5.13}$$

Subtracting the above two equations and taking into account (7.5.4), we can obtain

$$\begin{aligned}\left(k_n^2 - k_m^2\right)\int_0^1 \phi_n\phi_m dx &= \left(\phi_n'\phi_m - \phi_m'\phi_n'\right)\big|_0^1\\ &= \left[\phi_n'(1)\phi_m(1) - \phi_m'(1)\phi_n(1)\right]\\ &\quad - \left[\phi_n'(0)\phi_m(0) - \phi_m'(0)\phi_n(0)\right]\\ &= \left[-h_2\phi_n'(1)\phi_m'(1) + h_2\phi_m'(1)\phi_n'(1)\right]\\ &\quad - \left[h_1\phi_n'(0)\phi_m'(0) - h_1\phi_m'(0)\phi_n'(0)\right]\\ &= 0\end{aligned} \tag{7.5.14}$$

So, when $n \neq m$, we can obtain

$$\int_0^1 \phi_n\phi_m dx = 0, \int_0^1 \phi_n'\phi_m' dx = 0 \tag{7.5.15}$$

When $h_1 = 1$ and $h_2 = 1.52$, the Eq. (7.5.7) becomes

$$\left(1 - 1.52k^2\right)\tan k + 2.52k = 0 \tag{7.5.16}$$

The first five nonzero roots of this equation are:

$$k_1 = 1.198, \quad k_2 = 3.594, \quad k_3 = 6.535, \quad k_4 = 9.597, \quad k_5 = 12.697$$

It can be seen that

$$k_2 \approx 3k_1 \text{ so } \omega_2 \approx 3\omega_1$$

When $F(x, t) = \hat{F}(x)\cos\Omega t$, the governing differential equation of the system becomes

$$w_{tt} - c_2^2 w_{xx} = \frac{c_1^2}{2l} w_{xx} \int_0^l w_x^2 dx - 2\hat{\mu} w_t + \hat{F}(x)\cos\Omega t \tag{7.5.17}$$

Let the solution of the equation be

$$w(x, t) = \sum_{n=1}^{\infty} \phi_n(x)\psi_n(t) \tag{7.5.18}$$

Substituting (7.5.18) into (7.5.17) yields

$$\begin{aligned} &\sum_{n=1}^{\infty} \phi_n(x)\ddot{\psi}_n(t) + c_2^2 \sum_{n=1}^{\infty} k_n^2 \phi_n(x)\psi_n(t) \\ &= -\frac{c_1^2}{2l} \sum_{n=1}^{\infty} k_n^2 \phi_n(x)\psi_n(t) \int_0^l \left[\sum_{r=1}^{\infty} \phi_r'(x)\psi_r(t)\right]^2 dx \\ &\quad - 2\hat{\mu} \sum_{n=1}^{\infty} \phi_n(x)\dot{\psi}_n(t) + \hat{F}(x)\cos\Omega t \end{aligned} \tag{7.5.19}$$

Multiplying both sides of the above equation by $\phi_m(x)$, and integrating over [0, 1], we get

$$\begin{aligned} &\sum_{n=1}^{\infty} \ddot{\psi}_n \int_0^1 \phi_m\phi_n dx + c_2^2 \sum_{n=1}^{\infty} k_n^2 \psi_n \int_0^1 \phi_m\phi_n dx \\ &= -\frac{c_1^2}{2l} \sum_{n=1}^{\infty} k_n^2 \psi_n \left(\int_0^1 \phi_m\phi_n dx\right)\left(\sum_{p,q}^{\infty} \psi_p\psi_q \int_0^1 \phi'_p \phi'_q dx\right) \\ &\quad - 2\sum_{n=1}^{\infty} \dot{\psi}_n \int_0^1 \hat{\mu}\phi_m\phi_n dx + \cos\Omega t \int_0^1 \phi_m \hat{F}(x) dx \end{aligned} \tag{7.5.20}$$

Utilizing the orthogonality of the mode shape function, we can obtain

$$\ddot{\psi}_n + c_2^2 k_n^2 \psi_n = -\lambda_n^2 \psi_n \sum_m \alpha_m^2 \psi_m^2 - 2\hat{\mu}_n \dot{\psi}_n + K_n \cos\Omega t \tag{7.5.21}$$

where

$$\alpha_m^2 = \int_0^1 \phi\prime_m^2 dx,\ K_n = \int_0^1 \phi_n \hat{F}(x)dx,\ \lambda_n^2 = c_1^2 k_n^2/2l \tag{7.5.22}$$

Let the solution of (7.5.22) be

$$\psi_n = \varepsilon\psi_{n1}(T_0, T_1, T_2) + \varepsilon^2\psi_{n2}(T_0, T_1, T_2) + \varepsilon^3\psi_{n3}(T_0, T_1, T_2) + \cdots \tag{7.5.23}$$

In order to make both the damping and nonlinear terms appearing in the third order equation, let

$$\hat{\mu}_n = \varepsilon^2\tilde{\mu}_n \tag{7.5.24}$$

Substituting (7.4.57) and (7.4.58) into (7.4.55) and retaining to O(ε^3), we obtain

$$\begin{aligned}
0 &= \ddot{\psi}_n + c_2^2 k_n^2 \psi_n + \lambda_n^2 \psi_n \sum\nolimits_m^\infty \alpha_m^2 \psi_m^2 + 2\hat{\mu}_n \dot{\psi}_n - K_n \cos\Omega t \\
&= \left[D_0^2 + 2\varepsilon D_0 D_1 + \varepsilon^2\left(D_1^2 + 2D_0 D_2\right)\right]\left(\varepsilon\psi_{n1} + \varepsilon^2\psi_{n2} + \varepsilon^3\psi_{n3}\right) \\
&\quad + \omega_n^2\left(\varepsilon\psi_{n1} + \varepsilon^2\psi_{n2} + \varepsilon^3\psi_{n3}\right) \\
&\quad + 2\varepsilon^2\hat{\mu}_n\left(D_0 + \varepsilon D_1 + \varepsilon^2 D_2\right)\left(\varepsilon\psi_{n1} + \varepsilon^2\psi_{n2} + \varepsilon^3\psi_{n3}\right) \\
&\quad + \lambda_n^2\left(\varepsilon\psi_{n1} + \varepsilon^2\psi_{n2} + \varepsilon^3\psi_{n3}\right)\sum_{m=1}^{\infty}\alpha_m^2\left(\varepsilon\psi_{m1} + \varepsilon^2\psi_{m2} + \varepsilon^3\psi_{m3}\right)^2 - K_n\cos\Omega t + \cdots \\
&= \varepsilon\left(D_0^2\psi_{n1} + \omega_n^2\psi_{n1}\right) + \varepsilon^2\left(D_0^2\psi_{n2} + \omega_n^2\psi_{n2} + 2D_0 D_1\psi_{n1}\right) \\
&\quad + \varepsilon^3[D_0^2\psi_{n3} + \omega_n^2\psi_{n3} + \left(D_1^2 + 2D_0 D_2\right)\psi_{n1} + 2D_0 D_1\psi_{n2} \\
&\quad + 2\hat{\mu}_n D_0\psi_{n1} + \lambda_n^2\psi_{n1}\sum_{m=1}^{\infty}\alpha_m^2\psi_{m1}^2] - K_n\cos\Omega t + \cdots
\end{aligned} \tag{7.5.25}$$

We specify:

$$K_1 = 2\varepsilon^3 k_1;\quad K_n = 2\varepsilon k_n,\quad n \geq 2 \tag{7.5.26}$$

Let the coefficient of the same power of ε in the Eq. (7.5.25) be zero, we obtain.

Order ε:

$$D_0^2\psi_{11} + \omega_1^2\psi_{11} = 0 \tag{7.5.27}$$

$$D_0^2\psi_{n1} + \omega_n^2\psi_{n1} = 2k_n\cos\Omega T_0,\quad n \geq 2 \tag{7.5.28}$$

Order ε^2:

$$D_0^2\psi_{n2} + \omega_n^2\psi_{n2} = -2D_0 D_1\psi_{n1} \tag{7.5.29}$$

Order ε^3:

$$D_0^2\psi_{13}+\omega_1^2\psi_{13}=-\left(D_1^2+2D_0D_2\right)\psi_{11}-2D_0D_1\psi_{12}$$
$$-2\hat{\mu}_1D_0\psi_{11}-\lambda_1^2\psi_{11}\sum_{m=1}^{\infty}\alpha_m^2\psi_{m1}^2+2k_1\cos\Omega T_0 \tag{7.5.30}$$

$$D_0^2\psi_{n3}+\omega_n^2\psi_{n3}=-\left(D_1^2+2D_0D_2\right)\psi_{n1}-2D_0D_1\psi_{n2}$$
$$-2\hat{\mu}_nD_0\psi_{n1}-\lambda_n^2\psi_{n1}\sum_{m=1}^{\infty}\alpha_m^2\psi_{m1}^2,\quad n\geq 2 \tag{7.5.31}$$

(7.5.27) and (7.5.28) can be solved by

$$\psi_{11}=A_1e^{i\omega_1T_0}+cc$$
$$\psi_{n1}=A_ne^{i\omega_nT_0}+\hat{k}_ne^{i\Omega T_0}+cc,\quad n\geq 2 \tag{7.5.32}$$

where $A_n=A_n(T_1,T_2)$ and

$$\hat{k}_n=\frac{k_n}{\omega_n^2-\Omega^2},\quad n\geq 2 \tag{7.5.33}$$

Substituting (7.5.32) into (7.5.29), we get

$$D_0^2\psi_{n2}+\omega_n^2\psi_{n2}=-2i\omega_nD_1A_ne^{i\omega_nT_0}+cc \tag{7.5.34}$$

In order to eliminate secular terms from the above equation, there must be

$$D_1A_n=0\ \Rightarrow\ A_n=A_n(T_2) \tag{7.5.35}$$

Therefore, the solution to (7.5.34) is

$$\psi_{n2}=0 \tag{7.5.36}$$

Substituting (7.5.32), (7.5.35) and (7.5.36) into (7.5.31), and taking into account $\Omega\approx\omega_1,\ \omega_2\approx 3\omega_1$, we obtain

$$D_0^2\psi_{n3}+\omega_n^2\psi_{n3}=-\left(D_1^2+2D_0D_2\right)\psi_{n1}-2\hat{\mu}_nD_0\psi_{n1}-\lambda_n^2\psi_{n1}\sum_{m=1}^{\infty}\alpha_m^2\psi_{m1}^2$$
$$=-2i\omega_nA_n'e^{i\omega_nT_0}-2i\omega_n\hat{\mu}_nA_ne^{i\omega_nT_0}-2\lambda_n^2\alpha_1^2A_1\overline{A}_1A_ne^{i\omega_nT_0}$$
$$-\lambda_n^2\alpha_n^2A_n^2\overline{A}_ne^{i\omega_nT_0}-4\lambda_n^2\alpha_n^2\hat{k}_n^2A_ne^{i\omega_nT_0}$$
$$-2\lambda_n^2A_ne^{i\omega_nT_0}\sum_{m=1}^{\infty}\alpha_m^2\left(A_m\overline{A}_m+\hat{k}_m^2\right)-\lambda_n^2\alpha_1^2\hat{k}_nA_1^2e^{i(\Omega+2\omega_1)T_0}$$

$$-\lambda_n^2\hat{k}_n e^{3i\Omega T_0}\sum_{m=2}^{\infty}\alpha_m^2\hat{k}_m^2 + cc + NST, \quad n \geq 2 \tag{7.5.37}$$

$$\begin{aligned}\psi_{11} &= A_1 e^{i\omega_1 T_0} + cc\\ \psi_{n1} &= A_n e^{i\omega_n T_0} + \hat{k}_n e^{i\Omega T_0} + cc, \quad n \geq 2\end{aligned} \tag{7.5.38}$$

The prime denotes the derivative with respect to T_2. When $n \geq 3$ is used, in order to eliminate secular terms from the above equation, there must be

$$\begin{aligned}&2i\omega_n A_n' + 2i\omega_n\hat{\mu}_n A_n + 2\lambda_n^2\alpha_1^2 A_1\overline{A}_1 A_n + \lambda_n^2\alpha_n^2 A_n^2\overline{A}_n + 4\lambda_n^2\alpha_n^2\hat{k}_n^2 A_n\\ &+ 2\lambda_n^2 A_n\sum_{m=2}^{\infty}\alpha_m^2(A_m\overline{A}_m + \hat{k}_m^2) = 0, \quad n \geq 3\end{aligned} \tag{7.5.39}$$

Substituting $A_n = \frac{1}{2}a_n e^{i\beta_n}$ into the above equation yields

$$\begin{aligned}&2i\omega_n A_n' + 2i\omega_n\hat{\mu}_n A_n + 2\lambda_n^2\alpha_1^2 A_1\overline{A}_1 A_n + \lambda_n^2\alpha_n^2 A_n^2\overline{A}_n + 4\lambda_n^2\alpha_n^2\hat{k}_n^2 A_n\\ &+ 2\lambda_n^2 A_n\sum_{m=2}^{\infty}\alpha_m^2(A_m\overline{A}_m + \hat{k}_m^2) = 0, \quad n \geq 3\end{aligned} \tag{7.5.40}$$

From the imaginary part of the above equation, we have

$$a_n' = -\hat{\mu}_n a_n, \ n \geq 3 \tag{7.5.41}$$

Therefore,

$$a_n = a_{n0}e^{-\hat{\mu}_n T_2} \to 0, \ n \geq 3 \tag{7.5.42}$$

and

$$\psi_{n1} = \hat{k}_n e^{i\Omega T_0} + cc, \quad n \geq 3 \tag{7.5.43}$$

The first order approximate solution of the equation can be written as

$$\psi_n \approx \psi_{n1} = 2\hat{k}_n \cos\Omega t, \ n \geq 3 \tag{7.5.44}$$

From (7.5.32), we can obtain

$$\begin{aligned}\psi_{11} &= A_1 e^{i\omega_1 T_0} + cc\\ \psi_{21} &= A_2 e^{i\omega_2 T_0} + \hat{k}_2 e^{i\Omega T_0} + cc\end{aligned} \tag{7.5.45}$$

Equation (7.5.30) changes into

$$D_0^2\psi_{13}+\omega_1^2\psi_{13}=-2D_0D_2\psi_{11}-2\hat{\mu}_1D_0\psi_{11}-\lambda_1^2\alpha_1^2\psi_{11}^3$$
$$-\lambda_1^2\alpha_2^2\psi_{11}\psi_{21}^2-\lambda_1^2\psi_{11}\sum_{m=3}^{\infty}\alpha_m^2\psi_{m1}^2+2\hat{k}_1\cos\Omega T_0 \quad (7.5.46)$$

Substituting (7.5.45) and (7.5.43) into (7.5.46) yields

$$\begin{aligned}D_0^2\psi_{13}+\omega_1^2\psi_{13}=&-2i\omega_1A_1'e^{i\omega_1T_0}-2i\omega_1\hat{\mu}_1A_1e^{i\omega_1T_0}-3\lambda_1^2\alpha_1^2A_1^2\overline{A}_1e^{i\omega_1T_0}\\&-2\lambda_1^2\alpha_2^2A_1A_2\overline{A}_2e^{i\omega_1T_0}-2\lambda_1^2\alpha_2^2\hat{k}_2^2A_1e^{i\omega_1T_0}\\&-2\lambda_1^2h_2^2A_1e^{i\omega_1T_0}-2\lambda_1^2\alpha_2^2\hat{k}_2\overline{A}_1A_2e^{i(\omega_2-\omega_1-\Omega)T_0}\\&-\lambda_1^2h_2^2\overline{A}_1e^{i(2\Omega-\omega_1)T_0}+k_1e^{i\Omega T_0}+cc+NST\end{aligned} \quad (7.5.47)$$

where

$$h_2^2=\sum_{m=3}^{\infty}\alpha_m^2\hat{k}_m^2 \quad (7.5.48)$$

Let $n=2$ in (7.5.37), we get

$$\begin{aligned}D_0^2\psi_{23}+\omega_2^2\psi_{23}=&-2i\omega_2A_2'e^{i\omega_2T_0}-2i\omega_2\hat{\mu}_2A_2e^{i\omega_2T_0}-2\lambda_2^2\alpha_1^2A_1\overline{A}_1A_2e^{i\omega_2T_0}\\&-3\lambda_2^2A_2^2\overline{A}_2e^{i\omega_2T_0}-6\lambda_2^2\alpha_2^2\hat{k}_2^2A_2e^{i\omega_2T_0}-2\lambda_2^2h_2^2A_2e^{i\omega_2T_0}\\&-\lambda_2^2\alpha_1^2\hat{k}_2A_1^2e^{i(\Omega+2\omega_1)T_0}-\lambda_2^2\alpha_2^2\hat{k}_2^3e^{3i\Omega T_0}-\lambda_2^2h_2^2\hat{k}_2e^{3i\Omega T_0}\\&+cc+NST\end{aligned} \quad (7.5.49)$$

Introduce the detuning parameters σ_1 and σ_2 so that

$$\Omega=\omega_1+\varepsilon^2\sigma_1,\ \ \omega_2=3\omega_1+\varepsilon^2\sigma_2 \quad (7.5.50)$$

In order to eliminate secular terms from (7.5.47) and (7.5.49), we need

$$\begin{aligned}&2i\omega_1A_1'+2i\omega_1\hat{\mu}_1A_1+3\lambda_1^2\alpha_1^2A_1^2\overline{A}_1+2\lambda_1^2\alpha_2^2A_1A_2\overline{A}_2+2\lambda_1^2\alpha_2^2\hat{k}_2^2A_1\\&+2\lambda_1^2h_2^2A_1+2\lambda_1^2\alpha_2^2\hat{k}_2\overline{A}_1A_2e^{i(\sigma_2-\sigma_1)T_2}+\lambda_1^2h_2^2\overline{A}_1e^{2i\sigma_1T_2}-k_1e^{i\sigma_1T_2}=0\end{aligned} \quad (7.5.51)$$

$$\begin{aligned}&2i\omega_2A_2'+2i\omega_2\hat{\mu}_2A_2+2\lambda_2^2\alpha_1^2A_1\overline{A}_1A_2+3\lambda_2^2\alpha_2^2A_2^2\overline{A}_2\\&+6\lambda_2^2\alpha_2^2\hat{k}_2^2A_2+2\lambda_2^2h_2^2A_2+\lambda_2^2\alpha_1^2\hat{k}_2A_1^2e^{i(\sigma_1-\sigma_2)T_2}\\&+\lambda_2^2\alpha_2^2\hat{k}_2^3e^{i(3\sigma_1-\sigma_2)T_2}+\lambda_2^2h_2^2\hat{k}_2e^{i(3\sigma_1-\sigma_2)T_2}=0\end{aligned} \quad (7.5.52)$$

Substitute $A_1=\frac{1}{2}a_1e^{i\beta_1}$ and $A_2=\frac{1}{2}a_2e^{i\beta_2}$ into (7.5.51) and (7.5.52), we obtain

$$i\omega_1a_1'-\omega_1a_1\beta_1'+i\omega_1\hat{\mu}_1a_1+\frac{3}{8}\lambda_1^2\alpha_1^2a_1^3+\frac{1}{4}\lambda_1^2\alpha_2^2a_1a_2^2+\lambda_1^2\left(\alpha_2^2\hat{k}_2^2+h_2^2\right)a_1$$

$$+\frac{1}{2}\lambda_1^2\alpha_2^2\hat{k}_2a_1a_2e^{i(\sigma_2T_2-\sigma_1T_2-2\beta_1+\beta_2)}+\lambda_1^2h_2^2a_1e^{i(2\sigma_1T_2-2\beta_1)}$$
$$-k_1e^{i(\sigma_1T_2-\beta_1)}=0 \tag{7.5.53}$$

$$i\omega_2a_2'-\omega_2a_2\beta_2'+i\omega_2\hat{\mu}_2a_2+\frac{1}{4}\lambda_2^2\alpha_1^2a_1^2a_2+\frac{3}{8}\lambda_2^2\alpha_2^2a_2^3$$
$$+\lambda_2^2\left(3\alpha_2^2\hat{k}_2^2+h_2^2\right)+\lambda_2^2\alpha_1^2\hat{k}_2a_1^2e^{i(\sigma_1T_2-\sigma_2T_2+2\beta_1-\beta_2)}$$
$$+\lambda_2^2\hat{k}_2\left(\alpha_2^2\hat{k}_2^2+h_2^2\right)e^{i(3\sigma_1T_2-\sigma_2T_2-\beta_2)}=0 \tag{7.5.54}$$

Let

$$\gamma_1=\sigma_1T_2-\beta_1$$
$$\gamma_2=\sigma_2T_2-\sigma_1T_2-2\beta_1+\beta_2 \tag{7.5.55}$$

Separating the real and imaginary parts of (7.5.53) and (7.5.54) yields

$$a_1'=-\hat{\mu}_1a_1-\frac{1}{2}\omega_1^{-1}\lambda_1^2\alpha_2^2\hat{k}_2a_1a_2\sin\gamma_2-\omega_1^{-1}\lambda_1^2h_2^2a_1\sin2\gamma_1+\omega_1^{-1}k_1\sin\gamma_1$$
$$a_1\gamma_1'=a_1\sigma_1-\frac{3}{8}\omega_1^{-1}\lambda_1^2\alpha_1^2a_1^3-\frac{1}{4}\omega_1^{-1}\lambda_1^2\alpha_2^2a_1a_2^2-\omega_1^{-1}\lambda_1^2\left(\alpha_2^2\hat{k}_2^2+h_2^2\right)a_1$$
$$-\frac{1}{2}\omega_1^{-1}\lambda_1^2\alpha_2^2\hat{k}_2a_1a_2\cos\gamma_2-\omega_1^{-1}\lambda_1^2h_2^2a_1\cos2\gamma_1+\omega_1^{-1}k_1\cos\gamma_1 \tag{7.5.56}$$

$$a_2\prime=-\hat{\mu}_2a_2+\omega_2^{-1}\lambda_2^2\alpha_1^2\hat{k}_2a_1^2\sin\gamma_2$$
$$-\omega_2^{-1}\lambda_2^2\hat{k}_2\left(\alpha_2^2\hat{k}_2^2+h_2^2\right)\sin(2\gamma_1-\gamma_2)$$
$$a_2\left(\gamma_2'-2\gamma_1'\right)=(\sigma_2-3\sigma_1)a_2+\frac{1}{4}\omega_2^{-1}\lambda_2^2\alpha_1^2a_1^2a_2+\frac{3}{8}\omega_2^{-1}\lambda_2^2\alpha_2^2a_2^3$$
$$+\omega_2^{-1}\lambda_2^2\left(3\alpha_2^2\hat{k}_2^2+h_2^2\right)+\omega_2^{-1}\lambda_2^2\alpha_1^2\hat{k}_2a_1^2\cos\gamma_2$$
$$+\omega_2^{-1}\lambda_2^2\hat{k}_2\left(\alpha_2^2\hat{k}_2^2+h_2^2\right)\cos(2\gamma_1-\gamma_2) \tag{7.5.57}$$

The steady state solution of the above system of equations can be found and its stability can be analyzed.

It can be concluded that the first order approximate solution of the system for the case of $F(x,t)=\hat{F}(x)\cos\Omega t$, $\Omega\approx\omega_1$, $\omega_2\approx3\omega_1$ is

$$\psi_1=a_1\cos(\omega_1t+\beta_1)+O\left(\varepsilon^3\right)$$
$$\psi_2=a_2\cos(\omega_2t+\beta_2)+2\hat{k}_2\cos\Omega t+O\left(\varepsilon^3\right)$$
$$\psi_n=2\hat{k}_n\cos\Omega t+O\left(\varepsilon^3\right),\quad n\geq3 \tag{7.5.58}$$

7.6 Exercise 7.6 (Nonlinear Analysis of Planar Transverse Oscillations of Elastic Tensioned Strings Under Parametric Excitation Formed by Time-Varying Tension)

Solution: **(a)** Substituting

$$w = \sqrt{2}\sum_{n=1}^{\infty} u_n(t)\sin\frac{n\pi x}{l}$$

into the governing equation of the system yields

$$\sqrt{2}\sum_{n=1}^{\infty}\ddot{u}_n\sin\frac{n\pi x}{l} + c_0^2\big[1+p(t)\big]\sqrt{2}\frac{n^2\pi^2}{l^2}\sum_{n=1}^{\infty}u_n\sin\frac{n\pi x}{l}$$

$$= -\frac{c_1^2}{2l}\sqrt{2}\frac{n^2\pi^2}{l^2}\sum_{n=1}^{\infty}u_n\sin\frac{n\pi x}{l}\int_0^l 2\sum_{p=1}^{\infty}\left[u_p\frac{p\pi}{l}\cos\frac{p\pi x}{l}\right]^2 dx - 2\sqrt{2}\hat{\mu}\sum_{n=1}^{\infty}\dot{u}_n\sin\frac{n\pi x}{l}$$

Multiplying the above equation by $\sqrt{2}\sin(m\pi x/l)$ and making the integration of x over $[0, l]$ and taking into account the orthogonality of this mode shape function yields

$$\ddot{u}_n + \frac{c_0^2 n^2\pi^2}{l^2}\big[1+p(t)\big]u_n = -\frac{n^2c_1^2\pi^4}{2l^4}u_n\sum_{m=1}^{\infty}m^2u_m^2 - 2\hat{\mu}_n\dot{u}_n$$

i.e.

$$\ddot{u}_n + \omega_n^2\big[1+p(t)\big]u_n = -2\hat{\mu}_n\dot{u}_n - \alpha n^2 u_n\sum_{m=1}^{\infty}m^2u_m^2 \qquad (7.6.1)$$

where

$$\omega_n = \frac{n\pi c_0}{l}, \quad \alpha = \frac{c_1^2\pi^4}{2l^4} \qquad (7.6.2)$$

When $p(t) = 2\varepsilon\cos\Omega t$, let $\varepsilon = \hat{\varepsilon}^2$, then Eq. (7.6.1) becomes

$$\ddot{u}_n + \omega_n^2 u_n = -2\hat{\mu}_n\dot{u}_n - \alpha n^2 u_n\sum_{m=1}^{\infty}m^2u_m^2 - 2\hat{\varepsilon}^2\omega_n^2u_n\cos\Omega t \qquad (7.6.3)$$

Let the solution of this equation be

$$u_n = \hat{\varepsilon}u_{n1}(T_0, T_1, T_2) + \hat{\varepsilon}^2 u_{n2}(T_0, T_1, T_2) + \hat{\varepsilon}^3 u_{n3}(T_0, T_1, T_2) + \cdots \tag{7.6.4}$$

In order to make the damping, nonlinear, and the parametric excitation term appearing simultaneously in the third-order equation, we designate $\hat{\mu}_n$ as $\varepsilon^2\mu_n$. Substituting (7.6.4) into (7.6.3) and retaining to O($\hat{\varepsilon}^3$) yields

$$\begin{aligned}
0 &= \ddot{u}_n + \omega_n^2 u_n + 2\mu_n \dot{u}_n + \alpha n^2 u_n \sum_{m=1}^{\infty} m^2 u_m^2 + 2\hat{\varepsilon}^2 \omega_n^2 u_n \cos \Omega t \\
&= \left[D_0^2 + 2\hat{\varepsilon} D_0 D_1 + \hat{\varepsilon}^2 \left(D_1^2 + 2D_0 D_2\right)\right]\left(\hat{\varepsilon}u_{n1} + \hat{\varepsilon}^2 u_{n2} + \hat{\varepsilon}^3 u_{n3}\right) \\
&\quad + \omega_n^2 \left(\hat{\varepsilon}u_{n1} + \hat{\varepsilon}^2 u_{n2} + \hat{\varepsilon}^3 u_{n3}\right) \\
&\quad + 2\hat{\varepsilon}^2 u_n \left(D_0 + \hat{\varepsilon} D_1 + \hat{\varepsilon}^2 D_2\right)\left(\hat{\varepsilon}u_{n1} + \hat{\varepsilon}^2 u_{n2} + \hat{\varepsilon}^3 u_{n3}\right) \\
&\quad + \alpha n^2 \left(\hat{\varepsilon}u_{n1} + \hat{\varepsilon}^2 u_{n2} + \hat{\varepsilon}^3 u_{n3}\right) \sum_{m=1}^{\infty} m^2 \left(\hat{\varepsilon}u_{m1} + \hat{\varepsilon}^2 u_{m2} + \hat{\varepsilon}^3 u_{m3}\right)^2 \\
&\quad + 2\hat{\varepsilon}^2 \omega_n^2 \left(\hat{\varepsilon}u_{n1} + \hat{\varepsilon}^2 u_{n2} + \hat{\varepsilon}^3 u_{n3}\right) \cos \Omega t + \cdots \\
&= \hat{\varepsilon}\left(D_0^2 u_{n1} + \omega_n^2 u_{n1}\right) + \hat{\varepsilon}^2 \left(D_0^2 u_{n2} + \omega_n^2 u_{n2} + 2D_0 D_1 u_{n1}\right) \\
&\quad + \hat{\varepsilon}^3 [D_0^2 u_{n3} + \omega_n^2 u_{n3} + \left(D_1^2 + 2D_0 D_2\right) u_{n1} + 2D_0 D_1 u_{n2} \\
&\quad + 2\mu_n D_0 u_{n1} + \alpha n^2 u_{n1} \sum_{m=1}^{\infty} m^2 u_{m1}^2 + 2\omega_n^2 u_{n1} \cos \Omega t] + \cdots
\end{aligned} \tag{7.6.5}$$

Making the coefficient of the same power of ε zero in the above equation yields

$$D_0^2 u_{n1} + \omega_n^2 u_{n1} = 0 \tag{7.6.6}$$

$$D_0^2 u_{n2} + \omega_n^2 u_{n2} = -2D_0 D_1 u_{n1} \tag{7.6.7}$$

$$\begin{aligned}
D_0^2 u_{n3} + \omega_n^2 u_{n3} &= -\left(D_1^2 + 2D_0 D_2\right) u_{n1} - 2D_0 D_1 u_{n2} - 2\mu_n D_0 u_{n1} \\
&\quad - \alpha n^2 u_{n1} \sum_{m=1}^{\infty} m^2 u_{m1}^2 - 2\omega_n^2 u_{n1} \cos\Omega t
\end{aligned} \tag{7.6.8}$$

The solution of (7.6.6) is

$$u_{n1} = A_n e^{i\omega_n T_0} + cc \tag{7.6.9}$$

where $A_n = A_n(T_1, T_2)$. Substituting (7.6.9) into (7.6.7) yields

$$D_0^2 u_{n2} + \omega_n^2 u_{n2} = -2i\omega_n D_1 A_n e^{i\omega_n T_0} + cc \tag{7.6.10}$$

In order to eliminate secular terms from the above equation, there must be

$$D_1A_n = 0 \Rightarrow A_n = A_n(T_2) \tag{7.6.11}$$

Then the solution of (7.6.10) is

$$u_{n2} = 0 \tag{7.6.12}$$

Substituting (7.6.9), (7.6.11) and (7.6.12) into (7.6.8) yields

$$\begin{aligned} D_0^2u_{n3} + \omega_n^2u_{n3} &= -2D_0D_2u_{n1} - 2\hat{\mu}_nD_0u_{n1} - \alpha n^2u_{n1}\sum_{m=1}^{\infty} m^2u_{m1}^2 - 2\omega_n^2u_{n1}\cos\Omega T_0 \\ &= -2i\omega_nA_n'e^{i\omega_nT_0} - 2i\omega_n\mu_nA_ne^{i\omega_nT_0} - 2\alpha n^2A_ne^{i\omega_nT_0}\sum_{m=1}^{\infty} m^2A_m\overline{A}_m \\ &\quad - \alpha n^4A_n^2\overline{A}_ne^{i\omega_nT_0} - \omega_n^2\overline{A}_ne^{i(\Omega-\omega_n)T_0} + cc + NST \end{aligned} \tag{7.6.13}$$

When $\Omega = 2\omega_1 + \varepsilon\sigma = 2\omega_1 + \hat{\varepsilon}^2\sigma$, in order to eliminate secular terms from the above equation, we need

$$2i\omega_1A_1' + 2i\omega_1\mu_1A_1 + 3\alpha A_1^2\overline{A}_1 + 2\alpha A_1\sum_{m=2}^{\infty} m^2A_m\overline{A}_m + \omega_1^2\overline{A}_1e^{i\sigma T_2} = 0 \tag{7.6.14}$$

$$2i\omega_nA_n\prime + 2i\omega_n\mu_nA_n + 3\alpha n^4A_n^2\overline{A}_n + 2\alpha n^2A_n\sum_{m\neq n}^{\infty} m^2A_m\overline{A}_m = 0,\ n \geq 2 \tag{7.6.15}$$

Substituting $A_n = \frac{1}{2}a_ne^{i\beta_n}$ into (7.6.15) yields

$$i\omega_na_n' - \omega_na_n\beta_n' + i\omega_n\mu_na_n + \frac{3}{8}\alpha n^4a_n^3 + \frac{1}{4}\alpha n^2a_n\sum_{m\neq n}^{\infty} m^2a_m^2 = 0,\ n \geq 2 \tag{7.6.16}$$

From the imaginary part of (7.6.16), we have

$$a_n' + \mu_na_n = 0, n \geq 2 \tag{7.6.17}$$

therefore,

$$a_n = a_{n0}e^{-i\mu_n} \to 0, A_n \to 0, n \geq 2 \tag{7.6.18}$$

and

$$u_{n1} \to 0, n \geq 2 \tag{7.6.19}$$

Considering (7.6.18), we can rewrite Eq. (7.6.14) as

$$2i\omega_1 A_1' + 2i\omega_1\mu_1 A_1 + 3\alpha A_1^2\overline{A}_1 + \omega_1^2\overline{A}_1 e^{i\sigma T_2} = 0 \tag{7.6.20}$$

Substituting $A_1 = \frac{1}{2}a_1 e^{i\beta_1}$ into (7.6.20), we can obtain

$$i\omega_1 a_1' - \omega_1 a_1\beta_1' + i\omega_1\mu_1 a_1 + \frac{3}{8}\alpha a_1^3 + \frac{1}{2}\omega_1^2 a_1 e^{i(\sigma T_2 - 2\beta_1)} = 0 \tag{7.6.21}$$

Separating the real and imaginary parts of the above equation yields

$$\begin{aligned} a_1' &= -\mu_1 a_1 - \frac{1}{2}\omega_1 a_1 \sin\gamma_1 \\ \gamma_1' a_1 &= \sigma a_1 - \frac{3}{4}\omega_1^{-1}\alpha a_1^3 - \omega_1 a_1 \cos\gamma_1 \end{aligned} \tag{7.6.22}$$

where

$$\gamma_1 = \sigma T_2 - 2\beta_1 \tag{7.6.23}$$

Let $a_1' = \gamma_1' = 0$ in (7.6.22), we can obtain that the steady state solution should satisfy the following two equations:

$$\begin{aligned} 0 &= -\mu_1 a_1 - \frac{1}{2}\omega_1 a_1 \sin\gamma_1 \\ 0 &= \sigma a_1 - \frac{3}{4}\omega_1^{-1}\alpha a_1^3 - \omega_1 a_1 \cos\gamma_1 \end{aligned} \tag{7.6.24}$$

Obviously, $a_{10} = 0, \gamma_1 = \gamma_{10}$ is a steady state solution. In this case, the corresponding linear perturbation equation can be obtained from (7.6.22)

$$\begin{aligned} a_1' &= -\mu_1 a_1 - \frac{1}{2}\omega_1 a_1 \sin\gamma_{10} \\ 0 &= \sigma - \omega_1 \cos\gamma_{10} \end{aligned} \tag{7.6.25}$$

Generally speaking, (7.6.25) cannot be satisfied and hence the steady state solution $a_{10} = 0, \gamma_1 = \gamma_{10}$ is unstable. When ε, the frequency–response equation can be obtained from (7.6.24):

$$4\mu_1^2 + \left(\sigma - \frac{3}{4}\omega_1^{-1}\alpha a_1^2\right) = \omega_1^2 \tag{7.6.26}$$

For any set of non-trivial steady state solution a_{10}, γ_{10}, let

$$a_1 = a_{10} + \hat{a}_1, \gamma_1 = \gamma_{10} + \hat{\gamma}_1 \tag{7.6.27}$$

Substituting (7.6.27) into (7.6.22), we can obtain the linearized equations

$$\begin{Bmatrix} \widehat{a_1'} \\ \widehat{\gamma_1'} \end{Bmatrix} = \begin{bmatrix} -\mu_1 - \frac{1}{2}\omega_1 \sin\gamma_{10} & -\frac{1}{2}\omega_1 a_{10}\cos\gamma_{10} \\ \sigma a_{10}^{-1} - \frac{9}{4}\omega_1^{-1}\alpha a_{10} - \omega_1 a_{10}^{-1}\cos\gamma_{10} & \omega_1 \sin\gamma_{10} \end{bmatrix} \begin{Bmatrix} \hat{a}_1 \\ \hat{\gamma}_1 \end{Bmatrix} \quad (7.6.28)$$

Finding the two eigenvalues of Eq. (7.6.28), we can determine the stability of the non-trivial steady state solution a_{10}, γ_{10}.

Readers are invited to complete Exercise 7.6(d) and (e).

7.7 Exercise 7.7 (Transverse Oscillation of a Hinged-Hinged Beam Excited by First- and Second-Order Primary Resonances at $u = O(w^2)$)

Solution: (a) The governing equation of the beam in this Exercise is

$$r^2(\ddot{w} + 2\mu\dot{w} + w_{xxxx}) = \left[P(t) + \frac{1}{2l}\int_0^l w_x^2 dx \right] w_{xx} + F_z(x, t) \quad (7.7.1)$$

The solution to this governing equation is

$$w(x, t) = r^k \sum_{m=1}^{\infty} u_m(t)\phi_m(x) \quad (7.7.2)$$

where $\phi_m(x)$ is the solution to the following characteristic equation

$$\phi_m^{iv} - \omega_m^2 \phi_m = 0 \quad (7.7.3)$$

For the hinged-hinged beam

$$\phi_m = \phi_m'' = 0 \quad (7.7.4)$$

$\phi_m(x)$ is a family of orthogonal functions, i.e.,

$$\int_0^l \phi_i \phi_j dx = \delta_{ij} \quad (7.7.5)$$

Substituting (7.7.2) into (7.7.1), multiplying by $\phi_n(x)$, making the integration of x over $[0, l]$ and taking into account the orthogonality of this mode shape function yields

$$\ddot{u}_n + \omega_n^2 u_n = r^{-2} \sum_{m=1}^{\infty} u_m P(t) \int_0^l \phi''{}_m \phi_n dx$$
$$+ \frac{1}{2l} r^{2(k-1)} \sum_{m,p,q=1}^{\infty} u_m u_p u_q \left(\int_0^l \phi''_m \phi_n dx \right) \left(\int_0^l \phi'_p \phi'_q dx \right) \Bigg]$$
$$+ 2 \left(\int_0^l \mu \phi_n^2 dx \right) \dot{u}_n + r^{k-4} \int_0^l F_z(x,t) \phi_n dx \tag{7.7.6}$$

In order to make the damping, nonlinear, and the parametric excitation term appearing simultaneously in the same order equation, we let

$$\varepsilon = r^{2(k-1)}, \int_0^l \mu \phi_n^2 dx = \varepsilon \mu_n, r^{-2} P(t) \int_0^l \phi''_m \phi_n dx = \varepsilon p_{nm}(t)$$

$$\Gamma_{nmpq} = \frac{1}{2l} \left(\int_0^l \phi''_m \phi_n dx \right) \left(\int_0^l \phi'_p \phi'_q dx \right), f_{zn}(t) = r^{-(k+2)} \int_0^l F_z(x,t) \phi_n dx \tag{7.7.7}$$

Then (7.7.6) can be written as

$$\ddot{u}_n + \omega_n^2 u_n = \varepsilon \left[\sum_{m=1}^{\infty} p_{nm}(t) u_m + \sum_{m,p,q=1}^{\infty} \Gamma_{nmpq} u_m u_p u_q - 2\mu_n \dot{u}_n \right] + f_{zn}(t) \tag{7.7.8}$$

From the method of integration by part, Γ_{nmpq} can be written as

$$\Gamma_{nmpq} = -\frac{1}{2l} \left(\int_0^l \phi'_m \phi'_n dx \right) \left(\int_0^l \phi'_p \phi'_q dx \right) \tag{7.7.9}$$

Therefore,

$$\Gamma_{nmpq} = \Gamma_{mnpq} = \Gamma_{nmqp} = \Gamma_{pqnm} \tag{7.7.10}$$

Since $P(t) = 0$, $p_{nm}(t) = 0$. We take the dimensionless length of the beam $l = 1$; and, in turn, by Eq. (7.7.3) and (7.7.4), the natural frequency of the linear free oscillation of the hinged-hinged beam can be obtained as $\omega_n = n^2\pi^2$ and the corresponding normalized modes are

$$\phi_n(x) = \sqrt{2} \sin n\pi x, n = 1, 2, 3, \ldots \tag{7.7.11}$$

Taking into account again Eq. (7.7.9), we can rewrite (7.7.8) as

$$\ddot{u}_n + n^4\pi^4 u_n = -\varepsilon\left(2\mu_n\dot{u}_n + \frac{1}{2}\pi^4 n^2 u_n \sum_{m=1}^{\infty} m^2 u_m^2\right) + f_{zn}(t), \quad n = 1, 2, 3, \ldots \tag{7.7.12}$$

We use the method of multiple scales to solve (7.7.12). Let the solution of the equation be

$$u_n(t;\varepsilon) = u_{n0}(T_0, T_1) + \varepsilon u_{n1}(T_0, T_1) + \cdots \tag{7.7.13}$$

Substituting (7.7.13) into (7.7.12) and retaining to $O(\varepsilon)$ yields

$$\begin{aligned}
0 &= \ddot{u}_n + n^4\pi^4 u_n + \varepsilon\left(2\mu_n\dot{u}_n + \frac{1}{2}\pi^4 n^2 u_n \sum_{m=1}^{\infty} m^2 u_m^2\right) - f_{zn}(t) \\
&= \left(D_0^2 + 2\varepsilon D_0 D_1\right)(u_{n0} + \varepsilon u_{n1}) + n^4\pi^4(u_{n0} + \varepsilon u_{n1}) \\
&\quad + \varepsilon\left(2\mu_n D_0 u_{n0} + \frac{1}{2}\pi^4 n^2 u_{n0} \sum_{m=1}^{\infty} m^2 u_{m0}^2\right) - f_{zn}(t) \\
&= D_0^2 u_{n0} + n^4\pi^4 u_{n0} \\
&\quad + \varepsilon\left(D_0^2 u_{n1} + n^4\pi^4 u_{n1} + 2D_0 D_1 u_{n0} + 2\mu_n D_0 u_{n0} + \frac{1}{2}\pi^4 n^2 u_{n0} \sum_{m=1}^{\infty} m^2 u_{m0}^2\right) - f_{zn}(t)
\end{aligned} \tag{7.7.14}$$

Making the coefficient of the same power of ε zero in the above equation and considering the expression of $f_{zn}(t)$, we obtain

$$D_0^2 u_{n0} + n^4\pi^4 u_{n0} = 0 \tag{7.7.15}$$

$$D_0^2 u_{n1} + n^4\pi^4 u_{n1} = -2D_0 D_1 u_{n0} - 2\mu_n D_0 u_{n0} - \frac{1}{2}\pi^4 n^2 u_{n0} \sum_{m=1}^{\infty} m^2 u_{m0}^2 + \tilde{f}_{zn}(t) \tag{7.7.16}$$

where

$$\tilde{f}_{z1} = 2k_{11}\cos\Omega_1 t,\ \tilde{f}_{z2} = 2k_{21}\cos(\Omega_2 t + \theta),\ \tilde{f}_{zn}(t) = 0 \text{ for } n \geq 3 \tag{7.7.17}$$

The solution of the Eq. (7.7.15) is

$$u_{n0} = A_n(T_1)e^{in^2\pi^2 T_0} + cc \tag{7.7.18}$$

Substituting (7.7.18) into (7.7.16) yields

$$\begin{aligned}
D_0^2 u_{n1} + n^4\pi^4 u_{n1} &= -2in^2\pi^2 A_n' e^{in^2\pi^2 T_0} - 2in^2\pi^2\mu_n A_n e^{in^2\pi^2 T_0} \\
&\quad - \left(\pi^4 n^2 A_n \sum_{m=1}^{\infty} m^2 A_m \bar{A}_m\right) e^{in^2\pi^2 T_0}
\end{aligned}$$

$$-\frac{1}{2}\pi^4 n^4 A_n^2 \overline{A}_n e^{in^2\pi^2 T_0} + cc + NST + \tilde{f}_{zn}(t) \quad (7.7.19)$$

Here

$$\Omega_1 = \pi^2 + \varepsilon\sigma_1 \quad \text{and} \quad \Omega_2 = 4\pi^2 + \varepsilon\sigma_2 \quad (7.7.20)$$

In order to eliminate secular terms from (7.7.19), we need

$$2i\pi^2 A_1' + 2i\pi^2 \mu_1 A_1 + \frac{3}{2}\pi^4 A_1^2 \overline{A}_1 + \pi^4 A_1 \sum_{m=2}^{\infty} m^2 A_m \overline{A}_m - k_{11} e^{i\sigma_1 T_1} = 0 \quad (7.7.21)$$

$$8i\pi^2 A_2' + 8i\pi^2 \mu_2 A_2 + 24\pi^4 A_2^2 \overline{A}_2 + 4\pi^4 A_2 \sum_{m\neq 2}^{\infty} m^2 A_m \overline{A}_m - k_{21} e^{i(\sigma_2 T_1 + \theta)} = 0 \quad (7.7.22)$$

$$2in^2 A_n' + 2in^2 \mu_n A_n + \frac{3}{2} n^4 \pi^2 A_n^2 \bar{A}_n + n^2 \pi^2 A_n \sum_{m\neq n}^{\infty} m^2 A_m \bar{A}_m = 0 \quad \text{for } n \geq 3 \quad (7.7.23)$$

Therefore, the first order approximate solution of the governing equation is

$$u_n = A_n(T_1) e^{in^2\pi^2 T_0} + cc + O(\varepsilon) \quad (7.7.24)$$

$$A_n = \frac{1}{2} a_n e^{i\beta_n} A_n = \frac{1}{2} a_n e^{i\beta_n} \quad (7.7.25)$$

and substitute them into (7.7.23), we can obtain

$$in^2 a_n' - n^2 a_n \beta_n' + in^2 \mu_n a_n + \frac{3}{16} n^4 \pi^2 a_n^3 + \frac{1}{8} n^2 \pi^2 a_n \sum_{m\neq n}^{\infty} m^2 a_m^3 = 0, \quad n \geq 3 \quad (7.7.26)$$

Separating the real and imaginary parts of the above equation yields

$$a_n' + \mu_n a_n = 0, \ n \geq 3 \quad (7.7.27)$$

Therefore,

$$a_n = a_{n0} e^{-\mu_n T_1} \to 0 \Rightarrow A_n \to 0, \quad n \geq 3 \quad (7.7.28)$$

(c) Let

$$A_1 = \frac{1}{2}a_1 e^{i\beta_1}, A_2 = \frac{1}{2}a_2 e^{i\beta_2} \tag{7.7.29}$$

substitute them into (7.7.21) and (7.7.22), and take into account (7.7.28), we have

$$i\pi^2 a_1' - \pi^2 a_1 \beta_1' + i\pi^2 \mu_1 a_1 + \frac{3}{16}\pi^4 a_1^3 + \frac{1}{2}\pi^4 a_1 a_2^2 - k_{11} e^{i(\sigma_1 T_1 - \beta_1)} = 0 \tag{7.7.30}$$

$$4i\pi^2 a_2' - 4\pi^2 a_2 \beta_2' + 4i\pi^2 \mu_2 a_2 + 3\pi^4 a_2^3 + \frac{1}{2}\pi^4 a_1^2 a_2 - k_{21} e^{i(\sigma_2 T_1 - \beta_2 + \theta)} = 0 \tag{7.7.31}$$

Separating the real and imaginary parts of (7.7.30) and (7.7.31), we obtain

$$\begin{aligned} a_1' &= -\mu_1 a_1 + \pi^{-2} k_{11} \sin(\sigma_1 T_1 - \beta_1) \\ a_1 \beta_1' &= \frac{3}{16}\pi^2 a_1^3 + \frac{1}{2}\pi^2 a_1 a_2^2 - \pi^{-2} k_{11} \cos(\sigma_1 T_1 - \beta_1) \end{aligned} \tag{7.7.32}$$

$$\begin{aligned} a_2' &= -\mu_2 a_2 + \frac{1}{4}\pi^{-2} k_{21} \sin(\sigma_2 T_1 - \beta_2 + \theta) \\ a_2 \beta_2' &= \frac{3}{4}\pi^2 a_2^3 + \frac{1}{8}\pi^2 a_1^2 a_2 - \frac{1}{4}\pi^{-2} k_{21} \cos(\sigma_2 T_1 - \beta_2 + \theta) \end{aligned} \tag{7.7.33}$$

where

$$\gamma_1 = \sigma_1 T_1 - \beta_1, \gamma_2 = \sigma_2 T_1 - \beta_2 + \theta \tag{7.7.34}$$

Then (7.7.32) and (7.7.33) become

$$\begin{aligned} 20l a_1' &= -\mu_1 a_1 + \pi^{-2} k_{11} \sin\gamma_1 \\ a_1 \gamma_1' &= a_1 \sigma_1 - \frac{3}{16}\pi^2 a_1^3 - \frac{1}{2}\pi^2 a_1 a_2^2 + \pi^{-2} k_{11} \cos\gamma_1 \end{aligned} \tag{7.7.35}$$

$$\begin{aligned} a_2' &= -\mu_2 a_2 + \frac{1}{4}\pi^{-2} k_{21} \sin\gamma_2 \\ a_2 \gamma_2' &= a_2 \sigma_2 - \frac{3}{4}\pi^2 a_2^3 - \frac{1}{8}\pi^2 a_1^2 a_2 + \frac{1}{4}\pi^{-2} k_{21} \cos\gamma_2 \end{aligned} \tag{7.7.36}$$

Let $a_1' = \gamma_1' = a_2' = \gamma_2' = 0$, we can obtain the equation satisfied by the steady state solution:

$$\begin{aligned} \mu_1 a_1 &= \pi^{-2} k_{11} \sin\gamma_1 \\ \frac{3}{16}\pi^2 a_1^3 + \frac{1}{2}\pi^2 a_1 a_2^2 - a_1 \sigma_1 &= \pi^{-2} k_{11} \cos\gamma_1 \end{aligned} \tag{7.7.37}$$

$$\mu_2 a_2 = \frac{1}{4}\pi^{-2} k_{21} \sin\gamma_2$$
$$\frac{3}{4}\pi^2 a_2^3 + \frac{1}{8}\pi^2 a_1^2 a_2 - a_2\sigma_2 = \frac{1}{4}\pi^{-2} k_{21} \cos\gamma_2 \tag{7.7.38}$$

Eliminating γ_1 and γ_2 from the above two sets of equations, respectively, we can obtain the following system of frequency–response equations:

$$\mu_1^2 a_1^2 + \left(\frac{3}{16}\pi^2 a_1^3 + \frac{1}{2}\pi^2 a_1 a_2^2 - a_1\sigma_1\right)^2 = \pi^{-4} k_{11}^2$$
$$\mu_2^2 a_2^2 + \left(\frac{3}{4}\pi^2 a_2^3 + \frac{1}{8}\pi^2 a_1^2 a_2 - a_2\sigma_2\right)^2 = \frac{1}{16}\pi^{-4} k_{21}^2 \tag{7.7.39}$$

It can be seen that a_1 and a_2 are coupled with each other and are related to two detuning parameters σ_1 and σ_2. From this set of equations, the amplitude-frequency response curve can be plotted.

7.8 Exercise 7.8 (Forced Response of a Hinged-Hinged Beam with a Single Non-resonant Excitation at $u = O(w^2)$)

Solution: **(a)** We use the method of multiple scales to solve (7.7.12). Let the solution of the equation be

$$u_n(t;\varepsilon) = u_{n0}(T_0, T_1) + \varepsilon u_{n1}(T_0, T_1) + \cdots \tag{7.8.1}$$

Substituting (7.8.1) into (7.7.12) and retaining to $O(\varepsilon)$yields

$$\begin{aligned}
0 &= \ddot{u}_n + n^4\pi^4 u_n + \varepsilon\left(2\mu_n\dot{u}_n + \frac{1}{2}\pi^4 n^2 u_n \sum_{m=1}^{\infty} m^2 u_m^2\right) - f_{zn}(t)\\
&= \left(D_0^2 + 2\varepsilon D_0 D_1\right)(u_{n0} + \varepsilon u_{n1}) + n^4\pi^4(u_{n0} + \varepsilon u_{n1})\\
&\quad + \varepsilon\left(2\mu_n D_0 u_{n0} + \frac{1}{2}\pi^4 n^2 u_{n0}\sum_{m=1}^{\infty} m^2 u_{m0}^2\right) - 2k_n\cos(\Omega t + \theta_n)\\
&= D_0^2 u_{n0} + n^4\pi^4 u_{n0} - 2k_n\cos(\Omega t + \theta_n)\\
&\quad + \varepsilon\left(D_0^2 u_{n1} + n^4\pi^4 u_{n1} + 2D_0 D_1 u_{n0} + 2\mu_n D_0 u_{n0} + \frac{1}{2}\pi^4 n^2 u_{n0}\sum_{m=1}^{\infty} m^2 u_{m0}^2\right)
\end{aligned} \tag{7.8.2}$$

Making the coefficient of the same power of ε zero in the above equation yields

$$D_0^2 u_{n0} + n^4\pi^4 u_{n0} = 2k_n \cos(\Omega t + \theta_n) \tag{7.8.3}$$

$$D_0^2 u_{n1} + n^4\pi^4 u_{n1} = -2D_0 D_1 u_{n0} - 2\mu_n D_0 u_{n0} - \frac{1}{2}\pi^4 n^2 u_{n0} \sum_{m=1}^{\infty} m^2 u_{m0}^2 \tag{7.8.4}$$

The solution of the Eq. (7.8.3) is

$$u_{n0} = A_n(T_1) e^{in^2\pi^2 T_0} + \Lambda_n e^{i(\Omega T_0 + \theta_n)} + cc \tag{7.8.5}$$

where

$$\Lambda_n = \frac{K_n}{n^4\pi^4 - \Omega^2} \tag{7.8.6}$$

Substituting (7.8.6) into (7.8.4) and taking into account $\Omega = 2\pi^2 + \varepsilon\sigma\Omega = 2\pi^2 + \varepsilon\sigma$ and $\omega_2 - 2\omega_1 \approx \Omega$, we obtain

$$\begin{aligned} D_0^2 u_{n1} + n^4\pi^4 u_{n1} = & -2in^2\pi^2 A_n' e^{in^2\pi^2 T_0} - 2in^2\pi^2 \mu_n A_n e^{in^2\pi^2 T_0} \\ & - \pi^4 n^2 A_n e^{in^2\pi^2 T_0} \sum_{m=1}^{\infty} m^2 \left(A_m \overline{A}_m + \Lambda_m^2\right) \\ & - \pi^4 n^2 \overline{A}_n \sum_{m=1}^{\infty} m^2 \Lambda_m A_m e^{i\left(m^2\pi^2 T_0 - \Omega T_0 - \theta_m - n^2\pi^2 T\right)} \\ & - \frac{1}{2}\pi^4 n^2 \overline{A}_n \sum_{m=1}^{\infty} m^2 A_m^2 e^{i\left(2m^2\pi^2 - n^2\pi^2\right) T_0} \\ & - \pi^4 n^2 \Lambda_n \sum_{m=1}^{\infty} m^2 \Lambda_m A_m e^{i\left(m^2\pi^2 T_0 + \theta_n - \theta_m\right)} \\ & - \pi^4 n^2 \Lambda_n \sum_{m=1}^{\infty} m^2 \Lambda_m A_m e^{i\left(m^2\pi^2 T_0 + \theta_m - \theta_n\right)} \\ & - \frac{1}{2}\pi^4 n^2 \Lambda_n \sum_{m=1}^{\infty} m^2 A_m^2 e^{i\left(2m^2\pi^2 T_0 + \Omega T_0 + \theta_n\right)} + cc + NST \end{aligned} \tag{7.8.7}$$

In order to eliminate secular terms from (7.8.7) yields

$$\begin{aligned} & 2i\left(A_1' + \mu_1 A_1\right) + \frac{3}{2}\pi^2 A_1\left(A_1\overline{A}_1 + 2\Lambda_1^2\right) \\ & + \pi^2 A_1 \sum_{m=2}^{\infty} m^2\left(A_m\overline{A}_m + \Lambda_m^2\right) + 4\pi^2 \Lambda_2 A_2 \overline{A}_1 e^{-i(\sigma T_1 + \theta_2)} = 0 \end{aligned} \tag{7.8.8}$$

$$
\begin{aligned}
&8i\left(A_2' + \mu_2 A_2\right) + 4\pi^2 A_2 \sum_{m\neq 2}^{\infty} m^2\left(A_m\overline{A}_m + \Lambda_m^2\right) \\
&\quad + 24\pi^2 A_2\left(A_2\overline{A}_2 + 2\Lambda_2^2\right) + 2\pi^2\Lambda_2 A_1^2 e^{i(\sigma T_1+\theta_2)} = 0
\end{aligned} \tag{7.8.9}
$$

$$
\begin{aligned}
&2i\left(A_n' + \mu_n A_n\right) + \frac{3}{2}n^2\pi^2 A_n\left(A_n\bar{A}_n + 2\Lambda_n^2\right) \\
&\quad + \pi^2 A_n \sum_{m\neq n}^{\infty} m^2\left(A_m\bar{A}_m + \Lambda_m^3\right) = 0, \quad n \geq 3
\end{aligned} \tag{7.8.10}
$$

Therefore, the first order approximate solution of the governing equation is

$$
u_n = A_n(T_1)e^{in^2\pi^2 T_0} + \Lambda_n e^{i(\Omega T_0+\theta_n)} + cc + O(\varepsilon) \tag{7.8.11}
$$

(b) Let

$$
A_n = \frac{1}{2}a_n e^{i\beta_n} \tag{7.8.12}
$$

and substitute it into (7.8.10), we can obtain

$$
\begin{aligned}
&in^2 a_n' - n^2 a_n\beta_n' + in^2\mu_n a_n + \frac{3}{4}n^2\pi^2 a_n\left(\frac{1}{4}a_n^2 + 2\Lambda_n^2\right) \\
&\quad + \frac{1}{2}\pi^2 a_n \sum_{m\neq n}^{\infty} m^2\left(\frac{1}{4}a_m^2 + \Lambda_m^2\right) = 0, \quad n \geq 3
\end{aligned} \tag{7.8.13}
$$

Separating the real and imaginary parts of the above equation yields

$$
a_n' + \mu_n a_n = 0, \quad n \geq 3 \tag{7.8.14}
$$

Therefore,

$$
a_n = a_{n0}e^{-\mu_n T_1} \to 0 \Rightarrow \quad A_n \to 0, \ n \geq 3 \tag{7.8.15}
$$

Let

$$
A_1 = \frac{1}{2}a_1 e^{i\beta_1}, \quad A_2 = \frac{1}{2}a_2 e^{i\beta_2} \tag{7.8.16}
$$

substitute them into (7.8.8) and (7.8.9), and take into account (7.8.15), we have

$$
ia_1' - a_1\beta_1' + i\mu_1 a_1 + \frac{3}{16}\pi^2 a_1\left(a_1^2 + 8\Lambda_1^2\right)
$$

$$+\frac{1}{2}\pi^2 a_1\left(a_2^2+4\Lambda_2^2\right)+\pi^2\Lambda_2 a_1 a_2 e^{-i(\sigma T_1+2\beta_1-\beta_2+\theta_2)}=0 \tag{7.8.17}$$

$$\begin{aligned}&8ia_2'-8a_2\beta_2'+8i\mu_2 a_2+\frac{1}{2}\pi^2 a_2\left(a_1^2+4\Lambda_1^2\right)\\&\quad+3\pi^2 a_2\left(a_2^2+8\Lambda_2^2\right)+\frac{1}{2}\pi^2\Lambda_2 a_1^2 e^{i(\sigma T_1+2\beta_1-\beta_2+\theta_2)}=0\end{aligned} \tag{7.8.18}$$

Separating the real and imaginary parts of (7.8.17) and (7.8.18), we obtain

$$\begin{aligned}a_1'&=-\mu_1 a_1+\pi^2\Lambda_2 a_1 a_2\sin(\sigma T_1+2\beta_1-\beta_2+\theta_2)\\a_1\beta_1'&=\frac{3}{16}\pi^2 A_1\left(a_1^2+8\Lambda_1^2\right)+\frac{1}{2}\pi^2 a_1\left(a_2^2+4\Lambda_2^2\right)\\&\quad-\pi^2\Lambda_2 a_1 a_2\cos(\sigma T_1+2\beta_1-\beta_2+\theta_2)\end{aligned} \tag{7.8.19}$$

$$\begin{aligned}a_2'&=-\mu_2 a_2-\frac{1}{16}\pi^2\Lambda_2 a_1^2\sin(\sigma T_1+2\beta_1-\beta_2+\theta_2)\\a_2\beta_2'&=\frac{1}{16}\pi^2 a_2\left(a_1^2+4\Lambda_1^2\right)+\frac{3}{8}\pi^2 a_2\left(a_2^2+8\Lambda_2^2\right)\\&\quad+\frac{1}{16}\pi^2\Lambda_2 a_1^2\cos(\sigma T_1+2\beta_1-\beta_2+\theta_2)\end{aligned} \tag{7.8.20}$$

where

$$\gamma=\sigma T_1+2\beta_1-\beta_2+\theta_2 \tag{7.8.21}$$

Then (7.8.18) and (7.8.20) become

$$\begin{aligned}a_1'&=-\mu_1 a_1+\pi^2\Lambda_2 a_1 a_2\sin\gamma\\a_1\beta_1'&=\frac{3}{16}\pi^2 a_1\left(a_1^2+8\Lambda_1^2\right)+\frac{1}{2}\pi^2 a_1\left(a_2^2+4\Lambda_2^2\right)-\pi^2\Lambda_2 a_1 a_2\cos\gamma\end{aligned} \tag{7.8.22}$$

$$\begin{aligned}a_2'&=-\mu_2 a_2-\frac{1}{16}\pi^2\Lambda_2 a_1^2\sin\gamma\\a_2\gamma'-2a_2\beta_1'&=a_2\sigma_1-\frac{1}{16}\pi^2 a_2\left(a_1^2+4\Lambda_1^2\right)-\frac{3}{8}\pi^2 a_2\left(a_2^2+8\Lambda_2^2\right)\\&\quad-\frac{1}{16}\pi^2\Lambda_2 a_1^2\cos\gamma\end{aligned} \tag{7.8.23}$$

Let $a_1'=a_2'=\beta_1'=\gamma'=0$, we can obtain the equation satisfied by the steady state solution:

$$\begin{aligned}&\mu_1 a_1=\pi^2\Lambda_2 a_1 a_2\sin\gamma\\&\frac{3}{16}\pi^2 A_1\left(a_1^2+8\Lambda_1^2\right)+\frac{1}{2}\pi^2 a_1\left(a_2^2+4\Lambda_2^2\right)=\pi^2\Lambda_2 a_1 a_2\cos\gamma\end{aligned} \tag{7.8.24}$$

$$\mu_2 a_2 = \frac{1}{16}\pi^2\Lambda_2 a_1^2 \sin\gamma$$
$$a_2\sigma_1 - \frac{1}{16}\pi^2 a_2\left(a_1^2 + 4\Lambda_1^2\right) - \frac{3}{8}\pi^2 a_2\left(a_2^2 + 8\Lambda_2^2\right) = \frac{1}{16}\pi^2\Lambda_2 a_1^2 \cos\gamma \quad (7.8.25)$$

Eliminating γ from each of the above two sets of equations yields a system of frequency–response equations as

$$\mu_1^2 a_1^2 + \left[\frac{3}{16}\pi^2 A_1\left(a_1^2 + 8\Lambda_1^2\right) + \frac{1}{2}\pi^2 a_1\left(a_2^2 + 4\Lambda_2^2\right)\right]^2 = \pi^4\Lambda_2^2 a_1^2 a_2^2$$
$$\mu_2^2 a_2^2 + \left[a_2\sigma_1 - \frac{1}{16}\pi^2 a_2\left(a_1^2 + 4\Lambda_1^2\right) - \frac{3}{8}\pi^2 a_2\left(a_2^2 + 8\Lambda_2^2\right)\right]^2 = \frac{1}{256}\pi^4\Lambda_2^2 a_1^4 \quad (7.8.26)$$

From this set of equations, we can plot the amplitude-frequency response curve.

7.9 Exercise 7.9 (Combined Resonance Analysis of Hinged-Hinged Beams with Two Excitations at $u = O(w^2)$)

Solution: **(a)** We use the method of multiple scales to solve (7.7.12). Let the solution of the equation be

$$u_n(t;\varepsilon) = u_{n0}(T_0, T_1) + \varepsilon u_{n1}(T_0, T_1) + \cdots \quad (7.9.1)$$

Substituting (7.9.1) into (7.7.12) and retaining to $O(\varepsilon)$ yields

$$\begin{aligned}
0 &= \ddot{u}_n + n^4\pi^4 u_n + \varepsilon\left(2\mu_n\dot{u}_n + \frac{1}{2}\pi^4 n^2 u_n \sum_{m=1}^{\infty} m^2 u_m^2\right) - f_{zn}(t)\\
&= \left(D_0^2 + 2\varepsilon D_0 D_1\right)(u_{n0} + \varepsilon u_{n1}) + n^4\pi^4(u_{n0} + \varepsilon u_{n1})\\
&\quad + \varepsilon\left(2\mu_n D_0 u_{n0} + \frac{1}{2}\pi^4 n^2 u_{n0}\sum_{m=1}^{\infty} m^2 u_{m0}^2\right) - 2k_n\cos(\Omega t + \theta_n)\\
&= D_0^2 u_{n0} + n^4\pi^4 u_{n0} - 2k_n\cos(\Omega t + \theta_n)\\
&\quad + \varepsilon\left(D_0^2 u_{n1} + n^4\pi^4 u_{n1} + 2D_0 D_1 u_{n0} + 2\mu_n D_0 u_{n0} + \frac{1}{2}\pi^4 n^2 u_{n0}\sum_{m=1}^{\infty} m^2 u_{m0}^2\right)
\end{aligned} \quad (7.9.2)$$

Making the coefficient of the same power of ε zero in the above equation yields

$$D_0^2 u_{n0} + n^4\pi^4 u_{n0} = 2k_n \cos(\Omega t + \theta_n) \tag{7.9.3}$$

$$D_0^2 u_{n1} + n^4\pi^4 u_{n1} = -2D_0 D_1 u_{n0} - 2\mu_n D_0 u_{n0} - \frac{1}{2}\pi^4 n^2 u_{n0} \sum_{m=1}^{\infty} m^2 u_{m0}^2 \tag{7.9.4}$$

The solution of the Eq. (7.9.3) is

$$u_{n0} = A_n(T_1) e^{in^2\pi^2 T_0} + \Lambda_n e^{i(\Omega T_0 + \theta_n)} + cc \tag{7.9.5}$$

where

$$\Lambda_n = \frac{K_n}{n^4\pi^4 - \Omega^2} \tag{7.9.6}$$

Substituting (7.9.5) into (7.9.4) and taking into account $\Omega = 2\pi^2 + \varepsilon\sigma$ and $\omega_2 - 2\omega_1 \approx \Omega$, we obtain

$$\begin{aligned} D_0^2 u_{n1} + n^4\pi^4 u_{n1} = & -2in^2\pi^2 A' e^{in^2\pi^2 T_0} - 2in^2\pi^2 \mu_n A_n e^{in^2\pi^2 T_0} \\ & - \pi^4 n^2 A_n e^{in^2\pi^2 T_0} \sum_{m=1}^{\infty} m^2 \left(A_m \overline{A}_m + \Lambda_m^2\right) \\ & - \pi^4 n^2 \overline{A}_n \sum_{m=1}^{\infty} m^2 \Lambda_m A_m e^{i\left(m^2\pi^2 T_0 - \Omega T_0 - \theta_m - n^2\pi^2 T\right)} \\ & - \frac{1}{2}\pi^4 n^2 \overline{A}_n \sum_{m=1}^{\infty} m^2 A_m^2 e^{i\left(2m^2\pi^2 - n^2\pi^2\right) T_0} \\ & - \pi^4 n^2 \Lambda_n \sum_{m=1}^{\infty} m^2 \Lambda_m A_m e^{i\left(m^2\pi^2 T_0 + \theta_n - \theta_m\right)} \\ & - \pi^4 n^2 \Lambda_n \sum_{m=1}^{\infty} m^2 \Lambda_m A_m e^{i\left(m^2\pi^2 T_0 + \theta_m - \theta_n\right)} \\ & - \frac{1}{2}\pi^4 n^2 \Lambda_n \sum_{m=1}^{\infty} m^2 A_m^2 e^{i\left(2m^2\pi^2 T_0 + \Omega T_0 + \theta_n\right)} + cc + NST \end{aligned} \tag{7.9.7}$$

In order to eliminate secular terms from (7.9.7) yields

$$\begin{aligned} & 2i\left(A_1' + \mu_1 A_1\right) + \frac{3}{2}\pi^2 A_1 \left(A_1 \overline{A}_1 + 2\Lambda_1^2\right) \\ & + \pi^2 A_1 \sum_{m=2}^{\infty} m^2 \left(A_m \overline{A}_m + \Lambda_m^2\right) + 4\pi^2 \Lambda_2 A_2 \overline{A}_1 e^{-i(\sigma T_1 + \theta_2)} = 0 \end{aligned} \tag{7.9.8}$$

$$8i\left(A_2' + \mu_2 A_2\right) + 4\pi^2 A_2 \sum_{m\neq 2}^{\infty} m^2\left(A_m\overline{A}_m + \Lambda_m^2\right)$$
$$+ 24\pi^2 A_2\left(A_2\overline{A}_2 + 2\Lambda_2^2\right) + 2\pi^2\Lambda_2 A_1^2 e^{i(\sigma T_1+\theta_2)} = 0 \tag{7.9.9}$$

$$2i\left(A_n' + \mu_n A_n\right) + \frac{3}{2}n^2\pi^2 A_n\left(A_n\overline{A}_n + 2\Lambda_n^2\right)$$
$$+ \pi^2 A_n \sum_{m\neq n}^{\infty} m^2\left(A_m\overline{A}_m + \Lambda_m^2\right) = 0, \quad n \geq 3 \tag{7.9.10}$$

Therefore, the first order approximate solution of the equation is

$$u_n = A_n(T_1)e^{in^2\pi^2 T_0} + \Lambda_n e^{i(\Omega T_0+\theta_n)} + cc + O(\varepsilon) \tag{7.9.11}$$

(b) Let

$$A_n = \frac{1}{2}a_n e^{i\beta_n} \tag{7.9.12}$$

and substitute it into (7.9.10), we can obtain

$$in^2 a_n' - n^2 a_n\beta_n' + in^2\mu_n a_n + \frac{3}{4}n^2\pi^2 a_n\left(\frac{1}{4}a_n^2 + 2\Lambda_n^2\right)$$
$$+ \frac{1}{2}\pi^2 a_n \sum_{m\neq n}^{\infty} m^2\left(a_m^2 + \Lambda_m^2\right) = 0, \quad n \geq 3 \tag{7.9.13}$$

Separating the real and imaginary parts of the above equation gives

$$a_n' + \mu_n a_n = 0, \ n \geq 3 \tag{7.9.14}$$

Therefore,

$$a_n = a_{n0}e^{-\mu_n T_1} \to 0 \Rightarrow A_n \to 0, \quad n \geq 3 \tag{7.9.15}$$

Let

$$A_1 = \frac{1}{2}a_1 e^{i\beta_1}, \quad A_2 = \frac{1}{2}a_2 e^{i\beta_2} \tag{7.9.16}$$

substitute them into (7.9.8) and (7.9.9), and take into account (7.9.15), we have

$$ia_1' - a_1\beta_1' + i\mu_1 a_1 + \frac{3}{16}\pi^2 A_1\left(a_1^2 + 8\Lambda_1^2\right)$$

$$+\frac{1}{2}\pi^2 a_1\left(a_2^2+4\Lambda_2^2\right)+\pi^2\Lambda_2 a_1 a_2 e^{-i(\sigma T_1+2\beta_1-\beta_2+\theta_2)}=0 \tag{7.9.17}$$

$$8ia_2'-8a_2\beta_2'+8i\mu_2 a_2+\frac{1}{2}\pi^2 a_2\left(a_1^2+4\Lambda_1^2\right)$$
$$+3\pi^2 a_2\left(a_2^2+8\Lambda_2^2\right)+\frac{1}{2}\pi^2\Lambda_2 a_1^2 e^{i(\sigma T_1+2\beta_1-\beta_2+\theta_2)}=0 \tag{7.9.18}$$

Separating the real and imaginary parts of (7.9.17) and (7.9.18), we obtain

$$\begin{aligned}a_1'&=-\mu_1 a_1+\pi^2\Lambda_2 a_1 a_2\sin(\sigma T_1+2\beta_1-\beta_2+\theta_2)\\ a_1\beta_1'&=\frac{3}{16}\pi^2 A_1\left(a_1^2+8\Lambda_1^2\right)+\frac{1}{2}\pi^2 a_1\left(a_2^2+4\Lambda_2^2\right)\\ &\quad-\pi^2\Lambda_2 a_1 a_2\cos(\sigma T_1+2\beta_1-\beta_2+\theta_2)\end{aligned} \tag{7.9.19}$$

$$\begin{aligned}a_2'&=-\mu_2 a_2-\frac{1}{16}\pi^2\Lambda_2 a_1^2\sin(\sigma T_1+2\beta_1-\beta_2+\theta_2)\\ a_2\beta_2'&=\frac{1}{16}\pi^2 a_2\left(a_1^2+4\Lambda_1^2\right)+\frac{3}{8}\pi^2 a_2\left(a_2^2+8\Lambda_2^2\right)\\ &\quad+\frac{1}{16}\pi^2\Lambda_2 a_1^2\cos(\sigma T_1+2\beta_1-\beta_2+\theta_2)\end{aligned} \tag{7.9.20}$$

where

$$\gamma=\sigma T_1+2\beta_1-\beta_2+\theta_2 \tag{7.9.21}$$

Then (7.9.19) and (7.9.20) become

$$\begin{aligned}a_1'&=-\mu_1 a_1+\pi^2\Lambda_2 a_1 a_2\sin\gamma\\ a_1\beta_1'&=\frac{3}{16}\pi^2 a_1\left(a_1^2+8\Lambda_1^2\right)+\frac{1}{2}\pi^2 a_1\left(a_2^2+4\Lambda_2^2\right)-\pi^2\Lambda_2 a_1 a_2\cos\gamma\end{aligned} \tag{7.9.22}$$

$$\begin{aligned}a_2'&=-\mu_2 a_2-\frac{1}{16}\pi^2\Lambda_2 a_1^2\sin\gamma\\ a_2\gamma'-2a_2\beta_1'&=a_2\sigma_1-\frac{1}{16}\pi^2 a_2\left(a_1^2+4\Lambda_1^2\right)-\frac{3}{8}\pi^2 a_2\left(a_2^2+8\Lambda_2^2\right)\\ &\quad-\frac{1}{16}\pi^2\Lambda_2 a_1^2\cos\gamma\end{aligned} \tag{7.9.23}$$

Let $a_1'=a_2'=\beta_1'=\gamma'=0$, we can obtain the equation satisfied by the steady state solution

$$\begin{aligned}&\mu_1 a_1=\pi^2\Lambda_2 a_1 a_2\sin\gamma\\ &\frac{3}{16}\pi^2 A_1\left(a_1^2+8\Lambda_1^2\right)+\frac{1}{2}\pi^2 a_1\left(a_2^2+4\Lambda_2^2\right)=\pi^2\Lambda_2 a_1 a_2\cos\gamma\end{aligned} \tag{7.9.24}$$

$$\mu_2 a_2 = \frac{1}{16}\pi^2 \Lambda_2 a_1^2 \sin\gamma$$
$$a_2\sigma_1 - \frac{1}{16}\pi^2 a_2\left(a_1^2 + 4\Lambda_1^2\right) - \frac{3}{8}\pi^2 a_2\left(a_2^2 + 8\Lambda_2^2\right) = \frac{1}{16}\pi^2 \Lambda_2 a_1^2 \cos\gamma \quad (7.9.25)$$

Eliminating γ from the above two sets of equations, respectively, we can obtain a system of frequency–response equations

$$\mu_1^2 a_1^2 + \left[\frac{3}{16}\pi^2 A_1\left(a_1^2 + 8\Lambda_1^2\right) + \frac{1}{2}\pi^2 a_1\left(a_2^2 + 4\Lambda_2^2\right)\right]^2 = \pi^4 \Lambda_2^2 a_1^2 a_2^2$$
$$\mu_2^2 a_2^2 + \left[a_2\sigma_1 - \frac{1}{16}\pi^2 a_2\left(a_1^2 + 4\Lambda_1^2\right) - \frac{3}{8}\pi^2 a_2\left(a_2^2 + 8\Lambda_2^2\right)\right]^2 = \frac{1}{256}\pi^4 \Lambda_2^2 a_1^4 \quad (7.9.26)$$

From this set of equations, the amplitude-frequency response curve can be plotted.

7.10 Exercise 7.10 (Combined Resonance Analysis of a Hinged-Hinged Beam with Three Excitations at $u = O(w^2)$)

Solution: (a) We use the method of multiple scales to solve (7.7.12). Let the solution of the equation be

$$u_n(t;\varepsilon) = u_{n0}(T_0, T_1) + \varepsilon u_{n1}(T_0, T_1) + \cdots \quad (7.10.1)$$

Substituting (7.10.1) into (7.7.12) and retaining to $O(\varepsilon)$ yields

$$\begin{aligned}
0 &= \ddot{u}_n + n^4\pi^4 u_n + \varepsilon\left(2\mu_n \dot{u}_n + \frac{1}{2}\pi^4 n^2 u_n \sum_{m=1}^{\infty} m^2 u_m^2\right) - f_{zn}(t)\\
&= \left(D_0^2 + 2\varepsilon D_0 D_1\right)(u_{n0} + \varepsilon u_{n1}) + n^4\pi^4(u_{n0} + \varepsilon u_{n1})\\
&\quad + \varepsilon\left(2\mu_n D_0 u_{n0} + \frac{1}{2}\pi^4 n^2 u_{n0} \sum_{m=1}^{3} m^2 u_{m0}^2\right) - 2\sum_{m=1}^{3} K_{nm}\cos(\Omega_m t + \theta_{nm})\\
&= D_0^2 u_{n0} + n^4\pi^4 u_{n0} - 2\sum_{m=1}^{3} K_{nm}\cos(\Omega_m t + \theta_{nm})\\
&\quad + \varepsilon\left(D_0^2 u_{n1} + n^4\pi^4 u_{n1} + 2D_0 D_1 u_{n0} + 2\mu_n D_0 u_{n0} + \frac{1}{2}\pi^4 n^2 u_{n0} \sum_{m=1}^{\infty} m^2 u_{m0}^2\right)
\end{aligned} \quad (7.10.2)$$

Making the coefficient of the same power of ε zero in the above equation yields

$$D_0^2 u_{n0} + n^4\pi^4 u_{n0} = 2\sum_{m=1}^{3} K_{nm}\cos(\Omega_m t + \theta_{nm}) \tag{7.10.3}$$

$$D_0^2 u_{n1} + n^4\pi^4 u_{n1} = -2D_0D_1u_{n0} - 2\mu_n D_0 u_{n0} - \frac{1}{2}\pi^4 n^2 u_{n0}\sum_{m=1}^{\infty} m^2 u_{m0}^2 \tag{7.10.4}$$

The solution of (7.10.3) is

$$u_{n0} = A_n(T_1)e^{in^2\pi^2 T_0} + \sum_{m=1}^{3}\Lambda_{nm}e^{i\Omega_m T_0} + cc \tag{7.10.5}$$

where

$$\Lambda_{nm} = \frac{K_{nm}e^{i\theta_{nm}}}{n^4\pi^4 - \Omega_m^2} \tag{7.10.6}$$

Substituting (7.10.5) into (7.10.4) yields

$$\begin{aligned} D_0^2 u_{n1} + n^2\pi^2 u_{n1} &= -2D_0D_1u_{n0} - 2\mu_n D_0 u_{n0} - \frac{1}{2}\pi^4 n^2 u_{n0}\sum_{m=1}^{\infty} m^2 u_{m0}^2 \\ &= -2in^2\pi^2 A_n' e^{in^2\pi^2 T_0} - 2in^2\pi^2\mu_n A_n e^{in^2\pi^2 T_0} + cc + NST \\ &\quad - \frac{1}{2}\pi^4 n^2 u_{n0}\sum_{m=1}^{\infty} m^2 u_{m0}^2 \end{aligned} \tag{7.10.7}$$

Expanding the last term on the right-hand side of the above equation with (7.10.5) and taking into account $\Omega_1 + \Omega_2 + \Omega_3 = \pi^2 + \varepsilon\sigma$, we obtain

$$\begin{aligned} &-\frac{1}{2}\pi^4 n^2 u_{n0}\sum_{m=1}^{\infty} m^2 u_{m0}^2 \\ &= -\frac{1}{2}\pi^4 n^2\left(A_n e^{in^2\pi^2 T_0} + \sum_{p=1}^{3}\Lambda_{np}e^{i\Omega_p T_0}\right) \\ &\quad \times \sum_{m=1}^{\infty} m^2\left[A_m e^{im^2\pi^2 T_0} + \sum_{k=1}^{3}\Lambda_{mk}e^{i\Omega_k T_0} + \bar{A}_m e^{-im^2\pi^2 T_0} + \sum_{k=1}^{3}\bar{\Lambda}_{mk}e^{-i\Omega_k T_0}\right]^2 \\ &= -\pi^4 n^2 \sum_{m=1}^{\infty} m^2\left[\frac{1}{2}\bar{A}_n A_m^2 e^{-in^2\pi^2 T_0}e^{2im^2\pi^2 T_0} + A_n A_m \bar{A}_m e^{in^2\pi^2 T_0}\right. \\ &\quad + A_n e^{in^2\pi^2 T_0}\sum_{k=1}^{3}|\Lambda_{mk}|^2 + (\Lambda_{m1}\Lambda_{m2}\Lambda_{n3} + \Lambda_{m1}\Lambda_{n2}\Lambda_{m3} \end{aligned}$$

$$+ \Lambda_{n1}\Lambda_{m2}\Lambda_{m3})e^{i(\Omega_1+\Omega_2+\Omega_3)T_0} + A_m e^{im^2\pi^2 T_0}\sum_{k=1}^{3}\bar{\Lambda}_{mk}\Lambda_{nk}$$

$$+ A_m e^{im^2\pi^2 T_0}\sum_{k=1}^{3}\bar{\Lambda}_{nk}\Lambda_{mk}\Bigg] + cc + NST \tag{7.10.8}$$

Substituting (7.10.8) into (7.10.7) yields

$$\begin{aligned}
D_0^2 u_{n1} + n^4\pi^4 u_{n1} &= -2D_0D_1u_{n0} - 2\mu_n D_0 u_{n0} - \frac{1}{2}\pi^4 n^2 u_{n0}\sum_{m=1}^{\infty} m^2 u_{m0}^2 \\
&= -2in^2\pi^2 A_n' e^{in^2\pi^2 T_0} - 2in^2\pi^2\mu_n A_n e^{in^2\pi^2 T_0} \\
&\quad + e^{i\pi^2 T_0}\sum_{m=1}^{\infty} m^2(\Lambda_{m1}\Lambda_{m2}\Lambda_{n3} + \Lambda_{m1}\Lambda_{n2}\Lambda_{m3} + \Lambda_{n1}\Lambda_{m2}\Lambda_{m3})e^{i\upsilon T_1} \\
&\quad - \pi^4 n^2\sum_{m=1}^{\infty} m^2\Bigg[\frac{1}{2}\bar{A}_n A_m^2 e^{-in^2\pi^2 T_0} e^{2im^2\pi^2 T_0} + A_n A_m \bar{A}_m e^{in^2\pi^2 T_0} \\
&\quad + A_n e^{in^2\pi^2 T_0}\sum_{k=1}^{3}|\Lambda_{mk}|^2 + A_m e^{im^2\pi^2 T_0}\sum_{k=1}^{3}\bar{\Lambda}_{mk}\Lambda_{nk} \\
&\quad + A_m e^{im^2\pi^2 T_0}\sum_{k=1}^{3}\bar{\Lambda}_{nk}\Lambda_{mk}\Bigg] + cc + NST
\end{aligned} \tag{7.10.9}$$

In order to eliminate secular terms from (7.10.9), we need

$$\begin{aligned}
&2i\pi^2(A_1' + \mu_1 A_1) + \frac{3}{2}\pi^4 A_1\left(A_1\bar{A}_1 + 2\sum_{k=1}^{3}|\Lambda_{1k}|^2\right) \\
&+ \pi^4 A_1\sum_{m=2}^{\infty} m^2\left(A_m\bar{A}_m + \sum_{k=1}^{3}|\Lambda_{mk}|^2\right) \\
&+ \pi^4\sum_{m=1}^{\infty} m^2(\Lambda_{m1}\Lambda_{m2}\Lambda_{13} + \Lambda_{m1}\Lambda_{m3}\Lambda_{12} + \Lambda_{m2}\Lambda_{m3}\Lambda_{11})e^{i\sigma T_1} = 0
\end{aligned} \tag{7.10.10}$$

$$\begin{aligned}
&*20l2in^2\pi^2(A_n' + \mu_n A_n) + \frac{3}{2}\pi^4 n^4 A_n\left(A_n\bar{A}_n + 2\sum_{k=1}^{3}|\Lambda_{nk}|^2\right) \\
&\quad + \pi^4 n^2 A_n\sum_{m\neq n}^{\infty} m^2\left(A_m\bar{A}_m + \sum_{k=1}^{3}|\Lambda_{mk}|^2\right) = 0, \quad n \geq 2
\end{aligned} \tag{7.10.11}$$

Therefore, the first order approximate solution of the equation is

$$u_n = A_n(T_1)e^{in^2\pi^2 T_0} + \sum_{m=1}^{3}\Lambda_{nm}e^{i\Omega_m T_0} + cc + O(\varepsilon) \tag{7.10.12}$$

(**b**) Let

$$A_n = \frac{1}{2}a_n e^{i\beta_n} \tag{7.10.13}$$

Substituting (7.10.13) into (7.10.11) yields

$$\begin{aligned} & in^2 a_n' - n^2 a_n \beta_n' + in^2 \mu_n a_n + \frac{3}{4}n^2\pi^2 a_n\left(\frac{1}{4}a_n^2 + 2\sum_{k=1}^{3}|\Lambda_{nk}|^2\right) \\ & + \frac{1}{2}n^2\pi^2 a_n \sum_{m\neq n}^{\infty} m^2\left(\frac{1}{4}a_m^2 + 2\sum_{k=1}^{3}|\Lambda_{nk}|^2\right) = 0, \quad n \geq 2 \end{aligned} \tag{7.10.14}$$

Separating the real and imaginary parts of the above equation yields

$$a_n' + \mu_n a_n = 0, \; n \geq 2 \tag{7.10.15}$$

therefore

$$a_n = a_{n0}e^{-\mu_n T_1} \to 0 \Rightarrow A_n \to 0, \; n \geq 2 \tag{7.10.16}$$

Let

$$A_1 = \frac{1}{2}a_1 e^{i\beta_1} \tag{7.10.17}$$

Substitute it into (7.10.10) and take into account (7.10.16), we have

$$\begin{aligned} & ia_1' - a_1\beta_1' + i\mu_1 a_1 + \frac{3}{16}\pi^2 a_1\left(a_1^2 + 8\sum_{k=1}^{3}|\Lambda_{1k}|^2\right) \\ & + \pi^2\sum_{m=1}^{\infty} m^2(\Lambda_{m1}\Lambda_{m2}\Lambda_{13} + \Lambda_{m1}\Lambda_{m3}\Lambda_{12} + \Lambda_{m2}\Lambda_{m3}\Lambda_{11})e^{i(\sigma T_1 - \beta_1)} = 0 \end{aligned} \tag{7.10.18}$$

Separating the real and imaginary parts of Eq. (7.10.18) yields

$$\begin{aligned} a_1' &= -\mu_1 a_1 - \pi^2 E_2 \sin\gamma \\ a_1\gamma' &= a_1\sigma - \frac{3}{16}\pi^2 a_1\left(a_1^2 + E_1\right) - \pi^2 E_2 \cos\gamma \end{aligned} \tag{7.10.19}$$

where

$$\gamma = \sigma T_1 - \beta_1 \tag{7.10.20}$$

$$E_1 = 8\sum_{k=1}^{3} |\Lambda_{1k}|^2$$
$$E_2 = \sum_{m=1}^{\infty} m^2(\Lambda_{m1}\Lambda_{m2}\Lambda_{13} + \Lambda_{m1}\Lambda_{m3}\Lambda_{12} + \Lambda_{m2}\Lambda_{m3}\Lambda_{11}) \tag{7.10.21}$$

Let $a_1' = \gamma' = 0$, we can obtain the equation satisfied by the steady state solution

$$-\mu_1 a_1 = \pi^2 E_2 \sin\gamma$$
$$a_1\sigma - \frac{3}{16}\pi^2 a_1\left(a_1^2 + E_1\right) = \pi^2 E_2 \cos\gamma \tag{7.10.22}$$

Eliminating γ in each of the above two sets of equations yields a system of frequency–response equations as

$$\mu_1^2 a_1^2 + a_1^2\left[\sigma - \frac{3}{16}\pi^2\left(a_1^2 + E_1\right)\right]^2 = \pi^4 E_2^2 \tag{7.10.23}$$

From this set of equations, the amplitude-frequency response curve can be plotted.

7.11 Exercise 7.11 (Parametric Resonance Analysis on a Hinged-Hinged Beam at ($u = O(w^2)$))

Solution: **(a)** The governing equation of the present problem is still (7.7.1), and its modal discretization equation is (7.7.8), i.e.

$$\ddot{u}_n + \omega_n^2 u_n = \varepsilon\left[\sum_{m=1}^{\infty} p_{nm}(t)u_m + \sum_{m,p,q=1}^{\infty} \Gamma_{nmpq}u_m u_p u_q - 2\mu_n \dot{u}_n\right] + f_{zn}(t) \tag{7.11.1}$$

We take the dimensionless length of the beam $l = 1$; and, in turn, by Eq. (7.7.3) and (7.7.4), the natural frequency of the linear free oscillation of the hinged-hinged beam can be obtained as $\omega_n = n^2\pi^2$ and the corresponding normalized modes are

$$\phi_n(x) = \sqrt{2}\sin n\pi x, \quad n = 1, 2, 3, \ldots \tag{7.11.2}$$

Since $P(t) \neq 0$,

$$\varepsilon p_{nm}(t) = r^{-2}P(t)\int_0^l \phi_m'\phi_n dx$$

$$= -2n^2\pi^2 r^{-2} P(t) \int_0^l \sin(m\pi x)\sin(n\pi x)dx$$

$$= \begin{cases} -\varepsilon n^2\pi^2 p(t), & m = n \\ 0, & m \neq n \end{cases} \tag{7.11.3}$$

where

$$\varepsilon p(t) = r^{-2} P(t) \tag{7.11.4}$$

Considering again (7.7.9), we can obtain the modal discretization equation for the beam of this problem (7.11.1) can be rewritten as

$$\ddot{u}_n + n^4\pi^4 u_n = -\varepsilon\left[2\mu_n\dot{u}_n + \frac{1}{2}\pi^4 n^2 u_n \sum_{m=1}^{\infty} m^2 u_m^2 + \pi^2 n^2 p(t) u_n\right] + f_{zn}(t) \tag{7.11.5}$$

It can be seen that this is a set of parametric excitation equations.

(b) We use the method of multiple scales to solve (7.11.5). Let the solution of the equation be

$$u_n(t;\varepsilon) = u_{n0}(T_0, T_1) + \varepsilon u_{n1}(T_0, T_1) + \cdots \tag{7.11.6}$$

Substituting (7.11.6) into (7.11.5) and retaining to O(ε), we obtain

$$\begin{aligned} 0 &= \ddot{u}_n + n^4\pi^4 u_n + \varepsilon\left(2\mu_n\dot{u}_n + \frac{1}{2}\pi^4 n^2 u_n \sum_{m=1}^{\infty} m^2 u_m^2 + \pi^2 n^2 p u_n\right) - f_{zn}(t) \\ &= \left(D_0^2 + 2\varepsilon D_0 D_1\right)(u_{n0} + \varepsilon u_{n1}) + n^4\pi^4(u_{n0} + \varepsilon u_{n1}) \\ &\quad + \varepsilon\left(2\mu_n D_0 u_{n0} + \frac{1}{2}\pi^4 n^2 u_{n0} \sum_{m=1}^{\infty} m^2 u_{m0}^2\right) + \varepsilon\pi^2 n^2 p u_{n0} - f_{zn}(t) \\ &= D_0^2 u_{n0} + n^4\pi^4 u_{n0} \\ &\quad + \varepsilon\left(D_0^2 u_{n1} + n^4\pi^4 u_{n1} + 2D_0 D_1 u_{n0} + 2\mu_n D_0 u_{n0} + \frac{1}{2}\pi^4 n^2 u_{n0} \sum_{m=1}^{\infty} m^2 u_{m0}^2\right) \\ &\quad + \varepsilon n^2\pi^2 p u_{n0} - f_{zn}(t) \end{aligned} \tag{7.11.7}$$

When $f_{zn} = 0$, $p(t) = 2p_1\cos\Omega_1 t + 2p_2\cos(\Omega_2 t + \theta)$. Making the coefficient of the same power of ε zero in the above equation, we obtain

$$D_0^2 u_{n0} + n^4\pi^4 u_{n0} = 0 \tag{7.11.8}$$

$$D_0^2 u_{n1} + n^4\pi^4 u_{n1} = -2D_0D_1u_{n0} - 2\mu_n D_0 u_{n0} - \frac{1}{2}\pi^4 n^2 u_{n0}\sum_{m=1}^{\infty} m^2 u_{m0}^2$$
$$- \left[2n^2\pi^2 p_1 \cos\Omega_1 t + 2n^2\pi^2 p_2 \cos(\Omega_2 t + \theta)\right]u_{n0} \quad (7.11.9)$$

The solution of the Eq. (7.11.8) is

$$u_{n0} = A_n(T_1)e^{in^2\pi^2 T_0} + cc \quad (7.11.10)$$

Substituting (7.11.10) into (7.11.9) yields

$$D_0^2 u_{n1} + n^4\pi^4 u_{n1} = -2D_0D_1u_{n0} - 2\mu_n D_0 u_{n0} - \frac{1}{2}\pi^4 n^2 u_{n0}\sum_{m=1}^{\infty} m^2 u_{m0}^2$$
$$- \left[n^2\pi^2 p_1 \cos\Omega_2 t + 2n^2\pi^2 p_2 \cos(\Omega_2 t + \theta)\right]u_{n0}$$
$$= -2in^2\pi^2 A_n' e^{in^2\pi^2 T_0} - 2in^2\pi^2\mu_n A_n e^{in^2\pi^2 T_0}$$
$$- \frac{1}{2}\pi^4 n^2 \sum_{m=1}^{\infty} m^2\left(2A_m\bar{A}_m A_n e^{in^2\pi^2 T_0} + \bar{A}_n A_m^2 e^{i(2m^2\pi^2 - n^2\pi^2)T_0}\right)$$
$$- n^2\pi^2 p_1 \bar{A}_n e^{i(\Omega_1 - n^2\pi^2)T_0} - n^2\pi^2 p_2 \bar{A}_n e^{i(\Omega_2 T_0 - n^2\pi^2 T_0 + \theta)} + cc + NST \quad (7.11.11)$$

In order to eliminate secular terms from (7.11.11) and take into account $\Omega_1 = 2\pi^2 + \varepsilon\sigma_1$ and $\Omega_2 = 8\pi^2 + \varepsilon\sigma_2$, we need

$$2i\left(A_1' + \mu_1 A_1\right) + p_1\overline{A}_1 e^{i\sigma_1 T_1} + \frac{3}{2}\pi^2 A_1^2\overline{A}_1 + \pi^2 A_1\sum_{m=2}^{\infty} m^2 A_m\overline{A}_m = 0 \quad (7.11.12)$$

$$8i\left(A_2' + \mu_2 A_2\right) + 4p_2\overline{A}_2 e^{i(\sigma_2 T_1 + \theta)} + 24\pi^2 A_2^2\overline{A}_2 + 4\pi^2 A_2\sum_{m\neq 2}^{\infty} m^2 A_m\overline{A}_m = 0$$
$$(7.11.13)$$

$$2in^2\left(A_n' + \mu_n A_n\right) + \frac{3}{2}n^4\pi^2 A_n^2\overline{A}_n + n^2\pi^2 A_n\sum_{m\neq n}^{\infty} m^2 A_m\overline{A}_m = 0, \quad n \geq 3 \quad (7.11.14)$$

Therefore, the first order approximate solution of the equation is

$$u_n = A_n(T_1)e^{in^2\pi^2 T_0} + cc + O(\varepsilon) \quad (7.11.15)$$

(c) Let

$$A_n = \frac{1}{2}a_n e^{i\beta_n} \quad (7.11.16)$$

and substitute it into (7.11.14), we can obtain

$$in^2a_n' - n^2a_n\beta_n' + in^2\mu_n a_n + \frac{3}{16}n^4\pi^2a_n^3 + \frac{1}{8}n^2\pi^2a_n\sum_{m\neq n}^{\infty}m^2a_m^2 = 0 \quad (7.11.17)$$

Separating the real and imaginary parts of the above equation yields

$$a_n' + \mu_n a_n = 0, \quad n \geq 3 \quad (7.11.18)$$

therefore

$$a_n = a_{n0}e^{-\mu_n T_1} \to 0 \Rightarrow A_n \to 0, \; n \geq 3 \quad (7.11.19)$$

Let

$$A_1 = \frac{1}{2}a_1e^{i\beta_1}, \; A_2 = \frac{1}{2}a_2e^{i\beta_2} \quad (7.11.20)$$

substitute them into (7.11.12) and (7.11.13) and take into account that (7.11.19), we have

$$ia_1' - a_1\beta_1' + i\mu_1a_1 + \frac{1}{16}\pi^2a_1\left(3a_1^2 + 2a_2^2\right) + \frac{1}{2}p_1a_1e^{i(\sigma_1T_1 - 2\beta_1)} = 0 \quad (7.11.21)$$

$$8ia_2' - 8a_2\beta_2' + 8i\mu_2a_2 + 3\pi^2a_2^3 + \frac{1}{2}\pi^2a_1^2a_2 + 2p_2a_2e^{i(\sigma_2T_1 - 2\beta_2 + \theta)} = 0 \quad (7.11.22)$$

Separating the real and imaginary parts of Eq. (7.11.20) yields

$$\begin{aligned} a_1' &= -\mu_1a_1 - \frac{1}{2}p_1a_1\sin\gamma_1 \\ a_1\gamma_1' &= \sigma_1a_1 - \frac{1}{8}\pi^2a_1\left(3a_1^2 + 2a_2^2\right) - p_1a_1\cos\gamma_1 \end{aligned} \quad (7.11.23)$$

$$\begin{aligned} *20la_2' &= -\mu_2a_2 - \frac{1}{4}p_2a_2\sin\gamma_2 \\ a_2\gamma_2' &= \sigma_2a_2 - \frac{3}{4}\pi^2a_2^3 - \frac{1}{8}\pi^2a_1^2a_2 - \frac{1}{2}p_2a_2\cos\gamma_2 \end{aligned} \quad (7.11.24)$$

where

$$\gamma_1 = \sigma_1T_1 - 2\beta_1, \gamma_2 = \sigma_2T_1 - 2\beta_2 + \theta \quad (7.11.25)$$

Let $a_1' = \gamma_1' = a_2' = \gamma_2' = 0$, we can obtain the equation satisfied by the steady state solution

$$-2\mu_1a_1 = p_1a_1\sin\gamma_1$$

$$a_1\sigma_1 - \frac{1}{8}\pi^2 a_1\left(3a_1^2 + 2a_2^2\right) = p_1 a_1 \cos\gamma_1 \tag{7.11.26}$$

$$-2\mu_2 a_2 = \frac{1}{2}p_2 a_2 \sin\gamma_2$$
$$a_2\sigma_2 - \frac{3}{4}\pi^2 a_2^3 - \frac{1}{8}\pi^2 a_1^2 a_2 = \frac{1}{2}p_2 a_2 \cos\gamma_2 \tag{7.11.27}$$

Eliminating γ_1,γ_2 in the above two sets of equations, respectively, we can obtain the system of frequency–response equations

$$4\mu_1^2 a_1^2 + a_1^2\left[\sigma_1 - \frac{1}{8}\pi^2\left(3a_1^2 + 2a_2^2\right)\right]^2 = p_1^2 a_1^2$$
$$4\mu_2^2 a_2^2 + a_2^2\left(\sigma_2 - \frac{3}{4}\pi^2 a_2^2 - \frac{1}{8}\pi^2 a_1^2\right)^2 = \frac{1}{4}p_2^2 a_2^2 \tag{7.11.28}$$

From this set of equations, the amplitude-frequency response curve can be plotted.

The stability of the steady state solution is analyzed below. It can be seen from the system of frequency–response Eq. (7.11.28) that there exists a trivial steady state solution $a_{10} = a_{20} = 0$, and a non-zero steady state solution $a_{10} = a_{20} \neq 0$.

(1) **Stability analysis on the trivial steady state solution $a_{10} = a_{20} = 0$.**

Let

$$a_1 = 0 + \tilde{a}_1, \quad \gamma_1 = \gamma_{10} + \tilde{\gamma}_1$$
$$a_2 = 0 + \tilde{a}_2, \quad \gamma_2 = \gamma_{20} + \tilde{\gamma}_2 \tag{7.11.29}$$

and substitute it into (7.11.23) and (7.11.27), we can obtain the corresponding linearized equations:

$$\widetilde{a_1'} = -\left(\mu_1 + \frac{1}{2}p_1 \sin\gamma_{10}\right)\tilde{a}_1$$
$$0 = (\sigma_1 - p_1\cos\gamma_{10})\tilde{a}_1 \tag{7.11.30}$$

$$\widetilde{a_2'} = -\left(\mu_2 + \frac{1}{2}p_2 \sin\gamma_{20}\right)\tilde{a}_2$$
$$0 = (\sigma_2 - p_2\cos\gamma_{20})\tilde{a}_2 \tag{7.11.31}$$

It can be seen from (7.11.26) and (7.11.27) that γ_{10} and γ_{20} are arbitrary when $a_{10} = a_{20} = 0$, therefore, the trivial steady state solution is stable when $\mu_1 > \frac{1}{2}p_1$ and $\mu_2 > \frac{1}{2}p_2$ by (7.11.30) and (7.11.31). Otherwise, the trivial steady state solution is unstable.

(2) **Stability analysis of non-trivial steady state solution $a_{10} = a_{20} \neq 0$.**

Let

$$\begin{aligned} a_1 &= a_{10} + \tilde{a}_1, \quad \gamma_1 = \gamma_{10} + \tilde{\gamma}_1 \\ a_2 &= a_{20} + \tilde{a}_2, \quad \gamma_2 = \gamma_{20} + \tilde{\gamma}_2 \end{aligned} \tag{7.11.32}$$

and substitute it into (7.11.23) and (7.11.24), we can obtain the corresponding linearized equation:

$$\begin{Bmatrix} \tilde{a}_1' \\ \tilde{\gamma}_1' \\ \tilde{a}_2' \\ \tilde{\gamma}_2' \end{Bmatrix} = \begin{bmatrix} l_{11} & l_{12} & l_{13} & l_{14} \\ l_{21} & l_{22} & l_{23} & l_{24} \\ l_{31} & l_{32} & l_{33} & l_{34} \\ l_{41} & l_{42} & l_{43} & l_{44} \end{bmatrix} \begin{Bmatrix} \tilde{a}_1 \\ \tilde{\gamma}_1 \\ \tilde{a}_2 \\ \tilde{\gamma}_2 \end{Bmatrix} \tag{7.11.33}$$

where

$$\begin{aligned} & l_{11} = -\left(\mu_1 + \frac{1}{2}p_1 \sin\gamma_{10}\right), \quad l_{12} = -\frac{1}{2}p_1 a_{10}\cos\gamma_{10}, \quad l_{13} = l_{14} = 0 \\ & l_{21} = a_{10}^{-1}\sigma_1 - \frac{9}{8}\pi^2 a_{10} - \frac{1}{4}\pi^2 a_{10}^{-1} a_{20}^2 - p_1 a_{10}^{-1}\cos\gamma_{10}, \\ & l_{22} = p_1 \sin\gamma_{10}, \quad l_{23} = -\frac{1}{2}\pi^2 a_{20}, \quad l_{24} = 0 \\ & l_{31} = l_{32} = 0, \quad l_{33} = -\left(\mu_2 + \frac{1}{4}p_2 \sin\gamma_{20}\right), \quad l_{34} = -\frac{1}{4}p_2 a_{20}\cos\gamma_{20} \\ & l_{41} = -\frac{1}{4}\pi^2 a_{10}, \quad l_{42} = 0 \\ & l_{43} = a_{20}^{-1}\sigma_2 - \frac{9}{4}\pi^2 a_{20} - \frac{1}{8}\pi^2 a_{10}^2 a_{20}^{-1} - \frac{1}{2}p_2 a_{20}^{-1}\cos\gamma_{20}, \quad l_{44} = \frac{1}{2}p_2 \sin\gamma_{20} \end{aligned} \tag{7.11.34}$$

The stability of the steady state solution can be determined by the real parts of these eigenvalues.

7.12 Exercise 7.12 (Transverse Oscillation of a Hinged-Clamped Beam Under Internal Resonance and Non-resonant Excitation at $u = O(w^2)$)

Solution: The governing equation of the beam in this exercise is

$$r^2(\ddot{w} + 2\mu\dot{w} + w_{xxxx}) = \left[P(t) + \frac{1}{2l}\int_0^l w_x^2 dx\right] w_{xx} + F_z(x, t)$$

the modal discretization equation for the beam of this problem is

$$\ddot{u}_n + n^4\pi^4 u_n = -\varepsilon\left(2\mu_n\dot{u}_n + \frac{1}{2}\pi^4 n^2 u_n \sum_{m=1}^{\infty} m^2 u_m^2\right) + f_{zn}(t), \quad n = 1, 2, 3, \ldots$$

Since $P(t) = 0$, $p_{nm}(t) = 0$.

We take the characteristic length L of the beam to be half of the actual length of the beam, then the dimensionless length of the beam $l = 2$; and, in turn, the characteristic equation of linear free oscillation of a hinged-clamped beam can be obtained from

$$\phi_n^{iv} - \omega_n^2\phi_n = 0 \tag{7.12.1}$$

with boundary conditions:

$$\phi_n(0) = \phi_n''(0) = 0 \quad \phi_n(2) = \phi_n'(2) = 0 \tag{7.12.2}$$

where ω_n is the natural frequency. The generalized solution of (7.12.1) is

$$\phi_n(x) = D_1\cos\eta_n x + D_2\sin\eta_n x + D_3\cosh\eta_n x + D_4\sinh\eta_n x \tag{7.12.3}$$

where $\omega_n = \eta_n^2$. Applying the boundary conditions, we get

$$\begin{aligned} &D_1 + D_3 = 0 \\ &D_1 - D_3 = 0 \\ &D_1\cos2\eta_n + D_2\sin2\eta_n + D_3\cosh2\eta_n + D_4\sinh2\eta_n = 0 \\ &-D_1\sin2\eta_n + D_2\cos2\eta_n + D_3\sinh2\eta_n + D_4\cosh2\eta_n = 0 \end{aligned} \tag{7.12.4}$$

Therefore,

$$D_1 = D_3 = 0 \tag{7.12.5}$$

$$\begin{aligned} D_2\sin2\eta_n + D_4\sinh2\eta_n &= 0 \\ D_2\cos2\eta_n + D_4\cosh2\eta_n &= 0 \end{aligned} \tag{7.12.6}$$

From the non-trivial solution condition, we can obtain

$$\tan2\eta_n - \tanh2\eta_n = 0 \tag{7.12.7}$$

From (7.12.3), (7.12.5) and (7.12.6), we have

$$\phi_n(x) = D_2\sin\eta_n x + D_4\sinh\eta_n x$$

$$
\begin{aligned}
&= D_2\sin\eta_n x - D_2\frac{\sin 2\eta_n}{\sinh 2\eta_n}\sinh\eta_n x \\
&= \frac{D_2}{\sinh 2\eta_n}(\sinh 2\eta_n \sin\eta_n x - \sin 2\eta_n \sinh\eta_n x)
\end{aligned}
$$

i.e., the mode shape function is

$$\phi_n(x) = C_n(\sinh 2\eta_n \sin\eta_n x - \sin 2\eta_n \sinh\eta_n x) \tag{7.12.8}$$

where the value of C_n is taken to normalize the mode shape function $\phi_n(x)$.

The first five natural frequency given by (7.12.7) are

$$\omega_1 = 3.8545, \quad \omega_2 = 12.491, \quad \omega_3 = 26.062, \quad \omega_4 = 44.568, \quad \omega_5 = 68.007 \tag{7.12.9}$$

We note that $\omega_1 : \omega_3 \approx 1:3$, so there is internal resonance.

Combining the above results, we can rewrite the modal discretization equation for the beam of this problem as

$$\ddot{u}_n + \omega_n^2 u_n = \varepsilon\left(\sum_{m,\,p,\,q=1}^{\infty}\Gamma_{nmpq}u_m u_p u_q - 2\mu_n\dot{u}_n\right) + f_{zn}(t) \tag{7.12.10}$$

where

$$\Gamma_{nmpq} = -\frac{1}{2l}\left(\int_0^l \phi'_n\phi'_m dx\right)\left(\int_0^l \phi'_p\phi'_q dx\right), \quad l = 2 \tag{7.12.11}$$

and

$$\Gamma_{nmpq} = \Gamma_{mnpq} = \Gamma_{nmqp} = \Gamma_{pqnm} \tag{7.12.12}$$

We use the method of multiple scales to solve (7.12.10). Let the solution of the equation be

$$u_n(t;\varepsilon) = u_{n0}(T_0, T_1) + \varepsilon u_{n1}(T_0, T_1) + \cdots \tag{7.12.13}$$

Substituting (7.12.13) into (7.12.10) and retaining to $O(\varepsilon)$, we get

$$
\begin{aligned}
0 &= \ddot{u}_n + \omega_n^2 u_n + \varepsilon\left(2\mu_n\dot{u}_n - \sum_{m,p,q=1}^{\infty}\Gamma_{nmpq}u_m u_p u_q\right) - f_{zn}(t) \\
&= \left(D_0^2 + 2\varepsilon D_0 D_1\right)(u_{n0} + \varepsilon u_{n1}) + \omega_n^2(u_{n0} + \varepsilon u_{n1})
\end{aligned}
$$

$$+\varepsilon\left(2\mu_n D_0 u_{n0} - \sum_{m,p,q=1}^{\infty} \Gamma_{nmpq} u_{m0} u_{p0} u_{q0}\right) - 2K_n \cos\Omega t$$
$$= D_0^2 u_{n0} + \omega_n^2 u_{n0} - 2K_n \cos\Omega t$$
$$+\varepsilon\left(D_0^2 u_{n1} + \omega_n^2 u_{n1} + 2D_0 D_1 u_{n0} + 2\mu_n D_0 u_{n0} - \sum_{m,p,q=1}^{\infty} \Gamma_{nmpq} u_{m0} u_{p0} u_{q0}\right) \tag{7.12.14}$$

Making the coefficient of the same power of ε zero in the above equation yields

$$D_0^2 u_{n0} + \omega_n^2 u_{n0} = 2K_n \cos\Omega t \tag{7.12.15}$$

$$D_0^2 u_{n1} + \omega_n^2 u_{n1} = -2D_0 D_1 u_{n0} - 2\mu_n D_0 u_{n0} + \sum_{m,\,p,\,q=1}^{\infty} \Gamma_{nmpq} u_{m0} u_{p0} u_{q0} \tag{7.12.16}$$

The solution of the Eq. (7.12.15) is

$$u_{n0} = A_n(T_1) e^{i\omega_n T_0} + \Lambda_n e^{i\Omega T_0} + cc \tag{7.12.17}$$

where

$$\Lambda_n = \frac{K_n}{\omega_n^2 - \Omega^2} \tag{7.12.18}$$

Substituting (7.12.17) into (7.12.16) and taking into account $\varepsilon\sigma_1 = \omega_2 - 3\omega_1$, $\varepsilon\sigma_2 = 2\Omega - 4\omega_1$, we obtain

$$\begin{aligned}
D_0^2 u_{n1} + \omega_n^2 u_{n1} &= -2D_0 D_1 u_{n0} - 2\mu_n D_0 u_{n0} + \sum_{m,p,q=1}^{\infty} \Gamma_{nmpq} u_{m0} u_{p0} u_{q0} \\
&= -2i\omega_n A_n' e^{i\omega_n T_0} - 2i\omega_n \mu_n A_n e^{i\omega_n T_0} \\
&\quad + \sum_{m,p,q=1}^{\infty} \Gamma_{nmpq} \big\{ 2A_m \Lambda_p \Lambda_q e^{i\omega_m T_0} + 2\Lambda_m \Lambda_p A_q e^{i\omega_q T_0} + 2\Lambda_m A_p \Lambda_q e^{i\omega_p T_0} \\
&\quad + A_m \bar{A}_p A_q e^{i\omega_m T_0} e^{-i\omega_p T_0} e^{i\omega_q T_0} + A_m A_p \bar{A}_q e^{i\omega_m T_0} e^{i\omega_p T_0} e^{-i\omega_q T_0} \\
&\quad + \bar{A}_m A_p A_q e^{-i\omega_m T_0} e^{i\omega_p T_0} e^{i\omega_q T_0} + \bar{A}_m A_p \bar{A}_q e^{-i\omega_m T_0} e^{i\omega_p T_0} e^{-i\omega_q T_0} \\
&\quad + \bar{A}_m \bar{A}_p A_q e^{-i\omega_m T_0} e^{-i\omega_p T_0} e^{i\omega_q T_0} + A_m \bar{A}_p \bar{A}_q e^{i\omega_m T_0} e^{-i\omega_p T_0} e^{-i\omega_q T_0} \\
&\quad + A_m A_p A_q e^{i\omega_m T_0} e^{i\omega_p T_0} e^{i\omega_q T_0} + \bar{A}_m \Lambda_p \Lambda_q e^{i(2\Omega - \omega_m) T_0} \\
&\quad + \Lambda_m \bar{A}_p \Lambda_q e^{i(2\Omega - \omega_p) T_0} + \Lambda_m \Lambda_p \bar{A}_q e^{i(2\Omega - \omega_q) T_0} \big\} + cc + NST
\end{aligned} \tag{7.12.19}$$

In order to eliminate secular terms from the above equation, we need

$$\begin{aligned}&2i\omega_1 A_1' + 2i\omega_1\mu_1 A_1 - 2A_1 \sum_{p,q=1}^{\infty} \left(\Gamma_{11pq} + 2\Gamma_{1p1q}\right)\Lambda_p\Lambda_q \\ &- A_1 \sum_{m=1}^{\infty} 2(\Gamma_{11mm} + 2\Gamma_{1m1m})A_m\overline{A}_m - 3\Gamma_{1112}\overline{A}_1^2 A_2 e^{i\sigma_1 T_1} \\ &- \overline{A}_2 e^{i(\sigma_2-\sigma_1)T_1} \sum_{p,q=1}^{\infty} \left(\Gamma_{12pq} + 2\Gamma_{1p2q}\right)\Lambda_p\Lambda_q = 0 \end{aligned} \tag{7.12.20}$$

$$\begin{aligned}&2i\omega_2 A_2' + 2i\omega_2\mu_2 A_2 - 2A_2 \sum_{p,q=1}^{\infty} \left(\Gamma_{22pq} + 2\Gamma_{2p2q}\right)\Lambda_p\Lambda_q \\ &- A_2 \sum_{m=1}^{\infty} 2(\Gamma_{22mm} + 2\Gamma_{2m2m})A_m\overline{A}_m - \Gamma_{2111} A_1^3 e^{-i\sigma_1 T_1} \\ &- \overline{A}_1 e^{i(\sigma_2-\sigma_1)T_1} \sum_{p,q=1}^{\infty} \left(\Gamma_{12pq} + 2\Gamma_{1p2q}\right)\Lambda_p\Lambda_q = 0 \end{aligned} \tag{7.12.21}$$

$$\begin{aligned}&2i\omega_n A_n' + 2i\omega_n\mu_n A_n - 2A_n \sum_{p,q=1}^{\infty} \left(\Gamma_{nnpq} + 2\Gamma_{npnq}\right)\Lambda_p\Lambda_q \\ &- 2A_n \sum_{m=1}^{\infty} (\Gamma_{nnmm} + 2\Gamma_{nmnm})A_m\bar{A}_m = 0, \quad n \geq 3 \end{aligned} \tag{7.12.22}$$

Let

$$\begin{aligned}&Q_1 = 3\Gamma_{1112} = -2.311, \quad Q_2 = \Gamma_{2111} = \frac{1}{3}Q_1 = -0.7703 \\ &\alpha_{mn} = \alpha_{nm} = \begin{cases} 2(2\Gamma_{nmnm} + \Gamma_{nnmm}) \text{ for } & n \neq m \\ 6\Gamma_{nnnn} \end{cases} \\ &H_{nm} = \sum_{p,q} \left(\Gamma_{nmpq} + 2\Gamma_{npmq}\right)\Lambda_p\Lambda_q \end{aligned} \tag{7.12.23}$$

then (7.12.20)–(7.12.22) become

$$\begin{aligned}&2i\omega_1 A_1' + 2i\omega_1\mu_1 A_1 - 2H_{11}A_1 - A_1 \sum_{m=1}^{\infty} \alpha_{1m}A_m\overline{A}_m \\ &- Q_1\overline{A}_1^2 A_2 e^{i\sigma_1 T_1} - H_{12}\overline{A}_2 e^{i(\sigma_2-\sigma_1)T_1} = 0 \end{aligned} \tag{7.12.24}$$

$$2i\omega_2 A_2' + 2i\omega_2\mu_2 A_2 - Q_2 A_1^3 e^{-i\sigma_1 T_1} - 2H_{22}A_2$$

$$-A_2\sum_{m=1}^{\infty}\alpha_{2m}A_m\overline{A}_m - H_{12}\overline{A}_1e^{i(\sigma_2-\sigma_1)T_1} = 0 \tag{7.12.25}$$

$$2i\omega_nA_n' + 2i\omega_n\mu_nA_n - 2H_{nn}A_n - A_n\sum_{m=1}^{\infty}\alpha_{nm}A_m\overline{A}_m = 0, \quad n \geq 3 \tag{7.12.26}$$

Let

$$A_n = \frac{1}{2}a_ne^{i\beta_n} \tag{7.12.27}$$

and substitute it into (7.12.26), we can obtain

$$i\omega_na_n' - \omega_na_n\beta_n' + i\omega_n\mu_na_n - H_{nn}a_n - \frac{1}{8}a_n\sum_{m=1}^{\infty}\alpha_{nm}a_m^2 = 0, \quad n \geq 3 \tag{7.12.28}$$

Separating the real and imaginary parts of the above equation yields

$$a_n' + \mu_na_n = 0, \quad n \geq 3 \tag{7.12.29}$$

therefore

$$a_n = a_{n0}e^{-\mu_nT_1} \to 0 \Rightarrow A_n \to 0, \ n \geq 3 \tag{7.12.30}$$

Let

$$A_1 = \frac{1}{2}a_1e^{i\beta_1}, \quad A_2 = \frac{1}{2}a_2e^{i\beta_2} \tag{7.12.31}$$

substitute them into (7.12.24) and (7.12.25), and take into account that (7.12.30), we have

$$\begin{aligned} &i\omega_1a_1' - \omega_1a_1\beta_1' + i\omega_1\mu_1a_1 - H_{11}a_1 \\ &\quad - \frac{1}{8}a_1\sum_{m=1}^{2}\alpha_{1m}a_m^2 - \frac{1}{8}Q_1a_1^2a_2e^{i\gamma_1} - \frac{1}{2}H_{12}a_2e^{i\gamma_2} = 0 \end{aligned} \tag{7.12.32}$$

$$\begin{aligned} &i\omega_2a_2' - \omega_2a_2\beta_2' + i\omega_2\mu_2a_2 - H_{22}a_2 \\ &\quad - \frac{1}{8}a_2\sum_{m=1}^{2}\alpha_{2m}a_m^2 - \frac{1}{8}Q_2a_1^3e^{-i\gamma_1} - H_{12}a_1e^{i\gamma_2} = 0 \end{aligned} \tag{7.12.33}$$

where

$$\gamma_1 = \sigma_1T_1 - 3\beta_1 + \beta_2, \quad \gamma_2 = \sigma_2T_1 - \sigma_1T_1 - \beta_1 - \beta_2 \tag{7.12.34}$$

Separating the real and imaginary parts from (7.12.32) and (7.12.33), we obtain

$$20l\omega_1 a_1' = -\omega_1\mu_1 a_1 + \frac{1}{8}Q_1 a_1^2 a_2 \sin\gamma_1 + \frac{1}{2}H_{12}a_2\sin\gamma_2$$

$$\omega_1 a_1\left(\gamma_1' + \gamma_2'\right) = \omega_1\sigma_2 a_1 + 4H_{11}a_1 + \frac{1}{2}a_1\sum_{m=1}^{2}\alpha_{1m}a_m^2 + \frac{1}{2}Q_1 a_1^2 a_2\cos\gamma_1 + 2H_{12}a_2\cos\gamma_2 \tag{7.12.35}$$

$$\omega_2 a_2' = -\omega_2\mu_2 a_2 + \frac{1}{8}Q_2 a_1^3\sin\gamma_1 + H_{12}a_1\sin\gamma_2$$

$$\omega_2 a_2\left(3\gamma_2' - \gamma_1'\right) = \omega_2(3\sigma_2 - 4\sigma_1)a_2 + 4H_{22}a_2 + \frac{1}{2}a_2\sum_{m=1}^{2}\alpha_{2m}a_m^2 + \frac{1}{2}Q_2 a_1^3\cos\gamma_1 + 2H_{12}a_1\cos\gamma_2 \tag{7.12.36}$$

Let $a_1' = \gamma_1' = a_2' = \gamma_2' = 0$ in (7.12.35) and (7.12.36), we can obtain the equation satisfied by the steady state solution:

$$\begin{aligned} &-\omega_1\mu_1 a_1 + \frac{1}{8}Q_1 a_1^2 a_2\sin\gamma_1 + \frac{1}{2}H_{12}a_2\sin\gamma_2 = 0 \\ &-\omega_2\mu_2 a_2 - \frac{1}{8}Q_2 a_1^3\sin\gamma_1 + \frac{1}{2}H_{12}a_1\sin\gamma_2 = 0 \end{aligned} \tag{7.12.37}$$

$$\begin{aligned} &\omega_1\sigma_2 a_1 + 4H_{11}a_1 + \frac{1}{2}a_1\sum_{m=1}^{2}\alpha_{1m}a_m^2 + \frac{1}{2}Q_1 a_1^2 a_2\cos\gamma_1 + 2H_{12}a_2\cos\gamma_2 = 0 \\ &\omega_2(3\sigma_2 - 4\sigma_1)a_2 + 4H_{22}a_2 + \frac{1}{2}a_2\sum_{m=1}^{2}\alpha_{2m}a_m^2 + \frac{1}{2}Q_2 a_1^3\cos\gamma_1 + 2H_{12}a_1\cos\gamma_2 = 0 \end{aligned} \tag{7.12.38}$$

Solve the steady state solution from (7.12.37) and (7.12.38) and substitute it into (7.12.17), we can obtain the approximate solution of $u_n(t)$:

$$\begin{aligned} u_1 &\approx u_{10} = a_1\cos(\omega_1 t + \beta_1) + 2\Lambda_1\cos\Omega t + O(\varepsilon) \\ u_2 &\approx u_{20} = a_2\cos(\omega_2 t + \beta_2) + 2\Lambda_2\cos\Omega t + O(\varepsilon) \\ u_n &\approx u_{n0} = 2\Lambda_n\cos\Omega t + O(\varepsilon), \quad n \ge 3 \end{aligned} \tag{7.12.39}$$

Note that

$$\beta_1 = \frac{1}{4}\varepsilon\sigma_2 t - \frac{1}{4}(\gamma_1 + \gamma_2)$$

$$\beta_2 = \frac{1}{4}(\gamma_1 - 3\gamma_2) - \varepsilon\sigma_1 t + \frac{3}{4}\varepsilon\sigma_2 t$$
$$\varepsilon\sigma_1 = \omega_2 - 3\omega_1, \varepsilon\sigma_2 = 2\Omega - 4\omega_1$$

So

$$\omega_1 t + \beta_1 = \frac{1}{2}\Omega t - \frac{1}{4}(\gamma_1 + \gamma_2), \quad \omega_2 t + \beta_2 = \frac{3}{2}\Omega t + \frac{1}{4}(\gamma_1 - 3\gamma_2) \qquad (7.12.40)$$

Then the Eq. (7.12.39) becomes

$$u_1 \approx a_1 \cos\left[\frac{1}{2}\Omega t - \frac{1}{4}(\gamma_1 + \gamma_2)\right] + 2\Lambda_1 \cos\Omega t + O(\varepsilon)$$
$$u_2 \approx a_2 \cos\left[\frac{3}{2}\Omega t + \frac{1}{4}(\gamma_1 - 3\gamma_2)\right] + 2\Lambda_2 \cos\Omega t + O(\varepsilon)$$
$$u_n \approx 2\Lambda_n \cos\Omega t + O(\varepsilon), \quad n \geq 3 \qquad (7.12.41)$$

Since the deflection of the beam is $w(x, t) = \sum_{n=1}^{\infty} u_n \phi_n(x)$, the approximate solution for the beam deflection response is

$$w(x, t) = a_1 \phi_1(x) \cos\left[\frac{1}{2}\Omega t - \frac{1}{4}(\gamma_1 + \gamma_2)\right]$$
$$+ a_2 \phi_2(x) \cos\left[\frac{3}{2}\Omega t + \frac{1}{4}(\gamma_1 - 3\gamma_2)\right] + 2\left[\sum_{n=1}^{\infty} \Lambda_n \phi_n(x)\right] \cos\Omega t$$
$$(7.12.42)$$

7.13 Exercise 7.13 (Transverse Oscillation of a Hinged-Clamped Beam Under Internal Resonance and Non-resonance Excitation at $u = O(w^2)$)

Solution: All procedures before Eq. (7.12.18) in the previous exercise can still be applied in this problem. Substituting (7.12.17) into (7.12.16) and taking into account $\varepsilon\sigma_1 = \omega_2 - 3\omega_1, \quad \varepsilon\sigma_2 = \Omega - 5\omega_1$, we obtain

$$D_0^2 u_{n1} + \omega_n^2 u_{n1} = -2D_0 D_1 u_{n0} - 2\mu_n D_0 u_{n0} + \sum_{m,p,q=1}^{\infty} \Gamma_{nmpq} u_{m0} u_{p0} u_{q0}$$
$$= -2i\omega_n A_n' e^{i\omega_n T_0} - 2i\omega_n \mu_n A_n e^{i\omega_n T_0}$$
$$+ \sum_{m,p,q=1}^{\infty} \Gamma_{nmpq} \Big\{ 2A_m \Lambda_p \Lambda_q e^{i\omega_m T_0} + 2\Lambda_m \Lambda_p A_q e^{i\omega_q T_0} + 2\Lambda_m A_p \Lambda_q e^{i\omega_p T_0}$$

$$
\begin{aligned}
&+ A_m\bar{A}_pA_q e^{i\omega_m T_0}e^{-i\omega_p T_0}e^{i\omega_q T_0} + A_mA_p\bar{A}_q e^{i\omega_m T_0}e^{i\omega_p T_0}e^{-i\omega_q T_0}\\
&+ \bar{A}_mA_pA_q e^{-i\omega_m T_0}e^{i\omega_p T_0}e^{i\omega_q T_0} + \bar{A}_mA_p\bar{A}_q e^{-i\omega_m T_0}e^{i\omega_p T_0}e^{-i\omega_q T_0}\\
&+ \bar{A}_m\bar{A}_pA_q e^{-i\omega_m T_0}e^{-i\omega_p T_0}e^{i\omega_q T_0} + A_m\bar{A}_p\bar{A}_q e^{i\omega_m T_0}e^{-i\omega_p T_0}e^{-i\omega_q T_0}\\
&+ A_mA_pA_q e^{i\omega_m T_0}e^{i\omega_p T_0}e^{i\omega_q T_0}\\
&+ \bar{A}_m\Lambda_p\bar{A}_q e^{i\Omega T_0}e^{-i\omega_m T_0}e^{-i\omega_q T_0} + \bar{A}_m\bar{A}_p\Lambda_q e^{i\Omega T_0}e^{-i\omega_m T_0}e^{-i\omega_p T_0}\\
&+ \Lambda_m\bar{A}_A\bar{A}_q e^{i\Omega T_0}e^{-i\omega_p T_0}e^{-i\omega_q T_0}\\
&+ A_m\bar{A}_p\Lambda_q e^{i\Omega T_0}e^{i\omega_m T_0}e^{-i\omega_p T_0} + A_m\Lambda_p\bar{A}_q e^{i\Omega T_0}e^{i\omega_m T_0}e^{-i\omega_q T_0}\\
&+ \bar{A}_m\Lambda_pA_q e^{i\Omega T_0}e^{-i\omega_m T_0}e^{i\omega_q T_0} + \bar{A}_mA_p\Lambda_q e^{i\Omega T_0}e^{-i\omega_m T_0}e^{i\omega_p T_0}\\
&+ \Lambda_m\bar{A}_pA_q e^{i\Omega T_0}e^{-i\omega_p T_0}e^{i\omega_q T_0} + \Lambda_mA_p\bar{A}_q e^{i\Omega T_0}e^{i\omega_p T_0}e^{-i\omega_q T_0}\Big\} + cc + NST
\end{aligned}
\tag{7.13.1}
$$

In order to eliminate secular terms from the above equation, we need

$$
\begin{aligned}
&2i\omega_1A_1' + 2i\omega_1\mu_1A_1 - 2A_1\sum_{p,\,q=1}^{\infty}\left(\Gamma_{11pq} + 2\Gamma_{1p1q}\right)\Lambda_p\Lambda_q\\
&\quad - 2A_1\sum_{m=1}^{\infty}(\Gamma_{11mm} + 2\Gamma_{1m1m})A_m\bar{A}_m - 3\Gamma_{1112}\bar{A}_1^2A_2e^{i\sigma_1T_1}\\
&\quad - 2\bar{A}_1\bar{A}_2e^{i(\sigma_2-\sigma_1)T_1}\sum_{m=1}^{\infty}(\Gamma_{112m} + 2\Gamma_{121m})\Lambda_m = 0
\end{aligned}
\tag{7.13.2}
$$

$$
\begin{aligned}
&2i\omega_2A_2' + 2i\omega_2\mu_2A_2 - 2A_2\sum_{p,q=1}^{\infty}\left(\Gamma_{22pq} + 2\Gamma_{2p2q}\right)\Lambda_p\Lambda_q\\
&\quad - 2A_2\sum_{m=1}^{\infty}(\Gamma_{22mm} + 2\Gamma_{2m2m})A_m\bar{A}_m - \Gamma_{2111}A_1^3e^{-i\sigma_1T_1}\\
&\quad - \bar{A}_1^2e^{i(\sigma_2-\sigma_1)T_1}\sum_{m=1}^{\infty}(\Gamma_{112m} + 2\Gamma_{121m})\Lambda_m\\
&\quad - 2A_1\bar{A}_2e^{i(\sigma_2-2\sigma_1)T_1}\sum_{m=1}^{\infty}(\Gamma_{221m} + 2\Gamma_{212m})\Lambda_m = 0
\end{aligned}
\tag{7.13.3}
$$

$$
\begin{aligned}
&2i\omega_nA_n' + 2i\omega_n\mu_nA_n - 2A_n\sum_{p,\,q=1}^{\infty}\left(\Gamma_{nnpq} + 2\Gamma_{npnq}\right)\Lambda_p\Lambda_q\\
&\quad - 2A_n\sum_{m=1}^{\infty}(\Gamma_{nnmm} + 2\Gamma_{nmnm})A_m\bar{A}_m = 0, \quad n \geq 3
\end{aligned}
\tag{7.13.4}
$$

Let

$$
Q_1 = 3\Gamma_{1112} = -2.311, \quad Q_2 = \Gamma_{2111} = \frac{1}{3}Q_1 = -0.7703
$$

$$\alpha_{mn} = \alpha_{nm} = \begin{cases} 2(\Gamma_{nnmm} + 2\Gamma_{nmnm}) \text{ for } & n \neq m \\ 6\Gamma_{nnnn} \end{cases}$$

$$H_{nm} = \sum_{p,\ q} \left(\Gamma_{nmpq} + 2\Gamma_{npmq}\right) \Lambda_p \Lambda_q,$$

$$G_{nm} = \sum_{q=1}^{\infty} \left(\Gamma_{nnmq} + 2\Gamma_{nmnq}\right) \Lambda_q \tag{7.13.5}$$

then (7.13.2)–(7.13.4) become

$$\begin{aligned} &2i\omega_1 A_1' + 2i\omega_1 \mu_1 A_1 - 2H_{11} A_1 - A_1 \sum_{m=1}^{\infty} \alpha_{1m} A_m \overline{A}_m \\ &- Q_1 \overline{A}_1^2 A_2 e^{i\sigma_1 T_1} - 2G_{12} \overline{A}_1 \overline{A}_2 e^{i(\sigma_2 - \sigma_1) T_1} = 0 \end{aligned} \tag{7.13.6}$$

$$\begin{aligned} &2i\omega_2 A_2' + 2i\omega_2 \mu_2 A_2 - 2H_{22} A_2 - A_2 \sum_{m=1}^{\infty} \alpha_{2m} A_m \overline{A}_m \\ &- Q_2 A_1^3 e^{-i\sigma_1 T_1} - G_{12} \overline{A}_1^2 e^{i(\sigma_2 - \sigma_1) T_1} - 2G_{21} A_1 \overline{A}_2 e^{i(\sigma_2 - 2\sigma_1) T_1} = 0 \end{aligned} \tag{7.13.7}$$

$$2i\omega_n A_n' + 2i\omega_n \mu_n A_n - 2H_{nn} A_n - A_n \sum_{m=1}^{\infty} \alpha_{nm} A_m \overline{A}_m = 0, \quad n \geq 3 \tag{7.13.8}$$

Let

$$A_n = \frac{1}{2} a_n e^{i\beta_n} \tag{7.13.9}$$

and substitute it into (7.13.8), we can obtain

$$i\omega_n a_n' - \omega_n a_n \beta_n' + i\omega_n \mu_n a_n - H_{nn} a_n - \frac{1}{8} a_n \sum_{m=1}^{\infty} \alpha_{nm} a_m^2 = 0, \quad n \geq 3 \tag{7.13.10}$$

Separating the real and imaginary parts of the above equation yields

$$a_n' + \mu_n a_n = 0, \quad n \geq 3 \tag{7.13.11}$$

therefore

$$a_n = a_{n0} e^{-\mu_n T_1} \to 0 \Rightarrow A_n \to 0,\ n \geq 3 \tag{7.13.12}$$

Let

$$A_1 = \frac{1}{2}a_1 e^{i\beta_1}, \quad A_2 = \frac{1}{2}a_2 e^{i\beta_2} \tag{7.13.13}$$

substitute them into (7.13.6) and (7.13.7) and take into account that (7.13.12), we have

$$\begin{aligned} & i\omega_1 a_1' - \omega_1 a_1 \beta_1' + i\omega_1 \mu_1 a_1 - H_{11} a_1 \\ & \quad - \frac{1}{8} a_1 \sum_{m=1}^{2} \alpha_{1m} a_m^2 - \frac{1}{8} Q_1 a_1^2 a_2 e^{i\gamma_1} - \frac{1}{4} G_{12} a_1 a_2 e^{i\gamma_2} = 0 \end{aligned} \tag{7.13.14}$$

$$\begin{aligned} & i\omega_2 a_2' - \omega_2 a_2 \beta_2' + i\omega_2 \mu_2 a_2 - H_{22} a_2 - \frac{1}{8} a_2 \sum_{m=1}^{2} \alpha_{2m} a_m^2 \\ & \quad - \frac{1}{8} Q_2 a_1^3 e^{-i\gamma_1} - \frac{1}{4} G_{12} a_1^2 e^{i\gamma_2} - \frac{1}{2} G_{21} a_1 a_2 e^{i(\gamma_2 - \gamma_1)} = 0 \end{aligned} \tag{7.13.15}$$

where

$$\gamma_1 = \sigma_1 T_1 - 3\beta_1 + \beta_2, \gamma_2 = \sigma_2 T_1 - \sigma_1 T_1 - 2\beta_1 - \beta_2 \tag{7.13.16}$$

Separating the real and imaginary parts from (7.13.15) and (7.13.16), we obtain

$$\begin{aligned} \omega_1 a_1' &= -\omega_1 \mu_1 a_1 + \frac{1}{8} Q_1 a_1^2 a_2 \sin\gamma_1 + \frac{1}{4} G_{12} a_1 a_2 \sin\gamma_2 \\ \omega_1 a_1 (\gamma_1' + \gamma_2') &= -\omega_1 \sigma_2 a_1 + 5H_{11} a_1 + \frac{5}{8} a_1 \sum_{m=1}^{2} \alpha_{1m} a_m^2 \\ & \quad + \frac{5}{8} Q_1 a_1^2 a_2 \cos\gamma_1 + \frac{5}{4} G_{12} a_1 a_2 \cos\gamma_2 \end{aligned} \tag{7.13.17}$$

$$\begin{aligned} \omega_2 a_2' &= -\omega_2 \mu_2 a_2 + \frac{1}{8} Q_2 a_1^3 \sin\gamma_1 + \frac{1}{4} G_{12} a_1^2 \sin\gamma_2 \\ & \quad + \frac{1}{2} G_{21} a_1 a_2 \sin(\gamma_2 - \gamma_1) \\ \omega_2 a_2 (3\gamma_2' - 2\gamma_1') &= \omega_2 a_2 (3\sigma_2 - 5\sigma_1) + 5H_{22} a_2 + \frac{5}{8} a_2 \sum_{m=1}^{2} \alpha_{2m} a_m^2 \\ & \quad + \frac{5}{8} Q_2 a_1^3 \cos\gamma_1 + \frac{5}{4} G_{12} a_1^2 \cos\gamma_2 + \frac{5}{2} G_{21} a_1 a_2 \cos(\gamma_2 - \gamma_1) \end{aligned} \tag{7.13.18}$$

Let $a_1' = \gamma_1' = a_2' = \gamma_2' = 0$ in (7.13.17) and (7.13.18), we can obtain the equation satisfied by the steady state solution

$$-\omega_1 \mu_1 a_1 + \frac{1}{8} Q_1 a_1^2 a_2 \sin\gamma_1 + \frac{1}{4} G_{12} a_1 a_2 \sin\gamma_2 = 0$$

$$-\omega_2\mu_2 a_2 + \frac{1}{8}Q_2 a_1^3 \sin\gamma_1 + \frac{1}{4}G_{12}a_1^2 \sin\gamma_2 + \frac{1}{2}G_{21}a_1 a_2 \sin(\gamma_2 - \gamma_1) = 0 \tag{7.13.19}$$

$$-\omega_1\sigma_2 a_1 + 5H_{11}a_1 + \frac{5}{8}a_1\sum_{m=1}^{2}\alpha_{1m}a_m^2 + \frac{5}{8}Q_1 a_1^2 a_2 \cos\gamma_1 + \frac{5}{4}G_{12}a_1 a_2 \cos\gamma_2 = 0$$

$$\omega_2 a_2(3\sigma_2 - 5\sigma_1) + 5H_{22}a_2 + \frac{5}{8}a_2\sum_{m=1}^{2}\alpha_{2m}a_m^2 + \frac{5}{8}Q_2 a_1^3 \cos\gamma_1$$
$$+\frac{5}{4}G_{12}a_1^2 \cos\gamma_2 + \frac{5}{2}G_{21}a_1 a_2 \cos(\gamma_2 - \gamma_1) = 0 \tag{7.13.20}$$

Solving the steady state solution from (7.13.19) and (7.13.20) and substituting it into (7.13.18), we can obtain the approximate solution of $u_n(t)$

$$\begin{aligned} u_1 &\approx u_{10} = a_1\cos(\omega_1 t + \beta_1) + 2\Lambda_1\cos\Omega t + O(\varepsilon) \\ u_2 &\approx u_{20} = a_2\cos(\omega_2 t + \beta_2) + 2\Lambda_2\cos\Omega t + O(\varepsilon) \\ u_n &\approx u_{n0} = 2\Lambda_n\cos\Omega t + O(\varepsilon),\ \ n \geq 3 \end{aligned} \tag{7.13.21}$$

Note that

$$\begin{aligned} \beta_1 &= \frac{1}{5}\varepsilon\sigma_2 t - \frac{1}{5}(\gamma_1 + \gamma_2) \\ \beta_2 &= \frac{3}{5}\varepsilon\sigma_2 t - \varepsilon\sigma_1 t + \frac{1}{5}(2\gamma_1 - 3\gamma_2) \\ \varepsilon\sigma_1 &= \omega_2 - 3\omega_1, \quad \varepsilon\sigma_2 = \Omega - 5\omega_1 \end{aligned}$$

So

$$\omega_1 t + \beta_1 = \frac{1}{5}\Omega t - \frac{1}{5}(\gamma_1 + \gamma_2), \quad \omega_2 t + \beta_2 = \frac{3}{5}\Omega t + \frac{1}{5}(2\gamma_1 - 3\gamma_2) \tag{7.13.22}$$

In turn, Eq. (7.12.40) becomes

$$\begin{aligned} u_1 &\approx a_1\cos\left[\frac{1}{5}\Omega t - \frac{1}{5}(\gamma_1 + \gamma_2)\right] + 2\Lambda_1\cos\Omega t + O(\varepsilon) \\ u_2 &\approx a_2\cos\left[\frac{3}{5}\Omega t + \frac{1}{5}(2\gamma_1 - 3\gamma_2)\right] + 2\Lambda_2\cos\Omega t + O(\varepsilon) \\ u_n &\approx 2\Lambda_n\cos\Omega t + O(\varepsilon), \quad n \geq 3 \end{aligned} \tag{7.13.23}$$

Since the deflection of the beam is $w(x, t) = \sum_{n=1}^{\infty} u_n\phi_n(x)$, the approximate solution for the beam deflection response is

$$w(x,t) = a_1\phi_1(x)\cos\left[\frac{1}{5}\Omega t - \frac{1}{5}(\gamma_1 + \gamma_2)\right]$$
$$+ a_2\phi_2(x)\cos\left[\frac{3}{5}\Omega t + \frac{1}{5}(2\gamma_1 - 3\gamma_2)\right] + 2\left[\sum_{n=1}^{\infty}\Lambda_n\phi_n(x)\right]\cos\Omega t \tag{7.13.24}$$

7.14 Exercise 7.14 (Transverse Oscillation of a Clamped–Clamped Beam at Internal Resonance and Resonant Excitation at $u = O(w^2)$)

Solution: (a) The governing equation of the beam in this exercise is still

$$r^2(\ddot{w} + 2\mu\dot{w} + w_{xxxx}) = \left[P(t) + \frac{1}{2l}\int_0^l w_x^2 dx\right]w_{xx} + F_z(x,t) \tag{7.14.1}$$

and the modal discretization equation for the beam of this problem is also the same as previous exercises, i.e.,

$$\ddot{u}_n + n^4\pi^4 u_n = -\varepsilon\left(2\mu_n\dot{u}_n + \frac{1}{2}\pi^4 n^2 u_n\sum_{m=1}^{\infty}m^2u_m^2\right) + f_{zn}(t), \quad n = 1,2,3,\ldots$$

Since $P(t) = 0$, $p_{nm}(t) = 0$.

We take the characteristic length of the beam L to be the **actual** length of the beam, then the dimensionless length of the beam $l = 1$; and, in turn, the characteristic equation for the linear free oscillation of a clamped- clamped beam can be obtained from

$$\phi_n^{iv} - \omega_n^2\phi_n = 0 \tag{7.14.2}$$

with boundary conditions:

$$\phi_n(0) = \phi_n'(0) = 0, \quad \phi_n(1) = \phi_n'(1) = 0 \tag{7.14.3}$$

where ω_n is the natural frequency. The generalized solution of (7.14.3) is

$$\phi_n(x) = D_1\cos\eta_n x + D_2\sin\eta_n x + D_3\cosh\eta_n x + D_4\sinh\eta_n x \tag{7.14.4}$$

where.$\omega_n = \eta_n^2$ Applying the boundary conditions, we get

$$\begin{aligned}&D_1 + D_3 = 0\\&D_2 + D_4 = 0\\&D_1\cos\eta_n + D_2\sin\eta_n + D_3\cosh\eta_n + D_4\sinh\eta_n = 0\\&-D_1\sin\eta_n + D_2\cos\eta_n + D_3\sinh\eta_n + D_4\cosh\eta_n = 0\end{aligned} \tag{7.14.5}$$

Therefore,

$$D_1 = -D_3, \quad D_2 = -D_4 \tag{7.14.6}$$

$$\begin{aligned}&(\cos\eta_n - \cosh\eta_n)D_1 + (\sin\eta_n - \sinh\eta_n)D_2 = 0\\&(\sin\eta_n + \sinh\eta_n)D_1 - (\cos\eta_n - \cosh\eta_n)D_2 = 0\end{aligned} \tag{7.14.7}$$

From the non-trivial solution condition, we can obtain

$$\cos\eta_n\cosh\eta_n = 1 \tag{7.14.8}$$

From (7.14.4), (7.14.6) and (7.14.7), we have

$$\begin{aligned}&\phi_n(x) = D_1(\cos\eta_n x - \cosh\eta_n x) + D_2(\sin\eta_n x - \sinh\eta_n x)\\&= D_1\Big[(\cos\eta_n x - \cosh\eta_n x)\\&-\frac{\cos\eta_n - \cosh\eta_n}{\sin\eta_n - \sinh\eta_n}(\sin\eta_n x - \sinh\eta_n x)\Big]\\&= \frac{D_1}{\sin\eta_n - \sinh\eta_n}[(\sin\eta_n - \sinh\eta_n)(\cos\eta_n x - \cosh\eta_n x)\\&-(\cos\eta_n - \cosh\eta_n)(\sin\eta_n x - \sinh\eta_n x)]\end{aligned}$$

i.e., the mode shape function is

$$\begin{aligned}\phi_n(x) = C_n[&(\sin\eta_n - \sinh\eta_n)(\cos\eta_n x - \cosh\eta_n x)\\&- (\cos\eta_n - \cosh\eta_n)(\sin\eta_n x - \sinh\eta_n x)]\end{aligned} \tag{7.14.9}$$

where the value of C_n is taken to normalize the mode shape function $\phi_n(x)$.

The function $D(\omega) = \cos\sqrt{\omega}\cosh\sqrt{\omega} - 1$ obtained from (7.14.8) is shown in Fig. 7.3 and the first 5 natural frequencies are

22.4261.700120.89199.80298.43

The first 4 natural frequencies presented by the exercise are

22.37, 61.67, 120.91, 199.85

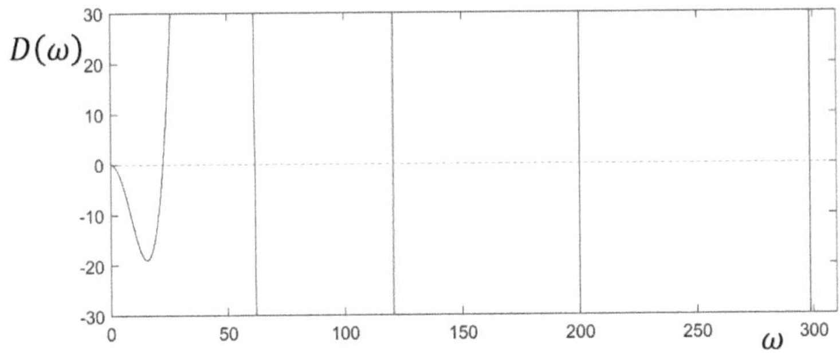

Fig. 7.3 The function of the frequency equation for Exercise 7.14

It can be seen that the calculated values of the first 4 natural frequencies are very close to the values given in the exercise; therefore, we adopt the values given in the exercise in the following analyses. We note that

$$\omega_1 + \omega_2 + \omega_3 = \omega_4 + \varepsilon\sigma_1 \tag{7.14.10}$$

Thus there is internal resonance.

Considering the above results, we can obtain the modal discretization equation for the beam of this problem

$$\ddot{u}_n + \omega_n^2 u_n = \varepsilon\left(\sum_{m,\,p,\,q=1}^{\infty} \Gamma_{nmpq} u_m u_p u_q - 2\mu_n \dot{u}_n\right) + f_{zn}(t) \tag{7.14.11}$$

where

$$\Gamma_{nmpq} = -\frac{1}{2l}\left(\int_0^l \phi'_n \phi'_m dx\right)\left(\int_0^l \phi'_p \phi'_q dx\right),\ l = 2 \tag{7.14.12}$$

and

$$\Gamma_{nmpq} = \Gamma_{mnpq} = \Gamma_{nmqp} = \Gamma_{pqnm} \tag{7.14.13}$$

We use the method of multiple scales to solve (7.14.11). Let the solution of the equation be

$$u_n(t;\varepsilon) = u_{n0}(T_0, T_1) + \varepsilon u_{n1}(T_0, T_1) + \cdots \tag{7.14.14}$$

Substituting (7.14.14) into (7.14.11) and retaining to $O(\varepsilon)$, we obtain

$$
\begin{aligned}
0 &= \ddot{u}_n + \omega_n^2 u_n + \varepsilon\left(2\mu_n \dot{u}_n - \sum_{m,p,q=1}^{\infty} \Gamma_{nmpq} u_m u_p u_q\right) - f_{zn}(t) \\
&= \left(D_0^2 + 2\varepsilon D_0 D_1\right)(u_{n0} + \varepsilon u_{n1}) + \omega_n^2 (u_{n0} + \varepsilon u_{n1}) \\
&+ \varepsilon\left(2\mu_n D_0 u_{n0} - \sum_{m,p,q=1}^{\infty} \Gamma_{nmpq} u_{m0} u_{p0} u_{q0}\right) - 2\varepsilon \delta_{ns} k_n \cos \Omega t \\
&= D_0^2 u_{n0} + \omega_n^2 u_{n0} + \varepsilon\big(D_0^2 u_{n1} + \omega_n^2 u_{n1} + 2 D_0 D_1 u_{n0} \\
&+ 2\mu_n D_0 u_{n0} - \sum_{m,p,q=1}^{\infty} \Gamma_{nmpq} u_{m0} u_{p0} u_{q0} - 2\delta_{ns} k_n \cos \Omega t\big)
\end{aligned} \tag{7.14.15}
$$

Making the coefficient of the same power of ε zero in the above equation yields

$$
D_0^2 u_{n0} + \omega_n^2 u_{n0} = 0 \tag{7.14.16}
$$

$$
\begin{aligned}
D_0^2 u_{n1} + \omega_n^2 u_{n1} &= -2 D_0 D_1 u_{n0} - 2\mu_n D_0 u_{n0} \\
&+ \sum_{m,\, p,\, q=1}^{\infty} \Gamma_{nmpq} u_{m0} u_{p0} u_{q0} \\
&+ 2\delta_{ns} k_n \cos \Omega t
\end{aligned} \tag{7.14.17}
$$

The solution of (7.14.16) is

$$
u_{n0} = A_n(T_1) e^{i\omega_n T_0} + cc \tag{7.14.18}
$$

Substituting (7.14.18) into (7.14.17) yields

$$
\begin{aligned}
D_0^2 u_{n1} + \omega_n^2 u_{n1} &= -2 D_0 D_1 u_{n0} - 2\mu_n D_0 u_{n0} \\
&+ \sum_{m,p,q=1}^{\infty} \Gamma_{nmpq} u_{m0} u_{p0} u_{q0} + 2\delta_{ns} k_n \cos \Omega t \\
&= -2i\omega_n A_n' e^{i\omega_n T_0} - 2i\omega_n \mu_n A_n e^{i\omega_n T_0} \\
&+ \sum_{m,p,q=1}^{\infty} \Gamma_{nmpq} \big\{A_m A_p A_q e^{i\omega_m T_0} e^{i\omega_p T_0} e^{i\omega_q T_0} \\
&+ A_m \overline{A}_p A_q e^{i\omega_m T_0} e^{-i\omega_p T_0} e^{i\omega_q T_0} + A_m A_p \overline{A}_q e^{i\omega_m T_0} e^{i\omega_p T_0} e^{-i\omega_q T_0} \\
&+ A_m \overline{A}_p \overline{A}_q e^{i\omega_m T_0} e^{-i\omega_p T_0} e^{-i\omega_q T_0}\big\} + \delta_{ns} k_n e^{i\Omega T_0} + cc + NST
\end{aligned} \tag{7.14.19}
$$

In order to eliminate secular terms from the above equation, and take into account $\omega_1 + \omega_2 + \omega_3 = \omega_4 + \varepsilon\sigma_1$ and $\Omega = \omega_s + \varepsilon\sigma_2$, we need

$$-2i\omega_1\left(A_1' + \mu_1 A_1\right) + A_1\sum_{m=1}^{\infty}\alpha_{1m}A_m\overline{A}_m + 8QA_4\overline{A}_3\overline{A}_2 e^{-i\sigma_1 T_1} + \delta_{1s}k_1 e^{i\sigma_2 T_1} = 0 \tag{7.14.20}$$

$$-2i\omega_2\left(A_2' + \mu_2 A_2\right) + A_2\sum_{m=1}^{\infty}\alpha_{2m}A_m\overline{A}_m + 8QA_4\overline{A}_3\overline{A}_1 e^{-i\sigma_1 T_1} + \delta_{2s}k_2 e^{i\sigma_2 T_1} = 0 \tag{7.14.21}$$

$$-2i\omega_3\left(A_3' + \mu_3 A_3\right) + A_3\sum_{m=1}^{\infty}\alpha_{3m}A_m\overline{A}_m + 8QA_4\overline{A}_2\overline{A}_1 e^{-i\sigma_1 T_1} + \delta_{3s}k_3 e^{i\sigma_2 T_1} = 0 \tag{7.14.22}$$

$$-2i\omega_4\left(A_4' + \mu_4 A_4\right) + A_4\sum_{m=1}^{\infty}\alpha_{4m}A_m\overline{A}_m + 8QA_3A_2A_1 e^{i\sigma_1 T_1} + \delta_{4s}k_1 e^{i\sigma_2 T_1} = 0 \tag{7.14.23}$$

$$-2i\omega_n\left(A_n' + \mu_n A_n\right) + A_n\sum_{m=1}^{\infty}\alpha_{nm}A_m\overline{A}_m = 0, \quad n \geq 5 \tag{7.14.24}$$

where

$$\begin{aligned} 8Q &= 2(\Gamma_{1234} + \Gamma_{1324} + \Gamma_{1423}) \\ \alpha_{mn} = \alpha_{nm} &= \begin{cases} 2(\Gamma_{nnmm} + 2\Gamma_{nmnm}) \text{ for } & n \neq m \\ 6\Gamma_{nnnn} \end{cases} \end{aligned} \tag{7.14.25}$$

Let

$$A_n = \frac{1}{2}a_n e^{i\beta_n} \tag{7.14.26}$$

Substituting it into (7.14.24) yields

$$i\omega_n a_n' - \omega_n a_n \beta_n' + i\omega_n \mu_n a_n - \frac{1}{8}a_n\sum_{m=1}^{\infty}\alpha_{nm}a_m^2 = 0, \ n \geq 5 \tag{7.14.27}$$

Separating the real and imaginary parts of the above equation yields

$$a_n' + \mu_n a_n = 0, \quad n \geq 5 \tag{7.14.28}$$

Therefore,

$$a_n = a_{n0}e^{-\mu_n T_1} \to 0 \Rightarrow A_n \to 0, \ n \geq 5 \tag{7.14.29}$$

When $s = 1$, the Eqs. (7.14.20)–(7.14.23) become

$$-2i\omega_1\left(A_1' + \mu_1 A_1\right) + A_1\sum_{m=1}^{4}\alpha_{1m}A_m\overline{A}_m + 8QA_4\overline{A}_3\overline{A}_2 e^{-i\sigma_1 T_1} + k_1 e^{i\sigma_2 T_1} = 0 \tag{7.14.30}$$

$$-2i\omega_2\left(A_2' + \mu_2 A_2\right) + A_2\sum_{m=1}^{4}\alpha_{2m}A_m\overline{A}_m + 8QA_4\overline{A}_3\overline{A}_1 e^{-i\sigma_1 T_1} = 0 \tag{7.14.31}$$

$$-2i\omega_3\left(A_3' + \mu_3 A_3\right) + A_3\sum_{m=1}^{4}\alpha_{3m}A_m\overline{A}_m + 8QA_4\overline{A}_2\overline{A}_1 e^{-i\sigma_1 T_1} = 0 \tag{7.14.32}$$

$$-2i\omega_4\left(A_4' + \mu_4 A_4\right) + A_4\sum_{m=1}^{4}\alpha_{4m}A_m\overline{A}_m + 8QA_3A_2A_1 e^{i\sigma_1 T_1} = 0 \tag{7.14.33}$$

where (7.14.29) is taken into account.

Substitute (7.14.26) into (7.14.30)–(7.14.33), we obtain

$$i\omega_1 a_1' - \omega_1 a_1\beta_1' + i\omega_1\mu_1 a_1 - \frac{1}{8}a_1\sum_{m=1}^{4}\alpha_{1m}a_m^2 - Qa_2a_3a_4e^{-i\gamma_1} - k_1e^{i\gamma_2} = 0 \tag{7.14.34}$$

$$i\omega_2 a_2' - \omega_2 a_2\beta_2' + i\omega_2\mu_2 a_2 - \frac{1}{8}a_2\sum_{m=1}^{4}\alpha_{2m}a_m^2 - Qa_1a_3a_4e^{-i\gamma_1} = 0 \tag{7.14.35}$$

$$i\omega_3 a_3' - \omega_3 a_3\beta_3' + i\omega_3\mu_3 a_3 - \frac{1}{8}a_3\sum_{m=1}^{4}\alpha_{3m}a_m^2 - Qa_1a_2a_4e^{-i\gamma_1} = 0 \tag{7.14.36}$$

$$i\omega_4 a_4' - \omega_4 a_4\beta_4' + i\omega_4\mu_4 a_4 - \frac{1}{8}a_4\sum_{m=1}^{4}\alpha_{4m}a_m^2 - Qa_1a_2a_3e^{i\gamma_1} = 0 \tag{7.14.37}$$

where

$$\gamma_1 = \sigma_1 T_1 + \beta_1 + \beta_2 + \beta_3 - \beta_4,\ \gamma_2 = \sigma_2 T_1 - \beta_1 \tag{7.14.38}$$

Separating the real and imaginary parts of (7.14.35) and (7.14.37), we obtain

$$\omega_2 a_2' = -\omega_2\mu_2 a_2 - Qa_1a_3a_4\sin\gamma_1 \tag{7.14.39}$$

$$\omega_4 a_4' = -\omega_4\mu_4 a_4 + Qa_1a_2a_3\sin\gamma_1 \tag{7.14.40}$$

The steady state solutions of a_1 and a_2 satisfy

$$\begin{aligned}\omega_2\mu_2 a_2 &= -Qa_1a_3a_4\sin\gamma_1\\ \omega_4\mu_4 a_4 &= Qa_1a_2a_3\sin\gamma_1\end{aligned}\tag{7.14.41}$$

Eliminating γ_1 from the above equation yields

$$\omega_2\mu_2 a_2^2 + \omega_4\mu_4 a_4^2 = 0\tag{7.14.42}$$

Therefore, $a_2 = a_4 = 0$. Separating the real and imaginary parts of the Eq. (7.14.36), we obtain

$$a_3' = -\mu_3 a_3 \Rightarrow a_3 = a_{30}e^{-\mu_3 T_1} \to 0\tag{7.14.43}$$

So

$$a_2 = a_3 = a_4 = 0\tag{7.14.44}$$

Thus, the Eq. (7.14.34) becomes

$$i\omega_1 a_1' - \omega_1 a_1\beta_1' + i\omega_1\mu_1 a_1 - \frac{1}{8}\alpha_{11}a_1^3 - k_1 e^{i\gamma_2} = 0\tag{7.14.45}$$

Separates the real part from the imaginary part of the above equation, we can obtain

$$\begin{aligned}\omega_1 a_1' &= -\omega_1\mu_1 a_1 + k_1\sin\gamma_2\\ \omega_1 a_1\gamma_2' &= \omega_1\sigma_2 a_1 + \frac{1}{8}\alpha_{11}a_1^3 + k_1\cos\gamma_2\end{aligned}\tag{7.14.46}$$

Let $a_1' = \gamma_2' = 0$ in (7.14.46), we can obtain the equation satisfied by the steady state solution:

$$\begin{aligned}\omega_1\mu_1 a_1 &= k_1\sin\gamma_2\\ -a_1\left(\omega_1\sigma_2 + \frac{1}{8}\alpha_{11}a_1^2\right) &= k_1\cos\gamma_2\end{aligned}\tag{7.14.47}$$

The frequency–response equation is

$$\omega_1^2\mu_1^2 a_1^2 + a_1^2\left(\omega_1\sigma_2 + \frac{1}{8}\alpha_{11}a_1^2\right)^2 = k_1^2\tag{7.14.48}$$

(f) and **(g)** Readers are invited to complete Exercise 7.14(f) and (g).
When $s = 4$, substituting (7.14.26) into (7.14.20)–(7.14.23), we can obtain

$$i\omega_1 a_1' - \omega_1 a_1 \beta_1' + i\omega_1 \mu_1 a_1 - \frac{1}{8} a_1 \sum_{m=1}^{4} \alpha_{1m} a_m^2 - Q a_2 a_3 a_4 e^{-i\gamma_1} = 0 \quad (7.14.49)$$

$$i\omega_2 a_2' - \omega_2 a_2 \beta_2' + i\omega_2 \mu_2 a_2 - \frac{1}{8} a_2 \sum_{m=1}^{4} \alpha_{2m} a_m^2 - Q a_1 a_3 a_4 e^{-i\gamma_1} = 0 \quad (7.14.50)$$

$$i\omega_3 a_3' - \omega_3 a_3 \beta_3' + i\omega_3 \mu_3 a_3 - \frac{1}{8} a_3 \sum_{m=1}^{4} \alpha_{3m} a_m^2 - Q a_1 a_2 a_4 e^{-i\gamma_1} = 0 \quad (7.14.51)$$

$$i\omega_4 a_4' - \omega_4 a_4 \beta_4' + i\omega_4 \mu_4 a_4 - \frac{1}{8} a_4 \sum_{m=1}^{4} \alpha_{4m} a_m^2 - Q a_1 a_2 a_3 e^{i\gamma_1} - k_4 e^{i\gamma_2} = 0 \quad (7.14.52)$$

where

$$\gamma_1 = \sigma_1 T_1 + \beta_1 + \beta_2 + \beta_3 - \beta_4, \gamma_2 = \sigma_2 T_1 - \beta_1 \quad (7.14.53)$$

and (7.14.29) is taken into account. Separating the real and imaginary parts of (7.14.35) and (7.14.37) and taking into account (7.14.53), we can obtain

$$\omega_1 a_1' = -\omega_1 \mu_1 a_1 - Q a_2 a_3 a_4 \sin\gamma_1 \quad (7.14.54)$$

$$\omega_2 a_2' = -\omega_2 \mu_2 a_2 - Q a_1 a_3 a_4 \sin\gamma_1 \quad (7.14.55)$$

$$\omega_3 a_3' = -\omega_3 \mu_3 a_3 - Q a_1 a_2 a_4 \sin\gamma_1 \quad (7.14.56)$$

$$\omega_4 a_4' = -\omega_4 \mu_4 a_4 + Q a_1 a_2 a_3 \sin\gamma_1 + k_4 \sin\gamma_2 \quad (7.14.57)$$

$$\omega_1 a_1 \gamma_2' = \omega_1 \sigma_2 a_1 + \frac{1}{8} a_1 \sum_{m=1}^{4} \alpha_{1m} a_m^2 + Q a_2 a_3 a_4 \cos\gamma_1 \quad (7.14.58)$$

$$\omega_2 a_2 \beta_2' = -\frac{1}{8} a_2 \sum_{m=1}^{4} \alpha_{2m} a_m^2 - Q a_1 a_3 a_4 \cos\gamma_1 \quad (7.14.59)$$

$$\omega_3 a_3 \beta_3' = -\frac{1}{8} a_3 \sum_{m=1}^{4} \alpha_{3m} a_m^2 - Q a_1 a_2 a_4 \cos\gamma_1 \quad (7.14.60)$$

$$-\omega_4 a_4 \gamma_1' - \omega_4 a_4 \gamma_2' + \omega_4 a_4 \beta_2' + \omega_4 a_4 \beta_3' = -\omega_4(\sigma_1 + \sigma_2) a_4 - \frac{1}{8} a_4 \sum_{m=1}^{4} \alpha_{4m} a_m^2 - Q a_1 a_2 a_3 \cos\gamma_1 - k_4 \cos\gamma_2 \quad (7.14.61)$$

The corresponding steady state solution must satisfy

$$\omega_1\mu_1 a_1 = -Qa_2a_3a_4\sin\gamma_1 \tag{7.14.62}$$

$$\omega_2\mu_2 a_2 = -Qa_1a_3a_4\sin\gamma_1 \tag{7.14.63}$$

$$\omega_3\mu_3 a_3 = -Qa_1a_2a_4\sin\gamma_1 \tag{7.14.64}$$

$$\omega_4\mu_4 a_4 - Qa_1a_2a_3\sin\gamma_1 - k_4\sin\gamma_2 = 0 \tag{7.14.65}$$

$$\omega_1\sigma_2 a_1 + \frac{1}{8}a_1\sum_{m=1}^{4}\alpha_{1m}a_m^2 + Qa_2a_3a_4\cos\gamma_1 = 0 \tag{7.14.66}$$

$$\frac{1}{8}a_2\sum_{m=1}^{4}\alpha_{2m}a_m^2 + Qa_1a_3a_4\cos\gamma_1 = 0 \tag{7.14.67}$$

$$\frac{1}{8}a_3\sum_{m=1}^{4}\alpha_{3m}a_m^2 + Qa_1a_2a_4\cos\gamma_1 = 0 \tag{7.14.68}$$

$$\omega_4(\sigma_1+\sigma_2)a_4 + \frac{1}{8}a_4\sum_{m=1}^{4}\alpha_{4m}a_m^2 + Qa_1a_2a_3\cos\gamma_1 + k_4\cos\gamma_2 = 0 \tag{7.14.69}$$

(7.14.62)–(7.14.69) have different steady state solutions, which are discussed separately as follows.

The Case of a Partially Zero Steady State Solution

From the direct observation on (7.14.62)–(7.14.69) we can find that there exists a steady state solution.

$$a_1 = a_2 = a_3 = 0 \text{ and } a_4 \text{ TBD} \tag{7.14.70}$$

Then (7.14.65) and (7.14.69) become

$$\begin{aligned}&\omega_4\mu_4 a_4 = k_4\sin\gamma_2\\ &a_4\left[\omega_4(\sigma_1+\sigma_2) + \frac{1}{8}\alpha_{44}a_4^2\right] = -k_4\cos\gamma_2\end{aligned} \tag{7.14.71}$$

The frequency–response equation is

$$\omega_4^2\mu_4^2a_4^2 + a_4^2[\omega_4(\sigma_1+\sigma_2) + \frac{1}{8}\alpha_{44}a_4^2]^2 = k_4^2 \tag{7.14.72}$$

Therefore, the steady state solution $a_4 \neq 0$.

The case of all non-zero steady state solutions

When $a_1 \neq 0, a_2 \neq 0, a_4 \neq 0$, we can solve for $\sin\gamma_1$ from (7.14.64) and substitute it into (7.14.65); solve for $\cos\gamma_1$ from (7.14.68) and substitute it into (7.14.69):

$$\omega_4\mu_4 a_4^2 + \omega_3\mu_3 a_3^2 = a_4 k_4 \sin\gamma_2 \tag{7.14.73}$$

$$\omega_4(\sigma_1+\sigma_2)a_4^2 + \frac{1}{8}a_4^2\sum_{m=1}^{4}\alpha_{4m}a_m^2 - \frac{1}{8}a_3^2\sum_{m=1}^{4}\alpha_{3m}a_m^2 = -a_4 k_4 \cos\gamma_2 \tag{7.14.74}$$

Eliminating γ_1 from three sets of Eqs. (7.14.62) and (7.14.66), (7.14.63) and (7.14.67), as well as (7.14.64) and (7.14.68); and eliminating γ_2 from (7.14.73) yield

$$\omega_1^2\mu_1^2 a_1^2 + a_1^2\left(\omega_1\sigma_2 + \frac{1}{8}\sum_{m=1}^{4}\alpha_{1m}a_m^2\right)^2 = Q^2 a_2^2 a_3^2 a_4^2 \tag{7.14.75}$$

$$\omega_2^2\mu_2^2 a_2^2 + a_2^2\left(\frac{1}{8}\sum_{m=1}^{4}\alpha_{2m}a_m^2\right)^2 = Q^2 a_1^2 a_3^2 a_4^2 \tag{7.14.76}$$

$$\omega_3^2\mu_3^2 a_3^2 + a_3^2\left(\frac{1}{8}\sum_{m=1}^{4}\alpha_{3m}a_m^2\right)^2 = Q^2 a_1^2 a_2^2 a_4^2 \tag{7.14.77}$$

$$\begin{aligned}&\left[\omega_4(\sigma_1+\sigma_2)a_4^2 + \frac{1}{8}a_4^2\sum_{m=1}^{4}\alpha_{4m}a_m^2 - \frac{1}{8}a_3^2\sum_{m=1}^{4}\alpha_{3m}a_m^2\right]^2 \\ &\quad + \left(\omega_4\mu_4 a_4^2 + \omega_3\mu_3 a_3^2\right)^2 = a_4^2 k_4^2\end{aligned} \tag{7.14.78}$$

This set of equations is the frequency–response equation for the steady-state amplitude. It can be seen that there is a non-trivial solution when $a_1 \neq 0,\ a_2 \neq 0,\ a_4 \neq 0$; for example, whenever $a_1 \neq 0$, then, by (7.14.75), we know that there must be $a_2 \neq 0,\ a_3 \neq 0,\ a_4 \neq 0$.

7.15 Exercise 7.15 (Transverse Oscillation of a Clamped–Clamped Beam Under Internal Resonance and Non-resonance Excitation at $u = O(w^2)$)

Solution: We take the characteristic length of the beam,L, to be the actual length of the beam, then the dimensionless length of the beam $l = 1$. In turn, the linear free oscillation mode function of the beam is

$$\phi_n(x) = C_n[(\sin\eta_n - \sinh\eta_n)(\cos\eta_n x - \cosh\eta_n x) \\ - (\cos\eta_n - \cosh\eta_n)(\sin\eta_n x - \sinh\eta_n x)] \tag{7.15.1}$$

where the value of C_n is taken to normalize the mode shape function $\phi_n(x)$. The first five natural frequencies of the beam are

22.42 61.700 120.89 199.80 298.43

We note that

$$\omega_1 + \omega_2 + \omega_3 = \omega_4 + \varepsilon\sigma_1 \tag{7.15.2}$$

Hence there is internal resonance. The discrete equation of the mode shape function of the beam in this problem is

$$\ddot{u}_n + \omega_n^2 u_n = \varepsilon\left(\sum_{m,p,q=1}^{\infty} \Gamma_{nmpq} u_m u_p u_q - 2\mu_n \dot{u}_n\right) + f_{zn}(t) \tag{7.15.3}$$

where

$$\Gamma_{nmpq} = -\frac{1}{2l}\left(\int_0^l \phi_n' \phi_m' dx\right)\left(\int_0^l \phi_p' \phi_q' dx\right), \quad l = 2 \tag{7.15.4}$$

and

$$\Gamma_{nmpq} = \Gamma_{mnpq} = \Gamma_{nmqp} = \Gamma_{pqnm} \tag{7.15.5}$$

We use the method of multiple scales to solve (7.15.3). Let the solution of the equation be

$$u_n(t;\varepsilon) = u_{n0}(T_0, T_1) + \varepsilon u_{n1}(T_0, T_1) + \cdots \tag{7.15.6}$$

Substituting (7.15.6) into (7.15.3) and retaining to $O(\varepsilon)$, we get

$$
\begin{aligned}
0 &= \ddot{u}_n + \omega_n^2 u_n + \varepsilon\left(2\mu_n \dot{u}_n - \sum_{m,p,q=1}^{\infty} \Gamma_{nmpq} u_m u_p u_q\right) - f_{zn}(t) \\
&= \left(D_0^2 + 2\varepsilon D_0 D_1\right)(u_{n0} + \varepsilon u_{n1}) + \omega_n^2 (u_{n0} + \varepsilon u_{n1}) \\
&+\varepsilon\left(2\mu_n D_0 u_{n0} - \sum_{m,p,q=1}^{\infty} \Gamma_{nmpq} u_{m0} u_{p0} u_{q0}\right) - 2K_n \cos \Omega t \\
&= D_0^2 u_{n0} + \omega_n^2 u_{n0} - 2K_n \cos \Omega t + \varepsilon\Big(D_0^2 u_{n1} + \omega_n^2 u_{n1} \\
&+2D_0 D_1 u_{n0} + 2\mu_n D_0 u_{n0} - \sum_{m,p,q=1}^{\infty} \Gamma_{nmpq} u_{m0} u_{p0} u_{q0}\Big)
\end{aligned}
\tag{7.15.7}
$$

Making the coefficient of the same power of ε zero in the above equation yields

$$
D_0^2 u_{n0} + \omega_n^2 u_{n0} = 2K_n \cos \Omega T_0 \tag{7.15.8}
$$

$$
D_0^2 u_{n1} + \omega_n^2 u_{n1} = -2D_0 D_1 u_{n0} - 2\mu_n D_0 u_{n0} + \sum_{m,\,p,\,q=1}^{\infty} \Gamma_{nmpq} u_{m0} u_{p0} u_{q0} \tag{7.15.9}
$$

The solution of (7.15.8) is

$$
u_{n0} = A_n(T_1) e^{i\omega_n T_0} + \Lambda_n e^{i\Omega T_0} + cc \tag{7.15.10}
$$

where

$$
\Lambda_n = \frac{K_n}{\omega_n^2 - \Omega^2} \tag{7.15.11}
$$

Substituting (7.15.10) into (7.15.9) and taking into account $\omega_1 + \omega_2 + \omega_3 = \omega_4 + \varepsilon\sigma_1$ and $\Omega = \omega_3 - \omega_1 - \omega_2 + \varepsilon\sigma_2$, we obtain

$$
\begin{aligned}
D_0^2 u_{n1} + \omega_n^2 u_{n1} &= -2D_0 D_1 u_{n0} - 2\mu_n D_0 u_{n0} + \sum_{m,p,q=1}^{\infty} \Gamma_{nmpq} u_{m0} u_{p0} u_{q0} \\
&= -2i\omega_n A_n' e^{i\omega_n T_0} - 2i\omega_n \mu_n A_n e^{i\omega_n T_0} + 2A_n e^{i\omega_n T_0} \sum_{p,q=1}^{\infty} \left(\Gamma_{nnpq} + 2\Gamma_{npnq}\right) \Lambda_p \Lambda_q \\
&+ A_4 \sum_{p,q=1}^{\infty} \left(\Gamma_{np4q} + \Gamma_{nqp4} + \Gamma_{n4pq}\right) \overline{A}_p \overline{A}_q e^{i\omega_4 T_0} e^{-i\omega_p T_0} e^{-i\omega_q T_0} \\
&+ \sum_{m,p,q=1}^{\infty} \Gamma_{nmpq} \left\{ A_m e^{i\omega_m T_0} \overline{A}_p e^{-i\omega_p T_0} \Lambda_q e^{i\Omega T_0} + A_m e^{i\omega_m T_0} \Lambda_p e^{i\Omega T_0} \overline{A}_q e^{-i\omega_q T_0} \right. \\
&+ A_m e^{i\omega_m T_0} \Lambda_p e^{-i\Omega T_0} \overline{A}_q e^{-i\omega_q T_0} + A_m e^{i\omega_m T_0} A_p e^{i\omega_p T_0} \Lambda_q e^{-i\Omega T_0} \\
&+ A_m e^{i\omega_m T_0} \overline{A}_p e^{-i\omega_p T_0} \Lambda_q e^{-i\Omega T_0} + \Lambda_m e^{i\Omega T_0} \overline{A}_p e^{-i\omega_p T_0} A_q e^{i\omega_q T_0} \\
&+ \Lambda_m e^{i\Omega T_0} A_p e^{i\omega_p T_0} \overline{A}_q e^{-i\omega_q T_0} + \Lambda_m e^{i\Omega T_0} \overline{A}_p e^{-i\omega_p T_0} \overline{A}_q e^{-i\omega_q T_0} \\
&+ A_m e^{i\omega_m T_0} A_p e^{i\omega_p T_0} \Lambda_q e^{i\Omega T_0} + A_m e^{i\omega_m T_0} \Lambda_p e^{i\Omega T_0} A_q e^{i\omega_q T_0} \\
&+ \Lambda_m e^{i\Omega T_0} A_p e^{i\omega_p T_0} A_q e^{i\omega_q T_0} + A_m e^{i\omega_m T_0} A_p e^{i\omega_p T_0} \overline{A}_q e^{-i\omega_q T_0} \\
&\left. + A_m e^{i\omega_m T_0} \overline{A}_p e^{-i\omega_p T_0} A_q e^{i\omega_q T_0} + \overline{A}_m e^{-i\omega_m T_0} A_p e^{i\omega_p T_0} A_q e^{i\omega_q T_0} \right\} \\
&+ 2(\Gamma_{1234} + \Gamma_{1324} + \Gamma_{1423}) A_1 A_2 A_3 e^{i\sigma_1 T_1} e^{i\omega_4 T_0} + cc + NST
\end{aligned}
\tag{7.15.12}
$$

In order to eliminate secular terms from the above equation, and taking into account $\omega_1 + \omega_2 + \omega_3 = \omega_4 + \varepsilon\sigma_1$ and $\Omega = \omega_3 - \omega_1 - \omega_2 + \varepsilon\sigma_2$, we need

$$
\begin{aligned}
&- 2i\omega_1 \left(A_1' + \mu_1 A_1\right) + 2H_{11} A_1 + A_1 \sum_{m=1}^{\infty} \alpha_{1m} A_m \bar{A}_m \\
&+ 8Q\bar{A}_2 \bar{A}_3 A_4 e^{-i\sigma_1 T_1} + 4F\bar{A}_2 A_3 e^{-i\sigma_2 T_1} = 0
\end{aligned}
\tag{7.15.13}
$$

$$
\begin{aligned}
&- 2i\omega_2 \left(A_2' + \mu_2 A_2\right) + 2H_{22} A_2 + A_2 \sum_{m=1}^{\infty} \alpha_{2m} A_m \bar{A}_m \\
&+ 8Q\bar{A}_1 \bar{A}_3 A_4 e^{-i\sigma_1 T_1} + 4F\bar{A}_1 A_3 e^{-i\sigma_2 T_1} = 0
\end{aligned}
\tag{7.15.14}
$$

$$
\begin{aligned}
&- 2i\omega_3 \left(A_3' + \mu_3 A_3\right) + 2H_{33} A_3 + A_3 \sum_{m=1}^{\infty} \alpha_{3m} A_m \overline{A}_m \\
&+ 8Q\overline{A}_1 \overline{A}_2 A_4 e^{-i\sigma_1 T_1} + 4F A_1 A_2 e^{i\sigma_2 T_1} = 0
\end{aligned}
\tag{7.15.15}
$$

$$
-2i\omega_4 \left(A_4' + \mu_4 A_4\right) + 2H_{44} A_4 + A_4 \sum_{m=1}^{\infty} \alpha_{4m} A_m \overline{A}_m + 8Q A_1 A_2 A_3 e^{i\sigma_1 T_1} = 0
\tag{7.15.16}
$$

$$-2i\omega_n\left(A_n' + \mu_n A_n\right) + 2H_{nn}A_n + A_n\sum_{m=1}^{\infty}\alpha_{nm}A_m\bar{A}_m = 0, \quad n \geq 5 \tag{7.15.17}$$

where

$$\begin{aligned}
&8Q = 2(\Gamma_{1234} + \Gamma_{1324} + \Gamma_{1423}), \quad 4F = 2\sum_{m=1}^{\infty}(\Gamma_{123m} + \Gamma_{231m} + \Gamma_{132m})\Lambda_m \\
&\alpha_{mn} = \alpha_{nm} = \begin{cases} 2(\Gamma_{nnmm} + 2\Gamma_{nmnm}) \text{ for } & n \neq m \\ 6\Gamma_{nnnn} \end{cases} \\
&H_{nm} = \sum_{p,q}\left(\Gamma_{nmpq} + 2\Gamma_{npmq}\right)\Lambda_p\Lambda_q
\end{aligned} \tag{7.15.18}$$

Let

$$A_n = \frac{1}{2}a_n e^{i\beta_n} \tag{7.15.19}$$

Substituting it into $2i\omega_n A_n' + 2i\omega_n\mu_n A_n - 2H_{nn}A_n - A_n\sum_{m=1}^{\infty}\alpha_{nm}A_m\bar{A}_m = 0, \quad n \geq 3$ yields

$$i\omega_n a_n' - \omega_n a_n\beta_n' + i\omega_n\mu_n a_n - H_{nn}a_n - \frac{1}{8}a_n\sum_{m=1}^{\infty}\alpha_{nm}a_m^2 = 0, \ n \geq 5 \tag{7.15.20}$$

Separating the real and imaginary parts of the above equation yields

$$a_n' + \mu_n a_n = 0, \quad n \geq 5 \tag{7.15.21}$$

therefore

$$a_n = a_{n0}e^{-\mu_n T_1} \to 0 \Rightarrow A_n \to 0, \ n \geq 5 \tag{7.15.22}$$

Substituting (7.15.19) into (7.15.13)–(7.15.16) and taking into account that (7.15.22), we have

$$\begin{aligned}
&-i\omega_1 a_1' + \omega_1 a_1\beta_1' - i\omega_1\mu_1 a_1 + H_{11}a_1 \\
&+\frac{1}{8}a_1\sum_{m=1}^{4}\alpha_{1m}a_m^2 + Qa_2a_3a_4e^{i\gamma_1} + Fa_2a_3e^{i\gamma_2} = 0
\end{aligned} \tag{7.15.23}$$

$$-i\omega_2 a_2' + \omega_2 a_2\beta_2' - i\omega_2\mu_2 a_2 + H_{22}a_2$$

$$+\frac{1}{8}a_2\sum_{m=1}^{4}\alpha_{2m}a_m^2 + Qa_1a_3a_4e^{i\gamma_1} + Fa_1a_3e^{i\gamma_2} = 0 \tag{7.15.24}$$

$$\begin{aligned} &- i\omega_3a_3' + \omega_3a_3\beta_3' - i\omega_3\mu_3a_3 + H_{33}a_3 \\ &+\frac{1}{8}a_3\sum_{m=1}^{4}\alpha_{3m}a_m^2 + Qa_1a_2a_4e^{i\gamma_1} + Fa_1a_2e^{-i\gamma_2} = 0 \end{aligned} \tag{7.15.25}$$

$$-i\omega_4a_4' + \omega_4a_4\beta_4' - i\omega_4\mu_4a_4 + H_{44}a_4 + \frac{1}{8}a_4\sum_{m=1}^{4}\alpha_{4m}a_m^2 + Qa_1a_2a_3e^{-i\gamma_1} = 0 \tag{7.15.26}$$

where

$$\gamma_1 = \beta_4 - \beta_3 - \beta_2 - \beta_1 - \sigma_1T_1, \quad \gamma_2 = \beta_3 - \beta_2 - \beta_1 - \sigma_2T_1 \tag{7.15.27}$$

Separating the real and imaginary parts of (7.15.23)–(7.15.26) and making $a_1' = a_2' = a_3' = a_4' = 0$, we obtain

$$-\omega_1\mu_1a_1 + Qa_2a_3a_4\sin\gamma_1 + Fa_2a_3\sin\gamma_2 = 0 \tag{7.15.28}$$

$$-\omega_2\mu_2a_2 + Qa_1a_3a_4\sin\gamma_1 + Fa_1a_3\sin\gamma_2 = 0 \tag{7.15.29}$$

$$-\omega_3\mu_3a_3 + Qa_1a_2a_4\sin\gamma_1 - Fa_1a_2\sin\gamma_2 = 0 \tag{7.15.30}$$

$$-\omega_4\mu_4a_4 - Qa_1a_2a_3\sin\gamma_1 = 0 \tag{7.15.31}$$

$$\omega_1a_1\beta_1' + H_{11}a_1 + \frac{1}{8}a_1\sum_{m=1}^{4}\alpha_{1m}a_m^2 + Qa_2a_3a_4\cos\gamma_1 + Fa_2a_3\cos\gamma_2 = 0 \tag{7.15.32}$$

$$\omega_2a_2\beta_2' + H_{22}a_2 + \frac{1}{8}a_2\sum_{m=1}^{4}\alpha_{2m}a_m^2 + Qa_1a_3a_4\cos\gamma_1 + Fa_1a_3\cos\gamma_2 = 0 \tag{7.15.33}$$

$$\omega_3a_3\beta_3' + H_{33}a_3 + \frac{1}{8}a_3\sum_{m=1}^{4}\alpha_{3m}a_m^2 + Qa_1a_2a_4\cos\gamma_1 + Fa_1a_2\cos\gamma_2 = 0 \tag{7.15.34}$$

$$\omega_4a_4\beta_4' + H_{44}a_4 + \frac{1}{8}a_4\sum_{m=1}^{4}\alpha_{4m}a_m^2 + Qa_1a_2a_3\cos\gamma_1 = 0 \tag{7.15.35}$$

(7.15.28)–(7.15.35) control the magnitude and phase of the mode shape function. For any set of amplitude steady state solutions, the phase is assumed to be

$$\beta_n = \beta_n' T_1 + \tau_n = \varepsilon \beta_n' t + \tau_n, \quad n = 1, 2, 3, 4 \tag{7.15.36}$$

Then by the Eqs. (7.15.10) and (7.15.6), we can obtain the approximate solution of $u_n(t)$:

$$\begin{aligned} u_n \approx u_{n0} &= a_n \cos\left[\left(\omega_n + \varepsilon \beta_n'\right)t + \tau_n\right] \\ &\quad + 2\Lambda_n \cos \Omega t + O(\varepsilon), \quad n = 1, 2, 3, 4 \\ u_n \approx u_{n0} &= 2\Lambda_n \cos \Omega t + O(\varepsilon), \quad n \geq 5 \end{aligned} \tag{7.15.37}$$

Since the deflection of the beam is $w(x, t) = \sum_{n=1}^{\infty} u_n \phi_n(x)$, the approximate solution for the beam deflection response is

$$w(x, t) = \sum_{n=1}^{4} a_n \phi_n(x) \cos\left[\left(\omega_n + \varepsilon \beta_n'\right)t + \tau_n\right] + 2\left[\sum_{n=1}^{4} \Lambda_n \phi_n(x)\right] \cos \Omega t \tag{7.15.38}$$

It is clear that the (7.15.28)–(7.15.35) has a zero amplitude steady state solution $a_1 = a_2 = a_3 = a_4 = 0$. For non-trivial amplitude solutions, (7.15.32)–(7.15.35) can be written as:

$$\begin{aligned} &\beta_1' + \omega_1^{-1} H_{11} + \frac{1}{8}\omega_1^{-1} \sum_{m=1}^{4} \alpha_{1m} a_m^2 \\ &\quad + \omega_1^{-1} Q a_1^{-1} a_2 a_3 a_4 \cos\gamma_1 + \omega_1^{-1} F a_1^{-1} a_2 a_3 \cos\gamma_2 = 0 \end{aligned} \tag{7.15.39}$$

$$\begin{aligned} &\beta_2' + \omega_2^{-1} H_{22} + \frac{1}{8}\omega_2^{-1} a_2^{-1} \sum_{m=1}^{4} \alpha_{2m} a_m^2 \\ &\quad + \omega_2^{-1} Q a_2^{-1} a_1 a_3 a_4 \cos\gamma_1 + \omega_2^{-1} F a_2^{-1} a_1 a_3 \cos\gamma_2 = 0 \end{aligned} \tag{7.15.40}$$

$$\begin{aligned} &\beta_3' + \omega_3^{-1} H_{33} + \frac{1}{8}\omega_3^{-1} \sum_{m=1}^{4} \alpha_{3m} a_m^2 \\ &\quad + \omega_3^{-1} Q a_3^{-1} a_1 a_2 a_4 \cos\gamma_1 + \omega_3^{-1} F a_3^{-1} a_1 a_2 \cos\gamma_2 = 0 \end{aligned} \tag{7.15.41}$$

$$\beta_4' + \omega_4^{-1} H_{44} + \frac{1}{8}\omega_4^{-1} \sum_{m=1}^{4} \alpha_{4m} a_m^2 + \omega_4^{-1} Q a_4^{-1} a_1 a_2 a_3 \cos\gamma_1 = 0 \tag{7.15.42}$$

From (7.15.27), we have

$$\gamma_1' + \sigma_1 = \beta_4' - \beta_3' - \beta_2' - \beta_1',\ \gamma_2' + \sigma_2 = \beta_3' - \beta_2' - \beta_1' \tag{7.15.43}$$

Combining (7.15.39)–(7.15.42) according to the right-hand side of the two equations of (7.15.43) yields

$$\begin{aligned}&\gamma_1' + \sigma_1 + \omega_4^{-1}H_{44} - \omega_3^{-1}H_{33} - \omega_2^{-1}H_{22} - \omega_1^{-1}H_{11}\\&+\frac{1}{8}\omega_4^{-1}\sum_{m=1}^{4}\alpha_{4m}a_m^2 - \frac{1}{8}\omega_3^{-1}\sum_{m=1}^{4}\alpha_{3m}a_m^2 - \frac{1}{8}\omega_2^{-1}a_2^{-1}\sum_{m=1}^{4}\alpha_{2m}a_m^2 - \frac{1}{8}\omega_1^{-1}\sum_{m=1}^{4}\alpha_{1m}a_m^2\\&+Q\left(\omega_4^{-1}a_4^{-1}a_1a_2a_3 - \omega_3^{-1}a_3^{-1}a_1a_2a_4 - \omega_2^{-1}a_2^{-1}a_1a_3a_4 - \omega_1^{-1}a_1^{-1}a_2a_3a_4\right)\cos\gamma_1\\&-F\left(\omega_3^{-1}a_3^{-1}a_1a_2 + \omega_2^{-1}a_2^{-1}a_1a_3 + \omega_1^{-1}a_1^{-1}a_2a_3\right)\cos\gamma_2 = 0\end{aligned} \tag{7.15.44}$$

$$\begin{aligned}&\gamma_2' + \sigma_2 + \omega_3^{-1}H_{33} - \omega_2^{-1}H_{22} - \omega_1^{-1}H_{11}\\&+\frac{1}{8}\omega_3^{-1}\sum_{m=1}^{4}\alpha_{3m}a_m^2 - \frac{1}{8}\omega_2^{-1}a_2^{-1}\sum_{m=1}^{4}\alpha_{2m}a_m^2 - \frac{1}{8}\omega_1^{-1}\sum_{m=1}^{4}\alpha_{1m}a_m^2\\&+Q\left(\omega_3^{-1}a_3^{-1}a_1a_2a_4 - \omega_2^{-1}a_2^{-1}a_1a_3a_4 - \omega_1^{-1}a_1^{-1}a_2a_3a_4\right)\cos\gamma_1\\&+F\left(\omega_3^{-1}a_3^{-1}a_1a_2 - \omega_2^{-1}a_2^{-1}a_1a_3 - \omega_1^{-1}a_1^{-1}a_2a_3\right)\cos\gamma_2 = 0\end{aligned} \tag{7.15.45}$$

Let $\gamma_1' = \gamma_2' = 0$, we obtain

$$\begin{aligned}&\sigma_1 + \omega_4^{-1}H_{44} - \omega_3^{-1}H_{33} - \omega_2^{-1}H_{22} - \omega_1^{-1}H_{11}\\&+\frac{1}{8}\omega_4^{-1}\sum_{m=1}^{4}\alpha_{4m}a_m^2 - \frac{1}{8}\omega_3^{-1}\sum_{m=1}^{4}\alpha_{3m}a_m^2 - \frac{1}{8}\omega_2^{-1}a_2^{-1}\sum_{m=1}^{4}\alpha_{2m}a_m^2 - \frac{1}{8}\omega_1^{-1}\sum_{m=1}^{4}\alpha_{1m}a_m^2\\&+Q\left(\omega_4^{-1}a_4^{-1}a_1a_2a_3 - \omega_3^{-1}a_3^{-1}a_1a_2a_4 - \omega_2^{-1}a_2^{-1}a_1a_3a_4 - \omega_1^{-1}a_1^{-1}a_2a_3a_4\right)\cos\gamma_1\\&-F\left(\omega_3^{-1}a_3^{-1}a_1a_2 + \omega_2^{-1}a_2^{-1}a_1a_3 + \omega_1^{-1}a_1^{-1}a_2a_3\right)\cos\gamma_2 = 0\end{aligned} \tag{7.15.46}$$

$$\begin{aligned}&\sigma_2 + \omega_3^{-1}H_{33} - \omega_2^{-1}H_{22} - \omega_1^{-1}H_{11}\\&+\frac{1}{8}\omega_3^{-1}\sum_{m=1}^{4}\alpha_{3m}a_m^2 - \frac{1}{8}\omega_2^{-1}a_2^{-1}\sum_{m=1}^{4}\alpha_{2m}a_m^2 - \frac{1}{8}\omega_1^{-1}\sum_{m=1}^{4}\alpha_{1m}a_m^2\\&+Q\left(\omega_3^{-1}a_3^{-1}a_1a_2a_4 - \omega_2^{-1}a_2^{-1}a_1a_3a_4 - \omega_1^{-1}a_1^{-1}a_2a_3a_4\right)\cos\gamma_1\\&+F\left(\omega_3^{-1}a_3^{-1}a_1a_2 - \omega_2^{-1}a_2^{-1}a_1a_3 - \omega_1^{-1}a_1^{-1}a_2a_3\right)\cos\gamma_2 = 0\end{aligned} \tag{7.15.47}$$

(7.15.46), (7.15.47) and (7.15.28)–(7.15.31) are the equations satisfied by the steady state solutions of $a_1, a_2, a_3, a_4, \gamma_1, \gamma_2$. After finding a set of steady state solutions, we substitute them into (7.15.39)–(7.15.42) and obtain the values for $\beta_1,\ \beta_2,\ \beta_3,\ \beta_4$ by integration. Obviously, these values contain integration constants, which need to be determined from the initial conditions.

7.16 Exercise 7.16 (Transverse Oscillation of a Clamped-Supported Beam Under Internal Resonance and Non-resonance Excitation at $u = O(w^2)$)

Solution: We take the characteristic length of the beam L to be the actual length of the beam, then the dimensionless length of the beam $l = 1$. In turn, the mode shape function of the linear free oscillation of the beam is

$$\phi_n(x) = C_n[(\sin \eta_n - \sinh \eta_n)(\cos \eta_n x - \cosh \eta_n x) - (\cos \eta_n - \cosh \eta_n)(\sin \eta_n x - \sinh \eta_n x)] \tag{7.16.1}$$

where the value of C_n C_n is taken to normalize the mode shape function $\phi_n(x)$. The first five natural frequencies of the beam are

$$22.42\,61.700\,120.89\,199.80\,298.43$$

We note that

$$\omega_1 + \omega_2 + \omega_3 = \omega_4 + \varepsilon\sigma_1 \tag{7.16.2}$$

Hence there is internal resonance. The discrete equation of the mode shape function of the beam of this problem is

$$\ddot{u}_n + \omega_n^2 u_n = \varepsilon\left(\sum_{m,p,q=1}^{\infty} \Gamma_{nmpq} u_m u_p u_q - 2\mu_n \dot{u}_n\right) + f_{zn}(t) \tag{7.16.3}$$

where

$$\Gamma_{nmpq} = -\frac{1}{2l}\left(\int_0^l \phi_n' \phi_m' dx\right)\left(\int_0^l \phi_p' \phi_q' dx\right), \quad l = 2 \tag{7.16.4}$$

and

$$\Gamma_{nmpq} = \Gamma_{mnpq} = \Gamma_{nmqp} = \Gamma_{pqnm} \tag{7.16.5}$$

We use the method of multiple scales to solve (7.16.3). Let the solution of the equation be

$$u_n(t;\varepsilon) = u_{n0}(T_0, T_1) + \varepsilon u_{n1}(T_0, T_1) + \cdots u_n(t;\varepsilon) = u_{n0}(T_0, T_1) + \varepsilon u_{n1}(T_0, T_1) + \cdots \tag{7.16.6}$$

Substituting (7.16.6) into (7.16.3) and retaining to $O(\varepsilon)$, we obtain

$$
\begin{aligned}
0 &= \ddot{u}_n + \omega_n^2 u_n + \varepsilon \left(2\mu_n \dot{u}_n - \sum_{m,p,q=1}^{\infty} \Gamma_{nmpq} u_m u_p u_q \right) - f_{zn}(t) \\
&= \left(D_0^2 + 2\varepsilon D_0 D_1\right)(u_{n0} + \varepsilon u_{n1}) + \omega_n^2 (u_{n0} + \varepsilon u_{n1}) \\
&\quad + \varepsilon \left(2\mu_n D_0 u_{n0} - \sum_{m,p,q=1}^{\infty} \Gamma_{nmpq} u_{m0} u_{p0} u_{q0} \right) - 2K_n \cos \Omega t \\
&= D_0^2 u_{n0} + \omega_n^2 u_{n0} - 2K_n \cos \Omega t + \varepsilon \big(D_0^2 u_{n1} + \omega_n^2 u_{n1} \\
&\quad + 2D_0 D_1 u_{n0} + 2\mu_n D_0 u_{n0} - \sum_{m,p,q=1}^{\infty} \Gamma_{nmpq} u_{m0} u_{p0} u_{q0} \big)
\end{aligned}
\tag{7.16.7}
$$

Making the coefficient of the same power of ε zero in the above equation yields

$$
D_0^2 u_{n0} + \omega_n^2 u_{n0} = 2K_n \cos \Omega T_0 \tag{7.16.8}
$$

$$
D_0^2 u_{n1} + \omega_n^2 u_{n1} = -2D_0 D_1 u_{n0} - 2\mu_n D_0 u_{n0} + \sum_{m,\,p,\,q=1}^{\infty} \Gamma_{nmpq} u_{m0} u_{p0} u_{q0} \tag{7.16.9}
$$

The solution of the Eq. (7.16.8) is

$$
u_{n0} = A_n(T_1) e^{i\omega_n T_0} + \Lambda_n e^{i\Omega T_0} + cc \tag{7.16.10}
$$

where

$$
\Lambda_n = \frac{K_n}{\omega_n^2 - \Omega^2} \tag{7.16.11}
$$

Substituting (7.6.10) into (7.16.9) and taking into account $\omega_1 + \omega_2 + \omega_3 = \omega_4 + \varepsilon\sigma_1 \omega_1 + \omega_2 + \omega_3 = \omega_4 + \varepsilon\sigma_1$ and $2\Omega = \omega_4 - \omega_1 + \varepsilon\sigma_2$, we obtain

$$
\begin{aligned}
&D_0^2 u_{n1} + \omega_n^2 u_{n1} = -2D_0 D_1 u_{n0} - 2\mu_n D_0 u_{n0} + \sum_{m,p,q=1}^{\infty} \Gamma_{nmpq} u_{m0} u_{p0} u_{q0} \\
&+ A_4 \sum_{p,q=1}^{\infty} (\Gamma_{np4q} + \Gamma_{nqp4} + \Gamma_{n4pq}) \overline{A}_p \overline{A}_q e^{i\omega_4 T_0} e^{-i\omega_p T_0} e^{-i\omega_q T_0} \\
&+ \sum_{m,p,q=1}^{\infty} \Gamma_{nmpq} \Big\{ A_m \Lambda_p \Lambda_q e^{-2i\Omega T_0} e^{i\omega_m T_0} + \Lambda_m A_p \Lambda_q e^{-2i\Omega T_0} e^{i\omega_p T_0} + \Lambda_m \Lambda_p A_q e^{-2i\Omega T_0} e^{i\omega_q T_0} + A_m \Lambda_p \Lambda_q e^{2i\Omega T_0} e^{i\omega_m T_0} \\
&+ \Lambda_m \Lambda_p A_q e^{-2i\Omega T_0} e^{i\omega_q T_0} + A_m \Lambda_p \Lambda_q e^{2i\Omega T_0} e^{i\omega_m T_0} \\
&+ \Lambda_m A_p \Lambda_q e^{2i\Omega T_0} e^{i\omega_p T_0} + \Lambda_m \Lambda_p A_q e^{2i\Omega T_0} e^{i\omega_q T_0} \Big\} \\
&+ 2(\Gamma_{1234} + \Gamma_{1324} + \Gamma_{1423}) A_1 A_2 A_3 e^{i\sigma_1 T_1} e^{i\omega_4 T_0} + cc + NST
\end{aligned}
\tag{7.16.12}
$$

In order to eliminate secular terms from the above equation, and taking into account $\omega_1 + \omega_2 + \omega_3 = \omega_4 + \varepsilon\sigma_1$ and $2\Omega = \omega_4 - \omega_1 + \varepsilon\sigma_2$, we need

$$- 2i\omega_1\left(A_1' + \mu_1 A_1\right) + 2H_{11}A_1 + A_1\sum_{m=1}^{\infty}\alpha_{1m}A_m\overline{A}_m + 8Q\overline{A}_2\overline{A}_3A_4e^{-i\sigma_1T_1} + H_{14}A_4e^{-i\sigma_2T_1} = 0 \quad (7.16.13)$$

$$-2i\omega_2\left(A_2' + \mu_2A_2\right) + 2H_{22}A_2 + A_2\sum_{m=1}^{\infty}\alpha_{2m}A_m\overline{A}_m + 8Q\overline{A}_1\overline{A}_3A_4e^{-i\sigma_1T_1} = 0 \quad (7.16.14)$$

$$-2i\omega_3(A_3\prime + \mu_3A_3) + 2H_{33}A_3 + A_3\sum_{m=1}^{\infty}\alpha_{3m}A_m\overline{A}_m + 8Q\overline{A}_1\overline{A}_2A_4e^{-i\sigma_1T_1} = 0 \quad (7.16.15)$$

$$-2i\omega_4\left(A_4' + \mu_4A_4\right) + 2H_{44}A_4 + A_4\sum_{m=1}^{\infty}\alpha_{4m}A_m\overline{A}_m + 8QA_1A_2A_3e^{i\sigma_1T_1} + H_{14}A_1e^{i\sigma_2T_1} = 0 \quad (7.16.16)$$

$$-2i\omega_n(A_n\prime + \mu_nA_n) + 2H_{nn}A_n + A_n\sum_{m=1}^{\infty}\alpha_{nm}A_m\overline{A}_m = 0, \quad n \geq 5 \quad (7.16.17)$$

where

$$8Q = 2(\Gamma_{1234} + \Gamma_{1324} + \Gamma_{1423})$$
$$\alpha_{mn} = \alpha_{nm} = \begin{cases} 2(\Gamma_{nnmm} + 2\Gamma_{nmnm})\ for \quad n \neq m \\ 6\Gamma_{nnnn} \end{cases}$$
$$H_{nm} = \sum\nolimits_{p,q}\left(\Gamma_{nmpq} + 2\Gamma_{npmq}\right)\Lambda_p\Lambda_q \quad (7.16.18)$$

Let

$$A_n = \frac{1}{2}a_ne^{i\beta_n} \quad (7.16.19)$$

Substituting it into (7.16.17) yields

$$i\omega_na_n' - \omega_na_n\beta_n\prime + i\omega_n\mu_na_n - H_{nn}a_n - \frac{1}{8}a_n\sum_{m=1}^{\infty}\alpha_{nm}a_m^2 = 0, \ n \geq 5 \quad (7.16.20)$$

Separating the real and imaginary parts of the above equation yields

$$a_n' + \mu_n a_n = 0, \quad n \geq 5 \tag{7.16.21}$$

therefore

$$a_n = a_{n0} e^{-\mu_n T_1} \to 0 \Rightarrow A_n \to 0, \; n \geq 5 \tag{7.16.22}$$

Substituting (7.16.19) into (7.16.13)–(7.16.16) and taking into account that (7.6.22), we have

$$\begin{aligned} & -i\omega_1 a_1' + \omega_1 a_1 \beta_1' - i\omega_1 \mu_1 a_1 + H_{11} a_1 \\ & + \frac{1}{8} a_1 \sum_{m=1}^{4} \alpha_{1m} a_m^2 + Q a_2 a_3 a_4 e^{i\gamma_1} + \frac{1}{2} H_{14} a_4 e^{i\gamma_2} = 0 \end{aligned} \tag{7.16.23}$$

$$-i\omega_2 a_2\prime + \omega_2 a_2 \beta_2\prime - i\omega_2 \mu_2 a_2 + H_{22} a_2 + \frac{1}{8} a_2 \sum_{m=1}^{4} \alpha_{2m} a_m^2 + Q a_1 a_3 a_4 e^{i\gamma_1} = 0 \tag{7.16.24}$$

$$-i\omega_3 a_3\prime + \omega_3 a_3 \beta_3\prime - i\omega_3 \mu_3 a_3 + H_{33} a_3 + \frac{1}{8} a_3 \sum_{m=1}^{4} \alpha_{3m} a_m^2 + Q a_1 a_2 a_4 e^{i\gamma_1} = 0 \tag{7.16.25}$$

$$\begin{aligned} & -i\omega_4 a_4\prime + \omega_4 a_4 \beta_4\prime - i\omega_4 \mu_4 a_4 + H_{44} a_4 \\ & + \frac{1}{8} a_4 \sum_{m=1}^{4} \alpha_{4m} a_m^2 + Q a_1 a_2 a_3 e^{-i\gamma_1} + \frac{1}{2} H_{14} a_1 e^{-i\gamma_2} = 0 \end{aligned} \tag{7.16.26}$$

where

$$\gamma_1 = \beta_4 - \beta_3 - \beta_2 - \beta_1 - \sigma_1 T_1, \quad \gamma_2 = \beta_4 - \beta_1 - \sigma_2 T_1 \tag{7.16.27}$$

Separating the real and imaginary parts of (7.16.23)–(7.16.26) and making $a_1' = a_2\prime = a_3\prime = a_4\prime = 0$, we obtain

$$-\omega_1 \mu_1 a_1 + Q a_2 a_3 a_4 \sin\gamma_1 + \frac{1}{2} H_{14} a_4 \sin\gamma_2 = 0 \tag{7.16.28}$$

$$-\omega_2 \mu_2 a_2 + Q a_1 a_3 a_4 \sin\gamma_1 = 0 \tag{7.16.29}$$

$$-\omega_3 \mu_3 a_3 + Q a_1 a_2 a_4 \sin\gamma_1 = 0 \tag{7.16.30}$$

$$-\omega_4 \mu_4 a_4 - Q a_1 a_2 a_3 \sin\gamma_1 - \frac{1}{2} H_{14} a_1 \sin\gamma_2 = 0 \tag{7.16.31}$$

$$\omega_1 a_1 \beta_1{}' + H_{11} a_1 + \frac{1}{8} a_1 \sum_{m=1}^{4} \alpha_{1m} a_m^2 + Q a_2 a_3 a_4 \cos\gamma_1 + \frac{1}{2} H_{14} a_4 \cos\gamma_2 = 0 \tag{7.16.32}$$

$$\omega_2 a_2 \beta_2{}' + H_{22} a_2 + \frac{1}{8} a_2 \sum_{m=1}^{4} \alpha_{2m} a_m^2 + Q a_1 a_3 a_4 \cos\gamma_1 = 0 \tag{7.16.33}$$

$$\omega_3 a_3 \beta_3' + H_{33} a_3 + \frac{1}{8} a_3 \sum_{m=1}^{4} \alpha_{3m} a_m^2 + Q a_1 a_2 a_4 \cos\gamma_1 = 0 \tag{7.16.34}$$

$$\omega_4 a_4 \beta_4' + H_{44} a_4 + \frac{1}{8} a_4 \sum_{m=1}^{4} \alpha_{4m} a_m^2 + Q a_1 a_2 a_3 \cos\gamma_1 + \frac{1}{2} H_{14} a_1 \cos\gamma_2 = 0 \tag{7.16.35}$$

Equations (7.16.28)–(7.16.35) control the *magnitude* and phase of the mode shape function. For any set of amplitude steady state solutions, the phase is assumed to be

$$\beta_n = \beta_n' T_1 + \tau_n = \varepsilon \beta_n' t + \tau_n, \quad n = 1, 2, 3, 4 \tag{7.16.36}$$

Then by (7.16.10) and (7.16.6), we can obtain the approximate solution of $u_n(t)$

$$\begin{aligned} u_n &\approx u_{n0} = a_n \cos\left[\left(\omega_n + \varepsilon \beta_n'\right)t + \tau_n\right] + 2\Lambda_n \cos\Omega t + O(\varepsilon), \quad n = 1, 2, 3, 4 \\ u_n &\approx u_{n0} = 2\Lambda_n \cos\Omega t + O(\varepsilon), \quad n \geq 5 \end{aligned} \tag{7.16.37}$$

Since the deflection of the beam is $w(x, t) = \sum_{n=1}^{\infty} u_n \phi_n(x)$, the approximate solution for the beam deflection response is

$$w(x, t) = \sum_{n=1}^{4} a_n \phi_n(x) \cos\left[\left(\omega_n + \varepsilon \beta_n'\right)t + \tau_n\right] + 2\left[\sum_{n=1}^{4} \Lambda_n \phi_n(x)\right] \cos\Omega t \tag{7.16.38}$$

It is clear that (7.16.28)–(7.16.35) has a zero amplitude steady state solution. For non-trivial amplitude solutions, (7.16.28)–(7.16.35) can be written as:

$$\begin{aligned} &\beta_1' + \omega_1^{-1} H_{11} + \frac{1}{8} \omega_1^{-1} \sum_{m=1}^{4} \alpha_{1m} a_m^2 \\ &\quad + \omega_1^{-1} Q a_1^{-1} a_2 a_3 a_4 \cos\gamma_1 + \frac{1}{2} \omega_1^{-1} H_{14} a_1^{-1} a_4 \cos\gamma_2 = 0 \end{aligned} \tag{7.16.39}$$

$$\beta_2' + \omega_2^{-1} H_{22} + \frac{1}{8} \omega_2^{-1} a_2^{-1} \sum_{m=1}^{4} \alpha_{2m} a_m^2 + \omega_2^{-1} Q a_2^{-1} a_1 a_3 a_4 \cos\gamma_1 = 0 \tag{7.16.40}$$

$$\beta_3' + \omega_3^{-1} H_{33} + \frac{1}{8}\omega_3^{-1} \sum_{m=1}^{4} \alpha_{3m} a_m^2 + \omega_3^{-1} Q a_3^{-1} a_1 a_2 a_4 \cos\gamma_1 = 0 \tag{7.16.41}$$

$$\begin{aligned} &\beta_4' + \omega_4^{-1} H_{44} + \frac{1}{8}\omega_4^{-1} \sum_{m=1}^{4} \alpha_{4m} a_m^2 \\ &\quad + \omega_4^{-1} Q a_4^{-1} a_1 a_2 a_3 \cos\gamma_1 + \frac{1}{2}\omega_4^{-1} H_{14} a_4^{-1} a_1 \cos\gamma_2 = 0 \end{aligned} \tag{7.16.42}$$

From (7.16.27), we have

$$\gamma_1' + \sigma_1 = \beta_4' - \beta_3' - \beta_2' - \beta_1', \quad \gamma_2' + \sigma_2 = \beta_4' - \beta_1' \tag{7.16.43}$$

Combining (7.16.39)–(7.16.42) according to the right-hand side of the two equations of (7.16.43) yields

$$\begin{aligned} &\gamma_1' + \sigma_1 + \omega_4^{-1} H_{44} - \omega_3^{-1} H_{33} - \omega_2^{-1} H_{22} - \omega_1^{-1} H_{11} \\ &+ \frac{1}{8}\omega_4^{-1} \sum_{m=1}^{4} \alpha_{4m} a_m^2 - \frac{1}{8}\omega_3^{-1} \sum_{m=1}^{4} \alpha_{3m} a_m^2 - \frac{1}{8}\omega_2^{-1} a_2^{-1} \sum_{m=1}^{4} \alpha_{2m} a_m^2 - \frac{1}{8}\omega_1^{-1} \sum_{m=1}^{4} \alpha_{1m} a_m^2 \\ &+ Q\left(\omega_4^{-1} a_4^{-1} a_1 a_2 a_3 - \omega_3^{-1} a_3^{-1} a_1 a_2 a_4 - \omega_2^{-1} a_2^{-1} a_1 a_3 a_4 - \omega_1^{-1} a_1^{-1} a_2 a_3 a_4\right) \cos\gamma_1 \\ &+ \frac{1}{2} H_{14}\left(\omega_4^{-1} a_4^{-1} a_1 - \omega_1^{-1} a_1^{-1} a_4\right) \cos\gamma_2 = 0 \end{aligned} \tag{7.16.44}$$

$$\begin{aligned} &\gamma_2' + \sigma_2 + \omega_4^{-1} H_{44} - \omega_1^{-1} H_{11} + \frac{1}{8}\omega_4^{-1} \sum_{m=1}^{4} \alpha_{4m} a_m^2 - \frac{1}{8}\omega_1^{-1} \sum_{m=1}^{4} \alpha_{1m} a_m^2 \\ &+ Q\left(\omega_4^{-1} a_4^{-1} a_1 - \omega_1^{-1} a_1^{-1} a_4\right) a_2 a_3 \cos\gamma_1 \\ &+ \frac{1}{2} H_{14}\left(\omega_4^{-1} a_4^{-1} a_1 - \omega_1^{-1} a_1^{-1} a_4\right) \cos\gamma_2 = 0 \end{aligned} \tag{7.16.45}$$

Let $\gamma_1' = \gamma_2' = 0$, we obtain

$$\begin{aligned} &\sigma_1 + \omega_4^{-1} H_{44} - \omega_3^{-1} H_{33} - \omega_2^{-1} H_{22} - \omega_1^{-1} H_{11} \\ &+ \frac{1}{8}\omega_4^{-1} \sum_{m=1}^{4} \alpha_{4m} a_m^2 - \frac{1}{8}\omega_3^{-1} \sum_{m=1}^{4} \alpha_{3m} a_m^2 - \frac{1}{8}\omega_2^{-1} a_2^{-1} \sum_{m=1}^{4} \alpha_{2m} a_m^2 - \frac{1}{8}\omega_1^{-1} \sum_{m=1}^{4} \alpha_{1m} a_m^2 \\ &+ Q\left(\omega_4^{-1} a_4^{-1} a_1 a_2 a_3 - \omega_3^{-1} a_3^{-1} a_1 a_2 a_4 - \omega_2^{-1} a_2^{-1} a_1 a_3 a_4 - \omega_1^{-1} a_1^{-1} a_2 a_3 a_4\right) \cos\gamma_1 \\ &+ \frac{1}{2} H_{14}\left(\omega_4^{-1} a_4^{-1} a_1 - \omega_1^{-1} a_1^{-1} a_4\right) \cos\gamma_2 = 0 \end{aligned} \tag{7.16.46}$$

$$\begin{aligned}&\sigma_2+\omega_4^{-1}H_{44}-\omega_1^{-1}H_{11}+\frac{1}{8}\omega_4^{-1}\sum_{m=1}^{4}\alpha_{4m}a_m^2-\frac{1}{8}\omega_1^{-1}\sum_{m=1}^{4}\alpha_{1m}a_m^2\\&+Q\left(\omega_4^{-1}a_4^{-1}a_1-\omega_1^{-1}a_1^{-1}a_4\right)a_2a_3\cos\gamma_1\\&+\frac{1}{2}H_{14}\left(\omega_4^{-1}a_4^{-1}a_1-\omega_1^{-1}a_1^{-1}a_4\right)\cos\gamma_2=0\end{aligned}\tag{7.16.47}$$

(7.16.46), (7.16.47) and (7.16.28)–(7.16.31) are the equations satisfied by the steady state solutions of a_1, a_2, a_3, a_4, γ_1, γ_2. After finding a set of steady state solutions, we substitute them into (7.16.39)–(7.16.42) and obtain the values of β_1, β_2, β_3, β_4 by integration. Obviously, these values contain integration constants, which need to be determined by the initial conditions.

7.17 Exercise 7.17 (Coupled Longitudinal and Transverse Oscillation Analysis of Hinged-Hinged Beams at $u = O(w)$)

Solution: Consider the coupling of longitudinal and transverse oscillations of a hinged-hinged beam, the equations describing the modal coordinates are

$$\begin{aligned}\ddot{\xi}_n+\lambda_n^2\xi_n&=\varepsilon\left[-2\nu_n\dot{\xi}_n-n\kappa\sum_{m=1}^{\infty}m\,\eta_m\left(p\eta_p+q\eta_q\right)\right]+f_{xn}(t)\\\ddot{\eta}_n+\omega_n^2\eta_n&=\varepsilon\left[-2\mu_n\dot{\eta}_n-n\kappa\sum_{m=1}^{\infty}m\,\xi_m\left(p\eta_p+q\eta_q\right)\right]+f_{yn}(t)\end{aligned}\tag{7.17.1}$$

$$\lambda_n=\frac{n\pi}{rl},\quad \omega_n=\frac{n\pi}{l}\left(\frac{n^2\pi^2}{l^2}+N\right)^{1/2},\quad p=|n-m|,\ q=n+m\tag{7.17.2}$$

$$f_{xn}(t)=2\varepsilon g_n\cos(\Omega_1t-\tau),\quad f_{yn}(t)=2\varepsilon\sum_{m=1}^{M}f_{nm}\cos(\Omega_{2m}t-\theta_m)\tag{7.17.3}$$

We use the method of multiple scales to solve (7.17.1). Let the solution of the equation be

$$\begin{aligned}\xi_n(t;\varepsilon)&=\xi_{n0}(T_0,T_1)+\varepsilon\xi_{n1}(T_0,T_1)+\cdots\\\eta_n(t;\varepsilon)&=\eta_{n0}(T_0,T_1)+\varepsilon\eta_{n1}(T_0,T_1)+\cdots\end{aligned}\tag{7.17.4}$$

Substituting (7.17.4) and (7.17.3) into (7.17.1) and retaining to $O(\varepsilon)$, we obtain

$$
\begin{aligned}
0 &= \ddot{\xi}_n + \lambda_n^2 \xi_n + \varepsilon \left[2\nu_n \dot{\xi}_n + n\kappa \sum_{m=1}^{\infty} m\eta_m \left(p\eta_p + q\eta_q \right) \right] - f_{xn}(t) \\
&= \left(D_0^2 + 2\varepsilon D_0 D_1 \right) (\xi_{n0} + \varepsilon \xi_{n1}) + \lambda_n^2 (\xi_{n0} + \varepsilon \xi_{n1}) \\
&+ \varepsilon \left[2\nu_n D_0 \xi_{n0} + n\kappa \sum_{m=1}^{\infty} m\eta_{m0} \left(p\eta_{p0} + q\eta_{q0} \right) \right] - 2\varepsilon g_n \cos(\Omega_1 t - \tau) \\
&= D_0^2 \xi_{n0} + \lambda_n^2 \xi_{n0} + \varepsilon \left[D_0^2 \xi_{n1} + \lambda_n^2 \xi_{n1} + 2D_0 D_1 \xi_{n0} + 2\nu_n D_0 \xi_{n0} \right. \\
&\left. + n\kappa \sum_{m=1}^{\infty} m\eta_{m0} \left(p\eta_{p0} + q\eta_{q0} \right) - 2g_n \cos(\Omega_1 t - \tau) \right]
\end{aligned} \tag{7.17.5}
$$

$$
\begin{aligned}
0 &= \ddot{\eta}_n + \omega_n^2 \eta_n + \varepsilon \left[2\mu_n \dot{\eta}_n + n\kappa \sum_{m=1}^{\infty} m\xi_m \left(p\eta_p + q\eta_q \right) \right] - f_{yn}(t) \\
&= \left(D_0^2 + 2\varepsilon D_0 D_1 \right) (\eta_{n0} + \varepsilon \eta_{n1}) + \omega_n^2 (\eta_{n0} + \varepsilon \eta_{n1}) \\
&+ \varepsilon \left[2\mu_n D_0 \eta_{n0} + n\kappa \sum_{m=1}^{\infty} m\xi_{m0} \left(p\eta_{p0} + q\eta_{q0} \right) \right] - 2\varepsilon \sum_{m=1}^{M} f_{nm} \cos(\Omega_{2m} t - \theta_m) \\
&= D_0^2 \eta_{n0} + \omega_n^2 \eta_{n0} + \varepsilon \left[D_0^2 \eta_{n1} + \omega_n^2 \eta_{n1} + 2D_0 D_1 \eta_{n0} + 2\mu_n D_0 \eta_{n0} \right. \\
&\left. + n\kappa \sum_{m=1}^{\infty} m\xi_{m0} \left(p\eta_{p0} + q\eta_{q0} \right) - 2 \sum_{m=1}^{M} f_{nm} \cos(\Omega_{2m} t - \theta_m) \right]
\end{aligned} \tag{7.17.6}
$$

Making the coefficient of the same power of ε zero in the above two equations, we obtain

$$
\begin{aligned}
\ddot{\xi}_{n0} + \lambda_n^2 \xi_{n0} &= 0 \\
\ddot{\eta}_{n0} + \omega_n^2 \eta_{n0} &= 0
\end{aligned} \tag{7.17.7}
$$

$$
\begin{aligned}
D_0^2 \xi_{n1} + \lambda_n^2 \xi_{n1} &= -2D_0 D_1 \xi_{n0} - 2\nu_n D_0 \xi_{n0} - n\kappa \sum_{m=1}^{\infty} m\eta_{m0} \left(p\eta_{p0} + q\eta_{q0} \right) \\
&\quad + 2g_n \cos(\Omega_1 t - \tau) \\
D_0^2 \eta_{n1} + \omega_n^2 \eta_{n1} &= -2D_0 D_1 \eta_{n0} - 2\mu_n D_0 \eta_{n0} - n\kappa \sum_{m=1}^{\infty} m\xi_{m0} \left(p\eta_{p0} + q\eta_{q0} \right) \\
&\quad + 2 \sum_{k=1}^{M} f_{nk} \cos(\Omega_{2k} t - \theta_k)
\end{aligned} \tag{7.17.8}
$$

The solution of the Eq. (7.17.7) is

$$
\xi_{n0} = A_n(T_1) e^{i\lambda_n T_0} + cc
$$

$$\eta_{n0} = B_n(T_1)e^{i\omega_n T_0} + cc \tag{7.17.9}$$

Substituting (7.17.9) into (7.17.8) yields

$$\begin{aligned} D_0^2\xi_{n1} + \lambda_n^2\xi_{n1} &= -2i\lambda_n A_n' e^{i\lambda_n T_0} - 2i\lambda_n \nu_n A_n e^{i\lambda_n T_0} \\ &- n\kappa \sum_{m=1}^{\infty} mB_m \\ &\times \left\{p\, B_p e^{i(\omega_m+\omega_p)T_0} + p\bar{B}_p e^{i(\omega_m-\omega_p)T_0} + qB_q e^{i(\omega_m+\omega_q)T_0} + q\bar{B}_q e^{i(\omega_m-\omega_q)T_0}\right\} \\ &+ g_n e^{i(\Omega_1 T_0-\tau)} + cc \end{aligned} \tag{7.17.10}$$

$$\begin{aligned} D_0^2\eta_{n1} + \omega_n^2\eta_{n1} &= -2i\omega_n B_n' e^{i\omega_n T_0} - 2i\omega_n \mu_n B_n e^{i\omega_n T_0} \\ &- n\kappa \sum_{m=1}^{\infty} mA_m \\ &\times \left\{pB_p e^{i(\lambda_m+\omega_p)T_0} + p\bar{B}_p e^{i(\lambda_m-\omega_p)T_0} + qB_q e^{i(\lambda_m+\omega_q)T_0} + q\bar{B}_q e^{i(\lambda_m-\omega_q)T_0}\right\} \\ &+ \sum_{k=1}^{M} f_{nk} e^{i(\Omega_{2k} T_0-\theta_k)} + cc \end{aligned} \tag{7.17.11}$$

By (7.17.10) and (7.17.11), we can see that the internal resonance occurs when

$$\lambda_n \approx \omega_m \pm \omega_p \text{ or } \lambda_n \approx \omega_m \pm \omega_p \tag{7.17.12}$$

With (7.17.11), we can write (7.17.12) as

$$\begin{aligned} &\frac{n}{r} \approx m\left(\frac{m^2\pi^2}{l^2} + N\right)^{1/2} \pm |n-m| \left[\frac{(n-m)^2\pi^2}{l^2} + N\right]^{1/2} \\ \text{or } &\frac{n}{r} \approx m\left(\frac{m^2\pi^2}{l^2} + N\right)^{1/2} \pm |n+m| \left[\frac{(n-m)^2\pi^2}{l^2} + N\right]^{1/2} \end{aligned} \tag{7.17.13}$$

This suggests that r is a parameter regulating internal resonance.

Since the beam theory requires a small slenderness ratio, our interest in longitudinal oscillations is mainly restricted to their fundamental frequency modes, i.e., $n = 1$. In addition, when the plus sign is taken, the condition (7.17.13) is satisfied first. Therefore, to obtain the value of the aspect ratio parameter l/r when internal resonance occurs, we let $N = 0$, which yields

$$\frac{l}{r} \approx \left(2m^2 + 2m + 1\right)\pi \tag{7.17.14}$$

From this, it is possible to estimate the value of l/r required for the internal resonance between the fundamental frequency mode of longitudinal oscillation and the mth order frequency mode of transverse oscillation.

Consider the following case:

$$\begin{aligned}\lambda_1 &= \omega_2 + \omega_3 + \varepsilon\sigma_I, \quad \Omega_1 = \lambda_1 + \varepsilon\sigma_1 \\ \Omega_{22} &= \omega_2 + \varepsilon\sigma_2, \quad \Omega_{23} = \omega_3 + \varepsilon\sigma_3\end{aligned} \tag{7.17.15}$$

In order to eliminate secular terms from (7.17.10) and (7.17.11), we need

$$2i\lambda_1 A_1' + 2i\lambda_1\nu_1 A_1 + 6\kappa B_2 B_3 e^{-i\sigma_I T_1} - g_1 e^{i(\sigma_1 T_1 - \tau)} = 0 \tag{7.17.16}$$

$$A_n' + \nu_n A_n = 0, \quad n \geq 2 \tag{7.17.17}$$

$$2i\omega_2 B_2' + 2i\omega_2\mu_2 B_2 + 6\kappa A_1 \overline{B}_3 e^{i\sigma_I T_1} - f_{22} e^{i(\sigma_2 T_1 - \theta_2)} = 0 \tag{7.17.18}$$

$$2i\omega_3 B_3' + 2i\omega_3\mu_3 B_3 + 6\kappa A_1 \overline{B}_2 e^{i\sigma_I T_1} - f_{33} e^{i(\sigma_3 T_1 - \theta_3)} = 0 \tag{7.17.19}$$

$$B_n' + \mu_n B_n = 0, \; n \neq 2 \text{ and } 3 \tag{7.17.20}$$

From (7.17.17) and (7.17.20), we have

$$\begin{aligned} A_n &= A_{n0} e^{-\nu_n T_1} \to 0, \; n \geq 2 \\ \text{and } B_n &= B_{n0} e^{-\mu_n T_1} \to 0, \; n \neq 2 \text{ and } 3 \end{aligned} \tag{7.17.21}$$

Let

$$A_n = \frac{1}{2} a_n e^{i\alpha_n}, \quad B_n = \frac{1}{2} b_n e^{i\beta_n} \tag{7.17.22}$$

and substitute them into (7.17.16), (7.17.18) and (7.17.19), we have

$$i\lambda_1 a_1' - \lambda_1 a_1 \alpha_1' + i\lambda_1\nu_1 a_1 + k b_2 b_3 e^{-i\gamma_I} - g_1 e^{i\gamma_1} = 0 \tag{7.17.23}$$

$$i\omega_2 b_2' - \omega_2 b_2 \beta_2' + i\omega_2\mu_2 b_2 + k a_1 b_3 e^{i\gamma_I} - f_{22} e^{i\gamma_2} = 0 \tag{7.17.24}$$

$$i\omega_3 b_3' - \omega_3 b_3 \beta_3' + i\omega_3\mu_3 b_3 + k a_1 b_2 e^{i\gamma_I} - f_{33} e^{i\gamma_3} = 0 \tag{7.17.25}$$

where

$$\begin{aligned} &k = \tfrac{3}{2}\kappa, \qquad \gamma_I = \sigma_I T_1 + \alpha_1 - \beta_2 - \beta_3, \\ &\gamma_1 = \sigma_1 T_1 - \alpha_1 - \tau, \; \gamma_2 = \sigma_2 T_1 - \beta_2 - \theta_2, \qquad \gamma_3 = \sigma_3 T_1 - \beta_3 - \theta_3 \end{aligned} \tag{7.17.26}$$

Separate the real and imaginary parts of (7.17.23)–(7.17.25) and let $a_1' = b_2' = b_3' = 0$, we can obtain the governing equation of the steady state solution:

$$\lambda_1\nu_1 a_1 - k b_2 b_3 \sin\gamma_I - g_1 \sin\gamma_1 = 0$$

$$\begin{aligned}\omega_2\mu_2b_2 + ka_1b_3\sin\gamma_I - f_{22}\sin\gamma_2 &= 0\\ \omega_3\mu_3b_3 + ka_1b_2\sin\gamma_I - f_{33}\sin\gamma_3 &= 0\end{aligned} \tag{7.17.27}$$

$$\begin{aligned}-\lambda_1a_1\alpha_1' + kb_2b_3\cos\gamma_I - g_1\cos\gamma_1 &= 0\\ -\omega_2b_2\beta_2' + ka_1b_3\cos\gamma_I - f_{22}\cos\gamma_2 &= 0\\ -\omega_3b_3\beta_3' + ka_1b_2\cos\gamma_I - f_{33}\cos\gamma_3 &= 0\end{aligned} \tag{7.17.28}$$

When $g_1 = 0,\ f_{22} = 0$, we can obtain from (7.17.27) that

$$\begin{aligned}\lambda_1\nu_1a_1 &= kb_2b_3\sin\gamma_1\\ \omega_2\mu_2b_2 &= -ka_1b_3\sin\gamma_1\end{aligned} \tag{7.17.29}$$

Eliminating γ_I from (7.17.29) yields

$$\lambda_1\nu_1a_1^2 + \omega_2\mu_2b_2^2 = 0 \tag{7.17.30}$$

So $a_1 = b_2 = 0$.
When $g_1 = 0,\ f_{33} = 0$, we can obtain from (7.17.27) that

$$\begin{aligned}\lambda_1\nu_1a_1 &= kb_2b_3\sin\gamma_I\\ \omega_3\mu_3b_3 &= -ka_1b_2\sin\gamma_I\end{aligned} \tag{7.17.31}$$

Eliminating γ_I from (7.17.30) yields

$$\lambda_1\nu_1a_1^2 + \omega_3\mu_3b_3^2 = 0 \tag{7.17.32}$$

So $a_1 = b_3 = 0$.
When $g_1 = 0, f_{22} \neq 0, f_{33} \neq 0$, we can obtain from (7.17.27) and (7.17.28) that

$$\begin{aligned}\lambda_1\nu_1a_1 - kb_2b_3\sin\gamma_I &= 0\\ \omega_2\mu_2b_2 + ka_1b_3\sin\gamma_I - f_{22}\sin\gamma_2 &= 0\\ \omega_3\mu_3b_3 + ka_1b_2\sin\gamma_I - f_{33}\sin\gamma_3 &= 0\end{aligned} \tag{7.17.33}$$

$$\begin{aligned}\alpha_1' &= \frac{kb_2b_3}{\lambda_1a_1}\cos\gamma_I\\ \beta_2' &= \frac{ka_1b_3}{\omega_2b_2}\cos\gamma_I - \frac{f_{22}}{\omega_2b_2}\cos\gamma_2\\ \beta_3' &= \frac{ka_1b_2}{\omega_3b_3}\cos\gamma_I - \frac{f_{33}}{\omega_3b_3}\cos\gamma_3\end{aligned} \tag{7.17.34}$$

From (7.17.26), we have

$$\gamma_I' - \sigma_I = \alpha_1' - \beta_2' - \beta_3', \quad -\gamma_2' + \sigma_2 = \beta_2', \quad -\gamma_3' + \sigma_3 = \beta_3' \tag{7.17.35}$$

Then from (7.17.34) and (7.17.35), we have

$$\begin{aligned}
\gamma_I' &= \sigma_I + \left(\frac{kb_2b_3}{\lambda_1 a_1} - \frac{ka_1b_2}{\omega_3 b_3} - \frac{ka_1b_3}{\omega_2 b_2}\right)\cos\gamma_I + \frac{f_{22}}{\omega_2 b_2}\cos\gamma_2 + \frac{f_{33}}{\omega_3 b_3}\cos\gamma_3 \\
\gamma_2' &= \sigma_2 - \frac{ka_1b_3}{\omega_2 b_2}\cos\gamma_I + \frac{f_{22}}{\omega_2 b_2}\cos\gamma_2 \\
\gamma_3' &= \sigma_3 - \frac{ka_1b_2}{\omega_3 b_3}\cos\gamma_I + \frac{f_{33}}{\omega_3 b_3}\cos\gamma_3
\end{aligned} \tag{7.17.36}$$

Let $\gamma_I' = \gamma_2' = \gamma_3' = 0$, we obtain

$$\begin{aligned}
&\sigma_I + \left(\frac{kb_2b_3}{\lambda_1 a_1} - \frac{ka_1b_2}{\omega_3 b_3} - \frac{ka_1b_3}{\omega_2 b_2}\right)\cos\gamma_I + \frac{f_{22}}{\omega_2 b_2}\cos\gamma_2 + \frac{f_{33}}{\omega_3 b_3}\cos\gamma_3 = 0 \\
&\sigma_2 - \frac{ka_1b_3}{\omega_2 b_2}\cos\gamma_I + \frac{f_{22}}{\omega_2 b_2}\cos\gamma_2 = 0 \\
&\sigma_3 - \frac{ka_1b_2}{\omega_3 b_3}\cos\gamma_I + \frac{f_{33}}{\omega_3 b_3}\cos\gamma_3 = 0
\end{aligned} \tag{7.17.37}$$

The above three equations are combined to give

$$\frac{kb_2b_3}{\lambda_1 a_1}\cos\gamma_I = \sigma_2 + \sigma_3 - \sigma_I \tag{7.17.38}$$

From the first equation of (7.17.33) and (7.17.38), we have

$$\tan\gamma_I = \frac{\nu_1}{\sigma_2 + \sigma_3 - \sigma_I} \tag{7.17.39}$$

Solving for $\sin\gamma_I$ from the first equation of (7.17.33), substituting it into the second and third equations of (7.17.33) and then substituting (7.17.38) into the second and third equations of (7.17.37) yields

$$\begin{aligned}
\omega_2\mu_2 b_2^2 + \lambda_1\nu_1 a_1^2 &= b_2 f_{22}\sin\gamma_2 \\
\omega_3\mu_3 b_3^2 + \lambda_1\nu_1 a_1^2 &= b_3 f_{33}\sin\gamma_3
\end{aligned} \tag{7.17.40}$$

$$\begin{aligned}
\lambda_1 a_1^2(\sigma_2 + \sigma_3 - \sigma_I) - \omega_2 b_2^2\sigma_2 &= b_2 f_{22}\cos\gamma_2 \\
\lambda_1 a_1^2(\sigma_2 + \sigma_3 - \sigma_I) - \omega_3 b_3^2\sigma_3 &= b_3 f_{33}\cos\gamma_3
\end{aligned} \tag{7.17.41}$$

Eliminating γ_2 and γ_3 from these two sets of equations yields

$$(\omega_2\mu_2 b_2^2 + \lambda_1\nu_1 a_1^2)^2 + [\lambda_1 a_1^2(\sigma_2 + \sigma_3 - \sigma_I) - \omega_2 b_2^2\sigma_2]^2 = b_2^2 f_{22}^2$$

$$(\omega_3\mu_3 b_3^2 + \lambda_1\nu_1 a_1^2)^2 + [\lambda_1 a_1^2(\sigma_2 + \sigma_3 - \sigma_I) - \omega_3 b_3^2\sigma_3]^2 = b_3^2 f_{33}^2 \quad (7.17.42)$$

7.18 Exercise 7.18 (Analysis of Internal and Primary Resonances of Cylindrical Shells, Saturation Phenomena)

Solution: (a) We use the method of multiple scales to solve the given equation by setting the solution of the equation as

$$u_n(t;\varepsilon) = \varepsilon u_{n0}(T_0, T_1) + \varepsilon^2 u_{n1}(T_0, T_1) + \cdots \quad (7.18.1)$$

Substituting (7.18.1) into the given equation and keeping to $O(\varepsilon^2)$, we obtain

$$\begin{aligned}
0 &= \ddot{u}_n + \omega_n^2 u_n + 2\varepsilon\mu_n\dot{u}_n - \varepsilon\sum_{p,q}^{\infty}\alpha_{npq}u_p u_q - f_n(t)\\
&= \left(D_0^2 + 2\varepsilon D_0 D_1\right)\left(\varepsilon u_{n0} + \varepsilon^2 u_{n1}\right) + \omega_n^2\left(\varepsilon u_{n0} + \varepsilon^2 u_{n1}\right)\\
&\quad +2\varepsilon\mu_n(D_0 + \varepsilon D_1)\left(\varepsilon u_{n0} + \varepsilon^2 u_{n1}\right)\\
&\quad -\sum_{p,q}^{\infty}\alpha_{npq}\left(\varepsilon u_{p0} + \varepsilon^2 u_{p1}\right)\left(\varepsilon u_{q0} + \varepsilon^2 u_{q1}\right) - 2\varepsilon^2\delta_{ns}k_n\cos\Omega T_0\\
&= \varepsilon\left(D_0^2 u_{n0} + \omega_n^2 u_{n0}\right) + \varepsilon^2\Big(D_0^2 u_{n1} + \omega_n^2 u_{n1} + 2D_0 D_1 u_{n0}\\
&\quad +2\mu_n D_0 u_{n0} - \sum_{p,q}^{\infty}\alpha_{npq}u_{p0}u_{q0} - 2\delta_{ns}k_n\cos\Omega T_0\Big)
\end{aligned} \quad (7.18.2)$$

Making the coefficient of the same power of ε zero in the above equation yields

$$D_0^2 u_{n0} + \omega_n^2 u_{n0} = 0 \quad (7.18.3)$$

$$D_0^2 u_{n1} + \omega_n^2 u_{n1} = -2D_0 D_1 u_{n0} - 2\mu_n D_0 u_{n0} + \sum_{p,q}^{\infty}\alpha_{npq}u_{p0}u_{q0} + 2\delta_{ns}k_n\cos\Omega T_0 \quad (7.18.4)$$

The solution of the Eq. (7.18.3) is

$$u_{n0} = A_n(T_1)e^{i\omega_n T_0} + cc \quad (7.18.5)$$

Substituting (7.18.5) into (7.18.4) yields

$$D_0^2 u_{n1} + \omega_n^2 u_{n1} = -2D_0 D_1 u_{n0} - 2\mu_n D_0 u_{n0} + \sum_{p,q}^{\infty} \alpha_{npq} u_{p0} u_{q0} + 2\delta_{ns} k_n \cos\Omega T_0$$
$$= -2i\omega_n A_n' e^{i\omega_n T_0} - 2i\omega_n \mu_n A_n e^{i\omega_n T_0}$$
$$+ \sum_{p,q}^{\infty} \alpha_{npq} \left(A_p A_q e^{i(\omega_p + \omega_q)T_0} + A_p \overline{A}_q e^{i(\omega_p - \omega_q)T_0} \right) + \delta_{ns} k_n e^{i\Omega T_0} + cc \tag{7.18.6}$$

In order to eliminate secular terms from the above equation and taking into account $\omega_3 \approx \omega_1 + \omega_2 + \varepsilon\sigma, \Omega = \omega_n + \varepsilon\sigma_n$, $n = 1$ or 2 or 3, we need

$$2i\omega_1 \left(A_1' + \mu_1 A_1 \right) = (\alpha_{132} + \alpha_{123}) A_3 \overline{A}_2 e^{i\sigma T_1} + \delta_{1s} k_1 e^{i\sigma_1 T_1} \tag{7.18.7}$$

$$2i\omega_2 \left(A_2' + \mu_2 A_2 \right) = (\alpha_{213} + \alpha_{231}) A_3 \overline{A}_1 e^{i\sigma T_1} + \delta_{2s} k_2 e^{i\sigma_2 T_1} \tag{7.18.8}$$

$$2i\omega_3 \left(A_3' + \mu_3 A_3 \right) = (\alpha_{312} + \alpha_{321}) A_1 A_2 e^{-i\sigma T_1} + \delta_{3s} k_3 e^{i\sigma_3 T_1} \tag{7.18.9}$$

$$A_n' + \mu_n A_n = 0, \quad n \geq 4 \tag{7.18.10}$$

From (7.18.10), we have

$$A_n = A_{n0} e^{-\mu_n T_1} \to 0, n \geq 4 \tag{7.18.11}$$

When $s = 1$, (7.18.7)–(7.18.9) become

$$2i\omega_1 \left(A_1' + \mu_1 A_1 \right) = (\alpha_{132} + \alpha_{123}) A_3 \overline{A}_2 e^{i\sigma T_1} + k_1 e^{i\sigma_1 T_1} \tag{7.18.12}$$

$$2i\omega_2 \left(A_2' + \mu_2 A_2 \right) = (\alpha_{213} + \alpha_{231}) A_3 \overline{A}_1 e^{i\sigma T_1} \tag{7.18.13}$$

$$2i\omega_3 \left(A_3' + \mu_3 A_3 \right) = (\alpha_{312} + \alpha_{321}) A_1 A_2 e^{-i\sigma T_1} \tag{7.18.14}$$

Let

$$A_n = \frac{1}{2} a_n e^{i\beta_n} \tag{7.18.15}$$

and substitute it into (7.18.9)–(7.18.12), we have

$$i\omega_1 a_1' - \omega_1 a_1 \beta_1' + i\omega_1 \mu_1 a_1 = \frac{1}{4} (\alpha_{132} + \alpha_{123}) a_2 a_3 e^{i\gamma_l} + k_1 e^{i\gamma_1}$$
$$i\omega_2 a_2' - \omega_2 a_2 \beta_2' + i\omega_2 \mu_2 a_2 = \frac{1}{4} (\alpha_{213} + \alpha_{231}) a_1 a_3 e^{i\gamma_l}$$

$$i\omega_3 a_3' - \omega_3 a_3 \beta_3' + i\omega_3 \mu_3 a_3 = \frac{1}{4}(\alpha_{312} + \alpha_{321}) a_1 a_2 e^{-i\gamma_I} \tag{7.18.16}$$

where

$$\gamma_I = \sigma T_1 - \beta_1 - \beta_2 + \beta_3,\ \gamma_1 = \sigma_1 T_1 - \beta_1 \tag{7.18.17}$$

Separating the real and imaginary parts of (7.18.16) yields

$$\begin{aligned} \omega_1 a_1' + \omega_1 \mu_1 a_1 &= \frac{1}{4}(\alpha_{132} + \alpha_{123}) a_2 a_3 \sin\gamma_I + k_1 \sin\gamma_1 \\ \omega_2 a_2' + \omega_2 \mu_2 a_2 &= \frac{1}{4}(\alpha_{213} + \alpha_{231}) a_1 a_3 \sin\gamma_I \\ \omega_3 a_3' + \omega_3 \mu_3 a_3 &= -\frac{1}{4}(\alpha_{312} + \alpha_{321}) a_1 a_2 \sin\gamma_I \end{aligned} \tag{7.18.18}$$

$$\begin{aligned} -\omega_1 a_1 \beta_1' &= \frac{1}{4}(\alpha_{132} + \alpha_{123}) a_2 a_3 \cos\gamma_I + k_1 \cos\gamma_1 \\ -\omega_2 a_2 \beta_2' &= \frac{1}{4}(\alpha_{213} + \alpha_{231}) a_1 a_3 \cos\gamma_I \\ -\omega_3 a_3 \beta_3' &= \frac{1}{4}(\alpha_{312} + \alpha_{321}) a_1 a_2 \cos\gamma_I \end{aligned} \tag{7.18.19}$$

From (7.18.17) and (7.18.19), we have

$$\begin{aligned} &\omega_1 a_1 \gamma_1' = \omega_1 \sigma_1 a_1 + \frac{1}{4}(\alpha_{132} + \alpha_{123}) a_2 a_3 \cos\gamma_I + k_1 \cos\gamma_1 \\ &a_1 a_2 a_3 \gamma_I' = \sigma a_1 a_2 a_3 + \frac{1}{4}[\omega_1^{-1}(\alpha_{132} + \alpha_{123}) a_2^2 a_3^2 + \omega_2^{-1}(\alpha_{213} + \alpha_{231}) a_1^2 a_3^2 \\ &-\omega_3^{-1}(\alpha_{312} + \alpha_{321}) a_1^2 a_2^2] \cos\gamma_I - a_2 a_3 \omega_1^{-1} k_1 \cos\gamma_1 \end{aligned} \tag{7.18.20}$$

Then from (7.18.18) and (7.18.20), we can obtain the corresponding equations for steady state solution

$$\begin{aligned} \omega_1 \mu_1 a_1 &= \frac{1}{4}(\alpha_{132} + \alpha_{123}) a_2 a_3 \sin\gamma_I + k_1 \sin\gamma_1 \\ \omega_2 \mu_2 a_2 &= \frac{1}{4}(\alpha_{213} + \alpha_{231}) a_1 a_3 \sin\gamma_I \\ \omega_3 \mu_3 a_3 &= -\frac{1}{4}(\alpha_{312} + \alpha_{321}) a_1 a_2 \sin\gamma_I \end{aligned} \tag{7.18.21}$$

$$\begin{aligned} &\omega_1 \sigma_1 a_1 + \frac{1}{4}(\alpha_{132} + \alpha_{123}) a_2 a_3 \cos\gamma_I + k_1 \cos\gamma_1 = 0 \\ &\sigma a_1 a_2 a_3 + \frac{1}{4}[\omega_1^{-1}(\alpha_{132} + \alpha_{123}) a_2^2 a_3^2 + \omega_2^{-1}(\alpha_{213} + \alpha_{231}) a_1^2 a_3^2 \end{aligned}$$

$$-\omega_3^{-1}(\alpha_{312}+\alpha_{321})a_1^2a_2^2]\cos\gamma_I - a_2a_3\omega_1^{-1}k_1\cos\gamma_1 = 0 \tag{7.18.22}$$

Eliminating γ_I from the second and third equations of (7.18.21) yields

$$(\alpha_{312}+\alpha_{321})\omega_2\mu_2a_2^2 + (\alpha_{213}+\alpha_{231})\omega_3\mu_3a_3^2 = 0 \tag{7.18.23}$$

Therefore, we need $a_2 = a_3 = 0$. In this case, the first equations of (7.18.21) and (7.18.22) become

$$\begin{aligned} \omega_1\mu_1a_1 &= k_1\sin\gamma_1 \\ \omega_1\sigma_1a_1 &= -k_1\cos\gamma_1 \end{aligned} \tag{7.18.24}$$

So

$$k_1^2 = \omega_1^2(\mu_1^2+\sigma_1^2)a_1^2 \tag{7.18.25}$$

It follows that the response in this case contains only first order modes.

When $s = 2$, the results are similar to the above result. In this case, the response contains only second-order mode.

When $s = 3$, (7.18.7)–(7.18.9) become

$$2i\omega_1(A_1' + \mu_1A_1) = (\alpha_{132}+\alpha_{123})A_3\overline{A}_2e^{i\sigma T_1} \tag{7.18.26}$$

$$2i\omega_2(A_2' + \mu_2A_2) = (\alpha_{213}+\alpha_{231})A_3\overline{A}_1e^{i\sigma T_1} \tag{7.18.27}$$

$$2i\omega_3(A_3' + \mu_3A_3) = (\alpha_{312}+\alpha_{321})A_1A_2e^{-i\sigma T_1} + k_3e^{i\sigma_3T_1} \tag{7.18.28}$$

Substituting (7.18.15) into (7.18.26)–(7.18.28) yields

$$\begin{aligned} i\omega_1a_1' - \omega_1a_1\beta_1' + i\omega_1\mu_1a_1 &= \frac{1}{4}(\alpha_{132}+\alpha_{123})a_2a_3e^{i\gamma_I} \\ i\omega_2a_2' - \omega_2a_2\beta_2' + i\omega_2\mu_2a_2 &= \frac{1}{4}(\alpha_{213}+\alpha_{231})a_1a_3e^{i\gamma_I} \\ i\omega_3a_3' - \omega_3a_3\beta_3' + i\omega_3\mu_3a_3 &= \frac{1}{4}(\alpha_{312}+\alpha_{321})a_1a_2e^{-i\gamma_I} + k_3e^{i\gamma_3} \end{aligned} \tag{7.18.29}$$

where

$$\gamma_I = \sigma T_1 - \beta_1 - \beta_2 + \beta_3, \gamma_3 = \sigma_3T_1 - \beta_3 \tag{7.18.30}$$

Separating the real and imaginary parts of (7.18.30) gives

$$\omega_1a_1\prime + \omega_1\mu_1a_1 = \frac{1}{4}(\alpha_{132}+\alpha_{123})a_2a_3\sin\gamma_I$$

$$\omega_2 a_2{}' + \omega_2\mu_2 a_2 = \frac{1}{4}(\alpha_{213} + \alpha_{231})a_1 a_3 \sin\gamma_I$$

$$\omega_3 a_3{}' + \omega_3\mu_3 a_3 = -\frac{1}{4}(\alpha_{312} + \alpha_{321})a_1 a_2 \sin\gamma_I + k_3 \sin\gamma_3 \qquad (7.18.31)$$

$$-\omega_1 a_1 \beta_1{}' = \frac{1}{4}(\alpha_{132} + \alpha_{123})a_2 a_3 \cos\gamma_I$$

$$-\omega_2 a_2 \beta_2{}' = \frac{1}{4}(\alpha_{213} + \alpha_{231})a_1 a_3 \cos\gamma_I$$

$$-\omega_3 a_3 \beta_3{}' = \frac{1}{4}(\alpha_{312} + \alpha_{321})a_1 a_2 \cos\gamma_I + k_3 \cos\gamma_3 \qquad (7.18.32)$$

From (7.18.30) and (7.18.32), we can obtain

$$a_1 a_2 a_3 \gamma_I' = \sigma a_1 a_2 a_3 + \frac{1}{4}\big[\omega_1^{-1}(\alpha_{132} + \alpha_{123})a_2^2 a_3^2 + \omega_2^{-1}(\alpha_{213} + \alpha_{231})a_1^2 a_3^2$$
$$-\omega_3^{-1}(\alpha_{312} + \alpha_{321})a_1^2 a_2^2\big]\cos\gamma_I - a_1 a_2 \omega_3^{-1} k_3 \cos\gamma_3$$
$$\omega_3 a_3 \gamma_3' = \omega_3 \sigma_3 a_3 + \frac{1}{4}(\alpha_{312} + \alpha_{321})a_1 a_2 \cos\gamma_I + k_3 \cos\gamma_3$$

$$(7.18.33)$$

Then from (7.18.31) and (7.18.33), we can obtain the corresponding equations for the steady state solution as follows:

$$\omega_1\mu_1 a_1 = \frac{1}{4}(\alpha_{132} + \alpha_{123})a_2 a_3 \sin\gamma_I$$

$$\omega_2\mu_2 a_2 = \frac{1}{4}(\alpha_{213} + \alpha_{231})a_1 a_3 \sin\gamma_I$$

$$\omega_3\mu_3 a_3 = -\frac{1}{4}(\alpha_{312} + \alpha_{321})a_1 a_2 \sin\gamma_I + k_3 \sin\gamma_3 \qquad (7.18.34)$$

$$\sigma a_1 a_2 a_3 + \frac{1}{4}[\omega_1^{-1}(\alpha_{132} + \alpha_{123})a_2^2 a_3^2 + \omega_2^{-1}(\alpha_{213} + \alpha_{231})a_1^2 a_3^2$$
$$-\omega_3^{-1}(\alpha_{312} + \alpha_{321})a_1^2 a_2^2]\cos\gamma_I - a_1 a_2 \omega_3^{-1} k_3 \cos\gamma_3 = 0$$
$$\omega_3 \sigma_3 a_3 + \frac{1}{4}(\alpha_{312} + \alpha_{321})a_1 a_2 \cos\gamma_I + k_3 \cos\gamma_3 = 0 \qquad (7.18.35)$$

Eliminating γ_I from the first and second equations of (7.18.34) yields

$$(\alpha_{213} + \alpha_{231})\omega_1\mu_1 a_1^2 - (\alpha_{132} + \alpha_{123})\omega_2\mu_2 a_2^2 = 0 \qquad (7.18.36)$$

As can be seen,a_1 and a_2 can be either a trivial solution or a non-trivial solution.
When $\boldsymbol{a}_1 = \boldsymbol{a}_2 = 0$.
From (7.18.34) and (7.18.35), we have

$$\begin{aligned}\omega_3\mu_3a_3 &= k_3\sin\gamma_3\\ \omega_3\sigma_3a_3 &= -k_3\cos\gamma_3\end{aligned} \tag{7.18.37}$$

So

$$k_3^2 = \omega_3^2\left(\mu_3^2+\sigma_3^2\right)a_3^2 \tag{7.18.38}$$

When $a_1 \neq 0,\ a_2 \neq 0$.
From (7.18.35), we can obtain

$$(\sigma+\sigma_3)a_1a_2a_3 + \frac{1}{4}\left[\omega_1^{-1}(\alpha_{132}+\alpha_{123})a_2^2a_3^2 + \omega_2^{-1}(\alpha_{213}+\alpha_{231})a_1^2a_3^2\right]\cos\gamma_I = 0 \tag{7.18.39}$$

The first equation of (7.18.34) and the Eq. (7.18.39) can be written as:

$$\begin{aligned}a_3\sin\gamma_I &= \frac{4\omega_1\mu_1}{(\alpha_{132}+\alpha_{123})}\frac{a_1}{a_2}\\ a_3\cos\gamma_I &= -\frac{4(\sigma+\sigma_3)}{\left[\omega_1^{-1}(\alpha_{132}+\alpha_{123})\frac{a_2}{a_1}+\omega_2^{-1}(\alpha_{213}+\alpha_{231})\frac{a_1}{a_2}\right]}\end{aligned} \tag{7.18.40}$$

Eliminating γ_I from (7.18.40), we obtain

$$a_3^2 = \frac{16\omega_1^2\mu_1^2}{(\alpha_{132}+\alpha_{123})^2}\frac{a_1^2}{a_2^2} + \frac{16(\sigma+\sigma_3)^2}{\left[\omega_1^{-1}(\alpha_{132}+\alpha_{123})\frac{a_2}{a_1}+\omega_2^{-1}(\alpha_{213}+\alpha_{231})\frac{a_1}{a_2}\right]^2} \tag{7.18.41}$$

With (7.18.36), the above equation becomes

$$a_3^2 = \frac{16\omega_1\omega_2\mu_1\mu_2}{(\alpha_{132}+\alpha_{123})(\alpha_{213}+\alpha_{231})}\left[1+\frac{(\sigma+\sigma_3)^2}{(\mu_1+\mu_2)^2}\right] \tag{7.18.42}$$

It can be seen that the steady state amplitude a_3 is independent of the excitation amplitude k_3, thus, u_3 mode is saturated. From (7.18.38) and (7.18.42), we can see that as the excitation amplitude k_3 increases from zero, so does the steady state amplitude a_3 until it reaches the saturation value given by (7.18.42), while $a_1 = a_2 = 0$. When the saturation phenomenon occurs, k_3 reaches the critical value k_ck_c. Substituting (7.18.42) into (7.18.38), we can obtain that critical value k_c:

$$k_c^2 = \frac{16\omega_1\omega_2\omega_3^2\mu_1\mu_2\left(\mu_3^2+\sigma_3^2\right)}{(\alpha_{132}+\alpha_{123})(\alpha_{213}+\alpha_{231})}\left[1+\frac{(\sigma+\sigma_3)^2}{(\mu_1+\mu_2)^2}\right] \tag{7.18.43}$$

When $k_3 > k_c$, the steady state amplitude a_3 is independent of the excitation amplitude k_3 and the steady state amplitude a_3 is constant at the saturation value given by (7.18.42). At this point,$a_1 \neq 0$,$a_2 \neq 0$ and, from (7.18.34) and (7.18.35), a_1 and a_2a_2 depend on k_3.

7.19 Exercise 7.19 (Primary Resonance Analysis of Transverse Oscillations of a Taut String)

Solution: From Sect. 7.5 of **the Book** or Exercise 7.4, we can write the governing equation for the transverse oscillation of a taut string as

$$v_{tt} - c_2^2 v_{xx} + 2\hat{\mu} v_t = c_1^2 v_{xx} P(t) + \frac{c_1^2}{2l} v_{xx} \int_0^l \left(v_x^2 + w_x^2\right) dx + G(x, t) \tag{7.19.1}$$

$$w_{tt} - c_2^2 w_{xx} + 2\hat{\mu} w_t = c_1^2 w_{xx} P(t) + \frac{c_1^2}{2l} w_{xx} \int_0^l \left(v_x^2 + w_x^2\right) dx + F(x, t) \tag{7.19.2}$$

Expanding the displacements v and w of the string in terms of linear modes with fixed ends, i.e.

$$v(x, t) = \sum_{n=1}^{\infty} \zeta_n(t) \sin\frac{n\pi x}{l},\ w(x, t) = \sum_{n=1}^{\infty} \eta_n(t) \sin\frac{n\pi x}{l} \tag{7.19.3}$$

Substituting (7.19.3) into (7.19.1) and (7.19.2), taking into account the orthogonality of the mode shape function, we can obtain the discrete modal coordinate equations

$$\begin{aligned} \ddot{\zeta}_n + \omega_n^2 \zeta_n &= -\left[2\hat{\mu}_n \dot{\zeta}_n + n^2 p(t) \zeta_n + n^2 \Gamma \zeta_n \sum_{m=1}^{\infty} m^2 \left(\zeta_m^2 + \eta_m^2\right)\right] + g_n(t) \\ \ddot{\eta}_n + \omega_n^2 \eta_n &= -\left[2\hat{\mu}_n \dot{\eta}_n + n^2 p(t) \eta_n + n^2 \Gamma \eta_n \sum_{m=1}^{\infty} m^2 \left(\zeta_m^2 + \eta_m^2\right)\right] + f_n(t) \end{aligned} \tag{7.19.4}$$

where

$$\omega_n = \frac{n\pi c_2}{l},\quad \Gamma = \frac{\pi^4 c_1^2}{4l^4},\quad p(t) = \frac{\pi^2 c_1^2 P(t)}{l^2} \tag{7.19.5}$$

$$g_n = \frac{2}{l} \int_0^l G(x, t) \sin\frac{n\pi x}{l} dx,\quad f_n = \frac{2}{l} \int_0^l F(x, t) \sin\frac{n\pi x}{l}\, dx \tag{7.19.6}$$

In the following case, we analyze a primary resonance case. Let

$$g_n(t) = 2K_n\cos\Omega t, \quad f_n = 0, \quad p(t) = 0, \quad \Omega \approx \omega_s \tag{7.19.7}$$

where s is a given positive integer. Since (7.19.4) contains only cubic nonlinearities, the solution of the equation is given as follows to find a first-order approximate solution

$$\begin{aligned}\zeta_n(t;\varepsilon) &= \varepsilon\zeta_{n0}(T_0,T_1,T_2) + \varepsilon^2\zeta_{n1}(T_0,T_1,T_2) + \varepsilon^3\zeta_{n2}(T_0,T_1,T_2) + \cdots \\ \eta_n(t;\varepsilon) &= \varepsilon\eta_{n0}(T_0,T_1,T_2) + \varepsilon^2\eta_{n1}(T_0,T_1,T_2) + \varepsilon^3\eta_{n2}(T_0,T_1,T_2) + \cdots\end{aligned} \tag{7.19.8}$$

In order to let the damping, the resonant excitation term and the nonlinear term appear simultaneously in the third order equation, we make

$$\hat{\mu}_n = \varepsilon^2\mu_n, K_n = \varepsilon^3 k_n \tag{7.19.9}$$

Substituting (7.19.7)–(7.19.9) into (7.19.4) and retaining to $O(\varepsilon^3)$, we obtain

$$\begin{aligned}0 &= \ddot{\zeta}_n + \omega_n^2\zeta_n + 2\hat{\mu}_n\dot{\zeta}_n + n^2\Gamma\zeta_n\sum_{m=1}^{\infty} m^2\left(\zeta_m^2 + \eta_m^2\right) - 2\delta_{ns}K_s\cos\Omega t \\ &= \left[D_0^2 + 2\varepsilon D_0D_1 + \varepsilon^2\left(D_1^2 + 2D_0D_2\right)\right]\left(\varepsilon\zeta_{n0} + \varepsilon^2\zeta_{n1} + \varepsilon^3\zeta_{n2}\right) \\ &\quad + \omega_n^2\left(\varepsilon\zeta_{n0} + \varepsilon^2\zeta_{n1} + \varepsilon^3\zeta_{n2}\right) \\ &\quad + 2\varepsilon^2\mu_n\left(D_0 + \varepsilon D_1 + \varepsilon^2 D_2\right)\left(\varepsilon\zeta_{n0} + \varepsilon^2\zeta_{n1} + \varepsilon^3\zeta_{n2}\right) \\ &\quad + n^2\Gamma\left(\varepsilon\zeta_{n0} + \varepsilon^2\zeta_{n1} + \varepsilon^3\zeta_{n2}\right)\sum_{m=1}^{\infty} m^2\Big[\left(\varepsilon\zeta_{n0} + \varepsilon^2\zeta_{n1} + \varepsilon^3\zeta_{n2}\right)^2 \\ &\quad + \left(\varepsilon\eta_{n0} + \varepsilon^2\eta_{n1} + \varepsilon^3\eta_{n2}\right)^2\Big] - 2\varepsilon^3\delta_{ns}k_s\cos\Omega T_0 \\ &= \varepsilon\left(D_0^2\zeta_{n0} + \omega_n^2\zeta_{n0}\right) + \varepsilon^2\left(D_0^2\zeta_{n1} + \omega_n^2\zeta_{n1} + 2D_0D_1\zeta_{n0}\right) \\ &\quad + \varepsilon^3\Big[D_0^2\zeta_{n2} + \omega_n^2\zeta_{n2} + \left(D_1^2 + 2D_0D_2\right)\zeta_{n0} + 2D_0D_1\zeta_{n1} \\ &\quad + 2\mu_nD_0\zeta_{n0} + n^2\Gamma\zeta_{n0}\sum_{m=1}^{\infty} m^2\left(\zeta_{m0}^2 + \eta_{m0}^2\right) - 2\delta_{ns}k_s\cos\Omega T_0\Big]\end{aligned} \tag{7.19.10}$$

$$\begin{aligned}0 &= \ddot{\eta}_n + \omega_n^2\eta_n + \left[2\hat{\mu}_n\dot{\eta}_n + n^2\Gamma\eta_n\sum_{m=1}^{\infty} m^2\left(\zeta_m^2 + \eta_m^2\right)\right] \\ &= \varepsilon\left(D_0^2\eta_{n0} + \omega_n^2\eta_{n0}\right) + \varepsilon^2\left(D_0^2\eta_{n1} + \omega_n^2\eta_{n1} + 2D_0D_1\eta_{n0}\right) \\ &\quad + \varepsilon^3\Big[D_0^2\eta_{n2} + \omega_n^2\eta_{n2} + \left(D_1^2 + 2D_0D_2\right)\eta_{n0} + 2D_0D_1\eta_{n1} \\ &\quad + 2\mu_nD_0\eta_{n0} + n^2\Gamma\eta_{n0}\sum_{m=1}^{\infty} m^2\left(\zeta_{m0}^2 + \eta_{m0}^2\right)\Big]\end{aligned} \tag{7.19.11}$$

Let the coefficient of the like power of ε be zero in (7.19.10) and (7.19.11), we obtain.

Order ε:

$$\begin{aligned} D_0^2\zeta_{n0} + \omega_n^2\zeta_{n0} = 0 \\ D_0^2\eta_{n0} + \omega_n^2\eta_{n0} = 0 \end{aligned} \tag{7.19.12}$$

Order ε^2:

$$\begin{aligned} D_0^2\zeta_{n1} + \omega_n^2\zeta_{n1} &= -2D_0D_1\zeta_{n0} \\ D_0^2\eta_{n1} + \omega_n^2\eta_{n1} &= -2D_0D_1\eta_{n0} \end{aligned} \tag{7.19.13}$$

Order ε^3:

$$\begin{aligned} D_0^2\zeta_{n2} + \omega_n^2\zeta_{n2} &= -\left(D_1^2 + 2D_0D_2\right)\zeta_{n0} - 2D_0D_1\zeta_{n1} \\ &\quad -2\mu_n D_0\zeta_{n0} - n^2\Gamma\zeta_{n0}\sum_{m=1}^{\infty} m^2\left(\zeta_{m0}^2 + \eta_{m0}^2\right) + 2\delta_{ns}k_s\cos\Omega T_0 \\ D_0^2\eta_{n2} + \omega_n^2\eta_{n2} &= -\left(D_1^2 + 2D_0D_2\right)\eta_{n0} - 2D_0D_1\eta_{n1} \\ &\quad -2\mu_n D_0\eta_{n0} - n^2\Gamma\eta_{n0}\sum_{m=1}^{\infty} m^2\left(\zeta_{m0}^2 + \eta_{m0}^2\right) \end{aligned} \tag{7.19.14}$$

The solution of (7.19.12) is

$$\zeta_{n0} = A_n e^{i\omega_n T_0} + cc,\ \eta_{n0} = B_n e^{i\omega_n T_0} + cc \tag{7.19.15}$$

where $A_n = A_n(T_1, T_2)$, $B_n = B_n(T_1, T_2)$. Substituting (7.19.15) into (7.19.13), we obtain

$$\begin{aligned} D_0^2\zeta_{n1} + \omega_n^2\zeta_{n1} &= -2i\omega_n D_1 A_n e^{i\omega_n T_0} + cc \\ D_0^2\eta_{n1} + \omega_n^2\eta_{n1} &= -2i\omega_n D_1 B_n e^{i\omega_n T_0} + cc \end{aligned} \tag{7.19.16}$$

In order to eliminate secular terms from the above equation, we need

$$\begin{aligned} D_1A_n &= 0 \Rightarrow A_n = A_n(T_2) \\ D_1B_n &= 0 \Rightarrow B_n = B_n(T_2) \end{aligned} \tag{7.19.17}$$

By (7.19.13), we can obtain

$$\zeta_{n1} = 0,\ \eta_{n1} = 0 \tag{7.19.18}$$

Substituting (7.19.15), (7.19.17) and (7.19.18) into (7.19.14), we obtain

$$
\begin{aligned}
&D_0^2\zeta_{n2} + \omega_n^2\zeta_{n2} = -2D_0D_2\zeta_{n0} - 2\mu_n D_0\zeta_{n0} \\
&-n^2\Gamma\zeta_{n0}\sum_{m=1}^{\infty} m^2\left(\zeta_{m0}^2 + \eta_{m0}^2\right) + 2\delta_{ns}k_s\cos\Omega T_0 \\
&= -2i\omega_n A_n' e^{i\omega_n T_0} - 2i\omega_n\mu_n A_n e^{i\omega_n T_0} - \Gamma n^4\overline{A}_n\left(A_n^2 + B_n^2\right) \\
&-2\Gamma n^2 A_n e^{i\omega_n T_0}\sum_{m=1}^{\infty} m^2\left(A_m\overline{A}_m + B_m\overline{B}_m\right) + \delta_{ns}k_s e^{i\Omega T_0} \\
&+cc + NST
\end{aligned}
\tag{7.19.19}
$$

$$
\begin{aligned}
&D_0^2\eta_{n2} + \omega_n^2\eta_{n2} = -2D_0D_2\eta_{n0} - 2\mu_n D_0\eta_{n0} - n^2\Gamma\eta_{n0}\sum_{m=1}^{\infty} m^2\left(\zeta_{m0}^2 + \eta_{m0}^2\right) \\
&= -2i\omega_n B_n' e^{i\omega_n T_0} - 2i\omega_n\mu_n B_n e^{i\omega_n T_0} - \Gamma n^4\overline{B}_n\left(A_n^2 + B_n^2\right) \\
&-2\Gamma n^2 B_n e^{i\omega_n T_0}\sum_{m=1}^{\infty} m^2\left(A_m\overline{A}_m + B_m\overline{B}_m\right) + cc + NST
\end{aligned}
\tag{7.19.20}
$$

The prime denotes the derivative with respect to T_2. Let

$$\Omega = \omega_s + \varepsilon^2\sigma \tag{7.19.21}$$

In order to eliminate secular terms from the equation, we need

$$
\begin{aligned}
&2i\omega_n A_n' + 2i\omega_n\mu_n A_n + \Gamma n^4\overline{A}_n\left(A_n^2 + B_n^2\right) \\
&\quad +2\Gamma n^2 A_n\sum_{m=1}^{\infty} m^2\left(A_m\overline{A}_m + B_m\overline{B}_m\right) - \delta_{ns}k_s e^{i\sigma T_2} = 0
\end{aligned}
\tag{7.19.22}
$$

$$
\begin{aligned}
&2i\omega_n B_n' + 2i\omega_n\mu_n B_n + \Gamma n^4\overline{B}_n\left(A_n^2 + B_n^2\right) \\
&\quad +2\Gamma n^2 B_n\sum_{m=1}^{\infty} m^2\left(A_m\overline{A}_m + B_m\overline{B}_m\right) = 0
\end{aligned}
\tag{7.19.23}
$$

Let

$$A_n = \frac{1}{2}a_n e^{i\alpha_n}, \quad B_n = \frac{1}{2}b_n e^{i\beta_n} \tag{7.19.24}$$

and substitute (7.19.24) into (7.19.22) and (7.19.23), we can obtain

$$
\begin{aligned}
&i\omega_n a_n' - \omega_n a_n\alpha_n' + i\omega_n\mu_n a_n + \tfrac{1}{8}\Gamma n^4\left(a_n^3 + a_n b_n^2 e^{i\gamma_n}\right) \\
&\quad +\tfrac{1}{4}\Gamma n^2 a_n\sum_{m=1}^{\infty} m^2\left(a_m^2 + b_m^2\right) - \delta_{ns}k_s e^{i\upsilon} = 0
\end{aligned}
\tag{7.19.25}
$$

$$
\begin{aligned}
i\omega_n b_{n\prime} - \omega_n b_n \beta_{n\prime} + i\omega_n \mu_n b_n + \tfrac{1}{8}\Gamma n^4\left(a_n^2 b_n e^{-i\gamma_n} + b_n^3\right) \\
+\tfrac{1}{4}\Gamma n^2 b_n \sum_{m=1}^{\infty} m^2\left(a_m^2 + b_m^2\right) = 0
\end{aligned}
\tag{7.19.26}
$$

where

$$
\gamma_n = 2(\beta_n - \alpha_n), \quad \nu = \sigma T_2 - \alpha_s \tag{7.19.27}
$$

Separating the real and imaginary parts of (7.19.25) and (7.19.26), we obtain

$$
\omega_n a_{n\prime} + \omega_n \mu_n a_n + \frac{1}{8}\Gamma n^4 a_n b_n^2 \sin\gamma_n - \delta_{ns} k_s \sin\nu = 0 \tag{7.19.28}
$$

$$
\begin{aligned}
-\omega_n a_n \alpha_n' + \tfrac{1}{8}\Gamma n^4 a_n^3 + \tfrac{1}{4}\Gamma n^2 a_n \sum_{m=1}^{\infty} m^2\left(a_m^2 + b_m^2\right) \\
+\tfrac{1}{8}\Gamma n^4 a_n b_n^2 \cos\gamma_n - \delta_{ns} k_s \cos\nu = 0
\end{aligned}
\tag{7.19.29}
$$

$$
\omega_n b_n' + \omega_n \mu_n b_n - \frac{1}{8}\Gamma n^4 a_n^2 b_n \sin\gamma_n = 0 \tag{7.19.30}
$$

$$
-\omega_n b_n \beta_{n\prime} + \frac{1}{8}\Gamma n^4 b_n^3 + \frac{1}{4}\Gamma n^2 b_n \sum_{m=1}^{\infty} m^2\left(a_m^2 + b_m^2\right) + \frac{1}{8}\Gamma n^4 a_n^2 b_n \cos\gamma_n = 0 \tag{7.19.31}
$$

When $n \neq s$, for the steady state solutions of a_n and b_n, (7.19.28) and (7.19.30) degenerate into

$$
\begin{aligned}
\omega_n \mu_n a_n + \tfrac{1}{8}\Gamma n^4 a_n b_n^2 \sin\gamma_n &= 0 \\
\omega_n \mu_n b_n - \tfrac{1}{8}\Gamma n^4 a_n^2 b_n \sin\gamma_n &= 0
\end{aligned}
\tag{7.19.32}
$$

This set of equations has only trivial solution $a_n = 0, \ b_n = 0$. Otherwise, if $a_n \neq 0, \ b_n \neq 0$, eliminating γ_n from (7.19.32) yields

$$
a_n^2 + b_n^2 = 0 \tag{7.19.33}
$$

which contradicts with $a_n \neq 0, \ b_n \neq 0$. Therefore,

$$
a_n = 0, \ b_n = 0, \quad \text{for all } n \neq s \tag{7.19.34}
$$

When $n = s$, (7.19.28)–(7.19.31) can be written as

$$
\omega_s a_s' + \omega_s \mu_s a_s + \frac{1}{8}\Gamma s^4 a_s b_s^2 \sin\gamma_s - k_s \sin\nu = 0 \tag{7.19.35}
$$

$$-\omega_s a_s \alpha_s\prime + \tfrac{1}{8}\Gamma s^4 a_s^3 + \tfrac{1}{4}\Gamma s^4 a_s(a_s^2 + b_s^2) \\ + \tfrac{1}{8}\Gamma s^4 a_s b_s^2 \cos\gamma_s - k_s \cos\nu = 0 \tag{7.19.36}$$

$$\omega_s b_s' + \omega_s \mu_s b_s - \frac{1}{8}\Gamma s^4 a_s^2 b_s \sin\gamma_s = 0 \tag{7.19.37}$$

$$-\omega_s b_s \beta_s' + \frac{1}{8}\Gamma s^4 b_s^3 + \frac{1}{4}\Gamma s^4 b_s(a_s^2 + b_s^2) + \frac{1}{8}\Gamma s^4 a_s^2 b_s \cos\gamma_s = 0 \tag{7.19.38}$$

Let

$$\mu = \mu_s,\ (a, b) = \sqrt{\frac{\Gamma s^4}{8\omega_s}}(a_s, b_s),\ k = \sqrt{\frac{\Gamma s^4}{8\omega_s^3}}k_s,\ \gamma = \gamma_s \tag{7.19.39}$$

and also consider Eq. (7.19.27), we can write (7.19.35)–(7.19.38) can be written as

$$a\prime + \mu a + ab^2 \sin\gamma - k\sin\nu = 0 \tag{7.19.40}$$

$$\nu\prime - \sigma + 3a^2 + 2b^2 + b^2\cos\gamma - a^{-1}k\cos\nu = 0 \tag{7.19.41}$$

$$b\prime + \mu b - a^2 b\sin\gamma = 0 \tag{7.19.42}$$

$$b\left[\frac{1}{2}\gamma\prime - \nu\prime + \sigma - 3b^2 - 2a^2 - a^2\cos\gamma\right] = 0 \tag{7.19.43}$$

There are two possible steady state solutions for the amplitude: (1) $b = 0, a \neq 0$; (2) $a \neq 0, b \neq 0$. In the first case, the oscillation of the string degenerates into plane oscillation.

For the case of steady state plane oscillation of a string $b = 0, a \neq 0$, (7.19.42) and (7.19.43) are automatically satisfied, the Eqs. (7.19.40) and (7.19.41) become

$$\mu a - k\sin\nu = 0 \tag{7.19.44}$$

$$-\sigma + 3a^2 - a^{-1}k\cos\nu = 0 \tag{7.19.45}$$

Eliminating ν from the above equations, we can obtain the frequency–response equation

$$\sigma = 3a^2 \pm \left(\frac{k^2}{a^2} - \mu^2\right)^{1/2} \tag{7.19.46}$$

For the steady state solution $b = 0,\ a \neq 0$, we can linearize (7.19.40)–(7.19.43) near the steady state solution and obtain

$$\begin{aligned}&\Delta a\prime + \mu\Delta a - (k\cos\nu)\Delta\nu = 0\\&\Delta\nu\prime + \left(6a + a^{-2}k\cos\nu\right)\Delta a + \left(a^{-1}k\sin\nu\right)\Delta\nu = 0\end{aligned} \tag{7.19.47}$$

$$\begin{aligned}&\Delta b' + \left(\mu - a^2\sin\gamma\right)\Delta b = 0\\&\left(\sigma - 2a^2 - a^2\cos\gamma\right)\Delta b = 0\end{aligned} \tag{7.19.48}$$

Their characteristic equations can be written as

$$\begin{aligned}&(\lambda + \mu)\Delta a - (k\cos\nu)\Delta\nu = 0\\&\left(6a + a^{-2}k\cos\nu\right)\Delta a + \left(\lambda + a^{-1}k\sin\nu\right)\Delta\nu = 0\end{aligned} \tag{7.19.49}$$

$$\begin{aligned}&\left(\lambda + \mu - a^2\sin\gamma\right)\Delta b = 0\\&\left(\sigma - 2a^2 - a^2\cos\gamma\right)\Delta b = 0\end{aligned} \tag{7.19.50}$$

(7.19.47) and (7.19.48) are two independent sets of differential equations and, of course, the characteristic Eqs. (7.19.49) and (7.19.50) are also independent; therefore, the value of λ obtained from (7.19.49) can be different from (7.19.50). In another word, (7.19.47) and (7.19.48) can have different exponential forms.

From (7.19.49) and non-trivial solution condition, we have

$$\begin{aligned}\lambda = &\frac{-\left(\mu + a^{-1}k\sin\nu\right)}{2}\\&\pm\frac{\left\{\left(\mu + a^{-1}k\sin\nu\right)^2 - 4\left[\mu a^{-1}k\sin\nu + k\cos\nu\left(6a + a^{-2}k\cos\nu\right)\right]\right\}^{1/2}}{2}\end{aligned} \tag{7.19.51}$$

Taking into account (7.19.44) and (7.19.45), we can write the above equation as

$$\lambda = -\mu \pm \left[\left(\sigma - 3a^2\right)\left(9a^2 - \sigma\right)\right]^{1/2} \tag{7.19.52}$$

It can be seen that when $\sigma \le 3a^2$ or $\sigma \ge 9a^2$, λ is a negative real root or a complex root with a negative real part; therefore, Δa and $\Delta\nu$ decay. When $3a^2 < \sigma < 9a^2$, λ is a real root.

The frequency–response curve, which is obtained from Eq. (7.19.46), is shown in Fig. 7.4. The spine curve equation is

$$\sigma = 3a^2 \tag{7.19.53}$$

Taking the positive and negative signs, respectively, in the frequency–response Eq. (7.19.46), we can obtain branch 1 and branch 2. When $\sigma \le 3a^2$, the points are on branch 1; and when $\sigma \ge 3a^2$, the points are on branch 2. When $\left(k^2/a^2\right) - \mu^2 = 0$, the two branches intersect at point A *on the* spine curve; at this point $\sigma = 3a^2$, so the horizontal coordinate of the intersection point A, σ_A, is

$$a_A^2 = \frac{k^2}{\mu^2},\ \sigma_A = 3a_A^2 = \frac{3k^2}{\mu^2} \tag{7.19.54}$$

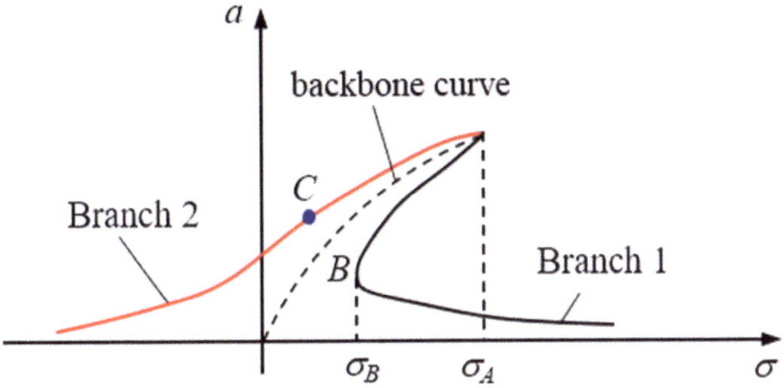

Fig. 7.4 Frequency–response curve for Exercise 7.19

Clearly, a_A is the maximum value of.a For branch 1, we have

$$\sigma = 3a^2 + \left(\frac{k^2}{a^2} - \mu^2\right)^{1/2} \tag{7.19.55}$$

From this, we have

$$a^2\left(\sigma - 3a^2\right)^2 = k^2 - \mu^2 a^2$$

Taking the derivative of the above equation with respect to a yields

$$\left(\sigma - 3a^2\right)\left(\sigma - 9a^2 + a\frac{d\sigma}{da}\right) = -\mu^2 \tag{7.19.56}$$

Let $d\sigma/da = 0$, the extreme point on branch 1 is

$$\sigma^2 - 12a^2\sigma + 27a^4 + \mu^2 = 0$$

therefore

$$\sigma = 6a^2 + 3a^2\sqrt{1 - \frac{\mu^2}{9a^4}} \tag{7.19.57}$$

It can be seen that the extreme point, B, on branch 1 is unique if it exists on branch 1. From (7.19.55), we know that as a decreases from a_A, $\sigma \to +\infty$ on branch 1; therefore, the value of point B can be the minimum. From (7.19.57), we know that as long as there exists an extreme point B with minimum value, the coordinates of point B satisfy $\sigma < 9a^2$. Therefore, the coordinates of each point of the AB segment on branch 1 satisfy $3a^2 < \sigma < 9a^2$.

From the above result, we have $\sigma_B < \sigma_A$, $a_B < a_A$ and there is only unique minimal point B on the segment of the curve AB. Therefore, on the segment of the curve AB (without points A and B), we have $d\sigma/da > 0$ and, from (7.19.56), we can obtain

$$\left(\sigma - 3a^2\right)\left(9a^2 - \sigma\right) - \mu^2 = a\left(\sigma - 3a^2\right)\frac{d\sigma}{da} > 0$$

i.e.,

$$\left(\sigma - 3a^2\right)\left(9a^2 - \sigma\right) > \mu^2 \tag{7.19.58}$$

Therefore, from the eigenvalue obtained by (7.19.52), we know that there must be a positive eigenvalue on the curve AB segment. Consequently, the perturbations of the steady state solutions a and ν on the curve AB segment diverge with the growth of time, i.e., the steady state solutions on the curve AB segment are unstable. It is already clear that points A and B are two points on the frequency–response curve, and their tangents are perpendicular to the σ axis.

From (7.19.50) and the non-trivial solution condition, we have

$$\begin{aligned} \lambda + \mu - a^2\sin\gamma &= 0 \\ \sigma - 2a^2 - a^2\cos\gamma &= 0 \end{aligned} \tag{7.19.59}$$

Eliminating γ from the above equation yields

$$(\lambda + \mu)^2 + \left(\sigma - 2a^2\right)^2 = a^4 \tag{7.19.60}$$

therefore

$$\lambda = -\mu \pm \left[\left(3a^2 - \sigma\right)\left(\sigma - a^2\right)\right]^{1/2} \tag{7.19.61}$$

It can be seen that when $\sigma \leq a^2$ or $\sigma \geq 3a^2$, λ is a negative real root or a complex root with a negative real part, the perturbation system (7.19.48) is stable. When $a^2 \leq \sigma \leq 3a^2$, λ is a real root; in this case, the steady state solution or the point representing the motion on the branch 2 or the upper half branch of the frequency–response curve, as shown in Fig. 7.4.

Therefore, the steady state solution on branch 2 might make $\lambda \geq 0$, which makes the perturbed system (7.19.48) unstable. However, from the result of Exercise 7.19(c), we know that the steady state solutions a and σ are inherently stable; therefore, it is only possible that the steady state solution $b = 0$ is unstable, i.e., the branching of the out-of-plane motion will diverge with the growth of time. Obviously, the steady state solutions a and σ, which makes the steady state solution $b = 0$ unstable, can only be on the CA segment of branch 2. The following is to determine the equation that should be satisfied by the coordinate of point C.

For branch 2 of the frequency–response curve, we have

$$\sigma = 3a^2 - \left(\frac{k^2}{a^2} - \mu^2\right)^{1/2} \tag{7.19.62}$$

The coordinate of point C makes $\lambda = 0$. Therefore, by (7.19.61), we have

$$-\mu + \left[\left(3a^2 - \sigma\right)\left(\sigma - a^2\right)\right]^{1/2} = 0 \tag{7.19.63}$$

Eliminating σ from (7.19.62) and (7.19.63) yields

$$4\mu^2 a^8 - 4k^2 a^6 + k^4 = 0 \tag{7.19.64}$$

This is the equation that the coordinates a_C of point C should satisfy.

When,$\mu = 0$ from (7.19.64), we can obtain $a_C = \left(\frac{1}{2}k\right)^{1/3}$. When $\mu \neq 0$ but μ/k is very small, we let a_C be

$$a_C = \left(\frac{1}{2}k\right)^{1/3} + c\mu^2 + \cdots \tag{7.19.65}$$

Substituting (7.19.65) into (7.19.64) and retaining to $O(\mu^2)$, we obtain

$$\begin{aligned} 0 &= 4\mu^2 a^8 - 4k^2 a^6 + k^4 \\ &= 4\mu^2\left(\frac{1}{2}k\right)^{8/3} - 4k^2\left[\left(\frac{1}{2}k\right)^2 + 6\left(\frac{1}{2}k^{5/3}\right)c\mu^2\right] + k^4 + O\left(\mu^3\right) \\ &= 4\left(\frac{1}{2}k\right)^{5/3}\left[\frac{1}{2}k - 6k^2 c\right]\mu^2 + O\left(\mu^3\right) \end{aligned} \tag{7.19.66}$$

Let the coefficient of μ^2 be zero, we obtain

$$c = \frac{1}{12k} \tag{7.19.67}$$

therefore

$$a_A = \left(\frac{1}{2}k\right)^{1/3} + \frac{\mu^2}{12k} + O\left(\mu^3\right) \tag{7.19.68}$$

From (7.19.68), we know that when the out-of-plane motion is unstable, damping makes the amplitude of the out-of-plane motion larger.

7.20 Exercise 7.20 (Oscillation Analysis of a Relief Valve with Boundary Nonlinearities)

Solution: The dynamic behavior of a relief valve which is used to protect a fluid system from overpressure, held against a seat by a helical spring and excited by a sinusoidal motion, is considered. The valve system under consideration is shown in Fig. 7.5(a). It consists of a valve having a mass, *m*, resting on a seat having nonlinear spring characteristics and retained by a helical spring which is considered as a continuous system. The mechanical model of this relief valve system is shown in Fig. 7.5(b). The linear spring is considered as a distributed parameter system whose motion is governed by the longitudinal wave equation and its boundary conditions, and the left end of the spring is fixed. Assume that the total mass of the linear spring is M, the length is L, and the stiffness coefficient is K. Let the fixed end of the linear spring be the origin of the coordinates; the displacement of the micro-segment at its coordinates x at the instant of t is denoted by.u The ball valve, with mass m, connects with the linear spring at the coordinate $x = L + u(L, t)$. The seat restricts the ball movement and is described by a massless nonlinear spring.

Here we analyze the longitudinal oscillation of the linear spring, which is regarded as a uniform continuum (or uniform elastic rod). Its governing differential equations and boundary conditions are:

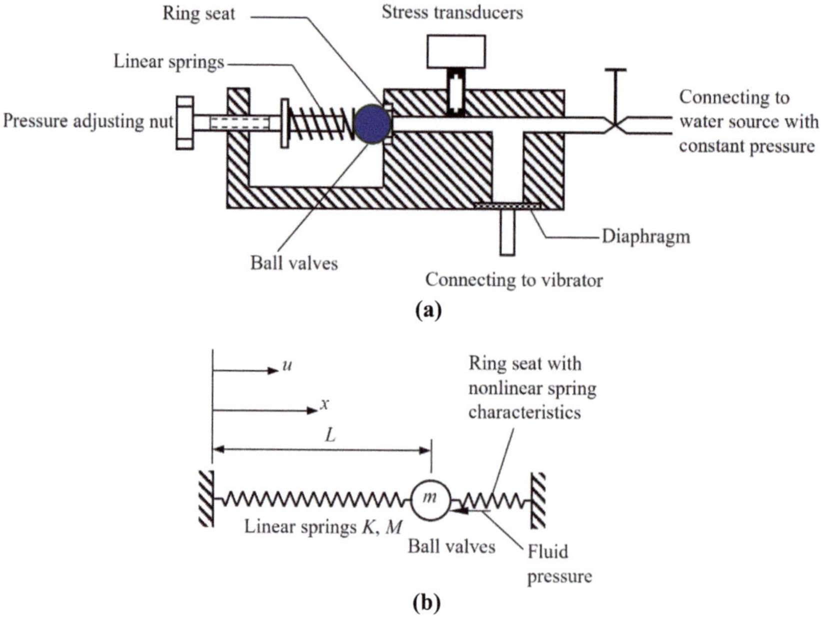

Fig. 7.5 **a** Schematic diagram of a relief valve system; **b** mechanical model of a relief valve for Exercise 7.20

$$\frac{\partial^2 u}{\partial t^2} = \frac{KL^2}{M}\frac{\partial^2 u}{\partial x^2}\frac{\partial^2 u}{\partial t^2} = \frac{KL^2}{M}\frac{\partial^2 u}{\partial x^2} \tag{7.20.1}$$

$$u = 0 \text{ at } x = 0 \tag{7.20.2}$$

$$m\frac{\partial^2 u}{\partial t^2} + KL\frac{\partial u}{\partial x} + \alpha u + \beta u^3 = -\left(\hat{F}_0 + \hat{F}_1\cos\Omega t\right) \text{ at } x = L \tag{7.20.3}$$

where α,β are the elasticity coefficients of the nonlinear ring seat,$\hat{F}_0$,$\hat{F}_1$ are the static and dynamic pressure amplitudes of the fluid, respectively. The following dimensionless variables are introduced

$$u^* = \frac{u}{L},\ t^* = \sqrt{\frac{K}{M}}t,\ x^* = \frac{x}{L},\ \Omega^* = \sqrt{\frac{M}{K}}\Omega \tag{7.20.4}$$

Substituting (7.20.4) into (7.20.1)–(7.20.3) and removing the asterisk from the variable in the result for brevity yields

$$\begin{aligned} &u_{tt} = u_{xx} \\ &u = 0, && \text{at}\, x = 0 \\ &u_{tt} + \alpha_1 u_x + \alpha_2 u + \alpha_3 u^3 = F_0 + F_1\cos\Omega t,\ \text{at}\ \ x = 1 \end{aligned} \tag{7.20.5}$$

where

$$\alpha_1 = \frac{M}{m},\ \alpha_2 = \frac{M\alpha}{mK},\ \alpha_3 = \frac{\beta L^2 M}{mK},\ F_0 = -\frac{\hat{F}_0 M}{mKL},\ F_1 = -\frac{\hat{F}_1 M}{mKL} \tag{7.20.6}$$

Let $F_1 = 0$ and all terms with time derivatives are zeros in (7.20.5), it becomes:

$$\begin{aligned} &u_{xx} = 0 \\ &u = 0, && \text{at}\ \ x = 0 \\ &\alpha_1 u_x + \alpha_2 u + \alpha_3 u^3 = F_0,\ \text{at}\ \ x = 1 \end{aligned} \tag{7.20.7}$$

The solution to (7.20.7) is

$$u = bx \tag{7.20.8}$$

where

$$\alpha_1 b + \alpha_2 b + \alpha_3 b^3 = F_0 \tag{7.20.9}$$

To determine the nonlinear oscillation of the system near the static equilibrium position, let

$$u(x, t) = bx + v(x, t) \tag{7.20.10}$$

Substituting (7.20.10) into (7.20.5) and applying (7.20.9) yields

$$\begin{aligned} &v_{tt} = v_{xx} \\ &v = 0, \quad \text{at } x = 0 \\ &v_{tt} + \alpha_1 v_x + \left(\alpha_2 + 3b^2\alpha_3\right)v + 3b\alpha_3 v^2 + \alpha_3 v^3 = F_1\cos\Omega t, \quad \text{at } x = 1 \end{aligned} \tag{7.20.11}$$

The linearized equation of (7.20.11) is

$$\begin{aligned} &v_{tt} = v_{xx} \\ &v = 0, \quad \text{at } x = 0 \\ &v_{tt} + \alpha_1 v_x + \left(\alpha_2 + 3b^2\alpha_3\right)v = F_1\cos\Omega t, \quad \text{at } x = 1 \end{aligned} \tag{7.20.12}$$

The solution of (7.20.12) can be represented by adding the forced oscillation solution v_1 and the free oscillation solution v_2, i.e.,

$$v(x, t) = v_1(x, t) + v_2(x, t) \tag{7.20.13}$$

where v_1 and v_2 satisfy the following equations, respectively:

$$\begin{aligned} &v_{1tt} = v_{1xx} \\ &v_1 = 0, \quad \text{at } x = 0 \\ &v_{1tt} + \alpha_1 v_{1x} + \left(\alpha_2 + 3b^2\alpha_3\right)v_1 = F_1\cos\Omega t, \quad \text{at } x = 1 \end{aligned} \tag{7.20.14}$$

$$\begin{aligned} &v_{2tt} = v_{2xx} \\ &v_2 = 0, \quad \text{at } x = 0 \\ &v_{2tt} + \alpha_1 v_{2x} + \left(\alpha_2 + 3b^2\alpha_3\right)v_2 = 0, \quad \text{at } x = 1 \end{aligned} \tag{7.20.15}$$

Let the solution to Eq. (7.20.14) be

$$v_1 = g(x)\cos\Omega t \tag{7.20.16}$$

Substituting this into (7.20.14), we obtain

$$\begin{aligned} &-\Omega^2 g = g_{xx} \\ &g = 0, \quad \text{at } x = 0 \\ &-\Omega^2 g + \alpha_1 g_x + \left(\alpha_2 + 3b^2\alpha_3\right)g = F_1 \quad \text{at } x = 1 \end{aligned} \tag{7.20.17}$$

The generalized solution of this equation is

$$g(x) = B\sin\Omega x + C\cos\Omega x \tag{7.20.18}$$

Substituting (7.20.18) into the boundary conditions, we obtain

$$\begin{aligned} & C = 0 \\ & -\Omega^2 B\sin\Omega + \alpha_1 \Omega B\cos\Omega + \left(\alpha_2 + 3b^2\alpha_3\right)B\sin\Omega = F_1 \end{aligned} \tag{7.20.19}$$

therefore

$$B \triangleq \Lambda = F_1\left[\alpha_1\Omega\cos\Omega + \left(\alpha_2 + 3b^2\alpha_3 - \Omega^2\right)\sin\Omega\right]^{-1} \tag{7.20.20}$$

$$v_1 = \Omega\sin\Omega x\cos\Omega t \tag{7.20.21}$$

Let the solution to Eq. (7.20.15) be

$$v_2 = \psi(x)e^{i\omega t} + cc \tag{7.20.22}$$

Substituting this into (7.20.15), we get

$$\begin{aligned} & \psi_{xx} = -\omega^2\psi \\ & \psi = 0, \quad \text{at } x = 0 \\ & -\omega^2\psi + \alpha_1\psi_x + \left(\alpha_2 + 3b^2\alpha_3\right)\psi = 0, \ \text{at } x = 1 \end{aligned} \tag{7.20.23}$$

The generalized solution of this equation is

$$\psi(x) = A\sin\omega x + D\cos\omega x \tag{7.20.24}$$

Substituting (7.20.24) into the boundary conditions, we get

$$\begin{aligned} & D = 0 \\ & \left[-\omega^2\sin\omega + \alpha_1\omega\cos\omega + \left(\alpha_2 + 3b^2\alpha_3\right)\sin\omega\right]A = 0 \end{aligned} \tag{7.20.25}$$

From the non-trivial solution condition, we can obtain the characteristic equation:

$$\left(\alpha_2 + 3b^2\alpha_3 - \omega^2\right)\tan\omega + \alpha_1\omega = 0 \tag{7.20.26}$$

From this, we can obtain natural frequencies ω_m and the corresponding mode shape functions

$$\psi_m(x) = A_m\sin\omega_m x \tag{7.20.27}$$

Substitute (7.20.27) into (7.20.22) and let $A_m = \frac{1}{2}a_m e^{i\beta_m}$, we get

$$v_2 = \sum_{m=1}^{\infty} A_m\sin\omega_m x\cos(\omega_m t + \beta_m) \tag{7.20.28}$$

Substituting (7.20.28) and (7.20.21) into (7.20.13), we obtain the linearized Eq. (7.20.12). Let the solution to the linearized equation be

$$v = \sum_{m=1}^{\infty} a_m \sin\omega_m x \cos(\omega_m t + \beta_m) + \Lambda \sin\Omega x \cos\Omega t \tag{7.20.29}$$

When $\Omega = \omega_n$, we can obtain $\Lambda \to \infty$ by (7.20.20) and (7.20.16); therefore, the primary resonance will occur when $\Omega \approx \omega_n$. The orthogonality condition for the mode shape function is given by (7.20.23), i.e.,

$$\begin{aligned} \omega_n^2 \psi_n &= -\frac{d^2\psi_n}{dx^2} \\ \omega_m^2 \psi_m &= -\frac{d^2\psi_m}{dx^2} \end{aligned} \tag{7.20.30}$$

Boundary conditions:

$$\begin{aligned} &\psi_k(0) = 0 \\ -\omega_k^2 \psi_k(1) + \alpha_1 &\tfrac{d\psi_k(1)}{dx} + (\alpha_2 + 3b^2\alpha_3)\psi_k(1) = 0 \end{aligned} \tag{7.20.31}$$

The two equations of (7.20.30) are multiplied by $\psi_m(x)$ and $\psi_n(x)$, and then integrated over [0, 1]. By integration by parts, we obtain

$$\omega_n^2 \int_0^1 \psi_n \psi_m dx = -\int_0^1 \psi_m \frac{d^2\psi_n}{dx^2} dx = -\left[\psi_m \frac{d\psi_n}{dx}\right]_0^1 + \int_0^1 \frac{d\psi_n}{dx}\frac{d\psi_m}{dx} dx \tag{7.20.32}$$

$$\omega_m^2 \int_0^1 \psi_n \psi_m dx = -\int_0^1 \psi_n \frac{d^2\psi_m}{dx^2} dx = -\left[\psi_n \frac{d\psi_m}{dx}\right]_0^1 + \int_0^1 \frac{d\psi_n}{dx}\frac{d\psi_m}{dx} dx \tag{7.20.33}$$

Subtracting these two equations and taking into account the boundary conditions yields

$$\begin{aligned} &\left(\omega_n^2 - \omega_m^2\right) \int_0^1 \psi_n \psi_m dx = -\psi_m(1) \tfrac{d\psi_n(1)}{dx} + \psi_n(1) \tfrac{d\psi_m(1)}{dx} \\ &= -\alpha_1^{-1}\left(\omega_n^2 - \omega_m^2\right)\psi_n(1)\psi_m(1) \end{aligned}$$

Thus, when $n \neq m, \omega_n \neq \omega_m$, there are

$$\alpha_1 \int_0^1 \psi_n \psi_m dx = -\psi_n(1)\psi_m(1) \tag{7.20.34}$$

When $n \neq m$ and $\omega_n \neq \omega_m$, by (7.20.32) and the boundary conditions, we have

$$\int_0^1 \frac{d\psi_n}{dx}\frac{d\psi_m}{dx}dx = -\alpha_1^{-1}\left(\alpha_2 + 3b^2\alpha_3\right)\psi_n(1)\psi_m(1) \tag{7.20.35}$$

Therefore, there is no regular orthogonality of the mode shape function of the system of this exercise. When $n = m$ is used by (7.20.31) and (7.20.32), we can obtain

$$\omega_n^2\left[\alpha_1\int_0^1 \psi_n^2(x)dx + \psi_n^2(1)\right] = \alpha_1\int_0^1\left[\frac{d\psi_n(x)}{dx}\right]^2 dx + \left(\alpha_2 + 3b^2\alpha_3\right)\psi_n^2(1) \tag{7.20.36}$$

To obtain the governing equation in discrete, modal coordinate form, let

$$v(x,t) = \sum_{n=1}^{\infty}\eta_n(t)\psi_n(x), \quad \psi_n(x) = \sin\omega_n x \tag{7.20.37}$$

and use the Lagrange equation to obtain the governing equation for.η_n

The kinetic energy of the system is

$$\begin{aligned} T &= \tfrac{1}{2}\int_0^L \tfrac{M}{L}\dot{u}^2dx + \tfrac{1}{2}m[\dot{u}(L)]^2 \\ &= \tfrac{1}{2}\int_0^1 \tfrac{M}{L}\left(L\tfrac{\partial v}{\partial t^*}\tfrac{dt^*}{dt}\right)^2 Ldx^* + \tfrac{1}{2}m\left[L\tfrac{\partial v(1)}{\partial t^*}\tfrac{dt^*}{dt}\right]^2 \\ &= \tfrac{1}{2}KL^2\int_0^1\left(\tfrac{\partial v}{\partial t}\right)^2dx + \tfrac{1}{2}\tfrac{mKL^2}{M}\left[\tfrac{\partial v(1)}{\partial t}\right]^2 \end{aligned} \tag{7.20.38}$$

Equation (7.20.4) is adopted to obtain the above equation. In addition, we remove the asterisk from the variable after the third equal sign.

The elastic potential energy of the spring and the potential energy of the fluid with constant pressure are

$$\begin{aligned}
\Pi &= \frac{1}{2}\int_0^L KL\left(\frac{\partial u}{\partial x}\right)^2 dx + \frac{1}{2}\alpha u^2(L) + \frac{1}{4}\beta u^4(L) + \hat{F}_0 u(L) \\
&= \frac{1}{2}KL^2\int_0^1 \left(\frac{\partial u^*}{\partial x^*}\right)^2 dx^* + \frac{1}{2}\alpha\cdot\left[Lu^*(1)\right]^2 + \frac{1}{4}\beta\cdot\left[Lu^*(1)\right]^4 + L\hat{F}_0 u^*(1) \\
&= \frac{1}{2}KL^2\int_0^1 \left(\frac{\partial u}{\partial x}\right)^2 dx + \frac{1}{2}\alpha L^2 u^2(1) + \frac{1}{4}\beta L^4 u^4(1) + L\hat{F}_0 u(1) \\
&= \frac{1}{2}KL^2\left[b^2 + 2bv(1)\right] + \frac{1}{2}KL^2\int_0^1 \left(\frac{\partial v}{\partial x}\right)^2 dx + \frac{1}{2}\alpha L^2\left[b^2 + 2bv(1)v^2(1)\right] \\
&+\frac{1}{4}\beta L^4\left[b^4 + 4b^3 v(1) + 6b^2 v^2(1) + 4bv^3(1) + v^4(1)\right] + L\hat{F}_0[b + v(1)] \\
&= \frac{1}{2}KL^2\int_0^1 \left(\frac{\partial v}{\partial x}\right)^2 dx + \frac{1}{2}L^2\left(\alpha + 3\beta L^2 b^2\right)v^2(1) + \beta L^4 bv^3(1) + \frac{1}{4}\beta L^4 v^4(1)
\end{aligned} \tag{7.20.39}$$

Equation (7.20.9) is adopted to obtain the above equation. In addition, we remove the constant part of the potential energy and the asterisk from the variable after the third equal sign.

Thus, the Lagrange function $\mathcal{L}$ is

$$\begin{aligned}
\mathcal{L} &= T - \Pi \\
&= \frac{1}{2}\sum_{p,q=1}^{\infty}\left[KL^2\int_0^1 \psi_p(x)\psi_q(x)dx + \frac{mKL^2}{M}\psi_p(1)\psi_q(1)\right]\dot{\eta}_p\dot{\eta}_q \\
&-\frac{1}{2}\sum_{p,q=1}^{\infty}\left[KL^2\int_0^1 \frac{d\psi_p}{dx}\frac{d\psi_q}{dx}dx + L^2\left(\alpha + 3\beta L^2 b^2\right)\psi_p(1)\psi_q(1)\right]\eta_p\eta_q \\
&-b\beta L^4\sum_{p,q,r=1}^{\infty}\psi_p(1)\psi_q(1)\psi_r(1)\eta_p\eta_q\eta_r \\
&-\frac{1}{4}\beta L^4\sum_{p,\,q,\,r,\,s=1}^{\infty}\psi_p(1)\psi_q(1)\psi_r(1)\psi_s(1)\eta_p\eta_q\eta_r\eta_s
\end{aligned} \tag{7.20.40}$$

The virtual work done by the dynamic pressure of the fluid is

$$\begin{aligned}
\delta W &= \left(-\hat{F}_1\cos\Omega t\right)\delta u(1) = \left(-L\hat{F}_1\cos\Omega^* t^*\right)\delta v^*(1) \\
&= \left(-L\hat{F}_1\cos\Omega t\right)\delta v(1) = \left(-L\hat{F}_1\cos\Omega t\right)\sum_{p=1}^{\infty}\psi_p(1)\delta\eta_p
\end{aligned} \tag{7.20.41}$$

Therefore, the corresponding generalized force is

$$Q_p = -L\hat{F}_1\psi_p(1)\cos\Omega t \tag{7.20.42}$$

From the Lagrange equation, we have

$$\frac{d}{dt}\frac{\partial \mathcal{L}}{\partial \dot{\eta}_n} - \frac{\partial \mathcal{L}}{\partial \eta_n} = Q_n, \ n = 1, 2, \ldots \tag{7.20.43}$$

Substituting the Lagrange function and the generalized force into (7.20.43) and taking into account (7.20.6), we obtain

$$\begin{aligned}
&\sum_{p=1}^{\infty}\left[\psi_p(1)\psi_n(1) + \alpha_1 \int_0^1 \psi_p(x)\psi_n(x)dx\right]\ddot{\eta}_p \\
&\quad + \sum_{p=1}^{\infty}\left[\left(\alpha_2 + 3\alpha_3 b^2\right)\psi_p(1)\psi_n(1) + \alpha_1 \int_0^1 \frac{d\psi_p}{dx}\frac{d\psi_n}{dx}dx\right]\eta_p \\
&\quad + 3b\alpha_3 \sum_{p,q=1}^{\infty} \psi_p(1)\psi_q(1)\psi_n(1)\eta_p\eta_q \\
&\quad + \alpha_3 \sum_{p,q,r=1}^{\infty} \psi_p(1)\psi_q(1)\psi_r(1)\psi_n(1)\eta_p\eta_q\eta_r = \psi_n(1)F_1 \cos \Omega t,
\end{aligned} \tag{7.20.44}$$

Taking into account (7.20.34), (7.20.35) and (7.20.36) and adding linear damping artificially, we can write this set of equations as

$$\ddot{\eta}_n + \omega_n^2\eta_n + 2\mu_n\dot{\eta}_n + \sum_{p,\ q=1}^{\infty} G_{npq}\eta_p\eta_q + \sum_{p,\ q,\ r=1}^{\infty} H_{npqr}\eta_p\eta_q\eta_r = K_n\cos\Omega t \tag{7.20.45}$$

where

$$\begin{aligned}
G_{npq} &= 3b\alpha_3\Gamma_n\psi_p(1)\psi_q(1), \qquad H_{npqr} = \alpha_3\Gamma_n\psi_p(1)\psi_q(1)\psi_r(1) \\
K_n &= \Gamma_n F_1, \ \Gamma_n = \psi_n(1)\Big/\left[\psi_n^2(1) + \alpha_1 \int_0^1 \psi_n^2(x)dx\right]
\end{aligned} \tag{7.20.46}$$

Noted that (7.20.44) can also be obtained from the governing differential equation, (7.20.11), of the system directly. Substituting (7.20.37) into the boundary conditions of $x = 1$ of (7.20.11), we obtain

$$\begin{aligned}
&\sum_{p=1}^{\infty} \psi_p(1)\ddot{\eta}_p + \alpha_1 \sum_{p=1}^{\infty} \tfrac{d\psi_p(1)}{dx}\eta_p + \left(\alpha_2 + 3b^2\alpha_3\right) \sum_{p=1}^{\infty} \psi_p(1)\eta_p \\
&\quad + 3b\alpha_3 \sum_{p,\ q=1}^{\infty} \psi_p(1)\psi_q(1)\eta_p\eta_q \\
&\quad + \alpha_3 \sum_{p,\ q,\ r=1}^{\infty} \psi_p(1)\psi_q(1)\psi_r(1)\eta_p\eta_q\eta_r = F_1\cos\Omega t
\end{aligned} \tag{7.20.47}$$

Substituting (7.20.37) into the first equation of (7.20.11), multiplying both sides by $\psi_n(x)$ and then integrating over [0, 1] yields

$$\sum_{p=1}^{\infty} \ddot{\eta}_p \int_0^1 \psi_n(x)\psi_p(x)dx = \sum_{p=1}^{\infty} \eta_p \int_0^1 \psi_n(x)\frac{d^2\psi_p(x)}{d^2x}dx$$

Utilizing the method of integration by parts and the boundary conditions at $x = 0$ yields

$$\sum_{p=1}^{\infty} \eta_p \psi_n(1)\frac{d\psi_p(1)}{dx} = \sum_{p=1}^{\infty} \ddot{\eta}_p \int_0^1 \psi_n(x)\psi_p(x)dx + \sum_{p=1}^{\infty} \eta_p \int_0^1 \frac{d\psi_n}{dx}\frac{d\psi_p}{dx}dx \tag{7.20.48}$$

Multiplying Eq. (7.20.47) by $\psi_n(1)$ and then substituting (7.20.48) into (7.20.47), we can obtain the Eq. (7.20.44).

In the following, we find the primary resonance solution of (7.20.45) when $\Omega \approx \omega_1$. Let

$$\Omega = \omega_1 + \varepsilon^2\sigma \tag{7.20.49}$$

$$\eta_n(t;\varepsilon) = \varepsilon\eta_{n0}(T_0, T_1, T_2) + \varepsilon^2\eta_{n1}(T_0, T_1, T_2) + \varepsilon^3\eta_{n2}(T_0, T_1, T_2) + \cdots \tag{7.20.50}$$

In order to make the nonlinear term appear in the third order equation along with the damping term and the excitation term, we need

$$\mu_n = 2\varepsilon^2\hat{\mu}_n,\ K_n = \begin{cases} 2\varepsilon^3 k_1,\ n = 1 \\ 2\varepsilon k_n,\ n \neq 1 \end{cases} \tag{7.20.51}$$

Substituting (7.20.50) and (7.20.51) into (7.20.45) and retaining to $O(\varepsilon^3)$, we obtain

$$
\begin{aligned}
0 &= \ddot{\eta}_n + \omega_n^2\eta_n + 2\hat{\mu}_n\dot{\eta}_n + \sum_{p,q=1}^{\infty} G_{npq}\eta_p\eta_q + \sum_{p,q,r=1}^{\infty} H_{npqr}\eta_p\eta_q\eta_r - K_n\cos\Omega t \\
&= \left[D_0^2 + 2\varepsilon D_0D_1 + \varepsilon^2\left(D_1^2 + 2D_0D_2\right)\right]\left(\varepsilon\eta_{n0} + \varepsilon^2\eta_{n1} + \varepsilon^3\eta_{n2}\right) \\
&\quad +\omega_n^2\left(\varepsilon\eta_{n0} + \varepsilon^2\eta_{n1} + \varepsilon^3\eta_{n2}\right) \\
&\quad +2\varepsilon^2\hat{\mu}_n\left(D_0 + \varepsilon D_1 + \varepsilon^2 D_2\right)\left(\varepsilon\eta_{n0} + \varepsilon^2\eta_{n1} + \varepsilon^3\eta_{n2}\right) \\
&\quad + \sum_{p,q=1}^{\infty} G_{npq}\left(\varepsilon\eta_{p0} + \varepsilon^2\eta_{p1} + \varepsilon^3\eta_{p2}\right)\left(\varepsilon\eta_{q0} + \varepsilon^2\eta_{q1} + \varepsilon^3\eta_{q2}\right) \\
&\quad +\varepsilon^3\sum_{p,q,r=1}^{\infty} H_{npqr}\eta_{p0}\eta_{q0}\eta_{r0} - 2\varepsilon^3 k_n\cos\Omega T_0 \\
&= \varepsilon D_0^2\eta_{n0} + \varepsilon\omega_n^2\eta_{n0} + \varepsilon^2\left[D_0^2\eta_{n1} + \omega_n^2\eta_{n1} + 2D_0D_1\eta_{n0} + \sum_{p,q=1}^{\infty} G_{npq}\eta_{p0}\eta_{q0}\right] \\
&\quad \left. + \sum_{p,q=1}^{\infty} G_{npq}\left(\eta_{p0}\eta_{q1} + \eta_{p1}\eta_{q0}\right) + \sum_{p,q,r=1}^{\infty} H_{npqr}\eta_{p0}\eta_{q0}\eta_{r0}\right] - K_n\cos\Omega T_0
\end{aligned}
\tag{7.20.52}
$$

Let the coefficient of the same power of ε in (7.20.52) be zero, we get

$$
D_0^2\eta_{10} + \omega_1^2\eta_{10} = 0 \tag{7.20.53}
$$

$$
D_0^2\eta_{n0} + \omega_n^2\eta_{n0} = 2k_n\cos\Omega T_0,\ n \neq 1 \tag{7.20.54}
$$

$$
D_0^2\eta_{n1} + \omega_n^2\eta_{n1} = -2D_0D_1\eta_{n0} - \sum_{p,\,q=1}^{\infty} G_{npq}\eta_{p0}\eta_{q0} \tag{7.20.55}
$$

$$
\begin{aligned}
&D_0^2\eta_{12} + \omega_1^2\eta_{12} = -\left(D_1^2 + 2D_0D_2\right)\eta_{10} - 2\hat{\mu}_1 D_0\eta_{10} - 2D_0D_1\eta_{11} \\
&- \sum_{p,\,q=1}^{\infty} G_{1pq}\left(\eta_{p0}\eta_{q1} + \eta_{p1}\eta_{q0}\right) - \sum_{p,\,q,\,r=1}^{\infty} H_{1pqr}\eta_{p0}\eta_{q0}\eta_{r0} \\
&+2k_1\cos\Omega T_0
\end{aligned}
\tag{7.20.56}
$$

$$
\begin{aligned}
&D_0^2\eta_{n2} + \omega_n^2\eta_{n2} = -\left(D_1^2 + 2D_0D_2\right)\eta_{n0} - 2\hat{\mu}_n D_0\eta_{n0} - 2D_0D_1\eta_{n1} \\
&\quad - \sum_{p,\,q=1}^{\infty} G_{npq}\left(\eta_{p0}\eta_{q1} + \eta_{p1}\eta_{q0}\right) - \sum_{p,\,q,\,r=1}^{\infty} H_{npqr}\eta_{p0}\eta_{q0}\eta_{r0},\ n \neq 1
\end{aligned}
\tag{7.20.57}
$$

The solutions to (7.20.53) and (7.20.54) are

$$
\eta_{n0} = A_n e^{i\omega_n T_0} + (1 - \delta_{1n})\hat{\Lambda}_n e^{i\Omega T_0} + cc \tag{7.20.58}
$$

where $A_n = A_n(T_1, T_2)$. Substituting (7.20.58) into (7.20.55), we get

$$\begin{aligned} D_0^2\eta_{n1} + \omega_n^2\eta_{n1} &= -2i\omega_n D_1 A_n e^{i\omega_n T_0} \\ &- \sum_{p,q=1}^{\infty} G_{npq}\Big[A_p A_q e^{i(\omega_p+\omega_q)T_0} + A_p\overline{A}_q e^{i(\omega_p-\omega_q)T_0} \\ &+(1-\delta_{1q})\hat{\Lambda}_q A_p e^{i(\omega_p+\Omega)T_0} + (1-\delta_{1q})\hat{\Lambda}_q A_p e^{i(\omega_p-\Omega)T_0} \\ &+(1-\delta_{1p})\hat{\Lambda}_p A_q e^{i(\Omega+\omega_q)T_0} + (1-\delta_{1p})\hat{\Lambda}_p \overline{A}_q e^{i(\Omega-\omega_q)T_0} \\ &+(1-\delta_{1p})(1-\delta_{1q})\hat{\Lambda}_p\hat{\Lambda}_q e^{2i\Omega T_0} \\ &+(1-\delta_{1p})(1-\delta_{1q})\hat{\Lambda}_p\hat{\Lambda}_q\Big] + cc \end{aligned} \tag{7.20.59}$$

In order to eliminate secular terms from the above equation, we need

$$D_1 A_n = 0 \;\Rightarrow\; A_n = A_n(T_2) \tag{7.20.60}$$

Further, by (7.20.59), we can obtain.

$$\begin{aligned} \eta_{n1} = &- \sum_{p,q=1}^{\infty} \Big[G_{npq}^{(1)} A_p A_q e^{i(\omega_p+\omega_q)T_0} + G_{npq}^{(2)} A_p\overline{A}_q e^{i(\omega_p-\omega_q)T_0} \\ &+G_{npq}^{(3)}(1-\delta_{1q})\hat{\Lambda}_q A_p e^{i(\omega_p+\Omega)T_0} + G_{npq}^{(4)}(1-\delta_{1q})\hat{\Lambda}_q A_p e^{i(\omega_p-\Omega)T_0} \\ &+G_{npq}^{(5)}(1-\delta_{1p})\hat{\Lambda}_p A_q e^{i(\Omega+\omega_q)T_0} + G_{npq}^{(6)}(1-\delta_{1p})\hat{\Lambda}_p \overline{A}_q e^{i(\Omega-\omega_q)T_0} \\ &+G_{npq}^{(7)}(1-\delta_{1p})(1-\delta_{1q})\hat{\Lambda}_p\hat{\Lambda}_q e^{2i\Omega T_0} \\ &+G_{npq}^{(8)}(1-\delta_{1p})(1-\delta_{1q})\hat{\Lambda}_p\hat{\Lambda}_q\Big] + cc \end{aligned} \tag{7.20.61}$$

where $G_{npq}^{(k)}$ is

$$\begin{aligned} G_{npq}^{(1)} &= \frac{G_{npq}}{\omega_n^2 - (\omega_p+\omega_q)^2}, \quad G_{npq}^{(2)} = \frac{G_{npq}}{\omega_n^2 - (\omega_p-\omega_q)^2} \\ G_{npq}^{(3)} &= \frac{G_{npq}}{\omega_n^2 - (\omega_p+\Omega)^2}, \quad G_{npq}^{(4)} = \frac{G_{npq}}{\omega_n^2 - (\omega_p-\Omega)^2} \\ G_{npq}^{(5)} &= \frac{G_{npq}}{\omega_n^2 - (\Omega+\omega_q)^2}, \quad G_{npq}^{(6)} = \frac{G_{npq}}{\omega_n^2 - (\Omega-\omega_q)^2} \\ G_{npq}^{(7)} &= \frac{G_{npq}}{\omega_n^2 - 4\Omega^2}, \quad G_{npq}^{(8)} = \frac{G_{npq}}{\omega_n^2} \end{aligned} \tag{7.20.62}$$

Substituting (7.20.58) and (7.20.60) into (7.20.57) yields

$$
\begin{aligned}
&D_0^2\eta_{n2} + \omega_n^2\eta_{n2} = -2D_0D_2\eta_{n0} - 2\hat{\mu}_n D_0\eta_{n0} \\
&- \sum_{p,q=1}^{\infty} G_{npq}\left(\eta_{p0}\eta_{q1} + \eta_{p1}\eta_{q0}\right) - \sum_{p,q,r=1}^{\infty} H_{npqr}\eta_{p0}\eta_{q0}\eta_{r0} \\
&= -2i\omega_n A_n' e^{i\omega_n T_0} - 2i\omega_n\hat{\mu}_n A_n e^{i\omega_n T_0} + ST \\
&+cc + NST, \quad n \neq 1
\end{aligned}
\tag{7.20.63}
$$

where the prime denotes the derivative with respect to T_2;ST denotes the resonance excitation term generated by $\left[-\sum_{p,q=1}^{\infty} G_{npq}\left(\eta_{p0}\eta_{q1} + \eta_{p1}\eta_{q0}\right) - \sum_{p,q,r=1}^{\infty} H_{npqr}\eta_{p0}\eta_{q0}\eta_{r0}\right]$. A closer inspection shows that after substituting $A_n = \frac{1}{2}a_n e^{i\beta_n}$ into ST and removing the factor $e^{i\beta_n}e^{i\omega_n T_0}$, the sum of the imaginary parts of all coefficients is zero. For example, the coefficients of the terms with non-zero imaginary parts in $\sum_{p,q=1}^{\infty} G_{npq}\eta_{p0}\eta_{q1}$ are:

$$
\begin{aligned}
&2\sum_{p,q=1}^{\infty}\Big[G_{npq}G_{qn1}^{(1)}\left(1-\delta_{p1}\right)\hat{\Lambda}_p A_n A_1 e^{-i\sigma T_2} + G_{n1q}G_{qnp}^{(3)}\left(1-\delta_{p1}\right)\hat{\Lambda}_p A_n \overline{A}_1 e^{i\sigma T_2} \\
&+G_{npq}G_{qn1}^{(2)}\left(1-\delta_{p1}\right)\hat{\Lambda}_p A_n \overline{A}_1 e^{i\sigma T_2} + G_{n1q}G_{qnp}^{(4)}\left(1-\delta_{p1}\right)\hat{\Lambda}_p A_n A_1 e^{-i\sigma T_2}\Big]e^{i\omega_n T_0} \\
&= \frac{1}{2}\sum_{p,q=1}^{\infty}\Big[G_{npq}G_{qn1}^{(1)}\left(1-\delta_{p1}\right)\hat{\Lambda}_p a_n a_1 e^{-(i\sigma T_2 - \beta_1)} \\
&+G_{n1q}G_{qnp}^{(3)}\left(1-\delta_{p1}\right)\hat{\Lambda}_p a_n a_1 e^{i\sigma T_2 - \beta_1} \\
&+G_{npq}G_{qn1}^{(2)}\left(1-\delta_{p1}\right)\hat{\Lambda}_p a_n a_1 e^{i\sigma T_2 - \beta_1} \\
&+G_{n1q}G_{qnp}^{(4)}\left(1-\delta_{p1}\right)\hat{\Lambda}_p a_n a_1 e^{-(i\sigma T_2 - \beta_1)}\Big]e^{i\beta_n}e^{i\omega_n T_0}
\end{aligned}
\tag{7.20.64}
$$

where

$$
\begin{aligned}
&G_{npq}G_{qn1}^{(1)} = \frac{G_{npq}G_{qn1}}{\omega_q^2 - (\omega_n + \omega_1)^2}, \quad G_{n1q}G_{qnp}^{(3)} = \frac{G_{n1q}G_{qnp}}{\omega_q^2 - (\omega_n + \Omega)^2} \\
&G_{npq}G_{qn1} = (3b\alpha_3)^2\Gamma_n\Gamma_q\psi_n(1)\psi_p(1)\psi_q(1)\psi_1(1) \\
&G_{n1q}G_{qnp} = (3b\alpha_3)^2\Gamma_n\Gamma_q\psi_n(1)\psi_q(1)\psi_p(1)\psi_1(1)
\end{aligned}
\tag{7.20.65}
$$

Apparently.

$$
G_{npq}G_{qn1}^{(1)} = G_{n1q}G_{qnp}^{(3)}
\tag{7.20.66}
$$

Therefore, the sum of the imaginary parts of the first and second terms on the right-hand side of (7.20.64) is zero. Similarly, the sum of the imaginary parts of the third and fourth terms on the right-hand side of (7.20.64). Thus, in order to eliminate secular terms in (7.20.63), we substitute $A_n = \frac{1}{2}a_n e^{i\beta_n}$ into the excitation term in (7.20.63), the sum of the imaginary parts of all the coefficient terms must be zero after removing the factor $e^{i\beta_n}e^{i\omega_n T_0}$, i.e.

$$a_n\prime + \hat{\mu}_n a_n = 0,\ n \neq 1 \tag{7.20.67}$$

or

$$a_n \to 0,\ n \neq 1 \ \text{ or } \ A_n \to 0,\ n \neq 1 \tag{7.20.68}$$

Then (7.20.53)–(7.20.57) degenerate to a unimodal (first order modal) equation, which can be rewritten as follows:

$$D_0^2\eta_{10} + \omega_1^2\eta_{10} = 0 \tag{7.20.69}$$

$$D_0^2\eta_{11} + \omega_1^2\eta_{11} = -2D_0D_1\eta_{10} - G_{111}\eta_{10}^2 \tag{7.20.70}$$

$$\begin{aligned} &D_0^2\eta_{12} + \omega_1^2\eta_{12} = -2D_0D_2\eta_{10} - 2\hat{\mu}_1 D_0\eta_{10} - 2G_{111}\eta_{10}\eta_{11} \\ &-H_{1111}\eta_{10}^3 + 2k_1\cos\Omega T_0 \end{aligned} \tag{7.20.71}$$

The solution of (7.20.69) is

$$\eta_{10} = A_1 e^{i\omega_1 T_0} + cc \tag{7.20.72}$$

Substituting (7.20.72) into (7.20.70) yields

$$D_0^2\eta_{11} + \omega_1^2\eta_{11} = -2i\omega_1 D_1 A_1 e^{i\omega_1 T_0} - G_{111}A_1^2 e^{2i\omega_1 T_0} - A_1\overline{A}_1 + cc \tag{7.20.73}$$

In order to eliminate secular terms from the above equation, we need

$$D_1 A_1 = 0 \Rightarrow A_1 = A_1(T_2) \tag{7.20.74}$$

Further, by (7.20.73), we can obtain

$$\eta_{11} = -\omega_1^{-2}A_1\overline{A}_1 + \frac{1}{3}\omega_1^{-2}G_{111}A_1^2 e^{2i\omega_1 T_0} + cc \tag{7.20.75}$$

Substituting (7.20.72) and (7.20.74) into (7.20.71) yields

$$\begin{aligned} &D_0^2\eta_{12} + \omega_1^2\eta_{12} = -2D_0D_2\eta_{10} - 2\hat{\mu}_1 D_0\eta_{10} - 2G_{111}\eta_{10}\eta_{11} \\ &-H_{1111}\eta_{10}^3 + 2k_1\cos\Omega T_0 \\ &= -2i\omega_1 A_1\prime e^{i\omega_1 T_0} - 2i\omega_1\hat{\mu}_1 A_1 e^{i\omega_1 T_0} - P_1 A_1^2\overline{A}_1 e^{i\omega_1 T_0} \\ &+k_1 e^{i\sigma T_2} e^{i\omega_1 T_0} + cc + NST \end{aligned} \tag{7.20.76}$$

where

$$P_1 = -\left(4\omega_1^{-2}G_{111} - \frac{2}{3}\omega_1^{-2}G_{111}^2 - 3H_{1111}\right) \tag{7.20.77}$$

In order to remove secular terms from (7.20.76), we need

$$2i\omega_1 A_1' + 2i\omega_1\hat{\mu}_1 A_1 + P_1 A_1^2\overline{A}_1 - k_1 e^{i\sigma T_2} = 0 \tag{7.20.78}$$

Substituting $A_1 = \frac{1}{2}a_1 e^{i\beta_1}$ into (7.20.78) yields

$$i\omega_1 a_1\prime - \omega_1 a_1\beta_1\prime + i\omega_1\hat{\mu}_1 a_1 + \frac{1}{8}P_1 a_1^3 - k_1 e^{i(\sigma T_2 - \beta_1)} = 0 \tag{7.20.79}$$

Separating the real and imaginary parts of the above equation yields

$$\begin{aligned} \omega_1 a_1\prime &= -\omega_1\hat{\mu}_1 a_1 + k_1\sin\gamma_1 \\ \omega_1 a_1\gamma_1\prime &= \omega_1 a_1\sigma - \tfrac{1}{8}P_1 a_1^3 + k_1\cos\gamma_1 \end{aligned} \tag{7.20.80}$$

Therefore, the first order modal response of the system is

$$\begin{aligned} \eta_1 &= \varepsilon\eta_{10} + O(\varepsilon^2) = \varepsilon a_1\cos(\omega_1 T_0 + \beta_1) + O(\varepsilon^2) \\ &= \varepsilon a_1\cos[\Omega T_0 - (\sigma T_2 - \beta_1)] + O(\varepsilon^2) \\ &= \varepsilon a_1\cos(\Omega t - \gamma_1) + O(\varepsilon^2) \end{aligned} \tag{7.20.81}$$

From the above results, the primary resonance response of the system can be written as

$$\eta_n \approx \varepsilon\eta_{n0} + O(\varepsilon^2) = \begin{cases} a_1\cos(\Omega t - \gamma_1) + O(\varepsilon^2), & n = 1 \\ O(\varepsilon^2), & n \neq 1 \end{cases} \tag{7.20.82}$$

where εa_1 has been written as a_1;a_1 and γ_1 are controlled by (7.20.80). Let $a_1\prime = \gamma_1\prime = 0$ in (7.20.80), we can obtain the equation satisfied by the steady state solution

$$\begin{aligned} \omega_1\hat{\mu}_1 a_1 &= k_1\sin\gamma_1 \\ \omega_1 a_1\sigma - \tfrac{1}{8}P_1 a_1^3 &= -k_1\cos\gamma_1 \end{aligned} \tag{7.20.83}$$

Therefore, the frequency–response equation is

$$\omega_1^2\hat{\mu}_1^2 a_1^2 + \left(\omega_1\sigma - \frac{1}{8}P_1 a_1^2\right)^2 a_1^2 = k_1^2 \tag{7.20.84}$$

In order to solve (7.20.45) with subharmonic ($\Omega \approx 2\omega_1$ and $\Omega \approx \frac{1}{2}\omega_1$) and combined ($\Omega \approx \omega_1 + \omega_2$) resonance, we need to specify the excitation term as a non-resonant hard excitation, and consequently, the resonance excitation term will appear in the second order equation. Therefore, in order to let the nonlinear term appear in the second order equation along with the damping and the excitation term, we let

$$\mu_n = 2\varepsilon\hat{\mu}_n, \quad K_n = 2\varepsilon k_n \tag{7.20.85}$$

and the solution of (7.20.45) be

$$\eta_n(t;\varepsilon) = \varepsilon\eta_{n0}(T_0, T_1) + \varepsilon^2\eta_{n1}(T_0, T_1) + \cdots \tag{7.20.86}$$

Substituting (7.20.85) and (7.20.86) into (7.20.45) and retaining to $O(\varepsilon^2)$, we obtain

$$\ddot{\eta}_n + \omega_n^2\eta_n + 2\mu_n\dot{\eta}_n + \sum_{p,\,q=1}^{\infty} G_{npq}\eta_p\eta_q + \sum_{p,\,q,\,r=1}^{\infty} H_{npqr}\eta_p\eta_q\eta_r = K_n\cos\Omega t \tag{7.20.87}$$

Substituting (7.20.50) and (7.20.51) into (7.20.45) yields

$$\begin{aligned}
0 &= \ddot{\eta}_n + \omega_n^2\eta_n + 2a_n\dot{\eta}_n + \sum_{p,q=1}^{\infty} G_{npq}\eta_p\eta_q + \sum_{p,q,r=1}^{\infty} H_{npqq}\eta_p\eta_q\eta_r - K_n\cos\Omega t \\
&= \left(D_0^2 + 2\varepsilon D_0D_1\right)\left(\varepsilon\eta_{n0} + \varepsilon^2\eta_{n1}\right) + \omega_n^2\left(\varepsilon\eta_{n0} + \varepsilon^2\eta_{n1}\right) \\
&\quad 2\varepsilon\hat{\mu}_n(D_0 + \varepsilon D_1)\left(\varepsilon\eta_{n0} + \varepsilon^2\eta_{n1}\right) \\
&\quad + \sum_{p,q=1}^{\infty} G_{npq}\left(\varepsilon\eta_{p0} + \varepsilon^2\eta_{p1}\right)\left(\varepsilon\eta_{q0} + \varepsilon^2\eta_{q1}\right) - 2\varepsilon k_n\cos\Omega T_0 \\
&= \varepsilon\left(D_0^2\eta_{n0} + \omega_n^2\eta_{n0} - 2k_n\cos\Omega T_0\right) \\
&\quad + \varepsilon^2\left[D_0^2\eta_{n1} + \omega_n^2\eta_{n1} + 2D_0D_1\eta_{n0} + 2\hat{\mu}_nD_0\eta_{n0} + \sum_{p,q=1}^{\infty} G_{npq}\eta_{p0}\eta_{q0}\right]
\end{aligned} \tag{7.20.88}$$

Let the coefficient of the same power of ε in the Eq. (7.20.88) be zero, we obtain

$$D_0^2\eta_{n0} + \omega_n^2\eta_{n0} = 2k_n\cos\Omega T_0 \tag{7.20.89}$$

$$D_0^2\eta_{n1} + \omega_n^2\eta_{n1} = -2D_0D_1\eta_{n0} - 2\hat{\mu}_nD_0\eta_{n0} - \sum_{p,\,q=1}^{\infty} G_{npq}\eta_{p0}\eta_{q0} \tag{7.20.90}$$

The solution of (7.20.89) is

$$\eta_{n0} = A_n e^{i\omega_n T_0} + \hat{\Lambda}_n e^{i\Omega T_0} + cc \tag{7.20.91}$$

where $A_n = A_n(T_1)$. Substituting (7.20.91) into (7.20.90), we obtain

$$\begin{aligned}
&D_0^2\eta_{n1} + \omega_n^2\eta_{n1} = -2D_0D_1\eta_{n0} - 2\hat{\mu}_n D_0\eta_{n0} - \sum_{p,q=1}^{\infty} G_{npq}\eta_{p0}\eta_{q0} \\
&= -2i\omega_n D_1 A_n e^{i\omega_n T_0} - 2i\omega_n A_n A_n e^{i\omega_n T_0} \\
&- \sum_{p,q=1}^{\infty} G_{npq}\Big[A_p A_q e^{i(\omega_p+\omega_q)T_0} + A_p\overline{A}_q e^{i(\omega_p-\omega_q)T_0} \\
&+\hat{\Lambda}_q A_p e^{i(\omega_p+\Omega)T_0} + \hat{\Lambda}_q A_p e^{i(\omega_p-\Omega)T_0} \\
&+\hat{\Lambda}_p A_q e^{i(\Omega+\omega_q)T_0} + \hat{\Lambda}_p\overline{A}_q e^{i(\Omega-\omega_q)T_0} + \hat{\Lambda}_p\hat{\Lambda}_q e^{2i\Omega T_0}\Big] + cc
\end{aligned} \tag{7.20.92}$$

In order to eliminate secular terms from the above equation, we need to consider the values of Ω,ω_n and their combinations. It is clear that different resonance cases such as $\Omega \approx 2\omega_1$, $\Omega \approx \frac{1}{2}\omega_1$ and $\Omega \approx \omega_1 + \omega_2$ will occur. Readers are invited to choose one of these cases to complete the solution.

7.21 Exercise 7.21 (First-Order Subharmonic Resonance Analysis of a Uniform Circular Plate Clamped Along Its Edge)

Solution: The dimensionless governing equation for the symmetric responses of a uniform circular plate is (7.21.1) to (7.21.4) (Sect. 7.6 of **the Book**).

$$\frac{\partial^2 w}{\partial t^2} + \nabla^4 w = \varepsilon\left[\frac{1}{r}\frac{\partial}{\partial r}\left(\frac{\partial F}{\partial r}\frac{\partial w}{\partial r}\right) - 2\mu\frac{\partial w}{\partial t}\right] + f(r,t) \tag{7.21.1}$$

$$\nabla^4 F = -\frac{1}{2r}\frac{\partial}{\partial r}\left[\left(\frac{\partial w}{\partial r}\right)^2\right] \tag{7.21.2}$$

$$\frac{\partial u}{\partial r} + \frac{1}{2}\left(\frac{\partial w}{\partial r}\right)^2 = \frac{1}{r}\frac{\partial F}{\partial r} - \nu\frac{\partial^2 F}{\partial r^2} \tag{7.21.3}$$

$$\frac{u}{r} = \frac{\partial^2 F}{\partial r^2} - \frac{\nu}{r}\frac{\partial F}{\partial r} \tag{7.21.4}$$

where

$$\varepsilon = 12h^2/R^2$$

The boundary conditions for clamped edges are

$$w = 0, \quad u = 0, \quad \partial w/\partial r = 0 \text{ at } r = 1 \tag{7.21.5}$$

Equation (7.21.3) and (7.21.4) can be combined to yield the following equation for F:

$$\frac{1}{2}\left(\frac{\partial w}{\partial r}\right)^2 = \frac{1}{r}\frac{\partial F}{\partial r} - \frac{\partial^2 F}{\partial r^2} - r\frac{\partial^3 F}{\partial r^3} \tag{7.21.6}$$

Then it follows from (7.21.4) and (7.21.5) that the boundary conditions for F is

$$\frac{\partial^2 F}{\partial r^2} - \nu\frac{\partial F}{\partial r} = 0 \text{ at } r = 1 \tag{7.21.7}$$

In addition,w and F are required to be finite at $r = 0$.
Let the deflection of the circular plate be

$$w(r, t; \varepsilon) = \sum_{m=1}^{\infty} \psi_m(t; \varepsilon)\phi_m(r) \tag{7.21.8}$$

where the ϕ_m are the linear, free-oscillation modes. Thus the ϕ_m are the solutions of the following eigenvalue problem:

$$\nabla^4\phi_m - \omega_m^2\phi_m = 0 \tag{7.21.9}$$

$$\phi_m(1) = 0, \quad \phi_m{}'(1) = 0, \quad \phi_m(0) < \infty \tag{7.21.10}$$

The eigenvalues ω_m are the natural frequencies of the plate. The ϕ_m are orthogonal with respect to the weighting function r. The amplitude of each mode is chosen such that

$$\int_0^1 r\phi_n\phi_m dr = \delta_{nm} \tag{7.21.11}$$

Next we obtain the solution of the eigenvalue problem defined above. We rewrite (7.21.9) in the following convenient form:

$$\left(\frac{\partial^2}{\partial r^2} + \frac{1}{r}\frac{\partial}{\partial r} - \kappa_m^2\right)\left(\frac{\partial^2}{\partial r^2} + \frac{1}{r}\frac{\partial}{\partial r} + \kappa_m^2\right)\phi_m = 0 \tag{7.21.12}$$

where $\kappa_m^4 = \omega_m^2$. Thus we can obtain the four linearly independent solutions of (7.21.12) from the following two equations:

$$\left(\frac{\partial^2}{\partial r^2} + \frac{1}{r}\frac{\partial}{\partial r} - \kappa_m^2\right)\phi_m = 0 \tag{7.21.13}$$

$$\left(\frac{\partial^2}{\partial r^2} + \frac{1}{r}\frac{\partial}{\partial r} + \kappa_m^2\right)\phi_m = 0 \tag{7.21.14}$$

From (7.21.13) and (7.21.14), we obtain

$$\phi_m^{(1)} = E_1 I_0(\kappa_m r) + E_2 K_0(\kappa_m r) \tag{7.21.15}$$

$$\phi_m^{(2)} = E_3 J_0(\kappa_m r) + E_4 Y_0(\kappa_m r) \tag{7.21.16}$$

where the E_n are constants of integration; J_n,Y_n are the Bessel functions of the first and second kind, respectively;I_n,K_n are the modified Bessel functions of the first and second kind, respectively. The complete solution is

$$\phi_m = \phi_m^{(1)} + \phi_m^{(2)} \tag{7.21.17}$$

The condition that $\phi_m(0)$ be bounded demands that $E_2 = E_4 = 0$ because both K_0 and Y_0 have logarithmic singularities at the origin. Thus it follows from (7.21.10) that

$$\phi_m = C_m[J_0(\kappa_m r)I_0(\kappa_m) - J_0(\kappa_m)I_0(\kappa_m r)] \tag{7.21.18}$$

where the κ_m are the roots of

$$I_0(\kappa)J_0\prime(\kappa) - J_0(\kappa)I_0\prime(\kappa) = 0 \tag{7.21.19}$$

and the C_m are obtained from (7.21.11).

The first five natural frequencies obtained from (7.21.19) are

$$\omega_m = 10.2158, \quad 39.7710, \quad 89.1040, \quad 158.1830, \quad 247.0050$$

We note that

$$\omega_1 + 2\omega_2 = 89.7578 \approx \omega_3 \tag{7.21.20}$$

Introduce the detuning parameter σ_1 such that

$$\omega_1 + 2\omega_2 = \omega_3 + \varepsilon\sigma_1 \tag{7.21.21}$$

Hence there is an internal resonance involving three modes.

Equations (7.21.6) and (7.21.7) suggest that it may be more convenient to solve $\partial F/\partial r$ instead of F. Thus we let

$$G = \partial F / \partial r \tag{7.21.22}$$

Substituting (7.21.22) into (7.21.7), we obtain the following boundary condition for G:

$$\partial G / \partial r - \nu G = 0 \text{ at } r = 1 \tag{7.21.23}$$

Substituting (7.21.22) and (7.21.8) into (7.21.6), we obtain

$$r^2 \frac{\partial^2 G}{\partial r^2} + r \frac{\partial G}{\partial r} - G = -\frac{1}{2} r \left(\sum_{m=1}^{\infty} \psi_m \phi_{m'} \right)^2 \tag{7.21.24}$$

The function G can be represented by an expansion in terms of a complete set of orthogonal eigenfunctions. Because

$$\left(r^2 \frac{d^2}{dr^2} + r \frac{d}{dr} - 1 \right) J_1(\zeta_m r) = -\zeta_m^2 r^2 J_1(\zeta_m r) \tag{7.21.25}$$

it is convenient to express G as follows:

$$G(r, t) = \sum_{m=1}^{\infty} \eta_m(t) J_1(\zeta_m r) \tag{7.21.26}$$

where the ζ_m are the roots of the following equation

$$\zeta J_0(\zeta) - (1 + \nu) J_1(\zeta) = 0 \tag{7.21.27}$$

For $\nu = \frac{1}{3}$, the first 12 roots of (7.21.27) are:

$$\begin{aligned}
&\zeta_1 = 1.545, \quad \zeta_2 = 5.266, \quad \zeta_3 = 9.497, \quad \zeta_4 = 11.68 \\
&\zeta_5 = 14.84, \quad \zeta_6 = 18.00, \quad \zeta_7 = 21.15, \quad \zeta_8 = 24.30 \\
&\zeta_9 = 30.59, \quad \zeta_{10} = 33.74, \quad \zeta_{11} = 36.88, \quad \zeta_{12} = 40.03
\end{aligned}$$

To obtain the functions $\eta_m(t)$, we substitute (7.21.26) into (7.21.25), multiply by $r^{-1} J_1(\zeta_n r)$ and integrate from $r = 0$ to $r = 1$. The result is

$$\eta_n(t) = \sum_{p,q=1}^{\infty} S_{npq} \psi_p(t) \psi_q(t) \tag{7.21.28}$$

where

$$S_{npq} = \left[\left(\zeta_n^2 - 1 + \nu^2\right)J_1^2(\zeta_n)\right]^{-1} \int_0^1 \phi_q{}'\phi_p{}'J_1(\zeta_n r)dr \tag{7.21.29}$$

Using (7.21.28) we can now rewrite (7.21.26) in the following form:

$$G(r,t) = \sum_{m,n,p=1}^{\infty} S_{mnp}\psi_n(t)\psi_p(t)J_1(\zeta_m r) \tag{7.21.30}$$

To obtain the equations governing the ψ_n, we substitute (7.21.30) and (7.21.8) into (7.21.21), multiply by $r\phi_n$ from $r = 0$ to $r = 1$. The result is

$$\ddot{\psi}_n + \omega_n^2\psi_n = -\varepsilon\left(2\mu_n\dot{\psi}_n - \sum_{m,p,q=1}^{\infty} \Gamma_{nmpq}\psi_m\psi_p\psi_q\right) + f_n(t) \tag{7.21.31}$$

where

$$f_n(t) = \int_0^1 r\phi_n f(r,t)dr,\ \mu_n = \int_0^1 \mu r\phi_n dr \tag{7.21.32}$$

To obtain the expression for the Γ_{nmpq}, we consider

$$\int_0^1 \frac{\partial}{\partial r}\left(G\frac{\partial w}{\partial r}\right)\phi_n dr = G\phi_n\frac{\partial w}{\partial r}\Big|_0^1 - \int_0^1 G\phi_n{}'\frac{\partial w}{\partial r}dr \tag{7.21.33}$$

The first term vanishes as a result of the boundary conditions and the symmetry of the deflection. Substituting (7.21.8) and (7.21.30) into (7.21.33) leads to

$$\int_0^1 \frac{\partial}{\partial r}\left(G\frac{\partial w}{\partial r}\right)\phi_n dr = -\sum_{m,n,p=1}^{\infty}\left[\sum_{k=1}^{\infty} S_{kpq}\int_0^1 J_1(\zeta_k r)\phi_n{}'\phi_m{}'dr\right]\psi_m\psi_n\psi_p \tag{7.21.34}$$

Using (7.21.29) leads to

$$\Gamma_{nmpq} = \sum_{k=1}^{\infty} \frac{\int_0^1 J_1(\zeta_k r)\phi_p'\phi_q'dr \int_0^1 J_1(\zeta_k r)\phi_m'\phi_m'dr}{\left(\zeta_k^2 - 1 + \nu^2\right)J_1^2(\zeta_k)} \tag{7.21.35}$$

Let

$$\Omega = 3\omega_1 + \varepsilon\sigma \tag{7.21.36}$$

Following the method of multiple scales, we let

$$\psi_n(t;\varepsilon) = \psi_{n0}(T_0, T_1) + \varepsilon\psi_{n1}(T_0, T_1) + \cdots \tag{7.21.37}$$

Substituting (7.21.37) into (7.21.31) and retaining to $O(\varepsilon)$), we obtain

$$\begin{aligned}
0 &= \ddot{\psi}_n + \omega_n^2\psi_n + \varepsilon\left(2\mu_n\dot{\psi}_n - \sum_{m,p,q=1}^{\infty}\Gamma_{nmpq}\psi_m\psi_p\psi_q\right) - f_n(t)\\
&= D_0^2\psi_{n0} + \omega_n^2\psi_{n0} - 2K_n\cos\Omega T_0 + \varepsilon\big(D_0^2\psi_{n1} + \omega_n^2\psi_{n1}\\
&\quad +2D_0D_1\psi_{n0} + 2\mu_n D_0\psi_{n0} - \sum_{m,p,q=1}^{\infty}\Gamma_{nmpq}\psi_{m0}\psi_{p0}\psi_{q0}\Big)
\end{aligned} \tag{7.21.38}$$

Equating coefficients of like powers of ε in the above equation yields

$$D_0^2\psi_{n0} + \omega_n^2\psi_{n0} = 2K_n\cos\Omega T_0 \tag{7.21.39}$$

$$D_0^2\psi_{n1} + \omega_n^2\psi_{n1} = -2D_0D_1\psi_{n0} - 2\mu_n D_0\psi_{n0} + \sum_{m,p,q=1}^{\infty}\Gamma_{nmpq}\psi_{m0}\psi_{p0}\psi_{q0} \tag{7.21.40}$$

The solution of (7.21.39) is

$$\psi_{n0} = A_n(T_1)e^{i\omega_n T_0} + \Lambda_n e^{i\Omega T_0} + cc, \quad \Lambda_n = \frac{K_n}{\omega_n^2 - \Omega^2} \tag{7.21.41}$$

Substituting (7.21.41) into (7.21.40) and taking into account $\Omega = 3\omega_1 + \varepsilon\sigma$ and $\omega_1 + 2\omega_2 = \omega_3 + \varepsilon\sigma_1$ yields

$$\begin{aligned}
&D_0^2\psi_{n1} + \omega_n^2\psi_{n1} = -2D_0D_1\psi_{n0} - 2\mu_n D_0\psi_{n0} + \sum_{m,p,q=1}^{\infty}\Gamma_{nmpq}\psi_{m0}\psi_{p0}\psi_{q0}\\
&= -2i\omega_n A_n' e^{i\omega_n T_0} - 2i\omega_n\mu_n A_n e^{i\omega_n T_0} + \sum_{m,p,q=1}^{\infty}\Gamma_{nmpq}\Big\{\overline{A}_m A_p A_q e^{-i\omega_m T_0}e^{i\omega_p T_0}e^{i\omega_q T_0}\\
&+A_m\overline{A}_p A_q e^{i\omega_m T_0}e^{-i\omega_p T_0}e^{i\omega_q T_0} + A_m A_p\overline{A}_q e^{i\omega_m T_0}e^{i\omega_p T_0}e^{-i\omega_q T_0}\\
&+\Lambda_m\overline{A}_p\overline{A}_q e^{i\Omega T_0}e^{-i\omega_p T_0}e^{-i\omega_q T_0} + \Lambda_p\overline{A}_m\overline{A}_q e^{-i\omega_m T_0}e^{i\Omega T_0}e^{-i\omega_q T_0}\\
&+\overline{A}_m\overline{A}_p\Lambda_q e^{-i\omega_m T_0}e^{-i\omega_p T_0}e^{i\Omega T_0} + 2\Lambda_m\Lambda_p A_q e^{i\omega_q T_0} + 2\Lambda_m A_p\Lambda_q e^{i\omega_p T_0}\\
&+2A_m\Lambda_p\Lambda_q e^{i\omega_m T_0}\Big\} + cc + NST
\end{aligned} \tag{7.21.42}$$

In order to eliminate secular terms from the above equation, we need

$$-2i\omega_1 A_1{}' - 2i\omega_1\mu_1 A_1 + 3\Gamma_{1111}A_1^2\overline{A}_1 + A_1\sum_{m\neq 1}^{\infty} 2(2\Gamma_{1m1m} + \Gamma_{11mm})A_m\overline{A}_m \\ +2A_1\sum_{m,\,p=1}^{\infty}\left(\Gamma_{11mp} + 2\Gamma_{1m1p}\right)\Lambda_m\Lambda_p + 3\overline{A}_1^2 e^{i\sigma T_1}\sum_{m=1}^{\infty}\Gamma_{111m}\Lambda_m = 0 \tag{7.21.43}$$

$$-2i\omega_n A_n{}' - 2i\omega_n\mu_n A_n + 3\Gamma_{nnnn}A_n^2\overline{A}_n + A_n\sum_{m\neq n}^{\infty} 2(2\Gamma_{nmnm} + \Gamma_{nnmm})A_m\overline{A}_m \\ +2A_n\sum_{m,\,p=1}^{\infty}\left(\Gamma_{nnmp} + 2\Gamma_{nmnp}\right)\Lambda_m\Lambda_p = 0,\ n\neq 1 \tag{7.21.44}$$

The above two equations can be written as

$$-2i\omega_1 A_1{}' - 2i\omega_1\mu_1 A_1 + A_1\sum_{m=1}^{\infty}\alpha_{1m}A_m\overline{A}_m + 2H_{11}A_1 + 4F\overline{A}_1^2 e^{i\sigma T_1} = 0 \tag{7.21.45}$$

$$-2i\omega_n A_n{}' - 2i\omega_n\mu_n A_n + A_n\sum_{m=1}^{\infty}\alpha_{nm}A_m\overline{A}_m + 2H_{nn}A_n = 0,\quad n\neq 1 \tag{7.21.46}$$

where

$$H_{nk} = \sum_{m,j}\left(\Gamma_{nkmj} + 2\Gamma_{nmkj}\right)\Lambda_m\Lambda_j,\ F = \frac{3}{4}\sum_{n=1}^{\infty}\Gamma_{111n}\Lambda_n$$

$$\alpha_{nj} = \alpha_{jn} = \begin{cases} 2\left(2\Gamma_{njnj} + \Gamma_{nnjj}\right) \text{ for } n\neq j \\ 3\Gamma_{nnnn} \end{cases}$$

Let

$$A_n = \frac{1}{2}a_n e^{i\beta_n} \tag{7.21.47}$$

Substituting (7.21.47) into (7.21.46) yields

$$-i\omega_n a_n{}' + \omega_n a_n\beta_n{}' - i\omega_n\mu_n a_n + \frac{1}{8}a_n\sum_{m=1}^{\infty}\alpha_{nm}a_m^2 + H_{nn}a_n = 0,\ n\neq 1 \tag{7.21.48}$$

Separating the real and imaginary parts of the above equation yields

$$a_n{}' + \mu_n a_n = 0,\ n\neq 1 \Rightarrow a_n = a_{n0}e^{-\mu_n T_1}\to 0,\ n\neq 1 \tag{7.21.49}$$

Substituting (7.21.47) into (7.21.45) and taking into account that (7.21.49), we obtain

$$-i\omega_1 a_1{}' + \omega_1 a_1 \beta_1{}' - i\omega_1 \mu_1 a_1 + \frac{1}{8}\alpha_{11} a_1^3 + H_{11} a_1 + F a_1^2 e^{i(\sigma T_1 - 3\beta)} = 0 \quad (7.21.50)$$

Separating the real and imaginary parts of the above equation yields

$$\begin{aligned} &\omega_1 a_1' + \omega_1 \mu_1 a_1 - F a_1^2 \sin\gamma = 0 \\ &-\omega_1 a_1 \gamma' + \omega_1 a_1 \sigma + \tfrac{3}{8}\alpha_{11} a_1^3 + 3H_{11} a_1 + 3F a_1^2 \cos\gamma = 0 \end{aligned} \quad (7.21.51)$$

where

$$\gamma = \sigma T_1 - 3\beta \quad (7.21.52)$$

From Eq. (7.21.51), we know that the steady state solution satisfies the following equations:

$$\begin{aligned} &a_1(\omega_1 \mu_1 - F a_1 \sin\gamma) = 0 \\ &a_1\left(\omega_1 \sigma + \tfrac{3}{8}\alpha_{11} a_1^2 + 3H_{11} + 3F a_1 \cos\gamma\right) = 0 \end{aligned} \quad (7.21.53)$$

Combining the above results, we can obtain the steady-state deflection of subharmonic resonance of the axisymmetric circular plate when $\Omega \approx 3\omega_1$ is

$$w(r,t) = a_1 \phi_1(r) \cos\left(\frac{1}{3}\Omega t - \frac{1}{3}\gamma\right) + \left[2\sum_{n=1}^{\infty} \Lambda_n \phi_n(r)\right] \cos\Omega t + O(\varepsilon) \quad (7.21.54)$$

where a_1 and γ are given by (7.21.53).

7.22 Exercise 7.22 (First-Order Superharmonic Resonance Analysis of a Uniform Circular Plate Clamped Along Its Edge)

Solution: Let

$$3\Omega = \omega_1 + \varepsilon\sigma \quad (7.22.1)$$

and

$$\psi_n(t;\varepsilon) = \psi_{n0}(T_0, T_1) + \varepsilon \psi_{n1}(T_0, T_1) + \cdots \quad (7.22.2)$$

Substituting (7.22.2) into (7.21.31) and retaining to $O(\varepsilon)$, we obtain

$$\begin{aligned}0 &= \ddot{\psi}_n + \omega_n^2\psi_n + \varepsilon\left(2\mu_n\dot{\psi}_n - \sum_{m,\,p,\,q=1}^{\infty}\Gamma_{nmpq}\psi_m\psi_p\psi_q\right) - f_n(t)\\ &= D_0^2\psi_{n0} + \omega_n^2\psi_{n0} - 2K_n\cos\Omega T_0 + \varepsilon\big(D_0^2\psi_{n1} + \omega_n^2\psi_{n1}\\ &+2D_0D_1\psi_{n0} + 2\mu_n D_0\psi_{n0} - \sum_{m,\,p,\,q=1}^{\infty}\Gamma_{nmpq}\psi_{m0}\psi_{p0}\psi_{q0}\Bigg)\end{aligned} \tag{7.22.3}$$

Equating coefficients of like powers of ε in the above equation yields

$$D_0^2\psi_{n0} + \omega_n^2\psi_{n0} = 2K_n\cos\Omega T_0 \tag{7.22.4}$$

$$D_0^2\psi_{n1} + \omega_n^2\psi_{n1} = -2D_0D_1\psi_{n0} - 2\mu_n D_0\psi_{n0} + \sum_{m,\,p,\,q=1}^{\infty}\Gamma_{nmpq}\psi_{m0}\psi_{p0}\psi_{q0} \tag{7.22.5}$$

The solution of (7.22.4) is

$$\psi_{n0} = A_n(T_1)e^{i\omega_n T_0} + \Lambda_n e^{i\Omega T_0} + cc,\ \ \Lambda_n = \frac{K_n}{\omega_n^2 - \Omega^2} \tag{7.22.6}$$

Substituting (7.22.6) into (7.22.5) and taking into account $3\Omega = \omega_1 + \varepsilon\sigma$ and $\omega_1 + 2\omega_2 = \omega_3 + \varepsilon\sigma_1$, we obtain

$$\begin{aligned}&D_0^2\psi_{n1} + \omega_n^2\psi_{n1} = -2D_0D_1\psi_{n0} - 2\mu_n D_0\psi_{n0} + \sum_{m,p,q=1}^{\infty}\Gamma_{nmpq}\psi_{m0}\psi_{p0}\psi_{q0}\\ &= -2i\omega_n A' e^{i\omega_n T_0} - 2i\omega_n\mu_n A_n e^{i\omega_n T_0} + \sum_{m,p,q=1}^{\infty}\Gamma_{nmpq}\big\{\overline{A}_m A_p A_q e^{-i\omega_m T_0}e^{i\omega_p T_0}e^{i\omega_q T_0}\\ &+A_m\overline{A}_p A_q e^{i\omega_m T_0}e^{-i\omega_p T_0}e^{i\omega_q T_0} + A_m A_p\overline{A}_q e^{i\omega_m T_0}e^{i\omega_p T_0}e^{-i\omega_q T_0}\\ &+2\Lambda_m\Lambda_p A_q e^{i\omega_q T_0} + 2\Lambda_m A_p\Lambda_q e^{i\omega_p T_0} + 2A_m\Lambda_p\Lambda_q e^{i\omega_m T_0}\\ &+\Lambda_m\Lambda_p\Lambda_q e^{3i\Omega T_0}\big\} + cc + NST\end{aligned} \tag{7.22.7}$$

In order to eliminate secular terms from the above equation, we need

$$\begin{aligned}&-2i\omega_1 A_1\prime - 2i\omega_1\mu_1 A_1 + 3\Gamma_{1111}A_1^2\overline{A}_1 + A_1\sum_{m\neq 1}^{\infty}2(2\Gamma_{1m1m} + \Gamma_{11mm})A_m\overline{A}_m\\ &+2A_1\sum_{m,\,p=1}^{\infty}\big(\Gamma_{11mp} + 2\Gamma_{1m1p}\big)\Lambda_m\Lambda_p + e^{i\sigma T_1}\sum_{m,\,p,\,q=1}^{\infty}\Gamma_{1mpq}\Lambda_m\Lambda_p\Lambda_q = 0\end{aligned} \tag{7.22.8}$$

$$-2i\omega_n A_n\prime - 2i\omega_n\mu_n A_n + 3\Gamma_{nnnn}A_n^2\overline{A}_n + A_n \sum_{m\neq n}^{\infty} 2(2\Gamma_{nmnm} + \Gamma_{nnmm})A_m\overline{A}_m$$
$$+2A_n \sum_{m,\,p=1}^{\infty} \left(\Gamma_{nnmp} + 2\Gamma_{nmnp}\right)\Lambda_m\Lambda_p = 0,\ \ n \neq 1 \tag{7.22.9}$$

The above two equations can be written as

$$-2i\omega_1 A_1\prime - 2i\omega_1\mu_1 A_1 + A_1\sum_{m=1}^{\infty}\alpha_{1m}A_m\overline{A}_m + 2H_{11}A_1 + Fe^{i\sigma T_1} = 0 \tag{7.22.10}$$

$$-2i\omega_n A_n\prime - 2i\omega_n\mu_n A_n + A_n\sum_{m=1}^{\infty}\alpha_{nm}A_m\overline{A}_m + 2H_{nn}A_n = 0,\ \ n \neq 1 \tag{7.22.11}$$

where

$$H_{nk} = \sum_{m,\,j}\left(\Gamma_{nkmj} + 2\Gamma_{nmkj}\right)\Lambda_m\Lambda_j,\ \ F = \sum_{m,\,p,\,q=1}^{\infty}\Gamma_{1mpq}\Lambda_m\Lambda_p\Lambda_q$$

$$\alpha_{nj} = \alpha_{jn} = \begin{cases} 2\left(2\Gamma_{njnj} + \Gamma_{nnjj}\right) \text{ for } n \neq j \\ 3\Gamma_{nnnn} \end{cases}$$

Let

$$A_n = \frac{1}{2}a_n e^{i\beta_n} \tag{7.22.12}$$

Substituting (7.22.12) into (7.22.11) yields

$$-i\omega_n a_n\prime + \omega_n a_n\beta_n\prime - i\omega_n\mu_n a_n + \frac{1}{8}a_n\sum_{m=1}^{\infty}\alpha_{nm}a_m^2 + H_{nn}a_n = 0,\quad n \neq 1 \tag{7.22.13}$$

Separating the real and imaginary parts of the above equation yields

$$a_n\prime + \mu_n a_n = 0,\ \ n \neq 1 \ \Rightarrow\ a_n = a_{n0}e^{-\mu_n T_1} \rightarrow 0,\ \ n \neq 1 \tag{7.22.14}$$

Substituting (7.22.12) into (7.22.10) and taking into account that (7.22.14), we obtain

$$-i\omega_1 a_1\prime + \omega_1 a_1\beta_1\prime - i\omega_1\mu_1 a_1 + \frac{1}{8}\alpha_{11}a_1^3 + H_{11}a_1 + Fe^{i(\sigma T_1 - \beta)} = 0 \tag{7.22.15}$$

Separating the real and imaginary parts of the above equation yields

$$\begin{aligned} &\omega_1 a_1' + \omega_1 \mu_1 a_1 - F\sin\gamma = 0 \\ &-\omega_1 a_1 \gamma' + \omega_1 \sigma a_1 + \tfrac{1}{8}\alpha_{11} a_1^3 + H_{11} a_1 + F\cos\gamma = 0 \end{aligned} \tag{7.22.16}$$

where

$$\gamma = \sigma T_1 - \beta \tag{7.22.17}$$

From Eq. (7.22.16), we know that the steady state solution satisfies the following equations:

$$\begin{aligned} &\omega_1 \mu_1 a_1 - F\sin\gamma = 0 \\ &\omega_1 \sigma a_1 + \tfrac{1}{8}\alpha_{11} a_1^3 + H_{11} a_1 + F\cos\gamma = 0 \end{aligned} \tag{7.22.18}$$

Combining the above results, we can obtain the steady-state deflection of subharmonic resonance of the axisymmetric circular plate when $\omega_1 \approx 3\Omega$ is

$$w(r,t) = a_1 \phi_1(r)\cos(3\Omega t - \gamma) + \left[2\sum_{n=1}^{\infty} \Lambda_n \phi_n(r)\right]\cos\Omega t + O(\varepsilon) \tag{7.22.19}$$

where a_1 and γ are given by (7.22.18).

7.23 Exercise 7.23 (Second-Order Superharmonic Resonance Analysis of a Uniform Circular Plate Clamped Along Its Edge)

Solution: Let

$$3\Omega = \omega_2 + \varepsilon\sigma \tag{7.23.1}$$

and

$$\psi_n(t;\varepsilon) = \psi_{n0}(T_0, T_1) + \varepsilon\psi_{n1}(T_0, T_1) + \cdots \tag{7.23.2}$$

Substituting (7.23.2) into (7.21.31) and retaining to $O(\varepsilon)$, we obtain

$$\begin{aligned} 0 &= \ddot{\psi}_n + \omega_n^2 \psi_n + \varepsilon\left(2\mu_n \dot{\psi}_n - \sum_{m,p,q=1}^{\infty} \Gamma_{nmpq} \psi_m \psi_p \psi_q\right) - f_n(t) \\ &= D_0^2 \psi_{n0} + \omega_n^2 \psi_{n0} - 2K_n \cos\Omega T_0 + \varepsilon\big(D_0^2 \psi_{n1} + \omega_n^2 \psi_{n1} \\ &\quad + 2D_0 D_1 \psi_{n0} + 2\mu_n D_0 \psi_{n0} - \sum_{m,p,q=1}^{\infty} \Gamma_{nmpq} \psi_{m0} \psi_{p0} \psi_{q0}\Big) \end{aligned} \tag{7.23.3}$$

Equating coefficients of like powers of ε in the above equation yields

$$D_0^2\psi_{n0} + \omega_n^2\psi_{n0} = 2K_n\cos\Omega T_0 \tag{7.23.4}$$

$$D_0^2\psi_{n1} + \omega_n^2\psi_{n1} = -2D_0D_1\psi_{n0} - 2\mu_n D_0\psi_{n0} + \sum_{m,\,p,\,q=1}^{\infty}\Gamma_{nmpq}\psi_{m0}\psi_{p0}\psi_{q0} \tag{7.23.5}$$

The solution of (7.23.4) is

$$\psi_{n0} = A_n(T_1)e^{i\omega_n T_0} + \Lambda_n e^{i\Omega T_0} + cc,\ \ \Lambda_n = \frac{K_n}{\omega_n^2 - \Omega^2} \tag{7.23.6}$$

Substituting (7.23.6) into (7.23.5) and taking into account $3\Omega = \omega_2 + \varepsilon\sigma$ and $\omega_1 + 2\omega_2 = \omega_3 + \varepsilon\sigma_1$ yields

$$\begin{aligned}
D_0^2\psi_{n1} + \omega_n^2\psi_{n1} &= -2D_0D_1\psi_{n0} - 2\mu_n D_0\psi_{n0} + \sum_{m,p,q=1}^{\infty}\Gamma_{nmpq}\psi_{m0}\psi_{p0}\psi_{q0}\\
&= -2i\omega_n A_n' e^{i\omega_n T_0} - 2i\omega_n\mu_n A_n e^{i\omega_n T_0} + \sum_{m,p,q=1}^{\infty}\Gamma_{nmpq}\{\overline{A}_m A_p A_q e^{-i\omega_m T_0}e^{i\omega_p T_0}e^{i\omega_q T_0}\\
&+A_m\overline{A}_p A_q e^{i\omega_m T_0}e^{-i\omega_p T_0}e^{i\omega_q T_0} + A_m A_p\overline{A}_q e^{i\omega_m T_0}e^{i\omega_p T_0}e^{-i\omega_q T_0}\\
&+2\Lambda_m\Lambda_p A_q e^{i\omega_q T_0} + 2\Lambda_m A_p\Lambda_q e^{i\omega_p T_0} + 2A_m\Lambda_p\Lambda_q e^{i\omega_m T_0}\\
&+\Lambda_m\Lambda_p\Lambda_q e^{3i\Omega T_0}\} + cc + NST
\end{aligned} \tag{7.23.7}$$

In order to eliminate secular terms from the above equation, we need

$$\begin{aligned}
&-2i\omega_2 A_2\prime - 2i\omega_2\mu_2 A_2 + 3\Gamma_{2222}A_2^2\overline{A}_2 + A_2\sum_{m\neq 2}^{\infty}2(2\Gamma_{2m2m} + \Gamma_{22mm})A_m\overline{A}_m\\
&\quad+2A_2\sum_{m,\,p=1}^{\infty}\left(\Gamma_{22mp} + 2\Gamma_{1m1p}\right)\Lambda_m\Lambda_p + e^{i\sigma T_1}\sum_{m,\,p,\,q=1}^{\infty}\Gamma_{2mpq}\Lambda_m\Lambda_p\Lambda_q = 0
\end{aligned} \tag{7.23.8}$$

$$\begin{aligned}
&-2i\omega_n A_n\prime - 2i\omega_n\mu_n A_n + 3\Gamma_{nnnn}A_n^2\overline{A}_n + A_n\sum_{m\neq n}^{\infty}2(2\Gamma_{nmnm} + \Gamma_{nnmm})A_m\overline{A}_m\\
&\quad+2A_n\sum_{m,p=1}^{\infty}\left(\Gamma_{nnmp} + 2\Gamma_{nmnp}\right)\Lambda_m\Lambda_p = 0,\ \ n\neq 2
\end{aligned} \tag{7.23.9}$$

The above two equations can be written as

$$-2i\omega_2 A_2\prime - 2i\omega_2\mu_2 A_2 + A_2\sum_{m=1}^{\infty}\alpha_{2m}A_m\overline{A}_m + 2H_{22}A_2 + Fe^{i\sigma T_1} = 0 \qquad (7.23.10)$$

$$-2i\omega_n A_n\prime - 2i\omega_n\mu_n A_n + A_n\sum_{m=1}^{\infty}\alpha_{nm}A_m\overline{A}_m + 2H_{nn}A_n = 0,\quad n\neq 2 \qquad (7.23.11)$$

where

$$H_{nk} = \sum_{m,j}\left(\Gamma_{nkmj} + 2\Gamma_{nmkj}\right)\Lambda_m\Lambda_j,\quad F = \sum_{m,p,q=1}^{\infty}\Gamma_{2mpq}\Lambda_m\Lambda_p\Lambda_q$$

$$\alpha_{nj} = \alpha_{jn} = \begin{cases} 2\left(2\Gamma_{njnj} + \Gamma_{nnjj}\right) & \text{for } n\neq j \\ 3\Gamma_{nnnn} \end{cases}$$

Let

$$A_n = \frac{1}{2}a_n e^{i\beta_n} \qquad (7.23.12)$$

Substituting (7.23.12) into (7.23.11) yields

$$-i\omega_n a_n\prime + \omega_n a_n\beta_n\prime - i\omega_n\mu_n a_n + \frac{1}{8}a_n\sum_{m=1}^{\infty}\alpha_{nm}a_m^2 + H_{nn}a_n = 0,\quad n\neq 2 \qquad (7.23.13)$$

Separating the real and imaginary parts of the above equation yields

$$a_n\prime + \mu_n a_n = 0,\ n\neq 1 \quad \Rightarrow a_n = a_{n0}e^{-\mu_n T_1}\to 0,\ n\neq 2 \qquad (7.23.14)$$

Substituting (7.23.12) into (7.23.10) and taking into account that (7.23.14), we obtain

$$-i\omega_2 a_2\prime + \omega_2 a_2\beta_2\prime - i\omega_2\mu_2 a_2 + \frac{1}{8}\alpha_{22}a_2^3 + H_{22}a_2 + Fe^{i(\sigma T_1 - \beta)} = 0 \qquad (7.23.15)$$

Separating the real part from the imaginary part of the above equation yields

$$\begin{aligned} &\omega_2 a_2\prime + \omega_2\mu_2 a_2 - F\sin\gamma = 0 \\ &-\omega_2 a_2\gamma\prime + \omega_2\sigma a_2 + \tfrac{1}{8}\alpha_{22}a_2^3 + H_{22}a_2 + F\cos\gamma = 0 \end{aligned} \qquad (7.23.16)$$

where

$$\gamma = \sigma T_1 - \beta \qquad (7.23.17)$$

From Eq. (7.23.16), we know that the steady state solution satisfies the following equations:

$$\begin{aligned} &\omega_2\mu_2 a_2 - F\sin\gamma = 0 \\ &\omega_2\sigma a_2 + \tfrac{1}{8}\alpha_{22}a_2^3 + H_{22}a_2 + F\cos\gamma = 0 \end{aligned} \tag{7.23.18}$$

Combining the above results, we can obtain the steady-state deflection of subharmonic resonance of the axisymmetric circular plate when $\omega_2 \approx 3\Omega$ is

$$w(r,t) = a_2\phi_2(r)\cos(3\Omega t - \gamma) + \left[2\sum_{n=1}^{\infty}\Lambda_n\phi_n(r)\right]\cos\Omega t + O(\varepsilon) \tag{7.23.19}$$

where a_1 and γ are given by (7.23.18).

7.24 Exercise 7.24 (Combined Resonance Analysis of a Uniform Circular Plate Clamped Along Its Edge)

Solution: Let

$$\varepsilon\sigma_1 = \omega_1 + 2\omega_2 - \omega_3, \quad \varepsilon\sigma_2 = 2\Omega - \omega_1 - \omega_2 \tag{7.24.1}$$

and

$$\psi_n(t;\varepsilon) = \psi_{n0}(T_0, T_1) + \varepsilon\psi_{n1}(T_0, T_1) + \cdots \tag{7.24.2}$$

Substituting (7.24.2) into (7.21.31) and retaining to $O(\varepsilon)$, we obtain

$$\begin{aligned} 0 &= \ddot{\psi}_n + \omega_n^2\psi_n + \varepsilon\left(2\mu_n\dot{\psi}_n - \sum_{m,p,q=1}^{\infty}\Gamma_{nmpq}\psi_m\psi_p\psi_q\right) - f_n(t) \\ &= D_0^2\psi_{n0} + \omega_n^2\psi_{n0} - 2K_n\cos\Omega T_0 + \varepsilon\big(D_0^2\psi_{n1} + \omega_n^2\psi_{n1} \\ &\quad + 2D_0D_1\psi_{n0} + 2\mu_n D_0\psi_{n0} - \sum_{m,p,q=1}^{\infty}\Gamma_{nmpq}\psi_{m0}\psi_{p0}\psi_{q0}\Big) \end{aligned} \tag{7.24.3}$$

Equating coefficients of like powers of ε in the above equation yields

$$D_0^2\psi_{n0} + \omega_n^2\psi_{n0} = 2K_n\cos\Omega T_0 \tag{7.24.4}$$

$$D_0^2\psi_{n1} + \omega_n^2\psi_{n1} = -2D_0D_1\psi_{n0} - 2\mu_n D_0\psi_{n0} + \sum_{m,p,q=1}^{\infty}\Gamma_{nmpq}\psi_{m0}\psi_{p0}\psi_{q0} \tag{7.24.5}$$

The solution of (7.24.4) is

$$\psi_{n0} = A_n(T_1)e^{i\omega_n T_0} + \Lambda_n e^{i\Omega T_0} + cc, \quad \Lambda_n = \frac{K_n}{\omega_n^2 - \Omega^2} \tag{7.24.6}$$

Substituting (7.24.6) into (7.24.5) and taking into account $\varepsilon\sigma_1 = \omega_1 + 2\omega_2 - \omega_3 \text{and} \varepsilon\sigma_2 = 2\Omega - \omega_1 - \omega_2$ yields

$$\begin{aligned}
D_0^2\psi_{n1} + \omega_n^2\psi_{n1} &= -2D_0D_1\psi_{n0} - 2\mu_n D_0\psi_{n0} + \sum_{m,p,q=1}^{\infty} \Gamma_{nmpq}\psi_{m0}\psi_{p0}\psi_{q0} \\
&= -2i\omega_n A_n' e^{i\omega_n T_0} - 2i\omega_n\mu_n A_n e^{i\omega_n T_0} \\
&+ \sum_{m,p,q=1}^{\infty} \Gamma_{nmpq}\{A_m A_p A_q e^{i\omega_m T_0}e^{i\omega_p T_0}e^{i\omega_q T_0} + \overline{A}_m A_p A_q e^{-i\omega_m T_0}e^{i\omega_p T_0}e^{i\omega_q T_0} \\
&+ A_m\overline{A}_p\overline{A}_q e^{i\omega_m T_0}e^{-i\omega_p T_0}e^{-i\omega_q T_0} + A_m\overline{A}_p A_q e^{i\omega_m T_0}e^{-i\omega_p T_0}e^{i\omega_q T_0} \\
&+ \overline{A}_m A_p\overline{A}_q e^{-i\omega_m T_0}e^{i\omega_p T_0}e^{-i\omega_q T_0} + \overline{A}_m\overline{A}_p A_q e^{-i\omega_m T_0}e^{-i\omega_p T_0}e^{i\omega_q T_0} \\
&+ A_m A_p\overline{A}_q e^{i\omega_m T_0}e^{i\omega_p T_0}e^{-i\omega_q T_0} + \Lambda_m\Lambda_p A_q e^{2i\Omega T_0}e^{i\omega_q T_0} \\
&+ \Lambda_m A_p\Lambda_q e^{2i\Omega T_0}e^{i\omega_p T_0} + A_m\Lambda_p\Lambda_q e^{2i\Omega T_0}e^{i\omega_m T_0} \\
&+ \overline{A}_m\Lambda_p\Lambda_q e^{2i\Omega T_0}e^{-i\omega_m T_0} + \Lambda_m\Lambda_p\overline{A}_q e^{2i\Omega T_0}e^{-i\omega_q T_0} + \Lambda_m\overline{A}_p\Lambda_q e^{2i\Omega T_0}e^{-i\omega_p T_0} \\
&+ 2\Lambda_m\Lambda_p A_q e^{i\omega_q T_0} + \Lambda_m A_p\Lambda_q e^{i\omega_p T_0} + \Lambda_m A_p\Lambda_q e^{i\omega_p T_0} \\
&+ A_m\Lambda_p\Lambda_q e^{i\omega_m T_0} + A_m\Lambda_p\Lambda_q e^{i\omega_m T_0}\} + cc + NST
\end{aligned} \tag{7.24.7}$$

In order to eliminate secular terms from the above equation, we need

$$\begin{aligned}
&-2i\omega_1 A_1\prime - 2i\omega_1\mu_1 A_1 + A_1\sum_{m=1}^{\infty}\alpha_{1m}A_m\overline{A}_m \\
&+2H_{11}A_1 + H_{12}\overline{A}_2 e^{i\sigma_2 T_1} + 8QA_3\overline{A}_2^2 e^{-i\sigma_1 T_1} = 0
\end{aligned} \tag{7.24.8}$$

$$\begin{aligned}
&-2i\omega_2 A\prime - 2i\omega_2\mu_2 A_2 + A_2\sum_{m=1}^{\infty}\alpha_{2m}A_m\overline{A}_m + 2H_{22}A_2 \\
&+16Q\overline{A}_1\overline{A}_2 A_3 e^{-i\sigma_1 T_1} + H_{12}\overline{A}_1 e^{i\sigma_2 T_1} + H_{23}A_3 e^{-i(\sigma_1+\sigma_2)T_1} = 0
\end{aligned} \tag{7.24.9}$$

$$\begin{aligned}
&-2i\omega_3 A_3\prime - 2i\omega_3\mu_3 A_3 + A_3\sum_{m=1}^{\infty}\alpha_{3m}A_m\overline{A}_m \\
&+2H_{33}A_3 + 8QA_1A_2^2 e^{i\sigma_1 T_1} + H_{23}A_2 e^{i(\sigma_1+\sigma_2)T_1} = 0
\end{aligned} \tag{7.24.10}$$

$$-2i\omega_n A_n\prime - 2i\omega_n\mu_n A_n + A_n\sum_{m=1}^{\infty}\alpha_{nm}A_m\overline{A}_m + 2H_{nn}A_n = 0, \; n \geq 4 \tag{7.24.11}$$

Let

$$A_n = \frac{1}{2} a_n e^{i\beta_n} \tag{7.24.12}$$

Substituting (7.24.12) into (7.24.11) yields

$$-i\omega_n a_n\prime + \omega_n a_n \beta_n\prime - i\omega_n \mu_n a_n + \frac{1}{8} a_n \sum_{m=1}^{\infty} \alpha_{nm} a_m^2 + H_{nn} a_n = 0, \quad n \geq 4 \tag{7.24.13}$$

Separating the real and imaginary parts of the above equation gives

$$a_n\prime + \mu_n a_n = 0,\ n \geq 4 \Rightarrow a_n = a_{n0} e^{-\mu_n T_1} \to 0, \quad n \geq 4 \tag{7.24.14}$$

Substituting (7.24.12) into (7.24.8)–(7.24.10) and taking into account that (7.24.14), we obtain

$$\begin{aligned} &-i\omega_1 a_1\prime + \omega_1 a_1 \beta_1\prime - i\omega_1 \mu_1 a_1 + \tfrac{1}{8} a_1 \sum_{m=1}^{3} \alpha_{1m} a_m^2 + H_{11} a_1 \\ &\quad + \tfrac{1}{2} H_{12} a_2 e^{i(\sigma_2 T_1 - \beta_1 - \beta_2)} + Q a_2^2 a_3 e^{i(-\sigma_1 T_1 - \beta_1 - 2\beta_2 + \beta_3)} = 0 \end{aligned} \tag{7.24.15}$$

$$\begin{aligned} &-i\omega_2 a_2\prime + \omega_2 a_2 \beta_2\prime - i\omega_2 \mu_2 a_2 + \tfrac{1}{8} a_2 \sum_{m=1}^{3} \alpha_{2m} a_m^2 + H_{22} a_2 \\ &+2Q a_1 a_2 a_3 e^{i(-\sigma_1 T_1 - \beta_1 - 2\beta_2 + \beta_3)} + \tfrac{1}{2} H_{12} a_1 e^{i(\sigma_2 T_1 - \beta_1 - \beta_2)} \\ &+\tfrac{1}{2} H_{23} a_3 e^{i(-\sigma_1 T_1 - \sigma_2 T_1 - \beta_2 + \beta_3)} = 0 \end{aligned} \tag{7.24.16}$$

$$\begin{aligned} &-i\omega_3 a_3\prime + \omega_3 a_3 \beta_3\prime - i\omega_3 \mu_3 a_3 + \tfrac{1}{8} a_3 \sum_{m=1}^{3} \alpha_{3m} a_m^2 + H_{33} a_3 \\ &+Q a_1 a_2^2 e^{i(\sigma_1 T_1 + \beta_1 + 2\beta_2 - \beta_3)} + \tfrac{1}{2} H_{23} a_2 e^{i(\sigma_1 T_1 + \sigma_2 T_1 + \beta_2 - \beta_3)} = 0 \end{aligned} \tag{7.24.17}$$

Separating the real part from the imaginary part of the above equation yields

$$\omega_1 a_1\prime + \omega_1 \mu_1 a_1 + Q a_2^2 a_3 \sin\gamma_1 + \frac{1}{2} H_{12} a_2 \sin\gamma_2 = 0 \tag{7.24.18}$$

$$\omega_2 a_2\prime + \omega_2 \mu_2 a_2 + 2Q a_1 a_2 a_3 \sin\gamma_1 + \frac{1}{2} H_{12} a_1 \sin\gamma_2 + \frac{1}{2} H_{23} a_3 \sin(\gamma_1 - \gamma_2) = 0 \tag{7.24.19}$$

$$\omega_3 a_3\prime + \omega_3 \mu_3 a_3 - Q a_1 a_2^2 \sin\gamma_1 - \frac{1}{2} H_{23} a_2 \sin(\gamma_1 - \gamma_2) = 0 \tag{7.24.20}$$

$$\omega_1 a_1 \beta_1\prime + \frac{1}{8} a_1 \sum_{m=1}^{3} \alpha_{1m} a_m^2 + H_{11} a_1 + Q a_2^2 a_3 \cos\gamma_1 + \frac{1}{2} H_{12} a_2 \cos\gamma_2 = 0 \tag{7.24.21}$$

$$\omega_2 a_2 \beta_2' + \tfrac{1}{8} a_2 \sum_{m=1}^{3} \alpha_{2m} a_m^2 + H_{22} a_2 + 2Q a_1 a_2 a_3 \cos\gamma_1 + \tfrac{1}{2} H_{12} a_1 \sin\gamma_2 + \tfrac{1}{2} H_{23} a_3 \cos(\gamma_1 - \gamma_2) = 0 \quad (7.24.22)$$

$$\omega_3 a_3 \beta_3' + \frac{1}{8} a_3 \sum_{m=1}^{3} \alpha_{3m} a_m^2 + H_{33} a_3 + Q a_1 a_2^2 \cos\gamma_1 + \frac{1}{2} H_{23} a_2 \cos(\gamma_1 - \gamma_2) = 0 \quad (7.24.23)$$

where

$$-\gamma_1 = \beta_3 - 2\beta_2 - \beta_1 - \sigma_1 T_1, \; -\gamma_2 = \sigma_2 T_1 - \beta_1 - \beta_2 \quad (7.24.24)$$

(7.24.24) can be rewritten as

$$-\gamma_2' = \sigma_2 - \beta_1' - \beta_2', \quad \gamma_1' - \gamma_2' = \sigma_1 + \sigma_2 - \beta_3' + \beta_2' \quad (7.24.25)$$

Clearly there exists a trivial solution $a_1 = a_2 = a_3 = 0$ in (7.24.18)–(7.24.23). When $a_1 \neq 0, a_2 \neq 0, a_3 \neq 0$, we can obtain the following equations from (7.24.21)–(7.24.23) considering (7.24.25):

$$-\gamma_2' = \sigma_2 + \tfrac{1}{8} \sum_{m=1}^{3} \left(\tfrac{\alpha_{1m} a_m^2}{\omega_1} + \tfrac{\alpha_{2m} a_m^2}{\omega_2} \right) + \tfrac{H_{11}}{\omega_1} + \tfrac{H_{22}}{\omega_2} + Q\left(\tfrac{a_2^2 a_3}{\omega_1 a_1} + \tfrac{2 a_1 a_3}{\omega_2} \right) \cos\gamma_1 + \tfrac{1}{2} H_{12} \left(\tfrac{a_2}{\omega_1 a_1} + \tfrac{a_1}{\omega_2 a_2} \right) \sin\gamma_2 + \tfrac{1}{2} H_{23} \tfrac{a_3}{\omega_2 a_2} \cos(\gamma_1 - \gamma_2) \quad (7.24.26)$$

$$\gamma_1' - \gamma_2' = \sigma_1 + \sigma_2 + \tfrac{1}{8} \sum_{m=1}^{3} \left(\tfrac{\alpha_{3m} a_m^2}{\omega_3} - \tfrac{\alpha_{2m} a_m^2}{\omega_2} \right) + \tfrac{H_{33}}{\omega_3} - \tfrac{H_{22}}{\omega_2} + Q\left(\tfrac{a_1 a_2^2}{\omega_3 a_3} - \tfrac{2 a_1 a_3}{\omega_2} \right) \cos\gamma_1 - \tfrac{1}{2} H_{12} \tfrac{a_1}{\omega_2 a_2} \sin\gamma_2 + \tfrac{1}{2} H_{23} \left(\tfrac{a_2}{\omega_3 a_3} - \tfrac{a_3}{\omega_2 a_2} \right) \cos(\gamma_1 - \gamma_2) \quad (7.24.27)$$

Let the time derivatives in (7.24.18)–(7.24.20), (7.24.26) and (7.24.27) be zero, we can obtain the equations satisfied by the non-trivial steady state solutions of a_1, a_2, a_3 and γ_1, γ_2:

$$\omega_1 \mu_1 a_1 + Q a_2^2 a_3 \sin\gamma_1 + \frac{1}{2} H_{12} a_2 \sin\gamma_2 = 0 \quad (7.24.28)$$

$$\omega_2 \mu_2 a_2 + 2Q a_1 a_2 a_3 \sin\gamma_1 + \frac{1}{2} H_{12} a_1 \sin\gamma_2 + \frac{1}{2} H_{23} a_3 \sin(\gamma_1 - \gamma_2) = 0 \quad (7.24.29)$$

$$\omega_3 \mu_3 a_3 - Q a_1 a_2^2 \sin\gamma_1 - \frac{1}{2} H_{23} a_2 \sin(\gamma_1 - \gamma_2) = 0 \quad (7.24.30)$$

$$0=\sigma_2+\frac{1}{8}\sum_{m=1}^{3}\left(\frac{\alpha_{1m}a_m^2}{\omega_1}+\frac{\alpha_{2m}a_m^2}{\omega_2}\right)+\frac{H_{11}}{\omega_1}+\frac{H_{22}}{\omega_2}+Q\left(\frac{a_2^2a_3}{\omega_1 a_1}+\frac{2a_1a_3}{\omega_2}\right)\cos\gamma_1$$
$$+\frac{1}{2}H_{12}\left(\frac{a_2}{\omega_1 a_1}+\frac{a_1}{\omega_2 a_2}\right)\sin\gamma_2+\frac{1}{2}H_{23}\frac{a_3}{\omega_2 a_2}\cos(\gamma_1-\gamma_2) \tag{7.24.31}$$

$$0=\sigma_1+\sigma_2+\frac{1}{8}\sum_{m=1}^{3}\left(\frac{\alpha_{3m}a_m^2}{\omega_3}-\frac{\alpha_{2m}a_m^2}{\omega_2}\right)+\frac{H_{33}}{\omega_3}-\frac{H_{22}}{\omega_2}$$
$$+Q\left(\frac{a_1a_2^2}{\omega_3 a_3}-\frac{2a_1a_3}{\omega_2}\right)\cos\gamma_1-\frac{1}{2}H_{12}\frac{a_1}{\omega_2 a_2}\sin\gamma_2$$
$$+\frac{1}{2}H_{23}\left(\frac{a_2}{\omega_3 a_3}-\frac{a_3}{\omega_2 a_2}\right)\cos(\gamma_1-\gamma_2) \tag{7.24.32}$$

Combining the above results, we can obtain the steady-state deflection of subharmonic resonance of the axisymmetric circular plate when $2\Omega\approx\omega_1+\omega_2$ is

$$w(r,t)=\sum_{n=1}^{3}a_n\phi_n(r)\cos[(\omega_n+\varepsilon\beta_n{'})t+\tau_n]+\left[2\sum_{n=1}^{\infty}\Lambda_n\phi_n(r)\right]\cos\Omega t+O(\varepsilon) \tag{7.24.33}$$

where $a_m$and $\beta_m{'}$ are determined by (7.24.18)–(7.24.27).

Readers are invited to complete Exercises 7.25 to 7.27 by referring to Exercises 7.21 to 7.24.

7.25 Exercise 7.25 (Combined Resonance Analysis of a Uniform Circular Plate Clamped Along Its Edge I)

7.26 Exercise 7.26 (Combined Resonance Analysis of a Uniform Circular Plate Clamped Along Its Edge II)

7.27 Exercise 7.27 (Axisymmetric Response of a Uniform Circular Plate Clamped Along Its Edge with 1st to 3rd Order Modes Subjecting To Corresponding Resonant Excitation III)

7.28 Exercise 7.28 (Derivation of Modal Discretization Equations for Berger's Equation for Axisymmetric Oscillation Analysis of a Clamped Circular Plates)

Solution: **(a)** In the Berger approximation, the nonlinear forced oscillations of a clamped plate are given by

$$D\nabla^4 w - \rho h e c_p^2 \nabla^2 w + \rho h w_{tt} = -2\mu w_t + f(\mathbf{r}, t) \tag{7.28.1}$$

$$Ae(t) = -\frac{1}{2}\iint_A w\nabla^2 w dxdy \tag{7.28.2}$$

where A is the area of the plate, and

$$D = \frac{Eh^3}{12(1-\nu^2)},\ c_p^2 = \frac{E}{\rho(1-\nu^2)} \tag{7.28.3}$$

The following dimensionless variables with asterisks are induced:

$$r = Rr^*,\ t = R^2\sqrt{\frac{\rho h}{D}}t^*,\ w = \frac{h^2}{R}w^*,\ \mu = \frac{12}{R^4}\sqrt{\rho h^5 D}\mu^*,\ f = \frac{Dh^2}{R^5}f^* \tag{7.28.4}$$

Taking into account

$$\nabla^4 = \left(\frac{\partial^2}{\partial r^2} + \frac{1}{r}\frac{\partial}{\partial r}\right)^2 \tag{7.28.5}$$

Substituting this into (7.28.1) and (7.28.2), and then dropping the asterisks, we obtain

$$\nabla^4 w - \alpha\frac{R^4}{h^4}e\nabla^2 w + w_{tt} = -2\alpha\mu w_t + f(r, t) \tag{7.28.6}$$

$$\begin{aligned}
e(t) &= -\frac{h^4}{R^4}\int_0^1 rw\nabla^2 w dr = -\frac{h^4}{R^4}\int_0^1 rw\left(\frac{d^2 w}{dr^2} + \frac{1}{r}\frac{dw}{dr}\right)dr \\
&= -\frac{h^4}{R^4}\left[\int_0^1 rw\frac{d^2 w}{dr^2}dr + \int_0^1 w\frac{dw}{dr}dr\right] \\
&= -\frac{h^4}{R^4}\left[rw\frac{dw}{dr}\bigg|_0^1 - \int_0^1 \frac{dw}{dr}d(rw) + \int_0^1 w\frac{dw}{dr}dr\right] \\
&= \frac{h^4}{R^4}\int_0^1 r\left(\frac{dw}{dr}\right)^2 dr
\end{aligned} \tag{7.28.7}$$

where $\alpha = 12h^2/R^2$. The above two equations are combined to give

$$\nabla^4 w - \alpha\nabla^2 w \int_0^1 r\left(\frac{dw}{dr}\right)^2 dr + w_{tt} = -2\alpha\mu w_t + f(r,t) \tag{7.28.8}$$

Let the deflection of the circular plate be

$$w(r,t) = \sum_{m=1}^{\infty} \psi_m(t)\phi_m(r) \tag{7.28.9}$$

where the ϕ_m are the linear, free oscillation modes of the plate. The undamped free oscillation equation of the plate is

$$\nabla^4 w + w_{tt} = 0 \tag{7.28.10}$$

Let $w(r,t) = \psi_{m0}e^{\omega_m t}\phi_m(r)$, the corresponding eigenvalue problem can be obtained as:

$$\nabla^4\phi_m - \omega_m^2\phi_m = 0 \tag{7.28.11}$$

$$\phi_m(1) = 0,\ \ \phi_m{}'(1) = 0,\ \ \phi_m(0) < \infty \tag{7.28.12}$$

The eigenvalue ω_m are the natural frequencies of the plate. The ϕ_m are orthogonal with respect to the weighting function r. The amplitude of each mode is chosen such that

$$\int_0^1 r\phi_n\phi_m dr = \delta_{nm} \tag{7.28.13}$$

The eigenvalue problem is solved below. Write (7.28.11) in a more convenient form below:

$$\left(\frac{\partial^2}{\partial r^2} + \frac{1}{r}\frac{\partial}{\partial r} - \kappa_m^2\right)\left(\frac{\partial^2}{\partial r^2} + \frac{1}{r}\frac{\partial}{\partial r} + \kappa_m^2\right)\phi_m = 0 \tag{7.28.14}$$

where $\kappa_m^4 = \omega_m^2$. Thus we can obtain the four linearly independent solutions of (7.28.14) from the following two equations:

$$\left(\frac{\partial^2}{\partial r^2} + \frac{1}{r}\frac{\partial}{\partial r} - \kappa_m^2\right)\phi_m = 0 \tag{7.28.15}$$

$$\left(\frac{\partial^2}{\partial r^2}+\frac{1}{r}\frac{\partial}{\partial r}+\kappa_m^2\right)\phi_m=0 \tag{7.28.16}$$

From (7.28.15) and (7.28.16), we obtain

$$\phi_m^{(1)}=E_1I_0(\kappa_m r)+E_2K_0(\kappa_m r) \tag{7.28.17}$$

$$\phi_m^{(2)}=E_3J_0(\kappa_m r)+E_4Y_0(\kappa_m r) \tag{7.28.18}$$

where the E_n are constants of integration; J_n, Y_n are the Bessel functions of the first and second kind, respectively;I_n,K_n are the modified Bessel functions of the first and second kind, respectively. The complete solution is

$$\phi_m=\phi_m^{(1)}+\phi_m^{(2)} \tag{7.28.19}$$

The condition that $\phi_m(0)$ be bounded demands that $E_2=E_4=0$ because both K_0 and Y_0 have logarithmic singularities at the origin. Thus it follows from (7.28.12) and (7.28.19) that

$$\phi_m=C_m[J_0(\kappa_m r)I_0(\kappa_m)-J_0(\kappa_m)I_0(\kappa_m r)] \tag{7.28.20}$$

where the κ_m are the roots of

$$I_0(\kappa)J_0{}'(\kappa)-J_0(\kappa)I_0{}'(\kappa)=0 \tag{7.28.21}$$

and the C_m are obtained from (7.28.13).

The first five natural frequencies obtained from (7.28.21) are

$$\omega_m=10.2158,\quad 39.7710,\quad 89.1040,\quad 158.1830,\quad 247.0050$$

We note that

$$\omega_1+2\omega_2=89.7578\approx\omega_3 \tag{7.28.22}$$

Hence there is an internal resonance involving three modes.

Substituting (7.28.9) into (7.28.8) and taking into account the eigenvalue Eq. (7.28.11) yields

$$\begin{aligned}\sum_{m=1}^{\infty}\omega_m^2\psi_m\phi_m-\alpha\sum_{m=1}^{\infty}\psi_m\nabla^2\phi_m\sum_{p,q=1}^{\infty}\psi_p\psi_q\int_0^1 r\phi_p{}'\phi_q{}'dr\\+\sum_{m=1}^{\infty}\ddot{\psi}_m\phi_m=-2\alpha\mu\sum_{m=1}^{\infty}\dot{\psi}_m\phi_m+f(r,t)\end{aligned} \tag{7.28.23}$$

where the prime denotes the derivative with respect to r. Multiply (7.28.23) by $r\phi_n(r)$ from $r = 0$ to $r = 1$. The result is

$$\sum_{m=1}^{\infty} \omega_m^2 \psi_m \int_0^1 r\phi_n\phi_m dr - \alpha \sum_{m=1}^{\infty} \psi_m \int_0^1 r\phi_n \nabla^2\phi_m dr \sum_{p,q=1}^{\infty} \psi_p\psi_q \int_0^1 r\phi_{p'}\phi_{q'} dr$$
$$+ \sum_{m=1}^{\infty} \ddot{\psi}_m \int_0^1 r\phi_n\phi_m dr = -2\alpha \sum_{m=1}^{\infty} \dot{\psi}_m \int_0^1 \mu r\phi_n\phi_m dr + \int_0^1 r\phi_n f(r,t) dr \tag{7.28.24}$$

Considering the orthogonality of the ϕ_m, we obtain

$$\ddot{\psi}_n + \omega_n^2\psi_n = \alpha \sum_{m=1}^{\infty} \psi_m \int_0^1 r\phi_n \nabla^2\phi_m dr \sum_{p,q=1}^{\infty} \psi_P\psi_q \int_0^1 r\phi_{p'}\phi_{q'} dr$$
$$-2\alpha\mu_n\dot{\psi}_n + f_n(t) \tag{7.28.25}$$

where

$$\mu_n = \int_0^1 \mu r\phi_n^2 dr, f_n(t) = \int_0^1 r\phi_n(r) f(r,t) dr \tag{7.28.26}$$

Dealing with the Eq. (7.28.25) again. The first integral on the right-hand side of (7.28.25) is

$$\int_0^1 r\phi_n\nabla^2\phi_m dr = \int_0^1 r\phi_n\left(\frac{d^2\phi_m}{dr^2} + \frac{1}{r}\frac{d\phi_m}{dr}\right)dr = \int_0^1 r\phi_n \frac{d^2\phi_m}{dr^2} dr + \int_0^1 \phi_n \frac{d\phi_m}{dr} dr$$
$$= r\phi_n\big|_0^1 - \int_0^1 \frac{d\phi_m}{dr} d(r\phi_n) + \int_0^1 \phi_n \frac{\partial\phi_m}{\partial r} dr \tag{7.28.27}$$
$$= -\int_0^1 r \frac{d\phi_n}{dr}\frac{d\phi_m}{dr} dr$$

Substituting (7.28.27) into (7.28.25) yields

$$\ddot{\psi}_n + \omega_n^2\psi_n = \alpha \sum_{m,p,q}^{\infty} \Gamma_{nmpq}\psi_m\psi_p\psi_q - 2\alpha\mu_n\dot{\psi}_n + f_n(t) \tag{7.28.28}$$

where

$$\Gamma_{nmpq} = -\int_0^1 r\phi_n'\phi_m' dr \int_0^1 r\phi_p'\phi_q' dr \tag{7.28.29}$$

7.29 Exercise 7.29 (Berger Equation for Nonlinear Oscillation Analysis of Simply Supported Rectangular Plates)

Solution: (a) In the Berger approximation, the nonlinear forced oscillations of a simply-supported rectangular plate are given by

$$D\nabla^4 w - \rho hec_p^2\nabla^2 w + \rho h w_{tt} = f(\mathbf{r}, t) \tag{7.29.1}$$

$$Ae(t) = \frac{1}{2}\iint\limits_A \left[\left(\frac{\partial w}{\partial x}\right)^2 + \left(\frac{\partial w}{\partial y}\right)^2\right] dxdy \tag{7.29.2}$$

where A is the area of the plate; and

$$D = \frac{Eh^3}{12\left(1-\nu^2\right)},\ c_p^2 = \frac{E}{\rho\left(1-\nu^2\right)},\ \nabla^2 = \frac{\partial}{\partial x^2} + \frac{\partial}{\partial y^2} \tag{7.29.3}$$

Let the deflection of the simply supported plate be $w(x, y) = \phi(x, y)e^{i\omega t}$, then the corresponding eigenvalue problem can be obtained as:

$$D\nabla^4\phi - \rho h\omega^2\phi = 0 \tag{7.29.4}$$

$$\phi(0, y) = \phi(a, y) = 0,\ \phi(x, 0) = \phi(x, b) = 0 \tag{7.29.5}$$

The oscillation mode function for a simply supported rectangular plate is given by

$$\phi_{nm}(x, y) = \sin\frac{n\pi x}{a}\sin\frac{m\pi y}{b} \tag{7.29.6}$$

Natural frequencies are

$$\omega_{nm}^2 = \frac{D}{\rho h}\left(\frac{n^2}{a^2} + \frac{m^2}{b^2}\right)^2 \tag{7.29.7}$$

Therefore, let the deflection of the plate be

$$w(x, y, t) = \sum_{n,\,m=1}^{\infty} \psi_{nm}(t)\phi_{nm} \sum_{n,\,m=1}^{\infty} \psi_{nm}(t)\sin\frac{n\pi x}{a}\sin\frac{m\pi y}{b} \tag{7.29.8}$$

Substituting (7.29.8) into (7.29.2) yields

$$
\begin{aligned}
e(t) &= \frac{1}{2A}\iint_A \left[\left(\frac{\partial w}{\partial x}\right)^2 + \left(\frac{\partial w}{\partial y}\right)^2\right] dxdy \\
&= \frac{1}{2ab}\left\{\iint_A \left[\sum_{n,m}^{\infty} \psi_{nm}\frac{n\pi}{a}\cos\frac{n\pi x}{a}\sin\frac{m\pi y}{b}\right]^2 dxdy\right. \\
&\quad \left. + \iint_A \left[\sum_{n,m}^{\infty} \psi_{nm}\frac{m\pi}{b}\sin\frac{n\pi x}{a}\cos\frac{m\pi y}{b}\right]^2 dxdy\right\} \\
&= \frac{1}{2ab}\left\{\iint_A \left[\sum_{n,m,p,q}^{\infty} \psi_{nm}\psi_{pq}\frac{n\pi}{a}\frac{p\pi}{a}\cos\frac{n\pi x}{a}\cos\frac{p\pi x}{a}\sin\frac{m\pi y}{b}\sin\frac{q\pi y}{b}\right] dxdy\right. \\
&\quad \left. + \iint_A \left[\sum_{n,m,p,q}^{\infty} \psi_{nm}\psi_{pq}\frac{m\pi}{b}\frac{q\pi}{b}\sin\frac{n\pi x}{a}\sin\frac{p\pi x}{a}\cos\frac{m\pi y}{b}\cos\frac{q\pi y}{b}\right] dxdy\right\} \\
&= \frac{\pi^2}{2ab}\left\{\sum_{n,m}^{\infty} \psi_{nm}^2\frac{n^2}{a^2}\int_0^a \cos^2\frac{n\pi x}{a}dx\int_0^b \sin^2\frac{m\pi y}{b}dy\right. \\
&\quad \left. + \sum_{n,m}^{\infty} \psi_{nm}^2\frac{m^2}{b^2}\int_0^a \sin^2\frac{n\pi x}{a}dx\int_0^b \cos^2\frac{m\pi y}{b}dy\right\} \\
&= \frac{\pi^2}{2ab}\sum_{n,m}^{\infty}\frac{ab}{4}\left(\frac{n^2}{a^2}+\frac{m^2}{b^2}\right)\psi_{nm}^2
\end{aligned}
$$

i.e.,

$$
e(t) = \frac{\pi^2}{8}\sum_{n,\,m=1}^{\infty}\left(\frac{n^2}{a^2}+\frac{m^2}{b^2}\right)\psi_{nm}^2 \tag{7.29.9}
$$

Substituting (7.29.8) and (7.29.9) into (7.29.1) and considering (7.29.4) yields

$$
\begin{aligned}
D\sum_{n,\,m}^{\infty} \omega_{nm}^2\psi_{nm}\phi_{nm} - \rho h c_p^2 \tfrac{\pi^2}{8}\sum_{p,\,q=1}^{\infty}\left(\tfrac{p^2}{a^2}+\tfrac{q^2}{b^2}\right)\psi_{nm}^2\sum_{n,\,m}^{\infty}\psi_{nm}\nabla^2\phi_{nm} \\
+\rho h\sum_{n,\,m}^{\infty}\ddot{\psi}_{nm}\phi_{nm} = f(x, y, t)
\end{aligned} \tag{7.29.10}
$$

Multiplying the above equation by $\phi_{rs}(x, y)$ and integrating over the area of the plate, we obtain

$$\begin{aligned}
&\rho h \sum_{n,m}^{\infty} \omega_{nm}^2 \psi_{nm} \iint_A \phi_{nm}\phi_{rs} dxdy \\
&-\rho h c_p^2 \frac{\pi^2}{8} \sum_{p,q=1}^{\infty} \left(\frac{p^2}{a^2} + \frac{q^2}{b^2}\right) \psi_{pq}^2 \sum_{n,m}^{\infty} \omega_{nm}^2 \iint_A \phi_{rs} \nabla^2 \phi_{nm} dxdy \\
&+\rho h \sum_{n,m}^{\infty} \ddot{\psi}_{nm} \iint_A \phi_{nm}\phi_{rs} dxdy = \iint_A \phi_{rs} f(x, y, t) dxdy
\end{aligned} \tag{7.29.11}$$

The oscillation mode function have orthogonality, i.e.

$$\iint_A \phi_{nm}\phi_{rs} dxdy = \begin{cases} \frac{1}{4}ab, & n = r, \quad m = s \\ 0, & n \neq r, \quad m \neq s \end{cases} \tag{7.29.12}$$

Dealing with (7.29.11) again. The integral in the second term on the right-hand side of (7.29.11) is

$$\begin{aligned}
\iint_A \phi_{rs} \nabla^2 \phi_{nm} dxdy &= \iint_A \phi_{rs} \left[\frac{\partial^2 \phi_{nm}}{\partial x^2} + \frac{\partial^2 \phi_{nm}}{\partial y^2}\right] dxdy \\
&= -\pi^2 \left(\frac{n^2}{a^2} + \frac{m^2}{b^2}\right) \int_0^a \sin\frac{r\pi x}{a} \sin\frac{n\pi x}{a} dx \int_0^b \sin\frac{s\pi y}{b} \sin\frac{m\pi y}{b} dy \\
&= \begin{cases} -\frac{\pi^2 ab}{4}\left(\frac{n^2}{a^2} + \frac{m^2}{b^2}\right), & n = r, m = s \\ 0, & n \neq r, m \neq s \end{cases}
\end{aligned} \tag{7.29.13}$$

Substituting (7.29.9), (7.29.12) and (7.29.13) into (7.29.11) yields

$$\begin{aligned}
\ddot{\psi}_{nm} + \omega_{nm}^2 \psi_{nm} + \frac{\pi^4 c_p^2}{8} \left(\frac{n^2}{a^2} + \frac{m^2}{b^2}\right) \sum_{p,q=1}^{\infty} \left(\frac{p^2}{a^2} + \frac{q^2}{b^2}\right) \psi_{pq}^2 \psi_{nm} \\
= \frac{4}{\rho hab} \iint_A \phi_{nm} f(x, y, t) dxdy
\end{aligned} \tag{7.29.14}$$

The following dimensionless variables with asterisks are induced:

$$\begin{aligned}
&\psi_{nm} = l\psi_{nm}^*, x = lx^*, y = ly^*, a = la^*, b = lb^* \\
&t = l^2 \sqrt{\frac{\rho h}{D}} t^*, \omega_{nm} = \frac{1}{l^2} \sqrt{\frac{D}{\rho h}} \omega_{nm}^*, f = \frac{Dab}{4l^5} f^*
\end{aligned} \tag{7.29.15}$$

where l is the characteristic width of the plate; for a square plate, $l = a$; for a rectangular plate, l can be taken as the radius of gyration of the rectangular plate.

Substituting (7.29.15) into (7.29.14), and then dropping the asterisk for brevity yields

$$\ddot{\psi}_{nm} + \omega_{nm}^2 \psi_{nm} = \alpha \left(\frac{n^2}{a^2} + \frac{m^2}{b^2} \right) \sum_{p,q=1}^{\infty} \left(\frac{p^2}{a^2} + \frac{q^2}{b^2} \right) \psi_{pq}^2 \psi_{nm} + f_{nm}(t) \quad (7.29.16)$$

where

$$\alpha = -\frac{\pi^4 \rho h l^2 c_p^2}{8D}, \; f_{nm}(t) = \iint_A \phi_{nm} f(x, y, t) dxdy \quad (7.29.17)$$

Note that ω_{nm} in (7.29.16) and f in (7.29.17) have been nondimensionalized by (7.29.15). The dimensionless natural frequency ω_{nm} is defined as:

$$\omega_{nm}^2 = \left(\frac{n^2}{a^2} + \frac{m^2}{b^2} \right)^2 \quad (7.29.18)$$

Adding dimensionless linear damping to (7.29.16) yields

$$\ddot{\psi}_{nm} + \omega_{nm}^2 \psi_{nm} = \alpha \left(\frac{n^2}{a^2} + \frac{m^2}{b^2} \right) \sum_{p,q=1}^{\infty} \left(\frac{p^2}{a^2} + \frac{q^2}{b^2} \right) \psi_{pq}^2 \psi_{nm} - 2\mu_{nm} \dot{\psi}_{nm} + f_{nm}(t) \quad (7.29.19)$$

This is the dimensionless governing equation of discrete modes of a simply-supported rectangular plate with the Berger approximation.

Following the method of multiple scales, we let

$$\psi_{nm}(t; \varepsilon) = \varepsilon \psi_{nm0}(T_0, T_1, T_2) + \varepsilon^2 \psi_{nm1}(T_0, T_1, T_2) + \varepsilon^3 \psi_{nm2}(T_0, T_1, T_2) + \cdots \quad (7.29.20)$$

In order to let the damping, excitation and the cubic nonlinearity terms appear in the same order equation:

$$\mu_{nm} = \varepsilon^2 \hat{\mu}_{nm}, K_{nm} = 2\varepsilon^3 k_{nm} \quad (7.29.21)$$

where $k_{nm} = 0$ when $n \neq 1, m \neq 1$.

Substituting (7.29.20) and (7.29.21) into (7.29.19) and retaining to $O(\varepsilon^3)$, we obtain

$$
\begin{aligned}
0 &= \ddot{\psi}_{nm} + \omega_{nm}^2\psi_{nm} - \alpha\Gamma_{nm}\sum_{p,q=1}^{\infty}\Gamma_{pq}\psi_{pq}^2\psi_{nm} + 2\mu_{nm}\dot{\psi}_{nm} - f_{nm}(t) \\
&= \varepsilon\left(D_0^2\psi_{nm0} + \omega_{nm}^2\psi_{nm0}\right) + \varepsilon^2\left(D_0^2\psi_{nm1} + \omega_{nm}^2\psi_{nm1} + 2D_0D_1\psi_{nm0}\right) \\
&+\varepsilon^3\Big[D_0^2\psi_{nm2} + \omega_{nm}^2\psi_{nm2} + \left(D_1^2 + 2D_0D_2\right)\psi_{nm0} + 2D_0D_1\psi_{nm1} \\
&+2\hat{\mu}_{nm}D_0\psi_{nm0} - \alpha\Gamma_{nm}\sum_{p,q=1}^{\infty}\Gamma_{pq}\psi_{pq0}^2\psi_{nm0} - 2k_{nm}\cos\Omega T_0\Big]
\end{aligned} \tag{7.29.22}
$$

where

$$\Gamma_{nm} = \frac{n^2}{a^2} + \frac{m^2}{b^2} \tag{7.29.23}$$

Let the coefficients of the like powers of ε in (7.29.22) be zero, we obtain

$$D_0^2\psi_{nm0} + \omega_{nm}^2\psi_{nm0} = 0 \tag{7.29.24}$$

$$D_0^2\psi_{nm1} + \omega_{nm}^2\psi_{nm1} = -2D_0D_1\psi_{nm0} \tag{7.29.25}$$

$$
\begin{aligned}
D_0^2\psi_{nm2} + \omega_{nm}^2\psi_{nm2} &= -\left(D_1^2 + 2D_0D_2\right)\psi_{nm0} - 2D_0D_1\psi_{nm1} - 2\hat{\mu}_{nm}D_0\psi_{nm0} \\
&+\alpha\Gamma_{nm}\sum_{p,q=1}^{\infty}\Gamma_{pq}\psi_{pq0}^2\psi_{nm0} + 2k_{nm}\cos\Omega T_0
\end{aligned} \tag{7.29.26}
$$

The solution of (7.29.24) is

$$\psi_{nm0} = A_{nm}(T_1, T_2)e^{i\omega_{nm}T_0} + cc \tag{7.29.27}$$

Substituting (7.29.27) into (7.29.25) yields

$$D_0^2\psi_{nm1} + \omega_{nm}^2\psi_{nm1} = -2i\omega_{nm}D_1A_{nm}e^{i\omega_{nm}T_0} + cc \tag{7.29.28}$$

In order to eliminate secular terms from the above equation, we need

$$D_1A_{nm} = 0 \Rightarrow A_{nm} = A_{nm}(T_2) \tag{7.29.29}$$

Further, from (7.29.28), we have

$$\psi_{nm1} = 0 \tag{7.29.30}$$

Let

$$\Omega = \omega_{11} + \varepsilon^2 \sigma \tag{7.29.31}$$

and substitute them into (7.29.26), we obtain

$$\begin{aligned} &D_0^2 \psi_{112} + \omega_{11}^2 \psi_{112} = -2i\omega_{11} A_{11}' e^{i\omega_{11} T_0} - 2i\omega_{11} \hat{\mu}_{11} A_{11} e^{i\omega_{11} T_0} \\ &+3\alpha \Gamma_{11}^2 A_{11}^2 \overline{A}_{11} e^{i\omega_{11} T_0} + k_{11} e^{i\sigma T_2} e^{i\omega_{11} T_0} + cc + NST \end{aligned} \tag{7.29.32}$$

$$\begin{aligned} &D_0^2 \psi_{nm2} + \omega_{nm}^2 \psi_{nm2} = -2i\omega_{nm} A_{nm}' e^{i\omega_{nm} T_0} - 2i\omega_{nm} \hat{\mu}_{nm} A_{nm} e^{i\omega_{nm} T_0} \\ &+3\alpha \Gamma_{nm}^2 A_{mm}^2 \overline{A}_{mm} e^{i\omega_{nm} T_0} + cc + NST, \quad n \neq 1, \quad m \neq 1 \end{aligned} \tag{7.29.33}$$

In order to eliminate secular terms from these two equations, we need

$$-2i\omega_{11} A_{11}' - 2i\omega_{11} \hat{\mu}_{11} D_0 A_{11} + 3\alpha \Gamma_{11}^2 A_{11}^2 \overline{A}_{11} + k_{11} e^{i\sigma T_2} = 0 \tag{7.29.34}$$

$$\begin{gathered} -2i\omega_{nm} A_{nm}' - 2i\omega_{nm} \hat{\mu}_{nm} A_{nm} + 3\alpha \Gamma_{nm}^2 A_{mm}^2 \overline{A}_{mm} = 0 \\ n \neq 1, \quad m \neq 1 \end{gathered} \tag{7.29.35}$$

Let

$$A_{nm} = a_{nm} e^{i\beta_{nm}} \tag{7.29.36}$$

and substitute this into (7.29.35), we can obtain

$$\begin{gathered} -i\omega_{nm} a_{nm}' + \omega_{nm} a_{nm} \beta_{nm}' - i\omega_{nm} \hat{\mu}_{nm} a_{nm} + \tfrac{3}{8} \alpha \Gamma_{nm}^2 a_{mm}^3 = 0 \\ n \neq 1, \quad m \neq 1 \end{gathered} \tag{7.29.37}$$

Separating the real and imaginary parts of the above equation yields

$$a_{nm}' + \hat{\mu}_{nm} a_{nm} = 0, \ n \neq 1, \ m \neq 1 \tag{7.29.38}$$

So

$$a_{nm} = a_{nm0} e^{-\hat{\mu}_{nm} T_2} \to 0, \ n \neq 1, \ m \neq 1 \tag{7.29.39}$$

Substituting (7.29.36) into (7.29.34) yields

$$-i\omega_{11} a_{11}' + \omega_{11} a_{11} \beta_{11}' - i\omega_{11} \hat{\mu}_{11} a_{11} + \frac{3}{8} \alpha \Gamma_{11}^2 a_{11}^3 + k_{11} e^{i(\sigma T_2 - \beta_{11})} = 0 \tag{7.29.40}$$

Separating the real part from the imaginary part of the above equation yields

$$\begin{aligned} &-\omega_{11} a_{11}' - \omega_{11} \hat{\mu}_{11} a_{11} + k_{11} \sin\gamma_{11} = 0 \\ &-\omega_{11} a_{11} \gamma_{11}' + \omega_{11} \sigma a_{11} + \tfrac{3}{8} \alpha \Gamma_{11}^2 a_{11}^3 + k_{11} \cos\gamma_{11} = 0 \end{aligned} \tag{7.29.41}$$

where

$$\gamma_{11} = \sigma T_2 - \beta_{11} \tag{7.29.42}$$

From Eq. (7.29.41), we know that the steady state solution satisfies the following equations:

$$\begin{aligned} &\omega_{11}\hat{\mu}_{11}a_{11} = k_{11}\sin\gamma_{11} \\ &-a_{11}\left(\omega_{11}\sigma + \tfrac{3}{8}\alpha\Gamma_{11}^2 a_{11}^2\right) = k_{11}\cos\gamma_{11} \end{aligned} \tag{7.29.43}$$

From this, the frequency–response equation is

$$\omega_{11}^2\hat{\mu}_{11}^2 a_{11}^2 + a_{11}^2(\omega_{11}\sigma + \frac{3}{8}\alpha\Gamma_{11}^2 a_{11}^2)^2 = k_{11}^2 \tag{7.29.44}$$

As seen from the above results, only 1 ~ 1 mode is excited under the above resonant excitation.

When $a = b,\omega_{12} = \omega_{21}$. Since this is still a primary resonance of the plate, the solution to Exercise 7.29(b), (7.29.20)–(7.29.30), still holds.

When $k_{12} \neq k_{21}$, Eqs. (7.29.32) and (7.29.33) change to:

$$\begin{aligned} &D_0^2\psi_{122} + \omega_{12}^2\psi_{122} = -2i\omega_{12}A_{12}' e^{i\omega_{12}T_0} - 2i\omega_{12}\hat{\mu}_{12}A_{12}e^{i\omega_{12}T_0} \\ &+3\alpha\Gamma_{12}^2 A_{12}^2\overline{A}_{12}e^{i\omega_{12}T_0} + k_{12}e^{i\sigma T_2}e^{i\omega_{12}T_0} + cc + NST \end{aligned} \tag{7.29.45}$$

$$\begin{aligned} &D_0^2\psi_{nm2} + \omega_{nm}^2\psi_{nm2} = -2i\omega_{nm}A_{nm}' e^{i\omega_{nm}T_0} - 2i\omega_{nm}\hat{\mu}_{nm}A_{nm}e^{i\omega_{nm}T_0} \\ &+3\alpha\Gamma_{nm}^2 A_{mm}^2\overline{A}_{mm}e^{i\omega_{nm}T_0} + cc + NST, \quad n \neq 1, \ m \neq 2 \end{aligned} \tag{7.29.46}$$

From this, we can obtain

$$a_{nm} = a_{nm0}e^{-\hat{\mu}_{nm}T_2} \to 0, \quad n \neq 1, \quad m \neq 2 \tag{7.29.47}$$

Hence only 1 ~ 2 mode is excited in this case.

When $k_{12} \neq k_{21}$, Eqs. (7.29.32) and (7.29.33) change to:

$$\begin{aligned} &D_0^2\psi_{122} + \omega_{12}^2\psi_{122} = -2i\omega_{12}A_{12}' e^{i\omega_{12}T_0} - 2i\omega_{12}\hat{\mu}_{12}A_{12}e^{i\omega_{12}T_0} \\ &+3\alpha\Gamma_{12}^2 A_{12}^2\overline{A}_{12}e^{i\omega_{12}T_0} + k_{12}e^{i\sigma T_2}e^{i\omega_{12}T_0} + cc + NST \end{aligned} \tag{7.29.48}$$

$$\begin{aligned} &D_0^2\psi_{212} + \omega_{21}^2\psi_{212} = -2i\omega_{21}A_{21}' e^{i\omega_{21}T_0} - 2i\omega_{21}\hat{\mu}_{21}A_{21}e^{i\omega_{21}T_0} \\ &+3\alpha\Gamma_{21}^2 A_{21}^2\overline{A}_{21}e^{i\omega_{21}T_0} + k_{21}e^{i\sigma T_2}e^{i\omega_{21}T_0} + cc + NST \end{aligned} \tag{7.29.49}$$

$$\begin{aligned} &D_0^2\psi_{nm2} + \omega_{nm}^2\psi_{nm2} = -2i\omega_{nm}A_{nm}' e^{i\omega_{nm}T_0} - 2i\omega_{nm}\hat{\mu}_{nm}A_{nm}e^{i\omega_{nm}T_0} \\ &+3\alpha\Gamma_{nm}^2 A_{mm}^2\overline{A}_{mm}e^{i\omega_{nm}T_0} + cc + NST, \quad n \neq 1, \ m \neq 2 \end{aligned} \tag{7.29.50}$$

From this, we can obtain

$$a_{nm} = a_{nm0}e^{-\beta_{nm}T_2} \to 0,\ n \neq 1,\ m \neq 2 \text{ and } n \neq 2,\ m \neq 1 \tag{7.29.51}$$

Hence both 1 ~ 2 and 2 ~ 1 modes are excited.

7.30 Exercise 7.30 (Nonlinear Oscillation Analysis of Rotating Circular Films)

Solution: In order to have a more thorough understanding of the problem, let us first derive the nonlinear governing equations for the rotating film in the problem. Neglecting the inertial force of the longitudinal motion of the plate, we can obtain the equation of motion of the plate in the Cartesian coordinate system is (Sect. 7.6.1 of **the Book**)

$$\begin{aligned}&\frac{\partial^2 M_x}{\partial x^2}+\frac{\partial^2 M_y}{\partial y^2}+2\frac{\partial^2 M_{xy}}{\partial x\partial y}+\frac{\partial}{\partial x}\left(N_x\frac{\partial w}{\partial x}\right)+\frac{\partial}{\partial y}\left(N_y\frac{\partial w}{\partial y}\right)\\&+\frac{\partial}{\partial x}\left(N_{xy}\frac{\partial w}{\partial y}\right)+\frac{\partial}{\partial y}\left(N_{xy}\frac{\partial w}{\partial x}\right)+f=\rho h\frac{\partial^2 w}{\partial t^2}\end{aligned} \tag{7.30.1}$$

$$\frac{\partial N_x}{\partial x}+\frac{\partial N_{xy}}{\partial y}=-q_x \tag{7.30.2}$$

$$\frac{\partial N_{xy}}{\partial x}+\frac{\partial N_y}{\partial y}=-q_y \tag{7.30.3}$$

where q_x and q_y are the centrifugal loads in the rotating surface. Assuming that the film does not resist bending, i.e., the bending moment and torque in (7.30.1) are zeros, yields

$$\frac{\partial}{\partial x}\left(N_x\frac{\partial w}{\partial x}\right)+\frac{\partial}{\partial y}\left(N_y\frac{\partial w}{\partial y}\right)+\frac{\partial}{\partial x}\left(N_{xy}\frac{\partial w}{\partial y}\right)+\frac{\partial}{\partial y}\left(N_{xy}\frac{\partial w}{\partial x}\right)+f=\rho h\frac{\partial^2 w}{\partial t^2} \tag{7.30.4}$$

Let's rewrite Eqs. (7.30.2)–(7.30.4) in the polar coordinate form. As shown in Fig. 7.6a, b, we have

$$q_r=\rho(hrdrd\theta)\Omega^2 r/rdrd\theta=\rho h\Omega^2 r \tag{7.30.5}$$

The force equilibrium equations for the film element in radial and tangential directions are:

$$\left(N_r+\frac{\partial N_r}{\partial r}dr\right)(r+dr)d\theta-N_r rd\theta+\left(N_{r\theta}+\frac{\partial N_{r\theta}}{\partial\theta}d\theta\right)dr-N_{r\theta}dr$$

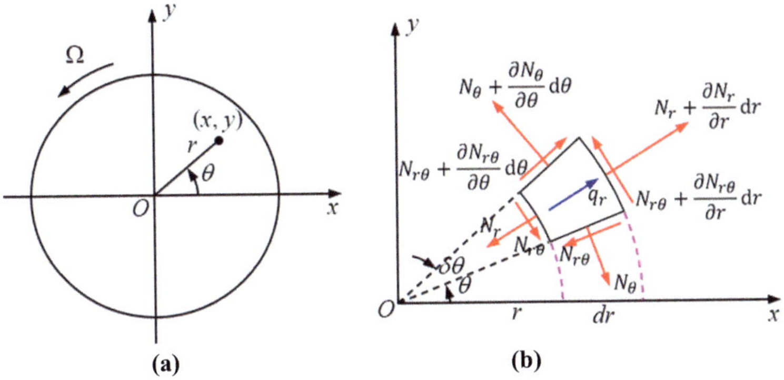

Fig. 7.6 **a** Polar coordinate, **b** force analysis on thin film elements for Exercise 7.30

$$- \left(N_\theta + \frac{\partial N_\theta}{\partial \theta} d\theta \right) drd\theta + q_r r\, drd\theta = 0$$

$$\left(N_\theta + \frac{\partial N_\theta}{\partial \theta} d\theta \right) dr - N_\theta dr + \left(N_{r\theta} + \frac{\partial N_{r\theta}}{\partial r} dr \right) (r + dr) d\theta - N_{r\theta} r d\theta$$
$$+ \left(N_{r\,\theta} + \frac{\partial N_{r\,\theta}}{\partial \theta} d\theta \right) drd\theta = 0$$

The above two equations are reorganized to give

$$\begin{aligned} \frac{\partial (N_r r)}{\partial r} + \frac{\partial N_{r\theta}}{\partial \theta} - N_\theta &= -\rho h \Omega^2 r^2 \\ \frac{\partial (N_{r\theta} r)}{\partial r} + \frac{\partial N_\theta}{\partial \theta} + N^r_{r\theta} &= 0 \end{aligned} \tag{7.30.6}$$

Define the stress function according to the following equation F:

$$\begin{aligned} N_r &= \frac{1}{r} \frac{\partial F}{\partial r} + \frac{1}{r^2} \frac{\partial^2 F}{\partial \theta^2} - \frac{1}{2} \rho h \Omega^2 r^2 \\ N_\theta &= \frac{\partial^2 F}{\partial r^2} - \frac{1}{2} \rho h \Omega^2 r^2, \quad N_{r\theta} = -\frac{\partial}{\partial r} \left(\frac{1}{r} \frac{\partial F}{\partial \theta} \right) \end{aligned} \tag{7.30.7}$$

It is easy to verify that the N_r, N_θ, $N_{r\theta}$ given by the above equation automatically satisfy the equations of equilibrium shown in (7.30.6).

In order to convert (7.30.4) into polar coordinate form, we consider the relationship between the derivatives in two coordinate systems. First, we have

$$r^2 = x^2 + y^2, \ \tan\theta = \frac{y}{x} \tag{7.30.8}$$

From this, we have

$$\frac{\partial r}{\partial x}=\frac{x}{r}=\cos\theta,\quad \frac{\partial r}{\partial y}=\frac{y}{r}=\sin\theta$$
$$\frac{\partial\theta}{\partial x}=-\frac{y}{r^2}=-\frac{\sin\theta}{r},\quad \frac{\partial\theta}{\partial y}=\frac{x}{r^2}=\frac{\cos\theta}{r} \tag{7.30.9}$$

Furthermore, for any function $G(x, y)$ in the xy plane, or $G(r\cos\theta, r\sin\theta)$ in the $r\theta$ plane, there are

$$\frac{\partial G}{\partial x}=\frac{\partial G}{\partial r}\frac{\partial r}{\partial x}+\frac{\partial G}{\partial\theta}\frac{\partial\theta}{\partial x}=\cos\theta\frac{\partial G}{\partial r}-\frac{\sin\theta}{r}\frac{\partial G}{\partial\theta}$$
$$\frac{\partial G}{\partial y}=\frac{\partial G}{\partial r}\frac{\partial r}{\partial y}+\frac{\partial G}{\partial\theta}\frac{\partial\theta}{\partial y}=\sin\theta\frac{\partial G}{\partial r}+\frac{\cos\theta}{r}\frac{\partial G}{\partial\theta} \tag{7.30.10}$$

$$\frac{\partial^2 G}{\partial x^2}=\cos^2\theta\frac{\partial^2 G}{\partial r^2}+\sin^2\theta\left(\frac{1}{r}\frac{\partial G}{\partial r}+\frac{1}{r^2}\frac{\partial^2 G}{\partial\theta^2}\right)-2\sin\theta\cos\theta\frac{\partial}{\partial r}\left(\frac{1}{r}\frac{\partial G}{\partial\theta}\right)$$
$$\frac{\partial^2 G}{\partial y^2}=\sin^2\theta\frac{\partial^2 G}{\partial r^2}+\cos^2\theta\left(\frac{1}{r}\frac{\partial G}{\partial r}+\frac{1}{r^2}\frac{\partial^2 G}{\partial\theta^2}\right)+2\sin\theta\cos\theta\frac{\partial}{\partial r}\left(\frac{1}{r}\frac{\partial G}{\partial\theta}\right)$$
$$\frac{\partial^2 G}{\partial x\partial y}=-\sin\theta\cos\theta\left(\frac{1}{r}\frac{\partial G}{\partial r}+\frac{1}{r^2}\frac{\partial^2 G}{\partial\theta^2}-\frac{\partial^2 G}{\partial r^2}\right)+\left(\cos^2\theta-\sin^2\theta\right)\frac{\partial}{\partial r}\left(\frac{1}{r}\frac{\partial G}{\partial\theta}\right) \tag{7.30.11}$$

$$\nabla^2 G=\left(\frac{\partial^2}{\partial x^2}+\frac{\partial^2}{\partial y^2}\right)G=\left(\frac{\partial^2}{\partial r^2}+\frac{1}{r}\frac{\partial}{\partial r}+\frac{1}{r^2}\frac{\partial^2}{\partial\theta^2}\right)G \tag{7.30.12}$$

The relationship between the internal forces in the midplane N_x, N_y, N_{xy} and $N_r, N_\theta, N_{r\theta}$ in both coordinate systems are

$$N_x=N_r\cos^2\theta+N_\theta\sin^2\theta-2N_{r\theta}\sin\theta\cos\theta$$
$$N_y=N_r\sin^2\theta+N_\theta\cos^2\theta+2N_{r\theta}\sin\theta\cos\theta$$
$$N_{xy}=(N_r-N_\theta)\sin\theta\cos\theta+N_{r\theta}\left(\cos^2\theta-\sin^2\theta\right) \tag{7.30.13}$$

$$N_r=N_x\cos^2\theta+N_y\sin^2\theta+2N_{xy}\sin\theta\cos\theta$$
$$N_\theta=N_x\sin^2\theta+N_y\cos^2\theta-2N_{xy}\sin\theta\cos\theta$$
$$N_{r\theta}=\left(N_y-N_x\right)\sin\theta\cos\theta+N_{xy}\left(\cos^2\theta-\sin^2\theta\right) \tag{7.30.14}$$

Substituting (7.30.7) into (7.30.13) yields

$$N_x=\frac{1}{r}\frac{\partial F}{\partial r}\cos^2\theta+\frac{1}{r^2}\frac{\partial^2 F}{\partial\theta^2}\cos^2\theta+\frac{\partial^2 F}{\partial r^2}\sin^2\theta$$
$$+2\frac{\partial}{\partial r}\left(\frac{1}{r}\frac{\partial F}{\partial\theta}\right)\sin\theta\cos\theta-\frac{1}{2}\rho h\Omega^2 r^2$$

$$
\begin{aligned}
N_y &= \frac{1}{r}\frac{\partial F}{\partial r}\sin^2\theta + \frac{1}{r^2}\frac{\partial^2 F}{\partial \theta^2}\sin^2\theta + \frac{\partial^2 F}{\partial r^2}\cos^2\theta \\
&\quad - 2\frac{\partial}{\partial r}\left(\frac{1}{r}\frac{\partial F}{\partial \theta}\right)\sin\theta\cos\theta - \frac{1}{2}\rho h\Omega^2 r^2 \\
N_{xy} &= \left(\frac{1}{r}\frac{\partial F}{\partial r} + \frac{1}{r^2}\frac{\partial^2 F}{\partial \theta^2} - \frac{\partial^2 F}{\partial r^2}\right)\sin\theta\cos\theta \\
&\quad - \frac{\partial}{\partial r}\left(\frac{1}{r}\frac{\partial F}{\partial \theta}\right)\left(\cos^2\theta - \sin^2\theta\right)
\end{aligned}
\tag{7.30.15}
$$

Remove the external load f from Eq. (7.30.4) and take into account (7.30.2) and (7.30.3), we obtain

$$
\begin{aligned}
\rho h\frac{\partial^2 w}{\partial t^2} &= \frac{\partial}{\partial x}\left(N_x\frac{\partial w}{\partial x}\right) + \frac{\partial}{\partial y}\left(N_y\frac{\partial w}{\partial y}\right) + \frac{\partial}{\partial x}\left(N_{xy}\frac{\partial w}{\partial y}\right) + \frac{\partial}{\partial y}\left(N_{xy}\frac{\partial w}{\partial x}\right) \\
&= N_x\frac{\partial^2 w}{\partial x^2} + N_y\frac{\partial^2 w}{\partial y^2} + 2N_{xy}\frac{\partial^2 w}{\partial x\partial y} + \left(\frac{\partial N_x}{\partial x} + \frac{\partial N_{xy}}{\partial y}\right)\frac{\partial w}{\partial x} + \left(\frac{\partial N_{xy}}{\partial x} + \frac{\partial N_y}{\partial y}\right)\frac{\partial w}{\partial y} \\
&= N_x\frac{\partial^2 w}{\partial x^2} + N_y\frac{\partial^2 w}{\partial y^2} + 2N_{xy}\frac{\partial^2 w}{\partial x\partial y} - q_x\frac{\partial w}{\partial x} - q_y\frac{\partial w}{\partial y}
\end{aligned}
$$

i.e.,

$$
\rho h\frac{\partial^2 w}{\partial t^2} = N_x\frac{\partial^2 w}{\partial x^2} + N_y\frac{\partial^2 w}{\partial y^2} + 2N_{xy}\frac{\partial^2 w}{\partial x\partial y} - q_x\frac{\partial w}{\partial x} - q_y\frac{\partial w}{\partial y} \tag{7.30.16}
$$

Let's deal with the terms on the right-hand side of (7.30.16). First, we know

$$
q_x = q_r\cos\theta, \quad q_y = q_r\sin\theta \tag{7.30.17}
$$

So

$$
q_x\frac{\partial w}{\partial x} + q_y\frac{\partial w}{\partial y} = q_r\cos\theta\frac{\partial w}{\partial x} + q_r\sin\theta\frac{\partial w}{\partial y} = \rho h\Omega^2 r\frac{\partial w}{\partial r} \tag{7.30.18}
$$

Using (7.30.15) and (7.30.11), we can write the first term on the right-hand side of (7.30.16) as

$$
\begin{aligned}
N_x\frac{\partial^2 w}{\partial x^2} &= \left[\frac{1}{r}\frac{\partial F}{\partial r}\cos^2\theta + \frac{1}{r^2}\frac{\partial^2 F}{\partial \theta^2}\cos^2\theta + \frac{\partial^2 F}{\partial r^2}\sin^2\theta + 2\frac{\partial}{\partial r}\left(\frac{1}{r}\frac{\partial F}{\partial \theta}\right)\sin\theta\cos\theta\right] \\
&\quad \times\left[\cos^2\theta\frac{\partial^2 w}{\partial r^2} + \sin^2\theta\left(\frac{1}{r}\frac{\partial w}{\partial r} + \frac{1}{r^2}\frac{\partial^2 w}{\partial \theta^2}\right) - 2\sin\theta\cos\theta\frac{\partial}{\partial r}\left(\frac{1}{r}\frac{\partial w}{\partial \theta}\right)\right] \\
&\quad - \frac{1}{2}\rho h\Omega^2 r^2\frac{\partial^2 w}{\partial x^2}
\end{aligned}
\tag{7.30.19}
$$

The second term on the right-hand side of (7.30.16) is

$$N_y\frac{\partial^2 w}{\partial y^2} = \left[\frac{1}{r}\frac{\partial F}{\partial r}\sin^2\theta + \frac{1}{r^2}\frac{\partial^2 F}{\partial \theta^2}\sin^2\theta + \frac{\partial^2 F}{\partial r^2}\cos^2\theta - 2\frac{\partial}{\partial r}\left(\frac{1}{r}\frac{\partial F}{\partial \theta}\right)\sin\theta\cos\theta\right]$$
$$\times\left[\sin^2\theta\frac{\partial^2 w}{\partial r^2} + \cos^2\theta\left(\frac{1}{r}\frac{\partial w}{\partial r} + \frac{1}{r^2}\frac{\partial^2 w}{\partial \theta^2}\right) + 2\sin\theta\cos\theta\frac{\partial}{\partial r}\left(\frac{1}{r}\frac{\partial w}{\partial \theta}\right)\right]$$
$$-\frac{1}{2}\rho h\Omega^2 r^2\frac{\partial^2 w}{\partial y^2} \tag{7.30.20}$$

The third term on the right-hand side of (7.30.16) is

$$2N_{xy}\frac{\partial^2 w}{\partial x\partial y} = 2\left[\left(\frac{1}{r}\frac{\partial F}{\partial r} + \frac{1}{r^2}\frac{\partial^2 F}{\partial \theta^2} - \frac{\partial^2 F}{\partial r^2}\right)\sin\theta\cos\theta - \frac{\partial}{\partial r}\left(\frac{1}{r}\frac{\partial F}{\partial \theta}\right)(\cos^2\theta - \sin^2\theta)\right]$$
$$\times\left[-\sin\theta\cos\theta\left(\frac{1}{r}\frac{\partial w}{\partial r} + \frac{1}{r^2}\frac{\partial^2 w}{\partial \theta^2} - \frac{\partial^2 w}{\partial r^2}\right) + (\cos^2\theta - \sin^2\theta)\frac{\partial}{\partial r}\left(\frac{1}{r}\frac{\partial w}{\partial \theta}\right)\right] \tag{7.30.21}$$

After reorganization, we obtain

$$N_x\frac{\partial^2 w}{\partial x^2} + N_y\frac{\partial^2 w}{\partial y^2} + 2N_{xy}\frac{\partial^2 w}{\partial x\partial y}$$
$$= -\frac{1}{2}\rho h\Omega^2 r^2\nabla^2 w + \frac{\partial^2 w}{\partial r^2}\left(\frac{1}{r}\frac{\partial F}{\partial r} + \frac{1}{r^2}\frac{\partial^2 F}{\partial \theta^2}\right)$$
$$+\left(\frac{1}{r}\frac{\partial w}{\partial r} + \frac{1}{r^2}\frac{\partial^2 w}{\partial \theta^2}\right)\frac{\partial^2 F}{\partial r^2} - 2\frac{\partial}{\partial r}\left(\frac{1}{r}\frac{\partial w}{\partial \theta}\right)\frac{\partial}{\partial r}\left(\frac{1}{r}\frac{\partial F}{\partial \theta}\right) \tag{7.30.22}$$

Substituting (7.30.18) and (7.30.22) into (7.30.16), we can obtain the governing equation for the rotating film

$$\rho h\left(\frac{\partial^2 w}{\partial t^2} + r\,\Omega^2\frac{\partial w}{\partial r} + \frac{1}{2}r^2\Omega^2\nabla^2 w\right) = \frac{\partial^2 w}{\partial r^2}\left(\frac{1}{r}\frac{\partial F}{\partial r} + \frac{1}{r^2}\frac{\partial^2 F}{\partial \theta^2}\right)$$
$$+\left(\frac{1}{r}\frac{\partial w}{\partial r} + \frac{1}{r^2}\frac{\partial^2 w}{\partial \theta^2}\right)\frac{\partial^2 F}{\partial r^2}$$
$$-2\frac{\partial}{\partial r}\left(\frac{1}{r}\frac{\partial w}{\partial \theta}\right)\frac{\partial}{\partial r}\left(\frac{1}{r}\frac{\partial F}{\partial \theta}\right) \tag{7.30.23}$$

The following is to derive the relation between the stress function F and the transverse displacement w. From the definition of the stress function (7.30.7), we obtain

$$\nabla^4 F = 4\rho h\Omega^2 + \nabla^2(N_r + N_\theta) \tag{7.30.24}$$

From (7.30.14), we can see that the Eq. (7.30.24) becomes

$$\nabla^4 F = 4\rho h \Omega^2 + \nabla^2 (N_x + N_y) \tag{7.30.25}$$

By (7.30.2) and (7.30.3), we have

$$\frac{\partial^2 N_x}{\partial x^2} + \frac{\partial^2 N_y}{\partial y^2} + 2\frac{\partial^2 N_{xy}}{\partial x \partial y} = -\left(\frac{\partial q_x}{\partial x} + \frac{\partial q_y}{\partial y}\right) \tag{7.30.26}$$

Substituting (7.30.6) into (7.30.25) yields

$$\nabla^4 F = 4\rho h \Omega^2 - \left(\frac{\partial q_x}{\partial x} + \frac{\partial q_y}{\partial y}\right) + \frac{\partial^2 N_y}{\partial x^2} + \frac{\partial^2 N_x}{\partial y^2} - 2\frac{\partial^2 N_{xy}}{\partial x \partial y} \tag{7.30.27}$$

The midplane strain is

$$\varepsilon_x = \frac{\partial u}{\partial x} + \frac{1}{2}\left(\frac{\partial w}{\partial x}\right)^2, \ \varepsilon_y = \frac{\partial v}{\partial y} + \frac{1}{2}\left(\frac{\partial w}{\partial y}\right)^2, \ \gamma_{xy} = \frac{\partial v}{\partial x} + \frac{\partial u}{\partial y} + \frac{\partial w}{\partial x}\frac{\partial w}{\partial y} \tag{7.30.28}$$

The midplane internal force can be expressed in terms of midplane strain as

$$N_x = \frac{Eh}{1-\nu^2}(\varepsilon_x + \nu \varepsilon_x) \tag{7.30.29}$$

$$N_y = \frac{Eh}{1-\nu^2}(\varepsilon_y + \nu \varepsilon_y) \tag{7.30.30}$$

$$N_{xy} = Gh\gamma_{xy} = \frac{Eh}{2(1-\nu^2)}(1-\nu)\gamma_{xy} \tag{7.30.31}$$

which adopts the relationship between the elastic constants of isotropic materials $G = E/2(1+\nu)$.

Substituting (7.30.29)–(7.30.31) into (7.30.26) yields

$$\frac{\partial^2 \varepsilon_x}{\partial x^2} + \frac{\partial^2 \varepsilon_y}{\partial y^2} + \frac{\partial^2 \varepsilon_{xy}}{\partial x \partial y} = -\nu\left(\frac{\partial^2 \varepsilon_x}{\partial x^2} + \frac{\partial^2 \varepsilon_y}{\partial y^2} - \frac{\partial^2 \varepsilon_{xy}}{\partial x \partial y}\right) - \frac{(1-\nu^2)}{Eh}\left(\frac{\partial q_x}{\partial x} + \frac{\partial q_y}{\partial y}\right) \tag{7.30.32}$$

From the expression of the midplane strain (7.30.28), we can obtain the compatibility equation for midplane strain

$$\frac{\partial^2 \varepsilon_x}{\partial y^2} + \frac{\partial^2 \varepsilon_y}{\partial x^2} - \frac{\partial^2 \gamma_{xy}}{\partial x \partial y} = \left(\frac{\partial^2 w}{\partial x \partial y}\right)^2 - \frac{\partial^2 w}{\partial x^2}\frac{\partial^2 w}{\partial y^2} \tag{7.30.33}$$

Substituting (7.30.29)–(7.30.30) into (7.30.27) yields

$$\nabla^4 F = 4\rho h\Omega^2 - \left(\frac{\partial q_x}{\partial x} + \frac{\partial q_y}{\partial y}\right) + \frac{Eh}{1-\nu^2}\left[\frac{\partial^2 \varepsilon_x}{\partial y^2} + \frac{\partial^2 \varepsilon_y}{\partial x^2} - \frac{\partial^2 \gamma_{xy}}{\partial x \partial y} + \nu\left(\frac{\partial^2 \varepsilon_x}{\partial x^2} + \frac{\partial^2 \varepsilon_y}{\partial y^2} + \frac{\partial^2 \gamma_{xy}}{\partial x \partial y}\right)\right] \tag{7.30.34}$$

Substituting (7.30.32) into (7.30.34) yields

$$\nabla^4 F = 4\rho h\Omega^2 - (1+\nu)\left(\frac{\partial q_x}{\partial x} + \frac{\partial q_y}{\partial y}\right) + Eh\left(\frac{\partial^2 \varepsilon_x}{\partial y^2} + \frac{\partial^2 \varepsilon_y}{\partial x^2} - \frac{\partial^2 \gamma_{xy}}{\partial x \partial y}\right) \tag{7.30.35}$$

Substituting the compatibility Eq. (7.30.33) into the above equation, we obtain

$$\nabla^4 F = 4\rho h\Omega^2 - (1+\nu)\left(\frac{\partial q_x}{\partial x} + \frac{\partial q_y}{\partial y}\right) + Eh[\left(\frac{\partial^2 w}{\partial x \partial y}\right)^2 - \frac{\partial^2 w}{\partial x^2}\frac{\partial^2 w}{\partial y^2}\Bigg] \tag{7.30.36}$$

Now, let's return the Eq. (7.30.36) back into the polar coordinate system. We have

$$\begin{aligned}\frac{\partial q_x}{\partial x} + \frac{\partial q_y}{\partial y} &= \cos\theta\frac{\partial(q_r\cos\theta)}{\partial r} - \frac{\sin\theta}{r}\frac{\partial(q_r\cos\theta)}{\partial\theta} \\ &\quad + \sin\theta\frac{\partial(q_r\sin\theta)}{\partial r} + \frac{\cos\theta}{r}\frac{\partial(q_r\sin\theta)}{\partial\theta} \\ &= 2\rho h\Omega^2\end{aligned} \tag{7.30.37}$$

$$\begin{aligned}\frac{\partial^2 w}{\partial x^2}\frac{\partial^2 w}{\partial y^2} &= \sin^2\theta\cos^2\theta\left(\frac{\partial^2 w}{\partial r^2}\right)^2 + \sin^2\theta\sin^2\theta\frac{\partial^2 w}{\partial r^2}\left(\frac{1}{r}\frac{\partial w}{\partial r} + \frac{1}{r^2}\frac{\partial^2 w}{\partial\theta^2}\right) \\ &\quad + \sin^2\theta\cos^2\theta\left(\frac{1}{r}\frac{\partial w}{\partial r} + \frac{1}{r^2}\frac{\partial^2 w}{\partial\theta^2}\right)^2 - 4\sin^2\theta\cos^2\theta\left[\frac{\partial}{\partial r}\left(\frac{1}{r}\frac{\partial w}{\partial\theta}\right)\right]^2 \\ &\quad + \cos^4\theta\left(\frac{1}{r}\frac{\partial w}{\partial r} + \frac{1}{r^2}\frac{\partial^2 w}{\partial\theta^2}\right)\frac{\partial^2 w}{\partial r^2} + 2\sin\theta\cos\theta\left(\cos^2\theta - \sin^2\theta\right)\frac{\partial^2 w}{\partial r^2}\frac{\partial}{\partial r}\left(\frac{1}{r}\frac{\partial w}{\partial\theta}\right) \\ &\quad - 2\sin\theta\cos\theta\left(\cos^2\theta - \sin^2\theta\right)\left(\frac{1}{r}\frac{\partial w}{\partial r} + \frac{1}{r^2}\frac{\partial^2 w}{\partial\theta^2}\right)\frac{\partial}{\partial r}\left(\frac{1}{r}\frac{\partial w}{\partial\theta}\right)\end{aligned}$$

$$\begin{aligned}\left(\frac{\partial^2 w}{\partial x \partial y}\right)^2 &= \sin^2\theta\cos^2\theta\left(\frac{1}{r}\frac{\partial w}{\partial r} + \frac{1}{r^2}\frac{\partial^2 w}{\partial\theta^2} - \frac{\partial^2 w}{\partial r^2}\right)^2 \\ &\quad + \left(\cos^2\theta - \sin^2\theta\right)^2\left[\frac{\partial}{\partial r}\left(\frac{1}{r}\frac{\partial w}{\partial\theta}\right)\right]^2 \\ &\quad - 2\sin\theta\cos\theta\left(\cos^2\theta - \sin^2\theta\right)\frac{\partial}{\partial r}\left(\frac{1}{r}\frac{\partial w}{\partial\theta}\right)\left(\frac{1}{r}\frac{\partial w}{\partial r} + \frac{1}{r^2}\frac{\partial^2 w}{\partial\theta^2} - \frac{\partial^2 w}{\partial r^2}\right)\end{aligned}$$

So

$$\left(\frac{\partial^2 w}{\partial x \partial y}\right)^2 - \frac{\partial^2 w}{\partial x^2}\frac{\partial^2 w}{\partial y^2} = -\left(\frac{1}{r}\frac{\partial w}{\partial r} + \frac{1}{r^2}\frac{\partial^2 w}{\partial \theta^2}\right)\frac{\partial^2 w}{\partial r^2} + \left[\frac{\partial}{\partial r}\left(\frac{1}{r}\frac{\partial w}{\partial \theta}\right)\right]^2$$
$$= -\left(\frac{1}{r}\frac{\partial w}{\partial r} + \frac{1}{r^2}\frac{\partial^2 w}{\partial \theta^2}\right)\frac{\partial^2 w}{\partial r^2} + \frac{1}{r^2}\left(\frac{\partial^2 w}{\partial r \partial \theta}\right)^2$$
$$- \frac{2}{r^3}\frac{\partial^2 w}{\partial r \partial \theta}\frac{\partial w}{\partial \theta} + \frac{1}{r^4}\left(\frac{\partial w}{\partial \theta}\right)^2 \tag{7.30.38}$$

Substituting (7.30.37) and (7.30.38) into (7.30.36) yields

$$\nabla^4 F = 2\rho h(1-\nu)\Omega^2$$
$$+ Eh\left[-\left(\frac{1}{r}\frac{\partial w}{\partial r} + \frac{1}{r^2}\frac{\partial^2 w}{\partial \theta^2}\right)\frac{\partial^2 w}{\partial r^2} + \frac{1}{r^2}\left(\frac{\partial^2 w}{\partial r \partial \theta}\right)^2 - \frac{2}{r^3}\frac{\partial^2 w}{\partial r \partial \theta}\frac{\partial w}{\partial \theta} + \frac{1}{r^4}\left(\frac{\partial w}{\partial \theta}\right)^2\right] \tag{7.30.39}$$

Now we have deduced all governing equations for a rotating film expressed in terms of the transverse displacement w and the stress function F.

Let

$$w = Ar^2 \sin 2(\theta \pm ct)$$
$$F = Br^4 \cos 4(\theta \pm ct) + \frac{1}{32}\left[2hEA^2 + (1-\nu)\rho h\Omega^2\right]r^4 - Cr^2 \tag{7.30.40}$$

Substituting (7.30.40) into the governing Eq. (7.30.23) yields

$$2\rho h\left(-2c^2 + \Omega^2\right) = -\frac{1}{2}\left[2hEA^2 + (1-\nu)\rho h\Omega^2\right] + 48B$$

The above equation is reorganized to give

$$c^2 - \frac{1}{8}(5-\nu)\Omega^2 = \frac{E}{4\rho}A^2 - \frac{12B}{\rho h} \tag{7.30.41}$$

Substituting (7.30.40) into (7.30.39), we can find that the equation is automatically satisfied.

For a free rotating circular film with radius R, the boundary conditions can be written as

$$N_r(R,\theta) = 0, \;\; N_{r\theta}(R,\theta) = 0 \tag{7.30.42}$$

Combining (7.30.7) and (7.30.40), we have

$$N_r(R,\theta) = \left[\frac{1}{r}\frac{\partial F}{\partial r} + \frac{1}{r^2}\frac{\partial^2 F}{\partial \theta^2} - \frac{1}{2}\rho h\Omega^2 r^2\right]\Big|_{r=R}$$

$$
\begin{aligned}
&= -12BR^2 \cos 4(\theta \pm ct) - \frac{1}{2}\rho h\Omega^2 R^2 \\
&\quad + \frac{1}{8}\left[2hEA^2 + (1-\nu)\rho h\Omega^2\right] R^2 - 2C = 0 \\
N_{r\theta}(R,\theta) &= \left[-\frac{\partial}{\partial r}\left(\frac{1}{r}\frac{\partial F}{\partial \theta}\right)\right]\Big|_{r=R} \\
&= -12BR^2 \sin 4(\theta \pm ct) = 0
\end{aligned} \tag{7.30.43}
$$

So

$$
B = 0,\ 2C = \left[\frac{1}{4}hEA^2 - \frac{1}{8}(3+\nu)\rho h\Omega^2\right]R^2 \tag{7.30.44}
$$

The displacement solution $w = Ar^2\sin 2(\theta \pm ct)$ given in **(c)** is a wave propagating in the circumferential direction with a velocity c. The wave does not change with time along the axial direction; therefore, this wave, with a definite wave form, propagates around the central axis with an angular velocity c. When this wave form rotates at the same speed and in the same direction as the film, it becomes a stationary wave; i.e., the transverse displacement, as well as the stress and strain, at any definite point on the film does not change with time. Therefore, for stationary wave,$c = \Omega$. Further, from (7.30.41) and (7.30.44), we have

$$
\Omega^2 = 2EA^2/\rho(3+\nu) \tag{7.30.45}
$$

For a free spinning membrane with radius R, we assume that the separated variable solution to (7.30.39) be

$$
w = \left[A\left(\frac{r}{R}\right)^2 \cos 2\theta\right]\psi(t) \tag{7.30.46}
$$

Substituting (7.30.46) into (7.30.39) yields

$$
\nabla^4 F = 2\rho h(1-\nu)\Omega^2 + \frac{4EhA^2}{R^4}\psi^2
$$

i.e.,

$$
\left(\frac{\partial^2}{\partial r^2} + \frac{1}{r}\frac{\partial}{\partial r} + \frac{1}{r^2}\frac{\partial^2}{\partial \theta^2}\right)^2 F = 2\rho h(1-\nu)\Omega^2 + \frac{4EhA^2}{R^4}\psi^2 \tag{7.30.47}
$$

The right-hand side of Eq. (7.30.47) is independent of the spatial variables r and θ, and the lowest order of the derivative of the left-hand side of the equation F with respect to θ is second order; therefore, if we wish to obtain a polynomial solution for F, then F can only take the quadratic and linear terms of θ. Since θ is a rotational variable in the whole circumference, F would be a multivalued function of θ, which might cause the film to break up along any radial direction and is not allowed (or a case that is outside of the scope of the present study). In summary, we can only take F to be independent of θ. Thus, the Eq. (7.30.47) becomes

$$\left(\frac{\partial^2}{\partial r^2}+\frac{1}{r}\frac{\partial}{\partial r}\right)^2 F=2\rho h(1-\nu)\Omega^2+\frac{4EhA^2}{R^4}\psi^2 \tag{7.30.48}$$

The solution to (7.30.48) can be assumed to be

$$F=a_4r^4+a_2r^2 \tag{7.30.49}$$

Substituting (7.30.49) into (7.30.48) yields

$$64a_4=2\rho h(1-\nu)\Omega^2+\frac{4EhA^2}{R^4}\psi^2 \tag{7.30.50}$$

Note that the Eq. (7.30.49) cannot contain the term r^3; otherwise, the Eq. (7.30.50) would have a term containing $1/r$ on the right-hand side.

Then, from the boundary condition (7.30.42), we have

$$\begin{aligned} N_r(R,\theta) &= \left[\frac{1}{r}\frac{\partial F}{\partial r}+\frac{1}{r^2}\frac{\partial^2 F}{\partial \theta^2}-\frac{1}{2}\rho h\Omega^2 r^2\right]\Big|_{r=R} \\ &= 4a_4R^2+2a_2-\frac{1}{2}\rho h\Omega^2R^2=0 \end{aligned} \tag{7.30.51}$$

The boundary condition $N_{r\theta}(R,\theta)=0$ is automatically satisfied.

From (7.30.50) and (7.30.51), we have

$$\begin{aligned} a_4 &= \frac{1}{32}\rho h(1-\nu)\Omega^2+\frac{EhA^2}{16R^4}\psi^2 \\ a_2 &= \frac{1}{16}\rho h(3+\nu)\Omega^2R^2-\frac{EhA^2}{8R^2}\psi^2 \end{aligned} \tag{7.30.52}$$

Therefore, the stress function F is

$$\begin{aligned} F=&\frac{EhA^2}{16}\left[\left(\frac{r}{R}\right)^4-2\left(\frac{r}{R}\right)^2\right] \\ &\left[\psi^2+\frac{1}{32}\rho h(1-\nu)\Omega^2r^4+\frac{1}{16}\rho h(3+\nu)\Omega^2R^2r^2\right] \end{aligned} \tag{7.30.53}$$

Substitute the expression for the displacement w (7.30.46) and the stress function F (7.30.53) into the governing Eq. (7.30.23), we obtain

$$\ddot{\psi}+\alpha_1\psi+\alpha_3\psi^3=0 \tag{7.30.54}$$

where

$$\alpha_1 = \frac{1}{2}(5 - \nu)\Omega^2, \quad \alpha_3 = \frac{EA^2}{\rho R^4} \tag{7.30.55}$$

Chapter 8
Traveling Waves

8.1 Exercise 8.1 (Determination on Dispersive Wave)

Solution: (a) Let the solution of the equation be

$$u(x, t) = ae^{i(kx-\omega t)} \tag{8.1.1}$$

Substituting this into the given equation gives

$$k - \omega = 0 \Rightarrow c_0 = \frac{\omega}{k} = 1 \tag{8.1.2}$$

It can be seen that the phase speed c_0 is constant and hence the solution to the given equation is a non-dispersive wave.

(b) Substituting (8.1.1) into the given equation, we get

$$\omega - k + ik^2 = 0 \Rightarrow c_0 = \frac{\omega}{k} = 1 - ik \tag{8.1.3}$$

It can be seen that the phase speed c_0 is a function of the wave number k and therefore the solution to the given equation is a dispersive wave.

(c) Substituting (8.1.1) into the given equation, we get

$$\omega = k - k^3 \Rightarrow c_0 = \frac{\omega}{k} = 1 - k^2 \tag{8.1.4}$$

It can be seen that the phase speed c_0 is a function of the wave number k and therefore the solution to the given equation is a dispersive wave.

(d) Substituting (8.1.1) into the given equation, we get

Z. He et al., *Solved Problems in Nonlinear Oscillations*,
https://doi.org/10.1007/978-981-97-6113-5_8

$$\omega^2 = k^2 + 1 \Rightarrow c_0 = \frac{\omega}{k} = \sqrt{1 + k^{-2}} \tag{8.1.5}$$

It can be seen that the phase speed c_0 is a function of the wave number k and therefore the solution to the given equation is a dispersive wave.

8.2 Exercise 8.2 (Direct Expansion and Reformulation Analysis of Longitudinal Waves of Semi-infinite Uniform Rods with Material Nonlinearity)

Solution: (a) The given equation is a nonlinear control equation for a uniform semi-infinite bar with material nonlinearity. Let the solution of the equation can be expressed as

$$u(x, t) = \varepsilon u_1(x, t) + \varepsilon^2 u_2(x, t) + \cdots \tag{8.2.1}$$

Substituting this into the control equation and retaining to $O(\varepsilon^2)$ yields

$$\begin{aligned} 0 &= \frac{\partial^2 u}{\partial x^2} - \frac{\partial^2 u}{\partial t^2} + 2E_1 \frac{\partial u}{\partial x} \frac{\partial^2 u}{\partial x^2} \\ &= \varepsilon \left(\frac{\partial^2 u_1}{\partial x^2} - \frac{\partial^2 u_1}{\partial t^2} \right) + \varepsilon^2 \left(\frac{\partial^2 u_2}{\partial x^2} - \frac{\partial^2 u_2}{\partial t^2} + 2E_1 \frac{\partial u_1}{\partial x} \frac{\partial^2 u_1}{\partial x^2} \right) \end{aligned} \tag{8.2.2}$$

Equating coefficients of like powers of ε in (8.2.2) yields

$$\frac{\partial^2 u_1}{\partial x^2} - \frac{\partial^2 u_1}{\partial t^2} = 0 \tag{8.2.3}$$

$$\frac{\partial^2 u_2}{\partial x^2} - \frac{\partial^2 u_2}{\partial t^2} = -2E_1 \frac{\partial u_1}{\partial x} \frac{\partial^2 u_1}{\partial x^2} \tag{8.2.4}$$

Equation (8.2.3) is a one-dimensional wave equation whose general solution for a right-running wave is

$$u_1(x, t) = f(t - x) = f(s_1), s_1 = t - x \tag{8.2.5}$$

Substituting (8.2.5) into (8.2.4) yields

$$\frac{\partial^2 u_2}{\partial x^2} - \frac{\partial^2 u_2}{\partial t^2} = 2E_1 \frac{df}{ds_1} \frac{d^2 f}{ds_1^2} \tag{8.2.6}$$

Let

$$u_2(x,t) = u_2(x,s_1) = X(x)S(s_1) \tag{8.2.7}$$

Substituting this into (8.2.6) yields

$$S\frac{d^2X}{dx^2} - 2\frac{dX}{dx}\frac{dS}{ds_1} = 2E_1\frac{df}{ds_1}\frac{d^2f}{ds_1^2} \tag{8.2.8}$$

where $\partial^2 S/\partial x^2 - \partial^2 S/\partial t^2 = 0$ has been considered. In order to make the left side of the above equation to be a function of s_1, we need

$$\frac{dX}{dx} = \text{constant} \Rightarrow X(x) = Cx + B \tag{8.2.9}$$

where C, B is the constant of integration. Since $BS(s_1)$ is the homogeneous solution of (8.2.6) and it need not be considered, hence $X(x) = Cx$. The Eq. (8.2.8) becomes

$$-2C\frac{dS}{ds_1} = E_1\frac{df'^2}{ds_1} \tag{8.2.10}$$

therefore

$$S = -\frac{1}{2C}E_1 f'^2 \tag{8.2.11}$$

and

$$u_2 = -\frac{1}{2}E_1 x f'^2 \tag{8.2.12}$$

Finally,

$$u = \varepsilon f(s_1) - \frac{1}{2}\varepsilon^2 E_1 x f'^2(s_1) + \cdots \tag{8.2.13}$$

Applying boundary conditions to the above equation yields

$$\phi(t) = f(t) \Rightarrow f(s_1) = \phi(s_1) \tag{8.2.14}$$

Therefore, the approximate solution of the original equation is

$$u = \varepsilon\phi(s_1) - \frac{1}{2}\varepsilon^2 E_1 x \phi'^2(s_1) + \cdots \tag{8.2.15}$$

(b) The longitudinal strain of the rod e is

$$e = \frac{\partial u}{\partial x} = -\varepsilon\phi'(s_1) - \frac{1}{2}\varepsilon^2 E_1 \phi'^2(s_1) + \varepsilon^2 E_1 x \phi'(s_1)\phi''(s_1) + \cdots$$

i.e.,

$$e = -\varepsilon\phi'(s_1) - \frac{1}{2}\varepsilon^2 E_1\left[\phi'^2(s_1) - 2x\phi'(s_1)\phi''(s_1)\right] + \cdots \tag{8.2.16}$$

The second term on the right-hand side of the above equation diverges as x increases. When $x = O(\varepsilon^{-1})$ or larger, the second term is of the same order as the first term or becomes of lower order. Therefore, the above expansion is not uniformly valid.

(c) Apply the reformulation method such that

$$s_1 = \xi + \varepsilon\xi_1(x, \xi) + \cdots \tag{8.2.17}$$

Substituting this into the expression of strain yields

$$e = -\varepsilon\phi'(\xi) - \frac{1}{2}\varepsilon^2 E_1\phi'^2(\xi) - \varepsilon^2\left[E_1 x\phi'(\xi)\phi''(\xi) - \xi_1\phi''(\xi)\right] + \cdots \tag{8.2.18}$$

In order to eliminate secular terms from the above equation, we need

$$\xi_1 = E_1 x\phi'(\xi) \tag{8.2.19}$$

So, the approximate expressions for e and s_1 become

$$e = -\varepsilon\phi'(\xi) + \cdots \tag{8.2.20}$$

$$s_1 = \xi + \varepsilon E_1 x\phi'(\xi) + \cdots \tag{8.2.21}$$

(d) Substituting (8.2.17) into (8.2.15) yields

$$\begin{aligned} u &= \varepsilon\phi(s_1) - \frac{1}{2}\varepsilon^2 E_1 x\phi'^2(s_1) + \cdots \\ &= \varepsilon\phi(\xi) + \varepsilon^2\left[\xi_1\phi'(\xi) - \frac{1}{2}E_1 x\phi'^2(\xi)\right] + \cdots \end{aligned} \tag{8.2.22}$$

In order to eliminate secular terms from the above equation, we need

$$\xi_1 = \frac{1}{2}E_1 x\phi'(\xi) \tag{8.2.23}$$

So, the approximate expressions for u and s_1 become

$$u = \varepsilon\phi(\xi) + \cdots \tag{8.2.24}$$

$$s_1 = \xi + \frac{1}{2}\varepsilon E_1 x\phi'(\xi) + \cdots \tag{8.2.25}$$

In this case, the approximate expression for the strain e becomes

$$\begin{aligned} e = \frac{\partial u}{\partial x} &= -\varepsilon\phi'(s_1) - \frac{1}{2}\varepsilon^2 E_1 \phi'^2(s_1) + \varepsilon^2 E_1 x\phi'(s_1)\phi''(s_1) + \cdots \\ &= -\varepsilon\phi'(\xi) - \frac{1}{2}\varepsilon^2 E_1 x\phi'(\xi)\phi''(\xi) - \frac{1}{2}\varepsilon^2 E_1 \phi'^2(\xi) + \varepsilon^2 E_1 x\phi'(\xi)\phi''(\xi) + \cdots \\ &= -\varepsilon\phi'(\xi) + \frac{1}{2}\varepsilon^2 E_1 x\phi'(\xi)\phi''(\xi) - \frac{1}{2}\varepsilon^2 E_1 \phi'^2(\xi) + \cdots \end{aligned}$$

i.e.,

$$e = -\varepsilon\phi'(s_1) - \frac{1}{2}\varepsilon^2 E_1\left[\phi'^2(s_1) - x\phi'(s_1)\phi''(s_1)\right] + \cdots \tag{8.2.26}$$

Clearly, such results for strain diverge with x and are not uniformly valid.

8.3 Exercise 8.3 (Direct Expansion and Reformulation Analysis of a Nonlinear Acoustic Equation)

Solution: (a) The given equation is the governing equation for the nonlinear acoustic wave of the pipe. Since the given edge value $\phi_x(0, t)$ contains a constant term, let its direct expansion be

$$\phi(x, t) = Mx + \varepsilon\phi_1(x, t) + \varepsilon^2\phi_2(x, t) + \cdots \tag{8.3.1}$$

Substituting this into the control equation and retaining to $O(\varepsilon^2)$ yields

$$\begin{aligned} 0 &= \phi_{tt} - \phi_{xx} + 2\phi_x\phi_{xt} - (1-\gamma)\left[\phi_t + \frac{1}{2}\phi_x^2 - \frac{1}{2}M^2\right]\phi_{xx} + \phi_x^2\phi_{xx} \\ &= \varepsilon\frac{\partial^2\phi_1}{\partial t^2} + \varepsilon^2\frac{\partial^2\phi_2}{\partial t^2} - \varepsilon\frac{\partial^2\phi_1}{\partial x^2} - \varepsilon^2\frac{\partial^2\phi_2}{\partial x^2} \\ &+ 2\left(M + \varepsilon\frac{\partial\phi_1}{\partial x}\right)\left(\varepsilon\frac{\partial^2\phi_1}{\partial x\partial t} + \varepsilon^2\frac{\partial^2\phi_2}{\partial x\partial t}\right) \\ &- (1-\gamma)\left[\varepsilon\frac{\partial\phi_1}{\partial t} + \frac{1}{2}\left(M + \varepsilon\frac{\partial\phi_1}{\partial x}\right)^2 - \frac{1}{2}M^2\right]\left(\varepsilon\frac{\partial^2\phi_1}{\partial x^2} + \varepsilon^2\frac{\partial^2\phi_2}{\partial x^2}\right) \\ &+ \left(M + \varepsilon\frac{\partial\phi_1}{\partial x}\right)^2\left(\varepsilon\frac{\partial^2\phi_1}{\partial x^2} + \varepsilon^2\frac{\partial^2\phi_2}{\partial x^2}\right) \end{aligned}$$

$$
\begin{aligned}
&= \varepsilon\left(\frac{\partial^2\phi_1}{\partial t^2} + 2M\frac{\partial^2\phi_1}{\partial x\partial t} + M^2\frac{\partial^2\phi_1}{\partial x^2} - \frac{\partial^2\phi_1}{\partial x^2}\right) \\
&+ \varepsilon^2\left[\begin{array}{l} \frac{\partial^2\phi_2}{\partial t^2} + 2M\frac{\partial^2\phi_2}{\partial x\partial t} + M^2\frac{\partial^2\phi_2}{\partial x^2} - \frac{\partial^2\phi_2}{\partial x^2} + 2\frac{\partial\phi_1}{\partial x}\frac{\partial^2\phi_1}{\partial x\partial t} \\ -(1-\gamma)\frac{\partial\phi_1}{\partial t}\frac{\partial^2\phi_1}{\partial x^2} - (1-\gamma)M\frac{\partial\phi_1}{\partial x}\frac{\partial^2\phi_1}{\partial x^2} + 2M\frac{\partial\phi_1}{\partial x}\frac{\partial^2\phi_1}{\partial x^2} \end{array}\right]
\end{aligned}
\quad (8.3.2)
$$

Equating coefficients of like powers of ε in (8.3.2) yields

$$
\begin{aligned}
&\frac{\partial^2\phi_1}{\partial t^2} + 2M\frac{\partial^2\phi_1}{\partial x\partial t} + M^2\frac{\partial^2\phi_1}{\partial x^2} - \frac{\partial^2\phi_1}{\partial x^2} = 0 \\
&\frac{\partial^2\phi_2}{\partial t^2} + 2M\frac{\partial^2\phi_2}{\partial x\partial t} + M^2\frac{\partial^2\phi_2}{\partial x^2} - \frac{\partial^2\phi_2}{\partial x^2} = -2\frac{\partial\phi_1}{\partial x}\frac{\partial^2\phi_1}{\partial x\partial t} \\
&+(1-\gamma)\tfrac{\partial\phi_1}{\partial t}\tfrac{\partial^2\phi_1}{\partial x^2} + (1-\gamma)M\tfrac{\partial\phi_1}{\partial x}\tfrac{\partial^2\phi_1}{\partial x^2} - 2M\tfrac{\partial\phi_1}{\partial x}\tfrac{\partial^2\phi_1}{\partial x^2}
\end{aligned}
$$

These two equations can be written as

$$
\left(\frac{\partial}{\partial t} + M\frac{\partial}{\partial x}\right)^2\phi_1 - \frac{\partial^2\phi_1}{\partial x^2} = 0 \quad (8.3.3)
$$

$$
\begin{aligned}
&\left(\tfrac{\partial}{\partial t} + M\tfrac{\partial}{\partial x}\right)^2\phi_2 - \tfrac{\partial^2\phi_2}{\partial x^2} = -2\tfrac{\partial\phi_1}{\partial x}\tfrac{\partial^2\phi_1}{\partial x\partial t} + (1-\gamma)\tfrac{\partial\phi_1}{\partial t}\tfrac{\partial^2\phi_1}{\partial x^2} \\
&+(1-\gamma)\tfrac{\partial\phi_1}{\partial x}\tfrac{\partial^2\phi_1}{\partial x^2} - 2M\tfrac{\partial\phi_1}{\partial x}\tfrac{\partial^2\phi_1}{\partial x^2}
\end{aligned}
\quad (8.3.4)
$$

If we make $\eta = (1+M)t - x$, $\phi_1 = \phi_1(\eta)$, the general solution of the Eq. (8.3.3) is

$$
\phi_1 = g(\eta)
$$

or

$$
\phi_1 = f(\xi), \xi = t - (1+M)^{-1}x \quad (8.3.5)
$$

Substituting (8.3.5) into (8.3.4) yields

$$
\begin{aligned}
&\left(\tfrac{\partial}{\partial t} + M\tfrac{\partial}{\partial x}\right)^2\phi_2 - \tfrac{\partial^2\phi_2}{\partial x^2} = -2(1+M)^{-2}\tfrac{dg}{d\xi}\tfrac{d^2g}{d\xi^2} + (1-\gamma)(1+M)^{-2}\tfrac{dg}{d\xi}\tfrac{d^2g}{d\xi^2} \\
&- (1-\gamma)M(1+M)^{-3}\tfrac{dg}{d\xi}\tfrac{d^2g}{d\xi^2} \\
&+ 2M(1+M)^{-3}\tfrac{dg}{d\xi}\tfrac{d^2g}{d\xi^2} \\
&= -(1+M)^{-3}(1+\gamma)\tfrac{dg}{d\xi}\tfrac{d^2g}{d\xi^2}.
\end{aligned}
\quad (8.3.6)
$$

Let

$$
\phi_2(x,t) = \phi_2(x,\xi) = X(x)H(\xi) \quad (8.3.7)
$$

Substituting (8.3.7) into (8.3.6) yields

$$\left(M^2-1\right)\frac{d^2X}{dx^2}H+2\frac{dX}{dx}\frac{dH}{d\xi}=-\frac{1}{2}(1+M)^{-3}(1+\gamma)\frac{dg'^2}{d\xi} \tag{8.3.8}$$

The prime represents the derivative with respect to ξ. In order to make the left side of the above equation to be a function of ξ, one must take

$$\frac{dX}{dx}=cons\tan t\Rightarrow X(x)=Cx+B \tag{8.3.9}$$

where C, B are the constants of integration. Since $BH(\xi)$ is the homogeneous solution of (8.3.3) it need not be considered. Hence $X(x)=Cx$. The equation (8.3.8) becomes

$$\frac{dH}{d\xi}=-\frac{1}{4C}(1+M)^{-3}(1+\gamma)\frac{dg'^2}{d\xi} \tag{8.3.10}$$

Therefore

$$H(\xi)=-\frac{1}{4C}(1+M)^{-3}(1+\gamma)g'^2 \tag{8.3.11}$$

And

$$\phi_2=-\frac{1}{4}(1+M)^{-3}(1+\gamma)xg'^2 \tag{8.3.12}$$

thereby

$$\phi(x,t)=Mx+\varepsilon g(\xi)-\frac{1}{4}\varepsilon^2(1+M)^{-3}(1+\gamma)xg'^2(\xi) \tag{8.3.13}$$

From all these, we can obtain

$$\begin{aligned}u=\tfrac{\partial\phi}{\partial x}=M&-\varepsilon(1+M)^{-1}g'(\xi)-\tfrac{1}{4}\varepsilon^2(1+M)^{-3}(1+\gamma)g'^2(\xi)\\+\tfrac{1}{2}\varepsilon^2(1+M)^{-4}(1+\gamma)xg'(\xi)g''(\xi)\end{aligned} \tag{8.3.14}$$

Applying boundary conditions to the above equation yields

$$f(t)=-(1+M)^{-1}g'(t)\Rightarrow g'(\xi)=-(1+M)f(\xi) \tag{8.3.15}$$

Substituting (8.3.15) into (8.3.14), with only secular terms are retained for higher-order terms, yields

$$u=M+\varepsilon f(\xi)+\frac{1}{2}\varepsilon^2(\gamma+1)(1+M)^{-2}xf(\xi)f\prime(\xi)+O\left(\varepsilon^3\right) \tag{8.3.16}$$

Clearly, this solution diverges with x and is a non-uniform expansion.

(b) Apply the reformulation method to solve the problem. Let $\xi = s + \varepsilon s_1(s, x) + \cdots$, and substitute this into (8.3.16), we can obtain

$$u = M + \varepsilon f(s) + \varepsilon^2 f'(s)\left[s_1 + \frac{1}{2}(\gamma + 1)(1 + M)^{-2} x f(s)\right] + \cdots \tag{8.3.17}$$

In order to eliminate secular terms in the above equation, we need

$$s_1 = -\frac{1}{2}(\gamma + 1)(1 + M)^{-2} x f(s) \tag{8.3.18}$$

Therefore, the solution after reformulation is

$$u = M + \varepsilon f(s) + \cdots \tag{8.3.19}$$

where

$$t - (1 + M)^{-1} x = s - \frac{1}{2}\varepsilon(\gamma + 1)(1 + M)^{-2} x f(s) + \cdots \tag{8.3.20}$$

8.4 Exercise 8.4 (Multiscale Analysis on a One-Dimensional Wave Equation with Cubic Nonlinearity)

Solution: (a) We use the method of multiple scales to find the solution. The corresponding linear homogeneous equation of the given equation is a one-dimensional wave equation with a right-running wave solution as a function of $s_1 = t - x$. The nonlinear term in the equation is the cube of the partial derivatives of u with respect to time t. In order to take this nonlinear effect into account, we introduce the slow-varying time variable $T_1 = \varepsilon t$; therefore, the solution of the equation is set as a function of s_1 and T_1. The space and time derivatives are transformed according to:

$$\begin{aligned} &\frac{\partial}{\partial t} = \frac{\partial}{\partial s_1} + \varepsilon\frac{\partial}{\partial T_1}, \ \frac{\partial^2}{\partial t^2} = \frac{\partial^2}{\partial s_1^2} + 2\varepsilon\frac{\partial^2}{\partial s_1 \partial T_1} + \varepsilon^2\frac{\partial^2}{\partial T_1^2} \\ &\frac{\partial^2}{\partial x^2} = \frac{\partial^2}{\partial s_1^2} \end{aligned} \tag{8.4.1}$$

Let the solution of the given equation be

$$u(x, t) = u_0(s_1, T_1) + \varepsilon u_1(s_1, T_1) + \cdots \tag{8.4.2}$$

Substituting (8.4.1) and (8.4.2) into the given equation and keeping to $O(\varepsilon)$, we get

$$\begin{aligned} 0 &= \frac{\partial^2 u}{\partial t^2} - \frac{\partial^2 u}{\partial x^2} + \varepsilon\left(\frac{\partial u}{\partial t}\right)^3 \\ &= \varepsilon\left[2\frac{\partial^2 u_0}{\partial s_1 \partial T_1} + \left(\frac{\partial u_0}{\partial s_1}\right)^3\right] \end{aligned} \tag{8.4.3}$$

Equating coefficients of like powers of ε in (8.4.3) yields

$$2\frac{\partial^2 u_0}{\partial s_1 \partial T_1} + \left(\frac{\partial u_0}{\partial s_1}\right)^3 = 0 \tag{8.4.4}$$

Let the solution of the Eq. (8.4.4) is

$$u_0 = f(s_1, T_1) \tag{8.4.5}$$

then from (8.4.2), we can obtain

$$u(x, t) = f(s_1, T_1) + O(\varepsilon) \tag{8.4.6}$$

f satisfies Eq. (8.4.4), i.e.,

$$2\frac{\partial^2 f}{\partial s_1 \partial T_1} + \left(\frac{\partial f}{\partial s_1}\right)^3 = 0 \tag{8.4.7}$$

Write Eq. (8.4.7) as

$$2\frac{\partial g}{\partial T_1} = -g^3,\ g = \frac{\partial f}{\partial s_1} \tag{8.4.8}$$

Separating the variables of the above equation yields

$$2\frac{dg}{g^3} = -dT_1 \tag{8.4.9}$$

After integration,

$$g^{-2} = T_1 + F(s_1) \tag{8.4.10}$$

i.e.,

$$\frac{\partial f}{\partial s_1} = [T_1 + F(s_1)]^{-1/2} \tag{8.4.11}$$

(c) From the given initial conditions as well as Eqs. (8.4.6) and (8.4.1), we have

$$f(-x,0) = a\sin\omega x \tag{8.4.12}$$

$$\frac{\partial}{\partial s_1} f(-x,0) + \varepsilon \frac{\partial}{\partial T_1} f(-x,0) = -a\omega\cos\omega x \tag{8.4.13}$$

Substituting (8.4.11) into (8.4.13) yields

$$\frac{1}{\sqrt{F(-x)}} + \varepsilon \frac{\partial}{\partial T_1} f(-x,0) = -a\omega\cos\omega x \tag{8.4.14}$$

Equating the coefficients of like powers of ε in the above equation yields

$$\frac{1}{\sqrt{F(-x)}} = -a\omega\cos\omega x \Rightarrow \frac{1}{\sqrt{F(s_1)}} = -a\omega\cos\omega s_1 \tag{8.4.15}$$

So

$$F(s_1) = \frac{\sec^2 \omega s_1}{a^2\omega^2} \tag{8.4.16}$$

8.5 Exercise 8.5 (A Simplified Model for Wind-Induced Oscillation of Overhead Power Lines)

Solution: (a) We use the method of multiple scales to find the solution. The corresponding linear homogeneous equation of the given equation is a one-dimensional wave equation with a right-running wave solution as a function of $s_1 = t - x$. The nonlinear term in the equation is the cube of the partial derivatives of u with respect to time t. In order to take this nonlinear effect into account, we introduce the slow-varying time variable $T_1 = \varepsilon t$; therefore, the solution of the equation is set as a function of s_1 *and* T_1. The space and time derivatives are transformed according to:

$$\begin{aligned} &\frac{\partial}{\partial t} = \frac{\partial}{\partial s_1} + \varepsilon \frac{\partial}{\partial T_1}, \frac{\partial^2}{\partial t^2} = \frac{\partial^2}{\partial s_1^2} + 2\varepsilon \frac{\partial^2}{\partial s_1 \partial T_1} + \varepsilon^2 \frac{\partial^2}{\partial T_1^2} \\ &\frac{\partial^2}{\partial x^2} = \frac{\partial^2}{\partial s_1^2} \end{aligned} \tag{8.5.1}$$

Let the solution of the given equation be

$$u(x,t) = u_0(s_1, T_1) + \varepsilon u_1(s_1, T_1) + \cdots \tag{8.5.2}$$

Substituting (8.5.1) and (8.5.2) into the given equation and keeping to O(ε), we get

$$0 = \frac{\partial^2 u}{\partial t^2} - \frac{\partial^2 u}{\partial x^2} - \varepsilon\left[\beta\frac{\partial u}{\partial t} - \alpha\left(\frac{\partial u}{\partial t}\right)^3\right]$$
$$= \varepsilon\left[2\frac{\partial^2 u_0}{\partial s_1 \partial T_1} - \beta\frac{\partial u_0}{\partial s_1} + \alpha\left(\frac{\partial u_0}{\partial s_1}\right)^3\right] \tag{8.5.3}$$

Equating coefficients of like powers of ε in the above equation yields

$$2\frac{\partial^2 u_0}{\partial s_1 \partial T_1} - \beta\frac{\partial u_0}{\partial s_1} + \alpha\left(\frac{\partial u_0}{\partial s_1}\right)^3 = 0 \tag{8.5.4}$$

Let the solution of the Eq. (8.5.4) be

$$u_0 = f(s_1, T_1) \tag{8.5.5}$$

then from (8.5.2), we can obtain

$$u(x, t) = f(s_1, T_1) + O(\varepsilon) \tag{8.5.6}$$

f satisfies the Eq. (8.5.4), i.e.,

$$2\frac{\partial^2 f}{\partial s_1 \partial T_1} - \beta\frac{\partial f}{\partial s_1} + \alpha\left(\frac{\partial f}{\partial s_1}\right)^3 = 0 \tag{8.5.7}$$

(b) Write the Eq. (8.5.7) as

$$2\frac{\partial g}{\partial T_1} = \beta g - \alpha g^3, g = \frac{\partial f}{\partial s_1}$$

Separating the variables of the above equation yields

$$\frac{2dg}{\beta g - \alpha g^3} = dT_1$$

After integration,

$$\ln\frac{g^2}{\alpha g^2 - \beta} = \beta T_1 + C(s_1) \Rightarrow \frac{g^2}{\alpha g^2 - \beta} = B(s_1)e^{\beta T_1}$$

i.e.,

$$\frac{\partial f}{\partial s_1} = g = \left[\frac{\alpha}{\beta} + F(s_1)e^{-\beta T_1}\right]^{-1/2} \tag{8.5.8}$$

(c) From the given initial conditions as well as Eqs. (8.5.6) and (8.5.1), we have

$$f(-x, 0) = -a\sin\omega x \tag{8.5.9}$$

$$\frac{\partial}{\partial s_1}f(-x, 0) + \varepsilon\frac{\partial}{\partial T_1}f(-x, 0) = -a\omega\cos\omega x \tag{8.5.10}$$

Substituting (8.5.8) into (8.5.10) yields

$$\frac{1}{\sqrt{\alpha/\beta + F(-x)}} + \varepsilon\frac{\partial}{\partial T_1}f(-x, 0) = -a\omega\cos\omega x \tag{8.5.11}$$

Equating the coefficients of like powers of ε in the above equation yields

$$\frac{1}{\sqrt{\alpha/\beta + F(-x)}} = -a\omega\cos\omega x \Rightarrow \frac{1}{\sqrt{\alpha/\beta + F(s_1)}} = -a\omega\cos\omega s_1 \tag{8.5.12}$$

So

$$F(s_1) = \frac{\sec^2\omega s_1}{a^2\omega^2} - \frac{\alpha}{\beta} \tag{8.5.13}$$

And then, by (8.5.8), we can get

$$\frac{\partial f}{\partial s_1} = \frac{a\omega\cos\omega s_1}{\left[\frac{\alpha}{\beta}a^2\omega^2\left(1 - e^{-\beta T_1}\right)\cos^2\omega s_1 + e^{-\beta T_1}\right]^{1/2}} \tag{8.5.14}$$

In order to make the integration of (8.5.14), we write it as

$$\begin{aligned} df &= \frac{a\cos\omega s_1 d(\omega s_1)}{\left\{\left[\frac{\alpha}{\beta}a^2\omega^2\left(1 - e^{-\beta T_1}\right) + e^{-\beta T_1}\right] - \frac{\alpha}{\beta}a^2\omega^2\left(1 - e^{-\beta T_1}\right)\cos^2\omega s_1\right\}^{1/2}} \\ &= \frac{a d(\sin\omega s_1)}{\left\{\left[\frac{\alpha}{\beta}a^2\omega^2\left(1 - e^{-\beta T_1}\right) + e^{-\beta T_1}\right] - \frac{\alpha}{\beta}a^2\omega^2\left(1 - e^{-\beta T_1}\right)\sin^2\omega s_1\right\}^{1/2}} \\ &= \frac{1}{\omega}\left(\frac{\beta/\alpha}{1 - e^{-\beta T_1}}\right)^{1/2}\frac{dz}{\sqrt{\Gamma^2 - z^2}}. \end{aligned} \tag{8.5.15}$$

where

$$\Gamma^2 = \frac{e^{-\beta T_1} + \frac{\alpha}{\beta}a^2\omega^2\left(1 - e^{-\beta T_1}\right)}{\frac{\alpha}{\beta}a^2\omega^2\left(1 - e^{-\beta T_1}\right)}, z = \sin\omega s_1 \tag{8.5.16}$$

Make the integration of Eq. (8.5.15), we can obtain

$$\begin{aligned} f &= \frac{1}{\omega}(\frac{\beta/\alpha}{1 - e^{-\beta T_1}})^{1/2}\arcsin\frac{z}{\Gamma} + h(T_1) \\ &= \frac{1}{\omega}(\frac{\beta/\alpha}{1 - e^{-\beta T_1}})^{1/2}\arcsin\left\{\left[\frac{\frac{\alpha}{\beta}a^2\omega^2\left(1 - e^{-\beta T_1}\right)}{e^{-\beta T_1} + \frac{\alpha}{\beta}a^2\omega^2\left(1 - e^{-\beta T_1}\right)}\right]^{1/2}\sin\omega s_1\right\} + h(T_1) \end{aligned}$$

Assume $h(T_1) = 0$, then

$$f = \frac{1}{\omega}\left(\frac{\beta/\alpha}{1 - e^{-\beta T_1}}\right)^{1/2}\arcsin\left\{\left[\frac{\frac{\alpha}{\beta}a^2\omega^2\left(1 - e^{-\beta T_1}\right)}{e^{-\beta T_1} + \frac{\alpha}{\beta}a^2\omega^2\left(1 - e^{-\beta T_1}\right)}\right]^{1/2}\sin\omega s_1\right\} \tag{8.5.17}$$

When $T_1 \to 0$, the above equation becomes

$$\begin{aligned} f|_{T_1\to 0} &= \lim_{T_1\to 0}\left\{\frac{1}{\omega}\left(\frac{\beta/\alpha}{1 - e^{-\beta T_1}}\right)^{1/2}\left[\frac{\frac{\alpha}{\beta}a^2\omega^2\left(1 - e^{-\beta T_1}\right)}{e^{-\beta T_1} + \frac{\alpha}{\beta}a^2\omega^2\left(1 - e^{-\beta T_1}\right)}\right]^{1/2}\right\}\sin(-\omega x) \\ &= -a\sin\omega x \end{aligned} \tag{8.5.18}$$

Therefore, the solution (8.5.17) satisfies the initial condition (8.5.9). Substituting (8.5.17) into (8.5.6) yields

$$u = \frac{1}{\omega}\left(\frac{\beta/\alpha}{1 - e^{-\beta T_1}}\right)^{1/2}\arcsin\left\{\left[\frac{\frac{\alpha}{\beta}a^2\omega^2\left(1 - e^{-\beta T_1}\right)}{e^{-\beta T_1} + \frac{\alpha}{\beta}a^2\omega^2\left(1 - e^{-\beta T_1}\right)}\right]^{1/2}\sin\omega s_1\right\} \tag{8.5.19}$$

8.6 Exercise 8.6 (Wave Propagating Along a Uniform, Initially Undeformed Nonlinear Elastic bar with Linear Damping)

Solution: (a) We use the method of multiple scales to find the solution. The corresponding linear homogeneous equation of the given equation is a one-dimensional wave equation with a right-running wave solution as a function of $s_1 = t - x/c$. The nonlinear term in the equation is the square of the spatial partial derivatives of u.

And the damping term is the temporal partial derivatives of u. In order to take these nonlinear and damping effects into account, we introduce a slow-varying spatial variable, $x_1 = \varepsilon x$, and a slow-varying temporal variable, $T_1 = \varepsilon t$; therefore, the solution of the equation is set to be a function of $s_1, x_1 \text{and} T_1$. The space and time derivatives are transformed according to:

$$\begin{aligned}\frac{\partial}{\partial t} &= \frac{\partial}{\partial s_1} + \varepsilon\frac{\partial}{\partial T_1}, \frac{\partial^2}{\partial t^2} = \frac{\partial^2}{\partial s_1^2} + 2\varepsilon\frac{\partial^2}{\partial s_1 \partial T_1} + \varepsilon^2\frac{\partial^2}{\partial T_1^2} \\ \frac{\partial}{\partial x} &= -\frac{1}{c}\frac{\partial}{\partial s_1} + \varepsilon\frac{\partial}{\partial x_1}, \frac{\partial^2}{\partial x^2} = \frac{1}{c^2}\frac{\partial^2}{\partial s_1^2} - 2\varepsilon\frac{1}{c}\frac{\partial^2}{\partial s_1 \partial x_1} + \varepsilon^2\frac{\partial^2}{\partial x_1^2}\end{aligned} \tag{8.6.1}$$

Let the solution of the given equation be

$$u(x, t) = \varepsilon u_1(s_1, x_1, T_1) + \varepsilon^2 u_2(s_1, x_1, T_1) + \cdots \tag{8.6.2}$$

Substituting (8.6.1) and (8.6.2) into the given equation and keeping to $O(\varepsilon^2)$, we get

$$\begin{aligned}0 &= \frac{\partial^2 u}{\partial x^2} - \frac{1}{c^2}\frac{\partial^2 u}{\partial t^2} + 2E_1\frac{\partial u}{\partial x}\frac{\partial^2 u}{\partial x^2} - 2\varepsilon\mu\frac{\partial u}{\partial t} \\ &= -2\varepsilon^2\left(\frac{1}{c}\frac{\partial^2 u_1}{\partial s_1 \partial x_1} + \frac{1}{c^2}\frac{\partial^2 u_1}{\partial s_1 \partial T_1} + \frac{E_1}{c^3}\frac{\partial u_1}{\partial s_1}\frac{\partial^2 u_1}{\partial s_1^2} + \mu\frac{\partial u_1}{\partial s_1}\right)\end{aligned} \tag{8.6.3}$$

Equating coefficients of like powers of ε in (8.6.3) yields

$$\frac{1}{c}\frac{\partial^2 u_1}{\partial s_1 \partial x_1} + \frac{1}{c^2}\frac{\partial^2 u_1}{\partial s_1 \partial T_1} + \frac{E_1}{c^3}\frac{\partial u_1}{\partial s_1}\frac{\partial^2 u_1}{\partial s_1^2} + \mu\frac{\partial u_1}{\partial s_1} = 0 \tag{8.6.4}$$

The strain of the rod e is

$$e = \frac{\partial u}{\partial x} = -\varepsilon\frac{1}{c}\frac{\partial u_1}{\partial s_1} + O(\varepsilon^2) = \varepsilon f(s_1, x_1, T_1) + O(\varepsilon^2) \tag{8.6.5}$$

where

$$f(s_1, x_1, T_1) = -\frac{1}{c}\frac{\partial u_1}{\partial s_1} \tag{8.6.6}$$

Substituting (8.6.6) into (8.6.4) yields

$$\frac{\partial f}{\partial x_1} + \frac{1}{c}\frac{\partial f}{\partial T_1} + \mu c f + \frac{E_1}{c} f\frac{\partial f}{\partial s_1} = 0 \tag{8.6.7}$$

(b) In order to examine the waves that are "nonlinearly distorted with distance", we let $\partial f/\partial T_1 = 0$, hence $f = f(s_1, x_1)$, (8.6.7) becomes

$$\frac{\partial f}{\partial x_1} + \mu c f + \frac{E_1}{c} f \frac{\partial f}{\partial s_1} = 0 \tag{8.6.8}$$

Let

$$f = Q(x_1)h(s_1, z), z = Z(x_1) \tag{8.6.9}$$

and substitute (8.6.9) into (8.6.8), we can obtain

$$h\left(Q' + \mu c Q\right) + cQ^2\left(\frac{Z\prime}{cQ}\frac{\partial h}{\partial z} + \frac{E_1}{c^2} h \frac{\partial h}{\partial s_1}\right) = 0 \tag{8.6.10}$$

The prime denotes the derivative with respect to x_1. In order to obtain the constant coefficient equations of Q and h from the above equation, we select Q, Z so that they satisfy.

$$Q' + \mu c Q = 0, Z' = cQ \tag{8.6.11}$$

Thus, from (8.6.10), we have

$$\frac{\partial h}{\partial z} + \frac{E_1}{c^2} h \frac{\partial h}{\partial s_1} = 0 \tag{8.6.12}$$

The solutions of (8.6.11) are

$$Q = A\exp(-\mu c x_1), Z = A\mu^{-1}\left[B - \exp(-\mu c x_1)\right]$$

where A and B are constants of integration. We take the variable z to have the same origin as x_1, i.e. $Z(0) = 0$ and hence $B = 1$. So

$$Q = A\exp(-\mu c x_1), Z = A\mu^{-1}\left[1 - \exp(-\mu c x_1)\right] \tag{8.6.13}$$

In order to solve (8.6.12), we consider the following linear equation in advance:

$$\frac{\partial h}{\partial z} + \frac{E_1}{c^2}\frac{\partial h}{\partial s_1} = 0$$

The solution to this equation is

$$h = \phi(\xi), \xi = s_1 - Dz$$

where D is a constant. Inspired by this, we use the parameter variation method to solve the nonlinear Eq. (8.6.12). For this purpose, we assume that the solution of (8.6.12) is

$$h = \phi(\xi), \xi = s_1 - D(\xi)z \tag{8.6.14}$$

In this way, we can treat ϕ as a function of s_1 and ξ. Substituting (8.6.14) into (8.6.12), we get

$$\frac{\partial \xi}{\partial z} + \frac{E_1}{c^2}\phi\frac{\partial \xi}{\partial s_1} = 0$$

From (8.6.14), we can obtain

$$\left(\frac{\partial \xi}{\partial z} = -D(\xi)\left(1 + z\frac{\partial D}{\partial \xi}\right)^{-1}, \frac{\partial \xi}{\partial s_1} = \left(1 + z\frac{\partial D}{\partial \xi}\right)^{-1}\right)$$

So

$$-D(\xi) + \frac{E_1}{c^2}\phi(\xi) = 0 \tag{8.6.15}$$

i.e.,

$$D(\xi) = E_1 c^{-2}\phi(\xi) \tag{8.6.16}$$

Combining the above results and taking into account $x_1 = \varepsilon x$ yields

$$f = \exp(-\varepsilon\mu c x)\phi(\xi) \tag{8.6.17}$$

$$s_1 = \xi + E_1\mu^{-1}c^{-2}\left[1 - \exp(-\varepsilon\mu c x)\right]\phi(\xi) \tag{8.6.18}$$

(c) In order to examine waves that are "nonlinearly distorted with time", we let $\partial f/\partial x_1 = 0$, hence, $f = f(s_1, T_1)$ and (8.6.7) becomes

$$\frac{\partial f}{\partial T_1} + \mu c^2 f + E_1 f\frac{\partial f}{\partial s_1} = 0 \tag{8.6.19}$$

Comparing (8.6.19) with (8.6.8), we know that by replacing μc, E_1/c and x_1 with μc^2, E_1 and T_1, respectively, (8.6.8) can be changed to (8.6.19). Therefore, the solution of (8.6.19) can be obtained just by making these substitutions in the solution of (8.6.8), i.e.,

$$s_1 = \xi + E_1\mu^{-1}c^{-2}\left[1 - \exp\left(-\varepsilon\mu c^2 t\right)\right]\phi(\xi) \tag{8.6.20}$$

$$e = \varepsilon\exp\left(-\varepsilon\mu c^2 t\right)\phi(\xi) + O\left(\varepsilon^2\right) \tag{8.6.21}$$

(d) For the nonlinear wave problem of a rod, the rightward and leftward waves in the rod cannot be superimposed in general. The conditions for their superposition are derived here.

The governing equations and solutions for the right-running wave have been given by (a). Similarly, let the left-running wave solution be a function of s_2, x_1 and T_1, $s_2 = t + x/c$ The space and time derivatives are transformed according to:

$$\frac{\partial}{\partial t} = \frac{\partial}{\partial s_2} + \varepsilon\frac{\partial}{\partial T_1}, \frac{\partial^2}{\partial t^2} = \frac{\partial^2}{\partial s_2^2} + 2\varepsilon\frac{\partial^2}{\partial s_2 \partial T_1} + \varepsilon^2\frac{\partial^2}{\partial T_1^2}$$
$$\frac{\partial}{\partial x} = \frac{1}{c}\frac{\partial}{\partial s_2} + \varepsilon\frac{\partial}{\partial x_1}, \frac{\partial^2}{\partial x^2} = \frac{1}{c^2}\frac{\partial^2}{\partial s_2^2} + 2\varepsilon\frac{1}{c}\frac{\partial^2}{\partial s_2 \partial x_1} + \varepsilon^2\frac{\partial^2}{\partial x_1^2} \tag{8.6.22}$$

Let the solution of the given equation be

$$u(x,t) = \varepsilon u_{1r}(s_1, x_1, T_1) + \varepsilon u_{1l}(s_2, x_1, T_1)$$
$$+ \varepsilon^2 u_{2r}(s_1, x_1, T_1) + \varepsilon^2 u_{2l}(s_2, x_1, T_1) + \cdots \tag{8.6.23}$$

Substituting (8.6.23), (8.6.22) and (8.6.2) into the nonlinear control equation of the rod and retaining to $O(\varepsilon^2)$ yields:

$$0 = \frac{\partial^2 u}{\partial x^2} - \frac{1}{c^2}\frac{\partial^2 u}{\partial t^2} - 2\varepsilon\mu\frac{\partial u}{\partial t} + 2E_1\frac{\partial u}{\partial x}\frac{\partial^2 u}{\partial x^2}$$
$$= -2\varepsilon^2\left(\frac{1}{c}\frac{\partial^2 u_{1r}}{\partial s_1 \partial x_1} + \frac{1}{c^2}\frac{\partial^2 u_{1r}}{\partial s_1 \partial T_1} + \mu\frac{\partial u_{1r}}{\partial s_1} + E_1\frac{1}{c^3}\frac{\partial u_{1r}}{\partial s_1}\frac{\partial^2 u_{1r}}{\partial s_1^2}\right)$$
$$+ 2\varepsilon^2\left(\frac{1}{c}\frac{\partial^2 u_{1l}}{\partial s_2 \partial x_1} - \frac{1}{c^2}\frac{\partial^2 u_{1l}}{\partial s_2 \partial T_1} - \mu\frac{\partial u_{1l}}{\partial s_2} + E_1\frac{1}{c^3}\frac{\partial u_{1l}}{\partial s_2}\frac{\partial^2 u_{1l}}{\partial s_2^2}\right)$$
$$+ 2\varepsilon^2 E_1\frac{1}{c^3}\left(\frac{\partial u_{1l}}{\partial s_2}\frac{\partial^2 u_{1r}}{\partial s_1^2} - \frac{\partial u_{1r}}{\partial s_1}\frac{\partial^2 u_{1l}}{\partial s_2^2}\right) \tag{8.6.24}$$

Equating coefficients of like powers of ε in (8.6.24) yields

$$\left(\frac{1}{c}\frac{\partial^2 u_{1r}}{\partial s_1 \partial x_1} + \frac{1}{c^2}\frac{\partial^2 u_{1r}}{\partial s_1 \partial T_1} + \mu\frac{\partial u_{1r}}{\partial s_1} + E_1\frac{1}{c^3}\frac{\partial u_{1r}}{\partial s_1}\frac{\partial^2 u_{1r}}{\partial s_1^2}\right)$$
$$+\left(\frac{1}{c}\frac{\partial^2 u_{1l}}{\partial s_2 \partial x_1} - \frac{1}{c^2}\frac{\partial^2 u_{1l}}{\partial s_2 \partial T_1} - \mu\frac{\partial u_{1l}}{\partial s_2} + E_1\frac{1}{c^3}\frac{\partial u_{1l}}{\partial s_2}\frac{\partial^2 u_{1l}}{\partial s_2^2}\right) \tag{8.6.25}$$
$$+E_1\frac{1}{c^3}\left(\frac{\partial u_{1l}}{\partial s_2}\frac{\partial^2 u_{1r}}{\partial s_1^2} - \frac{\partial u_{1r}}{\partial s_1}\frac{\partial^2 u_{1l}}{\partial s_2^2}\right) = 0$$

The strain of the rod e is

$$e = \frac{\partial u}{\partial x} = -\varepsilon\frac{1}{c}\frac{\partial u_{1r}}{\partial s_1} + \varepsilon\frac{1}{c}\frac{\partial u_{1l}}{\partial s_2} + O(\varepsilon^2)$$
$$= \varepsilon f(s_1, x_1, T_1) + \varepsilon g(s_2, x_1, T_1) + O(\varepsilon^2) \tag{8.6.26}$$

where

$$f(s_1, x_1, T_1) = -\frac{1}{c}\frac{\partial u_{1r}}{\partial s_1},\; g(s_2, x_1, T_1) = \frac{1}{c}\frac{\partial u_{1l}}{\partial s_2} \tag{8.6.27}$$

Substituting (8.6.27) into (8.6.25), we obtain

$$\begin{aligned}&\left(\frac{\partial f}{\partial x_1} + \frac{1}{c}\frac{\partial f}{\partial T_1} + E_1\frac{1}{c}f\frac{\partial f}{\partial s_1} + \mu c f\right)\\ &+\left(\frac{\partial g}{\partial x_1} - \frac{1}{c}\frac{\partial g}{\partial T_1} + E_1\frac{1}{c}g\frac{\partial g}{\partial s_2} - \mu c g\right)\\ &-\frac{E_1}{c}\left(g\frac{\partial f}{\partial s_1} - f\frac{\partial g}{\partial s_2}\right) = 0\end{aligned} \tag{8.6.28}$$

From the previous analyses, we know that the right-running wave f and the left-running wave g satisfy the following two equations, respectively:

$$\frac{\partial f}{\partial x_1} + \frac{1}{c}\frac{\partial f}{\partial T_1} + E_1\frac{1}{c}f\frac{\partial f}{\partial s_1} + \mu c f = 0 \tag{8.6.29}$$

$$\frac{\partial g}{\partial x_1} - \frac{1}{c}\frac{\partial g}{\partial T_1} + E_1\frac{1}{c}g\frac{\partial g}{\partial s_2} - \mu c g = 0 \tag{8.6.30}$$

Generally speaking, the superposition of right-running waves f and left-running waves g that satisfy these equations fail to satisfy (8.6.28) unless:

$$g\frac{\partial f}{\partial s_1} - f\frac{\partial g}{\partial s_2} = 0 \tag{8.6.31}$$

8.7 Exercise 8.7 (Modeling and Multiscale Analysis on the High-Frequency Oscillation of a Homogenous Visco-Elastic Rod)

Solution: **(a)** Applying Newton's second law to the bar control volume yields

$$\rho A\frac{\partial^2 u}{\partial t^2} = \frac{\partial (A\sigma)}{\partial x} \tag{8.7.1}$$

For a uniform bar, ρ and A are constants and the above equation becomes

$$\rho\frac{\partial^2 u}{\partial t^2} = \frac{\partial \sigma}{\partial x} \tag{8.7.2}$$

The longitudinal strain of the bar $e = \partial u/\partial x$. Therefore, from the approximate constitutive relation for a homogenous visco-elastic material with high frequencies, we can obtain:

$$\frac{\partial \sigma}{\partial t} = E\left(1 + 2E_1\frac{\partial u}{\partial x}\right)\frac{\partial^2 u}{\partial x \partial t} - 2\hat{\tau}\frac{\partial u}{\partial x} \tag{8.7.3}$$

where the coefficients of the linear correction term, τ, are denoted as $\hat{\tau}$ to facilitate different treatments depending on different situations. (8.7.2) and (8.7.3) are the governing equations for waves propagating along a uniform bar made of a homogenous visco-elastic material with high-frequencies.

(b) For small but finite-amplitude waves, let $\hat{\tau} = \varepsilon\tau$ and seek an expansion in the form

$$\begin{aligned} u &= \varepsilon u_1(s_1, s_2, x_1, T_1) + \varepsilon^2 u_2(s_1, s_2, x_1, T_1) + \cdots \\ \sigma &= \varepsilon \sigma_1(s_1, s_2, x_1, T_1) + \varepsilon^2 \sigma_2(s_1, s_2, x_1, T_1) + \cdots \end{aligned} \tag{8.7.4}$$

The space and time derivatives are transformed according to:

$$\begin{aligned}
&\frac{\partial}{\partial t} = \frac{\partial}{\partial s_1} + \frac{\partial}{\partial s_2} + \varepsilon\frac{\partial}{\partial T_1} \\
&\frac{\partial}{\partial x} = -\frac{1}{c}\frac{\partial}{\partial s_1} + \frac{1}{c}\frac{\partial}{\partial s_2} + \varepsilon\frac{\partial}{\partial x_1} \\
&\frac{\partial^2}{\partial t^2} = \frac{\partial^2}{\partial s_1^2} + \frac{\partial^2}{\partial s_2^2} + 2\frac{\partial^2}{\partial s_1 \partial s_2} + 2\varepsilon\frac{\partial^2}{\partial s_1 \partial T_1} + 2\varepsilon\frac{\partial^2}{\partial s_2 \partial T_1} + \varepsilon^2\frac{\partial^2}{\partial T_1^2} \\
&\frac{\partial^2}{\partial x^2} = \frac{1}{c^2}\frac{\partial^2}{\partial s_1^2} + \frac{1}{c^2}\frac{\partial^2}{\partial s_2^2} - 2\frac{1}{c^2}\frac{\partial^2}{\partial s_1 \partial s_2} - 2\varepsilon\frac{1}{c}\frac{\partial^2}{\partial s_1 \partial x_1} + 2\varepsilon\frac{1}{c}\frac{\partial^2}{\partial s_2 \partial x_1} + \varepsilon^2\frac{\partial^2}{\partial x_1^2} \\
&\frac{\partial^2}{\partial x \partial t} = -\frac{1}{c}\frac{\partial^2}{\partial s_1^2} + \frac{1}{c}\frac{\partial^2}{\partial s_2^2} + \varepsilon\frac{\partial^2}{\partial s_1 \partial x_1} + \varepsilon\frac{\partial^2}{\partial s_2 \partial x_1} - \varepsilon\frac{1}{c}\frac{\partial^2}{\partial s_1 \partial T_1} \\
&+\varepsilon\frac{1}{c}\frac{\partial^2}{\partial s_2 \partial T_1} + \varepsilon^2\frac{\partial^2}{\partial x_1 \partial T_1}
\end{aligned} \tag{8.7.5}$$

Substituting (8.7.4) and (8.7.5) into (8.7.2) and (8.7.3), we obtain

$$\begin{aligned}
0 = {} & \varepsilon\left(\rho\frac{\partial^2 u_1}{\partial s_1^2} + 2\rho\frac{\partial^2 u_1}{\partial s_1 \partial s_2} + \rho\frac{\partial^2 u_1}{\partial s_2^2} + \frac{1}{c}\frac{\partial \sigma_1}{\partial s_1} - \frac{1}{c}\frac{\partial \sigma_1}{\partial s_2}\right) \\
& + \varepsilon^2\left(\begin{array}{l} \rho\dfrac{\partial^2 u_2}{\partial s_1^2} + \rho\dfrac{\partial^2 u_2}{\partial s_2^2} + 2\rho\dfrac{\partial^2 u_2}{\partial s_1 \partial s_2} + \dfrac{1}{c}\dfrac{\partial \sigma_2}{\partial s_1} - \dfrac{1}{c}\dfrac{\partial \sigma_2}{\partial s_2} \\ -\dfrac{\partial \sigma_1}{\partial x_1} + 2\rho\dfrac{\partial^2 u_1}{\partial s_1 \partial T_1} + 2\rho\dfrac{\partial^2 u_1}{\partial s_2 \partial T_1} \end{array}\right)
\end{aligned} \tag{8.7.6}$$

$$
\begin{aligned}
0 = \varepsilon\left(\frac{\partial \sigma_1}{\partial s_1} + \frac{\partial \sigma_1}{\partial s_2} + \frac{E}{c}\frac{\partial^2 u_1}{\partial s_1^2} - \frac{E}{c}\frac{\partial^2 u_1}{\partial s_2^2}\right) \\
+ \varepsilon^2 \left[\begin{aligned}
&\frac{\partial \sigma_2}{\partial s_1} + \frac{\partial \sigma_2}{\partial s_2} + \frac{E}{c}\frac{\partial^2 u_2}{\partial s_1^2} - \frac{E}{c}\frac{\partial^2 u_2}{\partial s_2^2} \\
&+\frac{\partial \sigma_1}{\partial T_1} - E\frac{\partial^2 u_1}{\partial s_1 \partial x_1} - E\frac{\partial^2 u_1}{\partial s_2 \partial x_1} + \frac{E}{c}\frac{\partial^2 u_1}{\partial s_1 \partial T_1} - \frac{E}{c}\frac{\partial^2 u_1}{\partial s_2 \partial T_1} \\
&-2EE_1\frac{1}{c^2}\frac{\partial u_1}{\partial s_1}\frac{\partial^2 u_1}{\partial s_1^2} + 2EE_1\frac{1}{c^2}\frac{\partial u_1}{\partial s_1}\frac{\partial^2 u_1}{\partial s_2^2} + 2EE_1\frac{1}{c^2}\frac{\partial u_1}{\partial s_2}\frac{\partial^2 u_1}{\partial s_1^2} \\
&-2EE_1\frac{1}{c^2}\frac{\partial u_1}{\partial s_2}\frac{\partial^2 u_1}{\partial s_2^2} + 2\tau\left(-\frac{1}{c}\frac{\partial u_1}{\partial s_1} + \frac{1}{c}\frac{\partial u_1}{\partial s_2}\right)
\end{aligned}\right]
\end{aligned} \tag{8.7.7}
$$

Equating the coefficients of the like power of ε in the above two equations, we obtain

$$
\rho c\left(\frac{\partial}{\partial s_2} + \frac{\partial}{\partial s_1}\right)^2 u_1 - \left(\frac{\partial}{\partial s_2} - \frac{\partial}{\partial s_1}\right)\sigma_1 = 0 \tag{8.7.8}
$$

$$
\left(\frac{\partial}{\partial s_2} + \frac{\partial}{\partial s_1}\right)\sigma_1 - \frac{E}{c}\left(\frac{\partial^2}{\partial s_2^2} - \frac{\partial^2}{\partial s_1^2}\right)u_1 = 0 \tag{8.7.9}
$$

$$
\rho c\left(\frac{\partial}{\partial s_2} + \frac{\partial}{\partial s_1}\right)^2 u_2 - \left(\frac{\partial}{\partial s_2} - \frac{\partial}{\partial s_1}\right)\sigma_2 = c\frac{\partial \sigma_1}{\partial x_1} - 2\rho c\frac{\partial^2 u_1}{\partial s_1 \partial T_1} - 2\rho c\frac{\partial^2 u_1}{\partial s_2 \partial T_1} \tag{8.7.10}
$$

$$
\begin{aligned}
&\left(\frac{\partial}{\partial s_2} + \frac{\partial}{\partial s_1}\right)\sigma_2 - \frac{E}{c}\left(\frac{\partial^2}{\partial s_2^2} - \frac{\partial^2}{\partial s_1^2}\right)u_2 = -\frac{\partial \sigma_1}{\partial T_1} + E\frac{\partial^2 u_1}{\partial s_1 \partial x_1} + E\frac{\partial^2 u_1}{\partial s_2 \partial x_1} \\
&-\frac{E}{c}\frac{\partial^2 u_1}{\partial s_1 \partial T_1} + \frac{E}{c}\frac{\partial^2 u_1}{\partial s_2 \partial T_1} + 2EE_1\frac{1}{c^2}\frac{\partial u_1}{\partial s_1}\frac{\partial^2 u_1}{\partial s_1^2} \\
&-2EE_1\frac{1}{c^2}\frac{\partial u_1}{\partial s_1}\frac{\partial^2 u_1}{\partial s_2^2} - 2EE_1\frac{1}{c^2}\frac{\partial u_1}{\partial s_2}\frac{\partial^2 u_1}{\partial s_1^2} \\
&+2EE_1\frac{1}{c^2}\frac{\partial u_1}{\partial s_2}\frac{\partial^2 u_1}{\partial s_2^2} - 2\tau\left(-\frac{1}{c}\frac{\partial u_1}{\partial s_1} + \frac{1}{c}\frac{\partial u_1}{\partial s_2}\right)
\end{aligned} \tag{8.7.11}
$$

For a right-running wave, let $u_1 = f(s_1, x_1, T_1).$, $\sigma_1 = \sigma_1(s_1, x_1, T_1)$. and then from (8.7.8) or (8.7.9), we have

$$
\rho c\frac{\partial^2 f}{\partial s_1^2} + \frac{\partial \sigma_1}{\partial s_1} = 0 \tag{8.7.12}
$$

i.e.,

$$\sigma_1 = -\rho c \frac{\partial f}{\partial s_1} \tag{8.7.13}$$

in which we take $\sigma_1 = 0$ when the approximate strain $\varepsilon \partial f / \partial s_1 = 0$.

(c) For the right-running wave,$u_1 = f(s_1, x_1, T_1)$. Considering (8.7.13), we can change (8.7.10) and (8.7.11) into

$$\rho c \left(\frac{\partial}{\partial s_2} + \frac{\partial}{\partial s_1} \right)^2 u_2 - \left(\frac{\partial}{\partial s_2} - \frac{\partial}{\partial s_1} \right) \sigma_2 = -2\rho c \frac{\partial^2 f}{\partial s_1 \partial T_1} - \rho c^2 \frac{\partial^2 f}{\partial s_1 \partial x_1} \tag{8.7.14}$$

$$\left(\frac{\partial}{\partial s_2} + \frac{\partial}{\partial s_1} \right) \sigma_2 - \frac{E}{c} \left(\frac{\partial^2}{\partial s_2^2} - \frac{\partial^2}{\partial s_1^2} \right) u_2 = E \frac{\partial^2 f}{\partial s_1 \partial x_1} + \frac{2EE_1}{c^2} \frac{\partial f}{\partial s_1} \frac{\partial^2 f}{\partial s_1^2} + \frac{2\tau}{c} \frac{\partial f}{\partial s_1} \tag{8.7.15}$$

Subtracting these two equations yields

$$\begin{aligned} \frac{\partial}{\partial s_2} \left[\rho c \left(\frac{\partial}{\partial s_2} + \frac{\partial}{\partial s_1} \right) u_2 - \sigma_2 \right] &= -\rho c \frac{\partial^2 f}{\partial s_1 \partial T_1} - \rho c^2 \frac{\partial^2 f}{\partial s_1 \partial x_1} \\ -\rho E_1 \frac{\partial f}{\partial s_1} \frac{\partial^2 f}{\partial s_1^2} - \frac{\tau}{c} \frac{\partial f}{\partial s_1} \end{aligned} \tag{8.7.16}$$

After integration

$$\rho c \left(\frac{\partial}{\partial s_2} + \frac{\partial}{\partial s_1} \right) u_2 - \sigma_2 = \begin{bmatrix} -\rho c \dfrac{\partial^2 f}{\partial s_1 \partial T_1} - \rho c^2 \dfrac{\partial^2 f}{\partial s_1 \partial x_1} \\ -\rho E_1 \dfrac{\partial f}{\partial s_1} \dfrac{\partial^2 f}{\partial s_1^2} - \dfrac{\tau}{c} \dfrac{\partial f}{\partial s_1} \end{bmatrix} s_2 + A(s_1, x_1, T_1) \tag{8.7.17}$$

where $A(s_1, x_1, T_1)$ is a function to be determined. In order to eliminate secular terms from the above equation, we need

$$\frac{\partial^2 f}{\partial s_1 \partial x_1} + \frac{1}{c} \frac{\partial^2 f}{\partial s_1 \partial T_1} + \frac{E_1}{c^2} \frac{\partial f}{\partial s_1} \frac{\partial^2 f}{\partial s_1^2} + \frac{\tau}{\rho c^3} \frac{\partial f}{\partial s_1} = 0 \tag{8.7.18}$$

(d) For waves propagating along this uniform bar, the right-running and left-running waves in the bar cannot be superimposed in general. The conditions for their superposition are derived here.

For a left-running wave $u_1 = g(s_2, x_1, T_1), \sigma_1 = \sigma_1(s_2, x_1, T_1)$, by (8.7.8) or (8.7.9), we have

$$\rho c \left(\frac{\partial}{\partial s_2} + \frac{\partial}{\partial s_1} \right)^2 u_1 - \left(\frac{\partial}{\partial s_2} - \frac{\partial}{\partial s_1} \right) \sigma_1 = 0 \tag{8.7.19}$$

$$\rho c\frac{\partial^2 g}{\partial s_2^2}-\frac{\partial \sigma_1}{\partial s_2}=0 \tag{8.7.20}$$

i.e.,

$$\sigma_1=\rho c\frac{\partial g}{\partial s_2} \tag{8.7.21}$$

in which we take $\sigma_1=0$ when the approximate strain $\varepsilon\partial f/\partial s_1=0$.

For a left-running wave, by the Eqs. (8.7.10), (8.7.11) and (8.7.21), we have

$$\left(\rho c\left(\frac{\partial}{\partial s_2}+\frac{\partial}{\partial s_1}\right)^2 u_2-\left(\frac{\partial}{\partial s_2}-\frac{\partial}{\partial s_1}\right)\sigma_2=-2\rho c\frac{\partial^2 g}{\partial s_2\partial T_1}+\rho c^2\frac{\partial g}{\partial s_2\partial x_1}\right) \tag{8.7.22}$$

$$\left(\frac{\partial}{\partial s_2}+\frac{\partial}{\partial s_1}\right)\sigma_2-\frac{E}{c}\left(\frac{\partial^2}{\partial s_2^2}-\frac{\partial^2}{\partial s_1^2}\right)u_2=E\frac{\partial^2 g}{\partial s_2\partial x_1}+\frac{2EE_1}{c^2}\frac{\partial g}{\partial s_2}\frac{\partial^2 g}{\partial s_2^2}-\frac{2\tau}{c}\frac{\partial g}{\partial s_2} \tag{8.7.23}$$

Adding these two equations together yields

$$\begin{aligned}\frac{\partial}{\partial s_1}\left[\rho c\left(\frac{\partial}{\partial s_2}+\frac{\partial^2}{\partial s_1}\right)u_2+\sigma_2\right]&=\rho c^2\frac{\partial^2 g}{\partial s_2\partial x_1}-\rho c\frac{\partial^2 g}{\partial s_2\partial T_1}\\&+\frac{EE_1}{c^2}\frac{\partial g}{\partial s_2}\frac{\partial^2 g}{\partial s_2^2}-\frac{\tau}{c}\frac{\partial g}{\partial s_2}\end{aligned} \tag{8.7.24}$$

In order to eliminate secular terms from the above equation, we need

$$\frac{\partial^2 g}{\partial s_2\partial x_1}-\frac{1}{c}\frac{\partial^2 g}{\partial s_2\partial T_1}+\frac{E_1}{c^2}\frac{\partial g}{\partial s_2}\frac{\partial^2 g}{\partial s_2^2}-\frac{\tau}{\rho c^3}\frac{\partial g}{\partial s_2}=0 \tag{8.7.25}$$

Assuming that the right-running wave f and the left-running wave g controlled by (8.7.18) and (8.7.25) can be superimposed, we can obtain the solution to the original equation

$$\begin{aligned}u&=\varepsilon u_1+\cdots=\varepsilon f(s_1,x_1,T_1)+\varepsilon g(s_2,x_1,T_1)+\cdots\\\sigma&=\varepsilon\sigma_1+\cdots=-\varepsilon\rho c\left[\tfrac{\partial}{\partial s_1}f(s_1,x_1,T_1)-\tfrac{\partial}{\partial s_2}g(s_2,x_1,T_1)\right]+\cdots\end{aligned} \tag{8.7.26}$$

which satisfy (8.7.10) and (8.7.11). Substituting them into (8.7.10) and (8.7.11) yields

$$\begin{aligned}\rho c\left(\tfrac{\partial}{\partial s_2}+\tfrac{\partial}{\partial s_1}\right)^2u_2-\left(\tfrac{\partial}{\partial s_2}-\tfrac{\partial}{\partial s_1}\right)\sigma_2&=-\rho c^2\tfrac{\partial^2 f}{\partial s_1\partial x_1}+\rho c^2\tfrac{\partial^2 g}{\partial s_2\partial x_1}\\&-2\rho c\tfrac{\partial^2 f}{\partial s_1\partial T_1}-2\rho c\tfrac{\partial^2 g}{\partial s_1\partial T_1}-2\rho c\tfrac{\partial^2 f}{\partial s_2\partial T_1}-2\rho c\tfrac{\partial^2 g}{\partial s_2\partial T_1}\end{aligned} \tag{8.7.27}$$

$$
\begin{aligned}
&\left(\frac{\partial}{\partial s_2}+\frac{\partial}{\partial s_1}\right)\sigma_2-\frac{E}{c}\left(\frac{\partial^2}{\partial s_2^2}-\frac{\partial^2}{\partial s_1^2}\right)u_2=E\frac{\partial^2 f}{\partial s_1\partial x_1}+2EE_1\frac{1}{c^2}\frac{\partial f}{\partial s_1}\frac{\partial^2 f}{\partial s_1^2}\\
&+2\tau\frac{1}{c}\frac{\partial f}{\partial s_1}+E\frac{\partial^2 g}{\partial s_2\partial x_1}+2EE_1\frac{1}{c^2}\frac{\partial g}{\partial s_2}\frac{\partial^2 g}{\partial s_2^2}-2\tau\frac{1}{c}\frac{\partial g}{\partial s_2}\\
&-2EE_1\frac{1}{c^2}\frac{\partial f}{\partial s_1}\frac{\partial^2 g}{\partial s_2^2}-2EE_1\frac{1}{c^2}\frac{\partial g}{\partial s_2}\frac{\partial^2 f}{\partial s_1^2}
\end{aligned}
\tag{8.7.28}
$$

Adding up and subtracting the above two equations, respectively, yields

$$
\begin{aligned}
&\frac{\partial}{\partial s_2}\left[\rho c\left(\frac{\partial}{\partial s_2}+\frac{\partial}{\partial s_1}\right)u_2-\sigma_2\right]=-\rho c\frac{\partial^2 f}{\partial s_1\partial T_1}-\rho c^2\frac{\partial^2 f}{\partial s_1\partial x_1}\\
&-\rho E_1\frac{\partial f}{\partial s_1}\frac{\partial^2 f}{\partial s_1^2}-\frac{\tau}{c}\frac{\partial f}{\partial s_1}+\rho E_1\frac{1}{c^2}\frac{\partial f}{\partial s_1}\frac{\partial^2 g}{\partial s_2^2}+\rho E_1\frac{1}{c^2}\frac{\partial g}{\partial s_2}\frac{\partial^2 f}{\partial s_1^2}
\end{aligned}
\tag{8.7.29}
$$

$$
\begin{aligned}
&\frac{\partial}{\partial s_1}\left[\rho c\left(\frac{\partial}{\partial s_2}+\frac{\partial^2}{\partial s_1}\right)u_2+\sigma_2\right]=\rho c^2\frac{\partial^2 g}{\partial s_2\partial x_1}-\rho c\frac{\partial^2 g}{\partial s_2\partial T_1}\\
&+\frac{EE_1}{c^2}\frac{\partial g}{\partial s_2}\frac{\partial^2 g}{\partial s_2^2}-\frac{\tau}{c}\frac{\partial g}{\partial s_2}-\rho E_1\frac{1}{c^2}\frac{\partial f}{\partial s_1}\frac{\partial^2 g}{\partial s_2^2}-\rho E_1\frac{1}{c^2}\frac{\partial g}{\partial s_2}\frac{\partial^2 f}{\partial s_1^2}
\end{aligned}
\tag{8.7.30}
$$

From the above two equations, we can see that the right-running wave f and the left-running wave g are coupled with each other. The sufficient conditions that f and g are governed by (8.7.18) and (8.7.25), respectively, is

$$
\frac{\partial f}{\partial s_1}\frac{\partial^2 g}{\partial s_2^2}+\frac{\partial g}{\partial s_2}\frac{\partial^2 f}{\partial s_1^2}=0 \tag{8.7.31}
$$

(e) In order to examine the right-running wave with nonlinear distortion with distance, let $\partial f/\partial T_1=0$, therefore, $f=f(s_1,x_1)$ and (8.7.18) becomes

$$
\frac{\partial F}{\partial x_1}+\frac{E_1}{c^2}F\frac{\partial F}{\partial s_1}+\frac{\tau}{\rho c^3}F=0 \tag{8.7.32}
$$

in which $F=\partial f/\partial s_1$. Let

$$
F=Q(x_1)H(s_1,z),\ z=Z(x_1) \tag{8.7.33}
$$

and substitute them into (8.7.32) yields

$$
H\left(Q'+\frac{\tau}{\rho c^3}Q\right)+Q^2\left(\frac{Z'}{Q}\frac{\partial H}{\partial z}+\frac{E_1}{c^2}H\frac{\partial H}{\partial s_1}\right)=0 \tag{8.7.34}
$$

The prime denotes the derivative with respect to x_1. In order to obtain the differential equations of Q, H with constant coefficients, we make Q and Z satisfy

$$Q' + \frac{\tau}{\rho c^3} Q = 0, Z' = Q \tag{8.7.35}$$

Thus, from (8.7.34) we have

$$\frac{\partial H}{\partial z} + \frac{E_1}{c^2} H \frac{\partial H}{\partial s_1} = 0 \tag{8.7.36}$$

The solution for (8.7.35) is

$$Q = A \exp\left(-\frac{\tau}{\rho c^3} x_1\right), Z = A\left(\frac{\tau}{\rho c^3}\right)^{-1}\left[B - \exp\left(-\frac{\tau}{\rho c^3} x_1\right)\right]$$

where A and B are constants of integration. We take the variable z to have the same origin as x_1, i.e., $Z(0) = 0$, and hence $B = 1$. So

$$Q = A \exp\left(-\frac{\tau}{\rho c^3} x_1\right), Z = A\left(\frac{\tau}{\rho c^3}\right)^{-1}\left[1 - \exp\left(-\frac{\tau}{\rho c^3} x_1\right)\right] \tag{8.7.37}$$

Equation (8.7.36) is the same as Eq. (8.6.12), and its solution is

$$H = \psi(\xi), \xi = s_1 - D(\xi) z \tag{8.7.38}$$

$$D(\xi) = E_1 c^{-2} \psi(\xi) \tag{8.7.39}$$

Combining above results and taking $x_1 = \varepsilon x$ into account yields

$$F = \exp\left(-\frac{\varepsilon\tau}{\rho c^3} x\right)\psi(\xi) \tag{8.7.40}$$

$$s_1 = \xi + \frac{E_1 \rho c}{\tau}\left[1 - \exp\left(-\frac{\tau}{\rho c^3} x_1\right)\right]\psi(\xi) \tag{8.7.41}$$

where the undetermined A and function ψ are combined as ψ.

Combine (8.7.4) and (8.7.13), we have

$$\sigma = \varepsilon\sigma_1 = -\varepsilon\rho c \frac{\partial f}{\partial s_1} = -\varepsilon\rho c F = -\varepsilon\rho c \exp\left(-\frac{\varepsilon\tau}{\rho c^3} x\right)\psi(\xi)$$

i.e.

$$\left(\sigma = -\varepsilon\rho c \exp\left(-\frac{\varepsilon\tau}{\rho c^3}x\right)\psi(\xi)\right) \tag{8.7.42}$$

8.8 Exercise 8.8 (Modeling and Multiscale Analysis on Low-Frequency Oscillation of a Homogeneous Visco-Elastic Rod)

Solution: **(a)** Applying Newton's second law to the bar control volume yields

$$\rho A\frac{\partial^2 u}{\partial t^2} = \frac{\partial(A\sigma)}{\partial x} \tag{8.8.1}$$

For a uniform bar, ρ and A are constants and the above equation becomes

$$\rho\frac{\partial^2 u}{\partial t^2} = \frac{\partial\sigma}{\partial x} \tag{8.8.2}$$

The longitudinal strain of the bar $e = \partial u/\partial x$. Therefore, from the approximate constitutive relation for a homogenous visco-elastic material with low frequencies, we can obtain

$$\frac{\partial^2 u}{\partial x^2} - \frac{1}{c^2}\frac{\partial^2 u}{\partial t^2} = -2E_1\frac{\partial u}{\partial x}\frac{\partial^2 u}{\partial x^2} - 2\frac{\hat{\mu}}{E}\frac{\partial^3 u}{\partial x^2\partial t} \tag{8.8.3}$$

where $E = \rho c^2$. (8.8.3) is the governing equation for waves propagating along a uniform bar made of a homogenous visco-elastic material with high-frequency.

(b) For a small but finite amplitude right-running wave, let $\hat{\mu} = \varepsilon\mu E$ and set the solution of Eq. (8.8.3) as

$$u = \varepsilon f(s_1, x_1, T_1) + \cdots \tag{8.8.4}$$

The space and time derivatives are transformed according to:

$$\begin{aligned}
&\frac{\partial}{\partial t} = \frac{\partial}{\partial s_1} + \varepsilon\frac{\partial}{\partial T_1}, \ \frac{\partial^2}{\partial t^2} = \frac{\partial^2}{\partial s_1^2} + 2\varepsilon\frac{\partial^2}{\partial s_1\partial T_1} + \varepsilon^2\frac{\partial^2}{\partial T_1^2} \\
&\frac{\partial}{\partial x} = -\frac{1}{c}\frac{\partial}{\partial s_1} + \varepsilon\frac{\partial}{\partial x_1}, \ \frac{\partial^2}{\partial x^2} = \frac{1}{c^2}\frac{\partial^2}{\partial s_1^2} - 2\varepsilon\frac{1}{c}\frac{\partial^2}{\partial s_1\partial x_1} + \varepsilon^2\frac{\partial^2}{\partial x_1^2}
\end{aligned} \tag{8.8.5}$$

Substituting (8.8.4) and (8.8.5) into (8.8.3) yields

$$
\begin{aligned}
0 &= \tfrac{\partial^2 u}{\partial x^2} - \tfrac{1}{c^2}\tfrac{\partial^2 u}{\partial t^2} + 2E_1 \tfrac{\partial u}{\partial x}\tfrac{\partial^2 u}{\partial x^2} + 2\varepsilon\mu \tfrac{\partial^3 u}{\partial x^2 \partial t} \\
&= -\varepsilon^2 \tfrac{2}{c}\left(\tfrac{\partial^2 f}{\partial s_1 \partial x_1} + \tfrac{1}{c}\tfrac{\partial^2 f}{\partial s_1 \partial T_1} + \tfrac{E_1}{c^2}\tfrac{\partial f}{\partial s_1}\tfrac{\partial^2 f}{\partial s_1^2} - \tfrac{\mu}{c}\tfrac{\partial^3 f}{\partial s_1^3}\right)
\end{aligned} \tag{8.8.6}
$$

Equating the coefficients of the like power of ε in the above equation, we obtain

$$
\frac{\partial^2 f}{\partial s_1 \partial x_1} + \frac{1}{c}\frac{\partial^2 f}{\partial s_1 \partial T_1} + \frac{E_1}{c^2}\frac{\partial f}{\partial s_1}\frac{\partial^2 f}{\partial s_1^2} - \frac{\mu}{c}\frac{\partial^3 f}{\partial s_1^3} = 0 \tag{8.8.7}
$$

Or

$$
\frac{\partial F}{\partial x_1} + \frac{1}{c}\frac{\partial F}{\partial T_1} + \frac{E_1}{c^2}F\frac{\partial F}{\partial s_1} - \frac{\mu}{c}\frac{\partial^2 F}{\partial s_1^2} = 0 \tag{8.8.8}
$$

where

$$
F = \frac{\partial f}{\partial s_1} \tag{8.8.9}
$$

For a right-running wave that is nonlinearly distorted with time, let $\partial f/\partial x_1 = 0$, hence, $f = f(s_1, T_1)$, and (8.8.8) becomes

$$
\frac{\partial F}{\partial T_1} + \frac{E_1}{c}F\frac{\partial F}{\partial s_1} - \mu\frac{\partial^2 F}{\partial s_1^2} = 0 \tag{8.8.10}
$$

(c) For waves propagating along this uniform bar, the simple right-running and left-running waves in the bar cannot be superimposed in general. The conditions for their superposition are derived here.

Let the simple left-running wave be

$$
u = \varepsilon g(s_2, x_1, T_1) + \cdots \tag{8.8.11}
$$

The space and time derivatives are transformed according to:

$$
\begin{aligned}
&\frac{\partial}{\partial t} = \frac{\partial}{\partial s_2} + \varepsilon\frac{\partial}{\partial T_1}, \quad \frac{\partial^2}{\partial t^2} = \frac{\partial^2}{\partial s_2^2} + 2\varepsilon\frac{\partial^2}{\partial s_2 \partial T_1} + \varepsilon^2\frac{\partial^2}{\partial T_1^2} \\
&\frac{\partial}{\partial x} = \frac{1}{c}\frac{\partial}{\partial s_2} + \varepsilon\frac{\partial}{\partial x_1}, \quad \frac{\partial^2}{\partial x^2} = \frac{1}{c^2}\frac{\partial^2}{\partial s_2^2} + 2\varepsilon\frac{1}{c}\frac{\partial^2}{\partial s_2 \partial x_1} + \varepsilon^2\frac{\partial^2}{\partial x_1^2}
\end{aligned} \tag{8.8.12}
$$

Substituting (8.8.12) into (8.8.3) yields

$$
\begin{aligned}
0 &= \tfrac{\partial^2 u}{\partial x^2} - \tfrac{1}{c^2}\tfrac{\partial^2 u}{\partial t^2} + 2E_1 \tfrac{\partial u}{\partial x}\tfrac{\partial^2 u}{\partial x^2} + 2\varepsilon\mu \tfrac{\partial^3 u}{\partial x^2 \partial t} \\
&= 2\varepsilon^2 \tfrac{1}{c}\left(\tfrac{\partial^2 g}{\partial s_2 \partial x_1} - \tfrac{1}{c}\tfrac{\partial^2 g}{\partial s_2 \partial T_1} + \tfrac{E_1}{c^2}\tfrac{\partial g}{\partial s_2}\tfrac{\partial^2 g}{\partial s_1^2} + \tfrac{\mu}{c}\tfrac{\partial^3 g}{\partial s_2^3}\right)
\end{aligned} \tag{8.8.13}
$$

Equating the coefficients of the like power of ε in the above two equations, we obtain

$$\frac{\partial^2 g}{\partial s_2 \partial x_1} - \frac{1}{c}\frac{\partial^2 g}{\partial s_2 \partial T_1} + \frac{E_1}{c^2}\frac{\partial g}{\partial s_2}\frac{\partial^2 g}{\partial s_1^2} + \frac{\mu}{c}\frac{\partial^3 g}{\partial s_2^3} = 0 \tag{8.8.14}$$

Assuming that the right-running wave f and the left-running wave g controlled by (8.8.7) and (8.8.14) can be superimposed, we can obtain the solution to the original equation

$$u = \varepsilon f(s_1, x_1, T_1) + \varepsilon g(s_2, x_1, T_1) + \cdots \tag{8.8.15}$$

which satisfies Eq. (8.8.3). Substituting them into (8.8.3) yields

$$\begin{aligned}
0 &= \frac{\partial^2 u}{\partial x^2} - \frac{1}{c^2}\frac{\partial^2 u}{\partial t^2} + 2E_1\frac{\partial u}{\partial x}\frac{\partial^2 u}{\partial x^2} + 2\varepsilon\mu\frac{\partial^3 u}{\partial x^2 \partial t} \\
&= -\varepsilon^2\frac{2}{c}\left(\frac{\partial^2 f}{\partial s_1 \partial x_1} + \frac{1}{c}\frac{\partial^2 f}{\partial s_1 \partial T_1} + \frac{E_1}{c^2}\frac{\partial f}{\partial s_1}\frac{\partial^2 f}{\partial s_1^2} - \frac{\mu}{c}\frac{\partial^3 f}{\partial s_1^3}\right) \\
&\quad + 2\varepsilon^2\frac{1}{c}\left(\frac{\partial^2 g}{\partial s_2 \partial x_1} - \frac{1}{c}\frac{\partial^2 g}{\partial s_2 \partial T_1} + \frac{E_1}{c^2}\frac{\partial g}{\partial s_2}\frac{\partial^2 g}{\partial s_1^2} + \frac{\mu}{c}\frac{\partial^3 g}{\partial s_2^3}\right) \\
&\quad - 2\varepsilon^2\frac{E_1}{c^3}\left(\frac{\partial f}{\partial s_1}\frac{\partial^2 g}{\partial s_2^2} - \frac{\partial g}{\partial s_2}\frac{\partial^2 f}{\partial s_1^2}\right)
\end{aligned} \tag{8.8.16}$$

From the above equation, we can see that the right-running wave f and the left-running wave g are coupled with each other. The sufficient conditions that f and g are governed by (8.8.7) and (8.8.14), respectively, is

$$\frac{\partial f}{\partial s_1}\frac{\partial^2 g}{\partial s_2^2} - \frac{\partial g}{\partial s_2}\frac{\partial^2 f}{\partial s_1^2} = 0 \tag{8.8.17}$$

8.9 Exercise 8.9 (Transform the Burgers' Equation into the Heat Equation)

Solution: Let $u = -\frac{\delta}{\psi}\frac{\partial \psi}{\partial x}$ and $u = \frac{\partial \phi}{\partial x}$, then the Burgers' equation becomes

$$\phi_{xt} + \phi_x \phi_{xx} = \frac{1}{2}\delta\phi_{xxx}$$

Integrate over x, we can obtain

$$\phi_t + \frac{1}{2}\phi_x^2 = \frac{1}{2}\delta\phi_{xx} \tag{8.9.1}$$

Since $\phi = -\delta \log \psi$, we have

$$\phi_t = -\frac{\delta}{\psi}\psi_t, \phi_x = -\frac{\delta}{\psi}\psi_x, \phi_{xx} = -\frac{\delta}{\psi}\psi_{xx} + \frac{\delta}{\psi^2}\psi_x^2 \tag{8.9.2}$$

Substituting the above equations into (8.9.1) yields

$$\frac{\partial \psi}{\partial t} = \frac{1}{2}\delta\frac{\partial^2 \psi}{\partial x^2} \tag{8.9.3}$$

8.10 Exercise 8.10 (Stationary Solutions of the Burgers' Equation)

Solution: Let $\xi = x - ct$, the original equation can be written as

$$-c\frac{df}{d\xi} + f\frac{df}{d\xi} - \frac{1}{2}\delta\frac{d^2 f}{d\xi^2} = 0 \tag{8.10.1}$$

Integrate over ξ, we can obtain

$$-cf + \frac{1}{2}f^2 - \frac{1}{2}\delta\frac{df}{d\xi} = A \tag{8.10.2}$$

where A is the constant of integration. (8.10.2) can also be written as

$$\frac{df}{d\xi} = \frac{1}{\delta}\left(f^2 - 2cf - 2A\right) = \frac{1}{\delta}\left[(f - c)^2 - \left(c^2 + 2A\right)\right] \tag{8.10.3}$$

Assuming that $df/d\xi = 0$ is available at $f \neq c$, then $c^2 + 2A > 0$. Further, (8.10.3) can be rewritten as

$$\frac{df}{d\xi} = \frac{1}{\delta}(f - f_1)(f - f_2) \tag{8.10.4}$$

where

$$f_1 = c + \sqrt{c^2 + 2A}, f_2 = c - \sqrt{c^2 + 2A} \tag{8.10.5}$$

Make the integration of (8.10.4) over ξ, we can obtain

$$f = c - \frac{1}{2}(f_1 - f_2)\tanh\left[\frac{(f_1 - f_2)}{2\delta}(\xi - \xi_0)\right]$$

or

$$u = f = c - \sqrt{c^2 + 2A}\tanh\left[\frac{\sqrt{c^2 + 2A}}{\delta}(\xi - \xi_0)\right] \tag{8.10.6}$$

where ξ_0 is the constant of integration. This is the traveling wave solution (stationary solution) of the Burgers equation.

8.11 Exercise 8.11 (Steady-State Solution of the Burgers' Equation and Their Stability Analysis)

Solution: (a) Let the traveling wave solution of the Burgers equation be

$$u = u(\xi), \xi = x - ct \tag{8.11.1}$$

Substituting this into the Burgers' equation yields

$$-c\frac{du}{d\xi} + u\frac{du}{d\xi} - \nu\frac{d^2u}{d\xi^2} = 0 \tag{8.11.2}$$

Integrating the above equation over ξ, we obtain

$$-cu + \frac{1}{2}u^2 - \nu\frac{du}{d\xi} = A \tag{8.11.3}$$

where A is the constant of integration. (8.11.3) can be written as

$$\frac{du}{d\xi} = \frac{1}{2\nu}\left(u^2 - 2cu - 2A\right) = \frac{1}{2\nu}\left[(u - c)^2 - \left(c^2 + 2A\right)^2\right] \tag{8.11.4}$$

Assuming that $du/d\xi = 0$ is available at $u \neq c$, then $c^2 + 2A > 0$. Further, the Eq. (8.11.3) can be rewritten as

$$\frac{du}{d\xi} = \frac{1}{2\nu}(u - u_1)(u - u_2) \tag{8.11.5}$$

where

$$u_1 = c + \sqrt{c^2 + 2A}, u_2 = c - \sqrt{c^2 + 2A} \tag{8.11.6}$$

Make the integration of (8.11.5) over ξ, we can obtain

$$u = c - \frac{1}{2}(u_1 - u_2)\tanh\left[\frac{(u_1 - u_2)}{4\nu}(\xi - \xi_0)\right]$$

or

$$u = c - \sqrt{c^2 + 2A}\tanh\left[\frac{\sqrt{c^2 + 2A}}{2\nu}(\xi - \xi_0)\right] \tag{8.11.7}$$

where ξ_0 is the constant of integration. This is the traveling wave solution (steady state solution) of the Burgers equation.

Notice that the traveling wave solution (8.11.7) is the constant solution of the Burgers equation when $c = 0$. Therefore, let $c = 0$ and $\xi_0 = 0$, then $\xi = x$ and (8.11.7) becomes

$$u = -\sqrt{2A}\tanh\left[\frac{\sqrt{2A}}{2\nu}x\right] \tag{8.11.8}$$

From the above equation, $u(\pm\infty) = \mp\sqrt{2A}$. Write

$$u_\infty = \sqrt{2A} \tag{8.11.9}$$

Then the constant solution of the Burgers equation u_0 is

$$u_0 = -u_\infty\tanh\left(\frac{u_\infty x}{2\nu}\right) \tag{8.11.10}$$

(b) Let $u = u_0 + v$, where $|v|\langle\langle|u_0|$. Substituting this into the Burgers equation and retaining the linear term only yields

$$\frac{\partial v}{\partial t} + v\frac{du_0}{dx} + u_0\frac{\partial v}{\partial x} + u_0\frac{du_0}{dx} = \nu\frac{\partial^2 v}{\partial x^2} + \nu\frac{d^2 u_0}{dx^2} \tag{8.11.11}$$

For a constant solution u_0, (8.11.2) becomes

$$u_0\frac{du_0}{dx} - \nu\frac{d^2 u_0}{dx^2} = 0 \tag{8.11.12}$$

From (8.11.11) and (8.11.12), the control equation for the perturbation $v(x, t)$ is

$$\frac{\partial v}{\partial t} + u_0\frac{\partial v}{\partial x} + \frac{du_0}{dx}v = \nu\frac{\partial^2 v}{\partial x^2} \tag{8.11.13}$$

(c) Let the general solution of (8.11.13) be

$$v = v(\xi), \xi = x - ct \tag{8.11.14}$$

Substituting this into (8.11.13) yields

$$v\frac{d^2v}{d\xi^2} + c\frac{dv}{d\xi} - u_0\frac{dv}{d\xi} - \frac{du_0}{dx}v = 0 \tag{8.11.15}$$

Write this equation as a system of first order equations:

$$\begin{aligned} \frac{dv}{d\xi} &= w \\ \frac{dw}{d\xi} &= \frac{1}{v}\frac{du_0}{dx}v + \frac{1}{v}(u_0 - c)w \end{aligned} \tag{8.11.16}$$

For the constant solution u_0, we can obtain from (8.11.5), (8.11.6) and (8.11.9) that

$$\frac{du_0}{d\xi} = \frac{1}{2v}(u_0 - u_\infty)^2 \tag{8.11.17}$$

Substituting (8.11.17) into (8.11.16), we obtain

$$\begin{aligned} \frac{dv}{d\xi} &= w \\ \frac{dw}{d\xi} &= \frac{1}{2v^2}(u_0 - u_\infty)^2 v + \frac{1}{v}(u_0 - c)w \end{aligned} \tag{8.11.18}$$

The corresponding eigenvalues are

$$\lambda_{1,2} = \frac{(u_0 - c) \pm \sqrt{(u_0 - c)^2 + 2(u_0 - u_\infty)^2}}{2v} \tag{8.11.19}$$

It can be seen that one of the eigenvalues is positive and hence the constant solution u_0 is unstable.

8.12 Exercise 8.12 (Constant Solutions of the KdV Equation and Their Stability Analysis)

Solution: (a) The Kottweg-de Vries equation is also named as the KdV equation. Let the traveling wave solution of the KdV equation be given by

$$u = u(\xi), \xi = x - ct \tag{8.12.1}$$

Substituting this into the KdV equation yields

$$-c\frac{du}{d\xi}+u\frac{du}{d\xi}-\beta\frac{d^3u}{d\xi^3}=0 \tag{8.12.2}$$

Integrating the above equation over ξ, we obtain

$$-cu+\frac{1}{2}u^2-\beta\frac{d^2u}{d\xi^2}=A \tag{8.12.3}$$

where A is the constant of integration. (8.12.3) can be written as

$$\frac{d^2u}{d\xi^2}=\frac{1}{2\beta}\left(u^2-2cu-2A\right) \tag{8.12.4}$$

(8.12.4) is multiplied by $du/d\xi$. After integration, we have

$$\left(\frac{du}{d\xi}\right)^2=\frac{1}{3\beta}\left(u^3-3cu^2-6Au-6B\right) \tag{8.12.5}$$

where B is the constant of integration.

Assuming that the equation

$$u^3-3cu^2-6Au-6B=0 \tag{8.12.6}$$

has three real roots u_1, u_2, u_3, and let $u_1 \geq u_2 \geq u_3$. From the relation between roots and coefficients we have

$$\begin{aligned} c &= \tfrac{1}{3}(u_1+u_2+u_3) \\ A &= -\tfrac{1}{6}(u_1u_2+u_2u_3+u_3u_1) \\ B &= \tfrac{1}{6}u_1u_2u_3 \end{aligned} \tag{8.12.7}$$

(8.12.5) can be rewritten as

$$\left(\frac{du}{d\xi}\right)^2=\frac{1}{3\beta}(u-u_1)(u-u_2)(u-u_3) \tag{8.12.8}$$

For the case of $\beta > 0$, the solution of (8.12.8) is

$$u=u_2-(u_2-u_3)\mathrm{cn}^2\left[\sqrt{\frac{u_1-u_3}{12\beta}}(\xi-\xi_0), k\right], u_3 \leq u \leq u_2 \tag{8.12.9}$$

where

$$k=\sqrt{\frac{u_2-u_3}{u_1-u_3}} \tag{8.12.10}$$

(8.12.9) is the traveling wave solution of the KdV equation when $\varepsilon > 0$ and is an elliptical cosine wave, which is also called cnoidal waves, as shown in Fig. 8.1a.

This is a periodic function with amplitude

$$U = u_2 - u_3 \tag{8.12.11}$$

The period of the function $\mathrm{cn}^2(x)$ is $2K(k)$, so the wavelength of the elliptical cosine wave is

$$L = 2K(k)/\sqrt{\frac{u_1 - u_3}{12\beta}} \tag{8.12.12}$$

where

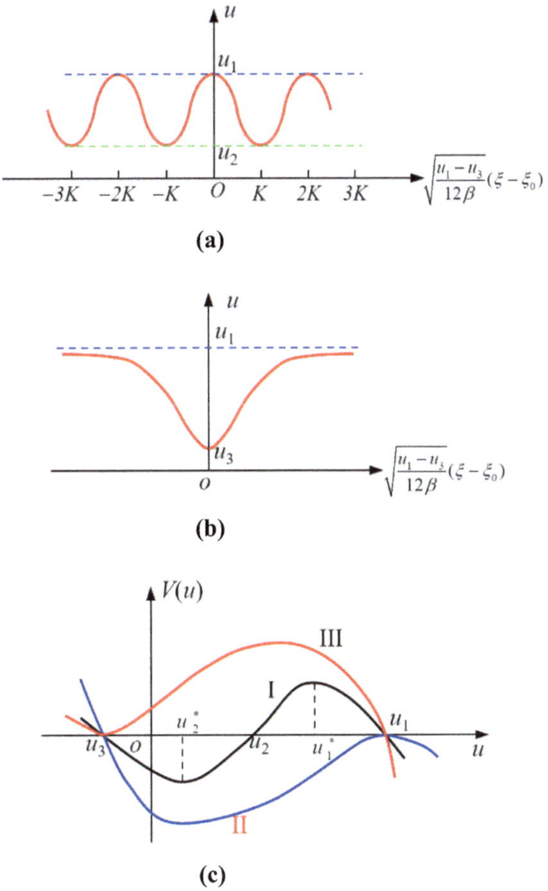

Fig. 8.1 **a** Cnoidal wave solution of the KdV equation, **b** isolator of the KdV equation, **c** schematic of $V(u)$ for Exercise 8.12

$$K(k) = \int_0^{\pi/2} \frac{1}{\sqrt{1 - k^2 \sin^2 \varphi}} d\varphi = \int_0^1 \frac{1}{\sqrt{(1 - x^2)(1 - k^2 x^2)}} dx \tag{8.12.13}$$

is Legendre's complete elliptic integral of the first kind.

When $u_2 \to u_3$, $k \to 0$. In this case, cn$x \to$ cos$x$; therefore, (8.12.9) becomes

$$u = \frac{1}{2}(u_2 + u_3) - \frac{1}{2}(u_2 - u_3)\cos\left[\sqrt{\frac{u_1 - u_3}{3\beta}}(\xi - \xi_0)\right] \tag{8.12.14}$$

This is a simple harmonic solution.

When $u_2 \to u_1$, $k \to 1$. In this case, cn$x \to$ sech$x$; therefore, (8.12.9) becomes

$$u = u_1 - (u_1 - u_3)\text{sech}^2\left[\sqrt{\frac{u_1 - u_3}{12\beta}}(\xi - \xi_0)\right] \tag{8.12.15}$$

This is a solitary wave (Fig. 8.1b), which is often called **a soliton** since it remains unchanged as it moves.

Notice that the traveling wave solution (8.12.15) is the constant solution of the KdV equation when $c = 0$. In this case,

$$u_3 = -2u_1 \tag{8.12.16}$$

Let $\xi_0 = 0$, (8.12.15) can be obtained by

$$u = u_1\left[1 - 3\,\text{sech}^2\left(\frac{u_1}{4\beta}\right)^{1/2} x\right] \tag{8.12.17}$$

From the above equation, $u(\pm\infty) = u_1 > 0$, therefore, let

$$u_\infty = u_1 \tag{8.12.18}$$

The constant solution of the KdV equation u_0 is

$$u_0 = u_\infty\left[1 - 3\,\text{sech}^2\left(\frac{u_\infty}{4\beta}\right)^{1/2} x\right],\ u_\infty > 0 \tag{8.12.19}$$

(b) Let $w = du/d\xi$ in (8.12.4), we obtain

$$\begin{aligned} \frac{du}{d\xi} &= w \\ \frac{dw}{d\xi} &= F(u) = -V'(u) \end{aligned} \tag{8.12.20}$$

where a prime denotes the derivative with respect to u; functions F and V are given below:

$$\begin{aligned} F(u) &= \frac{1}{2\beta}\left(u^2 - 2cu - 2A\right) = \frac{1}{2\beta}\left(u - u_1^*\right)\left(u - u_2^*\right) \\ u_1^* &= c + \sqrt{c^2 + 2A},\, u_2^* = c - \sqrt{c^2 + 2A} \end{aligned} \tag{8.12.21}$$

$$V(u) = -\frac{1}{6\beta}\left(u^3 - 3cu^2 - 6Au - 6B\right) \tag{8.12.22}$$

The situation with $c^2 + 2A > 0$ has been described in Exercise 8.11 (a). Furthermore, (8.12.5) can be written as

$$\frac{1}{2}w^2 + V(u) = 0 \tag{8.12.23}$$

which requires

$$V(u) \leq 0 \tag{8.12.24}$$

From (8.12.20), the equilibrium points of the system are

$$\left(\left(u_1^*, w_1^*\right) = \left(u_1^*, 0\right) \text{ and } \left(u_2^*, w_2^*\right) = \left(u_2^*, 0\right)\right) \tag{8.12.25}$$

Near the equilibrium position u^*, we let $u = u^* + v$, then (8.12.20) can be linearized as

$$\begin{Bmatrix} dv/d\xi \\ dw/d\xi \end{Bmatrix} = \begin{bmatrix} 0 & 1 \\ -V''(u^*) & 0 \end{bmatrix} \begin{Bmatrix} v \\ w \end{Bmatrix} \tag{8.12.26}$$

The corresponding characteristic equation is

$$\lambda^2 + V''\left(u^*\right) = 0 \tag{8.12.27}$$

where

$$V''(u) = -\frac{1}{\beta}(u - c) \tag{8.12.28}$$

From (8.12.21), (8.12.22) and (8.12.28), we can know that:

(1) For the equilibrium point $\left(u_1^*, 0\right)$, $V'\left(u_1^*\right) = 0$, $V''\left(u_1^*\right) < 0$, one of the eigenvalues is positive and another one is negative, so the equilibrium point $\left(u_1^*, 0\right)$ is the saddle point.

(2) For the equilibrium point $(u_2^*, 0)$, $V'(u_2^*) = 0$, $V''(u_2^*) > 0$, two eigenvalues are a pair of conjugate imaginary roots, so the equilibrium point $(u_2^*, 0)$ is the center.

A schematic of $V(u)$ is presented in Fig. 8.1c, where curve I is the potential energy curve for the general case. For a simple harmonic solution described by (8.12.14), there is $u_2 \to u_3$, so $u_2^* \to u_3$, the potential energy curve becomes curve II. For the soliton described by (8.12.15), there is $u_2 \to u_1$, so $u_1^* \to u_1$, the potential energy curve becomes curve III. Therefore, the equilibrium point $(u_1, 0)$ of the soliton (8.12.15) is the saddle point, which is unstable. Since the constant solution (8.12.19) is a stationary soliton, it is also unstable.

8.13 Exercise 8.13 (Traveling Wave Solutions of the Sine–Gordon Equation)

Solution: Substituting $\psi = \pi + \phi(s)$ into sine–Gordon equation, we obtain

$$\phi_{xx} - \phi_{tt} = -\sin\phi \tag{8.13.1}$$

Let $s = x - vt$, we get

$$\phi'' + \left(1 - v^2\right)^{-1} \sin\phi = 0 \tag{8.13.2}$$

where the prime denotes the derivative with respect to s. Equation (8.13.2) is a single pendulum equation.

(1) **When** $v \to 1$

In this case, $\left(1 - v^2\right)^{-1} \to \infty$, which corresponds to the situation that the torsional stiffness of the pendulum is infinity or the pendulum length is zero. Then Eq. (8.13.2) is solved by

$$\phi = 0 \tag{8.13.3}$$

Therefore, the solution to the sine–Gordon equation is

$$\psi = \pi \tag{8.13.4}$$

(2) **When** $v \neq 1$

In this case, $\left(1 - v^2\right)^{-1} \neq 0$. Make the integration of (8.13.2), we obtain

$$\frac{1}{2}\phi\prime^2 - \left(1 - v^2\right)^{-1} \cos\phi = E \tag{8.13.5}$$

Or write

$$\phi\prime = \pm\sqrt{2\left[E - \left(1 - v^2\right)^{-1}\cos\phi\right]} \tag{8.13.6}$$

Therefore, the solution to the sine–Gordon equation is

$$\psi\prime = \pm\sqrt{2\left[E - \left(1 - v^2\right)^{-1}\cos\psi\right]} \tag{8.13.7}$$

8.14 Exercise 8.14 (Characteristic Transformation and Straightforward Expansion for Waves Propagating Along a Uniform Elastic bar with Material Nonlinearity)

Solution: **(a)** The given nonlinear governing equation for the rod can be written as

$$\frac{1}{c^2}\frac{\partial^2 u}{\partial t^2} = (1 + 2E_1 e)\frac{\partial^2 u}{\partial x^2},\ e = \frac{\partial u}{\partial x} \tag{8.14.1}$$

Or

$$\frac{\partial^2 u}{\partial t^2} = \hat{c}^2(e)\frac{\partial^2 u}{\partial x^2},\ \hat{c}^2(e) = c^2(1 + 2E_1 e) \tag{8.14.2}$$

Let

$$v = \frac{\partial u}{\partial t} \tag{8.14.3}$$

then the following system of first order equations can be obtained by (8.14.3) and (8.14.2):

$$\frac{\partial e}{\partial t} - \frac{\partial v}{\partial x} = 0 \tag{8.14.4}$$

$$\frac{\partial v}{\partial t} - \hat{c}^2(e)\frac{\partial e}{\partial x} = 0 \tag{8.14.5}$$

A linear combination of these two equations yields

$$p_1\frac{\partial e}{\partial t} - p_2\hat{c}^2\frac{\partial e}{\partial x} + p_2\frac{\partial v}{\partial t} - p_1\frac{\partial v}{\partial x} = 0 \tag{8.14.6}$$

Let the coefficients p_1, p_2 satisfy the following conditions:

$$\frac{p_1}{p_2\hat{c}^2} = \frac{p_2}{p_1} or\, p_1 = \pm\hat{c}p_2 \tag{8.14.7}$$

Taking the positive and negative signs, respectively, and substituting them into (8.14.6) yields

$$\hat{c}\left(\frac{\partial e}{\partial t} - \hat{c}\frac{\partial e}{\partial x}\right) + \left(\frac{\partial v}{\partial t} - \hat{c}\frac{\partial v}{\partial x}\right) = 0 \tag{8.14.8}$$

$$\hat{c}\left(\frac{\partial e}{\partial t} + \hat{c}\frac{\partial e}{\partial x}\right) - \left(\frac{\partial v}{\partial t} + \hat{c}\frac{\partial v}{\partial x}\right) = 0 \tag{8.14.9}$$

In the $t \sim x$ plane, we specify a family of curves by the following equation:

$$\frac{dx}{dt} = -\hat{c} \tag{8.14.10}$$

Such curves are called characteristics. Thus, (8.14.8) along the characteristics becomes

$$\hat{c}\frac{\partial e}{\partial x} + \frac{\partial v}{\partial x} = 0 \text{or} \hat{c}\frac{\partial e}{\partial t} + \frac{\partial v}{\partial t} = 0 \tag{8.14.11}$$

Similarly, the characteristics of (8.14.9) are specified by the following equations:

$$\frac{dx}{dt} = \hat{c} \tag{8.14.12}$$

(8.14.11) and (8.14.12) are called the characteristics or canonical forms of the original problem. Squaring each of the above two equations yields the following equation:

$$(\frac{dx}{dt})^2 = \hat{c}^2 \tag{8.14.13}$$

Substituting the expressions of $\hat{c}^2$ and e into the above equation, we obtain

$$\left(1 + 2E_1\frac{\partial u}{\partial x}\right)\left(\frac{dt}{dx}\right)^2 = \frac{1}{c^2} \Rightarrow \frac{dt}{dx} = \pm\frac{1}{c}\left(1 + 2E_1\frac{\partial u}{\partial x}\right)^{-1/2} \tag{8.14.14}$$

Since $\partial u/\partial x$ is a small quantity, the right-hand side of the above equation is expanded to give

$$\frac{dt}{dx} = \pm\frac{1}{c}\left(1 - E_1\frac{\partial u}{\partial x}\right) + \text{higher} - \text{order terms} \tag{8.14.15}$$

(b) Omitting the higher order terms in (8.14.15), we obtain

$$\frac{dt}{dx} = -\frac{1}{c}(1 - E_1 e), \frac{dt}{dx} = \frac{1}{c}(1 - E_1 e) \tag{8.14.16}$$

The two equations in (8.14.16) define two consecutive families of characteristics in the $t \sim x$ plane, which are the integral curves (or solution curves) of two differential equations, and they are denoted as the ξ and η axes, respectively. ξ and η axes form a curve coordinate system in the $t \sim x$ plane, where the curve coordinates are denoted as (ξ, η). Thus, there are

$$Along\ \xi\ axis: \frac{dt}{dx} = -\frac{1}{c}(1 - E_1 e) \triangleq c_2 \tag{8.14.17}$$

$$Along\ \eta\ axis: \frac{dt}{dx} = \frac{1}{c}(1 - E_1 e) \triangleq c_1 \tag{8.14.18}$$

which are shown in Fig. 8.2a. Considering (8.14.18), we have

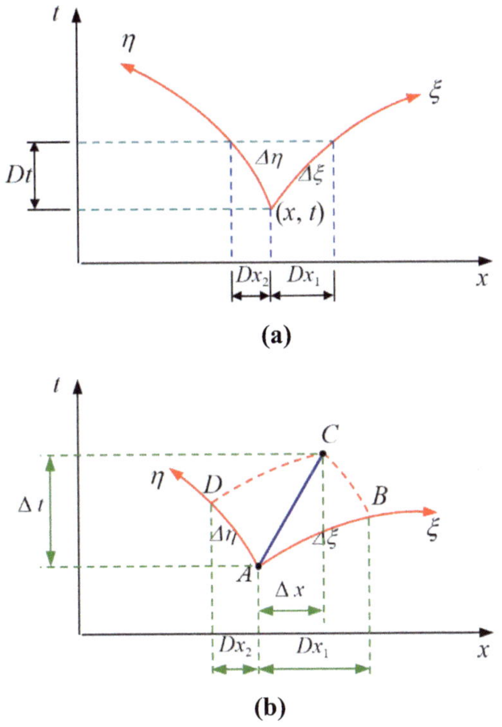

Fig. 8.2 **a** Line element involved in ∂t/∂x and ∂t/∂h, **b** line element relations formed by arbitrary two adjacent points *A* and *C* for Exercise 8.14

$$\frac{\partial t}{\partial \eta} = \frac{\Delta t}{\Delta \eta} = \frac{c_1 \Delta x_2}{\Delta \eta} = c_1 \frac{\partial x}{\partial \eta}$$

i.e.,

$$\frac{\partial t}{\partial \eta} = \left(\frac{1}{c} - \frac{E_1 e}{c}\right)\frac{\partial x}{\partial \eta} \triangleq c_1 \frac{\partial x}{\partial \eta} \tag{8.14.19}$$

Similarly, we can obtain

$$\frac{\partial t}{\partial \xi} = \left(-\frac{1}{c} + \frac{E_1 e}{c}\right)\frac{\partial x}{\partial \xi} \triangleq c_2 \frac{\partial x}{\partial \xi} \tag{8.14.20}$$

(c) For any function $F(x, t) = F(x(\xi, \eta), t(\xi, \eta))$, applying the chain rule yields

$$\frac{\partial F}{\partial x} = \frac{\partial F}{\partial \xi}\frac{\partial \xi}{\partial x} + \frac{\partial F}{\partial \eta}\frac{\partial \eta}{\partial x} \tag{8.14.21}$$

Referring to Fig. 8.2b, noting that quadrilateral $ABCD$ is an approximate parallelogram, and considering (8.14.17), (8.14.18), (8.14.19) and (8.14.20), we have

$$\frac{\partial \xi}{\partial x} = \frac{\Delta \xi}{\Delta x} = \frac{\Delta \xi}{\Delta x_1 - \Delta x_2} = \frac{1}{\left(\frac{1}{c_2} - \frac{1}{c_1}\right)\frac{\Delta t}{\Delta \xi}} = \frac{1}{\frac{c_1 - c_2}{c_1 c_2} c_2 x_\xi} = \frac{1}{c_1 - c_2}\frac{c_1}{x_\xi} \tag{8.14.22}$$

$$\frac{\partial \eta}{\partial x} = -\frac{\Delta \eta}{\Delta x} = \frac{\Delta \eta}{\Delta x_2 - \Delta x_1} = \frac{1}{\left(\frac{1}{c_1} - \frac{1}{c_2}\right)\frac{\Delta t}{\Delta \eta}} = \frac{1}{\frac{c_2 - c_1}{c_1 c_2} c_1 x_\eta} = \frac{1}{c_2 - c_1}\frac{c_2}{x_\eta} \tag{8.14.23}$$

It should be noted that (8.14.22) implies that $\partial \xi / \partial x$ is positive and $\Delta \xi$ increases (or decreases) as Δx increases (or decreases) (Fig. 8.2b); hence, $\partial \xi / \partial x = \Delta \xi / \Delta x$. (8.14.23) implies that $\partial \eta / \partial x$ is positive and $\Delta \eta$ decreases (or increases) as Δx increases (or decreases) (Fig. 8.2b); hence, $\partial \eta / \partial x = -\Delta \eta / \Delta x$. Substituting (8.14.22) and (8.14.23) into (8.14.21), we obtain

$$\frac{\partial F}{\partial x} = -\frac{1}{c_2 - c_1}\left(\frac{c_1}{x_\xi}\frac{\partial F}{\partial \xi} - \frac{c_2}{x_\eta}\frac{\partial F}{\partial \eta}\right)$$

i.e.,

$$\frac{\partial}{\partial x} = -\frac{1}{c_2 - c_1}\left(\frac{c_1}{x_\xi}\frac{\partial}{\partial \xi} - \frac{c_2}{x_\eta}\frac{\partial}{\partial \eta}\right) \tag{8.14.24}$$

With the chain rule and taking $c_2 = -c_1$ into account, we can obtain

$$\frac{\partial F}{\partial t} = \frac{\partial F}{\partial \xi}\frac{\partial \xi}{\partial t} + \frac{\partial F}{\partial \eta}\frac{\partial \eta}{\partial t} = \frac{\partial F}{\partial \xi}\frac{\partial \xi}{\partial x}\frac{dx}{dt}\Big|_{\text{on}\xi} + \frac{\partial F}{\partial \eta}\frac{\partial \eta}{\partial x}\frac{dx}{dt}\Big|_{\text{on}\eta}$$
$$= \frac{\partial F}{\partial \xi}\frac{1}{c_1 - c_2}\frac{c_1}{x_\xi}\frac{1}{c_2} + \frac{\partial F}{\partial \eta}\frac{1}{c_2 - c_1}\frac{c_2}{x_\eta}\frac{1}{c_1}$$
$$= \frac{1}{c_2 - c_1}\frac{\partial F}{\partial \xi}\frac{1}{x_\xi} - \frac{1}{c_2 - c_1}\frac{\partial F}{\partial \eta}\frac{1}{x_\eta}$$

i.e.,

$$\frac{\partial}{\partial t} = \frac{1}{c_2 - c_1}\left(\frac{1}{x_\xi}\frac{\partial}{\partial \xi} - \frac{1}{x_\eta}\frac{\partial}{\partial \eta}\right) \tag{8.14.25}$$

Using (8.14.24) and (8.14.25), we can replace the partial derivatives of x, t with the partial derivatives of ξ, η in (8.14.4) and (8.14.5) and obtain

$$x_\xi\left(e_\eta + c_2 v_\eta\right) - x_\eta\left(e_\xi + c_1 v_\xi\right) = 0 \tag{8.14.26}$$

$$x_\xi\left[v_\eta + c^2 c_2(1 + 2E_1 e)e_\eta\right] - x_\eta\left[v_\xi + c^2 c_1(1 + 2E_1 e)e_\xi\right] = 0 \tag{8.14.27}$$

which assumes $x_\xi \neq 0, x_\eta \neq 0$.

(d) (8.14.26) and (8.14.27) can be solved by the method of straightforward expansion. Let the solution of the equation be

$$\begin{aligned} e(\xi, \eta) &= \varepsilon e_1(\xi, \eta) + \varepsilon^2 e_2(\xi, \eta) + \cdots \\ v(\xi, \eta) &= \varepsilon v_1(\xi, \eta) + \varepsilon^2 v_2(\xi, \eta) + \cdots \end{aligned} \tag{8.14.28}$$

Substituting (8.14.28) into (8.14.26) and (8.14.27) and keeping to $O(\varepsilon^2)$, we obtain

$$\begin{aligned} 0 = \varepsilon&\left[x_\xi\left(\frac{\partial e_1}{\partial \eta} - \frac{1}{c}\frac{\partial v_1}{\partial \eta}\right) - x_\eta\left(\frac{\partial e_1}{\partial \xi} + \frac{1}{c}\frac{\partial v_1}{\partial \xi}\right)\right] \\ + \varepsilon^2&\left[x_\xi\left(\frac{\partial e_2}{\partial \eta} - \frac{1}{c}\frac{\partial v_2}{\partial \eta}\right) - x_\eta\left(\frac{\partial e_2}{\partial \xi} + \frac{1}{c}\frac{\partial v_2}{\partial \xi}\right) + E_1 e_1\left(x_\xi\frac{\partial v_1}{\partial \eta} + x_\eta\frac{\partial v_1}{\partial \xi}\right)\right] \end{aligned} \tag{8.14.29}$$

$$\begin{aligned} 0 = \varepsilon&\left[x_\xi\left(\frac{\partial e_1}{\partial \eta} - \frac{1}{c}\frac{\partial v_1}{\partial \eta}\right) + x_\eta\left(\frac{\partial e_1}{\partial \xi} + \frac{1}{c}\frac{\partial v_1}{\partial \xi}\right)\right] \\ + \varepsilon^2&\left[x_\xi\left(\frac{\partial e_2}{\partial \eta} - \frac{1}{c}\frac{\partial v_2}{\partial \eta}\right) + x_\eta\left(\frac{\partial e_2}{\partial \xi} - \frac{1}{c}\frac{\partial v_2}{\partial \xi}\right) + E_1 e_1\left(x_\xi\frac{\partial e_1}{\partial \eta} + x_\eta\frac{\partial e_1}{\partial \xi}\right)\right] \end{aligned} \tag{8.14.30}$$

Let the coefficient of the same power of ε be zero, we get

$$x_\xi\left(\frac{\partial e_1}{\partial\eta}-\frac{1}{c}\frac{\partial v_1}{\partial\eta}\right)-x_\eta\left(\frac{\partial e_1}{\partial\xi}+\frac{1}{c}\frac{\partial v_1}{\partial\xi}\right)=0 \tag{8.14.31}$$

$$x_\xi\left(\frac{\partial e_1}{\partial\eta}-\frac{1}{c}\frac{\partial v_1}{\partial\eta}\right)+x_\eta\left(\frac{\partial e_1}{\partial\xi}+\frac{1}{c}\frac{\partial v_1}{\partial\xi}\right)=0 \tag{8.14.32}$$

$$x_\xi\left(\frac{\partial e_2}{\partial\eta}-\frac{1}{c}\frac{\partial v_2}{\partial\eta}\right)-x_\eta\left(\frac{\partial e_2}{\partial\xi}+\frac{1}{c}\frac{\partial v_2}{\partial\xi}\right)=-E_1e_1\left(x_\xi\frac{\partial v_1}{\partial\eta}+x_\eta\frac{\partial v_1}{\partial\xi}\right) \tag{8.14.33}$$

$$x_\xi\left(\frac{\partial e_2}{\partial\eta}-\frac{1}{c}\frac{\partial v_2}{\partial\eta}\right)+x_\eta\left(\frac{\partial e_2}{\partial\xi}+\frac{1}{c}\frac{\partial v_2}{\partial\xi}\right)=-E_1e_1\left(x_\xi\frac{\partial e_1}{\partial\eta}+x_\eta\frac{\partial e_1}{\partial\xi}\right) \tag{8.14.34}$$

From (8.14.31) and (8.14.32), we have

$$\frac{\partial e_1}{\partial\eta}-\frac{1}{c}\frac{\partial v_1}{\partial\eta}=0,\ \frac{\partial e_1}{\partial\xi}+\frac{1}{c}\frac{\partial v_1}{\partial\xi}=0 \tag{8.14.35}$$

Integrating these two equations, respectively, yields

$$e_1-\frac{1}{c}v_1=2f(\xi),\ e_1+\frac{1}{c}v_1=2g(\eta) \tag{8.14.36}$$

From this

$$e_1=f(\xi)+g(\eta),\ v_1=-cf(\xi)+cg(\eta) \tag{8.14.37}$$

therefore,

$$\begin{aligned} e&=\varepsilon f(\xi)+\varepsilon g(\eta)+\cdots \\ v&=-\varepsilon cf(\xi)+\varepsilon cg(\eta)+\cdots \end{aligned} \tag{8.14.38}$$

Also make the expansion of t and x, we obtain

$$\begin{aligned} t(\xi,\eta)&=t_0(\xi,\eta)+\varepsilon t_1(\xi,\eta)+\cdots \\ x(\xi,\eta)&=x_0(\xi,\eta)+\varepsilon x_1(\xi,\eta)+\cdots \end{aligned} \tag{8.14.39}$$

Substituting (8.14.39) and (8.14.38) into (8.14.19) and (8.14.20), respectively, yields

$$\frac{\partial t_0}{\partial\eta}+\varepsilon\frac{\partial t_1}{\partial\eta}=\frac{1}{c}\{1-\varepsilon E_1[f(\xi)+g(\eta)]\}\left(\frac{\partial x_0}{\partial\eta}+\varepsilon\frac{\partial x_1}{\partial\eta}\right) \tag{8.14.40}$$

$$\frac{\partial t_0}{\partial\xi}+\varepsilon\frac{\partial t_1}{\partial\xi}=-\frac{1}{c}\{1-\varepsilon E_1[f(\xi)+g(\eta)]\}\left(\frac{\partial x_0}{\partial\xi}+\varepsilon\frac{\partial x_1}{\partial\xi}\right) \tag{8.14.41}$$

$$\frac{\partial t}{\partial \eta} = \left(\frac{1}{c} - \frac{E_1 e}{c}\right)\frac{\partial x}{\partial \eta} \triangleq c_1 \frac{\partial x}{\partial \eta} \tag{8.14.42}$$

$$\frac{\partial t}{\partial \xi} = \left(-\frac{1}{c} + \frac{E_1 e}{c}\right)\frac{\partial x}{\partial \xi} \triangleq c_2 \frac{\partial x}{\partial \xi} \tag{8.14.43}$$

Equating the coefficients of the same powers of ε, we obtain

$$\frac{\partial t_0}{\partial \eta} - \frac{1}{c}\frac{\partial x_0}{\partial \eta} = 0 \tag{8.14.44}$$

$$\frac{\partial t_0}{\partial \xi} + \frac{1}{c}\frac{\partial x_0}{\partial \xi} = 0 \tag{8.14.45}$$

$$\frac{\partial t_1}{\partial \eta} - \frac{1}{c}\frac{\partial x_1}{\partial \eta} = -\frac{1}{c}E_1\left[f(\xi) + g(\eta)\right]\frac{\partial x_0}{\partial \eta} \tag{8.14.46}$$

$$\frac{\partial t_1}{\partial \xi} + \frac{1}{c}\frac{\partial x_1}{\partial \xi} = \frac{1}{c}E_1\left[f(\xi) + g(\eta)\right]\frac{\partial x_0}{\partial \xi} \tag{8.14.47}$$

Equation (8.14.44) and (8.14.45) can be solved by

$$t_0 = A(\eta + \xi), x_0 = cA(\eta - \xi) \tag{8.14.48}$$

where A is the constant of integration. Substituting (8.14.48) into (8.14.46) and (8.14.47), we obtain

$$\frac{\partial t_1}{\partial \eta} - \frac{1}{c}\frac{\partial x_1}{\partial \eta} = -AE_1\left[f(\xi) + g(\eta)\right] \tag{8.14.49}$$

$$\frac{\partial t_1}{\partial \xi} + \frac{1}{c}\frac{\partial x_1}{\partial \xi} = -AE_1\left[f(\xi) + g(\eta)\right] \tag{8.14.50}$$

Integrating these two equations, respectively, yields the special solution of the equation

$$t_1 - \frac{1}{c}x_1 = -AE_1\left[(\eta - \xi)f + \int_{\xi}^{\eta} g(s)ds\right] \tag{8.14.51}$$

$$t_1 + \frac{1}{c}x_1 = -AE_1\left[\int_{\eta}^{\xi} f(s)ds + (\xi - \eta)g\right] \tag{8.14.52}$$

From this

$$t_1 = -\tfrac{1}{2}AE_1\left[(\eta-\xi)f + (\xi-\eta)g + \int_\eta^\xi f(s)ds + \int_\xi^\eta g(s)ds\right]$$
$$x_1 = \tfrac{1}{2}cAE_1\left[(\eta-\xi)f - (\xi-\eta)g - \int_\eta^\xi f(s)ds + \int_\xi^\eta g(s)ds\right] \tag{8.14.53}$$

Substituting (8.14.48) and (8.14.53) into (8.14.39) yields

$$t = A(\eta+\xi) - \tfrac{1}{2}\varepsilon AE_1\left[(\eta-\xi)f + (\xi-\eta)g + \int_\eta^\xi f(s)ds + \int_\xi^\eta g(s)ds\right] + \cdots$$
$$x = cA(\eta-\xi) + \tfrac{1}{2}\varepsilon cAE_1\left[(\eta-\xi)f - (\xi-\eta)g - \int_\eta^\xi f(s)ds + \int_\xi^\eta g(s)ds\right] + \cdots \tag{8.14.54}$$

Considering $x(\xi,\xi) = 0$ and $t(\xi,\xi) = \xi$, we can obtain

$$\xi = 2A\xi \Rightarrow A = \frac{1}{2} \tag{8.14.55}$$

Therefore,

$$t = \tfrac{1}{2}(\eta+\xi) - \tfrac{1}{4}\varepsilon E_1\left[(\eta-\xi)f + (\xi-\eta)g + \int_\eta^\xi f(s)ds + \int_\xi^\eta g(s)ds\right] + \cdots$$
$$x = \tfrac{1}{2}c(\eta-\xi) + \tfrac{1}{4}\varepsilon cE_1\left[(\eta-\xi)f - (\xi-\eta)g - \int_\eta^\xi f(s)ds + \int_\xi^\eta g(s)ds\right] + \cdots \tag{8.14.56}$$

From the above two equations, we can obtain

$$t - \tfrac{x}{c} = \xi - \tfrac{1}{2}\varepsilon E_1\left[\tfrac{2x}{c}f(\xi) + \int_\xi^\eta g(s)ds\right] + \cdots$$
$$t + \tfrac{x}{c} = \eta + \tfrac{1}{2}\varepsilon E_1\left[\tfrac{2x}{c}g(\eta) - \int_\eta^\xi f(s)ds\right] + \cdots \tag{8.14.57}$$

8.15 Exercise 8.15 (Forced Excitations of a Nonlinear Finite Elastic bar)

Solution: **(a)** The given equation is the nonlinear governing equation of forced excitations of a nonlinear finite elastic bar. Let the straightforward expansion of the solution of that equation be

$$u(x,t) = \varepsilon u_1(x,t) + \varepsilon^2 u_2(x,t) + \cdots \tag{8.15.1}$$

Substituting this into the governing equation and retaining to $O\left(\varepsilon^2\right)$ yields

$$\begin{aligned} 0 &= \frac{\partial^2 u}{\partial x^2} - \frac{\partial^2 u}{\partial t^2} + 2E_1 \frac{\partial u}{\partial x}\frac{\partial^2 u}{\partial x^2} \\ &= \varepsilon\left(\frac{\partial^2 u_1}{\partial x^2} - \frac{\partial^2 u_1}{\partial t^2}\right) + \varepsilon^2\left(\frac{\partial^2 u_2}{\partial x^2} - \frac{\partial^2 u_2}{\partial t^2} + 2E_1 \frac{\partial u_1}{\partial x}\frac{\partial^2 u_1}{\partial x^2}\right) \end{aligned} \tag{8.15.2}$$

Let the coefficient of the same power of ε be zero, we obtain

$$\frac{\partial^2 u_1}{\partial x^2} - \frac{\partial^2 u_1}{\partial t^2} = 0 \tag{8.15.3}$$

$$\frac{\partial^2 u_2}{\partial x^2} - \frac{\partial^2 u_2}{\partial t^2} = -2E_1 \frac{\partial u_1}{\partial x}\frac{\partial^2 u_1}{\partial x^2} \tag{8.15.4}$$

(8.15.3) is a one-dimensional wave equation, which can be solved by the variable separation method. Let

$$u_1(x,t) = \phi(x)q(t) \tag{8.15.5}$$

and substitute (8.15.5) into (8.15.4), we can obtain

$$\frac{\phi''}{\phi} = \frac{\ddot{q}}{q} = -\omega^2 \tag{8.15.6}$$

hence

$$\phi'' + \omega^2\phi = 0, \ddot{q} + \omega^2 q = 0 \tag{8.15.7}$$

and

$$\phi(x) = A\sin\omega x + B\cos\omega x,\ q(t) = C\cos(\omega t + \theta) \tag{8.15.8}$$

$$u_1(x,t) = (A\mathrm{sin}\omega x + B\mathrm{cos}\omega x)\mathrm{cos}(\omega t + \theta) \tag{8.15.9}$$

$$u(x,t) = \varepsilon(A\mathrm{sin}\omega x + B\mathrm{cos}\omega x)\mathrm{cos}(\omega t + \theta) + \varepsilon^2 u_2 \tag{8.15.10}$$

Applying the boundary conditions $u_1(0,t) = 0$ and$u_1(1,t) = \varepsilon p\mathrm{cos}\omega t$, we can obtain

$$\begin{aligned} &B\cos(\omega t + \theta) = 0 \\ &(A\sin\omega + B\cos\omega)\cos(\omega t + \theta) = p\cos\omega t \end{aligned} \tag{8.15.11}$$

therefore

$$B = 0, \theta = 0, A = p(\sin\omega)^{-1} \tag{8.15.12}$$

From the above results, we have

$$u_1 = p(\sin\omega)^{-1}\cos\omega t\sin\omega x \tag{8.15.13}$$

Substituting (8.15.13) into (8.15.4) yields

$$\frac{\partial^2 u_2}{\partial x^2} - \frac{\partial^2 u_2}{\partial t^2} = \frac{E_1\omega^3 p^2}{2\sin^2\omega}(1 + \cos 2\omega t)\sin 2\omega x \tag{8.15.14}$$

The boundary conditions for u_2 are

$$u_2(0, t) = u_2(1, t) = 0 \tag{8.15.15}$$

Let the solution of (8.15.14) be

$$u_2 = u_{21} + u_{22} \tag{8.15.16}$$

where u_{21} and u_{22} satisfy, respectively, the following two equations:

$$\frac{\partial^2 u_{21}}{\partial x^2} - \frac{\partial^2 u_{21}}{\partial t^2} = \frac{E_1\omega^3 p^2}{2\sin^2\omega}\sin 2\omega x \tag{8.15.17}$$

$$u_{21}(0, t) = u_{21}(1, t) = 0 \tag{8.15.18}$$

$$\frac{\partial^2 u_{22}}{\partial x^2} - \frac{\partial^2 u_{22}}{\partial t^2} = \frac{E_1\omega^3 p^2}{2\sin^2\omega}\cos 2\omega t\sin 2\omega x \tag{8.15.19}$$

$$u_{22}(0, t) = u_{22}(1, t) = 0 \tag{8.15.20}$$

Clearly, the solution to (8.15.17) and (8.15.18) can be written as

$$u_{21} = \psi_1(x) \tag{8.15.21}$$

where

$$\psi_1'' = \frac{E_1\omega^3 p^2}{2\sin^2\omega}\sin 2\omega x \tag{8.15.22}$$

$$\psi_1(0) = \psi_1(1) = 0 \tag{8.15.23}$$

therefore

$$u_{21} = \psi_1(x) = \frac{E_1 \omega p^2}{8\sin^2\omega}(x - \sin 2\omega x) \tag{8.15.24}$$

(8.15.19) and (8.15.20) can be solved by

$$u_{22} = \psi_2(x)\cos 2\omega t \tag{8.15.25}$$

where

$$\psi_2'' + 4\omega^2 \psi_2 = \frac{E_1 \omega^3 p^2}{2\sin^2\omega}\sin 2\omega x \tag{8.15.26}$$

$$\psi_2(0) = \psi_2(1) = 0 \tag{8.15.27}$$

The special solution of (8.15.26) is

$$\psi_2(x) = -\frac{E_1 \omega^2 p^2}{8\sin^2\omega} x\cos 2\omega x$$

However, this special solution does not satisfy the boundary condition (8.15.27). Superimposing the general solution on it, we have

$$\psi_2(x) = A_1 \sin 2\omega x + A_2 \cos 2\omega x - \frac{E_1 \omega^2 p^2}{8\sin^2\omega} x\cos 2\omega x \tag{8.15.28}$$

Substituting (8.15.28) into (8.15.27) yields

$$\begin{aligned} &A_2 = 0 \\ &A_1 \sin 2\omega + A_2 \cos 2\omega = \tfrac{E_1 \omega^2 p^2}{8\sin^2\omega}\cos 2\omega \end{aligned} \tag{8.15.29}$$

therefore

$$A_2 = 0, A_1 = \frac{E_1 \omega^2 p^2 \cos 2\omega}{8 \sin 2\omega \sin^2\omega} \tag{8.15.30}$$

and

$$\psi_2(x) = \frac{E_1 \omega^2 p^2}{8\sin^2\omega}\left(\frac{\cos 2\omega}{8\sin 2\omega}\sin 2\omega x - x\cos 2\omega x\right) \tag{8.15.31}$$

$$u_{22} = \frac{E_1 \omega^2 p^2}{8\sin^2\omega}\left(\frac{\cos 2\omega}{8\sin 2\omega}\sin 2\omega x - x\cos 2\omega x\right)\cos 2\omega t \tag{8.15.32}$$

From (8.15.32), (8.15.24) and (8.15.16), we have

$$u_2 = \frac{E_1 p^2 \omega^2}{8 \sin^2 \omega}\left[\frac{1}{\omega}(x \sin 2\omega - \sin 2\omega x) - \left(x \cos 2\omega x - \frac{\cos 2\omega \sin 2\omega x}{\sin 2\omega}\right) \cos 2\omega t\right] \tag{8.15.33}$$

From the above result, the solution of the given problem can be obtained as

$$\begin{aligned} u = {} & \varepsilon \frac{p}{\sin \omega} \cos \omega t \sin \omega x \\ & + \varepsilon^2 \frac{E_1 p^2 \omega^2}{8 \sin^2 \omega}\left[\begin{array}{l} \frac{1}{\omega}(x \sin 2\omega - \sin 2\omega x) \\ -\left(x \cos 2\omega x - \frac{\cos 2\omega \sin 2\omega x}{\sin 2\omega}\right) \cos 2\omega t \end{array}\right] \end{aligned} \tag{8.15.34}$$

(**b**) When $\omega \approx n\pi$, all terms in (8.15.34) tend to infinity and this solution is invalid. The corresponding free oscillation equation to the governing equation of the rod given in the problem is

$$\begin{aligned} & \frac{\partial^2 u}{\partial x^2} - \frac{\partial^2 u}{\partial t^2} = 0 \\ & u(0, t) = 0, u(1, t) = 0 \end{aligned} \tag{8.15.35}$$

It is easy to find out that the natural frequency of this set of equations is $\omega_n = n\pi, n = 1, 2, \ldots$. Therefore, this case demonstrates a primary resonance.

When $\omega \approx n\pi$, only the nth order modes are excited and the response of the other modes decays very quickly due to damping (see Exercise 7.20). (8.15.34) becomes

$$u = \varepsilon \frac{p}{\sin\omega} \cos\omega t \sin\omega x \tag{8.15.36}$$

Detuning parameter σ is introduced such that

$$\omega = n\pi + \varepsilon\sigma \tag{8.15.37}$$

Substituting this into (8.15.36) yields

$$u \approx \frac{p}{\sigma \cos n\pi} \cos n\pi t \sin n\pi x \tag{8.15.38}$$

Readers are invited to complete Exercise 8.15 (c)–(f).

8.16 Exercise 8.16 (Derivation of the Lagrangian Form of the Wave Equation for an Inviscid Isentropic Gas)

Solution: (a) In order to derive the Lagrangian form of one-dimensional wave equation for an inviscid isentropic gas, we take the gas control volume shown in Fig. 8.3 and write the following equation according to the law of conservation of mass:

$$\rho_0 dxdydz = \rho\left[\left(1+\frac{\partial\eta}{\partial x}\right)dx\right]dydz$$

Therefore, the mass conservation equation (continuity equation) is

$$\rho_0 = \rho\left(1+\frac{\partial\eta}{\partial x}\right) \tag{8.16.1}$$

Applying Newton's second law along the x direction of the control volume shown in Fig. 8.3 yields

$$(\rho_0 dxdydz)\frac{\partial^2\eta}{\partial t^2} = -\left(p+\frac{\partial p}{\partial x}dx\right)dydz + pdydz$$

Therefore, the equation of conservation of momentum (or equation of motion) is

$$\rho_0\frac{\partial^2\eta}{\partial t^2} = -\frac{\partial p}{\partial x} \tag{8.16.2}$$

For an isentropic gas, the equation of state is:

$$\frac{p}{p_0} = \left(\frac{\rho}{\rho_0}\right)^{\gamma} \tag{8.16.3}$$

where $\gamma = C_p/C_V$ is the specific heat ratio or adiabatic exponent.

Fig. 8.3 The control volume of inviscid gas for Exercise 8.16 and 8.19

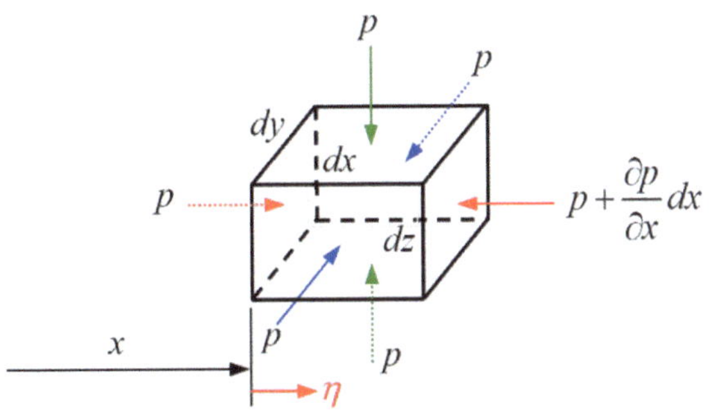

(b) Substituting (8.16.3) into (8.16.2) yields

$$\frac{\partial^2 \eta}{\partial t^2} = -\gamma \frac{p_0}{\rho_0} \left(\frac{\rho}{\rho_0} \right)^{\gamma-1} \frac{1}{\rho_0} \frac{\partial \rho}{\partial x} \tag{8.16.4}$$

Substituting (8.16.1) into (8.16.4) yields

$$\frac{\partial^2 \eta}{\partial t^2} = \gamma \frac{p_0}{\rho_0} \left(1 + \frac{\partial \eta}{\partial x} \right)^{-(\gamma+1)} \frac{\partial^2 \eta}{\partial x^2} \tag{8.16.5}$$

Let

$$c_0^2 = \gamma \frac{p_0}{\rho_0} \tag{8.16.6}$$

Therefore

$$\frac{\partial^2 \eta}{\partial t^2} = c_0^2 \left(1 + \frac{\partial \eta}{\partial x} \right)^{-(\gamma+1)} \frac{\partial^2 \eta}{\partial x^2} \tag{8.16.7}$$

8.17 Exercise 8.17 (Exact Solution of the One-Dimensional Wave Equation for an Inviscid Isentropic Gas)

Solution: **(a)** Substituting the given form of solution into the wave equation of the gas yields

$$\frac{\partial f}{\partial t} = c_0^2 \left(1 + \frac{\partial \eta}{\partial x} \right)^{-(\gamma+1)} \frac{\partial^2 \eta}{\partial x^2} \tag{8.17.1}$$

Let

$$g = \frac{\partial \eta}{\partial x} \tag{8.17.2}$$

then

$$\frac{\partial f}{\partial x} = \frac{\partial u}{\partial x} = \frac{\partial g}{\partial t} \tag{8.17.3}$$

therefore,

$$\frac{\partial f}{\partial t} = \frac{df}{dg} \frac{\partial g}{\partial t} = \frac{df}{dg} \frac{\partial f}{\partial x} = \left(\frac{df}{dg} \right)^2 \frac{\partial g}{\partial x}, \ \frac{\partial^2 \eta}{\partial x^2} = \frac{\partial g}{\partial x} \tag{8.17.4}$$

Substituting (8.17.4) into (8.17.1) yields

$$\left(\frac{df}{dg}\right)^2 = c_0^2(1+g)^{-(\gamma+1)} \tag{8.17.5}$$

or

$$f = c_0^2\left(1+\frac{\partial\eta}{\partial x}\right)^{-(\gamma+1)} \tag{8.17.6}$$

The prime denotes the derivative of f with respect to $\partial\eta/\partial x$.

(b) From (8.17.5), we have

$$\frac{df}{dg} = \pm c_0(1+g)^{-(\gamma+1)/2} \tag{8.17.7}$$

After integration, we have

$$f = u = \pm\frac{2c_0}{1-\gamma}(1+g)^{(1-\gamma)/2} + A \tag{8.17.8}$$

where A is the constant of integration. When the gas is undisturbed, $u = 0$. Assuming that the gas is homogeneous, then $g = \partial\eta/\partial x = 0$, and hence $u(0) = 0$. This gives

$$A = \mp\frac{2c_0}{1-\gamma} \tag{8.17.9}$$

Therefore, Eq. (8.17.8) becomes

$$u = \pm\frac{2c_0}{1-\gamma}\left[1-\left(1+\frac{\partial\eta}{\partial x}\right)^{(1-\gamma)/2}\right] \tag{8.17.10}$$

(c) Considering Eq. (8.17.3), we can obtain

$$\frac{df}{dg} = \frac{\partial f}{\partial x}\frac{\partial x}{\partial g} = \frac{\partial f}{\partial x}\frac{\partial t}{\partial g}\frac{dx}{dt} = \frac{\partial f/\partial x}{\partial g/\partial t}\frac{dx}{dt} = \frac{dx}{dt} \tag{8.17.11}$$

Substituting it into (8.17.7), we can obtain the characteristic line equation

$$\frac{dt}{dx} = \pm\frac{1}{c},\ c = c_0\left(1+\frac{\partial\eta}{\partial x}\right)^{-(\gamma+1)/2} \tag{8.17.12}$$

(d) From (8.17.10), we have

$$1+\frac{\partial\eta}{\partial x}=\left(1\mp\frac{1-\gamma}{2c_0}u\right)^{2/(1-\gamma)} \tag{8.17.13}$$

Substituting (8.17.13) into (8.17.12) yields

$$c=c_0\left(1\mp\frac{1-\gamma}{2c_0}u\right)^{(\gamma+1)/(\gamma-1)} \tag{8.17.14}$$

(e) From the above results, we assume that the right-running wave solution is

$$u=F\left(t-\frac{x}{c}\right) \tag{8.17.15}$$

For the right-running wave

$$c=c_0\left(1+\frac{\gamma-1}{2c_0}u\right)^{(\gamma+1)/(\gamma-1)} \tag{8.17.16}$$

Therefore, the solution can also be written as

$$u=F\left[t-\frac{x}{c_0}\left(1+\frac{\gamma-1}{2c_0}u\right)^{-(\gamma+1)/(\gamma-1)}\right] \tag{8.17.17}$$

Below we verify that the solution (8.17.15) satisfies the wave equation for a gas, i.e., it satisfies the following equation:

$$\frac{\partial^2\eta}{\partial t^2}-c_0^2\left(1+\frac{\partial\eta}{\partial x}\right)^{-(\gamma+1)}\frac{\partial^2\eta}{\partial x^2}=0 \tag{8.17.18}$$

We have.

$$\frac{\partial\eta}{\partial x}=\frac{\partial\eta}{\partial t}\frac{dt}{dx}=\frac{u}{c},\ \frac{\partial^2\eta}{\partial x^2}=\frac{1}{c}\frac{\partial u}{\partial x}=\frac{1}{c^2}\frac{\partial u}{\partial t} \tag{8.17.19}$$

For a right-running wave, we can obtain from (8.17.13) that

$$1+\frac{\partial\eta}{\partial x}=\left(1+\frac{\gamma-1}{2c_0}u\right)^{2/(1-\gamma)} \tag{8.17.20}$$

Substituting (8.17.20) and the second equation of (8.17.19) into the left side of (8.17.18), we obtain

$$
\begin{aligned}
&\frac{\partial^2 \eta}{\partial t^2} - c_0^2\left(1+\frac{\partial \eta}{\partial x}\right)^{-(\gamma+1)}\frac{\partial^2 \eta}{\partial x^2} \\
&= \frac{\partial u}{\partial t}\left[1-\frac{c_0^2}{c^2}\left(1+\frac{\gamma-1}{2c_0}u\right)^{2(\gamma+1)/(\gamma-1)}\right] \\
&= \frac{\partial u}{\partial t}(1-1) = 0
\end{aligned}
\tag{8.17.21}
$$

It can be seen that the solution (8.17.15) satisfies the wave equation for the gas.

(f) For small and finite propagation, u is a finite small quantity, so

$$
\left(1+\frac{\gamma-1}{2c_0}u\right)^{-(\gamma+1)/(\gamma-1)} \approx 1-\frac{\gamma+1}{2c_0}u \tag{8.17.22}
$$

Substituting (8.17.22) into (8.17.17) yields

$$
u = F\left[t-\frac{x}{c_0}\left(1-\frac{\gamma+1}{2c_0}u\right)\right] \tag{8.17.23}
$$

8.18 Exercise 8.18 (Approximate Solutions of the One-Dimensional Wave Equation for a Viscous Isentropic Gas and Its Fourier Expansion)

Solution: (a) Because $u(0,t) = u_0 \sin\omega t$, we replace the variable in this equation with the function F and obtain

$$
u = u_0 \sin\left[\omega t - \frac{\omega x}{c_0} + \frac{\gamma+1}{2c_0^2}\omega x u\right] \triangleq u_0 \sin\xi \tag{8.18.1}
$$

(b) Because

$$
\xi = \omega t - \frac{\omega x}{c_0} + \frac{\gamma+1}{2c_0^2}\omega x u = s + \frac{(\gamma+1)\omega u_0 x}{2c_0^2}\sin\xi,\ s = \omega t - \frac{\omega x}{c_0} \tag{8.18.2}
$$

Rewrite (8.18.1) as

$$
\frac{u}{u_0} = \sin\xi = \sin\left\{s\left[1+\frac{(\gamma+1)\omega u_0 x}{2c_0^2 s}\sin\xi\right]\right\} \tag{8.18.3}
$$

Since u_0 is a finite small quantity, u/u_0 is approximated as a periodic function of s by expanding it into the Fourier sine series of s:

$$\frac{u}{u_0} = \sin\xi = \sum_{n=1}^{\infty} b_n \sin ns \tag{8.18.4}$$

where

$$b_n = \frac{2}{\pi}\int_0^{2\pi} \sin\xi \sin ns ds \tag{8.18.5}$$

Integrating (8.18.5) by part, we obtain

$$\begin{aligned} b_n &= \frac{2}{\pi}\int_0^{2\pi} \sin\xi \sin ns ds = -\frac{2}{n\pi}\sin\xi\cos ns\Big|_0^{2\pi} + \frac{2}{n\pi}\int_0^{2\pi} \cos ns\cos\xi d\xi \\ &= \frac{2}{n\pi}\int_0^{2\pi} \cos ns\cos\xi d\xi \end{aligned} \tag{8.18.6}$$

From (8.18.2), we can obtain

$$s = \xi - \sigma \sin\xi, \sigma = \frac{(\gamma+1)\omega u_0 x}{2c_0^2} \tag{8.18.7}$$

Therefore,

$$\sigma\cos\xi d\xi = d\xi - ds \tag{8.18.8}$$

then

$$\begin{aligned} \frac{1}{\pi}\int_0^{2\pi} \cos ns \cos\xi d\xi &= \frac{1}{\sigma\pi}\int_0^{2\pi} \cos ns(d\xi - ds) = \frac{1}{\sigma\pi}\int_0^{2\pi} \cos ns d\xi \\ &= \frac{1}{\sigma\pi}\int_0^{2\pi} \cos ns(n\xi - n\sigma\sin\xi)d\xi \\ &= \frac{1}{\sigma}J_n(n\sigma) \end{aligned} \tag{8.18.9}$$

Substituting (8.18.9) into (8.18.6) yields

$$b_n = \frac{2}{n\sigma}J_n(n\sigma) \tag{8.18.10}$$

Substituting (8.18.10) and (8.18.7) into (8.18.4) yields

$$\frac{u}{u_0} = \frac{4c_0^2}{(\gamma+1)\omega u_0 x}\sum_{n=1}^{\infty}\frac{1}{n}J_n\left[\frac{n(\gamma+1)\omega u_0 x}{2c_0^2}\right]\sin[n\left(\omega t - \frac{\omega x}{c_0}\right)] \tag{8.18.11}$$

8.19 Exercise 8.19 (Derivation of the Eulerian Form of the One-Dimensional Wave Equation for an Inviscid Isentropic Gas)

Solution: (a) In order to derive the Eulerian form of one-dimensional wave equation for an inviscid isentropic gas, we take the gas control volume and write the following equation according to the law of conservation of mass:

$$\rho udtdydz - \left(\rho + \frac{\partial \rho}{\partial x}dx\right)\left(u + \frac{\partial u}{\partial x}dx\right)dtdydz = \frac{\partial \rho}{\partial t}dtdxdydz$$

Therefore, the equation of conservation of mass (or continuity equation) is

$$\frac{1}{\rho}\frac{\partial \rho}{\partial t} + \frac{u}{\rho}\frac{\partial \rho}{\partial x} = -\frac{\partial u}{\partial x} \tag{8.19.1}$$

Apply the momentum theorem to the control volume. The sum of the original momentum of the control volume and the momentum flowing into the control volume from t to $t + dt$ is

$$p(t) = \rho udxdydz + \rho u^2 dtdydz \tag{8.19.2}$$

At $t + dt$ instant, the sum of the momentum of the control volume and the momentum of the outgoing control volume is

$$\begin{aligned} p(t+dt) &= \left(\rho + \tfrac{\partial \rho}{\partial t}dt\right)\left(u + \tfrac{\partial u}{\partial t}dt\right)dxdydz + \left(\rho + \tfrac{\partial \rho}{\partial x}dx\right)\left(u + \tfrac{\partial u}{\partial x}dx\right)^2 dtdydz \\ &= \left(\rho + \tfrac{\partial \rho}{\partial t}dt\right)\left(u + \tfrac{\partial u}{\partial t}dt\right)dxdydz + \left(\rho + \tfrac{\partial \rho}{\partial x}dx\right)\left(u^2 + 2u\tfrac{\partial u}{\partial x}dx\right)dtdydz \end{aligned} \tag{8.19.3}$$

Then the increment of the momentum of the control volume during time interval dt is

$$\begin{aligned} dp &= p(t+dt) - p(t) \\ &= \rho\tfrac{\partial u}{\partial t}dtdxdydz + u\left[\tfrac{\partial \rho}{\partial t} + u\tfrac{\partial \rho}{\partial x} + 2\rho\tfrac{\partial u}{\partial x}\right]dtdxdydz \end{aligned} \tag{8.19.4}$$

Assuming that the mass in the control volume is conserved during time interval dt and applying the momentum theorem, we obtain

$$\begin{aligned} \tfrac{dp}{dt} &= \rho\tfrac{\partial u}{\partial t}dxdydz + u\left[\tfrac{\partial \rho}{\partial t} + u\tfrac{\partial \rho}{\partial x} + 2\rho\tfrac{\partial u}{\partial x}\right]dxdydz \\ &= pdydz - \left(p + \tfrac{\partial p}{\partial x}dx\right)dydz \end{aligned} \tag{8.19.5}$$

i.e.,

$$\rho\frac{\partial u}{\partial t}+u\left[\frac{\partial \rho}{\partial t}+u\frac{\partial \rho}{\partial x}+2\rho\frac{\partial u}{\partial x}\right]=-\frac{\partial p}{\partial x} \tag{8.19.6}$$

Applying the mass conservation Eq. (8.19.1) to the above equation yields

$$\frac{\partial u}{\partial t}+u\frac{\partial u}{\partial x}=-\frac{1}{\rho}\frac{\partial p}{\partial x} \tag{8.19.7}$$

This is the equation of conservation of momentum or the equation of motion. For an isentropic gas, the equation of state is

$$\frac{p}{p_0}=\left(\frac{\rho}{\rho_0}\right)^{\gamma} \tag{8.19.8}$$

where $\gamma = C_p/C_V$ is the specific heat ratio or adiabatic exponent.

(b) Multiplying (8.19.1) by c and $-c$, respectively, and adding them to (8.19.7) yields

$$\frac{c}{\rho}\frac{\partial \rho}{\partial t}+c\frac{u}{\rho}\frac{\partial \rho}{\partial x}+\frac{\partial u}{\partial t}+u\frac{\partial u}{\partial x}=-c\frac{\partial u}{\partial x}-\frac{1}{\rho}\frac{\partial p}{\partial x} \tag{8.19.9}$$

$$-\frac{c}{\rho}\frac{\partial \rho}{\partial t}-c\frac{u}{\rho}\frac{\partial \rho}{\partial x}+\frac{\partial u}{\partial t}+u\frac{\partial u}{\partial x}=c\frac{\partial u}{\partial x}-\frac{1}{\rho}\frac{\partial p}{\partial x} \tag{8.19.10}$$

where $c^2 = \partial p/\partial \rho$. (8.19.9) and (8.19.10) can be written as

$$\frac{\partial u}{\partial t}+\frac{c}{\rho}\frac{\partial \rho}{\partial t}=-(u+c)\left(\frac{\partial u}{\partial x}+\frac{c}{\rho}\frac{\partial \rho}{\partial x}\right) \tag{8.19.11}$$

$$\frac{\partial u}{\partial t}-\frac{c}{\rho}\frac{\partial \rho}{\partial t}=-(u-c)\left(\frac{\partial u}{\partial x}-\frac{c}{\rho}\frac{\partial \rho}{\partial x}\right) \tag{8.19.12}$$

Considering the equation of state (8.19.8), we can immediately rewrite these two equations as

$$\frac{\partial J_1}{\partial t}=-(u+c)\frac{\partial J_1}{\partial x} \tag{8.19.13}$$

$$\frac{\partial J_2}{\partial t}=-(u-c)\frac{\partial J_2}{\partial x} \tag{8.19.14}$$

where

$$J_1=u+\int_{\rho_0}^{\rho}\frac{c}{\rho}d\rho=u+\frac{2}{\gamma-1}(c-c_0) \tag{8.19.15}$$

$$J_2 = u - \int_{\rho_0}^{\rho} \frac{c}{\rho} d\rho = u - \frac{2}{\gamma - 1}(c - c_0) \tag{8.19.16}$$

$$c_0^2 = \gamma \frac{p_0}{\rho_0} \tag{8.19.17}$$

where c_0 is the speed of sound of a linear sound wave.

(c) Set up the characteristic line equation:

$$\frac{dx}{dt} = u + c, \ \frac{dx}{dt} = u - c \tag{8.19.18}$$

The coordinates of the curves, ξ, η are adopted. From (8.19.13) and (8.19.14), we can obtain

$$\frac{\partial J_1}{\partial \xi} \frac{\partial \xi}{\partial x} \frac{dx}{dt} = -(u + c) \frac{\partial J_1}{\partial \xi} \frac{\partial \xi}{\partial t} \frac{dt}{dx} = -\frac{\partial J_1}{\partial \xi} \frac{\partial \xi}{\partial x} \frac{dx}{dt}$$

i.e.,

$$(u + c) \frac{\partial \xi}{\partial x} \frac{\partial J_1}{\partial \xi} = 0 \tag{8.19.19}$$

therefore,

$$\frac{\partial J_1}{\partial \xi} = 0 \tag{8.19.20}$$

similarly, there are

$$\frac{\partial J_2}{\partial \eta} = 0 \tag{8.19.21}$$

Therefore, J_1 is a constant along the curve of ξ and J_2 is a constant along the curve of η. J_1 and J_2 are Riemann invariants.

(d) For the present one-dimensional wave propagation, a relatively simple exact solution to the nonlinear wave equation can be obtained.

From $c^2 = \partial p / \partial \rho$ and the equation of state (8.19.8), we have

$$c = c_0 \left(\frac{\rho}{\rho_0} \right)^{(\gamma-1)/2}, \rho = \rho_0 \left(\frac{c}{c_0} \right)^{2/(\gamma-1)}, \frac{\partial \rho}{\partial c} = \frac{2}{\gamma - 1} \frac{\rho_0}{c_0} \left(\frac{c}{c_0} \right)^{(3-\gamma)/(\gamma-1)} \tag{8.19.22}$$

Considering (8.19.22), we can rewrite (8.19.1) and (8.19.7) as follows:

$$\frac{\partial c}{\partial t} + u\frac{\partial c}{\partial x} + c\frac{\partial u}{\partial x} = 0 \tag{8.19.23}$$

$$\frac{\partial u}{\partial t} + u\frac{\partial u}{\partial x} = -\frac{2}{\gamma - 1}c\frac{\partial c}{\partial x} \tag{8.19.24}$$

Let there exist the relation $c = c(u)$, then

$$\frac{\partial c}{\partial t} = c'\frac{\partial u}{\partial t}, \frac{\partial c}{\partial x} = c'\frac{\partial u}{\partial x} \text{ where } c' = \frac{dc}{du} \tag{8.19.25}$$

Substituting (8.19.25) into (8.19.23) and (8.19.24) yields

$$c\prime\frac{\partial u}{\partial t} + c'u\frac{\partial u}{\partial x} + \frac{\gamma - 1}{2}c\frac{\partial u}{\partial x} = 0$$

$$\frac{\partial u}{\partial t} + u\frac{\partial u}{\partial x} + \frac{2}{\gamma - 1}cc'\frac{\partial u}{\partial x} = 0$$

Eliminating u form the above two equations, we can obtain

$$\frac{\gamma - 1}{2}\frac{c}{c\prime} = \frac{2}{\gamma - 1}cc'$$

or

$$\frac{dc}{du} = \pm\frac{\gamma - 1}{2} \tag{8.19.26}$$

The plus and minus signs of these determine the direction of propagation of the traveling wave. The positive sign indicates a right-running wave, therefore:

$$c = c_0 + \frac{\gamma - 1}{2}u \tag{8.19.27}$$

Substituting (8.19.27) into (8.19.24) yields

$$\frac{\partial u}{\partial t} + (c_0 + \beta u)\frac{\partial u}{\partial x} = 0 \tag{8.19.28}$$

where

$$\beta = \frac{1}{2}(\gamma + 1) \tag{8.19.29}$$

Now, (8.19.23) and (8.19.24) are reduced to a single first-order Eq. (8.19.28).

Notice that the general solution of the corresponding linear equation of this equation is

$$u = f\left(t - \frac{x}{c_0}\right) \tag{8.19.30}$$

Comparing (8.19.28) with the corresponding linear equation, we can suppose that the solution to this nonlinear equation is

$$u = F\left(t - \frac{x}{c_0 + \beta u}\right) \tag{8.19.31}$$

We can verify that (8.19.31) is exactly the solution of the Eq. (8.19.28).

8.20 Exercise 8.20 (Derivation of Linear Inviscid Acoustic Waves in a Hardwalled Duct)

Solution: (a) Let the velocity of the fluid mass be

$$\mathbf{u} = u\mathbf{i} + v\mathbf{j} + w\mathbf{k} \tag{8.20.1}$$

Take the control volume shown in Fig. 8.4a, which shows the mass inflow and outflow per unit time from the control volume. From the law of conservation of mass, we have

$$\begin{aligned}\frac{\partial}{\partial t}(\rho\Delta x\Delta y\Delta z) = \Delta y\Delta z\left[(\rho u)|_x - (\rho u)|_{x+\Delta x}\right] + \Delta z\Delta x\left[(\rho u)\big|_y - (\rho u)\big|_{y+\Delta y}\right] \\ + \Delta x\Delta y\left[(\rho u)|_z - (\rho u)|_{z+\Delta z}\right]\end{aligned} \tag{8.20.2}$$

i.e.,

$$\frac{\partial \rho}{\partial t} + \frac{\partial(\rho u)}{\partial x} + \frac{\partial(\rho u)}{\partial y} + \frac{\partial(\rho u)}{\partial z} = 0 \tag{8.20.3}$$

The above equation is written in vector form as

$$\frac{\partial \rho}{\partial t} + \nabla\cdot(\rho\mathbf{u}) = 0 \tag{8.20.4}$$

This is the continuity equation.

Based on the expression for the substantial derivative:

$$\frac{D(\cdot)}{Dt} = \frac{\partial(\cdot)}{\partial t} + u\cdot\nabla(\cdot) = \frac{\partial(\cdot)}{\partial t} + u\frac{\partial(\cdot)}{\partial x} + v\frac{\partial(\cdot)}{\partial y} + w\frac{\partial(\cdot)}{\partial z} \tag{8.20.5}$$

Equation (8.20.4) can be written as

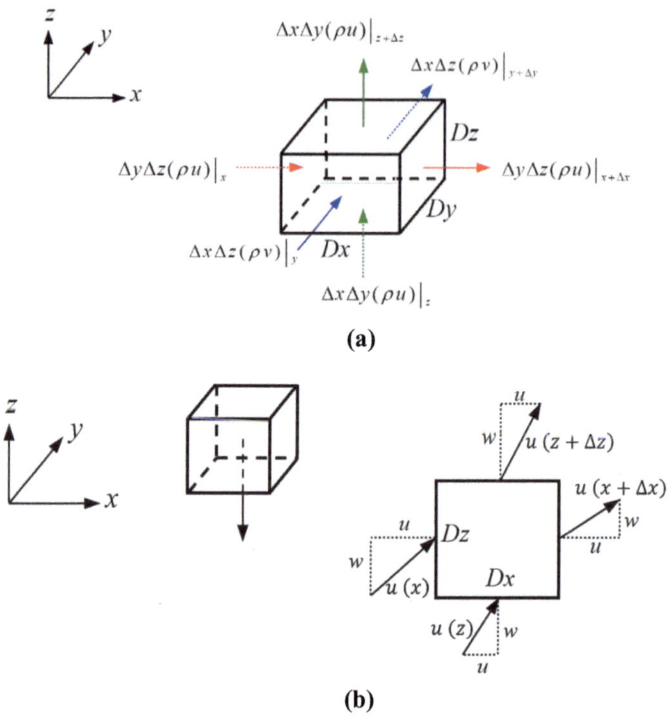

Fig. 8.4 **a** Mass flow of the control volume of inviscid fluid, **b** momentum flow of the control volume of inviscid fluid for Exercise 8.20

$$\frac{D\rho}{Dt} + \rho\nabla \cdot u = 0 \tag{8.20.6}$$

Applying the momentum theorem to the x direction of the control volume shown in Fig. 8.4b yields

$$\begin{aligned}
&\tfrac{\partial}{\partial t}(\rho u\Delta x\Delta y\Delta z) = P\Delta y\Delta z|_x - P\Delta y\Delta z|_{x+\Delta x} \\
&+\rho u^2\Delta y\Delta z|_x - \rho u^2\Delta y\Delta z|_{x+\Delta x} \\
&+\rho uv\Delta x\Delta z\big|_y - \rho uv\Delta x\Delta z\big|_{y+\Delta y} \\
&+\rho uw\Delta x\Delta y|_z - \rho uw\Delta x\Delta y|_{z+\Delta z}
\end{aligned}$$

where P is the pressure on each face. The above equation can be rearranged as

$$\frac{\partial(\rho u)}{\partial t} + \frac{\partial\left(\rho u^2\right)}{\partial x} + \frac{\partial(\rho uv)}{\partial y} + \frac{\partial(\rho uw)}{\partial z} + \frac{\partial P}{\partial x} = 0$$

Expanding this equation and applying the continuity equation yields the equation for the conservation of momentum along the x-direction

$$\rho\left(\frac{\partial u}{\partial t}+u\frac{\partial u}{\partial x}+v\frac{\partial u}{\partial y}+w\frac{\partial u}{\partial z}\right)+\frac{\partial P}{\partial x}=0$$

i.e.,

$$\rho\frac{Du}{Dt}+\frac{\partial P}{\partial x}=0 \tag{8.20.7}$$

Similarly, the equations of the conservation of momentum along the y- and z-direction can be obtained as

$$\rho\frac{Dv}{Dt}+\frac{\partial P}{\partial y}=0 \tag{8.20.8}$$

$$\rho\frac{Dw}{Dt}+\frac{\partial P}{\partial z}=-\rho g \tag{8.20.9}$$

The above three equations can be written in vector form:

$$\rho\frac{D\mathbf{u}}{Dt}+\nabla P=-\rho g\mathbf{k} \tag{8.20.10}$$

For an isentropic fluid, the equation of state is

$$\frac{p}{p_0}=\left(\frac{\rho}{\rho_0}\right)^{\gamma} \tag{8.20.11}$$

Ignoring gravity effects, the viscosity of the medium and heat conduction, we can degenerate the continuity and momentum equations into

$$\rho_t+\mathbf{u}\cdot\nabla\rho+\rho\nabla\cdot\mathbf{u}=0 \tag{8.20.12}$$

$$\rho[\mathbf{u}_t+(\mathbf{u}\cdot\nabla)\mathbf{u}]+\nabla P=0 \tag{8.20.13}$$

This set of equations is called Euler's equations. The equation of state can be written as

$$P-p_0=c_0^2(\rho-\rho_0)\left[1+\frac{B}{2!A}\left(\frac{\rho-\rho_0}{\rho_0}\right)+\frac{C}{3!A}\left(\frac{\rho-\rho_0}{\rho_0}\right)^2+\cdots\right] \tag{8.20.14}$$

where

$$c_0^2=\gamma\frac{p_0}{\rho_0} \tag{8.20.15}$$

When there are no sound waves in the medium, the solutions of the above three equations are

$$\rho = \rho_0,\ P = p_0,\ \mathbf{u} = 0 \tag{8.20.16}$$

This set of solutions is called the zeroth-order solutions, which describe the stationary state of the fluid.

Now suppose there is a small perturbation to the stationary state, i.e.,

$$\begin{aligned} &\rho = \rho_0 + \delta\rho,\ |\delta\rho| << \rho_0 \\ &P = p_0 + p,\ |\delta\rho| << \rho_0 c_0^2 \\ &\mathbf{u} = \mathbf{0} + \mathbf{u},\ |\mathbf{u}| << c_0 \end{aligned} \tag{8.20.17}$$

As a result, the continuity, momentum and equation of state become

$$\frac{\partial(\delta\rho)}{\partial t} + \underline{\mathbf{u}\cdot\nabla\delta\rho} + \rho_0\nabla\cdot\mathbf{u} + \underline{\delta\rho\nabla\cdot\mathbf{u}} = 0$$

$$\rho_0\frac{\partial\mathbf{u}}{\partial t} + \underline{\rho_0(\mathbf{u}\cdot\nabla)\mathbf{u}} + \underline{\delta\rho\frac{\partial\mathbf{u}}{\partial t}} + \underline{\delta\rho(\mathbf{u}\cdot\nabla)\mathbf{u}} + \nabla p = 0$$

$$p = c_0^2\delta\rho + \underline{\frac{B}{2!A}\frac{(\delta\rho)^2}{\rho_0}}$$

where the underlined terms are second-order or higher small quantities. Omitting these terms yields

$$\frac{\partial(\delta\rho)}{\partial t} + \rho_0\nabla\cdot\mathbf{u} = 0 \tag{8.20.18}$$

$$\rho_0\frac{\partial u}{\partial t} + \nabla p = 0 \tag{8.20.19}$$

$$c_0^2 = \frac{p}{\delta\rho} \tag{8.20.20}$$

This is the set of three-dimensional linearized wave equations for an inviscid, non-heat-transferring fluid that neglects gravity effects.

Eliminating $\delta\rho$ from (8.20.18) and (8.20.20), we obtain

$$\frac{\partial p}{\partial t} + \rho_0 c_0^2\nabla\cdot\mathbf{u} = 0$$

Then eliminate u from the above equation and (8.20.19), we can obtain

$$\frac{\partial^2 p}{\partial t^2} - c_0^2 \nabla^2 p = 0 \tag{8.20.21}$$

This is the three-dimensional wave equation for an inviscid fluid expressed in terms of the wave pressure p.

For a viscous fluid, the boundary condition is $\mathbf{u}_n = 0$. From (8.20.19), we can obtain

$$\frac{\partial p}{\partial n} = -\rho_0 \frac{\partial u_n}{\partial t} \tag{8.20.22}$$

Therefore, the boundary conditions are

$$\frac{\partial p}{\partial n} = 0 on\, \Gamma \tag{8.20.23}$$

(b) Since the fluid is inviscid, it can be conjectured that the pressure at the same cross-section of the pipe propagate along the pipe with the same velocity. Thus, for a right running wave in a pipe, we assume that the solution of Eq. (8.20.21) is

$$p = \psi(y, z)\exp[i(kx - \omega t)] \tag{8.20.24}$$

Substituting (8.20.24) into (8.20.21) and (8.20.23) yields

$$\frac{\partial^2 \psi}{\partial y^2} + \frac{\partial^2 \psi}{\partial z^2} + \lambda^2 \psi = 0 \tag{8.20.25}$$

$$\frac{\partial \psi}{\partial n} = 0 \text{ on } \Gamma \tag{8.20.26}$$

where

$$\omega^2 = c_0^2 \left(k^2 + \lambda^2\right) \tag{8.20.27}$$

(c) For a rectangular pipe, we assume that the cross section of the pipe is $0 \le y \le a$ and $0 \le z \le b$, and set the solution of Eq. (8.20.25) be

$$\psi = \Psi \cos \frac{n\pi y}{a} \cos \frac{m\pi z}{b}, n, m = 1, 2, 3, \ldots \tag{8.20.28}$$

so that the boundary conditions are automatically satisfied. Substituting (8.20.28) into (8.20.25), we obtain

$$\lambda^2 = \pi^2 \left(\frac{n^2}{a^2} + \frac{m^2}{b^2}\right) \tag{8.20.29}$$

Substituting (8.20.29) into (8.20.27) yields

$$\omega^2 = c_0^2\left[k^2 + \pi^2\left(\frac{n^2}{a^2} + \frac{m^2}{b^2}\right)\right] \tag{8.20.30}$$

It can be seen that the phase speed ω/k is not a constant and therefore the waves are dispersive.

The function ψ determined by the Eqs. (8.20.28) and (8.20.30) is the natural or acoustic mode of the traveling wave in a rectangular duct, and they are

$$\psi_{nm}(y, z) = \cos\frac{n\pi y}{a}\cos\frac{m\pi z}{b},\ n, m = 1, 2, 3, \ldots \tag{8.20.31}$$

It can be seen from (8.20.30) that for a given mode, the corresponding natural frequency ω_{nm} and wave number k_{nm} can be multivalued; while for a given frequency ω, there are

$$k_{nm}^2 = \frac{\omega^2}{c_0^2} - \pi^2\left(\frac{n^2}{a^2} + \frac{m^2}{b^2}\right) \tag{8.20.32}$$

When the right-hand side of the above equation is less than zero, k_{nm} is an imaginary number, and it can be seen from (8.20.24) that the wave is no longer a traveling wave. Therefore, the frequency ω must make the right-hand side of (8.20.32) greater than zero, otherwise it cannot propagate (precisely, it cannot propagate over long distances).

Write the dispersion relation as:

$$\left(\frac{\omega}{k}\right)^2 = c_0^2\left[1 + \frac{\pi^2}{k^2}\left(\frac{n^2}{a^2} + \frac{m^2}{b^2}\right)\right] \tag{8.20.33}$$

For the given a and b, it is assumed that the mode of $r \sim s$ order also satisfy (8.20.33), i.e.,

$$\left(\frac{\omega}{k}\right)^2 = c_0^2\left[1 + \frac{\pi^2}{k^2}\left(\frac{r^2}{a^2} + \frac{s^2}{b^2}\right)\right] \tag{8.20.34}$$

where $r \neq n, s \neq m$. In this case, the traveling wave mode of order $n \sim m$ has the same wave number and phase speed as the traveling wave mode of order $r \sim s$, and these two traveling waves interact strongly, i.e., they resonate harmonically.

Readers are invited to complete the analysis on Exercise 8.20 (d).

8.21 Exercise 8.21 (Analysis on the Linear Waves Propagating on the Surface of an Inviscid Liquid of Finite Depth)

Solution: (a) The governing equation of the problem is the two-dimensional Laplacian equation for the velocity potential function $\phi(x, y, t)$ in $x \sim y$ plane, and the boundary conditions are expressed in terms of the potential function $\phi(x, y, t)$ and the elevation of the interface above its undisturbed position $\eta(x, t)$. The propagation regions are $-\infty < x < \infty$ and $-h \le y \le 0$, as shown in Fig. 8.5.

Let the governing equation of the velocity potential function $\phi(x, y, t)$ have a separated variable solution:

$$\phi(x, y, t) = f(x)g(y)q(t) \tag{8.21.1}$$

Substituting this into the Laplacian equation of ϕ yields

$$\frac{\frac{d^2 f}{dx^2}}{f} = -\frac{\frac{d^2 g}{dy^2}}{g} = -k^2$$

So there's

$$\frac{d^2 f}{dx^2} + k^2 f = 0 \tag{8.21.2}$$

$$\frac{d^2 g}{dy^2} - k^2 g = 0 \tag{8.21.3}$$

Let the solution of Eq. (8.21.2) be

$$f(x) = \exp(ikx) \tag{8.21.4}$$

and the solution of Eq. (8.21.3) be

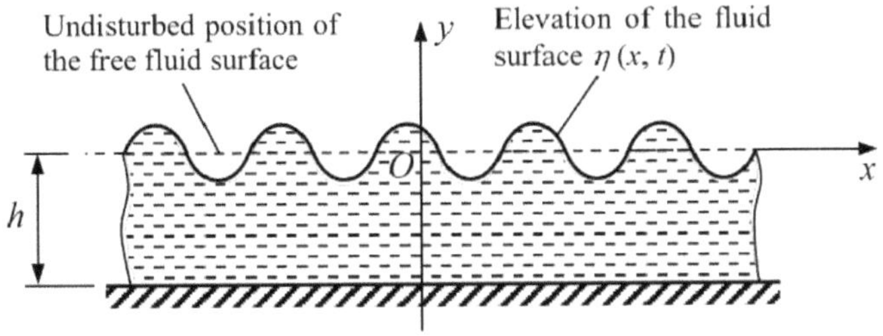

Fig. 8.5 Surface waves of a viscous-free fluid for Exercise 8.21

$$g(y) = A\cosh\left[k(y + y_1)\right] \tag{8.21.5}$$

where A and y_1 are the constants of integration. So

$$\phi(x, y, t) = A\cosh\left[k(y + y_1)\right]\exp(ikx)q(t)$$

Applying the boundary condition at $y = -h$ to the above equation yields

$$A\sinh\left[k(-h + y_1)\right] = 0 \Rightarrow y_1 = h \tag{8.21.6}$$

If we wish to obtain a simple harmonic solution for a right-running wave, we can take $q(t) = \exp(-i\omega t)$, and thus have

$$\phi(x, y, t) = A\cosh\left[k(y + h)\right]\exp\left[i(kx - \omega t)\right] \tag{8.21.7}$$

Substituting (8.21.7) into the joint boundary conditions of ϕ and η, we obtain

$$\frac{\partial \eta}{\partial t} = kA\sinh(kh)\exp[i(kx - \omega t)] \tag{8.21.8}$$

$$W\frac{\partial^2 \eta}{\partial x^2} - \eta = -i\omega A\cosh(kh)\exp[i(kx - \omega t)] \tag{8.21.9}$$

The solution of Eq. (8.21.8) can be set as

$$\eta(x, t) = \eta_0\exp[i(kx - \omega t)] \tag{8.21.10}$$

which leads to

$$A = -\frac{i\omega\eta_0}{k\sinh(kh)} \tag{8.21.11}$$

Substituting (8.21.10) and (8.21.11) into (8.21.9) yields

$$\omega^2 = k\tanh(kh)\left(Wk^2 + 1\right) \tag{8.21.12}$$

(b) For the case of small surface tension, $W \approx 0$, so the Eq. (8.21.12) becomes

$$\omega^2 = k\tanh(kh)$$

When $h \to 0$, the above equation becomes

$$\omega^2 \approx hk^2 \Rightarrow \frac{\omega}{k} \approx h \tag{8.21.13}$$

The phase speed is constant and hence the solution (8.21.12) represents a non-dispersive wave, i.e., the surface wave in the shallow water region is approximated as a non-dispersive wave.

(c) For deep water and nonnegligible surface tension,$h \to \infty$, the Eq. (8.21.12) becomes

$$\omega^2 = k\left(Wk^2 + 1\right) \tag{8.21.14}$$

For a positive integer n, we assume that $(n\omega, nk)$ also satisfies the dispersion relation (8.21.14) i.e.

$$(n\omega)^2 = nk\left[W(nk)^2 + 1\right] \tag{8.21.15}$$

then we need

$$k^2 = \frac{1}{nW} \tag{8.21.16}$$

If the condition (8.21.16) is satisfied, then the low-order traveling wave with wave number $k = 1/\sqrt{nW}$ has the same phase speed as the high-order traveling wave with wave number $k = \sqrt{n/W}$, and these two traveling waves will interact strongly, i.e., resonate harmonically.

8.22 Exercise 8.22 (The Method of Multiple Scales for Wave Group Propagation Governed by Klein–Gordon Equation)

Solution: (a) We consider the propagation of a group of waves whose frequencies and wave numbers are around a center frequency w and wave number k. We adopt the method of multiscale to complete this problem by introducing two slow-varying time scales T_1 and T_2, and two slow-varying space scales X_1 and X_2:

$$T_1 = \varepsilon t, T_2 = \varepsilon^2 t \, \text{and} \, X_1 = \varepsilon x, X_2 = \varepsilon^2 x \tag{8.22.1}$$

Therefore, the derivatives for the fast-varying scales $t = T_0$ and $x = X_0$ become

$$\begin{aligned} \frac{\partial}{\partial t} &= \frac{\partial}{\partial T_0} + \varepsilon \frac{\partial}{\partial T_1} + \varepsilon^2 \frac{\partial}{\partial T_2} \\ \frac{\partial}{\partial x} &= \frac{\partial}{\partial X_0} + \varepsilon \frac{\partial}{\partial X_1} + \varepsilon^2 \frac{\partial}{\partial X_2} \end{aligned} \tag{8.22.2}$$

Let the solution of the given Klein–Gordon equation be

$$u(x,t;\varepsilon)=\sum_{n=0}^{2}\varepsilon^n u_n(X_0,X_1,X_2,T_0,T_1,T_2)+O(\varepsilon^3) \tag{8.22.3}$$

Substituting (8.22.2) and (8.22.3) into the Klein–Gordon equation, we obtain

$$\begin{aligned}
0&=\frac{\partial^2 u}{\partial t^2}-\frac{\partial^2 u}{\partial x^2}-\gamma u-\varepsilon\alpha u^3\\
&=\left(\frac{\partial^2}{\partial T_0^2}+2\varepsilon\frac{\partial^2}{\partial T_0\partial T_1}+\varepsilon^2\frac{\partial^2}{\partial T_1^2}+2\varepsilon^2\frac{\partial^2}{\partial T_0\partial T_2}\right)(u_0+\varepsilon u_1+\varepsilon^2 u_2)\\
&-\left(\frac{\partial^2}{\partial X_0^2}+2\varepsilon\frac{\partial^2}{\partial X_0\partial X_1}+\varepsilon^2\frac{\partial^2}{\partial X_1^2}+2\varepsilon^2\frac{\partial^2}{\partial X_0\partial X_2}\right)(u_0+\varepsilon u_1+\varepsilon^2 u_2)\\
&-\gamma(u_0+\varepsilon u_1+\varepsilon^2 u_2)-\varepsilon^2\alpha(u_0+\varepsilon u_1+\varepsilon^2 u_2)^3\\
&=\frac{\partial^2 u_0}{\partial T_0^2}-\frac{\partial^2 u_0}{\partial X_0^2}-\gamma u_0\\
&+\varepsilon\left(\frac{\partial^2 u_1}{\partial T_0^2}-\frac{\partial^2 u_1}{\partial X_0^2}-\gamma u_1-2\frac{\partial^2 u_0}{\partial X_0\partial X_1}+2\frac{\partial^2 u_0}{\partial T_0\partial T_1}\right)\\
&+\varepsilon^2\left(\begin{array}{l}\frac{\partial^2 u_2}{\partial T_0^2}-\frac{\partial^2 u_2}{\partial X_0^2}-\gamma u_2+\frac{\partial^2 u_0}{\partial T_1^2}+2\frac{\partial^2 u_0}{\partial T_0\partial T_2}\\
-\frac{\partial^2 u_0}{\partial X_1^2}-2\frac{\partial^2 u_0}{\partial X_0\partial X_2}+2\frac{\partial^2 u_1}{\partial T_0\partial T_1}-2\frac{\partial^2 u_1}{\partial X_0\partial X_1}-\alpha u_0^3\end{array}\right)
\end{aligned} \tag{8.22.4}$$

Making the coefficient of the same power of ε be zero in the above equation yields

$$\frac{\partial^2 u_0}{\partial T_0^2}-\frac{\partial^2 u_0}{\partial X_0^2}-\gamma u_0=0 \tag{8.22.5}$$

$$\frac{\partial^2 u_1}{\partial T_0^2}-\frac{\partial^2 u_1}{\partial X_0^2}-\gamma u_1=2\frac{\partial^2 u_0}{\partial X_0\partial X_1}-2\frac{\partial^2 u_0}{\partial T_0\partial T_1} \tag{8.22.6}$$

$$\begin{aligned}
\frac{\partial^2 u_2}{\partial T_0^2}-\frac{\partial^2 u_2}{\partial X_0^2}-\gamma u_2&=-\frac{\partial^2 u_0}{\partial T_1^2}+\frac{\partial^2 u_0}{\partial X_1^2}-2\frac{\partial^2 u_0}{\partial T_0\partial T_2}+2\frac{\partial^2 u_0}{\partial X_0\partial X_2}\\
&-2\frac{\partial^2 u_1}{\partial T_0\partial T_1}+2\frac{\partial^2 u_1}{\partial X_0\partial X_1}+\alpha u_0^3
\end{aligned} \tag{8.22.7}$$

The solution to Eq. (8.22.5) can be written as

$$u_0=A(X_1,X_2,T_1,T_2)e^{i(kX_0-\omega T_0)}+cc \tag{8.22.8}$$

Substituting (8.22.8) into (8.22.6), we can obtain the dispersion relation:

$$\omega^2=k^2-\gamma \tag{8.22.9}$$

Substituting (8.22.8) into (8.22.6) yields

$$\frac{\partial^2 u_1}{\partial T_0^2} - \frac{\partial^2 u_1}{\partial X_0^2} - \gamma u_1 = \left(2ik\frac{\partial A}{\partial X_1} + 2i\omega\frac{\partial A}{\partial T_1}\right)e^{i(kX_0-\omega T_0)} + cc \tag{8.22.10}$$

In order to eliminate secular terms from the above equation, we need

$$k\frac{\partial A}{\partial X_1} + \omega\frac{\partial A}{\partial T_1} = 0 \tag{8.22.11}$$

Then the solution to (8.22.10) is

$$u_1 = 0 \tag{8.22.12}$$

Substituting (8.22.8) and (8.22.12) into (8.22.7) yields

$$\begin{gathered}\frac{\partial^2 u_2}{\partial T_0^2} - \frac{\partial^2 u_2}{\partial X_0^2} - \gamma u_2 = (-\frac{\partial^2 A}{\partial T_1^2} + \frac{\partial^2 A}{\partial X_1^2} + 2i\omega\frac{\partial A}{\partial T_2} + 2ik\frac{\partial A}{\partial X_2} \\ +3\alpha A^2\overline{A})e^{i(kX_0-\omega T_0)} + cc + NST\end{gathered} \tag{8.22.13}$$

In order to eliminate secular terms from the above equation, we need

$$-\frac{\partial^2 A}{\partial T_1^2} + \frac{\partial^2 A}{\partial X_1^2} + 2i\omega\frac{\partial A}{\partial T_2} + 2ik\frac{\partial A}{\partial X_2} + 3\alpha A^2\overline{A} = 0 \tag{8.22.14}$$

From Eq. (8.22.11), we have

$$\frac{\partial A}{\partial T_1} = -\frac{k}{\omega}\frac{\partial A}{\partial X_1} \tag{8.22.15}$$

Applying the partial derivative with respective to T_1 to (8.22.15) yields

$$\frac{\partial^2 A}{\partial T_1^2} = -\frac{k}{\omega}\frac{\partial^2 A}{\partial T_1 \partial X_1} = \frac{k^2}{\omega^2}\frac{\partial^2 A}{\partial X_1^2} \tag{8.22.16}$$

And then substituting (8.22.16) into (8.22.14) yields

$$\frac{\partial A}{\partial T_2} + \frac{k}{\omega}\frac{\partial A}{\partial X_2} - \frac{i}{2\omega}\left(1 - \frac{k^2}{\omega^2}\right)\frac{\partial^2 A}{\partial X_1^2} = \frac{3i\alpha}{2\omega}A^2\overline{A} \tag{8.22.17}$$

Using the dispersion relation (8.22.9), we can obtain

$$\omega' = \frac{k}{\omega}, \omega'' = \frac{\omega^2 - k^2}{\omega^3} \tag{8.22.18}$$

where $\omega' = d\omega/dk$ and $\omega'' = d^2\omega/dk^2$. Substituting (8.22.18) into (8.22.17) yields

$$\frac{\partial A}{\partial T_2}+\omega'\frac{\partial A}{\partial X_2}-\frac{1}{2}i\omega''\frac{\partial^2 A}{\partial X_1^2}=\frac{3i\alpha}{2\omega}A^2\overline{A} \tag{8.22.19}$$

Use (8.22.1) to replace the independent variables in the above equation with t and x, we obtain

$$\frac{\partial A}{\partial t}+\omega'\frac{\partial A}{\partial x}-\frac{1}{2}i\omega''\frac{\partial^2 A}{\partial x^2}=\frac{3i\varepsilon^2\alpha}{2\omega}A^2\overline{A} \tag{8.22.20}$$

From (8.22.11) and (8.22.14), we have

$$\frac{\partial A}{\partial x}+k'\frac{\partial A}{\partial t}+\frac{1}{2}ik''\frac{\partial^2 A}{\partial t^2}=\frac{3i\varepsilon^2\alpha}{2k}A^2\overline{A} \tag{8.22.21}$$

where $k'=dk/d\omega$ and $k''=d^2k/d\omega^2$. Therefore, the first order traveling wave solution of the Klein–Gordon equation is

$$U=a(x,t)\exp[i(kx-\omega t)]+cc+\cdots \tag{8.22.22}$$

(b) When ω is a real number, k is also a real number and the motion is a steady state fluctuation by (8.22.22). For a given k, the amplitude of the fluctuation A is independent of x and then (8.22.20) becomes

$$\frac{\partial A}{\partial t}=\frac{3i\varepsilon^2\alpha}{2\omega}A^2\overline{A} \tag{8.22.23}$$

Let

$$A=\frac{1}{2}ae^{i\beta} \tag{8.22.24}$$

and substitute (8.22.24) into (8.22.23), we can obtain

$$\dot{a}+ia\dot{\beta}=\frac{3i\varepsilon^2\alpha}{8\omega}a^3 \tag{8.22.25}$$

Then

$$\dot{a}=0,\ \dot{\beta}=\frac{3\varepsilon^2\alpha}{8\omega}a^2 \tag{8.22.26}$$

therefore,

$$a=a_0,\ \beta=\frac{3\varepsilon^2\alpha}{8\omega}a_0^2t+\beta_0 \tag{8.22.27}$$

$$A = \frac{1}{2}a_0\exp\left[i\left(\frac{3\varepsilon^2\alpha}{8\omega}a_0^2t + \beta_0\right)\right] \tag{8.22.28}$$

Substituting (8.22.28) into (8.22.22) yields

$$u = a_0\cos(kx - \hat{\omega}t + \beta_0) + \cdots \tag{8.22.29}$$

where

$$\hat{\omega} = \omega - \frac{3\varepsilon^2\alpha}{8\omega}a_0^2 \tag{8.22.30}$$

(c) In order to analyze the stability of steady state solution (8.22.29), we denote A as

$$A = \frac{1}{2}a(x,t)\exp[i\beta(x,t)] \tag{8.22.31}$$

Substituting (8.22.31) into (8.22.20) and then separating the real and imaginary parts yields

$$\frac{\partial a}{\partial t} + \omega'\frac{\partial a}{\partial x} + \omega''\left(\frac{\partial \beta}{\partial x}\frac{\partial a}{\partial x} + \frac{1}{2}a\frac{\partial^2\beta}{\partial x^2}\right) = 0 \tag{8.22.32}$$

$$a\frac{\partial \beta}{\partial t} + a\omega'\frac{\partial \beta}{\partial x} - \frac{1}{2}\omega''\left[\frac{\partial^2 a}{\partial x^2} - a\left(\frac{\partial \beta}{\partial x}\right)^2\right] - \frac{3\varepsilon^2\alpha}{8\omega}a^3 = 0 \tag{8.22.33}$$

We denote a and β as the superimposition of the steady state solution and the perturbation:

$$\begin{aligned} a &= a_0 + a_1(x,t) \\ \beta &= \frac{3\varepsilon^2\alpha}{8\omega}a_0^2t + \beta_0 + \beta_1(x,t) \end{aligned} \tag{8.22.34}$$

Substituting (8.22.34) into (8.22.32) and (8.22.33), we can obtain the linearized equation for the perturbation

$$\begin{aligned} &\frac{\partial a_1}{\partial t} + \omega'\frac{\partial a_1}{\partial x} + \frac{1}{2}\omega''a_0\frac{\partial^2\beta_1}{\partial x^2} = 0 \\ &a_0\frac{\partial \beta_1}{\partial t} + \omega'a_0\frac{\partial \beta_1}{\partial x} - \frac{1}{2}\omega''\frac{\partial^2 a_1}{\partial x^2} - \frac{3}{4\omega}\varepsilon^2\alpha a_0^2a_1 = 0 \end{aligned} \tag{8.22.35}$$

The solution to (8.22.35) is

$$\begin{aligned} a_1 &= a_{10}\exp[i(Kx - \Omega t)] \\ \beta_1 &= \beta_{10}\exp[i(Kx - \Omega t)] \end{aligned} \tag{8.22.36}$$

Substituting (8.22.36) into (8.22.35) yields

$$\begin{aligned} &i(\omega' K - \Omega)a_{10} - \frac{1}{2}\omega'' a_0 K^2 \beta_{10} = 0 \\ &\left(\frac{1}{2}\omega'' K^2 - \frac{3}{4\omega}\varepsilon^2 \alpha a_0^2\right)a_{10} + i a_0(\omega' K - \Omega)\beta_{10} = 0 \end{aligned} \tag{8.22.37}$$

From the non-trivial solution conditions for the system of Eq. (8.22.37), we have

$$(\omega' K - \Omega)^2 = \frac{1}{4}\omega''^2 K^4\left(1 - \frac{3\varepsilon^2 \alpha a_0^2}{2K^2 \omega\omega''}\right) \tag{8.22.38}$$

From (8.22.38), we can obtain:

(1) When α has the same sign as ω'' and $3\varepsilon^2\alpha a_0^2 > 2K^2\omega\omega''$, Ω is a complex number and a_1 and β_1 become unbounded.
(2) For any given real K and Ω, when $\alpha\omega'' < 0$, Ω is a real number, then a_1 and β_1 are bounded. Considering (8.22.18) and (8.22.9), we can obtain the sufficient condition for a_1 and β_1 to be bounded is $\alpha\gamma > 0$.

Readers are invited to complete the analyses on Exercise 8.22 (d) and (e).

MIX
Papier aus verantwortungsvollen Quellen
Paper from responsible sources
FSC® C105338

If you have any concerns about our products,
you can contact us on
ProductSafety@springernature.com

In case Publisher is established outside the EU,
the EU authorized representative is:
Springer Nature Customer Service Center GmbH
Europaplatz 3, 69115 Heidelberg, Germany

Printed by Libri Plureos GmbH
in Hamburg, Germany